Advanced Linear Algebra

Designed for advanced undergraduate and beginning graduate students in linear or abstract algebra, *Advanced Linear Algebra* covers theoretical aspects of the subject, along with examples, computations, and proofs. It explores a variety of advanced topics in linear algebra that highlight the rich interconnections of the subject to geometry, algebra, analysis, combinatorics, numerical computation, and many other areas of mathematics.

The author begins with chapters introducing basic notation for vector spaces, permutations, polynomials, and other algebraic structures. The following chapters are designed to be mostly independent of each other so that readers with different interests can jump directly to the topic they want. This is an unusual organization compared to many abstract algebra textbooks, which require readers to follow the order of chapters.

Each chapter consists of a mathematical vignette devoted to the development of one specific topic. Some chapters look at introductory material from a sophisticated or abstract viewpoint, while others provide elementary expositions of more theoretical concepts. Several chapters offer unusual perspectives or novel treatments of standard results.

A wide array of topics is included, ranging from concrete matrix theory (basic matrix computations, determinants, normal matrices, canonical forms, matrix factorizations, and numerical algorithms) to more abstract linear algebra (modules, Hilbert spaces, dual vector spaces, bilinear forms, principal ideal domains, universal mapping properties, and multilinear algebra).

The book provides a bridge from elementary computational linear algebra to more advanced, abstract aspects of linear algebra needed in many areas of pure and applied mathematics.

Nicholas A. Loehr received his Ph.D. in mathematics from the University of California at San Diego in 2003, studying algebraic combinatorics under the guidance of Professor Jeffrey Remmel. After spending two years at the University of Pennsylvania as an NSF postdoc, Dr. Loehr taught mathematics at the College of William and Mary, the United States Naval Academy, and Virginia Tech. Dr. Loehr has authored over sixty refereed journal articles and three textbooks on combinatorics, advanced linear algebra, and mathematical proofs. He teaches classes in these subjects and many others, including cryptography, vector calculus, modern algebra, real analysis, complex analysis, and number theory.

Textbooks in Mathematics
Series editors:
Al Boggess, Kenneth H. Rosen

https://www.routledge.com/Textbooks-in-Mathematics/book-series/CANDHTEXBOOMTH

Advanced Linear Algebra
Second Edition

Nicholas A. Loehr

CRC Press
Taylor & Francis Group
Boca Raton London New York

CRC Press is an imprint of the
Taylor & Francis Group, an **informa** business

A CHAPMAN & HALL BOOK

Second edition published 2024
by CRC Press
2385 Executive Center Drive, Suite 320, Boca Raton, FL 33431

and by CRC Press
4 Park Square, Milton Park, Abingdon, Oxon, OX14 4RN

CRC Press is an imprint of Taylor & Francis Group, LLC

© 2024 Nicholas A. Loehr

First edition published by Taylor and Francis 2014

Library of Congress Cataloging-in-Publication Data

Names: Loehr, Nicholas A, author.
Title: Advanced linear algebra / authored by Nicholas A Loehr.
Description: Second edition. | Boca Raton, FL : CRC Press, 2024. | Includes
bibliographical references and index.
Identifiers: LCCN 2023057206 | ISBN 9781032765723 (hbk) | ISBN
9781032777429 (pbk) | ISBN 9781003484561 (ebk)
Subjects: LCSH: Algebras, Linear. | BISAC: MATHEMATICS / Algebra / General.
| MATHEMATICS / Combinatorics. | SCIENCE / Physics.
Classification: LCC QA184.2 .L64 2024 | DDC 512/.5--dc23/eng/20240209
LC record available at https://lccn.loc.gov/2023057206

ISBN: 978-1-032-76572-3 (hbk)
ISBN: 978-1-032-77742-9 (pbk)
ISBN: 978-1-003-48456-1 (ebk)

DOI: 10.1201/9781003484561

Typeset in CMR10 font
by KnowledgeWorks Global Ltd.

Publisher's note: This book has been prepared from camera-ready copy provided by the authors.

*This book is dedicated to Dan Niehaus, T. S. Michael,
and Peter Linnell.*

Contents

Part II   Matrices   71

4   Basic Matrix Operations   73

5   Determinants via Calculations   105

Part IV The Interplay of Geometry and Linear Algebra   279

Preface

What is linear algebra, and how is it used? Upon examining almost any introductory text on linear algebra, we find a standard list of topics that seems to define the subject. On one hand, one part of linear algebra consists of computational techniques for solving linear equations, multiplying and inverting matrices, calculating and interpreting determinants, finding eigenvalues and eigenvectors, and so on. On the other hand, there is a theoretical side to linear algebra involving abstract vector spaces, subspaces, linear independence, spanning sets, bases, dimension, and linear transformations.

But there is much more to linear algebra than just vector spaces, matrices, and linear equations! The goal of this book is to explore a variety of advanced topics in linear algebra which highlight the rich interconnections linking this subject to geometry, algebra, analysis, combinatorics, numerical computation, and many other areas of mathematics. The book consists of twenty chapters, grouped into six main subject areas (algebraic structures, matrices, structured matrices, geometric aspects of linear algebra, modules, and multilinear algebra). Some chapters approach introductory material from a more sophisticated or abstract viewpoint, other chapters provide elementary expositions of more theoretical concepts, yet other chapters offer unusual perspectives or novel treatments of standard results.

Unlike some advanced mathematical texts, this book has been carefully designed to minimize the dependence of each chapter on material found in earlier chapters. Each chapter has been conceived as a "mathematical vignette" devoted to the development of one specific topic. If you need to learn about Jordan canonical forms, ruler and compass constructions, the singular value decomposition, Hilbert spaces, QR factorizations, convexity, normal matrices, or modules, you may turn immediately to the relevant chapter without first wading through ten chapters of algebraic background or worrying that you will need a theorem covered two hundred pages earlier. We do assume the reader has already encountered the basic linear algebra concepts described earlier (solving linear systems, computing with matrices and determinants, knowing elementary facts about vector spaces and linear maps). These topics are all revisited in a more sophisticated setting at various points throughout the book, but this is not the book you should read to learn the mechanics of Gaussian elimination or matrix multiplication for the first time! Chapter 1 provides a condensed review of some pertinent definitions from abstract algebra. But, the vast majority of the book requires very little knowledge of abstract algebra beyond the definitions of fields, vector spaces over a field, subspaces, linear transformations, linear independence, and bases. The last three chapters build on the material on modules covered in Chapter 17, and in a few places, we need some facts about permutations (covered in Chapter 2) and polynomials (discussed in Chapter 3).

Although this book focuses on theoretical aspects of linear algebra, giving complete proofs of all results, we supplement and explain the general theory with many specific examples and concrete computations. The level of abstraction gradually increases as the book progresses, as we move from matrices to vector spaces to modules. We have deliberately avoided presenting the mathematical content as a dry skeleton of itemized definitions, theorems, and proofs. Instead, the material is presented in a narrative format, providing motivation, examples, and informal discussions to help the reader understand the significance of the theorems and the intuition behind the proofs. Each chapter ends with a summary containing a structured list of the principal definitions and results covered in the chapter, followed by a set of exercises. Most exercises are not used in the main text and consist of examples, computations, and proofs intended to aid the reader in assimilating the material from the chapter. You are encouraged to use a computer algebra system to help solve computationally intensive exercises.

This text is designed for use by advanced undergraduates or beginning graduate students, or as a reference. A second course in linear algebra might cover portions of the chapters in Parts II, III, and IV, reviewing Part I as needed. A second course in abstract algebra, focusing on the role of linear-algebraic ideas, might cover Parts I, V, and VI, along with selections from the rest of the book (such as Chapter 12 or Chapter 13). The next section describes the contents of each chapter. For more details on what is covered in a given chapter, consult the summary for that chapter.

Synopsis of Topics Covered

Part I: Background on Algebraic Structures

Chapter 1: Overview of Algebraic Systems. This chapter contains a condensed reference for the abstract algebra background needed in some parts of the book. We review the definitions of groups, commutative groups, rings, integral domains, fields, vector spaces, algebras, and homomorphisms of these structures. We also quickly cover the constructions of subgroups, product groups, quotient groups, and their analogs for other algebraic systems. Finally, we recall some fundamental definitions and results from elementary linear algebra involving linear independence, spanning sets, bases, and dimension. Most of the later chapters in this book require only a small subset of the material covered here, such as the definitions of a field, vector space, linear map, subspace, linearly independent list, and ordered basis.

Chapter 2: Permutations. This chapter provides the basic definitions and facts about permutations that are required for the study of determinants and multilinear algebra. After describing the symmetric groups S_n, we show how to visualize functions on a finite set using directed graphs. This idea leads to factorizations of permutations into products of disjoint cycles. We can also write permutations as products of transpositions or basic transpositions. To understand the role of basic transpositions, we introduce inversions and prove that a list of numbers $f = [f(1), f(2), \ldots, f(n)]$ can be sorted into increasing order by interchanging adjacent elements $\mathrm{inv}(f)$ times. For permutations f, we define $\mathrm{sgn}(f) = (-1)^{\mathrm{inv}(f)}$ and use this formula to establish the fundamental properties of the sgn function.

Chapter 3: Polynomials. This chapter covers background on one-variable polynomials over a field. This material is used in chapters on canonical forms, ruler and compass constructions, primary decompositions, commuting matrices, etc. Topics include intuitive and formal definitions of polynomials, evaluation homomorphisms, the universal mapping property for polynomial rings, polynomial division with remainder, greatest common divisors, roots of polynomials, irreducible polynomials, existence and uniqueness of prime factorizations, minimal polynomials, and ways to test polynomials for irreducibility.

Part II: Matrices

Chapter 4: Basic Matrix Operations. This chapter revisits some basic material about matrices from a somewhat more sophisticated perspective. Topics studied include formal definitions of matrices and matrix operations, vector spaces of functions and matrices, matrix multiplication and its properties, invertible matrices, elementary row and column operations and elementary matrices, Smith canonical form, factorization of invertible matrices into elementary matrices, the equality of row rank and column rank, the theorem giving equivalent conditions for a matrix to be invertible, block matrix multiplication, and tensor products of matrices.

Chapter 5: Determinants via Calculations. The main properties of matrix determinants are often stated without full proofs in a first linear algebra course. This chapter establishes these properties starting from an explicit definition of $\det(A)$ as a sum, indexed by permutations, of signed products of entries of A. Topics include the multilinearity of the determinant as a function of the rows (or columns) of A, the alternating property, the effect of elementary row (or column) operations, the product formula, Laplace expansions, the classical adjoint of A, the explicit formula for A^{-1}, Cramer's Rule, the Cauchy–Binet Formula, the Cayley–Hamilton Theorem, and an introduction to permanents.

Chapter 6: Comparing Concrete Linear Algebra to Abstract Linear Algebra. In introductory linear algebra, we learn to execute computations involving concrete objects such as column vectors and matrices. In more advanced linear algebra, we study abstract vector spaces and linear transformations. Concrete concepts defined for matrices (such as matrix multiplication, matrix transpose, diagonalizability, idempotence) have abstract counterparts for linear transformations (such as composition of maps, dual maps, existence of a basis of eigenvectors, and being a projection). This chapter gives a thorough account of the relations between the concrete world of column vectors and matrices on the one hand, and the abstract world of vector spaces and linear maps on the other hand. We build a dictionary linking these two worlds, explaining the precise connection between each abstract concept and its concrete manifestation. In particular, we carefully describe how taking coordinates relative to an ordered basis yields a vector space isomorphism between an abstract vector space V and a concrete space F^n of column vectors. Similarly, computing the matrix of a linear map relative to an ordered basis gives an algebra isomorphism between an algebra of linear operators and an algebra of matrices. We also discuss how congruence, orthogonal similarity, and unitary similarity of matrices are related to bilinear maps, real inner product spaces, and complex inner product spaces.

Part III: Matrices with Special Structure

Chapter 7: Hermitian, Positive Definite, Unitary, and Normal Matrices. This chapter develops the basic facts about the special types of matrices mentioned in the title by building an analogy between properties of complex numbers (being real, being positive, having modulus 1) and properties of matrices (being Hermitian, being positive definite, being unitary). This analogy provides motivation for the central concept of a normal matrix. The chapter also includes a discussion of unitary similarity, triangularization, and diagonalization, the Spectral Theorem for Normal Matrices, simultaneous triangularization and diagonalization of commuting families, the Polar Decomposition of a matrix, and the Singular Value Decomposition.

Chapter 8: Jordan Canonical Forms. This chapter gives a novel and elementary derivation of the existence and uniqueness of the Jordan canonical form of a complex matrix. The first step in the proof is to classify nilpotent linear operators (with scalars in any field) by a visual analysis of partition diagrams. Once this classification is complete, we combine it with a version of Fitting's Lemma to obtain Jordan canonical forms with no explicit mention of generalized eigenspaces. The chapter concludes with remarks on how to compute the Jordan form, an application to systems of differential equations, and a discussion of the Jordan–Chevalley decomposition of a linear operator into a sum of commuting nilpotent and diagonalizable linear maps.

Chapter 9: Matrix Factorizations. There are many ways to factor a matrix into products of other matrices that have a special structure (such as triangular or unitary matrices). These factorizations often appear as building blocks in algorithms for the numerical solution of linear algebra problems. This chapter proves the existence and uniqueness of several matrix factorizations and explores the algebraic and geometric ideas leading to these factorizations. Topics covered include the Gram–Schmidt orthonormalization algorithm, QR-factorizations, Householder reflections, LU factorizations and their relation to Gaussian elimination, Cholesky's factorization of positive semidefinite matrices, the normal equations for solving least-squares problems, and the Singular Value Decomposition.

Chapter 10: Iterative Algorithms in Numerical Linear Algebra. This chapter studies methods for solving linear systems and computing eigenvalues that employ an iterative process to generate successive approximations converging to the true solution. Iterative algorithms for solving $A\mathbf{x} = \mathbf{b}$ include Richardson's method, the Jacobi method, and the Gauss–Seidel method. After developing some background on vector norms and matrix norms, we establish some theoretical results ensuring the convergence of these algorithms for certain classes of matrices. The chapter ends with an analysis of the power method for iteratively computing the largest eigenvalue of a matrix. Techniques for finding other eigenvalues (the shifted power method, inverse power method, and deflation) are also discussed.

Part IV: The Interplay of Geometry and Linear Algebra

Chapter 11: Affine Geometry and Convexity. Most introductions to linear algebra include an account of vector spaces, linear subspaces, the linear span of a subset of $\mathbb{R}^n$, linear independence, bases, dimension, and linear transformations. However, the parallel theory of affine sets and affine transformations is seldom covered in detail. This chapter starts by developing the basic concepts of affine geometry: affine sets, the affine span of a subset of $\mathbb{R}^n$, affine combinations, characterizations of affine sets (as translates of linear subspaces, as solution sets of linear equations, and as intersections of hyperplanes), affine independence, barycentric coordinates, and affine transformations. We continue with an introduction to the subject of convexity, discussing convex sets, convex hulls, convex combinations, Carathéodory's Theorem, simplexes, closed convex sets, convex cones, descriptions of convex sets as intersections of half-spaces, convex functions, epigraphs, Jensen's Inequality, and derivative tests for convex functions.

Chapter 12: Ruler and Compass Constructions. Why can a regular 17-gon be constructed with ruler and compass, while a regular 7-gon cannot? Which angles can be trisected with a ruler and compass? These questions and others are answered here in a linear-algebraic way by viewing field extensions as vector spaces and looking at dimensions. This material is often presented at the end of an abstract algebra course or is skipped altogether. Our development is quite elementary, requiring only basic facts about analytic geometry, plane geometry, fields, polynomials, and dimensions of vector spaces.

Chapter 13: Dual Vector Spaces. A fruitful idea in mathematics is the interplay between geometric spaces and structure-preserving functions defined on these spaces. This chapter studies the fundamental case of a finite-dimensional vector space V and the dual space V^* of linear functions from V to the field of scalars. We show how the concepts of *zero sets* and *annihilators* lead to inclusion-reversing bijections between the subspaces of V and the subspaces of V^*. Once this correspondence is understood, we explore how the choice of an inner product on V sets up an isomorphism between V and V^*. We also study the double dual V^{**} and its relation to V, along with dual maps and adjoint operators. The chapter concludes with a discussion of bilinear pairings and real and complex inner product spaces. We briefly comment on how the ideas in this chapter extend to Banach spaces and affine algebraic geometry.

Chapter 14: Bilinear Forms. A bilinear form on a vector space V generalizes the familiar dot product on $\mathbb{R}^n$. Just as the dot product enhances $\mathbb{R}^n$ with extra geometric information (such as the length of a vector, the distance between vectors, and the angle between vectors), a bilinear form confers similar geometric structure on the abstract space V. After defining bilinear forms and giving some examples, we study the matrix of a bilinear form, congruence of matrices, orthogonality, dual spaces, orthogonal complements, the radical of a bilinear form, totally isotropic subspaces, orthogonal maps, and reflections. We obtain structural results such as the representation of symmetric bilinear forms by diagonal matrices, the classification of alternating bilinear spaces as direct sums of hyperbolic spaces, Witt's Decomposition Theorem, Witt's Cancellation Theorem, and the generation of orthogonal groups by reflections.

Chapter 15: Metric Spaces and Hilbert Spaces. This chapter gives readers a glimpse of some aspects of infinite-dimensional linear algebra that play a fundamental role in modern analysis. We begin with a self-contained account of the material we need from analysis, including metric spaces, convergent sequences, closed sets, open sets, continuous functions, compactness, Cauchy sequences, and completeness. Next, we develop the basic theory of Hilbert spaces, exploring topics such as the Schwarz Inequality, the Parallelogram Law, orthogonal complements of closed subspaces, orthonormal sets, Bessel's Inequality, maximal orthonormal sets, Parseval's Equation, abstract Fourier expansions, the role of the Hilbert spaces $\ell_2(X)$, Banach spaces of continuous linear maps, the dual of a Hilbert space, and the adjoint of an operator. A major theme is the interaction between topological properties (especially completeness), algebraic computations, and geometric ideas such as convexity and orthogonality. Beyond a few brief allusions to L_2 spaces, this chapter does not assume any detailed knowledge of measure theory or Lebesgue integration.

Part V: Modules and Classification Theorems

Chapter 16: Finitely Generated Commutative Groups. A central theorem of group theory asserts that every finitely generated commutative group is isomorphic to a product of cyclic groups. This theorem is frequently stated, but not always proved, in first courses on abstract algebra. This chapter proves this fundamental result using linear-algebraic methods. We begin by reviewing basic concepts and definitions for commutative groups, stressing the analogy to corresponding concepts for vector spaces. The key idea in the proof of the classification theorem is to develop the analog of Gaussian elimination for integer-valued matrices and to use this algorithm to reduce such matrices to a certain canonical form. We also discuss elementary divisors and invariant factors and prove the uniqueness results for these objects. The uniqueness proof is facilitated by using partition diagrams to visualize finite groups whose size is a prime power. This chapter uses the very concrete setting of integer matrices to prepare the reader for more abstract algebraic ideas (universal mapping properties and the classification of finitely generated modules over principal ideal domains) covered in later chapters.

Chapter 17: Introduction to Modules. This chapter introduces the reader to some fundamental concepts in the theory of modules over arbitrary rings. We use the analogy to vector spaces (which are modules over fields) to motivate fundamental constructions for modules, while carefully pointing out that certain special facts about vector spaces fail to generalize to modules. In particular, the fact that not every module has a basis leads to a discussion of free modules and their properties. Specific topics covered include submodules, quotient modules, direct sums, generating sets, direct products, Hom modules, change of scalars, module homomorphisms, kernels, images, module isomorphism theorems, the Jordan–Hölder Theorem, length of a module, free modules, bases, Zorn's Lemma, existence of bases for modules over division rings, invariance of dimension for such modules, and an analogous result for modules using a commutative ring of scalars.

Chapter 18: Principal Ideal Domains, Modules over PIDs, and Canonical Forms. This chapter gives a detailed proof of one of the cornerstones of abstract algebra: the classification of all finitely generated modules over a PID. The chapter begins by proving the necessary algebraic properties of principal ideal domains, including the fact that every PID is a unique factorization domain. The classification proof is modeled upon the concrete case of commutative groups (covered in Chapter 16); a central idea is to develop a matrix reduction algorithm for matrices with entries in a PID. This algorithm changes any matrix into a diagonal matrix called a Smith normal form, which leads to the main structural results for modules. As special cases of the general theory, we deduce theorems on the rational canonical form and Jordan canonical form of linear operators and matrices. We also prove the uniqueness of the elementary divisors and invariant factors of a module, as well as the uniqueness of the various canonical forms for matrices.

Part VI: Universal Mapping Properties and Multilinear Algebra

Chapter 19: Introduction to Universal Mapping Properties. The concept of a universal mapping property (UMP) pervades linear and abstract algebra but is seldom mentioned in introductory treatments of these subjects. This chapter gives a careful and detailed introduction to this idea, starting with the UMP satisfied by a basis of a vector space. The UMP is formulated in several equivalent ways (as a diagram completion property, as a unique factorization property, and as a bijection between collections of functions). The concept is further developed by describing the UMPs characterizing free R-modules, quotient modules, direct products, and direct sums. This chapter serves as preparation for the next chapter on multilinear algebra.

Chapter 20: Universal Mapping Properties in Multilinear Algebra. The final chapter uses the idea of a universal mapping property (UMP) to organize the development of basic constructions in multilinear algebra. Topics covered include multilinear maps, alternating maps, symmetric maps, tensor products of modules, exterior powers, symmetric powers, and the homomorphisms of these structures induced by linear maps. We also discuss isomorphisms between tensor product modules, bases of tensor products and related modules, tensor products of matrices, the connection between exterior powers and determinants, and tensor algebras.

I wish you a rewarding journey through the boundless landscape of linear algebra!

Nicholas A. Loehr

Part I

Background on Algebraic Structures

1

Overview of Algebraic Systems

This chapter gives a rapid overview of the algebraic systems (such as groups, rings, fields, vector spaces, and algebras) that appear later in the book. After giving the axioms defining each of these systems and some basic examples, we describe some constructions (such as subspaces, product spaces, and quotient spaces) for building new systems from old ones. Then we discuss homomorphisms, which are structure-preserving maps between algebraic systems. The chapter concludes with a review of linear independence, spanning, basis, and dimension in the context of vector spaces over a field.

Unlike the rest of the book, this chapter is intended to be used as a review (for readers familiar with abstract algebra) or as a reference (for readers unfamiliar with the definitions), not as a leisurely introduction to the subject. To read the majority of the book, it suffices to know the meaning of the following terms defined in this chapter: commutative ring, field, vector space over a field, linear transformation, subspace, linearly independent list, basis, and dimension. Some further basic definitions regarding sets, functions, relations, and partially ordered sets appear in the Appendix.

1.1 Groups

Given any set S, a *binary operation* on S is a function $p : S \times S \to S$. We say that S is *closed under* this binary operation to emphasize that $p(a,b)$ is required to belong to S for all $a, b \in S$. There are many operation symbols that are used instead of the notation $p(a,b)$: for example, any of the expressions $a + b$, $a \cdot b$, $a \circ b$, $a \times b$, ab, or $[a,b]$ may be used to abbreviate $p(a,b)$ in different situations.

Next, we define some special properties that a binary operation p on S may or may not have. First, p is *commutative* iff[1] $p(a,b) = p(b,a)$ for all $a, b \in S$. Second, p is *associative* iff $p(a, p(b,c)) = p(p(a,b), c)$ for all $a, b, c \in S$. Third, S has an *identity element* relative to p iff there exists $e \in S$ (necessarily unique) such that $p(a,e) = a = p(e,a)$ for all $a \in S$. If such an identity e exists, we say $a \in S$ is *invertible* relative to p iff there exists $a' \in S$ with $p(a, a') = e = p(a', a)$. The element a' is called an *inverse* of a relative to p.

A *group* is a pair (G, p), where G is a set and p is an associative binary operation on G such that G has an identity element relative to p, and every element of G is invertible relative to p. Writing $p(a,b) = a \star b$, the group axioms take the form shown in Table 1.1.

For example, the set of all nonzero real numbers is a group under multiplication with identity $e = 1$. More generally, the set $\mathrm{GL}_n(\mathbb{R})$ of all $n \times n$ real-valued matrices A having nonzero determinant is a group under matrix multiplication, where the identity element is the $n \times n$ identity matrix. (To check the inverse axiom, we need the theorem from elementary linear algebra that says A is invertible iff $\det(A) \neq 0$. We prove a more general result in

---

[1]Throughout this text, the word *iff* is defined to mean "if and only if."

TABLE 1.1
Group axioms for $(G, \star)$.

> 1. For all $a, b \in G$, $a \star b$ is in G (closure).
> 2. For all $a, b, c \in G$, $a \star (b \star c) = (a \star b) \star c$ (associativity).
> 3. There exists $e \in G$ such that for all $a \in G$, $a \star e = a = e \star a$ (identity).
> 4. For all $a \in G$, there exists $a^{-1} \in G$ with $a \star a^{-1} = e = a^{-1} \star a$ (inverses).

Section 5.11.) For another example, let X be any set and $S(X)$ be the set of all bijections (one-to-one, onto functions) $f : X \to X$. Taking the binary operation to be composition of functions, we can verify that $(S(X), \circ)$ is a group. This group is discussed further in Chapter 2. The group operations in $\mathrm{GL}_n(\mathbb{R})$ and $S(X)$ are not commutative in general.

A *commutative group* (also called an *Abelian group*) is a group G in which the group operation does satisfy commutativity. Writing $p(a, b) = a + b$ for the group operation, the axioms for a commutative group (in additive notation) take the form shown in Table 1.2. We define *subtraction* in a commutative group $(G, +)$ by setting $a - b = a + (-b)$ for $a, b \in G$.

TABLE 1.2
Axioms for a commutative group $(G, +)$.

> 1. For all $a, b \in G$, $a + b$ is in G (closure).
> 2. For all $a, b, c \in G$, $a + (b + c) = (a + b) + c$ (associativity).
> 3. There exists $0_G \in G$ such that for all $a \in G$, $a + 0_G = a = 0_G + a$
>   (additive identity).
> 4. For all $a \in G$, there exists $-a \in G$ with $a + (-a) = 0_G = (-a) + a$
>   (additive inverses).
> 5. For all $a, b \in G$, $a + b = b + a$ (commutativity).

The familiar number systems (the integers $\mathbb{Z}$, the rational numbers $\mathbb{Q}$, the real numbers $\mathbb{R}$, and the complex numbers $\mathbb{C}$) are all commutative groups under addition. The set $\mathbb{R}^k$ of k-dimensional real vectors $\mathbf{v} = (v_1, \ldots, v_k)$ is a commutative group under vector addition; here, $(v_1, \ldots, v_k) + (w_1, \ldots, w_k) = (v_1 + w_1, \ldots, v_k + w_k)$ for all $v_i, w_i \in \mathbb{R}$. The identity element of this group is $\mathbf{0} = (0, \ldots, 0)$, and the inverse of $\mathbf{v} = (v_1, \ldots, v_k)$ is $-\mathbf{v} = (-v_1, \ldots, -v_k)$.

All of the commutative groups just mentioned are infinite. To give examples of finite groups, fix a positive integer n. Let $\mathbb{Z}_n = \{0, 1, 2, \ldots, n-1\}$ be the set of *integers modulo n*. Define *addition modulo n* as follows: given $a, b \in \mathbb{Z}_n$, let $a \oplus b = a + b$ if $a + b < n$ and $a \oplus b = a + b - n$ if $a + b \geq n$. Equivalently, $a \oplus b$ is the remainder when $a + b$ is divided by n, denoted $(a + b) \bmod n$. We may readily verify that $(\mathbb{Z}_n, \oplus)$ is a commutative group of size n.

1.2 Rings and Fields

A *ring* is a triple (R, p, q), where R is a set and p and q are binary operations on R (denoted by $p(a, b) = a + b$ and $q(a, b) = a \cdot b$) such that $(R, +)$ is a commutative group; the multiplication operation q is associative with an identity element 1_R; and the two *distributive laws* $a \cdot (b + c) = (a \cdot b) + (a \cdot c)$ and $(a + b) \cdot c = (a \cdot c) + (b \cdot c)$ hold for all $a, b, c \in R$. The ring axioms are written out in detail in Table 1.3. Note that some authors do not require, as we do, that all rings have a multiplicative identity.

TABLE 1.3
Ring axioms for $(R, +, \cdot)$.

1. $(R, +)$ is a commutative group. (See Table 1.2.)
2. For all $a, b \in R$, $a \cdot b$ is in R (closure under multiplication).
3. For all $a, b, c \in R$, $a \cdot (b \cdot c) = (a \cdot b) \cdot c$ (associativity of multiplication).
4. There is $1_R \in R$ so that for all $a \in R$, $a \cdot 1_R = a = 1_R \cdot a$ (multiplicative identity).
5. For all $a, b, c \in R$, $a \cdot (b + c) = (a \cdot b) + (a \cdot c)$ (left distributive law).
6. For all $a, b, c \in R$, $(a + b) \cdot c = (a \cdot c) + (b \cdot c)$ (right distributive law).

We say R is a *commutative ring* iff its multiplication operation is commutative (for all $a, b \in R$, $a \cdot b = b \cdot a$). We say R is an *integral domain* iff R is a commutative ring such that $0_R \neq 1_R$ and for all $a, b \in R$, if $a \neq 0_R$ and $b \neq 0_R$, then $a \cdot b \neq 0_R$. The last condition states that R has no *zero divisors*, which are nonzero ring elements whose product is zero. We say R is a *field* iff R is a commutative ring such that $0_R \neq 1_R$ and every nonzero element of R has a multiplicative inverse. When R is a field, the set of nonzero elements of R forms a commutative group under multiplication with identity 1_R. We may show that *every field is an integral domain.*

The number systems $\mathbb{Z}$, $\mathbb{Q}$, $\mathbb{R}$, and $\mathbb{C}$ are all commutative rings. $\mathbb{Q}$, $\mathbb{R}$, and $\mathbb{C}$ are fields, but $\mathbb{Z}$ is not, since most integers do not have multiplicative inverses in the set $\mathbb{Z}$ of integers. However, $\mathbb{Z}$ is an integral domain, since the product of two nonzero integers is never zero. For $n \geq 1$, we can make the set $\mathbb{Z}_n$ of integers modulo n into a commutative ring of size n. Addition here is addition mod n, and multiplication is the operation $\otimes$ defined (for $a, b \in \mathbb{Z}_n$) by letting $a \otimes b$ be the remainder when ab is divided by n, denoted $ab \bmod n$. We can verify that the ring $(\mathbb{Z}_n, \oplus, \otimes)$ is a field iff n is a prime number (an integer larger than 1 that is divisible only by itself and 1).

Another example of a commutative ring is the set $\mathbb{R}[x]$ of polynomials in one variable with real coefficients, with the ordinary rules for adding and multiplying polynomials. This ring is not a field, since the only polynomials whose inverses are also polynomials are the nonzero constants. This example can be generalized in several ways. We can replace the real coefficients with coefficients in an arbitrary field (or even an arbitrary ring). Or, we can allow polynomials in more than one variable. For more on polynomials, read Chapter 3.

An example of a non-commutative ring is provided by the set of all $n \times n$ real matrices under matrix addition and matrix multiplication, where $n \geq 2$ is a fixed integer. More generally, for any ring R, we can consider the set $M_n(R)$ of all $n \times n$ matrices with entries in R. Using the same rules for adding and multiplying matrices as in the real case, this set of matrices becomes a ring that is almost never commutative. For more details, read Chapter 4.

1.3   Vector Spaces

Most introductions to linear algebra study real vector spaces, where we can add two vectors or multiply a vector by a real number (also called a scalar). For more advanced investigations, it is helpful to replace the real scalars with elements of more general number systems (such as rings or fields). To ensure that the nice properties of real vector spaces carry over to the more general setting, it turns out that we need to have multiplicative inverses for all nonzero scalars. This leads to the following definition of a vector space over a field, which generalizes the concept of a real vector space.

Let $(F, +, \cdot)$ be a field, whose elements are called *scalars*. A *vector space over* F (also called an *F-vector space*) is a triple $(V, \boldsymbol{+}, s)$, where $(V, \boldsymbol{+})$ is a commutative group, and $s : F \times V \to V$ is a *scalar multiplication operation* satisfying the axioms listed in Table 1.4. Here, we use two different fonts for the addition $+$ in F and the addition $\boldsymbol{+}$ in V, but later, we use the same symbol $+$ for both. Similarly, using boldface letters for vectors $\mathbf{v}$ in V is not required. We almost always use juxtaposition to abbreviate various multiplication operations, writing $s(c, \mathbf{v}) = c\mathbf{v}$ and $c \cdot d = cd$ for $c, d \in F$ and $\mathbf{v} \in V$. If we replace the field F by any ring R, we obtain the definition of an *R-module*. Modules are studied later in the book, starting in Chapter 17.

TABLE 1.4
Axioms for a vector space $(V, \boldsymbol{+}, s)$ over a field $(F, +, \cdot)$.

---

1. $(V, \boldsymbol{+})$ is a commutative group. (See Table 1.2.)
2. For all $c \in F$ and $\mathbf{v} \in V$, $s(c, \mathbf{v}) = c\mathbf{v}$ is in V (closure under scalar multiplication).
3. For all $c, d \in F$ and $\mathbf{v} \in V$, $(c + d)\mathbf{v} = c\mathbf{v} \boldsymbol{+} d\mathbf{v}$
   (distributive law for scalar addition).
4. For all $c \in F$ and $\mathbf{v}, \mathbf{w} \in V$, $c(\mathbf{v} \boldsymbol{+} \mathbf{w}) = c\mathbf{v} \boldsymbol{+} c\mathbf{w}$
   (distributive law for vector addition).
5. For all $c, d \in F$ and $\mathbf{v} \in V$, $c(d\mathbf{v}) = (c \cdot d)\mathbf{v}$ (associativity of scalar multiplication).
6. For all $\mathbf{v} \in V$, $1_F \mathbf{v} = \mathbf{v}$ (identity for scalar multiplication).

The most well-known example of a real vector space is the set $\mathbb{R}^k$ of vectors $\mathbf{v} = (v_1, \ldots, v_k)$ with each $v_i \in \mathbb{R}$. As discussed earlier, $\mathbb{R}^k$ is a commutative group under vector addition. The scalar multiplication operation is given by $c\mathbf{v} = (cv_1, cv_2, \ldots, cv_k)$ for $c \in \mathbb{R}$ and $\mathbf{v} \in \mathbb{R}^k$. This example generalizes readily to the case where scalars come from an arbitrary field F. We define F^k to be the set of k-tuples $\mathbf{v} = (v_1, v_2, \ldots, v_k)$ with each $v_i \in F$. Given such a $\mathbf{v}$ and $\mathbf{w} = (w_1, w_2, \ldots, w_k) \in F^k$ and $c \in F$, define

$$\mathbf{v} \boldsymbol{+} \mathbf{w} = (v_1 + w_1, v_2 + w_2, \ldots, v_k + w_k);$$

$$c\mathbf{v} = (cv_1, cv_2, \ldots, cv_k).$$

The operations on the right sides of these equations are the given addition and multiplication operations in F. Using these definitions, it is tedious but straightforward to confirm that F^k is an F-vector space.

Some further examples of F-vector spaces, which we study in detail later in this book, are the set $M_{m,n}(F)$ of $m \times n$ matrices with entries in F (Chapter 4); the set of all functions from an arbitrary set X into F (Chapter 4); the set of all linear maps from one F-vector space to another (Chapters 6 and 13); any field K containing F as a subfield (Chapter 12); and the set of polynomials in one or more variables with coefficients in F (Chapter 3).

We conclude with the definition of an algebra over a field F, although this concept is only used in a few places in the book. For a field F, an *F-algebra* is a structure $(A, +, \star, s)$ such that $(A, +, \star)$ is a ring, $(A, +, s)$ is an F-vector space, and the ring multiplication and scalar multiplication are related by the identities

$$c(\mathbf{v} \star \mathbf{w}) = (c\mathbf{v}) \star \mathbf{w} = \mathbf{v} \star (c\mathbf{w}) \qquad \text{for all } c \in F \text{ and } \mathbf{v}, \mathbf{w} \in A.$$

(More precisely, A is an *associative F-algebra with identity*.) Some examples of F-algebras are the set $M_n(F)$ of $n \times n$ matrices with entries in F; the set $\operatorname{End}_F(V)$ of linear maps from a fixed F-vector space V to itself; and the set $F[x_1, \ldots, x_n]$ of polynomials in n variables with coefficients in F. Chapter 6 explores the close relationship between the first two of these algebras.

1.4 Subsystems

For each of the algebraic systems mentioned so far, we can construct new algebraic systems of the same kind using a variety of algebraic constructions. Some recurring constructions in abstract algebra are subsystems, direct products, and quotient systems. The next few sections review how these constructions are defined for groups, rings, fields, vector spaces, and algebras. For a more extensive discussion covering the case of modules, read Chapter 17.

Generally speaking, a *subsystem* of a given algebraic system is a subset that is closed under the relevant operations. For instance, a *subgroup* of a group $(G, \star)$ is a subset H of G such that for all $a, b \in H$, $a \star b \in H$; the identity e of G is in H; and for all $a \in H$, $a^{-1} \in H$. The first condition says that the subset H of G is *closed under the group operation* $\star$; the second condition says that H is *closed with respect to the identity*; and the third condition says that H is *closed under inverses*. When these closure conditions hold, it follows that the set H becomes a group if we restrict the binary operation $\star : G \times G \to G$ to the domain $H \times H$. A similar comment applies to the constructions of other types of subsystems below. H is a *normal* subgroup of G iff H is a subgroup such that $a \star h \star a^{-1} \in H$ for all $a \in G$ and all $h \in H$. The quantity $a \star h \star a^{-1}$ is called a *conjugate of h in G*. So the condition for normality can be stated by saying that H is *closed under conjugation*. Every subgroup of a commutative group G is automatically normal in G.

A *subring* of a ring $(R, +, \cdot)$ is a subset S of R such that $0_R \in S$, $1_R \in S$, and for all $a, b \in S$, $a + b \in S$, $-a \in S$, and $a \cdot b \in S$. In other words, S is a subring of R iff S is an additive subgroup of $(R, +)$ that is closed under multiplication and under the multiplicative identity. S is a *left ideal* of R iff S is an additive subgroup such that $a \cdot s \in S$ for all $a \in R$ and $s \in S$; we say that S is *closed under left multiplication by elements of R*. Similarly, S is a *right ideal* of R iff S is an additive subgroup such that $s \cdot a \in S$ for all $a \in R$ and $s \in S$ (i.e., S is *closed under right multiplication by elements of R*). S is an *ideal* of R (also called a *two-sided ideal* for emphasis) iff S is both a left ideal and a right ideal of R. Under our conventions, the different types of ideals in R need not be subrings, since the ideals are not required to contain the multiplicative identity 1_R. In fact, any left, right, or two-sided ideal of R containing 1_R must be the entire ring.

If F is a field, a *subfield* of F is a subring E of F such that for all nonzero $a \in E$, a^{-1} is in E. So, a subfield must contain the 0 and 1 of the original field and be closed under addition, additive inverses, multiplication, and multiplicative inverses for nonzero elements.

Let V be a vector space over a field F. A subset W of V is called a *subspace* of V iff: $\mathbf{0}_V \in W$; for all $\mathbf{v}, \mathbf{w} \in W$, $\mathbf{v} + \mathbf{w} \in W$; and for all $c \in F$ and $\mathbf{w} \in W$, $c\mathbf{w} \in W$. So, a subset W of V is a vector subspace of V iff W is closed under zero, vector addition, and scalar multiplication. The same definition applies if V is a module over a ring R, but now subspaces are called *submodules* or *R-submodules*.

Let A be an algebra over a field F. A subset B of A is a *subalgebra* iff B is both a subring of the ring A and a subspace of the vector space A. So, B contains 0_A and 1_A, and B is closed under addition, additive inverses, ring multiplication, and scalar multiplication.

Here are some examples to illustrate the preceding definitions. For each fixed integer n, the set $n\mathbb{Z} = \{nk : k \in \mathbb{Z}\}$ is a subgroup of the additive group $(\mathbb{Z}, +)$, which is a normal subgroup of $\mathbb{Z}$ because $\mathbb{Z}$ is commutative. Each set $n\mathbb{Z}$ is also an ideal of the ring $\mathbb{Z}$, but not a subring except when $n = \pm 1$. We can check, using the division algorithm for integers, that all subgroups of $\mathbb{Z}$ (and hence all ideals of $\mathbb{Z}$) are of this form. Given any multiplicative group $(G, \star)$ and any fixed $g \in G$, the set $\langle g \rangle = \{g^n : n \in \mathbb{Z}\}$ of powers of g is a subgroup of G, called the *cyclic subgroup generated by g*. This subgroup need not be normal in G. Given a commutative ring R and a fixed $b \in R$, the set $Rb = \{r \cdot b : r \in R\}$ is an ideal of R, called the *principal ideal generated by b*. In general, this does not coincide with $\mathbb{Z}b = \{nb : n \in \mathbb{Z}\}$, which is the additive subgroup of $(R, +)$ generated by b. $\mathbb{Z}$ is a subring of $\mathbb{R}$ that is not a subfield, since $\mathbb{Z}$ is not closed under multiplicative inverses. $\mathbb{Q}$ is a subfield of $\mathbb{R}$, as is the set $\mathbb{Q}(\sqrt{2})$ of all real numbers of the form $a + b\sqrt{2}$, where $a, b \in \mathbb{Q}$. The set of polynomials of degree at most 3 (together with zero) is a subspace of the vector space of all polynomials with real coefficients. For any field F, the set $\{(t, t + u, -u) : t, u \in F\}$ is a subspace of the F-vector space F^3. The set of upper-triangular $n \times n$ matrices is a subalgebra of the $\mathbb{R}$-algebra $M_n(\mathbb{R})$ of real $n \times n$ matrices.

1.5 Product Systems

Next, we consider the direct product construction for vector spaces. Suppose $V_1, V_2, \ldots, V_n$ are given vector spaces over the same field F. The *product set* $V = V_1 \times V_2 \times \cdots \times V_n$ consists of all ordered n-tuples $\mathbf{v} = (\mathbf{v}_1, \mathbf{v}_2, \ldots, \mathbf{v}_n)$ with each $\mathbf{v}_i \in V_i$. We can turn this product set into an F-vector space by defining addition and scalar multiplication as follows. Given $\mathbf{v} = (\mathbf{v}_1, \ldots, \mathbf{v}_n)$ and $\mathbf{w} = (\mathbf{w}_1, \ldots, \mathbf{w}_n)$ in V and given $c \in F$, define $\mathbf{v} + \mathbf{w} = (\mathbf{v}_1 + \mathbf{w}_1, \ldots, \mathbf{v}_n + \mathbf{w}_n)$ and $c\mathbf{v} = (c\mathbf{v}_1, \ldots, c\mathbf{v}_n)$. In these definitions, the operations in position i are the sum and scalar multiplication operations in the vector space V_i. It is tedious but routine to check the vector space axioms for V. In particular, the additive identity of V is $\mathbf{0}_V = (\mathbf{0}_{V_1}, \mathbf{0}_{V_2}, \ldots, \mathbf{0}_{V_n})$, and the additive inverse of $\mathbf{v} = (\mathbf{v}_1, \ldots, \mathbf{v}_n) \in V$ is $-\mathbf{v} = (-\mathbf{v}_1, \ldots, -\mathbf{v}_n)$. The set V with these operations is called the *direct product* of the F-vector spaces $V_1, V_2, \ldots, V_n$. Note that the F-vector space F^n is a special case of this construction obtained by taking every V_i to be F.

There are analogous constructions for the direct products of groups, rings, modules, and algebras. In each case, all algebraic operations are defined one component at a time. For example, given positive integers $n_1, n_2, \ldots, n_k$, the direct product $G = \mathbb{Z}_{n_1} \times \mathbb{Z}_{n_2} \times \cdots \times \mathbb{Z}_{n_k}$ is a commutative group of size $n_1 n_2 \cdots n_k$, with operation

$$(a_1, a_2, \ldots, a_k) + (b_1, b_2, \ldots, b_k) = ((a_1 + b_1) \bmod n_1, (a_2 + b_2) \bmod n_2, \ldots, (a_k + b_k) \bmod n_k)$$

for $a_i, b_i \in \mathbb{Z}_{n_i}$. In Chapter 16 we prove that every finite commutative group is isomorphic to a direct product of this form. (Isomorphisms are defined in §1.7.)

1.6   Quotient Systems

This section describes the notion of a quotient system of a given algebraic system. Quotient constructions are more subtle and less intuitive than the constructions of subsystems and product systems, but they play a prominent role in abstract algebra and other parts of mathematics.

The simplest instance of the quotient construction occurs when $(G, +)$ is a commutative group with subgroup H. We build a new commutative group, denoted G/H, and called the *quotient group of G by H*. Intuitively, the group G/H is a simplification of G obtained by discarding information contained in H. For instance, if G is the additive group $\mathbb{Z}$ of integers and H is the subgroup $n\mathbb{Z}$ of multiples of n, then $\mathbb{Z}/n\mathbb{Z}$ is isomorphic to the group $\mathbb{Z}_n$ of integers mod n. Intuitively, the formation of the quotient group discards multiples of n and focuses attention on the remainders when various integers are divided by n.

We now give the details of the construction of G/H, where $(G, +)$ is a commutative group with subgroup H. For each $a \in G$, define the *coset of H represented by a* to be $a + H = \{a + h : h \in H\}$. Each coset $a + H$ is a certain subset of G obtained by translating the subset H by adding the fixed element a to each h in H. Define the *quotient set G/H* to be $\{a + H : a \in G\}$, so G/H is the set of all cosets of H in G.

A critical point is the fact that each coset (element of G/H) can have several different names. In other words, there are often multiple ways of writing a particular coset in the form $a+H$ for some $a \in G$. The next theorem gives a precise criterion for when two elements of G give us different names for the same coset.

Coset Equality Theorem. Let $(G, +)$ be a commutative group with subgroup H. For all $a, b \in G$, $a + H = b + H$ if and only if $a - b \in H$.

Proof. Fix $a, b \in G$ satisfying $a + H = b + H$. Since $0_G \in H$, we have $a = a + 0_G \in a + H$. Since the set $a + H$ equals the set $b + H$ by assumption, we have $a \in b + H$, which means $a = b + h$ for some $h \in H$. So $a - b = h$ belongs to H, as needed.

For the converse, fix $a, b \in G$ satisfying $a - b \in H$; we must prove $a + H = b + H$. First we prove the set inclusion $a + H \subseteq b + H$. Fix $x \in a + H$. We can write $x = a + h_1$ for some $h_1 \in H$. Observe that $x = a + h_1 = b + (a - b) + h_1 = b + k$, where $k = (a - b) + h_1$ is the sum of two elements of H, hence is in H. So $x \in b + H$. To prove the opposite set inclusion $b + H \subseteq a + H$, fix $y \in b + H$. Write $y = b + h_2$ for some $h_2 \in H$. Then notice that $y = a + [-(a - b) + h_2] \in a + H$, where $a - b \in H$, $-(a - b) \in H$, and $-(a - b) + h_2 \in H$ because H is a subgroup. So $y \in a + H$ as required. $\square$

Next, we prove that G *is the disjoint union of the distinct cosets of H.* On one hand, every coset of H is a subset of G (by the closure axiom of G). On the other hand, for any $a \in G$, a is in the coset $a + H$ since $a = a + 0_G$ with $0_G \in H$. To see that any two distinct cosets of H must be disjoint, suppose $a + H$ and $b + H$ are two cosets of H such that some element c is in both cosets (here $a, b, c \in G$). Write $c = a + h = b + k$ for some $h, k \in H$. Then $a - b = k - h \in H$, so the Coset Equality Theorem gives $a + H = b + H$.

Here is an example. In the group $(\mathbb{Z}_{12}, \oplus)$, consider the subgroup $H = \{0, 3, 6, 9\}$. Three of the cosets of H are $0 + H = \{0, 3, 6, 9\}$, $1 + H = \{1, 4, 7, 10\}$, and $2 + H = \{2, 5, 8, 11\}$. In fact, these are all of the distinct cosets of H, since (for instance) $3 + H = \{3, 6, 9, 0\} = 0 + H$, $7 + H = 1 + H$ because $7 - 1 = 6 \in H$, and so on. Note that any element b in a coset $a + H$ can be used to give a new name $b + H$ for that coset. This holds since $b = a + h$ for some $h \in H$, so $b - a = h \in H$, so $a + H = b + H$ by the Coset Equality Theorem. For instance, the four possible names of the coset $\{2, 5, 8, 11\}$ are $2 + H$, $5 + H$, $8 + H$, and $11 + H$. The quotient set $\mathbb{Z}_{12}/H = \{a + H : a \in \mathbb{Z}_{12}\} = \{\{0, 3, 6, 9\}, \{1, 4, 7, 10\}, \{2, 5, 8, 11\}\}$ consists of three distinct cosets. The set $\mathbb{Z}_{12}$ is indeed the disjoint union of these cosets.

Returning to the general case, the next step is to define a binary addition operation p on G/H by setting $p(a + H, b + H) = (a + b) + H$ for all $a, b \in G$. If we use a circled plus symbol $\oplus$ to denote the new operation p, the definition reads: $(a + H) \oplus (b + H) = (a + b) + H$. Now a new subtlety emerges: our definition of the sum of two elements of G/H depends on the particular names $a + H$ and $b + H$ that we have chosen to represent these elements. If we change from these names to other names of the same two cosets, how can we be sure that the output is not affected? To answer this question, suppose $a, a_1, b, b_1 \in G$ satisfy $a + H = a_1 + H$ and $b + H = b_1 + H$. Using the original names $a + H$ and $b + H$, the sum of these cosets is defined to be $(a+b)+H$. Using the new names a_1+H and b_1+H for these same cosets, the sum of the cosets is calculated as $(a_1 + b_1) + H$. These two answers might appear to be different, but we can prove they are, in fact, different names for the same coset. To see this, note from the Coset Equality Theorem that $a - a_1 \in H$ and $b - b_1 \in H$. Therefore, since the subgroup H is closed under addition, $(a + b) - (a_1 + b_1) = (a - a_1) + (b - b_1) \in H$. (This calculation also requires commutativity of G.) Another application of the Coset Equality Theorem gives $(a + b) + H = (a_1 + b_1) + H$, as needed. We summarize this calculation by saying that the binary operation $\oplus$ is *well-defined* (or *single-valued*). We must perform such a check every time a binary operation or function is defined by a formula involving the name of an object that has several possible names.

With this technical issue out of the way, we verify the commutative group axioms for $(G/H, \oplus)$ as follows. Consider three arbitrary elements in G/H, which we can name as $a + H$, $b + H$, and $c + H$ for some $a, b, c \in G$. To check the closure of G/H under $\oplus$, note $(a + H) \oplus (b + H) = (a + b) + H$ does belong to G/H, since $a + b \in G$ by the closure axiom for G. To check the associativity of $\oplus$, compute

$$(a + H) \oplus ((b + H) \oplus (c + H)) = (a + H) \oplus ((b + c) + H) = (a + (b + c)) + H$$

$$= ((a + b) + c) + H = ((a + b) + H) \oplus (c + H) = ((a + H) \oplus (b + H)) \oplus (c + H).$$

The third equality holds by associativity of $+$ in G, and the other equalities hold by definition of $\oplus$. Similarly, $\oplus$ is commutative because $(a + H) \oplus (b + H) = (a + b) + H = (b + a) + H = (b + H) \oplus (a + H)$. This calculation uses commutativity of $+$ in G. The identity element of G/H is the coset $0 + H = H$, where 0 is the additive identity element in G, because

$$(a + H) \oplus (0 + H) = (a + 0) + H = a + H = (0 + a) + H = (0 + H) \oplus (a + H)$$

by the identity axiom for $(G, +)$. Finally, the additive inverse of $a + H$ relative to $\oplus$ is $(-a) + H$, since the inverse axiom in $(G, +)$ gives $(a + H) \oplus ((-a) + H) = (a + (-a)) + H = 0 + H = ((-a) + a) + H = ((-a) + H) \oplus (a + H)$. Note that $-a \in G$, and therefore the proposed inverse $(-a) + H$ does belong to the set G/H.

Continuing our earlier example where $G = \mathbb{Z}_{12}$ and $H = \{0, 3, 6, 9\}$, recall that $G/H = \{0 + H, 1 + H, 2 + H\}$. We have $(1 + H) + (2 + H) = (1 + 2) + H = 3 + H$, and another name for this answer is $0 + H$. Furthermore, $(2 + H) + (2 + H) = 4 + H = 1 + H$. Comparing the addition table for G/H to the addition table for the group $(\mathbb{Z}_3, \oplus)$, we can check that the groups $\mathbb{Z}_{12}/H$ and $\mathbb{Z}_3$ are isomorphic (as defined in §1.7). Similarly, it can be shown that for all $n \geq 1$, the quotient group $\mathbb{Z}/n\mathbb{Z}$ is isomorphic to the group $(\mathbb{Z}_n, \oplus)$ defined in §1.1.

The quotient construction can be generalized to other types of algebraic structures. First, we can replace the commutative group $(G, +)$ and its subgroup H by an arbitrary group $(G, \star)$ and a subgroup H that is normal in G. For each $a \in G$, we define the *left coset* $a \star H = \{a \star h : h \in H\}$, and we let $G/H = \{a \star H : a \in G\}$. In this setting, the *Left Coset Equality Theorem* states that for all $a, b \in G$, $a \star H = b \star H$ iff $a^{-1} \star b \in H$

iff $b^{-1} \star a \in H$ (note the inverse occurs to the left of the star in each case). The binary operation p on the set G/H of left cosets of H is defined by $p(a \star H, b \star H) = (a \star b) \star H$ for all $a, b \in G$. The verification that p is well-defined (Exercise 49) depends critically on H being a normal subgroup of G. Once this has been done, it is straightforward to prove (as above) that $(G/H, p)$ is a group. A similar construction could be executed using *right cosets* $H \star a = \{h \star a : h \in H\}$, which satisfy $H \star a = H \star b$ iff $a \star b^{-1} \in H$. However, the normality of H implies that every left coset $a \star H$ equals the right coset $H \star a$ (Exercise 50), so that the resulting quotient group is the same as before.

Second, we can replace G and H by a ring $(R, +, \cdot)$ and an ideal I of R. The set of (additive) cosets $R/I = \{a + I : a \in R\}$ is already known to be a commutative group, since $(R, +)$ is a commutative group with subgroup I. We introduce a second binary operation $\bullet$ on R/I by setting $(a + I) \bullet (b + I) = (a \cdot b) + I$ for all $a, b \in R$. We can check that $\bullet$ is well-defined using the assumption that I is a (two-sided) ideal. It is then routine to verify that $(R/I, \oplus, \bullet)$ is a ring with additive identity $0_R + I$ and multiplicative identity $1_R + I$. Also, R/I is commutative if R is commutative. The ring R/I is called the *quotient ring* of R by the ideal I.

Third, we can replace G and H by an F-vector space $(V, +, \cdot)$ and a subspace W, where F is a field. We construct the *quotient vector space* $(V/W, \oplus, \bullet)$ as follows. Since $(V, +)$ is a commutative group with subgroup W, we already have an additive commutative group $(V/W, \oplus)$, where $(\mathbf{x} + W) \oplus (\mathbf{y} + W) = (\mathbf{x} + \mathbf{y}) + W$ for all $\mathbf{x}, \mathbf{y} \in V$. We introduce a new scalar multiplication $\bullet$ (which is a function from $F \times V/W$ to V/W) by setting $c \bullet (\mathbf{v} + W) = (c \cdot \mathbf{v}) + W$ for all $c \in F$ and all $\mathbf{v} \in V$. We must check that $\bullet$ is well-defined. In the formula defining $\bullet$, elements of F do not have multiple names, but elements of V/W do. So, we fix $c \in F$ and $\mathbf{x}, \mathbf{y} \in V$, assume $\mathbf{x} + W = \mathbf{y} + W$, and prove $c \bullet (\mathbf{x} + W) = c \bullet (\mathbf{y} + W)$. In other words, we must prove $(c \cdot \mathbf{x}) + W = (c \cdot \mathbf{y}) + W$. By the Coset Equality Theorem, $\mathbf{x} - \mathbf{y} \in W$. Since W is a subspace of V, we get $c \cdot (\mathbf{x} - \mathbf{y}) \in W$. So $(c \cdot \mathbf{x}) - (c \cdot \mathbf{y}) \in W$. We deduce $(c \cdot \mathbf{x}) + W = (c \cdot \mathbf{y}) + W$ by the Coset Equality Theorem.

We can now verify the vector space axioms for $(V/W, \oplus, \bullet)$. We already proved that $(V/W, \oplus)$ is a commutative group. To verify the distributive law for vector addition in this space, fix $c \in F$ and $\mathbf{v}, \mathbf{x} \in V$, and calculate

$$c \bullet ((\mathbf{v} + W) \oplus (\mathbf{x} + W)) = c \bullet ((\mathbf{v} + \mathbf{x}) + W) = (c \cdot (\mathbf{v} + \mathbf{x})) + W.$$

By the known distributive law for vector addition in the original space V, the calculation continues:

$$(c \cdot (\mathbf{v} + \mathbf{x})) + W = (c \cdot \mathbf{v} + c \cdot \mathbf{x}) + W = (c \cdot \mathbf{v} + W) \oplus (c \cdot \mathbf{x} + W) = c \bullet (\mathbf{v} + W) \oplus c \bullet (\mathbf{x} + W),$$

and the axiom is verified. The remaining axioms are checked similarly. The construction of the quotient vector space generalizes at once to the case where R is a ring and V is an R-module with submodule W (see Chapter 17).

1.7 Homomorphisms

The concept of homomorphism allows us to compare algebraic structures. There are many different kinds of homomorphisms, one for each kind of algebraic system. Intuitively, a homomorphism of an algebraic system is a function from one system to another system of the same kind that preserves all relevant operations and structure.

For instance, if $(G, \star)$ and $(K, *)$ are groups, a *group homomorphism* from G to K is a function $T : G \to K$ such that $T(x \star y) = T(x) * T(y)$ for all $x, y \in G$. It follows from this condition that $T(e_G) = e_K$, where e_G is the identity element of G and e_K is the identity element of K. We can also show $T(x^{-1}) = T(x)^{-1}$ for all $x \in G$, where x^{-1} is the inverse of x in G and $T(x)^{-1}$ is the inverse of $T(x)$ in K. More generally, $T(x^n) = T(x)^n$ for all $x \in G$ and all $n \in \mathbb{Z}$. We say that the group homomorphism T *preserves* the group operations, the identity, inverses, and powers of group elements. In the case where G and K are commutative groups with operations written in additive notation, the definition of a group homomorphism becomes $T(x + y) = T(x) + T(y)$ for all $x, y \in G$. Now $T(0_G) = 0_K$, $T(-x) = -T(x)$, and $T(nx) = nT(x)$ for all $x \in G$ and all $n \in \mathbb{Z}$. (The notation nx denotes the sum of n copies of x for $n > 0$, the sum of $|n|$ copies of $-x$ for $n < 0$, or 0_G for $n = 0$.)

Analogously, given two rings R and S, a *ring homomorphism* is a function $T : R \to S$ such that for all $x, y \in R$, $T(x + y) = T(x) + T(y)$, $T(xy) = T(x)T(y)$, and $T(1_R) = 1_S$. It follows from these conditions that $T(x^n) = T(x)^n$ for all $x \in R$ and all integers $n \geq 0$. This formula also holds for negative integers n when x is an invertible element of R.

Next, suppose V and W are vector spaces over a field F. An *F-linear map* (also called a *vector space homomorphism* or *linear transformation*) is a function $T : V \to W$ such that $T(\mathbf{v} + \mathbf{z}) = T(\mathbf{v}) + T(\mathbf{z})$ and $T(c\mathbf{v}) = cT(\mathbf{v})$ for all $\mathbf{v}, \mathbf{z} \in V$ and all $c \in F$. The same definition applies if V and W are modules over a ring R; in this case, we call T an *R-linear map* or *R-module homomorphism*. Finally, an *algebra homomorphism* is a map between two F-algebras that is both a ring homomorphism and a vector space homomorphism.

Let $T : X \to Y$ be a homomorphism of any of the types defined above. We call T an *isomorphism* iff T is a bijective (one-to-one and onto) function from X to Y. In this case, T has a two-sided inverse function $T^{-1} : Y \to X$. We may check that T^{-1} is always a homomorphism of the same type as T and is also an isomorphism. Furthermore, the composition of homomorphisms (resp. isomorphisms) is a homomorphism (resp. isomorphism), and the identity map on a given algebraic structure X is an isomorphism for that type of structure. We write $X \cong Y$ if there exists an isomorphism between X and Y. The preceding remarks show that $\cong$ is an equivalence relation on any set of algebraic structures of a given kind. Specifically, for any X, Y, Z in the given set, $X \cong X$ (reflexivity); if $X \cong Y$, then $Y \cong X$ (symmetry); and if $X \cong Y$ and $Y \cong Z$, then $X \cong Z$ (transitivity).

Here are some examples of homomorphisms. If H is a subgroup of G, the inclusion map $i : H \to G$ given by $i(h) = h$ for all $h \in H$ is an injective group homomorphism; i is an isomorphism iff $H = G$. If H is normal in G, the projection map $p : G \to G/H$, given by $p(x) = x \star H$ for all $x \in G$, is a surjective group homomorphism; p is an isomorphism iff $H = \{e_G\}$. Similarly, if R is a ring with ideal I, the projection $p : R \to R/I$ given by $p(x) = x + I$ for $x \in R$ is a surjective ring homomorphism. If V is an F-vector space with subspace W, we obtain an analogous projection $p : V \to V/W$, which is F-linear. In the case of groups, vector spaces, and modules, the injective maps $j_i : V_i \to V_1 \times \cdots \times V_i \times \cdots \times V_n$ sending $x_i \in V_i$ to $(0, \ldots, x_i, \ldots, 0)$ are homomorphisms. However, this statement does not hold for products of rings, since $j_i(1_{V_i})$ is almost never the multiplicative identity of the product ring. On the other hand, for all the types of algebraic systems discussed, the projections $q_i : V_1 \times \cdots \times V_i \times \cdots \times V_n \to V_i$ sending $(x_1, \ldots, x_n)$ to x_i are surjective homomorphisms.

We now discuss kernels, images, and the Fundamental Homomorphism Theorem. Suppose $T : X \to Y$ is a function mapping a set X into an additive group Y with identity element 0_Y. The *kernel* of T is $\ker(T) = \{x \in X : T(x) = 0_Y\} \subseteq X$, and the *image* of T is $\operatorname{img}(T) = \{T(x) : x \in X\} \subseteq Y$. When T is a linear map between vector spaces, $\ker(T)$ is also called the *null space* of T, and $\operatorname{img}(T)$ is also called the *range* of T. The following facts are readily checked. (i) If T is a group homomorphism, then $\ker(T)$ is a normal subgroup of X, and $\operatorname{img}(T)$ is a subgroup of Y. (ii) If T is a ring homomorphism, then $\ker(T)$ is an ideal of X and img(T) is a subring of Y. (iii) If T is a linear transformation of F-vector spaces, then ker(T) is a subspace of X, and img(T) is a subspace of Y. (iv) For all types of homomorphisms considered here, T is injective iff ker(T) = $\{0_X\}$. (v) T is surjective iff img(T) = Y.

Fundamental Homomorphism Theorem for Commutative Groups. Let X and Y be commutative groups with operations written in additive notation. Suppose $T : X \to Y$ is a group homomorphism with kernel K and image I. There is a group isomorphism $T' : X/K \to I$ given by $T'(x + K) = T(x)$ for all $x \in X$.

Proof. Step 1. We check that T' is well-defined. Let a given element of X/K have two names $x + K = w + K$, where $x, w \in X$. We must check that $T'(x + K) = T'(w + K)$, which (by definition of T') is the same as proving $T(x) = T(w)$. By the Coset Equality Theorem, $x + K = w + K$ implies $x - w \in K$. Now $0_Y = T(x - w) = T(x) - T(w)$ by definition of the kernel and since the homomorphism T preserves subtraction. So $T(x) = T(w)$, as needed.

Step 2. We check that T' is a group homomorphism. Fix two elements of X/K, say $u + K$ and $x + K$ with $u, x \in X$. Compute

$$T'((u+K)+(x+K)) = T'((u+x)+K) = T(u+x) = T(u)+T(x) = T'(u+K)+T'(x+K).$$

From left to right, the equalities are true by definition of $+$ in X/K, by definition of T', by T being a homomorphism, and by definition of T'.

Step 3. We check that T' is surjective (onto). Note first that T' does map *into* the codomain I. This holds since each input to T' has the form $x + K$ for some $x \in X$, and $T'(x + K) = T(x)$ is in the image of T, which is the set I. Conversely, given any $y \in I$, the definition of image shows that $y = T(u)$ for some $u \in X$, hence $y = T'(u + K)$ for some $u + K \in X/K$.

Step 4. We check that T' is injective (one-to-one). Fix $u, x \in X$ satisfying $T'(u + K) = T'(x + K)$; we must prove $u + K = x + K$. By definition of T', we have $T(u) = T(x)$. So $T(u) - T(x) = 0_Y$, and $T(u - x) = 0_Y$ (homomorphisms preserve subtraction). This means $u - x$ belongs to the kernel of T, namely $u - x \in K$. By the Coset Equality Theorem, $u + K = x + K$, as needed. □

There are analogous Fundamental Homomorphism Theorems for groups, rings, vector spaces, modules, and algebras. Here is the version of the theorem for rings.

Fundamental Homomorphism Theorem for Rings. Suppose X and Y are rings, and $T : X \to Y$ is a ring homomorphism with kernel K and image I. There is a ring isomorphism $T' : X/K \to I$ given by $T'(x + K) = T(x)$ for all $x \in X$.

Proof. First, T is a group homomorphism from X to Y regarded as additive groups. By the Fundamental Homomorphism Theorem for Commutative Groups, we already know that T' is a bijective, well-defined homomorphism of additive groups. We need only check that T' also preserves the ring multiplication and identity. To do so, fix two elements of X/K, say $u + K$ and $x + K$ with $u, x \in X$. Using the definition of the product of cosets, the definition of T', and the fact that T preserves multiplication, we get

$$T'((u + K)(x + K)) = T'((ux) + K) = T(ux) = T(u)T(x) = T'(u + K)T'(x + K).$$

For similar reasons, we have $T'(1_{X/K}) = T'(1_X + K) = T(1_X) = 1_Y = 1_I$. □

The reader can formulate and prove the analogous theorems for other algebraic structures. Chapter 17 discusses this theorem and other isomorphism theorems for modules, which include commutative groups and vector spaces as special cases.

1.8 Spanning, Linear Independence, Basis, and Dimension

Introductions to linear algebra often discuss the concepts of linear independence, spanning sets, bases, and dimension in the case of real vector spaces. The same ideas occur in the more general setting of vector spaces over a field. In this section, after reviewing the basic definitions, we state without proof some theorems concerning linear independence and bases of vector spaces. We prove these theorems in a more general setting in Chapter 17.

Let V be a vector space over a field F. Let $L = (\mathbf{v}_1, \mathbf{v}_2, \ldots, \mathbf{v}_k)$ be a finite list of vectors with each $\mathbf{v}_i \in V$. Any vector of the form $c_1\mathbf{v}_1 + c_2\mathbf{v}_2 + \cdots + c_k\mathbf{v}_k$ with each $c_i \in F$ is called an *F-linear combination* of the vectors in L. The list L *spans* V iff every $\mathbf{w} \in V$ can be written in **at least one** way as an F-linear combination of the vectors in L. The list L is *linearly dependent over F* (or *F-linearly dependent*) iff there exist scalars $c_1, c_2, \ldots, c_k \in F$ such that at least one c_i is nonzero, and $\mathbf{0}_V = c_1\mathbf{v}_1 + c_2\mathbf{v}_2 + \cdots + c_k\mathbf{v}_k$. In other words, saying that L is linearly dependent means the zero vector can be written as a linear combination of the vectors in L where at least one coefficient is not the zero scalar. L is *linearly independent over F* iff L is not linearly dependent over F. Spelling this out, saying that L is F-linearly independent means: for all $c_1, \ldots, c_k \in F$, if $c_1\mathbf{v}_1 + \cdots + c_k\mathbf{v}_k = 0$, then $c_1 = c_2 = \cdots = c_k = 0$.

The definition of linear independence can be rephrased to resemble more closely the definition of spanning. We claim L is F-linearly independent iff every $\mathbf{w} \in V$ can be written in **at most one** way as an F-linear combination of the vectors in L. On one hand, if L is linearly dependent, then the vector $\mathbf{w} = \mathbf{0}$ can be written in at least two ways as a linear combination of vectors in L: one way is $\mathbf{0} = 0\mathbf{v}_1 + 0\mathbf{v}_2 + \cdots + 0\mathbf{v}_k$, and the other way is the linear combination appearing in the definition of linear dependence. On the other hand, assume L is linearly independent. Suppose a given $\mathbf{w} \in V$ can be written as $\mathbf{w} = \sum_{i=1}^{k} a_i\mathbf{v}_i$ and also as $\mathbf{w} = \sum_{i=1}^{k} b_i\mathbf{v}_i$ for some $a_i, b_i \in F$. Subtracting these equations gives $\mathbf{0} = \sum_{i=1}^{k}(a_i - b_i)\mathbf{v}_i$. The assumed linear independence of L then gives $a_i - b_i = 0$ for all i, hence $a_i = b_i$ for all i. So there is at most one way of writing $\mathbf{w}$ as a linear combination of the vectors in L.

Continuing the definitions, we say that the list $L = (\mathbf{v}_1, \ldots, \mathbf{v}_k)$ is an *ordered basis* of V iff L spans V and L is F-linearly independent. This means that every $\mathbf{v} \in V$ can be written in **exactly one way** as an F-linear combination of vectors in L. The vector space V is called *finite-dimensional* iff V has a finite ordered basis. The Basis Cardinality Theorem (proved in §17.15) states that any two ordered bases of a finite-dimensional vector space V must have the same size, where the size of a list L is its length (number of entries). The *dimension* of a finite-dimensional vector space V, denoted $\dim(V)$ or $\dim_F(V)$, is the size of any ordered basis of V.

For vector spaces that might be infinite-dimensional, we adapt the definitions of spanning, linear independence, and bases to apply to sets of vectors as well as lists of vectors. Note that a *set* of vectors is unordered and could be infinite, while a *list* of vectors (in this setting) is ordered and finite. Let S be any subset of an F-vector space V. The set S *spans* V iff every $\mathbf{w} \in V$ can be written as an F-linear combination of a finite list of vectors $(\mathbf{v}_1, \ldots, \mathbf{v}_n)$ with each $\mathbf{v}_i \in S$ (the list used depends on $\mathbf{w}$). The set S is *linearly independent* iff every finite list of distinct elements of S is linearly independent (as defined earlier). The set S is a *basis* of V iff S spans V and S is linearly independent. By adapting the proof for lists, we can show that S is a basis of V iff every $\mathbf{w} \in V$ can be written in exactly one way as a linear combination of elements of S. More precisely, this means that for each $\mathbf{w} \in V$, there is exactly one function $c : S \to F$ that is zero for all but finitely many inputs and satisfies $\mathbf{w} = \displaystyle\sum_{\mathbf{v} \in S : c(\mathbf{v}) \neq 0} c(\mathbf{v})\mathbf{v}$.

We state some facts about these concepts, which are proved in a more general setting in Chapter 17. Let V be a vector space over the field F, and let S and T be any subsets of V. Then:

(i) If S spans V and $S \subseteq T$, then T spans V.
(ii) If T is linearly independent and $S \subseteq T$, then S is linearly independent.
(iii) V spans V.
(iv) If T spans V, then there is a basis B of V with $B \subseteq T$.
(v) The empty set $\emptyset$ is linearly independent.
(vi) If S is linearly independent, then there is a basis C of V with $S \subseteq C$.
(vii) V has a basis (by (iii) and (iv), or (v) and (vi)).
(viii) If S is linearly independent and T spans V, then $|S| \leq |T|$.
(ix) **Basis Cardinality Theorem:** If S and T are any two bases of V, then $|S| = |T|$.

Here, $|S|$ denotes the cardinality of the set S, which is the number of elements in S if S is finite. In general, $|S| \leq |T|$ means there exists a one-to-one function $g : S \to T$, while $|S| = |T|$ means there exists a bijective (one-to-one and onto) function $h : S \to T$. Analogs of facts (i) through (ix) hold for ordered lists in a finite-dimensional vector space V. For instance, any linearly independent ordered list in V can be extended to an ordered basis of V by appending zero or more (appropriately chosen) vectors.

Because of facts (vii) and (ix), we can define the *dimension* of V, denoted $\dim(V)$ or $\dim_F(V)$, to be the unique cardinality of any basis of V. Fact (viii) says that a linearly independent set in V cannot have strictly larger cardinality than a spanning set in V. A basis is both a spanning set and a linearly independent set. Therefore:

(x) Any $S \subseteq V$ with $|S| > \dim(V)$ must be linearly dependent.
(xi) Any $T \subseteq V$ with $|T| < \dim(V)$ cannot span V.

Note also that an (ordered) list $L = (\mathbf{v}_1, \ldots, \mathbf{v}_k)$ spans V iff the (unordered) set $\{\mathbf{v}_1, \ldots, \mathbf{v}_k\}$ spans V. On the other hand, L is a linearly independent list iff L contains no repeated vectors and the set $\{\mathbf{v}_1, \ldots, \mathbf{v}_k\}$ is linearly independent. So in the finite-dimensional case, any ordered basis of V has the same size as any basis of V, and the two definitions of dimension (one using lists, the other using sets) are consistent.

We illustrate these ideas by proving the Rank–Nullity Theorem.

Rank–Nullity Theorem. Let V and W be F-vector spaces with V finite-dimensional. For any linear map $T : V \to W$ with null space $N = \ker(T)$ and image $R = \mathrm{img}(T)$,

$$\dim(N) + \dim(R) = \dim(V).$$

Here, $\dim(N)$ is called the *nullity* of T and $\dim(R)$ is called the *rank* of T. The theorem says that the nullity plus the rank equals the dimension of the domain of T.

Proof. Let $n = \dim(V)$ (which is finite) and $k = \dim(N)$. Note that k must be finite by fact (viii) or fact (x), since a basis of N is also a linearly independent subset of V. Let $\mathcal{B}_N = (\mathbf{v}_1, \ldots, \mathbf{v}_k)$ be an ordered basis of N. Since $\mathcal{B}_N$ is a linearly independent list, it can be extended (by fact (vi) for lists) to an ordered basis of V, say $\mathcal{B}_V = (\mathbf{v}_1, \ldots, \mathbf{v}_k, \mathbf{v}_{k+1}, \ldots, \mathbf{v}_n)$. Define $\mathbf{y}_i = T(\mathbf{v}_{k+i})$ for $1 \leq i \leq n-k$. We will show $\mathcal{B}_R = (\mathbf{y}_1, \ldots, \mathbf{y}_{n-k})$ is an ordered basis of R. We can then conclude $\dim(R) = n-k$, which proves the theorem since $k + (n-k) = n$.

First, we check that the list $\mathcal{B}_R$ spans the image R of T. Given an arbitrary $\mathbf{z} \in R$, we have $\mathbf{z} = T(\mathbf{x})$ for some $\mathbf{x} \in V$. Expressing $\mathbf{x}$ in terms of the basis $\mathcal{B}_V$, there are scalars $c_1, \ldots, c_n \in F$ with $\mathbf{x} = \sum_{i=1}^{n} c_i \mathbf{v}_i$. Applying the linear map T to this expression gives

$T(\mathbf{x}) = \sum_{i=1}^{n} c_i T(\mathbf{v}_i)$. Now $T(\mathbf{v}_i) = 0$ for $1 \leq i \leq k$ since these $\mathbf{v}_i$ are in the kernel of T, while $T(\mathbf{v}_i) = \mathbf{y}_{i-k}$ for $k < i \leq n$. We therefore get $\mathbf{z} = T(\mathbf{x}) = \sum_{i=k+1}^{n} c_i \mathbf{y}_{i-k}$, which expresses $\mathbf{z}$ as a linear combination of vectors in the list $\mathcal{B}_R$.

Second, we check that $\mathcal{B}_R$ is a linearly independent list. Assume $d_1, \ldots, d_{n-k} \in F$ satisfy $\sum_{j=1}^{n-k} d_j \mathbf{y}_j = \mathbf{0}_W$; we must show every d_j is zero. Since $\mathbf{y}_j = T(\mathbf{v}_{k+j})$, linearity of T gives $T(\sum_{j=1}^{n-k} d_j \mathbf{v}_{k+j}) = \mathbf{0}_W$. So the vector $\mathbf{u} = d_1 \mathbf{v}_{k+1} + \cdots + d_{n-k} \mathbf{v}_n$ is in the kernel N of T. As $\mathcal{B}_N$ is a basis of this kernel, $\mathbf{u}$ can be expressed as some linear combination of $\mathbf{v}_1, \ldots, \mathbf{v}_k$, say $\mathbf{u} = f_1 \mathbf{v}_1 + \cdots + f_k \mathbf{v}_k$ for some $f_i \in F$. Equating the two expressions for $\mathbf{u}$, we get

$$-f_1 \mathbf{v}_1 + \cdots + (-f_k) \mathbf{v}_k + d_1 \mathbf{v}_{k+1} + \cdots + d_{n-k} \mathbf{v}_n = \mathbf{0}_V.$$

Since $(\mathbf{v}_1, \ldots, \mathbf{v}_n)$ is a linearly independent list, we conclude that $-f_1 = \cdots = -f_k = d_1 = \cdots = d_{n-k} = 0$. In particular, every d_j is zero, as needed. $\qquad\square$

The Rank–Nullity Theorem can also be deduced from the Fundamental Homomorphism Theorem for Vector Spaces (see Exercise 66).

1.9   Summary

Table 1.5 summarizes some of the definitions of algebraic axioms, algebraic systems, subsystems, homomorphisms, and linear algebra concepts discussed in this chapter. We also recall the following points.

1. *Examples and Notation for Algebraic Structures.* $\mathbb{Z}$, $\mathbb{Q}$, $\mathbb{R}$, and $\mathbb{C}$ are commutative rings and additive groups. $\mathbb{Z}_n = \{0, 1, \ldots, n\}$ is a finite commutative ring with operations $a \oplus b = (a + b) \bmod n$ and $a \otimes b = (ab) \bmod n$. $\mathbb{Q}$, $\mathbb{R}$, $\mathbb{C}$, and $\mathbb{Z}_p$ (with p prime) are fields. The set $M_n(F)$ of $n \times n$ matrices with entries in a field F is a ring and an F-algebra. The set $\mathrm{GL}_n(F) = \{A \in M_n(F) : \det(A) \neq 0\}$ is a group under matrix multiplication. The set $S(X)$ of bijections $f : X \to X$ is a group under function composition. The set $F[x]$ of polynomials in one variable with coefficients in a field F is a commutative ring, F-vector space, and F-algebra. Examples of F-vector spaces include F^k with componentwise operations, the set of F-linear maps from one vector space to another, and the set $M_{m,n}(F)$ of $m \times n$ matrices with entries in F.

2. *Direct Products.* Given F-vector spaces $V_1, \ldots, V_k$, the set $V_1 \times \cdots \times V_k = \{(\mathbf{v}_1, \ldots, \mathbf{v}_k) : \mathbf{v}_i \in V_i\}$ becomes an F-vector space by defining addition and scalar multiplication one component at a time. A similar construction works for other algebraic systems.

3. *Cosets.* If $(G, +)$ is a commutative group with subgroup H, the quotient set G/H is the set of cosets $x + H = \{x + h : h \in H\}$ for $x \in G$. The Coset Equality Theorem states that for all $x, y \in G$, $x + H = y + H$ iff $x - y \in H$. G is the disjoint union of the distinct cosets of H. When defining functions or operations that act on cosets, we must check that the output is well-defined (independent of the names used for the input cosets).

4. *Quotient Systems.* Given a commutative group $(G, +)$ with subgroup H, the set G/H becomes a commutative group with operation $(x+H) \oplus (y+H) = (x+y)+H$ for $x, y \in G$. The identity of G/H is $0_G + H = H$, and the inverse of $x + H$ is $(-x) + H$. A similar construction is possible when G is a group with normal

TABLE 1.5
Summary of definitions of algebraic concepts.

Axiom for Set S	Meaning
closure of S under $\star$	$\forall a, b \in S,\ a \star b$ is in S
associativity	$\forall a, b, c \in S, (a \star b) \star c = a \star (b \star c)$
identity	$\exists e \in S, \forall a \in S, a \star e = a = e \star a$
inverses	$\forall a \in S, \exists b \in S,\ a \star b = e = b \star a$ (where e is the identity)
commutativity	$\forall a, b \in S, a \star b = b \star a$
left distributive law	$\forall a, b, c \in S, a \cdot (b + c) = (a \cdot b) + (a \cdot c)$
right distributive law	$\forall a, b, c \in S, (a + b) \cdot c = (a \cdot c) + (b \cdot c)$
Algebraic System	**Definition**
group $(G, \star)$	operation $\star$ has closure, associativity, identity, and inverses
commutative group	group where operation is commutative
ring $(R, +, \cdot)$	$(R, +)$ is comm. group; $\cdot$ has closure, associativity, identity, left distributive law, and right distributive law
commutative ring	ring where multiplication is commutative
integral domain	comm. ring with $1 \neq 0$ and no zero divisors
field F	comm. ring where $1 \neq 0$ and every $x \neq 0$ has a mult. inverse
F-vector space $(V, +, \cdot)$	$(V, +)$ is comm. group; scalar mult. obeys closure, scalar associativity, two distributive laws, and scalar identity law
R-module	same as vector space, but scalars come from ring R
F-algebra	ring and F-vector space with compatible multiplications
Subsystem	**Required Closure Properties**
subgroup H of $(G, \star)$	H closed under $\star$, identity, and inverses
normal subgroup of G	subgroup closed under conjugation by elements of G
subring S of $(R, +, \cdot)$	S closed under $+$, $-$, $\cdot$, 0, and 1
ideal I of $(R, +, \cdot)$	I closed under $+$, $-$, 0, and left/right mult. by elements of R
subfield of field F	subring closed under inverses of nonzero elements
subspace of vector space V	closed under 0, $+$, and scalar multiplication
submodule of module	same as subspace, but scalars come from a ring
subalgebra of A	subring and subspace of A
Homomorphism	**What Must Be Preserved**
group homomorphism	group operation (hence also identity, inverses, powers)
ring homomorphism	ring $+$, ring $\cdot$, and 1 (multiplicative identity)
F-linear map	vector addition, scalar multiplication by scalars in F
R-module homomorphism	vector addition, scalar multiplication by scalars in R
algebra homomorphism	ring $+$, ring $\cdot$, 1, scalar multiplication
Linear Algebra Term	**Definition**
linear combination of $\mathbf{v}_i$	a sum $c_1 \mathbf{v}_1 + \cdots + c_k \mathbf{v}_k$ with all $c_i \in F$
list L spans V	every $\mathbf{w} \in V$ is a linear combination of vectors in L
linearly independent list L	for all $c_i \in F$, if $\sum_{i=1}^{k} c_i \mathbf{v}_i = \mathbf{0}$, then $c_1 = \cdots = c_k = 0$
ordered basis of V	linearly independent list that spans V
spanning set in V	every $\mathbf{v} \in V$ is finite linear combination of vectors from set
linearly independent set	all finite lists of distinct elements of set are lin. independent
basis of V	linearly independent set that spans V
dimension of V	size of any basis (or ordered basis) of V

subgroup H; when R is a ring with ideal I; when V is a vector space with subspace W; and when V is an R-module with submodule W. In a quotient vector space V/W, scalar multiplication is given by $c(\mathbf{x}+W) = (c\mathbf{x})+W$ for $c \in F$ and $\mathbf{x} \in V$.

5. *Facts about Independence, Spanning, and Bases.* Every F-vector space V has a basis (possibly infinite). Any two bases of V have the same cardinality. Any linearly independent subset of V can be extended to a basis of V. Any spanning subset of V contains a basis of V. Subsets of linearly independent sets are also linearly independent, and supersets of spanning sets still span. A linearly independent subset of V can never be larger than a spanning set. For a vector space homomorphism $T : V \to W$ where $\dim(V)$ is finite, the Rank–Nullity Theorem says $\dim(V) = \dim(\ker(T)) + \dim(\mathrm{img}(T))$.

1.10   Exercises

1. Prove that a binary operation $\star$ on a set S can have at most one identity element.

2. Let $\star$ be an associative binary operation on a set S with identity element e. Show that each $a \in S$ can have at most one inverse relative to $\star$.

3. Explain why the following sets and binary operations are not groups by pointing out a group axiom that fails to hold.
   (a) S is the set of nonnegative integers, $p(x,y) = x + y$ for $x, y \in S$.
   (b) $S = \{-1, 0, 1\}$, $p(x,y) = x + y$ for $x, y \in S$.
   (c) $S = \mathbb{Z}$, $p(x,y) = 4$ for all $x, y \in S$.
   (d) $S = \mathbb{Z}$, $p(x,y) = (x + y) \bmod 10$ for all $x, y \in S$.
   (e) $S = \{0, 1, 2, 3, 4\}$, $p(0, x) = x = p(x, 0)$ for all $x \in S$, and $p(x, y) = 0$ for all nonzero $x, y \in S$.

4. Let X be a fixed subset of $\mathbb{R}$ and $x \star y = \min(x, y)$ for all $x, y \in X$. For each of the axioms for a commutative group, find conditions on X that are necessary and sufficient for $(X, \star)$ to satisfy that axiom. For which X is $(X, \star)$ a commutative group?

5. (a) For fixed $n \geq 1$, prove that the set of $n \times n$ matrices with real entries is a commutative group under the operation of matrix addition.
   (b) Is the set of matrices in (a) a group under matrix multiplication? Explain.

6. Decide (with proof) which sets are groups under function composition.
   (a) S is the set of injective (one-to-one) functions $f : \mathbb{R} \to \mathbb{R}$.
   (b) S is the set of surjective (onto) functions $g : \mathbb{R} \to \mathbb{R}$.
   (c) S is the set of one-to-one functions $h : \mathbb{Z}_5 \to \mathbb{Z}_5$.

7. Prove $\mathbb{Z}_n$ is a commutative group under $\oplus$ (addition mod n).

8. **Cancellation Laws for Groups.** Let $(G, \star)$ be a group. Prove: for all $a, x, y \in G$, if $a \star x = a \star y$, then $x = y$; and if $x \star a = y \star a$, then $x = y$. Point out where each of the four group axioms is used in the proof.

9. (a) Give a specific example of a ring $(G, +, \star)$ and nonzero $a, x, y \in G$ for which the cancellation laws in the previous exercise are not true.
   (b) If $(G, +, \star)$ is a field, are the cancellation laws for $\star$ true? Explain.
   (c) Find and prove a version of the multiplicative cancellation law that is valid in an integral domain.

10. **Sudoku Theorem for Groups.** Let $(G, \star)$ be a group consisting of the n distinct elements $x_1, \ldots, x_n$. Prove: for all $a \in G$, the list $a \star x_1, \ldots, a \star x_n$ is a rearrangement of the list $x_1, \ldots, x_n$. (A similar result holds for $x_1 \star a, \ldots, x_n \star a$.)

11. **Exponent Theorem for Groups.** Let $(G, \star)$ be an n-element commutative group with identity e. Prove: for all $a \in G$, $a^n = e$. Do this by letting $G = \{x_1, \ldots, x_n\}$ and evaluating the product $(a \star x_1) \star (a \star x_2) \star \cdots \star (a \star x_n)$ in two ways. (The theorem holds for non-commutative groups too, but the proof is harder.)

12. Let A and B be normal subgroups of a group G.
    (a) Show: if $A \cap B = \{e_G\}$, then $\forall a \in A, \forall b \in B, ab = ba$. (Study $aba^{-1}b^{-1}$.)
    (b) Prove $AB = \{ab : a \in A, b \in B\}$ is a subgroup of G.
    (c) Prove or disprove: AB must be a normal subgroup of G.

13. Fix $n \geq 1$. Let $\oplus$ and $\otimes$ denote addition mod n and multiplication mod n.
    (a) Prove: for all $a, b \in \mathbb{Z}_n$, $a \oplus b \in \mathbb{Z}_n$ and $a \oplus b = a + b - qn$ for some $q \in \mathbb{Z}$.
    (b) Prove: for all $a, b \in \mathbb{Z}_n$, $a \otimes b \in \mathbb{Z}_n$ and $a \otimes b = ab - sn$ for some $s \in \mathbb{Z}$.
    (c) Prove: for all $c, d \in \mathbb{Z}_n$, $c = d$ iff $c - d = tn$ for some $t \in \mathbb{Z}$.
    (d) Prove $(\mathbb{Z}_n, \oplus, \otimes)$ is a commutative ring. (*Hint:* Use (a) and (b) to eliminate $\oplus$ and $\otimes$ from each side of each ring axiom, then use (c) to see that the two sides must be equal.)

14. For any ring R, define R^* (the set of *units* of R) to be the set of invertible elements in R. So $x \in R$ belongs to R^* iff $\exists y \in R, xy = 1_R = yx$.
    (a) Prove that R^* is a group under the multiplication operation in R.
    (b) Show: If R is a commutative ring, then R^* is a commutative group.

15. Find R^* (see the previous exercise) for each ring R:
    (a) $\mathbb{Z}$ (b) $\mathbb{R}$ (c) a field R (d) $\mathbb{Z}_{12}$ (e) $M_3(\mathbb{R})$ (f) $\mathbb{R}[x]$

16. Prove that *every field is an integral domain*.

17. Prove that *every finite integral domain R is a field*. (For nonzero $a \in R$, define $L_a : R \to R$ by $L_a(x) = a \cdot x$ for $x \in R$. Prove L_a is one-to-one, hence onto.)

18. (a) Prove: if p is prime, then $\mathbb{Z}_p$ is a field.
    (b) Prove: if n is not prime, then $\mathbb{Z}_n$ is not an integral domain.

19. Let S be the set of all functions $f : \mathbb{R} \to \mathbb{R}$. For $f, g \in S$, define $f + g : \mathbb{R} \to \mathbb{R}$ by $(f + g)(x) = f(x) + g(x)$ for $x \in \mathbb{R}$. Prove $(S, +)$ is a commutative group.

20. Let S be the set of all functions $f : \mathbb{R} \to \mathbb{R}$. For $f, g \in S$, define $f \bullet g : \mathbb{R} \to \mathbb{R}$ by $(f \bullet g)(x) = f(x)g(x)$ for $x \in \mathbb{R}$. Is $(S, \bullet)$ a commutative group? Explain.

21. Let S be the set of all functions $f : \mathbb{R} \to \mathbb{R}$, with addition $+$ and multiplication $\bullet$ defined in the previous two exercises.
    (a) Prove $(S, +, \bullet)$ is a commutative ring.
    (b) Is S a field? Is S an integral domain? Explain.
    (c) Define a scalar multiplication $\cdot$ on S and prove $(S, +, \bullet, \cdot)$ is an $\mathbb{R}$-algebra.

22. Let S be the set of all functions $f : \mathbb{R} \to \mathbb{R}$, let $+$ be the sum defined in Exercise 19, and let $\circ$ be function composition. Which axioms for a commutative ring are true for $(S, +, \circ)$? Explain.

23. Let F be a field. Verify the vector space axioms for the direct product $V_1 \times V_2 \times \cdots \times V_k$ of F-vector spaces $V_1, V_2, \ldots, V_k$. Deduce that F^k is an F-vector space.

24. For $c \in \mathbb{R}$ and $\mathbf{v} \in \mathbb{Z}$, let $s(c, \mathbf{v}) = c\mathbf{v}$ be the ordinary product of the real number c and the integer $\mathbf{v}$. Which axioms for a real vector space hold for $(\mathbb{Z}, +, s)$?

25. Define scalar multiplication by $s(c, \mathbf{v}) = 0$ for all $c, \mathbf{v} \in \mathbb{R}$. Show that $(\mathbb{R}, +, s)$ satisfies all the axioms for a real vector space except the identity axiom for scalar multiplication.

26. Define scalar multiplication by $s(c, \mathbf{v}) = \mathbf{v}$ for all $c, \mathbf{v} \in \mathbb{R}$. Show that $(\mathbb{R}, +, s)$ satisfies all the axioms for a real vector space except for the distributive law for scalar addition.

27. Define scalar multiplication $s : \mathbb{C} \times \mathbb{R} \to \mathbb{R}$ by $s(a + ib, \mathbf{v}) = a\mathbf{v}$ for all $a, b, \mathbf{v} \in \mathbb{R}$. Show that $(\mathbb{R}, +, s)$ satisfies all the axioms for a complex vector space except for associativity of scalar multiplication.

28. Define scalar multiplication by $s(c, \mathbf{v}) = c^2\mathbf{v}$ for all $c, \mathbf{v} \in \mathbb{R}$. Which axioms for a real vector space hold for $(\mathbb{R}, +, s)$?

29. Define scalar multiplication s by $s(c, \mathbf{v}) = c$ for all $c, \mathbf{v} \in \mathbb{R}$. Which axioms for a real vector space hold for $(\mathbb{R}, +, s)$?

30. Let $V = \mathbb{R}_{>0}$ be the set of positive real numbers. Define the non-standard addition $p(\mathbf{v}, \mathbf{w}) = \mathbf{v}\mathbf{w}$ (the ordinary product of $\mathbf{v}$ and $\mathbf{w}$) for $\mathbf{v}, \mathbf{w} \in V$. Define scalar multiplication by $s(c, \mathbf{v}) = \mathbf{v}^c$ for $c \in \mathbb{R}$ and $\mathbf{v} \in V$. Which axioms for a real vector space hold for (V, p, s)?

31. Prove that for the field $\mathbb{Z}_p$ (where p is prime), the distributive law for vector addition follows from the other axioms for a $\mathbb{Z}_p$-vector space.

32. Prove that for the field $\mathbb{Q}$ of rational numbers, the distributive law for vector addition follows from the other axioms for a $\mathbb{Q}$-vector space.

33. †Can you find an example of a field F and a structure $(V, +, s)$ satisfying all the F-vector space axioms except the distributive law for vector addition?[2]

34. Let W be the set of matrices of the form $\begin{bmatrix} a & 0 \\ b & 0 \end{bmatrix}$ for some $a, b \in \mathbb{R}$. Is W an additive subgroup of $M_2(\mathbb{R})$? a subring? a left ideal? a right ideal? an ideal? a subspace? a subalgebra? Explain.

35. Give an example of a ring R and a right ideal I in R that is not an ideal of R.

36. If possible, give an example of a subset of $\mathbb{Z}$ that is closed under addition and inverses, yet is not a subgroup of $(\mathbb{Z}, +)$.

37. For any field F and $k \geq 1$, show that $W = \{(t, t, \ldots, t) : t \in F\}$ is a subspace of the F-vector space F^k. Is an analogous result true for groups or for rings?

38. Let G be the group $\mathrm{GL}_2(\mathbb{R})$ of invertible 2×2 matrices with real entries.
    (a) Give three different examples of normal subgroups of G.
    (b) Give an example of a subgroup of G that is not normal in G.

39. Prove that $\mathbb{Z}[i] = \{a + bi : a, b \in \mathbb{Z}\}$ is a subring of $\mathbb{C}$. Is this a subfield of $\mathbb{C}$?

40. *Quaternions.* Let $\mathbf{1} = \begin{bmatrix} 1 & 0 \\ 0 & 1 \end{bmatrix}$, $\mathbf{i} = \begin{bmatrix} i & 0 \\ 0 & -i \end{bmatrix}$, $\mathbf{j} = \begin{bmatrix} 0 & i \\ i & 0 \end{bmatrix}$, $\mathbf{k} = \begin{bmatrix} 0 & -1 \\ 1 & 0 \end{bmatrix}$.
    (a) Prove $Q = \{\pm\mathbf{1}, \pm\mathbf{i}, \pm\mathbf{j}, \pm\mathbf{k}\}$ is a non-commutative subgroup of $\mathrm{GL}_2(\mathbb{C})$.
    (b) Prove $\mathbb{H} = \{a\mathbf{1} + b\mathbf{i} + c\mathbf{j} + d\mathbf{k} : a, b, c, d \in \mathbb{R}\}$ is a real subalgebra of $M_2(\mathbb{C})$.
    (c) Prove every nonzero element of $\mathbb{H}$ has a multiplicative inverse in $\mathbb{H}$.
    (So $\mathbb{H}$ satisfies all field axioms except commutativity of multiplication.)

41. Given rings $R_1, \ldots, R_n$, carefully prove that the direct product $R = R_1 \times \cdots \times R_n$ is a ring (with componentwise operations). In particular, what are 0_R and 1_R?

---

[2]The mark † signals an exercise that may be especially challenging.

42. Prove that the direct product of two or more fields is never a field.

43. Suppose G_1 and G_2 are groups, H_1 is a subgroup of G_1, and H_2 is a subgroup of G_2. Prove $H_1 \times H_2$ is a subgroup of $G_1 \times G_2$. Prove if H_1 is normal in G_1 and H_2 is normal in G_2, then $H_1 \times H_2$ is normal in $G_1 \times G_2$.

44. State and prove results analogous to the previous exercise for subrings and ideals in a product ring and for subspaces in a product vector space.

45. Give an example of vector spaces V_1 and V_2 and a subspace W of $V_1 \times V_2$ such that W is not the direct product of a subspace of V_1 and a subspace of V_2.

46. Let $(G, \star)$ be a group. Assume H is a finite nonempty subset of G that is closed under $\star$. Prove H is a subgroup of G. (Given $a \in H$, show the list $a^1, a^2, a^3, \ldots$ has repeated entries. Deduce this list must contain e_G and a^{-1}.)

47. **Left Coset Equality Theorem.** Let $(G, \star)$ be a group with subgroup H. Prove: for all $a, b \in G$, the following conditions are equivalent: $a \star H = b \star H$; $a \in b \star H$; $b \in a \star H$; $a = b \star h$ for some $h \in H$; $b = a \star k$ for some $k \in H$; $b^{-1} \star a \in H$; $a^{-1} \star b \in H$.

48. State and prove the Right Coset Equality Theorem for groups.

49. Let H be a normal subgroup of a group $(G, \star)$. Verify that the binary operation $p : G/H \times G/H \to G/H$, given by $p(a \star H, b \star H) = (a \star b) \star H$ for all $a, b \in G$, is well-defined. Indicate where your proof uses normality of H. Then prove the group axioms for $(G/H, p)$.

50. Let $(G, \star)$ be a group with subgroup H. Prove: H is a normal subgroup of G iff for all $a \in G$, $a \star H = H \star a$.

51. Let I be an ideal in a ring $(R, +, \cdot)$. For $a, b \in R$, set $(a + I) \bullet (b + I) = (a \cdot b) + I$.
    (a) Prove $\bullet$ is a well-defined binary operation on R/I.
    (b) Prove the ring axioms for $(R/I, +, \bullet)$.
    (c) Prove: if R is commutative, then R/I is commutative.

52. Let $W = \{(t, -t) : t \in \mathbb{R}\}$, which is a subspace of the real vector space $\mathbb{R}^2$. Draw a picture of the set $\mathbb{R}^2/W$. Calculate $[(1, 2) + W] \oplus [(0, -1) + W]$, illustrate this calculation in your picture, and give three different names for the answer.

53. Suppose G and K are groups, and $T : G \to K$ is a group homomorphism. Prove: $\forall x \in G, \forall n \in \mathbb{Z}, T(x^n) = T(x)^n$. (Treat $n = 0$ and $n = -1$ separately.)

54. Let $T : X \to Y$ and $U : Y \to Z$ be F-linear maps of F-vector spaces.
    (a) Prove $U \circ T : X \to Z$ is F-linear.
    (b) Prove if T is a vector space isomorphism, then so is T^{-1}.
    (c) Prove $\ker(T)$ is a subspace of X, and $\text{img}(T)$ is a subspace of Y.
    (d) Prove $\ker(T) \subseteq \ker(U \circ T)$.
    (e) Prove $\text{img}(U \circ T) \subseteq \text{img}(U)$.
    (f) Find and prove a characterization of when $\ker(T) = \ker(U \circ T)$.

55. State and prove the Fundamental Homomorphism Theorem for Vector Spaces.

56. Given a field F and an integer $k > 0$, define vectors $\mathbf{e}_1, \ldots, \mathbf{e}_k \in F^k$ by letting $\mathbf{e}_i$ have 1_F in position i and 0_F elsewhere. Define $\mathbf{f}_i = \mathbf{e}_1 + \mathbf{e}_2 + \cdots + \mathbf{e}_i$ for $1 \le i \le k$.
    (a) Prove that $(\mathbf{e}_1, \mathbf{e}_2, \ldots, \mathbf{e}_k)$ is an ordered basis of F^k.
    (b) Prove that $(\mathbf{f}_1, \mathbf{f}_2, \ldots, \mathbf{f}_k)$ is an ordered basis of F^k.

57. For each real vector space V, find an ordered basis for V and compute $\dim(V)$.
    (a) the set of all real $n \times n$ matrices
    (b) the set of all complex $n \times n$ matrices

(c) the set of all real upper-triangular $n \times n$ matrices

(d) the set of all real symmetric $n \times n$ matrices

58. Prove facts (i), (ii), (iii) and (v) stated in Section 1.8 (page 15).

59. Assume V is a vector space over a field F, the list $T = (\mathbf{w}_1, \ldots, \mathbf{w}_n)$ spans V, and $L = (\mathbf{v}_1, \ldots, \mathbf{v}_k)$ is a linearly independent list in V. Show that if L does not span V, then there exists j such that $L' = (\mathbf{v}_1, \ldots, \mathbf{v}_k, \mathbf{w}_j)$ is linearly independent. Deduce that L can be extended to a basis of V by appending zero or more vectors from T.

60. Assume V is an F-vector space, S is a linearly independent subset of V, and T is a finite spanning set for V. Show that if S is not a subset of T, then there is a linearly independent set S' with $|S'| = |S|$ and $|S' \cap T| = |S \cap T| + 1$. Iterate this result to prove $|S| \leq |T|$.

61. Let $T : V \to W$ be a linear map between finite-dimensional vector spaces.
    (a) Prove: if U is a subspace of V, then $\dim(U) \leq \dim(V)$.
    (b) Prove: if U is a subspace of V and $\dim(U) = \dim(V)$, then $U = V$.
    (c) Prove: $\dim(\mathrm{img}(T)) \leq \min(\dim(V), \dim(W))$.
    (d) Find and prove inequalities relating $\dim(\ker(T))$ to $\dim(V)$ and $\dim(W)$.

62. Let V be an F-vector space and S be a list or set of vectors in V. Prove that the set W of all finite F-linear combinations of vectors in S is a subspace of V. Prove also that for every subspace Z of V that contains S, $W \subseteq Z$. W is called the *subspace of V generated by S*.

63. Let V and W be F-vector spaces, let $T : V \to W$ be an F-linear map and U be a subspace of V. Prove that $T[U] = \{T(\mathbf{u}) : \mathbf{u} \in U\}$ is a subspace of W. Show that if a list $(\mathbf{u}_1, \ldots, \mathbf{u}_k)$ spans U, then the list $(T(\mathbf{u}_1), \ldots, T(\mathbf{u}_k))$ spans $T[U]$.

64. Let $T : V \to W$ be a linear map of finite-dimensional vector spaces. For any list $L = (\mathbf{v}_1, \ldots, \mathbf{v}_k)$ of vectors in V, let $T[L]$ be the list $(T(\mathbf{v}_1), \ldots, T(\mathbf{v}_k))$.
    (a) Prove T is one-to-one iff for every linearly independent list L in V, $T[L]$ is a linearly independent list in W.
    (b) Prove T is onto iff for every list L in V that spans V, $T[L]$ spans W.
    (c) Prove T is an isomorphism iff for every ordered basis L of V, $T[L]$ is an ordered basis of W.

65. Let V be a finite-dimensional F-vector space with subspace W. Prove that $\dim(V/W) = \dim(V) - \dim(W)$. (*Hint:* Extend a basis of W to a basis of V, and prove that the cosets of the new basis vectors form a basis of V/W.)

66. Use Exercises 55 and 65 to reprove the Rank–Nullity Theorem.

67. **Endomorphism Ring of a Commutative Group.** Given a commutative group $(M, +)$, define $\mathrm{End}(M)$ to be the set of all group homomorphisms $f : M \to M$. Define the *sum* of $f, g \in \mathrm{End}(M)$ to be the function $f + g : M \to M$ given by $(f + g)(x) = f(x) + g(x)$ for all $x \in M$. Define the *product* of $f, g \in \mathrm{End}(M)$ to be the composition $f \circ g$, given by $(f \circ g)(x) = f(g(x))$ for all $x \in M$. Prove the ring axioms for $(\mathrm{End}(M), +, \circ)$. $\mathrm{End}(M)$ is called the *endomorphism ring of M*.

2

Permutations

This chapter gives the basic definitions and facts about permutations we need to discuss determinants and multilinear algebra. First we discuss how composition of functions confers a group structure on the set S_n of permutations of $\{1, 2, \ldots, n\}$. Next, we introduce a way of visualizing a function using a directed graph, which leads to a description of permutations in terms of disjoint directed cycles. We use this description to obtain some algebraic factorizations of permutations in the group S_n. The chapter concludes by studying inversions of functions and permutations, which give information about how many steps it takes to sort a list into increasing order. We use inversions to define the sign of a permutation, which plays a critical role in our subsequent treatment of determinants (Chapter 5).

2.1 Symmetric Groups

This section assumes familiarity with the definition of a group (§1.1). For each positive integer n, let $[n]$ denote the finite set $\{1, 2, \ldots, n\}$. The *symmetric group* S_n is defined to be the set of all bijective functions $f : [n] \to [n]$, with composition of functions as the binary operation. Recall that a function $f : X \to Y$ is a bijection iff f is *one-to-one* (for all $x, z \in X$, $f(x) = f(z)$ implies $x = z$) and f is *onto* (for each $y \in Y$ there is $x \in X$ with $f(x) = y$). For a function $f : X \to X$ mapping a finite set X into itself, we may check that f is one-to-one iff f is onto.

Recall that the *composition* of two functions $f : X \to Y$ and $g : Y \to Z$ is defined to be the function $g \circ f : X \to Z$ given by $(g \circ f)(x) = g(f(x))$ for $x \in X$. The composition of bijections is a bijection, so S_n is closed under the composition operation. Composition of functions is always associative, so associativity holds. The *identity function* on $[n]$, given by $\mathrm{id}(x) = x$ for $x \in [n]$, is a bijection and hence belongs to S_n. Since $f \circ \mathrm{id} = f = \mathrm{id} \circ f$ for all $f \in S_n$, S_n has an identity element relative to $\circ$. Finally, each $f \in S_n$ has an inverse function $f^{-1} : [n] \to [n]$, which is also a bijection and satisfies $f \circ f^{-1} = \mathrm{id} = f^{-1} \circ f$. Therefore, $(S_n, \circ)$ is a group. For all $n \geq 3$, S_n is not a commutative group.

Elements of S_n are called *permutations* of n objects. To explain this terminology, note that any function $f : [n] \to [n]$ is completely determined by the list $[f(1), f(2), \ldots, f(n)]$. We refer to this list as the *one-line form* for f. If f belongs to S_n, so that $i \neq j$ implies $f(i) \neq f(j)$, then this list is a rearrangement (or "permutation") of the list $[1, 2, \ldots, n]$.

For example, the function $f : [4] \to [4]$ given by $f(1) = 3$, $f(2) = 2$, $f(3) = 4$, and $f(4) = 1$ has one-line form $f = [3, 2, 4, 1]$. Given the one-line form $g = [4, 1, 3, 2] \in S_4$, we must have $g(1) = 4$, $g(2) = 1$, $g(3) = 3$, and $g(4) = 2$. Then $f \circ g = [1, 3, 4, 2]$ since $f \circ g(1) = f(g(1)) = f(4) = 1$, $f \circ g(2) = f(g(2)) = f(1) = 3$, etc. On the other hand, $g \circ f = [3, 1, 2, 4]$.

Using one-line forms, we now show that *the group S_n has exactly $n!$ elements.* We can build each $f \in S_n$ by making the following choices. There are n ways to choose $f(1)$. There are $n - 1$ ways to choose $f(2)$, which must be something in $[n]$ unequal to $f(1)$. Since $f(3)$ must be different from $f(1)$ and $f(2)$, there are $n - 2$ ways to choose $f(3)$. We continue

DOI: 10.1201/9781003484561-2

similarly. At the end there is one choice for $f(n)$. Combining all these choices by the Product Rule for Counting, we see there are $n \times (n-1) \times (n-2) \times \cdots \times 1 = n!$ permutations in S_n.

2.2   Representing Functions as Directed Graphs

Let $f : [n] \to [n]$ be any function. We create a graphical representation of f by drawing n dots labeled 1 through n, then drawing an arrow from i to $f(i)$ for all $i \in [n]$. If $i = f(i)$, this arrow is a loop pointing from i to itself. See Figure 2.1 for an example where $n = 23$ and

$$f = [10, 22, 7, 11, 15, \quad 19, 19, 12, 22, 12, \quad 11, 1, 6, 11, 5, \quad 8, 22, 9, 21, 11, \quad 3, 3, 2].$$

Note that we can recover the function f from its directed graph.

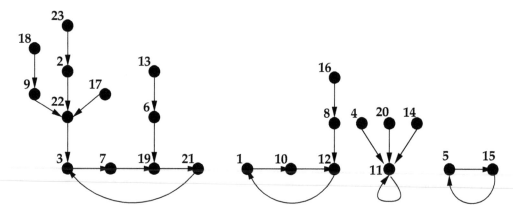

FIGURE 2.1
Directed graph for a function.

Each node in the graph of f has exactly one arrow leaving it, since f is a function. Thus, starting from any given node x_0, there is a well-defined path through the graph obtained by repeatedly following the arrows for f. This path visits the nodes x_0, $x_1 = f(x_0)$, $x_2 = f(f(x_0))$, $x_3 = f(f(f(x_0)))$, and so on. Since there are only finitely many nodes in the graph, there must exist integers i, j with $0 \le i < j$ and $x_i = x_j$. Pick the least i and then the least j (for this i) satisfying these conditions. Then the path starting at x_0 ends in a *directed cycle* (consisting of the nodes $x_i, x_{i+1}, \ldots, x_j = x_i$) and begins in a *tail* $x_0, \ldots, x_{i-1}$ that feeds into the cycle at position x_i. If $i = 0$, then there is no tail, and the path we are following lies completely on the cycle. If $j = i + 1$, the cycle consists of a single loop edge based at the node x_i.

For example, consider the f shown in Figure 2.1 and $x_0 = 2$. The sequence $(x_i : i \ge 0)$ is $(2, 22, 3, 7, 19, 21, 3, 7, 19, 21, \ldots)$, so $i = 2$, $j = 6$, the tail part is $2, 22$, and the cycle part is $3, 7, 19, 21$ with 21 leading back to 3. For $x_0 = 15$, we get a directed cycle $5, 15$ involving two vertices. For $x_0 = 20$, the tail is node 20 and the cycle is node 11 with its loop edge.

2.3 Cycle Decompositions of Permutations

Suppose that $f : [n] \to [n]$ is a permutation, so f is one-to-one and onto. What does the directed graph for f look like in this case? Since f is onto, there is at least one arrow entering each node in the graph. Since f is one-to-one, there is at most one arrow entering each node in the graph. Consider the path starting at x_0 with tail part $x_0, \ldots, x_{i-1}$ and cycle part $x_i, x_{i+1}, \ldots, x_{j-1}, x_j = x_i$. In this path, the tail part must be absent (meaning that $i = 0$), since otherwise the arrows starting at x_{i-1} and x_{j-1} both lead to x_i. This is impossible, since $x_{i-1} \neq x_{j-1}$ (by minimality of i) and f is one-to-one. We conclude that every path obtained by following arrows through the graph of f is a directed cycle, possibly involving a single node. Every node belongs to such a cycle, since every node has an arrow leaving it. Different cycles visit disjoint sets of nodes, since otherwise there would be two arrows entering the same node. So, *the directed graph representing a permutation is a disjoint union of directed cycles.*

It follows that we can describe a permutation f of $[n]$ by listing all the cycles in the directed graph of f, writing the elements within each cycle in the order they appear along the cycle. The cycles themselves can be listed in any order, and we can start at any position on each cycle when listing the nodes in that cycle. Any such listing of the directed cycles of f is called a *cycle decomposition* of f. The notation $(i_1, i_2, \ldots, i_k)$ represents a cycle that visits the k distinct nodes $i_1, i_2, \ldots, i_k$ in this order and then returns to i_1. Observe that $(i_2, \ldots, i_k, i_1)$ and $(i_3, \ldots, i_k, i_1, i_2)$ are alternate notations for the cycle just mentioned, but $(i_k, \ldots, i_2, i_1)$ is a different cycle (for $k \geq 3$) since the direction of the edges matters. If the graph of f has multiple cycles, we juxtapose the notation for each individual cycle to get a cycle decomposition of f. For example, let $n = 6$ and $f = [4, 2, 5, 6, 3, 1]$ in one-line form. The directed graph for f is shown in Figure 2.2, and a cycle decomposition of f is $(1, 4, 6)(2)(3, 5)$. Two other cycle decompositions for the same f are $(5, 3)(6, 1, 4)(2)$ and $(2)(3, 5)(4, 6, 1)$.

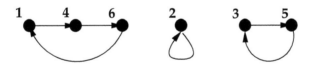

FIGURE 2.2
Directed graph of a permutation.

A cycle containing k nodes is called a *k-cycle*. When the domain $[n]$ has been specified, we are permitted to omit some or all 1-cycles from the cycle decomposition of a permutation. For example, the f in Figure 2.2 could also be written $(1, 4, 6)(3, 5)$. The identity function in S_6 has cycle decomposition $(1)(2)(3)(4)(5)(6)$, which we may abbreviate as (1). Given $h = (1, 3)(2, 5)(4)(6) \in S_6$, we can write $h = (1, 3)(2, 5) = (5, 2)(4)(3, 1)$. If we did not know $n = 6$, then the notation $(1, 3)(2, 5)$ is ambiguous: it could represent a permutation in S_n for any $n \geq 5$. Given that n is known, individual cycles of any length define permutations of $[n]$ by restoring the omitted 1-cycles. For instance, we have permutations $f_1 = (1, 4, 6)$, $f_2 = (2) = \mathrm{id}$, and $f_3 = (3, 5)$, which are given in one-line form by $f_1 = [4, 2, 3, 6, 5, 1]$, $f_2 = [1, 2, 3, 4, 5, 6]$, and $f_3 = [1, 2, 5, 4, 3, 6]$.

The one-line form for a function encloses the function values in square brackets, while the cycle notation for a k-cycle uses round parentheses around the nodes in the cycle. For example, $f = [4, 2, 5, 6, 3, 1]$ is different from the 6-cycle $(4, 2, 5, 6, 3, 1) = (1, 4, 2, 5, 6, 3)$. The one-line form of this 6-cycle is $[4, 5, 1, 2, 6, 3] \neq f$.

We can compose two permutations using their cycle decompositions. Given $f = (1, 4, 6)(2)(3, 5)$ and $g = (2, 6, 4, 3)(1, 5)$, let us find $f \circ g$ and $g \circ f$. For $f \circ g$, we could compute $f(g(1)) = f(5) = 3$, $f(g(2)) = f(6) = 1$, $f(g(3)) = f(2) = 2$, etc., leading to the one-line form $f \circ g = [3, 1, 2, 5, 4, 6]$. Converting this to a cycle decomposition, we have $f \circ g = (1, 3, 2)(4, 5)(6)$. Next, we see how to compute a cycle decomposition of $g \circ f$ without finding the one-line form first. Node 1 belongs to some cycle of $g \circ f$; indeed, $g(f(1)) = g(4) = 3$, so 1 is followed on its cycle by 3. Next, $g(f(3)) = g(5) = 1$, so 3 is followed on the cycle by 1, and we see that the 2-cycle $(1, 3)$ is part of the directed graph of $g \circ f$. Next, we consider node 2, and note $g(f(2)) = g(2) = 6$, then $g(f(6)) = g(1) = 5$, then $g(f(5)) = g(3) = 2$, so we have the cycle $(2, 6, 5)$. Finally, we find the cycle involving node 4 by noting $g(f(4)) = g(6) = 4$. Thus, one possible cycle decomposition of $g \circ f$ is $(1, 3)(2, 6, 5)(4)$. If needed, we can now find the one-line form $g \circ f = [3, 6, 1, 4, 2, 5]$.

Let $f = (1, 4, 6)(2)(3, 5)$, $f_1 = (1, 4, 6)$, $f_2 = (2)$, and $f_3 = (3, 5)$. By a computation similar to the one given in the last paragraph, we see that

$$f = (1, 4, 6)(2)(3, 5) = (1, 4, 6) \circ (2) \circ (3, 5) = f_1 \circ f_2 \circ f_3.$$

Note here that the single permutation f on the left side has been written as a product (in the group S_6) of three permutations f_1, f_2, f_3 on the right side. Intuitively, the action of f on $[n]$ can be accomplished by first applying f_3, which moves only the elements on the cycle $(3, 5)$; then applying f_2, which happens to be the identity map; and then applying f_1, which moves only the elements on the cycle $(1, 4, 6)$. Similarly, $f = f_1 \circ f_3 \circ f_2 = f_2 \circ f_1 \circ f_3 = f_3 \circ f_1 \circ f_2$, etc. We achieve the same net effect no matter what order we apply the functions f_1, f_2, and f_3, which translates into the algebraic statement that these three elements of S_n commute with each other.

2.4   Composition of Cycles

We now generalize the example at the end of the last section. Fix n, and consider two cycles $g = (a_1, \ldots, a_k)$ and $h = (b_1, \ldots, b_m)$ in the group S_n. We say g and h are *disjoint* cycles iff $\{a_1, \ldots, a_k\}$ and $\{b_1, \ldots, b_m\}$ are disjoint sets. We prove that if g and h are disjoint cycles, then $g \circ h = h \circ g$. We say that *disjoint cycles commute in S_n*. For the proof, recall that two functions $p, q : [n] \to [n]$ are *equal* iff $p(x) = q(x)$ for all $x \in [n]$. In the situation at hand, we have $p = g \circ h$ and $q = h \circ g$. Take an arbitrary $x \in [n]$, and consider three cases. First, if $x = a_i$ for some i, then

$$g \circ h(x) = g(h(a_i)) = g(a_i) = h(g(a_i)) = h \circ g(x)$$

since $h(a_j) = a_j$ for all j, and $g(a_i)$ equals some a_j. Second, if $x = b_i$ for some i, then

$$g \circ h(x) = g(h(b_i)) = h(b_i) = h(g(b_i)) = h \circ g(x)$$

since $g(b_j) = b_j$ for all j, and $g(b_i)$ equals some b_j. Third, if $x \notin \{a_1, \ldots, a_k, b_1, \ldots, b_m\}$, then

$$g \circ h(x) = x = h \circ g(x).$$

This completes the proof that $g \circ h = h \circ g$. We remark that the 1-cycle id commutes with every element of S_n, without needing any assumption of disjointness.

We now verify a few more identities involving composition of cycles. Suppose $f = C_1 C_2 \cdots C_k$ is any cycle decomposition of $f \in S_n$. Let us check carefully that $f = C_1 \circ C_2 \circ \cdots \circ C_k$. The notation C_i refers both to a cycle in the directed graph of f, and the function in S_n corresponding to that cycle. We must show that the function $C_1 \circ C_2 \circ \cdots \circ C_k$ applied to any $x \in [n]$ yields $f(x)$. If x does not appear in any of the cycles C_i, then all of the permutations $C_k, C_{k-1}, \ldots, C_1$ leave x fixed. (We say a function g *fixes* an element x iff $g(x) = x$.) Also, f itself must fix x since we are only allowed to omit 1-cycles in a cycle decomposition. Thus, $f(x) = x = C_1 \circ C_2 \circ \cdots \circ C_k(x)$. The other possibility is that x appears in a cycle C_i for some (necessarily unique) i. If $f(x) = y$, then $C_i(x) = y$ because of the way we constructed the cycles from the directed graph of f. All the other cycles C_j do not overlap with C_i, and therefore the functions associated with these other cycles must fix x and y. Starting at x and applying the functions C_k through C_1 in this order, x is always fixed until C_i sends x to y, and then y is always fixed. Hence, $C_1 \circ \cdots \circ C_k(x) = y = f(x)$, as needed.

Next, consider a general k-cycle $(a_1, a_2, \ldots, a_k)$, where $k \geq 3$. We show that

$$(a_1, a_2, \ldots, a_k) = (a_1, a_k) \circ (a_1, a_{k-1}) \circ \cdots \circ (a_1, a_3) \circ (a_1, a_2). \tag{2.1}$$

If x does not equal any a_j, then both sides of (2.1) fix x. Both sides send a_1 to a_2 and a_k to a_1. (Remember that the maps on the right side are applied from right to left.) For $1 < j < k$, the left side sends a_j to a_{j+1}. Reading the other side from right to left, the first $j - 2$ factors leave a_j fixed. The next 2-cycle, namely (a_1, a_j), sends a_j to a_1. The next 2-cycle, namely (a_1, a_{j+1}), sends a_1 to a_{j+1}. The remaining 2-cycles fix a_{j+1}, so the net effect is that a_j goes to a_{j+1}.

By similar arguments, we can verify the identity

$$(j, i) = (i, j) = (i, i+1) \circ (i+1, i+2) \circ \cdots \circ (j-2, j-1) \circ (j-1, j)$$
$$\circ (j-2, j-1) \circ \cdots \circ (i+1, i+2) \circ (i, i+1), \tag{2.2}$$

which holds for all $i, j \in [n]$ with $i < j$. For example, $(2, 5) = (2, 3) \circ (3, 4) \circ (4, 5) \circ (3, 4) \circ (2, 3)$.

2.5 Factorizations of Permutations

We now consider various ways of factoring permutations in the group S_n into products of simpler permutations. Here, "product" refers to the group operation in S_n, which is composition of functions. These results are analogous to the Fundamental Theorem of Arithmetic, which states that any positive integer can be written as a product of prime integers.

Factorization into Disjoint Cycles. Given any permutation $f \in S_n$, our analysis of the directed graph of f shows that f has a cycle decomposition $f = C_1 C_2 \cdots C_k$. We proved in Section 2.4 that this cycle decomposition implies the factorization $f = C_1 \circ C_2 \circ \cdots \circ C_k$. So, *every $f \in S_n$ can be written as a product of k pairwise disjoint cycles in S_n.*

To what extent is this factorization unique? On one hand, the cycles $C_1, \ldots, C_k$ are pairwise disjoint, so they all commute with each other. On the other hand, we are free to add or remove 1-cycles from this factorization, since all 1-cycles represent the identity element of S_n. If we do not consider these minor variations (reordering disjoint cycles, omitting 1-cycles) to be different from one another, then *the factorization of f into disjoint cycles is unique.* We omit a formal proof of uniqueness. The idea of the proof is to show that the non-identity cycles in any factorization of f into a product of disjoint cycles must correspond (in some order) to the cycles in the directed graph of f involving more than one node. These directed cycles are uniquely determined by f, so the cycles in the factorization of f are uniquely determined as well.

Factorizations into Transpositions. A *transposition* is a 2-cycle (i, j). The formula (2.1) shows that any k-cycle with $k \geq 3$ can be factored in S_n as a product of transpositions. Applying this formula to each cycle in a disjoint cycle factorization of f, we see that *every $f \in S_n$ can be written as a product of zero or more transpositions.* Here, an empty product represents the identity permutation, which can also be factored as $\text{id} = (1, 2) \circ (1, 2)$ for $n \geq 2$. Note that the transpositions used in such a factorization often are not disjoint. Furthermore, the factorization of a permutation into transpositions is not unique. For example, given any factorization of any $g \in S_n$ into transpositions, we can always create new factorizations of g by appending pairs of factors of the form $(i, j) \circ (i, j)$. There can also be essentially different factorizations involving the same number of transposition factors. For instance,

$$(1, 4, 3, 2) = [4, 1, 2, 3] = (1, 3) \circ (1, 4) \circ (2, 3) = (3, 4) \circ (2, 3) \circ (1, 2).$$

Factorizations into Basic Transpositions. A *basic transposition* in S_n is a transposition of the form $(i, i+1)$ for some $i \in \{1, 2, \ldots, n-1\}$. Formula (2.2) shows that any transposition can be factored in S_n into a product of basic transpositions. Applying this formula to each transposition in a factorization of $f \in S_n$ into transpositions, we see that *every $f \in S_n$ can be written as a product of zero or more basic transpositions.* These factorizations are not unique; for instance,

$$(1, 3, 4) = [3, 2, 4, 1] = (1, 2) \circ (2, 3) \circ (3, 4) \circ (1, 2) = (2, 3) \circ (1, 2) \circ (2, 3) \circ (3, 4).$$

The next section gives an algorithm (based on sorting the one-line form of f) for factoring a given $f \in S_n$ into a product of basic transpositions.

2.6 Inversions and Sorting

Consider a function $f : [n] \to [m]$, where n and m are positive integers. An *inversion* of f is a pair (i, j) such that $1 \leq i < j \leq n$ and $f(i) > f(j)$. In terms of the one-line form $[f(1), f(2), \ldots, f(n)]$, an inversion of f is a pair of positions (i, j), not necessarily adjacent, such that $f(i)$ and $f(j)$ appear out of order. Let $\text{Inv}(f)$ be the set of inversions of f and $\text{inv}(f) = |\text{Inv}(f)|$ be the number of inversions of f. These concepts apply, in particular, to permutations $f \in S_n$. For example, given $f = [3, 1, 4, 2]$, one inversion of f is $(1, 2)$, since $f(1) > f(2)$ $(3 > 1)$. Listing all the inversions, we find that $\text{Inv}(f) = \{(1, 2), (1, 4), (3, 4)\}$ and $\text{inv}(f) = 3$. For another example, $g = [1, 1, 2, 1, 1, 2]$ has $\text{Inv}(g) = \{(3, 4), (3, 5)\}$ and $\text{inv}(g) = 2$.

Given $f : [n] \to [m]$, we can sort the list $[f(1), f(2), \ldots, f(n)]$ into weakly increasing order by a sequence of basic transposition moves. By definition, a *basic transposition move* consists of switching two adjacent list elements that are strictly out of order; in other words, we are permitted to switch $f(i)$ and $f(i+1)$ iff $f(i) > f(i+1)$. How many such moves does it take to sort the list? We claim that, regardless of which moves are made in what order, the list is always sorted in exactly $\text{inv}(f)$ moves. This claim follows from the following three assertions. (1) The list is fully sorted iff $\text{inv}(f) = 0$. (2) If the list is not fully sorted, then there exists at least one permissible basic transposition move. (3) If g is obtained from f by applying any one basic transposition move, then $\text{inv}(g) = \text{inv}(f) - 1$. Assertions (1) and (2) are readily verified. To check (3), suppose $f(i) > f(i+1)$, and we get g from f by switching these two values in the one-line form of f. By definition of inversions, we see that the set $\text{Inv}(f)$ can be transformed into $\text{Inv}(g)$ by deleting the inversion $(i, i+1)$, and then replacing i by $i+1$ and $i+1$ by i in all other inversion pairs in which i or $i+1$ appears. We see from this that $\text{Inv}(g)$ has one less element than $\text{Inv}(f)$, so that (3) holds. For example, if $f = [3, 1, 4, 2]$ is transformed into $g = [3, 1, 2, 4]$ by switching the last two elements, then $\text{Inv}(f) = \{(1, 2), (1, 4), (3, 4)\}$ is transformed into $\text{Inv}(g) = \{(1, 2), (1, 3)\}$ by the process just described.

This sorting method leads to an algorithm for factoring permutations $f \in S_n$ into products of basic transpositions. Suppose g is obtained from f by switching $f(i)$ and $f(i+1)$ in the one-line form of f. We may check that $g = f \circ (i, i+1)$, so that (in the case $f(i) > f(i+1)$) *the basic transposition move switching the entries in positions i and $i+1$ can be accomplished algebraically in S_n by multiplying f on the right by $(i, i+1)$.* Sorting f to $\text{id} = [1, 2, \ldots, n]$ by a sequence of such moves therefore gives a formula of the form

$$f \circ (i_1, i_1 + 1) \circ (i_2, i_2 + 1) \circ \cdots \circ (i_k, i_k + 1) = \text{id},$$

where $k = \text{inv}(f)$. Solving for f, we find that $f = (i_k, i_k + 1) \circ \cdots \circ (i_2, i_2 + 1) \circ (i_1, i_1 + 1)$. In our running example where $f = [3, 1, 4, 2]$, we can write $f \circ (3, 4) \circ (1, 2) \circ (2, 3) = \text{id}$, and hence $f = (2, 3) \circ (1, 2) \circ (3, 4)$. To summarize, we have proved that *any $f \in S_n$ can be factored into the product of $\text{inv}(f)$ basic transpositions.*

We can also prove that *each $f \in S_n$ cannot be factored into a product of fewer than $\text{inv}(f)$ basic transpositions.* Suppose $f = t_1 \circ t_2 \circ \cdots \circ t_k$ where each t_j is a basic transposition. Then $f \circ t_k \circ \cdots \circ t_2 \circ t_1 = \text{id}$. This equation says that the one-line form for f can be sorted to the one-line form for id by k steps, where each step switches two adjacent entries. Each such step either raises or lowers the inversion count by 1. Since the starting list has $\text{inv}(f)$ inversions and the final list has zero inversions, we must have $k \geq \text{inv}(f)$, as needed.

2.7   Signs of Permutations

For any function $f : [n] \to [n]$, define the *sign* of f to be $\text{sgn}(f) = (-1)^{\text{inv}(f)}$ if f is a bijection, and $\text{sgn}(f) = 0$ otherwise. For example, if $f = \text{id} = [1, 2, \ldots, n]$, then $\text{inv}(f) = 0$ and $\text{sgn}(f) = 1$. If h is the basic transposition $(i, i+1)$, then the one-line form for h is

$$[1, \ldots, i-1, i+1, i, i+2, \ldots, n],$$

so that $\text{Inv}(h) = \{(i, i+1)\}$, $\text{inv}(h) = 1$, and $\text{sgn}(h) = -1$.

Sign Homomorphism Theorem. For all $n \geq 1$, the function $\text{sgn} : S_n \to \{-1, 1\}$ is a group homomorphism, meaning:

$$\text{for all } f, h \in S_n, \ \text{sgn}(f \circ h) = \text{sgn}(f) \cdot \text{sgn}(h). \tag{2.3}$$

Proof. First, (2.3) holds when $h = \text{id}$, since $f \circ \text{id} = f$ and $\text{sgn}(\text{id}) = 1$. Next, consider the special case where h is a basic transposition $(i, i+1)$. Fix $f \in S_n$, and set $g = f \circ h = f \circ (i, i+1)$. We know that the one-line form for g is obtained from the one-line form for f by switching the entries $f(i)$ and $f(i+1)$ in positions i and $i+1$. Hence, if $f(i) > f(i+1)$, then g is obtained from f by applying a basic transposition move. By assertion (3) in §2.6, $\mathrm{inv}(g) = \mathrm{inv}(f) - 1$, and so

$$\mathrm{sgn}(g) = -\mathrm{sgn}(f) = \mathrm{sgn}(f) \cdot \mathrm{sgn}((i, i+1)) = \mathrm{sgn}(f) \cdot \mathrm{sgn}(h).$$

In the case where $f(i) < f(i+1)$, we see that f is obtained from g by applying a basic transposition move. Applying assertion (3) with f and g interchanged, we see that $\mathrm{inv}(g) = \mathrm{inv}(f) + 1$, and again

$$\mathrm{sgn}(g) = -\mathrm{sgn}(f) = \mathrm{sgn}(f) \cdot \mathrm{sgn}((i, i+1)) = \mathrm{sgn}(f) \cdot \mathrm{sgn}(h).$$

Therefore, (2.3) holds when h is any basic transposition.

To prove (2.3) for arbitrary $h \in S_n$, write $h = h_1 \circ \cdots \circ h_s$ where each h_j is a basic transposition. We now use induction on s. If $s = 0$ or $s = 1$, then we already know that (2.3) holds for all $f \in S_n$. For the induction step, assume $s \geq 2$ and that (2.3) holds whenever h is a product of fewer than s basic transpositions. Using associativity, we get

$$
\begin{aligned}
\mathrm{sgn}(f \circ h) &= \mathrm{sgn}((f \circ h_1 \circ \cdots \circ h_{s-1}) \circ h_s) \\
&= \mathrm{sgn}(f \circ h_1 \circ \cdots \circ h_{s-1}) \cdot \mathrm{sgn}(h_s) \\
&= \mathrm{sgn}(f) \cdot \mathrm{sgn}(h_1 \circ \cdots \circ h_{s-1}) \cdot \mathrm{sgn}(h_s) \\
&= \mathrm{sgn}(f) \cdot \mathrm{sgn}(h_1 \circ \cdots \circ h_{s-1} \circ h_s) \\
&= \mathrm{sgn}(f) \cdot \mathrm{sgn}(h). \quad \square
\end{aligned}
$$

We note five consequences of the Sign Homomorphism Theorem. First, since $f \circ f^{-1} = \mathrm{id}$, we have $1 = \mathrm{sgn}(\mathrm{id}) = \mathrm{sgn}(f) \cdot \mathrm{sgn}(f^{-1})$. So $\mathrm{sgn}(f^{-1}) = 1/\mathrm{sgn}(f) = \mathrm{sgn}(f)^{-1}$. Since $1^{-1} = 1$ and $(-1)^{-1} = -1$, we conclude: $\mathrm{sgn}(f^{-1}) = \mathrm{sgn}(f)$ *for all* $f \in S_n$.

Second, if $h = h_1 \circ \cdots \circ h_s$ has been factored as a product of s basic transpositions h_i, then $\mathrm{sgn}(h) = \prod_{i=1}^{s} \mathrm{sgn}(h_i) = (-1)^s$. For $h = (i, j)$, (2.2) shows that we can write h as the product of an odd number of basic transpositions. Therefore, $\mathrm{sgn}((i, j)) = -1$ *for all transpositions* (not just basic transpositions).

Third, if $h = h_1 \circ \cdots \circ h_s$ has been factored as a product of s transpositions h_i, then $\mathrm{sgn}(h) = \prod_{i=1}^{s} \mathrm{sgn}(h_i) = (-1)^s$. In particular, if h can be written in some other way as a product of s' transpositions, then we must have $(-1)^s = (-1)^{s'}$, so that s and s' are both even or both odd. Thus, *while factorizations of a given permutation into transpositions are not unique, the parity of the number of transpositions in such factorizations is unique.*

Fourth, if h is a k-cycle, (2.1) shows that h can be written as a product of $k - 1$ transpositions. Therefore, $\mathrm{sgn}(h) = (-1)^{k-1}$. So *even-length cycles have negative sign, whereas odd-length cycles have positive sign.*

Fifth, suppose $h \in S_n$ is an arbitrary permutation whose disjoint cycle decomposition (including all 1-cycles) consists of c cycles with respective lengths $k_1, \ldots, k_c$. Then $\mathrm{sgn}(h) = \prod_{i=1}^{c} (-1)^{k_i - 1}$. Since $k_1 + \cdots + k_c = n$, the net power of -1 is $n - c$. So for $h \in S_n$, $\mathrm{sgn}(h) = (-1)^{n-c}$, *where c is the number of cycles in any cycle decomposition of h that includes all 1-cycles.*

2.8 Summary

1. *Symmetric Groups.* For each positive integer n, let $[n] = \{1, 2, \ldots, n\}$. S_n consists of all bijections (permutations) $f : [n] \to [n]$ with composition of functions as the group operation. S_n is a group of size $n!$, which is non-commutative for $n \geq 3$.

2. *Directed Graphs of Functions.* A function $f : [n] \to [n]$ can be represented by a directed graph with vertex set $[n]$ and a directed edge from i to $f(i)$ for each $i \in [n]$ (which is a loop if $f(i) = i$). For general f, the directed graph of f consists of directed trees feeding into directed cycles. For bijections $f \in S_n$, the directed graph of f consists of one or more directed cycles involving pairwise disjoint sets of vertices. We obtain cycle decompositions of f by traversing all these cycles (with the possible omission of length-1 cycles) and listing the elements encountered.

3. *Factorizations of Permutations.* Every $f \in S_n$ can be factored in S_n into a product of disjoint cycles. This factorization is unique except for the order of the factors and the possible omission of 1-cycles. Disjoint cycles commute. Every $f \in S_n$ can be factored as a product of transpositions (i, j) and as a product of basic transpositions $(i, i+1)$. These factorizations are not unique, but the parity (even or odd) of the number of transpositions used is unique.

4. *Facts about Inversions.* For any function $f : [n] \to [m]$, inv(f) is the number of pairs $i < j$ such that $f(i) > f(j)$. If we sort the list $[f(1), \ldots, f(n)]$ by interchanging adjacent elements that are out of order, then the sorting finishes after exactly inv(f) interchanges. Any $f \in S_n$ can be written as the product of inv(f) basic transpositions, but no fewer.

5. *Facts about Sign.* For $f \in S_n$, define sgn$(f) = (-1)^{\text{inv}(f)}$. For all $f, g \in S_n$, sgn$(f \circ g) = $ sgn$(f) \cdot$ sgn(g) (Sign Homomorphism Theorem). We have sgn$(\text{id}) = 1$, sgn$((i, j)) = -1$ for all $i \neq j$, and sgn$(g^{-1}) = $ sgn(g) for all $g \in S_n$. For $f \in S_n$, sgn$(f) = (-1)^{n-c}$, where c is the number of cycles in any cycle decomposition of f that includes all 1-cycles. If f can be factored into a product of s transpositions, then sgn$(f) = (-1)^s$. The sign of any k-cycle is $(-1)^{k-1}$.

2.9 Exercises

1. Let $f = [2, 2, 5, 1, 3]$ and $g = [3, 4, 2, 5, 1]$. Find $f \circ g$, $g \circ f$, f^3 (defined to be $f \circ f \circ f$), and g^{-1}, giving all answers in one-line form.

2. Let $f = [3, 1, 2, 4]$ and $g = [4, 3, 2, 1]$. Find $f \circ g$, $g \circ f$, $f \circ g \circ f^{-1}$, f^3, and f^{1001}, giving all answers in one-line form. Explain how you found f^{1001}.

3. Suppose $f : X \to Y$ and $g : Y \to Z$ are functions.
   (a) Prove: if f and g are one-to-one, then $g \circ f$ is one-to-one.
   (b) Prove: if f and g are onto, then $g \circ f$ is onto.
   (c) Prove: if f and g are bijections, then $g \circ f$ is a bijection.

4. Suppose $f : X \to Y$ and $g : Y \to Z$ are functions.
   (a) Prove: if $g \circ f$ is one-to-one, then f is one-to-one.
   (b) Prove: if $g \circ f$ is onto, then g is onto.
   (c) Give an example where $g \circ f$ is one-to-one but g is not one-to-one.
   (d) Give an example where $g \circ f$ is onto but f is not onto.

5. Prove that function composition is associative: for all functions $h : W \to X$, $g : X \to Y$, and $f : Y \to Z$, $f \circ (g \circ h) = (f \circ g) \circ h$.

6. Show that the group S_n is not commutative for $n \geq 3$.

7. (a) Given $f : [n] \to [n]$, explain informally why f is one-to-one iff f is onto.

(b) Give an example of a function $f : \mathbb{Z} \to \mathbb{Z}$ that is one-to-one but not onto.

(c) Give an example of a function $g : \mathbb{Z} \to \mathbb{Z}$ that is onto but not one-to-one.

8. For any set X, let $S(X)$ be the set of all bijections $f : X \to X$.
    (a) Prove that $(S(X), \circ)$ is a group.
    (b) For which sets X is $S(X)$ a commutative group?
    (c) Suppose $X = \{x_1, \ldots, x_n\}$ is an n-element set. Prove that the groups $S(X)$ and S_n are isomorphic.

9. Draw the directed graphs for each of these functions given in one-line form:
    (a) $[2, 3, 4, 5, 5]$ (b) $[5, 4, 3, 2, 1]$ (c) $[2, 2, 2, 2, 2]$ (d) $[1, 2, 3, 4, 5]$
    (e) $[3, 4, 5, 1, 2]$ (f) $[3, 3, 7, 12, 12, 11, 12, 9, 8, 7, 11, 10]$

10. Let f be the function whose directed graph is drawn in Figure 2.1. Find the one-line forms of $f \circ f$ and $f \circ f \circ f$, and draw the associated directed graphs.

11. Suppose X is an infinite set. Must the directed graph of a bijection $f : X \to X$ be a disjoint union of directed cycles? If not, can you describe the general structure of this graph?

12. Let $f = [4, 2, 1, 3] \in S_4$. Find all possible cycle decompositions of f.

13. Compute each of the following, giving a cycle decomposition for each answer.
    (a) $(1, 3, 6, 2) \circ (2, 4, 5)(1, 6)$
    (b) the inverse of $(1, 3, 5, 7, 9)(2)(4, 8, 6)$
    (c) $[3, 1, 2, 4] \circ (3, 1, 2, 4)$
    (d) $(3, 1, 2, 4) \circ [3, 1, 2, 4]$
    (e) $(3, 1, 2, 4) \circ (3, 1, 2, 4)$
    (f) $(1, 3, 4) \circ (2, 3)(1, 5) \circ (2, 5, 3, 1)^{-1} \circ (2, 3, 4) \circ (1, 2)(3, 4)$
    (g) $(2, 3) \circ (1, 2) \circ (2, 3) \circ (3, 4) \circ (2, 3) \circ (4, 5) \circ (3, 4) \circ (2, 3) \circ (1, 2)$

14. Let $f = (3, 1, 6)(2, 4)$ and $g = (5, 2, 1, 4, 3)$ in S_6.
    (a) Compute cycle decompositions of $f \circ g$ and $g \circ f$.
    (b) Compute cycle decompositions of $f \circ g \circ f^{-1}$ and $g \circ f \circ g^{-1}$.
    (c) Compute f^k for $k = 2, 3, 4, 5, 6$.
    (d) Find (with explanation) f^{670}.

15. List all $f \in S_4$ with $f \circ f = \mathrm{id}$. Explain why these f, and no others, work.

16. Given a cycle decomposition of $f \in S_n$, how can one quickly obtain a cycle decomposition of f^{-1}? Prove your answer.

17. Given a cycle decomposition of $f \in S_n$ and a large positive integer m, describe how one can find a cycle decomposition of $f^m = f \circ f \circ \cdots \circ f$ (m copies of f).

18. Find six permutations $h \in S_8$ such that $h \circ h = (1, 3)(2, 5)(4, 8)(6, 7)$.

19. How many $h \in S_{4n}$ satisfy $h \circ h = (1, 2)(3, 4)(5, 6) \cdots (4n - 1, 4n)$? Explain.

20. For which $f \in S_n$ does there exist $h \in S_n$ with $h \circ h = f$? Find and prove a necessary and sufficient condition involving the cycle decomposition of f.

21. Prove $(b_1, b_2, \ldots, b_m) = (b_1, b_2) \circ (b_2, b_3) \circ \cdots \circ (b_{m-1}, b_m)$.

22. Prove Formula (2.2) in §2.4.

23. Prove that the map $C_g : S_n \to S_n$, given by $C_g(f) = g \circ f \circ g^{-1}$ for $f \in S_n$, is a group isomorphism.

24. **Conjugation Rule for Permutations.** Suppose $g \in S_n$ and $f = (i_1, i_2, \ldots, i_k)$ is a k-cycle in S_n. Prove that $g \circ f \circ g^{-1} = (g(i_1), g(i_2), \ldots, g(i_k))$. Deduce that for all $f, g \in S_n$, a cycle decomposition for $g \circ f \circ g^{-1}$ can be obtained from a cycle

decomposition for f by applying g to each number in the cycle decomposition, leaving all parentheses unchanged.

25. Find all $f \in S_5$ that commute with $g = (1,2,4)(3,5)$.

26. Suppose the directed graph of $f \in S_n$ consists of k_1 1-cycles, k_2 2-cycles, and so on. Count the cycle decompositions of f in which no 1-cycles are omitted, and shorter cycles are always listed before longer cycles. Use the Conjugation Rule (Exercise 24) to prove that this count is the number of $g \in S_n$ that commute with f.

27. **Uniqueness of Disjoint Cycle Factorizations.** Let $f \in S_n$ have a cycle decomposition $C_1 C_2 \cdots C_k$ in which all 1-cycles have been omitted. View each C_i as an element of S_n. Prove: for any factorization $f = D_1 \circ D_2 \circ \cdots \circ D_j$, where the D_i are pairwise disjoint non-identity cycles in S_n, we have $j = k$ and $D_1, \ldots, D_k$ is a reordering of $C_1, \ldots, C_k$ of S_n). (Study the cycles of the directed graph of $D_1 \circ \cdots \circ D_j$.)

28. Make a table listing all permutations $f \in S_4$. In column 1, write f in one-line form. In column 2, draw the directed graph for f. In column 3, give a cycle decomposition of f. In column 4, compute $\mathrm{inv}(f)$. In column 5, compute $\mathrm{sgn}(f)$.

29. Let $g = [2,4,7,1,5,3,6] \in S_7$. Compute $\mathrm{Inv}(g)$, $\mathrm{inv}(g)$, $\mathrm{sgn}(g)$, $\mathrm{Inv}(g^{-1})$, $\mathrm{inv}(g^{-1})$, and $\mathrm{sgn}(g^{-1})$.

30. Let $h = (2,4,7)(1,5)(3,6) \in S_7$. Compute $\mathrm{Inv}(h)$, $\mathrm{inv}(h)$, $\mathrm{sgn}(h)$, $\mathrm{Inv}(h^{-1})$, $\mathrm{inv}(h^{-1})$, and $\mathrm{sgn}(h^{-1})$.

31. Suppose $f \in S_{12}$ has a cycle decomposition $f = (a,b,c)(d,e,g,h)(j,k)$. What is $\mathrm{sgn}(f)$?

32. Find and prove a simple relation between $\mathrm{inv}(f)$ and $\mathrm{inv}(f^{-1})$, for $f \in S_n$. Deduce that $\mathrm{sgn}(f) = \mathrm{sgn}(f^{-1})$.

33. Write each permutation as a product of basic transpositions.
    (a) $[4,1,3,5,2]$ (b) $(3,6)$ (c) $(1,2,4)(3,5)$ (d) $(1,4,2,5,6,3)$

34. Find, with explanation:
    (a) the number of basic transpositions in S_n;
    (b) the number of transpositions in S_n;
    (c) the number of n-cycles in S_n.

35. For even $n \geq 2$, how many $f \in S_n$ satisfy $f(1) < f(3) < f(5) < \cdots < f(n-1)$ and $f(i) < f(i+1)$ for all odd $i \in [n]$?

36. Write $f = [3,7,4,1,6,8,2,5]$ as a product of transpositions in three different ways. What is the least number of transpositions appearing in any such factorization for f?

37. Find all $f \in S_5$ with $\mathrm{inv}(f) = 8$.

38. Verify assertions (1) and (2) in §2.6.

39. Prove: for all functions $f, g : [n] \to [n]$, $\mathrm{sgn}(f \circ g) = \mathrm{sgn}(f) \cdot \mathrm{sgn}(g)$. (Recall that $\mathrm{sgn}(h) = 0$ when h is not a bijection.)

40. Fix $i < j$ in $[n]$. Use the definitions to find $\mathrm{Inv}((i,j))$, $\mathrm{inv}((i,j))$, and $\mathrm{sgn}((i,j))$.

41. Which permutation in S_n has the most inversions? Give the one-line form and a cycle decomposition for this permutation.

42. Suppose $f : [n] \to [n]$ has one-line form $[f(1), \ldots, f(n)]$, and g is obtained from this one-line form by switching $f(i)$ and $f(j)$ (the entries in positions i and j). Prove $g = f \circ (i,j)$.

43. Suppose $f : [n] \to [n]$ and $h = (i, j) \circ f$. Describe (with proof) how the one-line form for h can be obtained from the one-line form for f.

44. Use one of the previous two exercises and a sorting algorithm for one-line forms to prove that any $f \in S_n$ can be written as a product of transpositions.

45. Show that $A_n = \{f \in S_n : \text{sgn}(f) = +1\}$ is a normal subgroup of S_n. Find $|A_n|$.

46. How many permutations in S_n have sign -1?

47. Show that $V = \{(1), (1, 2)(3, 4), (1, 3)(2, 4), (1, 4)(2, 3)\}$ is a normal subgroup of S_4. Show that $H = \{(1), (1, 2, 3, 4), (1, 3)(2, 4), (1, 4, 3, 2)\}$ is a subgroup of S_4 that is not normal in S_4.

48. Prove: for all $n \geq 3$ and all $f \in S_n$ with $\text{sgn}(f) = +1$, f can be factored into a product of (not necessarily disjoint) 3-cycles.

49. (a) Prove that any $f \in S_n$ can be written as a product of factors, each of which is either $(1, 2)$ or $(1, 2, \ldots, n)$. Illustrate your proof with $f = [3, 5, 1, 4, 2]$.
    (b) What is the relation between the one-line forms of f and $f \circ (1, 2)$? What is the relation between the one-line forms of f and $f \circ (1, 2, \ldots, n)$?
    (c) Interpret parts (a) and (b) as a result about the ability to sort a list into increasing order using two particular sorting moves.

50. What can you say about the minimum number of steps needed to sort a list using the two sorting moves in the previous exercise? What if right-multiplication by $(n, \ldots, 2, 1)$ is also allowed as a move?

51. True or false? Explain each answer.
    (a) $(3, 1, 2, 4)$ is a 3-cycle.
    (b) $[3, 1, 2, 4]$ is a 3-cycle.
    (c) There exists $f \in S_4$ with $[f(1), f(2), f(3), f(4)] = (f(1), f(2), f(3), f(4))$.
    (d) For all functions $f : [n] \to [n]$, if $f \circ f \in S_n$ then $f \in S_n$.
    (e) For all $f \in S_n$, if $f^{-1} = f$ then $f = \text{id}$ or f is a 2-cycle.
    (f) For all distinct cycles $f, g \in S_n$, $f \circ g = g \circ f$ iff f and g are disjoint cycles.
    (g) Any product of c disjoint cycles in S_n has sign $(-1)^{n-c}$.
    (h) Every k-cycle in S_n $(k > 1)$ is a product of $k - 1$ basic transpositions.
    (i) Every $f \in S_n$ $(n > 1)$ can be written as a product of $\text{inv}(f) + 8$ transpositions.
    (j) For all $f, g \in S_n$ and $1 \leq i < n$, if $g = (i, i + 1) \circ f$, then $\text{inv}(g) = \text{inv}(f) \pm 1$.
    (k) Every $g \in S_n$ can be written as a product of disjoint transpositions.
    (l) For all $f \in S_n$, if $f = h \circ h$ for some $h \in S_n$, then $\text{sgn}(f) = +1$.
    (m) For all $f \in S_n$, if $\text{sgn}(f) = +1$, then $f = h \circ h$ for some $h \in S_n$.

3

Polynomials

Polynomials appear everywhere in linear algebra and throughout mathematics. Most of us learn about polynomial functions in calculus. In algebraic settings, we need a more formal concept of polynomial as a symbolic expression $a_0 + a_1 x + a_2 x^2 + \cdots + a_n x^n$ that is not necessarily a function of x. We begin this chapter with an intuitive description of these formal polynomials, which form a ring using the familiar algebraic rules for adding and multiplying polynomials. Then we make this intuitive discussion more precise by giving a rigorous definition of polynomials in terms of formal power series and connecting this definition with the idea of a polynomial function.

The next part of the chapter studies the divisibility structure of the ring of one-variable polynomials with coefficients in a field. There is a close analogy between factorization of such polynomials and corresponding results on factorization of integers. Recall that for any integers a, b with $b \neq 0$, there is a *division algorithm* that produces a unique quotient $q \in \mathbb{Z}$ and remainder $r \in \mathbb{Z}$ satisfying $a = bq + r$ and $0 \leq r < |b|$. We develop a similar division algorithm for polynomials, where the remainder must now have smaller degree than b. This algorithm gives useful information on divisors, greatest common divisors, and least common multiples of polynomials. To understand how polynomials factor, we introduce irreducible polynomials, which are non-constant polynomials that cannot be factored into products of two other polynomials of smaller degree. Irreducible polynomials are the analogs of prime integers. We show that every nonzero polynomial has a unique factorization into a product of irreducible polynomials, just as every nonzero positive integer can be written uniquely as a product of primes.

The chapter concludes with more facts on irreducible polynomials. We present some theorems and algorithms for testing whether polynomials with coefficients in certain fields are irreducible. In particular, Kronecker's factoring algorithm gives a laborious but definitive method for testing a polynomial with rational coefficients for irreducibility, or finding the irreducible factorization of such a polynomial. We also discuss minimal polynomials, which are polynomials that help us understand the structure of finite-dimensional F-algebras.

3.1 Intuitive Definition of Polynomials

We begin by presenting an informal definition of polynomials, which is made precise in §3.3. Let R be any ring. Intuitively, a *one-variable polynomial with coefficients in R* is a symbolic expression of the form
$$a_0 + a_1 x + a_2 x^2 + \cdots + a_n x^n,$$
where $n \geq 0$ is an integer, each a_i is an element of R, and the letter x is an *indeterminate* or *formal variable*. We often display such a polynomial using the summation notation $\sum_{i=0}^{n} a_i x^i$. It is convenient to define a_i for all integers $i \geq 0$ by setting $a_i = 0_R$ for $i > n$. By definition, two polynomials $\sum_{i \geq 0} a_i x^i$ and $\sum_{i \geq 0} b_i x^i$ are *equal* iff $a_i = b_i$ for all $i \geq 0$.

DOI: 10.1201/9781003484561-3

So, two polynomials are equal iff all corresponding coefficients are equal. A polynomial is only permitted to have finitely many nonzero coefficients. The symbol $R[x]$ denotes the set of all polynomials with coefficients in R. For example, $1 + 3x - \pi x^2 + \sqrt{7}x^4 \in \mathbb{R}[x]$, but $\sum_{n=0}^{\infty}(1/n!)x^n$ is not a polynomial.

The unique polynomial whose coefficients are all zero is called the *zero polynomial*. The *degree* of a nonzero polynomial $p = \sum_{i \geq 0} a_i x^i$ is the largest integer n such that $a_n \neq 0_R$. The degree of the zero polynomial is undefined. Given a polynomial $p = \sum_{i=0}^{n} a_i x^i$ of degree n, we write $\deg(p) = n$ and introduce the following terminology: $a_n x^n$ is called the *leading term* of p; x^n is called the *leading monomial* of p; and a_n is called the *leading coefficient* of p. We call p a *monic* polynomial iff $a_n = 1$. Polynomials of degree zero, one, two, three, four, and five are respectively called *constant* polynomials, *linear* polynomials, *quadratic* polynomials, *cubic* polynomials, *quartic* polynomials, and *quintic* polynomials. The zero polynomial is also considered to be a constant polynomial.

3.2 Algebraic Operations on Polynomials

Given any ring R, we can give $R[x]$ a ring structure by introducing the following addition and multiplication operations on polynomials. Suppose $p = \sum_{i \geq 0} a_i x^i$ and $q = \sum_{j \geq 0} b_j x^j$ are two elements of $R[x]$, where $a_i, b_j \in R$. We define the *sum* of p and q by

$$p + q = \sum_{k \geq 0}(a_k + b_k)x^k.$$

If the distributive law is to hold for polynomials, the product pq must be given by

$$pq = \sum_{i \geq 0}\sum_{j \geq 0} a_i b_j x^{i+j}.$$

To present this answer in standard form, we need to collect like powers of x. In the preceding expression, the coefficient of x^k is the sum of all products $a_i b_j$ such that $i + j = k$. So we define the *product* of p and q by

$$pq = \sum_{k \geq 0}\left(\sum_{i+j=k} a_i b_j\right) x^k = \sum_{k \geq 0}\left(\sum_{i=0}^{k} a_i b_{k-i}\right) x^k.$$

(In summations like this, it is always understood that the summation variables only take nonnegative integer values.) With these definitions, it is now tedious but straightforward to check that the ring axioms do hold for $R[x]$. We must confirm that addition of polynomials is closed, associative, commutative, has an identity element (the zero polynomial), and additive inverses exist. Similarly, multiplication of polynomials is closed, associative, has an identity element (the constant polynomial $1_{R[x]} = 1_R + 0_R x + 0_R x^2 + \cdots$), and distributes over addition. When the coefficient ring R is commutative, $R[x]$ is also commutative. We verify a few of the ring axioms now, leaving the others as exercises.

First, consider the additive inverse axiom. Assume we already have checked that the zero polynomial is the additive identity of $R[x]$. Given $p = \sum_{i=0}^{n} a_i x^i \in R[x]$, does p have an additive inverse in $R[x]$? Since R is a ring, each coefficient $a_i \in R$ has an additive inverse $-a_i \in R$, so that $q = \sum_{i=0}^{n}(-a_i)x^i$ is an element of $R[x]$. By definition of addition of polynomials, we compute

$$p + q = \sum_{i=0}^{n}(a_i + (-a_i))x^i = \sum_{i=0}^{n} 0_R x^i = 0_{R[x]} = \sum_{i=0}^{n}((-a_i) + a_i)x^i = q + p.$$

Therefore q is an additive inverse for p in $R[x]$.

Second, consider associativity of multiplication. Given polynomials $p, q, r \in R[x]$, write $p = \sum_{i \geq 0} a_i x^i$, $q = \sum_{j \geq 0} b_j x^j$, and $r = \sum_{k \geq 0} c_k x^k$, where all $a_i, b_j, c_k \in R$. We have $pq = \sum_{t \geq 0} d_t x^t$ where $d_t = \sum_{i+j=t} a_i b_j$, and $qr = \sum_{u \geq 0} e_u x^u$ where $e_u = \sum_{j+k=u} b_j c_k$. Applying the definition of product again, we see that $(pq)r = \sum_{v \geq 0} f_v x^v$ where

$$f_v = \sum_{t+k=v} d_t c_k = \sum_{(t,k):\ t+k=v} \left(\sum_{(i,j):\ i+j=t} a_i b_j \right) c_k = \sum_{(i,j,k):\ (i+j)+k=v} (a_i b_j) c_k.$$

The last step uses the distributive law in the ring R, as well as commutativity and associativity of addition in R. Likewise, $p(qr) = \sum_{v \geq 0} g_v x^v$ where

$$g_v = \sum_{i+u=v} a_i e_u = \sum_{(i,u):\ i+u=v} a_i \left(\sum_{(j,k):\ j+k=u} b_j c_k \right) = \sum_{(i,j,k):\ i+(j+k)=v} a_i (b_j c_k).$$

Since $(a_i b_j)c_k = a_i(b_j c_k)$ holds for all i, j, k by associativity in R, we see that $f_v = g_v$ for all $v \geq 0$. Therefore, by definition of equality of polynomials, $(pq)r = p(qr)$.

Now assume that F is a field. We define *scalar multiplication* on polynomials in $F[x]$ by setting $c \cdot \left(\sum_{i=0}^{n} a_i x^i \right) = \sum_{i=0}^{n}(ca_i)x^i$ for $c, a_i \in F$. Routine verifications show that $F[x]$, using polynomial addition and this scalar multiplication, is a vector space over F. Moreover, $B = \{1, x, x^2, x^3, \ldots, x^n, \ldots\}$ is a basis for $F[x]$. So $F[x]$ is an infinite-dimensional F-vector space. We can check that for all $c \in F$ and $p, q \in F[x]$, $c \cdot (pq) = (c \cdot p)q = p(c \cdot q)$. Thus, $F[x]$ is an F-algebra as defined in §1.3. In the case of an arbitrary ring R, we see similarly that $R[x]$ is an R-module.

3.3   Formal Power Series and Polynomials

In §3.1, we defined a polynomial $p \in R[x]$ to be a "symbolic expression" of the form $a_0 + a_1 x + \cdots + a_n x^n$, with each $a_i \in R$. However, we were rather vague about the meaning of the letter x in this expression. In order to build a sound theory, we need to have a more rigorous definition of what a polynomial is. The key to the formal definition of polynomials is our earlier stipulation that two polynomials are equal iff they have the same sequence of coefficients. This means we can identify a polynomial $\sum_{i=0}^{n} a_i x^i$ with its sequence of coefficients $(a_0, a_1, \ldots, a_n)$. For many purposes (such as adding two polynomials of different degrees), it is technically more convenient to use an infinite coefficient sequence $(a_0, a_1, \ldots, a_n, 0_R, 0_R, \ldots)$ that ends with an infinite list of zero coefficients.

Using this idea of a coefficient sequence, we define a *formal power series with coefficients in the ring R* to be an infinite sequence $f = (f_0, f_1, f_2, \ldots, f_i, \ldots) = (f_i : i \geq 0)$ with every $f_i \in R$. Two formal power series $f = (f_i : i \geq 0)$ and $g = (g_i : i \geq 0)$ are *equal* iff $f_i = g_i$ for all $i \geq 0$; this is the ordinary meaning of equality of two sequences. A formal power series f is called a *formal polynomial* iff there exists $N \in \mathbb{N}$ with $f_n = 0_R$ for all $n > N$ (which means f ends in an infinite string of zero terms). We write $R[[x]]$ to denote the set of all formal power series with coefficients in R, and (as above) we write $R[x]$ to denote the subset of $R[[x]]$ consisting of the formal polynomials. By analogy with our intuitive notation for polynomials, we often write $f = \sum_{i \geq 0} f_i x^i = \sum_{i=0}^{\infty} f_i x^i$ to denote a formal power series, and we call f_i *the coefficient of x^i in f*. However, note carefully that (for now) the summation symbol here is only a notational device — no summation (finite or infinite) is being performed! Similarly, the powers x^i appearing here are (for now) only symbols that form part of the notation for a formal power series or a formal polynomial.

Extending the earlier formulas for polynomial addition and multiplication to formal power series, we arrive at the following definitions. Suppose R is a ring and $f = (f_i : i \geq 0)$, $g = (g_i : i \geq 0)$ are two elements of $R[[x]]$. Define $f + g = (f_i + g_i : i \geq 0)$. Define $fg = (h_k : k \geq 0)$, where $h_k = \sum_{i+j=k} f_i g_j = \sum_{i=0}^{k} f_i g_{k-i}$ for each $k \geq 0$. Note that any particular coefficient in $f + g$ or fg can be found by computing finitely many sums and products in the underlying ring R. So our definitions of $f + g$ and fg do not rely on limits or any other infinite process. As in the case of polynomials, we can verify that $R[[x]]$ with these operations is a ring, which is commutative if R is commutative. For example, the proofs of the additive inverse axiom and associativity of multiplication, which we gave earlier for polynomials, carry over almost verbatim to the present situation. As another example, we check that $R[[x]]$ is closed under multiplication. Suppose $f, g \in R[[x]]$, and write $f = (f_i : i \geq 0)$, $g = (g_i : i \geq 0)$ for some $f_i, g_i \in R$. By definition, $fg = (h_k : k \geq 0)$ where $h_k = \sum_{i=0}^{k} f_i g_{k-i}$. To see that $fg \in R[[x]]$, we must prove that $h_k \in R$ for all $k \geq 0$. For fixed $k \geq 0$ and $0 \leq i \leq k$, $f_i g_{k-i}$ is in R because $f_i \in R$, $g_{k-i} \in R$, and R is closed under multiplication. Since h_k is a sum of finitely many elements of this form, h_k is also in R because R is closed under addition. So $fg \in R[[x]]$.

Once we prove all the ring axioms for $R[[x]]$, we can reprove that $R[x]$ is a ring by verifying that $R[x]$ is a subring of $R[[x]]$. Since the formal series $0 = (0, 0, 0, \ldots)$ and $1 = (1, 0, 0, \ldots)$ are polynomials, it suffices to check that for all $f, g \in R[x]$, $f + g \in R[x]$ and $-f \in R[x]$ and $fg \in R[x]$. In other words, if the sequences f and g each end in infinitely many zeroes, the same holds for $f + g$ and $-f$ and fg. We let the reader check this assertion in the case of $f + g$ and $-f$. To check the assertion for fg, fix $N, M \in \mathbb{N}$ such that $f_n = 0$ for all $n > N$ and $g_m = 0$ for all $m > M$. For any fixed $k > N + M$, the coefficient of x^k in fg is $\sum_{n+m=k} f_n g_m$. Since $k > N + M$, the condition $n + m = k$ forces $n > N$ or $m > M$, so $f_n = 0$ or $g_m = 0$, hence $f_n g_m = 0$. Thus, the coefficient of x^k is a sum of zeroes, so is 0.

We can now give a rigorous meaning to our original notation for polynomials as "symbolic expressions" involving x. *Define* the formal variable :w x to be the particular sequence $(0, 1, 0, 0, \ldots)$, which is both a polynomial in $R[x]$ and a formal series in $R[[x]]$. To motivate this definition, recall that we designed this coefficient sequence as the formal model for the informal polynomial expression $0 + 1x^1 + 0x^2 + 0x^3 + \cdots$. Iterating the definition of multiplication, we can check by induction that for all $i \geq 0$, x^i (the product of i copies of the sequence x in $R[[x]]$) is the sequence $(e_j : j \geq 0)$, where $e_i = 1$ and $e_j = 0$ for all $j \neq i$. Next, we can identify any ring element $a \in R$ with the constant polynomial $(a, 0, 0, \ldots)$ in $R[[x]]$. More precisely, the map $f(a) = (a, 0, 0, \ldots)$ is an injective ring homomorphism of R into $R[[x]]$. We use this injection to regard R as a subset of $R[[x]]$ (or $R[x]$), writing $a = (a, 0, 0, \ldots)$. Using this convention, we can check that

$$a_0 + a_1 x + a_2 x^2 + \cdots + a_n x^n = (a_0, a_1, a_2, \ldots, a_n, 0, 0, \ldots),$$

where the left side is built up by performing sum and product operations in the ring $R[x]$. Similarly, we can prove that the notation $\sum_{i=0}^{\infty} f_i x^i$ for a formal power series really is an infinite sum of terms, each of which is a product of f_i and i copies of x. But in this case, we need to introduce a limit concept for formal power series to define what the infinite sum means. We omit a detailed discussion of this point.

For a field F, $F[[x]]$ is an F-vector space and F-algebra with scalar multiplication defined by $c \cdot (f_i : i \geq 0) = (cf_i : i \geq 0)$ for all $c, f_i \in F$. $F[x]$ is a subspace and subalgebra of $F[[x]]$. Similarly, for a ring R, $R[[x]]$ is an R-module with submodule $R[x]$. Using the explicit description of the sequences x^i given in the previous paragraph, we can check that the set $B = \{1, x, x^2, \ldots, x^n, \ldots\}$ is an F-linearly independent subset of $F[[x]]$. However, B is not a basis of $F[[x]]$. The reason is that the span of B consists of all **finite** F-linear combinations of the powers of x, which are precisely the elements of $F[x]$. The formal series $(1_F, 1_F, 1_F, \ldots) = \sum_{n=0}^{\infty} x^n$ is an example of an element of $F[[x]]$ outside the span of B. It is true that B can be extended to a basis of $F[[x]]$, but it does not seem possible to give an explicit description of any specific basis of $F[[x]]$.

3.4 Properties of Degree

In this section, we study how the algebraic operations in $R[x]$ affect degree. Let R be any ring, and suppose p, q are nonzero polynomials in $R[x]$. Write $p = \sum_{i=0}^{N} a_i x^i$ and $q = \sum_{j=0}^{M} b_j x^j$ where $N = \deg(p)$, $M = \deg(q)$, and $a_i, b_j \in R$. By definition of addition, we see that $\deg(p+q) = N$ if $N > M$, while $\deg(p+q) = M$ if $M > N$. On the other hand, when $N = M$, it is possible for the leading terms to cancel. More precisely, when $N = M$, we have: $\deg(p+q) = N$ if $b_N \neq -a_N$; $\deg(p+q) < N$ if $b_N = -a_N$ and $p+q \neq 0$; and $\deg(p+q)$ is undefined if $p+q = 0$. To summarize, we always have

$$\deg(p+q) \leq \max(\deg(p), \deg(q))$$

assuming that p, q, and $p+q$ are nonzero; and strict inequality occurs iff $\deg(p) = \deg(q)$ and the leading terms of p and q cancel.

Next, consider the degree of the product pq, where $\deg(p) = N$ and $\deg(q) = M$ as above. In our earlier proof that $R[x]$ is a subring of $R[[x]]$, we showed that the coefficient of x^k in pq is zero for all $k > N + M$. We use a similar argument to find the coefficient of x^{N+M} in pq. By definition, this coefficient is $\sum_{i+j=N+M} a_i b_j$. One term in this sum is $a_N b_M$. For all the other terms, either $i > N$ and $a_i = 0$, or $j > M$ and $b_j = 0$. Hence,

$$pq = a_N b_M x^{N+M} + (\text{terms involving lower powers of } x).$$

We can deduce several consequences from this computation. First, *for all nonzero $p, q \in R[x]$, if the leading coefficients of p and q do not multiply to zero, then*

$$pq \neq 0 \text{ and } \deg(pq) = \deg(p) + \deg(q).$$

Let us call the displayed conclusion the *Degree Addition Formula*. Note that the Degree Addition Formula always holds if R is an integral domain (which has no zero divisors) or a field (which is a special kind of integral domain). In fact, *R is an integral domain iff $R[x]$ is an integral domain*. The forward implication follows from the Degree Addition Formula ($p, q \neq 0$ implies $pq \neq 0$), and the converse follows since R is isomorphic to a subring of $R[x]$, namely the subring of constant polynomials.

On the other hand, we can check that for any ring R, $R[x]$ *is never a field*. We now prove: *for an integral domain R, $p \in R[x]$ has a multiplicative inverse in $R[x]$ iff $\deg(p) = 0$ and p is invertible viewed as an element of R.* For the forward implication, assume $p \in R[x]$ has an inverse $q \in R[x]$. Applying the Degree Addition Formula to $pq = 1$ shows $\deg(p) + \deg(q) = \deg(1) = 0$. Since degrees are nonnegative integers, this forces $\deg(p) = \deg(q) = 0$. We also see that q (being a constant polynomial) can be regarded as an inverse of p in the ring R. For the converse implication, assume $\deg(p) = 0$ (so $p \in R$) and $pq = 1$ for some $q \in R$. The relation $pq = 1$ still holds when we view p and q as constant polynomials in $R[x]$. So p is invertible in $R[x]$. As a special case of this result, *for any field F, the invertible elements of $F[x]$ are precisely the nonzero constant polynomials.*

Given a field F, we can use degree to define some finite-dimensional subspaces of the F-vector space $F[x]$. Specifically, for each $n \geq 0$, let V_n be the set consisting of the zero polynomial and all $p \in F[x]$ of degree at most n. Each V_n is an $(n+1)$-dimensional subspace of $F[x]$ with ordered basis $(1, x, x^2, \ldots, x^n)$.

3.5 Evaluating Polynomials

From the outset, we have stressed that a formal polynomial is a sequence of coefficients, not a function of an input value x. Nevertheless, we can use a formal polynomial to create a polynomial function, as follows. Given a ring R and a polynomial $p = \sum_{i=0}^{n} a_i x^i \in R[x]$, define the *polynomial function* $f_p : R \to R$ *associated with* p by setting $f_p(c) = \sum_{i=0}^{n} a_i c^i$ for each $c \in R$. Note that the expression defining $f_p(c)$ is a formula involving the addition and multiplication operations in the given ring R. We call $f_p(c)$ the *value of the polynomial p at the input $x = c$*. We often write $p(c)$ instead of $f_p(c)$, although the latter notation is more precise.

We must take care to distinguish the (formal) polynomial p from its associated polynomial function f_p. We reiterate that p is the sequence $(a_0, a_1, \ldots, a_n, 0, 0, \ldots)$, while f_p is a certain function from R to R. This distinction may at first seem pedantic or trivial, but it is essential. Indeed, it is quite possible for two unequal polynomials to produce the same function from R to R. For example, let $R = \mathbb{Z}_2 = \{0, 1\}$ be the ring of integers modulo 2. There are only four distinct functions $g : \mathbb{Z}_2 \to \mathbb{Z}_2$, but there are infinitely many polynomials $p \in \mathbb{Z}_2[x]$ (which are sequences of zeroes and ones terminating in an infinite string of zeroes). Thus, one of these four functions must arise from infinitely many distinct polynomials. For an explicit example, consider the function $g : \mathbb{Z}_2 \to \mathbb{Z}_2$ such that $g(0) = 0$ and $g(1) = 1$. For any $i \geq 1$, the polynomial $x^i \in \mathbb{Z}_2[x]$ has g as its associated polynomial function, since $f_{x^i}(0) = 0^i = 0 = g(0)$ and $f_{x^i}(1) = 1^i = 1 = g(1)$. However, the polynomials $x^1 = (0, 1, 0, 0, \ldots)$, $x^2 = (0, 0, 1, 0, 0, \ldots)$, etc., are all distinct elements of $\mathbb{Z}_2[x]$.

We obtained the function f_p from a given formal polynomial $p = \sum_{i \geq 0} a_i x^i$ by regarding x as an actual variable that ranges over the elements of the coefficient ring R. For further study of polynomial rings, it is helpful to reverse this viewpoint, holding the value of x fixed and letting p range over all polynomials in $R[x]$. More precisely, suppose we are given a fixed ring element $c \in R$. We define a map $E_c : R[x] \to R$ by setting $E_c(p) = f_p(c) = p(c)$ for $p \in R[x]$. The map E_c sends each polynomial in $R[x]$ to the value of that polynomial at the fixed input $x = c$. We call E_c *evaluation at $x = c$*. *If R is a commutative ring and $c \in R$, then the map E_c is a ring homomorphism*, meaning that $(p+q)(c) = p(c) + q(c)$ and $(pq)(c) = p(c)q(c)$ for all $p, q \in R[x]$, and $1_{R[x]}(c) = 1_R$. The verification that E_c preserves multiplication requires commutativity of R.

The fact that E_c is a ring homomorphism is a special case of the following more general fact, called a *universal mapping property (UMP)*. Universal mapping properties appear throughout algebra and provide a unifying framework for understanding various algebraic constructions. Chapter 19 gives a systematic introduction to universal mapping properties.

Universal Mapping Property (UMP) for Polynomial Rings. Suppose R and S are commutative rings and $h : R \to S$ is a ring homomorphism. For every $c \in S$, there exists a unique ring homomorphism $H : R[x] \to S$ such that $H(x) = c$ and $H(r) = h(r)$ for all $r \in R$ (we say H *extends* h).

Proof. We first prove the uniqueness of H. Suppose $H : R[x] \to S$ is a ring homomorphism extending h such that $H(x) = c$. Let $p = \sum_{i=0}^{n} a_i x^i$ be an arbitrary polynomial in $R[x]$. As explained in §3.3, the displayed expression for p can be regarded as a sum in $R[x]$ of certain products of polynomials in $R[x]$ (namely, a_i times several copies of x). Since H is a ring homomorphism, we must have

$$H(p) = \sum_{i=0}^{n} H(a_i x^i) = \sum_{i=0}^{n} H(a_i)H(x)^i = \sum_{i=0}^{n} h(a_i)c^i.$$

Thus the value of $H(p)$ is completely determined by h and c, proving uniqueness.

To prove existence, use the formula just displayed to define a function $H : R[x] \to S$. Identifying each $a \in R$ with the constant polynomial $(a, 0, 0, \ldots)$, we see that $H(a) = h(a)c^0 = h(a)$ for all $a \in R$, so that H does extend h. In particular, $H(1_{R[x]}) = h(1_R) = 1_S$. Recalling that $x = (0, 1, 0, 0, \ldots)$, we see similarly that $H(x) = h(1_R)c^1 = c$. All that remains is to check that H preserves sums and products. Let $p = \sum_{i \geq 0} a_i x^i$ and $q = \sum_{i \geq 0} b_i x^i$ be two polynomials in $R[x]$, where $a_i, b_i \in R$. Since $p + q = \sum_{i \geq 0}(a_i + b_i)x^i$, the definition of H gives

$$H(p+q) = \sum_{i \geq 0} h(a_i + b_i)c^i = \sum_{i \geq 0}(h(a_i) + h(b_i))c^i = \sum_{i \geq 0} h(a_i)c^i + \sum_{i \geq 0} h(b_i)c^i = H(p) + H(q).$$

Next, since $pq = \sum_{i \geq 0}(\sum_{u+v=i} a_u b_v)x^i$, we find that

$$
\begin{aligned}
H(pq) &= \sum_{i \geq 0} h\left(\sum_{u+v=i} a_u b_v\right)c^i = \sum_{i \geq 0}\left(\sum_{u+v=i} h(a_u)h(b_v)\right)c^i \\
&= \sum_{i \geq 0}\sum_{u+v=i} h(a_u)c^u h(b_v)c^v \quad \text{(since S is commutative and $c^u c^v = c^i$)} \\
&= \left(\sum_{u \geq 0} h(a_u)c^u\right)\left(\sum_{v \geq 0} h(b_v)c^v\right) \quad \text{(using the distributive law in S)} \\
&= H(p)H(q). \quad \square
\end{aligned}
$$

The evaluation map E_c is the special case of the map H in the UMP obtained by taking $S = R$ and $h = \text{id}_R$. We can also apply the UMP in the situation where S is a commutative ring containing R as a subring, and $h : R \to S$ is the inclusion map. In this case, for every $c \in S$, we obtain a ring homomorphism $H_c : R[x] \to S$ such that, for each $p = \sum_{i \geq 0} a_i x^i$ in $R[x]$, $H_c(p) = p(c) = \sum_{i \geq 0} a_i c^i$. The image of this ring homomorphism is written $R[\{c\}]$ or sometimes $R[c]$. This image is a subring of S, and it is the smallest subring of S containing both R and c. You can check that this construction still works if, instead of assuming S is commutative, we only assume $rs = sr$ for all $r \in R$ and $s \in S$.

3.6   Polynomial Division with Remainder

Our next topic is the theory of divisibility in polynomial rings. The cornerstone of this theory is the following result, which precisely describes the process of polynomial long division with remainder.

Division Theorem for Polynomials. Suppose $f, g \in F[x]$, where F is a field and g is nonzero. There exist unique polynomials $q, r \in F[x]$ such that $f = gq + r$ and either $r = 0$ or $\deg(r) < \deg(g)$.

We call q the *quotient* and r the *remainder* when f is divided by g.

Proof. First we prove the existence of q and r. If $f = 0$, then we take $q = r = 0$. For nonzero f, we proceed by strong induction on $n = \deg(f)$. The base case occurs when $n < \deg(g)$; here we take $q = 0$ and $r = f$. For the induction step, fix a polynomial f with $\deg(f) = n \geq \deg(g)$, and assume that the existence assertion is already known for all polynomials of smaller degree than f. Write

$$f = ax^n + (\text{lower terms}) \quad \text{and} \quad g = bx^m + (\text{lower terms}),$$

where a and b are the leading coefficients of f and g, and $n \geq m$. *Since b is a nonzero element of the field F, its inverse b^{-1} exists in F.* So we can define $h = f - (ab^{-1}x^{n-m})g$, which is in $F[x]$. The polynomial h is the difference of two polynomials, whose leading terms both equal ax^n. It follows that $h = 0$ or $\deg(h) < n$. If $h = 0$, then we take $q = ab^{-1}x^{n-m}$ and $r = 0$. Otherwise, the induction hypothesis is applicable to h and yields polynomials $q_0, r_0 \in F[x]$ such that $h = q_0 g + r_0$ and either $r_0 = 0$ or $\deg(r_0) < \deg(g)$. Since $f = h + (ab^{-1}x^{n-m})g$, $f = qg + r$ holds if we take $q = q_0 + ab^{-1}x^{n-m}$ and $r = r_0$. Since $r = r_0$, we have $r = 0$ or $\deg(r) < \deg(g)$.

We now prove that the quotient and remainder associated with a given f and g are unique. Suppose $f = q_1 g + r_1$ and $f = q_2 g + r_2$ where: $q_1, q_2, r_1, r_2 \in F[x]$; $r_1 = 0$ or $\deg(r_1) < \deg(g)$; and $r_2 = 0$ or $\deg(r_2) < \deg(g)$. We need to prove that $q_1 = q_2$ and $r_1 = r_2$. Equating the given expressions for f and rearranging, we find that $(q_1 - q_2)g = r_2 - r_1$. The polynomial $r_2 - r_1$ is either zero or nonzero. In the case where $r_2 - r_1 = 0$, we deduce $r_1 = r_2$ and $(q_1 - q_2)g = 0$. *Since F is a field, $F[x]$ is an integral domain.* Then, since $g \neq 0$, it follows that $q_1 - q_2 = 0$ and $q_1 = q_2$. Thus the uniqueness result holds. In the other case, where $r_2 - r_1 \neq 0$, it follows that $q_1 - q_2 \neq 0$. So we may compute degrees, obtaining

$$\deg(r_2 - r_1) = \deg((q_1 - q_2)g) = \deg(q_1 - q_2) + \deg(g) \geq \deg(g).$$

This calculation uses the Degree Addition Formula in $F[x]$, which is valid *because the field F is an integral domain.* But the assumptions on r_1 and r_2 ensure that $r_2 - r_1$ has degree strictly less than $\deg(g)$. This contradiction shows that the second case never occurs.   □

We close this section with two further remarks, one computational and one theoretical. First, observe that the induction step in the existence proof can be unravelled into the iterative algorithm for long division of polynomials learned in elementary algebra. For example, suppose we are dividing $f = 6x^3 - 7x^2 + 4$ by $g = 3x + 1$ in the ring $\mathbb{Q}[x]$. The first step in the long division is to divide the leading terms of f and g, obtaining $6x^3/3x = 2x^2$. This monomial corresponds to the expression $ab^{-1}x^{n-m}$ in the proof above. As prescribed in the proof, we now multiply all of g by this monomial, subtract the result from f, and continue the division process. Usually the computations are presented in a tabular form as shown in Figure 3.1. The process ends when the leftover part is either zero

$$
\begin{array}{r}
2x^2 - 3x + 1 \\
\hline
3x+1) 6x^3 - 7x^2 + 4 \\
-6x^3 - 2x^2 \\
\hline
-9x^2 \\
9x^2 + 3x \\
\hline
3x + 4 \\
-3x - 1 \\
\hline
3
\end{array}
$$

FIGURE 3.1
Example of polynomial division with remainder.

or has degree less than the degree of g. In this example, once the division is complete, we obtain the quotient $q = 2x^2 - 3x + 1$ and the remainder $r = 3$.

The second remark concerns the possibility of extending the Division Theorem to polynomial rings $R[x]$ where R may not be a field. We restrict attention to commutative rings R. We have italicized the steps in the proof above that used the assumption that F was a field. We see that the uniqueness proof remains valid if $R[x]$ is an integral domain, which holds iff R itself is an integral domain. On the other hand, to make the existence proof work, we needed the leading coefficient of g (called b in the proof) to be an invertible element of the coefficient ring. We therefore obtain the following generalization of the Division Theorem.

Generalized Division Theorem for Polynomials. Given a commutative ring R and polynomials $f, g \in R[x]$ such that $g \neq 0$ and the leading coefficient of g is invertible in R, there exist $q, r \in R[x]$ with $f = qg + r$ and $r = 0$ or $\deg(r) < \deg(g)$. If R is an integral domain, then q and r are unique.

This theorem applies, in particular, when we divide an arbitrary polynomial f in $\mathbb{Z}[x]$ by a monic polynomial g in $\mathbb{Z}[x]$. The resulting quotient and remainder are uniquely determined polynomials in $\mathbb{Z}[x]$.

3.7 Divisibility and Associates

For the rest of this chapter, we focus attention on polynomial rings $F[x]$ where the ring F of coefficients is a field. Given $f, g \in F[x]$, we say that g *divides* f (written $g|f$) iff there exists $q \in F[x]$ with $f = gq$. We also say that g is a *divisor* of f, and that f is a *multiple* of g in $F[x]$. For $f = 0$, taking $q = 0$ shows that $g|0$ for every $g \in F[x]$. If f is nonzero and $g|f$, applying the Degree Addition Formula to the equation $f = gq$ shows that g is nonzero and $\deg(g) \leq \deg(f)$. For nonzero g, g divides f iff the remainder when we divide f by g is zero. This observation provides an algorithm for deciding whether or not one polynomial divides another.

Observe that $f|f$ for all $f \in F[x]$, since $f = f \cdot 1$. For all $f, g, h \in F[x]$, if $f|g$ and $g|h$, then $f|h$. To prove this, fix $f, g, h \in F[x]$ with $f|g$ and $g|h$. There exist $q, s \in F[x]$ with $g = fq$ and $h = gs$. So $h = gs = (fq)s = f(qs)$ with $qs \in F[x]$, hence $f|h$. Thus, divisibility is a reflexive and transitive relation on $F[x]$. Suppose $f|g$ and $g|f$; does it follow that $f = g$?

In general, the answer is no. The hypotheses $f|g$ and $g|f$ ensure that $g = fs$ and $f = gt$ for some $s, t \in F[x]$, and therefore $f1 = f = gt = f(st)$. Assuming f is nonzero, we can cancel f in the integral domain $F[x]$ to conclude that $st = 1$. By calculating degrees, we see that s and t are nonzero elements of the field F, and $t = s^{-1}$. Thus, $f = gt$ is a scalar multiple of g in the vector space $F[x]$, but f need not equal g. On the other hand, if f and g are both known to be monic, then $f|g$ and $g|f$ do imply $f = g$, since comparison of leading terms in the equation $f = gt$ forces $t = 1$.

In general, we can always multiply a polynomial in $F[x]$ by a nonzero constant from F (a polynomial of degree zero) without affecting divisibility properties. More precisely, suppose $f, g \in F[x]$ and $c \in F$ where $c \neq 0$. We can check that $f|g$ iff $(cf)|g$ iff $f|(cg)$. For any $f, h \in F[x]$, let us write $f \sim h$ iff there exists a nonzero $c \in F$ with $h = cf$. In this situation, we say that f and h are *associates* in the ring $F[x]$. We can verify that the relation $\sim$ is reflexive, symmetric, and transitive (symmetry requires the hypothesis that F is a field), so $\sim$ is an equivalence relation. Furthermore, the equivalence class of any nonzero f (namely, the set of associates of f in $F[x]$) contains exactly one monic polynomial, which can be found by dividing f by its leading coefficient.

Suppose $f, p, q, a, b \in F[x]$ are polynomials such that $f|p$ and $f|q$. We claim that $f|(ap + bq)$. To prove this, write $p = fc$ and $q = fd$ for some $c, d \in F[x]$. We compute $ap + bq = a(fc) + b(fd) = f(ac + bd)$. Since $ac + bd \in F[x]$, $f|(ap + bq)$. It follows by induction that for all $n \geq 1$ and all $f, p_1, \ldots, p_n, a_1, \ldots, a_n \in F[x]$, if f divides every p_i, then f also divides $a_1 p_1 + \cdots + a_n p_n$. We call $a_1 p_1 + \cdots + a_n p_n$ an $F[x]$-*linear combination* of $p_1, \ldots, p_n$, since the coefficients a_i are polynomials in $F[x]$.

3.8   Greatest Common Divisors of Polynomials

Suppose f and g are polynomials in $F[x]$ that are not both zero. A *common divisor* of f and g is a polynomial $p \in F[x]$ such that $p|f$ and $p|g$. A *greatest common divisor* (gcd) of f and g is a common divisor of f and g whose degree is as large as possible. We claim that gcds of f and g always exist, although they may not be unique. To see this, let $\mathrm{CDiv}(f, g)$ denote the set of all common divisors of f and g in $F[x]$. This set is nonempty, since $1|f$ and $1|g$, and the set does not contain zero, since $f \neq 0$ or $g \neq 0$. If $f \neq 0$, then every common divisor has degree at most $\deg(f)$, and similarly if $g \neq 0$. Thus, the set of degrees of polynomials in $\mathrm{CDiv}(f, g)$ is a finite nonempty subset of $\mathbb{Z}_{>0}$, so this set has a greatest element. Any polynomial in $\mathrm{CDiv}(f, g)$ having this maximum degree is a gcd of f and g. The next theorem gives a more precise result.

Theorem on GCDs of Two Polynomials. Suppose F is a field and $f, g \in F[x]$ are not both zero. There exists exactly one monic gcd of f and g in $F[x]$, which is written $\gcd(f, g)$. The set of all gcds of f and g consists of the associates of $\gcd(f, g)$ in $F[x]$. The set of common divisors of f and g in $F[x]$ equals the set of divisors of $\gcd(f, g)$ in $F[x]$. There exist polynomials $s, t \in F[x]$ with $\gcd(f, g) = sf + tg$.

The proof we give, called *Euclid's Algorithm*, gives a specific computational procedure for finding $\gcd(f, g)$ and polynomials s, t such that $\gcd(f, g) = sf + tg$.

Proof. We begin with the following observation: *if $a, b, q, r \in F[x]$, $b \neq 0$, and $a = bq + r$, then* $\mathrm{CDiv}(a, b) = \mathrm{CDiv}(b, r)$. To prove this, first assume $p \in F[x]$ satisfies $p|a$ and $p|b$; then p divides $1a + (-q)b = r$. So $\mathrm{CDiv}(a, b) \subseteq \mathrm{CDiv}(b, r)$. To prove the reverse inclusion, assume $p \in F[x]$ satisfies $p|b$ and $p|r$. Then p divides $qb + 1r = a$. We have now proved

$\mathrm{CDiv}(a, b) = \mathrm{CDiv}(b, r)$. We may conclude that when $a = bq + r$ with $b \neq 0$, $h \in F[x]$ is a gcd of a and b iff h is a gcd of b and r.

We now describe the algorithm for computing a gcd of f and g. Switching f and g if needed, we may assume that $g \neq 0$. We construct two finite sequences of polynomials $r_0, r_1, r_2, \ldots, r_{N+1}$ and $q_1, q_2, \ldots, q_N$ by repeatedly performing polynomial long division. To start, let $r_0 = f$ and $r_1 = g \neq 0$. Divide r_0 by r_1 to obtain

$$r_0 = q_1 r_1 + r_2, \qquad \text{where } r_2 = 0 \text{ or } \deg(r_2) < \deg(r_1).$$

If $r_2 = 0$, the construction ends with $N = 1$. If $r_2 \neq 0$, we continue by dividing r_1 by r_2, obtaining

$$r_1 = q_2 r_2 + r_3, \qquad \text{where } r_3 = 0 \text{ or } \deg(r_3) < \deg(r_2).$$

If $r_3 = 0$, the construction ends with $N = 2$. If $r_3 \neq 0$, we continue similarly. At step i, assuming that r_i is nonzero, we divide r_{i-1} by r_i to obtain

$$r_{i-1} = q_i r_i + r_{i+1}, \qquad \text{where } r_{i+1} = 0 \text{ or } \deg(r_{i+1}) < \deg(r_i).$$

If $r_{i+1} = 0$, we end the construction with $N = i$. Otherwise, we proceed to step $i+1$ of the computation.

We claim first that the computation must end after a finite number of steps. We prove this claim by a contradiction argument. If the algorithm never terminates, then r_i is never zero for any i, so we obtain an infinite decreasing sequence of degrees

$$\deg(r_1) > \deg(r_2) > \deg(r_3) > \cdots > \deg(r_i) > \cdots. \tag{3.1}$$

Then the set $S = \{\deg(r_i) : i \geq 1\}$ has no least element. But S is a nonempty subset of the well-ordered set $\mathbb{Z}_{>0}$, so S must have a least element. This contradiction shows that the algorithm must terminate. In fact, we see from (3.1) that $N \leq \deg(r_1) + 1$.

We now analyze the common divisors of f and g. By the observation preceding the description of the algorithm, the relation $r_{i-1} = q_i r_i + r_{i+1}$ implies that $\mathrm{CDiv}(r_{i-1}, r_i) = \mathrm{CDiv}(r_i, r_{i+1})$ for $1 \leq i \leq N$. Accordingly,

$$\mathrm{CDiv}(f, g) = \mathrm{CDiv}(r_0, r_1) = \mathrm{CDiv}(r_1, r_2) = \cdots = \mathrm{CDiv}(r_N, r_{N+1}) = \mathrm{CDiv}(r_N, 0).$$

Thus, the common divisors of f and g are the same as the common divisors of r_N and 0. Since everything divides 0, the common divisors of f and g are the same as the divisors of r_N. Now, the divisors of r_N of maximum degree are r_N and its associates. Therefore, a polynomial h is a gcd of f and g iff h is an associate of r_N. As pointed out earlier, r_N has a unique monic associate, which is $\gcd(f, g)$. Our argument has shown that every common divisor of f and g divides r_N, hence divides every associate of r_N. In particular, all common divisors of f and g divide $\gcd(f, g)$. Conversely, any divisor of $\gcd(f, g)$ is a common divisor of f and g.

To finish the proof, we must find polynomials $s, t \in F[x]$ such that $\gcd(f, g) = sf + tg$. It suffices to construct polynomials $s_i, t_i \in F[x]$ such that $r_i = s_i f + t_i g$ for $0 \leq i \leq N$; we can then find s and t by dividing u_N and v_N by the leading coefficient of r_N. We use strong induction on i. If $i = 0$, choose $s_i = 1$ and $t_i = 0$ (recalling that $r_0 = f$). If $i = 1$, choose $s_i = 0$ and $t_i = 1$ (recalling that $r_1 = g$). For the induction step, assume $1 \leq i \leq N$ and the polynomials $s_{i-1}, t_{i-1}, s_i, t_i$ with the required properties have already been computed. Then

$$
\begin{aligned}
r_{i+1} &= r_{i-1} - q_i r_i \\
&= (s_{i-1} f + t_{i-1} g) - q_i (s_i f + t_i g) \\
&= (s_{i-1} - q_i s_i) f + (t_{i-1} - q_i t_i) g.
\end{aligned}
$$

Therefore, $r_{i+1} = s_{i+1}f + t_{i+1}g$ holds if we choose

$$s_{i+1} = s_{i-1} - q_i s_i \in F[x] \quad \text{and} \quad t_{i+1} = t_{i-1} - q_i t_i \in F[x].$$

These equations give an explicit recursive algorithm for computing s and t from f and g. $\square$

The theorem does not say that the polynomials s and t such that $\gcd(f, g) = sf + tg$ are unique. Indeed, for any polynomial $z \in F[x]$, $sf + tg = (s + gz)f + (t - fz)g$, so that there are many possible choices for s and t.

3.9   GCDs of Lists of Polynomials

We now extend the discussion of gcds to the case of more than two polynomials. Suppose $n \geq 1$ and $f_1, \ldots, f_n$ are polynomials in $F[x]$. There is no loss of generality in assuming that all f_i are nonzero. We call $g \in F[x]$ a *common divisor* of $f_1, \ldots, f_n$ iff g divides every f_i; let $\mathrm{CDiv}(f_1, \ldots, f_n)$ be the set of these polynomials. A *greatest common divisor* of $f_1, \ldots, f_n$ is a polynomial of maximum degree in $\mathrm{CDiv}(f_1, \ldots, f_n)$. As in the case $n = 2$, greatest common divisors of $f_1, \ldots, f_n$ exist, although they need not be unique. In the case $n = 1$, $\mathrm{CDiv}(f_1)$ is the set of all divisors of f_1, the greatest common divisors of the collection $\{f_1\}$ are the associates of f_1, and we let $\gcd(f_1)$ be the unique monic associate of f_1.

Theorem on GCDs of n Polynomials. For all $n \geq 1$ and all nonzero $f_1, \ldots, f_n \in F[x]$, there exists a unique monic greatest common divisor of $f_1, \ldots, f_n$, written $\gcd(f_1, \ldots, f_n)$. The set of common divisors of $f_1, \ldots, f_n$ equals the set of divisors of $\gcd(f_1, \ldots, f_n)$. There exist polynomials $s_1, \ldots, s_n \in F[x]$ such that $\gcd(f_1, \ldots, f_n) = s_1 f_1 + \cdots + s_n f_n$. If $n \geq 2$, then

$$\gcd(f_1, \ldots, f_{n-1}, f_n) = \gcd(\gcd(f_1, \ldots, f_{n-1}), f_n). \tag{3.2}$$

Proof. All assertions in the theorem are already known when $n = 1$ or $n = 2$. Proceeding by induction, assume $n > 2$ and the theorem holds for all lists of $n - 1$ nonzero polynomials in $F[x]$. We first prove (3.2) as follows. For fixed $h \in F[x]$, the following conditions are logically equivalent:
(a) h divides each of $f_1, \ldots, f_{n-1}, f_n$;
(b) h divides each of $f_1, \ldots, f_{n-1}$, and h divides f_n;
(c) h divides $\gcd(f_1, \ldots, f_{n-1})$, and h divides f_n;
(d) h divides $\gcd(\gcd(f_1, \ldots, f_{n-1}), f_n)$.
The equivalence of (b) and (c) follows from part of the induction hypothesis for lists of $n-1$ polynomials, and the equivalence of (c) and (d) follows similarly using the Theorem on GCDs of Two Polynomials. From the equivalence of (a) and (d), we see that $\mathrm{CDiv}(f_1, \ldots, f_n)$ is exactly the set of polynomials dividing $g = \gcd(\gcd(f_1, \ldots, f_{n-1}), f_n)$. It now follows from the definition that the greatest common divisors of $f_1, \ldots, f_n$ consist of g and its associates. In particular, g is the unique monic gcd of $f_1, \ldots, f_n$.

To prove the remaining statement in the theorem, use the induction hypothesis to write

$$\gcd(f_1, \ldots, f_{n-1}) = t_1 f_1 + \cdots + t_{n-1} f_{n-1}$$

for some polynomials $t_1, \ldots, t_{n-1} \in F[x]$. Setting $g = \gcd(\gcd(f_1, \ldots, f_{n-1}), f_n)$ as above, we can also write

$$g = c \gcd(f_1, \ldots, f_{n-1}) + d f_n$$

for some polynomials $c, d \in F[x]$. Combining these expressions, we see that

$$\gcd(f_1, \ldots, f_n) = g = (ct_1)f_1 + \cdots + (ct_{n-1})f_{n-1} + df_n.$$

Thus, $\gcd(f_1, \ldots, f_n)$ is expressible as an $F[x]$-linear combination of $f_1, \ldots, f_n$. $\qquad\square$

Using Euclid's Algorithm for computing the gcd of two polynomials, we may convert the preceding proof into an algorithm that computes $\gcd(f_1, \ldots, f_n)$ and the polynomials $s_1, \ldots, s_n$ satisfying $\gcd(f_1, \ldots, f_n) = \sum_{i=1}^{n} s_i f_i$. The next section presents an alternative algorithm to compute this information.

3.10 Matrix Reduction Algorithm for GCDs

Given polynomials $f, g \in F[x]$, we can compute $d, a, b \in F[x]$ with $d = \gcd(f, g) = af + bg$ using a matrix reduction technique similar to the Gaussian elimination algorithm learned in elementary linear algebra. We begin with the 2×3 matrix $\begin{bmatrix} 1 & 0 & f \\ 0 & 1 & g \end{bmatrix}$. We then perform a sequence of elementary row operations (described below) to convert this matrix into a new matrix of the form $\begin{bmatrix} a & b & d \\ r & s & 0 \end{bmatrix}$, where all entries are elements of $F[x]$ and one entry in the rightmost column is 0. We will show that d is a gcd of f and g, and $d = af + bg$.

The allowable row operations are as follows. First, we can multiply a row by a nonzero constant in F. Second, we can switch the two rows. Third, for any polynomial $z \in F[x]$, we can add z times row 1 to row 2, or add z times row 2 to row 1.

We illustrate the algorithm by computing $\gcd(x^2 + 2x - 3, x^3 - 2x^2 + 2x - 1)$ in $\mathbb{Q}[x]$. We begin by adding $-x$ times row 1 to row 2 to eliminate the x^3 term:

$$\begin{bmatrix} 1 & 0 & x^2 + 2x - 3 \\ 0 & 1 & x^3 - 2x^2 + 2x - 1 \end{bmatrix} \longrightarrow \begin{bmatrix} 1 & 0 & x^2 + 2x - 3 \\ -x & 1 & -4x^2 + 5x - 1 \end{bmatrix}.$$

Next, we add 4 times row 1 to row 2:

$$\begin{bmatrix} 1 & 0 & x^2 + 2x - 3 \\ -x & 1 & -4x^2 + 5x - 1 \end{bmatrix} \longrightarrow \begin{bmatrix} 1 & 0 & x^2 + 2x - 3 \\ -x + 4 & 1 & 13x - 13 \end{bmatrix}.$$

Now multiply row 2 by $1/13$, then add $-x$ times row 2 to row 1:

$$\longrightarrow \begin{bmatrix} 1 & 0 & x^2 + 2x - 3 \\ (-x+4)/13 & 1/13 & x - 1 \end{bmatrix} \longrightarrow \begin{bmatrix} (1/13)x^2 - (4/13)x + 1 & -x/13 & 3x - 3 \\ (-x+4)/13 & 1/13 & x - 1 \end{bmatrix}.$$

Finally, add -3 times row 2 to row 1 to obtain the matrix

$$\begin{bmatrix} (1/13)x^2 - (1/13)x + (1/13) & (-x-3)/13 & 0 \\ (-x+4)/13 & 1/13 & x - 1 \end{bmatrix}.$$

Switching rows 1 and 2, we conclude that the gcd is $x - 1$, and

$$((-x+4)/13) \cdot (x^2 + 2x - 3) + (1/13) \cdot (x^3 - 2x^2 + 2x - 1) = x - 1.$$

Why does the matrix reduction algorithm work correctly? First, we show that the algorithm always produces a zero in the rightmost column in finitely many steps if we use appropriate row operations. Suppose the current matrix looks like $\begin{bmatrix} a & b & c \\ r & s & t \end{bmatrix}$. If c and t are both nonzero, we may proceed by induction on $\deg(c) + \deg(t)$. As in the proof of the Division Theorem for Polynomials, if $\deg(c) \leq \deg(t)$, we can add an appropriate polynomial multiple of row 1 to row 2 to lower the degree of t. Similarly, if $\deg(c) > \deg(t)$, we can add a multiple of row 2 to row 1 to lower the degree of c. In either case, the nonnegative integer $\deg(c) + \deg(t)$ strictly decreases as a result of the row operation (or one of c or t becomes zero, causing termination). Since there is no infinite strictly decreasing sequence of nonnegative integers, the algorithm does terminate in finitely many steps.

To see why the algorithm gives correct results, we prove the following statement by induction on the number of row operations performed: if the current matrix is $A = \begin{bmatrix} a & b & c \\ r & s & t \end{bmatrix}$, then $af + bg = c$, $rf + sg = t$, and there exist $u, v, w, y \in F[x]$ with $f = uc + vt$ and $g = wc + yt$. When the algorithm starts, we have $a = s = 1$, $b = r = 0$, $c = f$, and $t = g$. So the given statement holds, taking $u = y = 1$ and $v = w = 0$. For the induction step, assume the statement holds for the matrix A shown above. Suppose we multiply row 1 by a nonzero constant $k \in F$, producing the new matrix $B = \begin{bmatrix} ka & kb & kc \\ r & s & t \end{bmatrix}$. Then $rf + sg = t$ still holds; $(ka)f + (kb)g = (kc)$ follows from $af + bg = c$; and $f = (uk^{-1})(kc) + vt$, $g = (wk^{-1})(kc) + yt$. We argue similarly if row 2 is multiplied by k. Now suppose we modify the matrix A by adding z times row 2 to row 1, for some $z \in F[x]$. The new matrix is $C = \begin{bmatrix} a + zr & b + zs & c + zt \\ r & s & t \end{bmatrix}$. Adding z times $rf + sg = t$ to $af + bg = c$, we deduce $(a + zr)f + (b + zs)g = c + zt$, and the equation $rf + sg = t$ is still true. Furthermore, given that $f = uc + vt$ and $g = wc + yt$, we have $f = u(c + zt) + (v - uz)t$ and $g = w(c + zt) + (y - wz)t$ where $u, v - uz, w, y - wz \in F[x]$. Thus, the given statement holds for the new matrix C. We argue similarly in the case where we add z times row 1 to row 2. It is routine to check that the given statement remains true if we interchange the two rows of A.

Now consider what happens at the end of the algorithm, when one of the entries in the rightmost column of A (say t) becomes zero. Since $af + bg = c$, we see that any common divisor of f and g must also divide c. On the other hand, since $f = uc + v0 = uc$ and $g = wc + y0 = wc$, c is a common divisor of f and g. Thus, c must be a gcd of f and g. We can use one more row operation to make c monic.

We can generalize the matrix reduction algorithm to take given nonzero inputs $f_1, \ldots, f_n$ in $F[x]$ (where $n \geq 1$) and compute outputs $d, s_1, \ldots, s_n$ in $F[x]$ with $d = \gcd(f_1, \ldots, f_n) = s_1 f_1 + \cdots + s_n f_n$. We start with an $n \times n$ identity matrix, augmented by an extra column with entries $f_1, \ldots, f_n$. Performing elementary row operations, we reduce this matrix to a new matrix with only one nonzero entry d in the last column. Then d is a gcd of $f_1, \ldots, f_n$, and the remaining entries in the row containing d give coefficients s_i expressing d as a $F[x]$-linear combination of the $f_1, \ldots, f_n$. We ask the reader to provide a detailed proof of this assertion (Exercise 50), which is similar to the proof given above for $n = 2$.

The matrix reduction algorithm can also be adapted to the computation of gcds of two or more integers. In this case, we are only allowed to multiply a particular row by 1 or -1, but we can add any integer multiple of a row to a different row. To prove termination, we can use induction on the sum of the absolute values of the entries in the rightmost column. Examples and proofs for the integer version of this algorithm appear in the exercises.

3.11 Roots of Polynomials

Let F be a field. Given $p \in F[x]$ and $c \in F$, we call c a *root* or *zero* of p iff $p(c) = f_p(c) = 0$. We now show that c *is a root of p iff* $(x - c)$ *divides p in the ring $F[x]$*. To see this, fix $p \in F[x]$ and $c \in F$. Dividing p by $x - c$ gives $p = (x - c)q + r$ for some $q, r \in F[x]$ with $r = 0$ or $\deg(r) < \deg(x - c) = 1$. This means that the remainder r is a constant polynomial (possibly zero). Evaluating both sides of $p = (x - c)q + r$ at $x = c$ shows that $p(c) = (c - c)q(c) + r = r$. Now, c is a root of p iff $p(c) = 0$ iff the remainder r is 0 iff $(x - c)|p$ in $F[x]$.

More generally, if K is a field with subfield F, $p \in F[x]$, and $c \in K$, then c is called a *root* of p iff $p(c) = 0$. If $c \notin F$, the previous theorem does not apply directly, since $x - c$ is not an element of $F[x]$. However, since $F \subseteq K$, we can regard p as an element of the polynomial ring $K[x]$. Then c *is a root of p in K iff* $(x - c)|p$ *in the ring $K[x]$*.

We can now establish a bound on the number of roots of a polynomial.

Theorem on Number of Roots of Polynomials. Suppose F is a field and $p \in F[x]$ has degree $n > 0$. Then p has at most n roots in F.

Proof. We use induction on n. For $n = 1$, write $p = ax + b$ for some $a, b \in F$ with $a \neq 0$. We can check that $x = -a^{-1}b$ is the unique root of this degree 1 polynomial. Now suppose $n > 1$ and the result is known for polynomials of degree $n - 1$. Given p of degree n, consider two cases. If p has no root in F, then the result certainly holds. Otherwise, let c be a root of p in F. Then we can write $p = (x - c)q$ for some polynomial q of degree $n - 1$. If $d \in F$ is any root of p different from c, then $0 = p(d) = (d - c)q(d)$. Since $d - c \neq 0$ and F is a field, we see that $q(d) = 0$. This means that every root of p besides c is a root of q; conversely, every root of q is also a root of p. Now, by the induction hypothesis, q has at most $n - 1$ roots in F (possibly including c). Thus, p has at most n roots in F, namely all the roots of q together with c. $\qquad\square$

The preceding theorem also holds for polynomials p in $R[x]$, where R is an integral domain. But the theorem can fail dramatically for general commutative rings R. For instance, the polynomial $p = x^2 + 7 \in \mathbb{Z}_8[x]$ has four roots (namely 1, 3, 5, and 7) in the ring $R = \mathbb{Z}_8$. Since $2 \cdot 4 = 0$ in $\mathbb{Z}_8$, this ring is neither an integral domain nor a field.

Next, let F be any field, and consider the question of finding the roots of a given $p \in F[x]$ in the field F. If F is a finite field, we can find the roots of p (if any) by evaluating p at all the elements of F and seeing when the answer is zero. Now let $F = \mathbb{Q}$ be the field of rational numbers. Given a non-constant $p \in \mathbb{Q}[x]$, we can replace p by an associate cp (for some nonzero $c \in \mathbb{Q}$) such that cp is in $\mathbb{Z}[x]$, and cp has the same rational roots as p. To accomplish this, we could pick c to be the product (or least common multiple) of all denominators appearing in coefficients of p. The next theorem gives a finite list of possibilities for the rational roots of a polynomial in $\mathbb{Z}[x]$.

Rational Root Theorem. Let $f = a_n x^n + a_{n-1} x^{n-1} + \cdots + a_1 x + a_0$ be in $\mathbb{Z}[x]$, where $a_0, a_1, \ldots, a_n$ are integers and $a_n \neq 0$. Every rational root of f must have the form r/s for some integers r, s such that r divides a_0 and s divides a_n.

Proof. Assume r, s are relatively prime integers such that r/s is a root of f. Evaluating f at this root gives the equation

$$0 = a_n(r^n/s^n) + a_{n-1}(r^{n-1}/s^{n-1}) + \cdots + a_1(r^1/s^1) + a_0.$$

Now multiply both sides by the integer s^n to remove the fractions, giving

$$0 = a_n r^n + a_{n-1} r^{n-1} s^1 + \cdots + a_1 r^1 s^{n-1} + a_0 s^n.$$

Isolating $a_0 s^n$ on one side, we obtain

$$a_0 s^n = -(a_n r^n + a_{n-1} r^{n-1} s^1 + \cdots + a_1 r^1 s^{n-1}).$$

The right side is a multiple of r in $\mathbb{Z}$, so r divides $a_0 s^n$ in $\mathbb{Z}$. As r and s are relatively prime, we deduce that r divides a_0. (This can be shown, for instance, by considering prime factorizations; see Exercise 37.) Similarly, isolating $a_n r^n$ in the earlier equation shows that s divides $a_n r^n$, so s divides a_n because $\gcd(r, s) = 1$. $\qquad\square$

To illustrate the Rational Root Theorem, consider the polynomial $p = x^4 + (13/3)x^3 + (22/3)x^2 + 6x + (4/3)$ in $\mathbb{Q}[x]$. Taking $c = 3$, we have $cp = 3x^4 + 13x^3 + 22x^2 + 18x + 4 \in \mathbb{Z}[x]$. We find $a_4 = 3$ and $a_0 = 4$, so the possible rational roots of p are ± 1, $\pm 1/3$, ± 2, $\pm 2/3$, ± 4, and $\pm 4/3$. Evaluating p at each of these points, we find that p has rational roots $-1/3$ and -2. We can then factor $p = (x + 1/3)(x + 2)(x^2 + 2x + 2)$ in $\mathbb{Q}[x]$. As another example, let $q = x^2 - 3 \in \mathbb{Q}[x]$. By the Rational Root Theorem, the only possible roots of q in $\mathbb{Q}$ are ± 1 and ± 3. Since none of these four numbers is a root of q, we deduce that $\sqrt{3}$ is not rational. For a generalization of this example, see Exercise 80.

3.12 Irreducible Polynomials

Prime numbers and factorizations of numbers into products of primes play a pivotal role in the theory of integers. Our next goal is to develop analogous concepts for the theory of polynomials.

Let F be a field and $p \in F[x]$ have positive degree. We call p an *irreducible polynomial over F* iff whenever $p = fg$ for some polynomials f and g in $F[x]$, either f or g has degree zero. In this case, if p has degree $n > 0$ and f (say) has degree zero, then g has degree n and is an associate of p. Conversely, for any nonzero scalar $c \in F$, we have the factorization $p = c(c^{-1}p)$ where $c^{-1}p$ is an associate of p. It follows that the divisors of an irreducible polynomial p are precisely the nonzero constant polynomials and the associates of p. We can check that an associate of p is irreducible iff p is irreducible.

Negating the definition of irreducibility, we see that p is *reducible* in $F[x]$ iff there exists a *proper* factorization $p = fg$, meaning that neither f nor g is constant. Computing degrees, we see that we must have $0 < \deg(f) < \deg(p)$ and $0 < \deg(g) < \deg(p)$ in this situation. This leads to a criterion for detecting irreducibility of polynomials p of degree at most 3. First, every polynomial of degree 1 is irreducible, since otherwise $\deg(f)$ would be an integer strictly between 0 and 1. Second, if $p \in F[x]$ is such that $\deg(p) > 1$ and $p(c) = 0$ for some $c \in F$, then $(x - c)|p$ in $F[x]$ and hence p is reducible. Conversely, if $\deg(p)$ is 2 or 3 and p is reducible, then one of the factors in a proper factorization $p = fg$ must have degree 1. Moving a constant from f to g if needed, we can assume p has a monic factor $x - c$, and then $c \in F$ is a root of p. Thus, *a polynomial p of degree 2 or 3 is irreducible in $F[x]$ if and only if p has no root in F. An irreducible polynomial of degree at least 4 has no root in F, but the converse does not hold in general.* For example, $p = (x^2 - 3)^2$ is reducible in $\mathbb{Q}[x]$, yet p has no root in $\mathbb{Q}$.

Note that the field of coefficients F must be considered when determining irreducibility, since we only allow factorizations $p = fg$ where the factors f and g belong to $F[x]$. For example, consider the polynomial $p = x^2 - 2$. Applying the criterion in the last paragraph, we see that p is irreducible in $\mathbb{Q}[x]$ (since there is no rational number c with $c^2 = 2$). But p is reducible in $\mathbb{R}[x]$, with factorization $x^2 - 2 = (x - \sqrt{2})(x + \sqrt{2})$. Similarly, $x^2 + 1$ is irreducible in $\mathbb{Q}[x]$ and in $\mathbb{R}[x]$, but becomes reducible in the ring $\mathbb{C}[x]$, with factorization $x^2 + 1 = (x - i)(x + i)$.

For some fields F, we can give specific algorithms for testing the irreducibility of polynomials $p \in F[x]$. For example, if F is any finite field, such as $\mathbb{Z}_q$ for prime q, then a given $p \in F[x]$ has only a finite number of possible divisors. We can use polynomial long division to test each non-constant $f \in F[x]$ of degree less than $n = \deg(p)$ to see if $f|p$; p is irreducible in $F[x]$ iff no such divisor f exists. In fact, we can use the Degree Addition Formula to verify that if $p = fg$ is reducible, then one of its factors must have degree at most $\lfloor n/2 \rfloor$, so we need only test divisors up to this degree. The same process gives a factoring algorithm to find all possible divisors of a reducible $f \in F[x]$.

Irreducibility over infinite fields is harder to check, with a few exceptions. For the field $\mathbb{C}$, the *Fundamental Theorem of Algebra* asserts that every non-constant $p \in \mathbb{C}[x]$ has a root in $\mathbb{C}$. We omit the proof of this theorem. Combining this theorem with our earlier observations relating roots to irreducibility, we see that $p \in \mathbb{C}[x]$ *is irreducible in* $\mathbb{C}[x]$ *iff* $\deg(p) = 1$. With a little more work, we can also deduce from the Fundamental Theorem of Algebra that $p \in \mathbb{R}[x]$ *is irreducible in* $\mathbb{R}[x]$ *iff* $\deg(p) = 1$ *or* $p = ax^2 + bx + c$ *for some* $a, b, c \in \mathbb{R}$ *with* $a \neq 0$ *and* $b^2 - 4ac < 0$. Later in this chapter, we give an algorithm for testing irreducibility of polynomials in $\mathbb{Q}[x]$.

Let us return to the case of a general field F. We now establish a crucial property of irreducible polynomials, which is an analog of a corresponding property of prime integers: *for all* $p, r, s \in F[x]$, *if* p *is irreducible and* $p|rs$, *then* $p|r$ *or* $p|s$. To prove this, fix $p, r, s \in F[x]$ with p irreducible, and assume $p|rs$ but p does not divide r. Since $p|rs$, we may write $rs = pt$ for some $t \in F[x]$. What is $\gcd(p, r)$? This gcd must be a divisor of the irreducible polynomial p, so it is either 1 or the unique monic associate of p. The latter possibility is ruled out since p does not divide r, so $\gcd(p, r) = 1$. We may therefore write $1 = ap + br$ for some $a, b \in F[x]$. Multiplying by s gives $s = (ap + br)s = aps + brs = aps + bpt = p(as + bt)$. We see from this that $p|s$. More generally, an induction argument using the result just proved shows that *if an irreducible polynomial* $p \in F[x]$ *divides a product* $f_1 f_2 \cdots f_n$, *then* p *divides* f_i *for some* i. Conversely, if $p \in F[x]$ is any non-constant polynomial such that for all $r, s \in F[x]$, $p|rs$ implies $p|r$ or $p|s$, then p must be irreducible in $F[x]$. To see this, suppose p is reducible, with proper factorization $p = fg$ for some $f, g \in F[x]$. Then $p|fg$ (as p divides itself), but p cannot divide f or g since these polynomials have degree less than p.

3.13   Unique Factorization of Polynomials

We can now prove the following fundamental result on factorizations of polynomials with coefficients in a field.

Unique Factorization Theorem for Polynomials. Let F be a field. Every non-constant polynomial $g \in F[x]$ can be written in the form $g = cp_1 p_2 \cdots p_r$, where $c \in F$ is a nonzero constant and each p_i is a monic irreducible polynomial in $F[x]$. This factorization is unique except for reordering $p_1, p_2, \ldots, p_r$.

In more detail, the uniqueness assertion means: if $g = dq_1 q_2 \cdots q_s$, where $d \in F$ is constant and each q_j is a monic irreducible polynomial in $F[x]$, then $c = d$, $r = s$, and (after reordering factors if needed) $p_i = q_i$ for $1 \leq i \leq r$.

Proof. To prove the existence of the asserted factorization for g, we use strong induction on $\deg(g)$. Suppose $n = \deg(g) \geq 1$ and the result is already known for all non-constant polynomials of degree less than n. If g happens to be irreducible (which always occurs when $n = 1$), then we have the factorization $g = c(c^{-1}g)$ where $c \in F$ is the leading coefficient of g. If g is reducible, then we may write $g = fh$, where $f, h \in F[x]$ satisfy $0 < \deg(f) < \deg(g)$ and $0 < \deg(h) < \deg(g)$. The induction hypothesis tells us that f and h are constants times products of monic irreducible polynomials. Multiplying these expressions, we obtain the needed factorization for g by combining the two constants.

To prove uniqueness, we also use strong induction on the degree of g. Suppose $g = cp_1p_2\cdots p_r = dq_1q_2\cdots q_s$ where $c, d \in F$ and each p_i and q_j is monic and irreducible in $F[x]$. Assume the uniqueness result is already known to hold for factorizations of polynomials of degree less than $\deg(g)$. Since the p_i and q_j are all monic, comparison of leading coefficients immediately gives $c = d$. Multiplying by c^{-1} gives $p_1p_2\cdots p_r = q_1q_2\cdots q_s$. We deduce from this equality that p_r divides the product $q_1q_2\cdots q_s$. Since p_r is irreducible, p_r must divide some q_j. By reordering $q_1, \ldots, q_s$, we may assume that p_r divides q_s. Since q_s is irreducible and p_r is non-constant, p_r must be an associate of q_s. But p_r and q_s are both monic, so $p_r = q_s$ is forced. Since $F[x]$ is an integral domain, we can cancel this nonzero factor from $p_1p_2\cdots p_r = q_1q_2\cdots q_s$ to get $p_1p_2\cdots p_{r-1} = q_1q_2\cdots q_{s-1}$. If $r = 1$, then the empty product on the left side is 1, which forces $s = 1 = r$. Similarly, $s = 1$ forces $r = 1 = s$. In the remaining case, we have a nonconstant polynomial $h = p_1p_2\cdots p_{r-1} = q_1q_2\cdots q_{s-1}$. Since $\deg(h) < \deg(g)$, the induction hypothesis is applicable to these two factorizations of h. We conclude that $r - 1 = s - 1$ and (after reordering) $p_i = q_i$ for $1 \leq i \leq r - 1$. Therefore, $r = s$, $p_i = q_i$ for $1 \leq i \leq r - 1$, and $p_r = q_s = q_r$. This completes the induction. □

We can rephrase the preceding result in the following way. Given a field F, let $\{p_i : i \in I\}$ be the collection of all monic irreducible polynomials in $F[x]$, where I is some indexing set. *Every non-constant $f \in F[x]$ can be written uniquely in the form*

$$f = c\prod_{i \in I} p_i^{e_i}, \tag{3.3}$$

where c is a nonzero element of F, the e_i are nonnegative integers, and all but a finite number of e_i are zero. This expression for f is obtained from the previous theorem by writing $f = cq_1q_2\cdots q_s$ (where the q_j are monic and irreducible in $F[x]$) and then letting e_i be the number of j with $q_j = p_i$. Uniqueness of the e_i follows from the uniqueness of the factorization of f, because reordering the q_j does not affect the exponents e_i. If we allow all e_i to be zero, we see that every nonzero polynomial (including constant polynomials) can be written uniquely in the form (3.3). The zero polynomial can also be written in this form if we allow $c = 0$, but the e_i are not unique in this case (typically we would take all e_i to be zero). The expression (3.3) is often called the *prime factorization of the polynomial* f. When needed, we may write $c = c_f$ and $e_i = e_i(f)$ to indicate the dependence of these quantities on f.

3.14 Prime Factorizations and Divisibility

Now that unique prime factorizations of polynomials are available, we can approach the theory of divisibility and polynomial gcds from a new viewpoint. Let F be a field and f, g be nonzero polynomials in $F[x]$. Assume first that $g \mid f$ in $F[x]$, so that $f = gh$ for some polynomial $h \in F[x]$. Writing the prime factorizations (3.3) of f, g, and h, the relation $f = gh$ becomes

$$c_f \prod_{i \in I} p_i^{e_i(f)} = \left(c_g \prod_{i \in I} p_i^{e_i(g)} \right) \cdot \left(c_h \prod_{i \in I} p_i^{e_i(h)} \right) = (c_g c_h) \prod_{i \in I} p_i^{e_i(g) + e_i(h)},$$

where $c_f, c_g, c_h \in F$ and $e_i(f), e_i(g), e_i(h) \in \mathbb{Z}_{\geq 0}$. By uniqueness of the prime factorization of f, it follows that $c_f = c_g c_h$ and $e_i(f) = e_i(g) + e_i(h)$ for all $i \in I$. Since $e_i(h) \geq 0$, we see that $e_i(g) \leq e_i(f)$ for all $i \in I$. Conversely, if $e_i(g) \leq e_i(f)$ for all $i \in I$, then $g | f$ since

$$f = g \cdot (c_f / c_g) \prod_{i \in I} p_i^{e_i(f) - e_i(g)}.$$

So: *for all nonzero $f, g \in F[x]$, g divides f in $F[x]$ iff $e_i(g) \leq e_i(f)$ for all $i \in I$.*

Next, consider nonzero polynomials $f_1, \ldots, f_n, g \in F[x]$. By the preceding paragraph, we see that g is a common divisor of $f_1, \ldots, f_n$ iff $e_i(g) \leq e_i(f_j)$ for all $i \in I$ and all $j \in [n]$ iff $e_i(g) \leq \min(e_i(f_1), \ldots, e_i(f_n))$ for all $i \in I$. To obtain a common divisor g of the largest possible degree, we need to take $e_i(g) = \min(e_i(f_1), \ldots, e_i(f_n))$ for all $i \in I$. This means that

$$\gcd(f_1, \ldots, f_n) = \prod_{i \in I} p_i^{\min(e_i(f_1), \ldots, e_i(f_n))}.$$

The previous discussion also gives a new proof that the set of common divisors of $f_1, \ldots, f_n$ equals the set of divisors of $\gcd(f_1, \ldots, f_n)$.

A *common multiple* of $f_1, \ldots, f_n \in F[x]$ is a polynomial $h \in F[x]$ such that every f_j divides h. A *least common multiple* of $f_1, \ldots, f_n$ is a common multiple of $f_1, \ldots, f_n$ having minimum degree. Note that h is a common multiple of $f_1, \ldots, f_n$ iff $e_i(f_j) \leq e_i(h)$ for all $j \in [n]$ and all $i \in I$ iff $\max(e_i(f_1), \ldots, e_i(f_n)) \leq e_i(h)$ for all $i \in I$. To keep the degree of h as small as possible, we need to take $e_i(h) = \max(e_i(f_1), \ldots, e_i(f_n))$ for all $i \in I$. It is now evident that the least common multiples of $f_1, \ldots, f_n$ are precisely the associates of the monic polynomial

$$\mathrm{lcm}(f_1, \ldots, f_n) = \prod_{i \in I} p_i^{\max(e_i(f_1), \ldots, e_i(f_n))}.$$

Furthermore, the common multiples of $f_1, \ldots, f_n$ are precisely the multiples of $\mathrm{lcm}(f_1, \ldots, f_n)$. In the case $n = 2$, we have $\max(e_i(f_1), e_i(f_2)) + \min(e_i(f_1), e_i(f_2)) = e_i(f_1) + e_i(f_2)$ for all $i \in I$. Comparing prime factorizations, we deduce the identity

$$\mathrm{lcm}(f_1, f_2) \gcd(f_1, f_2) = c f_1 f_2,$$

where $c \in F$ is chosen to make the right side monic.

3.15 Irreducible Polynomials in $\mathbb{Q}[x]$

Let us now return to the question of how to decide whether a specific polynomial $f \in \mathbb{Q}[x]$ is irreducible, where $\mathbb{Q}$ is the field of rational numbers. We remarked earlier that f is irreducible iff any associate of f is irreducible. By replacing f by one of its associates, we can assume that f belongs to $\mathbb{Z}[x]$, meaning that all the coefficients of f are integers. For example, we can accomplish this by multiplying the original f by the product of all the denominators of the coefficients appearing in f. For the rest of our discussion, assume f is in $\mathbb{Z}[x]$. We seek a method for deciding if f is irreducible in $\mathbb{Q}[x]$.

Recall that all degree 1 polynomials are irreducible. Also, for f of degree 2 or 3, we know f is irreducible in $\mathbb{Q}[x]$ iff f has a rational root. For $f \in \mathbb{Z}[x]$ of arbitrary degree, we can find all rational roots of f (if any) by checking the finitely many rational numbers of the form described in the Rational Root Theorem (§3.11).

To continue, we need the following technical lemma. *Assume $f \in \mathbb{Z}[x]$ has degree $n > 0$ and is reducible in $\mathbb{Q}[x]$. Then there exist g, h in $\mathbb{Z}[x]$ with $f = gh$ and $\deg(g), \deg(h) < n$.* Note that the conclusion would follow immediately from the definition of reducible polynomials if we allowed g, h to come from $\mathbb{Q}[x]$, but we are demanding the stronger conclusion that g and h belong to $\mathbb{Z}[x]$. Obtaining this stronger result is surprisingly tricky, so we proceed in several steps.

Step 1: We show that for all prime integers p and all $g, h \in \mathbb{Z}[x]$, if p divides gh in the ring $\mathbb{Z}[x]$, then $p|g$ or $p|h$ in the ring $\mathbb{Z}[x]$. We prove the contrapositive. First, note that $p|g$ in $\mathbb{Z}[x]$ iff $g = pq$ for some $q \in \mathbb{Z}[x]$ iff every coefficient in g is divisible (in $\mathbb{Z}$) by p. Now, fix a prime $p \in \mathbb{Z}$ and $g, h \in \mathbb{Z}[x]$, and assume p does not divide g and p does not divide h in $\mathbb{Z}[x]$. We must prove that p does not divide gh in $\mathbb{Z}[x]$. Write $g = \sum_{i \geq 0} g_i x^i$, $h = \sum_{j \geq 0} h_j x^j$ for some $g_i, h_j \in \mathbb{Z}$. By our assumption on g and h, at least one g_i and at least one h_j are not divisible (in $\mathbb{Z}$) by p. Let r be maximal such that p does not divide g_r, and let s be maximal such that p does not divide h_s. Now, observe that the coefficient of x^{r+s} in gh is

$$g_r h_s + \sum_{i,j:\ i+j=r+s,\ i>r} g_i h_j + \sum_{i,j:\ i+j=r+s,\ j>s} g_i h_j.$$

By maximality of r, each term in the first sum is an integer multiple of p. By maximality of s, each term in the second sum is an integer multiple of p. But, since p is prime and neither g_r nor h_s is a multiple of p, $g_r h_s$ is not an integer multiple of p. We conclude that the coefficient of x^{r+s} in gh is not a multiple of p, and hence p does not divide gh in $\mathbb{Z}[x]$.

Step 2: We show that if $f \in \mathbb{Z}[x]$ has degree $n > 0$ and is reducible in $\mathbb{Q}[x]$, then there exist $g, h \in \mathbb{Z}[x]$ and $d \in \mathbb{Z}_{>0}$ with $df = gh$ and $\deg(g), \deg(h) < n$. By reducibility of f in $\mathbb{Q}[x]$, there exist $u, v \in \mathbb{Q}[x]$ with $f = uv$ and $\deg(u), \deg(v) < n$. The idea is to remove the denominators in u and v. Say $u = \sum_{i=0}^{s} (a_i/b_i) x^i$ and $v = \sum_{j=0}^{t} (c_j/d_j) x^j$ for some $s, t \in \mathbb{Z}_{\geq 0}$, $a_i, c_j \in \mathbb{Z}$, and $b_i, d_j \in \mathbb{Z}_{>0}$. Choose $d = b_0 b_1 \cdots b_s d_0 d_1 \cdots d_t \in \mathbb{Z}_{>0}$, $g = (b_0 \cdots b_s) u$, and $h = (d_0 \cdots d_t) v$. We see that $g, h \in \mathbb{Z}[x]$, $df = gh$, and $\deg(g), \deg(h) < n$.

Step 3: Among all possible choices of d, g, h satisfying the conclusion of Step 2, choose one with d minimal (which can be done, by the Well-Ordering Axiom for $\mathbb{Z}_{>0}$). We finish the proof of the lemma by showing $d = 1$. To get a contradiction, assume that $d > 1$. Then some prime integer p divides d in $\mathbb{Z}$, say $d = pd_1$ for some $d_1 \in \mathbb{Z}_{>0}$. Now, p divides df in $\mathbb{Z}[x]$, and $df = gh$, so $p|gh$ in $\mathbb{Z}[x]$. By Step 1, $p|g$ or $p|h$ in $\mathbb{Z}[x]$. Say $p|g$ (the other case is similar). Then $g = pg^*$ for some $g^* \in \mathbb{Z}[x]$. Now $df = gh$ becomes $pd_1 f = pg^* h$. As $\mathbb{Z}[x]$ is an integral domain, we can cancel p to obtain $d_1 f = g^* h$. Now d_1, g^*, and h satisfy the conclusion of Step 2. But since $d_1 < d$, this contradicts the minimality of d. The proof is now complete.

3.16   Testing Irreducibility in $\mathbb{Q}[x]$ via Reduction Modulo a Prime

Using the lemma from the last section, we can prove two theorems for testing irreducibility of certain polynomials in $\mathbb{Q}[x]$. The first theorem involves reducing a polynomial modulo p, where p is a prime integer. Given $f = \sum_{i \geq 0} f_i x^i \in \mathbb{Z}[x]$ and a prime $p \in \mathbb{Z}$, define

$\nu_p(f) = \sum_{i \geq 0} (f_i \bmod p) x^i \in \mathbb{Z}_p[x]$; we call $\nu_p(f)$ the *reduction of f modulo p*. Using the definitions of the algebraic operations on polynomials, you can check that $\nu_p : \mathbb{Z}[x] \to \mathbb{Z}_p[x]$ is a ring homomorphism. In particular, $\nu_p(gh) = \nu_p(g)\nu_p(h)$ for all $g, h \in \mathbb{Z}[x]$.

Testing Irreducibility in $\mathbb{Q}[x]$ via Reduction Modulo a Prime. Suppose $f \in \mathbb{Z}[x]$ has degree $n > 0$ and $p \in \mathbb{Z}$ is a prime not dividing the leading coefficient of f. If $\nu_p(f)$ is irreducible in $\mathbb{Z}_p[x]$, then f is irreducible in $\mathbb{Q}[x]$.

Proof. We prove the contrapositive. Fix $f \in \mathbb{Z}[x]$ of degree $n > 0$ and a prime p not dividing the leading coefficient of f, and assume f is reducible in $\mathbb{Q}[x]$. By the lemma of §3.15, we can write $f = gh$ for some g, h in $\mathbb{Z}[x]$ with $\deg(g), \deg(h) < n$. Reducing $f = gh$ modulo p gives $\nu_p(f) = \nu_p(g)\nu_p(h)$ in $\mathbb{Z}_p[x]$. By hypothesis, the leading coefficient of f does not become zero when we pass to $\nu_p(f)$ by reducing coefficients mod p. So $\nu_p(f)$ still has degree n in $\mathbb{Z}_p[x]$, and $\nu_p(g), \nu_p(h)$ have lower degree than n. So $\nu_p(f)$ is reducible in $\mathbb{Z}_p[x]$. $\square$

For example, consider $f = x^4 + 6x^2 + 3x + 7 \in \mathbb{Z}[x]$. Reducing mod 2 gives $\nu_2(f) = x^4 + x + 1 \in \mathbb{Z}_2[x]$. The irreducible polynomials of degree at most 2 in $\mathbb{Z}_2[x]$ are x, $x+1$, and $x^2 + x + 1$. None of these divide $x^4 + x + 1$ in $\mathbb{Z}_2[x]$, so the latter polynomial is irreducible in $\mathbb{Z}_2[x]$. As f is monic, our criterion applies to show f is irreducible in $\mathbb{Q}[x]$. Similarly, we can check that $x^5 - 4x + 2$ is irreducible in $\mathbb{Q}[x]$ by verifying that its reduction mod 3, namely $x^5 + 2x + 2$, is irreducible in $\mathbb{Z}_3[x]$. The latter fact follows by exhaustively checking the finitely many possible factors of degree 2 or less. On the other hand, note that $\nu_2(x^5 - 4x + 2) = x^5$ is reducible in $\mathbb{Z}_2[x]$. This shows that for a fixed choice of p, the converse of the irreducibility criterion need not hold. In fact, there exist monic polynomials $f \in \mathbb{Z}[x]$ that are irreducible in $\mathbb{Q}[x]$, and yet $\nu_p(f)$ is reducible in $\mathbb{Z}_p[x]$ for every prime integer p. It can be shown that $x^4 - 10x^2 + 1$ has this property (see [52, Example 26, p. 66]).

3.17 Eisenstein's Irreducibility Criterion for $\mathbb{Q}[x]$

Here is a second method for testing irreducibility in $\mathbb{Q}[x]$.

Eisenstein's Irreducibility Criterion for $\mathbb{Q}[x]$. Suppose $f = a_0 + a_1 x + \cdots + a_n x^n \in \mathbb{Z}[x]$ with $n > 0$ and $a_n \neq 0$. If some prime $p \in \mathbb{Z}$ divides a_i for $0 \leq i < n$, but p does not divide a_n and p^2 does not divide a_0, then f is irreducible in $\mathbb{Q}[x]$.

Proof. The proof is by contradiction. Assume f and p satisfy the given hypotheses, but f is reducible in $\mathbb{Q}[x]$. By the lemma of §3.15, there exist g, h in $\mathbb{Z}[x]$ with $f = gh$ and $\deg(g), \deg(h) < n$. Reducing $f = gh$ modulo p, we get $\nu_p(f) = \nu_p(g)\nu_p(h)$. Let $b = a_n \bmod p$. By hypothesis, b is nonzero, but all other coefficients of f reduce to 0 mod p. So $\nu_p(f) = bx^n \in \mathbb{Z}_p[x]$ with $b \neq 0$. Since p is prime, we know $\mathbb{Z}_p$ is a field, and therefore we can apply the Unique Factorization Theorem for Polynomials (§3.13) in the ring $\mathbb{Z}_p[x]$. As x is evidently irreducible in $\mathbb{Z}_p[x]$, the only way that $\nu_p(f) = bx^n$ can factor into a product $\nu_p(g)\nu_p(h)$ is to have $\nu_p(g) = cx^i$ and $\nu_p(h) = dx^j$ for some $c, d \in \mathbb{Z}_p$ with $cd = b$ and some integers i, j with $i + j = n$ and $0 < i, j < n$. For the reductions of g and h to have this form, we must have

$$g = c'x^i + \cdots + ps, \qquad h = d'x^j + \cdots + pt$$

for some $c', d', s, t \in \mathbb{Z}$, where $c' \bmod p = c$, $d' \bmod p = d$, and all coefficients not shown are integer multiples of p. Then the constant term of $f = gh$ is $a_0 = p^2 st$, contradicting the assumption that p^2 does not divide a_0. $\square$

For example, Eisenstein's Criterion applies (with $p = 2$) to prove once again that $x^5 - 4x + 2$ is irreducible in $\mathbb{Q}[x]$. We can see that $x^{10} - 2100$ is irreducible in $\mathbb{Q}[x]$ by applying Eisenstein's Criterion with $p = 3$ or $p = 7$, but not with $p = 2$ or $p = 5$. More generally, for all $n \geq 1$, $a \in \mathbb{Z}$, and prime $p \in \mathbb{Z}$ such that p divides a but p^2 does not divide a, Eisenstein's Criterion shows that $x^n - a$ is irreducible in $\mathbb{Q}[x]$. With some extra work (see Exercise 83), Eisenstein's Criterion can be used to show that for every prime $p \in \mathbb{Z}$, the polynomial $x^{p-1} + x^{p-2} + \cdots + x^2 + x + 1$ is irreducible in $\mathbb{Q}[x]$.

3.18   Lagrange's Interpolation Formula

In linear algebra and other settings, we often need to construct a polynomial that takes prescribed output values at a given list of input values. Lagrange's Interpolation Formula gives an explicit expression for such a polynomial.

Lagrange's Interpolation Formula. Given a field F, an integer $n \geq 0$, distinct elements $a_0, a_1, \ldots, a_n \in F$, and (not necessarily distinct) $b_0, b_1, \ldots, b_n \in F$, there exists a unique polynomial $g \in F[x]$ such that $g(a_i) = b_i$ for $0 \leq i \leq n$, and either $g = 0$ or $\deg(g) \leq n$.

Proof. First we prove existence. For each i between 0 and n, let

$$p_i = \prod_{\substack{j=0 \\ j \neq i}}^{n} (x - a_j) \Bigg/ \prod_{\substack{j=0 \\ j \neq i}}^{n} (a_i - a_j)$$

Each p_i is a polynomial of degree n in $F[x]$ satisfying $p_i(a_i) = 1_F$ and $p_i(a_j) = 0_F$ for all $j \neq i$. Note that the denominator in p_i is a nonzero element of F, hence is invertible, because $a_0, \ldots, a_n$ are distinct. Define $g = \sum_{i=0}^{n} b_i p_i$. Then g is in $F[x]$, $\deg(g) \leq n$ or $g = 0$, and $g(a_i) = b_i$ for $0 \leq i \leq n$.

To prove uniqueness of g, suppose $h \in F[x]$ also satisfies $h(a_i) = b_i$ for $0 \leq i \leq n$ and either $h = 0$ or $\deg(h) \leq n$. Then $f = g - h$ is a polynomial in $F[x]$ with $f(a_i) = 0$ for $0 \leq i \leq n$. This f has degree at most n or is the zero polynomial. But f has at least $n + 1$ distinct roots $a_0, a_1, \ldots, a_n$ in F. By the Theorem on Number of Roots of Polynomials, $\deg(f) \leq n$ is impossible. So f must be 0, which means $g = h$. $\qquad\square$

3.19   Kronecker's Algorithm for Factoring in $\mathbb{Q}[x]$

This section describes *Kronecker's Algorithm*, which takes as input an arbitrary polynomial $f \in \mathbb{Q}[x]$ and returns as output the factorization of f into irreducible factors in $\mathbb{Q}[x]$. In particular, we can use this algorithm to test if a given $f \in \mathbb{Q}[x]$ is irreducible in $\mathbb{Q}[x]$.

By clearing denominators, we can assume the input f is a polynomial of degree $n > 1$ with integer coefficients. If f is reducible in $\mathbb{Q}[x]$, then the lemma of §3.15 shows that $f = gh$ for some $g, h \in \mathbb{Z}[x]$ of smaller degree than n. In fact, the Degree Addition Formula ensures that one of g or h (say g) has degree at most $m = \lfloor n/2 \rfloor$. It suffices to find g (or, to prove irreducibility of f, to show that g must be a constant). Then we can find h by polynomial division, and apply the algorithm recursively to g and to h until the complete factorization of f into irreducible polynomials is found.

To proceed, pick arbitrary distinct integers $a_0, a_1, \ldots, a_m$. If $f(a_i) = 0$ for some i, then a_i is a root of f in $\mathbb{Z}$, and hence $g = x - a_i$ is a divisor of f in $\mathbb{Q}[x]$. Assume henceforth that $f(a_i) \neq 0$ for every i. The hypothesized factorization $f = gh$ implies $f(a_i) = g(a_i)h(a_i)$ for all i. In these equations, everything is an integer, and each $f(a_i)$ is a nonzero integer with only finitely many divisors. The algorithm tries to discover g by looping over all possible sequences of integers $(b_0, b_1, \ldots, b_m)$ with b_i a divisor of $f(a_i)$ in $\mathbb{Z}$ for all i; note that there are only finitely many such sequences. By Lagrange's Interpolation Formula, each sequence of this type yields exactly one $g \in \mathbb{Q}[x]$ such that $g(a_i) = b_i$ for $0 \leq i \leq m$, and we can explicitly calculate g knowing all a_i and b_i. Construct this finite list of possibilities for g and see if anything on the list divides f, belongs to $\mathbb{Z}[x]$, and has positive degree at most m. If so, then we have found a proper factor of f. If no g on this list works, then we know that no other g can possibly work, because of the equations $f(a_i) = g(a_i)h(a_i)$ that must hold in $\mathbb{Z}$ if $f = gh$. In this latter case, the algorithm has proved irreducibility of f in $\mathbb{Q}[x]$.

For example, we apply Kronecker's Algorithm to prove that $f = x^4 + x^3 + x^2 + x + 1$ is irreducible in $\mathbb{Q}[x]$. Here, $n = \deg(f) = 5$ and $m = 2$. For convenience of calculation, we choose $a_0 = 0$, $a_1 = 1$, and $a_2 = -1$. Now $f(a_0) = 1$ forces $b_0 \in \{1, -1\}$; $f(a_1) = 5$ forces $b_1 \in \{1, -1, 5, -5\}$; and $f(a_2) = 1$ forces $b_2 \in \{1, -1\}$. So there are $2 \cdot 4 \cdot 2 = 16$ sequences (b_0, b_1, b_2) to be tested. Now, if g is the unique polynomial of degree at most 2 such that $g(a_i) = b_i$ for $i = 0, 1, 2$, then $-g$ is evidently the unique polynomial of degree at most 2 such that $g(a_i) = -b_i$ for $i = 0, 1, 2$. Because of this sign symmetry, we really only have 8 sequences to test. We examine two of these sequences explicitly and let the reader (or the reader's computer) take care of the others. First, if $b_0 = b_1 = b_2 = 1$, the associated polynomial g is the constant polynomial 1. This g is, of course, a divisor of f, but we need to examine the other sequences to see if one of them yields a non-constant divisor. Second, say $b_0 = 1$, $b_1 = 5$, and $b_2 = -1$. From this data, Lagrange's Interpolation Formula produces

$$g = 1 \cdot \frac{(x-1)(x+1)}{(0-1)(0+1)} + 5 \cdot \frac{(x-0)(x+1)}{(1-0)(1+1)} + (-1) \cdot \frac{(x-0)(x-1)}{(-1-0)(-1-1)} = x^2 + 3x + 1.$$

Dividing f by g in $\mathbb{Q}[x]$ gives quotient $x^2 - 2x + 6$ and remainder $-15x - 5$, so this g is not a divisor of f. Checking all the other possible sequences (b_0, b_1, b_2), we never find a divisor of f other than ± 1. This proves that f is, indeed, irreducible in $\mathbb{Q}[x]$.

It is nice that we can use Kronecker's Algorithm to factor any specific $f \in \mathbb{Q}[x]$ or prove that f is irreducible. But, this algorithm is quite time-consuming if $\deg(f)$ is large or if the integers $f(a_i)$ have many prime divisors. Another drawback is that the algorithm cannot be used to prove irreducibility of collections of polynomials that depend on one or more parameters (like $x^n - 2$ as n varies). This is one reason why it is also helpful to have general theorems such as Eisenstein's Criterion that can prove the irreducibility of polynomials in $\mathbb{Q}[x]$ with special structure.

3.20 Algebraic Elements and Minimal Polynomials

Let F be a field and V be an F-algebra. So, V is an F-vector space and a ring (with multiplication operation $\star$, say) satisfying $c(x \star y) = (cx) \star y = x \star (cy)$ for all $x, y \in V$ and all $c \in F$. Fix an element $z \in V$. We are interested in the F-*algebra generated by z*, which is the intersection of all subalgebras of V containing z. We write $F[\{z\}]$ to denote this subalgebra (some authors write $F[z]$ instead, but this could be confused with a polynomial ring with formal indeterminate z). You may check that $F[\{z\}] = \{g(z) : g \in F[x]\}$, which is the set of all elements in the algebra V that can be written as polynomial expressions in z using coefficients from F. It follows that the set $S = \{1_V, z, z^2, \ldots, z^n, \ldots\}$ spans the subspace $F[\{z\}]$ of the F-vector space V.

Consider the case where $W = F[\{z\}]$ is finite-dimensional over F. Letting $\dim(W) = n$, the list $(1, z, z^2, \ldots, z^n)$ of $n + 1$ elements of W must be F-linearly dependent. So, there exist scalars $c_0, c_1, \ldots, c_n \in F$, not all zero, with $\sum_{i=0}^n c_i z^i = 0_V$. This means that z is a root in V of the nonzero polynomial $f = \sum_{i=0}^n c_i x^i \in F[x]$. By definition, we say that $v \in V$ is *algebraic over F* iff there exists a nonzero $f \in F[x]$ with $f(v) = 0$. We just proved that *for $z \in V$, if $F[\{z\}]$ is finite-dimensional, then z is algebraic over F.* The next theorem gives more specific information about algebraic elements of V.

Theorem on Minimal Polynomials. Suppose F is a field, V is any F-algebra, and $z \in V$ is algebraic over F. Let I be the set of all nonzero polynomials in $F[x]$ having z as a root. There exists a unique monic polynomial $m \in I$ of minimum degree, and I equals the set of all nonzero multiples of m in $F[x]$.

The polynomial m is called the *minimal polynomial of z over F*. The theorem says $m(z) = 0$, m is monic, and for any $g \in F[x]$ such that $g(z) = 0$, m divides g.

Proof. Note that I is nonempty because z is algebraic over F. Since the degrees of polynomials in I constitute a nonempty subset of $\mathbb{Z}_{\geq 0}$, there exists a polynomial $m \in I$ of minimum degree. Dividing by the leading coefficient if needed, we can assume that m is monic. Now, let $g \in I$ be any nonzero polynomial having z as a root. Dividing g by m in $F[x]$ gives $g = mq + r$ for some $q, r \in F[x]$ with $r = 0$ or $\deg(r) < \deg(m)$. Evaluating at z gives $0 = g(z) = m(z)q(z) + r(z) = 0q(z) + r(z) = r(z)$. If r were nonzero, then $r \in I$ and $\deg(r) < \deg(m)$, which contradicts the minimality of $\deg(m)$. So, in fact, $r = 0$ and m divides g in $F[x]$. Conversely, any multiple of m in $F[x]$ has z as a root. We have now proved everything except uniqueness of m. If m_1 is another monic polynomial of minimum degree in I, then seen that m divides m_1. Since the degrees are the same, m_1 must be a constant multiple of m. The constant must be 1, since m and m_1 are both monic. This proves that m is unique. $\square$

The minimal polynomial m of an algebraic $z \in V$ need not be irreducible in $F[x]$. But m must be irreducible if the F-algebra V is an integral domain (in particular, if V is a field). To prove this, suppose $m = fg$ for some $f, g \in F[x]$ that both have lower degree than m. Evaluating at z gives $0 = m(z) = f(z)g(z)$. If V is an integral domain, then $f(z) = 0$ or $g(z) = 0$. Both possibilities contradict the minimality of the degree of m.

Let z be an arbitrary element of any F-algebra V. We proved that if $W = F[\{z\}]$ is finite-dimensional, then z is algebraic over F and has a minimal polynomial. Note that the hypothesis on W automatically holds if V itself is finite-dimensional. The converse is also true, as we now prove.

Theorem on a Basis of $F[\{z\}]$. Suppose F is a field, V is an F-algebra, $z \in V$ is algebraic over F, and z has minimal polynomial m over F with $\deg(m) = d$. Then $F[\{z\}]$ is a d-dimensional F-vector space with ordered basis $B = (1_V, z, z^2, \ldots, z^{d-1})$.

Proof. We prove F-linear independence of B. Assume $a_0, \ldots, a_{d-1} \in F$ satisfy $a_0 1 + a_1 z + \cdots + a_{d-1} z^{d-1} = 0_V$. Then z is a root of the polynomial $\sum_{i=0}^{d-1} a_i x^i \in F[x]$. If this polynomial were nonzero, the minimality of $\deg(m)$ would be contradicted. So this polynomial is zero, which means $a_0, \ldots, a_{d-1}$ are all zero.

Next, we show B spans W. It suffices to show that every element in the spanning set $\{z^k : k \geq 0\}$ for W is an F-linear combination of the vectors in B. This is certainly the case for $0 \leq k < d$, since these powers of z are in the list B. Now fix $k \geq d$, and assume by induction that $z^0, z^1, \ldots, z^{k-1}$ are already known to belong to the subspace spanned by B. Write $m = x^d + \sum_{i=0}^{d-1} c_i x^i$ for some $c_i \in F$. Evaluating at z, multiplying by z^{k-d}, and solving for z^k gives $z^k = \sum_{i=0}^{d-1} -c_i z^{i+k-d}$, so that z^k is an F-linear combination of lower powers of z. All these powers belong to the subspace spanned by B, by induction hypothesis. So z^k is also in this subspace, completing the induction step. $\square$

How can we find the minimal polynomial m of a specific algebraic z in a specific F-algebra V? The key observation is that if $\deg(m) = d$, then z^d is the lowest power of z that is an F-linear combination of preceding powers of z, as we saw in the proof above. If m is not known in advance, we look at $d = 1, 2, 3, \ldots$ in turn and check if the list $(1, z, z^2, \ldots, z^d)$ in V is F-linearly dependent. If it is, the linear dependence relation $c_0 1 + c_1 z + c_2 z^2 + \cdots + c_d z^d$ with $c_d = 1_F$ gives us the minimal polynomial $m = \sum_{i=0}^{d} c_i x^i$ of z.

For example, consider $z = \sqrt{2} + \sqrt{3}$, which is an element of the $\mathbb{Q}$-algebra $\mathbb{R}$. We compute $z^0 = 1$, $z^1 = \sqrt{2} + \sqrt{3}$, $z^2 = 5 + 2\sqrt{6}$, and $z^3 = 11\sqrt{2} + 9\sqrt{3}$. These powers of z are linearly independent over $\mathbb{Q}$ (Exercise 92). But $z^4 = 49 + 20\sqrt{6} = 10z^2 - 1$, so $x^4 - 10x^2 + 1$ must be the minimal polynomial of z over $\mathbb{Q}$. Because the $\mathbb{Q}$-algebra $\mathbb{R}$ is a field, it follows that $x^4 - 10x^2 + 1$ (being a minimal polynomial) must be irreducible in $\mathbb{Q}[x]$.

Now consider the F-algebra $V = M_s(F)$ of all $s \times s$ matrices with entries in F. This algebra has finite dimension (namely s^2) over F, so every matrix $A \in V$ is algebraic over F and has a minimal polynomial $m_A \in F[x]$. The degree of m_A is the dimension of the subspace of matrices spanned by $\{I_s, A, A^2, \ldots, A^n, \ldots\}$, where I_s denotes the $s \times s$ identity matrix. This subspace has dimension at most s^2, so $\deg(m_A) \leq s^2$ for every $s \times s$ matrix A. For a specific example, let $s = 2$, $F = \mathbb{R}$, and $A = \begin{bmatrix} 1 & 2 \\ 3 & 4 \end{bmatrix}$. The matrices I_2 and A are linearly independent, but $A^2 = \begin{bmatrix} 7 & 10 \\ 15 & 22 \end{bmatrix} = 5A + 2I_2$, so the minimal polynomial of A over $\mathbb{R}$ is $m_A = x^2 - 5x - 2$. Observe that this polynomial is reducible in $\mathbb{R}[x]$, since it has real roots $(5 \pm \sqrt{33})/2$. A has the same minimal polynomial over $\mathbb{Q}$, but m_A is irreducible in $\mathbb{Q}[x]$ since m_A has no rational roots.

In the case of matrices A in $M_s(F)$, we can say more about the minimal polynomial of A. Define the *characteristic polynomial of A* to be $\chi_A = \det(xI_s - A)$, which is a monic polynomial of degree s in $F[x]$.

Cayley–Hamilton Theorem. For all $A \in M_s(F)$ with characteristic polynomial χ_A, $\chi_A(A) = 0$.

In other words, if $\chi_A = \sum_{i=0}^{s} c_i x^i$ with $c_i \in F$, then $\sum_{i=0}^{s} c_i A^i$ is the zero matrix. It follows that the minimal polynomial m_A of A must divide χ_A in $F[x]$, giving the improved bound $\deg(m_A) \leq s$.

For the 2×2 matrix A considered above, m_A is equal to the characteristic polynomial $\chi_A = \det \begin{bmatrix} x - 1 & -2 \\ -3 & x - 4 \end{bmatrix}$, but this equality does not hold for all matrices. In later chapters, we prove the Cayley–Hamilton Theorem and use canonical forms to obtain a deeper understanding of the relationship between the characteristic polynomial and the minimal polynomial of a square matrix (see §5.15, §8.14, and §18.19).

3.21 Multivariable Polynomials

We conclude this chapter with an overview of formal power series and polynomials that involve more than one variable. Intuitively, a polynomial in variables $x_1, \ldots, x_m$ with coefficients in a given ring R is a finite sum of monomials $cx_1^{e_1} \cdots x_m^{e_m}$, where $c \in R$

and $e_1, \ldots, e_m$ are nonnegative integers. To specify the polynomial, we must indicate the coefficient c associated with each possible m-tuple of exponents $(e_1, \ldots, e_m) \in \mathbb{Z}_{\geq 0}^m$. This informal description motivates the following precise definitions.

Given a ring R and a positive integer m, let $R[[x_1, \ldots, x_m]]$ be the set of all functions from $\mathbb{Z}_{\geq 0}^m$ to R. Such a function is called a *formal power series in $x_1, \ldots, x_m$ with coefficients in R*. Given $f \in R[[x_1, \ldots, x_m]]$, we introduce the formal summation notation

$$f = \sum_{e_1 \geq 0} \cdots \sum_{e_m \geq 0} f(e_1, \ldots, e_m) x_1^{e_1} \cdots x_m^{e_m}$$

to represent f. We also define $R[x_1, \ldots, x_m]$ to be the subset of $R[[x_1, \ldots, x_m]]$ consisting of all f where $f(e_1, \ldots, e_m) = 0_R$ for all but finitely many inputs $(e_1, \ldots, e_m) \in \mathbb{Z}_{\geq 0}^m$. Elements of $R[x_1, \ldots, x_m]$ are called *formal polynomials in the variables $x_1, \ldots, x_m$*.

Generalizing the $m = 1$ case, we can define addition and multiplication operations on $R[[x_1, \ldots, x_m]]$ that make this set a ring with subring $R[x_1, \ldots, x_m]$. For all $f, g \in R[[x_1, \ldots, x_m]]$ and all $v \in \mathbb{Z}_{\geq 0}^m$, define $(f + g)(v) = f(v) + g(v)$ and

$$(f \cdot g)(v) = \sum_{\substack{w, y \in \mathbb{Z}_{\geq 0}^m: \\ w + y = v}} f(w) g(y).$$

We invite the reader to execute the tedious computations needed to verify the ring axioms. As a hint for the proof of associativity of multiplication, you should show that $(fg)h$ and $f(gh)$ are functions from $\mathbb{Z}_{\geq 0}^m$ to R that both send each $v \in \mathbb{Z}_{\geq 0}^n$ to

$$\sum_{\substack{w, y, z \in \mathbb{Z}_{\geq 0}^m: \\ w + y + z = v}} f(w) g(y) h(z).$$

If R is commutative, then $R[[x_1, \ldots, x_m]]$ and its subring $R[x_1, \ldots, x_m]$ are commutative.

We can identify a ring element $c \in R$ with the constant polynomial (and formal series) that sends $(0, 0, \ldots, 0) \in \mathbb{Z}_{\geq 0}^m$ to c and every other element of $\mathbb{Z}_{\geq 0}^m$ to zero. This identification allows us to view R as a subring of $R[x_1, \ldots, x_m]$ and $R[[x_1, \ldots, x_m]]$. Next, define the formal variable x_i to be the function from $\mathbb{Z}_{\geq 0}^m$ to R that sends $(0, \ldots, 0, 1, 0, \ldots, 0)$ to 1_R (where the 1 appears in position i) and sends every other element of $\mathbb{Z}_{\geq 0}^m$ to 0_R. Using the definition of multiplication, you can then check that the ring element $cx_1^{e_1} \cdots x_m^{e_m}$ (formed by multiplying together c, then e_1 copies of x_1, etc.) is the function from $\mathbb{Z}_{\geq 0}^m$ to R that sends $(e_1, \ldots, e_m)$ to c and everything else to 0_R. It follows that our formal notation

$$f = \sum_{(e_1, \ldots, e_m) \in \mathbb{Z}_{\geq 0}^m} f(e_1, \ldots, e_m) x_1^{e_1} \cdots x_m^{e_m} \tag{3.4}$$

for the polynomial f can also be viewed as a valid algebraic identity in the ring $R[x_1, \ldots, x_m]$, in which f is built up from various ring elements c and $x_1, \ldots, x_m$ by addition and multiplication in the ring. (This comment can be generalized to formal power series, but only after defining what an infinite sum of ring elements means in $F[[x_1, \ldots, x_m]]$.)

As in the case $m = 1$, the multivariable polynomial ring $R[x_1, \ldots, x_m]$ satisfies a universal mapping property.

Universal Mapping Property (UMP) for $R[x_1, \ldots, x_m]$. Suppose R and S are commutative rings, $h : R \to S$ is a ring homomorphism, and $c_1, \ldots, c_m$ is any list of m elements of S. There exists a unique ring homomorphism $H : R[x_1, \ldots, x_m] \to S$ such that H extends h and $H(x_i) = c_i$ for $1 \leq i \leq m$.

Proof. We outline the proof. Given $f = \sum f(e_1, \ldots, e_m) x_1^{e_1} \cdots x_m^{e_m}$ in $R[x_1, \ldots, x_m]$, if a ring homomorphism H exists with the stated properties, then H must send f to $\sum h(f(e_1, \ldots, e_m)) c_1^{e_1} \cdots c_m^{e_m}$. This proves the uniqueness of H. To prove existence, take the preceding formula as the definition of $H(f)$, then check that H is a well-defined ring homomorphism extending h and sending x_i to c_i for all i between 1 and m. (Another approach to the existence proof uses induction on m and the UMP for one-variable polynomial rings. This approach relies on the following recursive characterization of multivariable polynomial rings: for $m > 1$, $R[x_1, \ldots, x_m]$ is isomorphic to the one-variable polynomial ring $T[x_m]$, where $T = R[x_1, \ldots, x_{m-1}]$. See Exercise 11.) $\qquad\square$

We now state without proof some of the basic facts about divisibility for multivariable polynomials with coefficients in a field F. Given $f, g \in F[x_1, \ldots, x_m]$, we say f *divides* g (written $f|g$, as before) iff $f = qg$ for some $q \in F[x_1, \ldots, x_m]$; and we say f and g are *associates* iff $f = cg$ for some nonzero $c \in F$. A polynomial $p \in F[x_1, \ldots, x_m]$ is *irreducible in* $F[x_1, \ldots, x_m]$ iff the only divisors of p in $F[x_1, \ldots, x_m]$ are nonzero constants and associates of p. It can be proved that every nonzero $f \in F[x_1, \ldots, x_m]$ can be factored uniquely into the form $c_f \prod_{i \in I} p_i^{e_i(f)}$, where $c_f \in F$ is nonzero, the p_i range over the set of all monic irreducible polynomials in $F[x_1, \ldots, x_m]$, and each $e_i(f) \in \mathbb{Z}_{\geq 0}$. Then divisibility, gcds, and lcms of multivariable polynomials can be analyzed in terms of these irreducible factorizations, as we did in §3.14 for $m = 1$. However, we warn the reader that some techniques and results used in the one-variable case do not extend to polynomials in more than one variable. For instance, given $f, g \in F[x_1, \ldots, x_m]$ with $m > 1$, it need not be true that $d = \gcd(f, g)$ is a polynomial combination of f and g. For a specific example, you can check that $\gcd(x_1, x_2) = 1$, but 1 cannot be written as $ax_1 + bx_2$ for any choice of $a, b \in F[x_1, \ldots, x_m]$. There is a division algorithm for multivariable polynomials, but it is much more subtle than the one-variable version. We do not give any details here, but instead refer the reader to the outstanding book by Cox, Little, and O'Shea [12].

3.22 Summary

1. *Polynomials and Formal Power Series.* Informally, a polynomial with coefficients in a ring R is an expression $\sum_{n=0}^{d} f_n x^n$ with $d \in \mathbb{Z}_{\geq 0}$ and each $f_n \in R$. A formal power series is an expression $\sum_{n=0}^{\infty} f_n x^n$ with each $f_n \in R$. More precisely, the set $R[[x]]$ of formal power series consists of all sequences $f = (f_n : n \in \mathbb{Z}_{\geq 0})$ with each $f_n \in R$. The set $R[x]$ of formal polynomials consists of all $(f_n : n \in \mathbb{Z}_{\geq 0})$ in $R[[x]]$ such that $f_n = 0$ for all but finitely many n. The set $R[[x]]$ becomes a ring using the operations

$$\sum_{n \geq 0} f_n x^n + \sum_{n \geq 0} g_n x^n = \sum_{n \geq 0} (f_n + g_n) x^n;$$

$$\left(\sum_{n \geq 0} f_n x^n \right) \cdot \left(\sum_{n \geq 0} g_n x^n \right) = \sum_{n \geq 0} \left(\sum_{j,k \in \mathbb{Z}_{\geq 0}: \ j+k=n} f_j g_k \right) x^n.$$

$R[x]$ is a subring of $R[[x]]$ containing an isomorphic copy of R (the constant polynomials). R is commutative iff $R[x]$ is commutative, and R is an integral domain iff $R[x]$ is an integral domain; similarly for $R[[x]]$. $R[x]$ is never a field. If F is a field, then $F[x]$ and $F[[x]]$ are infinite-dimensional F-vector spaces and

F-algebras. There are analogous definitions and results for formal power series and polynomials in m variables $x_1, \ldots, x_m$.

2. *Degree.* The degree of a nonzero polynomial $f = \sum_{n \geq 0} f_n x^n$ is the largest n with $f_n \neq 0$. The degree of the zero polynomial is undefined. For nonzero polynomials p, q such that $p + q \neq 0$, $\deg(p + q) \leq \max(\deg(p), \deg(q))$, and strict inequality occurs iff $\deg(p) = \deg(q)$ and the leading terms of p and q sum to zero. Similarly, if the leading terms of p and q do not multiply to zero, then the Degree Addition Formula $\deg(pq) = \deg(p) + \deg(q)$ is valid.

3. *Polynomial Functions, Evaluation Homomorphisms, and the UMP for $R[x]$.* Given a ring R and $p = \sum_{i=0}^{n} a_i x^i \in R[x]$, the associated polynomial function $f_p : R \to R$ is given by $f_p(c) = p(c) = \sum_{i=0}^{n} a_i c^i$ for $c \in R$. Polynomial rings satisfy the following universal mapping property (UMP): for any commutative rings R and S and any ring homomorphism $h : R \to S$ and any $c \in S$, there exists a unique ring homomorphism $H : R[x] \to S$ extending h such that $H(x) = c$. The special case $h = \mathrm{id}_R$ yields the evaluation homomorphisms $E_c : R[x] \to R$ given by $E_c(p) = f_p(c) = p(c)$ for all $p \in R[x]$.

4. *Divisibility Definitions.* For a field F and $f, g \in F[x]$, f divides g in $F[x]$ (written $f|g$) iff $g = qf$ for some $q \in F[x]$. Polynomials $f, g \in F[x]$ are associates iff $g = cf$ for some nonzero $c \in F$; every nonzero polynomial has a unique monic associate. We say $f \in F[x]$ is a greatest common divisor (gcd) of nonzero $g_1, \ldots, g_k \in F[x]$ iff f divides all g_i and has maximum degree among all common divisors of $g_1, \ldots, g_k$. The notation $\gcd(g_1, \ldots, g_k)$ refers to the unique monic gcd of $g_1, \ldots, g_k$. Similarly, $h \in F[x]$ is a least common multiple (lcm) of $g_1, \ldots, g_k$ iff every g_i divides h and h has minimum degree among all common multiples of $g_1, \ldots, g_k$. The notation $\mathrm{lcm}(g_1, \ldots, g_k)$ refers to the unique monic lcm of $g_1, \ldots, g_k$. A polynomial $p \in F[x]$ is irreducible in $F[x]$ iff its only divisors are constants and associates of p. Whether p is irreducible depends on the field F of coefficients.

5. *Polynomial Division with Remainder.* For a field F and $f, g \in F[x]$ with $g \neq 0$, there exist a unique quotient $q \in F[x]$ and remainder $r \in F[x]$ with $f = qg + r$ and either $r = 0$ or $\deg(r) < \deg(g)$. There is an algorithm to compute q and r from f and g. Replacing F by a commutative ring R, q and r still exist if the leading coefficient of g is invertible in R; q and r are unique if R is an integral domain.

6. *Greatest Common Divisors.* Given a field F and nonzero $f_1, \ldots, f_n \in F[x]$, there exists a unique monic gcd g of $f_1, \ldots, f_n$, and the set of common divisors of $f_1, \ldots, f_n$ equals the set of divisors of g. There exist $s_1, \ldots, s_n \in F[x]$ with $g = s_1 f_1 + \cdots + s_n f_n$; this fact does not extend to polynomials in more than one variable. When $n = 2$, we can find g by repeatedly dividing the previous remainder by the current remainder (starting with f_1 divided by f_2), letting g be the last nonzero remainder, then working backwards to find s_1 and s_2. For any $n \geq 1$, we can find g and $s_1, \ldots, s_n$ by starting with an $n \times n$ identity matrix augmented with a column containing $f_1, \ldots, f_n$, row-reducing this matrix until a single nonzero entry g remains in the extra column, and looking at the other entries in g's row to find $s_1, \ldots, s_n$.

7. *Roots of Polynomials.* Given a field F, $p \in F[x]$, and $c \in F$, c is a root of p iff $p(c) = 0$ iff $(x - c)|p$ in $F[x]$. If $\deg(p) = n$, then p has at most n distinct roots in F. For finite F, we can find the roots of p in F by exhaustive search. For $F = \mathbb{Q}$,

we can replace p by an associate $\sum_{i=0}^{n} a_i x^i$ in $\mathbb{Z}[x]$. The possible rational roots of p have the form r/s, where r divides a_0 and s divides a_n in $\mathbb{Z}$.

8. *Irreducible Polynomials.* Let F be a field. Every $p \in F[x]$ of degree 1 is irreducible. Given $p \in F[x]$ with $\deg(p) > 1$, if p is irreducible then p has no root in F. If p has no root in F and p has degree 2 or 3, then p is irreducible in $F[x]$. If F is finite, then we can factor $p \in F[x]$ or prove p is irreducible by dividing by finitely many potential divisors of degree at most $\deg(p)/2$. For $F = \mathbb{C}$, p is irreducible iff $\deg(p) = 1$. For $F = \mathbb{R}$, p is irreducible iff $\deg(p) = 1$ or $p = ax^2 + bx + c$ with $a, b, c \in \mathbb{R}$, $a \neq 0$, and $b^2 - 4ac < 0$. If an irreducible $p \in F[x]$ divides a product $f_1 f_2 \cdots f_n$ with all $f_i \in F[x]$, then $p | f_i$ for some i; this property characterizes irreducible polynomials.

9. *Unique Factorization in $F[x]$.* Let F be a field. Every non-constant $g \in F[x]$ can be written $g = c q_1 q_2 \cdots q_r$ where $c \in F$, $c \neq 0$, and each q_i is monic and irreducible in $F[x]$; this factorization is unique except for reordering $q_1, \ldots, q_r$. We can also write the factorization as $g = c_g \prod_{i \in I} p_i^{e_i(g)}$ where $c_g \in F$ is nonzero, $\{p_i : i \in I\}$ is an indexed set of all monic irreducible polynomials in $F[x]$, and each $e_i(g) \in \mathbb{Z}_{\geq 0}$. For all $f, g \in F[x]$, $f | g$ iff $e_i(f) \leq e_i(g)$ for all $i \in I$. So we can compute gcds and lcms by the formulas $e_i(\gcd(f_1, \ldots, f_n)) = \min(e_i(f_1), \ldots, e_i(f_n))$ and $e_i(\mathrm{lcm}(f_1, \ldots, f_n)) = \max(e_i(f_1), \ldots, e_i(f_n))$ for all $i \in I$. In particular, $\mathrm{lcm}(f, g) \gcd(f, g) = fg$.

10. *Criteria for Irreducibility in $\mathbb{Q}[x]$.* If $u \in \mathbb{Z}[x]$ has degree $n > 0$ and is reducible in $\mathbb{Q}[x]$, then $u = gh$ for some $g, h \in \mathbb{Z}[x]$ with $\deg(g), \deg(h) < n$. Given $f = \sum_{i=0}^{n} f_i x^i \in \mathbb{Z}[x]$ of degree n, if the reduction of f mod p is irreducible in $\mathbb{Z}_p[x]$ for some prime integer p not dividing f_n, then f is irreducible in $\mathbb{Q}[x]$. Eisenstein's Criterion says that if there is a prime $p \in \mathbb{Z}$ such that $p | f_i$ for all $i < n$, p does not divide f_n, and p^2 does not divide f_0, then f is irreducible in $\mathbb{Q}[x]$.

11. *Lagrange's Interpolation Formula.* Given a field F, distinct $a_0, \ldots, a_n \in F$, and arbitrary $b_0, \ldots, b_n \in F$, there exists a unique $g \in F[x]$ with $g = 0$ or $\deg(g) \leq n$ such that $g(a_i) = b_i$ for all i. Explicitly,

$$g = \sum_{i=0}^{n} b_i p_i, \text{ where } p_i = \prod_{j \neq i} (x - a_j) \Big/ \prod_{j \neq i} (a_i - a_j) \text{ for } 0 \leq i \leq n.$$

12. *Kronecker's Factoring Algorithm in $\mathbb{Q}[x]$.* Given $f \in \mathbb{Z}[x]$ of degree n, Kronecker's Algorithm produces a finite list of potential divisors of f of degree at most $m = \lfloor n/2 \rfloor$ as follows. Pick distinct $a_0, \ldots, a_m \in \mathbb{Z}$. Loop over all $(b_0, \ldots, b_m) \in \mathbb{Z}^{m+1}$ such that b_i divides $f(a_i)$ in $\mathbb{Z}$ for all i. Use Lagrange's Interpolation Formula to build $g \in \mathbb{Q}[x]$ with $g(a_i) = b_i$ for all i. If any such g is non-constant in $\mathbb{Z}[x]$ and divides f, then we have partially factored f. If no g works, then f must be irreducible in $\mathbb{Q}[x]$. As a special case, if some $f(a_i) = 0$, then $(x - a_i) | f$ in $\mathbb{Q}[x]$.

13. *Minimal Polynomials.* Let F be a field, V an F-algebra, and $z \in V$. We say z is algebraic over F iff $g(z) = 0$ for some nonzero $g \in F[x]$. For algebraic z, there exists a unique monic polynomial $m \in F[x]$ with $m(z) = 0$ such that m divides every other polynomial in $F[x]$ having z as a root; m is called the minimal polynomial of z over F. The subalgebra $F[\{z\}]$ spanned by all powers of z is finite-dimensional iff z is algebraic over F. In this case, a basis for this subalgebra (viewed as an F-vector space) is $(1, z, z^2, \ldots, z^{d-1})$ where d is the degree of the minimal polynomial of z. The minimal polynomial of z must be irreducible for an

integral domain or field V, but can be reducible in other algebras. The Cayley–Hamilton Theorem says that every $s \times s$ matrix $A \in M_s(F)$ has a minimal polynomial that divides the characteristic polynomial of A.

3.23   Exercises

1. Let R be a ring. Prove carefully that $R[[x]]$ is a commutative group under addition, indicating which ring axioms for R are needed at each point in the proof. Prove that $R[x]$ is an additive subgroup of $R[[x]]$.

2. Let R be a ring. Carefully prove the ring axioms for $R[[x]]$ involving multiplication, indicating which ring axioms for R are needed at each point in the proof. Prove $R[[x]]$ is commutative iff R is commutative. Prove $R[x]$ is a subring of $R[[x]]$.

3. Repeat the previous two exercises for $R[[x_1, \ldots, x_m]]$ and $R[x_1, \ldots, x_m]$.

4. Let F be a field. Prove $F[[x]]$ is an F-vector space and F-algebra using the scalar multiplication $c \cdot (f_i : i \geq 0) = (cf_i : i \geq 0)$ for $c, f_i \in F$. Prove $F[x]$ is a subspace and subalgebra of $F[[x]]$. Prove $\{1, x, x^2, \ldots, x^n, \ldots\}$ is a basis of $F[x]$.

5. Repeat the previous exercise for formal series and polynomials in m variables, replacing the set of powers of x with $\{x_1^{e_1} x_2^{e_2} \cdots x_m^{e_m} : (e_1, \ldots, e_m) \in \mathbb{Z}_{\geq 0}^m\}$.

6. Let R be a ring. (a) Show that $j : R \to R[x]$ given by $j(a) = (a, 0, 0, \ldots)$ for $a \in R$ is an injective ring homomorphism. (b) For $m \geq 2$, define a similar map $j_m : R \to R[x_1, \ldots, x_m]$ and verify that it is an injective ring homomorphism.

7. Let R be a ring with subring S. (a) Prove $S[[x]]$ is a subring of $R[[x]]$.
   (b) Prove $S[x]$ is a subring of $R[x]$.

8. Let R be a ring with ideal I. Write $I[[x]]$ (resp. $I[x]$) for the subset of formal series (resp. polynomials) having all coefficients in I. Prove there are ring isomorphisms $R[[x]]/I[[x]] \cong (R/I)[[x]]$ and $R[x]/I[x] \cong (R/I)[x]$.

9. Let R be a ring. Let $x = (0_R, 1_R, 0_R, 0_R, \ldots) \in R[x]$.
   (a) Prove carefully that for $i \geq 1$, the product of i copies of x in $R[x]$ is the sequence $(e_j : j \geq 0)$ with $e_i = 1_R$ and $e_j = 0_R$ for all $j \neq i$.
   (b) For $m > 1$ and $1 \leq k \leq m$, define $x_k : \mathbb{Z}_{\geq 0}^m \to R$ by letting x_k map $(0, \ldots, 1, \ldots, 0)$ to 1_R (the 1 is in position k) and letting x_k map everything else in $\mathbb{Z}_{\geq 0}^m$ to 0_R. Prove carefully that $x_1^{e_1} x_2^{e_2} \cdots x_m^{e_m}$ (the product of $e_1 + \cdots + e_m$ elements of $R[x_1, \ldots, x_m]$) is the function mapping $(e_1, \ldots, e_m)$ to 1_R and everything else in $\mathbb{Z}_{\geq 0}^m$ to 0_R. (This result justifies the monomial notation used for multivariable formal series and polynomials.)

10. **Binomial Theorem.** Let R be a ring. (a) Prove: for all $r \in R$ and $n \in \mathbb{Z}_{>0}$,

$$(r + x)^n = \sum_{k=0}^{n} \binom{n}{k} r^{n-k} x^k$$

in $R[x]$. Here, $\binom{n}{k}$ is the integer $\frac{n!}{k!(n-k)!}$, and the notation js (for $j \in \mathbb{Z}_{\geq 0}$, $s \in R$) denotes the sum of j copies of s in R.
   (b) Give a specific example showing that the identity in (a) can be false if the formal variable $x \in R[x]$ is replaced by an element of R.

11. Let R be a ring. Prove there are ring isomorphisms $R[[x_1, \ldots, x_{m-1}]][[x_m]] \cong R[[x_1, \ldots, x_{m-1}, x_m]]$ and $R[x_1, \ldots, x_{m-1}][x_m] \cong R[x_1, \ldots, x_m]$ for each $m > 1$. (This gives a recursive description of multivariable formal series and polynomials.)

12. Let R be a ring. Define the *order* of a nonzero formal series $f \in R[[x]]$, denoted $\operatorname{ord}(f)$, to be the smallest $n \geq 0$ with $f_n \neq 0_R$. (a) Suppose f, g are nonzero formal series with $f + g \neq 0$. Find and prove an inequality relating $\operatorname{ord}(f + g)$ to $\operatorname{ord}(f)$ and $\operatorname{ord}(g)$, and determine when strict inequality holds. (b) Find (with proof) a condition on $f, g \in R[[x]]$ ensuring $\operatorname{ord}(fg) = \operatorname{ord}(f) + \operatorname{ord}(g)$.

13. (a) Prove that a ring R is an integral domain iff $R[[x]]$ is an integral domain. (b) Fix $m > 1$. Prove R is an integral domain iff $R[x_1, \ldots, x_m]$ is an integral domain iff $R[[x_1, \ldots, x_m]]$ is an integral domain.

14. Give a specific example of a commutative ring R and polynomials of all degrees in $R[x]$ that have multiplicative inverses in $R[x]$.

15. Let F be a field. Prove that $g \in F[[x]]$ is invertible in $F[[x]]$ iff the constant coefficient in g is nonzero. Is $F[[x]]$ ever a field?

16. Let F be a field and $n > 0$. Is the set of $p \in F[x]$ with $\deg(p) = n$ or $p = 0$ a subspace of $F[x]$? What about the set of p with $\deg(p) > n$ or $p = 0$? What about the set of $p \in F[[x]]$ with $\operatorname{ord}(p) > n$ or $p = 0$? (See Exercise 12.)

17. Suppose F is a field and, for all $n \in \mathbb{Z}_{\geq 0}$, $p_n \in F[x]$ satisfies $\deg(p_n) = n$. Prove $\{p_n : n \geq 0\}$ is a basis of the F-vector space $F[x]$.

18. Let $p = x^4 + 3x^2 + 4x + 1 \in \mathbb{Z}_5[x]$. Find the polynomial function $f_p : \mathbb{Z}_5 \to \mathbb{Z}_5$.

19. (a) Explicitly describe all functions $g : \mathbb{Z}_2 \to \mathbb{Z}_2$.
    (b) For each g found in (a), find all polynomials $p \in \mathbb{Z}_2[x]$ such that $f_p = g$.

20. Let F be a finite field. Prove that for every function $g : F \to F$, there exists at least one $p \in F[x]$ such that $g = f_p$. Is it true that for each $g : F \to F$, there exist infinitely many $p \in F[x]$ with $g = f_p$?

21. Give a specific example of an infinite field F and a function $g : F \to F$ such that $g \neq f_p$ for all polynomials $p \in F[x]$.

22. Given a field F, let S be the set of all functions from F to F.
    (a) Verify that S is a commutative ring under pointwise operations on functions: $(f + g)(c) = f(c) + g(c)$ and $(f \cdot g)(c) = f(c) \cdot g(c)$ for all $f, g \in S$ and all $c \in F$.
    (b) Define $\phi : F[x] \to S$ by $\phi(p) = f_p$, where f_p is the polynomial function sending $c \in F$ to $p(c)$. Prove ϕ is a ring homomorphism.
    (c) Prove: ϕ is surjective iff F is finite.
    (d) Prove: ϕ is injective iff F is infinite. (So, we may only identify formal polynomials with polynomial functions when the field of coefficients is infinite.)

23. Let R be a commutative ring and $c \in R$. Prove that the map $\phi_c : R[x] \to R[x]$ given by $\phi_c(p) = p(x - c)$ for $p \in R[x]$ is a ring isomorphism. (Use the UMP.)

24. Fill in the details of the two proofs of the universal mapping property for $R[x_1, \ldots, x_m]$ outlined in §3.21.

25. Let R and S be commutative rings and $f : R \to S$ a ring homomorphism. Use the UMP to show that $F : R[x] \to S[x]$ defined by $F(\sum_{i \geq 0} a_i x^i) = \sum_{i \geq 0} f(a_i) x^i$ is a ring homomorphism. Prove if f is one-to-one, then F is one-to-one. Prove if f is onto, then F is onto.

26. Let R be a subring of a commutative ring S, and let $c \in S$. Prove that $R[\{c\}] = \{f_p(c) : p \in R[x]\}$ is the smallest subring of S containing R and c.

27. Let F be a field and V a (not necessarily commutative) F-algebra. For $z \in V$, prove that $F[\{z\}] = \{f_p(z) : p \in F[x]\}$ is an F-subalgebra of V. Prove also that $F[\{z\}]$ is spanned (as an F-vector space) by $\{z^n : n \geq 0\}$ and is the smallest subalgebra of V containing F and z.

28. Let $h : M_2(\mathbb{R}) \to M_2(\mathbb{R})$ be the identity map. Show there does not exist any extension of h to a ring homomorphism $H : M_2(\mathbb{R})[x] \to M_2(\mathbb{R})$ such that $H(x) = \begin{bmatrix} 1 & 1 \\ 0 & 1 \end{bmatrix}$. (So the UMP for $R[x]$ can fail if R is not commutative.)

29. For each $f, g \in \mathbb{Q}[x]$, find the quotient q and remainder r when f is divided by g.
    (a) $f = x^8 + x^4 + 1$, $g = x^4 - x^2 + 1$ (b) $f = 3x^3 - 5x^2 + 11x$, $g = x^2 - 2x + 3$
    (c) $f = x^4 - 5x^2 - x + 1$, $g = x - 2$ (d) $f = x^5 - 1$, $g = 2x^2 - 1$

30. For each $f, g \in F[x]$, find the quotient q and remainder r when f is divided by g.
    (a) $f = x^5 + x^4 + x + 1$, $g = x^3 + x$ in $\mathbb{Z}_2[x]$
    (b) $f = x^5 + x^4 + x + 1$, $g = x^3 + x$ in $\mathbb{Z}_3[x]$
    (c) $f = 3x^4 + x^3 + 2x + 4$, $g = 2x^2 + 3x + 1$ in $\mathbb{Z}_5[x]$
    (d) $f = 3x^4 + x^3 + 2x + 4$, $g = 2x^2 + 3x + 1$ in $\mathbb{Z}_7[x]$

31. (a) Give a specific example of an integral domain R and nonzero $f, g \in R[x]$ such that there are no $q, r \in R[x]$ satisfying $f = qg + r$ and $r = 0$ or $\deg(r) < \deg(g)$.
    (b) Give a specific example of a commutative ring R and $f, g \in R[x]$ such that the quotient q when f is divided by g (with remainder) is not unique.
    (c) Give a specific example of a commutative ring R and $f, g \in R[x]$ such that the remainder r when f is divided by g is not unique.

32. **Division Theorem for $\mathbb{Z}$.** Prove that for all $a, b \in \mathbb{Z}$ with b nonzero, there exist unique $q, r \in \mathbb{Z}$ with $a = bq + r$ and $0 \leq r < |b|$.

33. **Euclid's Algorithm for Integer GCDs.** Prove: for all nonzero integers $a_1, \ldots, a_n$, there exists a unique positive gcd d of $a_1, \ldots, a_n$, and there exist integers $s_1, \ldots, s_n$ with $d = s_1 a_1 + \cdots + s_n s_n$. Moreover, the set of common divisors of $a_1, \ldots, a_n$ equals the set of divisors of d. For $n = 2$, describe an algorithm (involving repeated integer division) for finding d and s_1, s_2 from a_1, a_2.

34. Prove: for all integers p, r, s, if p is prime and $p|rs$, then $p|r$ or $p|s$.

35. **Unique Factorization Theorem for Integers.** Prove: every integer $n \neq 0$ can be factored as $n = cp_1 p_2 \cdots p_k$, where $c \in \{1, -1\}$, $k \geq 0$, and each p_i is a prime positive integer. Prove the factorization is unique except for reordering $p_1, \ldots, p_k$.

36. **Criterion for Invertibility Modulo n.** Prove: for all $n, t \in \mathbb{Z}_{>0}$, $\gcd(n, t) = 1$ iff there exists $s \in \mathbb{Z}_n$ with $st \bmod n = 1$; and s is unique when it exists.

37. Let F be a field and $f, g, h \in F[x]$.
    (a) Prove: if $f|gh$ and $\gcd(f, g) = 1$, then $f|h$.
    (b) Prove: if $\gcd(f, g) = 1$ and $f|h$ and $g|h$, then $fg|h$.
    (c) Prove: if $d \in F[x]$ is a gcd of f and g, then hd is a gcd of hf and hg.
    (d) Give examples to show that (a) and (b) can fail if $\gcd(f, g) \neq 1$.

38. Repeat the previous exercise replacing $F[x]$ by $\mathbb{Z}$.

39. Assume F is a field and $f, f_1 \in F[x]$ are associates. Prove for all $h \in F[x]$, $f|h$ iff $f_1|h$; and $h|f$ iff $h|f_1$. Prove f is irreducible iff f_1 is irreducible.

40. Let R be any commutative ring. For $r, s \in R$, define $r|s$ iff $s = qr$ for some $q \in R$. Define $r, s \in R$ to be *associates in R*, denoted $r \sim s$, iff $r|s$ and $s|r$.
    (a) Prove that the divisibility relation $|$ on R is reflexive and transitive.
    (b) Prove that the association relation $\sim$ is an equivalence relation on R.

(c) Prove that for all integral domains R and all $r, s \in R$, $r \sim s$ iff $s = ur$ for some invertible element $u \in R$.

41. For the equivalence relation $\sim$ in Exercise 40, describe the equivalence classes of $\sim$ for these rings: $R = \mathbb{Z}$; R is a field; $R = F[x]$, where F is a field; $R = \mathbb{Z}_{12}$.

42. For each $f, g \in \mathbb{Q}[x]$, compute $d = \gcd(f, g)$ and $s, t \in \mathbb{Q}[x]$ with $d = sf + tg$.
    (a) $f = x^6 - 1$, $g = x^4 - 1$ (b) $f = x^3 - 5$, $g = x^2 - 2$
    (c) $f = x^3 + x + 1$, $g = 2x^2 - 3x + 1$ (d) $f = x^5 + x^4 - 3x^2 - 2x - 2$, $g = x^4 - x^3 - x^2 + 6$

43. Let $f_1 = x^2 - 3x + 2$, $f_2 = x^2 - 5x + 6$, and $f_3 = x^2 - 4x + 3$ in $\mathbb{Q}[x]$. Follow the proof in §3.9 to find $d = \gcd(f_1, f_2, f_3)$ and $s_1, s_2, s_3 \in \mathbb{Q}[x]$ with $d = s_1 f_1 + s_2 f_2 + s_3 f_3$.

44. Let K be a field with subfield F. Prove that for all nonzero $f_1, \ldots, f_n \in F[x]$, $\gcd(f_1, \ldots, f_n)$ in the ring $F[x]$ equals $\gcd(f_1, \ldots, f_n)$ in the ring $K[x]$ (even though the sets of common divisors of $f_1, \ldots, f_n$ and the irreducible factorizations of $f_1, \ldots, f_n$ may be different in these two rings).

45. Let F be a field. Use the Division Theorem to prove that for every ideal I of $F[x]$, there exists $g \in F[x]$ such that $I = F[x]g = \{hg : h \in F[x]\}$. (This says that $F[x]$ is a *principal ideal domain*, or *PID*. We study PIDs in Chapter 18.)

46. Let F be a field. Given ideals $I = F[x]g$ and $J = F[x]h$ in $F[x]$ (where $g, h \in F[x]$ are generators of the ideals), show that $I \subseteq J$ iff $h | g$ in $F[x]$. Deduce that $I = J$ iff h and g are associates in $F[x]$.

47. Let F be a field. Given nonzero $f_1, \ldots, f_n \in F[x]$, reprove the theorem that these polynomials have a unique monic gcd g, and that g has the form $\sum_{i=1}^{n} s_i f_i$ for some $s_i \in F[x]$, by studying the ideal $J = \{s_1 f_1 + \cdots + s_n f_n : s_i \in F[x]\}$.

48. Let F be a field. Given nonzero $f_1, \ldots, f_n \in F[x]$, show the set $K = \bigcap_{i=1}^{n} F[x]f_i$ is an ideal of $F[x]$, so K is generated by some $h \in F[x]$. Find (with proof) the specific relation between h and $f_1, \ldots, f_n$.

49. For each field F and $f, g \in F[x]$, use the matrix reduction algorithm in §3.10 to find $d = \gcd(f, g)$ and $s, t \in F[x]$ with $d = sf + tg$.
    (a) $F = \mathbb{Q}$, $f = x^{12} - 1$, $g = x^8 - 1$
    (b) $F = \mathbb{Z}_5$, $f = x^3 + x^2 + 3x + 2$, $g = 2x^4 + 2x^2 + x + 3$
    (c) $F = \mathbb{Z}_3$, $f = x^3 + 2x^2 + 2$, $g = x^3 + x^2 + 2x + 2$
    (d) $F = \mathbb{Z}_7$, $f = x^3 + 2x^2 + 4x + 1$, $g = x^2 + 5x + 6$
    (e) $F = \mathbb{Z}_2$, $f = x^8 + x^4 + x^3 + x + 1$, $g = x^5 + x^2 + x$

50. Give a careful proof that the matrix reduction algorithm in §3.10 for finding the gcd of a list of n polynomials terminates and gives correct results.

51. Let $f_1 = x^6 + x^4 + x + 1$, $f_2 = x^8 + x^6 + x^5 + x^4 + x^3 + x^2 + 1$, and $f_3 = x^6 + x^5 + x^2 + 1$ in $\mathbb{Z}_2[x]$. Use matrix reduction to find $d = \gcd(f_1, f_2, f_3)$ and $s_1, s_2, s_3 \in \mathbb{Z}_2[x]$ with $d = s_1 f_1 + s_2 f_2 + s_3 f_3$.

52. **Matrix Reduction Algorithm for Integer GCDs.** Describe a version of the matrix reduction algorithm in §3.10 that takes as input nonzero integers $a_1, \ldots, a_n$ and returns as output $d, s_1, \ldots, s_n \in \mathbb{Z}$ with $d = \gcd(a_1, \ldots, a_n) = \sum_{i=1}^{n} s_i a_i$. Prove that your algorithm terminates and returns a correct answer.

53. For each a, b, use the algorithm in the previous problem to compute $d, s, t \in \mathbb{Z}$ with $d = \gcd(a, b) = sa + tb$. (a) $a = 101$, $b = 57$ (b) $a = 516$, $b = 215$
    (c) $a = 1300$, $b = 967$ (d) $a = 1702$, $b = 483$

54. Let F be a field. Reprove the theorem that a polynomial $p \in F[x]$ of degree n has at most n roots in F using the uniqueness of the irreducible factorization of p.

55. Prove: for all $n \in \mathbb{Z}_{>0}$, there exists a commutative ring R such that $x^2 - 1_R \in R[x]$ has more than n roots in R.

56. Given $f = x^5 + 6x^4 + 9x^3 + 3x^2 + 10x + 11 \in \mathbb{Z}_{13}[x]$, find all roots of f in $\mathbb{Z}_{13}$.

57. Prove: for any field F and any finite subset S of F, there exist infinitely many $p \in F[x]$ such that S is the set of roots of p in F.

58. Prove: for any finite field F of size n, $\prod_{c \in F}(x - c) = x^n - x$ in $F[x]$. (Use Exercise 11 in Chapter 1.)

59. Given a polynomial $f = 7x^5 + 10x^4 + \cdots - 9 \in \mathbb{Z}[x]$ of degree 5 (where the middle coefficients are unknown integers), list all possible roots of f in $\mathbb{Q}$.

60. Find all rational roots of $f = 10x^5 + 11x^4 - 41x^3 + 29x^2 - 51x + 18$, and then factor f into irreducible polynomials in $\mathbb{Q}[x]$ and in $\mathbb{C}[x]$.

61. Find all $x \in \mathbb{Q}$ solving $0 = 6x^6 + 18.6x^5 + 12.6x^4 - 5.4x^3 - 25.2x^2 + 24.6x - 6$.

62. Use the Fundamental Theorem of Algebra to prove that $f \in \mathbb{R}[x]$ is irreducible iff $\deg(f) = 1$ or $f = ax^2 + bx + c$ for some $a, b, c \in \mathbb{R}$ with $a \neq 0$ and $b^2 - 4ac < 0$.

63. Find all irreducible polynomials of degree 5 or less in $\mathbb{Z}_2[x]$. Explain how you know your answers are irreducible, and how you know you have found all of them.

64. Find all monic irreducible polynomials of degree 2 in $\mathbb{Z}_3[x]$. Explain how you know your answers are irreducible, and how you know you have found all of them.

65. For prime p, how many $f \in \mathbb{Z}_p[x]$ are monic irreducible polynomials of degree 2?

66. (a) Prove that $x^4 + x^2 + 2x + 1$ is irreducible in $\mathbb{Z}_3[x]$.
    (b) Describe all polynomials in $\mathbb{Q}[x]$ whose irreducibility in $\mathbb{Q}[x]$ can be deduced directly from (a) by reduction mod 3.
    (c) Use (a) and Exercise 23 to show $x^4 + 8x^3 + 25x^2 + 38x + 25$ is irreducible in $\mathbb{Q}[x]$.

67. For each field F, decide (with proof) whether $x^4 + x^3 + x^2 + x + 1$ is irreducible in $F[x]$. (a) $\mathbb{C}$ (b) $\mathbb{R}$ (c) $\mathbb{Q}$ (d) $\mathbb{Z}_2$ (e) $\mathbb{Z}_3$ (f) $\mathbb{Z}_5$

68. For each field F, decide (with proof) whether $(1_F, 0_F, -1_F, 0_F, 1_F, 0_F, 0_F, \ldots)$ is irreducible in $F[x]$. (a) $\mathbb{Z}_2$ (b) $\mathbb{Z}_3$ (c) $\mathbb{Z}_5$ (d) $\mathbb{Z}_7$ (e) $\mathbb{Z}_{13}$ (f) $\mathbb{Q}$ (g) $\mathbb{C}$

69. Prove that $f = x^4 + 3x^2 + 1$ is irreducible in $\mathbb{Q}[x]$ as follows. Show f has no rational root. Explain why the reducibility of f implies the existence of monic quadratics $g, h \in \mathbb{Z}[x]$ with $f = gh$. Write $g = x^2 + bx + c$, $h = x^2 + dx + e$ for $b, c, d, e \in \mathbb{Z}$. By comparing the coefficients of gh to the coefficients of f, show that g, h cannot exist.

70. Prove $x^5 - 4x^2 + 2$ is irreducible in $\mathbb{Q}[x]$ by reducing modulo a prime.

71. Prove $x^7 + 20x^5 - 8x^4 + 11x^3 + 14x - 9$ is irreducible in $\mathbb{Q}[x]$ (reduce mod 2).

72. Prove $x^7 + 30x^5 - 6x^2 + 12$ is irreducible in $\mathbb{Q}[x]$ using Eisenstein's Criterion.

73. Prove $x^4 - 5x^2 + 1$ is irreducible in $\mathbb{Q}[x]$.

74. Prove $x^3 + ax + 1$ is irreducible in $\mathbb{Q}[x]$, where $a \in \mathbb{Z}$, $a \neq -2$, and $a \neq 0$.

75. Prove $x^5 + 30x^4 + 210x^3 + 300$ is irreducible in $\mathbb{Q}[x]$.

76. Prove $x^3 + 2x + 1$ is irreducible in $\mathbb{Z}_5[x]$.

77. Prove $x^4 - 8$ is irreducible in $\mathbb{Q}[x]$.

78. Find the factorization of $x^8 - 1$ into a product of irreducible polynomials in $\mathbb{C}[x]$, $\mathbb{R}[x]$, $\mathbb{Q}[x]$, and $\mathbb{Z}_2[x]$. Explain how you know that the factors are irreducible.

79. Find the factorization of $x^{12}-1$ into a product of irreducible polynomials in $\mathbb{C}[x]$, $\mathbb{R}[x]$, $\mathbb{Q}[x]$, $\mathbb{Z}_2[x]$, $\mathbb{Z}_3[x]$, and $\mathbb{Z}_5[x]$. Explain why the factors are irreducible.

80. Use the Rational Root Theorem to prove that for all $a, n \in \mathbb{Z}_{>0}$, $\sqrt[n]{a}$ is rational iff $a = b^n$ for some integer b.

81. Use unique prime factorizations in $\mathbb{Z}$ to prove: for all $a, n \in \mathbb{Z}_{>0}$, $\sqrt[n]{a} \in \mathbb{Q}$ iff a has prime factorization $\prod_i p_i^{e_i}$ with n dividing every e_i iff $\sqrt[n]{a} \in \mathbb{Z}$.

82. Let F be a field, and assume $p = a_0 + a_1 x + a_2 x^2 + \cdots + a_{n-1}x^{n-1} + a_n x^n$ is irreducible in $F[x]$, where $a_i \in F$ and $a_n \neq 0$. Let $q = a_n + a_{n-1}x + a_{n-2}x^2 + \cdots + a_1 x^{n-1} + a_0 x^n$. Prove q is irreducible in $F[x]$.

83. Let p be a prime integer and $g = x^{p-1}+x^{p-2}+\cdots+x^2+x+1$. This problem proves that g is irreducible in $\mathbb{Q}[x]$. Recall (Exercise 23) that the map $\phi : \mathbb{Q}[x] \to \mathbb{Q}[x]$, given by $\phi(f) = f(x + 1)$ for $f \in \mathbb{Q}[x]$, is a ring isomorphism. Compute the coefficients of the polynomial $g^* = g(x+1)$. (Start by applying ϕ to the equation $x^p - 1 = (x-1)g$.) Use a theorem from the text to prove g^* is irreducible in $\mathbb{Q}[x]$. Explain why irreducibility of g follows.

84. Does Eisenstein's Criterion hold if p is a positive integer (not necessarily prime)? Prove or give a justified counterexample. What happens if p is a product of two or more distinct primes?

85. Find $a, b, c, d \in \mathbb{Q}$ such that $f = a + bx + cx^2 + dx^3 \in \mathbb{Q}[x]$ satisfies $f(0) = -4$, $f(1) = -2$, $f(2) = 6$, and $f(1/2) = -3$.

86. Find $a, b, c \in \mathbb{Z}_7$ such that $g = a + bx + cx^2 \in \mathbb{Z}_7[x]$ satisfies $g(1) = 5$, $g(2) = 3$, and $g(3) = 0$.

87. Find $a, b, c \in \mathbb{Z}_{13}$ such that $h = a+bx+cx^2 \in \mathbb{Z}_{13}[x]$ satisfies $h(2) = 12$, $h(4) = 10$, and $h(6) = 9$.

88. **Secret Sharing.** A set of n people wish to share information about a master secret in such a way that any k of the n people can pool their knowledge to recover the secret, but no subset of fewer than k people can gain any knowledge about the secret by working together. This goal can be achieved using polynomials, as follows. Pick a prime $p > n$, and encode the master secret as a value $a_0 \in \mathbb{Z}_p$. Choose random $a_1, \ldots, a_{k-1} \in \mathbb{Z}_p$, and let $g = \sum_{i=0}^{k-1} a_i x^i \in \mathbb{Z}_p[x]$. Number the people 1 to n arbitrarily, and give person j the pair $(j, g(j))$. (a) Show how any subset of k or more people can use their collective knowledge to recover a_0. (b) Prove that the collective knowledge of any subset of $k - 1$ or fewer people gives no information about the secret.

89. **Chinese Remainder Theorem.** There is an analog of Lagrange's Interpolation Formula that lets us solve systems of congruences with pairwise relatively prime moduli. Assume $n_1, \ldots, n_k$ are positive integers with $\gcd(n_i, n_j) = 1$ for all $i \neq j$. Set $N = n_1 n_2 \cdots n_k$. Prove: for every list $(b_1, \ldots, b_k)$ with each $b_i \in \mathbb{Z}_{n_i}$, there exists a unique $x \in \mathbb{Z}_N$ such that $x \bmod n_i = b_i$ for $1 \leq i \leq k$. (To prove existence, define $q_i = \prod_{j \neq i} n_j$ for $1 \leq i \leq k$. Note $\gcd(q_i, n_i) = 1$, so Exercise 36 gives an integer $r_i \in \mathbb{Z}_{n_i}$ with $q_i r_i \bmod n_i = 1$. Define $p_i = q_i r_i \in \mathbb{Z}_N$; what is $p_i \bmod n_i$ and $p_i \bmod n_j$ for $j \neq i$? Define x in terms of $p_1, \ldots, p_n, b_1, \ldots, b_n$.)

90. Find $x \in \mathbb{Z}_{1001}$ solving $x \bmod 7 = 5$, $x \bmod 11 = 4$, and $x \bmod 13 = 1$.

91. Use Kronecker's Algorithm to find the irreducible factorization of each polynomial in $\mathbb{Q}[x]$. (a) $x^4 - x^2 + 2x - 1$ (b) $x^5 - 4x + 2$ (c) $x^5 - 3x^4 + 10x^3 - 10x^2 + 24x - 7$

92. (a) Prove that $(1, \sqrt{2}, \sqrt{3})$ is a $\mathbb{Q}$-linearly independent list.
    (b) Deduce that $(1, \sqrt{2}, \sqrt{3}, \sqrt{6})$ is a $\mathbb{Q}$-linearly independent list.
    (c) Let $z = \sqrt{2} + \sqrt{3}$. Prove that $(1, z, z^2, z^3)$ is $\mathbb{Q}$-linearly independent.

93. Viewing $\mathbb{C}$ as a $\mathbb{Q}$-algebra, find (with proof) the minimal polynomials over $\mathbb{Q}$ of each of these complex numbers: $-i$, $\sqrt[3]{7}$, $e^{\pi i/4}$, $\sqrt{5} + 3i$, $\sqrt{2} + \sqrt[3]{5}$.

94. Find the minimal polynomial of each matrix over $\mathbb{R}$ (where $a, b, c, d \in \mathbb{R}$):
$$\begin{bmatrix} 0 & 0 \\ 0 & 0 \end{bmatrix}, \begin{bmatrix} 2 & 0 & 0 \\ 0 & 2 & 0 \\ 0 & 0 & 2 \end{bmatrix}, \begin{bmatrix} 1 & 1 & 1 \\ 1 & 1 & 1 \\ 1 & 1 & 1 \end{bmatrix}, \begin{bmatrix} 0 & 1 & 0 \\ 0 & 0 & 1 \\ -d & -c & -b \end{bmatrix}, \begin{bmatrix} a & b \\ 0 & d \end{bmatrix}.$$

95. Let F be a field. Show $g \in F[x]$ is algebraic over F iff g is a constant.

96. **Formal Derivatives.** Given a commutative ring R and $f = \sum_{n \geq 0} f_n x^n \in R[[x]]$, define the *formal derivative of f* to be $D(f) = \sum_{n \geq 1} n f_n x^{n-1}$, where $n f_n$ denotes the sum of n copies of f_n in R. Prove the following formal derivative rules.
    (a) For all $f, g \in R[[x]]$ and $c \in R$, $D(f + g) = D(f) + D(g)$ and $D(cf) = cD(f)$.
    (b) For all $f, g \in R[[x]]$, $D(fg) = D(f)g + fD(g)$.
    (c) For all $f \in R[x]$ and $g \in R[[x]]$, $D(f(g)) = (D(f))(g) \cdot D(g)$. Here, $f(g)$ is the evaluation of the polynomial f at $x = g$, and similarly for $(D(f))(g)$.

97. **Repeated Roots.** Let F be a field, $p \in F[x]$, and $c \in F$. We say c is a *repeated root* of p iff $(x-c)^2 | p$ in $F[x]$. Prove c is a repeated root of p iff $p(c) = 0 = D(p)(c)$ (see Exercise 96). Prove if p has a repeated root in some field K containing F as a subfield, then $\gcd(p, D(p))$ (computed in $F[x]$) is not 1. (The converse is also true, but you may need Exercise 44 in Chapter 12 to prove it.)

98. True or false? Explain each answer.
    (a) For all rings R and all $g \in R[x]$ of degree $n > 0$, g has at most n roots in R.
    (b) For any field F, an irreducible $p \in F[x]$ has no root in F.
    (c) For any field F, any $p \in F[x]$ with no root in F is irreducible in $F[x]$.
    (d) For all fields F and all $p \in F[x]$, if $p(c) = 0$ for all $c \in F$, then $p = 0$.
    (e) For all irreducible $f \in \mathbb{Q}[x]$, the only divisors of f in $\mathbb{Q}[x]$ are 1 and f.
    (f) The polynomial $x^5 + x^4 + x^3 + x^2 + x + 1$ is irreducible in $\mathbb{Q}[x]$.
    (g) For all $b, c \in \mathbb{Z}$, every rational root of $2x^3 + bx^2 + cx + 3$ must be in the set $\{\pm 1, \pm 2, \pm 1/3, \pm 2/3\}$.
    (h) For all fields F and $f, g \in F[x]$, if $1 = sf + tg$ for some $s, t \in F[x]$, then $\gcd(f, g) = 1$.
    (i) For all rings R and all nonzero $f, g \in R[x]$, $\deg(fg) = \deg(f) + \deg(g)$.
    (j) For all $f, g \in \mathbb{Z}_2[x]$, if $f|g$ and $g|f$ then $f = g$.
    (k) For all even $a \in \mathbb{Z}$ and odd $b, c \in \mathbb{Z}$, $x^3 + ax^2 + bx + c$ is irreducible in $\mathbb{Q}[x]$.
    (l) Every monic $f \in \mathbb{Q}[x]$ with $\deg(f) \geq 1$ can be factored as $f = p_1 p_2 \cdots p_k$, where each p_i is irreducible in $\mathbb{Q}[x]$, and the only other factorizations of f into irreducible factors are obtained by reordering $p_1, \ldots, p_k$.
    (m) For all $p, q \in \mathbb{Q}[x]$, if $p(m) = q(m)$ for all $m \in \mathbb{Z}$, then $p = q$.
    (n) The minimal polynomial of an algebraic element in a commutative F-algebra must be irreducible in $F[x]$.
    (o) For all fields F and all monic $d, f, g \in F[x]$, $\gcd(f, g) = d$ iff $d = sf + tg$ for some $s, t \in F[x]$.

Part II
Matrices

4

Basic Matrix Operations

Our goal in this chapter is to study some basic material about matrices from a somewhat more advanced viewpoint. We begin by giving formal definitions of matrices, ordered n-tuples, row vectors, and column vectors. We then give several equivalent formulations of the basic matrix operations (addition, scalar multiplication, matrix multiplication, transpose, etc.) and derive their main properties. We also show how to execute elementary row and column operations by multiplying a matrix on the left or right by certain elementary matrices.

Throughout this chapter, the letter F denotes an arbitrary field, and elements of F are called *scalars*. However, most of our results on matrices extend to the case where F is a ring. In this more general setting, vector spaces are replaced by modules (see Chapter 17), and nonzero scalars in the field F are replaced by units (invertible elements) in the ring F. For those readers who need this level of generality, we include annotations in square brackets at any places in the chapter that explicitly require the assumption that F is a field, or that multiplication in F is commutative.

4.1 Formal Definition of Matrices and Vectors

For each positive integer n, write $[n]$ for the set $\{1, 2, \ldots, n\}$. Informally, an n-tuple is an ordered list $x = (x_1, x_2, \ldots, x_n)$ of n scalars $x_i \in F$. Formally, we define an *n-tuple with entries in F* to be a function $x : [n] \to F$ given by $x(i) = x_i$ for $i \in [n]$. We frequently use either notation (lists or functions) to describe n-tuples. Let F^n be the set of all n-tuples with entries in F.

Informally, an $m \times n$ matrix A is an array of scalars consisting of m rows and n columns. The element of F in row i and column j of A is called the i, j-*entry of A*. This entry is often written $A_{i,j}$ or A_{ij} or a_{ij}. Formally, let $[m] \times [n]$ be the set of ordered pairs $\{(i, j) : i \in [m], j \in [n]\}$. We define an *$m \times n$ matrix with entries in F* to be a function $A : [m] \times [n] \to F$, where $A(i, j)$ is the i, j-entry of A. Let $M_{m,n}(F)$ be the set of all $m \times n$ matrices over F and $M_n(F)$ be the set of all $n \times n$ (square) matrices over F.

A *column vector* of length n is an $n \times 1$ matrix. A *row vector* of length n is a $1 \times n$ matrix. Formally, a column vector is a function $x_c : [n] \times [1] \to F$, while a row vector is a function $x_r : [1] \times [n] \to F$. There is a bijection (one-to-one correspondence) between column vectors and n-tuples that sends x_c to the n-tuple $(x_c(1, 1), x_c(2, 1), \ldots, x_c(n, 1))$. Similarly, there is a bijection between row vectors and n-tuples that sends x_r to the n-tuple $(x_r(1, 1), x_r(1, 2), \ldots, x_r(1, n))$. Using these bijections, we can regard column vectors and row vectors as n-tuples.

Given a matrix $A \in M_{m,n}(F)$ and j in $[n]$, define $A^{[j]}$ to be the jth column of A, which can be viewed as the m-tuple $(A(1, j), A(2, j), \ldots, A(m, j))$. For i in $[m]$, define $A_{[i]}$ to be the ith row of A, which can be viewed as the n-tuple $(A(i, 1), A(i, 2), \ldots, A(i, n))$. Any matrix A is completely determined by the ordered list of its columns $(A^{[1]}, \ldots, A^{[n]})$. A is

DOI: 10.1201/9781003484561-4

also determined by the ordered list of its rows $(A_{[1]}, \ldots, A_{[m]})$. More formally, we have a bijection from $M_{m,n}(F)$ to $(F^m)^n$ and another bijection from $M_{m,n}(F)$ to $(F^n)^m$.

Next, we define some special matrices. The rectangular *zero matrix* $0_{m,n} \in M_{m,n}(F)$ is defined by $0_{m,n}(i,j) = 0$ for $1 \leq i \leq m$ and $1 \leq j \leq n$. We write 0_n for $0_{n,n}$. The square *identity matrix* $I_n \in M_n(F)$ is defined by $I_n(i,i) = 1$ and $I_n(i,j) = 0$ for $i \neq j$ (where $1 \leq i, j \leq n$). The *unit matrix* $J_{m,n} \in M_{m,n}(F)$ is defined by $J_{m,n}(i,j) = 1$ for $1 \leq i \leq m$ and $1 \leq j \leq n$. We omit the subscripts from the notation whenever they are understood from context.

Recall that two functions f and g are *equal* iff they have the same domain, the same codomain, and $f(x) = g(x)$ for all x in the common domain. Applying this remark to the formal definition of matrices, we see that an $m \times n$ matrix A equals an $m' \times n'$ matrix A' iff $m = m'$ and $n = n'$ and $A(i,j) = A'(i,j)$ for all i,j with $1 \leq i \leq m$ and $1 \leq j \leq n$. Similarly, $v \in F^n$ equals $v' \in F^{n'}$ iff $n = n'$ and $v(i) = v'(i)$ for all i between 1 and n.

In most cases, we label the rows and columns of a matrix by positive integers. Specifically, for an $m \times n$ matrix A, we use the set $[m] = \{1, 2, \ldots, m\}$ to index the rows of A, and we use the set $[n] = \{1, 2, \ldots, n\}$ to index the columns of A. Sometimes, it is more convenient to use other indexing sets to name the rows and columns (see §4.15 for one prominent example). Let I and J be any finite, nonempty sets. Define an *$I \times J$ matrix with entries in F* to be a function $A : I \times J \to F$. For $i \in I$ and $j \in J$, $A(i,j)$ is the *entry* of A in row i, column j. We call I the *row index set* of A and J the *column index set* of A. To display an $I \times J$ matrix as a rectangular array, we order the sets I and J, say $I = \{i_1 < i_2 < \ldots < i_m\}$ and $J = \{j_1 < j_2 < \cdots < j_n\}$. We then place $A(i_s, j_t)$ in row s, column t of the array.

4.2   Vector Spaces of Functions

We are about to give entry-by-entry definitions of algebraic operations on matrices and vectors. Before doing so, it is helpful to consider a more general situation. Let F be the field of scalars, let S be an arbitrary set, and V be the set of all functions $g : S \to F$. By introducing *pointwise operations* on functions, we can turn the set V into an F-vector space. [When F is a ring, V is a left F-module.] First, we define an addition operation $\boldsymbol{+} : V \times V \to V$. Let $g, h : S \to F$ be two elements of V. Define $g \boldsymbol{+} h : S \to F$ by setting $(g \boldsymbol{+} h)(x) = g(x) + h(x)$ for all $x \in S$. Note that the plus symbol on the right side is addition in the given field F, while the bold plus symbol on the left side is the new addition of functions that is being defined. Since F is closed under addition, $g(x) + h(x)$ always belongs to F, so that $g \boldsymbol{+} h$ is a function from S to F. In other words, we have the closure condition: for all $g, h \in V$, $g \boldsymbol{+} h \in V$. Similarly, the other additive axioms in the definition of a vector space [or module] follow from the corresponding axioms for the field [or ring] F. For instance, to confirm that $(g \boldsymbol{+} h) \boldsymbol{+} k = g \boldsymbol{+} (h \boldsymbol{+} k)$ for all $g, h, k \in V$, we check that both sides (which are functions from S to S) agree at every x in the domain S:

$$[(g \boldsymbol{+} h) \boldsymbol{+} k](x) = (g \boldsymbol{+} h)(x) + k(x) = (g(x) + h(x)) + k(x)$$
$$= g(x) + (h(x) + k(x)) = g(x) + (h \boldsymbol{+} k)(x) = [g \boldsymbol{+} (h \boldsymbol{+} k)](x). \quad (4.1)$$

This calculation uses associativity of addition in F. A similar computation shows that $g \boldsymbol{+} h = h \boldsymbol{+} g$ for all $g, h \in V$. You can check that the additive identity of V is the *zero function* $0_V : S \to F$ given by $0_V(x) = 0_F$ for all $x \in S$. The additive inverse of $f \in V$ is the function $(-f) : S \to F$ defined by $(-f)(x) = -(f(x))$ for $x \in X$. This equation specifies that the value of the function $-f$ at input x is the additive inverse (in F) of the scalar $f(x)$.

Next, we define a scalar multiplication operation $\cdot : F \times V \to V$. Given any $g : S \to F$ in V and any scalar $c \in F$, define the function $c \cdot g : S \to F$ by the equation $(c \cdot g)(x) = c \cdot (g(x))$ for $x \in S$. The $\cdot$ on the right side is multiplication in the given field [or ring] F. You can check the remaining axioms for V to be a vector space [or module], using the corresponding properties of the field [or ring] F. We can now consider finite linear combinations of elements of V, as in any vector space [or module]. Given $f_i \in V$ and $c_i \in F$ for $i \in [n]$, $\sum_{i=1}^{n} c_i f_i \in V$ is the function from S to F sending input $x \in S$ to output $\sum_{i=1}^{n} c_i f_i(x) \in F$.

Now assume S is a finite set. For each $x \in S$, define a function $e_x : S \to F$ by letting $e_x(x) = 1_F$ and $e_x(y) = 0_F$ for all $y \in S$ with $y \neq x$. Each e_x is an element of V, so the set $X = \{e_x : x \in S\}$ is a subset of V. The map from S to X sending $x \in S$ to $e_x \in X$ is a bijection, because of the field axiom $1_F \neq 0_F$ [which also holds in any nonzero ring].

Theorem on Bases of Function Spaces. Suppose S is a finite set and V is the F-vector space of functions $g : S \to V$. The set $X = \{e_x : x \in S\}$ is a basis for V, so $\dim_F(V) = |S|$.

Proof. First we prove X is a linearly independent set. Assume $\sum_{x \in S} c_x e_x = 0_V$, where each $c_x \in F$. Both sides of this equation are functions with domain S. Evaluate each side at a fixed $y \in S$. Using the fact that $e_x(y) = 0$ for $x \neq y$ and $e_y(y) = 1_F$, the left side evaluates to $c_y 1_F = c_y$. The right side evaluates to $0_V(y) = 0_F$. So $c_y = 0_F$ for all $y \in S$, proving the linear independence of X.

Next, let $f : S \to F$ be any element of V. To prove X spans V, we prove $\sum_{x \in X} f(x) e_x = f$, which exhibits f as an F-linear combination of elements of X. To prove this equality between two functions, we must show that both sides agree at any $y \in S$. We compute

$$\left(\sum_{x \in X} f(x) e_x \right)(y) = \sum_{x \in X} f(x) e_x(y) = f(y),$$

as needed. So X is a basis of V. By definition of dimension, $\dim(V) = |X| = |S|$. $\qquad\square$

4.3 Matrix Operations via Entries

We define vector space operations on matrices and n-tuples as special cases of the constructions in the previous section. First consider the set F^n of n-tuples of scalars from F. Formally, F^n is the set of all functions from the set $[n]$ to F. Taking the set S in the previous section to be $[n]$, we see that F^n is an F-vector space under pointwise operations on functions. Rewriting the definitions of these operations using the list notation for n-tuples, we obtain:

$$(x_1, x_2, \ldots, x_n) + (y_1, y_2, \ldots, y_n) = (x_1 + y_1, x_2 + y_2, \ldots, x_n + y_n) \qquad \text{for all } x_i, y_i \in F;$$

$$c(x_1, x_2, \ldots, x_n) = (cx_1, cx_2, \ldots, cx_n) \qquad \text{for all } c, x_i \in F.$$

The basis vector e_i (where $1 \leq i \leq n$) is the function such that $e_i(i) = 1_F$ and $e_i(j) = 0_F$ for all $j \neq i$. Using list notation, e_i is a list that has 1_F in position i and 0_F elsewhere: $e_i = (0, \ldots, 1, \ldots, 0)$. By the Theorem on Bases of Function Spaces, $X = (e_1, \ldots, e_n)$ is an ordered basis for F^n. We call X the *standard ordered basis for F^n*. In list notation, we have

$$(x_1, \ldots, x_n) = x_1 e_1 + \cdots + x_n e_n \text{ for all } x_i \in F.$$

Note that the vector e_i depends on n (the domain of e_i is $[n]$), and this dependence is not indicated in the notation e_i. In most cases, we can infer n from context, so no ambiguity occurs. We have $\dim(F^n) = n$.

Next, consider the set $M_{m,n}(F)$ of $m \times n$ matrices over F. Letting $S = [m] \times [n]$, the formal definition of matrices shows that $M_{m,n}(F)$ is the set of all functions from S to F. We know this set is a vector space using pointwise operations on functions. Writing the operations using matrix entries, we get:

$$(A + B)(i, j) = A(i, j) + B(i, j), \quad (cA)(i, j) = c(A(i, j)),$$

where $A, B \in M_{m,n}(F)$, $c \in F$, $i \in [m]$, and $j \in [n]$ are arbitrary. The basis vector $e_{(i,j)}$, which we abbreviate e_{ij}, is the function from $[m] \times [n]$ to F such that $e_{ij}(i, j) = 1_F$ and $e_{ij}(i', j') = 0_F$ if $i' \neq i$ or $j' \neq j$. Using the array notation for matrices, e_{ij} is the array that has 1_F in row i and column j and has 0_F in all other positions. By the Theorem on Bases of Function Spaces, $X = \{e_{ij} : 1 \leq i \leq m, 1 \leq j \leq n\}$ is a basis for $M_{m,n}(F)$. X is called the *standard basis for* $M_{m,n}(F)$. If A is a matrix with i, j-entry a_{ij}, we have the matrix identity $A = \sum_{i=1}^{m} \sum_{j=1}^{n} a_{ij} e_{ij}$. For example,

$$\begin{bmatrix} 2 & -1 \\ 0 & 6 \end{bmatrix} = 2 \begin{bmatrix} 1 & 0 \\ 0 & 0 \end{bmatrix} + (-1) \begin{bmatrix} 0 & 1 \\ 0 & 0 \end{bmatrix} + 0 \begin{bmatrix} 0 & 0 \\ 1 & 0 \end{bmatrix} + 6 \begin{bmatrix} 0 & 0 \\ 0 & 1 \end{bmatrix}.$$

As before, the meaning of e_{ij} does depend on m and n. We have $\dim(M_{m,n}(F)) = |X| = mn$.

Similar comments apply to $I \times J$ matrices for any index sets I and J. The set of all such matrices is a vector space with basis $X = \{e_{ij} : i \in I, j \in J\}$ and dimension $|X| = |I| \cdot |J|$.

For each positive integer n, we have n-dimensional vector spaces $M_{n,1}(F)$, $M_{1,n}(F)$, and F^n, consisting (respectively) of column vectors of length n, row vectors of length n, and n-tuples. You can check that the bijections between these sets (described earlier) are vector space isomorphisms. For $i \in [n]$, the standard basis vectors $e_{i1} \in M_{n,1}(F)$, $e_{1i} \in M_{1,n}(F)$, and $e_i \in F^n$ correspond to each other under these isomorphisms.

We now define the operations of matrix multiplication, matrix-vector multiplication, matrix transpose, and conjugate-transpose one entry at a time. Let $A \in M_{m,n}(F)$ and $B \in M_{n,p}(F)$. The *matrix product* AB is the $m \times p$ matrix defined by

$$(AB)(i, j) = \sum_{k=1}^{n} A(i, k) B(k, j), \quad \text{where } 1 \leq i \leq m, 1 \leq j \leq p.$$

By letting $m = 1$ or $p = 1$, we get the formulas for multiplying a matrix on the left or right by a row vector or column vector. Specifically, if $A \in M_{m,n}(F)$ and $v \in F^n$, then $Av \in F^m$ is defined by

$$(Av)(i) = \sum_{k=1}^{n} A(i, k) v(k) \quad \text{for } i \in [m].$$

Here, we identify lists in F^n and F^m with column vectors, writing $(Av)(i) = (Av)(i, 1)$ and $v(k) = v(k, 1)$. Alternatively, if $w \in F^m$ and $A \in M_{m,n}(F)$, then $wA \in F^n$ is given by

$$(wA)(j) = \sum_{k=1}^{m} w(k) A(k, j) \quad \text{for } j \in [n],$$

where we now identify lists in F^m and F^n with row vectors.

For example, if $A = \begin{bmatrix} 2 & -1 & 4 \\ 0 & 1 & 3 \end{bmatrix} \in M_{2,3}(\mathbb{R})$, $B = \begin{bmatrix} 3 & 2 \\ 1 & 1 \\ -2 & 5 \end{bmatrix} \in M_{3,2}(\mathbb{R})$, and $v = (4, 0, 1) \in \mathbb{R}^3$, then

$$AB = \begin{bmatrix} -3 & 23 \\ -5 & 16 \end{bmatrix}, \quad BA = \begin{bmatrix} 6 & -1 & 18 \\ 2 & 0 & 7 \\ -4 & 7 & 7 \end{bmatrix}, \quad vB = [10 \ 13], \quad \text{and } Av = \begin{bmatrix} 12 \\ 3 \end{bmatrix}.$$

Note that we view the 3-tuple v as a row vector when computing vB, but we view v as a column vector when computing Av.

For $A \in M_{m,n}(F)$, the *transpose* of A, written A^{T}, is the $n \times m$ matrix with i, j-entry

$$A^{\mathrm{T}}(i, j) = A(j, i) \quad \text{for } 1 \leq i \leq n, \, 1 \leq j \leq m.$$

When $F = \mathbb{C}$, the *conjugate-transpose* of A, written A^*, is the $n \times m$ complex matrix with i, j-entry

$$A^*(i, j) = \overline{A(j, i)} \quad \text{for } 1 \leq i \leq n, \, 1 \leq j \leq m.$$

The bar denotes *complex conjugation*: $\overline{x + iy} = x - iy$ for all $x, y \in \mathbb{R}$. For example, if $A = \begin{bmatrix} 2 + i & -3i & 4 \\ 1 - 2i & 0 & e^{\pi i / 4} \end{bmatrix} \in M_{2,3}(\mathbb{C})$, then

$$A^{\mathrm{T}} = \begin{bmatrix} 2 + i & 1 - 2i \\ -3i & 0 \\ 4 & e^{\pi i / 4} \end{bmatrix} \quad \text{and } A^* = \begin{bmatrix} 2 - i & 1 + 2i \\ 3i & 0 \\ 4 & e^{-\pi i / 4} \end{bmatrix}.$$

We make analogous definitions for matrices with rows and columns indexed by arbitrary finite sets. The *product* of an $I \times K$ matrix A and a $K \times J$ matrix B is the $I \times J$ matrix $C = AB$ with entries $C(i, j) = \sum_{k \in K} A(i, k)B(k, j)$ for all $i \in I$ and $j \in J$. The *transpose* of an $I \times J$ matrix D is the $J \times I$ matrix D^{T} with entries $D^{\mathrm{T}}(j, i) = D(i, j)$ for all $j \in J$ and $i \in I$. All properties proved below generalize at once to this setting.

4.4 Properties of Matrix Multiplication

We use the definition of the product of two matrices to prove some fundamental algebraic properties of matrix multiplication. Let $A, A' \in M_{m,n}(F)$ and $B, B' \in M_{n,p}(F)$. First, consider the two *distributive laws* $A(B + B') = AB + AB'$ and $(A + A')B = AB + A'B$. To prove the first of these laws, compute the i, j-entry of each side, for $i \in [m]$ and $j \in [p]$:

$$\begin{aligned}
[A(B + B')](i, j) &= \sum_{k=1}^{n} A(i, k)(B + B')(k, j) = \sum_{k=1}^{n} A(i, k)[B(k, j) + B'(k, j)] \\
&= \sum_{k=1}^{n} [A(i, k)B(k, j) + A(i, k)B'(k, j)] \\
&= \sum_{k=1}^{n} A(i, k)B(k, j) + \sum_{k=1}^{n} A(i, k)B'(k, j) \\
&= (AB)(i, j) + (AB')(i, j) = (AB + AB')(i, j).
\end{aligned}$$

The second distributive law is proved similarly. Next, for $A \in M_{m,n}(F)$ and $B \in M_{n,p}(F)$ and $c \in F$ [and F commutative], we have $c(AB) = (cA)B = A(cB)$. This is proved by showing that the i,j-entry of all three expressions is $\sum_{k=1}^{n} cA(i,k)B(k,j)$. We also have $AI_n = A = I_m A$. For instance, we prove $I_m A = A$ by computing

$$(I_m A)(i,j) = \sum_{k=1}^{m} I_m(i,k)A(k,j) = A(i,j) \quad \text{for all } i \in [m] \text{ and } j \in [n],$$

since $I_m(i,k)$ is 1 for $k = i$ and 0 otherwise.

Given $A \in M_{m,n}(F)$, $B \in M_{n,p}(F)$, and $C \in M_{p,q}(F)$, we prove the *associative law* $(AB)C = A(BC)$. Note first that $AB \in M_{m,p}(F)$ and $BC \in M_{n,q}(F)$. So both products $(AB)C$ and $A(BC)$ are defined and give outputs in $M_{m,q}(F)$. To prove $(AB)C = A(BC)$, fix arbitrary $i \in [m]$ and $j \in [q]$. On one hand,

$$
\begin{aligned}
[(AB)C](i,j) &= \sum_{k'=1}^{p}(AB)(i,k')C(k',j) = \sum_{k'=1}^{p}\left(\sum_{k=1}^{n} A(i,k)B(k,k')\right)C(k',j) \\
&= \sum_{k'=1}^{p}\sum_{k=1}^{n}(A(i,k)B(k,k'))C(k',j).
\end{aligned}
$$

On the other hand,

$$
\begin{aligned}
[A(BC)](i,j) &= \sum_{k=1}^{n} A(i,k)(BC)(k,j) = \sum_{k=1}^{n} A(i,k)\left(\sum_{k'=1}^{p} B(k,k')C(k',j)\right) \\
&= \sum_{k=1}^{n}\sum_{k'=1}^{p} A(i,k)(B(k,k')C(k',j)).
\end{aligned}
$$

The two final expressions are equal, since the finite sums (indexed by k and k') can be interchanged and since multiplication is associative in the field F. [The proof works for any ring F, possibly non-commutative.]

Letting $m = n = p = q$, the results proved above show that the F-vector space $M_n(F)$ is an associative F-algebra with identity (see §1.3). This means that $M_n(F)$ is an F-vector space and a ring (with identity I_n) such that $c(AB) = (cA)B = A(cB)$ for all $c \in F$ and all $A, B \in M_n(F)$.

We conclude this section with some properties of the transpose and conjugate-transpose operations. Let $A, B \in M_{m,n}(F)$ and $C \in M_{n,p}(F)$. We have $(A + B)^T = A^T + B^T$ since the i,j-entry of both sides is $A(j,i) + B(j,i)$ for all $i \in [n]$ and $j \in [m]$. For $c \in F$, we have $(cA)^T = c(A^T)$ since the i,j-entry of both sides is $cA(j,i)$. We have $(AC)^T = C^T A^T$ since, for all $i \in [p]$ and $j \in [m]$,

$$(AC)^T(i,j) = (AC)(j,i) = \sum_{k=1}^{n} A(j,k)C(k,i) = \sum_{k=1}^{n} C^T(i,k)A^T(k,j) = (C^T A^T)(i,j).$$

[Commutativity of multiplication in F is needed in this proof.] When $F = \mathbb{C}$, the conjugate-transpose operation has similar properties: $(A + B)^* = A^* + B^*$; $(cA)^* = \overline{c}(A^*)$; and $(AC)^* = C^* A^*$. The proofs are similar to those just given. For instance, the second property is true because, for all $i \in [n]$ and $j \in [m]$,

$$(cA)^*(i,j) = \overline{(cA)(j,i)} = \overline{c(A(j,i))} = \overline{c} \cdot \overline{A(j,i)} = \overline{c}(A^*(i,j)) = (\overline{c}A^*)(i,j).$$

We also have $(A^T)^T = A$ and, for $F = \mathbb{C}$, $(A^*)^* = A$.

4.5 Generalized Associativity

We now prove a general associativity result that applies to products of three or more matrices whose dimensions match. Informally, the result says that no parentheses are required to evaluate such a product unambiguously (although the order of the factors certainly matters in general). Formally, suppose we are given s matrices $A_1, \ldots, A_s$ and integers $n_0, \ldots, n_s$ such that $A_i \in M_{n_{i-1}, n_i}(F)$ for each i. Suppose also that we are given any complete parenthesization of the sequence $A_1 A_2 \cdots A_s$, which specifies exactly how these matrices are to be combined via the binary operation of matrix multiplication. For example, if $s = 4$, there are five possible complete parenthesizations:

$$A_1(A_2(A_3 A_4)), \ ((A_1 A_2)A_3)A_4, \ (A_1 A_2)(A_3 A_4), \ A_1((A_2 A_3)A_4), \ (A_1(A_2 A_3))A_4.$$

We define the *standard* complete parenthesization to be the $n_0 \times n_s$ matrix

$$\prod_{i=1}^{s} A_i = (\cdots (((A_1 A_2)A_3)A_4) \cdots)A_s.$$

Formally, the symbol $\prod_{i=1}^{s} A_i$ is defined recursively by $\prod_{i=1}^{1} A_i = A_1$ and $\prod_{i=1}^{s} A_i = (\prod_{i=1}^{s-1} A_i)A_s$ for all $s > 1$.

Generalized Associative Law for Matrix Multiplication. Suppose $A_1, \ldots, A_s$ are matrices with $A_i \in M_{n_{i-1}, n_i}(F)$ for each i. Any complete parenthesization of $A_1 A_2 \cdots A_s$ equals the standard parenthesization $\prod_{i=1}^{s} A_i$.

Proof. We prove the result by induction on s. The result holds for $s = 1$ and $s = 2$ since there is only one possible parenthesization. The result holds for the case $s = 3$ by the associative law proved in the previous section. Fix $s > 3$, and assume the result is known to hold for smaller values of s. Given any complete parenthesization of $A_1 \cdots A_s$, there exists a unique index $t < s$ such that the last binary product operation involved in the computation involves multiplying some complete parenthesization of $A_1 \cdots A_t$ by some complete parenthesization of $A_{t+1} \cdots A_s$. For example, for the five parenthesizations listed above in the case $s = 4$, we have $t = 1$, $t = 3$, $t = 2$, $t = 1$, and $t = 3$, respectively. By induction, the complete parenthesization of the first t factors evaluates to the $n_0 \times n_t$ matrix $B = \prod_{i=1}^{t} A_i$, while the complete parenthesization of the last $s - t$ factors evaluates to the $n_t \times n_s$ matrix $C = \prod_{j=t+1}^{s} A_j$. If $t = s-1$, this second product is A_s, so the given complete parenthesization of $A_1 \cdots A_s$ evaluates to $BA_s = \prod_{i=1}^{s} A_i$ by definition. Otherwise, in the case $t < s - 1$, set $D = \prod_{j=t+1}^{s-1} A_j$. We have $C = DA_s$ by definition, so that the given complete parenthesization of $A_1 \cdots A_s$ evaluates to $BC = B(DA_s)$. By associativity for three factors and the induction hypothesis, $B(DA_s) = (BD)A_s = (\prod_{i=1}^{s-1} A_i)A_s = \prod_{i=1}^{s} A_i$. $\square$

There is a formula for the entries of $\prod_{u=1}^{s} A_u$ that extends the formula for a product of two matrices. Specifically, the i, j-entry of $\prod_{u=1}^{s} A_u$ is

$$\sum_{k_1=1}^{n_1} \sum_{k_2=1}^{n_2} \cdots \sum_{k_{s-1}=1}^{n_{s-1}} A_1(i, k_1) A_2(k_1, k_2) A_3(k_2, k_3) \cdots A_{s-1}(k_{s-2}, k_{s-1}) A_s(k_{s-1}, j).$$

We prove this formula by induction on s. The base case $s = 2$ holds by definition of matrix multiplication. Fix $s > 2$, and assume the formula holds for products of $s - 1$ matrices. By definition of matrix multiplication, the i, j-entry of $\prod_{u=1}^{s} A_u = (\prod_{u=1}^{s-1} A_u) A_s$ is

$$\sum_{k=1}^{n_{s-1}} \left[\prod_{u=1}^{s-1} A_u \right] (i, k) A_s(k, j).$$

By induction hypothesis, this entry equals

$$\sum_{k=1}^{n_{s-1}} \left(\sum_{k_1=1}^{n_1} \cdots \sum_{k_{s-2}=1}^{n_{s-2}} A_1(i, k_1) A_2(k_1, k_2) \cdots A_{s-1}(k_{s-2}, k) \right) A_s(k, j).$$

Renaming the summation variable k to be k_{s-1}, using the distributive law in F, and reordering the finite summations, we obtain the required formula for the i, j-entry of a product of s factors.

Now that we know the products are unambiguously defined, we can prove that $(A_1 A_2 \cdots A_s)^{\mathrm{T}} = A_s^{\mathrm{T}} \cdots A_2^{\mathrm{T}} A_1^{\mathrm{T}}$ and, when $F = \mathbb{C}$, $(A_1 A_2 \cdots A_s)^{*} = A_s^{*} \cdots A_2^{*} A_1^{*}$. We already proved these identities for $s = 2$, and the general case follows by induction using generalized associativity. With generalized associativity in hand, we can also define the positive powers of a square matrix $A \in M_n(F)$. For each integer $s \geq 1$, we let $A^s = \prod_{i=1}^{s} A_i$ where every $A_i = A$. Informally, A^s is the product of s copies of A. We also define $A^0 = I_n$.

4.6   Invertible Matrices

A matrix A with entries in F is *invertible* iff A is square (say $n \times n$) and there exists a matrix $B \in M_n(F)$ with $AB = I_n = BA$. B is called an *inverse* of A. If such a matrix B exists, then B is unique. To prove this, suppose $B_1 \in M_n(F)$ also satisfies $AB_1 = I_n = B_1 A$. Then associativity of matrix multiplication gives

$$B_1 = B_1 I_n = B_1(AB) = (B_1 A)B = I_n B = B.$$

Given an invertible matrix A, define A^{-1} to be the unique inverse of A. For particular matrices $C, D \in M_n(F)$, we can prove that C^{-1} exists and that $C^{-1} = D$ by verifying the defining condition $CD = I_n = DC$. The next theorem illustrates how to use this proof technique.

Theorem on Closure Properties of Matrix Inverses.
(a) For all $A \in M_n(F)$, if A is invertible, then A^{-1} is invertible and $(A^{-1})^{-1} = A$.
(b) The identity matrix I_n is invertible, with $I_n^{-1} = I_n$.
(c) For all invertible $U, V \in M_n(F)$, UV is invertible, and $(UV)^{-1} = V^{-1}U^{-1}$.

Proof. To prove (a), suppose $A \in M_n(F)$ is invertible. Replace C by A^{-1} and D by A in the remark preceding the theorem. Since $A^{-1}A = I_n = AA^{-1}$ by definition of A^{-1}, the remark tells us that $(A^{-1})^{-1}$ exists and $(A^{-1})^{-1} = A$. To prove (b), replace C by I_n and D by I_n and observe that $I_n I_n = I_n = I_n I_n$. To prove (c), fix invertible matrices $U, V \in M_n(F)$. Compute

$$(UV)(V^{-1}U^{-1}) = U(VV^{-1})U^{-1} = UI_nU^{-1} = UU^{-1} = I_n.$$

A similar computation shows that $(V^{-1}U^{-1})(UV) = I_n$. The required conclusion follows by taking $C = UV$ and $D = V^{-1}U^{-1}$ in the remark. $\square$

Given k invertible matrices $U_1, U_2, \ldots, U_k \in M_n(F)$, the product $U_1 U_2 \cdots U_k$ is also invertible, and $(U_1 U_2 \cdots U_k)^{-1} = U_k^{-1} \cdots U_2^{-1} U_1^{-1}$. You can prove this by induction on k, using the case $k = 2$ that we just proved. (Note that generalized associativity is tacitly needed here and whenever we write unparenthesized products of matrices.) If we take all U_i equal to a given invertible matrix A, then we get $(A^k)^{-1} = (A^{-1})^k$ for all $k \geq 1$. For any invertible $A \in M_n(F)$ and positive integer k, define the *negative power* A^{-k} to be $(A^{-1})^k = (A^k)^{-1}$.

For all fields F and all positive integers n, let $\mathrm{GL}_n(F)$ be the set of all invertible matrices in $M_n(F)$. The set $\mathrm{GL}_n(F)$ is a group under matrix multiplication, which is almost always non-commutative. The group axioms (closure, associativity, identity, and inverses) follow from the preceding theorem. $\mathrm{GL}_n(F)$ is called the *general linear group of degree n*.

For any matrix $A \in M_n(F)$, A is invertible iff the transpose A^{T} is invertible, in which case $(A^{\mathrm{T}})^{-1} = (A^{-1})^{\mathrm{T}}$. To see this, first assume A^{-1} exists. Then

$$(A^{\mathrm{T}})(A^{-1})^{\mathrm{T}} = (A^{-1}A)^{\mathrm{T}} = (I_n)^{\mathrm{T}} = I_n = (I_n)^{\mathrm{T}} = (AA^{-1})^{\mathrm{T}} = (A^{-1})^{\mathrm{T}}(A^{\mathrm{T}}),$$

so the conclusion follows by taking $C = A^{\mathrm{T}}$ and $D = (A^{-1})^{\mathrm{T}}$ in the remark preceding the theorem. So the forward implication holds for all matrices. Applying the forward implication with A replaced by A^{T}, we conclude that the invertibility of A^{T} implies the invertibility of $(A^{\mathrm{T}})^{\mathrm{T}} = A$, which gives the reverse implication for the original A. An entirely analogous proof, using the fact that $(I_n)^* = I_n$, shows that *for all $A \in M_n(\mathbb{C})$, A is invertible iff A^* is invertible, in which case $(A^*)^{-1} = (A^{-1})^*$.*

[The rest of this section applies to fields and commutative rings F.] The next few paragraphs assume familiarity with determinants, which we study in detail in Chapter 5. Recall that for every square matrix $A \in M_n(F)$, there is an associated scalar $\det(A) \in F$ called the *determinant of A*. A famous theorem of linear algebra states that *for any field F and any $A \in M_n(F)$, A^{-1} exists iff $\det(A) \neq 0_F$.* In §5.11, we prove the more general fact that for any commutative ring R and any $A \in M_n(R)$, A^{-1} exists in $M_n(R)$ iff $\det(A)$ is an invertible element of the ring R. The proof provides explicit formulas (involving determinants) for every entry of A^{-1}, when the inverse exists. We also prove in §5.13 that for any commutative ring R and all $A, B \in M_n(R)$, $\det(AB) = \det(A)\det(B)$.

In our definition of an invertible matrix A, we require that A be square and that the inverse matrix B satisfy both $AB = I_n$ and $BA = I_n$. In fact, *given that A and B are square, either of the conditions $AB = I_n$ and $BA = I_n$ implies the other one.* For example, assume $A, B \in M_n(F)$ and $AB = I_n$. Taking determinants gives $\det(A)\det(B) = \det(I_n) = 1_F$. Since F is commutative, we also have $1_F = \det(B)\det(A)$, so that $\det(A)$ is an invertible element of F. By one of the theorems quoted in the last paragraph, A^{-1} (the unique *two-sided* inverse of A) exists. Multiplying both sides of $AB = I_n$ on the left by A^{-1}, we get $B = A^{-1}$, so that $BA = A^{-1}A = I_n$. By a similar proof, you can show that $BA = I_n$ implies $AB = I_n$. However, our proof (which uses determinants) relies crucially on the fact that the matrices in question are square. For $A \in M_{m,n}(F)$, we say $B \in M_{n,m}(F)$ is a *left inverse* of A iff $BA = I_n$. We say $B \in M_{n,m}(F)$ is a *right inverse* of A iff $AB = I_m$. When $m \neq n$, you can give examples where B is a left inverse but not a right inverse of A, and where C is a right inverse but not a left inverse of A.

We now prove that *for all $U, V \in M_n(F)$, U and V are invertible iff UV is invertible.* We proved the forward implication already by checking that $(UV)^{-1} = V^{-1}U^{-1}$. Conversely, assume UV is invertible with inverse $W \in M_n(F)$. Then $U(VW) = (UV)W = I_n$, so the result proved in the last paragraph shows that VW is the inverse of U. Similarly, $(WU)V = W(UV) = I_n$ shows that WU is the inverse of V. More generally, induction on k shows that for $U_1, \ldots, U_k \in M_n(F)$, every U_i is invertible iff the product $U_1 U_2 \cdots U_k$ is invertible. These results also follow quickly from the determinant criterion for invertibility.

4.7 Matrix Operations via Columns

We have seen that any $m \times n$ matrix A can be identified with the list of its columns in $(F^m)^n$: $A = (A^{[1]}, A^{[2]}, \ldots, A^{[n]})$, where $A^{[j]} = (A(1, j), A(2, j), \ldots, A(m, j))$ is column j of A. In symbols, $A^{[j]}(i) = A(i, j)$ for $1 \le i \le m$ and $1 \le j \le n$. It is fruitful to recast the definitions of the matrix operations in terms of the columns of the matrices involved.

First, assume $A, B \in M_{m,n}(F)$ and $c \in F$. For each column j, $(A + B)^{[j]} = A^{[j]} + B^{[j]}$ since, for $1 \le i \le m$, $(A + B)^{[j]}(i) = (A + B)(i, j) = A(i, j) + B(i, j) = A^{[j]}(i) + B^{[j]}(i) = (A^{[j]} + B^{[j]})(i)$. Similarly, you can check that $(cA)^{[j]} = c \cdot A^{[j]}$. This means that we can add two matrices one column at a time, and multiplying a matrix by a scalar multiplies each column by that scalar.

Next, assume $A \in M_{m,n}(F)$ and $B \in M_{n,p}(F)$, and let $C = AB \in M_{m,p}(F)$. How are the columns of A and B related to the columns of C? To answer this, fix i, j with $1 \le i \le m$ and $1 \le j \le p$. Compute

$$C^{[j]}(i) = (AB)(i, j) = \sum_{k=1}^{n} A(i, k)B(k, j) = \sum_{k=1}^{n} A(i, k)B^{[j]}(k) = (A(B^{[j]}))(i).$$

The last equality is the definition of multiplying the matrix A on the right by the column vector $B^{[j]}$. So $(AB)^{[j]} = A(B^{[j]})$ for all j, which means that *column j of AB is found by multiplying matrix A on the right by column j of B.* For example, the second column of the product

$$\begin{bmatrix} 2 & 0 \\ -1 & 3 \\ 0 & 4 \end{bmatrix} \begin{bmatrix} \pi & 2 & e & 3i & \sqrt{37} \\ 9! & 3 & \sqrt[3]{-11} & 10^9 & (2 + 3i)^7 \end{bmatrix}$$

is

$$\begin{bmatrix} 2 & 0 \\ -1 & 3 \\ 0 & 4 \end{bmatrix} \begin{bmatrix} 2 \\ 3 \end{bmatrix} = \begin{bmatrix} 4 \\ 7 \\ 12 \end{bmatrix}.$$

Now consider the product Av, where A is an $m \times n$ matrix and v is an $n \times 1$ column vector. For all $i \in [m]$,

$$(Av)(i) = \sum_{j=1}^{n} A(i, j)v(j) = \sum_{j=1}^{n} A^{[j]}(i)v(j) = \left(\sum_{j=1}^{n} A^{[j]}v(j) \right)(i).$$

So $Av = \sum_{j=1}^{n} A^{[j]}v(j)$ in F^m [or $Av = \sum_{j=1}^{n} v(j)A^{[j]}$ for *commutative* rings F], which means that *the matrix-vector product Av is a linear combination of the columns of A, and the coefficients in this linear combination are the entries in v.* For example,

$$\begin{bmatrix} 1 & 2 & 0 & -1 \\ 3 & i & -2 & 4 \\ 1/2 & 0 & 0 & \pi \end{bmatrix} \begin{bmatrix} 2 \\ -1 \\ 5 \\ 0 \end{bmatrix} = \begin{bmatrix} 0 \\ -4 - i \\ 1 \end{bmatrix} = 2 \begin{bmatrix} 1 \\ 3 \\ 1/2 \end{bmatrix} - 1 \begin{bmatrix} 2 \\ i \\ 0 \end{bmatrix} + 5 \begin{bmatrix} 0 \\ -2 \\ 0 \end{bmatrix} + 0 \begin{bmatrix} -1 \\ 4 \\ \pi \end{bmatrix}.$$

[The rest of this section assumes F is a field.] Let us deduce some consequences of the fact that Av is a linear combination of the columns of A. Given $A \in M_{m,n}(F)$, define the *range* or *image* of A to be the set $R(A) = \{Av : v \in F^n\} \subseteq F^m$, which is readily seen to be a subspace of F^m. Our formula for Av shows that

$$R(A) = \{v(1)A^{[1]} + \cdots + v(n)A^{[n]} : v(i) \in F\}.$$

So the range of A coincides with the subspace of all linear combinations of the columns of A in F^m. This subspace is also called the *column space* of A and may be written $\text{Col}(A)$. The dimension of this subspace is called the *column rank* of A, written $\text{colrk}(A)$.

For another application, consider a general matrix product AB, where $A \in M_{m,n}(F)$ and $B \in M_{n,p}(F)$. We prove that $\text{colrk}(AB) \leq \min(\text{colrk}(A), \text{colrk}(B))$, or (equivalently) $\text{colrk}(AB) \leq \text{colrk}(A)$ and $\text{colrk}(AB) \leq \text{colrk}(B)$. First, since $(AB)^{[j]} = A(B^{[j]})$, where $B^{[j]}$ is a column vector, we see that column j of AB is a linear combination of the columns of A, where the coefficients come from column j of B. In particular, each column $(AB)^{[j]}$ belongs to the column space of A. Knowing that $(AB)^{[j]} \in \text{Col}(A)$ for all j, it follows that $\text{Col}(AB) \subseteq \text{Col}(A)$ since the columns of AB generate the column space of AB. Taking dimensions, we see that $\text{colrk}(AB) \leq \text{colrk}(A)$. Next, we show that $\text{colrk}(AB) \leq \text{colrk}(B)$. Define a map $L_A : F^n \to F^m$ by sending each column vector $v \in F^n$ to the column vector $L_A(v) = Av \in F^m$. You can check that L_A is a linear map. Let $W = \text{Col}(B) \subseteq F^n$, and recall that W is spanned by the columns $B^{[1]}, \ldots, B^{[p]}$ of B. Applying the linear map L_A, we see (using Exercise 63 of Chapter 1) that $L_A[W]$ is spanned by the vectors $L_A(B^{[j]}) = A(B^{[j]}) = (AB)^{[j]}$, for $j = 1, \ldots, p$. The latter vectors are exactly the columns of AB. We conclude that $\text{Col}(AB) = L_A[W]$. Since the column space of AB is the image of the column space of B under a linear map, we must have $\dim(\text{Col}(AB)) \leq \dim(\text{Col}(B))$ (see Exercise 61 in Chapter 1). In other words, $\text{colrk}(AB) \leq \text{colrk}(B)$, as needed.

Suppose $A \in M_{m,n}(F)$. If P is any invertible $m \times m$ matrix, then we have inequalities

$$\text{colrk}(A) \geq \text{colrk}(PA) \geq \text{colrk}(P^{-1}(PA)) = \text{colrk}(A),$$

so that $\text{colrk}(PA) = \text{colrk}(A)$. If Q is any invertible $n \times n$ matrix, then the inequalities

$$\text{colrk}(A) \geq \text{colrk}(AQ) \geq \text{colrk}((AQ)Q^{-1}) = \text{colrk}(A)$$

show that $\text{colrk}(AQ) = \text{colrk}(A)$. Thus, *multiplication on the left or right by an invertible matrix does not change the column rank*. In particular, let A be the identity matrix I_n, which has column rank n since its columns form the basis $(e_1, \ldots, e_n)$ of F^n. We deduce that *every invertible $n \times n$ matrix has column rank n*.

4.8   Matrix Operations via Rows

We now rewrite the definitions of matrix operations in terms of the rows of the matrices and vectors involved. This time, we start with the fact that any $m \times n$ matrix A can be identified with the list of its rows in $(F^n)^m$: $A = (A_{[1]}, A_{[2]}, \ldots, A_{[m]})$, where $A_{[i]} = (A(i,1), A(i,2), \ldots, A(i,n))$ is row i of A. In symbols, $A_{[i]}(j) = A(i,j)$ for $1 \leq i \leq m$ and $1 \leq j \leq n$.

As in the case of columns, a short calculation shows that $(A + B)_{[i]} = A_{[i]} + B_{[i]}$ and $(cA)_{[i]} = c(A_{[i]})$ for all $A, B \in M_{m,n}(F)$, all $c \in F$, and all $i \in [m]$. This means that we can add two matrices of the same size one row at a time, and multiplying a matrix by a scalar multiplies each row by that scalar.

Next, assume $A \in M_{m,n}(F)$ and $B \in M_{n,p}(F)$, and let $C = AB \in M_{m,p}(F)$. How are the rows of A and B related to the rows of C? To answer this, fix i, j with $1 \leq i \leq m$ and $1 \leq j \leq p$. Compute

$$C_{[i]}(j) = (AB)(i,j) = \sum_{k=1}^{n} A(i,k)B(k,j) = \sum_{k=1}^{n} A_{[i]}(k)B(k,j) = (A_{[i]}B)(j).$$

So $(AB)_{[i]} = (A_{[i]})B$ for all $i \in [m]$, which means that *row i of AB is found by multiplying matrix B on the left by row i of A*. For example, the third row of the product

$$\begin{bmatrix} 2 & 0 \\ -1 & 3 \\ 0 & 4 \end{bmatrix} \begin{bmatrix} 5 & 7 & 9 & 11 & 13 \\ 1 & 1 & 1 & 1 & 10 \end{bmatrix}$$

is

$$\begin{bmatrix} 0 & 4 \end{bmatrix} \begin{bmatrix} 5 & 7 & 9 & 11 & 13 \\ 1 & 1 & 1 & 1 & 10 \end{bmatrix} = \begin{bmatrix} 4 & 4 & 4 & 4 & 40 \end{bmatrix}.$$

Now consider the product wA, where A is an $m \times n$ matrix and w is a $1 \times m$ row vector. For all $j \in [n]$,

$$(wA)(j) = \sum_{i=1}^{m} w(i)A(i,j) = \sum_{i=1}^{m} w(i)A_{[i]}(j) = \left(\sum_{i=1}^{m} w(i)A_{[i]} \right)(j).$$

So $wA = \sum_{i=1}^{m} w(i)A_{[i]}$ in F^n [and the scalars $w(i)$ must appear on the left for non-commutative rings F]. This says that *the vector-matrix product wA is a linear combination of the rows of A, and the coefficients in this linear combination are the entries in w*. For example,

$$\begin{bmatrix} 3 & 0 & -2 \end{bmatrix} \begin{bmatrix} 1 & 2 & 0 \\ 3 & i & -2 \\ 1/2 & 0 & 1 \end{bmatrix} = 3 \begin{bmatrix} 1 & 2 & 0 \end{bmatrix} + 0 \begin{bmatrix} 3 & i & -2 \end{bmatrix} - 2 \begin{bmatrix} 1/2 & 0 & 1 \end{bmatrix}.$$

[The rest of this section assumes F is a field.] Given $A \in M_{m,n}(F)$, define the *row space* of A to be the subspace $\mathrm{Row}(A) \subseteq F^n$ spanned by the m rows of A, which consists of all F-linear combinations of the rows of A. Define the *row rank* of A, written $\mathrm{rowrk}(A)$, to be the dimension of the row space of A. Reasoning as we did earlier for columns, our formula for wA shows that $\mathrm{Row}(A) = \{wA : w \in F^m\} \subseteq F^n$. Furthermore, for all $B \in M_{n,p}(F)$, each row i of the matrix product AB has the form $(AB)_{[i]} = (A_{[i]})B$. So row i of AB is a linear combination of the rows of B with coefficients coming from row i of A. Since the rows of AB generate $\mathrm{Row}(AB)$, we conclude that $\mathrm{Row}(AB) \subseteq \mathrm{Row}(B)$ and $\mathrm{rowrk}(AB) \leq \mathrm{rowrk}(B)$. On the other hand, consider the linear map $R_B : F^n \rightarrow F^p$ that sends a row vector $w \in F^n$ to the row vector $R_B(w) = wB \in F^p$. Let $V = R_B[\mathrm{Row}(A)]$ be the image of $\mathrm{Row}(A)$ under this linear map. Since $\mathrm{Row}(A)$ is spanned by $A_{[1]}, \ldots, A_{[m]}$, V is spanned by $R_B(A_{[1]}), \ldots, R_B(A_{[m]})$. In other words, V is spanned by $A_{[1]}B = (AB)_{[1]}, \ldots, A_{[m]}B = (AB)_{[m]}$. But these are exactly the rows of AB, and hence $V = \mathrm{Row}(AB)$. Since $V = R_B[\mathrm{Row}(A)]$, we must have $\dim(V) \leq \dim(\mathrm{Row}(A))$, and hence $\mathrm{rowrk}(AB) \leq \mathrm{rowrk}(A)$. In summary,

$$\mathrm{rowrk}(AB) \leq \min(\mathrm{rowrk}(A), \mathrm{rowrk}(B)).$$

Let $A \in M_{m,n}(F)$ be any matrix. If P is any invertible $m \times m$ matrix, then

$$\mathrm{rowrk}(A) \geq \mathrm{rowrk}(PA) \geq \mathrm{rowrk}(P^{-1}(PA)) = \mathrm{rowrk}(A),$$

so $\mathrm{rowrk}(PA) = \mathrm{rowrk}(A)$. Similarly, if Q is any invertible $n \times n$ matrix, then

$$\mathrm{rowrk}(A) \geq \mathrm{rowrk}(AQ) \geq \mathrm{rowrk}((AQ)Q^{-1}) = \mathrm{rowrk}(A),$$

so $\mathrm{rowrk}(AQ) = \mathrm{rowrk}(A)$. Thus, *multiplication on the left or right by an invertible matrix does not change the row rank*. In particular, let A be the identity matrix I_n, which has row rank n since its rows form the basis $(e_1, \ldots, e_n)$ of F^n. We deduce that *every invertible*

$n \times n$ matrix has row rank n. In §4.12, we prove that $\text{rowrk}(A) = \text{colrk}(A)$ for every matrix $A \in M_{m,n}(F)$.

Before leaving this section, we mention interpretations for the transpose and conjugate-transpose operations in terms of rows and columns. Suppose A is an $m \times n$ matrix. From the definition $A^{\mathrm{T}}(i, j) = A(j, i)$, we see immediately that $(A^{\mathrm{T}})_{[i]}(j) = A^{\mathrm{T}}(i, j) = A(j, i) = A^{[i]}(j)$ for $1 \leq j \leq m$, so that $(A^{\mathrm{T}})_{[i]} = A^{[i]}$ for all $i \in [n]$. Similarly, $(A^{\mathrm{T}})^{[j]}(i) = A^{\mathrm{T}}(i, j) = A(j, i) = A_{[j]}(i)$ for all $i \in [n]$ implies that $(A^{\mathrm{T}})^{[j]} = A_{[j]}$ for all $j \in [m]$. Translating these equations into words, we see that: row i of A^{T} is column i of A (for all $i \in [n]$), while column j of A^{T} is row j of A (for all $j \in [m]$). Thus, the matrix transpose operation interchanges the rows and columns of A. An analogous calculation when $F = \mathbb{C}$ shows that $(A^{*})_{[i]} = \overline{A^{[i]}}$ and $(A^{*})^{[j]} = \overline{A_{[j]}}$, where the complex conjugate of a row or column vector is found by conjugating each entry. So, we obtain A^{*} from A by interchanging rows and columns and taking the complex conjugate of every entry.

4.9   Elementary Operations and Elementary Matrices

This section describes the link between elementary row and column operations on matrices and multiplication by certain matrices called elementary matrices. There are three types of elementary row operations that we can apply to an $m \times n$ matrix A. First, for any $i \in [m]$, we can multiply row i of A by a nonzero scalar c in the field F. [In the case of rings, we require c to be an invertible element (unit) in the ring F. For non-commutative rings, each entry of the row is multiplied on the left by c.] Second, for any i, j in $[m]$ with $i \neq j$, we can interchange row i and row j of A. Third, for any i, j in $[m]$ with $i \neq j$ and any scalar $c \in F$, we can add c times row j to row i of A [multiplying by c on the left for non-commutative rings]. There are three analogous elementary column operations that act on the columns of A. [For non-commutative rings, we multiply column entries by c on the right.] These elementary operations are helpful for solving systems of linear equations, for computing normal forms and other invariants of matrices, and for other applications.

We now define three kinds of *elementary matrices*. First, for $i \in [p]$ and nonzero $c \in F$ [or units c in the case of rings], let $E_p^1[c; i]$ be the $p \times p$ matrix with i, i-entry equal to c, j, j-entries equal to 1_F for $j \neq i$, and all other entries equal to 0_F. Second, for $i \neq j$ in $[p]$, let $E_p^2[i, j]$ be the $p \times p$ matrix with 1_F in positions (i, j), (j, i), and (k, k) for all $k \neq i, j$, and 0_F elsewhere. Third, for $c \in F$ and $i \neq j$ in $[p]$, let $E_p^3[c; i, j]$ be the $p \times p$ matrix having 1_F in each diagonal position (k, k), c in position (i, j), and 0_F elsewhere. We prove below that the three types of elementary matrices encode the three types of elementary row or column operations defined above. As a mnemonic aid, note that we can obtain the elementary matrix for a given elementary operation by performing that operation on the $p \times p$ identity matrix I_p. For example,

$$E_3^1[4; 2] = \begin{bmatrix} 1 & 0 & 0 \\ 0 & 4 & 0 \\ 0 & 0 & 1 \end{bmatrix} ; \quad E_3^2[2, 3] = \begin{bmatrix} 1 & 0 & 0 \\ 0 & 0 & 1 \\ 0 & 1 & 0 \end{bmatrix} ; \quad E_3^3[3; 2, 1] = \begin{bmatrix} 1 & 0 & 0 \\ 3 & 1 & 0 \\ 0 & 0 & 1 \end{bmatrix} .$$

Fix an $m \times n$ matrix A, and take $p = m$. We claim that *multiplying A on the **left** by an elementary matrix has the same effect as applying the corresponding elementary **row** operation to A*. This is not hard to check by writing formulas for all the entries, but it is more instructive to give a conceptual proof using the ideas from §4.8. For example, consider the matrix $B = E_m^3[c; i, j]A$. Row k of B is $E_m^3[c; i, j]_{[k]}A$. This row is the linear combination of the m rows of A with coefficients given by the entries in the row vector $E_m^3[c; i, j]_{[k]}$. If $k \neq i$, then the relevant row of the elementary matrix is e_k, so row $B_{[k]}$ is 1 times row $A_{[k]}$ plus 0 times every other row of A. This means row k row of B equals row k of A for all $k \neq i$. If $k = i$, then $E_m^3[c; i, j]_{[i]} = ce_j + e_i$ by definition, so row $B_{[i]}$ is c times $A_{[j]}$ plus 1 times $A_{[i]}$. This means row i of B is c times row j of A plus row i of A. This proves that multiplication on the left by $E_m^3[c; i, j]$ has the same effect as the elementary row operation of the third type. The assertions for $E_m^1[c; i]$ and $E_m^2[i, j]$ are checked in the same way.

On the other hand, *multiplying $A \in M_{m,n}(F)$ on the **right** by an $n \times n$ elementary matrix has the same effect as applying the corresponding elementary **column** operation to A.* For example, consider the matrix $C = AE_n^2[i, j]$. For $k \in [n]$, column k of C is $A(E_n^2[i, j]^{[k]})$, which is the linear combination of the columns of A with coefficients given by the entries in the column vector $E_n^2[i, j]^{[k]}$. For $k \neq i, j$, this column vector is e_k, so $C^{[k]} = A^{[k]}$. For $k = i$, this column vector is e_j, so $C^{[i]} = A^{[j]}$. For $k = j$, this column vector is e_i, so $C^{[j]} = A^{[i]}$. So the columns of C are precisely the columns of A with column i and column j interchanged. The assertions for the other elementary matrices are checked in the same way. Note carefully that to add c times row j to row i, we left-multiply by $E_m^3[c; i, j]$; but to add c times column j to column i, we right-multiply by $E_n^3[c; j, i]$.

The inverse of an elementary matrix exists and is an elementary matrix of the same type. Specifically, $E_p^1[c; i]^{-1} = E_p^1[c^{-1}; i]$, $E_p^2[i, j]^{-1} = E_p^2[i, j]$, and $E_p^3[c; i, j]^{-1} = E_p^3[-c; i, j]$. You can verify these formulas by computing with entries, but they can also be checked by applying the remarks in the previous two paragraphs. For example, the product $E_p^3[c; i, j]E_p^3[-c; i, j]$ is the matrix obtained from $E_p^3[-c; i, j]$ by adding c times row j to row i. Row j of $E_p^3[-c; i, j]$ is e_j, and row i is $e_i + (-c)e_j$, so the new row i after the row operation is e_i. For all $k \neq i$, row k of $E_p^3[-c; i, j]$ is e_k, and this is unchanged by the row operation. So $E_p^3[c; i, j]E_p^3[-c; i, j] = I_p$, and the product in the other order is the identity for analogous reasons. The formulas for the inverses of the other types of elementary matrices can be verified similarly.

4.10   Elementary Matrices and Gaussian Elimination

[This section assumes F is a field.] Elementary matrices can help us prove many facts about general matrices, such as the determinant product formula $\det(AB) = \det(A)\det(B)$ (see §5.9). We begin here by reviewing some basic facts concerning the Gaussian elimination algorithm and relating this algorithm to elementary matrices. We assume readers have already seen some version of this algorithm in the context of solving systems of linear equations. For a more detailed study of Gaussian elimination and its connection to matrix factorizations, see Chapter 9.

Starting with any $m \times n$ matrix A, the Gaussian elimination algorithm applies a sequence of elementary row operations to transform A to its *reduced row-echelon form*, denoted A_{ech}. To describe A_{ech} precisely, let us call the leftmost nonzero entry in a given nonzero row of A_{ech} the *leading entry* of that row. Then A_{ech} is characterized by the following properties: any rows of A_{ech} with all entries zero (if any such rows exist) occur at the bottom of the matrix; the leading entry of each nonzero row is 1_F; for all $i \in [m - 1]$, the leading entry in row i is strictly left of the leading entry in row $i + 1$ (if any); and for each leading entry, the other elements in the column containing that entry are all zero. It can be shown that *for every $A \in M_{m,n}(F)$, there exists a unique reduced row-echelon form A_{ech} that can be reached from A by applying finitely many elementary row operations.* We can prove existence using a version of Gaussian elimination called Gauss–Jordan elimination, which produces the zeroes in the columns of the leading entries. The uniqueness of A_{ech} is more subtle, but we do not need to invoke uniqueness in what follows.

We have seen that each elementary row operation used in the passage from A to A_{ech} can be accomplished by multiplying on the left by an appropriate elementary matrix. Thus, $A_{\text{ech}} = E_k E_{k-1} \cdots E_1 A$, where E_s is the elementary matrix corresponding to the row operation applied at step s of the Gaussian elimination algorithm.

For example, let us compute the reduced row-echelon form for the matrix

$$A = \begin{bmatrix} 0 & 3 & -4 & 1 \\ 2 & 4 & -2 & 6 \\ 4 & -1 & 8 & 9 \end{bmatrix}$$

and find a specific factorization $A_{\text{ech}} = E_k \cdots E_1 A$. To obtain a leading 1 in the $1,1$-position, we switch row 1 and row 2 of A and then multiply the new row 1 by $1/2$. This can be accomplished by left-multiplying A first by $E_1 = E_3^2[1,2]$, and then by $E_2 = E_3^1[1/2;1]$:

$$E_1 A = \begin{bmatrix} 2 & 4 & -2 & 6 \\ 0 & 3 & -4 & 1 \\ 4 & -1 & 8 & 9 \end{bmatrix}; \qquad E_2 E_1 A = \begin{bmatrix} 1 & 2 & -1 & 3 \\ 0 & 3 & -4 & 1 \\ 4 & -1 & 8 & 9 \end{bmatrix}.$$

Next, we add -4 times row 1 to row 3, by left-multiplying by the elementary matrix $E_3 = E_3^3[-4;3,1]$. We continue by adding 3 times row 2 to row 3 (left-multiply by $E_4 = E_3^3[3;3,2]$):

$$E_3 E_2 E_1 A = \begin{bmatrix} 1 & 2 & -1 & 3 \\ 0 & 3 & -4 & 1 \\ 0 & -9 & 12 & -3 \end{bmatrix}; \qquad E_4 E_3 E_2 E_1 A = \begin{bmatrix} 1 & 2 & -1 & 3 \\ 0 & 3 & -4 & 1 \\ 0 & 0 & 0 & 0 \end{bmatrix}.$$

To finish, we multiply row 2 by $1/3$ (left-multiply by $E_5 = E_3^1[1/3;2]$) and then add -2 times the new row 2 to row 1 (left-multiply by $E_6 = E_3^3[-2;1,2]$):

$$E_5 E_4 E_3 E_2 E_1 A = \begin{bmatrix} 1 & 2 & -1 & 3 \\ 0 & 1 & -4/3 & 1/3 \\ 0 & 0 & 0 & 0 \end{bmatrix};$$

$$A_{\text{ech}} = E_6 E_5 E_4 E_3 E_2 E_1 A = \begin{bmatrix} 1 & 0 & 5/3 & 7/3 \\ 0 & 1 & -4/3 & 1/3 \\ 0 & 0 & 0 & 0 \end{bmatrix}.$$

4.11 Elementary Matrices and Invertibility

[This section assumes F is a field.] Our first theorem involving elementary matrices states that *a square matrix A is invertible iff A can be written as a product of elementary matrices.* To prove this, assume first that A is a product of elementary matrices. We know every elementary matrix is invertible, so any product of elementary matrices is also invertible (§4.6). Conversely, assume that $A \in M_n(F)$ is invertible. Write $A_{\text{ech}} = E_k E_{k-1} \cdots E_1 A$ for certain elementary matrices E_j. Since A and each elementary matrix E_j is invertible, so is their product A_{ech}. It readily follows that A_{ech} cannot have any zero rows. Since A_{ech} is square, the definition of reduced row-echelon form forces A_{ech} to be the identity matrix I_n. Consequently, $E_k E_{k-1} \cdots E_1 A = I_n$. Solving for A, we find that $A = E_1^{-1} E_2^{-1} \cdots E_k^{-1}$. Since the inverse of an elementary matrix is elementary, A has been expressed as a product of elementary matrices.

For example, let us express the invertible matrix $A = \begin{bmatrix} 2 & 1 \\ 4 & 3 \end{bmatrix}$ as a product of elementary matrices. We reduce A to $A_{\text{ech}} = I_2$ by the following sequence of elementary row operations: add -2 times row 1 to row 2, producing $\begin{bmatrix} 2 & 1 \\ 0 & 1 \end{bmatrix}$; multiply row 1 by $1/2$, producing $\begin{bmatrix} 1 & 1/2 \\ 0 & 1 \end{bmatrix}$; add $-1/2$ times row 2 to row 1, producing I_2. In terms of matrices, we have

$$I_2 = E_2^3[-\frac{1}{2};1,2]E_2^1[1/2;1]E_2^3[-2;2,1]A = \begin{bmatrix} 1 & -1/2 \\ 0 & 1 \end{bmatrix}\begin{bmatrix} 1/2 & 0 \\ 0 & 1 \end{bmatrix}\begin{bmatrix} 1 & 0 \\ -2 & 1 \end{bmatrix}\begin{bmatrix} 2 & 1 \\ 4 & 3 \end{bmatrix}.$$

Solving for A gives

$$\begin{bmatrix} 2 & 1 \\ 4 & 3 \end{bmatrix} = A = E_2^3[2;2,1]E_2^1[2;1]E_2^3[1/2;1,2] = \begin{bmatrix} 1 & 0 \\ 2 & 1 \end{bmatrix}\begin{bmatrix} 2 & 0 \\ 0 & 1 \end{bmatrix}\begin{bmatrix} 1 & 1/2 \\ 0 & 1 \end{bmatrix}.$$

4.12   Row Rank and Column Rank

[This section assumes F is a field.] As our second application of elementary matrices, we prove that *for any $m \times n$ matrix A*, $\text{rowrk}(A) = \text{colrk}(A)$. We can therefore define the *rank of A*, written $\text{rank}(A)$, as the common value of the row rank of A and the column rank of A. Recall that we have already shown (§4.8) that multiplying a matrix on the left or right by an invertible matrix does not change the row rank. These operations do not affect the column rank either (§4.7). Moreover, we have seen (§4.10) that $A_{\text{ech}} = PA$ for some matrix P that must be invertible (as P is a product of elementary matrices).

Having transformed A to A_{ech} by left-multiplying by P, let us continue to simplify the matrix A_{ech} by applying elementary column operations. These operations are achieved by multiplying on the right by certain elementary matrices. The leading 1 in each nonzero row of A_{ech} is already the only nonzero entry in its column. Adding appropriately chosen constant multiples of this column to the columns to its right, we can arrange that the leading 1 in each nonzero row is the only nonzero entry in its row. The matrix now consists of some number r of nonzero rows, each of which has a single 1 in it, and these 1s appear in distinct columns. Using column interchanges, we can arrange the matrix so that the 1s appear in the first r positions along the main diagonal. (The *main diagonal* of a rectangular $m \times n$ matrix consists of the i,i-entries for $1 \leq i \leq \min(m,n)$.) This final matrix is obtained from $A_{\text{ech}} = PA$ by right-multiplying by some matrix Q, which is a product of elementary matrices and is therefore invertible.

In summary, given $A \in M_{m,n}(F)$, we can find invertible matrices $P \in M_m(F)$ and $Q \in M_n(F)$ and an integer $r \in \{0,1,\ldots,\min(m,n)\}$ such that $(PAQ)_{i,i} = 1$ for all i between 1 and r, and all other entries of PAQ are 0. P is the product of the elementary matrices implementing the elementary row operations used to transform A, while Q is the product of the elementary matrices implementing the elementary column operations. The matrix PAQ is called the *Smith canonical form* or *Smith normal form* for A. [See Chapter 18 for a discussion of the more general case in which F is a principal ideal domain (PID).] Now we can prove our main result. We know that $\text{rowrk}(A) = \text{rowrk}(PAQ)$ and $\text{colrk}(A) = \text{colrk}(PAQ)$. It is evident from the form of PAQ that $\text{rowrk}(PAQ) = r = \text{colrk}(PAQ)$. Thus, $\text{rowrk}(A) = \text{colrk}(A)$. This proof provides an algorithm for computing the rank of A. You can also show that the row rank (and hence the column rank) of A equals the number of nonzero rows in the reduced row-echelon form A_{ech}. This observation gives a faster way to compute the rank of A.

For example, the matrix $A = \begin{bmatrix} 0 & 3 & -4 & 1 \\ 2 & 4 & -2 & 6 \\ 4 & -1 & 8 & 9 \end{bmatrix}$ has reduced row-echelon form

$$A_{\text{ech}} = PA = \begin{bmatrix} 1 & 0 & 5/3 & 7/3 \\ 0 & 1 & -4/3 & 1/3 \\ 0 & 0 & 0 & 0 \end{bmatrix},$$

where $P = \begin{bmatrix} -2/3 & 1/2 & 0 \\ 1/3 & 0 & 0 \\ 3 & -2 & 1 \end{bmatrix}$ is the product of the six elementary matrices used to reduce A to A_{ech} in §4.10. We see at once that $\text{rowrk}(A) = \text{rowrk}(A_{\text{ech}}) = 2$, and you can also check quickly that $\text{colrk}(A) = \text{colrk}(A_{\text{ech}}) = 2$. Let us continue transforming A_{ech} to reach the Smith canonical form of A. We use the following elementary column operations to remove the nonzero entries to the right of the leading 1s in A_{ech}: add $-5/3$ times column 1 to column 3; add $-7/3$ times column 1 to column 4; add $4/3$ times column 2 to column 3; add $-1/3$ times column 2 to column 4. In terms of matrices, let

$$Q = E_4^3[-5/3; 1, 3] E_4^3[-7/3; 1, 4] E_4^3[4/3; 2, 3] E_4^3[-1/3; 2, 4] = \begin{bmatrix} 1 & 0 & -5/3 & -7/3 \\ 0 & 1 & 4/3 & -1/3 \\ 0 & 0 & 1 & 0 \\ 0 & 0 & 0 & 1 \end{bmatrix}.$$

Then $PAQ = \begin{bmatrix} 1 & 0 & 0 & 0 \\ 0 & 1 & 0 & 0 \\ 0 & 0 & 0 & 0 \end{bmatrix}$ is the Smith normal form of A. The matrix PAQ has $r = 2$ nonzero entries on the main diagonal, confirming that $\text{rank}(A) = \text{rowrk}(A) = \text{colrk}(A) = 2$.

4.13 Conditions for Invertibility of a Matrix

[This section assumes F is a field.] Table 4.1 displays one of the central theorems of matrix algebra, which gives a long list of conditions, each of which is logically equivalent to the invertibility of a square matrix A. Condition (8b) says that the system of linear equations $Ax = b$ always has a solution x for any right side b. Condition (10b) says the system $Ax = b$ always has a unique solution x for any right side b. Condition (9b) says the homogeneous system $Ax = 0$ has no solutions other than the zero solution $x = 0$. The proof shows that these conditions can be reformulated by saying a certain linear map L_A (left multiplication by A) is surjective, or bijective, or injective. To detect when all these conditions hold, it is sufficient to check that A has nonzero determinant, or to test the rows (or columns) of A for linear independence, or to see if the reduced row-echelon form of A is I_n. We also see from conditions (2a) and (2b) that A is invertible (meaning A has a two-sided inverse) if A has a left inverse or if A has a right inverse.

We now prove the theorem in Table 4.1. We must verify many conditional and biconditional statements.

(1)$\Rightarrow$(2a): If A is invertible, then (2a) follows by choosing $B = A^{-1}$.
(1)$\Rightarrow$(2b): If A is invertible, then (2b) follows by choosing $C = A^{-1}$.

TABLE 4.1

Theorem on Equivalent Conditions for Invertibility of a Matrix.

Assume F is a field, $A \in M_n(F)$, and $L_A : F^n \to F^n$ is the linear map given by $L_A(x) = Ax$ for all column vectors $x \in F^n$. The following conditions are all equivalent:

(1)	A is invertible.
(2a)	$BA = I_n$ for some $B \in M_n(F)$.
(2b)	$AC = I_n$ for some $C \in M_n(F)$.
(3)	$\det(A) \neq 0_F$.
(4)	A is a product of finitely many elementary matrices.
(5)	The reduced row-echelon form A_{ech} is I_n.
(6a)	$\text{rowrk}(A) = n$ (the row space of A has dimension n).
(6b)	The n rows of A form a linearly independent list in F^n.
(7a)	$\text{colrk}(A) = n$ (the column space of A has dimension n).
(7b)	The n columns of A form a linearly independent list in F^n.
(7c)	The range of A has dimension n.
(8a)	The map L_A is surjective (onto).
(8b)	For all $b \in F^n$, there exists $x \in F^n$ with $Ax = b$.
(9a)	The map L_A is injective (one-to-one).
(9b)	For all $x \in F^n$, if $Ax = 0$ then $x = 0$.
(9c)	The null space of A (the kernel of L_A) is $\{0\}$.
(10a)	The map L_A is bijective (and hence an isomorphism).
(10b)	For all $b \in F^n$, there exists a unique $x \in F^n$ with $Ax = b$.
(11)	A^{T} is invertible.
(12)	[when $F = \mathbb{C}$] A^* is invertible.

(2a)$\Rightarrow$(3): Given $BA = I_n$, take determinants and use the product formula to get $\det(B)\det(A) = \det(I_n) = 1_F$. Since $1_F \neq 0_F$ in a field F, this forces $\det(A) \neq 0_F$.

(2b)$\Rightarrow$(3): The proof is analogous to (2a)$\Rightarrow$(3). (Similarly, (1) implies (3).)

(3)$\Rightarrow$(1): If $\det(A) \neq 0_F$, then we can use the explicit formula for A^{-1} (see §5.11) to confirm that A has a two-sided inverse. This formula involves a division by $\det(A)$, which is why we need A to have nonzero determinant.

(1)$\Leftrightarrow$(4): We proved this in §4.11.

(1)$\Leftrightarrow$(5): For any matrix A, we saw in §4.10 that $A_{\text{ech}} = PA$, where $P \in M_n(F)$ is a product of elementary matrices, hence is invertible. So, A is invertible iff A_{ech} is invertible (§4.6). Let $B \in M_n(F)$ be any square matrix in reduced row-echelon form. Using the definition of reduced row-echelon form, we see that B has no zero rows iff every row of B contains a leading 1 iff every column of B contains a leading 1 iff $B = I_n$. Since I_n is invertible and no matrix with a row of zeroes is invertible, we see that B is invertible iff $B = I_n$. Applying this remark to $B = A_{\text{ech}}$, we see that A is invertible iff $A_{\text{ech}} = I_n$.

(1)$\Leftrightarrow$(6a): As above, $A_{\text{ech}} = PA$ for some invertible matrix P. Multiplying by an invertible matrix does not change the row rank, so $\text{rowrk}(A_{\text{ech}}) = \text{rowrk}(A)$. If A is invertible, then $A_{\text{ech}} = I_n$ (since (1) implies (5)), so $\text{rowrk}(A_{\text{ech}}) = n$, so (6a) is true. Conversely, if (6a) holds, then $\text{rowrk}(A_{\text{ech}}) = n$, so A_{ech} cannot have any zero rows. As shown in the last paragraph, this forces $A_{\text{ech}} = I_n$. Since (5) implies (1), A is invertible.

(6a)$\Leftrightarrow$(6b): By definition, the row space of A is spanned by the n rows of A. If (6b) holds, then the list of n rows of A is a basis for the row space of A, so that $\text{rowrk}(A) = n$. Conversely, if (6b) fails, then we can delete one or more vectors from the list of rows to get a basis for the row space, forcing $\text{rowrk}(A) < n$.

(6a)$\Leftrightarrow$(7a): This follows from the identity $\mathrm{rowrk}(A) = \mathrm{colrk}(A)$, valid for all matrices A. We proved this identity in §4.12.

(7a)$\Leftrightarrow$(7b)$\Leftrightarrow$(7c): We prove the equivalence of (7a) and (7b) in the same way we proved the equivalence of (6a) and (6b). The range of A is the same as the column space of A (§4.7), so (7a) is equivalent to (7c) by definition.

(7c)$\Leftrightarrow$(8a): The range of the matrix A is the subspace $\{Ax : x \in F^n\} = \{L_A(x) : x \in F^n\}$, which is precisely the image of the map L_A. Note $L_A : F^n \to F^n$ is surjective iff the image of L_A is all of F^n iff the dimension of the image is n (since $\dim(F^n) = n$, but all proper subspaces of F^n have dimension smaller than n) iff the range of A has dimension n.

(8a)$\Leftrightarrow$(8b): Once we recall that $Ax = L_A(x)$, we see that (8b) is the very definition of L_A being surjective.

(8a)$\Leftrightarrow$(9c): Let the linear map $L_A : F^n \to F^n$ have image R and kernel N. We know R is the range of A. Moreover, N is the null space of A, since $x \in N$ iff $L_A(x) = 0$ iff $Ax = 0$. By the Rank–Nullity Theorem (see §1.8), $\dim(R) + \dim(N) = \dim(F^n) = n$. Now (8a) is equivalent to (7c), which says $\dim(R) = n$. In turn, this is equivalent to $\dim(N) = 0$, which is equivalent to $N = \{0\}$.

(9a)$\Leftrightarrow$(9c): The linear map L_A is injective iff its kernel is $\{0\}$. As observed above, the kernel of L_A is the same as the null space of A.

(9b)$\Leftrightarrow$(9c): The null space of A equals $\{0\}$ iff the null space of A is a subset of $\{0\}$. Expanding the definition of the latter condition, we obtain (9b).

(9a)$\Leftrightarrow$(10a): Assume (9a) holds. Then (8a) holds also, so L_A is injective and surjective, so (10a) holds. The converse is immediate, since all bijections are one-to-one.

(10a)$\Leftrightarrow$(10b): Recalling that $Ax = L_A(x)$, we see that (10b) is none other than the definition of L_A being a bijection.

(1)$\Leftrightarrow$(11),(1)$\Leftrightarrow$(12): We proved these equivalences in §4.6.

4.14 Block Matrix Multiplication

Block matrix multiplication is a useful method of computing the product of matrices that are built from smaller submatrices. As an example of this method, suppose X and Y are matrices built from smaller matrices (with subscripts indicating their sizes) as follows:

$$X = \left[\begin{array}{c|c} P_{3\times 2} & Q_{3\times 3} \\ \hline R_{5\times 2} & S_{5\times 3} \end{array}\right], \qquad Y = \left[\begin{array}{c|c|c} A_{2\times 3} & B_{2\times 4} & C_{2\times 1} \\ \hline D_{3\times 3} & E_{3\times 4} & F_{3\times 1} \end{array}\right]. \tag{4.2}$$

Block matrix multiplication lets us compute XY (in block form) by treating each block as if it were an individual matrix entry. In this example, the answer is

$$XY = \left[\begin{array}{c|c|c} PA + QD & PB + QE & PC + QF \\ \hline RA + SD & RB + SE & RC + SF \end{array}\right],$$

where the upper-left block is 3×3, the block below it is 5×3, and so on.

To state and prove the general rule for block matrix multplication, we need to use general index sets for the rows and columns of matrices. Suppose X is an $I \times J$ matrix and Y is a $J \times K$ matrix. Recall that a *set partition* of a set S is a collection of nonempty, pairwise disjoint sets S_i whose union is S. Suppose we are given a set partition $\{I_1, \ldots, I_m\}$ of I, a set partition $\{J_1, \ldots, J_n\}$ of J, and a set partition $\{K_1, \ldots, K_p\}$ of K. You can check that $\{I_i \times J_j : i \in [m], j \in [n]\}$ is a set partition of $I \times J$, and $\{J_j \times K_k : j \in [n], k \in [p]\}$ is a set partition of $J \times K$. Recall that the matrix X is a function from $I \times J$ to the field F. The restriction of the function X to each subset $I_i \times J_j$ gives an $I_i \times J_j$ *submatrix* of X, written $X_{i,j} : I_i \times J_j \to F$. Similarly, Y decomposes into submatrices $Y_{j,k} : J_j \times K_k \to F$.

This formal description captures the visual decomposition of matrices into smaller pieces as shown in (4.2). In that example, we have $I = [8]$, $J = [5]$, $K = [8]$, $m = n = 2$, $p = 3$, $I_1 = \{1,2,3\}$, $I_2 = \{4,5,6,7,8\}$, $J_1 = \{1,2\}$, $J_2 = \{3,4,5\}$, $K_1 = \{1,2,3\}$, $K_2 = \{4,5,6,7\}$, $K_3 = \{8\}$, $X_{1,1} = P$, $X_{1,2} = Q$, $X_{2,1} = R$, $X_{2,2} = S$, $Y_{1,1} = A$, and so on. In these set partitions, each subset in the partition consists of a block of *consecutive* integers, but this is not required in general.

Block Matrix Multiplication Rule. With the above notation, the matrix product $Z = XY : I \times K \to F$ consists of submatrices $Z_{i,k} : I_i \times K_k \to F$ given by

$$Z_{i,k} = \sum_{j \in [n]} X_{i,j} Y_{j,k} \qquad \text{for all } i \in [m],\, k \in [p]. \tag{4.3}$$

Proof. Given $r \in I$ and $c \in K$, the entry $Z(r,c)$ in row r and column c of $Z = XY$ is part of exactly one submatrix $Z_{i,k}$, namely the one where $r \in I_i$ and $c \in K_k$. We need only check that the two sides of (4.3) have the same r, c-entry. The left side $Z_{i,k}$ is a submatrix of $Z = XY$, so the r, c-entry is

$$Z_{i,k}(r,c) = Z(r,c) = (XY)(r,c) = \sum_{s \in J} X(r,s) Y(s,c).$$

For the right side, we first compute the r, c-entry of $X_{i,j} Y_{j,k}$ for a fixed $j \in [n]$. Since $X_{i,j}$ is an $I_i \times J_j$ submatrix of X and $Y_{j,k}$ is a $J_j \times K_k$ submatrix of Y, this entry is

$$(X_{i,j} Y_{j,k})(r,c) = \sum_{s \in J_j} X_{i,j}(r,s) Y_{j,k}(s,c) = \sum_{s \in J_j} X(r,s) Y(s,c).$$

Summing over $j \in [n]$, the r, c-entry of the right side of (4.3) is

$$\sum_{j \in [n]} \sum_{s \in J_j} X(r,s) Y(s,c) = \sum_{s \in J} X(r,s) Y(s,c),$$

since J is the disjoint union of its subsets $J_1, \ldots, J_n$. Thus, the left and right sides of (4.3) have the same r, c-entry. $\square$

4.15   Tensor Product of Matrices

This section introduces the tensor product operation on matrices, which has applications in quantum computing, multilinear algebra, and other settings. Given an $m \times n$ matrix A and an $s \times t$ matrix B, the tensor product $A \otimes B$ is a certain $ms \times nt$ matrix. To define this matrix, it is helpful to use general indexing sets for the rows and columns of the matrices.

Let I, J, K, L be finite, nonempty index sets. Let $A : I \times J \to F$ be an $I \times J$ matrix and $B : K \times L \to F$ be a $K \times L$ matrix, where A and B have entries in a field [or ring] F. The *tensor product matrix* $A \otimes B$ is the $(I \times K) \times (J \times L)$ matrix defined by setting

$$(A \otimes B)((i,k),(j,\ell)) = A(i,j)B(k,\ell) \qquad \text{for all } i \in I, k \in K, j \in J, \ell \in L. \tag{4.4}$$

Note that the rows of $A \otimes B$ are indexed by ordered pairs (i, k) with $i \in I$ and $k \in K$, while the columns of $A \otimes B$ are indexed by ordered pairs (j, ℓ) with $j \in J$ and $\ell \in L$. The defining equation (4.4) says that the entry of $A \otimes B$ in row (i, k), column (j, ℓ) is the product of the i, j-entry of A and the k, ℓ-entry of B.

We can visualize the tensor product $A \otimes B$ as a block matrix, where each block is a copy of B scaled by some entry of A. For example, let $I = J = K = \{0, 1\}$, $L = \{1, 2, 3\}$,

$$A = \begin{matrix} & 0 & 1 \\ \begin{matrix} 0 \\ 1 \end{matrix} & \left[\begin{matrix} 2 & 5 \\ -1 & 3 \end{matrix}\right] \end{matrix}, \quad \text{and} \quad B = \begin{matrix} & 1 & 2 & 3 \\ \begin{matrix} 0 \\ 1 \end{matrix} & \left[\begin{matrix} 4 & 1 & -3 \\ 2 & 0 & 7 \end{matrix}\right] \end{matrix}.$$

Writing bc as an abbreviation for the ordered pair (b, c), the rows of $A \otimes B$ are labeled by the 4 bit strings $00, 01, 10, 11$, while the columns of $A \otimes B$ are labeled by $01, 02, 03, 11, 12, 13$. Using this ordering for the row and column labels, we draw $A \otimes B$ as follows:

$$A \otimes B = \begin{matrix} & \begin{matrix} 01 & 02 & 03 & 11 & 12 & 13 \end{matrix} \\ \begin{matrix} 00 \\ 01 \\ 10 \\ 11 \end{matrix} & \left[\begin{matrix} 8 & 2 & -6 & 20 & 5 & -15 \\ 4 & 0 & 14 & 10 & 0 & 35 \\ -4 & -1 & 3 & 12 & 3 & -9 \\ -2 & 0 & -7 & 6 & 0 & 21 \end{matrix}\right] \end{matrix} = \left[\begin{array}{c|c} 2B & 5B \\ \hline -1B & 3B \end{array}\right].$$

We see the same pattern in the general case. Suppose each index set I, J, K, L is totally ordered, say $I = \{i_1 < i_2 < \cdots < i_m\}$, $J = \{j_1 < j_2 < \cdots < j_n\}$, etc. We introduce *lexicographic order* on the set $I \times K$ by letting $(i, k) < (i', k')$ mean that $i < i'$, or $i = i'$ and $k < k'$, for all $i, i' \in I$ and $k, k' \in K$. We use the analogous lexicographic order on $J \times L$. Displaying $A \otimes B$ by listing rows and columns in lexicographic order, we find that $A \otimes B$ has the block form

$$A \otimes B = \left[\begin{array}{c|c|c|c} A(i_1, j_1)B & A(i_1, j_2)B & \cdots & A(i_1, j_n)B \\ \hline A(i_2, j_1)B & A(i_2, j_2)B & \cdots & A(i_2, j_n)B \\ \hline \cdots & \cdots & \cdots & \cdots \\ \hline A(i_m, j_1)B & A(i_m, j_2)B & \cdots & A(i_m, j_n)B \end{array}\right].$$

We could have instead defined $(i, k) < (i', k')$ to mean $k < k'$, or $k = k'$ and $i < i'$, with a similar ordering on $J \times L$. Using these orderings, $A \otimes B$ has a block form where a typical block is $B(k, \ell)A$.

Theorem on Bilinearity of Matrix Tensor Products. Suppose A and A' are $I \times J$ matrices, B and B' are $K \times L$ matrices, and $c \in F$. Then:
(a) $(A + A') \otimes B = A \otimes B + A' \otimes B$ (left distributive law).
(b) $A \otimes (B + B') = A \otimes B + A \otimes B'$ (right distributive law).
(c) $c(A \otimes B) = (cA) \otimes B = A \otimes (cB)$ (scalar property).

Proof. Note that both sides of (a) are $(I \times K) \times (J \times L)$ matrices. Fix $i \in I$, $j \in J$, $k \in K$, and $\ell \in L$. The entry in row (i, k) and column (j, ℓ) of $(A + A') \otimes B$ is

$$(A + A')(i, j)B(k, \ell) = (A(i, j) + A'(i, j))B(k, \ell) = A(i, j)B(k, \ell) + A'(i, j)B(k, \ell).$$

The entry in row (i, k) and column (j, ℓ) of $A \otimes B + A' \otimes B$ is

$$(A \otimes B)((i, k), (j, \ell)) + (A' \otimes B)((i, k), (j, \ell)) = A(i, j)B(k, \ell) + A'(i, j)B(k, \ell).$$

The entries are equal, so (a) is true. Parts (b) and (c) are proved similarly (Exercise 96). $\square$

Theorem on Matrix Multiplication and Tensor Products. Suppose A, B, C, D are matrices where A is $I \times J$, B is $K \times L$, C is $J \times M$, and D is $L \times P$ for index sets I, J, K, L, M, P. If multiplication in F is commutative, then

$$(A \otimes B)(C \otimes D) = (AC) \otimes (BD). \tag{4.5}$$

Proof. Note that $A \otimes B$ is a $(I \times K) \times (J \times L)$ matrix, $C \otimes D$ is a $(J \times L) \times (M \times P)$ matrix, AC is $I \times M$, and BD is $K \times P$. So both sides of (4.5) are $(I \times K) \times (M \times P)$ matrices. Fix $i \in I$, $k \in K$, $m \in M$, and $p \in P$. By definition of matrix multiplication, the $((i,k),(m,p))$-entry of $(A \otimes B)(C \otimes D)$ is

$$\sum_{(j,\ell) \in J \times L} (A \otimes B)((i,k),(j,\ell))(C \otimes D)((j,\ell),(m,p)) = \sum_{j \in J} \sum_{\ell \in L} A(i,j)B(k,\ell)C(j,m)D(\ell,p).$$

The $((i,k),(m,p))$-entry of $(AC) \otimes (BD)$, namely $(AC)(i,m)(BD)(k,p)$, expands to

$$\left[\sum_{j \in J} A(i,j)C(j,m) \right] \cdot \left[\sum_{\ell \in L} B(k,\ell)D(\ell,p) \right] = \sum_{j \in J} \sum_{\ell \in L} A(i,j)C(j,m)B(k,\ell)D(\ell,p).$$

Because multiplication in F is commutative, this entry agrees with the previously computed entry. So $(A \otimes B)(C \otimes D) = (AC) \otimes (BD)$. $\square$

4.16 Summary

1. *Formal Definitions for Matrices.* An element of F^n is a function $x : [n] \to F$, which is often presented as a list $(x(1), \ldots, x(n))$. An $m \times n$ matrix is a function $A : [m] \times [n] \to F$. Special cases are column vectors ($n = 1$) and row vectors ($m = 1$), which can be identified with elements in F^m and F^n, respectively. A matrix can be identified with the list of its columns $(A^{[1]}, \ldots, A^{[n]})$ or with the list of its rows $(A_{[1]}, \ldots, A_{[m]})$; by definition, $A^{[j]}(i) = A(i,j) = A_{[i]}(j)$ for $i \in [m]$ and $j \in [n]$. An $I \times J$ matrix is a function $A : I \times J \to F$, where I and J are finite index sets for the rows and columns.

2. *Vector Spaces of Functions.* If S is any set and F is a field, the set V of all functions from S to F becomes an F-vector space under pointwise addition and scalar multiplication of functions. For $x \in S$, let $e_x \in V$ be the function that sends x to 1_F and everything else in S to 0_F. When S is finite, the set $X = \{e_x : x \in S\}$ is a basis for V such that $|X| = |S|$. In particular, $\dim(V) = |S|$ in this case.

3. *Spaces of Vectors and Matrices.* The sets F^n and $M_{m,n}(F)$ are F-vector spaces under componentwise operations. Viewing vectors and matrices as functions, these operations are the same as pointwise addition and scalar multiplication of functions from $[n]$ or $[m] \times [n]$ into F. A basis for F^n consists of the vectors e_i defined by $e_i(i) = 1$ and $e_i(j) = 0$ for $j \neq i$. A basis for $M_{m,n}(F)$ consists of the matrices e_{ij} defined by $e_{ij}(i,j) = 1$ and $e_{ij}(i',j') = 0$ whenever $i \neq i'$ or $j \neq j'$. We have $\dim(F^n) = n$ and $\dim(M_{m,n}(F)) = mn$.

4. *Matrix Multiplication.* For $A \in M_{m,n}(F)$ and $B \in M_{n,p}(F)$, $AB \in M_{m,p}(F)$ is the matrix whose i,j-entry is $\sum_{k=1}^n A(i,k)B(k,j)$. We have $AI_n = A = I_m A$ and $c(AB) = (cA)B = A(cB)$ for $c \in F$ [and F commutative]. Matrix multiplication

is distributive and associative. When $m = n$ [and F is commutative], the product operation turns the vector space $M_n(F)$ into an F-algebra. By generalized associativity, we can omit parentheses when writing a product of s matrices. If A_u is an $n_{u-1} \times n_u$ matrix for $1 \leq u \leq s$, then the i, j-entry of $\prod_{u=1}^s A_u$ is

$$\sum_{k_1=1}^{n_1} \sum_{k_2=1}^{n_2} \cdots \sum_{k_{s-1}=1}^{n_{s-1}} A_1(i, k_1) A_2(k_1, k_2) \cdots A_{s-1}(k_{s-2}, k_{s-1}) A_s(k_{s-1}, j).$$

Block matrix multiplication lets us compute AB by a formula that treats blocks (submatrices of A and B) as if they were individual matrix entries.

5. *Transpose and Conjugate-Transpose.* For $A \in M_{m,n}(F)$, we define $A^T \in M_{n,m}(F)$ and [when $F = \mathbb{C}$] $A^* \in M_{n,m}(\mathbb{C})$ by $A^T(i, j) = A(j, i)$ and $A^*(i, j) = \overline{A(j, i)}$ for $i \in [n]$ and $j \in [m]$. These operations satisfy the following identities: $(A + B)^T = A^T + B^T$, $(A^T)^T = A$, and $(cA)^T = c(A^T)$ for $c \in F$; $(A_1 \cdots A_s)^T = A_s^T \cdots A_1^T$ [for F commutative]; $(A + B)^* = A^* + B^*$, $(A^*)^* = A$, and $(cA)^* = \overline{c}(A^*)$ for $c \in \mathbb{C}$; $(A_1 \cdots A_s)^* = A_s^* \cdots A_1^*$.

6. *Computing with Rows and Columns.* On one hand, $(AB)^{[j]} = A(B^{[j]})$, which says that column j of AB is A times column j of B. On the other hand, $(AB)_{[i]} = (A_{[i]})B$, which says that row i of AB is row i of A times B. For a matrix-vector product Av, $Av = \sum_j A^{[j]} v(j)$, so that Av is a linear combination of the columns of A. For a vector-matrix product wA, $wA = \sum_i w(i) A_{[i]}$, so that wA is a linear combination of the rows of A. Some consequences [for fields F] are:

$$\mathrm{colrk}(AB) \leq \min(\mathrm{colrk}(A), \mathrm{colrk}(B)); \quad \mathrm{rowrk}(AB) \leq \min(\mathrm{rowrk}(A), \mathrm{rowrk}(B));$$

left or right multiplication by an invertible matrix does not change the row rank or the column rank; and the rank of an invertible $n \times n$ matrix is equal to n. Matrix transposition interchanges the rows and columns of A. The conjugate-transpose operation interchanges rows and columns and conjugates every entry.

7. *Elementary Operations and Elementary Matrices.* The three elementary row operations are interchange two rows; multiply a row by an invertible scalar; add any scalar multiple of one row to a different row. To apply an elementary row operation to A, we multiply A on the left by the corresponding elementary matrix (formed by applying the same row operation to the identity matrix). There are analogous elementary column operations, which can be achieved by multiplying A on the right by the corresponding elementary matrix. Elementary matrices are invertible, and their inverses are also elementary.

8. *Applications of Elementary Matrices.* [This item assumes F is a field.] A square matrix is invertible iff it is a product of elementary matrices. For any A (possibly rectangular), there exists an invertible matrix P such that PA is in reduced row-echelon form. Furthermore, there exists another invertible matrix Q such that PAQ is a matrix in Smith canonical form, with $\mathrm{rank}(A)$ copies of 1 on the main diagonal and all other entries 0. In particular, the column rank of A is always equal to the row rank of A. Elementary matrices can be used to prove the product formula $\det(AB) = \det(A)\det(B)$ for determinants.

9. *Invertible Matrices.* [This item assumes F is a field.] A matrix $A \in M_n(F)$ is invertible iff there exists a (necessarily unique) matrix $A^{-1} \in M_n(F)$ with $AA^{-1} = I_n = A^{-1}A$. The following conditions are equivalent to the existence of

A^{-1}: A has a left inverse; A has a right inverse; A has nonzero determinant; A is a product of elementary matrices; the reduced row-echelon form of A is I_n; A has row rank n; A has column rank n; the n rows of A are linearly independent; the n columns of A are linearly independent; the range of A has dimension n; the null space of A is $\{0\}$; the map $L_A : F^n \to F^n$ given by $L_A(x) = Ax$ for $x \in F^n$ is injective; L_A is surjective; L_A is bijective; the system $Ax = b$ has a solution $x \in F^n$ for every $b \in F^n$; $Ax = b$ has a *unique* solution x for every b; $Ax = 0$ has only the zero solution $x = 0$; A^{T} is invertible; [when $F = \mathbb{C}$] A^* is invertible. If A^{-1} exists, then $(A^{-1})^{-1} = A$, $(A^{\mathrm{T}})^{-1} = (A^{-1})^{\mathrm{T}}$, and [when $F = \mathbb{C}$] $(A^*)^{-1} = (A^{-1})^*$. A product $A = A_1 A_2 \cdots A_s$ is invertible iff every A_i is invertible, in which case $A^{-1} = A_s^{-1} \cdots A_2^{-1} A_1^{-1}$. The set $\mathrm{GL}_n(F)$ of invertible matrices in $M_n(F)$ is a group under matrix multiplication.

10. *Tensor Products of Matrices.* Given an $I \times J$ matrix A and a $K \times L$ matrix B, $A \otimes B$ is the $(I \times K) \times (J \times L)$ matrix where the entry in row (i, k), column (j, ℓ) is $A(i, j)B(k, \ell)$. Using lexicographic order for the index sets, $A \otimes B$ is a block matrix with blocks $A(i, j)B$. The matrix tensor product is F-bilinear. For commutative F, $(A \otimes B)(C \otimes D) = (AC) \otimes (BD)$ when these products are defined.

4.17 Exercises

Unless otherwise stated, assume F is a field in these exercises.

1. State a precise formula for a map $I : F^n \to M_{n,1}(F)$ that identifies n-tuples with column vectors, and check that I is a vector space isomorphism. (Note that the inputs and outputs for I are functions.)

2. Define a map $J : F^n \to M_{1,n}(F)$ that identifies n-tuples with row vectors, and prove J is a vector space isomorphism.

3. (a) Define a vector space isomorphism from F to F^1.
   (b) Define a vector space isomorphism from F to $M_{1,1}(F)$.
   (This problem lets us identify 1-tuples and 1×1 matrices with scalars in F.)

4. Consider the map $C : M_{m,n}(F) \to (F^m)^n$ sending $A \in M_{m,n}(F)$ to its list of columns $(A^{[1]}, \ldots, A^{[n]})$. Give a precise formula for C^{-1}, and check that C is F-linear. Do the same for the map $R : M_{m,n}(F) \to (F^n)^m$ sending $A \in M_{m,n}(F)$ to its list of rows $(A_{[1]}, \ldots, A_{[m]})$.

5. Given finite index sets I and J, choose bijections $f : I \to [m]$ and $g : J \to [n]$. Use f and g to define an isomorphism from the vector space of $I \times J$ matrices with entries in F to the vector space $M_{m,n}(F)$.

6. Let S be a set and V be the set of all functions $g : S \to F$ with pointwise addition and scalar multiplication.
   (a) Prove: addition on V is commutative.
   (b) Prove: the zero function from S to F is the additive identity element of V.
   (c) Prove: the additive inverse of $f \in V$ is the function sending $x \in S$ to $-f(x)$.

7. Let V be the set of functions $g : S \to F$. Prove that scalar multiplication on V satisfies the five axioms in the definition of a vector space.

8. Let S be any set and E be any F-vector space. Let V be the set of all functions $g : S \to E$. Define pointwise addition and scalar multiplication operations on V, and prove that V with these operations is an F-vector space.

9. Define S, E, and V as in the previous exercise. Suppose $S = \{x_1, \ldots, x_k\}$ is finite, and E is finite-dimensional with ordered basis $(v_1, \ldots, v_m)$. Describe an explicit basis for V, and compute $\dim_F(V)$.

10. (a) Cite a theorem from the text to explain why the set V of all functions $f : \mathbb{R} \to \mathbb{R}$ (under pointwise operations) is a real vector space.
    (b) For $x \in \mathbb{R}$, let $e_x : \mathbb{R} \to \mathbb{R}$ send x to 1 and all other inputs to 0. Which functions in V are in the span of the set $\{e_x : x \in \mathbb{R}\}$? Is this set a basis for V?
    (c) Explain why the set of continuous $f : \mathbb{R} \to \mathbb{R}$ is a subspace of V. Find the intersection of this subspace and the span of $\{e_x : x \in \mathbb{R}\}$.

11. Let $A = \begin{bmatrix} 5 & 2 & 1 \\ 1/2 & -3 & 0 \\ -1 & -4 & -3/2 \end{bmatrix}$, $B = \begin{bmatrix} 1 & -1 & 0 \\ 0 & 2 & -2 \\ 3 & -3 & 0 \end{bmatrix}$, $C = \begin{bmatrix} 4 & 2 & -7 \\ 1 & 1/3 & 3 \end{bmatrix}$,
    $w = [2\ 4\ 1]$, and $v = [1\ -3\ -1]^{\mathrm{T}}$. Compute $A + B$, AB, BA, CA, CB, Av, wB, wv, and vw.

12. For the matrices A, B, C in the previous exercise, compute $B \otimes C$, $C \otimes B$, and $A \otimes A$.

13. Let $A = \begin{bmatrix} 5 - i & 2i + 3 \\ 0 & -1 - 3i \end{bmatrix}$, $B = \begin{bmatrix} 3 + 4i & -2i \\ e^{5\pi i/3} & \sqrt{2} + 2i \end{bmatrix}$, and $v = \begin{bmatrix} i \\ 1 \end{bmatrix}$.
    Compute $A + B$, AB, A^{T}, B^{T}, A^*, B^*, Bv, $v^{\mathrm{T}}v$, v^*v, v^*Av, and v^*A^*Av.

14. Prove: for all $A, B \in M_n(F)$, A and B commute iff A^{T} and B^{T} commute.

15. Prove: for all invertible $A, B \in M_n(F)$, A and B commute iff A^{-1} and B^{-1} commute.

16. Let $A = \begin{bmatrix} 1 & 4 \\ 2 & 3 \end{bmatrix} \in M_2(F)$. For each choice of F, find A^2, A^3, AA^{T}, and (if possible) A^{-1}. (a) $F = \mathbb{R}$ (b) $F = \mathbb{Z}_5$ (c) $F = \mathbb{Z}_7$

17. Let $A = \begin{bmatrix} 1 & 0 & 0 \\ 0 & 0 & 1 \\ 0 & 1 & 0 \end{bmatrix}$, $B = \begin{bmatrix} 0 & 0 & 1 \\ 1 & 0 & 1 \\ 0 & 1 & 0 \end{bmatrix}$, and $C = \begin{bmatrix} 1 & 0 & 0 \\ 0 & 0 & 1 \\ 0 & 1 & 1 \end{bmatrix}$ in $M_2(\mathbb{Z}_2)$.
    Compute A^k, B^k, and C^k for all $k \in \mathbb{Z}$.

18. Let $A, A' \in M_{m,n}(F)$, $B \in M_{n,p}(F)$, and $c \in F$. Prove in detail:
    $(A + A')B = AB + A'B$, $c(AB) = (cA)B$, $c(AB) = A(cB)$, and $AI_n = A$.
    Which properties must be true if F is a non-commutative ring?

19. Let $A, B \in M_{m,n}(\mathbb{C})$ and $C \in M_{n,p}(\mathbb{C})$. Prove in detail:
    $(A + B)^* = A^* + B^*$, $(AC)^* = C^*A^*$, $(A^*)^* = A$,
    A is invertible iff A^* is invertible, in which case $(A^*)^{-1} = (A^{-1})^*$.

20. **Hadamard Product of Matrices.** For matrices $A, B \in M_{m,n}(F)$, define the *Hadamard product* of A and B, written $A \odot B$, to be the matrix in $M_{m,n}(F)$ with entries $(A \odot B)(i, j) = A(i, j)B(i, j)$ for $i \in [m]$ and $j \in [n]$. We obtain $A \odot B$ by multiplying corresponding entries of A and B. Prove that $\odot$ is commutative, associative, satisfies the left and right distributive laws, and has an identity element. Which matrices in $M_{m,n}(F)$ have inverses with respect to $\odot$?

21. Give a specific example of matrices $A, B \in M_n(F)$ with $AB = 0$ and $BA \neq 0$.

22. Give an example of matrices $A, B \in M_{m,n}(F)$ with $m \neq n$, $BA = 0$, and $AB \neq 0$.

23. Give an example of $A, B \in M_4(\mathbb{R})$ such that every entry of A and B is nonzero, yet $AB = BA = 0$.

24. A matrix $A \in M_n(F)$ is called *diagonal* iff $A(i, j) = 0_F$ for all $i \neq j$ in $[n]$. This means that the only nonzero entries of A occur on the main diagonal. Prove that the set of diagonal matrices is a commutative subalgebra of the F-algebra $M_n(F)$. Find a basis for this subalgebra and compute its dimension.

25. A matrix $A \in M_n(F)$ is called *upper-triangular* iff $A(i, j) = 0_F$ for all $i, j \in [n]$ with $i > j$. This means that all entries of A below the main diagonal are zero. Prove the set U of upper-triangular matrices is a subalgebra of $M_n(F)$. Find a basis for U and compute $\dim(U)$.

26. A matrix $A \in M_n(F)$ is called *lower-triangular* iff $A(i, j) = 0_F$ for all $i, j \in [n]$ with $i < j$. This means that all entries of A above the main diagonal are zero. Prove A is lower-triangular iff A^{T} is upper-triangular. Use this and the previous exercise to deduce that the set L of lower-triangular matrices is a subalgebra of $M_n(F)$. Find a basis for L and compute $\dim(L)$.

27. A matrix $A \in M_n(F)$ is called *strictly upper-triangular* iff $A(i, j) = 0_F$ for all $i, j \in [n]$ with $i \geq j$. This means that the only nonzero entries of A occur strictly above the main diagonal. Prove: if A is strictly upper-triangular, then $A^n = 0_n$. (Use the formula for $[\prod_{u=1}^{s} A_u](i, j)$ in §4.5.) Deduce that A is not invertible.

28. Prove: For all $A, B \in M_n(F)$, if A is strictly upper-triangular and B is upper-triangular, then AB and BA are strictly upper-triangular.

29. Show that the set S of strictly upper-triangular matrices is a subalgebra of $M_n(F)$. Exhibit a basis for S and compute $\dim(S)$.

30. A matrix A in $M_n(F)$ is called *unitriangular* iff A is upper-triangular and $A(i, i) = 1_F$ for all $i \in [n]$. Let T be the set of unitriangular matrices in $M_n(F)$. Is T closed under addition? additive inverses? additive identity? multiplication? multiplicative identity? scalar multiplication? multiplicative inverses? (For the last question, see Exercise 65.)

31. A matrix $A \in M_n(F)$ is called *nilpotent* iff $A^k = 0$ for some integer $k > 0$. Exercise 27 shows that any strictly upper-triangular matrix is nilpotent. Give examples of the following:
    (a) a nilpotent matrix that is neither upper-triangular nor lower-triangular
    (b) two nilpotent matrices whose sum is not nilpotent
    (c) two nilpotent matrices whose product is not nilpotent

32. Prove: for all $A, B \in M_n(F)$, if $A, B \in M_n(F)$ are nilpotent and $AB = BA$, then $A + B$ and AB are nilpotent.

33. A matrix $A \in M_n(F)$ is called *skew-symmetric* iff $A^{\mathrm{T}} = -A$. Prove that the set of skew-symmetric matrices is a subspace of $M_n(F)$. Find a basis for this subspace and compute its dimension.

34. Give an example of a nonzero matrix that is both symmetric and skew-symmetric, or explain why no such matrix exists.

35. Prove: for all $A, B \in M_n(F)$, if A and B are skew-symmetric and commute, then AB is symmetric. Does the result hold without assuming $AB = BA$?

36. **Trace of a Matrix.** The *trace* of a square matrix $A \in M_n(F)$ is defined by $\mathrm{tr}(A) = \sum_{i=1}^{n} A(i, i)$, which is the sum of the main diagonal entries of A.
    (a) Compute $\mathrm{tr}(0_n)$, $\mathrm{tr}(I_n)$, and $\mathrm{tr}(J_n)$.
    (b) Prove that $\mathrm{tr} : M_n(F) \to F$ is an F-linear map.

(c) Prove: for all $A, B \in M_n(F)$, $\mathrm{tr}(AB) = \mathrm{tr}(BA)$.

(d) Does part (c) hold for all $A \in M_{m,n}(F)$ and $B \in M_{n,m}(F)$ with $n \neq m$?

37. (a) Prove: for all $A \in M_n(F)$, $\mathrm{tr}(A^{\mathrm{T}}) = \mathrm{tr}(A)$.

(b) Find and prove an identity relating $\mathrm{tr}(A^*)$ and $\mathrm{tr}(A)$ for $A \in M_n(\mathbb{C})$.

38. Prove or disprove: for all invertible $A \in M_n(F)$, $\mathrm{tr}(A^{-1}) = \mathrm{tr}(A)^{-1}$.

39. Give an example of a field F and a matrix $A \in M_n(F)$ with all entries of A nonzero and $\mathrm{tr}(A) = 0_F$.

40. Given $A \in M_m(F)$ and $B \in M_n(F)$, prove that $\mathrm{tr}(A \otimes B) = \mathrm{tr}(A)\,\mathrm{tr}(B)$.

41. **Commutator of Matrices.** Given $A, B \in M_n(F)$, define the *commutator* of A and B to be $[A, B] = AB - BA$. Prove the following hold for all $A, B, C \in M_n(F)$ and all $d \in F$:

(a) $[A, B] = 0$ iff A and B commute.

(b) $[A, B] = -[B, A]$ (anticommutativity).

(c) $[A + B, C] = [A, C] + [B, C]$, $[A, B + C] = [A, B] + [A, C]$, and $[dA, B] = d[A, B] = [A, dB]$ (bilinearity).

(d) $[A, [B, C]] + [B, [C, A]] + [C, [A, B]] = 0$ (Jacobi Identity).

(e) The trace of $[A, B]$ is zero.

42. Is $[A, [B, C]] = [[A, B], C]$ true for all $A, B, C \in M_n(F)$? Either prove it, or find sufficient conditions on A, B, C for this identity to hold.

43. Write all 14 complete parenthesizations of a product $ABCDE$ of five matrices.

44. For $n \geq 1$, let p_n be the number of complete parenthesizations of a product of n matrices. Show $p_1 = p_2 = 1$, $p_3 = 2$, $p_4 = 5$, and $p_n = \sum_{k=1}^{n-1} p_k p_{n-k}$ for $n > 1$. (It follows from this recursion that p_n is the *Catalan number* $\frac{1}{n}\binom{2n-2}{n-1}$.)

45. **Laws of Exponents for Matrix Powers.** Let $A \in M_n(F)$. Prove:

(a) For all integers $r, s \geq 0$, $A^{r+s} = A^r A^s$.

(b) For all integers $r, s \geq 0$, $(A^r)^s = A^{rs}$.

(c) If A^{-1} exists, then (a) and (b) hold for all $r, s \in \mathbb{Z}$.

46. Let $(G, \star)$ be any group, $s \geq 1$, and $a_1, \ldots, a_s \in G$. State and prove a Generalized Associativity Law that justifies writing the product $a_1 \star a_2 \star \cdots \star a_s$ with no parentheses. (Imitate the proof given in the text for matrix products.)

47. (a) Given $U \in M_{m,n}(F)$ and $V \in M_{n,p}(F)$, how many additions and multiplications in F are needed to compute UV using the definition of matrix product?

(b) Suppose A is 2×10, B is 10×15, C is 15×4, and D is 4×8. For each complete parenthesization of $ABCD$, count the number of multiplications needed to evaluate this matrix product by multiplying matrices in the order determined by the parenthesization.

48. Prove: for all $a, b, c, d \in F$, if $ad - bc \neq 0_F$, then $A = \begin{bmatrix} a & b \\ c & d \end{bmatrix}$ is invertible, and $A^{-1} = (ad - bc)^{-1} \begin{bmatrix} d & -b \\ -c & a \end{bmatrix}$. Assuming $ad - bc \neq 0_F$, find an explicit factorization of A into a product of elementary matrices.

49. Find the inverse of $A = \begin{bmatrix} 0 & 1 & 2 & 1 \\ 1 & 0 & 1 & 1 \\ 0 & 2 & 2 & 1 \\ 1 & 2 & 2 & 1 \end{bmatrix}$ in $M_4(\mathbb{Z}_3)$.

50. Find necessary and sufficient conditions on the integer n and the field F for the group $\mathrm{GL}_n(F)$ to be commutative.

51. (a) Find the size of the group $GL_3(\mathbb{Z}_2)$. (Build $A \in GL_3(\mathbb{Z}_2)$ by choosing nonzero rows, one at a time, that are not in the span of the previous rows.)
    (b) Find the size of the group $GL_2(\mathbb{Z}_7)$.
    (c) For any $n \geq 1$ and any prime p, find the size of $GL_n(\mathbb{Z}_p)$.

52. Give an example of functions $f, g : \mathbb{R} \to \mathbb{R}$ with $f \circ g = \mathrm{id}_\mathbb{R}$ but $g \circ f \neq \mathrm{id}_\mathbb{R}$.

53. Prove: for all functions $f, g : \mathbb{R} \to \mathbb{R}$, if $f \circ g = \mathrm{id}_\mathbb{R}$ and $g \circ f \neq \mathrm{id}_\mathbb{R}$, then f is onto and not one-to-one, while g is one-to-one and not onto.

54. Show that $A = \begin{bmatrix} 1 & 0 & 0 \\ 1 & 0 & 1 \end{bmatrix} \in M_{2,3}(\mathbb{R})$ has a right inverse but no left inverse. Find all right inverses of A.

55. Give a specific example of a matrix $A \in M_{4,3}(\mathbb{R})$ that has a left inverse but no right inverse. Find all left inverses of A.

56. Prove: for all $A \in M_{m,n}(F)$ with $m > n$, if A has a left inverse, then the left inverse is not unique.

57. Prove: for all $A \in M_{m,n}(F)$, $\mathrm{rank}(A) \leq \min(m, n)$.

58. (a) Prove: for all $A, B \in M_{m,n}(F)$, $\mathrm{rank}(A + B) \leq \mathrm{rank}(A) + \mathrm{rank}(B)$.
    (b) Give an example where strict inequality holds in (a).

59. Let m, n be positive integers with $m < n$. By studying rank, show that no A in $M_{m,n}(F)$ has a left inverse, and no B in $M_{n,m}(F)$ has a right inverse.

60. Suppose $A \in M_{m,n}(F)$ and $B \in M_{n,p}(F)$. Prove $[(AB)(i,j)] = A_{[i]}B^{[j]}$ for all $i \in [m]$ and $j \in [p]$. Give a verbal description of what this equation means.

61. Define $A \in M_{3,5}(\mathbb{R})$ and $B \in M_{5,3}(\mathbb{R})$ by setting $A(i,j) = i^2 j$ and $B(j,i) = j + 2i$ for $i \in [3]$ and $j \in [5]$. Compute row 3 of AB. Compute column 2 of AB. Express column 4 of BA as a linear combination of the columns of B. Express row 1 of BA as a linear combination of the rows of A.

62. Let $A, B \in M_{m,n}(F)$ and $c \in F$. Prove $(A+B)_{[i]} = A_{[i]} + B_{[i]}$ and $(cA)_{[i]} = c(A_{[i]})$ for all $i \in [m]$.

63. Let $A \in M_n(F)$ be upper-triangular with $A(i,i) \neq 0$ for all $i \in [n]$ and $B \in M_{m,n}(F)$ be arbitrary. Prove: for all $i \in [n]$, the subspace of F^n spanned by the first i columns of B equals the subspace of F^n spanned by the first i columns of BA. Formulate and prove a similar result involving left multiplication by A.

64. Formulate and prove results similar to the previous exercise for multiplication on the left or right by lower-triangular matrices with no zeroes on the diagonal.

65. Let $A \in M_n(F)$ be unitriangular (see Exercise 30). Prove A is invertible, and find a recursive formula for computing the entries of A^{-1}. (Solve $BA = I_n$ for the columns of B from left to right.)

66. Illustrate the formula in Exercise 65 by finding the inverses of

$$A = \begin{bmatrix} 1 & 1 & 1 & 1 & 1 \\ 0 & 1 & 1 & 2 & 3 \\ 0 & 0 & 1 & 1 & 2 \\ 0 & 0 & 0 & 1 & 3 \\ 0 & 0 & 0 & 0 & 1 \end{bmatrix} \quad \text{and} \quad C = \begin{bmatrix} 1 & 1 & 1 & 1 & 1 & 1 & 1 & 1 \\ 0 & 1 & 0 & 1 & 0 & 1 & 0 & 1 \\ 0 & 0 & 1 & 1 & 0 & 0 & 1 & 1 \\ 0 & 0 & 0 & 1 & 0 & 0 & 0 & 1 \\ 0 & 0 & 0 & 0 & 1 & 1 & 1 & 1 \\ 0 & 0 & 0 & 0 & 0 & 1 & 0 & 1 \\ 0 & 0 & 0 & 0 & 0 & 0 & 1 & 1 \\ 0 & 0 & 0 & 0 & 0 & 0 & 0 & 1 \end{bmatrix}.$$

67. Which lower-triangular $A \in M_n(F)$ are invertible? Prove that A^{-1}, when it exists, is also lower-triangular.

68. Given $A \in M_n(F)$ and positive integers $r < s$, prove with minimal calculation that $\mathrm{Col}(A^s) \subseteq \mathrm{Col}(A^r)$. Similarly, prove $\mathrm{Row}(A^s) \subseteq \mathrm{Row}(A^r)$.

69. For each matrix below, determine the rank of the matrix by finding a subset of the columns that forms a basis for the column space:

(a) $\begin{bmatrix} 1 & -1 & 0 \\ 0 & 2 & -2 \\ 3 & -3 & 0 \end{bmatrix}$
(b) $\begin{bmatrix} 4 & 2 & -7 \\ 1 & 1/3 & 3 \end{bmatrix}$
(c) $\begin{bmatrix} 0 & 1 & 1 & 1 \\ 1 & 1 & 1 & 0 \\ 0 & 0 & 0 & 0 \\ 1 & 0 & 0 & 1 \end{bmatrix}$ in $M_4(\mathbb{Z}_2)$

(d) $J_{m,n}$, which has every entry equal to 1
(e) the matrix $A \in M_{m,n}(\mathbb{R})$ with $A(i, j) = ij$ for $i \in [m]$ and $j \in [n]$
(f) the matrix $A \in M_n(\mathbb{R})$ with $A(i, j) = (i - 1)n + j$ for $i, j \in [n]$

70. Repeat Exercise 69, but compute the rank by finding a subset of the rows that forms a basis for the row space.

71. Repeat Exercise 69, but compute the rank by finding the reduced row-echelon form of the matrix and counting the nonzero rows.

72. Let $A \in M_{m,n}(F)$, $i, \ell \in [m]$, $j, k \in [n]$, and $c, d \in F$ with $c \neq 0$. Prove:
(a) Left-multiplication by $E_m^1[c; i]$ multiplies row i of A by c.
(b) Right-multiplication by $E_n^1[c; j]$ multiplies column j of A by c.
(c) $E_m^1[c; i]$ is invertible with inverse $E_m^1[c^{-1}; i]$.
(d) For $i \neq \ell$, left-multiplication by $E_m^2[i, \ell]$ interchanges row i and row ℓ of A.
(e) For $i \neq \ell$, $E_m^2[i, \ell]$ is its own inverse.
(f) For $j \neq k$, right-multiplication by $E_n^3[d; j, k]$ adds d times column j of A to column k of A.

73. Prove that every elementary matrix $E_p^2[i, j]$ can be expressed as a product of several elementary matrices of the form $E_p^3[\pm 1; r, s]$. Conclude that the elementary operation of interchanging two rows of a matrix can be simulated by other elementary row operations.

74. List all elementary matrices in $M_3(\mathbb{Z}_2)$. Express each matrix as a product $D_1 D_2 \cdots D_k$, where each D_i is one of the matrices A, B, C in Exercise 17.

75. Draw pictures of all possible matrices $A \in M_{3,5}(F)$ in reduced row-echelon form. Each matrix entry should be a 0, a 1, or a $*$ to indicate an arbitrary scalar.

76. For each matrix A, compute the reduced row-echelon form A_{ech} and an invertible matrix P (a product of elementary matrices) with $A_{\mathrm{ech}} = PA$.
(a) $A \in M_{3,4}(\mathbb{R})$ given by $A(i, j) = i$ for $i \in [3]$ and $j \in [4]$
(b) $A \in M_4(\mathbb{R})$ given by $A(i, j) = (-1)^{i+j}$ for $i, j \in [4]$
(c) the matrix A in Exercise 13

(d) $A = \begin{bmatrix} 2 & 4 & -3 & 0 & -14 \\ 1 & 2 & 4 & -2 & 13 \\ 0 & 0 & 1 & 0 & 4 \\ 2 & 4 & 2 & 0 & 6 \end{bmatrix}$
(e) $A = \begin{bmatrix} 0 & 1 & 0 & 1 & 0 & 0 & 1 \\ 1 & 1 & 1 & 1 & 0 & 0 & 0 \\ 1 & 0 & 0 & 0 & 1 & 0 & 1 \\ 0 & 0 & 0 & 0 & 0 & 0 & 0 \\ 0 & 0 & 1 & 0 & 1 & 0 & 0 \\ 0 & 0 & 0 & 0 & 0 & 1 & 1 \\ 1 & 0 & 1 & 0 & 0 & 0 & 1 \\ 1 & 1 & 1 & 1 & 0 & 0 & 0 \end{bmatrix} \in M_{8,7}(\mathbb{Z}_2)$

77. For each matrix A in Exercise 76, find an invertible matrix Q (a product of elementary matrices) such that PAQ is the Smith canonical form of A.

78. If possible, write each matrix and its inverse as a product of elementary matrices.
    (a) A in Exercise 11 (b) B in Exercise 11 (c) A in Exercise 49
    (d) A in Exercise 66   (e) C in Exercise 66

79. Prove: for all $A, B \in M_{m,n}(F)$, rank$(A) = $ rank(B) iff there exist invertible matrices $U \in M_m(F)$ and $V \in M_n(F)$ with $A = UBV$.

80. Use the definitions to give a direct proof of (7b)$\Leftrightarrow$(9b) in Table 4.1.

81. Assume conditions (1), (6b), and (11) in Table 4.1 are known to be equivalent. Using this, prove (6b)$\Leftrightarrow$(7b).

82. Prove (8b)$\Rightarrow$(2b) in Table 4.1 by solving for the columns of C one at a time.

83. Let $A \in M_n(F)$, and define $R_A : F^n \to F^n$ by $R_A(x) = xA$ for each row vector $x \in F^n$. Prove the following conditions are equivalent:
    A is invertible; R_A is injective; R_A is surjective; R_A is bijective.

84. Let $T : V \to W$ be an F-linear map between two n-dimensional F-vector spaces V and W, where n is finite. Prove the following conditions are equivalent:
    T is injective; T is surjective; T is bijective.

85. Give an example of an F-vector space V and a linear map $T : V \to V$ that is one-to-one but not onto.

86. Give an example of an F-vector space V and a linear map $S : V \to V$ that is onto but not one-to-one.

87. Give an example of an F-vector space V and linear maps $T, S : V \to V$ with $S \circ T = \mathrm{id}_V$ but $T \circ S \neq \mathrm{id}_V$.

88. **Theorem on Right Invertibility.** Fix integers m, n with $0 < m < n$ and $A \in M_{m,n}(F)$. Prove that the following conditions on A are equivalent:
    (a) A has a right inverse.
    (b) A has at least two right inverses.
    (c) The reduced row-echelon form of A has no zero rows.
    (d) rowrk$(A) = m$.
    (e) The list of m rows of A is linearly independent in F^n.
    (f) colrk$(A) = m$.
    (g) The range of A has dimension m.
    (h) The map $L_A : F^n \to F^m$ sending each $x \in F^n$ to Ax is surjective.
    (i) For all $b \in F^m$, there exists $x \in F^n$ with $Ax = b$.
    (j) A^{T} has a left inverse.

89. **Theorem on Left Invertibility.** Fix integers m, n with $0 < n < m$ and $A \in M_{m,n}(F)$. Formulate and prove a list of conditions on A (analogous to those in Exercise 88), each of which is equivalent to A having a left inverse.

90. Consider the theorem: *for all $A, B \in M_n(F)$, if $BA = I_n$ then $AB = I_n$.*
    (a) Show that the theorem follows from this statement:
       (∗) for all $A, B \in M_n(F)$, if $BA = I_n$ then A is invertible.
    (b) We proved (∗) in the text using determinants (see the proof of (2a)$\Rightarrow$(3)$\Rightarrow$(1) in §4.13). Give several new proofs of (∗) by using the equivalences proved in §4.13 (excluding conditions (2a) and (2b)) and showing that (2a)$\Rightarrow$(6a), (2a)$\Rightarrow$(7a), and (2a)$\Rightarrow$(9b).
    (c) Similarly, give proofs that $AC = I_n$ implies A is invertible by showing (2b)$\Rightarrow$(6a), (2b)$\Rightarrow$(7a), (2b)$\Rightarrow$(8b), and by using (11) and (∗).
    (d) Prove, without using determinants, that for all $U, V \in M_n(F)$, if UV is invertible, then U and V are invertible.

91. Show that the formulas $(AB)^{[j]} = A(B^{[j]})$ and $(AB)_{[i]} = (A_{[i]})B$ (see §4.7 and §4.8) are special cases of the Block Matrix Multiplication Rule in §4.14.

92. Show that the formulas $Av = \sum_{j=1}^{n} A^{[j]}v(j)$ and $wA = \sum_{i=1}^{m} w(i)A_{[i]}$ (see §4.7 and §4.8) are special cases of the Block Matrix Multiplication Rule in §4.14.

93. **Block Matrix Addition Rule.** Suppose A and B are $I \times J$ matrices, $\{I_1, \ldots, I_m\}$ is a set partition of I, and $\{J_1, \ldots, J_n\}$ is a set partition of J. For $i \in [m]$ and $j \in [n]$, let $A_{i,j}$ and $B_{i,j}$ be the submatrices of A and B with domain $I_i \times J_j$. Prove $C = A + B$ consists of $I_i \times J_j$ submatrices $C_{i,j}$ where $C_{i,j} = A_{i,j} + B_{i,j}$ for $i \in [m]$ and $j \in [n]$.

94. Formally define the concept of a *block diagonal* matrix (cf. Exercise 24). State and prove a rule for multiplying two block diagonal matrices with compatible block sizes.

95. Formally define the concept of a *block upper-triangular* matrix (cf. Exercise 25). Prove that if A and B are block upper-triangular matrices with compatible block sizes, then AB is also block upper-triangular with the same block sizes.

96. Prove parts (b) and (c) of the Theorem on Bilinearity of Matrix Tensor Products.

97. For any index set K, define the $K \times K$ *identity matrix* I_K by setting $I_K(k, k) = 1$ and $I_K(k, k') = 0$ for all $k, k' \in K$ with $k \neq k'$. Prove: for all finite index sets K and L, $I_K \otimes I_L = I_{K \times L}$.

98. Suppose $A \in M_m(F)$ and $B \in M_n(F)$ are invertible matrices. Prove $A \otimes B$ is invertible, and $(A \otimes B)^{-1} = A^{-1} \otimes B^{-1}$.

99. Suppose λ is an eigenvalue of $A \in M_m(F)$ with eigenvector v, and μ is an eigenvalue of $B \in M_n(F)$ with eigenvector w. Prove $\lambda\mu$ is an eigenvalue of $A \otimes B \in M_{mn}(F)$ with eigenvector $v \otimes w$.

100. Let A be an $I \times I$ matrix, B be a $J \times J$ matrix, and C be a $K \times K$ matrix. Show that $(A \otimes B) \otimes C = A \otimes (B \otimes C)$ if we identify the index sets $(I \times J) \times K$ and $I \times (J \times K)$ using the bijection sending $((i, j), k)$ to $(i, (j, k))$.

101. Prove: for all $A \in M_m(F)$ and $B \in M_n(F)$, $A \otimes B$ can be transformed to $B \otimes A$ by permuting rows and columns appropriately.

102. This exercise outlines another proof [39] that $\text{rowrk}(A) = \text{colrk}(A)$ for all $A \in M_{m,n}(F)$. Call a row of A *extraneous* iff the row is an F-linear combination of the other rows of A. Define extraneous columns analogously.
(a) Prove: if A has no extraneous rows, then $\text{rowrk}(A) = m$ and $m \leq n$.
(b) Prove: if A has no extraneous columns, then $\text{colrk}(A) = n$ and $n \leq m$.
(c) Prove: if A has no extraneous rows or columns, then $\text{rowrk}(A) = \text{colrk}(A)$.
(d) Suppose A has an extraneous row. Show that deletion of this row produces a smaller matrix with the same row rank and column rank as A. (Say row i is extraneous. Consider the map $T : F^m \to F^{m-1}$ that deletes the ith coordinate. Show that the restriction of T to the column space of A is injective.)
(e) Prove a result similar to (d) if A has an extraneous column.
(f) Show that repeated use of (d) and (e) always leads to a smaller matrix B having no extraneous rows or columns, such that $\text{rowrk}(B) = \text{rowrk}(A)$ and $\text{colrk}(B) = \text{colrk}(A)$. Explain why $\text{rowrk}(A) = \text{colrk}(A)$ follows.

103. This exercise outlines a proof of Hans Liebeck [37] that $\text{colrk}(A) = \text{rowrk}(A)$ for all $A \in M_n(\mathbb{C})$. Prove:
(a) For all $y \in M_{n,1}(\mathbb{C})$, $y = 0$ iff $y^*y = 0$.
(b) For all $A \in M_n(\mathbb{C})$ and $x \in M_{n,1}(\mathbb{C})$, $Ax = 0$ iff $A^*Ax = 0$.

(c) For all $A \in M_n(\mathbb{C})$, $\mathrm{colrk}(A) = \mathrm{colrk}(A^*A)$.
(d) For all $A \in M_n(\mathbb{C})$, $\mathrm{colrk}(A^*) = \mathrm{rowrk}(A)$.
(e) For all $A \in M_n(\mathbb{C})$, $\mathrm{colrk}(A) \leq \mathrm{colrk}(A^*)$.
(f) $\mathrm{colrk}(A) = \mathrm{rowrk}(A)$ follows from (d) and (e).

5

Determinants via Calculations

Most introductions to linear algebra list the basic properties of determinants, including formulas for computing and manipulating determinants. However, complete justifications of these properties and formulas are not always provided. For example, the determinant of a general $n \times n$ matrix A is often defined recursively by Laplace expansion along the first row. We are told that using Laplace expansion along any row or column gives the same answer, but this fact is not always proved.

In this chapter, we rigorously define determinants by an explicit, non-recursive formula that involves a sum indexed by permutations. (See Chapter 2 for the background on permutations needed here.) We use the explicit formula to prove some familiar properties of determinants, such as the Laplace expansions. We give computational proofs whenever possible. These proofs work for matrices with entries in any commutative ring (as defined in §1.2), so we operate at that level of generality. Chapter 20 introduces a more abstract approach to determinants based on multilinear algebra. While that approach is conceptually harder, it yields many properties of determinants with a minimum of calculation.

5.1 Matrices with Entries in a Ring

Before defining determinants, we review the definitions of matrices with entries in a ring and operations on such matrices. (This material is developed more fully in Chapter 4.) Let R be a ring. For each positive integer m, let $[m]$ be the set $\{1, 2, \ldots, m\}$. An $m \times n$ *matrix* A *over* R is a function $A : [m] \times [n] \to R$. For $i \in [m]$ and $j \in [n]$, the ring element $A(i, j)$ is also written $A_{i,j}$ or A_{ij} or a_{ij}, and we often display the matrix A as an array of m rows and n columns such that $A(i, j)$ appears in row i and column j. Let $M_{m,n}(R)$ be the set of all $m \times n$ matrices with entries in R and $M_n(R)$ be the set of all $n \times n$ (square) matrices with entries in R. For each $n \geq 1$, we have a *zero matrix* $0_n \in M_n(R)$ defined by setting $0_n(i, j) = 0_R$ for all $i, j \in [n]$. For each $n \geq 1$, we also define the *identity matrix* I_n by setting $I_n(i, i) = 1_R$ for all $i \in [n]$ and $I_n(i, j) = 0_R$ for all $i \neq j$ in $[n]$. We write 0 for 0_n and I for I_n if n is understood from context.

A *row vector* is an element of $M_{1,n}(R)$. A *column vector* is an element of $M_{n,1}(R)$. We often identify row vectors and column vectors with n-tuples in R^n, which are defined to be functions $x : [n] \to R$. We often represent such a function by the ordered list $(x(1), \ldots, x(n))$. If x is a row vector, we may write $x(i)$ or x_i instead of $x(1, i)$; similarly if x is a column vector. We prefer using parentheses instead of subscripts here, since we often work with lists of row vectors. For instance, if $(x_1, \ldots, x_m)$ is a list of row vectors, then $x_3(5)$ is the fifth entry in the row vector x_3.

Table 5.1 defines some algebraic operations on matrices. The basic properties of these operations are discussed in Chapter 4. The operations $\overline{A}$ and A^* are defined only for matrices with complex entries.

DOI: 10.1201/9781003484561-5

TABLE 5.1

Definitions of matrix operations. Here, R is a ring, $A, B \in M_{m,n}(R)$, $C \in M_{n,p}(R)$, $d \in R$, and bars denote complex conjugation $(\overline{x + iy} = x - iy$ for $x, y \in \mathbb{R})$.

Matrix Operation	Defining Formula
1. matrix sum $A + B \in M_{m,n}(R)$	$(A + B)(i, j) = A(i, j) + B(i, j)$
2. scalar multiplication $dA \in M_{m,n}(R)$	$(dA)(i, j) = d \cdot A(i, j)$
3. matrix product $BC \in M_{m,p}(R)$	$(BC)(i, j) = \sum_{k=1}^{n} B(i, k) \cdot C(k, j)$
4. transpose $A^{\mathrm{T}} \in M_{n,m}(R)$	$A^{\mathrm{T}}(i, j) = A(j, i)$
5. matrix conjugate $\overline{A} \in M_{m,n}(\mathbb{C})$	$\overline{A}(i, j) = \overline{A(i, j)}$
6. conjugate-transpose $A^* \in M_{n,m}(\mathbb{C})$	$A^*(i, j) = \overline{A(j, i)}$

5.2 Explicit Definition of the Determinant

From now on, R is a fixed *commutative* ring. The *determinant* is a function that associates with each $n \times n$ square matrix A an element of R (called the determinant of A). We define the determinant function $\det : M_n(R) \to R$ by the formula

$$\det(A) = \sum_{f \in S_n} \mathrm{sgn}(f) A(f(1), 1) \cdot A(f(2), 2) \cdot \ldots \cdot A(f(n), n) \qquad \text{for } A \in M_n(R). \quad (5.1)$$

(See Chapter 2 for the definitions of S_n and $\mathrm{sgn}(f)$.) The sums and products on the right side of formula (5.1) are carried out in the ring R. Moreover, $\mathrm{sgn}(f)$ is interpreted as $+1_R$ or -1_R, where 1_R is the multiplicative identity of R.

One motivation for this mysterious formula comes from the theory of exterior powers in multilinear algebra, as we discuss in §20.17. For now, we describe a way of visualizing the formula in terms of the geometric layout of the matrix A. A typical term in the determinant formula arises by choosing one entry in each column of the matrix A and multiplying these n entries together. If $f(i)$ is the row of the entry chosen in column i, for each $i \in [n]$, we obtain the product $\prod_{i=1}^{n} A(f(i), i)$. In order for this product to contribute to the determinant, f is required to be a permutation. This means that that the entries f picks for two different columns cannot come from the same row. We attach the sign $\mathrm{sgn}(f)$ to each term in the formula, where $\mathrm{sgn}(f) = (-1)^{\mathrm{inv}(f)}$ is the number of transpositions of adjacent elements needed to sort the list $[f(1), \ldots, f(n)]$ into increasing order (see §2.6). The determinant of A is the sum of the signed terms coming from all $n!$ choices of the permutation f. We remark that we could also sum over all functions $f : [n] \to [n]$, if we use the convention that $\mathrm{sgn}(f) = 0$ when f is not a permutation.

For example, when $n = 3$,

$$\det \begin{bmatrix} a & b & c \\ d & e & f \\ g & h & i \end{bmatrix} = aei + gbf + dhc - gec - dbi - ahf.$$

The term $-dbi$ comes from the permutation $f = [f(1), f(2), f(3)] = [2, 1, 3]$. The sign is $-1 = (-1)^1$, since it takes 1 transposition to sort the list $[2, 1, 3]$. The term gbf comes from the permutation $f = [3, 1, 2]$. The sign is $+1 = (-1)^2$, since it takes two transpositions of adjacent elements to sort the list $[3, 1, 2]$.

When $n = 4$, $\det(A)$ is a sum of $4! = 24$ terms. When $n = 5$, $\det(A)$ is a sum of $5! = 120$ terms. When $n = 8$, $\det(A)$ is a sum of $8! = 40320$ terms. We see that the explicit formula for the determinant quickly becomes unwieldy for doing computations. However, it is very useful for theoretical purposes.

5.3 Diagonal and Triangular Matrices

If a matrix A has many entries equal to 0, then many of the $n!$ terms in the formula for $\det(A)$ are 0. This observation allows us to compute the determinants of certain special matrices quickly. For instance, consider the $n \times n$ identity matrix $I_n = I$. By definition,

$$\det(I) = \sum_{f \in S_n} \operatorname{sgn}(f) I(f(1), 1) I(f(2), 2) \cdots I(f(n), n).$$

Since $I(i, j) = 0_R$ for all $i \neq j$ in $[n]$, the only way to get a nonzero summand is to choose f so that $f(1) = 1$, $f(2) = 2$, ..., and $f(n) = n$. In other words, f must be the identity permutation of S_n. Since $\operatorname{sgn}(\operatorname{id}) = +1$ and $I(i, i) = 1_R$ for all $i \in [n]$, we see that $\det(I) = 1_R$.

More generally, suppose D is a *diagonal* $n \times n$ matrix, which means that $D(i, j) = 0_R$ for all $i \neq j$ in $[n]$. The argument given for I applies to D as well. So

$$\det(D) = D(1, 1) D(2, 2) \cdots D(n, n) = \prod_{i=1}^{n} D(i, i).$$

Thus, *the determinant of a diagonal matrix is the product of the entries on the diagonal.*

A slight adjustment of this technique allows us to compute the determinant of a triangular matrix. A matrix $A \in M_n(R)$ is *upper-triangular* iff $A(i, j) = 0$ for all $i > j$ in $[n]$. A is *lower-triangular* iff $A(i, j) = 0$ for all $i < j$ in $[n]$.

Determinant Rule for Triangular Matrices. The determinant of an upper-triangular or lower-triangular matrix A is the product of the diagonal entries of A.

Proof. We give the proof for a lower-triangular matrix A. We show that if $f \in S_n$ generates a nonzero summand $A(f(1), 1) A(f(2), 2) \cdots A(f(n), n)$ in (5.1), then f must be the identity permutation. First, note that $A(f(n), n) \neq 0$ forces $f(n) \geq n$ and hence $f(n) = n$. Second, note that $A(f(n-1), n-1) \neq 0$ forces $f(n-1) \geq n-1$. Since f is a permutation and $f(n) = n$, we must have $f(n-1) = n-1$. Continuing by backwards induction, fix $i \in [n]$ and assume that $f(j) = j$ for all $j > i$ in $[n]$. Then $A(f(i), i) \neq 0$ forces $f(i) \geq i$, and hence $f(i) = i$ since the values larger than i have already been used. After n steps, we see that $f = \operatorname{id}$ is the only permutation that gives a nonzero contribution to $\det(A)$. Hence, $\det(A) = \prod_{i=1}^{n} A(i, i)$ as claimed. The proof for upper-triangular matrices is analogous. $\square$

5.4 Changing Variables

To prove many properties of determinants, we often use a change of variable to simplify sums of the form $\sum_{y \in S} H(y)$. Here, S is a finite set (often, S is S_n when studying determinants), and $H : S \to R$ is a function that maps each summation index $y \in S$ to a summand $H(y)$. Suppose T is a set and $C : T \to S$ is a bijection (a function that is one-to-one and onto). Each $y \in S$ has the form $y = C(x)$ for a unique $x \in T$. Therefore, if a new summation variable x ranges over all elements of T exactly once, then the values $y = C(x)$ range over all elements of S exactly once. Using generalized associativity and commutativity of addition in the target ring R, it follows that

$$\sum_{y \in S} H(y) = \sum_{x \in T} H(C(x)). \tag{5.2}$$

In particular, if T and S are the same set, we have

$$\sum_{x \in S} H(x) = \sum_{y \in S} H(y) = \sum_{x \in S} H(C(x)). \tag{5.3}$$

For example, let $S = S_n$ and $C : S \to S$ be the inversion map given by $C(f) = f^{-1} \in S_n$ for every $f \in S_n$. We have $C(C(f)) = (f^{-1})^{-1} = f = \mathrm{id}_{S_n}(f)$ for all $f \in S_n$. So $C \circ C = \mathrm{id}_{S_n}$, which proves that C is a bijection on S_n and $C = C^{-1}$. Therefore,

$$\sum_{f \in S_n} H(f) = \sum_{f \in S_n} H(C(f)) = \sum_{f \in S_n} H(f^{-1}).$$

For another example, let $g \in S_n$ be fixed and $L_g : S_n \to S_n$ be the map given by $L_g(f) = g \circ f$ for all $f \in S_n$. The map L_g is called *left multiplication by g*. You can check that $L_{g^{-1}}$ (left multiplication by g^{-1}) is the two-sided inverse for L_g, so L_g is a bijection. The change-of-variable formula (5.3) therefore gives

$$\sum_{f \in S_n} H(f) = \sum_{f \in S_n} H(L_g(f)) = \sum_{f \in S_n} H(g \circ f).$$

The same technique applies to the simplification of finite products $\prod_{x \in S} H(x)$, assuming that the values $H(x)$ belong to a given commutative ring R. Specifically, if S is finite and $C : S \to S$ is any bijection, then

$$\prod_{x \in S} H(x) = \prod_{x \in S} H(C(x)). \tag{5.4}$$

5.5 Transposes and Determinants

In this section, we give an alternate formula for $\det(A)$ by applying a change of variable to the defining formula for the determinant. As a consequence, we obtain the identity $\det(A^{\mathrm{T}}) = \det(A)$.

Consider the definition (5.1) of the determinant of $A \in M_n(R)$. Writing the term $A(f(1), 1) \cdots A(f(n), n)$ as a product over $i \in [n]$, we see that

$$\det(A) = \sum_{f \in S_n} \mathrm{sgn}(f) \prod_{i \in [n]} A(f(i), i). \tag{5.5}$$

Since $f \in S_n$, the map $f^{-1} : [n] \to [n]$ is a bijection of the index set of the product operation in this formula. Therefore, applying (5.4) with x replaced by i, S replaced by $[n]$, $H(x)$ replaced by $A(f(i), i)$, and C replaced by f^{-1}, we see that

$$\prod_{i \in [n]} A(f(i), i) = \prod_{i \in [n]} A(f(f^{-1}(i)), f^{-1}(i)) = \prod_{i \in [n]} A(i, f^{-1}(i))$$

for each $f \in S_n$. Putting this into the determinant formula gives

$$\det(A) = \sum_{f \in S_n} \operatorname{sgn}(f) \prod_{i \in [n]} A(i, f^{-1}(i)).$$

We now perform a change of variable on the outer summation. Let $C : S_n \to S_n$ be the bijection $C(f) = f^{-1}$ for $f \in S_n$. Applying (5.3), we obtain

$$\sum_{f \in S_n} \operatorname{sgn}(f) \prod_{i \in [n]} A(i, f^{-1}(i)) = \sum_{f \in S_n} \operatorname{sgn}(f^{-1}) \prod_{i \in [n]} A(i, (f^{-1})^{-1}(i))$$

$$= \sum_{f \in S_n} \operatorname{sgn}(f) \prod_{i \in [n]} A(i, f(i)).$$

The last step uses the facts $\operatorname{sgn}(f) = \operatorname{sgn}(f^{-1})$ and $(f^{-1})^{-1} = f$. We conclude that

$$\det(A) = \sum_{f \in S_n} \operatorname{sgn}(f) \prod_{i \in [n]} A(i, f(i)). \tag{5.6}$$

Intuitively, this result means that we can compute the determinant as follows. Select one element from each row of A, say the element in column $f(i)$ of row i for each $i \in [n]$. Multiply these chosen elements together, and multiply by $\operatorname{sgn}(f)$ (which is zero if f is not a permutation). Then sum all terms obtained by making such choices. In contrast, our previous description chose one element from each column of A. The symmetric roles of the rows and columns in Formulas (5.5) and (5.6) let us deduce many column-oriented properties of determinants from the corresponding row-oriented properties. Examples of this symmetry between rows and columns appear in the coming sections.

Determinant Rules for A^{T}, $\overline{A}$, and A^*. For all $A \in M_n(R)$, $\det(A^{\mathrm{T}}) = \det(A)$. For all $A \in M_n(\mathbb{C})$, $\det(A^*) = \det(\overline{A}) = \overline{\det(A)}$.

Proof. Fix $A \in M_n(R)$. Using the original formula (5.1) for $\det(A^{\mathrm{T}})$, then the definition of A^{T}, then the new formula (5.6) for $\det(A)$, we compute:

$$\begin{aligned} \det(A^{\mathrm{T}}) &= \sum_{f \in S_n} \operatorname{sgn}(f) A^{\mathrm{T}}(f(1), 1) A^{\mathrm{T}}(f(2), 2) \cdots A^{\mathrm{T}}(f(n), n) \\ &= \sum_{f \in S_n} \operatorname{sgn}(f) A(1, f(1)) A(2, f(2)) \cdots A(n, f(n)) \\ &= \sum_{f \in S_n} \operatorname{sgn}(f) \prod_{i \in [n]} A(i, f(i)) = \det(A). \end{aligned}$$

For the rest of the proof, assume A has complex entries. Given $z \in \mathbb{C}$, we temporarily write $c(z) = \overline{z}$ for complex conjugation, so $c(x + iy) = x - iy$ for $x, y \in \mathbb{R}$. You can check that $c(z + z') = c(z) + c(z')$, $c(zz') = c(z)c(z')$, and $c(az) = ac(z)$ for all $z, z' \in \mathbb{C}$ and $a \in \mathbb{R}$.

The first two formulas extend by induction to finite sums and products. Hence,

$$
\begin{aligned}
\det(\overline{A}) &= \sum_{f \in S_n} \operatorname{sgn}(f) \prod_{i \in [n]} \overline{A}(f(i), i) = \sum_{f \in S_n} \operatorname{sgn}(f) \prod_{i \in [n]} c(A(f(i), i)) \\
&= \sum_{f \in S_n} \operatorname{sgn}(f) c \left(\prod_{i \in [n]} A(f(i), i) \right) \\
&= c \left(\sum_{f \in S_n} \operatorname{sgn}(f) \prod_{i \in [n]} A(f(i), i) \right) \\
&= c(\det(A)) = \overline{\det(A)}.
\end{aligned}
$$

So the determinant of the matrix conjugate of A is the complex conjugate of the determinant of A. Since $A^* = \overline{A^{\mathrm{T}}}$, we get $\det(A^*) = \overline{\det(A^{\mathrm{T}})} = \overline{\det(A)}$.   □

5.6   Multilinearity and the Alternating Property

We now investigate the relationship between the determinant of a square matrix A and the rows of A. Given $A \in M_n(R)$ and $i \in [n]$, we let $A_{[i]} \in R^n$ be row i of A, so that $A_{[i]}(j) = A(i, j)$ for all $j \in [n]$. In this section, we often identify the matrix A with the ordered list of rows $(A_{[1]}, A_{[2]}, \dots, A_{[n]}) \in (R^n)^n$. Thus, the determinant map can be regarded as a function $\det : (R^n)^n \to R$ of n input variables, where each variable $A_{[i]} = A_i$ is an element of R^n that corresponds to row i of A. Using the row notation, formulas (5.1) and (5.6) for the determinant of A become

$$
\det(A_1, \dots, A_n) = \sum_{f \in S_n} \operatorname{sgn}(f) \prod_{k=1}^{n} A_{f(k)}(k) = \sum_{f \in S_n} \operatorname{sgn}(f) \prod_{k=1}^{n} A_k(f(k)). \tag{5.7}
$$

Multilinearity Property of Determinants.   The determinant, viewed as a function $\det : (R^n)^n \to R$ of the n rows of A, is R-multilinear.

Multilinearity means that, if any index $i \in [n]$ and the rows A_j for all $j \neq i$ are held constant, then the determinant is an R-linear function of the remaining input argument A_i, which is an n-tuple in R^n. In more detail, fix $i \in [n]$ and n-tuples $A_j \in R^n$ for all $j \neq i$ in $[n]$, and define $D : R^n \to R$ by $D(v) = \det(A_1, \dots, A_{i-1}, v, A_{i+1}, \dots, A_n)$ for $v \in R^n$. The multilinearity property says that $D(v + w) = D(v) + D(w)$ and $D(rv) = rD(v)$ for all $v, w \in R^n$ and $r \in R$.

Proof. With the above notation, replace A_i by $v = (v(1), \dots, v(n))$ in (5.7). We obtain

$$
D(v) = \sum_{f \in S_n} \operatorname{sgn}(f) \prod_{k=1}^{i-1} A_k(f(k)) v(f(i)) \prod_{k=i+1}^{n} A_k(f(k)) = \sum_{f \in S_n} v(f(i)) \operatorname{sgn}(f) \prod_{\substack{k \in [n]: \\ k \neq i}} A_k(f(k)).
$$

Each summand here is one of the components of v multiplied by a scalar in R. Grouping together summands that involve the same component of v, we see that we can write $D(v) = \sum_{j=1}^{n} c_j v(j)$, where each c_j is a scalar in R that depends on the fixed A_k with $k \neq i$ but does not depend on v. Explicitly, for each $j \in [n]$, we have

$$c_j = \sum_{\substack{f \in S_n: \\ f(i)=j}} \text{sgn}(f) \prod_{\substack{k \in [n]: \\ k \neq i}} A_k(f(k)).$$

The R-linearity of the function D follows by computing

$$D(v+w) = \sum_{j=1}^{n} c_j(v+w)(j) = \sum_{j=1}^{n} c_j v(j) + \sum_{j=1}^{n} c_j w(j) = D(v) + D(w);$$

$$D(av) = \sum_{j=1}^{n} c_j(av)(j) = a \sum_{j=1}^{n} c_j v(j) = aD(v)$$

for all $v, w \in R^n$ and $a \in R$. $\qquad \square$

Since D is linear, $D(0_{R^n}) = 0_R$, which means that *a matrix containing a row of zeroes has determinant zero.*

Here is an example of the formula $D(v) = \sum_{j=1}^{n} c_j v(j)$ in the case where $A \in M_3(\mathbb{R})$ and $i = 3$. Fix rows $A_1 = (4, 1, -3)$ and $A_2 = (0, 2, 5)$. For $v = (v_1, v_2, v_3) \in \mathbb{R}^3$,

$$D(v) = \det \begin{bmatrix} 4 & 1 & -3 \\ 0 & 2 & 5 \\ v_1 & v_2 & v_3 \end{bmatrix} = 11v_1 - 20v_2 + 8v_3,$$

and this is an $\mathbb{R}$-linear function of the row vector v.

Alternating Property of Determinants. The determinant of a matrix with two equal rows is 0.

Proof. Let $A \in M_n(R)$ have rows $(A_1, A_2, \ldots, A_n)$. We assume $A_i = A_j$ for some $i \neq j$ and prove $\det(A_1, \ldots, A_n) = 0$. Using the assumption $A_i = A_j$, formula (5.7) becomes

$$\det(A) = \sum_{f \in S_n} \text{sgn}(f) \prod_{k \in [n]} A_k(f(k)) = \sum_{f \in S_n} \text{sgn}(f) A_i(f(i)) A_i(f(j)) \prod_{\substack{k \in [n]: \\ k \neq i,j}} A_k(f(k)).$$

Let us compare the summand coming from some $f \in S_n$ with the summand coming from $f' = f \circ (i, j)$. We have $f'(i) = f(j)$, $f'(j) = f(i)$, $f'(k) = f(k)$ for all $k \neq i, j$, and $\text{sgn}(f') = -\text{sgn}(f)$ by §2.7. It follows that

$$\text{sgn}(f') A_i(f'(i)) A_i(f'(j)) \prod_{\substack{k \in [n]: \\ k \neq i,j}} A_k(f'(k)) = -\text{sgn}(f) A_i(f(j)) A_i(f(i)) \prod_{\substack{k \in [n]: \\ k \neq i,j}} A_k(f(k)).$$

Since R is commutative, the summands for f and f' cancel each other. Pairing the $n!$ terms in the sum for $\det(A)$ in this way, we see that all terms cancel, hence $\det(A) = 0$. $\qquad \square$

5.7 Elementary Row Operations and Determinants

Using the fact that the determinant is alternating and multilinear in the rows of A, we can deduce some results on how elementary row operations affect determinants. Write

$A = (A_1, \ldots, A_n)$ (a list of n rows). The first type of elementary row operation multiplies row i of A by an invertible scalar $c \in R$. From the multilinearity property, we know

$$\det(A_1, \ldots, cA_i, \ldots, A_n) = c\det(A_1, \ldots, A_i, \ldots, A_n)$$

for any scalar $c \in R$ (invertible or not). So, *multiplying one row of a matrix by a scalar multiplies the determinant of the matrix by that scalar*. Multiplying the whole matrix by a given scalar c is the same as multiplying each of the n rows by c. Each row multiplication introduces a factor of c, so

$$\det(cA) = c^n \det(A) \quad \text{for all } A \in M_n(R).$$

The second type of elementary row operation interchanges two distinct rows of A, say row i and row j. We claim this operation changes the sign of $\det(A)$:

$$\det(A_1, \ldots, A_i, \ldots, A_j, \ldots, A_n) = -\det(A_1, \ldots, A_j, \ldots, A_i, \ldots, A_n). \tag{5.8}$$

To prove this, fix i, j in $[n]$ with $i \neq j$, fix the other rows A_k for all $k \neq i, j$ in $[n]$. Define $D : R^n \times R^n \to R$ by $D(v, w) = \det(A_1, \ldots, v, \ldots, w, \ldots, A_n)$ for $v, w \in R^n$, where v is in position i and w is in position j. By the alternating property already proved, $D(v, v) = 0$ for all $v \in R^n$. Furthermore, D is linear in each of its two inputs, since the determinant is multilinear. We therefore have

$$\begin{aligned}
0 &= D(A_i + A_j, A_i + A_j) = D(A_i + A_j, A_i) + D(A_i + A_j, A_j) \\
&= D(A_i, A_i) + D(A_j, A_i) + D(A_i, A_j) + D(A_j, A_j) \\
&= 0 + D(A_j, A_i) + D(A_i, A_j) + 0.
\end{aligned}$$

Rearranging, we get $D(A_i, A_j) = -D(A_j, A_i)$, which proves (5.8).

The third type of elementary row operation adds c times row j to row i of a matrix, where i, j in $[n]$, $i \neq j$, and $c \in R$. We claim this operation does not change $\det(A)$:

$$\det(A_1, \ldots, A_i + cA_j, \ldots, A_j, \ldots, A_n) = \det(A_1, \ldots, A_i, \ldots, A_j, \ldots, A_n).$$

We prove this using multilinearity and the alternating property. Defining D as in the last paragraph, we have

$$D(A_i + cA_j, A_j) = D(A_i, A_j) + cD(A_j, A_j) = D(A_i, A_j) + c \cdot 0 = D(A_i, A_j).$$

We can interpret these results as facts about elementary matrices (see §4.9). Suppose $A \in M_n(R)$ and A' results from A by performing some elementary row operation. Then $A' = EA$, where E is the elementary matrix obtained from I_n by applying the same row operation that took A to A'. We now show that $\det(A') = \det(E)\det(A)$ for all three types of elementary operations. First, suppose A' is obtained by multiplying row i of A by an invertible $c \in R$. Here, E is a diagonal matrix with a c in the i, i-position and other diagonal entries equal to 1. We know $\det(A') = c\det(A)$ and $\det(E) = c\det(I) = c$, so $\det(A') = \det(E)\det(A)$ in this case. Second, suppose A' is obtained from A by interchanging rows i and j, so E is obtained from I by interchanging rows i and j. Then $\det(A') = -\det(A)$ and $\det(E) = -\det(I) = -1_R$, so $\det(A') = \det(E)\det(A)$ in this case. Third, suppose A' is obtained from A by adding c times row j to row i. Then $\det(A') = \det(A)$ and $\det(E) = \det(I) = 1_R$, so $\det(A') = \det(E)\det(A)$ holds. To summarize, *for all square matrices A and all elementary matrices E of the same size as A, $\det(EA) = \det(E)\det(A)$*. When R is a field, we can use this fact to prove the Determinant Product Formula: $\det(BA) = \det(B)\det(A)$ for all $A, B \in M_n(R)$ (see §5.9).

5.8 Determinant Properties Involving Columns

The discussion in the last two sections, relating determinants to the rows of a matrix, also applies to the columns of a matrix. Given $A \in M_n(R)$ and $j \in [n]$, let $A^{[j]} \in R^n$ be column j of A, so that $A^{[j]}(i) = A(i, j)$ for all $i \in [n]$. We identify the matrix A with the ordered list of its columns, writing $A = (A^{[1]}, \ldots, A^{[n]}) \in (R^n)^n$. Using this column notation, our formulas (5.5) and (5.6) for the determinant become

$$\det(A) = \det(A^{[1]}, \ldots, A^{[n]}) = \sum_{f \in S_n} \operatorname{sgn}(f) \prod_{k=1}^{n} A^{[k]}(f(k)) = \sum_{f \in S_n} \operatorname{sgn}(f) \prod_{k=1}^{n} A^{[f(k)]}(k).$$

Starting from these formulas, we can repeat all the proofs in the last two sections verbatim, changing all subscripts (row indices) to superscripts (column indices). We thereby obtain the following facts.

(a) The determinant is an R-multilinear function of the n columns of A.
(b) If any two columns of A are equal or if A has a column of zeroes, then $\det(A) = 0$.
(c) Multiplying a column of A by $c \in R$ multiplies the determinant by c.
(d) Interchanging two columns of A flips the sign of the determinant.
(e) Adding c times column i to column $j \neq i$ does not change the determinant.

Since elementary column operations correspond to right multiplication by elementary matrices (§4.9), facts (c), (d), and (e) imply that $\det(AE) = \det(A) \det(E)$ for all matrices $A \in M_n(R)$ and all elementary matrices $E \in M_n(R)$.

All of these results can also be derived quickly from the corresponding results for rows by invoking the identity $\det(A) = \det(A^{\mathrm{T}})$. For instance, if A has two equal columns, then A^{T} has two equal rows, and so $\det(A) = \det(A^{\mathrm{T}}) = 0$. For another example of this proof method, assume A is arbitrary and E is elementary. Then E^{T} is also elementary (as is readily checked), so that

$$\det(AE) = \det((AE)^{\mathrm{T}}) = \det(E^{\mathrm{T}} A^{\mathrm{T}}) = \det(E^{\mathrm{T}}) \det(A^{\mathrm{T}}) = \det(A) \det(E).$$

This computation uses commutativity of the ring R (cf. Exercise 73).

5.9 Product Formula via Elementary Matrices

In this section only, we consider matrices with entries in a field F. We use facts about elementary matrices and row reduction to prove the Product Formula ($\det(AB) = \det(A) \det(B)$ for $A, B \in M_n(F)$). We also prove that A is invertible iff $\det(A) \neq 0_F$. Proofs of the corresponding results for matrices with entries in any commutative ring are given later (see §5.11 and §5.13).

First, we indicate a method for computing determinants of matrices in $M_n(F)$ that is much more efficient than computing all $n!$ terms in the defining formula. Given a square matrix A, perform the Gaussian Elimination Algorithm, using a finite sequence of elementary row operations to bring A into echelon form. Since A is square, the echelon form must be an upper-triangular matrix. We can use the results in §5.7 to keep track of how the determinant of the matrix changes as each row operation is applied. As we have seen, for every elementary row operation, the determinant is multiplied by some nonzero scalar in the field F (possibly the scalar 1). When a triangular matrix is reached, we can find the determinant by multiplying the diagonal entries.

For example, let the matrix $A = \begin{bmatrix} 0 & 2 & 1 \\ 3 & 6 & -3 \\ 2 & 5 & 1 \end{bmatrix}$ have unknown determinant $\det(A) = d$.

We perform Gaussian elimination steps on A as shown below, keeping track of the effect of each elementary row operation on the determinant:

$$\begin{bmatrix} 0 & 2 & 1 \\ 3 & 6 & -3 \\ 2 & 5 & 1 \end{bmatrix} \xrightarrow{\text{R1}\leftrightarrow\text{R2}} \begin{bmatrix} 3 & 6 & -3 \\ 0 & 2 & 1 \\ 2 & 5 & 1 \end{bmatrix} \xrightarrow{\text{R1}\times 1/3} \begin{bmatrix} 1 & 2 & -1 \\ 0 & 2 & 1 \\ 2 & 5 & 1 \end{bmatrix}$$
$$\det = d \qquad\qquad\qquad \det = -d \qquad\qquad\qquad \det = -d/3$$

$$\xrightarrow{\text{R3}-2\text{R1}} \begin{bmatrix} 1 & 2 & -1 \\ 0 & 2 & 1 \\ 0 & 1 & 3 \end{bmatrix} \xrightarrow{\text{R2}\leftrightarrow\text{R3}} \begin{bmatrix} 1 & 2 & -1 \\ 0 & 1 & 3 \\ 0 & 2 & 1 \end{bmatrix} \xrightarrow{\text{R3}-2\text{R2}} \begin{bmatrix} 1 & 2 & -1 \\ 0 & 1 & 3 \\ 0 & 0 & -5 \end{bmatrix}$$
$$\det = -d/3 \qquad\qquad\qquad \det = d/3 \qquad\qquad\qquad \det = d/3$$

The final matrix is triangular, with determinant $1 \cdot 1 \cdot (-5) = -5$. Since this also equals $d/3$, we see that $\det(A) = -15$.

Determinant Criterion for Matrix Invertibility. Let F be a field and $A \in M_n(F)$. The matrix A is invertible if and only if $\det(A) \neq 0_F$.

Proof. Given any $A \in M_n(F)$, perform row-reduction steps to convert A to a reduced row-echelon form matrix A_{ech}, as defined in §4.10. We know $A_{\text{ech}} = PA$ for some invertible matrix $P \in M_n(F)$, where P is the product of the elementary matrices associated with the elementary row operations used to reduce A. So, A is invertible iff A_{ech} is invertible. First suppose A is invertible. Then A_{ech} is invertible, so A_{ech} cannot have any zero rows. Since A_{ech} is square, we must have $A_{\text{ech}} = I_n$, which has determinant 1_F. The determinant of A differs from this by some factor that is a product of nonzero elements of F. Hence, A invertible implies $\det(A) \neq 0_F$. Conversely, suppose A is not invertible. Then A_{ech} is not invertible, hence A_{ech} must be an upper-triangular matrix with at least one zero on the main diagonal. So $\det(A_{\text{ech}}) = 0_F$. The determinant of A is a nonzero scalar multiple of this determinant, so A non-invertible implies $\det(A) = 0_F$. $\square$

Determinant Product Formula. Let F be a field. For all $A, B \in M_n(F)$,

$$\det(AB) = \det(A)\det(B).$$

Proof. We already proved the Product Formula when A is an elementary matrix and B is any matrix (§5.7). Next, we show that the formula holds when $A = E_1 \cdots E_k$ is a finite product of elementary matrices, and B is arbitrary. We prove this by induction on k, the case $k = 1$ already being known. Fix $k > 1$, and assume the formula holds when A is a product of $k-1$ elementary factors. For $A = E_1 \cdots E_k$, we compute:

$$\begin{aligned} \det(AB) &= \det(E_1(E_2 \cdots E_k B)) = \det(E_1)\det((E_2 \cdots E_k)B) \\ &= \det(E_1)\det(E_2 \cdots E_k)\det(B) = \det(E_1(E_2 \cdots E_k))\det(B) \\ &= \det(A)\det(B). \end{aligned}$$

The first and fifth equalities hold by associativity of matrix multiplication; the second and fourth equalities hold by the product formula where the left factor is elementary; and the third equality holds by induction hypothesis.

Using Gaussian elimination, you can show that $A \in M_n(R)$ is invertible iff A is a finite product of elementary matrices (see §4.11). Therefore, we have proved that the Product Formula $\det(AB) = \det(A)\det(B)$ holds when A is invertible and B is arbitrary. Similarly, the formula holds when B is invertible and A is arbitrary. Finally, consider the case where neither A nor B is invertible. As B is not invertible, there exists some nonzero $\mathbf{v} \in F^n$ with $B\mathbf{v} = \mathbf{0}$ (see §4.13). Hence $(AB)\mathbf{v} = A(B\mathbf{v}) = A\mathbf{0} = \mathbf{0}$, so AB is not invertible. It follows that $\det(AB) = 0_F = 0_F \cdot 0_F = \det(A)\det(B)$. □

5.10 Laplace Expansions

We now prove the Laplace expansions for determinants. Given $A \in M_n(R)$, let $A[i|j]$ be the matrix in $M_{n-1}(R)$ obtained by deleting row i and column j of A. Formally, the s,t-entry of $A[i|j]$ is $A(s,t)$ if $s < i$ and $t < j$; $A(s+1,t)$ if $i \le s \le n-1$ and $t < j$; $A(s,t+1)$ if $s < i$ and $j \le t \le n-1$; and $A(s+1,t+1)$ if $i \le s \le n-1$ and $j \le t \le n-1$. For any logical statement P, write $\chi(P) = 1$ if P is true and $\chi(P) = 0$ if P is false. Then we can define $A[i|j]$ more succinctly by the formula

$$A[i|j](s,t) = A(s + \chi(s \ge i), t + \chi(t \ge j)) \text{ for all } s, t \in [n-1].$$

Laplace Expansions for Determinants. For all $A \in M_n(R)$, $i \in [n]$, and $j \in [n]$,

$$\det(A) = \sum_{j=1}^{n} A(i,j)(-1)^{i+j}\det(A[i|j]) \quad \text{and} \quad \det(A) = \sum_{i=1}^{n} A(i,j)(-1)^{i+j}\det(A[i|j]).$$

The first formula, called *Laplace expansion along row i*, arises by grouping the $n!$ terms in the definition of $\det(A)$ based on the entry chosen from row i. The second formula, called *Laplace expansion along column j*, arises by grouping terms based on the entry chosen from column j.

Proof. Step 1: We prove the row expansion formula for $i = n$. Starting with the defining formula

$$\det(A) = \sum_{f \in S_n} \text{sgn}(f) \prod_{k \in [n]} A(f(k), k),$$

we group together the terms in this sum that involve $A(n,j)$ for each j between 1 and n. The entry $A(n,j)$ appears in the term indexed by $f \in S_n$ iff $f(j) = n$. Therefore,

$$\det(A) = \sum_{j=1}^{n} A(n,j)U_j, \text{ where } U_j = \sum_{\substack{f \in S_n: \\ f(j)=n}} \text{sgn}(f) \prod_{\substack{k \in [n]: \\ k \ne j}} A(f(k), k).$$

We complete Step 1 by showing that $U_j = (-1)^{n+j}\det(A[n|j])$ for $j \in [n]$.

From now on, keep j fixed. Introduce the notation $S = \{f \in S_n : f(j) = n\}$ and $T = S_{n-1}$. Define a bijection $C : T \to S$ as follows. View an element $g \in S_{n-1}$ as a list $[g(1), \ldots, g(n-1)]$ that uses the numbers 1 through $n-1$ once each. Define $C(g)$ to be the list

$$g' = [g(1), \ldots, g(j-1), n, g(j), \ldots, g(n-1)],$$

which is an element of S. C is a bijection; the inverse of C takes an input list $f \in S$ and deletes the entry in position j, which must be the integer n.

Applying the change-of-variable formula (5.2) and writing g' for $C(g)$, we see that

$$U_j = \sum_{f \in S} \text{sgn}(f) \prod_{\substack{k \in [n]: \\ k \neq j}} A(f(k), k) = \sum_{g \in T} \text{sgn}(g') \prod_{\substack{k \in [n]: \\ k \neq j}} A(g'(k), k). \tag{5.9}$$

To continue simplifying, let us first relate $\text{sgn}(g')$ to $\text{sgn}(g)$. Recall that $\text{sgn}(g) = (-1)^{\text{inv}(g)}$, where $\text{inv}(g)$ is the minimum number of transpositions of adjacent elements needed to sort the list $g = [g(1), \ldots, g(n-1)]$ (see §2.6). Similarly, $\text{sgn}(g') = (-1)^{\text{inv}(g')}$, where $\text{inv}(g')$ is the minimum number of transpositions of adjacent elements needed to sort the list

$$g' = [g(1), \ldots, g(j-1), n, g(j), \ldots, g(n-1)].$$

The latter list may be sorted by first moving the largest entry n from position j to position n (which requires $n - j$ transposition moves) and then sorting the first $n - 1$ elements of the resulting list. Since the resulting list is $[g(1), \ldots, g(n-1), n]$, the second stage of the sorting requires $\text{inv}(g)$ moves. Therefore, $\text{inv}(g') = \text{inv}(g) + n - j$. Raising -1_R to this power and noting that $(-1)^{n-j} = (-1)^{n+j}$ in any ring, we see that

$$\text{sgn}(g') = (-1)^{n+j} \text{sgn}(g). \tag{5.10}$$

Next, we find $A(g'(k), k)$ in terms of g and $A[n|j]$. If $1 \leq k < j$, then $g'(k) = g(k)$ by definition. Also $g(k) < n$, so $A[n|j](g(k), k) = A(g(k), k) = A(g'(k), k)$ in this case. On the other hand, if $j < k \leq n$, then $g'(k) = g(k-1) < n$ and $A[n|j](g(k-1), k-1) = A(g(k-1), k) = A(g'(k), k)$ by definition. We can therefore write

$$\prod_{\substack{k \in [n]: \\ k \neq j}} A(g'(k), k) = \prod_{1 \leq k < j} A[n|j](g(k), k) \prod_{j < k \leq n} A[n|j](g(k-1), k-1)$$

$$= \prod_{1 \leq k < j} A[n|j](g(k), k) \prod_{j \leq k \leq n-1} A[n|j](g(k), k)$$

$$= \prod_{k \in [n-1]} A[n|j](g(k), k).$$

Using this and (5.10) in (5.9), we see that

$$U_j = (-1)^{n+j} \sum_{g \in S_{n-1}} \text{sgn}(g) \prod_{k \in [n-1]} A[n|j](g(k), k) = (-1)^{n+j} \det(A[n|j]).$$

Step 2: We prove that if the Laplace expansion formula holds for a given row $i > 1$, then it holds for row $i - 1$. Combining this with Step 1, we can conclude that the Laplace expansion is valid for any row of A. To prove Step 2, assume that $1 < i \leq n$ is fixed and

$$\det(B) = \sum_{j=1}^{n} B(i, j)(-1)^{i+j} \det(B[i|j])$$

is known to be true for any $B \in M_n(R)$. Given A, let B be the matrix obtained from A by interchanging row $i - 1$ and row i. We know that $\det(B) = -\det(A)$. We see from the definitions that $B[i|j] = A[i-1|j]$ and $B(i, j) = A(i-1, j)$ for all $j \in [n]$. So,

$$\det(A) = -\det(B) = \sum_{j=1}^{n} B(i, j)(-1)^{i+j+1} \det(B[i|j])$$

$$= \sum_{j=1}^{n} A(i-1, j)(-1)^{(i-1)+j} \det(A[i-1|j]).$$

This is the Laplace expansion of $\det(A)$ along row $i - 1$.

Using Gaussian elimination, you can show that $A \in M_n(R)$ is invertible iff A is a finite product of elementary matrices (see §4.11). Therefore, we have proved that the Product Formula $\det(AB) = \det(A)\det(B)$ holds when A is invertible and B is arbitrary. Similarly, the formula holds when B is invertible and A is arbitrary. Finally, consider the case where neither A nor B is invertible. As B is not invertible, there exists some nonzero $\mathbf{v} \in F^n$ with $B\mathbf{v} = \mathbf{0}$ (see §4.13). Hence $(AB)\mathbf{v} = A(B\mathbf{v}) = A\mathbf{0} = \mathbf{0}$, so AB is not invertible. It follows that $\det(AB) = 0_F = 0_F \cdot 0_F = \det(A)\det(B)$. $\qquad \square$

5.10 Laplace Expansions

We now prove the Laplace expansions for determinants. Given $A \in M_n(R)$, let $A[i|j]$ be the matrix in $M_{n-1}(R)$ obtained by deleting row i and column j of A. Formally, the s, t-entry of $A[i|j]$ is $A(s, t)$ if $s < i$ and $t < j$; $A(s+1, t)$ if $i \leq s \leq n - 1$ and $t < j$; $A(s, t+1)$ if $s < i$ and $j \leq t \leq n - 1$; and $A(s+1, t+1)$ if $i \leq s \leq n - 1$ and $j \leq t \leq n - 1$. For any logical statement P, write $\chi(P) = 1$ if P is true and $\chi(P) = 0$ if P is false. Then we can define $A[i|j]$ more succinctly by the formula

$$A[i|j](s, t) = A(s + \chi(s \geq i), t + \chi(t \geq j)) \text{ for all } s, t \in [n - 1].$$

Laplace Expansions for Determinants. For all $A \in M_n(R)$, $i \in [n]$, and $j \in [n]$,

$$\det(A) = \sum_{j=1}^{n} A(i, j)(-1)^{i+j} \det(A[i|j]) \quad \text{and} \quad \det(A) = \sum_{i=1}^{n} A(i, j)(-1)^{i+j} \det(A[i|j]).$$

The first formula, called *Laplace expansion along row i*, arises by grouping the $n!$ terms in the definition of $\det(A)$ based on the entry chosen from row i. The second formula, called *Laplace expansion along column j*, arises by grouping terms based on the entry chosen from column j.

Proof. Step 1: We prove the row expansion formula for $i = n$. Starting with the defining formula

$$\det(A) = \sum_{f \in S_n} \text{sgn}(f) \prod_{k \in [n]} A(f(k), k),$$

we group together the terms in this sum that involve $A(n, j)$ for each j between 1 and n. The entry $A(n, j)$ appears in the term indexed by $f \in S_n$ iff $f(j) = n$. Therefore,

$$\det(A) = \sum_{j=1}^{n} A(n, j) U_j, \text{ where } U_j = \sum_{\substack{f \in S_n: \\ f(j)=n}} \text{sgn}(f) \prod_{\substack{k \in [n]: \\ k \neq j}} A(f(k), k).$$

We complete Step 1 by showing that $U_j = (-1)^{n+j} \det(A[n|j])$ for $j \in [n]$.

From now on, keep j fixed. Introduce the notation $S = \{f \in S_n : f(j) = n\}$ and $T = S_{n-1}$. Define a bijection $C : T \to S$ as follows. View an element $g \in S_{n-1}$ as a list $[g(1), \ldots, g(n-1)]$ that uses the numbers 1 through $n - 1$ once each. Define $C(g)$ to be the list

$$g' = [g(1), \ldots, g(j - 1), n, g(j), \ldots, g(n - 1)],$$

which is an element of S. C is a bijection; the inverse of C takes an input list $f \in S$ and deletes the entry in position j, which must be the integer n.

Applying the change-of-variable formula (5.2) and writing g' for $C(g)$, we see that

$$U_j = \sum_{f \in S} \text{sgn}(f) \prod_{\substack{k \in [n]: \\ k \neq j}} A(f(k), k) = \sum_{g \in T} \text{sgn}(g') \prod_{\substack{k \in [n]: \\ k \neq j}} A(g'(k), k). \qquad (5.9)$$

To continue simplifying, let us first relate $\text{sgn}(g')$ to $\text{sgn}(g)$. Recall that $\text{sgn}(g) = (-1)^{\text{inv}(g)}$, where $\text{inv}(g)$ is the minimum number of transpositions of adjacent elements needed to sort the list $g = [g(1), \ldots, g(n-1)]$ (see §2.6). Similarly, $\text{sgn}(g') = (-1)^{\text{inv}(g')}$, where $\text{inv}(g')$ is the minimum number of transpositions of adjacent elements needed to sort the list

$$g' = [g(1), \ldots, g(j-1), n, g(j), \ldots, g(n-1)].$$

The latter list may be sorted by first moving the largest entry n from position j to position n (which requires $n-j$ transposition moves) and then sorting the first $n-1$ elements of the resulting list. Since the resulting list is $[g(1), \ldots, g(n-1), n]$, the second stage of the sorting requires $\text{inv}(g)$ moves. Therefore, $\text{inv}(g') = \text{inv}(g) + n - j$. Raising -1_R to this power and noting that $(-1)^{n-j} = (-1)^{n+j}$ in any ring, we see that

$$\text{sgn}(g') = (-1)^{n+j} \text{sgn}(g). \qquad (5.10)$$

Next, we find $A(g'(k), k)$ in terms of g and $A[n|j]$. If $1 \leq k < j$, then $g'(k) = g(k)$ by definition. Also $g(k) < n$, so $A[n|j](g(k), k) = A(g(k), k) = A(g'(k), k)$ in this case. On the other hand, if $j < k \leq n$, then $g'(k) = g(k-1) < n$ and $A[n|j](g(k-1), k-1) = A(g(k-1), k) = A(g'(k), k)$ by definition. We can therefore write

$$\prod_{\substack{k \in [n]: \\ k \neq j}} A(g'(k), k) = \prod_{1 \leq k < j} A[n|j](g(k), k) \prod_{j < k \leq n} A[n|j](g(k-1), k-1)$$

$$= \prod_{1 \leq k < j} A[n|j](g(k), k) \prod_{j \leq k \leq n-1} A[n|j](g(k), k)$$

$$= \prod_{k \in [n-1]} A[n|j](g(k), k).$$

Using this and (5.10) in (5.9), we see that

$$U_j = (-1)^{n+j} \sum_{g \in S_{n-1}} \text{sgn}(g) \prod_{k \in [n-1]} A[n|j](g(k), k) = (-1)^{n+j} \det(A[n|j]).$$

Step 2: We prove that if the Laplace expansion formula holds for a given row $i > 1$, then it holds for row $i - 1$. Combining this with Step 1, we can conclude that the Laplace expansion is valid for any row of A. To prove Step 2, assume that $1 < i \leq n$ is fixed and

$$\det(B) = \sum_{j=1}^{n} B(i, j)(-1)^{i+j} \det(B[i|j])$$

is known to be true for any $B \in M_n(R)$. Given A, let B be the matrix obtained from A by interchanging row $i - 1$ and row i. We know that $\det(B) = -\det(A)$. We see from the definitions that $B[i|j] = A[i-1|j]$ and $B(i, j) = A(i-1, j)$ for all $j \in [n]$. So,

$$\det(A) = -\det(B) = \sum_{j=1}^{n} B(i, j)(-1)^{i+j+1} \det(B[i|j])$$

$$= \sum_{j=1}^{n} A(i-1, j)(-1)^{(i-1)+j} \det(A[i-1|j]).$$

This is the Laplace expansion of $\det(A)$ along row $i - 1$.

Step 3: We prove the Laplace expansion along column k, for each $k \in [n]$. From Steps 1 and 2, we already know that

$$\det(B) = \sum_{s=1}^{n} B(k,s)(-1)^{k+s} \det(B[k|s])$$

holds for any B and any row index k. Given A, take $B = A^{\mathrm{T}}$. The definitions show that $B(k,s) = A(s,k)$ and $B[k|s] = A[s|k]^{\mathrm{T}}$ (since deleting row k and column s of $B = A^{\mathrm{T}}$ has the same effect as deleting row s and column k of A, and then transposing). We now compute

$$
\begin{aligned}
\det(A) \quad &= \quad \det(B) = \sum_{s=1}^{n} B(k,s)(-1)^{k+s} \det(B[k|s]) \\
&= \quad \sum_{s=1}^{n} A(s,k)(-1)^{s+k} \det(A[s|k]^{\mathrm{T}}) = \sum_{i=1}^{n} A(i,k)(-1)^{i+k} \det(A[i|k]).
\end{aligned}
$$

This is the Laplace expansion of $\det(A)$ along column k. $\qquad\square$

5.11 Classical Adjoints and Inverses

Given a matrix $A \in M_n(R)$, we define an $n \times n$ matrix called the *classical adjoint* or *adjunct* of A, written adj(A) or A'. For $i,j \in [n]$, we set $A'(i,j) = (-1)^{i+j} \det(A[j|i])$. So the i,j-entry of A' is the determinant of the matrix obtained by deleting row j and column i of A, times a certain sign.

Classical Adjoint Formula. For all $A \in M_n(R)$, $AA' = A'A = \det(A)I_n$.

Proof. We show $AA' = \det(A)I_n$ in two steps. First, consider the i,j-entry of AA', where $i,j \in [n]$ and $i \neq j$. The definition of matrix multiplication gives

$$(AA')(i,j) = \sum_{k=1}^{n} A(i,k)A'(k,j) = \sum_{k=1}^{n} A(i,k)(-1)^{j+k} \det(A[j|k]).$$

Let C be the matrix obtained from A by replacing row j of A by row i. We see that $C[j|k] = A[j|k]$ and $C(j,k) = A(i,k)$ for all $k \in [n]$. Since C has two equal rows, $\det(C) = 0$. On the other hand, the Laplace expansion along row j of C gives

$$0 = \det(C) = \sum_{k=1}^{n} C(j,k)(-1)^{j+k} \det(C[j|k]) = \sum_{k=1}^{n} A(i,k)(-1)^{j+k} \det(A[j|k]).$$

Comparing to the previous formula, we see that $(AA')(i,j) = 0 = (\det(A)I_n)(i,j)$.

The second stage is to compute $(AA')(i,i)$, where $i \in [n]$. Here, we find that

$$(AA')(i,i) = \sum_{k=1}^{n} A(i,k)A'(k,i) = \sum_{k=1}^{n} A(i,k)(-1)^{i+k} \det(A[i|k]) = \det(A),$$

where the last equality is Laplace expansion along row i. Thus, $(AA')(i,i) = \det(A) = (\det(A)I_n)(i,i)$, as needed.

The companion formula $A'A = I_n$ can be proved similarly, using Laplace expansions along columns. Alternatively, it follows from the definitions that $(A')^{\mathrm{T}} = (A^{\mathrm{T}})'$. So the result already proved (applied to A^{T}) gives

$$(A'A)^{\mathrm{T}} = A^{\mathrm{T}}(A')^{\mathrm{T}} = A^{\mathrm{T}}(A^{\mathrm{T}})' = \det(A^{\mathrm{T}})I_n = \det(A)I_n.$$

Transposing both sides gives $A'A = \det(A)I_n^{\mathrm{T}} = \det(A)I_n$. $\square$

One consequence of the Classical Adjoint Formula is the following criterion for invertibility of a matrix.

Determinant Criterion for Matrix Invertibility. Suppose R is a commutative ring and $A \in M_n(R)$. The matrix A has an inverse in $M_n(R)$ if and only if $\det(A)$ has a multiplicative inverse in R.

Proof. First, assume $\det(A)$ is an invertible element of R. Multiplying the Classical Adjoint Formula by the inverse of $\det(A)$ we see that A has an inverse in $M_n(R)$ given by $A^{-1} = \det(A)^{-1}A'$. For $i, j \in [n]$, an explicit formula for the i, j-entry of A^{-1} is

$$A^{-1}(i, j) = (-1)^{i+j} \det(A[j|i]) \det(A)^{-1}. \tag{5.11}$$

For the converse, suppose A has an inverse B in $M_n(R)$, so $AB = I_n = BA$. Taking determinants and using the Product Formula, we get $\det(A)\det(B) = \det(I_n) = \det(B)\det(A)$. Since $\det(I_n) = 1_R$, we see that $\det(B)$ is a two-sided inverse for $\det(A)$ in the ring R. $\square$

For example, consider the matrix $A = \begin{bmatrix} 0 & 2 & 1 \\ 3 & 6 & -3 \\ 2 & 5 & 1 \end{bmatrix} \in M_3(\mathbb{Z})$. The classical adjoint

of A is $A' = \begin{bmatrix} 21 & 3 & -12 \\ -9 & -2 & 3 \\ 3 & 4 & -6 \end{bmatrix}$. For instance, $A'(2, 3) = (-1)^{2+3} \det \begin{bmatrix} 0 & 1 \\ 3 & -3 \end{bmatrix} = 3$. We

found earlier that $\det(A) = -15$, and you can check that $AA' = A'A = -15I_3$. Since -15 is not invertible in the ring $\mathbb{Z}$, A has no inverse in the ring $M_3(\mathbb{Z})$. On the other hand, A is invertible in the ring $M_3(\mathbb{R})$, with inverse $A^{-1} = (-1/15)A'$.

5.12   Cramer's Rule

One application of the formula for A^{-1} is Cramer's Rule for solving a nonsingular system of n linear equations in n unknowns (*nonsingular* means the system has a unique solution). Write this system in matrix notation as $A\mathbf{x} = \mathbf{b}$, where $A \in M_n(R)$ has invertible determinant, $\mathbf{x} \in R^n$ is a column vector of unknowns, and $\mathbf{b} \in R^n$ is a given column vector. To solve for a particular unknown x_i, multiply both sides by A^{-1} on the left, obtaining $\mathbf{x} = A^{-1}\mathbf{b}$, and take the ith component:

$$x_i = (A^{-1}\mathbf{b})_i = \sum_{k=1}^{n} A^{-1}(i, k)b_k.$$

Using the formula (5.11) for A^{-1}, we get

$$x_i = \det(A)^{-1} \sum_{k=1}^{n} b_k(-1)^{i+k} \det(A[k|i]).$$

To obtain Cramer's Rule, let C_i be the matrix whose columns are

$$(A^{[1]}, \ldots, A^{[i-1]}, \mathbf{b}, A^{[i+1]}, \ldots, A^{[n]}).$$

Evidently, $C_i[k|i] = A[k|i]$ and $C_i(k, i) = b_k$ for all $k \in [n]$. Laplace expansion along column i of C_i therefore gives

$$\det(C_i) = \sum_{k=1}^{n} C_i(k, i)(-1)^{k+i} \det(C_i[k|i]) = \sum_{k=1}^{n} b_k(-1)^{i+k} \det(A[k|i]).$$

Comparing this to the formula for x_i, we see that $x_i = \det(A)^{-1} \det(C_i)$ for all $i \in [n]$. To summarize, we have proved:

Cramer's Rule. Suppose $A \in M_n(R)$, $\det(A)$ is invertible in R, $\mathbf{b}$ is a given column vector, and $\mathbf{x}$ is the unique solution to $A\mathbf{x} = \mathbf{b}$. For all $i \in [n]$, we can compute x_i by replacing column i of A by $\mathbf{b}$, taking the determinant of this matrix, and dividing by $\det(A)$.

We stress that Cramer's Rule only applies when the coefficient matrix A is square with invertible determinant. Furthermore, when R is a field, Gaussian elimination is often a more efficient way to solve the linear system compared to the multiple determinant evaluations required by Cramer's Rule.

As an example of Cramer's Rule, let us solve the linear system of equations

$$\begin{cases} & 2x_2 & +x_3 & = & 9 \\ 3x_1 & +6x_2 & -3x_3 & = & -2 \\ 2x_1 & +5x_2 & +x_3 & = & 4. \end{cases}$$

This system has the form $A\mathbf{x} = \mathbf{b}$, where $A = \begin{bmatrix} 0 & 2 & 1 \\ 3 & 6 & -3 \\ 2 & 5 & 1 \end{bmatrix}$ and $\mathbf{b} = \begin{bmatrix} 9 \\ -2 \\ 4 \end{bmatrix}$. Earlier, we calculated $\det(A) = -15$. Hence, Cramer's Rule gives

$$x_1 = \frac{1}{-15} \det \begin{bmatrix} 9 & 2 & 1 \\ -2 & 6 & -3 \\ 4 & 5 & 1 \end{bmatrix} = \frac{135}{-15} = -9;$$

$$x_2 = \frac{1}{-15} \det \begin{bmatrix} 0 & 9 & 1 \\ 3 & -2 & -3 \\ 2 & 4 & 1 \end{bmatrix} = \frac{-65}{-15} = \frac{13}{3};$$

$$x_3 = \frac{1}{-15} \det \begin{bmatrix} 0 & 2 & 9 \\ 3 & 6 & -2 \\ 2 & 5 & 4 \end{bmatrix} = \frac{-5}{-15} = \frac{1}{3}.$$

5.13 Product Formula via Computations

Determinant Product Formula. Let R be a commutative ring. For all $A, B \in M_n(R)$, $\det(AB) = \det(A) \det(B)$.

Proof. Fix $A, B \in M_n(R)$. Replacing A by AB in the definition (5.1) gives:

$$\det(AB) = \sum_{f \in S_n} \text{sgn}(f)(AB)(f(1), 1)(AB)(f(2), 2) \cdots (AB)(f(n), n).$$

Next, use the definition of matrix multiplication to write each factor as

$$(AB)(f(j), j) = \sum_{k_j=1}^{n} A(f(j), k_j) B(k_j, j).$$

Putting these expressions into the previous formula and using the distributive law repeatedly (see Exercise 64), we get:

$$\det(AB) = \sum_{f \in S_n} \sum_{k_1=1}^{n} \cdots \sum_{k_n=1}^{n} \operatorname{sgn}(f) A(f(1), k_1) B(k_1, 1) \cdots A(f(n), k_n) B(k_n, n).$$

We reorder the sums and products, using commutativity of addition and multiplication in R:

$$\det(AB) = \sum_{k_1=1}^{n} \cdots \sum_{k_n=1}^{n} \sum_{f \in S_n} \operatorname{sgn}(f) A(f(1), k_1) \cdots A(f(n), k_n) B(k_1, 1) \cdots B(k_n, n).$$

We can regard the n outer summations over integers $k_j \in [n]$ as a single summation over lists $\mathbf{k} = [k_1, k_2, \ldots, k_n] \in [n]^n$. Also, we can factor out the values $B(k_j, j)$ that do not depend on the inner summation index f. We get:

$$\det(AB) = \sum_{\mathbf{k} \in [n]^n} B(k_1, 1) \cdots B(k_n, n) \left[\sum_{f \in S_n} \operatorname{sgn}(f) A(f(1), k_1) \cdots A(f(n), k_n) \right].$$

Given a fixed list $\mathbf{k} \in [n]^n$, note that the term in brackets is the determinant of the matrix C whose columns are $(A^{[k_1]}, \ldots, A^{[k_n]})$. To see why, note that $C(i, j) = A(i, k_j)$ for $i, j \in [n]$, and so

$$\det(C) = \sum_{f \in S_n} \operatorname{sgn}(f) \prod_{j \in [n]} C(f(j), j) = \sum_{f \in S_n} \operatorname{sgn}(f) \prod_{j \in [n]} A(f(j), k_j).$$

If $k_i = k_j$ for any $i \neq j$, then $\det(C) = 0$ because C has two equal columns. Therefore, we can discard all summands of the outer summation indexed by lists $\mathbf{k} = [k_1, \ldots, k_n]$ with a repeated entry. The surviving summands are indexed by permutations $\mathbf{k} = [k_1, \ldots, k_n] \in S_n$. Suppose we sort the list $[k_1, \ldots, k_n]$ into increasing order by interchanging adjacent elements, and at the same time interchanging the corresponding adjacent columns in C. Each such interchange multiplies $\det(C)$ by -1. We need $\operatorname{inv}(\mathbf{k})$ interchanges to reach the list $[1, 2, \ldots, n]$ (see §2.6), and at that point we have transformed C to the matrix A whose columns are in their original order. Therefore, $\det(C) = (-1)^{\operatorname{inv}(\mathbf{k})} \det(A) = \operatorname{sgn}(\mathbf{k}) \det(A)$. Replacing the term in square brackets by this formula, and factoring $\det(A)$ out of the main sum, we now see that

$$\det(AB) = \det(A) \sum_{\mathbf{k} \in S_n} \operatorname{sgn}(\mathbf{k}) B(k_1, 1) \cdots B(k_n, n).$$

The remaining sum is precisely the definition of $\det(B)$, so the proof is finished. $\square$

A short but abstract proof of the Determinant Product Formula is given in §20.17.

5.14 Cauchy–Binet Formula

The Cauchy–Binet formula is a generalization of the Determinant Product Formula to products of rectangular matrices.

Cauchy–Binet Formula. Suppose R is a commutative ring, $A \in M_{m,n}(R)$, $B \in M_{n,m}(R)$, and $m \leq n$. Let J be the set of all strictly increasing lists $\mathbf{j} = [j_1, j_2, \ldots, j_m] \in [n]^m$. Then

$$\det(AB) = \sum_{\mathbf{j} \in J} \det(A^{[j_1]}, A^{[j_2]}, \ldots, A^{[j_m]}) \det(B_{[j_1]}, B_{[j_2]}, \ldots, B_{[j_m]}).$$

Here, $A^{[j_k]}$ is the j_kth column of A, and $B_{[j_k]}$ is the j_kth row of B. So every determinant appearing in the formula is the determinant of an $m \times m$ matrix. If $m = n$, then J consists of the single list $[1, 2, \ldots, n]$, and the Cauchy–Binet Formula reduces to the Determinant Product Formula for square matrices.

Proof. We generalize the proof given in the last section. First, the definitions of determinants and matrix products give

$$\det(AB) = \sum_{f \in S_m} \text{sgn}(f) \prod_{i \in [m]} (AB)(f(i), i) = \sum_{f \in S_m} \text{sgn}(f) \prod_{i \in [m]} \sum_{k_i \in [n]} A(f(i), k_i) B(k_i, i).$$

Using the distributive law repeatedly (see Exercise 64), we get

$$\det(AB) = \sum_{f \in S_m} \sum_{k_1 \in [n]} \cdots \sum_{k_m \in [n]} \text{sgn}(f) \prod_{i \in [m]} A(f(i), k_i) \prod_{i \in [m]} B(k_i, i).$$

Now we move the summation over f inside the other sums, replace the multiple summations over indices $k_1, \ldots, k_m$ by a single summation over lists $\mathbf{k} = [k_1, \ldots, k_m] \in [n]^m$, and rearrange factors in the resulting summation:

$$\det(AB) = \sum_{\mathbf{k} \in [n]^m} \prod_{i \in [m]} B(k_i, i) \left[\sum_{f \in S_m} \text{sgn}(f) \prod_{i \in [m]} A(f(i), k_i) \right].$$

We recognize the term in brackets as the definition of $\det(A^{[k_1]}, \ldots, A^{[k_m]})$. If $k_i = k_j$ for some $i \neq j$, then the matrix $(A^{[k_1]}, \ldots, A^{[k_m]})$ has two equal columns, so its determinant is zero. Discarding these summands, we now have

$$\det(AB) = \sum_{\substack{\mathbf{k} \in [n]^m: \\ k_i \text{ distinct}}} \det(A^{[k_1]}, \ldots, A^{[k_m]}) \prod_{i \in [m]} B(k_i, i).$$

Given a list $\mathbf{k} = [k_1, \ldots, k_m]$ of distinct elements of $[n]$, let $\text{sort}(\mathbf{k})$ be the list obtained by sorting $\mathbf{k}$ into increasing order. Note that $\mathbf{j} = \text{sort}(\mathbf{k})$ belongs to J. Define $\text{inv}(\mathbf{k})$ and $\text{sgn}(\mathbf{k})$ as in the case $m = n$ (see §2.6). It takes $\text{inv}(\mathbf{k})$ interchanges of adjacent columns to turn the matrix $(A^{[k_1]}, \ldots, A^{[k_m]})$ into the matrix $(A^{[j_1]}, \ldots, A^{[j_m]})$. Therefore, $\det(A^{[k_1]}, \ldots, A^{[k_m]}) = \text{sgn}(\mathbf{k}) \det(A^{[j_1]}, \ldots, A^{[j_m]})$. Grouping summands in the formula for

$\det(AB)$ based on the value of $\mathbf{j} = \mathrm{sort}(\mathbf{k})$, we obtain

$$\det(AB) = \sum_{\substack{\mathbf{j}\in J}} \sum_{\substack{\mathbf{k}\in[n]^m: \\ \mathrm{sort}(\mathbf{k})=\mathbf{j}}} \det(A^{[k_1]},\ldots,A^{[k_m]}) \prod_{i\in[m]} B(k_i,i)$$

$$= \sum_{\mathbf{j}\in J} \det(A^{[j_1]},\ldots,A^{[j_m]}) \left[\sum_{\substack{\mathbf{k}\in[n]^m: \\ \mathrm{sort}(\mathbf{k})=\mathbf{j}}} \mathrm{sgn}(\mathbf{k}) \prod_{i\in[m]} B(k_i,i)\right].$$

To finish, compare the term in square brackets to the definition of $\det(B_{[j_1]},\ldots,B_{[j_m]})$. The formula in brackets looks like the original definition of the determinant of a matrix, except now we are indexing the rows of the matrix by the increasing sequence $j_1 < \cdots < j_m$ instead of the standard indexing sequence $1 < 2 < \cdots < m$. This renaming of indices makes no difference when we calculate $\mathrm{inv}(\mathbf{k})$ and $\mathrm{sgn}(\mathbf{k})$. Therefore, the term in brackets is none other than $\det(B_{[j_1]},\ldots,B_{[j_m]})$. (To prove this formally, change the summation variable from a sum over the set of bijections $\mathbf{k} : [m] \to \{j_1,\ldots,j_m\}$ to a sum over the set S_m of bijections from $[m]$ to $[m]$.) To summarize,

$$\det(AB) = \sum_{\mathbf{j}\in J} \det(A^{[j_1]}, A^{[j_2]},\ldots,A^{[j_m]}) \det(B_{[j_1]}, B_{[j_2]},\ldots,B_{[j_m]}). \qquad \square$$

For another proof, see Exercise 56 in Chapter 20.

5.15 Cayley–Hamilton Theorem

Given A in $M_n(R)$, the *characteristic polynomial of A* is defined as $\chi_A = \det(xI_n - A)$, which is a polynomial in $R[x]$.

Cayley–Hamilton Theorem. Suppose R is a commutative ring. For all $A \in M_n(R)$, $\chi_A(A) = 0$. More specifically, if $\chi_A = \sum_{i\geq 0} c_i x^i \in R[x]$ with $c_i \in R$, then $\sum_{i\geq 0} c_i A^i$ is the zero matrix.

We give a tricky, technical proof of this theorem here. For other proofs, see §8.14 and §18.19. We begin with a short, but incorrect, proof attempt. Fix $A \in M_n(R)$ and let $p(x) = \chi_A(x) = \det(xI_n - A)$ be the characteristic polynomial of A. Suppose we try to compute $p(A)$ by simply replacing the variable x (inside the determinant) by the matrix A. We apparently obtain $p(A) = \det(AI_n - A) = \det(0_n) = 0$, giving the conclusion of the Cayley–Hamilton Theorem. However, something must be wrong with this calculation, since $p(A)$ is an $n \times n$ matrix, but $\det(0_n) = 0_R$ is a scalar in R!

The correct proof needs to be much more careful about the evaluation of the formal variable x appearing in the characteristic polynomial. First recall the *universal mapping property (UMP) for polynomials* proved in §3.5: given two commutative rings R and S, a ring homomorphism $h : R \to S$, and $c \in S$, there exists a unique ring homomorphism $H : R[x] \to S$ such that $H(r) = h(r)$ for all $r \in R$, and $H(x) = c$.

We would like to apply this result taking c to be the given matrix A. Another technical problem immediately occurs, since A belongs to the non-commutative ring $M_n(R)$, but S (in the UMP) is required to be commutative. To address this issue, define S to be the set of all matrices in $M_n(R)$ of the form $r_0 I_n + r_1 A + r_2 A^2 + \cdots + r_k A^k$ for some $k \geq 0$ and $r_0, r_1, \ldots, r_k \in R$. By Exercise 61, S is a commutative subring of $M_n(R)$ containing I_n and

A. We define $h : R \to S$ by setting $h(r) = rI_n$ for $r \in R$; Exercise 61 shows that h is a ring homomorphism. The UMP says there is a ring homomorphism $H : R[x] \to S$ extending h and sending x to A. More specifically, for any $r_0, r_1, \ldots, r_k \in R$,

$$H(r_0 + r_1 x + \cdots + r_k x^k) = r_0 I_n + r_1 A + \cdots + r_k A^k \in S.$$

We now have a precise way of talking about $p(A)$: namely, $p(A) = H(p)$ is the image of p under the ring homomorphism H that extends h and sends x to A.

Since $p = \det(xI_n - A)$, we need to find $H(p) = H(\det(xI_n - A))$. For this, we need another technical digression. Suppose $F : T \to U$ is a ring homomorphism between commutative rings T and U. We claim that there is an associated ring homomorphism $F^* : M_n(T) \to M_n(U)$ such that $F^*(B)_{i,j} = F(B_{i,j})$ for all $B \in M_n(T)$ and all $i, j \in [n]$. This definition says that F^* acts on a matrix B by applying F to each entry $B_{i,j}$. We claim also that $F(\det(B)) = \det(F^*(B))$ for all $B \in M_n(T)$. See Exercise 62 for the proofs of these claims.

Applying this discussion to the ring homomorphism $H : R[x] \to S$, we obtain a ring homomorphism $H^* : M_n(R[x]) \to M_n(S)$ that acts by applying H to every entry of a matrix in $M_n(R[x])$. Let $C = H^*(xI_n - A)$. Then

$$\det(C) = \det(H^*(xI_n - A)) = H(\det(xI_n - A)) = H(p) = p(A). \tag{5.12}$$

Note that $C \in M_n(S)$ is an $n \times n$ matrix with entries in S, so each entry of C is itself an $n \times n$ matrix built up from powers of A. This is what makes this proof tricky!

For example, suppose $A = \begin{bmatrix} r & s \\ t & u \end{bmatrix} \in M_2(R)$. Then $xI_2 - A = \begin{bmatrix} x - r & -s \\ -t & x - u \end{bmatrix}$ is in $M_2(R[x])$ and $p = \chi_A = \det(xI_2 - A) = (x - r)(x - u) - st = x^2 - (r + u)x + (ru - st)$. So

$$p(A) = H(p) = A^2 - (r + u)A + (ru - st)I_2.$$

On the other hand, we get $C = H^*(xI_2 - A)$ by applying H to every entry of $xI_2 - A$. This produces the matrix of matrices

$$C = \begin{bmatrix} A - rI_2 & -sI_2 \\ -tI_2 & A - uI_2 \end{bmatrix} = \begin{bmatrix} \begin{bmatrix} 0 & s \\ t & u - r \end{bmatrix} & \begin{bmatrix} -s & 0 \\ 0 & -s \end{bmatrix} \\ \begin{bmatrix} -t & 0 \\ 0 & -t \end{bmatrix} & \begin{bmatrix} r - u & s \\ t & 0 \end{bmatrix} \end{bmatrix}.$$

Working in the commutative ring S, we verify (5.12) by computing

$$\det(C) = (A - rI_2)(A - uI_2) - (-sI_2)(-tI_2) = A^2 - (r + u)A + (ru - st)I_2 = p(A).$$

Return to the case of general n and A. By definition of C, for all $i, j \in [n]$ with $j \neq i$, $C(i, i) = A - A(i, i)I_n$ and $C(i, j) = -A(i, j)I_n$. We know that $p(A) = H(p) = \det(C)$ by (5.12), so our task is to prove that $\det(C)$ is the zero matrix in S. For reasons that emerge shortly, we instead prove the equivalent statement $\det(C^T) = 0$.

For a matrix $A \in M_n(R)$ with entries in R, we compute the vector-matrix product $\mathbf{w}A$ for any row vector $\mathbf{w} \in R^n$ via the formula $(\mathbf{w}A)_j = \sum_{i=1}^n w_i A(i, j)$ for $j \in [n]$. Since $C^T \in M_n(S)$ is a matrix of matrices, we instead compute an expression of the form $\mathbf{z}C^T$, where $\mathbf{z}$ is a row vector of row vectors. More precisely, suppose $U \in M_n(S)$ and $\mathbf{z} = [\mathbf{z}_1 \ \mathbf{z}_2 \ \cdots \ \mathbf{z}_n] \in (R^n)^n$, where each $\mathbf{z}_i$ is a row vector in R^n. Then $\mathbf{z}U \in (R^n)^n$ is the row vector whose jth component is the row vector $\sum_{i=1}^n \mathbf{z}_i U(i, j)$. You can verify the associativity property $(\mathbf{z}U)V = \mathbf{z}(UV)$, valid for all $\mathbf{z} \in (R^n)^n$ and $U, V \in M_n(S)$ (Exercise 61).

Continuing the proof, let us take $\mathbf{z} = [\mathbf{e}_1 \ \mathbf{e}_2 \ \cdots \ \mathbf{e}_n]$, where each $\mathbf{e}_i$ is the row vector of length n with 1_R in position i and 0_R in all other positions. We claim $\mathbf{z}C^\mathrm{T} = \mathbf{0}$, where $\mathbf{0}$ denotes the row vector consisting of n row vectors all equal to zero. To see this, note that the jth row vector appearing in $\mathbf{z}C^\mathrm{T}$ is

$$
\sum_{i=1}^{n} \mathbf{e}_i C^\mathrm{T}(i,j) = \mathbf{e}_j C(j,j) + \sum_{i:\, i\neq j} \mathbf{e}_i C(j,i)
$$

$$
= \mathbf{e}_j(A - A(j,j)I_n) + \sum_{i:\, i\neq j} \mathbf{e}_i(-A(j,i)I_n) = \mathbf{e}_j A - \sum_{i=1}^{n} A(j,i)\mathbf{e}_i.
$$

The row vectors $\mathbf{e}_j A$ and $\sum_{i=1}^{n} A(j,i)\mathbf{e}_i$ are both equal to the jth row of A, so their difference is indeed zero. This holds for all j, so $\mathbf{z}C^\mathrm{T} = \mathbf{0}$ as claimed.

To finish, multiply the equation $\mathbf{z}C^\mathrm{T} = \mathbf{0}$ on the right by the classical adjoint $\mathrm{adj}(C^\mathrm{T})$. The right side is still $\mathbf{0}$ in $(R^n)^n$. The left side becomes

$$
(\mathbf{z}C^\mathrm{T})\,\mathrm{adj}(C^\mathrm{T}) = \mathbf{z}(C^\mathrm{T}\,\mathrm{adj}(C^\mathrm{T})) = \mathbf{z}(\det(C^\mathrm{T})I_n).
$$

Using the definition of $\mathbf{z}U$ one more time, we see that the n row vectors appearing in $\mathbf{0} = \mathbf{z}(\det(C^\mathrm{T})I_n)$ are $\mathbf{e}_1 \det(C^\mathrm{T}), \ldots, \mathbf{e}_n \det(C^\mathrm{T})$, which are precisely the n rows of $\det(C^\mathrm{T})$. Since all these row vectors are zero, the matrix $\det(C^\mathrm{T}) = \det(C)$ must be zero, completing the proof of the Cayley–Hamilton Theorem.

Continuing the 2×2 example, we have

$$
\mathbf{z}C^\mathrm{T} = [[1\ 0]\ [0\ 1]]\begin{bmatrix} A - rI_2 & -tI_2 \\ -sI_2 & A - uI_2 \end{bmatrix}.
$$

The first component of $\mathbf{z}C^\mathrm{T}$ is

$$
[1\ 0](A - rI_2) + [0\ 1](-sI_2) = [r\ s] - [r\ 0] - [0\ s] = [0\ 0].
$$

Similarly, the second component of $\mathbf{z}C^\mathrm{T}$ is

$$
[1\ 0](-tI_2) + [0\ 1](A - uI_2) = -[t\ 0] + [t\ u] - [0\ u] = [0\ 0].
$$

Multiplying on the right by $\mathrm{adj}(C^\mathrm{T})$ gives

$$
[[0\ 0]\ [0\ 0]] = (\mathbf{z}C^\mathrm{T})\,\mathrm{adj}(C^\mathrm{T}) = \mathbf{z}(\det(C^\mathrm{T})I_2) = [[1\ 0]\ [0\ 1]]\begin{bmatrix} \det(C^\mathrm{T}) & 0 \\ 0 & \det(C^\mathrm{T}) \end{bmatrix}.
$$

Expanding the far right side shows that row 1 and row 2 of $\det(C^\mathrm{T})$ are $[0\ 0]$, so $\det(C^\mathrm{T}) = \det(C)$ is the 2×2 zero matrix.

5.16 Permanents

The *permanent* of a square matrix $A \in M_n(R)$ is defined by the formula

$$
\mathrm{per}(A) = \sum_{f \in S_n} \prod_{j \in [n]} A(f(j), j).
$$

This formula comes from the defining formula (5.1) for $\det(A)$ by omitting the sign factor $\operatorname{sgn}(f)$. The permanent of A is the sum of $n!$ terms, each of which is a product of n entries of A, which are chosen from distinct rows and columns in all possible ways.

Many of the results in this chapter extend without difficulty to permanents, by deleting all occurrences of $\operatorname{sgn}(f)$ in the relevant proofs. For example, the changes of variable $g = f^{-1}$ followed by $j = g(i)$ show that

$$\operatorname{per}(A) = \sum_{g \in S_n} \prod_{i \in [n]} A(i, g(i)) = \operatorname{per}(A^{\mathsf{T}}).$$

Similarly, you can prove the following facts about permanents.
(a) For $A \in M_n(\mathbb{C})$, $\operatorname{per}(\overline{A}) = \operatorname{per}(A^*) = \overline{\operatorname{per}(A)}$.
(b) $\operatorname{per} : (R^n)^n \to R$ is an R-multilinear function of the n rows of A.
(c) $\operatorname{per} : (R^n)^n \to R$ is an R-multilinear function of the n columns of A.
(d) The permanent of a matrix with a row (or column) of zeroes is zero.
(e) The permanent of a triangular or diagonal matrix is the product of the diagonal entries.
(f) Multiplying one row (or column) of A by $c \in R$ multiplies the permanent by c.
(g) Multiplying the entire $n \times n$ matrix by c multiplies the permanent by c^n.

Instead of the alternating property of the determinant, the permanent has the following *symmetry property*:
(h) Permuting the rows (or columns) of A does not change the permanent.
To prove (h), it suffices to show $\operatorname{per}(B) = \operatorname{per}(A)$ when B is obtained from A by interchanging row i and row j, since any permutation of rows can be achieved by a sequence of such interchanges. Change the summation variable from f to $g = f \circ (i, j)$. Since $g(i) = f(j)$ and $g(j) = f(i)$, we find that

$$\operatorname{per}(A) = \sum_{f \in S_n} \prod_{k \in [n]} A(k, f(k)) = \sum_{f \in S_n} A(i, f(i)) A(j, f(j)) \prod_{k \neq i, j} A(k, f(k))$$

$$= \sum_{g \in S_n} A(i, g(j)) A(j, g(i)) \prod_{k \neq i, j} A(k, g(k)) = \sum_{g \in S_n} B(j, g(j)) B(i, g(i)) \prod_{k \neq i, j} B(k, g(k))$$

$$= \sum_{g \in S_n} \prod_{k \in [n]} B(k, g(k)) = \operatorname{per}(B).$$

Unlike the case of determinants, it is not true that the permanent of a matrix with two equal rows (or columns) must be zero. For example, the permanent of $\begin{bmatrix} 1 & 1 \\ 1 & 1 \end{bmatrix}$ is $1+1 = 2$. In other words, the multilinear permanent function is not alternating. Consequently, those properties of determinants whose proofs invoked the alternating property are no longer valid for permanents. For instance, if B is obtained from A by adding c times row j of A to row i of A, we cannot conclude that $\operatorname{per}(B) = \operatorname{per}(A)$. Unfortunately, this means that we can no longer use Gaussian elimination as an efficient method for computing permanents as we did for determinants. The Laplace expansions for determinants work for permanents after omitting the sign factors. For instance, for any $i \in [n]$,

$$\operatorname{per}(A) = \sum_{k=1}^{n} A(i, k) \operatorname{per}(A[i|k]).$$

(The general case can be deduced from the special case $i = n$ by using the symmetry of the permanent to reorder rows.) However, the recursive Laplace expansion still requires the evaluation of $n!$ terms to compute the permanent of A. The formula $AA' = \det(A)I_n = A'A$ does not generalize to permanents, and the product formula $\operatorname{per}(AB) = \operatorname{per}(A) \operatorname{per}(B)$ does not hold in general.

5.17 Summary

Here are the main facts we proved about the determinant of a square $n \times n$ matrix A with entries in a commutative ring R.

1. *Defining Formulas for the Determinant:*
   $\det(A) = \sum_{f \in S_n} \operatorname{sgn}(f) \prod_{i=1}^{n} A(f(i), i) = \sum_{f \in S_n} \operatorname{sgn}(f) \prod_{i=1}^{n} A(i, f(i))$.

2. *Transpose and Conjugation:* $\det(A^{\mathrm{T}}) = \det(A)$.
   For complex matrices, $\det(\overline{A}) = \overline{\det(A)} = \det(A^*)$.

3. *Diagonal and Triangular Matrices:* If A is upper-triangular, lower-triangular, or diagonal, then $\det(A) = \prod_{i=1}^{n} A(i, i)$. So $\det(I_n) = 1_R$ and $\det(0_n) = 0_R$.

4. *Multilinearity and the Alternating Property:* We can view $\det : (R^n)^n \to R$ as a function of the n rows of A or the n columns of A: $\det(A) = \det(A_{[1]}, \ldots, A_{[n]}) = \det(A^{[1]}, \ldots, A^{[n]})$. Either way, $\det$ is an R-multilinear function of its n inputs. In other words, if all rows other than row i are fixed, then the function $D(v) = \det(A_1, \ldots, A_{i-1}, v, A_{i+1}, \ldots, A_n)$ (for $v \in R^n$) is R-linear; similarly for columns. If a matrix A has two equal rows, two equal columns, a zero row, or a zero column, then $\det(A) = 0$.

5. *Elementary Operations and Determinants:* There are three elementary row and column operations on matrices, which may be executed by multiplying on the left or right by appropriate elementary matrices. First, multiplying one row of a matrix by $c \in R$ multiplies the determinant by c; the corresponding elementary matrix has determinant c (this matrix is elementary only for c invertible). Second, interchanging any two distinct rows of a matrix multiplies the determinant by -1_R; the corresponding elementary matrix has determinant -1_R. Third, adding a scalar multiple of one row of a matrix to another row leaves the determinant unchanged; the corresponding elementary matrix has determinant 1_R. Combining these facts with row-reduction via Gaussian elimination, we obtain an efficient method for computing determinants when R is a field.

6. *Product Formula:* For all $A, B \in M_n(R)$, $\det(AB) = \det(A)\det(B)$.

7. *Laplace Expansions:* For all $i, j \in [n]$,

$$\det(A) = \sum_{j=1}^{n} A(i,j)(-1)^{i+j} \det(A[i|j]) = \sum_{i=1}^{n} A(i,j)(-1)^{i+j} \det(A[i|j]),$$

   where $A[i|j]$ is the matrix obtained by deleting row i and column j of A.

8. *Classical Adjoint:* The classical adjoint of $A \in M_n(R)$ is the matrix $A' = \operatorname{adj}(A)$ whose i, j-entry is $(-1)^{i+j} \det(A[j|i])$. We have $A'A = \det(A)I_n = AA'$.

9. *Explicit Formula for Inverses:* A matrix A is invertible in $M_n(R)$ iff $\det(A)$ is invertible in the commutative ring R. In this case, $A^{-1} = \det(A)^{-1} \operatorname{adj}(A)$, so

$$A^{-1}(i,j) = \det(A)^{-1}(-1)^{i+j} \det(A[j|i]).$$

10. *Cramer's Rule:* If $A\mathbf{x} = \mathbf{b}$ is a system of n linear equations in n unknowns whose coefficient matrix A is invertible, then the unique solution $\mathbf{x}$ has components $x_i = \det(A)^{-1} \det(C_i)$ for each $i \in [n]$, where C_i is the matrix obtained from A by substituting $\mathbf{b}$ for the ith column of A.

11. *Cauchy–Binet Formula:* If $A \in M_{m,n}(R)$, $B \in M_{n,m}(R)$, and $m \leq n$, then

$$\det(AB) = \sum_{1 \leq j_1 < j_2 < \cdots < j_m \leq n} \det(A^{[j_1]}, A^{[j_2]}, \ldots, A^{[j_m]}) \det(B_{[j_1]}, B_{[j_2]}, \ldots, B_{[j_m]}).$$

12. *Cayley–Hamilton Theorem:* The characteristic polynomial of A is $\chi_A(x) = \det(xI_n - A)$. If $\chi_A = \sum_{i \geq 0} c_i x^i$ with $c_i \in R$, then $\chi_A(A) = \sum_{i \geq 0} c_i A^i = 0$.

13. *Permanents:* For $A \in M_n(R)$, the permanent of A is

$$\mathrm{per}(A) = \sum_{f \in S_n} \prod_{j \in [n]} A(f(j), j) = \sum_{f \in S_n} \prod_{i \in [n]} A(i, f(i)) = \mathrm{per}(A^{\mathsf{T}}).$$

The permanent is a symmetric, R-multilinear function per $: (R^n)^n \to R$ whose n inputs are the rows (or columns) of A. We have $\mathrm{per}(A) = \prod_{i \in [n]} A(i, i)$ when A is triangular or diagonal. For each $k \in [n]$, the Laplace expansions

$$\mathrm{per}(A) = \sum_{j=1}^{n} A(k, j) \, \mathrm{per}(A[k|j]) = \sum_{i=1}^{n} A(i, k) \, \mathrm{per}(A[i|k])$$

are valid. However, the permanent is not alternating; we cannot use elementary row operations to evaluate permanents; and there is no product formula for permanents.

5.18   Exercises

Unless otherwise specified, assume R is a commutative ring in these exercises.

1. Use (5.1) to write a formula for $\det(A)$ when $A \in M_4(R)$.

2. Suppose $A \in M_5(R)$ satisfies $A(i, j) = 0_R$ for all $i, j \in [5]$ with $i+j$ odd. Use (5.1) to compute $\det(A)$.

3. Describe how to find the determinant of a matrix that has exactly one nonzero entry in every row and column. Illustrate your method by computing

$$\det \begin{bmatrix} 0 & 4 & 0 & 0 & 0 \\ 2 & 0 & 0 & 0 & 0 \\ 0 & 0 & 0 & 0 & 1 \\ 0 & 0 & 5 & 0 & 0 \\ 0 & 0 & 0 & 3 & 0 \end{bmatrix}.$$

4. Prove: for any upper-triangular matrix $A \in M_n(R)$, $\det(A) = \prod_{i=1}^{n} A(i, i)$.

5. Suppose $A \in M_n(R)$ satisfies $A(i, j) = 0_R$ for all $i, j \in [n]$ with $i + j > n + 1$. Find and prove a simple formula for $\det(A)$.

6. Suppose $A \in M_n(R)$ satisfies $A(i, j) = 0$ for all $i, j \in [n]$ with $i > j + 1$. In general, how many of the $n!$ terms in (5.1) might be nonzero for such an A?

7. Consider a matrix $U \in M_{n+m}(R)$ written in block form $U = \begin{bmatrix} A_{n \times n} & B_{n \times m} \\ 0_{m \times n} & D_{m \times m} \end{bmatrix}$, where A, B, C, D have the indicated sizes. Prove: $\det(U) = \det(A) \det(D)$.

8. Suppose $U \in M_{2k}(R)$ has the block form $\begin{bmatrix} A & B \\ C & D \end{bmatrix}$ where $A, B, C, D \in M_k(R)$ and $k \geq 2$. Is $\det(U) = \det(A)\det(D) - \det(C)\det(B)$ always true?

9. Call $U \in M_n(R)$ block *upper-triangular* iff there exist positive integers $s, k_1, \ldots, k_s$ such that $k_1 + \cdots + k_s = n$ and for all integers i, j, t with $k_1 + \cdots + k_{t-1} < i \leq k_1 + \cdots + k_t$ and all $j \leq k_1 + \cdots + k_{t-1}$, $U(i,j) = 0$. Draw a picture showing the block structure of U, and find a formula for $\det(U)$. (Use Exercise 7 and induction on s.)

10. Let $(G, \star)$ be any group. For $g \in G$, define maps L_g, R_g, and C_g from G to G by setting $L_g(x) = g \star x$, $R_g(x) = x \star g$, and $C_g(x) = g \star x \star g^{-1}$ for all $x \in G$.
(a) Prove L_g is a bijection by showing L_g is one-to-one and onto.
(b) Prove R_g is a bijection by showing $R_{g^{-1}}$ is a two-sided inverse of R_g.
(c) Prove C_g is a group isomorphism.
(d) Prove $I : G \to G$, given by $I(g) = g^{-1}$ for $g \in G$, is a bijection.
(e) If T is any ring, G is finite, and $f : G \to T$ is any function, use L_g to explain why $\sum_{x \in G} f(x) = \sum_{x \in G} f(g \star x)$ for each fixed $g \in G$. Write similar formulas based on the maps R_g, C_g, and I.

11. Assume $A \in M_n(\mathbb{R})$ satisfies $AA^T = I_n$. Prove $\det(A) \in \{+1, -1\}$.

12. Assume n is odd and $A \in M_n(\mathbb{R})$ satisfies $A^T = -A$. Prove $\det(A) = 0$.

13. Assume $A \in M_n(\mathbb{C})$ satisfies $AA^* = I_n$. Prove $|\det(A)| = 1$ and $A^*A = I_n$.

14. Assume $A \in M_n(\mathbb{R})$ satisfies $A^k = 0$ for some $k > 0$. Prove $\det(A) = 0$.

15. Give an example of a commutative ring R and $A \in M_2(R)$ where $AA^T = I_2$ but $\det(A) \notin \{+1, -1\}$.

16. Give an example of a commutative ring R and $A \in M_3(R)$ where $A^T = -A$ but $\det(A) \neq 0$.

17. Give an example of a commutative ring R and $A \in M_2(R)$ where $A \neq 0$ but $A^k = 0$ for some positive integer k.

18. Use Laplace expansions and induction on n to reprove that $\det(A) = \det(A^T)$ for all $A \in M_n(R)$.

19. Fix row vectors $A_1 = (2, 6, -1, -1)$, $A_2 = (3, 4, 0, -2)$, and $A_4 = (2, -3, 1, 5)$ in $\mathbb{R}^4$. Define $D : \mathbb{R}^4 \to \mathbb{R}$ by $D(v) = \det(A_1, A_2, v, A_4)$ for $v = (v_1, v_2, v_3, v_4) \in \mathbb{R}^4$. Write $D(v)$ explicitly in the form $c_1 v_1 + \cdots + c_4 v_4$ for some $c_i \in \mathbb{R}$, and confirm that D is an $\mathbb{R}$-linear function.

20. Repeat Exercise 19 using column vectors $A^{[2]} = [3\ 5\ 4]^T$, $A^{[3]} = [2\ 7\ -1]^T$, and $D : \mathbb{R}^3 \to R$ defined on column vectors $v = [v_1\ v_2\ v_3]^T \in \mathbb{R}^3$ by $D(v) = \det(v, A^{[2]}, A^{[3]})$.

21. In the formula $D(v) = \sum_{j=1}^n c_j v(j)$ from §5.6, find an expression for c_j involving a sign times a determinant of a submatrix of A.

22. Use elementary row operations to prove that if some row of $A \in M_n(R)$ is an R-linear combination of other rows of A, then $\det(A) = 0_R$. Deduce a corresponding result for columns.

23. Suppose the rows of a matrix $A \in M_n(R)$ are R-linearly dependent, so there exist $r_1, \ldots, r_n \in R$ (not all 0) with $r_1 A_{[1]} + \cdots + r_n A_{[n]} = 0$ in R^n. Give a specific example to show that $\det(A)$ need not be 0. (Compare to the previous exercise.)

24. Use Laplace expansions to reprove the fact that multiplying row i of $A \in M_n(R)$ by $c \in R$ replaces $\det(A)$ by $c\det(A)$.

25. Use Laplace expansions to reprove the fact that interchanging two rows of a matrix changes the sign of the determinant.

26. Use Laplace expansions and induction to reprove the fact that adding c times row i to row $j \neq i$ of $A \in M_n(R)$ does not change $\det(A)$.

27. Let $A_1, \ldots, A_n$ be row vectors in R^n and $c_{ij} \in R$ for each $i < j$ in $[n]$. Prove that $\det(A_1, \ldots, A_n)$ equals

$$\det\left(A_1 + \sum_{j>1} c_{1j}A_j, A_2 + \sum_{j>2} c_{2j}A_j, \ldots, A_i + \sum_{j>i} c_{ij}A_j, \ldots, A_n\right).$$

Give two proofs, one using using elementary row operations, and one using the Determinant Product Formula.

28. Prove: for all elementary matrices E, E^{T} is also an elementary matrix.

29. Prove or disprove: for all elementary matrices $E \in M_n(R)$ and all $k \in \mathbb{Z}$, E^k is also an elementary matrix.

30. Use the definition (5.1) to prove that $\det(A)$ is a multilinear function of the n columns of $A \in M_n(R)$. Prove that each elementary column operation has the expected effect on determinants, without invoking the corresponding facts about elementary row operations. Use this to prove that $\det(AE) = \det(A)\det(E)$ for all $A \in M_n(R)$ and all elementary matrices $E \in M_n(R)$.

31. Given $a, b \in R$, evaluate $\det(aI_n + bJ_n)$, where $J_n \in M_n(R)$ has every entry equal to 1_R.

32. Define $A \in M_n(\mathbb{R})$ by $A(i, j) = \min(i, j)$ for $i, j \in [n]$. Compute $\det(A)$.

33. Given distinct $r_1, r_2, \ldots, r_n$ in a field F, define $A \in M_{n+1}(F[x])$ by

$$\begin{aligned}
A(i, i) &= x & \text{for } i \in [n]; \\
A(i, n+1) &= 1 & \text{for } i \in [n+1]; \\
A(i, j) &= r_j & \text{for } i > j \text{ in } [n+1]; \\
A(i, j) &= r_{j-1} & \text{for } i < j \text{ in } [n].
\end{aligned}$$

Prove $\det(A) = (x - r_1)(x - r_2) \cdots (x - r_n)$ by showing that the determinant is a monic polynomial of degree n having every r_j as a root. (Apply Exercise 62 to certain evaluation homomorphisms.)

34. In the previous exercise, suppose $r_1, r_2, \ldots, r_n$ are not necessarily distinct elements of a commutative ring R. Does the formula for $\det(A)$ still hold?

35. Let $A \in M_n(F)$ satisfy $A(i, j) = 0$ for all $i \neq j$ in $[n-1]$. Compute $\det(A)$.

36. Compute the determinant of each matrix by using row operations to reduce the matrix to an upper-triangular matrix in the given ring.

(a) $\begin{bmatrix} 1 & 4 & 6 \\ 3 & 2 & 1 \\ -2 & 4 & 5 \end{bmatrix} \in M_3(\mathbb{R})$ (b) $\begin{bmatrix} 4 & 0 & 1 & 2 \\ 3 & 3 & 1 & 4 \\ 2 & 1 & 1 & 3 \\ 3 & 1 & 3 & 4 \end{bmatrix} \in M_4(\mathbb{Z}_5)$

$$(c) \quad \begin{bmatrix} 1 & 1 & 1 & 0 & 1 & 0 & 1 \\ 0 & 0 & 0 & 1 & 1 & 0 & 0 \\ 1 & 1 & 1 & 0 & 0 & 1 & 0 \\ 0 & 0 & 1 & 0 & 1 & 1 & 0 \\ 0 & 1 & 1 & 1 & 0 & 1 & 1 \\ 1 & 0 & 1 & 1 & 0 & 1 & 1 \\ 1 & 0 & 0 & 0 & 0 & 0 & 1 \end{bmatrix} \in M_7(\mathbb{Z}_2)$$

37. Compute the determinants of each matrix in Exercise 36 by repeated use of Laplace expansions along convenient rows and columns.

38. Find the characteristic polynomial of each matrix in Exercise 36, and verify by a direct computation that the Cayley–Hamilton Theorem holds for these matrices.

39. Compute the permanent of each matrix in Exercise 36.

40. **Vandermonde Determinant.** Given $x_1, \ldots, x_n \in R$, define $V \in M_n(R)$ by $V(i,j) = x_i^{n-j}$ for $i, j \in [n]$. Prove $\det(V) = \prod_{1 \le i < j \le n}(x_i - x_j)$. (Use elementary row and column operations and induction on n.)

41. Given a field F, distinct elements $a_0, \ldots, a_n$ in F, and $b_0, \ldots, b_n$ in F, prove there exist unique $c_0, \ldots, c_n \in F$ such that the polynomial $p = \sum_{i=0}^{n} c_i x^i \in F[x]$ satisfies $p(a_i) = b_i$ for $0 \le i \le n$. Do this by setting up a system of $n+1$ linear equations in $n+1$ unknowns, and using the Vandermonde matrix V and its determinant from Exercise 40. How is this result related to Lagrange's Interpolation Formula (§3.19)? Can you use that formula to describe the columns of V^{-1}?

42. Given $p = a_0 + a_1 x + a_2 x^2 + \cdots + a_{n-1} x^{n-1} + x^n \in R[x]$, define the *companion matrix* $C_p \in M_n(R)$ by setting $C_p(j+1, j) = 1$ for $j \in [n-1]$, $C_p(i, n) = -a_{i-1}$ for $i \in [n]$, and letting all other entries be zero. Prove that $p = \det(xI_n - C_p) = \chi_{C_p}$.

43. **Generalized Laplace Expansion.** Given $S \subseteq [n]$, let $\mathrm{inv}(S)$ be the number of ordered pairs (s, t) with $s \in S$, $t \in [n]$, $t \notin S$, and $s > t$. Let $\mathrm{sgn}(S) = (-1)^{\mathrm{inv}(S)}$. Fix a k-element subset I of $[n]$. Prove: for all $A \in M_n(R)$,

$$\det(A) = \mathrm{sgn}(I) \sum_{\substack{J \subseteq [n]: \\ |J|=k}} \mathrm{sgn}(J) \det(A[I|J]) \det(A[I^c|J^c]),$$

where $A[I|J]$ denotes the matrix in $M_{n-k}(R)$ obtained by erasing all rows in I and all columns in J, and $A[I^c|J^c]$ denotes the matrix in $M_k(R)$ obtained by erasing all rows not in I and all columns not in J. (Reduce to the case $I = [k]$.) Show that $\mathrm{sgn}(I)\,\mathrm{sgn}(J) = (-1)^{\sum_{i \in I} i + \sum_{j \in J} j}$. Show how to deduce the Laplace expansion of $\det(A)$ along row i as a special case.

44. Use Laplace expansions along columns to prove that $A'A = (\det A)I_n$ for all $A \in M_n(R)$.

45. Given $a, b, c, d \in R$, find the classical adjoint of $A = \begin{bmatrix} a & b \\ c & d \end{bmatrix}$. Use this to give an explicit formula for A^{-1}, when it exists.

46. In $M_3(\mathbb{R})$, dind the classical adjoint and the inverse of $\begin{bmatrix} 4 & 1 & -2 \\ 3 & 0 & 5 \\ 7 & -1 & 3 \end{bmatrix}$.

47. In $M_4(\mathbb{Z}_7)$, find the classical adjoint and (if possible) the inverse of

$$\begin{bmatrix} 2 & 3 & 1 & 0 \\ 1 & 5 & 3 & 1 \\ 2 & 2 & 6 & 4 \\ 3 & 4 & 5 & 1 \end{bmatrix}.$$

48. Use the explicit formula for A^{-1} to prove that if A is a unitriangular matrix (see Exercise 30 of Chapter 4), then A^{-1} exists and is also unitriangular.

49. Given $b_1, b_2, b_3 \in \mathbb{R}$, solve $\begin{bmatrix} 4 & 1 & -2 \\ 3 & 6 & -1 \\ 1 & 1 & 5 \end{bmatrix} \begin{bmatrix} x_1 \\ x_2 \\ x_3 \end{bmatrix} = \begin{bmatrix} b_1 \\ b_2 \\ b_3 \end{bmatrix}$ using Cramer's Rule.

50. Given the linear system of equations

$$\begin{cases} 2x_1 & +3x_2 & -x_3 & & -x_5 & = & 4 \\ 3x_1 & +5x_2 & +x_3 & +2x_4 & +x_5 & = & 0 \\ -x_1 & & & -3x_4 & +x_5 & = & 1 \\ & x_2 & +4x_3 & +x_4 & & = & -1 \\ & -2x_2 & -3x_3 & & +4x_5 & = & 6 \end{cases}$$

use Cramer's Rule to find x_3.

51. State and prove a version of Cramer's Rule for solving the system $\mathbf{x}A = \mathbf{b}$, where $A \in M_n(R)$ is invertible, $\mathbf{b} \in R^n$ is a given row vector, and $\mathbf{x} \in R^n$ is an unknown row vector.

52. Let $A \in M_n(R)$ have $\det(A) \neq 0_R$. Use Cramer's Rule and Laplace expansions to derive the explicit formula for the entries of A^{-1}.

53. Prove: if $A \in M_n(R)$ is invertible, then $\det(A^{-1}) = \det(A)^{-1}$ in R.

54. Matrices $A, B \in M_n(R)$ are called *similar* iff there exists an invertible $S \in M_n(R)$ with $B = S^{-1}AS$. Prove that similar matrices have the same trace, determinant, and characteristic polynomial.

55. For a field F, recall $\mathrm{GL}_n(F)$ is the group of invertible matrices $A \in M_n(F)$. Prove that $\mathrm{SL}_n(F) = \{A \in M_n(F) : \det(A) = 1_F\}$ is a normal subgroup of $\mathrm{GL}_n(F)$. $\mathrm{SL}_n(F)$ is called the *special linear group of degree n over F*.

56. Let $A = \begin{bmatrix} a & b \\ c & d \end{bmatrix}$ and $B = \begin{bmatrix} r & s \\ t & u \end{bmatrix}$. Expand the definition of $\det(AB)$ to get a sum of 8 terms, and show how terms cancel to produce a sum of 4 terms equal to $\det(A)\det(B)$. (This illustrates the proof in §5.13.)

57. Let $A, B \in M_n(R)$. Give direct proofs of the product formula $\det(AB) = \det(A)\det(B)$ for each of the following special cases:
    (a) A is diagonal;
    (b) A is triangular (cf. Exercise 27);
    (c) A is a permutation matrix (every row and column of A has exactly one 1, with all other entries 0);
    (d) A can be factored as $A = PLU$ for some $P, L, U \in M_n(R)$ with P a permutation matrix, L lower-triangular, and U upper-triangular.

58. Compute both sides of the Cauchy–Binet Formula for the matrices

$$A = \begin{bmatrix} 2 & 1 & 0 & 3 \\ 1 & -1 & 4 & 3 \\ 5 & -3 & -2 & 0 \end{bmatrix}, \qquad B = \begin{bmatrix} 3 & 1 & 2 \\ 0 & -2 & 3 \\ 1 & 1 & 2 \\ 3 & -3 & 1 \end{bmatrix}.$$

59. Use the Cauchy–Binet Formula to prove the *Cauchy–Schwarz inequality*: for all real $x_1, \ldots, x_n, y_1, \ldots, y_n$,

$$\left| \sum_{i=1}^{n} x_i y_i \right| \leq \left(\sum_{i=1}^{n} x_i^2 \right)^{1/2} \left(\sum_{i=1}^{n} y_i^2 \right)^{1/2}.$$

60. Illustrate the proof of the Cayley–Hamilton Theorem given in §5.15 for a general matrix $A \in M_3(R)$ by computing the entries of the matrices C, $\mathbf{z}C^{\mathrm{T}}$, and $\mathbf{z}C^{\mathrm{T}} \mathrm{adj}(C^{\mathrm{T}})$ appearing in that proof.

61. This exercise fills in some details in the proof given in §5.15. (a) Show that the set S of matrices defined in that proof is a subring of $M_n(R)$, and S is commutative. (b) Show that $h : R \to S$, defined by $h(r) = rI_n$ for $r \in R$, is a ring homomorphism. (c) Prove that $(\mathbf{z}U)V = \mathbf{z}(UV)$ for all $U, V \in M_n(S)$ and all $\mathbf{z} \in (R^n)^n$.

62. Suppose R and S are commutative rings, and $F : R \to S$ is a ring homomorphism. Define $F^* : M_n(R) \to M_n(S)$ by $F^*(A)_{i,j} = F(A_{i,j})$ for $A \in M_n(R)$ and $i, j \in [n]$.
    (a) Prove F^* is a ring homomorphism.
    (b) Prove: for all $A \in M_n(R)$, $\det(F^*(A)) = F(\det(A))$.
    (c) Deduce the formula $\det(\overline{A}) = \overline{\det(A)}$ as a special case of (b).

63. Suppose $A \in M_n(R[x])$ has the property that for all $i, j \in [n]$, $A(i, j)$ is either zero or has degree at most d_i. Prove $\det(A) \in R[x]$ is either zero or has degree at most $d_1 + d_2 + \cdots + d_n$.

64. **Generalized Distributive Law.** Let R be a ring and $a_{ij} \in R$ for $1 \leq i \leq m$ and $1 \leq j \leq n_i$. Use the distributive laws in R and induction to prove

$$\prod_{i=1}^{m} \left(\sum_{j_i=1}^{n_i} a_{ij_i} \right) = \sum_{j_1=1}^{n_1} \sum_{j_2=1}^{n_2} \cdots \sum_{j_m=1}^{n_m} \left(\prod_{i=1}^{m} a_{ij_i} \right).$$

(This formula was used several times in §5.13 and §5.14.)

65. Given $A \in M_n(R)$, use (5.1) to show that the characteristic polynomial χ_A is a monic polynomial of degree n.

66. Given $A \in M_n(R)$, show that the coefficient of x^{n-1} in χ_A is $-\sum_{i=1}^{n} A(i, i)$ (the negative of the trace of A). Prove that the constant coefficient in χ_A is $(-1)^n \det(A)$.

67. Suppose F is a field, $A \in M_n(F)$ has characteristic polynomial $p \in F[x]$, and $\lambda \in F$. Explain very carefully why $p(\lambda) = \det(\lambda I_n - A)$. Deduce that the set of roots of p in F equals the set of eigenvalues of A in F.

68. Given $A \in M_m(R)$ and $B \in M_n(R)$, prove $\det(A \otimes B) = \det(A)^n \det(B)^m$. ($A \otimes B$ is defined in §4.15. First prove the formula for $A \otimes I_n$ and $I_m \otimes B$.)

69. Prove facts (a) through (g) in §5.16.

70. Prove: for $A \in M_n(R)$ and $i \in [n]$, $\mathrm{per}(A) = \sum_{k=1}^{n} A(i, k) \, \mathrm{per}(A[i|k])$.

71. Give an example with $A, B \in M_2(\mathbb{R})$ to show that $\mathrm{per}(AB) = \mathrm{per}(A) \, \mathrm{per}(B)$ is false in general.

72. **Pfaffians.** For even $n > 0$, let SPf_n be the set of $f \in S_n$ with $f(1) < f(3) < f(5) < \cdots < f(n-1)$ and $f(i) < f(i+1)$ for all odd $i \in [n]$. Given $A \in M_n(R)$ with $A^{\mathrm{T}} = -A$, define the *Pfaffian of A* by the formula

$$\mathrm{Pf}(A) = \sum_{f \in \mathrm{SPf}_n} \mathrm{sgn}(f) A(f(1), f(2)) A(f(3), f(4)) \cdots A(f(n-1), f(n)).$$

(a) Compute $\mathrm{Pf}(A)$ for $n \in \{2, 4, 6\}$.
(b) Prove: for even $n > 0$, $\mathrm{Pf}(A)$ is a sum of $1 \cdot 3 \cdot 5 \cdot \ldots \cdot (n-1)$ terms.
(c) For $i < j$ in $[n]$, let $A[[i, j]]$ be the matrix obtained from A by deleting row i, row j, column i, and column j. Prove: for even $n \geq 4$,

$$\mathrm{Pf}(A) = \sum_{j=2}^{n} (-1)^j A(1, j) \, \mathrm{Pf}(A[[1, j]]).$$

(d) Prove that $\det(A) = \mathrm{Pf}(A)^2$ for $n \in \{2, 4, 6\}$.
(e) Can you prove $\det(A) = \mathrm{Pf}(A)^2$ for all even $n > 0$?

73. Let R be the non-commutative ring $M_2(\mathbb{R})$. Use formula (5.1) to define the determinant of any $A \in M_n(R)$. Give specific examples to show that all but two statements below are false. Prove the two true statements.
(a) For all $r, s, t, u \in R$, $\det \begin{bmatrix} r & s \\ t & u \end{bmatrix} = ru - st$.
(b) For all $A \in M_n(R)$, $\det(A) = \det(A^{\mathrm{T}})$.
(c) Switching two columns of $A \in M_n(R)$ replaces $\det(A)$ by $-\det(A)$.
(d) Switching two rows of $A \in M_n(R)$ replaces $\det(A)$ by $-\det(A)$.
(e) For all $A, B \in M_n(R)$, $\det(AB) = \det(A) \det(B)$.
(f) For all $A \in M_n(R)$, $\det(A) = \sum_{i=1}^{n} (-1)^{i+1} A(i, 1) \det(A[i|1])$.
(g) For all $A \in M_n(R)$, $\det(A) = \sum_{j=1}^{n} (-1)^{1+j} A(1, j) \det(A[1|j])$.

74. Let $A \in M_{k,n}(R)$ with $k \leq n$. Fix lists $(i_1, i_2, \ldots, i_k)$ and $(j_1, j_2, \ldots, j_k)$ in $[n]^k$ and fix $s \in [k]$. Prove

$$\det(A^{[i_1]}, A^{[i_2]}, \ldots, A^{[i_k]}) \det(A^{[j_1]}, A^{[j_2]}, \ldots, A^{[j_k]})$$

$$= \sum_{t=1}^{k} \det(A^{[i_1]}, \ldots, A^{[j_t]}, \ldots, A^{[i_k]}) \det(A^{[j_1]}, \ldots, A^{[i_s]}, \ldots, A^{[j_k]}),$$

where the matrices on the right side arise from the matrices on the left side by interchanging the positions of columns $A^{[i_s]}$ and $A^{[j_t]}$.

75. True or false? Explain each answer.
(a) For all $A, B \in M_n(R)$, $\det(A + B) = \det(A) + \det(B)$.
(b) For all $A \in M_n(R)$ and $c \in R$, $\det(cA) = c \det(A)$.
(c) For all $A \in M_n(\mathbb{C})$, if $A^k = I_n$ for some $k > 0$, then $|\det(A)| = 1$.
(d) For all upper-triangular $A \in M_n(R)$ and all strictly upper-triangular $B \in M_n(R)$, $\det(A + B) = \det(A) + \det(B)$.
(e) For all $n \times n$ matrices A with integer entries, if $\det(A) = \pm 1$, then every entry of A^{-1} is also an integer.
(f) For all $A \in M_{m,n}(\mathbb{R})$ and $B \in M_{n,m}(\mathbb{R})$, if $m > n$, then $\det(AB) = 0$.
(g) For all $A \in M_n(R)$, A^{-1} exists in $M_n(R)$ iff $\det(A) \neq 0_R$.
(h) For all $A \in M_n(\mathbb{R})$, if $|A(i, j)| \leq K$ for all $i, j \in [n]$, then $|\det(A)| \leq n! K^n$.
(i) For all $n \in \mathbb{Z}_{>0}$ and all $A, B \in M_n(\mathbb{Z}_2)$, $\mathrm{per}(AB) = \mathrm{per}(A) \mathrm{per}(B)$.

6

Comparing Concrete Linear Algebra to Abstract Linear Algebra

In introductions to linear algebra, we learn about column vectors, matrices, and the algebraic operations on these objects: vector addition, multiplication of a vector by a scalar, matrix addition, multiplication of a matrix by a scalar, matrix multiplication, matrix inversion, matrix transpose, and so on. Later in linear algebra, we study abstract vector spaces and linear maps between such spaces. The abstract setting provides a powerful tool for theoretical work. But for computational applications, it is often more convenient to work with column vectors and matrices. The goal of this chapter is to give a thorough explanation of the relation between the concrete world of column vectors and matrices on the one hand, and the abstract world of vector spaces and linear maps on the other hand. We build a dictionary translating between these two worlds, which explains the precise connection between each abstract concept and its concrete counterpart. The full dictionary linking matrix theory to linear algebra appears in the summary for this chapter.

We are mostly concerned with the following three abstract ideas: vector spaces, linear transformations, and bilinear maps. Given a vector v from an abstract n-dimensional vector space V over a field F, we show how to represent v concretely by a column vector (n-tuple) with entries from F. Given a linear transformation $T : V \to W$ mapping an abstract n-dimensional F-vector space V to an abstract m-dimensional F-vector space W, we show how to represent T concretely by an $m \times n$ matrix with entries from F. Given a bilinear form B on an abstract n-dimensional vector space V, we show how to represent B concretely by an $n \times n$ matrix with entries in F. For each of these constructions, we discuss how the explicit, concretely defined operations on column vectors and matrices correspond to abstract algebraic operations on vectors, linear maps, and bilinear maps.

All of the constructions for converting abstract entities to concrete entities depend heavily on choosing ordered bases for the abstract vector spaces involved. Different choices of bases lead to different concrete representations of a given vector, linear map, or bilinear form. We might ask if some bases are better than others for computational or theoretical purposes. For instance, since it is easier to compute with diagonal matrices compared to arbitrary matrices, we could ask if there is a basis such that a given linear operator is represented by a diagonal matrix. This leads us to a discussion of similarity, congruence, transition matrices, change of coordinates, diagonalization, and triangularization. We conclude the chapter by studying real and complex inner product spaces, orthogonal and unitary maps and matrices, and orthonormal bases.

6.1 Column Vectors versus Abstract Vectors

In this chapter, we assume the background material on matrices covered in Chapter 4. For each field F and each positive integer n, define a concrete vector space F^n whose elements

DOI: 10.1201/9781003484561-6

are n-tuples $(c_1, c_2, \ldots, c_n)$ with each $c_i \in F$. For example, if $F = \mathbb{C}$ and $n = 3$, some specific elements of $\mathbb{C}^3$ are $(i, \pi, e - 3i)$, $(7/2, 0, -i\sqrt{3})$, and $(0, 0, 1)$. We often identify the

n-tuple $(c_1, c_2, \ldots, c_n)$ with the column vector $\begin{bmatrix} c_1 \\ c_2 \\ \vdots \\ c_n \end{bmatrix}$.

On the other hand, let V be an abstract finite-dimensional vector space over F (as defined in Section 1.3). Without more specific information about V, it is not clear at the outset how to represent or describe particular elements in V. The best we can do is to introduce generic letters like v, w, etc., to denote vectors in V.

The concept of linear combinations allows us to relate the concrete set F^n to the abstract set V. Suppose $X = (x_1, \ldots, x_n)$ is any ordered list of vectors in V. Define the *linear combination function* $L_X : F^n \to V$ by setting

$$L_X(c_1, \ldots, c_n) = c_1 x_1 + c_2 x_2 + \cdots + c_n x_n \in V \qquad \text{for all } (c_1, \ldots, c_n) \in F^n.$$

The function L_X maps an n-tuple of scalars $(c_1, \ldots, c_n)$ to the linear combination of $x_1, \ldots, x_n$ having scalar coefficients $c_1, \ldots, c_n$.

Let us find conditions ensuring that the function L_X is surjective, or injective, or bijective. First, the following conditions are logically equivalent.
(a) L_X is surjective (onto).
(b) For every $v \in V$, there exists $w \in F^n$ with $v = L_X(w)$.
(c) For all $v \in V$, there exist $c_1, \ldots, c_n \in F$ with $v = \sum_{i=1}^{n} c_i x_i$.
(d) Every $v \in V$ can be written in *at least one way* as a linear combination of the list X.
(e) The list X spans V.

Second, we claim the following conditions are logically equivalent.
(a) L_X is injective (one-to-one).
(b) For all $w, y \in F^n$, if $L_X(w) = L_X(y)$, then $w = y$.
(c) For all $c_i, d_i \in F$, if $\sum_{i=1}^{n} c_i x_i = \sum_{i=1}^{n} d_i x_i$, then $(c_1, \ldots, c_n) = (d_1, \ldots, d_n)$.
(d) Each $v \in V$ can be written in *at most one way* as a linear combination of the list X.
(e) The list X is linearly independent.

To see why linear independence of X is equivalent to the preceding conditions, assume that X is linearly independent. Given $c_i, d_i \in F$ with $\sum_{i=1}^{n} c_i x_i = \sum_{i=1}^{n} d_i x_i$, note that $\sum_{i=1}^{n} (c_i - d_i) x_i = 0$. By the linear independence of X, we conclude $c_i - d_i = 0$ and $c_i = d_i$ for all i. Conversely, suppose L_X is injective and $\sum_{i=1}^{n} c_i x_i = 0$ for some $c_i \in F$. Then $L_X(c_1, \ldots, c_n) = 0_V = L_X(0, \ldots, 0)$ forces $(c_1, \ldots, c_n) = (0, \ldots, 0)$, so all $c_i = 0$. This means that the list X is linearly independent.

Combining the preceding remarks on spanning and linear independence, we see that the following conditions are logically equivalent.
(a) $L_X : F^n \to V$ is a bijection (one-to-one and onto).
(b) For every $v \in V$, there exists a unique $w \in F^n$ with $v = L_X(w)$.
(c) For all $v \in V$, there exist unique $c_1, \ldots, c_n \in F$ with $v = \sum_{i=1}^{n} c_i x_i$.
(d) Every $v \in V$ can be written in *exactly one way* as a linear combination of the list X.
(e) The list X is an ordered basis of V (meaning X spans V and X is linearly independent).

Assume that $X = (x_1, \ldots, x_n)$ is an ordered basis of V, so that L_X is a bijection and an invertible function. For all $v \in V$, define $[v]_X = L_X^{-1}(v) \in F^n$. So, for all $c_i \in F$ and all $v \in V$,

$$[v]_X = (c_1, \ldots, c_n) \quad \text{iff} \quad v = c_1 x_1 + c_2 x_2 + \cdots + c_n x_n.$$

For each $v \in V$, there exist unique scalars $c_1, \ldots, c_n \in F$ such that $v = \sum_{i=1}^{n} c_i x_i$. We call the scalars $c_1, \ldots, c_n$ the *coordinates of v relative to the ordered basis X*, and we call $[v]_X$

the *coordinate vector of v relative to X*. For example, take any index j in $\{1, 2, \ldots, n\}$. Let $\mathbf{e}_j \in F^n$ be the n-tuple $(0, \ldots, 1, \ldots, 0)$ that has a 1 in position j and 0s elsewhere. Since $L_X(\mathbf{e}_j) = 0x_1 + \cdots + 1x_j + \cdots + 0x_n = x_j$, we have $[x_j]_X = \mathbf{e}_j$.

Summary. For every ordered basis X of V, there is a bijection L_X mapping the concrete vector space F^n to the abstract vector space V. The function L_X maps an n-tuple $(c_1, \ldots, c_n)$ to the linear combination $c_1 x_1 + \cdots + c_n x_n$. The inverse function L_X^{-1} maps each abstract vector $v \in V$ to the n-tuple (coordinate vector) $[v]_X$.

> For all $c_i \in F$ and $v \in V$, $[v]_X = (c_1, \ldots, c_n)$ iff $v = c_1 x_1 + \cdots + c_n x_n$.

The notation reminds us that the coordinate vector $[v]_X \in F^n$ depends on both the vector v and the choice of ordered basis X. Changing to a different ordered basis Y replaces $[v]_X$ with a new n-tuple $[v]_Y$. The relationship between $[v]_X$ and $[v]_Y$ is explained in §6.10.

6.2   Examples of Computing Coordinates

Throughout this chapter, we use the real vector spaces and ordered bases shown in Table 6.1 as running examples of the computations discussed in the text. You may check that each list displayed in the table really is an ordered basis for the associated vector space.

How to Compute Coordinate Vectors.
Input: a vector $v \in V$ and an ordered basis $X = (x_1, \ldots, x_n)$ of V.
Output: the coordinate vector $[v]_X = (c_1, \ldots, c_n) \in F^n$.
Method: Solve the vector equation $v = c_1 x_1 + \cdots + c_n x_n$ for the unknown scalars $c_1, \ldots, c_n$. Upon expanding definitions of $v, x_1, \ldots, x_n$, we often obtain a system of n linear scalar equations in the n unknowns c_i. We may solve this system by Gaussian elimination.

Example 1. Let $v = (3, 3, 1) \in \mathbb{R}^3$. Since

$$v = 3(1, 0, 0) + 3(0, 1, 0) + 1(0, 0, 1) = 3\mathbf{e}_1 + 3\mathbf{e}_2 + 1\mathbf{e}_3,$$

we see that $[v]_{X_1} = (3, 3, 1) = v$. More generally, for any $w \in F^n$, $[w]_X = w$ when X is the standard ordered basis $(\mathbf{e}_1, \ldots, \mathbf{e}_n)$.

On the other hand, to find $[v]_{X_2}$, we must solve the vector equation

$$(3, 3, 1) = c_1(1, 1, 1) + c_2(1, 2, 4) + c_3(1, 3, 9).$$

Equating components gives a system of three linear equations

$$\begin{cases} 3 &= 1c_1 + 1c_2 + 1c_3 \\ 3 &= 1c_1 + 2c_2 + 3c_3 \\ 1 &= 1c_1 + 4c_2 + 9c_3. \end{cases}$$

Solving this system gives $c_1 = 2$, $c_2 = 2$, and $c_3 = -1$. Therefore $[v]_{X_2} = (2, 2, -1)$.

To find coordinates relative to X_3, we must solve the vector equation $(3, 3, 1) = c_1(0, 2, 1) + c_2(-1, 1, 3) + c_3(1, 0, 4)$. The new system is

$$\begin{cases} 3 &= 0c_1 - 1c_2 + 1c_3 \\ 3 &= 2c_1 + 1c_2 + 0c_3 \\ 1 &= 1c_1 + 3c_2 + 4c_3. \end{cases}$$

The solution is $[v]_{X_3} = (32/13, -25/13, 14/13)$.

TABLE 6.1
Real vector spaces and ordered bases used as examples in this chapter.

1. $\mathbb{R}^3 = \{(c_1, c_2, c_3) : c_i \in \mathbb{R}\}$ (3-tuples or column vectors).   Basis $X_1 = (\mathbf{e}_1, \mathbf{e}_2, \mathbf{e}_3) = ((1,0,0),(0,1,0),(0,0,1))$.   Basis $X_2 = ((1,1,1),(1,2,4),(1,3,9))$.   Basis $X_3 = ((0,2,1),(-1,1,3),(1,0,4))$.
2. $P_{\leq 3} = \{f \in \mathbb{R}[t] : f = 0 \text{ or } \deg(f) \leq 3\}$ (polynomials in t of degree at most 3).   Basis $X_1 = (1, t, t^2, t^3)$.   Basis $X_2 = ((t-2)^3, (t-2)^2, t-2, 1)$.   Basis $X_3 = (t+3, 2t^2-4, t^3-t^2, t^3+t^2)$.
3. $M_2(\mathbb{R}) = \left\{ \begin{bmatrix} a & b \\ c & d \end{bmatrix} : a, b, c, d \in \mathbb{R} \right\}$ (2 × 2 matrices).   Basis $X_1 = (\mathbf{e}_{11}, \mathbf{e}_{12}, \mathbf{e}_{21}, \mathbf{e}_{22}) = \left(\begin{bmatrix} 1 & 0 \\ 0 & 0 \end{bmatrix}, \begin{bmatrix} 0 & 1 \\ 0 & 0 \end{bmatrix}, \begin{bmatrix} 0 & 0 \\ 1 & 0 \end{bmatrix}, \begin{bmatrix} 0 & 0 \\ 0 & 1 \end{bmatrix} \right).$ Basis $X_2 = \left(\begin{bmatrix} 1 & 0 \\ 0 & 1 \end{bmatrix}, \begin{bmatrix} 0 & 1 \\ 1 & 0 \end{bmatrix}, \begin{bmatrix} 1 & 0 \\ 0 & -1 \end{bmatrix}, \begin{bmatrix} 0 & -1 \\ 1 & 0 \end{bmatrix} \right).$ Basis $X_3 = \left(\begin{bmatrix} 1 & 2 \\ 0 & 3 \end{bmatrix}, \begin{bmatrix} 1 & 0 \\ 2 & 3 \end{bmatrix}, \begin{bmatrix} 0 & 1 \\ 2 & 3 \end{bmatrix}, \begin{bmatrix} 1 & 2 \\ 3 & 0 \end{bmatrix} \right).$
4. $\mathbb{C} = \{a + ib : a, b \in \mathbb{R}\}$ (complex numbers).   Basis $X_1 = (1, i)$.   Basis $X_2 = (e^{\pi i/6}, e^{2\pi i/3})$.   Basis $X_3 = (3 + 4i, 2 - i)$.

Example 2. Let $v = (t-3)^3 \in P_{\leq 3}$. To find $[v]_{X_1}$, solve $(t-3)^3 = c_1 1 + c_2 t + c_3 t^2 + c_4 t^3$ by expanding the left side into monomials and comparing coefficients. Since $(t-3)^3 = t^3 - 9t^2 + 27t - 27$, we get $c_1 = -27$, $c_2 = 27$, $c_3 = -9$, $c_4 = 1$, and $[v]_{X_1} = (-27, 27, -9, 1)$.

On the other hand, to find $[v]_{X_2}$, we must solve

$$(t-3)^3 = d_1(t-2)^3 + d_2(t-2)^2 + d_3(t-2) + d_4 1$$

for unknowns $d_1, d_2, d_3, d_4 \in \mathbb{R}$. Expanding both sides gives

$$t^3 - 9t^2 + 27t - 27 = d_1 t^3 + (d_2 - 6d_1)t^2 + (d_3 - 4d_2 + 12d_1)t + (d_4 - 2d_3 + 4d_2 - 8d_1).$$

Equating coefficients leads to the triangular system of linear equations

$$\begin{cases} d_1 & & & = 1 \\ -6d_1 & +d_2 & & = -9 \\ 12d_1 & -4d_2 & +d_3 & = 27 \\ -8d_1 & +4d_2 & -2d_3 & +d_4 = -27. \end{cases}$$

Solving gives $d_1 = 1$, $d_2 = -3$, $d_3 = 3$, $d_4 = -1$, and $[v]_{X_2} = (1, -3, 3, -1)$. By solving another system, you can check that $[v]_{X_3} = (27, 27, 32, -31)$.

Example 3. Let $A = \begin{bmatrix} 1 & 2 \\ 3 & 4 \end{bmatrix} \in M_2(\mathbb{R})$. We have $A = 1\mathbf{e}_{11} + 2\mathbf{e}_{12} + 3\mathbf{e}_{21} + 4\mathbf{e}_{22}$, so that $[A]_{X_1} = (1, 2, 3, 4)$. To find $[A]_{X_2}$, write

$$\begin{bmatrix} 1 & 2 \\ 3 & 4 \end{bmatrix} = A = d_1 \begin{bmatrix} 1 & 0 \\ 0 & 1 \end{bmatrix} + d_2 \begin{bmatrix} 0 & 1 \\ 1 & 0 \end{bmatrix} + d_3 \begin{bmatrix} 1 & 0 \\ 0 & -1 \end{bmatrix} + d_4 \begin{bmatrix} 0 & -1 \\ 1 & 0 \end{bmatrix}$$

where d_1, d_2, d_3, d_4 are unknown scalars. Equating corresponding matrix entries on each side, we obtain a system of linear equations

$$\begin{cases} d_1 & & +d_3 & & = & 1 \\ & d_2 & & -d_4 & = & 2 \\ & d_2 & & +d_4 & = & 3 \\ d_1 & & -d_3 & & = & 4. \end{cases}$$

Solving this system gives $[A]_{X_2} = (d_1, d_2, d_3, d_4) = (5/2, 5/2, -3/2, 1/2)$. Similarly, by solving

$$\begin{bmatrix} 1 & 2 \\ 3 & 4 \end{bmatrix} = c_1 \begin{bmatrix} 1 & 2 \\ 0 & 3 \end{bmatrix} + c_2 \begin{bmatrix} 1 & 0 \\ 2 & 3 \end{bmatrix} + c_3 \begin{bmatrix} 0 & 1 \\ 2 & 3 \end{bmatrix} + c_4 \begin{bmatrix} 1 & 2 \\ 3 & 0 \end{bmatrix},$$

we find that $[A]_{X_3} = (1/3, 1/3, 2/3, 1/3)$.

Example 4. Let $v = e^{\pi i/3} \in \mathbb{C}$. By definition of complex exponentials, we have $v = \cos(\pi/3) + \sin(\pi/3)i = (1/2)1 + (\sqrt{3}/2)i$, and hence $[v]_{X_1} = (1/2, \sqrt{3}/2)$. To find $[v]_{X_2}$, we write $e^{\pi i/3} = d_1 e^{\pi i/6} + d_2 e^{2\pi i/3}$ for real unknowns d_1 and d_2. Equating real and imaginary parts of both sides leads to the 2×2 linear system

$$\begin{cases} 1/2 & = & (\sqrt{3}/2)d_1 & +(-1/2)d_2 \\ \sqrt{3}/2 & = & (1/2)d_1 & +(\sqrt{3}/2)d_2. \end{cases}$$

The solution is $d_1 = \sqrt{3}/2$, $d_2 = 1/2$, so $[v]_{X_2} = (\sqrt{3}/2, 1/2)$. As a side remark, note that reversing the order of the basis X_1 gives $[v]_{(i,1)} = (\sqrt{3}/2, 1/2) = [v]_{X_2}$. This shows that a vector can have the same coordinates relative to two different ordered bases. By solving one more 2×2 system, you can check that $[v]_{X_3} = ((1 + 2\sqrt{3})/22, (4 - 3\sqrt{3})/22)$.

6.3 Operations on Column Vectors versus Abstract Vectors

Vector addition and scalar multiplication in F^n are defined by the explicit formulas

$$(c_1, \ldots, c_n) + (d_1, \ldots, d_n) = (c_1 + d_1, \ldots, c_n + d_n), \qquad a(c_1, \ldots, c_n) = (ac_1, \ldots, ac_n)$$

for all $c_i, d_i, a \in F$. On the other hand, all we know initially about the addition and scalar multiplication operations in an abstract vector space V are the axioms listed in Table 1.4.

We can use the linear combination maps $L_X : F^n \to V$ to relate the concrete vector space operations in F^n to the abstract vector space operations in V. Let $X = (x_1, \ldots, x_n)$ be any ordered list in V. For $\mathbf{c} = (c_1, \ldots, c_n) \in F^n$, $\mathbf{d} = (d_1, \ldots, d_n) \in F^n$, and $a \in F$, we use the vector space axioms in V to compute:

$$L_X(\mathbf{c}+\mathbf{d}) = L_X(c_1+d_1, \ldots, c_n+d_n) = \sum_{i=1}^n (c_i+d_i)x_i = \sum_{i=1}^n c_i x_i + \sum_{i=1}^n d_i x_i = L_X(\mathbf{c}) + L_X(\mathbf{d});$$

$$L_X(a\mathbf{c}) = L_X(ac_1, \ldots, ac_n) = \sum_{i=1}^n (ac_i)x_i = a\sum_{i=1}^n c_i x_i = aL_X(\mathbf{c}).$$

This computation shows that $L_X : F^n \to V$ is always an F-linear map (vector space homomorphism) from F^n into V. Combining this fact with the results in §6.1, we see that L_X is a vector space isomorphism from F^n to V iff X is an ordered basis for V. In this case, the map $L_X^{-1} : V \to F^n$, which sends each $v \in V$ to its coordinate vector $[v]_X$ in F^n, is also a vector space isomorphism (see Exercise 54 of Chapter 1). Linearity of this inverse map means that

$$[v + w]_X = [v]_X + [w]_X \quad \text{and} \quad [cv]_X = c[v]_X \quad \text{for all } v, w \in V \text{ and } c \in F.$$

Example. Let us find $[\mathbf{e}_{22}]_{X_3}$ using computations from §6.2. Let $A = \begin{bmatrix} 1 & 2 \\ 3 & 4 \end{bmatrix}$ and $B = \begin{bmatrix} 1 & 2 \\ 3 & 0 \end{bmatrix}$, and note $\mathbf{e}_{22} = (1/4)A - (1/4)B$. We already calculated $[A]_{X_3} = (1/3, 1/3, 2/3, 1/3)$, while $[B]_{X_3} = (0, 0, 0, 1)$. Therefore,

$$[\mathbf{e}_{22}]_{X_3} = [(1/4)(A - B)]_{X_3} = (1/4)[A]_{X_3} - (1/4)[B]_{X_3} = (1/12, 1/12, 1/6, -1/6).$$

So far, we have seen that each ordered basis $X = (x_1, \ldots, x_n)$ of V gives us a vector space isomorphism from the abstract n-dimensional space V to the concrete vector space F^n. This isomorphism is given by the coordinate map $v \mapsto [v]_X$ for $v \in V$, and the inverse isomorphism L_X sends $(c_1, \ldots, c_n) \in F^n$ to $\sum_{i=1}^n c_i x_i \in V$. We have $L_X(\mathbf{e}_j) = x_j$ and $[x_j]_X = \mathbf{e}_j$ for all j. This says that the standard ordered basis $E = (\mathbf{e}_1, \ldots, \mathbf{e}_n)$ of F^n corresponds to the ordered basis X of V under the isomorphism. Note carefully that we have found not one, but many isomorphisms $V \cong F^n$, which are indexed by the ordered bases of V. Many problems in linear algebra involve finding a good ordered basis X for V, so that the associated isomorphism $V \cong F^n$ has additional properties useful for solving the given problem.

Recall the theorem that any finite-dimensional vector space V has a unique dimension, which is the length of any ordered basis of V. Using this, we show that *two finite-dimensional F-vector spaces are isomorphic iff they have the same dimension.* On one hand, if V and W both have dimension n, then we can compose the isomorphisms $V \cong F^n$ and $F^n \cong W$ to conclude that V and W are isomorphic. Conversely, if $V \cong W$, then any isomorphism from V to W is a bijection mapping each ordered basis of V to an ordered basis of W. Thus ordered bases of V and W must have the same length, so $\dim(V) = \dim(W)$. Although we omit the proof, similar results hold for infinite-dimensional spaces. Specifically, vector spaces V and W are isomorphic iff $\dim(V) = \dim(W)$, where the dimensions are viewed as infinite cardinals.

We have constructed a family of isomorphisms $V \cong F^n$ (along with their inverse isomorphisms $F^n \cong V$) parametrized by ordered bases of V. We may ask whether we have found all such isomorphisms, and whether our parametrization via ordered bases is unique. To state this question formally, note that we have a well-defined function ϕ, given by the formula $\phi(X) = L_X$, that maps the set $\mathcal{B}$ of all ordered bases of V into the set $\mathcal{I}$ of all vector space isomorphisms from F^n onto V. We show that $\phi : \mathcal{B} \to \mathcal{I}$ is a bijection, which means that each $L \in \mathcal{I}$ has the form $\phi(X) = L_X$ for a unique $X \in \mathcal{B}$. In other words, *every isomorphism $L : F^n \to V$ has the form L_X for a unique ordered basis X of V.*

First we prove that ϕ is one-to-one. Assume $X = (x_1, \ldots, x_n)$ and $Y = (y_1, \ldots, y_n)$ are ordered bases of V such that $\phi(X) = \phi(Y)$, meaning $L_X = L_Y$. We must show that $X = Y$. This follows because $x_j = L_X(\mathbf{e}_j) = L_Y(\mathbf{e}_j) = y_j$ for all j between 1 and n.

Next, we prove that ϕ is surjective. Let $L : F^n \to V$ be any vector space isomorphism. We must find an ordered basis X such that $L = L_X = \phi(X)$. We find X by applying L to the standard ordered basis $E = (\mathbf{e}_1, \ldots, \mathbf{e}_n)$ of F^n. Define $x_j = L(\mathbf{e}_j)$ for $j = 1, 2, \ldots, n$, and define $X = (x_1, \ldots, x_n)$. This list X is an ordered basis of V since X is the image of the ordered basis E of F^n under an isomorphism L. To prove that the functions L and L_X from F^n to V are equal, we check that they have the same effect on each input $\mathbf{c} = (c_1, \ldots, c_n)$ in the domain F^n. Since $\mathbf{c} = \sum_{j=1}^{n} c_j \mathbf{e}_j$, linearity of L gives

$$L(\mathbf{c}) = \sum_{j=1}^{n} c_j L(\mathbf{e}_j) = \sum_{j=1}^{n} c_j x_j = L_X(\mathbf{c}).$$

This completes the proof that ϕ is a bijection.

6.4 Matrices versus Linear Maps

The next step in our comparison of concrete matrix theory to abstract linear algebra is establishing a correspondence between matrices and linear transformations. This correspondence is fundamental and essential for all that follows. For any field F, let

$$M_{m,n}(F) = \text{the set of all } m \times n \text{ matrices with entries in } F.$$

For any F-vector spaces V and W, let

$$L(V, W) = \text{the set of all } F\text{-linear maps } T \text{ from } V \text{ to } W.$$

A function $T : V \to W$ belongs to $L(V, W)$ iff $T(v + v') = T(v) + T(v')$ and $T(cv) = cT(v)$ for all $v, v' \in V$ and all $c \in F$.

Suppose $\dim(V) = n$ and $\dim(W) = m$. We construct not one, but many bijections from $L(V, W)$ to $M_{m,n}(F)$. These bijections are parametrized by ordered bases of V and W. Let $X = (x_1, \ldots, x_n)$ be any ordered basis for V and $Y = (y_1, \ldots, y_m)$ be any ordered basis for W. We define a bijection $M_{X,Y} : L(V, W) \to M_{m,n}(F)$ that sends a linear transformation T to a certain matrix $M_{X,Y}(T)$. We introduce special notation $_Y[T]_X$ for the matrix $M_{X,Y}(T)$. The matrix $_Y[T]_X$ is called *the matrix of T using the input basis X and the output basis Y.*

Our plan is to build a correspondence between linear maps and matrices by composing three bijections

$$L(V, W) \longrightarrow W^n \longrightarrow (F^m)^n \longrightarrow M_{m,n}(F), \text{ where:} \qquad (6.1)$$

$$
\begin{array}{lll}
L(V, W) & = & \text{set of } F\text{-linear maps } T : V \to W; \\
W^n & = & \text{set of lists } (w_1, \ldots, w_n) \text{ with all } w_j \in W; \\
(F^m)^n & = & \text{set of lists } (\mathbf{z}_1, \ldots, \mathbf{z}_n) \text{ with all } \mathbf{z}_j \in F^m; \\
M_{m,n}(F) & = & \text{set of } m \times n \text{ matrices with entries in } F.
\end{array}
$$

The crucial observation for setting up this correspondence is that a linear map T with domain V is completely determined by the values of T on the basis vectors in X. The first map in (6.1) is the *restriction map* $R_X : L(V, W) \to W^n$, defined by setting $R_X(T) = (T(x_1), T(x_2), \ldots, T(x_n)) \in W^n$ for each $T \in L(V, W)$.

Let us check that R_X is one-to-one and onto. First, suppose $S, T \in L(V, W)$ are two linear maps such that $R_X(S) = R_X(T)$. This means $S(x_j) = T(x_j)$ for $j = 1, 2, \ldots, n$. We prove $S = T$ by showing $S(v) = T(v)$ for all $v \in V$. Given $v \in V$, write $v = \sum_{j=1}^{n} c_j x_j$ for some $c_j \in F$. Then, use linearity of S and T to compute

$$S(v) = S\left(\sum_{j=1}^{n} c_j x_j\right) = \sum_{j=1}^{n} c_j S(x_j) = \sum_{j=1}^{n} c_j T(x_j) = T\left(\sum_{j=1}^{n} c_j x_j\right) = T(v).$$

Second, given any $\mathbf{w} = (w_1, w_2, \ldots, w_n) \in W^n$, we must build $T \in L(V, W)$ such that $R_X(T) = \mathbf{w}$. We define

$$T\left(\sum_{j=1}^{n} c_j x_j\right) = \sum_{j=1}^{n} c_j w_j \quad \text{for all } c_j \in F. \tag{6.2}$$

It is routine to check that T is a well-defined linear map from V to W such that $T(x_j) = w_j$ for $j = 1, 2, \ldots, n$. Hence, $T \in L(V, W)$ and $R_X(T) = (T(x_1), \ldots, T(x_n)) = \mathbf{w}$. We now know R_X is a bijection. For each $\mathbf{w} \in W^n$, the proof of surjectivity shows that $R_X^{-1}(\mathbf{w})$ is the map $T \in L(V, W)$ defined by (6.2).

To continue, we define a bijection from W^n to $(F^m)^n$. Starting with $\mathbf{w} = (w_1, \ldots, w_n) \in W^n$, we apply the bijection $L_Y^{-1} : W \to F^m$ to each element of the list $\mathbf{w}$ to obtain a list $([w_1]_Y, \ldots, [w_n]_Y)$ of n column vectors in F^m. The inverse bijection takes a list $(\mathbf{z}_1, \ldots, \mathbf{z}_n) \in (F^m)^n$ and maps it to the list $(L_Y(\mathbf{z}_1), \ldots, L_Y(\mathbf{z}_n)) \in W^n$.

To finish, recall from §4.1 that there is a bijection from $(F^m)^n$ to $M_{m,n}(F)$. This bijection maps a list $(\mathbf{z}_1, \ldots, \mathbf{z}_n)$ of n column vectors in F^m to the $m \times n$ matrix A whose columns are $\mathbf{z}_1, \ldots, \mathbf{z}_n$ in this order. The inverse map sends each $A \in M_{m,n}(F)$ to the list $(A^{[1]}, \ldots, A^{[n]}) \in (F^m)^n$, where $A^{[j]} \in F^m$ is the jth column of A.

Composing the three bijections discussed above, we get a bijection $M_{X,Y}$ from $L(V, W)$ to $M_{m,n}(F)$. Applying the formulas for each map, we see that $T \in L(V, W)$ is first sent to the list $(T(x_1), \ldots, T(x_n)) \in W^n$, which is then sent to the list of coordinate vectors $([T(x_1)]_Y, \ldots, [T(x_n)]_Y) \in (F^m)^n$, which is finally sent to the matrix having these coordinate vectors as columns. Writing $_Y[T]_X = M_{X,Y}(T)$, we see that the jth column of this matrix is $[T(x_j)]_Y$ for $j = 1, 2, \ldots, n$.

The inverse bijection starts with a matrix $A \in M_{m,n}(F)$ and does three things. First, replace the matrix by its list of columns $(A^{[1]}, \ldots, A^{[n]}) \in (F^m)^n$. Second, replace this list by the list of vectors $(w_1, \ldots, w_n) \in W^n$, where $w_j = L_Y(A^{[j]}) = \sum_{i=1}^{m} A(i, j) y_i$ for $j = 1, 2, \ldots, n$. Third, replace this list by the linear map T defined in (6.2). The net effect is that A maps to T, where

$$T\left(\sum_{j=1}^{n} c_j x_j\right) = \sum_{j=1}^{n} c_j \sum_{i=1}^{m} A(i, j) y_i = \sum_{i=1}^{m} \left(\sum_{j=1}^{n} A(i, j) c_j\right) y_i \quad \text{for all } c_j \in F. \tag{6.3}$$

Summary. Given an ordered basis $X = (x_1, \ldots, x_n)$ for V and an ordered basis $Y = (y_1, \ldots, y_m)$ for W, we obtain a bijection from $L(V, W)$ to $M_{m,n}(F)$ sending the linear map T to the matrix $_Y[T]_X$. By definition,

> $_Y[T]_X$ is the matrix whose jth column is the coordinate vector $[T(x_j)]_Y$.

To say this another way,

> $A = {}_Y[T]_X$ means $T(x_j) = \sum_{i=1}^{m} A(i, j) y_i$ for $j = 1, 2, \ldots, n$.

Equivalently, $A = {}_Y[T]_X$ iff T satisfies (6.3). The notation $_Y[T]_X$ reminds us that the bijection constructed here depends on the choice of ordered bases X and Y.

6.5 Examples of Matrices Associated with Linear Maps

We give some examples of computing the matrix of a linear map.

How to Compute the Matrix of a Linear Map.
Input: A linear map $T : V \to W$, an ordered basis $X = (x_1, \ldots, x_n)$ of V, and an ordered basis $Y = (y_1, \ldots, y_m)$ of W.
Output: the matrix $_Y[T]_X$ of T using input basis X and output basis Y.
Method: Apply T to each input basis vector $x_j \in X$. Compute the coordinate vector of $T(x_j)$ relative to the basis Y. Write this vector as column j of the matrix.

Example 1. (See Table 6.1 on page 137 for the notation used in these examples.)
Let $T : P_{\leq 3} \to P_{\leq 3}$ be the differentiation operator $T(f) = df/dt$ for $f \in P_{\leq 3}$. By a theorem from calculus, T is a linear map. Let us find $A = {}_{X_1}[T]_{X_1}$. To get the first column of A, compute $T(1) = 0$, which has coordinates $(0, 0, 0, 0)$ relative to X_1. To get the second column of A, compute $T(t) = 1$, which has coordinates $(1, 0, 0, 0)$ relative to X_1. To get the third column of A, compute $T(t^2) = 2t$, which has coordinates $(0, 2, 0, 0)$ relative to X_1. To get the fourth column of A, compute $T(t^3) = 3t^2$, which has coordinates $(0, 0, 3, 0)$ relative to X_1. Arranging these columns in a matrix, we have

$$A = {}_{X_1}[T]_{X_1} = \begin{bmatrix} 0 & 1 & 0 & 0 \\ 0 & 0 & 2 & 0 \\ 0 & 0 & 0 & 3 \\ 0 & 0 & 0 & 0 \end{bmatrix}.$$

Suppose we change the input basis to X_2, but keep X_1 as the output basis. Applying T to the vectors $(t-2)^3$, $(t-2)^2$, $(t-2)^1$, 1 produces the vectors $3(t-2)^2 = 3t^2 - 12t + 12$, $2(t-2) = 2t-4$, 1, and 0, respectively. We find coordinates $[3t^2 - 12t + 12]_{X_1} = (12, -12, 3, 0)$, and so on, leading to the matrix

$$_{X_1}[T]_{X_2} = \begin{bmatrix} 12 & -4 & 1 & 0 \\ -12 & 2 & 0 & 0 \\ 3 & 0 & 0 & 0 \\ 0 & 0 & 0 & 0 \end{bmatrix}.$$

On the other hand, taking X_2 as both input basis and output basis gives

$$_{X_2}[T]_{X_2} = \begin{bmatrix} 0 & 0 & 0 & 0 \\ 3 & 0 & 0 & 0 \\ 0 & 2 & 0 & 0 \\ 0 & 0 & 1 & 0 \end{bmatrix}.$$

Example 2. Let $T : M_2(\mathbb{R}) \to M_2(\mathbb{R})$ be the transpose map given by $T(A) = A^{\mathrm{T}}$ for all $A \in M_2(\mathbb{R})$. This map is linear. We compute $T(\mathbf{e}_{11}) = \mathbf{e}_{11}$, $T(\mathbf{e}_{12}) = \mathbf{e}_{21}$, $T(\mathbf{e}_{21}) = \mathbf{e}_{12}$, and $T(\mathbf{e}_{22}) = \mathbf{e}_{22}$. Therefore,

$$_{X_1}[T]_{X_1} = \begin{bmatrix} 1 & 0 & 0 & 0 \\ 0 & 0 & 1 & 0 \\ 0 & 1 & 0 & 0 \\ 0 & 0 & 0 & 1 \end{bmatrix}.$$

Similar computations show that:

$$_{X_2}[T]_{X_2} = \begin{bmatrix} 1 & 0 & 0 & 0 \\ 0 & 1 & 0 & 0 \\ 0 & 0 & 1 & 0 \\ 0 & 0 & 0 & -1 \end{bmatrix}; \quad _{X_3}[T]_{X_3} = \begin{bmatrix} 0 & 1 & 1/2 & 1/2 \\ 1 & 0 & -1/2 & -1/2 \\ 0 & 0 & 1 & 0 \\ 0 & 0 & 0 & 1 \end{bmatrix};$$

$$X_1[T]_{X_3} = \begin{bmatrix} 1 & 1 & 0 & 1 \\ 0 & 2 & 2 & 3 \\ 2 & 0 & 1 & 2 \\ 3 & 3 & 3 & 0 \end{bmatrix} ; \quad X_2[T]_{X_1} = \begin{bmatrix} 1/2 & 0 & 0 & 1/2 \\ 0 & 1/2 & 1/2 & 0 \\ 1/2 & 0 & 0 & -1/2 \\ 0 & 1/2 & -1/2 & 0 \end{bmatrix}.$$

Example 3. For any vector spaces V and W, the zero map $0 \in L(V, W)$ is the linear map sending every $v \in V$ to 0_W. For all ordered bases X and Y, $_Y[0]_X$ is the zero matrix $0 \in M_{m,n}(F)$. On the other hand, for the identity map id_V on any vector space V, we have $_X[\mathrm{id}]_X = I_n$ (the identity matrix) for all ordered bases X of V. But if X and Y are distinct ordered bases for V, then $_Y[\mathrm{id}]_X$ is not the identity matrix. For example, using the bases of $\mathbb{R}^3$ in Table 6.1, we see that $_{X_1}[\mathrm{id}]_{X_2} = \begin{bmatrix} 1 & 1 & 1 \\ 1 & 2 & 3 \\ 1 & 4 & 9 \end{bmatrix}$. We study matrices of the form $_Y[\mathrm{id}]_X$ in more detail later (see §6.10).

Example 4. Let $z = a + ib$ be a fixed complex number. Define $T_z : \mathbb{C} \to \mathbb{C}$ by $T_z(w) = zw$ for all $w \in \mathbb{C}$. You can check that T_z is an $\mathbb{R}$-linear map (and also a $\mathbb{C}$-linear map). Let us compute $_{X_1}[T_z]_{X_1}$. For column 1, compute $T_z(1) = z1 = z = a + ib$, and note that $[z]_{X_1} = (a, b)$. For column 2, compute $T_z(i) = zi = -b + ia$, and note that $[zi]_{X_1} = (-b, a)$. So

$$X_1[T_z]_{X_1} = \begin{bmatrix} a & -b \\ b & a \end{bmatrix}.$$

Example 5. Define a linear map $T : P_{\leq 3} \to \mathbb{R}^3$ by $T(f) = (f(1), f(2), f(3))$ for $f \in P_{\leq 3}$. Let $X = (1, t, t^2, t^3)$, $Y = (\mathbf{e}_1, \mathbf{e}_2, \mathbf{e}_3)$, and $Z = ((1, 1, 1), (1, 2, 3), (1, 4, 9))$. Applying T to each vector in X, we get $T(1) = (1, 1, 1)$, $T(t) = (1, 2, 3)$, $T(t^2) = (1, 4, 9)$, and $T(t^3) = (1, 8, 27)$. Computing coordinates of these vectors relative to Y and Z, we find:

$$Y[T]_X = \begin{bmatrix} 1 & 1 & 1 & 1 \\ 1 & 2 & 4 & 8 \\ 1 & 3 & 9 & 27 \end{bmatrix} ; \quad Z[T]_X = \begin{bmatrix} 1 & 0 & 0 & 6 \\ 0 & 1 & 0 & -11 \\ 0 & 0 & 1 & 6 \end{bmatrix}.$$

If we change the input basis to $X' = (1, (t - 2), (t - 2)^2, (t - 2)^3)$, we get:

$$Y[T]_{X'} = \begin{bmatrix} 1 & -1 & 1 & -1 \\ 1 & 0 & 0 & 0 \\ 1 & 1 & 1 & 1 \end{bmatrix} ; \quad Z[T]_{X'} = \begin{bmatrix} 1 & -2 & 4 & -2 \\ 0 & 1 & -4 & 1 \\ 0 & 0 & 1 & 0 \end{bmatrix}.$$

Example 6. Suppose $V = F^n$, $W = F^m$, and A is a given matrix in $M_{m,n}(F)$. Define a linear map $L^A : F^n \to F^m$ by setting $L^A(\mathbf{v}) = A\mathbf{v} \in F^m$ for all column vectors $\mathbf{v}$ in F^n. The map L^A is called *left multiplication by the matrix A*. Let $E = (\mathbf{e}_1, \ldots, \mathbf{e}_n)$ be the standard ordered basis of F^n and $E' = (\mathbf{e}'_1, \ldots, \mathbf{e}'_m)$ be the standard ordered basis of F^m. We show that $_{E'}[L^A]_E = A$. To get the jth column of $_{E'}[L^A]_E$, we apply L^A to the jth vector in E, obtaining $L^A(\mathbf{e}_j) = A\mathbf{e}_j = A^{[j]}$. Next, we take the coordinate vector of this column vector relative to E'. For any column vector $\mathbf{w} \in F^m$, $[\mathbf{w}]_{E'} = \mathbf{w}$, so the coordinate vector is $A^{[j]}$. Thus, the jth column of $_{E'}[L^A]_E$ equals the jth column of A for all j, proving that $_{E'}[L^A]_E = A$.

We can restate the result of this example by saying that *the matrix of left multiplication by A (using the standard ordered bases) is A itself*. It follows that the map $A \mapsto L^A$ from $M_{m,n}(F)$ to $L(F^n, F^m)$ is the inverse of the general bijection $T \mapsto {}_{E'}[T]_E$ constructed earlier, so this map is also a bijection. Bijectivity of the map $A \mapsto L^A$ means that *every linear transformation from F^n to F^m is left multiplication by a uniquely determined $m \times n$ matrix*.

6.6   Vector Operations on Matrices and Linear Maps

The set of matrices $M_{m,n}(F)$ is a vector space with the following operations. Given $A, B \in M_{m,n}(F)$ and $c \in F$, $A+B$ is the matrix with i,j-entry $A(i,j)+B(i,j)$, and cA is the matrix with i,j-entry $cA(i,j)$. We can also describe these operations in terms of the columns of the matrices using the formulas $(A+B)^{[j]} = A^{[j]} + B^{[j]}$ and $(cA)^{[j]} = c(A^{[j]})$ (see §4.7). The operations on the right sides of the last two formulas are addition and scalar multiplication of column vectors in the vector space F^m.

Suppose V, W are given vector spaces, S and T are linear maps in $L(V,W)$, and $c \in F$. Define $S + T : V \to W$ by $(S+T)(v) = S(v) + T(v)$ for all $v \in V$. Define $cT : V \to W$ by $(cT)(v) = c(T(v))$ for all $v \in V$. The operations on the right sides of these definitions are the addition and scalar multiplication in the given abstract vector space W. You can check that $S + T$ and cT really are F-linear maps, so that we have the closure properties $S + T \in L(V,W)$ and $cT \in L(V,W)$. Also, the zero function from V to W belongs to $L(V,W)$. It follows that $L(V,W)$ is a subspace of the vector space of all functions from V to W (see Exercise 8 in Chapter 4), so that $L(V,W)$ is also a vector space using the operations defined here.

Let $X = (x_1, \ldots, x_n)$ be an ordered basis for V and $Y = (y_1, \ldots, y_m)$ be an ordered basis for W. We prove that the bijection $M_{X,Y} : L(V,W) \to M_{m,n}(F)$ given by $M_{X,Y}(T) = {}_Y[T]_X$ is F-linear, hence is a vector space isomorphism. Let $S, T \in L(V,W)$ and $c \in F$. For j between 1 and n, the jth column of ${}_Y[S+T]_X$ is

$$[(S+T)(x_j)]_Y = [S(x_j) + T(x_j)]_Y = [S(x_j)]_Y + [T(x_j)]_Y,$$

which is the sum of the jth column of ${}_Y[S]_X$ and the jth column of ${}_Y[T]_X$. This means that $M_{X,Y}(S+T) = M_{X,Y}(S) + M_{X,Y}(T)$. Similarly, the jth column of ${}_Y[cT]_X$ is

$$[(cT)(x_j)]_Y = [c(T(x_j))]_Y = c \cdot [T(x_j)]_Y.$$

This is c times the jth column of ${}_Y[T]_X$, and therefore $M_{X,Y}(cT) = cM_{X,Y}(T)$. To summarize: *given any ordered bases X of V and Y of W, the map $T \mapsto {}_Y[T]_X$ is a vector space isomorphism $L(V,W) \cong M_{m,n}(F)$.*

Since isomorphic vector spaces have the same dimension, we deduce $\dim(L(V,W)) = \dim(M_{m,n}(F)) = mn$. We can obtain an even stronger conclusion by recalling that vector space isomorphisms send bases to bases. We have seen (§4.3) that the set $E = \{\mathbf{e}_{ij} : 1 \leq i \leq m, 1 \leq j \leq n\}$ is a basis for $M_{m,n}(F)$. Note that $\mathbf{e}_{ij}$ can be described as the $m \times n$ matrix whose jth column is $\mathbf{e}_{ij}^{[j]} = \mathbf{e}_i \in F^m$, and whose kth column (for each $k \neq j$) is $\mathbf{e}_{ij}^{[k]} = \mathbf{0} \in F^m$. The image of E under $M_{X,Y}^{-1}$ is a basis for $L(V,W)$. Let us describe the linear map $T_{ij} = M_{X,Y}^{-1}(\mathbf{e}_{ij})$. Since ${}_Y[T_{ij}]_X = \mathbf{e}_{ij}$ by definition, we see that $[T_{ij}(x_k)]_Y = \mathbf{e}_{ij}^{[k]} = \mathbf{0}$ for all $k \neq j$, so that $T_{ij}(x_k) = L_Y(\mathbf{0}) = 0_W$ for all $k \neq j$. On the other hand, $[T_{ij}(x_j)]_Y = \mathbf{e}_{ij}^{[j]} = \mathbf{e}_i$ implies $T_{ij}(x_j) = L_Y(\mathbf{e}_i) = y_i$. Thus T_{ij} sends the basis vector x_j to y_i and all other basis vectors in X to zero. By linearity, $T_{ij} : V \to W$ must be given by the formula

$$T_{ij}\left(\sum_{k=1}^{n} c_k x_k\right) = c_j y_i \quad \text{for all } c_k \in F. \tag{6.4}$$

6.7 Matrix Transpose versus Dual Maps

Given a matrix $A \in M_{m,n}(F)$, the *transpose* of A is the matrix $A^{\mathrm{T}} \in M_{n,m}(F)$ defined by $A^{\mathrm{T}}(i,j) = A(j,i)$ for $1 \leq i \leq n$ and $1 \leq j \leq m$. To understand the significance of this operation in the world of abstract linear algebra, we must first briefly discuss dual vector spaces. (See Chapter 13 for a more complete treatment.)

Given an n-dimensional F-vector space V, the *dual space* V^* is defined to be $L(V,F)$, the vector space of all linear maps from V into the field F (viewed as a 1-dimensional F-vector space). Applying our general results on $L(V,W)$ with $W = F$, we see that $V^* = L(V,F)$ is isomorphic to the vector space $M_{1,n}(F)$ of $1 \times n$ matrices or *row vectors*. More specifically, let $X = (x_1, \ldots, x_n)$ be any ordered basis of V. The one-element list $Z = (1_F)$ is an ordered basis for the F-vector space F, as you can check. For $j = 1, 2, \ldots, n$, define linear maps $f_j \in V^* = L(V,F)$ by setting $f_j(x_j) = 1_F$, setting $f_j(x_k) = 0_F$ for each $k \neq j$, and extending by linearity. Explicitly, $f_j(\sum_{k=1}^{n} c_k x_k) = c_j$ for all $c_1, \ldots, c_n \in F$. Comparing this to the description of the maps T_{ij} at the end of the last section, we see that $X^* = (f_1, \ldots, f_n)$ is the ordered basis for $V^* = L(V,F)$ that corresponds to the standard ordered basis for $M_{1,n}(F)$ under the isomorphism $M_{X,Z}^{-1}$. X^* is called the *dual basis* of X. Note that $\dim(V^*) = \dim(V) = n$ (in the finite-dimensional case we are considering).

Next, let W be an m-dimensional F-vector space with ordered basis $Y = (y_1, \ldots, y_m)$ and $S \in L(V,W)$ be a fixed linear map. Given any map $g \in W^* = L(W,F)$, note that $g \circ S : V \to F$ is a linear map from V to F; that is, $g \circ S$ is an element of V^*. Thus, S induces a map $S^* : W^* \to V^*$ given by $S^*(g) = g \circ S$ for $g \in W^*$. The map S^* is itself linear, since for all $g, h \in W^*$ and all $c \in F$,

$$S^*(g + h) = (g + h) \circ S = (g \circ S) + (h \circ S) = S^*(g) + S^*(h);$$

$$S^*(cg) = (cg) \circ S = c(g \circ S) = cS^*(g).$$

So $S^* \in L(W^*, V^*)$. From our discussion above, we have the dual bases Y^* for W^* and X^* for V^*. We claim that

$$_{X^*}[S^*]_{Y^*} = (_Y[S]_X)^{\mathrm{T}}.$$

That is, transposing the matrix of S (computed using the input basis X and output basis Y) gives the matrix of S^* (computed using the input basis Y^* and output basis X^*).

To prove this claim, set $A = {}_Y[S]_X$ and $B = {}_{X^*}[S^*]_{Y^*}$. We must show that $B(i,j) = A(j,i)$ whenever $1 \leq i \leq n$ and $1 \leq j \leq m$. Write $X^* = (f_1, \ldots, f_n)$ and $Y^* = (g_1, \ldots, g_m)$. By definition, the jth column of B is

$$B^{[j]} = [S^*(g_j)]_{X^*} = [g_j \circ S]_{X^*}.$$

To proceed, recall that $[g_j \circ S]_{X^*}$ is the unique n-tuple $(c_1, \ldots, c_n) \in F^n$ such that

$$g_j \circ S = c_1 f_1 + \cdots + c_n f_n.$$

To find the ith coordinate c_i of $B^{[j]}$, evaluate both sides of the previous identity at x_i. The right side is c_i, while the left side is $g_j(S(x_i))$. Therefore, $B(i,j) = B^{[j]}(i) = c_i = g_j(S(x_i))$. On the other hand, since $A^{[i]} = [S(x_i)]_Y$, we have $S(x_i) = \sum_{k=1}^{m} A^{[i]}(k)y_k = \sum_{k=1}^{m} A(k,i)y_k$. Applying g_j to both sides gives $g_j(S(x_i)) = A(j,i)$. We conclude that $B(i,j) = A(j,i)$.

6.8 Matrix/Vector Multiplication versus Evaluation of Maps

Suppose A is an $m \times n$ matrix and $\mathbf{z} = (z_1, \ldots, z_n) \in F^n$ is an n-tuple, regarded as an $n \times 1$ column vector. We can form the matrix-vector product $A\mathbf{z}$, which is an m-tuple ($m \times 1$ column vector) whose ith component is $\sum_{j=1}^{n} A(i, j)z_j$. As shown in Chapter 4, we can also write this formula as $A\mathbf{z} = \sum_{j=1}^{n} z_j A^{[j]}$, which expresses $A\mathbf{z}$ as a linear combination of the columns $A^{[j]}$ of A.

On the other hand, given a linear map $T : V \to W$ and an abstract vector $v \in V$, we can apply the map T to v to obtain another vector $w = T(v) \in W$. Let $X = (x_1, \ldots, x_n)$ and $Y = (y_1, \ldots, y_m)$ be ordered bases for V and W, respectively. We prove that

$$[T(v)]_Y = {}_Y[T]_X \, [v]_X. \tag{6.5}$$

We can state this formula in words: the coordinate vector of the output of T at input v (using basis Y) is found by multiplying the matrix of T (using bases X and Y) by the coordinate vector of v (using basis X). The proof consists of three steps. First, writing $[v]_X = (c_1, \ldots, c_n)$, we know $v = \sum_{j=1}^{n} c_j x_j$ by definition of coordinates. Second, note that $T(v) = \sum_{j=1}^{n} c_j T(x_j)$ by linearity of T. Third, applying the formula at the end of the previous paragraph to $A = {}_Y[T]_X$ and $z = [v]_X$, we get

$$
{}_Y[T]_X \, [v]_X = A\mathbf{z} = \sum_{j=1}^{n} z_j A^{[j]} = \sum_{j=1}^{n} c_j [T(x_j)]_Y = \left[\sum_{j=1}^{n} c_j T(x_j) \right]_Y = [T(v)]_Y.
$$

6.9 Matrix Multiplication versus Composition of Linear Maps

Given matrices $B \in M_{p,m}(F)$ and $A \in M_{m,n}(F)$, the *matrix product* $BA \in M_{p,n}(F)$ is defined by

$$(BA)(i, j) = \sum_{k=1}^{m} B(i, k)A(k, j) \quad \text{for } i = 1, 2, \ldots, p \text{ and } j = 1, 2, \ldots, n.$$

For linear maps $T \in L(V, W)$ and $S \in L(W, U)$, the *composition* $S \circ T \in L(V, U)$ is defined by

$$(S \circ T)(v) = S(T(v)) \quad \text{for all } v \in V.$$

You can check that $S \circ T$ really is linear.

We now compare the concrete operation of matrix multiplication to the abstract operation of composition of linear maps. Let $X = (x_1, \ldots, x_n)$, $Y = (y_1, \ldots, y_m)$, and $Z = (z_1, \ldots, z_p)$ be ordered bases for V, W, and U, respectively. Given linear maps S and T as above, let $A = {}_Y[T]_X \in M_{m,n}(F)$, $B = {}_Z[S]_Y \in M_{p,m}(F)$, and $C = {}_Z[S \circ T]_X \in M_{p,n}(F)$. We claim that $C = BA$, or in other notation,

$$
{}_Z[S \circ T]_X = {}_Z[S]_Y \, {}_Y[T]_X. \tag{6.6}
$$

We know (§4.7) that the jth column of BA is $B(A^{[j]})$. So it suffices to show that $C^{[j]} = B(A^{[j]})$ for $j = 1, 2, \ldots, n$. This follows from the computation:

$$
C^{[j]} = [(S \circ T)(x_j)]_Z = [S(T(x_j))]_Z = {}_Z[S]_Y \, [T(x_j)]_Y = B(A^{[j]}),
$$

where we have used the definition of C, then the definition of $S \circ T$, then (6.5), then the definition of $A^{[j]}$. To summarize: *the matrix of the composition of two linear maps is the product of the matrices of the individual maps, whenever the output basis of the map acting first equals the input basis of the map acting second.*

As one consequence of this formula, suppose $U = V$, $m = n = p$, $X = Z$, and $T \in L(V, W)$ is an invertible linear map with inverse $T^{-1} \in L(W, V)$. Formula (6.6) gives

$$_Y[T]_X \; _X[T^{-1}]_Y = {}_Y[T \circ T^{-1}]_Y = {}_Y[\mathrm{id}_W]_Y = I_n;$$

$$_X[T^{-1}]_Y \; _Y[T]_X = {}_X[T^{-1} \circ T]_X = {}_X[\mathrm{id}_V]_X = I_n.$$

These equations show that

$$({}_Y[T]_X)^{-1} = {}_X[T^{-1}]_Y. \tag{6.7}$$

Thus, *matrix inversion corresponds to functional inversion of the associated linear map.*

6.10 Transition Matrices and Changing Coordinates

Given an element v of some vector space V, we sometimes need to compute the coordinates of v relative to more than one basis. Specifically, suppose $X = (x_1, \ldots, x_n)$ and $Y = (y_1, \ldots, y_n)$ are ordered bases for V, we are given $[v]_X$ for some $v \in V$, and we need to compute $[v]_Y$. Applying formula (6.5) to the identity map $\mathrm{id} : V \to V$, we see that

$$[v]_Y = {}_Y[\mathrm{id}]_X \, [v]_X.$$

Here, $_Y[\mathrm{id}]_X$ is the matrix whose jth column is $[\mathrm{id}(x_j)]_Y = [x_j]_Y$. We call this matrix *the transition matrix from X to Y*, since left-multiplication by this matrix transforms coordinates relative to X into coordinates relative to Y. Since $\mathrm{id}^{-1} = \mathrm{id}$, (6.7) shows that $({}_Y[\mathrm{id}]_X)^{-1} = {}_X[\mathrm{id}]_Y$. Thus, transition matrices between ordered bases are invertible, and the transition matrix from Y to X is the inverse of the transition matrix from X to Y.

To compute the transition matrix $_Y[\mathrm{id}]_X$: Compute the coordinates $[x_j]_Y$ of each x_j relative to Y, and put these coordinates in the jth column of the transition matrix.

To change coordinates from $[v]_X$ **to** $[v]_Y$: Compute the matrix-vector product $[v]_Y = {}_Y[\mathrm{id}]_X \, [v]_X$.

Example 1. For $V = P_{\leq 3}$ (see Table 6.1), we have

$$_{X_1}[\mathrm{id}]_{X_2} = \begin{bmatrix} -8 & 4 & -2 & 1 \\ 12 & -4 & 1 & 0 \\ -6 & 1 & 0 & 0 \\ 1 & 0 & 0 & 0 \end{bmatrix}.$$

For instance, the first column comes from the computation $(t-2)^3 = -8 + 12t - 6t^2 + 1t^3$. You may check directly that

$$_{X_2}[\mathrm{id}]_{X_1} = \begin{bmatrix} 0 & 0 & 0 & 1 \\ 0 & 0 & 1 & 6 \\ 0 & 1 & 4 & 12 \\ 1 & 2 & 4 & 8 \end{bmatrix},$$

and this matrix is the inverse of the preceding one. Suppose $v = 2t^3 - 4t + 1$, so that $[v]_{X_1} = (1, -4, 0, 2)$. Then

$$[v]_{X_2} = {}_{X_2}[\text{id}]_{X_1} [v]_{X_1} = (2, 12, 20, 9),$$

which says that $v = 2(t-2)^3 + 12(t-2)^2 + 20(t-2) + 9(1)$.

Example 2. For the three ordered bases of $\mathbb{R}^3$ in Table 6.1, it is immediate that

$$_{X_1}[\text{id}]_{X_2} = \begin{bmatrix} 1 & 1 & 1 \\ 1 & 2 & 3 \\ 1 & 4 & 9 \end{bmatrix}; \qquad _{X_1}[\text{id}]_{X_3} = \begin{bmatrix} 0 & -1 & 1 \\ 2 & 1 & 0 \\ 1 & 3 & 4 \end{bmatrix}.$$

So, for instance,

$$_{X_2}[\text{id}]_{X_3} = {}_{X_2}[\text{id}]_{X_1} {}_{X_1}[\text{id}]_{X_3} = (_{X_1}[\text{id}]_{X_2})^{-1} {}_{X_1}[\text{id}]_{X_3} = \begin{bmatrix} -4.5 & -4 & 5 \\ 7 & 4 & -7 \\ -2.5 & -1 & 3 \end{bmatrix}.$$

To check the third column of the answer, observe that $(1, 0, 4) = 5(1, 1, 1) - 7(1, 2, 4) + 3(1, 3, 9)$.

It turns out that all invertible matrices occur as transition matrices $_Y[\text{id}]_X$. More precisely, suppose $\dim(V) = n$ and Y is a fixed ordered basis for V. Let $\mathcal{B}$ be the set of all ordered bases for V and $\text{GL}_n(F)$ be the set of all invertible $n \times n$ matrices with entries in F. Define $\phi : \mathcal{B} \to \text{GL}_n(F)$ by setting $\phi(Z) = {}_Y[\text{id}]_Z$ for each ordered basis $Z \in \mathcal{B}$. We show that ϕ is a bijection.

For the proof, we define a map $\phi' : \text{GL}_n(F) \to \mathcal{B}$ that will be shown to be ϕ^{-1}. For any invertible matrix $A \in \text{GL}_n(F)$ with columns $A^{[1]}, \ldots, A^{[n]}$, let

$$\phi'(A) = (L_Y(A^{[1]}), \ldots, L_Y(A^{[n]})).$$

We first show that $\phi'(A)$ is an ordered basis for V, meaning that $\phi'(A) \in \mathcal{B}$. Since A is invertible, the column rank of A is n (see §4.13). Consequently, the list of columns $(A^{[1]}, \ldots, A^{[n]})$ is an ordered basis for F^n. Applying the isomorphism $L_Y : F^n \to V$ to this ordered basis, we see that $\phi'(A)$ is indeed an ordered basis for V. Next, let $A \in \text{GL}_n(F)$ and $Z = (z_1, \ldots, z_n) \in \mathcal{B}$ be arbitrary. Using the definition of $_Y[\text{id}]_Z$, we have the following list of equivalent statements.
(a) $\phi(Z) = A$.
(b) $[z_j]_Y = A^{[j]}$ for $j = 1, 2, \ldots, n$.
(c) $z_j = L_Y(A^{[j]})$ for $j = 1, 2, \ldots, n$.
(d) $Z = \phi'(A)$.
We conclude that ϕ' is the two-sided inverse to ϕ.

The bijectivity of ϕ means that *for any fixed ordered basis Y of V, every invertible matrix in $M_n(F)$ has the form $_Y[\text{id}]_Z$ for a unique ordered basis Z of V.* By similar methods, you can check that *for fixed Y, every invertible matrix in $M_n(F)$ has the form $_X[\text{id}]_Y$ for a unique ordered basis X of V.*

6.11 Changing Bases

Let V and W be vector spaces with $\dim(V) = n$ and $\dim(W) = m$. Suppose $T \in L(V, W)$ is represented by a matrix $A = {}_Y[T]_X$ relative to an ordered basis X for V and an ordered basis Y for W. What happens to A if we change the basis for the domain V from X to another ordered basis X'? Since $T = T \circ \mathrm{id}_V$, formula (6.6) gives the relation

$$_Y[T]_{X'} = {}_Y[T]_X \; {}_X[\mathrm{id}_V]_{X'}.$$

Writing $A' = {}_Y[T]_{X'}$ and $P = {}_X[\mathrm{id}_V]_{X'}$, this says that $A' = AP$. As X' ranges over all possible ordered bases for V (while X remains fixed), we know that P ranges over all invertible $n \times n$ matrices.

For any $A, A' \in M_{m,n}(F)$, say that A is *column-equivalent* to A' iff $A' = AP$ for some invertible $P \in M_n(F)$. You can check that this defines an equivalence relation on $M_{m,n}(F)$. Since P is invertible iff P is a finite product of elementary matrices, we see that A' is column-equivalent to A iff A' can be obtained from A by a finite sequence of elementary column operations, which are encoded by the matrix P (see Chapter 4). For a matrix $A = {}_Y[T]_X$, the equivalence class of A relative to column-equivalence is the set of all matrices ${}_Y[T]_{X'}$, as X' ranges over all ordered bases for V.

Next, we ask: what happens to $A = {}_Y[T]_X$ if we change the basis for the codomain W from Y to Y'? Since $T = \mathrm{id}_W \circ T$, formula (6.6) gives

$$_{Y'}[T]_X = {}_{Y'}[\mathrm{id}_W]_Y \; {}_Y[T]_X.$$

Writing $A' = {}_{Y'}[T]_X$ and $Q = {}_{Y'}[\mathrm{id}_W]_Y$, this says that $A' = QA$. As Y' ranges over all possible ordered bases for W (while Y remains fixed), we know that Q ranges over all invertible $m \times m$ matrices.

For all $A, A' \in M_{m,n}(F)$, say that A is *row-equivalent* to A' iff $A' = QA$ for some invertible $Q \in M_m(F)$. This defines another equivalence relation on $M_{m,n}(F)$. A' is row-equivalent to A iff A' can be obtained from A by a finite sequence of elementary row operations, which are encoded by the matrix Q. If $A = {}_Y[T]_X$, then the equivalence class of A relative to row-equivalence is the set of all matrices ${}_{Y'}[T]_X$, as Y' ranges over all ordered bases for W.

Suppose we change the input basis from X to X' and also change the output basis from Y to Y'. Then

$$_{Y'}[T]_{X'} = {}_{Y'}[\mathrm{id}_W]_Y \; {}_Y[T]_X \; {}_X[\mathrm{id}_V]_{X'}.$$

Accordingly, we say that $A, A' \in M_{m,n}(F)$ are *equivalent* (or, more precisely, *row/column-equivalent*) iff $A' = QAP$ for some invertible matrices $Q \in M_m(F)$ and $P \in M_n(F)$ iff A' can be obtained from A by a finite sequence of elementary row and column operations. The equivalence class of ${}_Y[T]_X$ relative to this equivalence relation is the set of all matrices ${}_{Y'}[T]_{X'}$ as X' and Y' range independently over all ordered bases for V and W, respectively. Given a field F and any matrix $A \in M_{m,n}(F)$, we can find invertible matrices P and Q such that QAP has i, i-entry equal to 1 for $1 \le i \le r = \mathrm{rank}(A)$, and all other entries of QAP are 0 (see §4.12). Taking $A = {}_Y[T]_X$, where $T \in L(V, W)$ is an arbitrary linear map, we obtain the following result.

Projection Theorem. For any $T \in L(V, W)$, there exist ordered bases $X' = (x_1', \ldots, x_n')$ for V and $Y' = (y_1', \ldots, y_m')$ for W such that $T(x_i') = y_i'$ for $1 \le i \le \mathrm{rank}(T)$, and $T(x_i') = 0$ for all $i > \mathrm{rank}(T)$.

We can also give an abstract proof of this theorem that makes no mention of matrices; see Exercise 46. In Chapter 18, we obtain a more general theorem by replacing the field F of scalars by a type of ring called a *principal ideal domain*.

6.12   Algebras of Matrices versus Algebras of Linear Operators

Let F be a field. As defined in §1.3, an F-*algebra* is a structure $(\mathcal{A}, +, \cdot, \star)$ such that $(\mathcal{A}, +, \cdot)$ is an F-vector space, $(\mathcal{A}, +, \star)$ is a ring, and $c \cdot (x \star y) = (c \cdot x) \star y = x \star (c \cdot y)$ for all $x, y \in \mathcal{A}$ and all $c \in F$. According to this definition, all F-algebras are assumed to be associative and have a multiplicative identity. There are more general versions of F-algebras, but they do not occur in this book. A map $g : \mathcal{A} \to \mathcal{B}$ between two F-algebras is called an F-*algebra homomorphism* iff g is F-linear and a ring homomorphism, meaning that

$$g(x + y) = g(x) + g(y), \quad g(c \cdot x) = c \cdot g(x), \; g(x \star y) = g(x) \star g(y), \; \text{and} \; g(1_{\mathcal{A}}) = 1_{\mathcal{B}}$$

for all $x, y \in \mathcal{A}$ and all $c \in F$. An F-*algebra isomorphism* is a bijective F-algebra homomorphism.

In linear algebra, there are two prominent examples of F-algebras. First, we have the concrete F-algebra $M_n(F) = M_{n,n}(F)$ consisting of all square $n \times n$ matrices over the field F. Using the standard matrix operations, $M_n(F)$ is both a vector space (of dimension n^2) and a ring, and $c(AB) = (cA)B = A(cB)$ for all $A, B \in M_n(F)$ and all $c \in F$. The multiplicative identity element is the identity matrix I_n. Second, we have the abstract F-algebra $L(V) = L(V, V)$ consisting of all F-linear operators $T : V \to V$ on an n-dimensional vector space V. We know $L(V)$ is a vector space of dimension n^2 under the standard pointwise operations on linear maps. This vector space becomes a ring (as you can verify) by defining the product of $S, T \in L(V)$ to be the composition of functions $S \circ T$. The identity function id_V is the identity element for $L(V)$.

We have already seen that the vector spaces $M_n(F)$ and $L(V)$ are isomorphic (§6.6). Indeed, for each pair of ordered bases X and Y of V, the map $T \mapsto {}_Y[T]_X$ is a vector space isomorphism $L(V) \cong M_n(F)$. Here, we make the stronger statement that the F-algebras $M_n(F)$ and $L(V)$ are isomorphic. We obtain the required algebra isomorphisms by letting $X = Y$. More specifically, for each ordered basis X of V, consider the vector space isomorphism $M_X = M_{X,X} : L(V) \to M_n(F)$ given by $M_X(T) = {}_X[T]_X$ for $T \in L(V)$. By (6.6), for all $S, T \in L(V)$,

$$M_X(S \circ T) = {}_X[S \circ T]_X = {}_X[S]_X \; {}_X[T]_X = M_X(S)M_X(T).$$

This says that the bijective linear map M_X preserves products. Also, $M_X(\mathrm{id}_V) = {}_X[\mathrm{id}_V]_X = I_n$, so M_X is an algebra isomorphism. Let us write $[T]_X$ as an abbreviation for ${}_X[T]_X$.

We review some previously established facts using the new notation. Suppose $S, T \in L(V)$, $v \in V$, $c \in F$, and $X = (x_1, \ldots, x_n)$ is an ordered basis for V. First, the following statements are equivalent.
(a) $A = [S]_X$.
(b) The jth column of A is $[S(x_j)]_X$ for $j = 1, 2, \ldots, n$.
(c) $S(x_j) = L_X(A^{[j]}) = \sum_{i=1}^n A(i, j) x_i$ for $j = 1, 2, \ldots, n$.
Second, M_X preserves the algebraic structure:

$$[S + T]_X = [S]_X + [T]_X; \quad [cS]_X = c[S]_X; \quad [S \circ T]_X = [S]_X \, [T]_X;$$

$$[S(v)]_X = [S]_X \, [v]_X; \quad [0_V]_X = 0_{n \times n}; \quad [\mathrm{id}_V]_X = I_n.$$

Third, S is an invertible operator in $L(V)$ iff $[S]_X$ is an invertible matrix in $M_n(F)$, in which case

$$[S^{-1}]_X = ([S]_X)^{-1}.$$

It follows from these remarks that $[S^k]_X = ([S]_X)^k$ holds for all integers $k \geq 0$, and for all negative integers k when S is invertible.

An element x in an F-algebra is called *nilpotent* iff $x^m = 0$ for some $m \geq 1$. For example, a square matrix is nilpotent if multiplying the matrix by itself enough times gives the zero matrix. A linear operator on V is nilpotent if applying the operator enough times in succession gives the zero operator on V. From the identity $[S^k]_X = ([S]_X)^k$, we see that $S \in L(V)$ is a nilpotent operator iff $[S]_X \in M_n(F)$ is a nilpotent matrix.

6.13 Similarity of Matrices versus Similarity of Linear Maps

Two $n \times n$ matrices A and A' are called *similar* iff there exists an invertible matrix $P \in M_n(F)$ such that $A' = P^{-1}AP$. Similarity of matrices is readily seen to be an equivalence relation on $M_n(F)$. So, $M_n(F)$ is the disjoint union of equivalence classes relative to the similarity relation.

To understand the abstract significance of similarity of matrices, suppose $T \in L(V)$ and X is an ordered basis for V. Let $A = [T]_X$. If we change the basis for V from X to X', what happens to the representing matrix A for T? Letting $A' = [T]_{X'}$ and $P = {}_X[\mathrm{id}_V]_{X'}$, formula (6.6) gives

$$A' = {}_{X'}[\mathrm{id}_V]_X \; {}_X[T]_X \; {}_X[\mathrm{id}_V]_{X'} = P^{-1}AP.$$

In other words, the matrix $[T]_{X'}$ is similar to the matrix $[T]_X$. Holding X fixed and letting X' vary over all ordered bases for V, we know that P ranges over all invertible $n \times n$ matrices (§6.10). Therefore, the set of all matrices similar to A, namely $\{P^{-1}AP : P \in \mathrm{GL}_n(F)\}$, equals the set of matrices $[T]_{X'}$ that represent the given linear operator T relative to all the possible ordered bases X' for V.

Compare this to the following abstract version of similarity. Two linear maps $T, T' : V \to V$ are called *similar* iff there exists an invertible linear map $S : V \to V$ such that $T' = S^{-1} \circ T \circ S$. Similarity of linear maps defines an equivalence relation on the F-algebra $L(V)$. Choosing a fixed ordered basis X for V and applying the isomorphism $M_X : L(V) \cong M_n(F)$, we see that the linear maps T' and T are similar iff the matrices $A' = [T']_X$ and $A = [T]_X$ are similar. As T' ranges over all linear maps similar to T, A' ranges over all matrices similar to A.

Given a matrix $A = [T]_X \in M_n(F)$, we now have two interpretations for its equivalence class $\{P^{-1}AP : P \in \mathrm{GL}_n(F)\}$ relative to similarity. First, this equivalence class equals the set of all matrices of the form $[T]_{X'}$ as X' ranges over ordered bases for V. Second, this equivalence class equals the set of all matrices $[T']_X$ where T' ranges over all linear maps similar to T. The first interpretation of equivalence classes is fundamental in linear algebra, for the following reason. If we can find a particularly simple matrix in the similarity class of $A = [T]_X$, then the action of the operator T is correspondingly simple for an appropriately chosen ordered basis X' for V. More specifically, if $A' = P^{-1}AP$ is nice (in some sense) when P is the matrix ${}_X[\mathrm{id}]_{X'}$, then the relation $A' = [T]_{X'}$ implies that the action of T on the basis X' is nice in the same way. The next section describes some possibilities for the "nice" matrices we might hope to find in the equivalence class of A under similarity.

6.14 Diagonalizability and Triangulability

Given a matrix $A \in M_n(F)$, we make the following definitions.
(a) A is *diagonal* iff $A(i, j) = 0$ for all $i \neq j$.
(b) A is *upper-triangular* iff $A(i, j) = 0$ for all $i > j$.

(c) A is *strictly upper-triangular* iff $A(i,j) = 0$ for all $i \geq j$.

(d) A is *lower-triangular* iff $A(i,j) = 0$ for all $i < j$.

(e) A is *strictly lower-triangular* iff $A(i,j) = 0$ for all $i \leq j$.

For example, the matrices

$$\begin{bmatrix} 2 & 0 \\ 0 & 3 \end{bmatrix}, \begin{bmatrix} 2 & 1 \\ 0 & 3 \end{bmatrix}, \begin{bmatrix} 0 & 1 \\ 0 & 0 \end{bmatrix}, \begin{bmatrix} 2 & 0 \\ 1 & 3 \end{bmatrix}, \begin{bmatrix} 0 & 0 \\ 1 & 0 \end{bmatrix},$$

are diagonal, upper-triangular, strictly upper-triangular, lower-triangular, and strictly lower-triangular, respectively.

Suppose V is an n-dimensional vector space and $T \in L(V)$ is a linear map. T is called *diagonalizable* iff there exists an ordered basis $X = (x_1, \ldots, x_n)$ for V such that $[T]_X$ is a diagonal matrix. In this case, writing $A = [T]_X$, we have

$$T(x_j) = \sum_{i=1}^{n} A(i,j)x_i = A(j,j)x_j \quad \text{for } j = 1, 2, \ldots, n.$$

This says that every x_j in X is an eigenvector for T with eigenvalue $A(j,j)$. Conversely, if X is an ordered basis such that $T(x_j) = c_j x_j$ for each $x_j \in X$, then $[T]_X$ is the diagonal matrix with $c_1, \ldots, c_n$ on the main diagonal. So far, we have shown: *T is diagonalizable iff there exists an ordered basis X for V consisting of eigenvectors for T.*

Next, define $T \in L(V)$ to be *triangulable* (resp. *strictly triangulable*) iff there exists an ordered basis $X = (x_1, \ldots, x_n)$ for V such that $A = [T]_X$ is an upper-triangular matrix (resp. strictly upper-triangular matrix). By definition, $T(x_j) = \sum_{i=1}^{n} A(i,j)x_i$ for $j = 1, 2, \ldots, n$. So the upper-triangularity condition ($A(i,j) = 0$ for all $i > j$) is equivalent to the identities $T(x_j) = \sum_{i=1}^{j} A(i,j)x_i$ for $j = 1, 2, \ldots, n$. In other words, $[T]_X$ is upper-triangular iff for $j = 1, 2, \ldots, n$, $T(x_j)$ is a linear combination of $x_1, \ldots, x_j$. Similarly, $[T]_X$ is strictly upper-triangular iff $T(x_1) = 0$ and each $T(x_j)$ for $j = 2, 3, \ldots, n$ is a linear combination of $x_1, \ldots, x_{j-1}$. Analogous considerations show that $[T]_X$ is lower-triangular iff each $T(x_j)$ is a linear combination of $x_j, \ldots, x_n$ (similarly for strict lower-triangularity).

We can also phrase these results in terms of similarity of matrices. Given $T \in L(V)$, define $A = [T]_Y$ where Y is any ordered basis for V (e.g., Y might be the standard ordered basis for $V = F^n$). These conditions are equivalent: T is diagonalizable; $A' = [T]_X$ is diagonal for some ordered basis X; the similarity equivalence class of A contains a diagonal matrix; $P^{-1}AP$ is diagonal for some invertible matrix $P \in M_n(F)$. Likewise, these conditions are equivalent: T is triangulable; $A' = [T]_X$ is upper-triangular for some ordered basis X; the similarity equivalence class of A contains an upper-triangular matrix; $P^{-1}AP$ is upper-triangular for some invertible $P \in M_n(F)$. We say that a matrix A is *diagonalizable* (resp. *triangulable*) iff $P^{-1}AP$ is diagonal (resp. upper-triangular) for some invertible $P \in M_n(F)$.

Let us look more closely at the matrix equation $P^{-1}AP = D$, where $A \in M_n(F)$ is given, P is an unknown invertible matrix in $M_n(F)$, and D is an unknown diagonal matrix in $M_n(F)$. Say P has columns $\mathbf{x}_1, \mathbf{x}_2, \ldots, \mathbf{x}_n$, and D has diagonal entries $c_1, \ldots, c_n$. The following conditions are equivalent: $P^{-1}AP = D$; $AP = PD$; $A(P^{[j]}) = P(D^{[j]})$ for $j = 1, 2, \ldots, n$; $A\mathbf{x}_j = c_j \mathbf{x}_j$ for $j = 1, 2, \ldots, n$. Also, invertibility of P is equivalent to requiring that $(\mathbf{x}_1, \ldots, \mathbf{x}_n)$ is an ordered basis for F^n. This calculation shows: *a matrix $A \in M_n(F)$ is diagonalizable iff F^n has an ordered basis consisting of eigenvectors of A.* In this case, we diagonalize A by arranging n linearly independent eigenvectors of A as the columns of a matrix P; then $P^{-1}AP = D$ is diagonal with the corresponding eigenvalues as diagonal entries. So, any algorithm for finding eigenvectors and eigenvalues of A leads to an algorithm for diagonalizing A, if possible. Some numerical methods for approximating the eigenvalues and eigenvectors of A are covered in Chapter 10. As a complement to this approach, Chapters 7 and 8 develop some easy-to-test sufficient conditions for a matrix to be diagonalizable or triangulable.

Next, we discuss an abstract formulation of triangulability. To do so, we introduce the notion of a flag of subspaces. Given an n-dimensional vector space V, a *(complete) flag of subspaces* is a list $(V_0, V_1, \ldots, V_n)$ of subspaces of V such that $\{0\} = V_0 \subseteq V_1 \subseteq \cdots \subseteq V_n = V$ and $\dim(V_i) = i$ for each i. For example, if $X = (x_1, \ldots, x_n)$ is an ordered basis for V and V_i is the subspace spanned by $x_1, \ldots, x_i$, then $(V_0, \ldots, V_n)$ is a flag of subspaces. Call this flag the flag of subspaces *associated with* the ordered basis X. Conversely, we can obtain an ordered basis X for V from a flag of subspaces by letting x_i be any vector in V_i but not in V_{i-1}. Note that X is not uniquely determined by the flag since we have choices for each x_i. We say that a linear map $T \in L(V)$ *stabilizes* a flag of subspaces iff $T[V_i] \subseteq V_i$ for $i = 0, 1, \ldots, n$. T is said to *strictly stabilize* the flag iff $T[V_i] \subseteq V_{i-1}$ for $i = 1, 2, \ldots, n$. From our earlier discussion of triangulability, we see that $[T]_X$ is (strictly) upper-triangular iff T (strictly) stabilizes the flag of subspaces associated with X. Conversely, if T (strictly) stabilizes some flag, and we form an ordered basis X from this flag as described above, then $[T]_X$ is (strictly) upper-triangular. To summarize, $T \in L(V)$ *is (strictly) triangulable iff there exists a flag of subspaces of V (strictly) stabilized by T.*

We claim that *a strictly triangulable operator T must be nilpotent.* Letting $V_0 \subseteq V_1 \subseteq \cdots \subseteq V_n = V$ be a flag strictly stabilized by T, and setting $V_i = \{0\}$ for all $i < 0$, we see by induction on $k \geq 1$ that $T^k[V_j] \subseteq V_{j-k}$ for all $j \leq n$. In particular, $T^n[V] = T^n[V_n] \subseteq V_0 = \{0\}$, so that $T^n = 0$ and T is nilpotent. Translating these comments into matrix terms, we see that a strictly upper-triangular matrix A is nilpotent, with $A^n = 0$.

6.15   Block-Triangular Matrices and Invariant Subspaces

We say a matrix A in $M_n(F)$ is *block-triangular with two blocks* iff there is a k between 1 and $n-1$ such that for all i, j with $k < i \leq n$ and $1 \leq j \leq k$, $A(i, j) = 0$. Such a matrix has the form $\begin{bmatrix} B & C \\ 0 & D \end{bmatrix}$, where $B \in M_k(F)$, $C \in M_{k, n-k}(F)$, 0 is the $(n-k) \times k$ zero matrix, and $D \in M_{n-k}(F)$.

Let W be a subspace of a vector space V. Given $T \in L(V)$, we say W is a *T-invariant subspace* of V iff $T[W] \subseteq W$, which means that $T(w) \in W$ for all $w \in W$. By linearity, it is equivalent to require that $T(x) \in W$ for all x in some ordered basis for W. If W is T-invariant, then the restricted linear map $T|W : W \to V$ can be viewed as a linear map with codomain W. In other words, the restriction $T|W$ belongs to the F-algebra $L(W)$. Furthermore, we have an induced map $T' : V/W \to V/W$ on the quotient vector space V/W, defined by $T'(v + W) = T(v) + W$ for all $v \in V$. This map is well-defined, because for $v, u \in V$, $v + W = u + W$ implies $v - u \in W$, hence $T(v) - T(u) \in W$ by linearity and T-invariance of W, hence $T(v) + W = T(u) + W$. Since T' is also linear, T' belongs to the F-algebra $L(V/W)$.

We can relate these abstract concepts to block-triangular matrices. Suppose that W is a T-invariant subspace of V with $\{0\} \neq W \neq V$, and $X = (x_1, \ldots, x_n)$ is any ordered basis for V such that $X_1 = (x_1, \ldots, x_k)$ is an ordered basis for W. You can check that $X_2 = (x_{k+1} + W, \ldots, x_n + W)$ is an ordered basis for V/W (Exercise 56). We claim: the matrix $A = [T]_X$ is block-triangular with block sizes k and $n-k$; the upper-left $k \times k$ block of A is $[T|W]_{X_1}$; and the lower-right $(n-k) \times (n-k)$ block of A is $[T']_{X_2}$. Pictorially,

$$A = [T]_X = \begin{bmatrix} [T|W]_{X_1} & C \\ 0 & [T']_{X_2} \end{bmatrix},$$

where C is some $k \times (n-k)$ matrix.

To prove this, let us first compute the column $A^{[j]}$ where $1 \le j \le k$. On one hand, since $A = [T]_X$, we have $T(x_j) = \sum_{i=1}^n A(i,j)x_i$, where this expression of $T(x_j)$ in terms of the basis X is unique. On the other hand, letting $B = [T|W]_{X_1}$, we have

$$T(x_j) = T|W(x_j) = \sum_{i=1}^k B(i,j)x_i = \sum_{i=1}^k B(i,j)x_i + 0x_{k+1} + \cdots + 0x_n.$$

We have just written two expressions for $T(x_j)$ as linear combinations of the elements in the basis X. Since the expansion in terms of a basis is unique, we conclude that $A(i,j) = B(i,j)$ for $1 \le i \le k$, and $A(i,j) = 0$ for $k < i \le n$. These conclusions hold for $j = 1, 2, \ldots, k$, so we have confirmed that the upper-left and lower-left blocks of A are given by $B = [T|W]_{X_1}$ and $0 \in M_{n-k,k}(F)$, respectively.

Next, consider a column $A^{[j]}$ with $k < j \le n$. Let $D = [T']_{X_2}$. On one hand, starting from $T(x_j) = \sum_{i=1}^n A(i,j)x_i$, the definition of T' shows that

$$T'(x_j + W) = \sum_{i=1}^n A(i,j)(x_i + W) = \sum_{i=k+1}^n A(i,j)(x_i + W).$$

The last equality holds because $x_i + W = 0 + W$ for $i = 1, 2, \ldots, k$, since $x_i \in W$. On the other hand, the definitions of D and X_2 show that

$$T'(x_j + W) = \sum_{i=1}^{n-k} D(i, j-k)(x_{i+k} + W) = \sum_{i=k+1}^n D(i-k, j-k)(x_i + W).$$

Comparing these expressions, we see that $A(i,j) = D(i-k, j-k)$, so that the lower-right block of A is the matrix $D = [T']_{X_2}$.

Conversely, suppose X is any ordered basis for V such that $A = [T]_X$ is block-triangular with diagonal blocks of size k and $n-k$, in this order. Let W be the subspace spanned by $X_1 = (x_1, \ldots, x_k)$. For $j = 1, 2, \ldots, k$, the block-triangularity of A gives $T(x_j) = \sum_{i=1}^n A(i,j)x_i = \sum_{i=1}^k A(i,j)x_i$, so that $T(x_j) \in W$ for all x_j in X_1. We conclude that W is a T-invariant subspace of V.

6.16 Block-Diagonal Matrices and Reducing Subspaces

A matrix $A \in M_n(F)$ is *block-diagonal with two blocks* iff for some k between 1 and $n-1$, some $B \in M_k(F)$, and some $D \in M_{n-k}(F)$, A has the form

$$\begin{bmatrix} B & 0_{k \times (n-k)} \\ 0_{(n-k) \times k} & D \end{bmatrix}.$$

To relate this concept to linear maps, we must first discuss direct sums. Given a vector space V with subspaces W and Z, we say V is the *direct sum* of W and Z, and we write

$V = W \oplus Z$, to mean that $V = W + Z$ and $W \cap Z = \{0\}$. Equivalently, $V = W \oplus Z$ means that for each $v \in V$, there is exactly one way to write $v = w + z$ where $w \in W$ and $z \in Z$. Suppose $V = W \oplus Z$, X_1 is an ordered basis for W, and X_2 is an ordered basis for Z. You can check that the list X formed by concatenating X_1 and X_2 is an ordered basis for V. This implies $\dim(V) = \dim(W \oplus Z) = \dim(W) + \dim(Z)$.

Given $T \in L(V)$, we say that T is *reducible* iff there exist nonzero T-invariant subspaces W and Z such that $V = W \oplus Z$. In this situation, consider ordered bases $X_1 = (x_1, \ldots, x_k)$ for W, $X_2 = (x_{k+1}, \ldots, x_n)$ for Z, and $X = (x_1, \ldots, x_n)$ for V. We claim that

$$A = [T]_X = \begin{bmatrix} B & 0 \\ 0 & D \end{bmatrix}, \quad \text{where } B = [T|W]_{X_1} \text{ and } D = [T|Z]_{X_2}.$$

On one hand, for each $j \in \{1, 2, \ldots, k\}$, x_j is in the T-invariant subspace W, so $T(x_j) = \sum_{i=1}^{n} A(i,j)x_i = \sum_{i=1}^{k} A(i,j)x_i$. Also, $T(x_j) = T|W(x_j) = \sum_{i=1}^{k} B(i,j)x_i$. On the other hand, for each $j \in \{k+1, \ldots, n\}$, x_j is in the T-invariant subspace Z, so $T(x_j) = \sum_{i=1}^{n} A(i,j)x_i = \sum_{i=k+1}^{n} A(i,j)x_i$. Also, $T(x_j) = T|Z(x_j) = \sum_{i=1}^{n-k} D(i, j-k)x_{i+k}$. The claim follows from these remarks.

Conversely, suppose $X = (x_1, \ldots, x_n)$ is any ordered basis of V such that $A = [T]_X$ is block-diagonal with two diagonal blocks of size k and $n-k$. Let W be the subspace spanned by $(x_1, \ldots, x_k)$ and Z be the subspace spanned by $(x_{k+1}, \ldots, x_n)$. Since X is an ordered basis, it follows that $V = W \oplus Z$. The block-diagonal form of A shows that W and Z are T-invariant subspaces. To summarize: *there exists a decomposition $V = W \oplus Z$ into two T-invariant subspaces with $\dim(W) = k$ iff there exists an ordered basis X of V such that $[T]_X$ is block-diagonal with blocks of size k and $n-k$.*

6.17 Idempotent Matrices and Projections

A matrix $A \in M_n(F)$ is called *idempotent* iff $A^2 = A$. For example, given $0 \leq k \leq n$, the block-diagonal matrix $I_{k,n} = \begin{bmatrix} I_k & 0 \\ 0 & 0 \end{bmatrix} \in M_n(F)$ is idempotent. Analogously, a linear map $T \in L(V)$ is called *idempotent* iff $T \circ T = T$. Applying the F-algebra isomorphism $T \mapsto [T]_X$, where X is an ordered basis of V, we see that *a linear map T is idempotent iff $[T]_X$ is an idempotent matrix.*

Suppose V is an F-vector space and W, Z are subspaces of V such that $V = W \oplus Z$. For all $v \in V$, there exist unique $w \in W$ and $z \in Z$ with $v = w + z$. So, we can define a map $P = P_{W,Z} : V \to V$ by letting $P(v)$ be the unique $w \in W$ appearing in the expression $v = w + z$. We call P the *projection of V onto W along Z*. We claim that P is F-linear and idempotent, with $\text{img}(P) = W$ and $\ker(P) = Z$. Given $v, v' \in V$ and $c \in F$, write $v = w + z$ and $v' = w' + z'$ for unique $w, w' \in W$ and unique $z, z' \in Z$. By definition, $P(v) = w$ and $P(v') = w'$. Note that $v + v' = (w + w') + (z + z')$ with $w + w' \in W$ and $z + z' \in Z$. So the definition of P gives $P(v + v') = w + w' = P(v) + P(v')$. Similarly, $cv = cw + cz$ with $cw \in W$ and $cz \in Z$, so $P(cv) = cw = cP(v)$. Thus P is F-linear. Given $w \in W$, we have $w = w + 0$ with $w \in W$ and $0 \in Z$. So $P(w) = w$ *for all $w \in W$*. Given $v \in V$ with $v = w + z$ as above, we see that $P^2(v) = P(P(v)) = P(w) = w = P(v)$. This holds for all $v \in V$, so $P^2 = P$. Turning to the image of P, the identity $w = P(w)$ for $w \in W$ shows that $W \subseteq \text{img}(P)$. The reverse inclusion $\text{img}(P) \subseteq W$ is immediate from the definition of P, so $\text{img}(P) = W$. As for the kernel, given $z \in Z$, we have $z = 0 + z$ with $0 \in W$ and $z \in Z$. So, $P(z) = 0$, which shows that $Z \subseteq \ker(P)$. Conversely, fix $v \in V$ with $v \in \ker(P)$.

Write $v = w + z$ with $w \in W$ and $z \in Z$. Then $w = P(v) = 0$ shows that $v = z \in Z$. So $\ker(P) \subseteq Z$.

Let us compute $[P]_X$, where X is the ordered basis of V obtained by concatenating an ordered basis $X_1 = (x_1, \ldots, x_k)$ for W with an ordered basis $X_2 = (x_{k+1}, \ldots, x_n)$ for Z. For $j = 1, 2, \ldots, k$, x_j is in W, so we know that $P(x_j) = x_j = 1 x_j + \sum_{i \neq j} 0 x_i$. For $j = k+1, \ldots, n$, x_j is in Z, so we know that $P(x_j) = 0 = \sum_{i=1}^{n} 0 x_i$. We conclude that $[P]_X = I_{k,n}$. It also follows that W and Z are P-invariant subspaces of V with $P|W = \mathrm{id}_W$ and $P|Z = 0_{L(Z)}$.

Next, we show that *for every idempotent $T \in L(V)$, there exist unique subspaces W and Z with $V = W \oplus Z$ and $T = P_{W,Z}$.* To prove uniqueness, suppose we had subspaces W, Z, W_1, Z_1 with $V = W \oplus Z = W_1 \oplus Z_1$ and $P_{W,Z} = T = P_{W_1, Z_1}$. Then $Z = \ker(P_{W,Z}) = \ker(T) = \ker(P_{W_1, Z_1}) = Z_1$ and $W = \mathrm{img}(P_{W,Z}) = \mathrm{img}(T) = \mathrm{img}(P_{W_1, Z_1}) = W_1$. To prove existence, assume $T \in L(V)$ is idempotent, and define $W = \mathrm{img}(T)$ and $Z = \ker(T)$. Let us first check that $V = W \oplus Z$. Given $z \in W \cap Z$, we have $z = T(v)$ for some $v \in V$ and also $T(z) = 0$. So $z = T(v) = T(T(v)) = T(z) = 0$, proving $W \cap Z = \{0\}$. Given $v \in V$, note $v = T(v) + (v - T(v))$ where $T(v) \in W = \mathrm{img}(T)$. Since $T(v - T(v)) = T(v) - T(T(v)) = T(v) - T(v) = 0$, we have $v - T(v) \in Z = \ker(T)$. So $v \in W + Z$, and $V = W \oplus Z$. To finish, we check that $T = P_{W,Z}$. Fix $v \in V$, and write $v = w + z$ with $w \in W$ and $z \in Z$. We have $T(z) = 0$ and $w = T(y)$ for some $y \in V$. Applying T to $v = T(y) + z$ gives

$$T(v) = T(T(y) + z) = T(T(y)) + T(z) = T(y) + 0 = T(y) = w = P_{W,Z}(v),$$

as needed. To summarize: *every idempotent $T \in L(V)$ is the projection $P_{W,Z}$ determined by unique subspaces $W = \mathrm{img}(T)$ and $Z = \ker(T)$, which satisfy $V = W \oplus Z$. T is idempotent iff for some ordered basis X of V and some $k \in \{0, \ldots, n\}$, $[T]_X = I_{k,n}$.*

6.18   Bilinear Maps and Matrices

Let V be an n-dimensional F-vector space. A function $B : V \times V \to F$ is called F-*bilinear* iff for all $v, v', w \in V$ and all $c \in F$, $B(v + v', w) = B(v, w) + B(v', w)$, $B(w, v + v') = B(w, v) + B(w, v')$, and $B(cv, w) = cB(v, w) = B(v, cw)$. It follows from this definition and induction that for all bilinear maps B and all $x_i, y_j \in V$ and all $c_i, d_j \in F$,

$$B\left(\sum_{i=1}^{r} c_i x_i, \sum_{j=1}^{s} d_j y_j\right) = \sum_{i=1}^{r} \sum_{j=1}^{s} c_i d_j B(x_i, y_j). \tag{6.8}$$

Let $\mathrm{BL}(V)$ be the set of all F-bilinear maps on V. You can check that the zero map is F-bilinear; the pointwise sum of two F-bilinear maps is F-bilinear; and any scalar multiple of an F-bilinear map is F-bilinear. So, $\mathrm{BL}(V)$ is a subspace of the vector space of all functions from $V \times V$ to F, and therefore $\mathrm{BL}(V)$ is a vector space.

For each ordered basis $X = (x_1, \ldots, x_n)$ of V, we define a map $N_X : \mathrm{BL}(V) \to M_n(F)$ that turns out to be a vector space isomorphism. Given $B \in \mathrm{BL}(V)$, we define $N_X(B)$ to be the matrix $[B]_X$ with i, j-entry $B(x_i, x_j)$ for $i, j = 1, 2, \ldots, n$. We call $[B]_X$ the *matrix of the bilinear map B relative to the ordered basis X.* You can check that N_X is an F-linear map. We show that N_X is one-to-one and onto.

To prove N_X is one-to-one, suppose $B, C \in \mathrm{BL}(V)$ are two bilinear maps with $N_X(B) = N_X(C)$; we must prove $B = C$. Comparing the entries of $[B]_X$ and $[C]_X$, our assumption tells us that $B(x_i, x_j) = C(x_i, x_j)$ for all $i, j \in \{1, 2, \ldots, n\}$. We must show $B(v, w) = C(v, w)$ for all $v, w \in V$. Fix $v, w \in V$, and write $v = \sum_{i=1}^n c_i x_i$ and $w = \sum_{j=1}^n d_j x_j$ for some $c_i, d_j \in F$. Since B and C satisfy (6.8),

$$B(v, w) = \sum_{i=1}^n \sum_{j=1}^n c_i d_j B(x_i, x_j) = \sum_{i=1}^n \sum_{j=1}^n c_i d_j C(x_i, x_j) = C(v, w).$$

To prove N_X is onto, let $A \in M_n(F)$ be a given matrix; we must find $B \in \mathrm{BL}(V)$ with $N_X(B) = A$. In other words, we must construct an F-bilinear map on V satisfying $B(x_i, x_j) = A(i, j)$ for all $i, j \in \{1, 2, \ldots, n\}$. To do so, take any $v, w \in V$ and write $v = \sum_{i=1}^n c_i x_i$ and $w = \sum_{j=1}^n d_j x_j$ for some $c_i, d_j \in F$. Let

$$B(v, w) = \sum_{i=1}^n \sum_{j=1}^n c_i d_j A(i, j),$$

which gives a well-defined function since expansions in terms of the ordered basis X are unique. We must now check that B is F-bilinear and $B(x_i, x_j) = A(i, j)$ for all $i, j \in \{1, 2, \ldots, n\}$. We prove one required identity and let you check the rest. Given $v, w \in W$ as above, and given $v' = \sum_{i=1}^n c_i' x_i$ with $c_i' \in F$, note that $v + v' = \sum_{i=1}^n (c_i + c_i') x_i$. Therefore, the definition of B gives

$$\begin{aligned}
B(v + v', w) &= \sum_{i=1}^n \sum_{j=1}^n (c_i + c_i') d_j A(i, j) \\
&= \sum_{i=1}^n \sum_{j=1}^n c_i d_j A(i, j) + \sum_{i=1}^n \sum_{j=1}^n c_i' d_j A(i, j) = B(v, w) + B(v', w).
\end{aligned}$$

To summarize: *for each ordered basis X of V, there is an F-vector space isomorphism* $\mathrm{BL}(V) \cong M_n(F)$ *that sends a bilinear map $B \in \mathrm{BL}(V)$ to the matrix $[B]_X$ with i, j-entry* $B(x_i, x_j)$. *Therefore*, $\dim(\mathrm{BL}(V)) = n^2$, *where* $n = \dim(V)$.

6.19 Congruence of Matrices

Two matrices $A, A' \in M_n(F)$ are called *congruent* iff there exists an invertible matrix $P \in M_n(F)$ with $A' = P^{\mathrm{T}} A P$. Congruence of matrices defines an equivalence relation on $M_n(F)$, as you can check. To understand the abstract significance of congruence of matrices, suppose $B \in \mathrm{BL}(V)$ and $X = (x_1, \ldots, x_n)$ and $Y = (y_1, \ldots, y_n)$ are ordered bases of V. Given that $A = [B]_X$, what happens to A if we change the basis from X to Y? We claim that A is replaced by a congruent matrix $A' = P^{\mathrm{T}} A P$, where $P = {}_X[\mathrm{id}]_Y$. In particular, as Y ranges over all ordered bases of V, we know that P ranges over all invertible matrices in $M_n(F)$. So the set of matrices representing a bilinear map B relative to some ordered basis of V equals the set of matrices congruent to A.

We prove the claim by computing $A' = [B]_Y$. We know $y_j = \sum_{i=1}^{n} P(i,j)x_i$ for $j = 1, 2, \ldots, n$. For fixed r, s between 1 and n, the r, s-entry of A' is

$$
\begin{aligned}
B(y_r, y_s) &= B\left(\sum_{i=1}^{n} P(i,r)x_i, \sum_{j=1}^{n} P(j,s)x_j\right) \\
&= \sum_{i=1}^{n}\sum_{j=1}^{n} P(i,r)P(j,s)B(x_i, x_j) \\
&= \sum_{i=1}^{n}\sum_{j=1}^{n} P^{\mathrm{T}}(r,i)A(i,j)P(j,s) = (P^{\mathrm{T}}AP)(r,s).
\end{aligned}
$$

So $A' = P^{\mathrm{T}}AP$ as claimed.

As in the case of similarity, for each congruence class in $M_n(F)$, we would like to find a matrix in that congruence class that is as simple as possible. The answer to this question depends heavily on the field F. See Chapter 14 for more details.

6.20 Real Inner Product Spaces and Orthogonal Matrices

In this section, we consider real vector spaces. A *real inner product space* consists of a real vector space V and a bilinear map $B \in \mathrm{BL}(V)$ such that $B(v, w) = B(w, v)$ for all $v, w \in V$, and $B(v, v) > 0$ for all nonzero $v \in V$. For $v, w \in V$, we write $\langle v, w \rangle_V = B(v, w)$. The most common inner product space is $\mathbb{R}^n$, taking the bilinear form B to be the *dot product*: for all $\mathbf{v} = (v_1, \ldots, v_n)$ and $\mathbf{w} = (w_1, \ldots, w_n)$ in $\mathbb{R}^n$,

$$
B(\mathbf{v}, \mathbf{w}) = \langle \mathbf{v}, \mathbf{w} \rangle_{\mathbb{R}^n} = \mathbf{v} \bullet \mathbf{w} = v_1 w_1 + v_2 w_2 + \cdots + v_n w_n.
$$

We see at once that $[B]_E = I_n$, where $E = (\mathbf{e}_1, \ldots, \mathbf{e}_n)$ is the standard ordered basis of $\mathbb{R}^n$. In any inner product space V, the *norm* or *length* of $v \in V$ is defined by $||v|| = \sqrt{B(v,v)}$. An ordered basis $X = (x_1, \ldots, x_n)$ of V is called an *orthonormal* basis iff for all i, j between 1 and n, $\langle x_i, x_j \rangle_V$ is 1 if $i = j$ and is 0 if $i \neq j$. For example, the standard ordered basis E of $\mathbb{R}^n$ is orthonormal.

Given inner product spaces V and W, a linear map $T \in L(V, W)$ is called *orthogonal* iff for all $v, v' \in V$, $\langle T(v), T(v') \rangle_W = \langle v, v' \rangle_V$. To check if a given linear map T is orthogonal, it suffices to verify that $\langle T(x_i), T(x_j) \rangle_W = \langle x_i, x_j \rangle_V$ for all x_i, x_j in a given ordered basis X of V. To see why, suppose the condition holds for pairs of basis elements. Let $v = \sum_{i=1}^{n} c_i x_i$ and $v' = \sum_{j=1}^{n} d_j x_j$ be any vectors in V, where $c_i, d_j \in \mathbb{R}$. By linearity of T and bilinearity of the inner product, we find that

$$
\begin{aligned}
\langle T(v), T(v') \rangle_W &= \left\langle T\left(\sum_{i=1}^{n} c_i x_i\right), T\left(\sum_{j=1}^{n} d_j x_j\right) \right\rangle_W = \left\langle \sum_{i=1}^{n} c_i T(x_i), \sum_{j=1}^{n} d_j T(x_j) \right\rangle_W \\
&= \sum_{i=1}^{n}\sum_{j=1}^{n} c_i d_j \langle T(x_i), T(x_j) \rangle_W = \sum_{i=1}^{n}\sum_{j=1}^{n} c_i d_j \langle x_i, x_j \rangle_V \\
&= \left\langle \sum_{i=1}^{n} c_i x_i, \sum_{j=1}^{n} d_j x_j \right\rangle_V = \langle v, v' \rangle_V.
\end{aligned}
$$

Let $O(V)$ denote the set of all orthogonal linear maps from V to V. $O(V)$ is not a subspace of the vector space $L(V)$. Instead, $O(V)$ is a subgroup of the group $\text{GL}(V)$ of all invertible linear maps $T \in L(V)$. $O(V)$ is called the *orthogonal group on V*. We check that $O(V) \subseteq \text{GL}(V)$ and ask you to confirm the three closure conditions in the definition of a subgroup (see §1.4). Given $T \in O(V)$, we know T is a linear map from V to V. To see that T is invertible, it suffices to check that $\ker(T) = \{0_V\}$. Given $v \in \ker(T)$, orthogonality of T gives $\langle v, v\rangle_V = \langle T(v), T(v)\rangle_V = \langle 0, 0\rangle_V = 0$, which forces $v = 0$ by the definition of an inner product space.

We have seen that each ordered basis $X = (x_1, \ldots, x_n)$ of V induces a vector space isomorphism $L_X : \mathbb{R}^n \to V$. We now seek conditions for L_X to be an orthogonal map (which can be regarded as an isomorphism of inner product spaces). The following conditions are equivalent.

(a) L_X is orthogonal.
(b) $\langle L_X(\mathbf{e}_i), L_X(\mathbf{e}_j)\rangle_V = \langle \mathbf{e}_i, \mathbf{e}_j\rangle_{\mathbb{R}^n}$ for all $i, j \in \{1, 2, \ldots, n\}$.
(c) For all $i, j \in \{1, 2, \ldots, n\}$, $\langle x_i, x_j\rangle_V$ is 1 for $i = j$ and is 0 for $i \neq j$.
(d) $X = (x_1, \ldots, x_n)$ is an orthonormal basis of V.

So, L_X *is an inner product space isomorphism* $\mathbb{R}^n \cong V$ *iff X is an orthonormal basis of V.*

Let $X = (x_1, \ldots, x_n)$ be an orthonormal basis for the inner product space V. Recall there is an $\mathbb{R}$-algebra isomorphism from $L(V)$ to $M_n(\mathbb{R})$ sending $T \in L(V)$ to $[T]_X \in M_n(\mathbb{R})$. Let us describe the concrete group of matrices that corresponds to the abstract orthogonal group $O(V)$ when we apply this isomorphism. The following conditions are equivalent, for a given $T \in L(V)$ with corresponding matrix $A = [T]_X$.

(a) T is in $O(V)$.
(b) $\langle T(x_i), T(x_j)\rangle_V = \langle x_i, x_j\rangle_V$ for all i, j between 1 and n.
(c) $\langle L_X(A^{[i]}), L_X(A^{[j]})\rangle_V$ is 1 for $i = j$ and is 0 for $i \neq j$.
(d) $\langle A^{[i]}, A^{[j]}\rangle_{\mathbb{R}^n}$ is 1 for $i = j$ and is 0 for $i \neq j$.

Conditions (c) and (d) are equivalent, since L_X is an orthogonal map by the preceding remarks. Call any matrix $A \in M_n(\mathbb{R})$ *orthogonal* iff A satisfies condition (d). Let $O_n(\mathbb{R})$ be the set of all orthogonal matrices in $M_n(\mathbb{R})$. We have shown that *the linear map T is orthogonal iff its matrix A (relative to an orthonormal basis) is orthogonal.*

Our definition of an orthogonal matrix A states that the column vectors $A^{[1]}, \ldots, A^{[n]}$ must be orthonormal. Geometrically, this condition says that each column of A is a unit vector (has norm 1), and any two distinct columns of A are perpendicular. We can restate this condition by noting that $\langle A^{[i]}, A^{[j]}\rangle_{\mathbb{R}^n} = A^{[i]} \bullet A^{[j]}$ is the i, j-entry of the matrix $A^{\mathrm{T}}A$. So, the following conditions are equivalent for a given $A \in M_n(\mathbb{R})$ with rows $A_{[i]}$ and columns $A^{[j]}$:

(a) A is an orthogonal matrix.
(b) The columns $A^{[1]}, \ldots, A^{[n]}$ form an orthonormal list in $\mathbb{R}^n$.
(c) $A^{[i]} \bullet A^{[j]}$ is 1 for $i = j$ and is 0 for $i \neq j$.
(d) $(A^{\mathrm{T}}A)(i, j)$ is 1 for $i = j$ and is 0 for $i \neq j$.
(e) $A^{\mathrm{T}}A = I_n$ (the identity matrix).
(f) A is invertible and $A^{-1} = A^{\mathrm{T}}$.
(g) $AA^{\mathrm{T}} = I_n$.
(h) $(AA^{\mathrm{T}})(i, j)$ is 1 for $i = j$ and is 0 for $i \neq j$.
(i) $A_{[i]} \bullet A_{[j]}$ is 1 for $i = j$ and is 0 for $i \neq j$.
(j) The rows $A_{[1]}, \ldots, A_{[n]}$ form an orthonormal list in $\mathbb{R}^n$.
(k) A^{T} is an orthogonal matrix.
(l) A is invertible and A^{-1} is an orthogonal matrix.

For more on why (e), (f), and (g) are equivalent, see §4.6.

Consider a transition matrix $P = {}_Y[\text{id}]_X$ between two orthonormal bases X and Y for an inner product space V. By orthonormality of Y, the map L_Y^{-1} sending $v \in V$ to $[v]_Y$ is orthogonal. By orthonormality of X, $\langle x_i, x_j \rangle_V$ is 1 for $i = j$ and is 0 for $i \neq j$. It follows that the columns $[x_1]_Y, \ldots, [x_n]_Y$ of P are orthonormal in $\mathbb{R}^n$, so that P is an orthogonal matrix. You can check that every orthogonal matrix in $O_n(\mathbb{R})$ has this form for some choice of orthonormal bases X and Y. Define two matrices $A, A' \in M_n(\mathbb{R})$ to be *orthogonally similar* iff $A' = P^{-1}AP = P^{\mathrm{T}}AP$ for some orthogonal matrix $P \in O_n(\mathbb{R})$. Orthogonal similarity is an equivalence relation on $M_n(\mathbb{R})$. Given $T \in L(V)$, the equivalence class of $A = [T]_X$ consists of all possible matrices that represent T relative to different choices of orthonormal bases for V.

6.21   Complex Inner Product Spaces and Unitary Matrices

In this section, we consider the complex version of inner product spaces. A *complex inner product space* consists of a complex vector space V and a map $B : V \times V \to \mathbb{C}$, written $B(v, w) = \langle v, w \rangle_V$, satisfying the following identities for all $v, v', w \in V$ and all $c \in \mathbb{C}$:

$$\langle v + v', w \rangle_V = \langle v, w \rangle_V + \langle v', w \rangle_V; \quad \langle cv, w \rangle_V = c\langle v, w \rangle_V; \quad \langle w, v \rangle_V = \overline{\langle v, w \rangle_V};$$
$$\langle v, v \rangle \in \mathbb{R}_{>0} \text{ for all } v \neq 0.$$

The notation $\overline{z}$ stands for the complex conjugate of the complex number z, namely, $\overline{a + ib} = a - ib$ for $a, b \in \mathbb{R}$. It follows from the preceding identities that

$$\langle w, v + v' \rangle_V = \langle w, v \rangle_V + \langle w, v' \rangle_V \quad \text{and} \quad \langle w, cv \rangle_V = \overline{c}\langle w, v \rangle_V.$$

An ordered basis $X = (x_1, \ldots, x_n)$ of V is called *orthonormal* iff $\langle x_i, x_j \rangle_V$ is 1 if $i = j$ and is 0 if $i \neq j$. For example, $\mathbb{C}^n$ is a complex inner product space with inner product

$$\langle \mathbf{v}, \mathbf{w} \rangle_{\mathbb{C}^n} = v_1\overline{w_1} + v_2\overline{w_2} + \cdots + v_n\overline{w_n} \quad \text{for } \mathbf{v} = (v_1, \ldots, v_n), \mathbf{w} = (w_1, \ldots, w_n) \in \mathbb{C}^n.$$

Using this inner product, the standard ordered basis $E = (\mathbf{e}_1, \ldots, \mathbf{e}_n)$ of $\mathbb{C}^n$ is orthonormal.

For complex inner product spaces V and W, $T \in L(V, W)$ is called a *unitary map* iff $\langle T(v), T(v') \rangle_W = \langle v, v' \rangle_V$ for all $v, v' \in V$. As in the real case, it suffices to check that $\langle T(x_i), T(x_j) \rangle_W = \langle x_i, x_j \rangle_V$ for all x_i, x_j in an ordered basis X of V. Let $U(V)$ be the set of all unitary maps in $L(V)$. You can check that $U(V)$ is a subgroup of $\mathrm{GL}(V)$.

As in the real case, we see that the isomorphism $L_X : \mathbb{C}^n \to V$ is a unitary map iff X is an orthonormal basis of V. Given an orthonormal basis X and a linear map $T \in L(V)$, T is in $U(V)$ iff $A = [T]_X$ is a matrix with orthonormal columns in $\mathbb{C}^n$. Call any matrix $A \in M_n(\mathbb{C})$ with orthonormal columns *unitary*, and let $U_n(\mathbb{C})$ be the set of all unitary matrices in $M_n(\mathbb{C})$. Recall that the conjugate-transpose of A is the matrix A^* with i, j-entry $A^*(i, j) = \overline{A(j, i)}$. The orthonormality of the columns of A in $\mathbb{C}^n$ is equivalent to the matrix identity $A^*A = I_n$, which is equivalent to $AA^* = I_n$ (see §4.6). So, A is unitary iff A is invertible with $A^{-1} = A^*$ iff the rows of A are orthonormal in $\mathbb{C}^n$.

As in the real case, a transition matrix $P = {}_Y[\mathrm{id}]_X$ between two orthonormal bases of a complex inner product space V must be unitary. Moreover, all unitary matrices have this form for some choice of orthonormal bases X and Y. Define two matrices $A, A' \in M_n(\mathbb{C})$ to be *unitarily similar* iff $A' = P^{-1}AP = P^*AP$ for some unitary $P \in U_n(\mathbb{C})$. Unitary similarity is an equivalence relation on $M_n(\mathbb{C})$. Given $T \in L(V)$, the equivalence class of $A = [T]_X$ consists of all possible matrices that represent T relative to different choices of orthonormal bases for V.

6.22   Summary

1. *Notation.* Table 6.2 recalls the notation used in this chapter for various vector spaces, algebras, and groups.

TABLE 6.2
Summary of notation.

Symbol	Meaning
F^n	vector space of n-tuples (column vectors) with entries in F
$M_{m,n}(F)$	vector space of $m \times n$ matrices with entries in F
$M_n(F)$	F-algebra of $n \times n$ matrices
$\mathrm{GL}_n(F)$	group of invertible $n \times n$ matrices with entries in F
$O_n(\mathbb{R})$	group of orthogonal $n \times n$ real matrices ($A^{-1} = A^{\mathrm{T}}$)
$U_n(\mathbb{C})$	group of unitary $n \times n$ complex matrices ($A^{-1} = A^*$)
V, W	abstract F-vector spaces
$L(V, W)$	vector space of F-linear maps $T : V \to W$
$L(V)$	F-algebra of F-linear maps $T : V \to V$
$\mathrm{BL}(V)$	vector space of bilinear maps $B : V \times V \to F$
$\mathrm{GL}(V)$	group of invertible F-linear maps $T : V \to V$
$O(V)$	group of orthogonal linear maps on real inner product space V
$U(V)$	group of unitary linear maps on complex inner product space V

2. *Main Isomorphisms.* Table 6.3 reviews the isomorphisms between abstract algebraic structures and concrete versions of these structures covered in this chapter.

3. *Linear Combination Maps.* Given any list $X = (x_1, \ldots, x_n)$ of vectors in some abstract F-vector space V, we have a linear map $L_X : F^n \to V$ given by $L_X(c_1, \ldots, c_n) = \sum_{i=1}^{n} c_i x_i$ for $c_i \in F$. L_X is surjective iff X spans V. L_X is injective iff X is linearly independent. L_X is bijective iff X is an ordered basis for V. An F-vector space V has dimension n iff there exists a vector space isomorphism $V \cong F^n$. Any two n-dimensional F-vector spaces are isomorphic. The function $X \mapsto L_X$ is a bijection from the set of ordered bases for V onto the set of vector space isomorphisms $L : F^n \to V$. The inverse bijection sends such

TABLE 6.3
Main isomorphisms between abstract linear-algebraic objects and concrete matrix-theoretic objects.

Isomorphism	Type of iso.	Isomorphism depends on:
$V \cong F^n$	vector space	ordered basis X of V
$L(V, W) \cong M_{m,n}(F)$	vector space	ordered bases X of V and Y of W
$L(V) \cong M_n(F)$	F-algebra	ordered basis X of V
$\mathrm{BL}(V) \cong M_n(F)$	vector space	ordered basis X of V
$\mathrm{GL}(V) \cong \mathrm{GL}_n(F)$	group	ordered basis X of V
$V \cong \mathbb{R}^n$ or $\mathbb{C}^n$	inner prod. space	orthonormal basis X of V
$O(V) \cong O_n(\mathbb{R})$	group	orthonormal basis X of V
$U(V) \cong U_n(\mathbb{C})$	group	orthonormal basis X of V

an isomorphism L to the ordered basis $(L(\mathbf{e}_1), \ldots, L(\mathbf{e}_n))$, where $\mathbf{e}_1, \ldots, \mathbf{e}_n$ are the standard basis vectors in F^n.

4. *Vector Spaces of n-tuples vs. Abstract Vectors.* If $X = (x_1, \ldots, x_n)$ is an ordered basis for an abstract F-vector space V, the coordinate map $v \mapsto [v]_X$ is an F-vector space isomorphism $V \cong F^n$. In detail, for all $c, c_i \in F$ and all $v, w \in V$:

$$(c_1, \ldots, c_n) = [v]_X \Leftrightarrow v = L_X(c_1, \ldots, c_n) \Leftrightarrow v = c_1 x_1 + \cdots + c_n x_n;$$

$$[v + w]_X = [v]_X + [w]_X; \quad [cv]_X = c[v]_X.$$

5. *Vector Spaces of Matrices and Linear Maps.* If $X = (x_1, \ldots, x_n)$ and $Y = (y_1, \ldots, y_m)$ are ordered bases for F-vector spaces V and W, then the map $T \mapsto {}_Y[T]_X$ is an F-vector space isomorphism $L(V, W) \cong M_{m,n}(F)$. In detail, for all $A \in M_{m,n}(F)$, $S, T \in L(V, W)$, $c \in F$:

$$A = {}_Y[T]_X \Leftrightarrow \forall j, A^{[j]} = [T(x_j)]_Y \Leftrightarrow \forall j, T(x_j) = L_Y(A^{[j]}) = \sum_{i=1}^{n} A(i,j) y_i;$$

$$_Y[S + T]_X = {}_Y[S]_X + {}_Y[T]_X; \quad {}_Y[cT]_X = c({}_Y[T]_X).$$

If $R \in L(W, U)$ where U has ordered basis Z, then

$$_Z[R \circ T]_X = {}_Z[R]_Y \, {}_Y[T]_X.$$

6. *Algebras of Matrices and Operators.* For each ordered basis $X = (x_1, \ldots, x_n)$ of an F-vector space V, the map $T \mapsto [T]_X$ is an F-algebra isomorphism $L(V) \cong M_n(F)$. In detail, for all $A \in M_n(F)$, $S, T \in L(V)$, $c \in F$, $v \in V$:

$$A = [T]_X \Leftrightarrow \forall j \, A^{[j]} = [T(x_j)]_X \Leftrightarrow \forall j, T(x_j) = L_X(A^{[j]}) = \sum_{i=1}^{n} A(i,j) x_i;$$

$$[S + T]_X = [S]_X + [T]_X; \quad [cS]_X = c[S]_X; \quad [S \circ T]_X = [S]_X \, [T]_X;$$

$$[S(v)]_X = [S]_X \, [v]_X; \quad [0_{L(V)}]_X = 0_{M_n(F)}; \quad [\mathrm{id}_V]_X = I_n;$$

$$\forall k \in \mathbb{Z}_{\geq 0}, [S^k]_X = ([S]_X)^k; \text{this holds for all } k \in \mathbb{Z} \text{ when } S \text{ is invertible.}$$

7. *Transition Matrices.* For ordered bases X and Y of V, the *transition matrix from X to Y* is the matrix $A = {}_Y[\mathrm{id}_V]_X$ defined by $A^{[j]} = [x_j]_Y$ for $j = 1, 2, \ldots, n$. We have $x_j = \sum_{i=1}^{n} A(i,j) y_i$, $A^{-1} = {}_X[\mathrm{id}]_Y$, and ${}_X[\mathrm{id}]_X = I_n$. For each fixed X, the map $Y \mapsto {}_Y[\mathrm{id}]_X$ is a bijection from the set of ordered bases of V to the set of invertible matrices $\mathrm{GL}_n(F)$. For each fixed Y, the map $X \mapsto {}_Y[\mathrm{id}]_X$ is also a bijection between these sets. If $P = {}_Y[\mathrm{id}]_X$ and $T \in L(V)$, then $[T]_X = P^{-1}[T]_Y P$.

8. *Bilinear Maps and Congruence of Matrices.* For each ordered basis X of V, there is a vector space isomorphism $\mathrm{BL}(V) \cong M_n(F)$ that sends a bilinear map B to the matrix $[B]_X$ with i, j-entry $B(x_i, x_j)$. Changing the basis from X to Y replaces $A = [B]_X$ by $A' = [B]_Y = P^T A P$, where $P = {}_X[\mathrm{id}]_Y$. So, the set of all matrices congruent to A is the set of matrices representing B relative to some ordered basis of V.

9. *Inner Product Spaces.* For each orthonormal basis X of a real inner product space V, the map $v \mapsto [v]_X$ (and its inverse L_X) are isomorphisms preserving the inner product. Orthogonal maps correspond to orthogonal matrices under the isomorphism $T \mapsto [T]_X$, for $T \in L(V)$. The transition matrix between two orthonormal bases is orthogonal. The equivalence class of a matrix $A = [T]_X$ under orthogonal similarity is the set of all matrices that represent T as we vary the orthonormal basis X. Similar facts holds for complex inner product spaces, replacing "orthogonal" by "unitary" throughout.

10. *Equivalence Relations on Matrices.* Table 6.4 summarizes some commonly used equivalence relations on rectangular and square matrices.

TABLE 6.4
Equivalence relations on matrices.

Equiv. relation	Definition of $A \sim A'$	Abstract significance
col. equivalence	$\exists P \in \mathrm{GL}_n(F), A' = AP$	changes input basis in $_Y[T]_X$
row equivalence	$\exists Q \in \mathrm{GL}_m(F), A' = QA$	changes output basis in $_Y[T]_X$
row/col equiv.	$\exists P \in \mathrm{GL}_n(F), \exists Q \in \mathrm{GL}_m(F),$ $A' = QAP$	changes input/output bases in $_Y[T]_X$
similarity	$\exists P \in \mathrm{GL}_n(F), A' = P^{-1}AP$	go from $[T]_X$ to $[T]_Y$
congruence	$\exists P \in \mathrm{GL}_n(F), A' = P^{\mathrm{T}}AP$	change basis for bilinear map
orthogonal sim.	$\exists P \in O_n(\mathbb{R}), A' = P^{-1}AP = P^{\mathrm{T}}AP$	orthonormal basis change $(F = \mathbb{R})$
unitary sim.	$\exists P \in U_n(\mathbb{C}), A' = P^{-1}AP = P^*AP$	orthonormal basis change $(F = \mathbb{C})$

11. *Diagonalizability.* These conditions on a linear map $T : V \to V$ are equivalent: T is diagonalizable; $[T]_X$ is diagonal for some ordered basis X of V; V has an ordered basis consisting of eigenvectors of T; for any ordered basis Y of V, $[T]_Y$ is similar to a diagonal matrix; for any ordered basis Y of V, $P^{-1}[T]_Y P$ is diagonal for some invertible $P \in M_n(F)$.

12. *Triangulability.* A *flag of subspaces* of V is a chain of subspaces $\{0\} = V_0 \subseteq V_1 \subseteq \cdots \subseteq V_n = V$ with $\dim(V_i) = i$ for all i. These conditions on a linear map $T : V \to V$ are equivalent: T is triangulable; $[T]_X$ is upper-triangular for some ordered basis X of V; T stabilizes some flag of subspaces of V; for any ordered basis Y of V, $[T]_Y$ is similar to an upper-triangular matrix; for any ordered basis Y of V, $P^{-1}[T]_Y P$ is upper-triangular for some invertible $P \in M_n(F)$.

13. *Reducibility and Invariant Subspaces.* Given $T \in L(V)$ and an ordered basis $X = (x_1, \ldots, x_n)$ of V, the matrix $A = [T]_X$ has the block-triangular form $\begin{bmatrix} B_{k \times k} & C \\ 0 & D_{(n-k) \times (n-k)} \end{bmatrix}$ iff $X_1 = (x_1, \ldots, x_k)$ spans a T-invariant subspace W of V (which is a subspace with $T(w) \in W$ for all $w \in W$). In this case, $B = [T|W]_{X_1}$ and $D = [T']_{X_2'}$, where $T|W \in L(W)$ is the restriction of T to W, $T' : V/W \to V/W$ in $L(V/W)$ is given by $T'(v + W) = T(v) + W$ for $v \in V$, and $X_2' = (x_{k+1} + W, \ldots, x_n + W)$. Moreover, $A = [T]_X$ is block-diagonal with two diagonal blocks of size k and $n - k$ iff $C = 0$ iff $X_1 = (x_1, \ldots, x_k)$ and $X_2 = (x_{k+1}, \ldots, x_n)$ both span T-invariant subspaces of V iff T is reduced by the pair of subspaces W and Z spanned by X_1 and X_2. In this case, $B = [T|W]_{X_1}$ and $D = [T|Z]_{X_2}$.

14. *Idempotent Matrices and Projections.* A linear map $T \in L(V)$ is *idempotent* iff $T \circ T = T$. Given $V = W \oplus Z$, the projection on W along Z is the map $P_{W,Z}$ that sends each $v \in V$ to the unique $w \in W$ such that $v = w + z$ for some $z \in Z$. The map $P_{W,Z}$ is idempotent with kernel Z and image W, and $[P_{W,Z}]_X = I_{k,n}$, where X is obtained by concatenating ordered bases for W and Z. Every idempotent $T \in L(V)$ has the form $P_{W,Z}$ for unique subspaces W, Z with $V = W \oplus Z$; here, $W = \text{img}(T)$ and $Z = \text{ker}(T)$.

15. *Computational Procedures.* Assume $X = (x_1, \dots, x_n)$ and $Y = (y_1, \dots, y_m)$ are ordered bases of F-vector spaces V and W.

  - To **compute coordinates** $[v]_X$ when given $v \in V$ and X: solve the equation $v = c_1 x_1 + \cdots + c_n x_n$ for the unknowns c_j in F; output the answer $[v]_X = (c_1, \dots, c_n)$.

  - To **compute the matrix** $_Y[T]_X$ **of a linear map** when given T and X and Y: for $j = 1, 2, \dots, n$, find $T(x_j)$; compute the coordinate vector of $T(x_j)$ relative to Y, namely $[T(x_j)]_Y$; and write these coordinates as the jth column of the matrix.

  - To **find** $[T]_X$ given T and X: perform the algorithm in the preceding item taking $Y = X$.

  - To **find the transition matrix from** X **to** Y: compute the coordinate vector $[x_j]_Y$ of each x_j relative to Y, and write this vector as the jth column of the transition matrix. Alternatively, compute the matrix inverse of the transition matrix from Y to X if the latter matrix is already available. (For instance, this occurs when X is the standard ordered basis for F^n and each $\mathbf{y}_j \in Y$ is presented as a column vector, so $[\mathbf{y}_j]_X = \mathbf{y}_j$ for all j.)

  - To **change coordinates** from $[v]_X$ to $[v]_Y$ where $[v]_X$ and X and Y are given: find the transition matrix $_Y[\text{id}]_X$ from X to Y; compute the matrix-vector product $[v]_Y = {_Y[\text{id}]_X} [v]_X$.

  - To **find the matrix of a linear map relative to a new basis** where $[T]_X$ and X and Y are given: find the transition matrix $P = {_X[\text{id}]_Y}$ and its inverse $P^{-1} = {_Y[\text{id}]_X}$; compute the matrix product $[T]_Y = P^{-1}[T]_X P$.

  - To **diagonalize a diagonalizable linear operator** T: compute an ordered basis X for V consisting of eigenvectors for T (for instance, by finding roots of the characteristic polynomial and solving linear equations); output $[T]_X$, which is the diagonal matrix with the eigenvalues corresponding to each $x_j \in X$ appearing on the main diagonal.

  - To **diagonalize a diagonalizable matrix** $A \in M_n(F)$: Find a linearly independent list of n eigenvectors of A in F^n, say $\mathbf{x}_1, \dots, \mathbf{x}_n$, with associated eigenvalues $c_1, \dots, c_n$. Let P be the matrix with columns $\mathbf{x}_1, \dots, \mathbf{x}_n$. Let D be the diagonal matrix with entries $c_1, \dots, c_n$ on the diagonal. Then A is similar to the diagonal matrix D via $P^{-1}AP = D$.

  - To **compute the matrix** $[B]_X$ **of a bilinear map** B: let the i, j-entry be $B(x_i, x_j)$ for all $i, j \in \{1, 2, \dots, n\}$.

  - To **find the matrix of a bilinear map relative to a new basis**: compute $P = {_X[\text{id}]_Y}$ and $[B]_Y = P^{\mathrm{T}}[B]_X P$.

Our dictionary for translating between matrix theory and linear algebra is presented in Tables 6.2, 6.3, 6.4, and 6.5.

TABLE 6.5

Dictionary connecting concrete matrix theory and abstract linear algebra.

Concept in Matrix Theory	Connecting Formula	Concept in Linear Algebra
concrete vector space F^n of n-tuples	isomorphisms $L_X : F^n \cong V$ $L_X(c_1, \ldots, c_n) = \sum_{i=1}^n c_i x_i$	abstract F-vector space V with ordered basis $X = (x_1, \ldots, x_n)$
n-tuple $(c_1, \ldots, c_n) \in F^n$	$[v]_X = (c_1, \ldots, c_n)$ iff $v = c_1 x_1 + \cdots + c_n x_n$	abstract vector $v \in V$
componentwise operations on n-tuples	$[v + w]_X = [v]_X + [w]_X,$ $[cv]_X = c[v]_X$	abstract vector addition and scalar multiplication in V
set of all isomorphisms $L : F^n \cong V$	bijections: $X \mapsto L_X$ and $L \mapsto (L(\mathbf{e}_1), \ldots, L(\mathbf{e}_n))$	set of all ordered bases X of V
vector space $M_{m,n}(F)$ of $m \times n$ matrices	isomorphisms $T \mapsto {}_Y[T]_X$ indexed by ordered bases	vector space $L(V, W)$ of linear maps from V to W
matrix $A \in M_{m,n}(F)$ with columns $A^{[j]}$ and entries $A(i, j)$	$A = {}_Y[T]_X$ iff $A^{[j]} = [T(x_j)]_Y$ iff $\forall j,\ T(x_j) = \sum_{i=1}^m A(i,j) y_i$	linear map $T \in L(V, W)$ with X an ordered basis for V and Y an ordered basis for W
matrix addition $(A + B)(i, j) = A(i, j)$ $+B(i, j)$	${}_Y[S + T]_X = {}_Y[S]_X + {}_Y[T]_X$	addition of linear maps $(S + T)(v) = S(v) + T(v)$
scalar multiple of a matrix $(cA)(i, j) = c(A(i, j))$	${}_Y[cT]_X = c({}_Y[T]_X)$	scalar multiple of linear map $(cT)(v) = c \cdot T(v)$
matrix/vector multiplication $(Av)_i = \sum_{j=1}^n A(i, j) v_j$	$[T(v)]_Y = {}_Y[T]_X\,[v]_X$	evaluation of linear map T on input v gives $T(v)$
matrix multiplication $(BA)(i, j) = \sum_k B(i, k)$ $A(k, j)$	${}_Z[S \circ T]_X = {}_Z[S]_Y\ {}_Y[T]_X$	composition of linear maps $(S \circ T)(v) = S(T(v))$
inverse of a matrix $AA^{-1} = I_n = A^{-1}A$	${}_X[T^{-1}]_Y = ({}_Y[T]_X)^{-1}$	inverse of a map $T \circ T^{-1} = \mathrm{id}_W,\ T^{-1} \circ T = \mathrm{id}_V$
matrix transpose $A^{\mathrm{T}}(i, j) = A(j, i)$	${}_{X^*}[T^*]_{Y^*} = ({}_Y[T]_X)^{\mathrm{T}}$ $(X^*, Y^*$ are dual bases$)$	dual map $T^* \in L(W^*, V^*)$ $T^*(g) = g \circ T : V \to F$
$m \times n$ matrix A	${}_{E'}[L^A]_E = A$ $(E, E'$ are standard bases$)$	left multiplication by A $L^A : F^n \to F^m,\ L^A(\mathbf{v}) = A\mathbf{v}$
F-algebra $M_n(F)$	isomorphisms $T \mapsto [T]_X$	F-algebra $L(V)$
identity matrix I_n	$[\mathrm{id}_V]_X = I_n$	identity map $\mathrm{id}_V : V \to V$ $\mathrm{id}_V(v) = v$ for $v \in V$
zero matrix 0_n	$[0_V]_X = 0_n$	zero map $0_V : V \to V$ $0_V(v) = 0 \in V$ for $v \in V$

6.23 Exercises

Unless otherwise stated, assume that F is a field, V is an F-vector space with ordered basis $X = (x_1, \ldots, x_n)$, and W is an F-vector space with ordered basis $Y = (y_1, \ldots, y_m)$. Exercises involving $\mathbb{R}^3$, $P_{\leq 3}$, $M_2(\mathbb{R})$, and $\mathbb{C}$ use the notation from Table 6.1 on page 137.

1. For each vector space V in Table 6.1, prove that X_1, X_2, and X_3 are ordered bases for V. Compute $\dim(V)$.

2. For each vector $\mathbf{v} \in \mathbb{R}^3$, compute $[\mathbf{v}]_{X_1}$, $[\mathbf{v}]_{X_2}$, and $[\mathbf{v}]_{X_3}$.
   (a) $(1, 2, 4)$ (b) $(-2, 5, 2)$ (c) $(0, 0, c)$

3. For each vector $v \in P_{\leq 3}$, compute $[v]_{X_1}$, $[v]_{X_2}$, and $[v]_{X_3}$.
   (a) $1 + t + t^2 + t^3$ (b) $(t+1)^2$ (c) $t(3t-4)(2t+1)$

4. For each vector $v \in M_2(\mathbb{R})$, compute $[v]_{X_1}$, $[v]_{X_2}$, and $[v]_{X_3}$.
   (a) $\begin{bmatrix} 1 & 1 \\ 1 & 0 \end{bmatrix}$ (b) $\begin{bmatrix} 4 & 4 \\ -1 & -1 \end{bmatrix}$ (c) $\begin{bmatrix} \cos\theta & -\sin\theta \\ \sin\theta & \cos\theta \end{bmatrix}$

5. For each vector $v \in \mathbb{C}$, compute $[v]_{X_1}$, $[v]_{X_2}$, and $[v]_{X_3}$.
   (a) $-i$ (b) $7 - 5i$ (c) $e^{\pi i/4}$

6. (a) Which vector $\mathbf{v} \in \mathbb{R}^3$ satisfies $[\mathbf{v}]_{X_2} = (3,2,1)$?
   (b) Which vector $v \in P_{\leq 3}$ satisfies $[v]_{X_3} = (2,-1,0,5)$?
   (c) Which vector $v \in \mathbb{C}$ satisfies $[v]_{X_3} = (2,-3)$?

7. For all $c \in \mathbb{R}$, find $[(t+c)^3]_{X_1}$ and $[(t+c)^3]_{X_2}$.

8. (a) Find a nonzero $A \in M_2(\mathbb{R})$ such that $[A]_{X_1} = 6[A]_{X_3}$.
   (b) Find all $c \in \mathbb{R}$ such that for some nonzero $A \in M_2(\mathbb{R})$, $[A]_{X_1} = c[A]_{X_3}$.

9. Let X' be obtained from X by switching the positions of x_i and x_j. Show X' is an ordered basis of V. For $v \in V$, how is $[v]_{X'}$ related to $[v]_X$?

10. Let X' be obtained from X by replacing x_i by bx_i, where $b \in F$ is nonzero. Show X' is an ordered basis of V. For $v \in V$, how is $[v]_{X'}$ related to $[v]_X$?

11. Let X' be obtained from X by replacing x_i by $x_i + ax_j$, where $a \in F$ and $j \neq i$. Show X' is an ordered basis of V. For $v \in V$, how is $[v]_{X'}$ related to $[v]_X$?

12. Define $T : \mathbb{R}^3 \to \mathbb{R}^2$ by $T(a,b,c) = (a+b-c, 3b+2c)$ for $a,b,c \in \mathbb{R}$.
    Let $Y = ((1,0),(0,1))$ and $Z = ((1,3),(-1,2))$.
    (a) Find $_Y[T]_{X_1}$ and $_Z[T]_{X_1}$.
    (b) Find $_Y[T]_{X_2}$ and $_Z[T]_{X_2}$.
    (c) There is an ordered basis W of $\mathbb{R}^2$ such that $_W[T]_{X_3} = \begin{bmatrix} 3 & 1 & b \\ -1 & 1 & c \end{bmatrix}$. Find W, b, and c.

13. Define $T : P_{\leq 3} \to \mathbb{R}^3$ by $T(f) = \left(f(2), f'(2), \int_0^2 f(t)\,dt \right)$ for $f \in P_{\leq 3}$. Check that T is $\mathbb{R}$-linear. For $Y = (\mathbf{e}_1, \mathbf{e}_2, \mathbf{e}_3)$, compute $_Y[T]_{X_1}$, $_Y[T]_{X_2}$, and $_Y[T]_{X_3}$. Do the same for $Y = ((1,1,1),(1,2,4),(1,3,9))$.

14. Let $\mathrm{tr} : M_2(\mathbb{R}) \to \mathbb{R}$ be the trace map, defined by $\mathrm{tr}(A) = A(1,1) + A(2,2)$ for $A \in M_2(\mathbb{R})$. Let $Z = (1)$. Explain why $\mathrm{tr} \in M_2(\mathbb{R})^*$. Find $_Z[\mathrm{tr}]_{X_1}$, $_Z[\mathrm{tr}]_{X_2}$, and $_Z[\mathrm{tr}]_{X_3}$. Let $T : M_2(\mathbb{R}) \to M_2(\mathbb{R})$ be the transpose map. Show that $T^*(\mathrm{tr}) = \mathrm{tr}$.

15. Define $T : \mathbb{C} \to \mathbb{C}$ by $T(z) = iz$ for $z \in \mathbb{C}$. (a) Compute $[T]_{X_1}$, $[T]_{X_2}$, and $[T]_{X_3}$. (b) Square each matrix found in (a), and discuss the answers. (c) Find $_{X_3}[T]_{X_1}$ and $_{X_1}[T]_{X_3}$.

16. Let $A = \begin{bmatrix} 2 & 1 \\ -1 & 3 \end{bmatrix}$. Define three maps $\lambda_A, \rho_A, \kappa_A : M_2(\mathbb{R}) \to M_2(\mathbb{R})$ by setting $\lambda_A(B) = AB$, $\rho_A(B) = BA$, and $\kappa_A(B) = AB - BA$ for all $B \in M_2(\mathbb{R})$. Confirm that λ_A, ρ_A, and κ_A are $\mathbb{R}$-linear. Compute $[\lambda_A]_{X_1}$, $[\rho_A]_{X_1}$, and $[\kappa_A]_{X_1}$. Compute $[\lambda_A]_{X_2}$, $[\rho_A]_{X_2}$, and $[\kappa_A]_{X_2}$.

17. Given $A \in M_{m,n}(F)$, define $\lambda_A : M_{n,p}(F) \to M_{m,p}(F)$ by $\lambda_A(B) = AB$ for $B \in M_{n,p}(F)$. Describe the entries of $_Y[\lambda_A]_X$, where X and Y are standard ordered bases for $M_{n,p}(F)$ and $M_{m,p}(F)$ (e.g., $X = (\mathbf{e}_{11}, \mathbf{e}_{12}, \ldots, \mathbf{e}_{1p}, \mathbf{e}_{21}, \mathbf{e}_{22}, \ldots, \mathbf{e}_{np})$).

18. Given $A \in M_{n,p}(F)$, define $\rho_A : M_{m,n}(F) \to M_{m,p}(F)$ by $\rho_A(B) = BA$ for $B \in M_{m,n}(F)$. Describe the entries of $_Y[\rho_A]_X$, where X and Y are standard ordered bases for $M_{m,n}(F)$ and $M_{m,p}(F)$.

19. Given $A \in M_n(F)$, define $\kappa_A : M_n(F) \to M_n(F)$ by $\kappa_A(B) = AB - BA$ for $B \in M_n(F)$. Let $X = (\mathbf{e}_{11}, \mathbf{e}_{12}, \ldots, \mathbf{e}_{nn})$ be the standard ordered basis of $M_n(F)$. For each i, j, compute $[\kappa_{\mathbf{e}_{ij}}]_X$. Illustrate by writing all entries of $[\kappa_{\mathbf{e}_{23}}]_X$ for $n = 3$.

20. Show that each of the three bijections in (6.1) is F-linear. (This gives another proof that the map $T \mapsto {}_Y[T]_X$ is F-linear.)

21. Using only the definitions of spanning and linear independence (not matrices), prove that the maps T_{ij} defined in (6.4) form a basis of the F-vector space $L(V, W)$.

22. (a) Check that the map T defined in (6.2) is F-linear, well-defined, and sends x_j to w_j for $j = 1, 2, \ldots, n$. (b) Reprove (a) by showing that $T = L_{\mathbf{w}} \circ L_X^{-1}$, where $\mathbf{w} = (w_1, \ldots, w_n) \in W^n$.

23. Define a map $M'_{X,Y} : L(V, W) \to M_{n,m}(F)$ by letting $M'_{X,Y}(T)$ be the $n \times m$ matrix whose jth row (for $j = 1, 2, \ldots, n$) is $[T(x_j)]_Y$ viewed as a row vector.
    (a) Is $M'_{X,Y}$ a vector space isomorphism? Explain.
    (b) If $V = W$, $X = Y$, and $m = n$, is $M'_{X,X}$ an F-algebra isomorphism? Explain.

24. Let $\mathcal{B}$ be the set of pairs (X, Y) where X is an ordered basis for V and Y is an ordered basis for W. Let $\mathcal{I}$ be the set of vector space isomorphisms $S : L(V, W) \to M_{m,n}(F)$. Define a map $\phi : \mathcal{B} \to \mathcal{I}$ by $\phi(X, Y) = M_{X,Y}$, where $M_{X,Y}(T) = {}_Y[T]_X$. Prove ϕ is not injective, in general. Prove ϕ is not surjective, in general.

25. Show that each list is an ordered basis of $P_{\leq 3}^*$.
    (a) (f_0, f_1, f_2, f_3), where $f_i(c_0 + c_1 t + c_2 t^2 + c_3 t^3) = c_i$ for $i = 0, 1, 2, 3$.
    (b) (g_0, g_1, g_2, g_3), where $g_i(p) = (d/dt)^i(p)|_{t=0}$.
    (c) (h_0, h_1, h_2, h_3), where $h_i(p) = p(i)$.

26. Let $D : P_{\leq 3} \to P_{\leq 3}$ be the differentiation operator. For each basis X of $P_{\leq 3}^*$ found in Exercise 25, compute $[D^*]_X$.

27. Suppose $S \in L(V, W)$ and $U \in L(W, Z)$. Prove $(U \circ S)^* = S^* \circ U^*$. Translate this fact into an identity involving the matrix transpose.

28. Define $T : \mathbb{C} \to M_2(\mathbb{R})$ by $T(a + bi) = \begin{bmatrix} a & -b \\ b & a \end{bmatrix}$ for all $a, b \in \mathbb{R}$. Show that T is a one-to-one $\mathbb{R}$-algebra homomorphism.

29. Prove (6.6) by expressing $(S \circ T)(x_j)$ as a specific linear combination of vectors in Z and showing that the coefficient of z_i in this combination is $(BA)(i, j)$.

30. Suppose $T : V \to W$ is an F-linear map. You are given ${}_Y[T]_X$ but do not know any other specific information about T, X, or Y.
    (a) Explain how to use matrix algorithms to compute a basis for $\ker(T)$.
    (b) Explain how to use matrix algorithms to compute a basis for $\mathrm{img}(T)$.

31. Let $X_4 = ((1, 1, 1), (1, 0, -1), (1, 0, 1))$, which is an ordered basis of $\mathbb{R}^3$. Compute ${}_{X_i}[\mathrm{id}]_{X_4}$ and ${}_{X_4}[\mathrm{id}]_{X_i}$ for $i = 1, 2, 3$.

32. For $V = P_{\leq 3}$, compute the transition matrix:
    (a) from X_3 to X_1; (b) from X_1 to X_3; (c) from X_3 to X_2; (d) from X_2 to X_3.

33. Compute transition matrices between all pairs of ordered bases for $\mathbb{C}$ listed in Table 6.1.

34. Compute transition matrices between all pairs of ordered bases for $M_2(\mathbb{R})$ listed in Table 6.1.

35. (a) Given $[\mathbf{v}]_{X_3} = (5, -1, -1)$ in $\mathbb{R}^3$, find $[\mathbf{v}]_{X_2}$.
     (b) Given $[v]_{X_2} = (2, 1, 1, -2)$ in $P_{\leq 3}$, find $[v]_{X_3}$.
     (c) Given $[v]_{X_2} = (4, -5, 0, 1)$ in $M_2(\mathbb{R})$, find $[v]_{X_1}$ and $[v]_{X_3}$.

36. Let $P_{<n} = \{f \in \mathbb{R}[t] : f = 0 \text{ or } \deg(f) < n\}$. $P_{<n}$ has ordered bases $X = (1, t, t^2, \ldots, t^{n-1})$ and $Y = (1, (t+c), (t+c)^2, \ldots, (t+c)^{n-1})$ for fixed $c \in \mathbb{R}$. Find $_X[\text{id}]_Y$ and $_Y[\text{id}]_X$.

37. Let V consist of polynomials in $\mathbb{R}[t]$ of degree at most 4. Let $X = (1, t, t^2, t^3, t^4)$ and $Y = (1, t, t(t-1), t(t-1)(t-2), t(t-1)(t-2)(t-3))$. Find $_X[\text{id}]_Y$ and $_Y[\text{id}]_X$.

38. Prove: for any given ordered basis Y of V, every invertible matrix in $M_n(F)$ has the form $_X[\text{id}]_Y$ for a unique ordered basis X of V.

39. Let $A = \begin{bmatrix} 2 & 1 & -1 \\ 2 & 3 & 0 \\ 4 & 1 & -2 \end{bmatrix}$. Find the unique ordered basis Z of $\mathbb{R}^3$ such that $_{X_3}[\text{id}]_Z = A$. Find the unique ordered basis Y of $\mathbb{R}^3$ such that $_Y[\text{id}]_{X_3} = A$.

40. Verify that each relation is an equivalence relation.
     (a) column-equivalence on $M_{m,n}(F)$
     (b) row-equivalence on $M_{m,n}(F)$
     (c) row/column-equivalence on $M_{m,n}(F)$
     (d) similarity on $M_n(F)$
     (e) congruence on $M_n(F)$
     (f) orthogonal similarity on $M_n(\mathbb{R})$
     (g) unitary similarity on $M_n(\mathbb{C})$

41. Find all the similarity equivalence classes in $M_2(\mathbb{Z}_2)$.

42. Find all the congruence equivalence classes in $M_2(\mathbb{Z}_2)$.

43. Find all similarity equivalence classes in $M_n(F)$ consisting of a single element.

44. Let $D : P_{\leq 3} \to P_{\leq 3}$ be given by $D(f) = f'$ for $f \in P_{\leq 3}$. Find ordered bases X and Y of $P_{\leq 3}$ such that $_Y[D]_X$ is diagonal with 0s and 1s on the diagonal. Find an ordered basis X of $P_{\leq 3}$ such that $[D]_X$ is diagonal, or explain why this cannot be done.

45. Let $A \in M_n(F)$ be an invertible matrix. Explicitly describe ordered bases X and Y of F^n such that $_Y[L^A]_X = I_n$. Prove: if $[L^A]_X = I_n$ for some ordered basis X, then $A = I_n$.

46. Prove the Projection Theorem (see §6.11) without using matrices.

47. Let $A \in M_n(F)$ and $P \in \text{GL}_n(F)$. Prove $\mathbf{x} \in F^n$ is an eigenvector of A with associated eigenvalue c iff $P^{-1}\mathbf{x}$ is an eigenvector of $P^{-1}AP$ with associated eigenvalue c. Conclude that similar matrices have the same eigenvalues.

48. Show that for all $A \in M_n(F)$ and all $c \in F$, A is diagonalizable iff $A + cI_n$ is diagonalizable.

49. Show that a strictly upper-triangular nonzero matrix is not diagonalizable.

50. Show that if $c \in F$ is the only eigenvalue of $A \in M_n(F)$, then $A = cI_n$ or A is not diagonalizable.

51. Show that if $A \in M_n(F)$ has no eigenvalues in F, then A is not triangulable.

52. Show that $A = \begin{bmatrix} 0 & -1 \\ 1 & 0 \end{bmatrix}$ is not triangulable in $M_2(\mathbb{R})$, but A is diagonalizable in $M_2(\mathbb{C})$.

53. Give an example of a field F and $A \in M_n(F)$ such that A has an eigenvalue in F, but A is not triangulable.

54. Decide if each matrix A is diagonalizable using the given field F of scalars. If it is, find a diagonal D and an invertible P with $P^{-1}AP = D$.

(a) $A = \frac{1}{7} \begin{bmatrix} 32 & 12 & -6 \\ 9 & 20 & -3 \\ 9 & 6 & 11 \end{bmatrix}$, $F = \mathbb{R}$.

(b) $A = \begin{bmatrix} 1 & 1 & 0 \\ -1 & 1 & 0 \\ 0 & 0 & -1 \end{bmatrix}$, $F = \mathbb{R}$.

(c) the matrix in (b), using $F = \mathbb{C}$.

(d) $A = \begin{bmatrix} 1 & 0 & 1 & 0 \\ 0 & 1 & 0 & 1 \\ 1 & 0 & 1 & 0 \\ 0 & 1 & 0 & 1 \end{bmatrix}$, $F = \mathbb{Q}$.

(e) the matrix in (d), using $F = \mathbb{Z}_2$ (the integers mod 2).

55. Given a complete flag of subspaces $(V_0, V_1, \ldots, V_n)$ of V, let x_i be any element in V_i but not in V_{i-1}, for $i = 1, 2, \ldots, n$. Prove $X = (x_1, \ldots, x_n)$ is an ordered basis of V.

56. Let $(x_1, \ldots, x_n)$ be an ordered basis for V such that $(x_1, \ldots, x_k)$ is an ordered basis for a subspace W. Prove $(x_{k+1} + W, \ldots, x_n + W)$ is an ordered basis for the quotient vector space V/W.

57. Give an example of a map $T \in L(\mathbb{R}^2)$ such that the only T-invariant subspaces of $\mathbb{R}^2$ are $\{0\}$ and $\mathbb{R}^2$.

58. Give an example of a map $T \in L(\mathbb{R}^2)$ such that $\mathbb{R}^2$ has exactly four T-invariant subspaces.

59. Does there exist $T \in L(\mathbb{R}^2)$ such that T has exactly three T-invariant subspaces? Explain.

60. Give an example of $T \in L(\mathbb{R}^3)$ such that $\mathbb{R}^3$ has infinitely many T-invariant subspaces, but T is not a scalar multiple of id.

61. Given positive integers $i_1, i_2, \ldots, i_b$ with sum n, let

$$I_1 = \{1, 2, \ldots, i_1\}, \ I_2 = \{i_1 + 1, i_1 + 2, \ldots, i_1 + i_2\}, \ \ldots,$$
$$I_b = \{i_1 + \cdots + i_{b-1} + 1, \ldots, n\}.$$

Call a matrix $A \in M_n(F)$ *block-triangular* with block sizes $i_1, i_2, \ldots, i_b$ iff for all $i \in I_r$ and $j \in I_s$ with $r > s$, $A(i, j) = 0$. Define a *partial flag* of type $(i_1, \ldots, i_b)$ to be a chain of subspaces $V_1 \subseteq V_2 \subseteq \cdots \subseteq V_b$ with $\dim(V_j) = i_1 + i_2 + \cdots + i_j$ for $j = 1, 2, \ldots, b$. Show that there exists an ordered basis X of V such that $[T]_X$ is block-triangular with block sizes $i_1, i_2, \ldots, i_b$ iff V has a partial flag of type $(i_1, \ldots, i_b)$ stabilized by T.

62. Define $I_1, \ldots, I_b$ as in Exercise 61. Call a matrix $A \in M_n(F)$ *block-diagonal* with block sizes $i_1, \ldots, i_b$ iff for all $i \in I_r$ and $j \in I_s$ with $r \neq s$, $A(i, j) = 0$. Find and prove an abstract criterion on $T \in L(V)$ that is equivalent to the existence of an ordered basis X of V such that $[T]_X$ is block-diagonal with block sizes $i_1, \ldots, i_b$.

63. Give an example of a 4×4 real matrix A with all entries nonzero such that A is similar to a block-triangular matrix, but A is not triangulable.

64. Let $A = \begin{bmatrix} 1/2 & -1/2 & -1/2 & 1/2 \\ 1/2 & 5/2 & 5/2 & 1/2 \\ -1/2 & -3/2 & -3/2 & -1/2 \\ 1/2 & 1/2 & 1/2 & 1/2 \end{bmatrix}$. Verify that A is idempotent. Find an ordered basis X of $\mathbb{R}^4$ with $[L^A]_X = I_{k,4}$ for some k. Find an invertible $P \in M_4(\mathbb{R})$ with $P^{-1}AP = I_{k,4}$.

65. Given an idempotent matrix $A \in M_n(F)$, describe an algorithm for finding k and $P \in \mathrm{GL}_n(F)$ with $P^{-1}AP = I_{k,n}$.

66. **Projections of a Direct Sum.** We say that V is the *direct sum* of subspaces $W_1, \ldots, W_k$, and we write $V = W_1 \oplus \cdots \oplus W_k$, iff every $v \in V$ can be written uniquely in the form $v = w_1 + \cdots + w_k$ with $w_i \in W_i$. (a) Given $V = W_1 \oplus \cdots \oplus W_k$, define maps $P_1, \ldots, P_k : V \to V$ by setting $P_i(v) = w_i$, where $v = w_1 + \cdots + w_k$ as above. Show that each P_i is linear and idempotent with image W_i, $P_i P_j = 0$ for all $i \neq j$, and $P_1 + \cdots + P_k = \mathrm{id}_V$. What is $\ker(P_i)$? (b) Let $X_1, \ldots, X_k$ be ordered bases for $W_1, \ldots, W_k$. Let X be the concatenation of $X_1, \ldots, X_k$. Show that X is an ordered basis of V, and find $[P_i]_X$ for each i.

67. Suppose $Q_1, \ldots, Q_k$ are idempotent elements of $L(V)$ such that $Q_i Q_j = 0$ for all $i \neq j$ and $Q_1 + \cdots + Q_k = \mathrm{id}_V$.
(a) Write $W_i = \mathrm{img}(Q_i)$ for $i = 1, 2, \ldots, k$. Show that $V = W_1 \oplus \cdots \oplus W_k$.
(b) Define P_i as in Exercise 66. Show that $P_i = Q_i$ for all i.

68. Suppose $V = W \oplus Z$, and let $P = P_{W,Z}$ be the associated projection.
(a) Prove: W is T-invariant iff $PTP = TP$ in $L(V)$.
(b) Prove: W and Z are both T-invariant iff $PT = TP$ in $L(V)$.

69. Verify the following facts stated in §6.18.
(a) $\mathrm{BL}(V)$ is a subspace of the vector space of all functions from $V \times V$ to F.
(b) The map N_X defined in §6.18 is F-linear.
(c) Equation (6.8) holds.
(d) The map B defined at the end of §6.18 is F-bilinear, and $B(x_i, x_j) = A(i, j)$.

70. Check that each map B is $\mathbb{R}$-bilinear, and compute $[B]_X$ for the given ordered basis X.
(a) $B\left(\begin{bmatrix} a \\ c \end{bmatrix}, \begin{bmatrix} b \\ d \end{bmatrix} \right) = \det \begin{bmatrix} a & b \\ c & d \end{bmatrix}$, using $X = (\mathbf{e}_1, \mathbf{e}_2)$.
(b) For $f, g \in P_{\leq 3}$, $B(f, g) = \int_0^1 f(t)g(t)\, dt$, using $X = (1, t, t^2, t^3)$.
(c) For $z, w \in \mathbb{C}$, $B(z, w) = $ the real part of zw, using $X = (1, i)$.

71. Use your answers to each part of Exercise 70 to find $[B]_Y$ for the new ordered basis Y. (a) $Y = ((1, 2), (3, 4))$. (b) $Y = (t + 3, 2t^2 - 4, t^3 - t^2, t^3 + t^2)$.
(c) $Y = (3 + 4i, 2 - i)$.

72. For $V = \mathbb{R}^3$, we are given $B \in \mathrm{BL}(V)$ with $[B]_{X_2} = \begin{bmatrix} 2 & 1 & 0 \\ -1 & 3 & 1 \\ 0 & 1 & -1 \end{bmatrix}$.
(a) Find $B((1, 0, -4), (1, 3, 7))$. (b) Find $[B]_{X_1}$. (c) Find $[B]_{X_3}$.

73. Given $B \in \mathrm{BL}(V)$ and $v, w \in V$, show that $[B(v, w)]_{1 \times 1} = ([v]_X)^{\mathrm{T}} [B]_X [w]_X$.

74. How does the matrix $[B]_X$ change if we switch the positions of x_i and x_j in X?

75. How does the matrix $[B]_X$ change if we multiply x_i by a nonzero $c \in F$?

76. Given $B \in \mathrm{BL}(V)$ and $v, w \in V$, define $L_v(w) = B(v, w)$ and $R_v(w) = B(w, v)$.
(a) Prove: for all $v \in V$, L_v and R_v are in V^*.
(b) How is the row vector $_{(1_F)}[L_{x_i}]_X$ related to the matrix $[B]_X$?

(c) How is the row vector $_{(1_F)}[R_{x_j}]_X$ related to the matrix $[B]_X$?

(d) Describe how to use $[B]_X$ to compute $_{(1_F)}[L_v]_X$ and $_{(1_F)}[R_v]_X$ for any $v \in V$.

77. Let $\mathcal{B}$ be the set of ordered bases for V and $\mathcal{I}$ be the set of vector space isomorphisms $S : \mathrm{BL}(V) \to M_n(F)$. Define $\phi : \mathcal{B} \to \mathcal{I}$ by $\phi(X) = N_X$, where $N_X(B) = [B]_X$ for $B \in \mathrm{BL}(V)$. Prove or disprove: ϕ is a bijection.

78. (a) For a real inner product space V, prove $O(V)$ is a subgroup of $\mathrm{GL}(V)$.

(b) Prove $O_n(\mathbb{R})$ is a subgroup of $\mathrm{GL}_n(\mathbb{R})$.

79. (a) For a complex inner product space V, prove $U(V)$ is a subgroup of $\mathrm{GL}(V)$.

(b) Prove $U_n(\mathbb{C})$ is a subgroup of $\mathrm{GL}_n(\mathbb{C})$.

80. Fix an orthonormal basis X of a real inner product space V. (a) Show that every orthogonal matrix $P \in O_n(\mathbb{R})$ has the form $_Y[\mathrm{id}]_X$ for a unique orthonormal basis Y of V. (b) Show that every orthogonal matrix $P \in O_n(\mathbb{R})$ has the form $_X[\mathrm{id}]_Y$ for a unique orthonormal basis Y of V. (c) Prove analogs of (a) and (b) for complex inner product spaces.

81. Let $\mathbf{v} = (v_1, v_2, v_3)$ and $\mathbf{w} = (w_1, w_2, w_3)$ be vectors in $\mathbb{R}^3$ with $\|\mathbf{v}\| = 1 = \|\mathbf{w}\|$ and $\mathbf{v} \bullet \mathbf{w} = 0$. Find two vectors $\mathbf{z} \in \mathbb{R}^3$ such that $(\mathbf{v}, \mathbf{w}, \mathbf{z})$ is an orthonormal basis of $\mathbb{R}^3$.

82. (a) Show that $P_{\leq 3}$ becomes a real inner product space if we define $\langle f, g \rangle = \int_0^1 f(t)g(t)\,dt$ for $f, g \in P_{\leq 3}$. (b) Find an orthonormal basis (f_0, f_1, f_2, f_3) of $P_{\leq 3}$ such that $\deg(f_i) = i$ for $0 \leq i \leq 3$.

83. (a) For each $t \in \mathbb{R}$, show that $A_t = \begin{bmatrix} \cos t & -\sin t \\ \sin t & \cos t \end{bmatrix}$ is an orthogonal matrix.

(b) Show that every orthogonal matrix in $O_2(\mathbb{R})$ has the form A_t or $A_t \begin{bmatrix} 1 & 0 \\ 0 & -1 \end{bmatrix}$ for some $t \in \mathbb{R}$.

84. Let $\mathcal{B}$ be the set of orthonormal bases for a real inner product space V. Let $\mathcal{I}$ be the set of orthogonal vector space isomorphisms $S : \mathbb{R}^n \to V$. Define $\phi : \mathcal{B} \to \mathcal{I}$ by $\phi(X) = L_X$ for all $X \in \mathcal{B}$. Prove or disprove: ϕ is a bijection.

85. Let $\mathcal{B}$ be the set of ordered bases for V and $\mathcal{I}$ be the set of F-algebra isomorphisms $S : L(V) \to M_n(F)$. Define $\phi : \mathcal{B} \to \mathcal{I}$ by $\phi(X) = M_X$, where $M_X(T) = [T]_X$ for $T \in L(V)$. Is ϕ one-to-one? Is ϕ onto? Explain.

Part III

Matrices with Special Structure

7

Hermitian, Positive Definite, Unitary, and Normal Matrices

We begin by recalling some fundamental facts about complex numbers.

- For every complex number z, there exist unique real numbers x and y such that $z = x + iy$. We call x the *real part* of z and y the *imaginary part* of z, writing $x = \mathrm{Re}(z)$ and $y = \mathrm{Im}(z)$. The formula $z = x + iy$ is called the *Cartesian decomposition* of the complex number z.

- For each complex number $z = x + iy$ with $x, y \in \mathbb{R}$, the *complex conjugate* of z is $\overline{z} = x - iy$. For all $z, w \in \mathbb{C}$, the following properties hold: $\overline{\overline{z}} = z$; $\overline{z + w} = \overline{z} + \overline{w}$; $\overline{zw} = \overline{z} \cdot \overline{w} = \overline{w} \cdot \overline{z}$; $\mathrm{Re}(z) = (z + \overline{z})/2$; $\mathrm{Im}(z) = (z - \overline{z})/2i$; z is real iff $\overline{z} = z$; z is pure imaginary (meaning $\mathrm{Re}(z) = 0$) iff $\overline{z} = -z$.

- The *magnitude* of a complex number $z = x + iy$ is the nonnegative real number $|z| = \sqrt{x^2 + y^2} = (\overline{z}z)^{1/2}$. Every nonzero complex number can be written uniquely in the form $z = rw$, where r is a positive real number and w is a complex number of magnitude 1. We often write $w = e^{i\theta} = \cos\theta + i\sin\theta$ for some (non-unique) real number θ. Then $z = re^{i\theta}$ is called the *polar decomposition* of z.

Now consider the set $M_n(\mathbb{C})$ of all $n \times n$ matrices with complex entries. If $n = 1$, we can identify $M_n(\mathbb{C})$ with $\mathbb{C}$. Our goal in this chapter is to build an analogy between $\mathbb{C}$ and $M_n(\mathbb{C})$ for $n > 1$. We define matrix versions of the concepts of real numbers, positive numbers, pure imaginary numbers, complex numbers of magnitude 1, complex conjugation, the Cartesian decomposition of a complex number, and the polar decomposition of a complex number. This approach allows us to unify a diverse collection of fundamental results in matrix theory. The analogy with $\mathbb{C}$ also provides some motivation for the introduction of certain special classes of matrices — namely Hermitian, positive definite, unitary, and normal matrices — that have many remarkable properties.

7.1 Conjugate-Transpose of a Matrix

The first step toward implementing the analogy between complex numbers and complex-valued matrices is to introduce a matrix version of complex conjugation. One possibility is to consider the operation on matrices that replaces each entry of a matrix by its complex conjugate. However, it turns out to be much more fruitful to combine this operation with the transpose map that interchanges the rows and columns of a matrix.

Formally, suppose A is an $m \times n$ matrix with complex entries $A(i, j)$, where $1 \le i \le m$ and $1 \le j \le n$. Define the *conjugate-transpose* of A to be the $n \times m$ matrix A^* such that

$$A^*(i, j) = \overline{A(j, i)} \qquad \text{for } 1 \le i \le n \text{ and } 1 \le j \le m.$$

DOI: 10.1201/9781003484561-7

For example, given $A = \begin{bmatrix} 2+i & 3 & 5i \\ 1-2i & 0 & -1+3i \end{bmatrix}$, we find $A^* = \begin{bmatrix} 2-i & 1+2i \\ 3 & 0 \\ -5i & -1-3i \end{bmatrix}$. In the case of a 1×1 matrix, the conjugate-transpose operation reduces to the complex conjugation operation on $\mathbb{C}$. Many algebraic properties of complex conjugation extend to the conjugate-transpose.

Theorem on the Conjugate-Transpose.
(a) [conjugate-linearity] For all $A, B \in M_{m,n}(\mathbb{C})$ and all $z \in \mathbb{C}$, $(A+B)^* = A^* + B^*$ and $(zA)^* = \overline{z}(A^*)$.
(b) [product reversal] For all $A \in M_{m,n}(\mathbb{C})$ and all $B \in M_{n,p}(\mathbb{C})$, $(AB)^* = B^*A^*$.
(c) [idempotence] For all $A \in M_{m,n}(\mathbb{C})$, $(A^*)^* = A$.
(d) [inverse rule] For all invertible $A \in M_n(\mathbb{C})$, A^* is invertible and $(A^*)^{-1} = (A^{-1})^*$.
(e) [trace rule] For all $A \in M_n(\mathbb{C})$, $\mathrm{tr}(A^*) = \overline{\mathrm{tr}(A)}$, where $\mathrm{tr}(A) = \sum_i A(i,i)$ is the sum of the diagonal entries of A.
(f) [determinant rule] For all $A \in M_n(\mathbb{C})$, $\det(A^*) = \overline{\det(A)}$.

Proof. We prove a matrix identity by showing that the left side and the right side have the same i,j-entry for all relevant i and j. The conclusions in (a) involve $n \times m$ matrices, so we fix i, j with $1 \le i \le n$ and $1 \le j \le m$. We compute

$$(A+B)^*(i,j) = \overline{(A+B)(j,i)} = \overline{A(j,i) + B(j,i)} = \overline{A(j,i)} + \overline{B(j,i)}$$
$$= A^*(i,j) + B^*(i,j) = (A^* + B^*)(i,j).$$

So $(A+B)^* = A^* + B^*$. For the second part of (a), compute

$$(zA)^*(i,j) = \overline{(zA)(j,i)} = \overline{z(A(j,i))} = \overline{z}\,\overline{A(j,i)} = \overline{z}(A^*(i,j)) = (\overline{z}A^*)(i,j).$$

So $(zA)^* = \overline{z}A^*$. In particular, if x is a real scalar, then $(xA)^* = x(A^*)$.

In part (b), both sides of the conclusion are $p \times m$ matrices. Fix i, j with $1 \le i \le p$ and $1 \le j \le m$. Compute

$$(AB)^*(i,j) = \overline{AB(j,i)} = \overline{\sum_{k=1}^{n} A(j,k)B(k,i)} = \sum_{k=1}^{n} \overline{A(j,k)} \cdot \overline{B(k,i)}$$
$$= \sum_{k=1}^{n} B^*(i,k)A^*(k,j) = (B^*A^*)(i,j).$$

So $(AB)^* = B^*A^*$. For (c), fix i, j with $1 \le i \le m$ and $1 \le j \le n$, and note that

$$(A^*)^*(i,j) = \overline{A^*(j,i)} = \overline{\overline{A(i,j)}} = A(i,j).$$

So $(A^*)^* = A$. To prove (d), let $A \in M_n(\mathbb{C})$ be invertible, $B = A^{-1}$, and I be the $n \times n$ identity matrix. Apply part (b) to the equations $AB = I = BA$ to obtain $B^*A^* = I^* = I = A^*B^*$. Thus, $B^* = (A^{-1})^*$ is the two-sided matrix inverse for A^*, as needed.

To prove (e), compute

$$\mathrm{tr}(A^*) = \sum_{i=1}^{n} A^*(i,i) = \sum_{i=1}^{n} \overline{A(i,i)} = \overline{\sum_{i=1}^{n} A(i,i)} = \overline{\mathrm{tr}(A)}.$$

To prove (f), use the definition of determinants (Chapter 5) to calculate

$$
\begin{aligned}
\det(A^*) &= \sum_{f \in S_n} \mathrm{sgn}(f) A^*(f(1),1) A^*(f(2),2) \cdots A^*(f(n),n) \\
&= \sum_{f \in S_n} \mathrm{sgn}(f) \overline{A(1,f(1))}\, \overline{A(2,f(2))} \cdots \overline{A(n,f(n))} \\
&= \overline{\sum_{f \in S_n} \mathrm{sgn}(f) A^{\mathrm{T}}(f(1),1) A^{\mathrm{T}}(f(2),2) \cdots A^{\mathrm{T}}(f(n),n)} \\
&= \overline{\det(A^{\mathrm{T}})} = \overline{\det(A)}. \quad \square
\end{aligned}
$$

The conjugate-transpose operation is related to the standard inner product on $\mathbb{C}^n$. For vectors $\mathbf{v} = (v_1, \dots, v_n)$ and $\mathbf{w} = (w_1, \dots, w_n)$ in $\mathbb{C}^n$, the *inner product* of $\mathbf{v}$ and $\mathbf{w}$ is defined to be $\langle \mathbf{v}, \mathbf{w} \rangle = \sum_{i=1}^n \overline{w_i} v_i$. We often identify $\mathbb{C}^n$ with the space of column vectors $M_{n,1}(\mathbb{C})$ and identify 1×1 matrices with complex numbers. Using this convention, the definition of matrix multiplication shows that $\langle \mathbf{v}, \mathbf{w} \rangle = \mathbf{w}^* \mathbf{v}$. Observe that $\mathbf{v}^* \mathbf{v} = \langle \mathbf{v}, \mathbf{v} \rangle = \sum_{i=1}^n \overline{v_i} v_i = \sum_{i=1}^n |v_i|^2$ is a nonnegative real number, which is zero iff $\mathbf{v} = \mathbf{0}$. We define the *Euclidean norm* of $\mathbf{v}$ to be $\|\mathbf{v}\| = (\mathbf{v}^* \mathbf{v})^{1/2}$ for all $\mathbf{v} \in \mathbb{C}^n$. If we require $\mathbf{v}$ to be in $\mathbb{R}^n$, then the complex conjugates disappear and we recover the standard inner product (the dot product) and the Euclidean norm on $\mathbb{R}^n$. We have the following *adjoint property* of the conjugate-transpose:

for all $A \in M_n(\mathbb{C})$ and all $\mathbf{v}, \mathbf{w} \in \mathbb{C}^n$, $\langle A\mathbf{v}, \mathbf{w} \rangle = \langle \mathbf{v}, A^* \mathbf{w} \rangle$.

This follows since $\langle A\mathbf{v}, \mathbf{w} \rangle = \mathbf{w}^* A \mathbf{v} = \mathbf{w}^* (A^*)^* \mathbf{v} = (A^* \mathbf{w})^* \mathbf{v} = \langle \mathbf{v}, A^* \mathbf{w} \rangle$.

Given a matrix $A \in M_n(\mathbb{C})$, we can use the conjugate-transpose operation to define a quadratic form on $\mathbb{C}^n$ associated with A. This quadratic form is the function $Q : \mathbb{C}^n \to \mathbb{C}$ defined by $Q(\mathbf{v}) = \mathbf{v}^* A \mathbf{v}$ for $\mathbf{v} \in \mathbb{C}^n$. To get a formula for $Q(\mathbf{v})$ in terms of the components of $\mathbf{v}$, write $\mathbf{v} = (v_1, \dots, v_n)$ with each $v_k \in \mathbb{C}$. The kth component of the column vector $A\mathbf{v}$ is $\sum_{j=1}^n A(k,j) v_j$. Multiplying this column vector on the left by the row vector $\mathbf{v}^*$, we see that

$$
Q(\mathbf{v}) = \sum_{k=1}^n \sum_{j=1}^n \overline{v_k} A(k,j) v_j \qquad \text{for all } \mathbf{v} \in \mathbb{C}^n. \tag{7.1}
$$

For example, if $A = \begin{bmatrix} 1 & 2-i \\ 3i & 0 \end{bmatrix}$ and $v = (3+i, -4i)$, then

$$
Q(v) = (3-i)1(3+i) + (3-i)(2-i)(-4i) + (4i)(3i)(3+i) + (4i)0(-4i) = -46 - 32i.
$$

7.2 Hermitian Matrices

A complex number z is real iff $\overline{z} = z$. By analogy, we make the following definition: a complex-valued matrix A is *Hermitian* iff $A^* = A$. This equality requires A to be a square matrix. We see from the definition of A^* that $A \in M_n(\mathbb{C})$ is Hermitian iff $A(i,j) = \overline{A(j,i)}$ for all i, j between 1 and n. For example,

$$
A = \begin{bmatrix}
2 & 1+8i & 0 & i \\
1-8i & -1 & 3-5i & 7+i/2 \\
0 & 3+5i & 0 & 1 \\
-i & 7-i/2 & 1 & \pi
\end{bmatrix}
$$

is a Hermitian matrix. A real-valued matrix A is Hermitian iff A is symmetric (meaning $A^{\mathrm{T}} = A$). For any Hermitian matrix A, the diagonal entries of A must all be real. This follows from the requirement $A(i,i) = \overline{A(i,i)}$ for all i. A 1×1 complex matrix is Hermitian iff the sole entry of that matrix is real. We regard Hermitian matrices as being matrix analogs of real numbers. This analogy is strengthened by results proved later in this section.

Suppose A and B are $n \times n$ Hermitian matrices and x is a real scalar. Then $A + B$ and xA are Hermitian, since $(A+B)^* = A^* + B^* = A + B$ and $(xA)^* = x(A^*) = xA$. More generally, any real linear combination of Hermitian matrices is also Hermitian. On the other hand, AB need not be Hermitian in general, since $(AB)^* = B^*A^* = BA$. This equation shows that $(AB)^* = AB$ holds iff $AB = BA$. Thus, the product of two commuting Hermitian matrices is also Hermitian. If $A \in M_n(\mathbb{C})$ is Hermitian and $B \in M_n(\mathbb{C})$ is arbitrary, then B^*AB is also Hermitian, because $(B^*AB)^* = B^*A^*(B^*)^* = B^*AB$.

If A is Hermitian and $\mathbf{v}, \mathbf{w} \in \mathbb{C}^n$, then $\langle A\mathbf{v}, \mathbf{w} \rangle = \langle \mathbf{v}, A^*\mathbf{w} \rangle = \langle \mathbf{v}, A\mathbf{w} \rangle$. This fact is sometimes expressed by saying that a Hermitian matrix is *self-adjoint*. Next, consider the quadratic form Q associated with a Hermitian matrix A. We claim that for all $\mathbf{v} \in \mathbb{C}^n$, $Q(\mathbf{v}) = \mathbf{v}^*A\mathbf{v}$ is real. To confirm this, we need only check that $\overline{Q(\mathbf{v})} = Q(\mathbf{v})$. Since $\mathbf{v}^*A\mathbf{v}$ is a 1×1 matrix and A is Hermitian,

$$\overline{Q(\mathbf{v})} = \overline{\mathbf{v}^*A\mathbf{v}} = (\mathbf{v}^*A\mathbf{v})^* = \mathbf{v}^*A^*(\mathbf{v}^*)^* = \mathbf{v}^*A\mathbf{v} = Q(\mathbf{v}).$$

Another key fact about Hermitian matrices is that *all eigenvalues of a Hermitian matrix A must be real*. To prove this, let $\lambda \in \mathbb{C}$ be an eigenvalue of A with an associated nonzero eigenvector $\mathbf{v} = (v_1, \ldots, v_n) \in \mathbb{C}^n$. The definition shows that $\mathbf{v}^*\mathbf{v} = \sum_{i=1}^{n} |v_i|^2$ is a positive real number. We have seen that $\mathbf{v}^*A\mathbf{v}$ is real. Also $\mathbf{v}^*A\mathbf{v} = \mathbf{v}^*(\lambda\mathbf{v}) = \lambda(\mathbf{v}^*\mathbf{v})$. Thus, $\lambda = (\mathbf{v}^*A\mathbf{v})/(\mathbf{v}^*\mathbf{v})$ is a quotient of two real numbers and is therefore real. In contrast, a real-valued matrix $B \in M_n(\mathbb{R})$ need not have all real eigenvalues. For example, the eigenvalues of $B = \begin{bmatrix} 0 & 1 \\ -1 & 0 \end{bmatrix}$ are i and $-i$. This gives more evidence that the Hermitian property (as opposed to the requirement of having all real entries) is a good generalization of the property of being real. Note here that iB is Hermitian although every entry of iB is pure imaginary.

The property of having a real-valued quadratic form characterizes the class of Hermitian matrices. More precisely, *if $A \in M_n(\mathbb{C})$ is a matrix such that $Q(\mathbf{v}) = \mathbf{v}^*A\mathbf{v}$ is real for all $\mathbf{v} \in \mathbb{C}^n$, then A must be Hermitian*. To prove this, consider the unit vectors $\mathbf{e}_k \in \mathbb{C}^n$, where $\mathbf{e}_k$ has a 1 in position k and 0 elsewhere. Using (7.1), we see that $Q(\mathbf{e}_k) = A(k,k)$ must be real for $1 \leq k \leq n$. Next, for all $j \neq k$, we see similarly that $Q(\mathbf{e}_j + \mathbf{e}_k) = A(j,j) + A(j,k) + A(k,j) + A(k,k)$ is real, and therefore $A(j,k) + A(k,j)$ is real for all $j \neq k$. Since we know $Q(\mathbf{e}_j + i\mathbf{e}_k) = A(j,j) + iA(j,k) + \overline{i}A(k,j) + A(k,k)$ is real, it follows that $i(A(j,k) - A(k,j))$ is real for all $j \neq k$. Fix $j \neq k$, and write $A(j,k) = a + ib$ and $A(k,j) = c + id$ with $a, b, c, d \in \mathbb{R}$. The preceding conditions mean that $b + d = 0$ and $a - c = 0$, so $A(k,j) = c + id = a - ib = \overline{A(j,k)}$ for all $j \neq k$. We also have $A(j,j) = \overline{A(j,j)}$ for all j, since $A(j,j)$ is real. Thus, $A = A^*$ as claimed.

On the other hand, a matrix $A \in M_n(\mathbb{R})$ with all real eigenvalues need not be Hermitian. For instance, $A = \begin{bmatrix} 1 & 2 \\ 0 & 3 \end{bmatrix}$ has eigenvalues 1 and 3 (the entries on the main diagonal), but $A \neq A^*$. Later, we introduce a matrix property called normality and prove that a normal matrix with all real eigenvalues must be Hermitian.

7.3   Hermitian Decomposition of a Matrix

The next theorem gives a matrix analog of the decomposition of a complex number z in the form $x + iy$ with x, y real.

Theorem on Hermitian Decomposition of a Matrix. For each $A \in M_n(\mathbb{C})$, there exist unique Hermitian matrices $X, Y \in M_n(\mathbb{C})$ such that $A = X + iY$.

Proof. To prove existence, recall that complex conjugation can be used to recover the real and imaginary parts of a complex number $z = x + iy$, since $x = (z + \overline{z})/2$ and $y = (z - \overline{z})/2i$. Given the matrix $A \in M_n(\mathbb{C})$, we are led by analogy to consider the matrices $X = (A + A^*)/2$ and $Y = (A - A^*)/2i$. We have $X^* = (A^* + A)/2 = X$ and $Y^* = \frac{1}{2i}(A^* - A) = \frac{1}{-2i}(A^* - A) = Y$, so X and Y are Hermitian. A calculation confirms that $X + iY = A$. To prove uniqueness, suppose that $A = X + iY = U + iV$ where X, Y, U, V are all Hermitian. We must prove $X = U$ and $Y = V$. Observe that $X - U = i(V - Y)$. Applying the conjugate-transpose operation to both sides of this matrix identity, we get $X - U = -i(V - Y)$. Thus, $i(V - Y) = -i(V - Y)$, which implies $V - Y = 0$ and hence $Y = V$. Then $X - U = i(V - Y) = 0$ leads to $X = U$, as needed. $\qquad\qquad\square$

The preceding theorem can also be rephrased in terms of skew-Hermitian matrices. A complex-valued matrix A is called *skew-Hermitian* iff $A^* = -A$. This concept is the analog of a pure imaginary number in $\mathbb{C}$. You can check that a real linear combination of skew-Hermitian matrices is also skew-Hermitian. The zero matrix is the only matrix that is both Hermitian and skew-Hermitian. A matrix B is Hermitian iff iB is skew-Hermitian, since $(iB)^* = -i(B^*)$. Combining this fact with the previous theorem, we see that any matrix $A \in M_n(\mathbb{C})$ can be written uniquely in the form $A = H + S$, where H is Hermitian and S is skew-Hermitian. More precisely, we have $H = X = (A + A^*)/2$ and $S = iY = (A - A^*)/2$. H is called the *Hermitian part of A*, and S is called the *skew-Hermitian part of A*. The decomposition $A = X + iY = H + S$ is called the *Hermitian decomposition* or *Cartesian decomposition* of the matrix A.

Consider a complex number $z \in \mathbb{C}$ and its Cartesian decomposition $z = x + iy$ with $x, y \in \mathbb{R}$. Then $xy = yx$, since multiplication of real numbers is commutative. The analogous property for complex matrices does not always hold. Suppose $A = X + iY$ is the Cartesian decomposition of $A \in M_n(\mathbb{C})$, so $X = (A + A^*)/2$ and $Y = (A - A^*)/2i$. A calculation with the distributive law shows that

$$XY = \frac{1}{4i}(AA + A^*A - AA^* - A^*A^*); \qquad YX = \frac{1}{4i}(AA - A^*A + AA^* - A^*A^*).$$

Comparing these expressions, we see that $XY = YX$ holds iff $A^*A - AA^* = -A^*A + AA^*$ iff $2A^*A = 2AA^*$ iff $A^*A = AA^*$. The latter condition does not hold for every matrix; for example, if $A = \begin{bmatrix} 1 & 2 \\ 0 & 3 \end{bmatrix}$, then $A^* = \begin{bmatrix} 1 & 0 \\ 2 & 3 \end{bmatrix}$, $AA^* = \begin{bmatrix} 5 & 6 \\ 6 & 9 \end{bmatrix}$, and $A^*A = \begin{bmatrix} 1 & 2 \\ 2 & 13 \end{bmatrix}$. We say that a matrix $A \in M_n(\mathbb{C})$ is a *normal matrix* iff $A^*A = AA^*$. We shall soon see that normal matrices have nicer properties than general complex matrices.

7.4 Positive Definite Matrices

Our next goal is to find the matrix analogs of positive and negative real numbers. To begin, recall that a matrix $A \in M_n(\mathbb{C})$ is Hermitian iff $A = A^*$ iff $Q(\mathbf{v}) = \mathbf{v}^* A \mathbf{v}$ is real for all $\mathbf{v} \in \mathbb{C}^n$ (§7.2). This motivates the following definition: we say a matrix $A \in M_n(\mathbb{C})$ is *positive definite* iff $\mathbf{v}^* A \mathbf{v}$ is a positive[1] real number for all nonzero $\mathbf{v} \in \mathbb{C}^n$. Say A is *positive semidefinite* iff $\mathbf{v}^* A \mathbf{v}$ is a nonnegative real number for all nonzero $\mathbf{v} \in \mathbb{C}^n$. Similarly, A is *negative definite* iff $\mathbf{v}^* A \mathbf{v}$ is in $\mathbb{R}_{<0}$ for all nonzero $\mathbf{v} \in \mathbb{C}^n$, and A is *negative semidefinite* iff $\mathbf{v}^* A \mathbf{v}$ is in $\mathbb{R}_{\leq 0}$ for all nonzero $\mathbf{v} \in \mathbb{C}^n$. Positive definite matrices are Hermitian, since positive real numbers are real; similarly for positive semidefinite, negative definite, and negative semidefinite matrices.

Suppose $A, B \in M_n(\mathbb{C})$ are positive definite and c is a positive real scalar. Then $A + B$ and cA are positive definite, since for nonzero $\mathbf{v} \in \mathbb{C}^n$, $\mathbf{v}^*(A + B)\mathbf{v} = \mathbf{v}^* A \mathbf{v} + \mathbf{v}^* B \mathbf{v}$ is a sum of two positive real numbers and $\mathbf{v}^*(cA)\mathbf{v} = c(\mathbf{v}^* A \mathbf{v})$ is a product of two positive real numbers. Similarly, any linear combination of positive definite matrices with positive real coefficients is positive definite. Any linear combination of negative definite matrices with positive real coefficients is negative definite. A is positive definite iff $-A$ is negative definite. Analogous properties hold for semidefinite matrices.

Another helpful fact about positive definite matrices is that *all eigenvalues of a positive definite matrix A are positive real numbers*. To prove this, let $\lambda \in \mathbb{C}$ be an eigenvalue of A and $\mathbf{v}$ be an associated nonzero eigenvector. We know $\mathbf{v}^* \mathbf{v} = \sum_{i=1}^n |v_i|^2$ is a positive real number. By positive definiteness, $\mathbf{v}^* A \mathbf{v} = \mathbf{v}^*(\lambda \mathbf{v}) = \lambda(\mathbf{v}^* \mathbf{v})$ is also in $\mathbb{R}_{>0}$. So $\lambda = (\mathbf{v}^* A \mathbf{v})/(\mathbf{v}^* \mathbf{v})$ is positive and real, being the ratio of two positive real numbers. The same argument shows that the eigenvalues of a positive semidefinite matrix are all positive or zero; the eigenvalues of a negative definite matrix are all negative real numbers; and the eigenvalues of a negative semidefinite matrix are all in $\mathbb{R}_{\leq 0}$. Since the trace of a matrix is the sum of its eigenvalues, we see that *the trace of a positive definite matrix is a positive real number*. Since the determinant of a matrix is the product of its eigenvalues, we see that *the determinant of a positive definite matrix is a positive real number*. In particular, *positive definite matrices are always invertible*. In the negative definite case, the trace is always negative, but the determinant is negative iff n (the size of the matrix) is odd; the determinant is positive for n even. We prove in §7.15 that a matrix $A \in M_n(\mathbb{C})$ is positive definite iff $A^* = A$ and for all k between 1 and n, the matrix consisting of the first k rows and k columns of A has positive determinant.

Every real number is either positive, negative, or zero. However, the corresponding property is not true for Hermitian matrices with size $n > 1$. It is certainly possible for a Hermitian matrix to have both positive and negative real eigenvalues; consider, for example, diagonal matrices with real diagonal entries. Such matrices are neither positive definite nor negative definite. A Hermitian matrix with both positive and negative eigenvalues is sometimes called *indefinite*.

A matrix with all positive eigenvalues need not be Hermitian and hence need not be positive definite; consider $A = \begin{bmatrix} 1 & 2 \\ 0 & 3 \end{bmatrix}$ for an example. We show in §7.9 that a normal matrix with all positive eigenvalues must be positive definite.

---

[1] In this section, "positive" always means "strictly positive" when applied to real numbers.

7.5 Unitary Matrices

Next, we develop a matrix analog of the unit circle in $\mathbb{C}$. Recall that a complex number $z \in \mathbb{C}$ is on the unit circle iff $|z| = 1$ iff $|z|^2 = 1$ iff $z\overline{z} = 1$ iff $z^{-1} = \overline{z}$. Accordingly, we define a matrix $U \in M_n(\mathbb{C})$ to be *unitary* iff $UU^* = I$. A 1×1 matrix is unitary iff its sole entry has modulus 1. Set $u = \det(U) \in \mathbb{C}$, so $\det(U^*) = \overline{u}$. Taking determinants in the formula $UU^* = I$, we see that $|u|^2 = u\overline{u} = 1$; thus, *the determinant of a unitary matrix is a complex number of modulus 1.* In particular, $\det(U) \neq 0$, so U^{-1} exists. Multiplying $UU^* = I$ by U^{-1}, we see that $U^{-1} = U^*$ and hence $U^*U = I$. Reversing this argument, we see that the following three conditions on $U \in M_n(\mathbb{C})$ are equivalent:

$$UU^* = I; \qquad U^*U = I; \qquad U \text{ is invertible and } U^{-1} = U^*.$$

We could have taken any of these conditions as the definition of a unitary matrix.

If U is unitary, then U^* is also unitary, since $U^*(U^*)^* = U^*U = I$. If U is unitary, then U^{-1} is also unitary, since $U^{-1} = U^*$. If U is unitary, then the transpose U^{T} is also unitary, since $U^{\mathrm{T}}(U^{\mathrm{T}})^* = U^{\mathrm{T}}(U^*)^{\mathrm{T}} = (U^*U)^{\mathrm{T}} = I^{\mathrm{T}} = I$. If U and V are unitary matrices of the same size, then UV is unitary, since $(UV)(UV)^* = UVV^*U^* = UIU^* = UU^* = I$. If U is a diagonal matrix such that each diagonal entry is a number $z_k \in \mathbb{C}$ of modulus 1, then U is unitary, because $z_k^{-1} = \overline{z_k}$ implies $U^{-1} = U^*$ for such a matrix U. In particular, the identity matrix I is unitary. Let $U_n(\mathbb{C})$ denote the set of all unitary matrices in $M_n(\mathbb{C})$. The preceding remarks show that $U_n(\mathbb{C})$ is closed under identity, matrix product, and inverses. Therefore, the set $U_n(\mathbb{C})$ of unitary matrices forms a subgroup of the multiplicative group $\mathrm{GL}_n(\mathbb{C})$ of invertible complex matrices.

Unitary matrices preserve inner products; more precisely, if U is unitary and $\mathbf{v}, \mathbf{w} \in \mathbb{C}^n$, then $\langle U\mathbf{v}, U\mathbf{w} \rangle = \langle \mathbf{v}, \mathbf{w} \rangle$. This follows since $\langle U\mathbf{v}, U\mathbf{w} \rangle = (U\mathbf{w})^*(U\mathbf{v}) = \mathbf{w}^*(U^*U)\mathbf{v} = \mathbf{w}^*I\mathbf{v} = \mathbf{w}^*\mathbf{v} = \langle \mathbf{v}, \mathbf{w} \rangle$. Similarly, *unitary matrices preserve norms,* meaning that $||U\mathbf{v}|| = ||\mathbf{v}||$ for all unitary U and all $\mathbf{v} \in \mathbb{C}^n$. This follows since $||U\mathbf{v}||^2 = \langle U\mathbf{v}, U\mathbf{v} \rangle = \langle \mathbf{v}, \mathbf{v} \rangle = ||\mathbf{v}||^2$.

All eigenvalues of a unitary matrix are complex numbers of modulus 1. To see this, let $U \in M_n(\mathbb{C})$ be unitary with eigenvalue $\lambda \in \mathbb{C}$ and associated nonzero eigenvector $\mathbf{v}$. We have $||\mathbf{v}|| = ||U\mathbf{v}|| = ||\lambda\mathbf{v}|| = |\lambda| \cdot ||\mathbf{v}||$. Dividing by the nonzero scalar $||\mathbf{v}||$ gives $|\lambda| = 1$, as needed. Taking the product of all the eigenvalues, we get another proof that $|\det(U)| = 1$ for U unitary.

A list of vectors $\mathbf{v}_1, \ldots, \mathbf{v}_m \in \mathbb{C}^n$ is called *orthonormal* iff $\langle \mathbf{v}_j, \mathbf{v}_k \rangle = 0$ for all $j \neq k$ and $\langle \mathbf{v}_j, \mathbf{v}_j \rangle = 1$ for all j. Consider the condition $U^*U = I$, which can be used to define unitary matrices. Let the columns of U be $\mathbf{v}_1, \ldots, \mathbf{v}_n \in \mathbb{C}^n$. The j, k-entry of the product U^*U is found by taking the product of row j of U^* and column k of U. Row j of U^* is $\mathbf{v}_j^*$, and column k of U is $\mathbf{v}_k$. We conclude that $(U^*U)(j, k) = \mathbf{v}_j^*\mathbf{v}_k = \langle \mathbf{v}_k, \mathbf{v}_j \rangle$ for all j, k. On the other hand, $I(j, k) = 1$ for $j = k$ and $I(j, k) = 0$ for $j \neq k$. Comparing to the definition of orthonormality, we see that *a matrix $U \in M_n(\mathbb{C})$ is unitary iff $U^*U = I$ iff the columns of U are orthonormal vectors in $\mathbb{C}^n$.* Applying similar reasoning to the condition $UU^* = I$ (or applying the result just stated to U^{T}), we see that *a matrix $U \in M_n(\mathbb{C})$ is unitary iff the rows of U are orthonormal vectors in $\mathbb{C}^n$.* Thus, the rows (resp. columns) of a unitary matrix U are mutually perpendicular unit vectors in $\mathbb{C}^n$, further strengthening the analogy between U and the unit circle in $\mathbb{C}$.

We have seen that a unitary matrix preserves inner products. Conversely, *if $U \in M_n(\mathbb{C})$ satisfies $\langle U\mathbf{v}, U\mathbf{w} \rangle = \langle \mathbf{v}, \mathbf{w} \rangle$ for all $\mathbf{v}, \mathbf{w} \in \mathbb{C}^n$, then U is unitary.* Proof: The condition on scalar products says that $\mathbf{w}^*(U^*U)\mathbf{v} = \mathbf{w}^*\mathbf{v}$ for all $\mathbf{v}, \mathbf{w} \in \mathbb{C}^n$. Choosing $\mathbf{v} = \mathbf{e}_j$ and $\mathbf{w} = \mathbf{e}_k$ (standard basis vectors) and carrying out the matrix multiplication on the left side, we see that $(U^*U)(k, j)$ is 0 for $k \neq j$ and is 1 for $k = j$. So $U^*U = I$ and U is unitary.

Similarly, *if $U \in M_n(\mathbb{C})$ satisfies $||U\mathbf{v}|| = ||\mathbf{v}||$ for all $\mathbf{v} \in \mathbb{C}^n$, then U is unitary.* Proof: The condition on norms means that $\langle U\mathbf{v}, U\mathbf{v} \rangle = \langle \mathbf{v}, \mathbf{v} \rangle$ for all $\mathbf{v} \in \mathbb{C}^n$. We show that U must preserve inner products, which implies that U is unitary by the previous result. Fix $\mathbf{v}, \mathbf{w} \in \mathbb{C}^n$, and apply the assumed condition to the vectors $\mathbf{v}$, $\mathbf{w}$, $\mathbf{v} + \mathbf{w}$, and $\mathbf{v} + i\mathbf{w}$. We conclude that:

$$\langle U\mathbf{v}, U\mathbf{v} \rangle = \langle \mathbf{v}, \mathbf{v} \rangle; \quad \langle U\mathbf{w}, U\mathbf{w} \rangle = \langle \mathbf{w}, \mathbf{w} \rangle;$$

$$\langle U(\mathbf{v} + \mathbf{w}), U(\mathbf{v} + \mathbf{w}) \rangle = \langle \mathbf{v} + \mathbf{w}, \mathbf{v} + \mathbf{w} \rangle; \quad \langle U(\mathbf{v} + i\mathbf{w}), U(\mathbf{v} + i\mathbf{w}) \rangle = \langle \mathbf{v} + i\mathbf{w}, \mathbf{v} + i\mathbf{w} \rangle.$$

For any complex scalars c, d and any vectors $\mathbf{v}, \mathbf{w} \in \mathbb{C}^n$, we have

$$\langle c\mathbf{v} + d\mathbf{w}, c\mathbf{v} + d\mathbf{w} \rangle = (c\mathbf{v} + d\mathbf{w})^*(c\mathbf{v} + d\mathbf{w}) = ||c\mathbf{v}||^2 + \overline{d}c\mathbf{w}^*\mathbf{v} + \overline{c}d\mathbf{v}^*\mathbf{w} + ||d\mathbf{w}||^2.$$

Applying this formula to the preceding expressions, we get:

$$\begin{aligned}
\langle \mathbf{v} + \mathbf{w}, \mathbf{v} + \mathbf{w} \rangle &= ||\mathbf{v}||^2 + \mathbf{w}^*\mathbf{v} + \mathbf{v}^*\mathbf{w} + ||\mathbf{w}||^2, \\
\langle U\mathbf{v} + U\mathbf{w}, U\mathbf{v} + U\mathbf{w} \rangle &= ||U\mathbf{v}||^2 + \mathbf{w}^*U^*U\mathbf{v} + \mathbf{v}^*U^*U\mathbf{w} + ||U\mathbf{w}||^2, \\
\langle \mathbf{v} + i\mathbf{w}, \mathbf{v} + i\mathbf{w} \rangle &= ||\mathbf{v}||^2 - i\mathbf{w}^*\mathbf{v} + i\mathbf{v}^*\mathbf{w} + ||\mathbf{w}||^2, \\
\langle U\mathbf{v} + iU\mathbf{w}, U\mathbf{v} + iU\mathbf{w} \rangle &= ||U\mathbf{v}||^2 - i\mathbf{w}^*U^*U\mathbf{v} + i\mathbf{v}^*U^*U\mathbf{w} + ||U\mathbf{w}||^2.
\end{aligned}$$

Set $a = \mathbf{w}^*\mathbf{v} = \langle \mathbf{v}, \mathbf{w} \rangle$, $b = \mathbf{w}^*(U^*U)\mathbf{v} = \langle U\mathbf{v}, U\mathbf{w} \rangle$, so $\mathbf{v}^*\mathbf{w} = \overline{a}$ and $\mathbf{v}^*(U^*U)\mathbf{w} = \overline{b}$. Putting these values into the preceding formulas and recalling that $||U\mathbf{v}||^2 = ||\mathbf{v}||^2$ and $||U\mathbf{w}||^2 = ||\mathbf{w}||^2$, we get $a + \overline{a} = b + \overline{b}$ and $-i(a - \overline{a}) = -i(b - \overline{b})$. These equations say that $2\operatorname{Re}(a) = 2\operatorname{Re}(b)$ and $2\operatorname{Im}(a) = 2\operatorname{Im}(b)$; so the complex numbers a and b are equal. This means that $\langle U\mathbf{v}, U\mathbf{w} \rangle = \langle \mathbf{v}, \mathbf{w} \rangle$, hence U is unitary.

To summarize, all of the following conditions on a matrix $U \in M_n(\mathbb{C})$ are equivalent: U is unitary; $UU^* = I$; $U^*U = I$; U is invertible and $U^{-1} = U^*$; the rows of U are orthonormal in $\mathbb{C}^n$; the columns of U are orthonormal in $\mathbb{C}^n$; U^* is unitary; U^{-1} is unitary; U^{T} is unitary; U preserves inner products in $\mathbb{C}^n$; U preserves norms in $\mathbb{C}^n$. Also, if U is unitary then the determinant and all eigenvalues of U have norm 1 in $\mathbb{C}$, but the converse does not hold in general. The failure of the converse can be demonstrated by considering an upper-triangular matrix V with numbers of modulus 1 on the main diagonal and at least one nonzero off-diagonal entry. Here, V^* is lower-triangular, V^{-1} is upper-triangular, and neither matrix is diagonal, so $V^* \neq V^{-1}$.

7.6   Unitary Similarity

Two square matrices $A, B \in M_n(\mathbb{C})$ are called *similar* iff there is an invertible matrix $S \in M_n(\mathbb{C})$ such that $B = S^{-1}AS$. We say A and B are *unitarily similar* iff there is a unitary matrix $U \in M_n(\mathbb{C})$ such that $B = U^{-1}AU = U^*AU$. You can check that similarity and unitary similarity are equivalence relations on $M_n(\mathbb{C})$. The next theorem shows that unitary similarity preserves the properties of being Hermitian, positive definite, or unitary.

Theorem on Unitary Similarity. Suppose B is unitarily similar to A.
(a) A is Hermitian iff B is Hermitian.
(b) A is positive definite iff B is positive definite.
(c) A is unitary iff B is unitary.

Proof. Let $B = U^*AU$ where U is unitary. If A is Hermitian, then $B^* = (U^*AU)^* = U^*A^*U = U^*AU = B$, so B is Hermitian. If A is positive definite and $\mathbf{v} \in \mathbb{C}^n$ is nonzero, then $U\mathbf{v} \neq 0$ since U is invertible. Hence $\mathbf{v}^*B\mathbf{v} = \mathbf{v}^*U^*AU\mathbf{v} = (U\mathbf{v})^*A(U\mathbf{v})$ is a positive real number, so B is positive definite. If A is unitary, then $B = U^*AU$ is a product of unitary matrices and is therefore unitary. The converse statements follow from the symmetry of unitary similarity. $\square$

Next, we discuss the geometric significance of similarity and unitary similarity. Suppose V is a finite-dimensional complex vector space (which we can take to be $\mathbb{C}^n$ with no loss of generality), and suppose $T : V \to V$ is a linear operator on V. Suppose A is the matrix representing T relative to some ordered basis of V. If we switch to some other ordered basis of V, then the matrix representing T relative to the new basis is similar to A. The columns of the similarity matrix S give the coordinates of the new basis relative to the old basis. Conversely, any matrix similar to A is the matrix of T relative to some ordered basis of V. For more details, see Chapter 6.

Unitary similarity has an analogous geometric interpretation. Here, we suppose V is a finite-dimensional complex inner product space (which we can take to be $\mathbb{C}^n$ with the standard inner product, without loss of generality), and we assume T is a linear operator on V. Let A be the matrix of T relative to an orthonormal basis of V, which might be the standard ordered basis $(\mathbf{e}_1, \ldots, \mathbf{e}_n)$ of $\mathbb{C}^n$. You can check that A is a length-preserving matrix (meaning $||A\mathbf{x}|| = ||\mathbf{x}||$ for all $\mathbf{x} \in \mathbb{C}^n$) iff T is a length-preserving map (meaning $||T(v)|| = ||v||$ for all $v \in V$). If we switch to some other orthonormal basis of V, we obtain a new matrix B such that $B = U^{-1}AU$ for some invertible matrix U. As pointed out above, the columns of U are the coordinates of the new basis vectors relative to the old ones. Since both bases are orthonormal, we see that the columns of U are orthonormal in $\mathbb{C}^n$, and hence U is unitary. Thus, B is unitarily similar to A. Conversely, any matrix that is unitarily similar to A is the matrix of T relative to some orthonormal basis of V. So *unitary similarity corresponds to orthonormal change of basis, while ordinary similarity corresponds to arbitrary change of basis.* For more details, see §6.21.

A matrix $A \in M_n(\mathbb{C})$ is called *unitarily diagonalizable* iff there exists a unitary matrix $U \in M_n(\mathbb{C})$ and a diagonal matrix $D \in M_n(\mathbb{C})$ such that $U^*AU = D$. A linear operator T on V is called *unitarily diagonalizable* iff the matrix of T relative to some orthonormal basis is diagonal iff the matrix of T relative to any orthonormal basis is unitarily diagonalizable. This means that there exists an orthonormal list of vectors $(u_1, \ldots, u_n)$ in V and scalars $c_1, \ldots, c_n \in \mathbb{C}$ such that $T(u_k) = c_k u_k$ for all k between 1 and n. In other words, *an operator T on V is unitarily diagonalizable iff the inner product space V has an orthonormal basis consisting of eigenvectors of T.*

In the case where $V = \mathbb{C}^n$ and A is the matrix of T relative to $(\mathbf{e}_1, \ldots, \mathbf{e}_n)$, the matrix equation $U^*AU = D$ is equivalent to $AU = UD$ (since $U^* = U^{-1}$). Let the columns of U be $\mathbf{u}_1, \ldots, \mathbf{u}_n \in \mathbb{C}^n$ and the diagonal entries of D be $c_1, \ldots, c_n$. The kth column of AU is $A\mathbf{u}_k$, while the kth column of UD is $c_k\mathbf{u}_k$. Thus, we get a matrix proof of the observation in the previous paragraph: a matrix $A \in M_n(\mathbb{C})$ is unitarily diagonalizable iff there exist n orthonormal vectors in $\mathbb{C}^n$ that are eigenvectors of A. In this case, the unitary matrix achieving diagonalization has columns consisting of the orthonormal eigenvectors of A, and the resulting diagonal matrix has the eigenvalues of A on its main diagonal.

7.7 Unitary Triangularization

Not every matrix can be unitarily diagonalized. For example, the matrix $A = \begin{bmatrix} 0 & 1 \\ 0 & 0 \end{bmatrix}$ has zero as its only eigenvalue. If A were similar (or unitarily similar) to a diagonal matrix, that matrix must have zeroes on its main diagonal, since similar matrices have the same eigenvalues. But then A would be similar to the zero matrix, forcing A itself to be zero. We prove below that normality of A is a necessary and sufficient condition for unitary diagonalizability of A.

On the other hand, if we weaken the concept of unitary diagonalizability, we can obtain a result valid for all square matrices. A matrix $A \in M_n(\mathbb{C})$ is called *triangulable* iff there is an invertible matrix $S \in M_n(\mathbb{C})$ such that $S^{-1}AS$ is an upper-triangular matrix. Call A *unitarily triangulable* iff there is a unitary $U \in M_n(\mathbb{C})$ such that $U^{-1}AU = U^*AU$ is an upper-triangular matrix. Geometrically, if $T : V \to V$ is the linear operator represented by the matrix A, then T is triangulable iff there exists an ordered basis $(x_1, \ldots, x_n)$ of V such that $T(x_i)$ is in the span of $(x_1, \ldots, x_i)$ for $i = 1, 2, \ldots, n$. T is unitarily triangulable iff there exists an orthonormal basis of V with this property. In this situation, x_1 must be an eigenvector of T.

Schur's Theorem on Unitary Triangularization. Every matrix $A \in M_n(\mathbb{C})$ can be unitarily triangularized. Equivalently, for every linear operator T on an n-dimensional complex inner product space V, there is an orthonormal basis X of V such that $[T]_X$ (the matrix of T relative to X) is upper-triangular.

Proof. We prove the result for linear operators by induction on $n \geq 1$. For $n = 1$, let x be any nonzero vector in V. Letting $X = (x/||x||)$, X is an orthonormal basis of V, and the 1×1 matrix $[T]_X$ is upper-triangular. Next, fix $n > 1$, and assume the theorem is already known for all inner product spaces of dimension less than n. Since we are working over the algebraically closed field $\mathbb{C}$, we know that T has an eigenvalue $c_1 \in \mathbb{C}$ with associated eigenvector x_1 (take c_1 to be any root of the characteristic polynomial of T). Dividing x_1 by a scalar, we can ensure that $||x_1|| = 1$. We claim $V = \mathbb{C}x_1 \oplus W$, where

$$W = x_1^\perp = \{w \in V : \langle w, x_1 \rangle = 0\}.$$

To see this, consider the map $S : V \to \mathbb{C}$ given by $S(v) = \langle v, x_1 \rangle$ for $v \in V$. S is a linear map with kernel W and image $\mathbb{C}$ (since $S(x_1) = 1$). So the Rank–Nullity Theorem shows that $n = \dim(V) = \dim(W) + \dim(\mathbb{C})$. As $\dim(\mathbb{C}) = 1$, we obtain $\dim(W) = n - 1$. On the other hand, given $dx_1 \in \mathbb{C}x_1 \cap W$ with $d \in \mathbb{C}$, we have $0 = S(dx_1) = \langle dx_1, x_1 \rangle = d||x_1||^2 = d$, so $\mathbb{C}x_1 \cap W = \{0\}$. This proves the claim.

It follows that every vector in V can be written uniquely in the form $ax_1 + w$, where $a \in \mathbb{C}$ and $w \in W$. Let $P : V \to W$ be the projection map defined by $P(ax_1 + w) = w$, and define a linear map $T' : W \to W$ by setting $T'(w) = P(T(w))$ for $w \in W$. By induction, we can find an orthonormal basis $X' = (x_2, \ldots, x_n)$ of W such that $[T']_{X'}$ is upper-triangular. Let $X = (x_1, x_2, \ldots, x_n)$, which is readily verified to be an orthonormal basis for V. We assert that

$$[T]_X = \left[\begin{array}{c|ccc} c_1 & * & \cdots & * \\ \hline 0 & & & \\ \vdots & & [T']_{X'} & \\ 0 & & & \end{array} \right], \tag{7.2}$$

which is an upper-triangular matrix. To verify this, recall that the jth column of $[T]_X$ consists of the coordinates of $T(x_j)$ relative to the basis X. Since $T(x_1) = c_1 x_1$, the first column of $[T]_X$ is $(c_1, 0, \ldots, 0)$. For $j > 1$, write $T(x_j) = c_{1,j} x_1 + c_{2,j} x_2 + \cdots + c_{n,j} x_n$ with $c_{i,j} \in \mathbb{C}$. Then

$$T'(x_j) = P(T(x_j)) = c_{2,j} x_2 + \cdots + c_{n,j} x_n.$$

Thus, the jth column of $[T]_X$ consists of some scalar $c_{1,j}$ followed by the $(j-1)$th column of $[T']_{X'}$. So $[T]_X$ has the form shown in (7.2). $\qquad\square$

7.8 Simultaneous Triangularization

Given two matrices $A, B \in M_n(\mathbb{C})$, Schur's Theorem guarantees that there are unitary matrices U and V such that $U^* A U$ and $V^* B V$ are upper-triangular matrices. However, there is no guarantee that one single matrix U can triangularize both A and B at the same time. If such a unitary matrix does exist, we say that A and B can be *simultaneously unitarily triangularized*.

Theorem on Simultaneous Triangularization of Two Commuting Matrices. If $A, B \in M_n(\mathbb{C})$ satisfy $AB = BA$, then A and B can be simultaneously unitarily triangularized. Equivalently, if T and U are two commuting linear operators on a complex inner product space V, then there exists an orthonormal basis X of V such that $[T]_X$ and $[U]_X$ are both upper-triangular matrices.

Proof. We use induction on $n = \dim(V)$, the case $n = 1$ being immediate. Assume $n > 1$ and that the theorem holds for linear maps on spaces of dimension less than n. If the required result is to be true, consideration of the first column of $[T]_X$ and $[U]_X$ shows that we need to find a unit vector $x_1 \in \mathbb{C}^n$ that is simultaneously an eigenvector for T and an eigenvector for U. To do this, fix an eigenvalue c_1 of T. Let $Z = \{v \in V : T(v) = c_1 v\}$, which is a nonzero subspace of V, namely, the eigenspace associated with the eigenvalue c_1. We claim that U maps Z into Z. Given $v \in Z$, commutativity of T and U gives

$$T(U(v)) = U(T(v)) = U(c_1 v) = c_1(U(v)),$$

so that $U(v) \in Z$. It follows that the restriction $U|Z : Z \to Z$ is a linear operator on Z. Thus, U has an eigenvector in Z (with associated eigenvalue d_1, say). Let $x_1 \in Z$ be a unit eigenvector of U. Since $x_1 \in Z$, x_1 is also an eigenvector for T.

We continue as in the proof of Schur's Theorem. Let $W = x_1^\perp$, so that $V = \mathbb{C}x_1 \oplus W$. Let $P : \mathbb{C}x_1 \oplus W \to W$ be the projection given by $P(ax_1 + w) = w$ for $a \in \mathbb{C}$ and $w \in W$. Consider the two linear operators on W defined by $T' = (P \circ T)|W$ and $U' = (P \circ U)|W$. We intend to apply our induction hypothesis to the $(n-1)$-dimensional space W. Before doing so, we must check that the new operators T' and U' on W still commute. To verify this, we first show that $P \circ T \circ P = P \circ T$. Given any $v = ax_1 + w$ in V (where $a \in \mathbb{C}$ and $w \in W$),

$$P(T(v)) = P(T(ax_1 + w)) = P(aT(x_1) + T(w)) =$$
$$P(ac_1 x_1) + P(T(w)) = P(T(w)) = P(T(P(v))). \quad (7.3)$$

Similarly, $P \circ U \circ P = P \circ U$. It follows that

$$\begin{aligned} T' \circ U' &= (P \circ T \circ P \circ U)|W = (P \circ T \circ U)|W \\ &= (P \circ U \circ T)|W = (P \circ U \circ P \circ T)|W = U' \circ T'. \end{aligned}$$

The induction hypothesis, applied to T' and U' acting on W, produces an orthonormal basis $X' = (x_2, \ldots, x_n)$ of W such that $[T']_{X'}$ and $[U']_{X'}$ are both upper-triangular. Letting $X = (x_1, x_2, \ldots, x_n)$, we see as in the proof of Schur's Theorem that X is an orthonormal basis of V such that $[T]_X$ and $[U]_X$ are both upper-triangular. $\square$

We can extend this result to an arbitrary (possibly infinite) collection of commuting matrices or operators.

Theorem on Simultaneous Triangularization of Commuting Operators. Suppose V is an n-dimensional complex inner product space and $\{T_i : i \in I\}$ is an indexed collection of linear operators on V such that $T_i \circ T_j = T_j \circ T_i$ for all $i, j \in I$. Then there exists an orthonormal basis X of V such that $[T_i]_X$ is upper-triangular for all $i \in I$.

Proof. We prove this result by repeating the previous induction proof. The key point is that we must find a unit vector in V that is simultaneously an eigenvector for every operator T_i. To do this, consider a nonzero subspace Z of V of minimum positive dimension such that $T_i[Z] \subseteq Z$ for all $i \in I$. Such a subspace must exist, since $T_i[V] \subseteq V$ and V is finite-dimensional. Let T be a fixed operator from the collection $\{T_i : i \in I\}$. Let Z' be a nonzero eigenspace for $T|Z$. The reasoning used before shows that $T_i[Z'] \subseteq Z'$ for all $i \in I$, since each T_i commutes with T. By minimality of $\dim(Z)$, we see that $Z' = Z$. Thus, Z itself is part of an eigenspace for T, so every nonzero $z \in Z$ is an eigenvector for T. But T was arbitrary, so every nonzero $z \in Z$ is an eigenvector for every T_i in the given collection. Choose any such z and normalize it to get a common unit eigenvector x_1 for all the T_i. The rest of the proof is the same as before. In particular, any two maps from the collection $\{T'_i = (P \circ T_i)|W : i \in I\}$ commute (by the same calculation done previously), so the induction hypothesis does apply to give us the basis X' needed to complete the induction step. $\square$

7.9   Normal Matrices and Unitary Diagonalization

Recall that a matrix $A \in M_n(\mathbb{C})$ is unitarily diagonalizable iff there exists a unitary matrix U such that U^*AU is a diagonal matrix. On the other hand, A is normal iff $AA^* = A^*A$. The next theorem states a surprising connection between these two concepts.

Spectral Theorem on Normal Matrices and Unitary Diagonalizability. For all $A \in M_n(\mathbb{C})$, A is unitarily diagonalizable if and only if A is normal.

Proof. First assume that A is unitarily diagonalizable, so there is a unitary matrix U and a diagonal matrix D with $D = U^*AU$. Since $D^* = U^*A^*U$ is also diagonal, we see that $DD^* = D^*D$. Then

$$AA^* = (UDU^*)(UD^*U^*) = U(DD^*)U^* = U(D^*D)U^* = (UD^*U^*)(UDU^*) = A^*A,$$

so A is normal.

Conversely, suppose $A \in M_n(\mathbb{C})$ is a normal matrix. Then $AA^* = A^*A$ by definition, so A and A^* can be simultaneously unitarily triangularized. Choose a unitary matrix U such that $C = U^*AU$ and $D = U^*A^*U$ are both upper-triangular. On one hand, C is an upper-triangular matrix. On the other hand, $C = U^*AU = (U^*A^*U)^* = D^*$ is the conjugate-transpose of an upper-triangular matrix, so C is a lower-triangular matrix. This means that C must be a diagonal matrix, so we have unitarily diagonalized A. $\square$

Here are some useful corollaries of the Spectral Theorem.

Theorem on Unitary Diagonalization of Special Matrices. Hermitian matrices, skew-Hermitian matrices, positive definite matrices, negative definite matrices, real symmetric matrices, and unitary matrices can always be unitarily diagonalized.

Proof. By the Spectral Theorem, it suffices to check that each type of matrix mentioned in the theorem must be normal. Any Hermitian matrix A is normal, since $A = A^*$ implies $AA^* = A^2 = A^*A$. In particular, positive definite, negative definite, and real symmetric matrices are normal. Any unitary matrix U is normal, since $UU^* = I = U^*U$. Any skew-Hermitian matrix S is normal, since $S^* = -S$ implies $SS^* = -S^2 = S^*S$. □

There are normal matrices that are neither Hermitian, skew-Hermitian, nor unitary. For example, consider $A = \begin{bmatrix} 1+i & 0 \\ 0 & 4 \end{bmatrix}$ or any matrix unitarily similar to A.

The nature of the eigenvalues of a normal matrix leads to a characterization of the properties of being Hermitian, positive definite, or unitary.

Theorem on Eigenvalues of Normal Matrices. Suppose $A \in M_n(\mathbb{C})$ is normal.
(a) A is Hermitian iff all eigenvalues of A are real.
(b) A is positive definite iff all eigenvalues of A are strictly positive real numbers.
(c) A is unitary iff all eigenvalues of A are complex numbers of modulus 1.
(d) A is invertible iff all eigenvalues of A are nonzero.

Proof. In each part, the forward implication holds with no restriction on A, as we saw earlier in the chapter. For normal matrices, we can prove both directions of each implication using the Spectral Theorem. Given a normal matrix A, write $D = U^*AU$ where U is unitary and D is a diagonal matrix with the eigenvalues of A on its diagonal. Unitary similarity preserves the properties of being Hermitian, positive definite, unitary, or invertible, so A has one of these properties iff D does. It is immediate from the definitions that a diagonal matrix D is Hermitian iff all diagonal entries of D are real; D is positive definite iff all diagonal entries of D are strictly positive; D is unitary iff all diagonal entries of D have norm 1 in $\mathbb{C}$; and D is invertible iff all diagonal entries of D are nonzero. □

We know that every positive real number has a positive square root. Analogy leads us to the following matrix result.

Theorem on Square Roots of Positive Definite Matrices. If $A \in M_n(\mathbb{C})$ is positive definite, then there exists a positive definite matrix B such that $B^2 = A$.

This positive definite square root of A is unique, as shown later (§7.11).

Proof. There is a unitary matrix U and a diagonal matrix D with $D = U^*AU$. Each diagonal entry of D is a positive real number; let E be the diagonal matrix obtained by replacing each such entry by its positive square root. Then $E^2 = D$. Setting $B = UEU^*$, we compute

$$B^2 = (UEU^*)(UEU^*) = UE(U^*U)EU^* = UE^2U^* = UDU^* = A.$$

Since E is positive definite and B is unitarily similar to E, B is also positive definite. □

A similar proof shows that every Hermitian matrix A has a Hermitian cube root, which is a Hermitian matrix C such that $C^3 = A$.

7.10   Polynomials and Commuting Matrices

We sometimes need to know which matrices in $M_n(\mathbb{C})$ commute with a given matrix A. For example, we proved that commutativity of A and B is a sufficient condition for A and B to be simultaneously triangulable. In general, it is difficult to characterize the set of all matrices that commute with A. However, we can describe a particular collection of matrices that always commute with A. Consider a polynomial $p \in \mathbb{C}[x]$ with complex coefficients, say $p = c_0 + c_1 x + c_2 x^2 + \cdots + c_d x^d$ where all $c_i \in \mathbb{C}$. We can replace x by A in this polynomial to obtain a matrix $p(A) = c_0 I + c_1 A + c_2 A^2 + \cdots + c_d A^d \in M_n(\mathbb{C})$. For any polynomial p, the matrix A commutes with $p(A)$, because

$$Ap(A) = c_0 A + c_1 A^2 + c_2 A^3 + \cdots + c_d A^{d+1} = p(A)A.$$

More generally, suppose C is any matrix that commutes with a given matrix A. We show by induction that C commutes with A^k for all integers $k \geq 0$. This holds for $k = 0$ and $k = 1$, since C commutes with I and A. Assuming the result holds for some k, note that

$$CA^{k+1} = C(A^k A) = (CA^k)A = (A^k C)A = A^k(CA) = A^k(AC) = A^{k+1}C.$$

Next, observe that if C commutes with U and V, then C commutes with $U + V$. If C commutes with U and r is a scalar, then C commutes with rU. It follows that if C commutes with matrices $U_0, \ldots, U_m$, then C commutes with any linear combination of $U_0, \ldots, U_m$. Taking $U_k = A^k$, we deduce that *if C commutes with A, then C commutes with $p(A)$ for every polynomial p.* For example, if A is invertible, then $C = A^{-1}$ commutes with A (since $AA^{-1} = I = A^{-1}A$), so A^{-1} commutes with every polynomial in A.

We have shown that for all matrices $A, C \in M_n(\mathbb{C})$, if C commutes with A, then $p(A)$ commutes with C for every polynomial p. We can apply this result with A replaced by C and C replaced by $p(A)$ to conclude: *if $AC = CA$, then $p(A)$ commutes with $q(C)$ for all polynomials $p, q \in \mathbb{C}[x]$.*

Here are a few more facts that can be proved by similar methods. First, suppose $B = U^*AU$ for some unitary matrix U. Then $p(B) = U^*p(A)U$ and $p(A) = Up(B)U^*$ for all polynomials $p \in \mathbb{C}[x]$. We prove this for the particular polynomials $p = x^k$ by induction on $k \geq 0$, then deduce the result for general p by linearity. Second, suppose D is a diagonal matrix with diagonal entries $d_1, \ldots, d_n$. The same proof technique shows that $p(D)$ is a diagonal matrix with diagonal entries $p(d_1), \ldots, p(d_n)$.

Theorem on Polynomial Characterization of Normality. For all $A \in M_n(\mathbb{C})$, A is normal iff $A^* = p(A)$ for some polynomial $p \in \mathbb{C}[x]$. If A is normal and invertible, then A^{-1} is a polynomial in A.

Proof. First assume $A^* = p(A)$ for some polynomial p. Every polynomial in A commutes with A, so A is normal. Conversely, assume A is normal. Choose a unitary U such that $D = U^*AU$ is diagonal. Let $d_1, \ldots, d_n$ be the diagonal entries of D. We know that $D^* = U^*A^*U$, where D^* has diagonal entries $\overline{d_1}, \ldots, \overline{d_n}$. Suppose we can find a polynomial p such that $p(D) = D^*$. Then a calculation shows that $p(A) = p(UDU^*) = Up(D)U^* = UD^*U^* = A^*$, so the proof is complete once p is found.

To construct p, we use *Lagrange's Interpolation Formula* (see §3.19): if $x_1, \ldots, x_m$ are any m distinct numbers in $\mathbb{C}$, and $y_1, \ldots, y_m$ are arbitrary elements of $\mathbb{C}$, then there exists a polynomial $p \in \mathbb{C}[x]$ such that $p(x_i) = y_i$ for all i. You can check that

$$p(x) = \sum_{j=1}^{m} y_j \frac{\prod_{i:i \neq j}(x - x_i)}{\prod_{i:i \neq j}(x_j - x_i)}$$

is a polynomial with the required values. To apply this formula in the current situation, note that $d_i = d_j$ iff $\overline{d_i} = \overline{d_j}$. We can take the x_i to be the distinct complex numbers that occur on the main diagonal of D, and we can take $y_i = \overline{x_i}$ for each i. Choosing p as above, it then follows that $p(d_i) = \overline{d_i}$ for all i. Accordingly, $p(D) = D^*$ and $p(A) = A^*$.

In the case where A is normal and invertible, D^{-1} is the diagonal matrix with diagonal entries $d_1^{-1}, \ldots, d_n^{-1}$. Using Lagrange's Interpolation Formula as before, we can find a polynomial $q \in \mathbb{C}[x]$ such that $q(d_i) = d_i^{-1}$ for all i. Then $q(D) = D^{-1}$, and hence $q(A) = A^{-1}$. $\square$

Using the Cayley–Hamilton Theorem, you can check that A^{-1} is a polynomial in A for any invertible matrix A, even if A is not normal.

7.11   Simultaneous Unitary Diagonalization

We say that a collection of matrices $\{A_i : i \in I\} \subseteq M_n(\mathbb{C})$ can be *simultaneously unitarily diagonalized* iff there exists a unitary matrix U (independent of i) such that $D_i = U^* A_i U$ is diagonal for all $i \in I$.

Theorem on Simultaneous Unitary Diagonalization. A collection $\{A_i : i \in I\}$ of matrices in $M_n(\mathbb{C})$ is simultaneously unitarily diagonalizable if and only if every A_i is a normal matrix and $A_i A_j = A_j A_i$ for all $i, j \in I$.

Proof. First assume $\{A_i : i \in I\}$ is simultaneously unitarily diagonalizable. Since each A_i is unitarily diagonalizable, we know each A_i is normal by a previous theorem. Next, choose a unitary U such that $D_i = U^* A_i U$ is diagonal for all $i \in I$. Diagonal matrices commute, so $D_i D_j = D_j D_i$ for all $i, j \in I$. Therefore,

$$A_i A_j = (U D_i U^*)(U D_j U^*) = U(D_i D_j)U^* = U(D_j D_i)U^* = (U D_j U^*)(U D_i U^*) = A_j A_i$$

for all $i, j \in I$.

Next, assume every A_i is normal and $A_i A_j = A_j A_i$ for all $i, j \in I$. We recycle some ideas from the earlier proof that a normal matrix is unitarily diagonalizable. The key idea in that proof was to simultaneously unitarily upper-triangularize A and A^*, and then observe that the resulting triangular matrices must actually be diagonal. To use this idea here, we need to invoke our previous theorem about simultaneous upper-triangularization of a collection of matrices (§7.8). Consider the collection of matrices $\mathcal{F} = \{A_i : i \in I\} \cup \{A_i^* : i \in I\}$. Each A_i is normal, so $A_i^* = p_i(A_i)$ for certain polynomials $p_i \in \mathbb{C}[x]$. To use our theorem, we need to know that $\mathcal{F}$ is a commuting family. For any $i, j \in I$, A_i commutes with A_j by hypothesis. So, by our results on polynomials, A_i commutes with $p_j(A_j) = A_j^*$; $p_i(A_i) = A_i^*$ commutes with A_j; and $p_i(A_i) = A_i^*$ commutes with $p_j(A_j) = A_j^*$. This covers all possible pairs of matrices in $\mathcal{F}$. So our theorem does apply, and we know there is a unitary matrix U such that $D_i = U^* A_i U$ and $E_i = U^* A_i^* U$ are upper-triangular matrices for all $i \in I$. We finish the proof as before: D_i is both upper-triangular and lower-triangular, since $D_i = E_i^*$; so D_i is a diagonal matrix for all $i \in I$. Thus the collection $\{A_i : i \in I\}$ has been simultaneously unitarily diagonalized. (In fact, we have diagonalized the larger collection $\mathcal{F}$.) $\square$

As an illustration of this theorem, we prove the uniqueness part of the Theorem on Square Roots of Positive Definite Matrices. Let $A \in M_n(\mathbb{C})$ be a positive definite matrix. We show that there exists a unique positive definite matrix B such that $B^2 = A$. Recall how existence of B follows from the Spectral Theorem. First, we find a unitary U such that $D = U^*AU$ is diagonal with positive real diagonal entries $d_1, \ldots, d_n$. Then we let E be diagonal with diagonal entries $\sqrt{d_1}, \ldots, \sqrt{d_n}$, and put $B = UEU^*$. By Lagrange interpolation, choose a polynomial $p \in \mathbb{C}[x]$ such that $p(d_i) = \sqrt{d_i}$ for all i. It follows that $p(D) = E$ and hence $p(A) = B$. Now, suppose C is any positive definite matrix such that $C^2 = A$. Since $B = p(C^2)$ is a polynomial in C, B and C commute. Both B and C are normal, so we can simultaneously unitarily diagonalize B and C. Let V be a unitary matrix such that $D_1 = V^*BV$ and $D_2 = V^*CV$ are both diagonal. D_1 and D_2 are positive definite (since B and C are), so all diagonal entries of D_1 and D_2 are positive. Furthermore, $D_1^2 = V^*AV = D_2^2$, so that $D_1(k,k)^2 = D_2(k,k)^2 \in \mathbb{R}$ for all k. Since every positive real number has a unique positive square root, it follows that $D_1(k,k) = D_2(k,k)$ for all k. So $D_1 = D_2$, which implies $B = VD_1V^* = VD_2V^* = C$. A similar proof shows that for all integers $k \geq 1$, every positive definite matrix in $M_n(\mathbb{C})$ has a unique positive definite $(2k)$th root, and every Hermitian matrix has a unique Hermitian $(2k-1)$th root. We use the notation $\sqrt{A}$ to denote the unique positive definite square root of a positive definite matrix A, $\sqrt[3]{B}$ to denote the unique Hermitian cube root of a Hermitian matrix B, and so on.

7.12 Polar Decomposition: Invertible Case

For every nonzero complex number z, there exist $r, u \in \mathbb{C}$ such that r is a positive real number, $|u| = 1$, and $z = ru = ur$. We call this the *polar decomposition of z*. Analogy suggests the following result: for every invertible normal matrix $A \in M_n(\mathbb{C})$, there exist $R, U \in M_n(\mathbb{C})$ such that R is positive definite, U is unitary, and $A = RU = UR$. The techniques developed so far lead to a quick proof, as follows. By normality of A, find a unitary V and a diagonal D such that $D = V^*AV$. Each diagonal entry $d_i = D(i,i)$ is a nonzero complex number (since A is invertible). Let $d_i = r_iu_i$ be the polar decomposition of d_i. Let R' and U' be diagonal matrices with diagonal entries r_i and u_i, so R' is positive definite and U' is unitary. Observe that $D = R'U' = U'R'$. Putting $R = VR'V^*$ and $U = VU'V^*$, we then have $A = RU = UR$ where R is positive definite and U is unitary.

The next theorem generalizes this result to the case where A is invertible but not necessarily normal.

Polar Decomposition Theorem for Invertible Matrices. If $A \in M_n(\mathbb{C})$ is invertible, then there exist unique matrices $R, U \in M_n(\mathbb{C})$ such that R is positive definite, U is unitary, and $A = RU$. This factorization is called the *polar decomposition of A*.

Proof. To motivate the proof, consider the polar decomposition $z = ru$ of a nonzero complex number z. To find r given z, we could calculate $z\bar{z} = |z|^2 = r^2|u|^2 = r^2$ and then take square roots to find r. Then u must be $r^{-1}z$. Given the invertible matrix A, we are therefore led to consider the matrix AA^*. This matrix is positive definite, since for nonzero $\mathbf{v} \in \mathbb{C}^n$, $\mathbf{v}^*AA^*\mathbf{v} = \|A^*\mathbf{v}\|^2$ is a positive real number. Let R be the unique positive definite square root of AA^* and $U = R^{-1}A$. Then $A = RU$. Since $R^* = R$ and $RR = AA^*$, the fact that U is unitary follows from the calculation

$$UU^* = R^{-1}AA^*(R^{-1})^* = R^{-1}(RR)(R^*)^{-1} = (R^{-1}R)(RR^{-1}) = I.$$

To see that R and U are uniquely determined by A, suppose $A = SW$ where S is positive definite and W is unitary. Then $AA^* = SWW^*S^* = SIS = S^2$. So $S = R$ since the positive definite matrix AA^* has a unique positive definite square root. It follows that $W = S^{-1}A = R^{-1}A = U$. $\square$

In the polar decomposition $A = RU$, R and U need not commute. In fact, $RU = UR$ holds iff A is a normal matrix (compare this to the corresponding fact for the Cartesian decomposition of A, which motivated the definition of normality). If A is normal, then $RU = UR$ follows from the unitary diagonalization proof given in the first paragraph of this section. Conversely, suppose R and U commute. Since R and U are both normal, it suffices to prove that *the product of two commuting normal matrices is also normal.* Suppose B and C are normal commuting matrices. By normality, B^* is a polynomial in B and C^* is a polynomial in C, so the four matrices B, B^*, C, C^* all commute with one another. Then

$$(BC)(BC)^* = BCC^*B^* = C^*B^*BC = (BC)^*(BC),$$

so BC is normal.

For A invertible but not normal, we can obtain a dual polar decomposition $A = U_1 R_1$, where R_1 is positive definite and U_1 is unitary. It suffices to take the conjugate-transpose of the polar decomposition $A^* = R_2 U_2$, which gives $A = U_2^* R_2^*$ where U_2^* is unitary and $R_2^* = R_2$ is positive definite. As above, U_1 and R_1 are uniquely determined by A.

7.13 Polar Decomposition: General Case

The complex number 0 can be written in polar form as $0 = ru = ur$, where $r = 0$ and u is any complex number of modulus 1. Analogy suggests the following result.

Polar Decomposition Theorem for Non-Invertible Matrices. For any non-invertible matrix $A \in M_n(\mathbb{C})$, there exist a positive semidefinite R and a unitary U such that $A = UR$, where R (but not U) is uniquely determined by A.

Proof. Finding R and proving its uniqueness is not difficult, given what we did earlier for invertible A: if $A = UR$ is to hold, we must have $A^*A = R^*U^*UR = R^2$, forcing R to be the unique positive semidefinite square root of the matrix A^*A (where A^*A is readily seen to be positive semidefinite). However, finding U becomes trickier, since neither A nor R is invertible in the present situation.

To proceed, we need the observation that R and A have the same null space. To see this, fix $\mathbf{v} \in \mathbb{C}^n$, and compute (using $R^2 = A^*A$ and $R = R^*$)

$$R\mathbf{v} = \mathbf{0} \Leftrightarrow ||R\mathbf{v}||^2 = 0 \Leftrightarrow \mathbf{v}^*R^*R\mathbf{v} = 0 \Leftrightarrow \mathbf{v}^*A^*A\mathbf{v} = 0 \Leftrightarrow ||A\mathbf{v}||^2 = 0 \Leftrightarrow A\mathbf{v} = \mathbf{0}.$$

By the Rank–Nullity Theorem, the image of R and the image of A have the same dimension. Let $V \subseteq \mathbb{C}^n$ be the image of R and $W \subseteq \mathbb{C}^n$ be the image of A. We define a map $\phi : V \to W$ as follows. Given $\mathbf{z} \in V$, write $\mathbf{z} = R\mathbf{x}$ for some $\mathbf{x} \in \mathbb{C}^n$, and set $\phi(\mathbf{z}) = A\mathbf{x}$. We must check that ϕ is well-defined, i.e., that the value $\phi(\mathbf{z})$ does not depend on the choice of $\mathbf{x}$ such that $R\mathbf{x} = \mathbf{z}$. If $\mathbf{x}'$ is another vector such that $R\mathbf{x}' = \mathbf{z}$, then $R(\mathbf{x} - \mathbf{x}') = \mathbf{0}$, hence $A(\mathbf{x} - \mathbf{x}') = \mathbf{0}$, so $A\mathbf{x} = A\mathbf{x}'$. Thus, ϕ is well-defined, and this map satisfies $\phi(R\mathbf{x}) = A\mathbf{x}$ for all $\mathbf{x} \in \mathbb{C}^n$. You can check that ϕ is a linear map. Furthermore, ϕ is length-preserving, since

$$||\phi(R\mathbf{x})||^2 = ||A\mathbf{x}||^2 = \mathbf{x}^*A^*A\mathbf{x} = \mathbf{x}^*R^2\mathbf{x} = \mathbf{x}^*R^*R\mathbf{x} = ||R\mathbf{x}||^2$$

for all $\mathbf{x} \in \mathbb{C}^n$. We can write $\mathbb{C}^n = V \oplus V^\perp = W \oplus W^\perp$ (see Exercise 26). Extend ϕ to a linear operator u on $\mathbb{C}^n$ by sending an orthonormal basis of $V^\perp$ to an orthonormal basis of $W^\perp$ and extending by linearity; the operator u is readily seen to be length-preserving. Let U be the matrix of u relative to the standard ordered basis of $\mathbb{C}^n$, so that $u(\mathbf{x}) = U\mathbf{x}$

(matrix-vector product) for all $\mathbf{x} \in \mathbb{C}^n$. Since u is a length-preserving map, U is a length-preserving (unitary) matrix. Furthermore, $UR\mathbf{x} = u(R\mathbf{x}) = \phi(R\mathbf{x}) = A\mathbf{x}$ for all $\mathbf{x} \in \mathbb{C}^n$. Thus, $A = UR$, and the proof is complete. Moreover, the proof reveals the extent of the non-uniqueness of U: the action of U is forced on the image V of $R = \sqrt{A^*A}$, but U can act as an arbitrary length-preserving map from $V^\perp$ to $W^\perp$. $\square$

Reasoning as in the invertible case, $UR = RU$ holds iff A is normal. By forming the conjugate-transpose of the polar decomposition $A^* = U_2 R_2$, we see that every $A \in M_n(\mathbb{C})$ can be written as a product $R_1 U_1$, where $R_1 = \sqrt{AA^*}$ is positive semidefinite and U_1 is unitary.

The polar decomposition can be written in the following equivalent form.

Singular Value Decomposition. Any matrix $A \in M_n(\mathbb{C})$ can be factored in the form $A = PDQ^*$, where Q and P are unitary, and D is a diagonal matrix with nonnegative real entries. This factorization is called the *singular value decomposition of A*.

Proof. Given $A \in M_n(\mathbb{C})$, start with a polar decomposition $A = UR$ with U unitary and R positive semidefinite. Unitarily diagonalize R, say $D = Q^*RQ$ for some unitary matrix Q. Then $R = QDQ^*$, $P = UQ$ is a unitary matrix, and $A = UR = UQDQ^* = PDQ^*$. $\square$

In the singular value decomposition $A = PDQ^*$, the entries on the diagonal of D are the eigenvalues of R (occurring with the correct multiplicities), and R is uniquely determined by A. It follows that the diagonal entries of D (disregarding order) are uniquely determined by the matrix A; these entries are called the *singular values of A*. By passing from matrices to the linear maps they represent, we obtain the following geometric interpretation of the matrix factorization $A = PDQ^*$. *Every linear transformation on $\mathbb{C}^n$ is the composition of an isometry (length-preserving linear map), followed by a rescaling of the coordinate axes by nonnegative real numbers, followed by another isometry.* The multiset of rescaling factors is uniquely determined by the linear map. We give another proof of the singular value decomposition, extended to rectangular matrices, in §9.12.

7.14 Interlacing Eigenvalues for Hermitian Matrices

Given a Hermitian matrix $B \in M_n(\mathbb{C})$ with $n > 1$, let $A \in M_{n-1}(\mathbb{C})$ be the submatrix of B obtained by deleting the last row and column of B. Then A is also Hermitian, so all eigenvalues of A and B are real numbers. The next theorem shows that each eigenvalue of A is interleaved (or "interlaced") between two consecutive eigenvalues of B.

Interlacing Theorem for Hermitian Matrices. Let $B \in M_n(\mathbb{C})$ be a Hermitian matrix with eigenvalues $b_1 \leq b_2 \leq \cdots \leq b_n$. Let A be B with the last row and column deleted. Let A have eigenvalues $a_1 \leq a_2 \leq \cdots \leq a_{n-1}$. Then

$$b_1 \leq a_1 \leq b_2 \leq a_2 \leq b_3 \leq a_3 \leq \cdots \leq b_{n-1} \leq a_{n-1} \leq b_n,$$

Proof. Step 1. We prove the theorem for all matrices B of the form

$$B = \begin{bmatrix} a_1 & & & & c_1 \\ & a_2 & \mathbf{0} & & c_2 \\ & \mathbf{0} & \ddots & & \vdots \\ & & & a_{n-1} & c_{n-1} \\ \hline \overline{c_1} & \overline{c_2} & \cdots & \overline{c_{n-1}} & a_n \end{bmatrix}, \tag{7.4}$$

where A is diagonal, $a_1, \ldots, a_{n-1}$ are distinct and in increasing order, $a_n \in \mathbb{R}$, and every $c_k \in \mathbb{C}$ is nonzero. Using the definition of determinants in §5.2, we see that any matrix of the form (7.4) has characteristic polynomial

$$\chi_B(t) = \det(tI_n - B) = \prod_{k=1}^{n}(t - a_k) - \sum_{k=1}^{n-1}(t - a_1)\cdots(t - a_{k-1})|c_k|^2(t - a_{k+1})\cdots(t - a_{n-1})$$

(7.5)

(see Exercise 66). We also know $\chi_B(t) = \prod_{k=1}^{n}(t - b_k)$ is a monic polynomial of degree n in $\mathbb{R}[t]$ whose roots are the eigenvalues of B (§5.15).

To proceed, we evaluate the polynomial $\chi_B(t)$ at $t = a_j$, where $1 \leq j \leq n - 1$. Putting $t = a_j$ in (7.5), the first product becomes 0, and all summands indexed by k with $k \neq j$ also become 0 because of the factor $(t - a_j)$. Thus,

$$\chi_B(a_j) = -(a_j - a_1)\cdots(a_j - a_{j-1})|c_j|^2(a_j - a_{j+1})\cdots(a_j - a_{n-1}).$$

Since the a_k are distinct and occur in increasing order, and since $|c_j|^2 \neq 0$, we see that $\chi_B(a_j)$ is negative for $j = n-1, n-3, n-5, \ldots$, and $\chi_B(a_j)$ is positive for $j = n-2, n-4, n-6, \ldots$. Since $\chi_B(t)$ is monic of degree n, $\lim_{x \to \infty} \chi_B(x) = +\infty$, while $\lim_{x \to -\infty} \chi_B(x)$ is $+\infty$ for n even and is $-\infty$ for n odd. We illustrate this situation in Figure 7.1 in the cases $n = 6$ and $n = 7$. Since polynomial functions are continuous, we can apply the Intermediate Value Theorem from calculus to conclude that each of the n intervals $(-\infty, a_1)$, (a_k, a_{k+1}) for $1 \leq k < n - 1$, and (a_{n-1}, ∞) contains at least one root of $\chi_B(t)$. Since $\chi_B(t)$ has exactly n real roots, we see that each interval contains exactly one eigenvalue of B. Therefore $b_1 < a_1 < b_2 < a_2 < \cdots < b_{n-1} < a_{n-1} < b_n$, as needed.

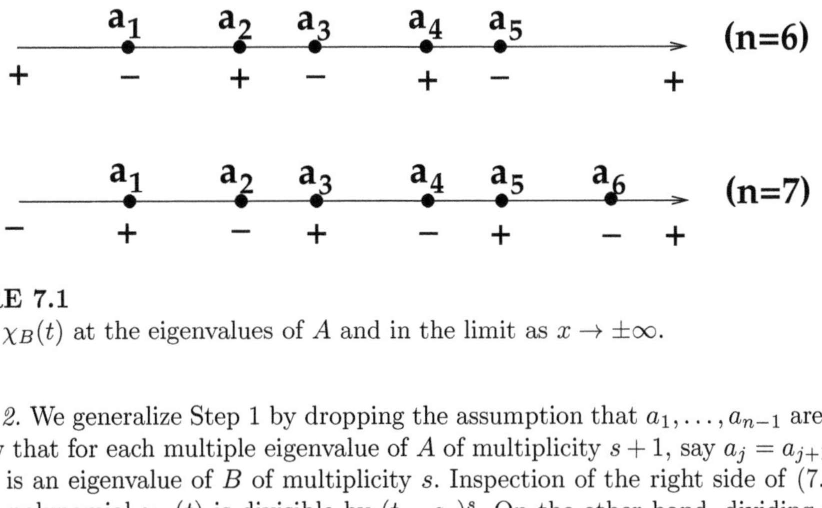

FIGURE 7.1
Signs of $\chi_B(t)$ at the eigenvalues of A and in the limit as $x \to \pm\infty$.

Step 2. We generalize Step 1 by dropping the assumption that $a_1, \ldots, a_{n-1}$ are distinct. We show that for each multiple eigenvalue of A of multiplicity $s + 1$, say $a_j = a_{j+1} = \cdots = a_{j+s}$, a_j is an eigenvalue of B of multiplicity s. Inspection of the right side of (7.5) shows that the polynomial $\chi_B(t)$ is divisible by $(t - a_j)^s$. On the other hand, dividing $\chi_B(t)$ by $(t - a_j)^s$ and then setting $t = a_j$, we obtain

$$-\sum_{k=j}^{j+s}(a_k - a_1)\cdots(a_k - a_{j-1})|c_k|^2(a_k - a_{j+s+1})\cdots(a_k - a_{n-1})$$

$$= -(a_j - a_1)\cdots(a_j - a_{j-1})(a_j - a_{j+s+1})\cdots(a_j - a_{n-1})(|c_j|^2 + \cdots + |c_{j+s}|^2) \neq 0.$$

Thus, a_j is a root of $\chi_B(t)$ of multiplicity s. The proof of Step 2 can now be completed by a sign-change analysis similar to Step 1 (see Exercise 67).

Step 3. We generalize Step 2 by dropping the assumption that every c_k be nonzero. Proceed by induction on $n \geq 2$. If $n = 2$ and $c_1 = 0$, the eigenvalues of B are a_1 and a_2 (in some order), and the interlacing inequalities immediately follow. If $n > 2$ and some $c_k = 0$, let B' (resp. A') be obtained from B (resp. A) by deleting row k and column k. Since $c_k = 0$, we see from (7.4) that $\chi_B(t) = (t - a_k)\chi_{B'}(t)$ and $\chi_A(t) = (t - a_k)\chi_{A'}(t)$. By induction, the eigenvalues of B' and A' satisfy the required interlacing property. To obtain the eigenvalues of B and A, we insert one more copy of a_k into both lists of eigenvalues. You can check that the interlacing property still holds for the new lists.

Step 4. We prove the result for general Hermitian $B \in M_n(\mathbb{C})$. As A is unitarily diagonalizable, there is a unitary matrix $U \in M_{n-1}(\mathbb{C})$ such that U^*AU is diagonal with entries $a_1 \leq \cdots \leq a_{n-1}$ on the diagonal. Add a zero row at the bottom of U, and then add the column $\mathbf{e}_n$ on the right end to get a unitary matrix $V \in M_n(\mathbb{C})$. The matrix V^*BV has the same eigenvalues as B and is of the form (7.4). By Step 3, the required interlacing property holds. $\qquad\square$

7.15   Determinant Criterion for Positive Definite Matrices

Given $A \in M_n(\mathbb{C})$, we can quickly decide whether A is Hermitian ($A^* = A$), unitary ($A^*A = I$), or normal ($A^*A = AA^*$) via matrix computations that do not require knowing the eigenvalues of A. On the other hand, the definition of a positive definite matrix ($\mathbf{v}^*A\mathbf{v} \in \mathbb{R}_{>0}$ for all nonzero $\mathbf{v} \in \mathbb{C}^n$) cannot be checked so easily. The next theorem gives a determinant condition on A that is necessary and sufficient for A to be positive definite. While computing determinants of large matrices is time-consuming, it may be even harder to find all the eigenvalues of A. The theorem uses the following notation. For k between 1 and n and $A \in M_n(\mathbb{C})$, let $A[k]$ be the matrix in $M_k(\mathbb{C})$ consisting of the first k rows and k columns of A.

Determinant Criterion for Positive Definite Matrices. For all $A \in M_n(\mathbb{C})$, A is positive definite if and only if $A^* = A$ and $\det(A[k])$ is a strictly positive real number for $k = 1, 2, \ldots, n$.

Proof. First assume $A \in M_n(\mathbb{C})$ is positive definite, so $A^* = A$ holds. Fix k between 1 and n. We show that $A[k]$ is positive definite, which implies $\det(A[k])$ is real and positive. Fix a nonzero $\mathbf{v} \in \mathbb{C}^k$. Let $\mathbf{w} \in \mathbb{C}^n$ be $\mathbf{v}$ with $n - k$ zeroes appended. Then $\mathbf{v}^*A[k]\mathbf{v} = \mathbf{w}^*A\mathbf{w}$ is a positive real number. So $A[k]$ is positive definite.

Next, assume $B \in M_n(\mathbb{C})$, $B^* = B$, and $\det(B[k]) \in \mathbb{R}_{>0}$ for $1 \leq k \leq n$. We prove B is positive definite by induction on n. In the base case $n = 1$, we see that $B \in M_1(\mathbb{C})$ is positive definite iff $B(1,1) \in \mathbb{R}_{>0}$ iff $\det(B[1]) \in \mathbb{R}_{>0}$. Fix $n > 1$, and assume the result is known for matrices in $M_{n-1}(\mathbb{C})$. Let $A = B[n-1] \in M_{n-1}(\mathbb{C})$ be obtained from B by deleting the last row and column. Note $A^* = A$ and $\det(A[k]) = \det(B[k]) \in \mathbb{R}_{>0}$ for $1 \leq k \leq n - 1$, so the induction hypothesis shows that A is positive definite. Let the eigenvalues of A be the positive real numbers $a_1 \leq a_2 \leq \cdots \leq a_{n-1}$ and the eigenvalues of B be the real numbers $b_1 \leq b_2 \leq \cdots \leq b_n$. By the Interlacing Theorem in §7.14, we know $b_1 \leq a_1 \leq b_2 \leq a_2 \leq \cdots \leq a_{n-1} \leq b_n$. Thus, every eigenvalue of B, except possibly b_1, is $\geq a_1$ and hence is positive. On the other hand, $\det(B) = b_1 b_2 \cdots b_n > 0$ by hypothesis. Since $b_2, \ldots, b_n > 0$, it follows that $b_1 > 0$ as well. Since B is normal and all its eigenvalues are positive real numbers, B is positive definite. This completes the induction step. $\qquad\square$

7.16 Summary

1. *Definitions.* Given any complex matrix A, the *conjugate-transpose* A^* is the matrix obtained by transposing A and replacing each entry by its complex conjugate. A matrix $A \in M_n(\mathbb{C})$ is *Hermitian* iff $A^* = A$; A is *unitary* iff $A^* = A^{-1}$; A is *normal* iff $AA^* = A^*A$; A is *positive definite* iff $\mathbf{v}^*A\mathbf{v}$ is a positive real number for all nonzero $\mathbf{v} \in \mathbb{C}^n$; A is *positive semidefinite* iff $\mathbf{v}^*A\mathbf{v}$ is a nonnegative real number for all $\mathbf{v} \in \mathbb{C}^n$.

2. *Properties of Conjugate-Transpose.* Whenever the matrix operations are defined, the following identities hold: $(A+B)^* = A^*+B^*$; $(cA)^* = \overline{c}(A^*)$; $(AB)^* = B^*A^*$; $(A^*)^* = A$; $(A^{-1})^* = (A^*)^{-1}$; $\text{tr}(A^*) = \overline{\text{tr}(A)}$; $\det(A^*) = \overline{\det(A)}$; $\langle A\mathbf{v}, \mathbf{w} \rangle = \mathbf{w}^*A\mathbf{v} = \langle \mathbf{v}, A^*\mathbf{w} \rangle$.

3. *Hermitian Matrices.* The following conditions on a matrix $A \in M_n(\mathbb{C})$ are equivalent: A is Hermitian; $A = A^*$; $\mathbf{v}^*A\mathbf{v}$ is real for all $\mathbf{v} \in \mathbb{C}^n$; A is normal with all real eigenvalues. Real linear combinations of Hermitian matrices are Hermitian, as are products of commuting Hermitian matrices. Every matrix $B \in M_n(\mathbb{C})$ can be written uniquely in the form $B = X + iY$, where X and Y are Hermitian matrices; X and Y commute iff B is normal. This is called the *Cartesian decomposition* or *Hermitian decomposition* of B.

4. *Positive Definite Matrices.* The following conditions on a matrix $A \in M_n(\mathbb{C})$ are equivalent: A is positive definite; $\mathbf{v}^*A\mathbf{v}$ is real and positive for all nonzero $\mathbf{v} \in \mathbb{C}^n$; A is normal with all positive real eigenvalues; $A^* = A$ and for all k between 1 and n, the matrix consisting of the first k rows and columns of A has positive determinant. Positive linear combinations of positive definite matrices are positive definite.

5. *Unitary Matrices.* The following conditions on a matrix U in $M_n(\mathbb{C})$ are equivalent: U is unitary; $UU^* = I$; $U^*U = I$; $U^{-1} = U^*$; U^* is unitary; U^{-1} is unitary; U^{T} is unitary; the rows of U are orthonormal vectors in $\mathbb{C}^n$; the columns of U are orthonormal vectors in $\mathbb{C}^n$; $\langle U\mathbf{v}, U\mathbf{w} \rangle = \langle \mathbf{v}, \mathbf{w} \rangle$ for all $\mathbf{v}, \mathbf{w} \in \mathbb{C}^n$; U preserves inner products; $||U\mathbf{v}|| = ||\mathbf{v}||$ for all $\mathbf{v} \in \mathbb{C}^n$; U is length-preserving; U is the matrix of a length-preserving linear map relative to an orthonormal basis; U is normal with all eigenvalues having modulus 1 in $\mathbb{C}$. For U unitary, we have $|\det(U)| = 1$ in $\mathbb{C}$. Unitary similarity is an equivalence relation on square matrices that corresponds to an orthonormal change of basis on the associated vector space.

6. *Unitary Triangularization and Diagonalization.* A matrix $A \in M_n(\mathbb{C})$ is *unitarily triangulable* iff $U^*AU = U^{-1}AU$ is upper-triangular for some unitary matrix U; A is *unitarily diagonalizable* iff U^*AU is diagonal for some unitary matrix U. *Schur's Theorem* says every matrix $A \in M_n(\mathbb{C})$ can be unitarily triangularized. The *Spectral Theorem* says A can be unitarily diagonalized iff A is normal iff there exist n orthonormal eigenvectors for A. In this case, the eigenvectors are the columns of the diagonalizing matrix U, and the diagonal entries of U^*AU are the eigenvalues of A. The Spectral Theorem applies, in particular, to Hermitian, positive definite, and unitary matrices (which are all normal).

7. *Simultaneous Triangularization and Diagonalization.* A collection $\{A_i : i \in I\}$ of $n \times n$ matrices can be *simultaneously* triangularized (resp. diagonalized) iff there exists a single invertible matrix S (independent of $i \in I$) such that $S^{-1}A_iS$ is upper-triangular (resp. diagonal) for all $i \in I$. A commuting collection of $n \times n$

complex matrices can be simultaneously unitarily triangularized. A collection of normal matrices can be simultaneously unitarily diagonalized iff the matrices in the collection commute.

8. *kth Roots of Matrices.* For any even $k \geq 1$, every positive definite matrix A has a unique positive definite kth root, which is a positive definite matrix B such that $B^k = A$. For any odd $k \geq 1$, every Hermitian matrix A has a unique Hermitian kth root. In each case, $B = p(A)$ for some polynomial p; we can find B by unitarily diagonalizing A and taking positive (resp. real) kth roots of the positive (resp. real) diagonal entries.

9. *Polar Decomposition.* For every matrix A in $M_n(\mathbb{C})$, there exist a positive semidefinite matrix R and a unitary matrix U such that $A = UR$. The matrix $R = \sqrt{A^*A}$ is always uniquely determined by A; U is unique if A is invertible. U and R commute iff A is normal. There is a dual decomposition $A = R'U'$ even when A is not normal.

10. *Singular Value Decomposition.* Every matrix $A \in M_n(\mathbb{C})$ can be written in the form $A = PDQ^*$, where P and Q are unitary and D is a diagonal matrix with nonnegative real entries. The entries of D are uniquely determined by A (up to rearrangement); they are called the *singular values of A*. This result says that every linear transformation of $\mathbb{C}^n$ is the composition of an isometry, then a nonnegative rescaling of orthonormal axes, then another isometry.

11. *Characterizations of Normality.* The following conditions on A in $M_n(\mathbb{C})$ are equivalent: A is normal; $AA^* = A^*A$; A is unitarily diagonalizable; A has n orthonormal eigenvectors; the unique Hermitian matrices X and Y in the Cartesian decomposition $A = X + iY$ commute; $A^* = p(A)$ for some polynomial p; the matrices U and R in the polar decomposition $A = UR$ commute.

12. *Interlacing Theorem.* Given $B \in M_n(\mathbb{C})$ with $B^* = B$, let A be B with the last row and column erased. If B has eigenvalues $b_1 \leq \cdots \leq b_n$ and A has eigenvalues $a_1 \leq \cdots \leq a_{n-1}$, then $b_1 \leq a_1 \leq b_2 \leq a_2 \leq \cdots \leq b_{n-1} \leq a_{n-1} \leq b_n$.

7.17   Exercises

1. Prove the properties of complex conjugation stated in the introduction to this chapter.

2. Let $A = \begin{bmatrix} 1-i & 0 & 2+2i \\ 3 & -1+3i & 5i \end{bmatrix}$ and $B = \begin{bmatrix} i & -2+i \\ 1 & 1+3i \end{bmatrix}$.
   Compute A^*, B^*, AA^*, A^*A, $(BA)^*$, A^*B^*, B^{-1}, $(B^*)^{-1}$, and $(B^{-1})^*$.

3. For $A \in M_{m,n}(\mathbb{C})$, define $\overline{A} \in M_{m,n}(\mathbb{C})$ to be the matrix with i,j-entry $\overline{A(i,j)}$. Which properties in the Theorem on the Conjugate-Transpose (§7.1) have analogs for $\overline{A}$? Prove your answers.

4. Let Q_A be the quadratic form associated with a matrix $A \in M_n(\mathbb{C})$ (see §7.1). (a) Prove: for all $A \in M_n(\mathbb{R})$, there exists a symmetric matrix $B \in M_n(\mathbb{R})$ with $Q_A = Q_B$. (b) Does part (a) hold if we replace $\mathbb{R}$ by $\mathbb{C}$? Explain. (c) Given $A \in M_n(\mathbb{C})$, must there exist a Hermitian $B \in M_n(\mathbb{C})$ with $Q_A = Q_B$?

5. Decide whether each matrix is Hermitian, unitary, positive definite, or normal (select all that apply, and explain).

(a) the $n \times n$ identity matrix   (b) the $n \times n$ zero matrix

(c) $\begin{bmatrix} 0 & i \\ -i & 0 \end{bmatrix}$ (d) $\begin{bmatrix} 5 & -\sqrt{3} \\ -\sqrt{3} & 7 \end{bmatrix}$ (e) $\begin{bmatrix} 7 & 1 \\ 3 & 3 \end{bmatrix}$ (f) $\begin{bmatrix} 1/3 & 2/3 & 2/3 \\ 2/3 & -2/3 & 1/3 \\ 2/3 & 1/3 & -2/3 \end{bmatrix}$

(g) $\begin{bmatrix} -1+i & -4-2i & 6 \\ -4-2i & 3 & 2-2i \\ 6 & 2-2i & -2-i \end{bmatrix}$ (h) $\begin{bmatrix} 3 & i & 0 \\ i & 3 & i \\ 0 & i & 3 \end{bmatrix}$

6. For a 2×2 matrix $A = \begin{bmatrix} a & b \\ c & d \end{bmatrix}$ with $a, b, c, d \in \mathbb{C}$, find algebraic conditions on a, b, c, d that are equivalent to A being: (a) Hermitian (b) unitary (c) positive definite (d) positive semidefinite (e) normal.

7. For a fixed $t \in \mathbb{C}$, call a matrix $A \in M_n(\mathbb{C})$ t-*Hermitian* iff $A^* = tA$.
   (a) Prove that the set of t-Hermitian matrices is a real subspace of $M_n(\mathbb{C})$.
   (b) Suppose A is t-Hermitian, B is s-Hermitian, and $AB = BA$. Prove AB is st-Hermitian.

8. (a) Prove: if $A \in M_n(\mathbb{C})$ is skew-Hermitian, then the associated quadratic form Q takes values in $\{ib : b \in \mathbb{R}\}$. (b) Is the converse of (a) true? Prove or give a counterexample. (c) Must a matrix with all pure imaginary eigenvalues be skew-Hermitian? Explain.

9. Find the Hermitian decomposition of each matrix.
   (a) $\begin{bmatrix} 2 & 5 \\ -1 & 3 \end{bmatrix}$ (b) $\begin{bmatrix} 1+3i & 2-5i \\ 3+4i & -1-2i \end{bmatrix}$ (c) $\begin{bmatrix} 3 & 7 \\ 7 & 3 \end{bmatrix}$
   (d) $\begin{bmatrix} 2i & 1+3i \\ -1+3i & -5i \end{bmatrix}$ (e) $\begin{bmatrix} 0 & i & 2i \\ -i & 0 & 3i \\ 2i & 3i & 4 \end{bmatrix}$

10. Given nonzero $\mathbf{v} \in \mathbb{C}^n$, let $A = I - (2/\mathbf{v}^*\mathbf{v})(\mathbf{v}\mathbf{v}^*) \in M_n(\mathbb{C})$.
    (a) Compute A for $\mathbf{v} = (1, 2)$ and $\mathbf{v} = (1, 7, 5, 5)$.
    (b) Show that A is Hermitian and unitary.
    (c) Show that $A\mathbf{v} = -\mathbf{v}$ and $A\mathbf{w} = \mathbf{w}$ for all $\mathbf{w} \in \mathbb{C}^n$ with $\langle \mathbf{w}, \mathbf{v} \rangle = 0$.
    (d) Describe all diagonal matrices unitarily similar to A.

11. Let F be a field in which $1_F + 1_F \neq 0_F$. Prove: for all $A \in M_n(F)$, there exist unique $B, C \in M_n(F)$ with $A = B + C$, $B^{\mathrm{T}} = B$, and $C^{\mathrm{T}} = -C$. Does this result hold when $1_F + 1_F = 0_F$? Explain.

12. Let $A \in M_n(\mathbb{C})$ be positive definite. (a) Show A^{T} is positive definite. (b) Show A^{-1} is positive definite. (c) For $k \in \mathbb{Z}_{>0}$, must A^k be positive definite? Explain.

13. Give an example of an $n \times n$ matrix A with $n \geq 2$ satisfying the indicated properties, or explain why no such example exists.
    (a) A is positive definite, but $A(i, j)$ is a negative real number for all i, j.
    (b) A is positive definite, but $A(i, j)$ is pure imaginary for all $i \neq j$.
    (c) A is positive semidefinite and negative semidefinite and nonzero.
    (d) A is negative definite, but $\det(A) > 0$.
    (e) All eigenvalues of A are real and negative, but A is not negative definite.

14. Suppose $A \in M_n(\mathbb{C})$ satisfies $\mathbf{v}^*A\mathbf{v} \in \mathbb{R}_{>0}$ for all $\mathbf{v}$ in a fixed basis of $\mathbb{C}^n$. Give an example to show that A need not be positive definite.

15. Suppose $A \in M_n(\mathbb{R})$ satisfies $A^{\mathrm{T}} = A$ and $\mathbf{v}^{\mathrm{T}}A\mathbf{v} \geq 0$ for all nonzero $\mathbf{v} \in \mathbb{R}^n$. Prove A is positive semidefinite. Give a specific example where this result fails if we omit the hypothesis $A^{\mathrm{T}} = A$.

16. Use the determinant criterion in §7.15 to find all $c \in \mathbb{C}$ such that the following matrices are positive definite.

(a) $\begin{bmatrix} 2 & c \\ c & 3 \end{bmatrix}$ (b) $\begin{bmatrix} 4 & c & 0 \\ c & 4 & c \\ 0 & c & 4 \end{bmatrix}$ (c) $\begin{bmatrix} c & 1 & 2 & 0 \\ 1 & c & 1 & 2 \\ 2 & 1 & c & 1 \\ 0 & 2 & 1 & c \end{bmatrix}$

17. Suppose $A \in M_n(\mathbb{C})$ satisfies $A^* = A$ and $\det(A[k]) \geq 0$ for $1 \leq k \leq n$. Give an example to show that A need not be positive semidefinite.

18. Suppose $A \in M_n(\mathbb{C})$ satisfies $A^* = A$, $\det(A[k]) > 0$ for $1 \leq k < n$, and $\det(A) \geq 0$. Prove A is positive semidefinite.

19. State and prove a determinant characterization of negative definite matrices.

20. Suppose $f : \mathbb{R}^2 \to \mathbb{R}$ has continuous second partial derivatives, and f has a local minimum at $(a, b) \in \mathbb{R}^2$. (a) Prove that $A = \begin{bmatrix} f_{xx}(a, b) & f_{xy}(a, b) \\ f_{yx}(a, b) & f_{yy}(a, b) \end{bmatrix}$ is positive semidefinite. Deduce that $f_{xx}f_{yy} - f_{xy}^2 \geq 0$ and $f_{xx}, f_{yy} \geq 0$ at (a, b). (For any nonzero $\mathbf{v} = (v_1, v_2) \in \mathbb{R}^2$, study $g(t) = f((a, b) + t(v_1, v_2))$ for $t \in \mathbb{R}$.) (b) State and prove a version of (a) for a local maximum of f.

21. Generalize the previous exercise to functions $f : \mathbb{R}^n \to \mathbb{R}$.

22. How many diagonal unitary matrices are there in $M_n(\mathbb{R})$?

23. Give an example of a unitary matrix U with each property, or explain why this cannot be done. (a) $U^{\mathrm{T}} = -U$. (b) $U(1, 1) = 2$. (c) $U^* = U \neq I$. (d) U is upper-triangular and not diagonal. (e) U is symmetric with all entries nonzero.

24. Let U be any 2×2 real unitary matrix. Show that the first row of U has the form $(\cos \theta, \sin \theta)$ for some $\theta \in [0, 2\pi)$. Given such a first row, find all possible second rows of U.

25. Let Z be any subspace of $\mathbb{C}^n$. Prove there exists an orthonormal basis of Z by induction on $\dim(Z)$. (Use the claim in §7.7.)

26. Let W be a subspace of $\mathbb{C}^n$ with orthonormal basis $(\mathbf{w}_1, \ldots, \mathbf{w}_k)$. Define the *orthogonal complement* $W^\perp$ to be the set of $\mathbf{z} \in \mathbb{C}^n$ such that $\langle \mathbf{z}, \mathbf{w} \rangle = 0$ for all $\mathbf{w} \in W$. Prove $\mathbf{z} \in W^\perp$ iff $\langle \mathbf{z}, \mathbf{w}_i \rangle = 0$ for $1 \leq i \leq k$. Prove $W^\perp$ is a subspace of $\mathbb{C}^n$. Prove $W \cap W^\perp = \{0\}$. Prove $\dim(W) + \dim(W^\perp) = n$. Conclude that $\mathbb{C}^n = W \oplus W^\perp$.

27. Let W be a subspace of $\mathbb{C}^n$ with orthonormal basis $(\mathbf{w}_1, \ldots, \mathbf{w}_k)$. We can extend this list to a basis $(\mathbf{w}_1, \ldots, \mathbf{w}_k, \mathbf{x}_{k+1}, \ldots, \mathbf{x}_n)$ of $\mathbb{C}^n$. For $k + 1 \leq j \leq n$, define

$$\mathbf{y}_j = \mathbf{x}_j - \sum_{r=1}^{j-1} (\mathbf{w}_r^* \mathbf{x}_j) \mathbf{w}_r, \qquad \mathbf{w}_j = \mathbf{y}_j / \|\mathbf{y}_j\|.$$

Use induction to prove these facts: (a) Each $\mathbf{y}_j$ is nonzero. (b) $\langle \mathbf{y}_j, \mathbf{w}_s \rangle = 0$ for all $s < j$. (c) $(\mathbf{w}_1, \ldots, \mathbf{w}_j)$ spans the same subspace as $(\mathbf{w}_1, \ldots, \mathbf{w}_k, \mathbf{x}_{k+1}, \ldots, \mathbf{x}_j)$ for $k + 1 \leq j \leq n$. (d) $(\mathbf{w}_1, \ldots, \mathbf{w}_n)$ is an orthonormal basis for $\mathbb{C}^n$. So, *every orthonormal list in $\mathbb{C}^n$ can be extended to an orthonormal basis of $\mathbb{C}^n$.*

28. Fix $A \in M_n(\mathbb{C})$. Prove: if there exists $U \in U_n(\mathbb{C})$ such that A is similar to U, then A^{-1} is similar to A^*.

29. Prove or disprove the converse of the statement in the previous exercise.

30. Find all $A \in M_n(\mathbb{C})$ such that A is the only matrix unitarily similar to A.

31. Given $A \in M_n(\mathbb{C})$ and $f \in S_n$, define $B \in M_n(\mathbb{C})$ by $B(i,j) = A(f(i), f(j))$ for all i, j between 1 and n. Prove A and B are unitarily similar.

32. Given any subgroup H of $\mathrm{GL}_n(\mathbb{C})$, call two matrices $A, B \in M_n(\mathbb{C})$ H-*similar* iff $B = S^{-1}AS$ for some $S \in H$. (a) Show that H-similarity is an equivalence relation on $M_n(\mathbb{C})$. (b) Give an example of subgroups $H \neq K$ in $\mathrm{GL}_n(\mathbb{C})$ such that H-similarity coincides with K-similarity.

33. With as little calculation as possible, explain why each pair of matrices cannot be unitarily similar. (a) $A = \begin{bmatrix} 2 & 1 \\ 1 & 2 \end{bmatrix}$, $B = \begin{bmatrix} -1 & -8 \\ 1 & 5 \end{bmatrix}$. (b) $A = \begin{bmatrix} 0.6 & -0.8 \\ 0.8 & 0.6 \end{bmatrix}$, $B = \begin{bmatrix} -3.4 & -1.6 \\ 10.4 & 4.6 \end{bmatrix}$. (c) $A = \begin{bmatrix} 2 & 2 \\ 2 & 5 \end{bmatrix}$, $B = \begin{bmatrix} 9 & 2 \\ -12 & -2 \end{bmatrix}$.

34. Suppose $e_1 \cdots e_k$ is any finite sequence of 1s and $*$s. Prove: for all $A, B \in M_n(\mathbb{C})$, if A is unitarily similar to B, then $\mathrm{tr}(A^{e_1} A^{e_2} \cdots A^{e_k}) = \mathrm{tr}(B^{e_1} B^{e_2} \cdots B^{e_k})$. For example, if $e_1 e_2 e_3 e_4 = *1 * *$, the condition states that $\mathrm{tr}(A^* A A^* A^*) = \mathrm{tr}(B^* B B^* B^*)$. (It can be shown that, conversely, if the equality of traces holds for all such sequences $e_1 \cdots e_k$, then A and B are unitarily similar.)

35. Prove that any normal matrix $A \in M_n(\mathbb{C})$ is unitarily similar to A^{T}. Is every $A \in M_n(\mathbb{C})$ unitarily similar to A^{T}?

36. For each matrix A, find a unitary U and an upper-triangular T with $T = U^* A U$.

(a) $\begin{bmatrix} 2.5 & -0.5 \\ 0.5 & 3.5 \end{bmatrix}$ (b) $\begin{bmatrix} 5 & -108 & -36 \\ 39 & -127 & -26 \\ -36 & 72 & -25 \end{bmatrix}$ (c) $\begin{bmatrix} 4 & 1 & 0 & 0 \\ 1 & 4 & 1 & 0 \\ 0 & 1 & 4 & 1 \\ 0 & 0 & 1 & 4 \end{bmatrix}$

37. Define $A \in M_5(\mathbb{C})$ by $A(i,j) = i$ for $1 \leq i, j \leq 5$.
(a) Find a matrix $S \in \mathrm{GL}_5(\mathbb{C})$ such that $S^{-1}AS$ is diagonal.
(b) Find a matrix $U \in U_5(\mathbb{C})$ such that $U^* A U$ is upper-triangular.
(c) Is there a matrix $V \in U_5(\mathbb{C})$ such that $V^* A V$ is diagonal? Why?

38. Give an example of a matrix $A \in M_2(\mathbb{R})$ for which there does not exist any unitary $U \in M_2(\mathbb{R})$ with $U^* A U$ upper-triangular.

39. Suppose $A, B \in M_n(\mathbb{C})$ are unitarily similar. Prove $\sum_{i,j=1}^n |A(i,j)|^2 = \sum_{i,j=1}^n |B(i,j)|^2$. (Look at the trace of AA^*.)

40. Suppose A is unitarily similar to two upper-triangular matrices T_1 and T_2. Prove $\sum_{i<j} |T_1(i,j)|^2 = \sum_{i<j} |T_2(i,j)|^2$.

41. Give an example of upper-triangular matrices T_1 and T_2 in $M_3(\mathbb{C})$ such that T_1 and T_2 are unitarily similar, $T_1(i,i) = T_2(i,i)$ for $i = 1, 2, 3$, but $T_1 \neq T_2$.

42. Give an example of upper-triangular matrices T_1 and T_2 with the same diagonal such that $\sum_{i<j} |T_1(i,j)|^2 = \sum_{i<j} |T_2(i,j)|^2$, but T_1 and T_2 are not unitarily similar.

43. Let $A \in M_n(\mathbb{C})$ have eigenvalues $c_1, \ldots, c_n \in \mathbb{C}$. Prove A is normal iff $\sum_{i,j=1}^n |A(i,j)|^2 = \sum_{i=1}^n |c_i|^2$.

44. Which pairs of matrices can be simultaneously unitarily triangularized? Which can be simultaneously unitarily diagonalized? Decide without calculating any eigenvalues.
(a) $A = \begin{bmatrix} 1 & 4 \\ 3 & 1 \end{bmatrix}$, $B = \begin{bmatrix} 0 & 12 \\ 9 & 0 \end{bmatrix}$.

(b) $A = \begin{bmatrix} 3 & -1 \\ 0 & 2 \end{bmatrix}$, $B = \begin{bmatrix} 1 & 1 \\ 1 & 0 \end{bmatrix}$.

(c) $A = \begin{bmatrix} a & b \\ b & a \end{bmatrix}$, $B = \begin{bmatrix} c & d \\ d & c \end{bmatrix}$, where $a, b, c, d \in \mathbb{R}$.

45. For even $k > 0$, show that for each positive definite (resp. semidefinite) $A \in M_n(\mathbb{C})$, there exists a unique positive definite (resp. semidefinite) $B \in M_n(\mathbb{C})$ with $B^k = A$.

46. For odd $k > 0$, show that for each Hermitian $A \in M_n(\mathbb{C})$, there exists a unique Hermitian $B \in M_n(\mathbb{C})$ with $B^k = A$.

47. Compute a numerical approximation to the unique positive definite square root of each positive definite matrix.

(a) $\begin{bmatrix} 5 & -1 \\ -1 & 5 \end{bmatrix}$ (b) $\begin{bmatrix} 3 & 2 & 1 \\ 2 & 3 & 2 \\ 1 & 2 & 3 \end{bmatrix}$ (c) $\begin{bmatrix} 2 & 1+i \\ 1-i & 2 \end{bmatrix}$

48. Prove $A \in M_n(\mathbb{C})$ is positive definite iff there is an invertible $B \in M_n(\mathbb{C})$ with $A = B^*B$.

49. (a) Give an example of a matrix $A \in M_2(\mathbb{C})$ such that $B^2 = A$ has no solution $B \in M_2(\mathbb{C})$. (b) For each $n > 2$, find $A \in M_n(\mathbb{C})$ with no square root in $M_n(\mathbb{C})$.

50. Fix $n > 1$. (a) Show that $I_n \in M_n(\mathbb{C})$ has infinitely many square roots in $M_n(\mathbb{C})$. (b) Show that $0 \in M_n(\mathbb{C})$ has infinitely many square roots in $M_n(\mathbb{C})$.

51. Let $T \in M_2(\mathbb{C})$ be upper-triangular. Find all upper-triangular $U \in M_2(\mathbb{C})$ with $U^2 = T$. Do the same for $M_3(\mathbb{C})$. Do there exist matrices T, U such that $U^2 = T$, T is upper-triangular, but U is not upper-triangular?

52. Let $T \in M_n(\mathbb{C})$ be upper-triangular with distinct nonzero diagonal entries. Prove there exist exactly 2^n upper-triangular $U \in M_n(\mathbb{C})$ with $U^2 = T$.

53. Prove that for $n > 1$, a matrix $A \in M_n(\mathbb{C})$ cannot have a finite odd number of square roots in $M_n(\mathbb{C})$.

54. Find (with proof) an example of a matrix $A \in M_2(\mathbb{C})$ that has exactly two square roots in $M_2(\mathbb{C})$.

55. (a) Suppose $A \in M_n(\mathbb{C})$ has eigenvalues $c_1, \ldots, c_n$, $B \in M_n(\mathbb{C})$ has eigenvalues $d_1, \ldots, d_n$, and $AB = BA$. Prove there exists $f \in S_n$ such that $A + B$ has eigenvalues $c_1 + d_{f(1)}, \ldots, c_n + d_{f(n)}$.
(b) With the setup in (a), what can you say about the eigenvalues of AB?
(c) Give an example to show that (a) can fail without the hypothesis $AB = BA$.

56. For each matrix A, find a polynomial $p \in \mathbb{C}[x]$ with $A^* = p(A)$, or explain why this is impossible. (a) $\begin{bmatrix} 2 & 0 & 0 \\ 0 & 3+4i & 0 \\ 0 & 0 & 2 \end{bmatrix}$ (b) $\begin{bmatrix} 5 & 5 \\ 2 & 2 \end{bmatrix}$ (c) $\begin{bmatrix} 5 & 2 \\ -2 & 5 \end{bmatrix}$

(d) $\begin{bmatrix} 1 & 1+2i & 1 & -1+2i \\ -1-2i & 1 & -1+2i & -1 \\ 1 & 1-2i & 1 & -1-2i \\ 1-2i & -1 & 1+2i & 1 \end{bmatrix}$

57. Prove: for any invertible $A \in M_n(\mathbb{C})$, $A^{-1} = p(A)$ for some $p \in \mathbb{C}[x]$.

58. Suppose $A \in M_n(\mathbb{C})$ is normal. Must $\overline{A} = p(A)$ for some $p \in \mathbb{C}[x]$?

59. Suppose $\overline{A} = p(A)$ for some $p \in \mathbb{C}[x]$. Must A be normal?

60. Suppose $A \in M_n(\mathbb{C})$ is normal. Must $A^T = p(A)$ for some $p \in \mathbb{C}[x]$?

61. Suppose $A^T = p(A)$ for some $p \in \mathbb{C}[x]$. Must A be normal?

62. Find a polar decomposition $A = UR$ for each matrix A.

(a) $\begin{bmatrix} 1 & 1 \\ 1 & -1 \end{bmatrix}$ (b) $\begin{bmatrix} 2 & 2 \\ 2 & 2 \end{bmatrix}$ (c) $\begin{bmatrix} -5 & 0 & 0 \\ 0 & 1 & 0 \\ 0 & 0 & -7 \end{bmatrix}$ (d) $\begin{bmatrix} 0 & 0 & 0 \\ 2 & 0 & 0 \\ 0 & 1 & 0 \end{bmatrix}$

63. Find a singular value decomposition and the singular values for each matrix in Exercise 62.

64. Prove the following abstract version of the singular value decomposition: given a linear map $T : V \to W$ between two n-dimensional complex inner product spaces, there exist an orthonormal basis $X = (x_1, \ldots, x_n)$ for V and an orthonormal basis $Y = (y_1, \ldots, y_n)$ for W and nonnegative real numbers $c_1, \ldots, c_n$ such that $T(x_i) = c_i y_i$ for $1 \le i \le n$.

65. Prove this singular value decomposition theorem for rectangular matrices: given $A \in M_{m,n}(\mathbb{C})$, there exist unitary matrices $P \in M_m(\mathbb{C})$ and $Q \in M_n(\mathbb{C})$ and a matrix $D \in M_{m,n}(\mathbb{C})$ with $D(i,i) \in \mathbb{R}_{\ge 0}$ for all i and $D(i,j) = 0$ for $i \ne j$, such that $A = PDQ^*$.

66. Verify (7.5) on page 193.

67. In Step 2 of §7.14, assume $a_{j-1} < a_j = \cdots = a_{j+s} < a_{j+s+1}$. Compute the sign of $\lim_{\epsilon \to 0^+} \chi_B(a_j - \epsilon)/\epsilon^s$ and the sign of $\lim_{\epsilon \to 0^+} \chi_B(a_{j+s} + \epsilon)/\epsilon^s$. Use this to complete the proof of Step 2.

68. For the matrix $A = \begin{bmatrix} 1 & 4 & 0 & 0 \\ 4 & 7 & -1 & -5 \\ 0 & -1 & 1 & 0 \\ 0 & -5 & 0 & 5 \end{bmatrix}$, verify the Interlacing Theorem in §7.14 by computing the eigenvalues of $A[k]$ for $k = 1, 2, 3, 4$.

69. True or false? Explain each answer. Assume matrices are in $M_n(\mathbb{C})$ unless otherwise stated.
(a) Every diagonal matrix is normal.
(b) A is positive semidefinite iff $-A$ is negative semidefinite.
(c) For all $A \in M_{m,n}(\mathbb{C})$, AA^* is positive definite.
(d) For all $A \in M_{m,n}(\mathbb{C})$, A^*A is positive semidefinite.
(e) Every real symmetric matrix is unitarily diagonalizable.
(f) The product of two Hermitian matrices is always Hermitian.
(g) The product of two positive definite matrices is always positive definite.
(h) The product of two unitary matrices is always unitary.
(i) The square of a Hermitian matrix is always positive semidefinite.
(j) Every negative definite matrix has a negative definite inverse.
(k) The inverse of an invertible normal matrix must be normal.
(l) If A is normal, then $p(A)$ is normal for all $p \in \mathbb{C}[x]$.
(m) Every positive definite matrix has a unique square root.
(n) For fixed $n > 1$, I_n is the only matrix that is both unitary and Hermitian.
(o) A normal matrix with all entries in $\mathbb{R}_{>0}$ must be positive definite.
(p) If every complex eigenvalue of A has modulus 1, then A must be unitary.
(q) If $A \in M_2(\mathbb{C})$ has $A(1,1) \ge 0$ and $\det(A) \ge 0$, then A must be positive semidefinite.

8

Jordan Canonical Forms

Let V be an n-dimensional vector space over a field F. Given any linear operator $T : V \to V$ and any ordered basis X of V, recall from Chapter 6 that there is a matrix $A = [T]_X$ in $M_n(F)$ that represents T relative to the basis X. If we switch from X to another ordered basis Y, the matrix A is replaced by a similar matrix of the form $[T]_Y = P^{-1}AP$, where $P \in M_n(F)$ is some invertible matrix (specifically, P is the matrix of id_V relative to the input basis Y and output basis X).

A fundamental question in linear algebra is how to pick the ordered basis Y to make the matrix $[T]_Y$ as simple as possible. From the viewpoint of matrix computations, the simplest $n \times n$ matrices are *diagonal* matrices D, which satisfy $D(i,j) = 0_F$ for all $i \neq j$. A linear map T is called *diagonalizable* iff there exists an ordered basis Y such that $[T]_Y$ is a diagonal matrix. Regrettably, for $n > 1$, not all linear maps on V can be diagonalized. For example, suppose T is a *nilpotent* linear operator, which means that $T^k = T \circ T \circ \cdots \circ T$ (k factors) $= 0$ for some positive integer k. Suppose $[T]_Y$ were a diagonal matrix D. On one hand, $[T^k]_Y = [0]_Y = 0$. On the other hand, $[T^k]_Y = [T]_Y^k = D^k$. Since D is diagonal, $D^k = 0$ forces every diagonal entry of D to be zero, so that $D = 0$ and $T = 0$. But for $n > 1$, there are many examples of nonzero nilpotent linear maps (as we see below). These maps cannot be diagonalized.

In this chapter, we prove that linear maps on a vector space using complex scalars ($F = \mathbb{C}$) can always be represented by certain nearly-diagonal matrices called *Jordan canonical forms*. More precisely, for any field F, $c \in F$, and $m \in \mathbb{Z}_{>0}$, define the *Jordan block*

$$
J(c; m) = \begin{bmatrix} c & 1 & 0 & \cdots & 0 \\ 0 & c & 1 & \cdots & 0 \\ 0 & 0 & c & \cdots & 0 \\ 0 & 0 & 0 & \cdots & 1 \\ 0 & 0 & 0 & \cdots & c \end{bmatrix}_{m \times m}.
$$

This matrix has m diagonal entries equal to c, $m-1$ copies of 1_F on the next higher diagonal, and all other entries zero. When $m = 1$, $J(c; 1)$ is the 1×1 matrix $[c]$. Next, define a *Jordan canonical form* to be any matrix $\mathbf{J}$ that has the block-diagonal structure

$$
\mathbf{J} = \begin{bmatrix} J(c_1; m_1) & 0 & \cdots & 0 \\ 0 & J(c_2; m_2) & \cdots & 0 \\ \vdots & & \ddots & \vdots \\ 0 & 0 & \cdots & J(c_s; m_s) \end{bmatrix} \tag{8.1}
$$

for some $m_1, \ldots, m_s$ in $\mathbb{Z}_{>0}$ and $c_1, \ldots, c_s$ in F (repetitions may occur here). Writing blk-diag$(A_1, \ldots, A_s)$ to denote a block-diagonal matrix with diagonal blocks $A_1, \ldots, A_s$, we have $\mathbf{J} = $ blk-diag$(J(c_1; m_1), \ldots, J(c_s; m_s))$. We can now state our main result.

Jordan Canonical Form Theorem for Linear Maps. For any linear map T on a finite-dimensional complex vector space V, there is an ordered basis X of V such that the matrix $\mathbf{J} = [T]_X$ is a Jordan canonical form. Furthermore, if Y is any ordered basis such that

DOI: 10.1201/9781003484561-8

$\mathbf{J}' = [T]_Y$ is also a Jordan canonical form, then $\mathbf{J}'$ is obtained from $\mathbf{J}$ by rearranging the Jordan blocks in (8.1).

Here is a reformulation of the theorem as a statement about matrices.

Jordan Canonical Form Theorem for Matrices. For any $A \in M_n(\mathbb{C})$, there exists an invertible $P \in M_n(\mathbb{C})$ such that $\mathbf{J} = P^{-1}AP$ is a Jordan canonical form. If $\mathbf{J}' = Q^{-1}AQ$ is any Jordan canonical form similar to A, then $\mathbf{J}'$ is obtained by rearranging the Jordan blocks of $\mathbf{J}$.

Our approach to proving the Jordan Canonical Form Theorems begins by proving the following classification result for nilpotent linear maps over any field.

Theorem Classifying Nilpotent Linear Maps. Suppose V is an n-dimensional vector space over any field F and $T : V \to V$ is a nilpotent linear map. There exist unique integers $m_1 \geq m_2 \geq \cdots \geq m_s > 0$ such that for some ordered basis X of V,

$$[T]_X = \text{blk-diag}(J(0_F; m_1), J(0_F; m_2), \ldots, J(0_F; m_s)).$$

We prove this theorem by analyzing partition diagrams that give a geometric representation of the sequence $(m_1, \ldots, m_s)$. The next proof ingredient is Fitting's Lemma, which (roughly speaking) takes an arbitrary linear map $T : V \to V$ and breaks V into two pieces, such that T restricted to one piece is nilpotent, and T restricted to the other piece is an isomorphism. Combining this result with the classification of nilpotent maps, we obtain the Jordan Canonical Form Theorems.

We continue by discussing methods for actually computing the Jordan canonical form of a specific linear map or matrix. Then we describe an application of this theory to the solution of systems of linear ordinary differential equations. We conclude by studying a more abstract version of the Jordan canonical form that is needed in the study of Lie algebras. Specifically, we prove that every linear map on a finite-dimensional complex vector space V can be written uniquely as the sum of a diagonalizable linear map and a nilpotent linear map that commute with each other.

8.1 Examples of Nilpotent Maps

We begin our analysis of nilpotent linear maps by constructing some specific examples of nonzero nilpotent maps. Let F be any field and V be an F-vector space with ordered basis $X = (x_1, \ldots, x_n)$. Recall that we can define an F-linear map $T : V \to V$ by choosing any elements $y_1, \ldots, y_n \in V$, declaring that $T(x_j) = y_j$ for $1 \leq j \leq n$, and then setting $T(\sum_{j=1}^n c_j x_j) = \sum_{j=1}^n c_j y_j$ for all $c_j \in F$. In other words, we can build a unique linear map by sending basis elements anywhere and then *extending by linearity*. Furthermore, writing $T(x_j) = y_j = \sum_{i=1}^n a_{ij} x_i$ for $a_{ij} \in F$, we know that the matrix $A = [T]_X$ has i,j-entry a_{ij}.

To illustrate this procedure, suppose $n = 5$ and we decide that $T(x_1) = 0$, $T(x_2) = x_1$, $T(x_3) = x_2$, $T(x_4) = x_3$, and $T(x_5) = x_4$. The following arrow diagram illustrates the action of T on the basis X:

$$0 \overset{T}{\leftarrow} x_1 \overset{T}{\leftarrow} x_2 \overset{T}{\leftarrow} x_3 \overset{T}{\leftarrow} x_4 \overset{T}{\leftarrow} x_5.$$

For a general vector $v = c_1 x_1 + c_2 x_2 + c_3 x_3 + c_4 x_4 + c_5 x_5$ with $c_j \in F$,

$$T(v) = c_1 \cdot 0 + c_2 x_1 + c_3 x_2 + c_4 x_3 + c_5 x_4.$$

Applying T again, we find that

$$
\begin{aligned}
T^2(v) &= T(T(v)) = 0 + c_3 x_1 + c_4 x_2 + c_5 x_3, \\
T^3(v) &= c_4 x_1 + c_5 x_2, \\
T^4(v) &= c_5 x_1, \\
T^5(v) &= 0.
\end{aligned}
$$

Since $T^5(v) = 0$ for all $v \in V$, $T^5 = 0$ and T is nilpotent. On the other hand, the powers T^k for $1 \le k < 5$ are nonzero maps. In general, we say a nilpotent map T has *index of nilpotence* m iff m is the least positive integer with $T^m = 0$. The matrix of T relative to the ordered basis X is

$$
\begin{bmatrix}
0 & 1 & 0 & 0 & 0 \\
0 & 0 & 1 & 0 & 0 \\
0 & 0 & 0 & 1 & 0 \\
0 & 0 & 0 & 0 & 1 \\
0 & 0 & 0 & 0 & 0
\end{bmatrix} = J(0; 5).
$$

Generalizing this example, for any ordered basis $X = (x_1, \ldots, x_n)$, we may define a linear map T on V by setting $T(x_1) = 0$ and $T(x_j) = x_{j-1}$ for $2 \le j \le n$:

$$
0 \overset{T}{\leftarrow} x_1 \overset{T}{\leftarrow} x_2 \overset{T}{\leftarrow} x_3 \overset{T}{\leftarrow} \cdots \overset{T}{\leftarrow} x_{n-1} \overset{T}{\leftarrow} x_n.
$$

Since $T(x_1) = 0$ and $T(x_j) = 1 x_{j-1} + \sum_{k \neq j-1} 0 x_k$ for $j > 1$, we see that the matrix $[T]_X$ has 1_F as its $(j-1, j)$-entry (for $2 \le j \le n$) and 0_F elsewhere. In other words, $[T]_X = J(0; n)$. Extending by linearity, we have

$$
T\left(\sum_{k=1}^n c_k x_k \right) = \sum_{k=2}^n c_k x_{k-1} \quad \text{for all } c_k \in F.
$$

Iterating the map T gives

$$
T^i\left(\sum_{k=1}^n c_k x_k \right) = \sum_{k=i+1}^n c_k x_{k-i} \quad \text{for all } i \ge 0.
$$

We see that $T^i = 0$ for all $i \ge n$, and T is nilpotent of index n.

Next, consider an example where $n = 9$ and T acts on basis vectors as shown here:

$$
\begin{aligned}
& 0 \overset{T}{\leftarrow} x_1 \overset{T}{\leftarrow} x_2 \overset{T}{\leftarrow} x_3 \\
& 0 \overset{T}{\leftarrow} x_4 \overset{T}{\leftarrow} x_5 \\
& 0 \overset{T}{\leftarrow} x_6 \overset{T}{\leftarrow} x_7 \\
& 0 \overset{T}{\leftarrow} x_8 \\
& 0 \overset{T}{\leftarrow} x_9
\end{aligned}
\tag{8.2}
$$

You can check that $[T]_X = \text{blk-diag}(J(0; 3), J(0; 2), J(0; 2), J(0; 1), J(0; 1))$, $T \neq 0$, $T^2 \neq 0$, but $T^3 = 0$. So T is a nilpotent map of index 3.

For our next example, let $n = 4$ and define T on the basis by $T(x_1) = 0$, $T(x_2) = x_1$, $T(x_3) = 2x_1$, and $T(x_4) = 3x_1$. We see that T is nilpotent of index 2, and

$$
[T]_X = \begin{bmatrix}
0 & 1 & 2 & 3 \\
0 & 0 & 0 & 0 \\
0 & 0 & 0 & 0 \\
0 & 0 & 0 & 0
\end{bmatrix}.
$$

This matrix is not a Jordan canonical form. However, if we let $Z = (x_1, x_2, x_3 - 2x_2, x_4 - 3x_2)$, then Z is an ordered basis of V with $[T]_Z = \text{blk-diag}(J(0; 2), J(0; 1), J(0; 1))$.

8.2 Partition Diagrams

We intend to show that every nilpotent linear map on V can be described by an arrow diagram like (8.2), if we choose the right ordered basis for V. To discuss these arrow diagrams more precisely, we introduce the idea of partition diagrams.

A *partition* of an integer $n \geq 0$ is a sequence $\mu = (\mu_1, \mu_2, \ldots, \mu_s)$, where each μ_i is a positive integer, $\mu_1 \geq \mu_2 \geq \cdots \geq \mu_s$, and $\mu_1 + \mu_2 + \cdots + \mu_s = n$. We let $\ell(\mu)$ be the length of the sequence μ. For example, $\mu = (3, 2, 2, 1, 1)$ is a partition of $n = 9$ with $\ell(\mu) = 5$. For any $c \in F$ and partition μ, define

$$J(c; \mu) = \text{blk-diag}(J(c; \mu_1), J(c; \mu_2), \ldots, J(c; \mu_s)).$$

Given any partition μ, the *diagram of μ* is the set

$$D(\mu) = \{(i, j) \in \mathbb{Z} \times \mathbb{Z} : 1 \leq i \leq \ell(\mu),\ 1 \leq j \leq \mu_i\}.$$

We visualize $D(\mu)$ by drawing a picture with a box in row i and column j for each (i, j) in $D(\mu)$. For example, the diagram of $(3, 2, 2, 1, 1)$ is the set

$$\{(1, 1), (1, 2), (1, 3), (2, 1), (2, 2), (3, 1), (3, 2), (4, 1), (5, 1)\},$$

which is drawn as follows:

For all $k \geq 1$, let μ'_k be the number of boxes in column k of the diagram of μ. In our example, $\mu'_1 = 5$, $\mu'_2 = 3$, $\mu'_3 = 1$, and $\mu'_k = 0$ for all $k > 3$. In general, $\mu'_1 = \ell(\mu)$ and $\mu'_k = 0$ for all $k > \mu_1$.

We can recover a partition μ from its diagram $D(\mu)$. A partition diagram $D(\mu)$ can be reconstructed from the sequence of column lengths $(\mu'_1, \mu'_2, \ldots)$. We can compute these column lengths if we are given the sequence of partial sums $(\mu'_1, \mu'_1 + \mu'_2, \mu'_1 + \mu'_2 + \mu'_3, \ldots)$, where the kth entry in the sequence counts the number of boxes in the first k columns. So, for any partitions μ and ν,

$$\mu = \nu \quad \text{iff} \quad \text{for all } k \geq 1,\ \mu'_1 + \cdots + \mu'_k = \nu'_1 + \cdots + \nu'_k. \tag{8.3}$$

This observation is the key fact needed to establish the uniqueness properties of Jordan canonical forms.

8.3 Partition Diagrams and Nilpotent Maps

Let V be an n-dimensional vector space over any field F. For each partition μ of n and each ordered basis $X = (x_1, \ldots, x_n)$ of V, we can build a nilpotent linear map $T = T_{X,\mu}$ as follows. Fill the boxes in the diagram of μ with the elements of X, row by row, working from left to right in each row. Define the map T by sending each basis vector in the leftmost column to zero, sending every other basis vector to the element to its immediate left, and extending by linearity. For example, given $\mu = (5, 5, 3, 1, 1, 1)$ and $X = (x_1, \ldots, x_{16})$, we first fill $D(\mu)$ as shown here:

x_1	x_2	x_3	x_4	x_5
x_6	x_7	x_8	x_9	x_{10}
x_{11}	x_{12}	x_{13}		
x_{14}				
x_{15}				
x_{16}				

Then $T_{X,\mu}$ is the unique linear map on V that sends basis vectors x_1, x_6, x_{11}, x_{14}, x_{15}, and x_{16} to 0, x_2 to x_1, x_7 to x_6, x_9 to x_8, and so on. You can check that T is nilpotent of index 5, and $[T]_X = J(0; (5, 5, 3, 1, 1, 1))$.

Let us describe the preceding construction more formally. Given a partition μ of n and an ordered basis $X = (x_1, \ldots, x_n)$, we can view the *ordered* basis X as an *indexed set* $X = \{x(i, j) : (i, j) \in D(\mu)\}$ by setting $x(i, j) = x_{j + \mu_1 + \mu_2 + \cdots + \mu_{i-1}}$ for all $(i, j) \in D(\mu)$. Here, $x(i, j)$ is the basis vector placed in cell (i, j) of the diagram. We define $T = T_{X,\mu}$ on the basis X by setting $T(x(i, 1)) = 0_V$ for $1 \le i \le \ell(\mu)$ and $T(x(i, j)) = x(i, j - 1)$ for all $(i, j) \in D(\mu)$ with $j > 1$. Extending by linearity, we see that

$$T\left(\sum_{(i,j) \in D(\mu)} c(i, j) x(i, j) \right) = \sum_{i=1}^{\ell(\mu)} \sum_{j=2}^{\mu_i} c(i, j) x(i, j - 1) \quad \text{for all } c(i, j) \in F. \tag{8.4}$$

Returning to the ordered basis $(x_1, \ldots, x_n)$, we see that T sends x_1, $x_{\mu_1 + 1}$, $x_{\mu_1 + \mu_2 + 1}$, etc., to zero, and T sends every other x_k to x_{k-1}. This observation is equivalent to the statement that $[T]_X = J(0; \mu)$.

Our goal is to prove that for every nilpotent linear map T on V, there exists a unique partition μ of $n = \dim(V)$ such that $[T]_X = J(0; \mu)$ for some ordered basis X of V. By the preceding remarks, this is equivalent to proving that for every nilpotent linear map T on V, there exists a unique partition μ and a (not necessarily unique) ordered basis X with $T = T_{X,\mu}$. Before giving this proof, we show how the partition diagram $D(\mu)$ encodes information about the image and null space of $T_{X,\mu}$ and its powers.

8.4 Computing Images via Partition Diagrams

Given any linear map $T : V \to V$, recall that the *image* of T (also called the *range* of T) is the subspace $\operatorname{img}(T) = T[V] = \{T(v) : v \in V\}$. The *null space* of T (also called the *kernel* of T) is the subspace $\operatorname{Null}(T) = \ker(T) = \{v \in V : T(v) = 0_V\}$. Given a nilpotent map of the special form $T = T_{X,\mu}$, let us see how to use the diagram of μ to compute the dimensions of $\operatorname{img}(T^k)$ and $\operatorname{Null}(T^k)$ for all $k \ge 1$. We use $\mu = (5, 5, 3, 1, 1, 1)$ as a running example to illustrate the general discussion.

We begin by giving a visual representation of the formula (8.4) defining $T_{X,\mu}$. Represent a vector $v = \sum_{(i,j) \in D(\mu)} c(i, j) x(i, j)$ in V by filling each cell $(i, j) \in D(\mu)$ with the scalar $c(i, j) = c_{i,j} \in F$ that multiplies the basis vector $x(i, j)$ indexed by that cell. Applying T shifts all these coefficients one cell to the left, with coefficients $c(i, 1)$ falling off the left edge and new zero coefficients coming into the right cell of each row. For example,

$$
T \left(
\begin{array}{|c|c|c|c|c|}
\hline
c_{1,1} & c_{1,2} & c_{1,3} & c_{1,4} & c_{1,5} \\
\hline
c_{2,1} & c_{2,2} & c_{2,3} & c_{2,4} & c_{2,5} \\
\hline
\end{array}
\right)
=
\begin{array}{|c|c|c|c|c|}
\hline
c_{1,2} & c_{1,3} & c_{1,4} & c_{1,5} & 0 \\
\hline
c_{2,2} & c_{2,3} & c_{2,4} & c_{2,5} & 0 \\
\hline
\end{array}
\tag{8.5}
$$

(The matrix diagram shows: left side — row 1: $c_{1,1}, c_{1,2}, c_{1,3}, c_{1,4}, c_{1,5}$; row 2: $c_{2,1}, c_{2,2}, c_{2,3}, c_{2,4}, c_{2,5}$; row 3: $c_{3,1}, c_{3,2}, c_{3,3}$; row 4: $c_{4,1}$; row 5: $c_{5,1}$; row 6: $c_{6,1}$. Right side — row 1: $c_{1,2}, c_{1,3}, c_{1,4}, c_{1,5}, 0$; row 2: $c_{2,2}, c_{2,3}, c_{2,4}, c_{2,5}, 0$; row 3: $c_{3,2}, c_{3,3}, 0$; row 4: 0; row 5: 0; row 6: 0.)

It is immediate from this picture that the image of T consists of all F-linear combinations of the linearly independent basis vectors

$$\{x(1,1), x(1,2), x(1,3), x(1,4), x(2,1), x(2,2), x(2,3), x(2,4), x(3,1), x(3,2)\},$$

so that this set is a basis of $\mathrm{img}(T)$. In general, we see pictorially or via (8.4) that the set

$$\{x(i, j-1) : 1 \le i \le \ell(\mu), 2 \le j \le \mu_i\} = \{x(i,j) : (i,j) \in D(\mu) \text{ and } (i, j+1) \in D(\mu)\}$$

is a basis for $\mathrm{img}(T_{X,\mu})$. These are the basis vectors occupying all cells of $D(\mu)$ excluding the rightmost cell in each row.

What happens if we apply T again? In our example,

$$
T^2 \left(
\begin{array}{|c|c|c|c|c|}
\hline
c_{1,1} & c_{1,2} & c_{1,3} & c_{1,4} & c_{1,5} \\
\hline
c_{2,1} & c_{2,2} & c_{2,3} & c_{2,4} & c_{2,5} \\
\hline
\end{array}
\right)
=
\begin{array}{|c|c|c|c|c|}
\hline
c_{1,3} & c_{1,4} & c_{1,5} & 0 & 0 \\
\hline
c_{2,3} & c_{2,4} & c_{2,5} & 0 & 0 \\
\hline
\end{array}
$$

(The matrix diagram shows: left side — row 1: $c_{1,1}, c_{1,2}, c_{1,3}, c_{1,4}, c_{1,5}$; row 2: $c_{2,1}, c_{2,2}, c_{2,3}, c_{2,4}, c_{2,5}$; row 3: $c_{3,1}, c_{3,2}, c_{3,3}$; row 4: $c_{4,1}$; row 5: $c_{5,1}$; row 6: $c_{6,1}$. Right side — row 1: $c_{1,3}, c_{1,4}, c_{1,5}, 0, 0$; row 2: $c_{2,3}, c_{2,4}, c_{2,5}, 0, 0$; row 3: $c_{3,3}, 0, 0$; row 4: 0; row 5: 0; row 6: 0.)

So $\mathrm{img}(T^2)$ consists of all F-linear combinations of $x(1,1)$, $x(1,2)$, $x(1,3)$, $x(2,1)$, $x(2,2)$, $x(2,3)$, and $x(3,1)$, and these vectors form a basis for this subspace. Similarly, for any X and μ, a basis for $\mathrm{img}(T^2_{X,\mu})$ is the set of all $x(i,j)$ occupying cells $(i,j) \in D(\mu)$ excluding the two rightmost cells in each row.

Continuing to iterate (8.4), we see that for each positive integer k,

$$
T^k_{X,\mu} \left(\sum_{(i,j) \in D(\mu)} c(i,j) x(i,j) \right) = \sum_{(i,j) \in D(\mu) : (i, j-k) \in D(\mu)} c(i,j) x(i, j-k)
$$

$$
= \sum_{(i,j) \in D(\mu) : (i, j+k) \in D(\mu)} c(i, j+k) x(i,j) \quad \text{for all } c(i,j) \in F, \tag{8.6}
$$

which shows that $\{x(i,j) : (i,j) \in D(\mu) \text{ and } (i, j+k) \in D(\mu)\}$ is an F-basis for $\mathrm{img}(T^k_{X,\mu})$. We get this basis by ignoring the k rightmost cells in each row of $D(\mu)$ and taking all remaining $x(i,j)$. In particular, taking $k = \mu_1 - 1$ and $k = \mu_1$ in (8.6) shows that $T_{X,\mu}$ *is a nilpotent map of index μ_1.*

8.5 Computing Null Spaces via Partition Diagrams

Next, let us see how to use partition diagrams to find bases for the null spaces $\mathrm{Null}(T^k_{X,\mu})$. For our running example of $\mu = (5,5,3,1,1,1)$, inspection of (8.5) reveals that $v = \sum_{(i,j)\in D(\mu)} c(i,j)x(i,j)$ is sent to zero by T iff

$$c(1,2) = c(1,3) = c(1,4) = c(1,5)$$
$$=c(2,2) = c(2,3) = c(2,4) = c(2,5)$$
$$=c(3,2) = c(3,3) = 0.$$

Pictorially,

$$\mathrm{Null}(T) = \left\{ \begin{array}{|c|c|c|c|c|} \hline c_{1,1} & 0 & 0 & 0 & 0 \\ \hline c_{2,1} & 0 & 0 & 0 & 0 \\ \hline c_{3,1} & 0 & 0 \\ \cline{1-3} c_{4,1} \\ \cline{1-1} c_{5,1} \\ \cline{1-1} c_{6,1} \\ \cline{1-1} \end{array} \quad : c(i,1) \in F \right\}. \tag{8.7}$$

In general, $\mathrm{Null}(T_{X,\mu})$ consists of all F-linear combinations of basis vectors residing in the leftmost column of $D(\mu)$, so that $\{x(i,1) : 1 \le i \le \ell(\mu) = \mu'_1\}$ is a basis for $\mathrm{Null}(T_{X,\mu})$.

Similarly, applying T^2 to v produces output zero iff $c(i,j) = 0$ for all $j > 2$, so that a basis of $\mathrm{Null}(T^2)$ consists of all basis vectors in the first two columns of $D(\mu)$. In general, for all $k \ge 1$, the set

$$\{x(i,j) : (i,j) \in D(\mu) \text{ and } 1 \le j \le k\} \tag{8.8}$$

is a basis for $\mathrm{Null}(T^k_{X,\mu})$, and therefore

$$\dim(\mathrm{Null}(T^k_{X,\mu})) = \mu'_1 + \mu'_2 + \cdots + \mu'_k \quad \text{for all } k \ge 1. \tag{8.9}$$

To prove this assertion without pictures, note from (8.6) that $T^k_{X,\mu}(v) = 0$ iff $c(i,j) = 0$ for all $(i,j) \in D(\mu)$ such that $(i, j-k) \in D(\mu)$, which holds iff v is a linear combination of those $x(i,j)$ satisfying $(i,j) \in D(\mu)$ and $1 \le j \le k$.

8.6 Classification of Nilpotent Maps (Stage 1)

We are now ready to prove the following classification theorem.

Theorem Classifying Nilpotent Linear Maps. Suppose V is an n-dimensional vector space over any field F and $T : V \to V$ is a nilpotent linear map. There exists a unique partition μ of n such that $[T]_X = J(0;\mu)$ for some ordered basis X of V.

As we saw at the end of §8.3, the conclusion is equivalent to the existence of a unique partition μ such that for some ordered basis X of V, $T = T_{X,\mu}$.

We can prove uniqueness of μ very quickly. Suppose $T : V \to V$ is nilpotent, and $T = T_{X,\mu} = T_{Y,\nu}$ for some partitions μ and ν and some ordered bases X and Y of V. For all $k \ge 1$, (8.9) shows that

$$\mu'_1 + \cdots + \mu'_k = \dim(\mathrm{Null}(T^k)) = \nu'_1 + \cdots + \nu'_k.$$

So $\mu = \nu$ follows from (8.3).

Proving existence of μ and X is more subtle. We use induction on $n = \dim(V)$. For the base case $n = 0$, we must have $V = \{0\}$ and $T = 0$. Then $T = T_{X,\mu}$ holds if we choose $X = \emptyset$ and $\mu = ()$, which is a partition of $n = 0$ of length zero. If you prefer to start the induction at $n = 1$, you can check that for $n = 1$ we have $T = T_{X,\mu}$, where $X = \{x\}$ is any nonzero vector in V and $\mu = (1)$ is the unique partition of 1. Incidentally, this shows that the ordered basis X is not unique, in general.

In the rest of the proof, we assume $n = \dim(V) > 1$ and that the existence assertion is already known to hold for all F-vector spaces of smaller dimension than n. Given a nilpotent linear map $T : V \to V$, we build μ and X such that $T = T_{X,\mu}$ in three stages. In Stage 1, we consider the subspace $V_1 = \mathrm{img}(T) = T[V]$. Since $\dim(V) > 0$, we cannot have $V_1 = V$; otherwise $T[V] = V$ implies $T^k[V] = V \neq \{0\}$ for all $k > 0$, contradicting the nilpotence of T. Furthermore, T maps $V_1 = T[V]$ into itself, so we can consider the restricted function $T|V_1 : V_1 \to V_1$, given by $T|V_1(v) = T(v)$ for all $v \in V_1$. The map $T|V_1$ is linear and nilpotent. Since V_1 is a proper subspace of V, we know $n_1 = \dim(V_1) < \dim(V)$, and the induction hypothesis gives us an ordered basis X_1 of V_1 and a partition $\nu = (\nu_1, \ldots, \nu_s)$ of n_1 such that $T|V_1 = T_{X_1, \nu}$. For example, $D(\nu)$ and X_1 might look like this:

$$D(\nu) = \qquad\qquad
\begin{aligned}
&0 \xleftarrow{T|V_1} x_{1,1} \xleftarrow{T|V_1} x_{1,2} \xleftarrow{T|V_1} x_{1,3} \xleftarrow{T|V_1} x_{1,4} \\
&0 \xleftarrow{T|V_1} x_{2,1} \xleftarrow{T|V_1} x_{2,2} \xleftarrow{T|V_1} x_{2,3} \xleftarrow{T|V_1} x_{2,4} \\
&0 \xleftarrow{T|V_1} x_{3,1} \xleftarrow{T|V_1} x_{3,2}
\end{aligned}$$

8.7 Classification of Nilpotent Maps (Stage 2)

Continuing the proof, our goal in the induction step is to prove that $T : V \to V$ has the form $T_{X,\mu}$ for some partition μ of n. If this conclusion were true, we know from §8.4 that $D(\nu)$ consists of all cells of $D(\mu)$ excluding the rightmost cell of each row. Stage 2 of the proof is to recover the rows of the unknown partition μ of length 1, if there are any.

To do this, we study $V_2 = V_1 + \mathrm{Null}(T) = \{u + v : u \in V_1, v \in \mathrm{Null}(T)\}$, which is a subspace of V containing $V_1 = \mathrm{img}(T)$. Write $n_2 = \dim(V_2)$. The subspace V_2 is spanned by $X_1 \cup \mathrm{Null}(T)$, so we can extend the ordered basis X_1 for V_1 to an ordered basis X_2 for V_2 by adding appropriate vectors from $\mathrm{Null}(T)$ to the end of the list X_1. Suppose t additional vectors are needed, so $n_2 = n_1 + t$. Let ρ be the partition of n_2 obtained by adding t parts of size 1 to the end of ν. Note that $T|V_2$ is nilpotent and maps V_2 to itself (in fact, it maps V_2 into V_1). We can see that $T|V_2 = T_{X_2, \rho}$ since both sides are linear maps having the same effect on basis vectors in X_2; in particular, both sides send all new vectors in $X_2 \setminus X_1$ to zero. For example, $D(\rho)$ and X_2 might look like this, where new boxes added to ν are marked with stars:

$$D(\rho) = \qquad\qquad
\begin{aligned}
&0 \xleftarrow{T|V_2} x_{1,1} \xleftarrow{T|V_2} x_{1,2} \xleftarrow{T|V_2} x_{1,3} \xleftarrow{T|V_2} x_{1,4} \\
&0 \xleftarrow{T|V_2} x_{2,1} \xleftarrow{T|V_2} x_{2,2} \xleftarrow{T|V_2} x_{2,3} \xleftarrow{T|V_2} x_{2,4} \\
&0 \xleftarrow{T|V_2} x_{3,1} \xleftarrow{T|V_2} x_{3,2} \\
&0 \xleftarrow{T|V_2} x_{4,1} \\
&0 \xleftarrow{T|V_2} x_{5,1} \\
&0 \xleftarrow{T|V_2} x_{6,1}
\end{aligned}$$

We claim that $n_2 = n - s$, where $s = \ell(\nu)$ is the number of rows in $D(\nu)$. To prove the claim, recall that $V_2 = \text{img}(T) + \text{Null}(T)$, so (using Exercise 31)

$$n_2 = \dim(V_2) = \dim(\text{img}(T)) + \dim(\text{Null}(T)) - \dim(\text{img}(T) \cap \text{Null}(T)). \qquad (8.10)$$

By the Rank–Nullity Theorem (see §1.8), $\dim(\text{img}(T)) + \dim(\text{Null}(T)) = \dim(V) = n$. On the other hand, $\text{Null}(T|_{V_1}) = V_1 \cap \text{Null}(T) = \text{img}(T) \cap \text{Null}(T)$. Applying (8.9) to the map $T|_{V_1} = T_{X_1,\nu}$ (with $k = 1$), we get

$$\dim(\text{img}(T) \cap \text{Null}(T)) = \dim(\text{Null}(T|_{V_1})) = \nu_1' = \ell(\nu) = s.$$

Putting these formulas into (8.10) gives $n_2 = n - s$, as claimed.

8.8   Classification of Nilpotent Maps (Stage 3)

We can now find a partition μ of n and an ordered basis X of V for which $T = T_{X,\mu}$. Define μ by adding 1 to each of the first s parts of ρ, so $\mu = (\nu_1 + 1, \nu_2 + 1, \ldots, \nu_s + 1, 1, \ldots, 1)$ and μ ends with t parts equal to 1. Extend the indexed basis $X_2 = \{x(i,j) : (i,j) \in D(\rho)\}$ to an indexed set $X = \{x(i,j) : (i,j) \in D(\mu)\}$ by letting $x(i, \mu_i)$ be any vector in V with $T(x(i, \mu_i)) = x(i, \mu_i - 1)$, for $1 \le i \le s$. Such vectors must exist, since $(i, \mu_i - 1) \in D(\nu)$ means that $x(i, \mu_i - 1) \in X_1 \subseteq \text{img}(T)$. For example, $D(\mu)$ and X might look like this, where new boxes added to ρ are marked with stars:

$$D(\mu) = \begin{array}{|c|c|c|c|c|} \hline & & & & \star \\ \hline & & & & \star \\ \hline & & & \star & \\ \hline & & & & \\ \hline & & & & \\ \hline & & & & \\ \hline \end{array}$$

$$0 \xleftarrow{T} x_{1,1} \xleftarrow{T} x_{1,2} \xleftarrow{T} x_{1,3} \xleftarrow{T} x_{1,4} \xleftarrow{T} x_{1,5}$$
$$0 \xleftarrow{T} x_{2,1} \xleftarrow{T} x_{2,2} \xleftarrow{T} x_{2,3} \xleftarrow{T} x_{2,4} \xleftarrow{T} x_{2,5}$$
$$0 \xleftarrow{T} x_{3,1} \xleftarrow{T} x_{3,2} \xleftarrow{T} x_{3,3}$$
$$0 \xleftarrow{T} x_{4,1}$$
$$0 \xleftarrow{T} x_{5,1}$$
$$0 \xleftarrow{T} x_{6,1}$$

If we can show that the indexed set X is a basis of V, then $T = T_{X,\mu}$ follows since both linear maps agree on all basis vectors in X. By the claim in the last section, μ is a partition of $n_2 + s = n = \dim(V)$. Hence, if we can show that the indexed set $X = \{x(i,j) : (i,j) \in D(\mu)\}$ is linearly independent, then the n vectors $x(i,j)$ must be distinct and form a basis of V.

To check linear independence, assume that for some scalars $c(i,j) \in F$,

$$0 = \sum_{(i,j) \in D(\mu)} c(i,j) x(i,j).$$

We must show every $c(i,j)$ is zero. Applying the linear map T gives

$$0 = \sum_{(i,j) \in D(\mu) : (i,j+1) \in D(\mu)} c(i,j+1) x(i,j) = \sum_{(i,j) \in D(\nu)} c(i,j+1) x(i,j)$$

(see (8.5) and (8.6)). Since $X_1 = \{x(i,j) : (i,j) \in D(\nu)\}$ is known to be linearly independent already (by induction), we see that $c(i, j+1) = 0$ for all $(i,j) \in D(\nu)$. Our original linear combination now reduces to

$$0 = \sum_{i=1}^{\ell(\mu)} c(i,1) x(i,1) = \sum_{(i,j) \in D(\rho) : j=1} c(i,j) x(i,j).$$

But $X_2 = \{x(i,j) : (i,j) \in D(\rho)\}$ is linearly independent by construction, so we deduce that $c(i,1) = 0$ for all i. Hence the indexed set X is linearly independent, and the existence proof is finally complete.

Here is one corollary of the classification of nilpotent linear maps. If V is n-dimensional and $T : V \to V$ is a nilpotent linear map, then $T^n = 0$ (so T has index of nilpotence at most n). To see why this holds, note $T = T_{X,\mu}$ for some ordered basis X and some partition μ of n. The diagram $D(\mu)$ can have at most n nonempty columns. By (8.9), $\mathrm{Null}(T^n)$ has dimension $\mu'_1 + \cdots + \mu'_n = n$, so $\mathrm{Null}(T^n)$ must be all of V. This means that $T^n = 0$, as claimed.

8.9   Fitting's Lemma

We prove the existence part of the Jordan Canonical Form Theorem by combining the classification of nilpotent linear maps with a result called Fitting's Lemma. Before stating this result, we need some preliminary concepts. Let V be a vector space over a field F. Given two subspaces W and Z of V, the *sum* of these subspaces is the subspace $W + Z = \{w + z : w \in W, z \in Z\}$. We say this sum is a *direct sum*, denoted $W \oplus Z$, iff $W \cap Z = \{0_V\}$. You can check that for a direct sum $W \oplus Z$, if X_1 is an ordered basis of W and X_2 is an ordered basis of Z, then the list consisting of the vectors in X_1 followed by the vectors in X_2 is an ordered basis of $W \oplus Z$. In particular, $\dim(W \oplus Z) = \dim(W) + \dim(Z)$ when W and Z are finite-dimensional.

Given any linear map $S : V \to V$, we say a subspace W of V is S-*invariant* iff $S(w) \in W$ for all $w \in W$. In this situation, the restriction of S to W, denoted $S|W$, maps W into itself and is a linear map from W to W. Now assume $n = \dim(V)$ is finite. Recall that $\mathrm{Null}(S)$ and $\mathrm{img}(S)$ are always subspaces of V, and by the Rank–Nullity Theorem in §1.8, $\dim(\mathrm{Null}(S)) + \dim(\mathrm{img}(S)) = n = \dim(V)$. If $\mathrm{Null}(S) \cap \mathrm{img}(S) = \{0\}$, then we have a direct sum $\mathrm{Null}(S) \oplus \mathrm{img}(S)$ of dimension n, which must therefore be the entire space V. However, there is no guarantee that $\mathrm{Null}(S)$ and $\mathrm{img}(S)$ have zero intersection. For example, if S is the nilpotent map defined in (8.5), then $\mathrm{Null}(S) \cap \mathrm{img}(S)$ is a 3-dimensional subspace spanned by $x(1,1)$, $x(2,1)$, and $x(3,1)$.

We can rectify this situation by imposing an appropriate hypothesis on S. Specifically, *if* $S : V \to V$ *is a linear map such that* $\mathrm{Null}(S) = \mathrm{Null}(S^2)$, *then* $V = \mathrm{Null}(S) \oplus \mathrm{img}(S)$. By the comments in the preceding paragraph, it is enough to show that $\mathrm{Null}(S) \cap \mathrm{img}(S) = \{0\}$. Suppose $z \in \mathrm{Null}(S) \cap \mathrm{img}(S)$. On one hand, $S(z) = 0$. On the other hand, $z = S(y)$ for some $y \in V$. Now, $S^2(y) = S(S(y)) = S(z) = 0$ means that $y \in \mathrm{Null}(S^2) = \mathrm{Null}(S)$. Then $z = S(y) = 0$, so $\mathrm{Null}(S) \cap \mathrm{img}(S) = \{0\}$.

Fitting's Lemma: Given an n-dimensional F-vector space V and a linear map $U : V \to V$, there exist U-invariant subspaces Z and W of V such that $V = Z \oplus W$, $U|Z$ is nilpotent, and $U|W$ is an isomorphism.

In the case where $\mathrm{Null}(U) = \{0\}$, we have $\dim(\mathrm{img}(U)) = n$ by the Rank–Nullity Theorem, so $\mathrm{img}(U) = V$ and U is an isomorphism. So we may take $Z = \{0\}$ and $W = V$ in this situation. Next, consider the case where $\mathrm{Null}(U)$ is a nonzero subspace of V. Since $\mathrm{Null}(U^k) \subseteq \mathrm{Null}(U^{k+1})$ for all $k \geq 1$, we have a chain of subspaces

$$\{0\} \neq \mathrm{Null}(U) \subseteq \mathrm{Null}(U^2) \subseteq \mathrm{Null}(U^3) \subseteq \cdots \subseteq \mathrm{Null}(U^k) \subseteq \cdots \subseteq V.$$

All these subspaces are finite-dimensional, so there is some $m \geq 1$ with $\mathrm{Null}(U^k) = \mathrm{Null}(U^m)$ for all $k \geq m$. In particular, taking $k = 2m$, we can apply our earlier result to the linear map $S = U^m$ to conclude that $V = \text{Null}(S) \oplus \text{img}(S) = \text{Null}(U^m) \oplus \text{img}(U^m)$. Let $Z = \text{Null}(U^m) \neq \{0\}$ and $W = \text{img}(U^m)$. Given $z \in Z$, we know $U^m(z) = 0$, so $U^m(U(z)) = U(U^m(z)) = U(0) = 0$, so $U(z) \in Z$. Given $w \in W$, we know $w = U^m(v)$ for some $v \in V$, so $U(w) = U(U^m(v)) = U^m(U(v)) \in W$. Thus Z and W are U-invariant subspaces. By the very definition of Z, $U|Z$ is a nilpotent linear map of index at most m. On the other hand, suppose $w \in W$ is in $\text{Null}(U|W)$. Then $U(w) = 0$, hence $U^m(w) = 0$ and $w \in \text{Null}(U^m) \cap \text{img}(U^m) = \{0\}$. So $\ker(U|W) = \{0\}$, which means $U|W : W \to W$ is injective and hence surjective. Thus, $U|W$ is an isomorphism.

8.10   Existence of Jordan Canonical Forms

In this section, we prove the existence assertion in the Jordan Canonical Form Theorem. Before doing so, we need some basic facts about eigenvalues. Given a field F and a matrix $A \in M_n(F)$, a scalar $c \in F$ is called an *eigenvalue* of A iff $Av = cv$ for some nonzero $n \times 1$ column vector v. Any such v is called an *eigenvector* of A associated with the eigenvalue c. The eigenvalues of A are the roots of the characteristic polynomial $\chi_A = \det(xI_n - A) \in F[x]$ in the field F. This polynomial has degree n, so A has at most n distinct eigenvalues in F. For the field $F = \mathbb{C}$ of complex numbers, the polynomial χ_A always has a complex root by the Fundamental Theorem of Algebra. So every $n \times n$ complex matrix A has between 1 and n complex eigenvalues. For $A \in M_n(F)$, let $\text{Spec}(A)$ (the *spectrum* of A) be the set of all eigenvalues of A. When A is triangular, $\text{Spec}(A)$ is the set of scalars appearing on the main diagonal of A.

Suppose V is an n-dimensional F-vector space, and $T : V \to V$ is a linear map. An *eigenvalue* of T is a scalar $c \in F$ such that $T(v) = cv$ for some nonzero $v \in V$. Any such v is called an *eigenvector* of T associated with the eigenvalue c. If X is any ordered basis of V and $A = [T]_X$ is the matrix of T relative to X, then T and A have the same eigenvalues. This follows since $T(v) = cv$ iff $A[v]_X = c[v]_X$, where $[v]_X$ is the column vector giving the coordinates of v relative to X. In particular, every linear map T on an n-dimensional complex vector space has between 1 and n complex eigenvalues. Let $\text{Spec}(T)$ (the *spectrum* of T) be the set of all eigenvalues of T.

Jordan Canonical Form Theorem (Existence Part).   For every finite-dimensional complex vector space V and every linear map $T : V \to V$, there exists an ordered basis X of V such that $[T]_X$ is a Jordan canonical form.

The proof is by induction on $n = \dim(V)$. For $n = 0$ and $n = 1$, the result follows immediately. Now assume $n > 1$ and the result is already known for all complex vector spaces of smaller dimension than n.

Pick a fixed eigenvalue c of the given linear map T, and let $v \neq 0$ be an associated eigenvector. Let $U = T - c\,\text{id}_V$, where id_V denotes the identity map on V. Applying Fitting's Lemma to the linear map U, we get a direct sum $V = Z \oplus W$ where Z and W are U-invariant (hence also T-invariant) subspaces of V such that $U|Z$ is nilpotent and $U|W$ is an isomorphism. On one hand, $U|Z$ is nilpotent, so we know there is an ordered basis X_1 of Z and a partition μ of the integer $k = \dim(Z)$ such that $[U|Z]_{X_1} = J(0; \mu)$. It follows that

$$[T|Z]_{X_1} = [U|Z + c\,\text{id}_Z]_{X_1} = [U|Z]_{X_1} + [c\,\text{id}_Z]_{X_1} = J(0; \mu) + cI_k = J(c; \mu).$$

On the other hand, v cannot belong to W, since otherwise $U|W(v) = U(v) = 0 = U|W(0)$ contradicts the fact that $U|W$ is an isomorphism. So $W \neq V$ and $\dim(W) < \dim(V)$. By the induction hypothesis, there exists an ordered basis X_2 of W such that $[U|W]_{X_2}$ is a

Jordan canonical form matrix J_1. By the same calculation used above, we see that $[T|W]_{X_2}$ is the matrix J_2 obtained from J_1 by adding c to every diagonal entry. This new matrix is also a Jordan canonical form. Finally, taking X to be the concatenation of X_1 and X_2, we know X is an ordered basis of V such that $[T]_X = \text{blk-diag}(J(c; \mu), J_2)$. This matrix is a Jordan canonical form, so the induction proof is complete.

The existence of Jordan canonical forms for linear maps implies the existence of Jordan canonical forms for matrices, as follows. Given $A \in M_n(\mathbb{C})$, let $T : \mathbb{C}^n \to \mathbb{C}^n$ be the linear map defined by $T(v) = Av$ for all column vectors $v \in \mathbb{C}^n$. Choose an ordered basis X of $\mathbb{C}^n$ such that $J = [T]_X$ is a Jordan canonical form. We know $J = P^{-1}AP$ for some invertible $P \in M_n(\mathbb{C})$, so A is similar to a Jordan canonical form.

The only special feature of the field $\mathbb{C}$ needed in this proof was that every linear map on a nonzero finite-dimensional $\mathbb{C}$-vector space has an eigenvalue in $\mathbb{C}$. This follows from the fact that for $F = \mathbb{C}$, all non-constant polynomials in $F[x]$ split into products of linear factors. Any field F with this property is called *algebraically closed*. The Jordan Canonical Form Theorem holds for vector spaces with scalars coming from any algebraically closed field.

8.11   Uniqueness of Jordan Canonical Forms

Jordan Canonical Form Theorem (Uniqueness Part). Suppose V is an n-dimensional complex vector space, $T : V \to V$ is a linear map, and X and Y are ordered bases of V such that $A = [T]_X$ and $B = [T]_Y$ are both Jordan canonical forms. Then B is obtained from A by rearranging the Jordan blocks in A.

Before proving this statement, we consider an example that conveys the idea of the proof. Suppose $X = (x_1, \ldots, x_{14})$ and $A = [T]_X$ is the Jordan canonical form

$$\text{blk-diag}(J(7; 2), J(7; 2), J(7; 1), J(-4; 5), J(i; 3), J(i; 1)).$$

This matrix is triangular, so listing entries on the main diagonal gives $\text{Spec}(T) = \text{Spec}(A) = \{7, -4, i\}$. Some eigenvectors of A and T associated with the eigenvalue 7 are x_1, x_3, and x_5. In fact, you can check that the set of all eigenvectors for this eigenvalue is the set of all nonzero $\mathbb{C}$-linear combinations of x_1, x_3, and x_5. On the other hand, x_2 and x_4 are not eigenvectors for the eigenvalue 7, even though these basis vectors are associated with Jordan blocks with 7 on the diagonal.

The key to the uniqueness proof is that the subspace of V spanned by x_1, x_2, x_3, x_4, x_5 can be described using just the linear map T, not the ordered basis X or the matrix A. To see how this is done, consider the linear map $U = T - 7\,\text{id}_V$. The matrix of U relative to X is

$$C = [U]_X = A - 7I_{14} = \text{blk-diag}(J(0; 2), J(0; 2), J(0; 1), J(-11; 5), J(-7+i; 3), J(-7+i; 1)).$$

We can also write $C = \text{blk-diag}(J(0; \mu), C_1)$, where $\mu = (2, 2, 1)$ and C_1 is a triangular matrix with all diagonal entries nonzero. What is the null space of C^{14}? We compute $C^{14} = \text{blk-diag}(0_{5 \times 5}, C_2)$, where $C_2 = C_1^{14}$ is also a triangular matrix with all diagonal entries nonzero. It follows that $\text{Null}(C^{14})$ consists of all column vectors $v \in \mathbb{C}^{14}$ with $v_1, \ldots, v_5$ arbitrary and $v_6 = v_7 = \cdots = v_{14} = 0$. Translating back to the linear map T, this means that $\text{Null}((T - 7\,\text{id}_V)^{14})$ is the subspace of V with basis $(x_1, x_2, x_3, x_4, x_5)$. Since the restriction of $T - 7\,\text{id}_V$ to this subspace is nilpotent, we can appeal to the known uniqueness result for nilpotent maps to see that the partition $\mu = (2, 2, 1)$ is unique.

We now turn to the general uniqueness proof. Assume the setup in the theorem statement. First, $\mathrm{Spec}(A) = \mathrm{Spec}(T) = \mathrm{Spec}(B)$, so that A and B have the same set of diagonal entries (ignoring multiplicities), namely the set of eigenvalues of the map T.

Let $c \in \mathrm{Spec}(T)$ be any fixed eigenvalue. Write $U = T - c\,\mathrm{id}_V$. Solely for notational convenience, we can reorder the ordered basis X and the ordered basis Y so that all the Jordan blocks for c in A and B occur first, with the block sizes weakly decreasing, say $A = \mathrm{blk\text{-}diag}(J(c;\mu), A')$ for some partition μ of some k and $B = \mathrm{blk\text{-}diag}(J(c;\nu), B')$ for some partition ν of some m. Since c is fixed but arbitrary, uniqueness follows if we can show $k = m$ and $\mu = \nu$. Writing the reordered bases as $X = (x_1, \ldots, x_n)$ and $Y = (y_1, \ldots, y_n)$, we claim that $X_1 = (x_1, \ldots, x_k)$ and $Y_1 = (y_1, \ldots, y_m)$ are both bases for the same subspace $Z = \mathrm{Null}(U^n)$ of V. Assuming this claim is true, it follows that $k = m$, $[U|Z]_{X_1} = J(0; \mu)$, and $[U|Z]_{Y_1} = J(0; \nu)$. Since $U|Z$ is a nilpotent linear map, our previously proved uniqueness result for nilpotent operators allows us to conclude that $\mu = \nu$.

So we need only prove the claim that $X_1 = (x_1, \ldots, x_k)$ is a basis for $Z = \mathrm{Null}(U^n)$ (the corresponding assertion for Y_1 is proved in the same way). Since $U = T - c\,\mathrm{id}_V$, $[U]_X = \mathrm{blk\text{-}diag}(J(0; \mu), A' - cI_{n-k})$ where $A' - cI_{n-k}$ is an upper-triangular matrix with no zeroes on its diagonal (because all Jordan blocks for c occur in the first k rows of A). Taking the nth power of U, we get $[U^n]_X = [U]_X^n = \mathrm{blk\text{-}diag}(0_{k \times k}, A'')$, where $A'' \in M_{n-k}(\mathbb{C})$ is triangular with nonzero entries on its diagonal. From the form of the matrix $[U^n]_X$, we see that U^n sends $x_1, \ldots, x_k$, and every linear combination of these vectors to zero. So the span of X_1 is contained in $\mathrm{Null}(U^n) = Z$. For the reverse inclusion, suppose $v = \sum_{i=1}^n c_i x_i$ (where $c_i \in \mathbb{C}$) is not in the span of X_1. This means there exists $s > k$ with $c_s \neq 0$; choose the largest index s with this property. Since $[U^n]_X$ is upper-triangular with a nonzero entry in the s, s-position, there are scalars d_i with $U^n(v) = \sum_{i<s} d_i x_i + c_s A''(s, s)x_s \neq 0$. Hence $v \notin Z$, and Z is the subspace spanned by the list X_1. Since this list is linearly independent (being a sublist of the ordered basis X), X_1 is an ordered basis of Z.

The uniqueness of Jordan canonical forms for linear maps implies the uniqueness result for matrices, as follows. Given $A \in M_n(\mathbb{C})$, let $T : \mathbb{C}^n \to \mathbb{C}^n$ be the linear map $T(v) = Av$ for $v \in \mathbb{C}^n$. The set of matrices similar to A is exactly the set of matrices $[T]_X$ as X ranges over all ordered bases of $\mathbb{C}^n$ (see Chapter 6). So all Jordan canonical forms in the similarity class of A represent the map T relative to appropriate ordered bases, and these forms therefore differ from one another only by rearranging the Jordan blocks.

8.12   Computing Jordan Canonical Forms

In this section, we discuss how to compute a Jordan canonical form of a specific matrix $A \in M_n(\mathbb{C})$. More specifically, we want to find a Jordan canonical form $\mathbf{J} \in M_n(\mathbb{C})$ and an invertible matrix P with $\mathbf{J} = P^{-1}AP$. If all we need is the matrix $\mathbf{J}$, we can proceed as follows. Examining the uniqueness proof given above, we observe first that the diagonal entries of $\mathbf{J}$ must be the eigenvalues of A. These eigenvalues can be found, in principle, by computing the roots of the characteristic polynomial of A. In reality, for large n, it may be difficult or impossible to find the exact eigenvalues of A. But for the present discussion, let us assume that $\mathrm{Spec}(A)$ can be found.

For each $c \in \mathrm{Spec}(A)$, we need to find the Jordan blocks in $\mathbf{J}$ of the form $J(c; m)$. Assuming these blocks occur in decreasing order of size, they collectively constitute a Jordan matrix of the form $J(c; \mu)$ for some unknown partition μ depending on c. Looking at the uniqueness proof again, we are led to consider the null space Z of the matrix $(A - cI_n)^n$. The proof shows that the nilpotent linear map $U : Z \to Z$ defined by $U(z) = Az - cz$ for

$z \in Z$ has matrix $J(0; \mu)$ relative to some ordered basis of Z. We can find μ by using (8.9), which states that $\mu'_1 + \cdots + \mu'_k = \dim(\mathrm{Null}(U^k))$ for all $k \geq 1$. Translating back to matrices, we need only compute bases for all the null spaces $\mathrm{Null}((A - cI_n)^k)$ for $1 \leq k \leq n$, which can be done by Gaussian elimination. Letting d_k be the dimension of the kth null space (with $d_0 = 0$), we compute $\mu'_k = d_k - d_{k-1}$ for all $k \geq 1$, from which $D(\mu)$ and hence μ are readily found.

Finding the transition matrix P requires more work. Since $\mathbf{J}$ is now known, one straightforward but inefficient approach is to treat the entries of P as unknowns, which can be found by solving the linear system $P\mathbf{J} = AP$. Another method first finds bases for the various null spaces $\mathrm{Null}((A - cI_n)^n)$. Using these basis vectors as the columns of a matrix Q, we have $Q^{-1}AQ = \mathrm{blk\text{-}diag}(A_1, \ldots, A_k)$, where each A_i has a single eigenvalue c_i and $A_i - c_iI$ is nilpotent. Considering each block separately, we are reduced to solving the following problem: given a nilpotent matrix $B \in M_m(\mathbb{C})$, find a specific invertible matrix R and a partition μ of m such that $R^{-1}BR = J(0; \mu)$.

This problem can be solved by a recursive algorithm that implements the three-stage inductive proof of the Classification Theorem for Nilpotent Linear Maps. First, we recursively find an ordered basis X_1 for $V_1 = \mathrm{img}(B)$ and a partition $\nu = (\nu_1, \ldots, \nu_s)$ such that the map $(x \mapsto Bx : x \in V_1)$ has matrix $J(0; \nu)$ relative to the basis X_1. Second, we compute a basis for $\mathrm{Null}(B)$ and use appropriate vectors from this basis to augment X_1 to a basis X_2 for $V_2 = \mathrm{img}(B) + \mathrm{Null}(B)$. If t new basis vectors are added, we define ρ by adding t new parts of size 1 to the end of ν. Third, writing $X_2 = \{x(i, j) : (i, j) \in D(\rho)\}$, we solve linear equations to find vectors $x(i, \nu_j + 1)$ such that $Bx(i, \nu_j + 1) = x(i, \nu_j)$ for $1 \leq i \leq s$. We define $\mu = (\nu_1 + 1, \ldots, \nu_s + 1, 1, \ldots, 1)$ (where there are t parts of size 1) and $X = \{x(i, j) : (i, j) \in D(\mu)\}$. Finally, we obtain R by placing the column vectors $x(i, j)$ into a matrix, starting with the first row of $D(\mu)$ and working left to right, top to bottom. You may check that every step in this paragraph can be implemented by solving an explicitly computable system of linear equations. Admittedly, the entire algorithm requires a very substantial amount of computation.

For example, consider the matrix

$$A = \begin{bmatrix} -2 & -7 & 2 & 1 & 7 & -8 \\ -3.5 & -1.5 & -0.5 & 0.5 & 5 & 0 \\ -1 & -4 & 1 & 1 & 4 & -5 \\ 6.5 & 18.5 & -10.5 & -0.5 & -14 & 31 \\ -0.5 & -0.5 & -3.5 & 0.5 & 4 & 3 \\ 3 & 3 & -3 & 0 & -3 & 5 \end{bmatrix}. \tag{8.11}$$

Using a computer algebra system, we compute $\chi_A = (x + 1)^2(x - 2)^4$, so $\mathrm{Spec}(A) = \{-1, 2\}$. Row reduction of $(A + I)^6$ shows that the null space of this matrix consists of column vectors

$$\{(s, s - t, s + t, -s - 2t, s, t) : s, t \in \mathbb{C}\},$$

so $v_1 = (1, 1, 1, -1, 1, 0)$ and $v_2 = (0, -1, 1, -2, 0, 1)$ form a basis for this null space. Similarly, row reduction of $(A - 2I)^6$ yields a null space

$$\{(a - b/4 + c/4 + 3d/4, b/4 + 3c/4 - 7d/4, a, b, c, d) : a, b, c, d \in \mathbb{C}\},$$

so

$$v_3 = (1, 0, 1, 0, 0, 0), \quad v_4 = (-1, 1, 0, 4, 0, 0), \quad v_5 = (1, 3, 0, 0, 4, 0), \quad v_6 = (3, -7, 0, 0, 0, 4)$$

form a basis for this null space. Letting P_1 be the matrix with columns $v_1, \ldots, v_6$, we find

$$A_1 = P_1^{-1} A P_1 = \begin{bmatrix} -1 & -1 & 0 & 0 & 0 & 0 \\ 0 & -1 & 0 & 0 & 0 & 0 \\ 0 & 0 & 0 & 1 & 3 & 5 \\ 0 & 0 & -1 & 2.5 & 1.5 & 3.5 \\ 0 & 0 & -1 & 0.5 & 3.5 & 3.5 \\ 0 & 0 & 0 & 0 & 0 & 2 \end{bmatrix}.$$

By inspection, replacing v_2 by $-v_2$ converts the upper 2×2 block to $J(-1; 2)$. To deal with the lower 4×4 block (corresponding to the eigenvalue 2 of A), we consider the nilpotent matrix

$$B = \begin{bmatrix} -2 & 1 & 3 & 5 \\ -1 & 0.5 & 1.5 & 3.5 \\ -1 & 0.5 & 1.5 & 3.5 \\ 0 & 0 & 0 & 0 \end{bmatrix}.$$

A recursive call to the algorithm produces a basis $x(1,1) = (2,1,1,0)$, $x(1,2) = (2.5, 1.75, 1.75, 0)$ for $\text{img}(B)$ and associated partition $\nu = (2)$. Row-reduction of B shows that $\text{Null}(B)$ has a basis $((1,2,0,0), (3,0,2,0))$. Since the first basis vector is not a linear combination of $x(1,1)$ and $x(1,2)$, we set $x(2,1) = (1,2,0,0)$ and $\rho = (2,1)$. Finally, we solve $Bv = x(1,2)$ to obtain $x(1,3) = (2,1,1,0.5)$. Now, taking

$$P_2 = \begin{bmatrix} 1 & 0 & 0 & 0 & 0 & 0 \\ 0 & -1 & 0 & 0 & 0 & 0 \\ 0 & 0 & 2 & 2.5 & 2 & 1 \\ 0 & 0 & 1 & 1.75 & 1 & 2 \\ 0 & 0 & 1 & 1.75 & 1 & 0 \\ 0 & 0 & 0 & 0 & 0.5 & 0 \end{bmatrix}, \quad P = P_1 P_2 = \begin{bmatrix} 1 & 0 & 2 & 2.5 & 3.5 & -1 \\ 1 & 1 & 4 & 7 & 0.5 & 2 \\ 1 & -1 & 2 & 2.5 & 2 & 1 \\ -1 & 2 & 4 & 7 & 4 & 8 \\ 1 & 0 & 4 & 7 & 4 & 0 \\ 0 & -1 & 0 & 0 & 2 & 0 \end{bmatrix},$$

we find that $P^{-1} A P = \text{blk-diag}(J(-1; 2), J(2; 3), J(2; 1))$.

8.13 Application to Differential Equations

This section describes an application of the Jordan Canonical Form Theorem to the solution of homogeneous systems of linear ordinary differential equations. We let x denote a column vector of unknown continuously differentiable functions $x_1, \ldots, x_n : \mathbb{R} \to \mathbb{C}$. Given a fixed matrix $A \in M_n(\mathbb{C})$, our goal is to find all solutions to $x' = Ax$, which is equivalent to the system of scalar differential equations

$$x_i'(t) = \sum_{j=1}^{n} A(i,j) x_j(t) \qquad \text{for } 1 \leq i \leq n \text{ and } t \in \mathbb{R}.$$

This system can be solved easily if A is a diagonal matrix. In this case, $x_i'(t) = A(i,i) x_i(t)$ has general solution $x_i(t) = b_i e^{A(i,i)t}$, where b_i is any constant. More generally, if A is a Jordan block $J(c; n)$ for some $c \in \mathbb{C}$, we can backsolve for $x_n(t), x_{n-1}(t), \ldots, x_1(t)$ as follows. First, $x_n'(t) = c x_n(t)$ implies $x_n(t) = b_n e^{ct}$ for some constant b_n. Second, $x_{n-1}'(t) = c x_{n-1}(t) + x_n(t) = c x_{n-1}(t) + b_n e^{ct}$ has general solution $x_{n-1}(t) = b_n t e^{ct} + b_{n-1} e^{ct}$ where b_{n-1} is any constant. Third, $x_{n-2}'(t) = c x_{n-2}(t) + x_{n-1}(t)$ has general solution $x_{n-2}(t) =$

$(b_n/2)t^2 e^{ct} + b_{n-1} t e^{ct} + b_{n-2} e^{ct}$ where b_{n-2} is any constant. Continuing backwards through the block, you can check by induction that

$$x_{n-k}(t) = \sum_{i=0}^{k} \frac{b_{n-i}}{(k-i)!} t^{k-i} e^{ct} \qquad \text{for } 0 \le k < n, \qquad (8.12)$$

where $b_n, \ldots, b_1 \in \mathbb{C}$ are arbitrary constants.

Given $A = \text{blk-diag}(J(c_1; n_1), \ldots, J(c_k; n_k))$, we can use (8.12) to solve for the first n_1 functions $x_1, \ldots, x_{n_1}$. The same formula applies to recover the functions $x_{n_1+1}, \ldots, x_{n_1+n_2}$, starting with $x_{n_1+n_2}$ and working backwards, and similarly for all later blocks.

Finally, consider the case where $A \in M_n(\mathbb{C})$ is an arbitrary matrix. Using the algorithms in the previous section, we can find a Jordan canonical form $\mathbf{J}$ and an invertible matrix P in $M_n(\mathbb{C})$ with $\mathbf{J} = P^{-1}AP$. Introduce a new column vector y of unknown functions $y_1, \ldots, y_n$ by the linear change of variable $y = P^{-1}x$, $x = Py$. By linearity of the derivative, $x' = Py'$, and we see that x solves $x' = Ax$ iff $Py' = A(Py)$ iff $y' = (P^{-1}AP)y$ iff y solves $y' = \mathbf{J}y$. So, once we compute P and $\mathbf{J}$, y is given by the formulas above, and then $x = Py$ is the solution to the original system. We must point out that there may be more efficient ways of solving $x' = Ax$, especially when A has special structure, but a full discussion of this point is beyond the scope of this text.

For example, let us solve $x' = Ax$, where A is the matrix in (8.11). We computed a Jordan canonical form of A to be $\text{blk-diag}(J(-1; 2), J(2; 3), J(2; 1))$. Let $y = P^{-1}x$, where P is the matrix found at the end of §8.12. Then $y' = \mathbf{J}y$ has general solution

$$
\begin{aligned}
&y_1(t) = c_1 t e^{-t} + c_2 e^{-t}, & &y_2(t) = c_1 e^{-t}, \\
&y_3(t) = (c_3/2)t^2 e^{2t} + c_4 t e^{2t} + c_5 e^{2t}, & &y_4(t) = c_3 t e^{2t} + c_4 e^{2t}, \quad y_5(t) = c_3 e^{2t}, \\
&y_6(t) = c_6 e^{2t}.
\end{aligned}
$$

where $c_1, \ldots, c_6 \in \mathbb{C}$ are arbitrary constants, and the original system has solution $x = Py$.

8.14 Minimal Polynomials

Let V be an n-dimensional vector space over any field F and $T : V \to V$ be a linear map. Recall that T is in the F-algebra $L(V)$ of all F-linear maps from V to V, which is finite-dimensional. Therefore, we know from §3.20 that there exists a unique monic polynomial $m_T \in F[x]$ of least degree such that $m_T(T) = 0$, and m_T divides all polynomials $g \in F[x]$ such that $g(T) = 0$. We can find m_T by searching the list of powers $(\text{id}_V, T, T^2, T^3, \ldots)$ for the lowest power T^k that is a linear combination of preceding powers of T. Similarly, any matrix $A \in M_n(F)$ has a minimal polynomial m_A. For any fixed ordered basis X of V, the map sending $T \in L(V)$ to $[T]_X \in M_n(F)$ is an algebra isomorphism (see Chapter 6). It follows that the linear map T has the same minimal polynomial as the matrix $[T]_X$, for any choice of X. By letting X vary over all ordered bases of V, we conclude from this that *similar matrices have the same minimal polynomial*.

Suppose A is a block-diagonal matrix $\text{blk-diag}(A_1, \ldots, A_s)$. For any $f \in F[x]$, we have $f(A) = \text{blk-diag}(f(A_1), \ldots, f(A_s))$. Therefore, $f(A) = 0$ iff every $f(A_i) = 0$ iff m_{A_i} divides f for $1 \le i \le s$. It follows that $m_A = \text{lcm}(m_{A_1}, \ldots, m_{A_s})$. In particular, suppose A is a diagonal matrix. Then we can take each A_i to be a 1×1 matrix. The minimal polynomial of the matrix $[c]$ is $x - c$, for any $c \in F$. Hence, for diagonal A, $m_A = \text{lcm}_{1 \le i \le n}(x - A(i, i))$. Since the diagonal entries of A are the eigenvalues of A, we can also write this as $m_A = \prod_{c \in \text{Spec}(A)}(x - c)$. More generally, any *diagonalizable* matrix A is similar to a diagonal matrix with the eigenvalues in $\text{Spec}(A)$ appearing on the main diagonal (each with a certain multiplicity). It follows that

$$\text{for diagonalizable } A \in M_n(F), \quad m_A = \prod_{c \in \text{Spec}(A)} (x - c).$$

So for all $A \in M_n(F)$, if A is diagonalizable, then m_A splits into distinct linear factors in $F[x]$.

The converse statement also holds. We use Jordan canonical forms to prove the converse for $F = \mathbb{C}$; a different proof is needed for general fields F (Exercise 58). Assume $A \in M_n(\mathbb{C})$ is not diagonalizable. We know A is similar to a Jordan canonical form $\mathbf{J} = \text{blk-diag}(J(c_1; n_1), \ldots, J(c_s; n_s))$ that has the same minimal polynomial as A. The matrix $\mathbf{J}$ cannot be diagonal, so some $n_i > 1$. We know $m_{\mathbf{J}} = \text{lcm}_{1 \le i \le s} m_{J(c_i; n_i)}$. So it suffices to show that $m_{J(c;k)} = (x - c)^k$.

Evaluating the polynomial $g = (x - c)^k$ at $x = J(c; k)$ gives

$$g(J(c; k)) = (J(c; k) - cI_k)^k = J(0; k)^k = 0.$$

The last equality can be seen by a matrix calculation, or by noting $J(0; k)$ is the matrix of the linear map $T_{X,(k)}$, which is nilpotent of index k. So $m_{J(c;k)}$ must divide g. By unique factorization in $\mathbb{C}[x]$, the monic divisors of g in $\mathbb{C}[x]$ are $(x - c)^i$ where $0 \le i \le k$. Since $J(0; k)^i \ne 0$ for $i < k$, we see that g itself must be the minimal polynomial of $J(c; k)$.

This proof tells us how to compute m_A from a Jordan canonical form of A. We have $m_A = \prod_{c \in \text{Spec}(A)} (x - c)^{k(c)}$, where $k(c)$ is the maximum size of any Jordan block $J(c; k)$ appearing in a Jordan canonical form similar to A. On the other hand, consider the characteristic polynomial $\chi_A = \det(xI_n - A)$. Similar matrices have the same characteristic polynomial, so $\chi_A = \chi_{\mathbf{J}}$. The characteristic polynomial of a Jordan block $J(c; k)$ is $(x - c)^k$, since $xI_n - J(c; k)$ is a $k \times k$ triangular matrix with all diagonal entries equal to $x - c$. For any block-diagonal matrix $B = \text{blk-diag}(B_1, \ldots, B_s)$, computing the determinant shows that $\chi_B = \prod_{i=1}^{s} \chi_{B_i}$. Taking $B = \mathbf{J}$ here, we see that $\chi_A = \prod_{i=1}^{s} (x - c_i)^{n_i}$. Comparing to the earlier formula for m_A, we deduce the Cayley–Hamilton Theorem for complex matrices: *for all $A \in M_n(\mathbb{C})$, the minimal polynomial m_A divides the characteristic polynomial χ_A in $\mathbb{C}[x]$.*

Analogous results hold for linear maps $T \in L(V)$. In particular, *T is diagonalizable iff m_T splits into distinct linear factors in $F[x]$, in which case $m_T = \prod_{c \in \text{Spec}(T)} (x - c)$.* We use this theorem to prove that *if $T \in L(V)$ is diagonalizable and W is a T-invariant subspace of V, then $T|W$ is diagonalizable.* To prove this, write $g = m_T$ and $f = m_{T|W}$ in $F[x]$. We know $g(T)$ is the zero operator on V. Restricting to W, $g(T|W) = g(T)|W$ is the zero operator on W. So f, the minimal polynomial of $T|W$, must divide g in $F[x]$. By diagonalizability of T, g splits into a product of distinct linear factors in $F[x]$. Since f divides g, the unique factorization of f in $F[x]$ must also be a product of distinct linear factors, which are a subset of the factors for g. By the theorem, $T|W$ is diagonalizable.

8.15   Jordan–Chevalley Decomposition of a Linear Operator

Here, we prove an abstract version of the Jordan Canonical Form Theorem that is needed in the theory of Lie algebras.

Jordan–Chevalley Decomposition Theorem. Suppose V is an m-dimensional complex vector space and $T : V \to V$ is a linear map. There exist unique linear maps T_d and T_n on V such that T_d is diagonalizable, T_n is nilpotent, $T = T_d + T_n$, and $T_d \circ T_n = T_n \circ T_d$.

T_d is called the *diagonalizable part* of T, T_n is called the *nilpotent part* of T, and $T = T_d + T_n$ is called the *Jordan–Chevalley decomposition of T*.

To prove existence of T_d and T_n with the stated properties, write $\text{Spec}(T) = \{c_1, \ldots, c_s\}$. By the Jordan Canonical Form Theorem, we can pick an ordered basis X of V such that $[T]_X = \text{blk-diag}(J(c_1; \mu^{(1)}), J(c_2; \mu^{(2)}), \ldots, J(c_s; \mu^{(s)}))$, where each $\mu^{(i)}$ is a partition of some positive integer m_i. Let X_1 consist of the first m_1 vectors in the list X, let X_2 consist of the next m_2 vectors in X, and so on. Define a linear map T_d on the basis X by letting $T_d(z_i) = c_i z_i$ for all $z_i \in X_i$, and extending by linearity. Define $T_n = T - T_d$, which is linear since it is a linear combination of two linear maps. Evidently $T = T_d + T_n$. Also, $[T_d]_X$ is the diagonal matrix $\text{blk-diag}(c_1 I_{m_1}, c_2 I_{m_2}, \ldots, c_s I_{m_s})$, so T_d is diagonalizable. On the other hand,

$$[T_n]_X = [T]_X - [T_d]_X = \text{blk-diag}(J(0; \mu^{(1)}), J(0; \mu^{(2)}), \ldots, J(0; \mu^{(s)})).$$

It follows that every basis vector in X gets sent to zero after applying T_n at most $\max(m_1, \ldots, m_s)$ times, so that T_n is nilpotent. Finally, for each i, the scalar multiple of the identity $c_i I_{m_i}$ commutes with $J(0; \mu^{(i)})$. So the matrices $[T_d]_X$ and $[T_n]_X$ commute since each diagonal block of the first matrix commutes with the corresponding block of the second matrix. It follows that the linear maps T_d and T_n commute, completing the existence proof.

Turning to the uniqueness proof, let $T = T_d' + T_n'$ be any decomposition of T into the sum of a diagonalizable linear map T_d' and a nilpotent linear map T_n' that commute with each other. We must prove $T_d' = T_d$ and $T_n' = T_n$. Keep the notation of the previous paragraph. We saw in the uniqueness proof in §8.11 that each X_i is a basis of the T-invariant subspace $Z_i = \text{Null}((T - c_i \, \text{id}_V)^m)$. Now, T_d' commutes with T, since

$$T_d' \circ T = T_d' \circ (T_d' + T_n') = T_d' \circ T_d' + T_d' \circ T_n' = T_d' \circ T_d' + T_n' \circ T_d' = (T_d' + T_n') \circ T_d' = T \circ T_d'.$$

Then T_d' also commutes with $T - c_i \, \text{id}_V$ and any power of this linear map. It follows that each Z_i is a T_d'-invariant subspace, since $z \in Z_i$ implies $(T - c_i \, \text{id}_V)^m (T_d'(z)) = T_d'((T - c_i \, \text{id}_V)^m (z)) = T_d'(0) = 0$. Similarly, T_n' commutes with T, and so each Z_i is a T_n'-invariant subspace.

It follows that $[T_d']_X = \text{blk-diag}(B_1, \ldots, B_s)$ and $[T_n']_X = \text{blk-diag}(C_1, \ldots, C_s)$ for certain matrices $B_i, C_i \in M_{m_i}(\mathbb{C})$. Because B_i commutes with $c_i I_{m_i}$ for all i, T_d' commutes with T_d. Since T_d' also commutes with T, T_d' commutes with $T_n = T - T_d$. Similarly, T_n' commutes with T_d and T_n. Now, $T_d' + T_n' = T = T_d + T_n$ gives $T_d' - T_d = T_n - T_n'$. The left side of this equation is a diagonalizable linear map, since it is the difference of two commuting diagonalizable linear maps (Exercise 60). The right side of this equation is a nilpotent linear map, since it is the difference of two commuting nilpotent linear maps (Exercise 61). The two sides of the equation are equal, so we are considering a diagonalizable nilpotent linear map. The only such map is zero (as we saw in the introduction to this chapter), so $T_d' - T_d = 0 = T_n - T_n'$. Thus, $T_d' = T_d$ and $T_n = T_n'$, proving uniqueness.

The result just proved leads to the following Jordan–Chevalley Decomposition Theorem for Matrices: *for any matrix $A \in M_n(\mathbb{C})$, there exist unique matrices $B, C \in M_n(\mathbb{C})$ such that $A = B + C$, $BC = CB$, B is diagonalizable, and C is nilpotent.*

8.16   Summary

1. *Definitions.* Given a vector space V and a linear map $T : V \to V$, T is *nilpotent* iff $T^k = 0$ for some positive integer k. The least such k is the *index of nilpotence* of T. T is *diagonalizable* iff for some ordered basis X of V, $[T]_X$ is a diagonal matrix. $\mathrm{Spec}(T)$ is the set of eigenvalues of T. The *Jordan block* $J(c; k)$ is a $k \times k$ matrix with all diagonal entries equal to c, all entries 1 on the next higher diagonal, and zeroes elsewhere. A *Jordan canonical form* is a block-diagonal matrix with diagonal blocks $J(c_1; k_1), \ldots, J(c_s; k_s)$. A *partition of n* is a weakly decreasing sequence $\mu = (\mu_1, \ldots, \mu_s)$ of positive integers with sum n. We write $J(c; \mu) = $ blk-diag$(J(c; \mu_1), \ldots, J(c; \mu_s))$. The *diagram of μ* consists of s rows of boxes, with μ_i boxes in row i; μ'_k is the height of column k in this diagram. Formally, the diagram of μ is $D(\mu) = \{(i, j) : 1 \le i \le s, 1 \le j \le \mu_i\}$.

2. *Nilpotent Maps and Partition Diagrams.* Given a partition μ and a basis $X = \{x(i, j) : (i, j) \in D(\mu)\}$ for V, we obtain a nilpotent linear map $T_{X,\mu}$ on V by sending $x(i, 1)$ to zero for all i, sending $x(i, j)$ to $x(i, j - 1)$ for all i and all $j > 1$, and extending by linearity. A basis for $\mathrm{Null}(T_{X,\mu}^k)$ consists of $x(i, j)$ with $(i, j) \in D(\mu)$ and $1 \le j \le k$; so $\dim(\mathrm{Null}(T_{X,\mu}^k)) = \mu'_1 + \cdots + \mu'_k$. A basis for $\mathrm{img}(T_{X,\mu}^k)$ consists of all $x(i, j)$ with $(i, j) \in D(\mu)$ and $(i, j + k) \in D(\mu)$.

3. *Classification of Nilpotent Maps.* Given any field F, any n-dimensional F-vector space V, and any nilpotent linear map $T : V \to V$, there exists a unique partition μ of n such that $T = T_{X,\mu}$ (equivalently, $[T]_X = J(0; \mu)$) for some ordered basis X of V.

4. *Jordan Canonical Form Theorem.* For all algebraically closed fields F (such as $\mathbb{C}$), all finite-dimensional F-vector spaces V, and all linear maps $T : V \to V$, there exists an ordered basis X of V such that $[T]_X$ is a Jordan canonical form blk-diag$(J(c_1; m_1), \ldots, J(c_s; m_s))$. For any other ordered basis Y of V, if $[T]_Y = $ blk-diag$(J(d_1; n_1), \ldots, J(d_t; n_t))$, then $s = t$ and the Jordan blocks $J(d_k; n_k)$ are a rearrangement of the Jordan blocks $J(c_i; n_i)$. Every matrix $A \in M_n(F)$ is similar to a Jordan canonical form $\mathbf{J}$, and any other Jordan form similar to A is obtained by reordering the Jordan blocks of $\mathbf{J}$.

5. *Computing Jordan Forms.* The diagonal entries in any Jordan form of a linear map T (or matrix A) are the eigenvalues of T (or A). For each eigenvalue c, let $U_c = T - c\,\mathrm{id}_V$ (or $U_c = A - cI_n$). We can find an ordered basis X_c for the subspace $Z_c = \mathrm{Null}(U_c^n)$ such that $[U_c|Z_c]_{X_c} = J(0; \mu(c))$. Letting X be the concatenation of the bases X_c, $[T]_X$ is a Jordan canonical form. Column k of $D(\mu(c))$ has size $\dim(\mathrm{Null}(U_c^k)) - \dim(\mathrm{Null}(U_c^{k-1}))$.

6. *Application to Differential Equations.* One way to solve $x' = Ax$ is to find an invertible P and a Jordan form $\mathbf{J}$ with $\mathbf{J} = P^{-1}AP$. The substitution $x = Py$, $y = P^{-1}x$ converts the system $x' = Ax$ to $y' = \mathbf{J}y$, which can be solved by backsolving each Jordan block.

7. *Minimal Polynomials and Jordan Forms.* A matrix $A \in M_n(F)$ is diagonalizable iff the minimal polynomial m_A splits into distinct linear factors in $F[x]$; similarly for a linear map on an n-dimensional vector space. Given $A \in M_n(\mathbb{C})$, $m_A = \prod_{c \in \mathrm{Spec}(A)} (x - c)^{k(c)}$, where $k(c)$ is the maximum size of any Jordan block $J(c; k)$ appearing in a Jordan form similar to A. The characteristic polynomial $\chi_A = \prod_{c \in \mathrm{Spec}(A)} (x - c)^{m(c)}$, where $m(c)$ is the total size of all Jordan blocks

$J(c; k)$ in a Jordan form of A. The Cayley–Hamilton Theorem follows: m_A divides χ_A. If $T \in L(V)$ is diagonalizable and W is a T-invariant subspace of V, then the restriction $T|W$ is also diagonalizable.

8. *Jordan–Chevalley Decomposition.* Given a linear map T on a finite-dimensional vector space V over an algebraically closed field F, there exist unique linear maps T_d and T_n on V such that T_d is diagonalizable, T_n is nilpotent, $T_d \circ T_n = T_n \circ T_d$, and $T = T_d + T_n$.

8.17   Exercises

Unless otherwise specified, assume in these exercises that F is a field, V is a finite-dimensional F-vector space, and $T : V \to V$ is a linear map.

1. Decide whether each linear map is nilpotent. If not, explain why not. If so, find the index of nilpotence.
   (a) $T : \mathbb{R}^3 \to \mathbb{R}^3$ given by $T(a, b, c) = (c, 0, b)$ for $a, b, c \in \mathbb{R}$.
   (b) $T : \mathbb{R}^3 \to \mathbb{R}^3$ given by $T(a, b, c) = (0, c, b)$ for $a, b, c \in \mathbb{R}$.
   (c) $D : V \to V$ given by $D(f) = df/dx$, where V is the vector space of real polynomials of degree at most 4.

2. Decide whether each linear map is nilpotent. If not, explain why not. If so, find the index of nilpotence.
   (a) $T : M_3(\mathbb{R}) \to M_3(\mathbb{R})$ given by $T(A) = A^{\mathrm{T}}$ (the transpose map).
   (b) $T : \mathbb{C}^2 \to \mathbb{C}^2$ given by $T(a, b) = (-3a + 9b, -a + 3b)$ for $a, b \in \mathbb{C}$.
   (c) $T : \mathbb{C}^2 \to \mathbb{C}^2$ given by $T(a, b) = (-a + 3b, -a + 3b)$ for $a, b \in \mathbb{C}$.
   (d) $T(x) = Ax$ for $x \in \mathbb{R}^{10}$, where $A = \mathrm{blk\text{-}diag}(J(0; 4), J(0; 3)^{\mathrm{T}}, J(0; 3))$.

3. Decide whether each linear map is diagonalizable. If not, explain why not. If so, find an ordered basis X such that $[T]_X$ is diagonal.
   (a) $T : \mathbb{R}^2 \to \mathbb{R}^2$ given by $T(a, b) = (a, a + b)$ for $a, b \in \mathbb{R}$.
   (b) $T : \mathbb{R}^2 \to \mathbb{R}^2$ given by $T(a, b) = (a + b, a + b)$ for $a, b \in \mathbb{R}$.
   (c) $T_\theta : \mathbb{R}^2 \to \mathbb{R}^2$ given by $T_\theta(a, b) = (a \cos\theta - b \sin\theta, a \sin\theta + b \cos\theta)$ for $a, b \in \mathbb{R}$, where $0 < \theta < \pi/2$ is fixed.
   (d) $T_\theta : \mathbb{C}^2 \to \mathbb{C}^2$ given by the same formula in (c) for $a, b \in \mathbb{C}$.

4. Decide whether each linear map is diagonalizable. If not, explain why not. If so, find an ordered basis X such that $[T]_X$ is diagonal.
   (a) $D(f) = df/dx$ for f a real polynomial of degree at most 4.
   (b) $T : M_2(\mathbb{R}) \to M_2(\mathbb{R})$ given by $T(A) = A^{\mathrm{T}}$ (the transpose map).

5. Prove or disprove: for all $A \in M_n(F)$, if A is nilpotent, then A is not invertible.

6. Prove or disprove: for all $A \in M_n(F)$, if A is not invertible, then A is nilpotent.

7. Use the uniqueness part of the Jordan Canonical Form Theorem to prove that $A \in M_n(\mathbb{C})$ is diagonalizable iff every Jordan block in any Jordan canonical form similar to A has size 1.

8. Suppose that $A \in M_n(F)$ has only one eigenvalue c and $A \neq cI_n$. Prove that A is not diagonalizable. Deduce that nonzero nilpotent maps are not diagonalizable.

9. For $A = J(0; (5, 5, 5, 5, 2, 2, 2, 2, 2, 2, 2, 1, 1, 1))$, how many Jordan canonical forms are similar to A?

10. For $A = \text{blk-diag}(J(2;(2,2,2)), J(3;3), J(4;(3,3,2,2)))$, how many Jordan canonical forms are similar to A?

11. Describe the Jordan canonical form matrices that are not similar to any Jordan canonical form besides themselves.

12. Suppose V is finite-dimensional and $T : V \to V$ is a linear map such that for all $v \in V$, there exists a positive integer $k(v)$ (depending on v) with $T^{k(v)}(v) = 0$. Prove that T is nilpotent.

13. Give an example to show that the result of the previous exercise can fail if V is infinite-dimensional.

14. In the example at the end of §8.1, verify that Z is an ordered basis of V and $[T]_Z = J(0;(2,1,1))$.

15. For each linear map T defined on the basis $(x_1, \ldots, x_n)$, decide if T is nilpotent. If it is, find an ordered basis Z and a partition μ such that $[T]_Z = J(0;\mu)$.
    (a) $n = 5$, $T(x_i) = x_{i+1}$ for $1 \leq i < 4$, $T(x_5) = 0$.
    (b) $n = 4$, $T(x_1) = x_3$, $T(x_2) = 0$, $T(x_3) = x_4$, $T(x_4) = x_1$.
    (c) $n = 8$, $T(x_i) = x_{\lfloor i/2 \rfloor}$ for all i, letting $x_0 = 0$.
    (d) $n = 11$, $T(x_i) = x_{(2i \bmod 12)}$ for $1 \leq i < 12$, letting $x_0 = 0$.

16. For each linear map T defined on the basis $(x_1, \ldots, x_n)$, decide if T is nilpotent. If it is, find an ordered basis Z and a partition μ such that $[T]_Z = J(0;\mu)$.
    (a) $n = 7$, $T(x_1) = T(x_5) = x_2$, $T(x_7) = x_4$, $T(x_2) = T(x_4) = T(x_6) = x_3$, $T(x_3) = 0$.
    (b) $n = 3$, $T(x_1) = x_2 + 2x_3$, $T(x_2) = x_1 + 2x_3$, $T(x_3) = x_1 + 2x_2$.
    (c) $n = 4$, $T(x_i) = x_1 + x_2 + x_3 - 3x_4$ for $1 \leq i \leq 4$.
    (d) $T(x_i) = x_{i+1} + x_{i+2} + \cdots + x_n$ for $1 \leq i < n$, $T(x_n) = 0$.
    (e) $n = 15$, $T(x_i) = x_{(2i \bmod 16)}$ for $1 \leq i < 15$, letting $x_0 = 0$.

17. (a) Suppose $X = (x_1, \ldots, x_n)$ is an ordered basis for V and T is a linear map on V such that for $1 \leq j \leq n$, either $T(x_j) = 0$ or there exists $i < j$ with $T(x_j) = x_i$. Prove T is nilpotent. (b) More generally, assume that for all j, $T(x_j) = \sum_{i=1}^{j-1} a_{ij} x_i$ for some $a_{ij} \in F$. Without using matrices or Jordan forms, prove T is nilpotent.

18. List all partitions of 5. For each partition μ, state the elements of the set $D(\mu)$, draw a picture of $D(\mu)$, and compute μ'_k for all $k \geq 1$.

19. Find the partition μ, given that
$$(\mu'_1 + \mu'_2 + \cdots + \mu'_k : k \geq 1) = (8, 12, 16, 19, 21, 23, 25, 26, 27, 28, 29, 30, 30, 30, \ldots).$$

20. For a certain linear map $T_{X,\mu}$, it is known that the dimensions of $\text{img}(T_{X,\mu}^k)$ for $k \geq 1$ are 15, 11, 8, 5, 3, 2, 1, 0. What can be said about μ?

21. Prove: for all partitions μ of n, μ'_k is the number of $\mu_i \geq k$. Conclude that $\mu' = (\mu'_1, \mu'_2, \ldots : \mu'_k > 0)$ is a partition of n, and $\mu'' = \mu$.

22. Let $X = (x_1, x_2, x_3, x_4)$ be an ordered basis of V. For each partition μ of 4, compute $T_{X,\mu}(x_i)$ for $1 \leq i \leq 4$ and $T_{X,\mu}(v)$, where $v = \sum_{i=1}^4 c_i x_i$.

23. Let $T = T_{X,\mu}$ where $\mu = (6,3,3,3,2,1,1)$. For all $k \geq 1$, describe a basis for $\text{Null}(T^k)$ and a basis for $\text{img}(T^k)$.

24. Let $T = T_{X,\mu}$ where $\mu = (11,8,6,6,3,2,2,2,1,1)$. For all $k \geq 1$, compute the dimensions of $\text{Null}(T^k)$ and $\text{img}(T^k)$.

25. Assume $\dim(V) = 1$. Check directly that for any nilpotent linear map $T : V \to V$, $T = T_{(x),(1)}$ for all nonzero $x \in V$.

26. Given a partition μ and $r, k > 0$, use a visual analysis of $D(\mu)$ to find a basis for $\text{Null}(T^r_{X,\mu}) \cap \text{img}(T^k_{X,\mu})$. Illustrate for $\mu = (5, 5, 3, 1, 1, 1)$, $r = 3$, $k = 2$.

27. Given a partition μ and $r, k > 0$, find a basis for $\text{Null}(T^r_{X,\mu}) + \text{img}(T^k_{X,\mu})$. Illustrate for $\mu = (5, 5, 3, 1, 1, 1)$, $r = 1$, $k = 3$.

28. Let $\mu = (5, 5, 3, 1, 1, 1)$ and $X = (x_1, \ldots, x_{16})$. For each $k \geq 2$, find a partition ν and an ordered basis Z such that $T^k_{X,\mu} = T_{Z,\nu}$.

29. For a partition μ and $k > 0$, describe how to find Z and ν such that $T^k_{X,\mu} = T_{Z,\nu}$.

30. Let $[T]_X = J(0; (2, 2, 1))$, where $X = (x_1, \ldots, x_5)$. Let $Z = (z_1, \ldots, z_5)$ be another ordered basis given by $z_j = \sum_{i=1}^{5} c_{ij} x_i$ for some $c_{ij} \in F$. Find necessary and sufficient conditions on the c_{ij} to ensure that $[T]_Z = J(0; (2, 2, 1))$.

31. Prove: for any two subspaces W and Z of a finite-dimensional vector space V, $\dim(W + Z) + \dim(W \cap Z) = \dim(W) + \dim(Z)$. (One approach is to apply the Rank–Nullity Theorem to the map $f : W \times Z \to W + Z$ given by $f(w, z) = w - z$ for $w \in W$ and $z \in Z$.)

32. Let $T(a, b, c) = (2a + b - 3c, 2a + b - 3c, 2a + b - 3c)$ for $a, b, c \in \mathbb{R}$.
    (a) Check that T is nilpotent.
    (b) Find a basis X_1 for $V_1 = \text{img}(T)$ and a partition ν such that $T|V_1 = T_{X_1,\nu}$.
    (c) Extend X_1 to a basis X_2 of $V_2 = \text{img}(T) + \text{Null}(T)$, and find a partition ρ with $T|V_2 = T_{X_2,\rho}$.
    (d) Extend X_2 to get an indexed basis X of $\mathbb{R}^3$ and a partition μ with $T = T_{X,\mu}$.
    (e) Check that $X'_2 = ((1, -2, 0), (0, 3, 1))$ is a basis of V_2 such that $T|V_2 = T_{X'_2,\rho}$. What happens if we try to solve part (d) starting from the basis X'_2?

33. Let $T(a, b, c, d) = (-a + b + c + d, a + b - c + d, -a + b + c + d, -a - b + c - d)$ for $a, b, c, d \in \mathbb{R}$. Follow the proof in the text to find X and μ such that $T = T_{X,\mu}$.

34. Let $T(a, b, c, d, e, f) = (0, a + 2b + d - 3f, 0, 2a + 2b + c + d - 3f, 0, 3a + 2b + 2c + d + e - 3f)$ for $a, b, c, d, e, f \in \mathbb{R}$. Follow the proof in the text to find X and μ such that $T = T_{X,\mu}$. (In Stage 1, you can use the results of Exercise 32.)

35. Let a linear map $T : \mathbb{R}^9 \to \mathbb{R}^9$ be defined on the standard basis by $T(\mathbf{e}_k) = \mathbf{e}_{\lfloor k/3 \rfloor}$ for $1 \leq k \leq 9$, with $\mathbf{e}_0 = 0$. Follow the proof in the text to find a partition μ and ordered basis X with $T = T_{X,\mu}$.

36. Let T be a nilpotent linear map on V, $n = \dim(V)$, and $j = \dim(\text{Null}(T))$, so $\dim(\text{img}(T)) = n - j$. What are the possible values of $k = \dim(\text{Null}(T) \cap \text{img}(T))$? For each feasible k, construct an explicit example of a map T achieving this k.

37. **Direct Sums.** Let $W_1, \ldots, W_k$ be subspaces of V and X_i be an ordered basis of W_i for $1 \leq i \leq k$. Let $W = W_1 + \cdots + W_k$ and X be the concatenation of the lists $X_1, \ldots, X_k$. Prove the following conditions are equivalent:
    (a) For each $w \in W$, there exist unique $w_i \in W_i$ with $w = w_1 + \cdots + w_k$.
    (b) For $2 \leq i \leq k$, $W_i \cap (W_1 + W_2 + \cdots + W_{i-1}) = \{0\}$.
    (c) X is an ordered basis for W.
    (d) X is a linearly independent list.
    When any of these conditions holds, we say W is the *direct* sum of the W_i, written $W = W_1 \oplus W_2 \oplus \cdots \oplus W_k$.

38. Give an example of three subspaces W_1, W_2, W_3 of a vector space V such that $W_1 \cap W_2 = W_1 \cap W_3 = W_2 \cap W_3 = \{0\}$, but the sum $W_1 + W_2 + W_3$ is not direct.

39. Assume $n = \dim(V) < \infty$, $S : V \to V$ is linear, and $\text{img}(S) = \text{img}(S^2)$. Must $V = \text{Null}(S) \oplus \text{img}(S)$ follow? Explain.

40. For each linear map U on $\mathbb{R}^4$, find explicit U-invariant subspaces Z and W satisfying the conclusions of Fitting's Lemma.
    (a) $U(a, b, c, d) = (a, b, 0, 0)$
    (b) $U(a, b, c, d) = (0, c, b, a)$
    (c) $U(a, b, c, d) = (a + b, c + d, c + d, a + b)$
    (d) $U(a, b, c, d) = (c - a, d - a, c - a, 2d - b - a)$

41. Let V be the infinite-dimensional real vector space of all sequences $(x_0, x_1, x_2, \ldots)$ under componentwise operations. For each linear map $U : V \to V$, prove that the conclusion of Fitting's Lemma does not hold for U.
    (a) $U((x_0, x_1, x_2, \ldots)) = (x_1, x_2, x_3, \ldots)$ for all $x_i \in \mathbb{R}$.
    (b) $U((x_0, x_1, x_2, \ldots)) = (0, x_0, x_1, x_2, \ldots)$ for all $x_i \in \mathbb{R}$.

42. *Eigenspaces.* For each eigenvalue c of a linear map T on V, define the *eigenspace* $E_c = \{v \in V : T(v) = cv\}$. Thus, E_c consists of zero and all eigenvectors of T associated with the eigenvalue c. Show each E_c is a subspace of V.

43. For a complex vector space V, let $X = (x_1, \ldots, x_n)$ be an ordered basis of V such that $[T]_X = \text{blk-diag}(J(c; \mu), B)$, where B is a triangular matrix with no diagonal entries equal to c. Show that $(x_1, x_{\mu_1+1}, x_{\mu_1+\mu_2+1}, \ldots)$ is an ordered basis of E_c. Deduce that for $\text{Spec}(T) = \{c_1, \ldots, c_k\}$, the sum $E_{c_1} + \cdots + E_{c_k}$ is a direct sum (see Exercise 37) in the complex case.

44. For any field F, show that the sum of the eigenspaces of T is a direct sum. (If not, study a relation $v_1 + \cdots + v_i = 0$ with $v_j \in E_{c_j}$, $v_i \neq 0$, and i minimal.)

45. Give an example to show that the sum of all eigenspaces of V may be a proper subspace W of V. If $\dim(V) = n$ and $|\text{Spec}(T)| = k$, what is the minimum possible dimension of W?

46. **Generalized Eigenspaces.** Given a linear map T on an n-dimensional vector space V and $c \in \text{Spec}(T)$, the *generalized eigenspace* for c is $G_c = \text{Null}((T - c\,\text{id}_V)^n)$. Prove: if F is algebraically closed, then $V = \bigoplus_{c \in \text{Spec}(T)} G_c$.

47. Prove: for all $A \in M_n(\mathbb{C})$, A and A^T are similar to the same Jordan canonical forms and hence are similar to each other.

48. For $c \neq 0$ and $k \geq 1$, compute $J(c; k)^{-1}$. What is a Jordan canonical form similar to this matrix?

49. Compute a Jordan canonical form similar to each matrix.
$$\begin{bmatrix} 1 & 1 & 1 & 1 \\ 0 & 1 & 1 & 1 \\ 0 & 0 & 1 & 1 \\ 0 & 0 & 0 & 1 \end{bmatrix}, \begin{bmatrix} 3 & 3 & 3 & 3 \\ 3 & 3 & 3 & 3 \\ 3 & 3 & 3 & 3 \\ 3 & 3 & 3 & 3 \end{bmatrix}, \begin{bmatrix} 1 & 0 & 1 & 0 \\ 0 & 1 & 0 & 1 \\ 1 & 0 & 1 & 0 \\ 0 & 1 & 0 & 1 \end{bmatrix}, \begin{bmatrix} 4 & -2 & 9 & -2 \\ 1 & 1 & 4 & -1 \\ 0 & 0 & 2 & 0 \\ 1 & -1 & 5 & 1 \end{bmatrix}.$$

50. For each matrix A in Exercise 49, find an invertible matrix P such that $P^{-1}AP$ is a Jordan canonical form.

51. Write a program to find the Jordan canonical form of a given complex matrix.

52. Solve the following systems of ordinary differential equations.
    (a) $x_1' = 3x_1$, $x_2' = x_1 + 3x_2$.
    (b) $x_1' = x_2$, $x_2' = -x_1$.
    (c) $x_1' = x_1 + 2x_2 + 3x_3$, $x_2' = x_2 + 2x_3$, $x_3' = x_3$.
    (d) $x_1' = x_1 + x_2$, $x_2' = x_2 + x_3$, $x_3' = x_3 + x_4$, $x_4' = x_1 + x_4$.
    (e) $x_1' = x_3' = x_1 + 2x_2 + 2x_3 + x_4$, $x_2' = x_4' = 2x_1 + x_2 + x_3 + 2x_4$.

53. Check by induction on k that the functions in (8.12) satisfy $x' = J(c; n)x$.

54. Find the minimal polynomial and characteristic polynomial of each matrix in Exercise 49.

55. (a) Prove or disprove: for all $A \in M_n(\mathbb{C})$, if χ_A splits into distinct linear factors in $\mathbb{C}[x]$ then A is diagonalizable. (b) Prove or disprove: for all $A \in M_n(\mathbb{C})$, if A is diagonalizable then χ_A splits into distinct linear factors in $\mathbb{C}[x]$.

56. Suppose W is a T-invariant subspace of V and $h \in F[x]$ is any polynomial. Prove W is also $h(T)$-invariant.

57. **Primary Decomposition Theorem.** For any field F and linear map $T : V \to V$, suppose $m_T = p_1^{e_1} \cdots p_s^{e_s} \in F[x]$, where $p_1, \ldots, p_s$ are distinct monic irreducible polynomials in $F[x]$. Define $W_i = \ker(p_i^{e_i}(T))$ for $1 \le i \le s$.
(a) Show that each W_i is a T-invariant subspace of V.
(b) Show that $V = W_1 + W_2 + \cdots + W_s$. (Define $g_i = m_T/p_i^{e_i} \in F[x]$ and explain why there exist $h_i \in F[x]$ with $\sum_{i=1}^s h_i g_i = 1$. For $v \in V$, show $v = \sum_{i=1}^s h_i(T) g_i(T)(v)$ where the ith summand is in W_i.)
(c) Show that $V = W_1 \oplus W_2 \oplus \cdots \oplus W_s$, a direct sum. (If the sum is not direct, choose t minimal such that $0 = w_1 + w_2 + \cdots + w_t$ with $w_i \in W_i$ and $w_t \ne 0$. Contradict minimality of t.)

58. Prove: for any field F, if m_T splits into a product of distinct linear factors in $F[x]$, then T is diagonalizable. (Use Exercise 57.)

59. Let S and T be diagonalizable linear maps on an F-vector space V. Prove there exists an ordered basis X such that $[S]_X$ and $[T]_X$ are both diagonal iff $S \circ T = T \circ S$. So, two diagonalizable operators can be *simultaneously diagonalized* iff the operators commute. (For the hard direction, choose an ordered basis Y such that $[S]_Y = \text{blk-diag}(c_1 I_{n_1}, \ldots, c_s I_{n_s})$, where $c_1, \ldots, c_s$ are the distinct eigenvalues of S. Break Y into sublists $Y_1, \ldots, Y_s$ of length $n_1, \ldots, n_s$, and let $W_1, \ldots, W_s$ be the subspaces spanned by these sublists. Explain why each W_i is T-invariant. Show that there is an ordered basis X_i for W_i such that $[S|W_i]_{X_i}$ and $[T|W_i]_{X_i}$ are both diagonal.)

60. (a) Prove: if S and T are diagonalizable linear maps on V and $S \circ T = T \circ S$, then $aS + bT$ is diagonalizable for any $a, b \in F$. (b) Give a specific example of diagonalizable linear maps S and T such that $S - T$ is not diagonalizable.

61. (a) Suppose S and T are commuting nilpotent linear maps on an F-vector space V. Prove $S - T$ is nilpotent. (b) Must (a) hold if S and T do not commute? Prove or give a counterexample.

62. Deduce the Jordan–Chevalley decomposition for matrices from the corresponding result for linear maps.

63. Find the Jordan–Chevalley decomposition of each matrix in Exercise 49.

64. Find the Jordan–Chevalley decomposition of the matrix A in (8.11).

65. Given $A = \begin{bmatrix} 0 & -1 \\ 1 & 0 \end{bmatrix}$, show that there exist infinitely many pairs of matrices B, C with $A = B + C$, B diagonalizable, and C nilpotent. Why doesn't this contradict the uniqueness assertion in the Jordan–Chevalley Theorem?

66. Given $T \in L(V)$, show that there exist polynomials $f, g \in F[x]$ with no constant term such that $T_d = f(T)$ and $T_n = g(T)$.

9

Matrix Factorizations

Matrices with special structure — such as diagonal matrices, triangular matrices, and unitary matrices — are simpler to work with than general matrices. Many algorithms have been developed in numerical linear algebra to convert an input matrix into a form with specified special structure by using a sequence of carefully chosen matrix operations. These algorithms can often be described mathematically as providing factorizations of matrices into products of structured matrices. This chapter proves the existence and uniqueness properties of several matrix factorizations and explores the algebraic and geometric ideas leading to these factorizations.

The first factorization, called the *QR factorization*, writes a complex matrix as a product of a unitary matrix Q and an upper-triangular matrix R. More generally, Q can be replaced by a rectangular matrix with orthonormal columns. To obtain such factorizations, we study the Gram–Schmidt algorithm for converting a linearly independent list of vectors into an orthonormal list of vectors. Another approach to QR factorizations involves Householder transformations, which generalize geometric reflections in $\mathbb{R}^2$ or $\mathbb{R}^3$. By cleverly choosing and composing Householder matrices, we can force the entries of a matrix below the main diagonal to become 0, one column at a time.

The second family of factorizations, called *LU decompositions*, writes certain square matrices as products LU with L lower-triangular and U upper-triangular. When such a product exists, we can arrange for one of the matrices (L or U) to have all 1s on its main diagonal. In general, a triangular matrix is called *unitriangular* iff all of its diagonal entries are 1. Not every matrix has an LU factorization, but an appropriate permutation of the rows can convert any invertible matrix to a matrix of the form LU. Similarly, by permuting both rows and columns, any matrix can be converted to a matrix with an LU factorization. These LU factorizations are closely connected to the Gaussian Elimination Algorithm, which reduces a matrix to a simpler form by systematically making entries become 0.

One motivation for finding QR and LU factorizations is the efficient solution of a linear system of equations $Ax = b$. If A is lower-triangular, this system is readily solved by *forward substitution*, in which we solve for the components $x_1, \ldots, x_n$ of x in this order. Similarly, if A is upper-triangular, the system can be solved quickly by *backward substitution*, in which we solve for $x_n, \ldots, x_1$ in this order. If $A = LU$ with L lower-triangular and U upper-triangular, we can solve $Ax = LUx = b$ by first solving $Ly = b$ for y via forward substitution, then solving $Ux = y$ for x via backward substitution. Given $A = QR$ with Q unitary and R upper-triangular, we can solve $Ax = QRx = b$ by noting that $Q^{-1} = Q^*$ (which can be computed quickly from Q), so $Ax = b$ iff $Rx = Q^*b$. The latter system can be solved by backward substitution.

The chapter concludes with two more matrix factorization results. The *Cholesky factorization* expresses a positive semidefinite matrix as a product LL^*, where L is lower-triangular. This factorization is closely related to the QR and LU factorizations, and it has applications to least-squares approximation problems. The *singular value decomposition* expresses a matrix A as a product of a unitary matrix, a diagonal matrix with nonnegative diagonal entries, and another unitary matrix. These factorizations all play a fundamental role in numerical linear algebra.

DOI: 10.1201/9781003484561-9

9.1 Approximation by Orthonormal Vectors

To prepare for our study of the QR factorization, we first discuss how to approximate a vector as a linear combination of orthonormal vectors. Let V be a complex inner product space. This means V is a complex vector space with an *inner product* (the analog of the dot product in $\mathbb{R}^n$) satisfying these conditions for all $x, y, z \in V$ and $c \in \mathbb{C}$: $\langle x + y, z \rangle = \langle x, z \rangle + \langle y, z \rangle$; $\langle cx, y \rangle = c \langle x, y \rangle$; $\langle y, x \rangle = \overline{\langle x, y \rangle}$; and for $x \neq 0$, $\langle x, x \rangle \in \mathbb{R}_{>0}$. The most frequently used complex inner product space is $\mathbb{C}^n$ with the *standard inner product* $\langle v, w \rangle = \sum_{k=1}^n v_k \overline{w_k}$ for $v, w \in \mathbb{C}^n$.

For any $x \in V$, the *length* of x is $||x|| = \sqrt{\langle x, x \rangle}$. The *distance* between vectors $x, y \in V$ is $||x - y||$. Two vectors $x, y \in V$ are called *orthogonal* iff $\langle x, y \rangle = 0$, or equivalently $\langle y, x \rangle = 0$. Orthogonal vectors x and y satisfy the *Pythagorean Identity* $||x + y||^2 = ||x||^2 + ||y||^2$, since

$$||x + y||^2 = \langle x + y, x + y \rangle = \langle x, x \rangle + \langle x, y \rangle + \langle y, x \rangle + \langle y, y \rangle = ||x||^2 + 0 + 0 + ||y||^2.$$

A list of vectors $(u_1, \ldots, u_k)$ in V is called *orthogonal* iff $\langle u_r, u_s \rangle = 0$ for all $r \neq s$ in $[k]$. This list is called *orthonormal* iff the list is orthogonal and $\langle u_r, u_r \rangle = 1$ for all r in $[k]$. Geometrically, an orthonormal list consists of mutually perpendicular unit vectors in V.

Theorem on Independence of Orthogonal Vectors. Any orthogonal list of nonzero vectors in an inner product space V is linearly independent.

In particular, every orthonormal list of vectors is linearly independent.

Proof. Let $(u_1, \ldots, u_k)$ be an orthogonal list of nonzero vectors in V. Suppose $c_1 u_1 + \cdots + c_k u_k = 0$ for given $c_1, \ldots, c_k \in \mathbb{C}$. Fix r between 1 and k and take the inner product of this linear combination with u_r. We get

$$0 = \langle 0, u_r \rangle = \langle c_1 u_1 + \cdots + c_k u_k, u_r \rangle = \sum_{j=1}^k c_j \langle u_j, u_r \rangle = c_r \langle u_r, u_r \rangle.$$

Since u_r is nonzero, $\langle u_r, u_r \rangle > 0$, and so $c_r = 0$. This holds for all r, so $(u_1, \ldots, u_k)$ is linearly independent. $\square$

Theorem on Approximation by Orthonormal Vectors. Suppose $(u_1, \ldots, u_k)$ is an orthonormal list in a complex inner product space V. Let U be the subspace spanned by $(u_1, \ldots, u_k)$. For each $v \in V$, there is a unique vector $x \in U$ minimizing the distance $||v - x||$ from x to v, namely $x = \sum_{r=1}^k \langle v, u_r \rangle u_r$. Also, $v - x$ is orthogonal to every $u \in U$.

We call x the *orthogonal projection of v onto U*.

Proof. Define $x = \sum_{r=1}^k \langle v, u_r \rangle u_r \in U$. Given any s between 1 and k, we compute

$$\langle v - x, u_s \rangle = \langle v, u_s \rangle - \langle x, u_s \rangle = \langle v, u_s \rangle - \sum_{r=1}^k \langle v, u_r \rangle \langle u_r, u_s \rangle = \langle v, u_s \rangle - \langle v, u_s \rangle = 0.$$

Given any $u \in U$, write $u = \sum_{s=1}^k d_s u_s$ with $d_s \in \mathbb{C}$, and note

$$\langle v - x, u \rangle = \sum_{s=1}^k \overline{d_s} \langle v - x, u_s \rangle = \sum_{s=1}^k \overline{d_s} \cdot 0 = 0.$$

Hence, $v - x$ is orthogonal to every $u \in U$. In particular, for any $y \in U$, $x - y$ is in the subspace U, so the Pythagorean Identity gives

$$||v - y||^2 = ||(v - x) + (x - y)||^2 = ||v - x||^2 + ||x - y||^2.$$

Now $||x - y||^2 = \langle x - y, x - y \rangle \geq 0$, with equality iff $x - y = 0$. So $||v - y||^2 \geq ||v - x||^2$ for all $y \in U$, with equality iff $y = x$. So x as defined here is the unique vector in U with minimum distance to v. $\qquad\square$

The preceding theorem shows that any $v \in V$ can be decomposed into a sum $v = x + (v - x)$, where $x = \sum_{r=1}^{k} \langle v, u_r \rangle u_r \in U$ and $v - x$ is orthogonal to every vector in U. Observe that v is in the subspace U spanned by $(u_1, \ldots, u_r)$ iff v itself is the unique vector in U with minimum distance to v iff $x = v$ iff $v - x = 0$ iff $v = \sum_{r=1}^{k} \langle v, u_r \rangle u_r$. This gives an algorithm for detecting when a given $v \in V$ is a linear combination of the orthonormal list $(u_1, \ldots, u_k)$ and finding the coefficients of the linear combination when it exists. For example, let $u_1 = \frac{1}{2}(1, 1, 1, 1)$, $u_2 = \frac{1}{2}(1, -1, 1, -1)$, and $u_3 = \frac{1}{2}(-1, -1, 1, 1)$ in $V = \mathbb{C}^4$. You can check that (u_1, u_2, u_3) is an orthonormal list. Given $v = (-5, 1, 3, 9)$, we compute $\langle v, u_1 \rangle = 4$, $\langle v, u_2 \rangle = -6$, and $\langle v, u_3 \rangle = 8$. Then $x = 4u_1 - 6u_2 + 8u_3 = v$, so $v \in U$ and we have written v as a specific linear combination of u_1, u_2, u_3. For $v = (2, -3, 0, 1)$, we compute $\langle v, u_1 \rangle = 0$, $\langle v, u_2 \rangle = 2$, and $\langle v, u_3 \rangle = 1$. Here, $x = 0u_1 + 2u_2 + 1u_3 = \frac{1}{2}(1, -3, 3, -1)$ is the orthogonal projection of v onto U. We can see directly that $v - x = \frac{1}{2}(3, -3, -3, 3)$ is orthogonal to x, u_1, u_2, u_3, and hence to every vector in U. Although $v - x$ is not a unit vector, we can change it into one by dividing by its length, obtaining $u_4 = \frac{1}{2}(1, -1, -1, 1)$. We now have a longer orthonormal list (u_1, u_2, u_3, u_4), which (being linearly independent) must be a basis of $\mathbb{C}^4$. This illustrates the basic step in the Gram–Schmidt Orthonormalization Algorithm, which we describe in the next section.

9.2 Gram–Schmidt Orthonormalization Algorithm

Given a complex inner product space V, a finite-dimensional subspace U, and a vector $v \in V$, our test for determining whether v is in the subspace U requires us to know an orthonormal basis of U. This raises the question of how to convert an arbitrary basis for a subspace U into an orthonormal basis of U.

Let $(v_1, \ldots, v_k)$ be any linearly independent list of vectors in V. We describe a process, called the *Gram–Schmidt Orthonormalization Algorithm*, that produces an orthonormal list $(u_1, \ldots, u_k)$ such that for all r between 1 and k, the sublist $(v_1, \ldots, v_r)$ spans the same subspace as the sublist $(u_1, \ldots, u_r)$. The algorithm executes the following steps for $s = 1, 2, \ldots, k$ in this order. Upon starting step s, vectors $u_1, \ldots, u_{s-1}$ have already been found. Calculate $x_s = \sum_{r=1}^{s-1} \langle v_s, u_r \rangle u_r$ (taking $x_1 = 0$), and set $u_s = (v_s - x_s)/||v_s - x_s||$. This algorithm computes the next vector u_s by subtracting from v_s its orthogonal projection x_s onto the span of the preceding vectors, then normalizing to get a unit vector.

To see that the algorithm has the required properties, we prove by induction on r that $(u_1, \ldots, u_r)$ is an orthonormal basis for the subspace V_r spanned by $(v_1, \ldots, v_r)$, for all r between 1 and k. In the base case ($r = 1$), $x_1 = 0$ and $v_1 \neq 0$, so $u_1 = v_1/||v_1||$ is a well-defined unit vector. Because u_1 is a nonzero scalar multiple of v_1, (u_1) is a basis of V_1.

For the induction step, fix r with $1 < r \leq k$, and assume $(u_1, \ldots, u_{r-1})$ is an orthonormal basis of V_{r-1}. On one hand, since $(v_1, \ldots, v_r)$ is linearly independent by hypothesis, v_r is not in the subspace V_{r-1} spanned by $(v_1, \ldots, v_{r-1})$. Since $(u_1, \ldots, u_{r-1})$ spans this subspace by assumption, the orthogonal projection x_r is not equal to v_r. Thus $v_r - x_r \neq 0$, and hence we may divide this vector by its length to get a unit vector u_r. On the other hand, we proved earlier that $v_r - x_r$ is orthogonal to every vector in the space V_{r-1}, and the same is true of the scalar multiple u_r of $v_r - x_r$. In particular, u_r is orthogonal to $u_1, \ldots, u_{r-1}$, proving orthonormality (and hence linear independence) of the list $(u_1, \ldots, u_r)$. Finally, since x_r is in V_{r-1} and $u_r = ||v_r - x_r||^{-1}(v_r - x_r)$, we see that u_r is in the subspace V_r spanned by $(v_1, \ldots, v_r)$. Since the r linearly independent vectors $u_1, \ldots, u_r$ all belong to the r-dimensional subspace V_r, these vectors must be an orthonormal basis for this subspace, completing the induction step.

For example, suppose we are given the linearly independent vectors

$$v_1 = (2, 2, 2, 2), \quad v_2 = (4, -1, 4, -1), \quad v_3 = (-4, -2, 0, 2), \quad v_4 = (2, -3, 0, 1)$$

in $V = \mathbb{C}^4$. We execute the Gram–Schmidt algorithm as follows:

$$x_1 = 0, \quad u_1 = v_1/||v_1|| = \frac{1}{2}(1, 1, 1, 1);$$

$$x_2 = \langle v_2, u_1 \rangle u_1 = 3u_1 = \frac{1}{2}(3, 3, 3, 3), \quad v_2 - x_2 = \frac{1}{2}(5, -5, 5, -5), \quad u_2 = \frac{1}{2}(1, -1, 1, -1);$$

$$x_3 = \langle v_3, u_1 \rangle u_1 + \langle v_3, u_2 \rangle u_2 = -2u_1 - 2u_2 = (-2, 0, -2, 0),$$

$$v_3 - x_3 = (-2, -2, 2, 2), \quad u_3 = \frac{1}{2}(-1, -1, 1, 1);$$

and (see the example at the end of §9.1)

$$x_4 = 0u_1 + 2u_2 + 1u_3, \quad v_4 - x_4 = \frac{1}{2}(3, -3, -3, 3), \quad u_4 = \frac{1}{2}(1, -1, -1, 1).$$

The output is the orthonormal basis (u_1, u_2, u_3, u_4) of $\mathbb{C}^4$.

We close with a few remarks about the Gram–Schmidt orthonormalization process. First, this algorithm provides a constructive proof of the theorem that *every subspace of a finite-dimensional complex inner product space has an orthonormal basis*. Second, we can use the algorithm to see that *every orthonormal list in a finite-dimensional inner product space V can be extended to an orthonormal ordered basis of V* (Exercise 15). Third, the algorithm can be applied to real inner product spaces by restricting to real scalars. Fourth, the formulas in the algorithm remain valid for an infinite-dimensional complex inner product space. In this setting, the formulas transform a countably infinite linearly independent sequence $(v_1, v_2, \ldots, v_n, \ldots)$ into an infinite orthonormal sequence $(u_1, u_2, \ldots, u_n, \ldots)$ such that $(v_1, \ldots, v_n)$ and $(u_1, \ldots, u_n)$ span the same subspace for all positive integers n. However, calling this process an "algorithm" is misleading, since the calculation does not terminate in finitely many steps.

Finally, we can apply the algorithm to a list of vectors $(v_1, \ldots, v_k)$ that may be linearly dependent. If the list is linearly dependent and r is minimal such that v_r is a linear combination of $v_1, \ldots, v_{r-1}$, then the algorithm detects this by computing that the orthogonal projection x_r is equal to v_r, hence $v_r - x_r = 0$. In this case, we leave u_r undefined, discard v_r, and continue processing v_{r+1}, v_{r+2}, and so on. If further linear dependencies exist, they are detected as they arise. In the end, the algorithm tells us exactly which vectors v_r depend linearly on preceding vectors and provides an orthonormal list (of length at most k) that spans the same subspace as the list $(v_1, \ldots, v_k)$.

9.3   Gram–Schmidt QR Factorization

We now recast our results on the Gram–Schmidt Orthonormalization Algorithm as a theorem about matrix factorizations.

Theorem on QR Factorizations (Full Rank Case). Suppose a matrix $A \in M_{n,k}(\mathbb{C})$ has k linearly independent columns. There exist unique matrices $Q \in M_{n,k}(\mathbb{C})$ and $R \in M_k(\mathbb{C})$ such that $A = QR$, Q has orthonormal columns, and R is upper-triangular with all diagonal entries in $\mathbb{R}_{>0}$. When A has real entries, Q and R are real as well.

Proof. To prove existence of the factorization, we consider the complex inner product space $\mathbb{C}^n$ (viewed as a set of column vectors) and look at the linearly independent list of vectors $(v_1, \ldots, v_k)$, where $v_j = A^{[j]}$ is the jth column of the given matrix A. The Gram–Schmidt Orthonormalization Algorithm produces an orthonormal list $(u_1, \ldots, u_k)$ such that, for $1 \leq j \leq k$, the subspace V_j spanned by $(v_1, \ldots, v_j)$ equals the subspace spanned by $(u_1, \ldots, u_j)$. Let $Q \in M_{n,k}(\mathbb{C})$ be the matrix with columns $Q^{[j]} = u_j \in \mathbb{C}^n$ for $1 \leq j \leq k$. Define an upper-triangular matrix $R \in M_k(\mathbb{C})$ by setting $R(i,j) = \langle v_j, u_i \rangle$ for $1 \leq i \leq j \leq k$, and $R(i,j) = 0$ for all $i > j$. Note that Q and R have real entries if A does.

To check that $A = QR$, we use facts established in §4.7 and §9.1. First, the jth column of QR is $Q(R^{[j]})$. Second, $Q(R^{[j]})$ is a linear combination of the columns of Q with coefficients given by the entries of $R^{[j]}$. Specifically, the jth column of QR is $R(1,j)Q^{[1]} + R(2,j)Q^{[2]} + \cdots + R(k,j)Q^{[k]}$. Using the definitions of Q and R, the jth column of QR is

$$\langle v_j, u_1 \rangle u_1 + \langle v_j, u_2 \rangle u_2 + \cdots + \langle v_j, u_j \rangle u_j.$$

Since v_j is in V_j, which is the span of $(u_1, \ldots, u_j)$, the linear combination just written must be equal to v_j (see §9.1). This means that the jth column of QR is $v_j = A^{[j]}$ for all j, so $A = QR$ as needed.

We see that the diagonal entries of R are strictly positive real numbers as follows. Recall from the description of the orthonormalization algorithm that $u_j = (v_j - x_j)/||v_j - x_j||$, where x_j is the orthogonal projection of v_j onto the subspace V_{j-1}. We know that $v_j - x_j$ is orthogonal to everything in V_{j-1}, hence is orthogonal to x_j. Then $||v_j - x_j||^2 = \langle v_j - x_j, v_j - x_j \rangle = \langle v_j, v_j - x_j \rangle - \langle x_j, v_j - x_j \rangle = \langle v_j, v_j - x_j \rangle$. So, the diagonal entry $R(j,j)$ is $\langle v_j, u_j \rangle = \langle v_j, v_j - x_j \rangle / ||v_j - x_j|| = \frac{||v_j - x_j||^2}{||v_j - x_j||} = ||v_j - x_j|| \in \mathbb{R}_{>0}$.

Before discussing uniqueness, we consider an example. Suppose $A = \begin{bmatrix} 2 & 4 & -4 \\ 2 & -1 & -2 \\ 2 & 4 & 0 \\ 2 & -1 & 2 \end{bmatrix}$.

By the calculations in §9.2 and the definitions above, $A = QR$ holds for

$$Q = \begin{bmatrix} 1/2 & 1/2 & -1/2 \\ 1/2 & -1/2 & -1/2 \\ 1/2 & 1/2 & 1/2 \\ 1/2 & -1/2 & 1/2 \end{bmatrix}, \qquad R = \begin{bmatrix} 4 & 3 & -2 \\ 0 & 5 & -2 \\ 0 & 0 & 4 \end{bmatrix}.$$

Returning to the general case, we now prove uniqueness of Q and R. Assume $A = QR = Q_1 R_1$, where $Q, Q_1 \in M_{n,k}(\mathbb{C})$ both have orthonormal columns and $R, R_1 \in M_k(\mathbb{C})$ are both upper-triangular with all diagonal entries in $\mathbb{R}_{>0}$. Let the columns of A be $v_1, \ldots, v_k$, the columns of Q be $u_1, \ldots, u_k$, and the columns of Q_1 be $z_1, \ldots, z_k$. We show by strong induction on j that $u_j = z_j$ and $R^{[j]} = R_1^{[j]}$ for $1 \leq j \leq k$. Fix j in this range, and assume $u_s = z_s$ and $R^{[s]} = R_1^{[s]}$ is already known for all s with $1 \le s < j$. On one hand, consideration of the jth column of $A = QR$ shows (as in the existence proof) that

$$v_j = R(1,j)u_1 + \cdots + R(j-1,j)u_{j-1} + R(j,j)u_j.$$

Taking the inner product of each side with u_i, we see that $R(i,j) = \langle v_j, u_i \rangle$ for $1 \le i \le j$. Since R is upper-triangular, $R(i,j) = 0$ for $j < i \le k$.

On the other hand, looking at the jth column of $A = Q_1 R_1$ gives

$$
\begin{aligned}
v_j &= R_1(1,j)z_1 + \cdots + R_1(j-1,j)z_{j-1} + R_1(j,j)z_j \\
&= R_1(1,j)u_1 + \cdots + R_1(j-1,j)u_{j-1} + R_1(j,j)z_j.
\end{aligned}
$$

Since the list $(z_1, \ldots, z_{j-1}, z_j) = (u_1, \ldots, u_{j-1}, z_j)$ is orthonormal, taking the inner product of each side with u_i gives $R_1(i,j) = \langle v_j, u_i \rangle = R(i,j)$ for $1 \le i < j$. Since R_1 is upper-triangular, $R_1(i,j) = 0$ for $j < i \le k$. Now, equating the two expressions for v_j and cancelling $\sum_{i=1}^{j-1} \langle v_j, u_i \rangle u_i$, we deduce that $R(j,j)u_j = R_1(j,j)z_j$. Taking the length of both sides gives $|R(j,j)| = |R_1(j,j)|$. Since both $R(j,j)$ and $R_1(j,j)$ are positive real numbers, we conclude finally that $R(j,j) = R_1(j,j)$ and $u_j = z_j$. $\qquad\square$

We can extend the QR factorization to arbitrary matrices, as follows. Define a rectangular matrix R to be *upper-triangular* iff $R(i,j) = 0$ for all $i > j$.

Theorem on QR Factorizations (General Case). Suppose a matrix $A \in M_{n,k}(\mathbb{C})$ has column rank r. There exist matrices $Q \in M_{n,r}(\mathbb{C})$ and $R \in M_{r,k}(\mathbb{C})$ such that $A = QR$, Q has orthonormal columns, and R is upper-triangular with all diagonal entries in $\mathbb{R}_{\ge 0}$.

Proof. Let the columns of A be $v_1, \ldots, v_k$. Let the columns of Q be the orthonormal list $(u_1, \ldots, u_r)$ produced when the Gram–Schmidt algorithm is applied to the list $(v_1, \ldots, v_k)$. We have $r < k$ if some vector v_j is a linear combination of previous vectors v_i. Let $R(i,j) = \langle v_j, u_i \rangle$ for $1 \le i \le j \le k$, and $R(i,j) = 0$ for $r \ge i > j \ge 1$. As in the earlier existence proof, $A = QR$ follows from the fact that each v_j is in the span of the orthonormal list $(u_1, \ldots, u_{\min(j,r)})$. If some v_j is in the span of $(u_1, \ldots, u_{j-1})$, then $R(j,j)$ is 0. Otherwise, $R(j,j) \in \mathbb{R}_{>0}$ by the reasoning used before. $\qquad\square$

By deleting the columns of A corresponding to those v_j that depend linearly on earlier v_i, and deleting the same columns of R, we obtain a factorization $A' = QR'$ of the type initially discussed, with $A' \in M_{n,r}(\mathbb{C})$ having full column rank and $R' \in M_{r,r}(\mathbb{C})$ being square and upper-triangular with all diagonal entries strictly positive.

9.4 Householder Reflections

Our next goal is to derive a version of the QR factorization in which an arbitrary matrix $A \in M_{n,k}(\mathbb{C})$ is factored as $A = QR$, where $Q \in M_n(\mathbb{C})$ is a unitary matrix and $R \in M_{n,k}(\mathbb{C})$ is an upper-triangular matrix. We can obtain such a factorization by modifying the proof given earlier based on Gram–Schmidt orthonormalization (Exercise 19). This section and the next one describe Householder's Algorithm for reaching this factorization, which has certain computational advantages over algorithms using the Gram–Schmidt Algorithm.

The key idea is that we can apply a sequence of reflections to transform A into an upper-triangular matrix by forcing the required zeroes to appear, one column at a time. In $\mathbb{R}^n$, a *reflection* is a linear map that sends a given nonzero vector v to $-v$ and sends w to

w for each vector w perpendicular to v. The subspace of all such vectors w, which is the orthogonal complement of $\{v\}$, forms the mirror through which v is being reflected.

Let us define reflections formally in the complex inner product space $\mathbb{C}^n$ with the standard inner product $\langle x, y \rangle = \sum_{k=1}^{n} x_k \overline{y_k} = y^* x$ for $x, y \in \mathbb{C}^n$. Fix a nonzero vector $v \in \mathbb{C}^n$. Starting with any basis $(v, v_2, \ldots, v_n)$ of $\mathbb{C}^n$, use the Gram–Schmidt Algorithm to obtain an orthonormal ordered basis $B = (u_1, u_2, \ldots, u_n)$ of $\mathbb{C}^n$, where $u_1 = v/\|v\|$. There is a unique linear map $T_v : \mathbb{C}^n \to \mathbb{C}^n$ defined by setting

$$T_v(c_1 u_1 + c_2 u_2 + \cdots + c_n u_n) = -c_1 u_1 + c_2 u_2 + \cdots + c_n u_n \tag{9.1}$$

for all $c_1, \ldots, c_n \in \mathbb{C}$. Observe that T_v sends any scalar multiple of u_1 to its negative, whereas T_v sends any linear combination of $u_2, \ldots, u_n$ to itself. In particular, $T_v(v) = -v$ and $T_v(w) = w$ for all w orthogonal to v. These formulas uniquely determine the linear map T_v and show that T_v does not depend on the choice of the orthonormal basis B. The matrix of T_v relative to the ordered basis B is diagonal with diagonal entries $-1, 1, \ldots, 1$. This matrix is unitary and equal to its own inverse, so $\|T_v(z)\| = \|z\|$ for all $z \in \mathbb{C}^n$, and $T_v^{-1} = T_v$.

Define a matrix $Q_v = I_n - (2/\|v\|^2) v v^* \in M_n(\mathbb{C})$. We claim that $T_v(y) = Q_v y$ for all $y \in \mathbb{C}^n$. By linearity, it suffices to verify this for all $y \in \{v, u_2, \ldots, u_n\}$. For $y = v$, recall that $\|v\|^2 = \langle v, v \rangle = v^* v$ and compute

$$Q_v v = I_n v - \left[\frac{2}{\|v\|^2} v v^* \right] v = v - \frac{2}{v^* v} v (v^* v) = v - 2v = -v = T_v(v).$$

For $y = u_j$ with $2 \le j \le n$, note $v^* u_j = \langle u_j, v \rangle = \langle u_j, \|v\| u_1 \rangle = 0$, so

$$Q_v u_j = I_n u_j - \left[\frac{2}{\|v\|^2} v v^* \right] u_j = u_j - \frac{2}{\|v\|^2} v (v^* u_j) = u_j = T_v(u_j).$$

So the claim holds. We call T_v the *Householder transformation determined by v*, and we call Q_v the *Householder matrix determined by v*. We also refer to T_v and Q_v as *Householder reflections* to emphasize the geometric character of these maps and matrices.

You can check from the explicit formula that Q_v is unitary ($Q_v^* Q_v = Q_v Q_v^* = I_n$) and self-inverse ($Q_v^{-1} = Q_v$); these facts also follow from the corresponding properties of T_v or $[T_v]_B$. It is also routine to confirm that for any nonzero scalar $c \in \mathbb{C}$, $T_{cv} = T_v$ and $Q_{cv} = Q_v$. Note that if v is a unit vector, the formula for Q_v simplifies to $Q_v = I_n - 2vv^*$, and we can always arrange this by replacing v by an appropriate scalar multiple of itself. We also define $T_0 = \mathrm{id}_{\mathbb{C}^n}$ and $Q_0 = I_n$.

The following lemma shows how to find Householder reflections sending a given vector x to certain other vectors having the same length as x.

Lemma on Householder Reflections. For any $x, y \in \mathbb{C}^n$ with $x \ne y$ and $\|x\| = \|y\|$ and $\langle x, y \rangle \in \mathbb{R}$, there exists $v \in \mathbb{C}^n$ with $Q_v x = y$. In fact, $Q_v x = y$ iff v is a nonzero scalar multiple of $x - y$.

Proof. Fix $x \ne y$ in $\mathbb{C}^n$ with $\|x\| = \|y\|$ and $\langle x, y \rangle \in \mathbb{R}$. Define $v = x - y$ and $w = x + y$; the geometric motivation for choosing v and w in this way is suggested in Figure 9.1. The hypothesis $\langle x, y \rangle \in \mathbb{R}$ implies $\langle x, y \rangle = \overline{\langle x, y \rangle} = \langle y, x \rangle$, so that

$$\langle v, w \rangle = \langle x - y, x + y \rangle = \langle x, x \rangle - \langle y, x \rangle + \langle x, y \rangle - \langle y, y \rangle = \|x\|^2 - \|y\|^2 = 0.$$

Since w is orthogonal to v, $T_v(w) = w$. Now $x = (x - y)/2 + (x + y)/2 = v/2 + w/2$, so

$$T_v(x) = T_v(v/2 + w/2) = \frac{1}{2} T_v(v) + \frac{1}{2} T_v(w) = (-v)/2 + w/2 = (y - x)/2 + (x + y)/2 = y$$

(cf. Figure 9.1). Thus, $Q_v x = y$, and hence $Q_{cv} x = y$ for any nonzero $c \in \mathbb{C}$.

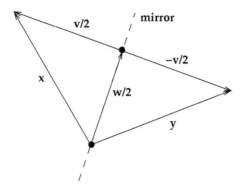

FIGURE 9.1
Finding a reflection that sends x to y.

To prove the converse, assume $z \in \mathbb{C}^n$ satisfies $T_z(x) = y$. Then $T_z(y) = T_z^{-1}(y) = x$, so $T_z(v) = T_z(x - y) = T_z(x) - T_z(y) = y - x = -v$. From the defining formula for T_z (see (9.1) with v there replaced by z), we find that the only vectors sent to their negatives by T_z are scalar multiples of z. So v is a nonzero scalar multiple of z and vice versa. $\square$

Observe that the v specified in the lemma is not unique even if we insist that v be a unit vector, since we can still replace v by $e^{i\theta}v$ for any $\theta \in [0, 2\pi)$. In order for v to exist, we must assume $||x|| = ||y||$ and $\langle x, y \rangle \in \mathbb{R}$ (see Exercise 27).

9.5 Householder QR Factorization

Given $A \in M_{n,k}(\mathbb{C})$, *Householder's QR algorithm* finds a factorization $A = QR$, where $Q \in M_n(\mathbb{C})$ is unitary and $R \in M_{n,k}(\mathbb{C})$ is upper-triangular with nonnegative diagonal entries, by determining a sequence of Householder reflections and a diagonal unitary matrix whose product is Q. The main idea is to use the lemma of §9.4 repeatedly to create the zero entries in R one column at a time.

To begin, let $x = A^{[1]} \in \mathbb{C}^n$ be the first column of A. Write $x(1) = A(1,1) = se^{i\theta}$ for some nonnegative real s and $\theta \in [0, 2\pi)$. Define $y = -||x||e^{i\theta}\mathbf{e}_1 \in \mathbb{C}^n$, where $\mathbf{e}_1$ is the standard basis vector $(1, 0, \ldots, 0)$. Note that $||y|| = ||x||$ and $\langle x, y \rangle = -se^{i\theta}||x||e^{-i\theta}$ is real. (Exercise 34 discusses why the negative sign is present in the definition of y.) Let $v_1 = x - y$, and consider the matrix $Q_{v_1}A$. The first column of this matrix is $Q_{v_1}x = y$, which is a multiple of $\mathbf{e}_1$ and hence has zeroes in rows 2 through n. Since $Q_0 = I_n$, this conclusion holds even if $x = 0$ or $x = y$.

At the next stage, let A_2 be the submatrix of $Q_{v_1}A$ obtained by deleting the first row and column. Repeat the construction in the previous paragraph, taking $x \in \mathbb{C}^{n-1}$ to be the first column of A_2, to obtain $v_2' \in \mathbb{C}^{n-1}$ such that $Q_{v_2'}A_2 \in M_{n-1,k-1}(\mathbb{C})$ has zeroes in the first column below row 1. Construct $v_2 \in \mathbb{C}^n$ by preceding v_2' with a zero. The first row and column of $v_2v_2^*$ contain all zeroes, so Q_{v_2} is a block-diagonal matrix with diagonal blocks I_1 and $Q_{v_2'}$. It follows that $Q_{v_2}Q_{v_1}A$ has all zeroes below the diagonal in the first two columns.

We continue similarly to process the remaining columns. After j steps, we have found a matrix $Q_{v_j} \cdots Q_{v_1}A$ with zeroes below the diagonal in the first j columns. Repeat the construction in step 1, taking $x \in \mathbb{C}^{n-j}$ to be the first column of the lower-right $(n-j) \times (k-j)$ submatrix of the current matrix. We thereby obtain $v_{j+1}' \in \mathbb{C}^{n-j}$ such that $Q_{v'_{j+1}}$ sends x to a multiple of $\mathbf{e}_1 \in \mathbb{C}^{n-j}$. Let $v_{j+1} \in \mathbb{C}^n$ be v'_{j+1} preceded by j zeroes, so that $Q_{v_{j+1}}$ is block-diagonal with diagonal blocks I_j and $Q_{v'_{j+1}}$. Then $Q_{v_{j+1}}Q_{v_j} \cdots Q_{v_1} A$ has all zeroes below the diagonal in the first $j+1$ columns.

After $s = \min(n-1, k)$ steps, we have a relation $Q_{v_s} \cdots Q_{v_1} A = R'$, where $R' \in M_{n,k}(\mathbb{C})$ is upper-triangular. Left-multiplying both sides by an appropriate diagonal matrix D with all diagonal entries of the form $e^{i\theta}$, we can obtain a factorization $DQ_{v_s} \cdots Q_{v_1} A = R$ where R is upper-triangular with nonnegative real diagonal entries. Solving for A and recalling that $Q_{v_j}^{-1} = Q_{v_j}$ for all j, we get $A = QR$ with $Q = Q_{v_1} Q_{v_2} \cdots Q_{v_s} D^{-1}$. The matrix Q is unitary, being a product of unitary matrices.

We illustrate Householder's QR algorithm on the matrix

$$A = \begin{bmatrix} -5 & 3 & -9 & 1 \\ -4 & 1 & 3 & 12 \\ 8 & 3 & -2 & 0 \\ 4 & 9 & 1 & -4 \end{bmatrix}.$$

In the first step, $x = (-5, -4, 8, 4)^{\mathrm{T}}$, $||x|| = 11$, $x(1) = -5 = 5e^{i\pi}$, so $y = -||x||e^{i\pi}\mathbf{e}_1 = (11, 0, 0, 0)$. Using $v_1 = x - y = (-16, -4, 8, 4)^{\mathrm{T}}$, we find that

$$Q_{v_1} = \frac{1}{11}\begin{bmatrix} -5 & -4 & 8 & 4 \\ -4 & 10 & 2 & 1 \\ 8 & 2 & 7 & -2 \\ 4 & 1 & -2 & 10 \end{bmatrix}, \quad Q_{v_1}A = \frac{1}{11}\begin{bmatrix} 121 & 41 & 21 & -69 \\ 0 & 13 & 63 & 112 \\ 0 & 29 & -82 & 40 \\ 0 & 97 & -19 & -24 \end{bmatrix}.$$

In the second step, we look at the lower-right 3×3 block and take $x = (13/11, 29/11, 97/11)^{\mathrm{T}}$, $||x|| = 9.27941$, $x(1) = (13/11)e^{i0}$, so $y = (-9.27941, 0, 0)^{\mathrm{T}}$. Taking $v'_2 = x - y$ gives

$$v_2 = (0, 10.4612, 2.63636, 8.81818)^{\mathrm{T}},$$

$$Q_{v_2} = \begin{bmatrix} 1 & 0 & 0 & 0 \\ 0 & -0.127 & -0.284 & -0.950 \\ 0 & -0.284 & 0.928 & -0.239 \\ 0 & -0.950 & -0.239 & 0.199 \end{bmatrix}, \quad Q_{v_2}Q_{v_1}A = \begin{bmatrix} 11 & 3.727 & 1.909 & -6.273 \\ 0 & -9.279 & 3.030 & -0.257 \\ 0 & 0 & -8.134 & 1.006 \\ 0 & 0 & -4.001 & -10.981 \end{bmatrix}.$$

In the third step, $x = (-8.134, -4.001)^{\mathrm{T}}$, $||x|| = 9.065$, $y = (9.065, 0)^{\mathrm{T}}$, and $v_3 = (0, 0, -17.199, -4.001)^{\mathrm{T}}$, so

$$Q_{v_3} = \begin{bmatrix} 1 & 0 & 0 & 0 \\ 0 & 1 & 0 & 0 \\ 0 & 0 & -0.897 & -0.441 \\ 0 & 0 & -0.441 & 0.897 \end{bmatrix}, \quad Q_{v_3}Q_{v_2}Q_{v_1}A = \begin{bmatrix} 11 & 3.727 & 1.909 & -6.273 \\ 0 & -9.279 & 3.030 & -0.257 \\ 0 & 0 & 9.065 & 3.944 \\ 0 & 0 & 0 & -10.297 \end{bmatrix}.$$

Finally, let D be diagonal with diagonal entries $1, -1, 1, -1$ and $Q = Q_{v_1} Q_{v_2} Q_{v_3} D^{-1}$. Then $A = QR$ holds with

$$Q = \begin{bmatrix} -0.455 & 0.506 & -0.728 & 0.086 \\ -0.364 & 0.254 & 0.492 & 0.749 \\ 0.727 & 0.031 & -0.363 & 0.581 \\ 0.363 & 0.824 & 0.309 & -0.306 \end{bmatrix}, \quad R = \begin{bmatrix} 11 & 3.727 & 1.909 & -6.273 \\ 0 & 9.279 & -3.030 & 0.257 \\ 0 & 0 & 9.065 & 3.944 \\ 0 & 0 & 0 & 10.297 \end{bmatrix}.$$

$$(9.2)$$

9.6 LU Factorization

In the next few sections, we discuss decompositions of a square matrix into the product of a lower-triangular matrix and an upper-triangular matrix. In order to assure the existence and uniqueness of such a decomposition, we must impose certain hypotheses on the given matrix A. For any field F and $A \in M_n(F)$, let $A[k]$ denote the matrix in $M_k(F)$ consisting of the first k rows and columns of A.

LU Factorization Theorem. For all $A \in M_n(F)$, there exist a lower-unitriangular matrix $L \in M_n(F)$ and an invertible upper-triangular matrix $U \in M_n(F)$ such that $A = LU$ iff for all k between 1 and n, $\det(A[k]) \neq 0_F$. L and U are unique if they exist.

Proof. On one hand, assume $A = LU$ for some matrices L and U with the stated properties. In particular, $0 \neq \det(U) = \prod_{i=1}^{n} U(i,i)$. Fix k between 1 and n. By writing $A = LU$ as a product of block matrices

$$\left[\begin{array}{c|c} A[k] & B \\ \hline C & D \end{array}\right] = \left[\begin{array}{c|c} L[k] & 0 \\ \hline M & N \end{array}\right] \left[\begin{array}{c|c} U[k] & V \\ \hline 0 & W \end{array}\right], \tag{9.3}$$

we see that $A[k] = L[k]U[k]$, where $L[k]$ is lower-unitriangular and $U[k]$ is upper-triangular. Taking determinants gives $\det(A[k]) = \det(L[k])\det(U[k]) = \prod_{i=1}^{k} L(i,i) \prod_{i=1}^{k} U(i,i) \neq 0$. So the determinant condition on A is necessary for the LU factorization to exist.

On the other hand, to prove uniqueness, assume $A = LU = L_1 U_1$ where both L, U and L_1, U_1 have the properties stated in the theorem. Since L_1 and U are invertible, we can write $L_1^{-1}L = U_1 U^{-1}$. The matrix $L_1^{-1}L$ is lower-unitriangular, and the matrix $U_1 U^{-1}$ is upper-triangular. These two matrices are equal, so $L_1^{-1}L$ is upper-triangular and lower-unitriangular. The only such matrix is I_n, which implies $L = L_1$ and $U = U_1$.

Next, we assume $\det(A[k]) \neq 0$ for $1 \leq k \leq n$ and prove existence of the factorization $A = LU$ by the following recursive computation. Define $L(i,i) = 1_F$ for $i \in [n]$ and $L(i,j) = U(j,i) = 0_F$ for $1 \leq i < j \leq n$. We determine the remaining entries of L one column at a time, and we simultaneously determine the remaining entries of U one row at a time. Fix k between 1 and n, and make the induction assumption that we have already determined all entries in the first $k-1$ columns of L and all entries in the first $k-1$ rows of U. Assume further that $A(i,j) = (LU)(i,j)$ holds for all $i, j \in [n]$ such that $i \leq k-1$ or $j \leq k-1$. For the base case $k = 1$, these assumptions are vacuously true.

To continue, we first recover $U(k,k)$ as follows. Comparing the k,k-entries of A and LU, we see that these entries are equal iff

$$A(k,k) = \sum_{r=1}^{n} L(k,r)U(r,k) = \sum_{r=1}^{k-1} L(k,r)U(r,k) + L(k,k)U(k,k). \tag{9.4}$$

By hypothesis, every entry in the sum from $r = 1$ to $k-1$ is already known. Since we chose $L(k,k) = 1$, equation (9.4) holds iff we define

$$U(k,k) = A(k,k) - \sum_{r=1}^{k-1} L(k,r)U(r,k). \tag{9.5}$$

We now know (among other things) that $A(i, j) = (LU)(i, j)$ for all $i, j \in [k]$ and that the zero blocks displayed in (9.3) are present in L and U. So the equation $A[k] = L[k]U[k]$ is valid. Taking determinants shows that $0 \neq \det(A[k]) = \det(L[k]) \det(U[k]) = \prod_{i=1}^{k} U(i, i)$. Since the field F has no zero divisors, we conclude that $U(k, k) \neq 0_F$.

The next step is to compute the unknown entries in column k of L, which are the entries $L(i, k)$ as i ranges from $k + 1$ to n. Comparing the i, k-entries of A and LU for each i in this range, we see that these entries agree iff

$$A(i, k) = \sum_{r=1}^{n} L(i, r)U(r, k) = \sum_{r=1}^{k-1} L(i, r)U(r, k) + L(i, k)U(k, k).$$

The entries $L(i, r)$ and $U(r, k)$ for $1 \leq r \leq k - 1$ are already known, and we have seen that $U(k, k) \neq 0_F$. Thus, $A(i, k) = (LU)(i, k)$ holds iff we define

$$L(i, k) = U(k, k)^{-1} \left[A(i, k) - \sum_{r=1}^{k-1} L(i, r)U(r, k) \right]. \tag{9.6}$$

We now know all of column k of L, and we have ensured that $A(i, j) = (LU)(i, j)$ whenever $i < k$ or $j \leq k$.

The last step is to compute the unknown entries in row k of U, which are the entries $U(k, j)$ as j ranges from $k + 1$ to n. Comparing the k, j-entries of A and LU for each j in this range, we see that these entries agree iff

$$A(k, j) = \sum_{r=1}^{n} L(k, r)U(r, j) = \sum_{r=1}^{k-1} L(k, r)U(r, j) + L(k, k)U(k, j).$$

The entries $L(k, r)$ and $U(r, j)$ for $1 \leq r \leq k - 1$ are already known, and we chose $L(k, k) = 1_F$. So $A(k, j) = (LU)(k, j)$ holds iff we define

$$U(k, j) = A(k, j) - \sum_{r=1}^{k-1} L(k, r)U(r, j). \tag{9.7}$$

We now know all of row k of U, and we have ensured that $A(i, j) = (LU)(i, j)$ whenever $i \leq k$ or $j \leq k$. This completes step k of the induction, and the existence proof is now finished. □

The factorization $A = LU$ just proved, where L is lower-unitriangular and U is upper-triangular and invertible, is sometimes called the *Doolittle factorization* of A. Some variants of this factorization can be found by changing the conditions on the diagonal entries of L and U. For example, assuming $\det(A[k]) \neq 0$ for $1 \leq k \leq n$, there exist unique L, D, U in $M_n(F)$ such that L is lower-unitriangular, D is diagonal and invertible, and U is upper-unitriangular. With the same assumption on A, there exist unique $L, U \in M_n(F)$ such that L is lower-triangular and invertible, and U is upper-unitriangular; this is called the *Crout factorization* of A. In the field $F = \mathbb{C}$ (or any field where every element has a square root), we can write $A = LU$ with $L(k, k) = U(k, k)$ for all k by setting $L(k, k) = U(k, k)$ in (9.4) and solving for $U(k, k)$. We ask the reader to derive these variant factorizations in Exercise 37.

9.7 Example of the LU Factorization

We illustrate the computation of an LU factorization using the matrix

$$A = \begin{bmatrix} 3 & 1 & -5 & 2 \\ -9 & -1 & 14 & -5 \\ 15 & 9 & -23 & 12 \\ 3 & 9 & -9 & 7 \end{bmatrix},$$

which does satisfy the determinant condition $\det(A[k]) \neq 0$ for $1 \leq k \leq 4$. Initially, the unknown factors L and U look like this:

$$L = \begin{bmatrix} 1 & 0 & 0 & 0 \\ a & 1 & 0 & 0 \\ b & d & 1 & 0 \\ c & e & f & 1 \end{bmatrix}, \quad U = \begin{bmatrix} q & r & s & t \\ 0 & u & v & w \\ 0 & 0 & x & y \\ 0 & 0 & 0 & z \end{bmatrix}.$$

Comparing the entries of the matrix equation $A = LU$ leads to the following sixteen scalar equations in the unknowns $a, b, c, \ldots, z$:

$$\begin{array}{llll}
3 = q & 1 = r & -5 = s & 2 = t \\
-9 = aq & -1 = ar + u & 14 = as + v & -5 = at + w \\
15 = bq & 9 = br + du & -23 = bs + dv + x & 12 = bt + dw + y \\
3 = cq & 9 = cr + eu & -9 = cs + ev + fx & 7 = ct + ew + fy + z
\end{array}$$

We see $q = 3$, then (working down the first column) $a = -3$, $b = 5$, $c = 1$, and (looking at the first row) $r = 1$, $s = -5$, $t = 2$. From the 2,2-entry, we now see $u = -1 - ar = 2$. Working down the second column, $9 = 5 + 2d$ gives $d = 2$, and $9 = 1 + 2e$ gives $e = 4$. Moving along the second row, $14 = 15 + v$ gives $v = -1$, and $-5 = -6 + w$ gives $w = 1$. The 3,3-entry gives $-23 = -25 - 2 + x$, so $x = 4$. Continuing similarly, we find $f = 0$, $y = 0$, and $z = 1$. So, the Doolittle version of the LU factorization is

$$A = \begin{bmatrix} 3 & 1 & -5 & 2 \\ -9 & -1 & 14 & -5 \\ 15 & 9 & -23 & 12 \\ 3 & 9 & -9 & 7 \end{bmatrix} = \begin{bmatrix} 1 & 0 & 0 & 0 \\ -3 & 1 & 0 & 0 \\ 5 & 2 & 1 & 0 \\ 1 & 4 & 0 & 1 \end{bmatrix} \cdot \begin{bmatrix} 3 & 1 & -5 & 2 \\ 0 & 2 & -1 & 1 \\ 0 & 0 & 4 & 0 \\ 0 & 0 & 0 & 1 \end{bmatrix}.$$

We can force U to be unitriangular by factoring out an appropriate diagonal matrix:

$$A = \begin{bmatrix} 1 & 0 & 0 & 0 \\ -3 & 1 & 0 & 0 \\ 5 & 2 & 1 & 0 \\ 1 & 4 & 0 & 1 \end{bmatrix} \cdot \begin{bmatrix} 3 & 0 & 0 & 0 \\ 0 & 2 & 0 & 0 \\ 0 & 0 & 4 & 0 \\ 0 & 0 & 0 & 1 \end{bmatrix} \cdot \begin{bmatrix} 1 & 1/3 & -5/3 & 2/3 \\ 0 & 1 & -1/2 & 1/2 \\ 0 & 0 & 1 & 0 \\ 0 & 0 & 0 & 1 \end{bmatrix}.$$

Then we can absorb the diagonal matrix into L to obtain the Crout factorization of A:

$$A = \begin{bmatrix} 3 & 0 & 0 & 0 \\ -9 & 2 & 0 & 0 \\ 15 & 4 & 4 & 0 \\ 3 & 8 & 0 & 1 \end{bmatrix} \cdot \begin{bmatrix} 1 & 1/3 & -5/3 & 2/3 \\ 0 & 1 & -1/2 & 1/2 \\ 0 & 0 & 1 & 0 \\ 0 & 0 & 0 & 1 \end{bmatrix}.$$

9.8 LU Factorizations and Gaussian Elimination

There is a close relationship between LU factorizations and Gaussian elimination. To see how this occurs, let us perform elementary row operations on the matrix A in §9.7 to create zeroes below the diagonal of A. The first step is to remove the -9 in the $2,1$-entry by adding 3 times row 1 of A to row 2 of A. Recall from §4.9 that this row operation can be achieved by multiplying A on the left by the elementary matrix E_1 obtained from I_4 by adding 3 times row 1 to row 2. The new matrix is

$$E_1 A = \begin{bmatrix} 1 & 0 & 0 & 0 \\ 3 & 1 & 0 & 0 \\ 0 & 0 & 1 & 0 \\ 0 & 0 & 0 & 1 \end{bmatrix} \cdot \begin{bmatrix} 3 & 1 & -5 & 2 \\ -9 & -1 & 14 & -5 \\ 15 & 9 & -23 & 12 \\ 3 & 9 & -9 & 7 \end{bmatrix} = \begin{bmatrix} 3 & 1 & -5 & 2 \\ 0 & 2 & -1 & 1 \\ 15 & 9 & -23 & 12 \\ 3 & 9 & -9 & 7 \end{bmatrix}.$$

We continue by adding -5 times row 1 to row 3 using an elementary matrix E_2, which produces a new matrix

$$E_2(E_1 A) = \begin{bmatrix} 1 & 0 & 0 & 0 \\ 0 & 1 & 0 & 0 \\ -5 & 0 & 1 & 0 \\ 0 & 0 & 0 & 1 \end{bmatrix} \cdot \begin{bmatrix} 3 & 1 & -5 & 2 \\ 0 & 2 & -1 & 1 \\ 15 & 9 & -23 & 12 \\ 3 & 9 & -9 & 7 \end{bmatrix} = \begin{bmatrix} 3 & 1 & -5 & 2 \\ 0 & 2 & -1 & 1 \\ 0 & 4 & 2 & 2 \\ 3 & 9 & -9 & 7 \end{bmatrix}.$$

Next, we add -1 times row 1 to row 4, obtaining

$$E_3(E_2 E_1 A) = \begin{bmatrix} 1 & 0 & 0 & 0 \\ 0 & 1 & 0 & 0 \\ 0 & 0 & 1 & 0 \\ -1 & 0 & 0 & 1 \end{bmatrix} \cdot \begin{bmatrix} 3 & 1 & -5 & 2 \\ 0 & 2 & -1 & 1 \\ 0 & 4 & 2 & 2 \\ 3 & 9 & -9 & 7 \end{bmatrix} = \begin{bmatrix} 3 & 1 & -5 & 2 \\ 0 & 2 & -1 & 1 \\ 0 & 4 & 2 & 2 \\ 0 & 8 & -4 & 5 \end{bmatrix}.$$

Moving to column 2, we add -2 times row 2 to row 3:

$$E_4(E_3 E_2 E_1 A) = \begin{bmatrix} 1 & 0 & 0 & 0 \\ 0 & 1 & 0 & 0 \\ 0 & -2 & 1 & 0 \\ 0 & 0 & 0 & 1 \end{bmatrix} \cdot \begin{bmatrix} 3 & 1 & -5 & 2 \\ 0 & 2 & -1 & 1 \\ 0 & 4 & 2 & 2 \\ 0 & 8 & -4 & 5 \end{bmatrix} = \begin{bmatrix} 3 & 1 & -5 & 2 \\ 0 & 2 & -1 & 1 \\ 0 & 0 & 4 & 0 \\ 0 & 8 & -4 & 5 \end{bmatrix}.$$

Then we add -4 times row 2 to row 4:

$$E_5(E_4 E_3 E_2 E_1 A) = \begin{bmatrix} 1 & 0 & 0 & 0 \\ 0 & 1 & 0 & 0 \\ 0 & 0 & 1 & 0 \\ 0 & -4 & 0 & 1 \end{bmatrix} \cdot \begin{bmatrix} 3 & 1 & -5 & 2 \\ 0 & 2 & -1 & 1 \\ 0 & 0 & 4 & 0 \\ 0 & 8 & -4 & 5 \end{bmatrix} = \begin{bmatrix} 3 & 1 & -5 & 2 \\ 0 & 2 & -1 & 1 \\ 0 & 0 & 4 & 0 \\ 0 & 0 & 0 & 1 \end{bmatrix}.$$

There is already a zero in the $4,3$-position, so we are finished.

Observe that the final matrix $(E_5 E_4 E_3 E_2 E_1)A$ is precisely the upper-triangular matrix U found in §9.7. Solving for A, we find that $A = LU$ where $L = E_1^{-1} E_2^{-1} E_3^{-1} E_4^{-1} E_5^{-1}$. The inverse of each elementary matrix E_i is the lower-triangular matrix obtained by negating the unique nonzero entry of E_i not on the diagonal. Using this observation, you can confirm that L as defined here agrees with the L found in §9.7. Note that for $i > j$, $L(i,j)$ is the negative of the multiple of row j that was added to row i to make the i,j-entry of A become zero.

Now consider what happens when we perform Gaussian elimination on a general matrix $A \in M_n(F)$ satisfying $\det(A[k]) \neq 0$ for k between 1 and n. The algorithm proceeds as follows, assuming no divisions by zero occur (we justify this assumption later):

```
for j=1 to n-1 do
  for i=j+1 to n do
    m(i,j) = A(i,j)/A(j,j);
    replace row i of A by row i minus m(i,j) times row j;
  end for;
end for.
```

To describe this algorithm using elementary matrices, let $E(i, j, c)$ be the elementary matrix with i, j-entry equal to c, diagonal entries equal to 1, and other entries equal to zero (for $i \neq j$ in $[n]$ and $c \in F$). For each column index j in the outer loop, the inner loop modifies the current version of the matrix A by left-multiplying by a product of elementary matrices

$$C_j = E(n, j, -m(n, j))E(n-1, j, -m(n-1, j)) \cdots E(j+1, j, -m(j+1, j)), \qquad (9.8)$$

which create zeroes in column j below the diagonal. The net effect of the algorithm is to replace A by the matrix

$$C_{n-1} \cdots C_2 C_1 A = U,$$

which is an upper-triangular matrix. Solving for A, we can write $A = LU$ with

$$L = C_1^{-1} C_2^{-1} \cdots C_{n-1}^{-1} I_n.$$

Observe that

$$C_j^{-1} = E(j+1, j, m(j+1, j)) \cdots E(n-1, j, m(n-1, j))E(n, j, m(n, j)).$$

To find the entries of L, we can start with I_n and work from right to left, applying the elementary row operations encoded by each elementary matrix $E(i, j, m(i, j))$ as j goes from $n-1$ down to 1 and (for each j) i goes from n down to $j+1$. Because the elementary operations are applied in this particular order, you can check that the final result is a lower-unitriangular matrix L with i, j-entry $L(i, j) = m(i, j)$ for all $i > j$ in $[n]$. To summarize, *in the Doolittle factorization $A = LU$, the entries of L encode the multipliers needed when we use Gaussian elimination to reduce A to the upper-triangular matrix U.*

To complete our analysis, we justify the assumption that the computation $m(i, j) = A(i, j)/A(j, j)$ in the algorithm never involves a division by zero. We show this by induction on the outer loop variable j. Fix j with $1 \leq j < n$, and assume no divisions by zero occurred in the first $j - 1$ iterations of the outer loop. At this point, the algorithm has transformed the original matrix A into a new matrix

$$C_{j-1} \cdots C_2 C_1 A = U_{j-1},$$

where $L_{j-1} = C_{j-1} \cdots C_2 C_1$ is lower-unitriangular, and U_{j-1} has zeroes below the diagonal in the first $j - 1$ columns. Write $L_{j-1}A = U_{j-1}$ as a product of block matrices

$$\left[\begin{array}{c|c} L_{j-1}[j] & 0 \\ \hline * & * \end{array} \right] \left[\begin{array}{c|c} A[j] & * \\ \hline * & * \end{array} \right] = \left[\begin{array}{c|c} U_{j-1}[j] & * \\ \hline * & * \end{array} \right], \qquad (9.9)$$

where each upper-left block is $j \times j$. We see that $L_{j-1}[j]A[j] = U_{j-1}[j]$. Taking determinants and noting that $U_{j-1}[j]$ is upper-triangular, we get

$$0 \neq \det(A[j]) = \det(L_{j-1}[j])\det(A[j]) = \det(U_{j-1}[j]) = \prod_{k=1}^{j} U_{j-1}(k, k).$$

So the value of $A(j, j)$ used in the jth iteration of the outer loop, namely $U_{j-1}(j, j)$, is nonzero, as claimed.

9.9 Permuted LU Factorizations

Our next goal is to generalize the analysis in the preceding section to the case where $A \in M_n(F)$ is an invertible matrix, but A does not satisfy the condition that all $\det(A[k])$ are nonzero. Such an A cannot have an LU factorization, but we prove there is a permutation matrix $P \in M_n(F)$ such that PA has an LU factorization. Recall that a *permutation matrix* P has exactly one 1 in each row and column, with all other entries of P equal to 0. For $F = \mathbb{C}$, the columns of P are orthonormal (relative to the standard inner product on $\mathbb{C}^n$), so that $P^*P = P^{\mathrm{T}}P = I_n$. It follows that P is invertible with $P^{-1} = P^{\mathrm{T}}$; the same result holds for any field F (Exercise 39). By §4.8, PA is obtained from A by reordering the rows of A. Specifically, for each i, j with $P(i, j) = 1$, row i of PA is row j of A.

Permuted LU Factorization Theorem (Invertible Case). For any invertible $A \in M_n(F)$, there exist a permutation matrix $P \in M_n(F)$, a lower-unitriangular matrix $L \in M_n(F)$, and an upper-triangular invertible matrix $U \in M_n(F)$ with $PA = LU$, or equivalently $A = P^{\mathrm{T}}LU$.

Proof. The idea of the proof is to reduce A to the upper-triangular matrix U via Gaussian elimination. The algorithm used before still works, with one modification. When we try to compute the multipliers $m(i, j)$ to process column j of the current matrix, we may find that $A(j, j)$ is zero. Recall that the current matrix (when the jth iteration of the outer loop begins) can be written $U_{j-1} = L_{j-1}A$, where A is the original input matrix, L_{j-1} is lower-unitriangular, and U_{j-1} has zeroes below the diagonal in the first $j-1$ columns. Since A and L_{j-1} are both invertible, U_{j-1} is invertible. Writing U_{j-1} in block form as

$$U_{j-1} = \left[\begin{array}{c|c} U_{j-1}[j-1] & * \\ \hline 0 & * \end{array} \right], \tag{9.10}$$

it follows by taking determinants that for some $k \geq j$, $U_{j-1}(k, j) \neq 0$. Choose k minimal with this property and interchange rows j and k at the start of the jth iteration of the outer loop. The rest of the algorithm proceeds as before.

In terms of elementary matrices, the modified elimination algorithm yields a factorization

$$C_{n-1}P_{n-1} \cdots C_3 P_3 C_2 P_2 C_1 P_1 A = U, \tag{9.11}$$

where C_j has the form (9.8), and each P_j is an elementary permutation matrix obtained from I_n by switching row j with some row $k_j \geq j$. To obtain the required factorization $PA = LU$, we need to move all the permutation matrices P_j to the right through all the matrices C_r. This manipulation requires some care, since P_j and C_r do not commute in general.

Consider a product $P_j C_r$ where $1 \leq r < j \leq n$. Each C_r is a product of elementary matrices of the form $E(s, r, m)$ with $s > r$ and $m \in F$. Suppose P_j is obtained by switching rows j and $k \geq j$ of I_n. We claim $P_j E(s, r, m) = E(s', r, m)P_j$, where $s' = k$ if $s = j$, $s' = j$ if $s = k$, and $s' = s$ otherwise. We verify that $P_j E(j, r, m) = E(k, r, m)P_j$, leaving the other assertions in the claim as exercises. It suffices to show that $P_j E(j, r, m)B = E(k, r, m)P_j B$ for any matrix $B \in M_n(F)$. The left side is the matrix obtained from B by first adding m times row r to row j, then switching rows j and k. The right side is the matrix obtained from B by first switching rows j and k, then adding m times row r to row k. Since r is different from j and k, the effect of these two operations on B is the same. The rest of the claim is verified similarly.

By using the claim repeatedly to move P_j to the right past each factor in C_r, we see that for all $r < j$, $P_j C_r = C'_r P_j$ where C'_r is also a product of elementary matrices $E(s', r, m)$ with s' ranging from $r+1$ to n in some order (cf. (9.8)). Using this in (9.11), we can first write $P_2 C_1 = C'_1 P_2$ and continue by rewriting $C_3 P_3 C_2 C'_1 P_2 P_1$ as $C_3 C'_2 C''_1 P_3 P_2 P_1$. Ultimately, we reach a new factorization

$$\tilde{C}_{n-1} \cdots \tilde{C}_2 \tilde{C}_1 (P_{n-1} \cdots P_2 P_1) A = U,$$

where each $\tilde{C}_j$ has the form (9.8) (with factors appearing in a different order). The product $P_{n-1} \cdots P_2 P_1$ is a permutation matrix P. We now have $PA = LU$, where $L = \tilde{C}_1^{-1} \cdots \tilde{C}_{n-1}^{-1}$ is lower-unitriangular. $\quad\square$

Finally, we extend our results to the case of a general square matrix A that need not be invertible.

Permuted LU Factorization Theorem (General Case). For any $A \in M_n(F)$, there exist permutation matrices $P, Q \in M_n(F)$, a lower-unitriangular $L \in M_n(F)$, and an upper-triangular $U \in M_n(F)$ with $\operatorname{rank}(A) = \operatorname{rank}(U)$, such that $PAQ = LU$.

Proof. Examining the preceding proof, we see that the only place we needed invertibility of A was when we used (9.10) to show $U_{j-1}(k, j) \neq 0$ for some $k \geq j$. Consider what happens when our algorithm reduces a general matrix A to the matrix U_{j-1} given in (9.10).

Case 1: There is a nonzero entry somewhere in the lower-right block shown in (9.10), say $U_{j-1}(k, p) \neq 0$ for some $k, p \geq j$. We can multiply the current matrix on the right by the permutation matrix Q_j that switches columns j and p, and then multiply on the left (as before) by the permutation matrix P_j that switches rows j and k, to create a matrix with a nonzero entry in the j, j-position. The algorithm proceeds to create zeroes below this entry, as before.

Case 2: All entries in the lower-right block of (9.10) are zero. Then the algorithm terminates at this point with a factorization

$$C_{j-1} P_{j-1} \cdots C_1 P_1 A Q_1 Q_2 \cdots Q_{j-1} = U_{j-1} = U,$$

where U is upper-triangular since $U_{j-1}[j-1]$ is upper-triangular and rows j through n of U_{j-1} contain all zeroes. Moving the matrices $P_1, \ldots, P_{j-1}$ to the right (as in the invertible case) and rearranging, we obtain the required factorization $PAQ = LU$. Since P, Q, and L are all invertible, and multiplying on the left or right by an invertible matrix does not change the rank, we have $\operatorname{rank}(A) = \operatorname{rank}(U)$. $\quad\square$

9.10 Cholesky Factorization

Cholesky Factorization Theorem. For every positive semidefinite matrix $A \in M_n(\mathbb{C})$, there exists a lower-triangular matrix L with nonnegative diagonal entries such that $A = LL^*$. If A is positive definite (hence invertible), then L is unique.

Proof. Recall from §7.11 that a positive semidefinite matrix $A \in M_n(\mathbb{C})$ has a unique positive semidefinite square root, which is a positive semidefinite matrix $B \in M_n(\mathbb{C})$ such that $A = B^2$. Since B must be Hermitian, we have $A = B^*B$. Now, B has a QR factorization $B = QR$, where $Q \in M_n(\mathbb{C})$ is unitary and $R \in M_n(\mathbb{C})$ is upper-triangular with nonnegative diagonal entries. Let $L = R^*$, which is lower-triangular with nonnegative diagonal entries. Since Q is unitary, we have $A = B^*B = (QR)^*QR = R^*Q^*QR = R^*R = LL^*$, as needed.

To prove uniqueness when A is invertible, suppose $A = LL^* = MM^*$ where both $L, M \in M_n(\mathbb{C})$ are lower-triangular with nonnegative diagonal entries. Taking determinants, we see that L and M are both invertible. Then $M^{-1}L = M^*(L^*)^{-1}$, where the left side is a lower-triangular matrix and the right side is an upper-triangular matrix. Since the two sides are equal, $M^{-1}L$ must be a diagonal matrix D with strictly positive diagonal entries. Then $L = MD$, so $MM^* = LL^* = MDD^*M^* = MD^2M^*$. Left-multiplying by M^{-1} and right-multiplying by $(M^*)^{-1}$ gives $I = D^2$. Since all diagonal entries of D are positive real numbers, this forces $D = I$ and $M = L$, completing the uniqueness proof.   □

One approach to computing the Cholesky factorization of a positive definite matrix A is to observe that $A = LL^*$ is one of the variations of the LU factorization from §9.6, in which L is lower-triangular, $L^* = U$ is upper-triangular, and $L(k, k) = L^*(k, k)$ for all $k \in [n]$. We can recover the columns of L (and hence the rows of $U = L^*$) one at a time using the formulas in §9.6. Specifically, for $k = 1$ to n, we first calculate

$$L(k, k) = \sqrt{A(k, k) - \sum_{r=1}^{k-1} L(k, r)L^*(r, k)}$$

and then compute

$$L(i, k) = L(k, k)^{-1}\left[A(i, k) - \sum_{r=1}^{k-1} L(i, r)L^*(r, k)\right]$$

for $i = k+1, \ldots, n$. Since we proved above that there is a factorization of the form $A = LL^*$, the uniqueness of the LU factorization ensures that the U computed by the LU algorithm must equal L^*. In particular, there is no need to use (9.7) to solve for the entries of U in this setting.

For example, consider $A = \begin{bmatrix} 4 & 2 & -1 \\ 2 & 5 & 1 \\ -1 & 1 & 3 \end{bmatrix}$. By the determinant test in §7.15, A is positive definite. We seek a factorization $A = LL^*$, say

$$\begin{bmatrix} 4 & 2 & -1 \\ 2 & 5 & 1 \\ -1 & 1 & 3 \end{bmatrix} = \begin{bmatrix} r & 0 & 0 \\ s & u & 0 \\ t & v & w \end{bmatrix} \begin{bmatrix} r & s & t \\ 0 & u & v \\ 0 & 0 & w \end{bmatrix}.$$

(Since A is real and positive definite, L must be real and $L^* = L^T$.) Comparing $1, 1$-entries, we find $r^2 = 4$ and $r = 2$. Working down the first column, $sr = 2$ and $tr = -1$, so $s = 1$ and $t = -1/2$. Moving to the $2, 2$-entry, $s^2 + u^2 = 5$ becomes $1 + u^2 = 5$, so $u = 2$. Looking at the $3, 2$-entry, $1 = st + vu$ implies $1 = -1/2 + 2v$, so $v = 3/4$. Finally, $t^2 + v^2 + w^2 = 3$ gives $1/4 + 9/16 + w^2 = 3$, so $w = \sqrt{35/16} = \sqrt{35}/4$. We can check the computation by confirming that

$$\begin{bmatrix} 4 & 2 & -1 \\ 2 & 5 & 1 \\ -1 & 1 & 3 \end{bmatrix} = \begin{bmatrix} 2 & 0 & 0 \\ 1 & 2 & 0 \\ -1/2 & 3/4 & \sqrt{35}/4 \end{bmatrix} \begin{bmatrix} 2 & 1 & -1/2 \\ 0 & 2 & 3/4 \\ 0 & 0 & \sqrt{35}/4 \end{bmatrix}.$$

9.11   Least Squares Approximation

This section describes an application of the Cholesky factorization to the solution of a least squares approximation problem. Our goal is to solve a linear system of the form $Ax = b$, where $A \in M_{m,n}(\mathbb{C})$ is a given matrix with $m \geq n$, $b \in \mathbb{C}^m$ is a given column vector, and $x \in \mathbb{C}^n$ is unknown. This linear system typically has many more equations than unknowns, so we would not expect there to be any exact solutions. Instead, we seek an approximate solution x that minimizes the length of the *error vector* $Ax - b$. The squared length of this error vector is $\sum_{i=1}^{m} |(Ax)_i - b_i|^2$, which is the sum of the squares of the errors in each individual equation.

Theorem on Least Squares Approximation. Suppose $m \geq n$ and $A \in M_{m,n}(\mathbb{C})$ has rank n (so the n columns of A are linearly independent). For each $b \in \mathbb{C}^m$, there exists a unique $x \in \mathbb{C}^n$ minimizing $||Ax - b||^2$, and this x is the unique solution to the square linear system $(A^*A)x = A^*b$.

The linear equations in the system $(A^*A)x = A^*b$ are called *normal equations* for x, and x is called the *least squares approximate solution* to $Ax = b$.

Proof. We first check that $A^*A \in M_n(\mathbb{C})$ is invertible. Fix $z \in \mathbb{C}^n$ with $A^*Az = 0$; it suffices to show $z = 0$ (see Table 4.1). Note that $||Az||^2 = \langle Az, Az \rangle = (Az)^*(Az) = z^*A^*Az = 0$, so $Az = 0$. If z were not zero, then writing $Az = 0$ as a linear combination of the n columns of A (as explained in §4.7) would show that these columns were linearly dependent, contrary to hypothesis. So $z = 0$, A^*A is invertible, and the normal equations have a unique solution $x = (A^*A)^{-1}(A^*b)$.

Now let $y \in \mathbb{C}^n$ be arbitrary. Since x solves the normal equations, the vectors $Ax - b$ and $A(y - x)$ are orthogonal: $\langle Ax - b, A(y - x) \rangle = (y - x)^*A^*(Ax - b) = (y - x)^*(A^*Ax - A^*b) = 0$. Using the Pythagorean Identity, we compute

$$||Ay - b||^2 = ||A(y - x) + (Ax - b)||^2 = ||A(y - x)||^2 + ||Ax - b||^2 \geq ||Ax - b||^2.$$

This inequality shows that among all vectors in $\mathbb{C}^n$, the x chosen above minimizes $||Ax - b||^2$. Furthermore, equality holds iff $||A(y - x)||^2 = 0$, which implies $A(y - x) = 0$, hence (as seen in the last paragraph) $y - x = 0$ and $y = x$. So the vector x minimizing the sum of the squares of the errors is unique, as needed. $\square$

To see how the Cholesky factorization can be used here, note that A^*A is positive definite, since for all nonzero $v \in \mathbb{C}^n$, $Av \neq 0$ and so $v^*A^*Av = \langle Av, Av \rangle = ||Av||^2$ is a positive real number. We can therefore write $A^*A = LL^*$ for a unique lower-triangular $L \in M_n(\mathbb{C})$. The normal equations characterizing the least squares solution x become $LL^*x = A^*b$. If we have found L, we can quickly solve the normal equations by first solving $Lw = A^*b$ using forward substitution, then solving $L^*x = w$ using backward substitution. We must point out, however, that solving least squares problems via the normal equations can lead to issues of numerical stability. Exercise 52 describes another approach to finding x based on a QR factorization of A.

9.12 Singular Value Decomposition

Singular Value Decomposition Theorem. For any $A \in M_{m,n}(\mathbb{C})$, there exist a unitary matrix $U \in M_n(\mathbb{C})$, a unitary matrix $V \in M_m(\mathbb{C})$, and a diagonal matrix $D \in M_{m,n}(\mathbb{R})$ with diagonal entries $d_1 \geq d_2 \geq \cdots \geq d_{\min(m,n)} \geq 0$, such that $A = VDU^*$. The numbers d_j are uniquely determined by A; the nonzero d_j are the positive square roots of the nonzero eigenvalues of A^*A.

We call $d_1, \ldots, d_{\min(m,n)}$ the *singular values* of A.

Proof. We proved the singular value decomposition for square matrices as a consequence of the polar decomposition in §7.13. This section gives an independent proof valid for rectangular matrices. Fix $A \in M_{m,n}(\mathbb{C})$. The matrix $A^*A \in M_n(\mathbb{C})$ is positive semidefinite, since for any $x \in \mathbb{C}^n$, $x^*A^*Ax = (Ax)^*(Ax) = \langle Ax, Ax \rangle \geq 0$. We saw in Chapter 7 that positive semidefinite matrices are normal, hence unitarily diagonalizable via an orthonormal basis of eigenvectors, and all their eigenvalues are nonnegative real numbers. Let the eigenvalues of A^*A (with multiplicity) be $c_1 \geq c_2 \geq \cdots \geq c_n \geq 0$. Let $u_1, u_2 \ldots, u_n \in \mathbb{C}^n$ be associated unit eigenvectors, so $(u_1, \ldots, u_n)$ is an orthonormal basis of $\mathbb{C}^n$ and $A^*Au_j = c_ju_j$ for $1 \leq j \leq n$. Let $d_j = \sqrt{c_j} \geq 0$ for $1 \leq j \leq n$ and r be the maximal index with $c_j > 0$. For $1 \leq j \leq r$, define $v_j = d_j^{-1}Au_j \in \mathbb{C}^m$. We claim $(v_1, \ldots, v_r)$ is an orthonormal list in $\mathbb{C}^m$. To see why, fix i, j between 1 and r and compute

$$\langle v_j, v_i \rangle = (d_i^{-1}Au_i)^*(d_j^{-1}Au_j) = d_i^{-1}d_j^{-1}u_i^*A^*Au_j = d_i^{-1}d_j^{-1}u_i^*(c_ju_j) = (d_j/d_i)\langle u_j, u_i \rangle.$$

Since the u_k are orthonormal, we get $\langle v_j, v_i \rangle = 0$ if $j \neq i$, and $\langle v_j, v_i \rangle = d_i/d_i = 1$ if $j = i$.

Using the Gram–Schmidt Orthonormalization Algorithm, we can extend the orthonormal list $(v_1, \ldots, v_r)$ to an orthonormal basis $(v_1, \ldots, v_r, \ldots, v_m)$ of $\mathbb{C}^n$. Let $U \in M_n(\mathbb{C})$ be the matrix with columns $u_1, \ldots, u_n$ and $V \in M_m(\mathbb{C})$ be the matrix with columns $v_1, \ldots, v_m$. Both U and V are unitary matrices, since they are square matrices with orthonormal columns (see §7.5). Let $D \in M_{m,n}(\mathbb{C})$ have diagonal entries $D(j,j) = d_j$ for $1 \leq j \leq \min(m,n)$, and all other entries 0. To prove that $A = VDU^*$, we prove the equivalent identity $AU = VD$ (recall $U^* = U^{-1}$ since U is unitary). Fix j between 1 and n. On the left side, the jth column of AU is $A(U^{[j]}) = Au_j$. On the right side, the jth column of VD is $V(D^{[j]}) = D(j,j)V^{[j]} = d_jv_j$. If $j \leq r$, then $d_jv_j = Au_j$ holds by definition of v_j. If $r < j \leq n$, then $c_j = d_j = 0$. So $\|Au_j\|^2 = u_j^*A^*Au_j = u_j^*(c_ju_j) = 0$, hence $Au_j = 0 = d_jv_j$. We conclude $AU = VD$ since all columns agree.

To prove the uniqueness of the d_j, let $A = VDU^*$ be any factorization with the properties given in the theorem statement (not necessarily using the matrices U, D, V constructed above). Since V is unitary, $A^*A = (VDU^*)^*(VDU^*) = UD^*V^*VDU^* = UD^*DU^*$. Since U is unitary, A^*A and D^*D are similar and so have the same eigenvalues (counted with multiplicity). Letting the nonzero diagonal entries of D be $d_1, \ldots, d_r$, you can check that D^*D has eigenvalues $d_1^2, \ldots, d_r^2$, together with $n-r$ eigenvalues equal to zero. So the nonzero diagonal entries of D are the positive square roots of the nonzero eigenvalues of A^*A, as claimed. $\qquad\square$

We remark that the matrices V and U appearing in the singular value decomposition of A are not unique in general. We can obtain different matrices U by picking different orthonormal bases for each eigenspace of A^*A when forming the list $(u_1, \ldots, u_n)$. Each choice of U leads to a unique list $(v_1, \ldots, v_r)$, but then we can get different matrices V by extending this list to an orthonormal basis of $\mathbb{C}^m$ in different ways.

In terms of linear transformations, the singular value decomposition can be phrased as follows. *For any linear map $T : \mathbb{C}^n \to \mathbb{C}^m$, there exist an orthonormal basis $(u_1, \ldots, u_n)$ of $\mathbb{C}^n$, an orthonormal basis $(v_1, \ldots, v_m)$ of $\mathbb{C}^n$, and nonnegative real numbers $d_1 \geq \cdots \geq d_{\min(m,n)}$ such that $T(u_j) = d_jv_j$ for $1 \leq j \leq \min(m,n)$, and $T(u_j) = 0$ for $m < j \leq n$.* We can also view the factorization $A = VDU^*$ geometrically, by saying that the linear transformation $x \mapsto Ax$ acts as the composition of an isometry on $\mathbb{C}^n$ encoded by the unitary matrix U^*, followed by a nonnegative rescaling of the coordinate axes encoded by D (which deletes the last $n - m$ coordinates when $n > m$, and adds $m - n$ zero coordinates when $m > n$), followed by an isometry on $\mathbb{C}^m$ encoded by the unitary matrix V.

9.13 Summary

1. *Orthogonality and Orthonormality.* A list $(u_1, \ldots, u_k)$ in an inner product space V is *orthogonal* iff $\langle u_i, u_j \rangle = 0$ for all $i \neq j$ in $[k]$. The list is *orthonormal* iff it is orthogonal and $\langle u_i, u_i \rangle = 1$ for all i in $[k]$. Orthogonal lists of nonzero vectors are linearly independent. Orthogonal vectors $x, y \in V$ satisfy the *Pythagorean Identity* $||x + y||^2 = ||x||^2 + ||y||^2$. A square matrix Q is unitary iff its columns are orthonormal iff $Q^*Q = I$ iff $Q^* = Q^{-1}$.

2. *Orthogonal Projections.* Let $(u_1, \ldots, u_k)$ be an orthonormal list in an inner product space V. Let U be the subspace spanned by this list. For any $v \in V$, there exists a unique $x \in U$ minimizing $||v - x||$, namely $x = \sum_{r=1}^{k} \langle v, u_r \rangle u_r$, and $v - x$ is orthogonal to every vector in U. We call x the *orthogonal projection of v onto U*. The vector v is in U iff $v = x$ iff $v = \sum_{r=1}^{k} \langle v, u_r \rangle u_r$.

3. *Gram–Schmidt Orthonormalization.* Given a finite or countably infinite linearly independent sequence $(v_1, v_2, \ldots)$ in an inner product space V, the *Gram–Schmidt Orthonormalization Algorithm* computes an orthonormal sequence $(u_1, u_2, \ldots)$ such that $(v_1, \ldots, v_r)$ and $(u_1, \ldots, u_r)$ span the same subspace V_r for all $r \geq 1$. Having found $u_1, \ldots, u_{s-1}$, we calculate $x_s = \sum_{r=1}^{s-1} \langle v_s, u_r \rangle u_r$ and $u_s = (v_s - x_s)/||v_s - x_s||$. When $\dim(V) < \infty$, we see that every subspace of V has an orthonormal basis, and every orthonormal list in V can be extended to an orthonormal basis of V. When v_s depends linearly on preceding vectors in the list, the algorithm detects this by computing $x_s = v_s$.

4. *QR Factorization via the Gram–Schmidt Algorithm.* Given $A \in M_{n,k}(\mathbb{C})$ with k linearly independent columns, there exist unique matrices $Q \in M_{n,k}(\mathbb{C})$ and $R \in M_k(\mathbb{C})$ such that $A = QR$, Q has orthonormal columns, and R is upper-triangular with strictly positive diagonal entries. When A is real, Q and R are real. If A has column rank $r < k$, we can write $A = QR$ where $Q \in M_{n,r}(\mathbb{C})$ has orthonormal columns and $R \in M_{r,k}(\mathbb{C})$ is upper-triangular with nonnegative diagonal entries. The columns of Q are the orthonormal vectors obtained by applying Gram–Schmidt orthonormalization to the columns of A. The entries of R are given by $R(i, j) = \langle v_j, u_i \rangle$ for $i \leq j$, and $R(i, j) = 0$ for $i > j$.

5. *Householder Reflections.* Given nonzero $v \in \mathbb{C}^n$, the *Householder matrix* for v is $Q_v = I_n - (2/||v||^2)vv^* \in M_n(\mathbb{C})$. These matrices are unitary and self-inverse $(Q_v^* = Q_v^{-1} = Q_v)$, $Q_v v = -v$, and $Q_v w = w$ for all $w \in \mathbb{C}^n$ orthogonal to v. For any $x \neq y$ in $\mathbb{C}^n$, there exists $v \in \mathbb{C}^n$ with $Q_v x = y$ iff $||x|| = ||y||$ and $\langle x, y \rangle \in \mathbb{R}$. Specifically, when x and y satisfy these conditions, $Q_v x = y$ iff v is a nonzero scalar multiple of $x - y$.

6. *Householder's QR Algorithm.* The Householder algorithm reduces $A \in M_{n,k}(\mathbb{C})$ to an upper-triangular matrix $R \in M_{n,k}(\mathbb{C})$ by applying $s = \min(n - 1, k)$ Householder reflections to create zeroes one column at a time. At the jth step, we find a Householder reflection sending the partial column vector $(A(j, j), A(j + 1, j), \ldots, A(n, j))^{\mathrm{T}}$ to a vector whose last $n - j$ entries are 0. The output of the algorithm is a factorization $A = QR$, where Q is the product of s Householder matrices and a diagonal unitary matrix.

7. *LU Factorizations.* For any field F and $A \in M_n(F)$, there exist a lower-unitriangular matrix $L \in M_n(F)$ and an invertible upper-triangular matrix $U \in M_n(F)$ such that $A = LU$ iff for all $k \in [n]$, $\det(A[k]) \neq 0$. When L

and U exist, they are unique. This is *Doolittle's LU factorization*; we obtain *Crout's LU factorization* by making U unitriangular instead of L. We can also write $A = LDU$ with both L and U unitriangular and D diagonal.

8. *Recursive Formula for LU Factorizations.* When the Doolittle factorization $A = LU$ exists, we can find L and U recursively as follows:

```
set L(i,i)=1 and L(i,j)=0=U(j,i) for all i<j in [n];
for k=1 to n do
  U(k,k)=A(k,k)-sum(r=1 to k-1) L(k,r)U(r,k);
  for i=k+1 to n do
    L(i,k)=(1/U(k,k))[A(i,k)-sum(r=1 to k-1) L(i,r)U(r,k)];
  end for;
  for j=k+1 to n do
    U(k,j)=A(k,j)-sum(r=1 to k-1) L(k,r)U(r,j);
  end for;
end for.
```

9. *LU Factorization via Gaussian Elimination.* Given $A \in M_n(F)$ with $\det(A[k]) \neq 0$ for all $k \in [n]$, the Gaussian elimination algorithm proceeds as follows:

```
for j=1 to n-1 do
  for i=j+1 to n do
    m(i,j) = A(i,j)/A(j,j);
    replace row i of A by row i minus m(i,j) times row j;
  end for;
end for.
```

In the Doolittle factorization $A = LU$, U is the output of the elimination algorithm, and the entries $L(i,j)$ below the diagonal are the multipliers $m(i,j)$ used in the algorithm.

10. *Permuted LU Factorizations.* For any $A \in M_n(F)$, there exist permutation matrices $P, Q \in M_n(F)$, a lower-unitriangular matrix $L \in M_n(F)$, and an upper-triangular matrix $U \in M_n(F)$ with $PAQ = LU$ and $\operatorname{rank}(A) = \operatorname{rank}(U)$. When A is invertible, we may take $Q = I_n$. These factorizations can be achieved by using Gaussian elimination with row and column interchanges.

11. *Cholesky Factorization.* Every positive semidefinite matrix $A \in M_n(\mathbb{C})$ can be factored as $A = LL^*$ for some lower-triangular $L \in M_n(\mathbb{C})$ with nonnegative diagonal entries. When A is positive definite, L is unique.

12. *Least Squares Approximation.* Given $b \in \mathbb{C}^m$ and $A \in M_{m,n}(\mathbb{C})$ of rank n with $m \geq n$, there exists a unique $x \in \mathbb{C}^n$ minimizing $\|Ax - b\|^2$, namely the unique solution to the normal equations $A^*Ax = A^*b$.

13. *Singular Value Decomposition.* Any $A \in M_{m,n}(\mathbb{C})$ can be factored as $A = VDU^*$, where $V \in M_m(\mathbb{C})$ and $U \in M_n(\mathbb{C})$ are unitary, and $D \in M_{m,n}(\mathbb{R})$ is diagonal with diagonal entries $d_1 \geq d_2 \geq \cdots \geq 0$. The d_j are called the *singular values* of A and are uniquely determined by A. The nonzero d_j are the positive square roots of the nonzero eigenvalues of A^*A.

9.14   Exercises

Assume V is a complex inner product space in these exercises.

1.   Prove: for all x, y in V, $||x + y||^2 + ||x - y||^2 = 2||x||^2 + 2||y||^2$.

2.   Give a specific example of vectors x, y with $||x+y||^2 = ||x||^2+||y||^2$ and $\langle x, y \rangle \neq 0$.

3.   Prove: for all $x, y \in V$, if $||x + y||^2 = ||x||^2 + ||y||^2 = ||x + iy||^2$, then $\langle x, y \rangle = 0$.

4.   Let $L = \begin{bmatrix} 5 & 0 & 0 & 0 \\ -3 & 2 & 0 & 0 \\ 7 & -1 & -4 & 0 \\ 8 & -3 & -5 & 2 \end{bmatrix}, U = \begin{bmatrix} 1 & 3 & 2 & -4 \\ 0 & 1 & 5 & -3 \\ 0 & 0 & 1 & 6 \\ 0 & 0 & 0 & 1 \end{bmatrix}, b = \begin{bmatrix} 10 \\ 6 \\ -12 \\ 4 \end{bmatrix}.$
     (a) Solve $Lx = b$ by forward substitution.
     (b) Solve $Ux = b$ by backward substitution.
     (c) Solve $Ax = b$, where $A = LU$.

5.   Let $Q = \begin{bmatrix} 0.28 & 0 & 0 & 0.96 \\ 0 & 0.6 & 0.8 & 0 \\ -0.96 & 0 & 0 & 0.28 \\ 0 & 0.8 & -0.6 & 0 \end{bmatrix}, R = \begin{bmatrix} 2 & 3 & -1 & -1 \\ 0 & 5 & 3 & -1 \\ 0 & 0 & 2 & -2 \\ 0 & 0 & 0 & 2 \end{bmatrix}, b = \begin{bmatrix} 1 \\ 3 \\ 2 \\ -2 \end{bmatrix}.$
     Confirm that Q is unitary. Solve $Ax = b$ efficiently, where $A = QR$.

6.   Suppose $A = LU$, where $A, L, U \in M_n(\mathbb{C})$ with L lower-unitriangular and U upper-triangular. If we know L and U, how many multiplications and divisions in $\mathbb{C}$ are required to solve $Ax = b$?

7.   Suppose $A = QR$, where $A, Q, R \in M_n(\mathbb{C})$ with Q unitary and R upper-triangular. If we know Q and R, how many multiplications and divisions in $\mathbb{C}$ are required to solve $Ax = b$?

8.   Apply the Gram–Schmidt algorithm to $v_1 = (1, 1, 1)$, $v_2 = (1, 2, 4)$, $v_3 = (1, 3, 9)$ to compute an orthonormal basis of $\mathbb{C}^3$.

9.   Let V be the subspace of $\mathbb{C}^4$ spanned by $v_1 = (3, 1, 5, -1)$, $v_2 = (2, 0, 1, 3)$, and $v_3 = (1, -1, -1, -1)$. Apply the Gram–Schmidt algorithm to the list (v_1, v_2, v_3) to compute an orthonormal basis of V. Find the orthogonal projection of each vector onto V: $\mathbf{e}_1$, $(-2, 2, 3, -2)$, $(1, 2, 3, 4)$, $(24, -2, 15, 7)$.

10.   Let V be the real inner product space of continuous functions $f : [0, 1] \to \mathbb{R}$, with inner product $\langle f, g \rangle = \int_0^1 f(x)g(x) \, dx$ for $f, g \in V$. Apply Gram–Schmidt orthonormalization to the list $(1, x, x^2, x^3)$ to obtain an orthonormal basis for the subspace W spanned by this list. Find the orthogonal projection of $f(x) = e^x$ onto W and the minimum distance from f to W. Do the same for $f(x) = \sin(2\pi x)$.

11.   Repeat Exercise 10, but let V consist of continuous functions $f : [-1, 1] \to \mathbb{R}$, with $\langle f, g \rangle = \int_{-1}^1 f(x)g(x) \, dx$ for $f, g \in V$.

12.   Given a list of orthonormal vectors $(u_1, \ldots, u_k)$ and $c_1, \ldots, c_k \in \mathbb{C}$, prove $||c_1 u_1 + \cdots + c_k u_k||^2 = |c_1|^2 + \cdots + |c_k|^2$. How does this formula change if we only assume $u_1, \ldots, u_k$ are orthogonal vectors?

13.   Let $(u_1, \ldots, u_k)$ be an orthonormal list in V and $v \in V$. Prove that $\sum_{j=1}^k |\langle v, u_j \rangle|^2 \leq ||v||^2$, with equality iff v is in the span of $u_1, \ldots, u_k$.

14.   Let $V = \mathbb{R}[x]$ be the real inner product space of polynomials, with inner product $\left\langle \sum_{i=0}^n a_i x^i, \sum_{j=0}^m b_j x^j \right\rangle = \sum_{i=0}^{\min(m,n)} a_i b_i$ for $a_i, b_j \in \mathbb{R}$. Suppose

$(f_0, f_1, f_2, \ldots, f_n, \ldots)$ is any infinite list of elements of V with $\deg(f_n) = n$ for all $n \geq 0$. Show that applying Gram–Schmidt orthonormalization to this list produces the sequence $(\pm 1, \pm x, \pm x^2, \ldots, \pm x^n, \ldots)$. Given a finite subset S of $\mathbb{Z}_{\geq 0}$, what is the orthogonal projection of $g \in \mathbb{R}[x]$ onto the subspace spanned by $(x^j : j \in S)$?

15. Use the Gram–Schmidt algorithm to prove that any orthonormal list $(u_1, \ldots, u_k)$ in a finite-dimensional inner product space V can be extended to an orthonormal ordered basis of V.

16. Show that if $R, R_1 \in M_k(\mathbb{C})$ are upper-triangular with positive real diagonal entries, then RR_1^{-1} is upper-triangular with positive real diagonal entries. Show that the only unitary upper-triangular matrix in $M_k(\mathbb{C})$ with positive real diagonal entries is I_k. Deduce the uniqueness of the factorization $A = QR$ in the case where $A \in M_k(\mathbb{C})$ is a square matrix.

17. For each matrix, use Gram–Schmidt orthonormalization to find the QR factorizations described in §9.3. (a) $\begin{bmatrix} 1 & 2 \\ 1 & 1 \end{bmatrix}$ (b) $\begin{bmatrix} 3 & 4 \\ 4 & 3 \end{bmatrix}$ (c) $\begin{bmatrix} 2 & 1 & 3 \\ 3 & 1 & 5 \\ 6 & 2 & -1 \end{bmatrix}$

(d) $\begin{bmatrix} 1 & 0 & 1 & 0 \\ 3 & 3 & 3 & 3 \\ 2 & 0 & 2 & 0 \end{bmatrix}$ (e) $\begin{bmatrix} 1 & 2 & 3 & 4 \\ 1 & 3 & 2 & 4 \\ 1 & 4 & 2 & 3 \\ 2 & 3 & 1 & 4 \end{bmatrix}$

18. Let $A \in M_{n,k}(\mathbb{R})$ have rank k. How many ways can we write $A = QR$ where $Q \in M_{n,k}(\mathbb{R})$ has orthonormal columns and $R \in M_k(\mathbb{R})$ is upper-triangular? (We do not assume R has positive diagonal entries here.)

19. Use the Gram–Schmidt algorithm to prove: any $A \in M_{n,k}(\mathbb{C})$ can be factored as $A = QR$, where $Q \in M_n(\mathbb{C})$ is unitary and $R \in M_{n,k}(\mathbb{C})$ is upper-triangular.

20. Formulate and prove a uniqueness result for the factorization $A = QR$ of an arbitrary matrix $A \in M_{n,k}(\mathbb{C})$ described in §9.3.

21. **Modified Gram–Schmidt Algorithm.** Given a linearly independent list $(v_1, \ldots, v_k)$ in an inner product space V, the Gram–Schmidt orthonormalization algorithm in §9.2 produces an orthonormal list $(u_1, \ldots, u_k)$. Show that the following algorithm also transforms $(v_1, \ldots, v_k)$ into $(u_1, \ldots, u_k)$. Loop from $j = 1$ to k. Replace the current vector v_j by $v_j/||v_j||$, and then for each r from $j+1$ to k, replace v_r by $v_r - \langle v_r, v_j \rangle v_j$. Output the final values of $(v_1, \ldots, v_k)$. (This version of Gram–Schmidt can be shown to be more stable numerically compared to the original version.)

22. Estimate the number of real multiplications, divisions, and square root extractions needed to compute the QR factorization of a matrix $A \in M_n(\mathbb{R})$ via the Gram–Schmidt Orthonormalization Algorithm.

23. Estimate the number of real multiplications, divisions, and square root extractions needed to compute the QR factorization of a matrix $A \in M_n(\mathbb{R})$ via Householder's Algorithm.

24. Describe how to adapt the Gram–Schmidt algorithm to transform a linearly independent list $(v_1, \ldots, v_k)$ into an orthogonal list $(u_1, \ldots, u_k)$ using a computation that avoids extractions of square roots.

25. Draw a picture in $\mathbb{R}^2$ illustrating the action of the reflection $T_{(-4,3)}$ on each of these vectors: $(-4, 3)$, $(3, 4)$, $(1, 0)$, $(0, 1)$, $(1, -2)$. Confirm your answers algebraically by multiplying each column vector by the matrix $Q_{(-4,3)}$.

26. Find a Householder matrix Q_v in $M_3(\mathbb{R})$ that fixes each point in the plane $2x_1 - 3x_2 + x_3 = 0$. Draw a sketch showing this plane and the vectors v, $w = (1, -1, 1)$, and $Q_v w$.

27. (a) Explain why the lemma in §9.4 fails to hold without the hypothesis $\|x\| = \|y\|$.
    (b) Does the lemma hold without the hypothesis $\langle x, y \rangle \in \mathbb{R}$?
    (c) Suppose $x = y$. Describe all $v \in \mathbb{C}^n$ with $Q_v x = y$.

28. For each $x, y \in \mathbb{C}^n$, find v such that $Q_v x = y$ or explain why this cannot be done.
    (a) $x = (5, 12)$, $y = (13, 0)$. (b) $x = (1, i)$, $y = (\sqrt{2}, 0)$. (c) $x = (i, 1)$, $y = (\sqrt{2}, 0)$.
    (d) $x = (3, 2, 6)$, $y = (2, 6, 3)$. (e) $x = (3, 2, 3)$, $y = (2, 3, 2)$.
    (f) $x = (3, 2, 6)$, $y = (b, 0, 0)$ for different choices of $b \in \mathbb{C}$.

29. For each matrix in Exercise 17, use Householder's Algorithm to compute the QR factorization. In each case, write Q as an explicit product of Householder reflections and a diagonal unitary matrix.

30. Apply Gram–Schmidt orthonormalization to the columns of the matrix A in §9.5 to obtain the matrices Q and R in (9.2).

31. **Givens Rotations.** For any $i \neq j$ in $[n]$ and $\theta \in [0, 2\pi)$, define the *Givens rotation matrix* $G = G(i, j; \theta) \in M_n(\mathbb{R})$ by letting $G(i, i) = G(j, j) = \cos\theta$, $G(j, i) = \sin\theta$, $G(i, j) = -\sin\theta$, $G(k, k) = 1$ for all $k \neq i, j$, and letting all other entries of G be zero. Left-multiplication by $G(i, j; \theta)$ rotates the plane spanned by $\mathbf{e}_i$ and $\mathbf{e}_j$ counterclockwise through an angle of θ.
    (a) Show that $G(i, j; \theta)$ is orthogonal with inverse $G(i, j; -\theta)$.
    (b) Let $A \in M_n(\mathbb{R})$ have $A(j, i) \neq 0$. Find θ so that $G(i, j; \theta)A$ has j, i-entry 0.

32. **Givens QR Algorithm.** Describe an algorithm for computing a factorization $A = QR$ of a matrix $A \in M_{n,k}(\mathbb{R})$ that reduces A to an upper-triangular matrix R by left-multiplying by a sequence of Givens rotation matrices (see Exercise 31). How many matrices are needed in general?

33. Show that every unitary matrix $Q \in M_n(\mathbb{C})$ has a factorization $Q = Q_{v_1} \cdots Q_{v_{n-1}} D$, where D is a diagonal unitary matrix and each Q_{v_j} is a Householder matrix. Can the matrix D be omitted? Explain.

34. In numerical computations, one source of inaccuracy is *subtractive cancellation*, in which two nearly equal real numbers are subtracted. Explain how the minus sign in the formula for y in §9.5 avoids subtractive cancellation in the computation of the first component of v_1.

35. Find Doolittle's LU factorization of each matrix using the recursive formulas in §9.6. (a) $\begin{bmatrix} 3 & -2 \\ -12 & 13 \end{bmatrix}$ (b) $\begin{bmatrix} 2 & -3 & 1 \\ -4 & 10 & 6 \\ 6 & -13 & -3 \end{bmatrix}$ (c) $\begin{bmatrix} 2 & 3 & -2 & 4 \\ 1 & 6.5 & -9 & 3 \\ -5 & 2.5 & -9 & -8 \\ -2 & -3 & 5 & -4 \end{bmatrix}$

36. Find an LU factorization of each matrix using Gaussian elimination. (a) $\begin{bmatrix} 2 & 8 \\ 6 & 19 \end{bmatrix}$ (b) $\begin{bmatrix} 3 & 2 & 1 \\ 2 & 10/3 & 5/3 \\ -5 & 8/3 & 7/3 \end{bmatrix}$ (c) $\begin{bmatrix} 2 & 10 & -13 & 8 \\ -10 & -35 & 58 & -46 \\ -6 & -24 & 39.2 & -35.4 \\ 0.8 & -41 & 16.4 & 21.4 \end{bmatrix}$

37. Justify the existence and uniqueness assertions for the variants of the LU factorization described in the last paragraph of §9.6.

38. State and prove a theorem regarding factorizations of square matrices of the form $A = UL$, where U is upper-triangular and L is lower-triangular. Give one proof by

modifying the formulas in §9.6, and a second proof that deduces UL factorizations from LU factorizations.

39. Show that for any field F and any permutation matrix $P \in M_n(F)$, $P^{\mathrm{T}} = P^{-1}$.

40. For each matrix A, find a factorization $PA = LU$ or $PAQ = LU$ as in §9.9.

(a) $\begin{bmatrix} 0 & 2 \\ 3 & 1 \end{bmatrix}$ (b) $\begin{bmatrix} 0 & 0 & 2 \\ 1 & -1 & 2 \\ 3 & 3 & 3 \end{bmatrix}$ (c) $\begin{bmatrix} 0 & 0 & 0 & 1 \\ 0 & 1 & 0 & 0 \\ 0 & 0 & 0 & 3 \\ 0 & 5 & 0 & 0 \end{bmatrix}$.

41. Extend the results on LU factorizations and permuted LU factorizations to rectangular matrices $A \in M_{m,n}(F)$.

42. Without using Gaussian elimination or permuted LU factorizations, prove by induction on n that for every invertible $A \in M_n(F)$, there exists a permutation matrix $P \in M_n(F)$ such that $\det((PA)[k]) \neq 0$ for $1 \leq k \leq n$.

43. Suppose $A \in M_n(\mathbb{C})$ has rank r and $\det(A[k]) \neq 0$ for $1 \leq k \leq r$. Prove that A has an LU factorization in which L is lower-unitriangular and U is upper-triangular with the last $n - r$ diagonal entries equal to zero.

44. Show that any $A \in M_n(F)$ can be factored as $A = LQU$, where L is lower-triangular and invertible, U is upper-triangular and invertible, and $Q \in M_n(F)$ has at most one 1 in each row and column, and all other entries of Q are 0. (Use elementary matrices to reduce A.)

45. Find a Cholesky factorization of each positive semidefinite matrix.

(a) $\begin{bmatrix} 4 & -2i \\ 2i & 2 \end{bmatrix}$ (b) $\begin{bmatrix} 9 & 15 & 0 \\ 15 & 26 & 1 \\ 0 & 1 & 5 \end{bmatrix}$ (c) $\begin{bmatrix} 16 & -4 & -12 & 4 \\ -4 & 1 & 3 & -1 \\ -12 & 3 & 19 & -1 \\ 4 & -1 & -1 & 4 \end{bmatrix}$

46. Give an example to show that the Cholesky factorization of a non-invertible positive semidefinite matrix need not be unique.

47. Count the number of multiplications, divisions, and square root extractions in $\mathbb{C}$ needed to compute the Cholesky factorization of an $n \times n$ positive definite matrix using the formulas in §9.10. How do these operation counts compare to those needed to find the LU decomposition of an $n \times n$ matrix using the formulas in §9.6?

48. Consider the inconsistent linear system $x + y = 3$, $2x - y = 0$, $x - 2y = 0$. Use the normal equations to find the least squares approximate solution to this system. Graph the equations in the system and the least squares solution in a single picture.

49. Find the least squares approximate solution to the system

$$\begin{cases} 3x & +7y & -2z & = & 4 \\ 5x & -3y & +2z & = & 8 \\ -x & & -5z & = & 2 \\ -2x & +y & +3z & = & -1 \\ 7x & -4y & +z & = & 0 \end{cases}$$

by setting up and solving the normal equations.

50. Given the system of two equations $ax = b$ and $cx = d$ with $(a, c) \neq (0, 0)$, what is the least squares approximate solution to this system?

51. Given n equations $a_j x = b_j$ with $a_1, \ldots, a_n, b_1, \ldots, b_n \in \mathbb{C}$ and not all a_j are 0, what is the least squares approximate solution to this system?

52. Suppose we need the least squares approximate solution to $Ax = b$, where $A \in M_{n,k}(\mathbb{C})$ has rank k, and we know a factorization $A = QR$ with the properties stated in §9.3. Show that this solution can be found efficiently by solving $Rx = Q^*b$.

53. Given n real data points $(x_1, y_1), \ldots, (x_n, y_n)$, consider the problem of finding real parameters m and b such that the line $y = mx + b$ best approximates the given data. Formulate this as a least squares problem, and solve the normal equations to obtain specific formulas for m and b. Specify exactly what quantity is being minimized by the optimal choice of m and b.

54. Follow the proof in §9.12 to find a singular value decomposition for each matrix.

$$\begin{bmatrix} 1.824 & 1.032 \\ -1.968 & 2.176 \end{bmatrix}, \quad \begin{bmatrix} 3 & 3 & 3 \\ -1 & -1 & -1 \end{bmatrix}, \quad \begin{bmatrix} 1 & 1 & 1 & 1 \\ 1 & -1 & 1 & -1 \\ 1 & 1 & -1 & -1 \end{bmatrix}, \quad \begin{bmatrix} 2 & 0 & 2 & 0 \\ 0 & 0 & 0 & 0 \\ 2 & 0 & 2 & 0 \\ 0 & 0 & 0 & i \end{bmatrix}.$$

55. **Penrose Properties.** (a) Prove that for every $A \in M_{m,n}(\mathbb{C})$, there exists at most one matrix $B \in M_{n,m}(\mathbb{C})$ with these four properties, called the *Penrose properties*.

I: $ABA = A$; II: $BAB = B$; III: $(AB)^* = AB$; IV: $(BA)^* = BA$.

(If B and C have these properties, show $B = (BA)^*(CA)^*C(AC)^*(AB)^* = C$.)
(b) Show that if $A \in M_n(\mathbb{C})$ is invertible, then $B = A^{-1}$ has the four Penrose properties.

56. **Pseudoinverses.** Suppose $A \in M_{m,n}(\mathbb{C})$ has singular value decomposition $A = VDU^*$. Let $D' \in M_{n,m}(\mathbb{C})$ be the matrix obtained from D^{T} by inverting the nonzero diagonal entries. The matrix $A^+ = UD'V^* \in M_{n,m}(\mathbb{C})$ is called the *pseudoinverse of A*. (a) Show that the pseudoinverse does not depend on the choice of singular value decomposition of A, by showing that A^+ has the four Penrose properties in Exercise 55. (b) Without using the Penrose properties, show that $A^+ = A^{-1}$ when $A \in M_n(\mathbb{C})$ is invertible.

57. For each matrix A in Exercise 54, find the pseudoinverse A^+ (see Exercise 56). Compute AA^+ and A^+A in each case.

58. Let $A = VDU^*$ be a singular value decomposition of $A \in M_{m,n}(\mathbb{C})$ such that D has k nonzero entries on the diagonal. (a) Show that the last $n - k$ columns of U are an orthonormal basis of the null space of A. (b) Show that the first k columns of V are an orthonormal basis of the image of A.

59. Let $A \in M_{m,n}(\mathbb{C})$. Use singular value decompositions to prove that the largest singular value of A is $\sup\{\|Ax\| : x \in \mathbb{C}^n, \|x\| = 1\}$.

10

Iterative Algorithms in Numerical Linear Algebra

In applications of linear algebra to science, engineering, and other areas, we often need to find numerical solutions to a huge linear system of equations or to approximate the eigenvalues of a large square matrix. The branch of numerical analysis called numerical linear algebra studies algorithms for solving such problems efficiently while minimizing the effects of rounding errors.

Consider the problem of solving the linear system $Ax = b$, where $A \in M_n(\mathbb{C})$ is a given invertible matrix, $b \in \mathbb{C}^n$ is a given column vector, and $x \in \mathbb{C}^n$ is a column vector of unknowns. Gaussian elimination and related algorithms solve this system by a finite sequence of steps, which (in the absence of roundoff errors) ultimately terminates with the exact solution. This chapter discusses a different class of algorithms for solving such problems, called iterative algorithms. The idea is to start with some initial approximation x_0 to the sought-for solution x, and then to apply some recursive formula $x_{k+1} = f(x_k)$ (depending on the problem instance) to compute an infinite sequence $x_0, x_1, x_2, \ldots, x_k, \ldots$ of approximate solutions. We can then ask if this sequence converges (under an appropriate definition of convergence) to the true solution x, and try to find bounds on how far away a given approximation x_k is to x.

In order to give a precise mathematical analysis of such algorithms, we need a formal way to measure the distance between two column vectors in $\mathbb{C}^n$. We define this distance using the idea of a vector norm. The related concept of a matrix norm helps us understand how multiplication by a fixed matrix affects the distances between vectors.

The analysis of iterative algorithms constitutes a vast subfield of numerical linear algebra, and we only have space to give a short introduction to this subject in this chapter. We discuss three basic iterative methods for solving linear systems (the algorithms of Richardson, Jacobi, and Gauss–Seidel) and analyze them using norms and the spectral radius of a matrix. We also study the power method, which can be used to approximate the largest eigenvalue of a given square matrix. To find other eigenvalues of the matrix, we can use variations of the power method or a technique called deflation.

10.1   Richardson's Algorithm

As a first illustration of an iterative method for solving the linear system $Ax = b$, we describe *Richardson's Algorithm*. The input to the algorithm is a matrix $A \in M_n(\mathbb{C})$, a column vector $b \in \mathbb{C}^n$, and a vector $x_0 \in \mathbb{C}^n$ that represents an initial guess or approximation to the true solution vector x. (We may take $x_0 = 0$ if we have no initial information about x.) The algorithm proceeds by forming a sequence of approximate solutions $x_0, x_1, x_2, \ldots, x_k, \ldots$, where for all $k \geq 0$,

$$x_{k+1} = x_k + (b - Ax_k).$$

After computing each vector x_k, we find the *output error vector* $b - Ax_k$ and add this vector to x_k to obtain the next input vector x_{k+1}. Ideally, we would reach the true solution x from

DOI: 10.1201/9781003484561-10

x_k by adding the *input error vector* $x - x_k$ to x_k, but we do not know this error vector. However, if the matrix A is close to the identity matrix in some sense, then we would expect the output error vector $b - Ax_k = A(x - x_k)$ to be close to the input error vector. In this case, we would expect the approximate solutions x_k to approach the true solution x. We make this informal analysis rigorous in §10.11, where we also derive a precise bound for the size of the error vector $x - x_k$.

Here is a small example to illustrate the execution of Richardson's Algorithm. Let $A = \begin{bmatrix} 1.1 & 0.3 \\ -0.4 & 0.8 \end{bmatrix}$ and $b = \begin{bmatrix} 2 \\ -3 \end{bmatrix}$. Starting with $x_0 = 0$, we successively compute

$$x_1 = x_0 + (b - Ax_0) = \begin{bmatrix} 2 \\ -3 \end{bmatrix}, \qquad x_2 = x_1 + (b - Ax_1) = \begin{bmatrix} 2.7 \\ -2.8 \end{bmatrix},$$

$$x_3 = x_2 + (b - Ax_2) = \begin{bmatrix} 2.57 \\ -2.48 \end{bmatrix}, \qquad x_4 = x_3 + (b - Ax_3) = \begin{bmatrix} 2.487 \\ -2.468 \end{bmatrix},$$

$$x_5 = x_4 + (b - Ax_4) = \begin{bmatrix} 2.4917 \\ -2.4988 \end{bmatrix}, \qquad x_6 = x_5 + (b - Ax_5) = \begin{bmatrix} 2.50047 \\ -2.50308 \end{bmatrix}.$$

On the other hand, by Gaussian elimination or Cramer's Rule, we find the exact solution is $x = \begin{bmatrix} 2.5 \\ -2.5 \end{bmatrix}$. We see that the sequence is converging rapidly to the exact solution.

However, suppose we take $A = \begin{bmatrix} 2 & 3 \\ 1 & 4 \end{bmatrix}$ and $b = \begin{bmatrix} 2 \\ -3 \end{bmatrix}$. Starting the algorithm at $x_0 = 0$, we compute the sequence of approximations

$$\begin{bmatrix} 2 \\ -3 \end{bmatrix}, \begin{bmatrix} 9 \\ 4 \end{bmatrix}, \begin{bmatrix} -19 \\ -24 \end{bmatrix}, \begin{bmatrix} 93 \\ 88 \end{bmatrix}, \begin{bmatrix} -355 \\ -360 \end{bmatrix}, \begin{bmatrix} 1437 \\ 1432 \end{bmatrix}, \dots,$$

which does not appear to be converging. The true solution here is $\begin{bmatrix} 3.4 \\ -1.6 \end{bmatrix}$.

10.2   Jacobi's Algorithm

We introduce the next iterative algorithm for solving $Ax = b$, called *Jacobi's Algorithm*, with a specific example. Suppose $A = \begin{bmatrix} 5 & 1 & -1 \\ -1 & 5 & 2 \\ 2 & 2 & 4 \end{bmatrix}$, $b = \begin{bmatrix} 4 \\ -1 \\ 3 \end{bmatrix}$, and $x = \begin{bmatrix} r \\ s \\ t \end{bmatrix}$. We are trying to solve a system of three linear equations in three unknowns. The basic idea is to solve the ith equation for the ith unknown, as shown here:

$$\begin{cases} 5r & +s & -t & = & 4 \\ -r & +5s & +2t & = & -1 \\ 2r & +2s & +4t & = & 3 \end{cases} \Rightarrow \begin{cases} r & = & (4 - s + t)/5 \\ s & = & (-1 + r - 2t)/5 \\ t & = & (3 - 2r - 2s)/4 \end{cases}$$

We use the rewritten equations to define a sequence of vectors $x_k = [r_k \ s_k \ t_k]^{\mathrm{T}}$ by choosing any initial vector $x_0 = [r_0 \ s_0 \ t_0]^{\mathrm{T}}$, then computing

$$\begin{cases} r_{k+1} & = & (4 - s_k + t_k)/5 \\ s_{k+1} & = & (-1 + r_k - 2t_k)/5 \\ t_{k+1} & = & (3 - 2r_k - 2s_k)/4 \end{cases} \qquad \text{for all } k \geq 0. \tag{10.1}$$

If we start with $x_0 = 0$, we compute the following sequence of approximate solutions:

$$\begin{bmatrix} 0.8 \\ -0.2 \\ 0.75 \end{bmatrix}, \begin{bmatrix} 0.99 \\ -0.34 \\ 0.45 \end{bmatrix}, \begin{bmatrix} 0.958 \\ -0.182 \\ 0.425 \end{bmatrix}, \begin{bmatrix} 0.921 \\ -0.178 \\ 0.362 \end{bmatrix}, \begin{bmatrix} 0.908 \\ -0.161 \\ 0.379 \end{bmatrix}, \begin{bmatrix} 0.908 \\ -0.170 \\ 0.376 \end{bmatrix}, \begin{bmatrix} 0.909 \\ -0.169 \\ 0.381 \end{bmatrix}, \ldots.$$

These vectors appear to be converging rapidly to the exact solution $x = [0.91 \; -0.17 \; 0.38]^{\mathrm{T}}$.

Now we describe how Jacobi's Algorithm works in general. The input to the algorithm is a matrix $A \in M_n(\mathbb{C})$, a column vector $b \in \mathbb{C}^n$, and an initial approximation $x_0 \in \mathbb{C}^n$. This algorithm requires that all diagonal entries $A(i,i)$ be nonzero. In fact, it is preferable for the diagonal entry in each row to be large in magnitude compared to the other entries in its row. If the initial matrix does not satisfy this requirement, we may adjust A by permuting rows and columns to make the requirement hold. Row permutations correspond to reordering the linear equations in the system $Ax = b$, and column permutations correspond to reordering the unknown components of x.

Write $A = Q + R$, where Q consists of the diagonal entries of A, and R consists of the off-diagonal entries of A. Formally, let $Q(i,i) = A(i,i)$ for $1 \leq i \leq n$, let all other entries of Q be zero, let $R(i,j) = A(i,j)$ for $1 \leq i,j \leq n$ with $i \neq j$, and let all other entries of R be zero. Since Q is a diagonal matrix with nonzero entries on the diagonal, we can quickly compute Q^{-1} by inverting each diagonal entry. We now compute the sequence of approximations $x_0, x_1, x_2, \ldots, x_k, \ldots$ by setting

$$x_{k+1} = Q^{-1}(b - Rx_k) \qquad \text{for all } k \geq 0. \tag{10.2}$$

To see that this matches the description of the algorithm given in the example above, note that the ith row of $Ax = b$ is the equation

$$b(i) = \sum_{j=1}^{n} A(i,j)x(j) = A(i,i)x(i) + \sum_{j \neq i} A(i,j)x(j) = Q(i,i)x(i) + \sum_{j=1}^{n} R(i,j)x(j).$$

Here, we write $x(i)$ for the ith component of the vector x to avoid confusion with the subscripts in the sequence of approximations $x_0, x_1, x_2, \ldots$. Solving the ith equation for $x(i)$ and using the resulting formula as the update rule for obtaining $x_{k+1}(i)$ from x_k, we get

$$x_{k+1}(i) = Q(i,i)^{-1} \left[b(i) - \sum_{j=1}^{n} R(i,j)x_k(j) \right]$$

for $1 \leq i \leq n$. These formulas are equivalent to the matrix equation (10.2).

We analyze the convergence properties of Jacobi's Algorithm in §10.12.

10.3   Gauss–Seidel Algorithm

The *Gauss–Seidel Algorithm* is a variation of Jacobi's Algorithm that uses the updated value of each unknown variable as soon as that value is computed. We illustrate the idea with the same 3×3 system $Ax = b$ considered in §10.2. The key modification to the previous example is that the update rules in (10.1) are replaced by

$$\begin{cases} r_{k+1} &= & (4 - s_k + t_k)/5 \\ s_{k+1} &= & (-1 + r_{k+1} - 2t_k)/5 \\ t_{k+1} &= & (3 - 2r_{k+1} - 2s_{k+1})/4 \end{cases} \qquad \text{for all } k \geq 0. \tag{10.3}$$

The update rule for r_{k+1} is the same as before. When we compute s_{k+1}, we have already computed the new value r_{k+1} but we have not yet computed t_{k+1}. So we use r_{k+1} and t_k on the right side of the update rule for s_{k+1}. When we compute t_{k+1}, we use the newly computed values r_{k+1} and s_{k+1} on the right side instead of the old values r_k and s_k.

Using the Gauss–Seidel formulas and starting with $x_0 = 0$, we compute the following sequence of approximate solutions:

$$\begin{bmatrix} 0.8 \\ -0.04 \\ 0.37 \end{bmatrix}, \begin{bmatrix} 0.882 \\ -0.172 \\ 0.395 \end{bmatrix}, \begin{bmatrix} 0.913 \\ -0.175 \\ 0.381 \end{bmatrix}, \begin{bmatrix} 0.911 \\ -0.170 \\ 0.379 \end{bmatrix}, \begin{bmatrix} 0.910 \\ -0.170 \\ 0.380 \end{bmatrix}, \ldots$$

This sequence appears to be converging to the exact solution $x = [0.91\ -0.17\ 0.38]^{\mathrm{T}}$ even faster than the sequence produced by the Jacobi method.

To solve a general system $Ax = b$ using the Gauss–Seidel Algorithm, we require that every diagonal entry of A be nonzero (which may require preprocessing A by permuting rows and columns). Starting with any initial vector x_0, we compute $x_1, x_2, \ldots$, via the update rules

$$x_{k+1}(i) = A(i,i)^{-1} \left[b(i) - \sum_{j<i} A(i,j)x_{k+1}(j) - \sum_{j>i} A(i,j)x_k(j) \right] \qquad \text{for all } k \geq 0,$$

(10.4)

which are computed in order as i increases from 1 to n.

Next, we give a matrix formulation of the Gauss–Seidel iterative procedure. We first rewrite (10.3) by moving all terms with subscript $k+1$ to the left side, obtaining

$$\begin{bmatrix} 5 & 0 & 0 \\ -1 & 5 & 0 \\ 2 & 2 & 4 \end{bmatrix} \begin{bmatrix} r_{k+1} \\ s_{k+1} \\ t_{k+1} \end{bmatrix} = \begin{bmatrix} 4 \\ -1 \\ 3 \end{bmatrix} + \begin{bmatrix} 0 & -1 & 1 \\ 0 & 0 & -2 \\ 0 & 0 & 0 \end{bmatrix} \begin{bmatrix} r_k \\ s_k \\ t_k \end{bmatrix}.$$

The matrix on the left consists of the entries in the original matrix A on or below the main diagonal. The matrix on the right is found by negating the entries of A above the main diagonal. For the general system $Ax = b$, define Q and R in $M_n(\mathbb{C})$ by setting $Q(i,j) = A(i,j)$ for all $i \geq j$, $Q(i,j) = 0$ for $i < j$, and $R = A - Q$. Multiplying (10.4) by $A(i,i)$ and adding $\sum_{j<i} A(i,j)x_{k+1}(j)$ to both sides, we see that the equations (10.4) (for i ranging from 1 to n) are equivalent to the matrix update rule

$$Qx_{k+1} = b - Rx_k \qquad \text{for all } k \geq 0.$$

When using this update rule, it should be kept in mind that we do not find x_{k+1} by explicitly computing Q^{-1} and applying this matrix to $b - Rx_k$. Rather, we compute $b - Rx_k$ and then backsolve for x_{k+1} as indicated in (10.4). Note that (10.2) can also be written in the form $Qx_{k+1} = b - Rx_k$, but with a different choice of Q and R. In both cases, a key property of Q is that $Qx_{k+1} = y$ can be quickly solved for x_{k+1} in much less time than the n^3 steps required for a general matrix inversion.

We study the convergence properties of the Gauss–Seidel Algorithm in §10.14.

10.4 Vector Norms

To analyze the convergence of iterative algorithms, we need to define the distance between two vectors in a vector space. To do so, we first introduce the concept of the length (or norm) of a vector. Given a real or complex vector space V, a *norm* on V is a function $N : V \to \mathbb{R}$, denoted $N(x) = ||x||$ for $x \in V$, which must satisfy the following axioms.

- *Nonnegativity:* For all $x \in V$, $0 \leq ||x|| < \infty$; and $||x|| = 0$ iff $x = 0$.

- *Scalar Rule:* For all $x \in V$ and all scalars c, $||cx|| = |c| \cdot ||x||$.

- *Triangle Inequality:* For all $x, y \in V$, $||x + y|| \leq ||x|| + ||y||$.

A vector space V together with a specific norm N on V is called a *normed vector space*.

For example, let V be the real vector space $\mathbb{R}^n$. For $x = (x_1, x_2, \ldots, x_n) \in \mathbb{R}^n$, define the *sup norm* by setting

$$||x||_\infty = \max(|x_1|, |x_2|, \ldots, |x_n|). \tag{10.5}$$

Let us check the axioms for a normed vector space. Fix $x, y \in \mathbb{R}^n$ and $c \in \mathbb{R}$. First, since $|x_i| \geq 0$ for all i, the maximum of these n numbers is also nonnegative and finite; and the maximum is zero iff all $x_i = 0$ iff $x = 0$. So $||x||_\infty \geq 0$ with equality iff $x = 0$. Second, $|cx_i| = |c| \cdot |x_i|$ for each i, so

$$||cx||_\infty = \max(|c| \cdot |x_1|, \ldots, |c| \cdot |x_n|) = |c| \max(|x_1|, \ldots, |x_n|) = |c| \cdot ||x||_\infty.$$

Third, the Triangle Inequality in $\mathbb{R}$ tells us that $|x_i + y_i| \leq |x_i| + |y_i|$ for $1 \leq i \leq n$. Now, $|x_i| \leq \max_{1 \leq j \leq n} |x_j| = ||x||_\infty$ and $|y_i| \leq \max_{1 \leq j \leq n} |y_j| = ||y||_\infty$, so

$$|x_i + y_i| \leq ||x||_\infty + ||y||_\infty.$$

This inequality holds for all i between 1 and n, so

$$||x + y||_\infty = \max(|x_1 + y_1|, \ldots, |x_n + y_n|) \leq ||x||_\infty + ||y||_\infty,$$

as needed. An identical proof shows that $\mathbb{C}^n$ is a complex normed vector space using the sup norm; in this case, $|x_j|$ denotes the magnitude of the complex number x_j.

Another norm on the real vector space $\mathbb{R}^n$ is the *1-norm*, defined by

$$||x||_1 = |x_1| + |x_2| + \cdots + |x_n| \tag{10.6}$$

for all $x = (x_1, \ldots, x_n) \in \mathbb{R}^n$. The first two axioms are readily verified; let us check the Triangle Inequality. Given $x, y \in \mathbb{R}^n$, we have $|x_i + y_i| \leq |x_i| + |y_i|$ in $\mathbb{R}$ for $1 \leq i \leq n$. Adding these n inequalities, we obtain

$$||x + y||_1 = \sum_{i=1}^{n} |x_i + y_i| \leq \sum_{i=1}^{n} (|x_i| + |y_i|) = \sum_{i=1}^{n} |x_i| + \sum_{i=1}^{n} |y_i| = ||x||_1 + ||y||_1.$$

A third example of a norm on $\mathbb{R}^n$ is the *Euclidean norm* or *2-norm*, defined by

$$||x||_2 = \sqrt{|x_1|^2 + |x_2|^2 + \cdots + |x_n|^2} \tag{10.7}$$

for $x = (x_1, \ldots, x_n) \in \mathbb{R}^n$. You can check that the norm axioms are satisfied; the proof of the Triangle Inequality follows from the Cauchy–Schwarz Inequality (see Exercise 29). Similarly, you can verify that formulas (10.6) and (10.7) define norms on the complex vector space $\mathbb{C}^n$.

A vector u in a normed vector space V is called a *unit vector* iff $||u|| = 1$. Given any nonzero $v \in V$, $||v||^{-1}v$ is a unit vector. To verify this, let $c = ||v||^{-1}$, and note $||cv|| = |c| \cdot ||v|| = ||v||^{-1}||v|| = 1$.

10.5   Metric Spaces

A *metric space* is a set X together with a *distance function* (or *metric*) $d : X \times X \to \mathbb{R}$, which must satisfy the following axioms.

- *Nonnegativity:* For all $x, y \in X$, $0 \le d(x, y) < \infty$, and $d(x, y) = 0$ iff $x = y$.

- *Symmetry:* For all $x, y \in X$, $d(x, y) = d(y, x)$.

- *Triangle Inequality:* For all $x, y, z \in X$, $d(x, z) \le d(x, y) + d(y, z)$.

The nonnegative real number $d(x, y)$ is the *distance* from x to y.

Any normed vector space V becomes a metric space if we define $d(x, y) = ||x - y||$ for $x, y \in V$. To check this, fix $x, y, z \in V$. We have $0 \le ||x - y|| < \infty$, with $||x - y|| = 0$ iff $x - y = 0$ iff $x = y$, so the first axiom for a metric space holds. We have

$$d(y, x) = ||y - x|| = ||(-1)(x - y)|| = |-1| \cdot ||x - y|| = ||x - y|| = d(x, y),$$

so the second axiom holds. We compute

$$d(x, z) = ||x - z|| = ||(x - y) + (y - z)|| \le ||x - y|| + ||y - z|| = d(x, y) + d(y, z),$$

so the third axiom holds. A metric defined from a norm in this way has two additional properties related to the vector space structure of V.

- *Translation Invariance:* For all $x, y, z \in V$, $d(x + z, y + z) = d(x, y)$.

- *Dilation Property:* For all $x, y \in V$ and all scalars c, $d(cx, cy) = |c| d(x, y)$.

Translation invariance holds since $d(x + z, y + z) = ||(x + z) - (y + z)|| = ||x - y|| = d(x, y)$. To check the dilation property, compute

$$d(cx, cy) = ||cx - cy|| = ||c(x - y)|| = |c| \cdot ||x - y|| = |c| d(x, y).$$

10.6   Convergence of Sequences

In any metric space (X, d), we can define what it means for a sequence to converge. Given $x \in X$ and given a sequence $(x_0, x_1, x_2, \ldots) = (x_k : k \ge 0)$ of points in X, we say that the sequence (x_k) *converges to* x iff for every $\epsilon > 0$, there exists $k_0 \in \mathbb{Z}_{\ge 0}$ such that for all $k \ge k_0$, $d(x_k, x) < \epsilon$. Informally, this definition says that the sequence gets arbitrarily close to x (and stays close) for terms far enough out in the sequence. When this condition holds, we write "$\lim_{k \to \infty} x_k = x$," and we also write "$x_k \to x$ as $k \to \infty$."

Not all sequences in a metric space (X, d) converge. But *if a sequence $(x_k : k \ge 0)$ converges to some limit x, then x is unique.* To prove this, suppose $x_k \to x$ and also $x_k \to y$ for some $x, y \in X$. Given $\epsilon > 0$, choose $k_1, k_2 \in \mathbb{Z}_{\ge 0}$ so that $d(x_k, x) < \epsilon/2$ for all $k \ge k_1$ and $d(x_k, y) < \epsilon/2$ for all $k \ge k_2$. Take $k = \max(k_1, k_2)$ to see that $d(x, y) \le d(x, x_k) + d(x_k, y) < \epsilon$. Since this holds for every positive ϵ, we must have $d(x, y) = 0$, so that $x = y$.

Now suppose V is a normed vector space; we always assume V has the metric $d(x, y) = ||x - y||$ defined using the norm. Given two sequences $(v_k : k \ge 0)$ and $(w_k : k \ge 0)$ in V, suppose $v_k \to v$ and $w_k \to w$ for some $v, w \in V$. We assert that $v_k + w_k \to v + w$.

To prove this, fix $\epsilon > 0$. Choose $k_1, k_2 \in \mathbb{Z}_{\geq 0}$ such that $d(v_k, v) < \epsilon/2$ for all $k \geq k_1$ and $d(w_k, w) < \epsilon/2$ for all $k \geq k_2$. For all $k \geq \max(k_1, k_2)$, the Triangle Inequality gives

$$d(v_k + w_k, v + w) = ||(v_k + w_k) - (v + w)|| = ||(v_k - v) + (w_k - w)||$$
$$\leq ||v_k - v|| + ||w_k - w|| < \epsilon/2 + \epsilon/2 = \epsilon.$$

So $\lim_{k \to \infty} (v_k + w_k) = v + w$.

Next, assume c is a scalar and $v_k \to v$; we assert that $cv_k \to cv$. This result is evident when $c = 0$, so assume $c \neq 0$. Given $\epsilon > 0$, choose k_0 so that $d(v_k, v) < \epsilon/|c|$ for all $k \geq k_0$. Then for all $k \geq k_0$, we compute $d(cv_k, cv) = |c|d(v_k, v) < \epsilon$. So $\lim_{k \to \infty}(cv_k) = cv$. More generally, you can show that if $c_k \to c$ in $\mathbb{R}$ or $\mathbb{C}$ and $v_k \to v$ in V, then $c_k v_k \to cv$ in V (Exercise 60). You can also show by induction that if $(v_k^{(1)} : k \geq 0), \ldots, (v_k^{(m)} : k \geq 0)$ are m sequences converging respectively to $v^{(1)}, \ldots, v^{(m)}$, and if $c_1, \ldots, c_m$ are fixed scalars, then

$$\lim_{k \to \infty} (c_1 v_k^{(1)} + \cdots + c_m v_k^{(m)}) = c_1 v^{(1)} + \cdots + c_m v^{(m)}.$$

We say that *limits preserve linear combinations of vectors.*

10.7 Comparable Norms

As we have seen, a given vector space V may possess many different norms and hence many different metrics. The definition of a convergent sequence depends on the metric used, so it is possible that a given sequence $(v_k : k \geq 0)$ might converge to some v relative to one norm, but not converge to v relative to some other norm. However, when V is finite-dimensional, we can show that this does not happen. In other words, for any two norms $|| \cdot ||$ and $|| \cdot ||'$ on V, we will prove that $v_k \to v$ relative to $|| \cdot ||$ iff $v_k \to v$ relative to $|| \cdot ||'$. So for studying questions of convergence, we can use whichever norm is most convenient for calculations.

We say that two norms $|| \cdot ||$ and $|| \cdot ||'$ on a vector space V are *comparable* iff there exist real constants C, D with $0 < C, D < \infty$ such that for all $x \in V$, $||x|| \leq C||x||'$ and $||x||' \leq D||x||$. For example, let us show that the sup norm and the 2-norm are comparable on $V = \mathbb{R}^n$. It suffices to prove that for all $x = (x_1, \ldots, x_n) \in \mathbb{R}^n$,

$$||x||_\infty \leq ||x||_2 \leq \sqrt{n}||x||_\infty$$

(here $C = 1$ and $D = \sqrt{n}$). To prove the first inequality, observe that $|x_i| = \sqrt{|x_i|^2} \leq \sqrt{\sum_{j=1}^n |x_j|^2} = ||x||_2$ for all i. Taking the maximum over all i gives $||x||_\infty \leq ||x||_2$. To prove the second inequality, fix an index k with $|x_k| = \max_i |x_i|$. Then $|x_i|^2 \leq |x_k|^2$ for all i, so

$$||x||_2 = \sqrt{\sum_{i=1}^n |x_i|^2} \leq \sqrt{\sum_{i=1}^n |x_k|^2} = \sqrt{n}|x_k| = \sqrt{n}||x||_\infty.$$

Similarly, you may prove that the sup norm and the 1-norm on $\mathbb{R}^n$ are comparable by showing that

$$||x||_\infty \leq ||x||_1 \leq n||x||_\infty$$

for all $x \in \mathbb{R}^n$. Comparability of norms is an equivalence relation (Exercise 65), so it follows that the 1-norm and the 2-norm are comparable as well.

We now sketch the proof that *any vector norm* $|| \cdot ||$ *on* $\mathbb{R}^n$ *is comparable to the* 2-*norm.* (A similar result holds for $\mathbb{C}^n$, but these results do not extend to infinite-dimensional spaces.) First we prove that there is a constant $C \in (0, \infty)$ such that for all $x = (x_1, \ldots, x_n) \in \mathbb{R}^n$, $||x|| \le C||x||_2$. Let $\mathbf{e}_1, \ldots, \mathbf{e}_n$ be the standard basis vectors in $\mathbb{R}^n$, so that $x = x_1\mathbf{e}_1 + \cdots + x_n\mathbf{e}_n$. Let $C = ||\mathbf{e}_1|| + ||\mathbf{e}_2|| + \cdots + ||\mathbf{e}_n||$, which is a positive finite constant. Use the norm axioms to compute

$$\begin{aligned} ||x|| &= ||x_1\mathbf{e}_1 + \cdots + x_n\mathbf{e}_n|| \le ||x_1\mathbf{e}_1|| + \cdots + ||x_n\mathbf{e}_n|| \\ &= |x_1| \cdot ||\mathbf{e}_1|| + \cdots + |x_n| \cdot ||\mathbf{e}_n|| \\ &\le ||x||_\infty ||\mathbf{e}_1|| + \cdots + ||x||_\infty ||\mathbf{e}_n|| = C||x||_\infty \le C||x||_2. \end{aligned}$$

Obtaining the inequality $||x||_2 \le D||x||$ requires a compactness argument. Here, we cite without proof some results from calculus on the metric space $\mathbb{R}^n$ with the Euclidean metric $d(x, y) = ||x - y||_2$. Let $S = \{x \in \mathbb{R}^n : ||x||_2 = 1\}$ be the unit sphere in this metric space. The set S is closed and bounded in $\mathbb{R}^n$, so S is a compact subset of $\mathbb{R}^n$. Define $f : S \to \mathbb{R}$ by $f(x) = ||x||$ for $x \in S$. We claim the function f is continuous. Given $\epsilon > 0$, let $\delta = \epsilon/C$. Then for $x, y \in S$ satisfying $d(x, y) = ||x - y||_2 < \delta$, we have

$$|f(x) - f(y)| = |\,||x|| - ||y||\,| \le ||x - y|| \le C||x - y||_2 < \epsilon.$$

(The first inequality follows from Exercise 38.) From calculus, we know that a continuous real-valued function on a compact set attains its minimum value at some point in the set. So, there exists $x_0 \in S$ such that $f(x_0) \le f(x)$ for all $x \in S$. Note that $x_0 \ne 0$ (since $||x_0||_2 = 1$), so $f(x_0) = ||x_0|| > 0$. Let $D = ||x_0||^{-1}$, which is a positive finite constant. Given any $x \in \mathbb{R}^n$, we can now prove that $||x||_2 \le D||x||$. This inequality certainly holds if $x = 0$. If $||x||_2 = 1$, then $x \in S$, so $f(x_0) \le f(x)$, which means $D^{-1} \le ||x||$. Multiplying by D, we get $||x||_2 = 1 \le D||x||$ as needed. Next, consider an arbitrary nonzero $y \in \mathbb{R}^n$. Let $c = ||y||_2$ and $x = c^{-1}y \in S$. Multiplying both sides of the known inequality $||x||_2 \le D||x||$ by c, we get $||cx||_2 \le D||cx||$, so $||y||_2 \le D||y||$. This completes the proof that $|| \cdot ||$ and $|| \cdot ||_2$ are comparable. By Exercise 65, any two norms on $\mathbb{R}^n$ (or $\mathbb{C}^n$) are comparable.

Given comparable norms $|| \cdot ||$ and $|| \cdot ||'$ on $\mathbb{R}^n$ or $\mathbb{C}^n$, we now show that $v_k \to v$ relative to the first norm iff $v_k \to v$ relative to the second norm. Choose constants C and D as in the definition of comparable norms. Suppose $v_k \to v$ relative to $|| \cdot ||$. Given $\epsilon > 0$, choose k_0 so that $k \ge k_0$ implies $||v_k - v|| < \epsilon/D$. Then $k \ge k_0$ implies $||v_k - v||' \le D||v_k - v|| < \epsilon$. So $v_k \to v$ relative to $|| \cdot ||'$. Conversely, suppose $v_k \to v$ relative to $|| \cdot ||'$. Given $\epsilon > 0$, choose k_1 so that $k \ge k_1$ implies $||v_k - v||' < \epsilon/C$. Then $k \ge k_1$ implies $||v_k - v|| \le C||v_k - v||' < \epsilon$. Combining this result with the preceding theorem, we conclude that *in* $\mathbb{R}^n$ *or* $\mathbb{C}^n$, $v_k \to v$ *relative to any given vector norm iff* $v_k \to v$ *relative to the 2-norm (or the 1-norm or the sup norm).*

10.8   Matrix Norms

Recall that $M_n(\mathbb{C})$ is the set of all $n \times n$ matrices with entries in $\mathbb{C}$. Since $M_n(\mathbb{C})$ is a complex vector space, we can consider vector norms defined on $M_n(\mathbb{C})$. All such norms satisfy $0 \le ||A|| < \infty$, $||A|| = 0$ iff $A = 0$, $||cA|| = |c| \cdot ||A||$, and $||A + B|| \le ||A|| + ||B||$ for all $A, B \in M_n(\mathbb{C})$ and all $c \in \mathbb{C}$. A *matrix norm* is a vector norm $|| \cdot ||$ on $M_n(\mathbb{C})$ satisfying this additional axiom:

- *Submultiplicativity:* For all $A, B \in M_n(\mathbb{C})$, $||AB|| \le ||A|| \cdot ||B||$.

For example, define $||A|| = n \cdot \max_{i,j} |A(i,j)|$ for $A \in M_n(\mathbb{C})$. The axioms for a vector norm can be checked as in §10.4. To prove submultiplicativity, fix $A, B \in M_n(\mathbb{C})$ and i, j between 1 and n. We know $(AB)(i,j) = \sum_{k=1}^{n} A(i,k)B(k,j)$, so

$$n|(AB)(i,j)| \leq \sum_{k=1}^{n} n|A(i,k)| \cdot |B(k,j)| \leq \sum_{k=1}^{n} n(n^{-1}||A||)(n^{-1}||B||) = ||A|| \cdot ||B||.$$

This holds for all i, j, so taking the maximum gives $||AB|| \leq ||A|| \cdot ||B||$ as needed. (The inequality would not hold if we omitted the n from the definition of $||A||$.)

Given any vector norm $||\cdot||$ on $\mathbb{C}^n$, we now construct a matrix norm on $M_n(\mathbb{C})$ called the matrix norm *induced by* the given vector norm. In addition to the properties listed above, this matrix norm satisfies $||I_n|| = 1$ and $||Av|| \leq ||A|| \cdot ||v||$ for all $A \in M_n(\mathbb{C})$ and $v \in \mathbb{C}^n$. To define the matrix norm, let $U = \{v \in \mathbb{C}^n : ||v|| = 1\}$ be the set of unit vectors in $\mathbb{C}^n$ relative to the given vector norm. For $A \in M_n(\mathbb{C})$, let

$$||A|| = \sup\{||Av|| : v \in U\}. \tag{10.8}$$

So the norm of A is the least upper bound of the set of lengths $||Av||$ as v ranges over all unit vectors in $\mathbb{C}^n$ (we view elements of $\mathbb{C}^n$ as column vectors here). For example, $||0|| = \sup\{||0v|| : v \in U\} = \sup\{0\} = 0$ and $||I_n|| = \sup\{||I_n v|| : v \in U\} = \sup\{1\} = 1$.

We now prove that our definition satisfies the requirements of a matrix norm. Fix A, B in $M_n(\mathbb{C})$ and $c \in \mathbb{C}$. For the nonnegativity axiom, note that $0 \leq ||A|| \leq \infty$ since $||A||$ is the least upper bound of a set of nonnegative real numbers. Proving that $||A||$ is finite is a bit tricky. We know $||\cdot||$ is comparable to $||\cdot||_\infty$, so there is a positive finite constant C with $||v||_\infty \leq C||v||$ for all $v \in \mathbb{C}^n$. Given any $v = (v_1, \ldots, v_n) \in U$, we have $v = v_1\mathbf{e}_1 + \cdots + v_n\mathbf{e}_n$, so $Av = A\sum_{i=1}^{n} v_i\mathbf{e}_i = \sum_{i=1}^{n} v_i(A\mathbf{e}_i)$. Taking norms gives

$$
\begin{aligned}
||Av|| &\leq \sum_{i=1}^{n} ||v_i(A\mathbf{e}_i)|| = \sum_{i=1}^{n} |v_i| \cdot ||A\mathbf{e}_i|| \\
&\leq \sum_{i=1}^{n} ||v||_\infty ||A\mathbf{e}_i|| \leq \sum_{i=1}^{n} C||v|| \cdot ||A\mathbf{e}_i|| = C\sum_{i=1}^{n} ||A\mathbf{e}_i||.
\end{aligned}
$$

Thus, $C\sum_{i=1}^{n} ||A\mathbf{e}_i||$ is a finite upper bound for the set $\{||Av|| : v \in U\}$, so the least upper bound $||A||$ of this set is indeed finite. We already saw that $||0|| = 0$. On the other hand, a nonzero matrix A must have some nonzero column $A^{[j]} = A\mathbf{e}_j$. Taking $v = ||\mathbf{e}_j||^{-1}\mathbf{e}_j$, which is a unit vector, we have $||Av|| = ||\mathbf{e}_j||^{-1}||A\mathbf{e}_j|| \neq 0$, so $||A|| > 0$.

For the scalar axiom, note that $||(cA)v|| = |c| \cdot ||Av||$ for all $v \in U$. So the least upper bound of the numbers $||(cA)v||$ over all $v \in U$ is $|c|$ times the least upper bound of the numbers $||Av||$, giving $||cA|| = |c| \cdot ||A||$. To verify the Triangle Inequality, fix $v \in U$ and compute $||(A+B)v|| = ||Av + Bv|| \leq ||Av|| + ||Bv|| \leq ||A|| + ||B||$. Thus $||A|| + ||B||$ is an upper bound for the set $\{||(A+B)v|| : v \in U\}$, and hence the least upper bound $||A+B||$ of this set satisfies $||A+B|| \leq ||A|| + ||B||$.

Next, we check that $||Av|| \leq ||A|| \cdot ||v||$ for $A \in M_n(\mathbb{C})$ and $v \in \mathbb{C}^n$. Both sides are zero when $v = 0$. For nonzero v, write $v = cu$ where $c = ||v||$ and $u \in U$. Then compute

$$||Av|| = ||A(cu)|| = ||c(Au)|| = |c| \cdot ||Au|| \leq |c| \cdot ||A|| = ||A|| \cdot ||v||.$$

To prove submultiplicativity, fix $A, B \in M_n(\mathbb{C})$. For any $v \in U$, the fact just proved gives

$$||(AB)v|| = ||A(Bv)|| \leq ||A|| \cdot ||Bv|| \leq ||A|| \cdot ||B|| \cdot ||v|| = ||A|| \cdot ||B||.$$

Thus $||A|| \cdot ||B||$ is an upper bound for the set $\{||(AB)v|| : v \in U\}$, and hence the least upper bound $||AB||$ of this set satisfies $||AB|| \leq ||A|| \cdot ||B||$.

Not all matrix norms are induced by vector norms via (10.8). For example, the norm $||A|| = n \cdot \max_{i,j} |A(i,j)|$ cannot be induced by any vector norm since $||I_n|| = n \neq 1$. Let us call a matrix norm $|| \cdot ||$ *induced* iff there exists a vector norm for which $|| \cdot ||$ is the matrix norm induced by this vector norm.

The proof of the following lemma uses the matrix norm induced by a given vector norm. Suppose $(v_k : k \geq 0)$ is a sequence in $\mathbb{C}^n$ converging to $v \in \mathbb{C}^n$ relative to a vector norm $||\cdot||$. Then for all $A \in M_n(\mathbb{C})$, $Av_k \to Av$. The result holds when $A = 0$, so assume $A \neq 0$. Fix $\epsilon > 0$, and choose k such that $k \geq k_0$ implies $||v_k - v|| < \epsilon/||A||$, where $||A||$ is the matrix norm induced by the given vector norm. Then $k \geq k_0$ implies $||Av_k - Av|| = ||A(v_k - v)|| \leq ||A|| \cdot ||v_k - v|| < \epsilon$. For this proof to work, it is critical to know (as proved above) that $||A||$ is finite.

10.9 Formulas for Matrix Norms

In this section, we develop some explicit formulas for computing the matrix norms induced by the sup norm, the 1-norm, and the 2-norm on $\mathbb{C}^n$. We start by showing that the matrix norm $||A||_\infty$ induced by the vector norm $||v||_\infty$ satisfies

$$||A||_\infty = \max_{1 \leq i \leq n} \sum_{j=1}^n |A(i,j)| = \max_{1 \leq i \leq n} ||A_{[i]}||_1, \qquad (10.9)$$

where $A_{[i]} = (A(i,1), \ldots, A(i,n))$ is row i of A. Let $U = \{v \in \mathbb{C}^n : ||v||_\infty = 1\}$. For any $v \in U$, we compute

$$||Av||_\infty = \max_{1 \leq i \leq n} |(Av)_i| = \max_{1 \leq i \leq n} \left| \sum_{j=1}^n A(i,j)v_j \right|$$
$$\leq \max_{1 \leq i \leq n} \sum_{j=1}^n |A(i,j)| \cdot |v_j| \leq \max_{1 \leq i \leq n} \sum_{j=1}^n |A(i,j)| \cdot ||v||_\infty = \max_{1 \leq i \leq n} \sum_{j=1}^n |A(i,j)|.$$

So the expression on the right side of (10.9) is an upper bound for $\{||Av||_\infty : v \in U\}$, hence $||A||_\infty \leq \max_{1 \leq i \leq n} \sum_{j=1}^n |A(i,j)|$. To establish the reverse inequality, fix an index k between 1 and n for which $\sum_{j=1}^n |A(k,j)|$ attains its maximum. For each j between 1 and n, choose v_j to be a complex number of modulus 1 such that $A(k,j)v_j = |A(k,j)|$. (If $A(k,j) = re^{i\theta}$ in polar form, we can take $v_j = e^{-i\theta}$.) Now observe that $v = (v_1, \ldots, v_n) \in \mathbb{C}^n$ has $||v||_\infty = 1$ since $|v_j| = 1$ for all j. So $v \in U$, and we conclude that

$$||A||_\infty \geq ||Av||_\infty = \max_{1 \leq i \leq n} |(Av)_i| \geq |(Av)_k| = \left| \sum_{j=1}^n A(k,j)v_j \right| = \sum_{j=1}^n |A(k,j)|$$
$$= \max_{1 \leq i \leq n} \sum_{j=1}^n |A(i,j)|.$$

Next, we prove that the matrix norm $||A||_1$ induced by the vector norm $||v||_1$ satisfies

$$||A||_1 = \max_{1 \leq j \leq n} \sum_{i=1}^n |A(i,j)| = \max_{1 \leq j \leq n} ||A^{[j]}||_1, \qquad (10.10)$$

where $A^{[j]} = (A(1,j), \ldots, A(n,j))^{\mathrm{T}}$ is column j of A. Let $U = \{v \in \mathbb{C}^n : ||v||_1 = 1\}$. For any $v \in U$,

$$
\begin{aligned}
||Av||_1 &= \sum_{i=1}^{n} |(Av)_i| = \sum_{i=1}^{n} \left| \sum_{j=1}^{n} A(i,j) v_j \right| \\
&\leq \sum_{i=1}^{n} \sum_{j=1}^{n} |A(i,j)| \cdot |v_j| = \sum_{j=1}^{n} |v_j| \sum_{i=1}^{n} |A(i,j)| \\
&\leq \sum_{j=1}^{n} |v_j| \left[\max_{1 \leq j \leq n} \sum_{i=1}^{n} |A(i,j)| \right] = ||v||_1 \max_{1 \leq j \leq n} \sum_{i=1}^{n} |A(i,j)| = \max_{1 \leq j \leq n} \sum_{i=1}^{n} |A(i,j)|.
\end{aligned}
$$

Since $v \in U$ is arbitrary, we conclude that $||A||_1 \leq \max_{1 \leq j \leq n} \sum_{i=1}^{n} |A(i,j)|$. For the reverse inequality, fix an index k for which $\sum_{i=1}^{n} |A(i,k)|$ is maximized. Since $||\mathbf{e}_k||_1 = 1$, $\mathbf{e}_k \in U$, and hence

$$
||A||_1 \geq ||A\mathbf{e}_k||_1 = ||A^{[k]}||_1 = \sum_{i=1}^{n} |A(i,k)| = \max_{1 \leq j \leq n} \sum_{i=1}^{n} |A(i,j)|.
$$

Let $||A||_2$ be the matrix norm induced by the vector norm $||v||_2$. For any $B \in M_n(\mathbb{C})$, define the *spectral radius* $\rho(B)$ to be the maximum magnitude of all the complex eigenvalues of B, i.e.,

$$
\rho(B) = \max\{|c| : c \in \mathbb{C} \text{ is an eigenvalue of } B\}.
$$

It can be shown (Exercise 72) that $||A||_2 = \sqrt{\rho(A^*A)}$ for $A \in M_n(\mathbb{C})$. You can also show that $||A||_2$ is the largest singular value of A (see Exercise 59 in Chapter 9).

10.10   Matrix Inversion via Geometric Series

The *Geometric Series Formula*

$$
\frac{1}{1-r} = 1 + r + r^2 + r^3 + \cdots + r^k + \cdots = \sum_{k=0}^{\infty} r^k
$$

is valid for all complex numbers r such that $|r| < 1$. In this section, we prove an analogous formula in which r is replaced by an $n \times n$ complex matrix C. We assume that $||C|| < 1$ for some induced matrix norm on $\mathbb{C}^n$; this is the analog of the hypothesis $|r| < 1$.

Under this assumption, we show that $I - C$ is an invertible matrix, and

$$
(I - C)^{-1} = I + C + C^2 + \cdots + C^k + \cdots = \sum_{k=0}^{\infty} C^k, \tag{10.11}
$$

where the infinite series of matrices denotes the limit of the sequence of partial sums $(I + C + C^2 + \cdots + C^k : k \geq 0)$. To get a contradiction, assume that $I - C$ is not invertible. Then there exists a nonzero vector x with $(I - C)x = 0$, so $x = Cx$. Taking norms gives $||x|| = ||Cx|| \leq ||C|| \cdot ||x||$. Dividing by the positive real number $||x||$ gives $1 \leq ||C||$, contradicting our hypothesis on C. So $B = (I - C)^{-1}$ does exist. B must be nonzero, so $||B|| > 0$.

Now fix $k \geq 0$, and compute

$$(I-C)(I+C+C^2+\cdots+C^k) = (I+C+C^2+\cdots+C^k)-(C+C^2+\cdots+C^k+C^{k+1}) = I-C^{k+1}.$$

Multiplying both sides by B gives $I + C + C^2 + \cdots + C^k = B - BC^{k+1}$. Given $\epsilon > 0$, we can choose k_0 so that $k \geq k_0$ implies $||C||^{k+1} < \epsilon/||B||$; this is possible since $||C|| < 1$. For any $k \geq k_0$, it follows that

$$||(I + C + C^2 + \cdots + C^k) - B|| = || - BC^{k+1}|| \leq ||B|| \cdot ||C||^{k+1} < \epsilon.$$

This proves that the sequence of partial sums converges to $B = (I - C)^{-1}$, as needed.

10.11 Affine Iteration and Richardson's Algorithm

In this section, we study the convergence of the iterative algorithm obtained by repeatedly applying an affine map to a given initial vector. The input to the algorithm consists of a fixed matrix $C \in M_n(\mathbb{C})$, a vector $d \in \mathbb{C}^n$, and an initial vector $x_0 \in \mathbb{C}^n$. We construct a sequence $(x_k : k \geq 0)$ in $\mathbb{C}^n$ by the recursive formula

$$x_{k+1} = Cx_k + d \qquad (k \geq 0). \tag{10.12}$$

We seek conditions under which this sequence of vectors converges to some limit $x \in \mathbb{C}^n$.

If the limit x exists, then we can find it by letting k go to infinity on both sides of (10.12). Assuming that the sequence (x_k) converges to x, the sequence (x_{k+1}) on the left side also converges to x. On the right side, (Cx_k) converges to Cx, so $(Cx_k + d)$ converges to $Cx + d$. The two sides converge to the same limit, so $x = Cx + d$. This means that the limit x must solve the linear system $(I - C)x = d$. If $I - C$ is invertible, then the only possible limit of the iterative algorithm is $x = (I - C)^{-1}d$.

To prove convergence, we assume that $||C|| < 1$ for some induced matrix norm on $\mathbb{C}^n$. As shown in the last section, this assumption ensures that $I - C$ is invertible. We prove that for any choice of initial vector $x_0 \in \mathbb{C}^n$, the sequence (x_k) defined by (10.12) converges to $x = (I - C)^{-1}d$, and we find an upper bound on the error $||x_k - x||$. We know $x_{k+1} = Cx_k + d$ for $k \geq 0$, and $x = Cx + d$. Subtracting these equations gives $x_{k+1} - x = (Cx_k + d) - (Cx + d) = Cx_k - Cx = C(x_k - x)$. Taking norms, we get

$$||x_{k+1} - x|| = ||C(x_k - x)|| \leq ||C|| \cdot ||x_k - x||$$

for all $k \geq 0$. If $k \geq 1$, we can write $||x_k - x|| \leq ||C|| \cdot ||x_{k-1} - x||$ to bound the right side, giving $||x_{k+1} - x|| \leq ||C||^2||x_{k-1} - x||$. Continuing in this way, we eventually obtain the bound

$$||x_{k+1} - x|| \leq ||C||^{k+1}||x_0 - x||,$$

valid for all $k \geq 0$. The right side goes to zero at an exponential rate as k goes to infinity, since $||C|| < 1$. Therefore, $\lim_{k \to \infty} ||x_k - x|| = 0$, proving that $x_k \to x$ as needed.

This iterative method finds the unique x solving $(I - C)x = d$. So, if we want to use this method to solve a linear system $Ax = b$ (for given $A \in M_n(\mathbb{C})$ and $b \in \mathbb{C}^n$), we can choose $C = I - A$ and $d = b$. As long as $r = ||I - A|| < 1$ for some induced matrix norm, the iterative method converges to the solution x, with the error in the kth term bounded by r^k times the initial error $||x_0 - x||$. Since $C = I - A$, the iteration formula (10.12) becomes

$$x_{k+1} = x_k + (b - Ax_k) \qquad (k \geq 0),$$

and we have recovered Richardson's Algorithm for solving a linear system.

We can use the formulas for matrix norms in §10.9 to find explicit sufficient conditions on A guaranteeing that Richardson's Algorithm converges. On one hand, if $\max_{1 \leq i \leq n} \sum_{j=1}^{n} |(I - A)(i, j)| < 1$, then (10.9) shows that $||I - A||_\infty < 1$, so the algorithm converges. On the other hand, if $\max_{1 \leq j \leq n} \sum_{i=1}^{n} |(I - A)(i, j)| < 1$, then (10.10) shows that $||I - A||_1 < 1$, so the algorithm converges. For example, the matrix $A = \begin{bmatrix} 1.1 & 0.3 \\ -0.4 & 0.8 \end{bmatrix}$ considered in §10.1 satisfies both conditions just mentioned, so Richardson's Algorithm for solving $Ax = b$ converges for any choice of b and x_0. In fact, since $||I - A||_1 = 1/2$, we know that $||x_k - x|| \leq 2^{-k}||x_0 - x||$ for all $k \geq 1$, so that the bound for the maximum error is cut in half in each successive iteration.

10.12   Splitting Matrices and Jacobi's Algorithm

We can apply the analysis in the preceding section to iterative algorithms that solve $Ax = b$ by the following procedure. The algorithm picks an invertible *splitting matrix* Q (depending on the given $A \in M_n(\mathbb{C})$) and computes $(x_k : k \geq 0)$ using the iteration formula

$$Qx_{k+1} = b - (A - Q)x_k \qquad (k \geq 0). \tag{10.13}$$

For example, Richardson's Algorithm takes $Q = I$; Jacobi's Algorithm takes Q to be the diagonal part of A; and the Gauss–Seidel Algorithm takes Q to be the lower-triangular part of A. Recall that an algorithm of this kind is only practical if $Qx_{k+1} = y$ can be solved quickly for x_{k+1} given y.

By multiplying both sides of (10.13) on the left by Q^{-1}, we see that (10.13) is algebraically (though not computationally) equivalent to

$$x_{k+1} = Q^{-1}b - (Q^{-1}A - Q^{-1}Q)x_k = (I - Q^{-1}A)x_k + Q^{-1}b.$$

Therefore, we can invoke the results of §10.11 taking $C = I - Q^{-1}A$ and $d = Q^{-1}b$. Note that $(I - C)x = d$ iff $Q^{-1}Ax = Q^{-1}b$ iff $Ax = b$ for this choice of C and d. We conclude that if $||I - Q^{-1}A|| < 1$ for some induced matrix norm, then the sequence defined by (10.13) converges to the correct solution for any starting vector x_0. We also have the error bound

$$||x_k - x|| \leq ||I - Q^{-1}A||^k ||x_0 - x|| \tag{10.14}$$

for all $k \geq 0$.

Using this analysis, we can prove a sufficient condition on A guaranteeing the convergence of the Jacobi Algorithm for solving $Ax = b$. Call $A \in M_n(\mathbb{C})$ *diagonally dominant* iff

$$|A(i, i)| > \sum_{\substack{j=1 \\ j \neq i}}^{n} |A(i, j)| \qquad \text{for all } i \text{ between 1 and } n.$$

(Such a matrix must have all diagonal entries nonzero.) We show that *the Jacobi method always converges given a diagonally dominant input matrix A.* In this case, Q^{-1} is a diagonal matrix with $Q^{-1}(i, i) = A(i, i)^{-1}$. Therefore, the i, j-entry of $I - Q^{-1}A$ is 0 if $i = j$, and $-A(i, i)^{-1}A(i, j)$ otherwise. So for each i between 1 and n,

$$\sum_{j=1}^{n} |(I - Q^{-1}A)(i, j)| = \sum_{\substack{j=1 \\ j \neq i}}^{n} \frac{|A(i, j)|}{|A(i, i)|} < 1.$$

Taking the maximum over all i, it follows from (10.9) that $||I - Q^{-1}A||_\infty < 1$, proving convergence.

10.13 Induced Matrix Norms and the Spectral Radius

In the previous sections, we proved that various iterative algorithms converge if a certain matrix $B \in M_n(\mathbb{C})$ satisfies $||B|| < 1$ for some induced matrix norm. It is possible that $||B||_1 \geq 1$, $||B||_\infty \geq 1$, and yet $||B|| < 1$ for some other matrix norm induced by some vector norm other than $||\cdot||_1$ or $||\cdot||_\infty$. To obtain a sufficient condition for convergence that is as powerful as possible, we would really like to know the quantity

$$f(B) = \inf\{||B|| : ||\cdot|| \text{ is an induced matrix norm on } \mathbb{C}^n\}, \qquad (10.15)$$

which is the greatest lower bound in $\mathbb{R}$ of the numbers $||B||$ as the norm varies over all possible induced matrix norms. Recall from §10.9 that the spectral radius of B is

$$\rho(B) = \max\{|c| : c \in \mathbb{C} \text{ is an eigenvalue of } B\}.$$

We prove that $f(B) = \rho(B)$ *for all* $B \in M_n(\mathbb{C})$.

To begin, let $c \in \mathbb{C}$ be an eigenvalue of B with $|c| = \rho(B)$ and $v \neq 0$ be an associated eigenvector. Fix an arbitrary vector norm $||\cdot||$ on $\mathbb{C}^n$, and note that $u = ||v||^{-1}v$ satisfies $||u|| = 1$ and $Bu = cu$. Therefore, using the matrix norm induced by this vector norm, we compute

$$||B|| \geq ||Bu|| = ||cu|| = |c| \cdot ||u|| = |c| = \rho(B).$$

So $\rho(B)$ is a lower bound for the set of numbers on the right side of (10.15). As $f(B)$ is the greatest lower bound of this set, $\rho(B) \leq f(B)$ follows.

Showing the opposite inequality $f(B) \leq \rho(B)$ requires more work. It is enough to prove that $f(B) \leq \rho(B) + \epsilon$ for each fixed $\epsilon > 0$. Given $\epsilon > 0$, we first show that there is an invertible matrix $S \in M_n(\mathbb{C})$ such that $S^{-1}BS$ is upper-triangular and $||S^{-1}BS||_\infty \leq \rho(B) + \epsilon$. We give an argument using Jordan canonical forms (Chapter 8); a different argument based on unitary triangularization of complex matrices is sketched in Exercise 73. Define $T : \mathbb{C}^n \to \mathbb{C}^n$ by $T(v) = Bv$ for $v \in \mathbb{C}^n$. We know there is an ordered basis $X = (x_1, x_2, \ldots, x_n)$ for $\mathbb{C}^n$ such that $[T]_X$ is a Jordan canonical form. This means that for certain scalars $c_1, \ldots, c_n \in \mathbb{C}$ and $d_1, \ldots, d_n \in \{0, 1\}$, $T(x_1) = c_1 x_1$ and $T(x_i) = c_i x_i + d_i x_{i-1}$ for $2 \leq i \leq n$. Replace the ordered basis X by the ordered basis $Y = (y_1, y_2, \ldots, y_n)$, where $y_i = \epsilon^i x_i$ for $1 \leq i \leq n$. We compute $T(y_1) = c_1 y_1$ and

$$T(y_i) = T(\epsilon^i x_i) = \epsilon^i T(x_i) = \epsilon^i(c_i x_i + d_i x_{i-1}) = c_i y_i + (d_i \epsilon)y_{i-1}$$

for $2 \leq i \leq n$. Letting $S = {}_E[\mathrm{id}]_Y$, where $E = (e_1, \ldots, e_n)$ is the standard ordered basis of $\mathbb{C}^n$, we see that $U = S^{-1}BS = [T]_Y$ is an upper-triangular matrix with main diagonal entries $c_1, \ldots, c_n$, entries equal to zero or ϵ on the next higher diagonal, and zeroes elsewhere. Since U is triangular, the eigenvalues of U are $c_1, \ldots, c_n$. Since B is similar to U, these are also the eigenvalues of B. For $1 \leq i \leq n$, $\sum_{j=1}^n |U(i,j)|$ is either $|c_i|$ or $|c_i| + \epsilon$. By (10.9), we see that

$$||U||_\infty \leq \max_{1 \leq i \leq n}(|c_i| + \epsilon) = \rho(B) + \epsilon.$$

To continue, we need a clever choice of a vector norm and its induced matrix norm. You can check (Exercise 39) that for any fixed invertible $S \in M_n(\mathbb{C})$, the formula $||v||_S =$

$||S^{-1}v||_\infty$ (for $v \in \mathbb{C}^n$) defines a vector norm on $\mathbb{C}^n$. Also, the induced matrix norm satisfies $||A||_S = ||S^{-1}AS||_\infty$ for all $A \in M_n(\mathbb{C})$. Taking S to be the matrix found in the previous paragraph, we find that

$$f(B) \leq ||B||_S = ||S^{-1}BS||_\infty = ||U||_\infty \leq \rho(B) + \epsilon,$$

as needed.

We have now proved that $f(B) = \rho(B)$ for all $B \in M_n(\mathbb{C})$. Using (10.15), we see that $\rho(B) < 1$ iff there exists an induced matrix norm such that $||B|| < 1$. Accordingly, we can restate the convergence result in §10.12 as follows. *An iterative algorithm based on (10.13) converges if the spectral radius $\rho(I - Q^{-1}A)$ is less than 1.*

10.14   Analysis of the Gauss–Seidel Algorithm

This section uses the convergence criterion $\rho(I - Q^{-1}A) < 1$ to prove that *the Gauss–Seidel Algorithm converges if the input matrix $A \in M_n(\mathbb{C})$ is diagonally dominant.* Recall that diagonal dominance means $|A(i,i)| > \sum_{j \neq i} |A(i,j)|$ for all i. Let $c \in \mathbb{C}$ be any eigenvalue of $I - Q^{-1}A$ with associated eigenvector $x = (x_1, \ldots, x_n) \neq 0$. It suffices to prove $|c| < 1$.

We know $(I - Q^{-1}A)x = cx$. Left-multiplying both sides by Q gives $(Q - A)x = c(Qx)$. Recall that, in the Gauss–Seidel Algorithm, Q contains the entries of A on or below the main diagonal, and $Q - A$ involves the entries of A above the main diagonal. So, taking the ith component of $(Q - A)x = c(Qx)$ gives

$$-\sum_{j:j>i} A(i,j)x_j = c \sum_{j:j\leq i} A(i,j)x_j$$

for all i. Isolating the term $cA(i,i)x_i$ and taking magnitudes gives

$$|c|\,|A(i,i)|\,|x_i| = \left| -\sum_{j:j>i} A(i,j)x_j - c\sum_{j:j<i} A(i,j)x_j \right| \leq \sum_{j:j>i} |A(i,j)|\,|x_j| + |c| \sum_{j:j<i} |A(i,j)|\,|x_j|.$$

Fix an index i such that $|x_i| = \max_{1 \leq k \leq n} |x_k| > 0$. Then $|x_j|/|x_i| \leq 1$ for all $j \neq i$, so dividing the preceding inequality by $|x_i|$ gives

$$|c|\,|A(i,i)| \leq \sum_{j:j>i} |A(i,j)| + |c| \sum_{j:j<i} |A(i,j)|. \tag{10.16}$$

Now, the diagonal dominance of A implies that $|A(i,i)| - \sum_{j:j<i} |A(i,j)| > \sum_{j:j>i} |A(i,j)| \geq 0$. Therefore, we can solve (10.16) for $|c|$ to conclude that

$$|c| \leq \frac{\sum_{j:j>i} |A(i,j)|}{|A(i,i)| - \sum_{j:j<i} |A(i,j)|} < 1,$$

as needed.

10.15   Power Method for Finding Eigenvalues

To compute the spectral radius of $A \in M_n(\mathbb{C})$ from the definition, we need to know the largest complex eigenvalue of A. We now discuss an iterative algorithm called the *power method* whose goal is to compute this largest eigenvalue and an associated eigenvector. The algorithm takes as input the matrix A and an arbitrary nonzero initial vector $x_0 \in \mathbb{C}^n$. The algorithm iteratively computes $y_{k+1} = Ax_k$ and $x_{k+1} = y_{k+1}/\|y_{k+1}\|_\infty$ for all $k \geq 0$. If we get $y_{k+1} = 0$, then the algorithm fails, and we try again with a different x_0. At stage k, x_k is the algorithm's approximation for the required eigenvector. The associated eigenvalue is estimated by choosing an index i and returning $c_k = (Ax_k)(i)/x_k(i)$. Any index i between 1 and n can be used here, as long as $|x_k(i)|$ is not too close to zero.

For example, let $A = \begin{bmatrix} 1 & 4 & -1 \\ 0 & 3 & 2 \\ 1 & -1 & -3 \end{bmatrix}$ and $x_0 = \begin{bmatrix} 1 \\ 0 \\ 0 \end{bmatrix}$. The vectors $x_1, x_2, \ldots, x_{12}$ are:

$$\begin{bmatrix} 1 \\ 0 \\ 1 \end{bmatrix}, \begin{bmatrix} 0 \\ 1 \\ -1 \end{bmatrix}, \begin{bmatrix} 1 \\ 0.2 \\ 0.4 \end{bmatrix}, \begin{bmatrix} 1 \\ 1 \\ -0.286 \end{bmatrix}, \begin{bmatrix} 1 \\ 0.459 \\ 0.162 \end{bmatrix}, \begin{bmatrix} 1 \\ 0.636 \\ 0.020 \end{bmatrix},$$

$$\begin{bmatrix} 1 \\ 0.553 \\ 0.086 \end{bmatrix}, \begin{bmatrix} 1 \\ 0.586 \\ 0.060 \end{bmatrix}, \begin{bmatrix} 1 \\ 0.572 \\ 0.071 \end{bmatrix}, \begin{bmatrix} 1 \\ 0.578 \\ 0.067 \end{bmatrix}, \begin{bmatrix} 1 \\ 0.575 \\ 0.068 \end{bmatrix}, \begin{bmatrix} 1 \\ 0.576 \\ 0.068 \end{bmatrix}.$$

At this stage, our estimate for the eigenvector is $x_{12} = (1, 0.576, 0.068)$, and $Ax_{12} = (3.236, 1.864, 0.22)$. Dividing each entry of Ax_{12} by the corresponding entry of x_{12} gives us three estimates for the largest eigenvalue of A: 3.236, $1.864/0.576 \approx 3.236$, and $0.22/0.068 \approx 3.235$. In fact, factoring the characteristic polynomial of A shows that the exact eigenvalues of A are -1, $1 - \sqrt{5}$, and $1 + \sqrt{5}$. Thus, $\rho(A) = 1 + \sqrt{5} \approx 3.23607$, so that we obtain quite good estimates for $\rho(A)$ after twelve iterations of the algorithm.

We now give a sufficient condition on A guaranteeing that the approximations produced by the power method converge to $\rho(A)$. The condition we impose on A is that A *is a diagonalizable matrix having a unique eigenvalue $c \in \mathbb{C}$ such that $|c| = \rho(A)$*. (Some of the exercises explore what happens if these conditions are not met.) Given these assumptions, we can choose an ordered basis $Z = (z_1, \ldots, z_n)$ of $\mathbb{C}^n$ consisting of n linearly independent eigenvectors of A. We have $Az_i = c_i z_i$ for the complex eigenvalues $c_1, \ldots, c_n$ of A. We can reorder the basis Z so that

$$\rho(A) = |c_1| > |c_2| \geq |c_3| \geq \cdots \geq |c_n|.$$

Given any $x_0 \in \mathbb{C}^n$, we obtain x_k by a sequence of k steps, each of which involves multiplying by A and then rescaling to get a unit vector. Letting $b_k \in \mathbb{R}_{>0}$ be the product of the rescaling factors used to reach x_k, we see that $x_k = b_k A^k x_0$ for all $k \geq 1$.

Write $x_0 = d_1 z_1 + d_2 z_2 + \cdots + d_n z_n$ for unique scalars $d_i \in \mathbb{C}$. We assume $d_1 \neq 0$, which happens almost surely if x_0 is chosen at random in $\mathbb{C}^n$. By replacing each z_i in the basis Z by $d_i z_i$, we can assume without loss of generality that every d_i is 1. We have $A^k z_i = c_i^k z_i$ for all i and all $k \geq 1$, so $x_k = b_k(c_1^k z_1 + c_2^k z_2 + \cdots + c_n^k z_n)$. Letting $b_k' = b_k c_1^k$ and $r_i = c_i/c_1$ for $2 \leq i \leq n$, we can rewrite this as

$$x_k = b_k'(z_1 + r_2^k z_2 + \cdots + r_n^k z_n).$$

We have $|r_i| = |c_i|/|c_1| < 1$ for $2 \leq i \leq n$, so for each such i, the sequence $r_i^k z_i$ goes to zero as k goes to infinity. Letting $x_k' = x_k/b_k'$, it follows that $x_k' = z_1 + r_2^k z_2 + \cdots + r_n^k z_n \to z_1$ as $k \to \infty$.

Next, let i be any index such that $z_1(i) \neq 0$. Since $x_k' \to z_1$ relative to $\|\cdot\|_\infty$, the inequality $0 \leq |x_k'(i) - z_1(i)| \leq \|x_k' - z_1\|_\infty$ shows that $x_k'(i) \to z_1(i)$ in $\mathbb{C}$ as k goes to infinity. Similarly, since $Ax_k' \to Az_1 = c_1 z_1$, we see that $(Ax_k')(i) \to c_1 z_1(i)$ as k goes to infinity. Finally, since $x_k = b'_k x'_k$,

$$\lim_{k \to \infty} \frac{(Ax_k)(i)}{x_k(i)} = \lim_{k \to \infty} \frac{(Ax'_k)(i)}{x'_k(i)} = \frac{c_1 z_1(i)}{z_1(i)} = c_1,$$

which says that the approximations produced by the power method do converge to the largest eigenvalue c_1 of A.

Although x'_k converges to the eigenvector z_1 associated with the eigenvalue c_1, we cannot conclude that the sequence x_k itself converges to any fixed eigenvector of A. The trouble is that the complex scaling constant $b'_k = b_k c_1^k$ may cause the sequence x_k to jump between several eigenvectors that differ by a complex scaling factor of modulus 1. For example, if $A = \begin{bmatrix} 0 & -1 \\ 1 & 0 \end{bmatrix}$ and x_0 is the eigenvector $\begin{bmatrix} i \\ 1 \end{bmatrix}$ associated with the eigenvalue i, the power method produces the periodic sequence

$$\begin{bmatrix} i \\ 1 \end{bmatrix}, \begin{bmatrix} -1 \\ i \end{bmatrix}, \begin{bmatrix} -i \\ -1 \end{bmatrix}, \begin{bmatrix} 1 \\ -i \end{bmatrix}, \begin{bmatrix} i \\ 1 \end{bmatrix}, \begin{bmatrix} -1 \\ i \end{bmatrix}, \begin{bmatrix} -i \\ -1 \end{bmatrix}, \begin{bmatrix} 1 \\ -i \end{bmatrix}, \dots$$

This sequence does not converge, but each vector in the sequence is an eigenvector associated with the eigenvalue i. On the other hand, you can adapt the analysis given above to show that $x_k - b'_k z_1 \to 0$ as $k \to \infty$ (Exercise 75). Informally, this means that for large enough k, x_k is guaranteed to come arbitrarily close to some eigenvector associated with the eigenvalue c_1, but this eigenvector may depend on k. Alternatively, since b_k is known and c_1 is being estimated by the algorithm, we can use an estimate for c_1 to compute $x_k/(b_k c_1^k)$ as an approximation of the eigenvector z_1.

10.16   Shifted and Inverse Power Method

The power method provides an iterative algorithm for computing the largest eigenvalue of $A \in M_n(\mathbb{C})$. What if we want to find one of the other eigenvalues of A? This section and the next one discuss several approaches to this question.

Observe first that the largest eigenvalue of A is the eigenvalue farthest from the origin in the complex plane. Suppose we replace the matrix A by a shifted matrix $A + bI$, where $b \in \mathbb{C}$ is a fixed constant. Let $c \in \mathbb{C}$ be an eigenvalue of A with associated eigenvector $x \in \mathbb{C}^n$. Since $Ax = cx$, we have $(A + bI)x = Ax + bx = (c + b)x$, so that $c + b \in \mathbb{C}$ is an eigenvalue of $A + bI$ with associated eigenvector x. Conversely, if c' is an eigenvalue of $A + bI$, a similar computation shows that $c' - b$ is an eigenvalue of A. Geometrically, the eigenvalues of $A + bI$ are obtained by shifting (or translating) all eigenvalues of A by the fixed scalar b. For any $c \in \mathbb{C}$, the distance from c to 0 in $\mathbb{C}$ is the same as the distance from $c + b$ to b in $\mathbb{C}$. So the eigenvalue of A farthest from the origin gets shifted to the eigenvalue of $A + bI$ farthest from the point b.

Suppose $C \in M_n(\mathbb{C})$ is a given matrix, $b \in \mathbb{C}$ is a given scalar, and we want to compute the complex eigenvalue of C farthest from b. Taking $A = C - bI$ in the previous paragraph, we first compute the eigenvalue c of A farthest from the origin using the power method. Then the required eigenvalue for C is $c + b$. This algorithm is called the *shifted power method*.

Next, suppose $A \in M_n(\mathbb{C})$ and we want to find the eigenvalue of A closest to the origin. If A is not invertible, this eigenvalue is zero. If A^{-1} exists and c is any eigenvalue of A with associated eigenvector x, then $Ax = cx$ implies $A^{-1}Ax = A^{-1}cx$, hence $x = cA^{-1}x$. As c must be nonzero, we see $A^{-1}x = c^{-1}x$, so that c^{-1} is an eigenvalue of A^{-1} with associated eigenvector x. The argument is reversible, so the eigenvalues of A^{-1} are the multiplicative inverses in $\mathbb{C}$ of the eigenvalues of A. The inverse of a complex number written in polar form is $(re^{i\theta})^{-1} = r^{-1}e^{-i\theta}$. It follows that for nonzero $c, d \in \mathbb{C}$, c is closer to the origin than d iff c^{-1} is farther from the origin than d^{-1}. Using the power method to find the eigenvalue c of A^{-1} that is farthest from the origin, it follows that c^{-1} is the eigenvalue of A closest to the origin. This algorithm is called the *inverse power method*.

We can use the *shifted inverse power method* to find the eigenvalue of a given $C \in M_n(\mathbb{C})$ that is closest to a given $b \in M_n(C)$. We apply the original power method to the matrix $(C - bI)^{-1}$ to obtain the largest eigenvalue c, and then return $c^{-1} + b$.

When implementing the inverse power method (or its shifted version), we need to compute $y_{k+1} = A^{-1}x_k$ given x_k. Equivalently, we must solve $Ay_{k+1} = x_k$ for the unknown vector y_{k+1}. In practice, it is often more efficient to obtain y_{k+1} by using Gaussian elimination algorithms to solve $Ay_{k+1} = x_k$ rather than computing A^{-1} explicitly. This is especially true when A has special structure such as sparsity.

10.17   Deflation

Suppose $A \in M_n(\mathbb{C})$ is a given matrix and we have found, by whatever method, one particular eigenvalue c of A and an associated eigenvector x. This section describes a general algorithm called *deflation* that produces a new matrix $B \in M_{n-1}(\mathbb{C})$ whose eigenvalues (counted by their algebraic multiplicities as roots of the characteristic polynomial) are precisely the eigenvalues of A with one copy of c removed. In particular, by repeatedly applying the power method (or its variants) followed by deflation to a given matrix A, we can eventually compute all the eigenvalues of A.

Recall Schur's Theorem from §7.7: for any matrix $A \in M_n(\mathbb{C})$, there exists a unitary matrix $U \in M_n(\mathbb{C})$ such that $U^{-1}AU = U^*AU$ is upper-triangular. We obtain the deflation algorithm by examining the computations implicit in the proof of this result. Assume $c_1 \in \mathbb{C}$ is the known eigenvalue of A with associated known eigenvector $x_1 \in \mathbb{C}^n$. Dividing x_1 by $||x_1||_2$, we can assume that $||x_1||_2 = 1$. Let $T : \mathbb{C}^n \to \mathbb{C}^n$ be defined by $T(v) = Av$ for $v \in \mathbb{C}^n$. The key step in the inductive proof of Schur's theorem was to extend the list (x_1) to an orthonormal basis $X = (x_1, x_2, \ldots, x_n)$ of $\mathbb{C}^n$. We can compute such an orthonormal basis X explicitly using the Gram–Schmidt Orthonormalization Algorithm (see §9.2 and Exercise 15 in Chapter 9).

Arrange the column vectors $x_1, x_2, \ldots, x_n$ as the columns of a matrix U, so $U = {}_E[\mathrm{id}]_X$ where $E = (\mathbf{e}_1, \ldots, \mathbf{e}_n)$ is the standard ordered basis of $\mathbb{C}^n$. Since the columns of U are orthonormal, U is a unitary matrix (see §7.5), so $U^{-1} = U^*$. We can compute $C = [T]_X = U^*AU$ explicitly. The first column of this matrix is $(c_1, 0, \ldots, 0)^T$. Let $B \in M_{n-1}(\mathbb{C})$ be obtained from C by deleting the first row and first column. Computing the characteristic polynomial of C by expanding the determinant along the first column, we see that $\chi_A(t) = \chi_C(t) = (t - c_1)\chi_B(t)$. Therefore, the matrix B has the required eigenvalues, and the deflation algorithm returns B as its output.

10.18   Summary

1. *Vector Norms.* A vector norm $||\cdot||$ on a real or complex vector space V satisfies

$$0 \leq ||x|| < \infty, \ ||x|| = 0 \Leftrightarrow x = 0, \ ||cx|| = |c| \cdot ||x||, \ ||x + y|| \leq ||x|| + ||y||$$

for all x, y in V and all scalars c. The pair $(V, ||\cdot||)$ is called a normed vector space. Examples of vector norms on $\mathbb{R}^n$ and $\mathbb{C}^n$ include

$$||x||_\infty = \max_{1 \leq i \leq n} |x_i|, \quad ||x||_1 = |x_1| + \cdots + |x_n|, \quad ||x||_2 = \sqrt{|x_1|^2 + \cdots + |x_n|^2}.$$

2. *Metric Spaces.* A metric space is a set X and a metric d satisfying

$$0 \leq d(x, y) < \infty, \ d(x, y) = 0 \Leftrightarrow x = y, \ d(x, y) = d(y, x), \ d(x, z) \leq d(x, y) + d(y, z)$$

for all $x, y, z \in X$. A normed vector space V has the metric $d(x, y) = ||x - y||$, which also satisfies $d(x + z, y + z) = d(x, y)$ and $d(cx, cy) = |c| d(x, y)$ for all $x, y, z \in V$ and all scalars c.

3. *Convergence Properties.* In a metric space (X, d), a sequence $(x_k)_{k \geq 0}$ converges to $x \in X$ iff for all $\epsilon > 0$, there exists $k_0 \in \mathbb{Z}_{\geq 0}$ such that for all $k \geq k_0$, $d(x_k, x) < \epsilon$. The limit of a convergent sequence is unique when it exists. In $\mathbb{C}^n$, if $v_k \to v$ and $w_k \to w$, then $v_k + w_k \to v + w$, $cv_k \to cv$ for any scalar c, and $Av_k \to Av$ for any matrix A.

4. *Comparable Norms.* Vector norms $||\cdot||$ and $||\cdot||'$ on a vector space V are comparable iff there exist real C, D with $0 < C, D < \infty$ such that for all $x \in V$, $||x|| \leq C||x||'$ and $||x||' \leq D||x||$. Any two vector norms on $\mathbb{R}^n$ or $\mathbb{C}^n$ are comparable. In particular, $||x||_\infty \leq ||x||_2 \leq \sqrt{n}||x||_\infty$ and $||x||_\infty \leq ||x||_1 \leq n||x||_\infty$.

5. *Matrix Norms.* A matrix norm $||\cdot|| : M_n(\mathbb{C}) \to \mathbb{R}_{\geq 0}$ satisfies

$$||A|| = 0 \Leftrightarrow A = 0, \ ||cA|| = |c| \cdot ||A||, \ ||A + B|| \leq ||A|| + ||B||, \ ||AB|| \leq ||A|| \cdot ||B||$$

for all $A, B \in M_n(\mathbb{C})$ and all $c \in \mathbb{C}$. Any vector norm $||\cdot||$ on $\mathbb{C}^n$ has an associated matrix norm $||A|| = \sup\{||Av|| : v \in \mathbb{C}^n, ||v|| = 1\}$, and $||Ax|| \leq ||A|| \cdot ||x||$ for all $x \in \mathbb{C}^n$. A matrix norm defined in this way from some vector norm is called an induced matrix norm.

6. *Formulas for Matrix Norms.* Fix $A \in M_n(\mathbb{C})$.

   (a) The matrix norm induced by the sup norm is $||A||_\infty = \max\limits_{1 \leq i \leq n} \sum\limits_{j=1}^{n} |A(i, j)|$.

   (b) The matrix norm induced by the 1-norm is $||A||_1 = \max\limits_{1 \leq j \leq n} \sum\limits_{i=1}^{n} |A(i, j)|$.

   (c) The spectral radius $\rho(A)$ is the maximum of the numbers $|c|$ as c ranges over all complex eigenvalues of A. The spectral radius of A also equals $\inf\{||A||\}$, where $||\cdot||$ varies over all induced matrix norms on $\mathbb{C}^n$.

   (d) The matrix norm induced by the 2-norm is $||A||_2 = \sqrt{\rho(A^*A)}$.

7. *Geometric Series for Matrix Inverses.* If $C \in M_n(\mathbb{C})$ satisfies $||C|| < 1$ for some induced matrix norm (or equivalently, if $\rho(C) < 1$), then $I - C$ is invertible and $(I - C)^{-1} = \sum_{k=0}^{\infty} C^k$. The error of the partial sum ending at C^k is at most $||(I - C)^{-1}|| \cdot ||C||^{k+1}$.

8. *Affine Iteration.* If $C \in M_n(\mathbb{C})$ satisfies $||C|| < 1$ for some induced matrix norm (or equivalently, if $\rho(C) < 1$), then for any $d, x_0 \in \mathbb{C}^n$, the sequence defined by $x_{k+1} = Cx_k + d$ (for $k \geq 0$) converges to the unique $x \in \mathbb{C}^n$ solving $(I - C)x = d$, and $||x_k - x|| \leq ||C||^k ||x_0 - x||$.

9. *Richardson's Algorithm.* This iterative method solves $Ax = b$ by the update rule $x_{k+1} = x_k + (b - Ax_k)$. It converges if $||I - A|| < 1$ for some induced matrix norm (i.e., $\rho(I - A) < 1$), in which case $||x_k - x|| \leq ||I - A||^k ||x_0 - x||$. Convergence is guaranteed if $\max_i \sum_j |(I - A)(i, j)| < 1$ or if $\max_j \sum_i |(I - A)(i, j)| < 1$.

10. *Jacobi's Algorithm.* This iterative method solves $Ax = b$ by the update rule $x_{k+1} = Q^{-1}(b - Rx_k)$ where Q is the diagonal part of A and $R = A - Q$, i.e.,

$$x_{k+1}(i) = A(i, i)^{-1}[b(i) - \sum_{j:j\neq i} A(i, j)x_k(j)].$$

If $\rho(I - Q^{-1}A) < 1$, convergence occurs with $||x_k - x|| \leq ||I - Q^{-1}A||^k ||x_0 - x||$. Convergence is guaranteed if $|A(i, i)| < \sum_{j:j\neq i} |A(i, j)|$ for all i.

11. *Gauss–Seidel Algorithm.* This iterative method solves $Ax = b$ by the update rule $Qx_{k+1} = b - Rx_k$ where Q is the lower-triangular part of A and $R = A - Q$, i.e.,

$$x_{k+1}(i) = A(i, i)^{-1}[b(i) - \sum_{j:j<i} A(i, j)x_{k+1}(j) - \sum_{j:j>i} A(i, j)x_k(j)].$$

If $\rho(I - Q^{-1}A) < 1$, convergence occurs with $||x_k - x|| \leq ||I - Q^{-1}A||^k ||x_0 - x||$. Convergence is guaranteed if $|A(i, i)| < \sum_{j:j\neq i} |A(i, j)|$ for all i.

12. *Power Method.* The power method approximates the largest eigenvalue c of $A \in M_n(\mathbb{C})$ by starting with $x_0 \neq 0$, computing $x_{k+1} = Ax_k / ||Ax_k||_\infty$ for $k \geq 0$, and estimating $c \approx (Ax_k)(i)/x_k(i)$ where i is chosen so $|x_k(i)|$ is not too close to zero. If A is diagonalizable with a unique eigenvalue c of magnitude $\rho(A)$, then for most choices of x_0, the power method approximations converge to c, and the x_k approach associated eigenvectors (depending on k). The requirement on x_0 for convergence is that the expansion of x_0 as a linear combination of eigenvectors of A must involve the eigenvector associated with c.

13. *Variations of the Power Method.* The shifted power method finds the eigenvalue of $A \in M_n(\mathbb{C})$ farthest from $b \in \mathbb{C}$ by applying the power method to $A - bI$ and adding b to the output. The inverse power method finds the eigenvalue of A closest to zero by applying the power method to A^{-1} and taking the reciprocal of the output. The shifted inverse power method finds the eigenvalue of A closest to b by using the power method to get the largest eigenvalue c of $(A - bI)^{-1}$, and returning $c^{-1} + b$.

14. *Deflation.* If we have found one eigenvalue c of $A \in M_n(\mathbb{C})$ and an associated unit eigenvector x, we can compute a matrix $B \in M_{n-1}(\mathbb{C})$ whose eigenvalues are the remaining eigenvalues of A as follows. Extend x to an orthonormal basis of $\mathbb{C}^n$ by the Gram–Schmidt algorithm. Let U be the unitary matrix having these basis vectors as columns, with x in column 1. Let B be the matrix U^*AU with the first row and column erased.

10.19   Exercises

1. Given $v = (2, -4, -1, 0, 3)$, compute $||v||_\infty$, $||v||_1$, and $||v||_2$.

2. Given $v = (3 + 4i, -2 - i, 5i)$, compute $||v||_\infty$, $||v||_1$, and $||v||_2$.

3. Sketch the set of unit vectors in $\mathbb{R}^2$ for the norms $||\cdot||_\infty$, $||\cdot||_1$, and $||\cdot||_2$.

4. Sketch the set of unit vectors in $\mathbb{R}^3$ for the norms $||\cdot||_\infty$, $||\cdot||_1$, and $||\cdot||_2$.

5. For each matrix A, execute Richardson's method (computing $x_1, \ldots, x_5$ by hand) to try to solve $Ax = b$ using $b = [1\ 2]^T$ and $x_0 = 0$. Compare x_5 to the exact solution x. Does the method appear to be converging?

    (a) $\begin{bmatrix} 1 & 1/2 \\ 0 & 1 \end{bmatrix}$ (b) $\begin{bmatrix} 1 & 2 \\ 2 & 1 \end{bmatrix}$ (c) $\begin{bmatrix} 1 & 0.1 \\ 0.1 & 1 \end{bmatrix}$ (d) $\begin{bmatrix} 1 & -1 \\ 1 & 1 \end{bmatrix}$

6. For each matrix A, predict whether Richardson's Algorithm converges. If so, give an estimate on the error $||x_k - x||$ for an appropriate vector norm.

    (a) $\begin{bmatrix} 1.3 & 0.2 & -0.3 \\ 0.1 & 0.9 & -0.2 \\ 0.2 & 0.2 & 1.2 \end{bmatrix}$ (b) $\begin{bmatrix} 1 & 1 & 2 \\ 0 & 1 & 2 \\ 2 & 1 & 1 \end{bmatrix}$ (c) $\begin{bmatrix} 3.5 & -4.5 \\ 1.8 & -2.2 \end{bmatrix}$

    (d) $A \in M_n(\mathbb{R})$ given by $A(i, j) = 2^{i-j}$ for $i \le j$, and $A(i, j) = 0$ for $i > j$.

7. Write a program in a computer algebra system to implement Richardson's Algorithm. Use the program to solve $Ax = b$, where

$$A = \begin{bmatrix} 1 & 0.1 & 0 & 0.2 & -0.1 & 0.1 \\ 0.1 & 1 & 0 & 0.1 & 0 & 0.2 \\ 0 & 0.2 & 1 & -0.1 & 0 & 0.1 \\ -0.3 & 0 & 0 & 1 & -0.1 & 0 \\ -0.1 & 0.1 & 0 & 0 & 1 & 0 \\ 0 & 0.2 & 0.1 & -0.1 & 0.3 & 1 \end{bmatrix}, \quad b = \begin{bmatrix} 2 \\ 1.3 \\ 0 \\ -1 \\ 0.6 \\ 1.7 \end{bmatrix}.$$

   Give an error estimate for your answer.

8. For each matrix A and column vector b, execute Jacobi's Algorithm (starting with $x_0 = 0$ and computing $x_1, \ldots, x_5$) to try to solve $Ax = b$. Compare x_5 to the exact solution x. Does the method appear to be converging?

    (a) $A = \begin{bmatrix} 2 & 1 \\ -1 & 2 \end{bmatrix}$, $b = \begin{bmatrix} 3 \\ 1 \end{bmatrix}$ (b) $A = \begin{bmatrix} 0 & 5 \\ -2 & 1 \end{bmatrix}$, $b = \begin{bmatrix} 2 \\ -2 \end{bmatrix}$

    (c) $A = \begin{bmatrix} 1 & 3 \\ 3 & 1 \end{bmatrix}$, $b = \begin{bmatrix} 1 \\ 2 \end{bmatrix}$ (d) $A = \begin{bmatrix} 3 & 1 & 1 \\ 2 & 4 & -1 \\ 0 & -1 & 2 \end{bmatrix}$, $b = \begin{bmatrix} 4 \\ -1 \\ 0 \end{bmatrix}$.

9. Repeat Exercise 8 using the Gauss–Seidel algorithm.

10. For each square matrix A, predict whether Jacobi's Algorithm converges. If so, give an estimate on the error $||x_k - x||$ for an appropriate vector norm.

    (a) $\begin{bmatrix} 1 & 2 & 3 \\ 2 & 3 & 5 \\ 3 & 4 & 5 \end{bmatrix}$ (b) $\begin{bmatrix} 8 & 4 & 2 \\ 3 & 9 & 1 \\ 1 & -2 & 4 \end{bmatrix}$ (c) $A = \begin{bmatrix} 2 & -5 \\ 1 & 3 \end{bmatrix}$

    (d) $A(i, i) = c > 2$, $A(i, j) = 1$ for $|i - j| = 1$, and $A(i, j) = 0$ otherwise.

11. Repeat Exercise 10 using the Gauss–Seidel algorithm.

12. Write a program in a computer algebra system to implement Jacobi's Algorithm.

Use the program to solve $Ax = b$, where

$$A = \begin{bmatrix} 3.2 & 1.1 & 0 & 0 & -1.2 & 0 \\ 1.3 & 4.1 & -1.2 & 1.5 & 0 & 0 \\ 0.8 & 0 & -1.7 & 0 & 0.5 & 0 \\ 0.7 & 2.1 & 0 & 5.3 & 0 & -1.3 \\ 0.3 & 0 & -0.6 & 1.1 & -2.4 & 0 \\ 1.4 & -0.8 & 0 & 0 & -1.0 & 4.9 \end{bmatrix}, \quad b = \begin{bmatrix} 4.1 \\ -2.2 \\ 1.3 \\ -0.8 \\ 0 \\ -3.2 \end{bmatrix}.$$

Give an error estimate for your answer.

13. Repeat Exercise 12 using the Gauss–Seidel algorithm.

14. For each matrix A, execute several iterations of the power method to approximate the largest eigenvalue of A and an associated eigenvector. Also find all eigenvalues of A exactly and compare to the algorithm's output.

(a) $\begin{bmatrix} 4 & -2 \\ -3 & -1 \end{bmatrix}$ (b) $\begin{bmatrix} 3 & 1 & 0 \\ 1 & 3 & 1 \\ 0 & 1 & 3 \end{bmatrix}$ (c) $\begin{bmatrix} 1 & 2 & i \\ 2 & 1 & -1 \\ -i & -1 & 1 \end{bmatrix}$

15. Repeat Exercise 14, but use the inverse power method to find the smallest eigenvalue of A and an associated eigenvector.

16. Repeat Exercise 14, but use the shifted power method to find the eigenvalue of A farthest from 4 and an associated eigenvector.

17. Let $A \in M_4(\mathbb{R})$ have $A(1,4) = A(3,2) = A(4,2) = A(4,4) = -1$ and all other entries equal to 1. Use the inverse shifted power method to find the eigenvalue of A closest to -3 and an associated eigenvector.

18. For each matrix A, find the exact eigenvalues of A (with multiplicities). Try executing the power method on A and x_0 for a few iterations, and explain what goes wrong (and why) in each case.

(a) $A = \begin{bmatrix} 2 & 0 \\ 0 & -2 \end{bmatrix}$, $x_0 = \begin{bmatrix} 1 \\ 1 \end{bmatrix}$ (b) $A = \begin{bmatrix} 1 & 2 & 3 \\ 4 & 5 & 6 \\ 7 & 8 & 9 \end{bmatrix}$, $x_0 = \begin{bmatrix} 1 \\ -2 \\ 1 \end{bmatrix}$

(c) $A = \begin{bmatrix} 3 & 2 & 1 \\ 0 & 2 & 1 \\ 0 & 0 & 1 \end{bmatrix}$, $x_0 = \begin{bmatrix} 0 \\ -3 \\ 4 \end{bmatrix}$ (d) $A = \begin{bmatrix} -1/2 & \sqrt{3}/2 \\ -\sqrt{3}/2 & -1/2 \end{bmatrix}$, $x_0 = \begin{bmatrix} 1 \\ 0 \end{bmatrix}$.

19. Write a program to implement the power method for calculating eigenvalues. Test your program on the matrices in Exercises 7 and 12.

20. For each matrix A in Exercise 5, try to approximate A^{-1} using the geometric series formula (10.11). For which matrices does the series converge?

21. Use the geometric series formula (10.11) to estimate A^{-1}, where A is the matrix in Exercise 7.

22. For fixed $b \in \mathbb{C}$, define $A \in M_n(\mathbb{C})$ by setting $A(i,j) = 1$ for $i = j$, $A(i,j) = b$ for $j = i+1$, and $A(i,j) = 0$ for all other i,j. Use (10.11) to find A^{-1}.

23. Prove: if C is *nilpotent* (meaning $C^k = 0$ for some positive integer k), then $I-C$ is invertible and $(I-C)^{-1}$ is given exactly by the finite sum $I+C+C^2+\cdots+C^{k-1}$.

24. Consider the affine iteration $x_{k+1} = Cx_k + d$ for fixed $C \in M_n(\mathbb{C})$ and $d \in \mathbb{C}^n$. Express x_k as a sum of terms involving powers of C, x_0, and d. Prove that if C is nilpotent with $C^r = 0$, then x_r is the exact solution to $(I - C)x = d$.

25. Suppose $C \in M_n(\mathbb{C})$ satisfies $\rho(C) > 1$ and $I - C$ is invertible. Show that for all $d \in \mathbb{C}^n$, there exists $x_0 \in \mathbb{C}^n$ such that the affine iteration (10.12) starting at x_0 converges. Show there exist $d, x_0 \in \mathbb{C}^n$ such that the affine iteration (10.12) starting at x_0 does not converge.

26. Prove that the 1-norm on $\mathbb{C}^n$ satisfies the axioms for a vector norm.

27. Define $||(x_1, x_2, x_3)|| = |x_1| + |x_3|$. Which axioms for a vector norm are satisfied?

28. Let $||(x_1, x_2)|| = (\sqrt{|x_1|} + \sqrt{|x_2|})^2$. Which axioms for a vector norm are satisfied?

29. **Cauchy–Schwarz Inequality.** Prove: for $x, y \in \mathbb{R}^n$, $|x \bullet y| \leq ||x||_2 ||y||_2$. (Compute $(ax + by) \bullet (ax + by)$ for $a, b \in \mathbb{R}$. For $x \neq 0 \neq y$, take $a = ||x||_2^{-1}$ and $b = \pm ||y||_2^{-1}$.)

30. Prove that the 2-norm on $\mathbb{R}^n$ satisfies the axioms for a vector norm. (For the triangle inequality, compute $||x + y||_2^2$.)

31. Let $||(x_1, x_2)|| = |x_1|^2 + |x_2|^2$. Which axioms for a vector norm are satisfied?

32. Let $(V, ||\cdot||_V)$ and $(W, ||\cdot||_W)$ be normed vector spaces. For $(v, w) \in V \times W$, define $||(v, w)||_1 = ||v||_V + ||w||_W$. Prove this defines a vector norm on $V \times W$.

33. Let $(V, ||\cdot||_V)$ and $(W, ||\cdot||_W)$ be normed vector spaces. For $(v, w) \in V \times W$, define $||(v, w)||_\infty = \max(||v||_V, ||w||_W)$. Prove this defines a vector norm on $V \times W$.

34. Explain why the 1-norm and sup norm on $\mathbb{R}^n$ are special cases of the norms defined in the previous two exercises.

35. For each real $p \in [1, \infty)$, the *p-norm* on $\mathbb{C}^n$ is defined on $x = (x_1, \ldots, x_n) \in \mathbb{C}^n$ by $||x||_p = (|x_1|^p + |x_2|^p + \cdots + |x_n|^p)^{1/p}$.
    (a) Show $||\cdot||_p$ satisfies the first two axioms for a vector norm.
    (b) **Hölder's Inequality.** Assume $1 < p, q < \infty$ and $p^{-1} + q^{-1} = 1$.
    Prove: for all $x, y \in \mathbb{C}^n$, $\displaystyle\sum_{i=1}^{n} |x_i y_i| \leq ||x||_p ||y||_q$.
    (First prove it assuming $||x||_p = ||y||_q = 1$, using Exercise 111 of Chapter 11.)
    (c) **Minkowski's Inequality.** Prove: for all $x, y \in \mathbb{C}^n$, $||x + y||_p \leq ||x||_p + ||y||_p$.
    (Start with $||x + y||_p^p$, note $|x_i + y_i|^p \leq (|x_i| + |y_i|) \cdot |x_i + y_i|^{p-1}$, and use (b).)

36. Prove: for $x \in \mathbb{R}^n$, $\lim_{p \to \infty} ||x||_p = ||x||_\infty$.

37. For fixed real numbers $p, r \geq 1$, find the minimum constant $C = C(p, r)$ such that for all $x \in \mathbb{C}^n$, $||x||_p \leq C ||x||_r$. (One approach is to use Lagrange multipliers.)

38. Prove: for all x, y in a normed vector space, $|\, ||x|| - ||y|| \,| \leq ||x - y||$.

39. For a fixed invertible $S \in M_n(\mathbb{C})$, define $||v||_S = ||S^{-1}v||_\infty$ for $v \in \mathbb{C}^n$. Prove $||\cdot||_S$ is a vector norm on $\mathbb{C}^n$. Prove that the matrix norm induced by $||\cdot||_S$ satisfies $||A||_S = ||S^{-1}AS||_\infty$ for all $A \in M_n(\mathbb{C})$.

40. Suppose $T : V \to W$ is a vector space isomorphism. Given a vector norm $||\cdot||_V$ on V, let $||w||_W = ||T^{-1}(w)||_V$ for $w \in W$. Show $||\cdot||_W$ is a vector norm on W. Prove that for the metrics associated with the norms on V and W, $d(T(v), T(v')) = d(v, v')$ for all $v, v' \in V$.

41. Explain why the vector norm $||\cdot||_S$ in §10.13 is a special case of the construction in the previous exercise.

42. (a) Prove directly from the definitions that for all $x = (x_1, \ldots, x_n) \in \mathbb{C}^n$ and all i, $|x_i| \leq ||x||$ where $||\cdot||$ is the 1-norm, the 2-norm, or the sup norm.
    (b) Use a theorem from the text to show that for any vector norm on $\mathbb{C}^n$ and all i, there is a constant K with $|x_i| \leq K ||x||$.

(c) Let V be a finite-dimensional real or complex normed vector space with ordered basis $Z = (z_1, \ldots, z_n)$. Each $v \in V$ has a unique expansion $v = p_1(v)z_1 + \cdots + p_n(v)z_n$ for certain scalars $p_i(v)$. Prove there are constants K_i such that for all $v \in V$, $|p_i(v)| \leq K_i\|v\|$.

43. Consider the $\mathbb{Q}$-vector space $V = \{a + b\sqrt{2} : a, b \in \mathbb{Q}\}$.
    (a) Define $\|a + b\sqrt{2}\|$ to be the absolute value (in $\mathbb{R}$) of $a + b\sqrt{2}$. Show that the axioms for a vector norm (using rational scalars) hold.
    (b) Show that there does not exist a constant K such that for all $x = a + b\sqrt{2} \in V$, $|a| \leq K\|x\|$. (Contrast to Exercise 42(c).)

44. Let V be the real vector space of continuous functions $f : [0, 1] \to \mathbb{R}$. Show that $\|f\|_\infty = \sup_{x \in [0,1]} |f(x)|$ is a vector norm on V.

45. Let V be the real vector space of continuous functions $f : [0, 1] \to \mathbb{R}$. Show that $\|f\|_I = \int_0^1 |f(x)|\, dx$ is a vector norm on V.

46. Let V, $\|\cdot\|_\infty$, and $\|\cdot\|_I$ be defined as in Exercises 44 and 45.
    (a) Show there is no constant C such that for all $f \in V$, $\|f\|_\infty \leq C\|f\|_I$.
    (b) Is there a constant D such that for all $f \in V$, $\|f\|_I \leq D\|f\|_\infty$? Explain.

47. For $A \in M_n(\mathbb{C})$ with $n > 1$, define $\|A\| = \max_{1 \leq i,j \leq n} |A(i, j)|$. Prove that this is not a matrix norm.

48. For $A \in M_n(\mathbb{C})$, define $\|A\| = \sum_{i,j=1}^n |A(i, j)|$. Is this a matrix norm? If so, is this an induced matrix norm?

49. **Frobenius Matrix Norm.** For $A \in M_n(\mathbb{C})$, define $\|A\|_F = \sqrt{\sum_{i,j=1}^n |A(i, j)|^2}$. Prove $\|\cdot\|_F$ is a matrix norm. Is $\|\cdot\|_F$ induced by any vector norm on $\mathbb{C}^n$? Explain. Prove $\|A\|_F^2$ is the trace (sum of diagonal entries) of A^*A.

50. Let U be a unitary matrix.
    (a) Prove: for all $v \in \mathbb{C}^n$, $\|Uv\|_2 = \|v\|_2$.
    (b) Must (a) hold for the 1-norm or the sup norm on $\mathbb{C}^n$?
    (c) Prove: for all $A \in M_n(\mathbb{C})$, $\|UA\|_2 = \|A\|_2 = \|AU\|_2$.
    (d) Does (c) hold for the matrix norms induced by the 1-norm or the sup norm on $\mathbb{C}^n$? Does (c) hold for the Frobenius matrix norm (Exercise 49)?

51. For any real or complex vector space V and $x, y \in V$, let $d(x, y) = 1$ if $x \neq y$, and $d(x, y) = 0$ if $x = y$. Prove that (V, d) is a metric space. Is there a vector norm on V such that d is the associated metric? Explain.

52. Suppose V is a real or complex vector space and d is a metric on V invariant under translations and respecting dilations. For $x \in V$, define $\|x\| = d(x, 0_V)$. Prove $\|\cdot\|$ is a vector norm whose associated metric is d.

53. Let (X, d_X) and (Y, d_Y) be metric spaces. Show $X \times Y$ is a metric space with metric given by $d((x_1, y_1), (x_2, y_2)) = d_X(x_1, x_2) + d_Y(y_1, y_2)$ for $x_1, x_2 \in X$ and $y_1, y_2 \in Y$.

54. Suppose (X, d) is a metric space and $(x_k)_{k \geq 0}$ is a sequence in X.
    (a) Suppose $x_k = x \in X$ for all k. Prove $\lim_{k \to \infty} x_k = x$.
    (b) Assume $x_k \to x \in X$. Prove: for all $j \in \mathbb{Z}_{>0}$, $\lim_{k \to \infty} x_{k+j} = x$.
    (c) Suppose $(y_k)_{k \geq 0}$ is a sequence such that $y_k = x_{j_k}$ for some indices $j_0 < j_1 < \cdots < j_k < \cdots$ (so (y_k) is a *subsequence* of (x_k)). Prove: if $x_k \to x$, then $y_k \to x$.
    (d) Give an example to show that the converse of the result in (c) can fail.

55. Suppose $(x_k)_{k \geq 0}$ is a sequence in a normed vector space V.
    (a) Prove: if $x_k \to x$ in V, then $\|x_k\| \to \|x\|$ in $\mathbb{R}$.

(b) Show that if $||x_k|| \to 0$ in $\mathbb{R}$, then $x_k \to 0_V$ in V.

(c) Give an example where $||x_k|| \to r > 0$ in $\mathbb{R}$ but (x_k) does not converge in V.

56. Let (X, d) be a metric space with $x_k \in X$ for $k \geq 0$. We say (x_k) is a *Cauchy sequence* iff for all $\epsilon > 0$, there exists $k_0 \in \mathbb{Z}_{\geq 0}$ such that for all $j, k \geq k_0$, $d(x_k, x_j) < \epsilon$. Prove that every convergent sequence is a Cauchy sequence. Give an example of a Cauchy sequence that does not converge.

57. Prove that if (x_k) is a Cauchy sequence in a metric space (X, d) and some subsequence $(y_k) = (x_{j_k})$ converges to x, then $x_k \to x$.

58. We say a metric space (X, d) is *complete* iff every Cauchy sequence in X converges to some point in X. Using the fact that $\mathbb{R}^1$ is complete, prove: for all $n > 1$, $\mathbb{R}^n$ is complete using the metric $d(x, y) = ||x - y||_\infty$ for $x, y \in \mathbb{R}^n$.

59. For any vector norm $|| \cdot ||$ on $\mathbb{R}^n$, use the previous exercise to prove that $\mathbb{R}^n$ is complete using the metric $d(x, y) = ||x - y||$ for $x, y \in \mathbb{R}^n$.

60. Prove: if $c_k \to c$ (in $\mathbb{R}$ or $\mathbb{C}$) and $v_k \to v$ in a normed vector space V, then $c_k v_k \to cv$.

61. Prove that $||x||_\infty \leq ||x||_1 \leq n||x||_\infty$ for all $x \in \mathbb{R}^n$.

62. Find a nonzero $x \in \mathbb{R}^n$ such that $||x||_1 = ||x||_2 = ||x||_\infty$.

63. Find all $y \in \mathbb{R}^n$ such that $||y||_2 = \sqrt{n}||y||_\infty$.

64. Find all $z \in \mathbb{R}^n$ such that $||z||_1 = n||z||_\infty$.

65. Let S be the set of all vector norms on a (possibly infinite-dimensional) vector space V. Show that comparability of norms defines an equivalence relation on S. How many equivalence classes are there when $V = \mathbb{R}^n$?

66. Let $||\cdot||$ be any vector norm on $\mathbb{R}^n$. Let $S = \{x \in \mathbb{R}^n : ||x|| = 1\}$ be the set of unit vectors relative to this norm. Show S is a closed and bounded (hence compact) subset of the metric space $(\mathbb{R}^n, d)$, where $d(y, z) = ||y - z||_2$ for $y, z \in \mathbb{R}^n$. For the matrix norm $||A||$ induced by $|| \cdot ||$, show there exists $x \in S$ with $||Ax|| = ||A||$.

67. Let $|| \cdot ||$ be any matrix norm. Prove: for all $A \in M_n(\mathbb{C})$ and all $k \in \mathbb{Z}_{\geq 0}$, $||A^k|| \leq ||A||^k$. If A is invertible, is there any relation between $||A^{-1}||$ and $||A||^{-1}$?

68. Let $||A||$ be the matrix norm induced by a vector norm $|| \cdot ||$ on $\mathbb{C}^n$.

(a) Prove: $||A|| = \sup\{||Ax||/||x|| : x \in \mathbb{C}^n, x \neq 0\}$.

(b) Prove: $||A|| = \inf\{C \in \mathbb{R} : ||Ax|| \leq C||x|| \text{ for all } x \in \mathbb{C}^n\}$.

69. For each matrix A in Exercise 6, compute $||A||_\infty$, $||A||_1$, $||A||_2$, and $||A||_F$ (see Exercise 49). Take $n = 4$ when computing $||A||_2$ in part (d).

70. (a) For $A \in M_n(\mathbb{C})$, show that $||A||_\infty = ||(||A_{[1]}||_1, \ldots, ||A_{[n]}||_1)||_\infty$.

(b) State a similar formula for $||A||_1$.

71. Let $S = \{u \in \mathbb{R}^n : ||u||_2 = 1\}$. For each matrix A below, define $f : S \to \mathbb{R}$ by $f(u) = ||Au||_2^2$ for $u \in S$. Find the maximum value of f on S and hence compute $||A||_2$. (Use Lagrange multipliers for (b) and (c).)

(a) $\begin{bmatrix} 3 & 1 \\ -1 & 3 \end{bmatrix}$
(b) $\begin{bmatrix} 1 & 1 \\ 1 & 1 \end{bmatrix}$
(c) $\begin{bmatrix} 1 & 1 & 0 \\ 0 & 1 & 1 \\ 1 & 0 & 1 \end{bmatrix}$

72. Prove: for all $A \in M_n(\mathbb{C})$, $||A||_2 = \sqrt{\rho(A^*A)}$. (Note A^*A is positive semidefinite.)

73. Use the theorem in §7.7 to show that given any $\epsilon > 0$ and any B in $M_n(\mathbb{C})$, there is an invertible S in $M_n(\mathbb{C})$ such that $S^{-1}BS$ is upper-triangular and $||S^{-1}BS||_\infty \leq \rho(B) + \epsilon$. (If T is triangular and D is diagonal with $D(i, i) = \delta^i$ for some $\delta > 0$, compare the entries of T and $D^{-1}TD$.)

74. Assume the setup in §10.15. (a) Suppose the largest eigenvalue c has several associated eigenvectors, say $c = c_1 = \cdots = c_m$ and $|c_m| > |c_{m+1}|$. Prove the power method still converges. (b) Suppose instead that $|c_1| = \cdots = |c_m| > |c_{m+1}|$ with $m > 1$. Must the power method converge? (c) Suppose we happen to pick x_0 with $d_1 = 0$. Assuming $|c_2| > |c_3|$, what does the power method do?

75. With the setup in §10.15, prove that $x_k - b'_k z_1 \to 0$ as $k \to \infty$.

76. Execute the deflation algorithm on the matrix $A = \begin{bmatrix} 1 & 4 & -1 \\ 0 & 3 & 2 \\ 1 & -1 & -3 \end{bmatrix}$, which has a known eigenvalue $c = -1$ and associated eigenvector $x = [3 \; -1 \; 2]^{\mathrm{T}}$.

77. Let $A = \begin{bmatrix} 0 & 1 & 1 \\ 1 & 1 & 0 \\ 1 & 0 & 1 \end{bmatrix}$. Find the largest eigenvalue of A and an associated eigenvector by the power method. Use deflation to recover the other two eigenvalues of A.

78. Let $A = \begin{bmatrix} 0 & 0 & 0 & 1 \\ 1 & 0 & 0 & 0 \\ 0 & 1 & 0 & 0 \\ 0 & 0 & 1 & 0 \end{bmatrix}$. By inspection, find an eigenvector of A associated with the eigenvalue 1. Use deflation to find $B \in M_3(\mathbb{C})$ such that $\chi_B(t) = (t-1)\chi_A(t)$.

79. (a) Give an example of a matrix A and a scalar $c \neq 0$ such that Richardson's Algorithm fails to converge when applied to $Ax = b$, but does converge when applied to $(cA)x = cb$. (b) For any $A \in M_n(\mathbb{C})$ and nonzero $c \in \mathbb{C}$, how are the eigenvalues of $I - A$ related to the eigenvalues of $I - cA$? (c) Find a condition on the eigenvalues of A guaranteeing the existence of $c \in \mathbb{C}$ such that Richardson's Algorithm converges for the system $(cA)x = cb$. Describe how to choose c if we know all the eigenvalues of A.

80. Let $A = \begin{bmatrix} a & b \\ c & d \end{bmatrix}$, $Q_1 = \begin{bmatrix} a & 0 \\ 0 & d \end{bmatrix}$, and $Q_2 = \begin{bmatrix} a & 0 \\ c & d \end{bmatrix}$ where $a, b, c, d \in \mathbb{C}$ with $a \neq 0 \neq d$. Find the eigenvalues of $I - Q_1^{-1}A$ and $I - Q_2^{-1}A$. Deduce sufficient conditions for the convergence of the Jacobi method and the Gauss–Seidel method on the system $Ax = y$.

81. Give a specific example of a matrix $A \in M_3(\mathbb{R})$ and $b \in \mathbb{R}^3$ such that the Jacobi method converges when applied to $Ax = b$, but the Gauss–Seidel method does not converge for this system.

82. Give a specific example of a matrix $A \in M_3(\mathbb{R})$ and $b \in \mathbb{R}^3$ such that the Gauss–Seidel method converges when applied to $Ax = b$, but the Jacobi method does not converge for this system.

83. Discuss how the shifted power method might be used to find the largest eigenvalue of $A \in M_n(\mathbb{C})$ in each of these cases: (a) A is not invertible; (b) two eigenvalues of A both have magnitude $\rho(A)$.

84. Suppose $A \in M_n(\mathbb{C})$ is lower-triangular with nonzero entries on the diagonal. (a) What happens if we use the Gauss–Seidel algorithm to solve $Ax = b$? (b) Must the Jacobi method converge for this system? Explain.

85. Given $A = \begin{bmatrix} a & b \\ c & d \end{bmatrix} \in M_2(\mathbb{R})$, find $\|A\|_2$ and $\rho(A)$ in terms of a, b, c, d.

86. Define $\|A\| = \rho(A)$ for $A \in M_n(\mathbb{C})$. Which axioms for a matrix norm are satisfied?

Part IV

The Interplay of Geometry and Linear Algebra

11

Affine Geometry and Convexity

A central concept in linear algebra is the idea of a subspace of a vector space. Subspaces provide an algebraic abstraction of uncurved geometric objects such as lines and planes through the origin in three-dimensional space. However, there are many other uncurved geometric figures that are not subspaces, including lines and planes not passing through the origin, individual points, line segments, triangles, quadrilaterals, polygons in the plane, tetrahedra, cubes, and solid polyhedra. This chapter discusses affine sets and convex sets, which provide an algebraic setting for studying such figures and their higher-dimensional analogs.

Geometrically, an affine subset of a vector space is a subspace that has been translated away from the origin by adding some fixed vector. Affine sets can also be described as intersections of hyperplanes, as solution sets of systems of linear equations, or as the set of affine combinations of a given set of vectors. A notion of affine independence, analogous to linear independence, leads to affine versions of bases, dimension, and coordinate systems for affine sets. Affine maps preserve the affine structure of affine sets, in the same way that linear maps preserve the linear structure of vector spaces and subspaces.

A convex set in $\mathbb{R}^n$ contains the line segment joining any two of its points. Given a set S of points in $\mathbb{R}^n$, there is a smallest convex set containing S, called the convex hull of S. We can build the convex hull from the generating set S, by taking all convex combinations of the generators, or we can obtain the convex hull by intersecting all convex sets that contain S. This fact is one instance of a family of theorems asserting the equivalence of a generative description of a class of sets and an intersectional description of that same class. We shall encounter several such theorems in this chapter, including the fundamental result that convex hulls of finite sets coincide with bounded intersections of finitely many closed half-spaces.

The chapter concludes with a discussion of convex real-valued functions, which are closely related to convex sets. Convex functions play a prominent role in analysis, linear algebra, and optimization theory, especially linear programming. We prove Jensen's Inequality for convex functions and describe how to test convexity by checking the first and second derivatives of a function.

11.1   Linear Subspaces

Before beginning our study of affine geometry, we discuss some facts about subspaces of vector spaces that are used in the development of the affine theory. Let F be a field, n be a positive integer, and V be an n-dimensional vector space over F. We know V is isomorphic to the vector space F^n consisting of n-tuples $v = (v_1, \ldots, v_n)$ with each v_i in F (see §6.3).

We often identify an n-tuple v with the $n \times 1$ column vector $\begin{bmatrix} v_1 \\ \vdots \\ v_n \end{bmatrix}$.

A *subspace* of V is a subset W of V satisfying these three closure conditions:
(i) 0_V belongs to W (closure under identity);
(ii) for all $v, w \in W$, $v + w$ is in W (closure under vector sum);
(iii) for all $v \in W$ and $c \in F$, cv is in W (closure under scalar multiplication).

In this chapter, we may refer to subspaces of V as *linear subspaces* to emphasize the distinction between subspaces and affine sets (defined later). Given a subspace W, an integer $k \geq 0$, vectors $v_1, \ldots, v_k$ in W, and scalars $c_1, \ldots, c_k$ in F, it follows by induction on k that $c_1 v_1 + \cdots + c_k v_k$ must belong to W. The expression $c_1 v_1 + \cdots + c_k v_k$ is called a *linear combination of* $v_1, \ldots, v_k$. Thus, subspaces are closed under taking linear combinations of their elements. Conversely, any subset of V that is closed under linear combinations must be a linear subspace of V. Here, we use the convention that a linear combination of $k = 0$ terms is interpreted as 0_V.

Geometrically, the condition that $0_V \in W$ means that all linear subspaces are required to pass through the origin. Every subspace W of V is itself an F-vector space, and the dimension of W is between 0 and n. One-dimensional subspaces of V are called *lines through the origin*; two-dimensional subspaces of V are called *planes through the origin*; and $(n-1)$-dimensional subspaces of the n-dimensional space V are called *linear hyperplanes*. Lines and planes in $\mathbb{R}^3$ that do not pass through $(0, 0, 0)$ are not linear subspaces; they are examples of the affine sets to be studied later.

11.2   Examples of Linear Subspaces

This section describes four ways of constructing linear subspaces, each of which leads to a different description of the points in the subspace.

First, suppose that V is F^n (viewed as a set of column vectors), and $A \in M_{m,n}(F)$ is an $m \times n$ matrix with entries in F. The *null space* of A is $W = \{x \in V : Ax = 0\}$. We check that W is a subspace of V as follows. W is a subset of V by definition. To check (i), note $A0 = 0$, so $0 \in W$. To check (ii), fix $v, w \in W$ and show that $v + w \in W$. We know $Av = 0 = Aw$, so $A(v + w) = Av + Aw = 0 + 0 = 0$, so $v + w$ is in W. To check (iii), fix $v \in W$ and $c \in F$ and show that $cv \in W$. We know $Av = 0$, so $A(cv) = c(Av) = c0 = 0$, so $cv \in W$. We can think of the null space of A more concretely as the solution set of the system of m homogeneous linear equations in n unknowns

$$\begin{cases} A(1,1)x_1 + A(1,2)x_2 + \cdots + A(1,n)x_n &= 0 \\ A(2,1)x_1 + A(2,2)x_2 + \cdots + A(2,n)x_n &= 0 \\ \qquad\qquad\qquad \cdots \\ A(m,1)x_1 + A(m,2)x_2 + \cdots + A(m,n)x_n &= 0. \end{cases} \tag{11.1}$$

The equations are called *homogeneous* because their right-hand sides are zero.

Second, suppose that V is F^m (viewed as a set of column vectors), and $A \in M_{m,n}(F)$ is a given matrix. The *range* of A is $W = \{v \in V : \exists x \in F^n, Ax = v\}$. It is routine to verify that W is a subspace of F^m. We can give a more abstract version of these first two constructions by considering an arbitrary F-linear map $T : V \to Z$ between two F-vector spaces V and Z. The *kernel of* T, defined to be $\ker(T) = \{v \in V : T(v) = 0_Z\}$, is a subspace of V. The *image of T*, defined to be $\text{img}(T) = \{z \in Z : \exists v \in V, T(v) = z\}$, is a subspace of Z. To recover the earlier examples involving the matrix A, take $T : F^n \to F^m$ to be $T(v) = Av$ for $v \in F^n$. This is a linear map whose kernel is the null space of A and whose image is the range of A.

Third, suppose $S = \{v_1, \ldots, v_k\}$ is a set of k vectors in V. Let W consist of all linear combinations of elements of S, so that a vector w is in W iff there exist scalars $c_1, \ldots, c_k \in F$ with $w = c_1 v_1 + \cdots + c_k v_k$. It is routine to check that W satisfies the closure conditions (i), (ii), and (iii), so that W is a subspace of V. We call W the *subspace spanned by S* or the *linear span of S*, writing $W = \text{Sp}(S)$ or $W = \text{Sp}_F(S)$. Observe that $S \subseteq W$, because each v_i can be written in the form $0v_1 + \cdots + 1v_i + \cdots + 0v_n$. For infinite $S \subseteq V$, let $\text{Sp}_F(S)$ consist of all linear combinations of finitely many elements of S. Again, we may check that this is a linear subspace of V containing S. For all linear subspaces Z of V, if $S \subseteq Z$ then $\text{Sp}_F(S) \subseteq Z$ since Z is closed under linear combinations. So, $\text{Sp}_F(S)$ is the smallest linear subspace of V containing S. The fact that the range of $A \in M_{m,n}(F)$ is a subspace is a special case of the spanning construction. This follows from the fact that the elements Ax in the range of A are precisely the linear combinations $x_1 A^{[1]} + x_2 A^{[2]} + \cdots + x_n A^{[n]}$ of the n columns of A (see §4.7).

Fourth, suppose $\{W_i : i \in I\}$ is any collection of linear subspaces of V. The intersection $W = \bigcap_{i \in I} W_i$ is readily seen to be a subspace of V. For example, any intersection of linear hyperplanes in V is a subspace of V. We claim that the null space of a matrix $A \in M_{m,n}(F)$ can be viewed as a special case of this construction. To see why, note that the solution set of the system of equations (11.1) is the intersection of the solution sets of the m individual equations considered separately. Consider the solution set Z of a particular equation $c_1 x_1 + \cdots + c_n x_n = 0$, where every $c_i \in F$. If all c_i are zero, then Z is all of F^n, and we can discard this equation from the system without changing the final solution set of the full system. If some $c_i \neq 0$, we solve the equation for x_i to get

$$x_i = c_i^{-1}(-c_1 x_1 - \cdots - c_{i-1} x_{i-1} - c_{i+1} x_{i+1} - \cdots - c_n x_n).$$

We obtain all solutions by choosing any scalar value for each x_j with $j \neq i$ and then using the displayed relation to compute x_i. For example, if $k \neq i$ is fixed and we choose $x_k = 1$ and $x_j = 0$ for all $j \neq k, i$, then $x_i = -c_i^{-1} c_k$. From this description, we can verify that the set $\{\mathbf{e}_k - c_i^{-1} c_k \mathbf{e}_i : 1 \leq k \leq n, k \neq i\}$ is a basis for Z of size $n - 1$, where $\mathbf{e}_i$ and $\mathbf{e}_k$ are standard basis vectors in F^n. This means that Z is a linear hyperplane in F^n. The null space of A is the intersection of the linear hyperplanes associated with each nonzero row of A. This result also holds when A is a zero matrix using the convention that the intersection of an empty collection of hyperplanes is all of F^n.

11.3 Characterizations of Linear Subspaces

We now show that every linear subspace can be described in one of the ways discussed in the last section.

Theorem Characterizing Linear Subspaces. Let W be any subspace of an n-dimensional F-vector space V.
(a) If $\dim(W) = k$, then W is the span of a list of k linearly independent vectors.
(b) If $V = F^n$ and $\dim(W) = n - m$, then for some $A \in M_{m,n}(F)$, W is the solution set of the system $Ax = 0$ of m homogeneous nonzero linear equations (W is the null space of A).
(c) If $\dim(W) = n - m$, then W is an intersection of m linear hyperplanes in V.

Proof. Part (a) follows from the fact that the subspace W is itself a vector space, so it has a basis S of size k (see §1.8 and §17.14). Since S spans W, W consists of all linear combinations of elements of S. In fact, the linear independence of S ensures that each $w \in W$ can be written in exactly one way as $c_1 v_1 + \cdots + c_k v_k$ with $c_i \in F$ and $v_i \in S$.

To prove (b), let W be an $(n - m)$-dimensional subspace of F^n with ordered basis $B = (v_{m+1}, v_{m+2}, \ldots, v_n)$. By adding appropriate vectors to the beginning of the list B, we can extend B to an ordered basis $C = (v_1, \ldots, v_m, v_{m+1}, \ldots, v_n)$ for F^n. Define $T : F^n \to F^m$ by setting

$$T(c_1 v_1 + \cdots + c_m v_m + c_{m+1} v_{m+1} + \cdots + c_n v_n) = \begin{bmatrix} c_1 \\ \vdots \\ c_m \end{bmatrix} \quad \text{for all } c_1, \ldots, c_n \in F.$$

Let $A \in M_{m,n}(F)$ be the matrix of T with respect to the standard ordered bases of F^n and F^m, so $T(x) = Ax$ for all $x \in F^n$ (see Chapter 6). On one hand, the kernel of T consists precisely of the linear combinations of vectors in B, which is the given subspace W. On the other hand, since $T(x) = Ax$, we see that the kernel of T is exactly the null space of A, which is the solution set of the system of equations $Ax = 0$. If some row i of A were zero, then the ith entry of Ax would be zero for every x. But the range of A equals the image of T, which is all of F^m. So none of the rows of A can be zero. We have now expressed W as the solution set of m homogeneous nonzero linear equations. As seen in the previous section, it follows that W is the intersection of the m linear hyperplanes determined by the m rows of A.

The previous paragraph establishes the special case of (c) where W is a subspace of F^n. For a general n-dimensional vector space V with $(n-m)$-dimensional subspace W, we know there is a vector space isomorphism $g : V \to F^n$. (For example, as seen in Chapter 6, we can obtain g by selecting an ordered basis of V and mapping each $v \in V$ to the coordinates of v relative to this basis.) Applying the isomorphism g to W gives a subspace $W' = g[W]$, which is an $(n - m)$-dimensional subspace of F^n. So we can find m linear hyperplanes $H'_1, \ldots, H'_m$ in F^n with $W' = H'_1 \cap \cdots \cap H'_m$. Let $H_i = g^{-1}[H'_i]$ for $1 \le i \le m$, which is an $(n - 1)$-dimensional subspace (linear hyperplane) in W since g^{-1} is an isomorphism. Then

$$W = g^{-1}[W'] = g^{-1}[H'_1 \cap \cdots \cap H'_m] = g^{-1}[H'_1] \cap \cdots \cap g^{-1}[H'_m] = H_1 \cap \cdots \cap H_m,$$

so W is an intersection of m linear hyperplanes in V. $\qquad\square$

11.4 Affine Combinations and Affine Sets

Let V be an n-dimensional vector space over a field F. We have seen that the linear subspaces of V are the subsets of V closed under linear combinations. Given $v_1, \ldots, v_k \in V$, define an *affine combination* of $v_1, \ldots, v_k$ to be a linear combination $c_1 v_1 + \cdots + c_k v_k$ where each $c_i \in F$ and $c_1 + c_2 + \cdots + c_k = 1_F$. An *affine set* in V is a subset W of V that is closed under affine combinations. In other words, $W \subseteq V$ is affine iff for all positive integers k, all $w_1, \ldots, w_k \in W$, and all $c_1, \ldots, c_k \in F$, if $c_1 + \cdots + c_k = 1$, then $c_1 w_1 + \cdots + c_k w_k \in W$. For the field $F = \mathbb{R}$, it can be shown that a set W is affine iff for all $w, z \in W$ and all $c \in \mathbb{R}$, $cw + (1-c)z$ is in W. Thus, in real vector spaces, it suffices to check the affine combinations with $k = 2$ terms. But this fact does not hold in every field; see Exercise 38.

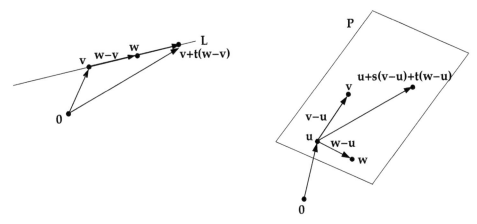

FIGURE 11.1
Vector description of lines and planes in $\mathbb{R}^3$.

To motivate the definition of affine combinations, we recall the vector equations of lines and planes in $\mathbb{R}^3$ that do not necessarily pass through the origin. Consider the line L in $\mathbb{R}^3$ through the distinct points $v, w \in \mathbb{R}^3$. Geometrically, we can reach each point on this line by starting at the origin, traveling to the tip of the vector v (viewed as an arrow starting at 0), then following some scalar multiple $t(w - v)$ of the direction vector $w - v$ for the line (where $t \in \mathbb{R}$). See Figure 11.1. Algebraically, the line consists of all vectors $v + t(w - v)$ as t ranges over $\mathbb{R}$. Restating this,

$$L = \{tw + (1 - t)v : t \in \mathbb{R}\} = \{tw + sv : t, s \in \mathbb{R} \text{ and } t + s = 1\}.$$

In other words, L consists of all affine combinations of v and w. You can check that L is an affine set in $\mathbb{R}^3$ (this is a special case of a general result proved in §11.6).

Similarly, consider the plane P in $\mathbb{R}^3$ through three non-collinear points $u, v, w \in \mathbb{R}^3$. As in the case of the line, we reach arbitrary points in P from the origin by going to the tip of u, then traveling within the plane some amount in the direction $v - u$ and some other amount in the direction $w - u$; see Figure 11.1. Algebraically,

$$P = \{u + s(v - u) + t(w - u) : s, t \in \mathbb{R}\} = \{(1 - s - t)u + sv + tw : s, t \in \mathbb{R}\}$$
$$= \{ru + sv + tw : r, s, t \in \mathbb{R}, r + s + t = 1\}.$$

So P is the set of affine combinations of u, v, w. You may check that P is an affine set in $\mathbb{R}^3$.

A plane in $\mathbb{R}^3$ can also be specified as the solution set of one linear (possibly non-homogeneous) equation in three unknowns. A line in $\mathbb{R}^3$ is the solution set of two linear equations in three unknowns. More generally, you may check that for any matrix $A \in M_{m,n}(F)$ and $b \in F^m$, the set of all solutions $x \in F^n$ to the non-homogeneous system of linear equations $Ax = b$ is an affine subset of F^n. We shall see later (§11.9) that all affine subsets of F^n have this form.

11.5 Affine Sets and Linear Subspaces

We can use linear subspaces of V to build more examples of affine sets in V. Let W be a fixed linear subspace of V. Since W is closed under all linear combinations of its elements, which include affine combinations as a special case, W is an affine subset of V. More generally, for any subset S of V and any vector $u \in V$, define the *translate of S by u* to be the set $u + S = \{u + x : x \in S\}$. We now prove that each translate $u + W$ is an affine set. Fix $v_1, \ldots, v_k \in u + W$ and $c_1, \ldots, c_k \in F$ with $c_1 + \cdots + c_k = 1$. We must check that $c_1 v_1 + \cdots + c_k v_k \in u + W$. Write each v_i as $u + w_i$ for some $w_i \in W$. Then

$$\sum_{i=1}^{k} c_i v_i = \sum_{i=1}^{k} c_i(u + w_i) = \sum_{i=1}^{k} c_i u + \sum_{i=1}^{k} c_i w_i = u \sum_{i=1}^{k} c_i + \sum_{i=1}^{k} c_i w_i.$$

We know $\sum_{i=1}^{k} c_i = 1$ and $\sum_{i=1}^{k} c_i w_i \in W$ (because W is a linear subspace), so our calculation shows that $\sum_{i=1}^{k} c_i v_i$ is in $u + W$, as needed. Exactly the same calculation proves that *any translate $u + X$ of an affine set X is also an affine set*.

Taking W to be the subspace V, we see that the whole space V is an affine set. Taking W to be the subspace $\{0\}$, we see that every one-point set $u + W = \{u\}$ is an affine set. By definition, an *affine line* is a translate $u + W$ where $\dim(W) = 1$; an *affine plane* is a translate $u + W$ where $\dim(W) = 2$; and an *affine hyperplane* is a translate $u + W$ where $\dim(W) = n - 1$ (meaning that W is a linear hyperplane).

We next prove that *for all $X \subseteq V$, X is a linear subspace iff X is an affine set and $0_V \in X$.* We already know the forward direction is true. Conversely, suppose X is an affine set containing 0_V. Given $v, w \in X$, $v + w$ is the affine combination $-1 \cdot 0_V + 1v + 1w$ of elements of X, so $v + w \in X$. Given $v \in X$ and $c \in F$, cv is the affine combination $(1 - c) \cdot 0_V + cv$ of elements of X, so $cv \in X$. We also know $0_V \in X$, so X is a linear subspace.

Theorem on Affine Sets and Linear Subspaces. For every nonempty affine set X in V, there exists a unique linear subspace W in V such that $X = u + W$ for some $u \in V$.

The vector u such that $X = u + W$ is not unique if $|X| > 1$. In fact, the proof will show that $X = u + W$ and $W = -u + X$ for any choice of $u \in X$. We call W the *direction subspace of X*. This theorem provides one geometric characterization of the nonempty affine sets in V: they are precisely the translates of the linear subspaces of V.

Proof. Fix a nonempty affine set X in V, and fix $u \in X$. Define $W = -u + X$. On one hand, W is an affine set, being a translate of the affine set X. On the other hand, $u \in X$ implies $0 = -u + u \in W$. By the result preceding the theorem, W is a linear subspace, and evidently $u + W = u + (-u + X) = X$. We now prove uniqueness of the linear subspace W. Say $X = u + W = v + W'$ for some $u, v \in V$ and linear subspaces W and W'. Since $0 \in W$ and $0 \in W'$, u and v belong to X. Then $v = u + w$ for some $w \in W$, hence $W' = -v + X = (-v + u) + W = -w + W = W$. (The last step uses the fact that $x + W = W$ for all x in a subspace W.) $\qquad\square$

Since the direction subspace of an affine set X is uniquely determined by X, we can define the *affine dimension* of X (written $\dim(X)$) to be the vector-space dimension of its direction subspace. For instance, points, affine lines, affine planes, and affine hyperplanes have respective affine dimensions 0, 1, 2, and $n - 1$. The affine dimension of $\emptyset$ is undefined.

11.6 The Affine Span of a Set

Let $S = \{v_1, \ldots, v_k\}$ be a finite subset of V. We know that the smallest linear subspace of V containing S is the linear span $\mathrm{Sp}(S)$ consisting of all linear combinations of $v_1, \ldots, v_k$. By analogy, define the *affine span* or *affine hull* of S, denoted $\mathrm{aff}(S)$, to be the set of all affine combinations of elements of S. In symbols,

$$\mathrm{aff}(\{v_1, \ldots, v_k\}) = \left\{ c_1 v_1 + \cdots + c_k v_k : c_i \in F, \sum_{i=1}^{k} c_i = 1 \right\}.$$

Similarly, for an infinite subset S of V, let $\mathrm{aff}(S)$ be the set of all affine combinations of finitely many vectors belonging to S. For example, in $V = \mathbb{R}^3$, our discussion of Figure 11.1 shows that $\mathrm{aff}(\{v, w\})$ is the line through v and w, and $\mathrm{aff}(\{u, v, w\})$ is the plane through u, v, and w. The next theorem shows that $\mathrm{aff}(S)$ is the smallest affine set containing S.

Theorem on Affine Spans. For all $S \subseteq V$, $\mathrm{aff}(S)$ is an affine set containing S. If $S \subseteq T$ and T is an affine set, then $\mathrm{aff}(S) \subseteq T$. The set $\mathrm{aff}(S)$ of all affine combinations of vectors in S equals the intersection of all affine sets in V containing S.

Proof. To check that $\mathrm{aff}(S)$ is affine, fix $w_1, \ldots, w_s \in \mathrm{aff}(S)$ and $c_1, \ldots, c_s \in F$ with $\sum_{j=1}^{s} c_j = 1$. We show $w = \sum_{j=1}^{s} c_j w_j$ belongs to $\mathrm{aff}(S)$. There are finitely many elements $v_1, \ldots, v_k \in S$ such that each w_j is some affine combination of these elements, say $w_j = \sum_{i=1}^{k} a_{ij} v_i$ with $a_{ij} \in F$ and $\sum_{i=1}^{k} a_{ij} = 1$ for $1 \le j \le s$. Compute

$$w = \sum_{j=1}^{s} c_j w_j = \sum_{j=1}^{s} c_j \left(\sum_{i=1}^{k} a_{ij} v_i \right) = \sum_{j=1}^{s} \sum_{i=1}^{k} c_j a_{ij} v_i = \sum_{i=1}^{k} \left(\sum_{j=1}^{s} c_j a_{ij} \right) v_i.$$

So w is a linear combination of the v_i, and the coefficients in this combination have sum

$$\sum_{i=1}^{k} \sum_{j=1}^{s} c_j a_{ij} = \sum_{j=1}^{s} c_j \sum_{i=1}^{k} a_{ij} = \sum_{j=1}^{s} c_j \cdot 1 = 1.$$

Hence w is an affine combination of $v_1, \ldots, v_k$, which proves that $w \in \mathrm{aff}(S)$. For each $x \in S$, $x = 1x$ is an affine combination of elements of S, so $S \subseteq \mathrm{aff}(S)$. Finally, suppose T is any affine subset of V with $S \subseteq T$. Since T is closed under affine combinations of any of its elements, affine combinations of elements taken from the subset S must be in T. So $\mathrm{aff}(S) \subseteq T$. This inclusion holds for all affine sets T containing S, so $\mathrm{aff}(S)$ is a subset of the intersection of all such sets. On the other hand, $\mathrm{aff}(S)$ is an affine set containing S, so $\mathrm{aff}(S)$ is one of the sets T involved in this intersection. Thus, the intersection is a subset of $\mathrm{aff}(S)$. $\qquad\square$

11.7 Affine Independence

In three-dimensional geometry, any two distinct points determine a line, and any three non-collinear points determine a plane. We would like similar descriptions of general affine sets A of the form $A = \mathrm{aff}(S)$, where the affine spanning set S is as small as possible. The examples of lines and planes indicate that the size of S exceeds the affine dimension of A by 1. To make this precise, we introduce the notion of affinely independent sets.

Recall that a list of vectors $L = (v_1, \ldots, v_k)$ in V is *linearly independent* iff the only linear combination of the v_i that gives 0_V is the combination with all zero coefficients. In other words, L is linearly independent iff for all $c_1, \ldots, c_k \in F$, if $c_1 v_1 + \cdots + c_k v_k = 0_V$, then $c_i = 0_F$ for $1 \le i \le k$. We now define L to be *affinely independent* iff for all $c_1, \ldots, c_k \in F$ such that $c_1 + c_2 + \cdots + c_k = 0_F$, $c_1 v_1 + \cdots + c_k v_k = 0_V$ implies $c_i = 0_F$ for $1 \le i \le k$. The list L is *affinely dependent* iff there exist $c_1, \ldots, c_k \in F$ with some $c_i \ne 0$, $\sum_{j=1}^{k} c_j = 0$, and $\sum_{j=1}^{k} c_j v_j = 0$. A set S of vectors (finite or not) is affinely independent iff every finite list of distinct vectors in S is affinely independent. The following result relates affine independence to linear independence.

Theorem Relating Affine Independence and Linear Independence. The list $L = (v_0, v_1, v_2, \ldots, v_k)$ is affinely independent iff the list $L' = (v_1 - v_0, v_2 - v_0, \ldots, v_k - v_0)$ is linearly independent.

Proof. We prove the contrapositive in both directions. Assuming L is affinely dependent, we have scalars $c_0, \ldots, c_k \in F$ summing to zero and not all zero, such that $\sum_{i=0}^{k} c_i v_i = 0$. We can rewrite the given combination of the v_i as

$$c_1(v_1 - v_0) + c_2(v_2 - v_0) + \cdots + c_k(v_k - v_0) + (c_0 + c_1 + c_2 + \cdots + c_k)v_0 = 0.$$

But the sum of all the c_i is zero, so we have expressed zero as a linear combination of the vectors in L' with coefficients $c_1, \ldots, c_k$. These coefficients cannot all be zero, since otherwise $\sum_{i=0}^{k} c_i = 0$ would give $c_0 = 0$ as well. This proves the linear dependence of L'. Conversely, assuming L' is linearly dependent, we have scalars $d_1, \ldots, d_k \in F$ (not all zero) with $d_1(v_1 - v_0) + \cdots + d_k(v_k - v_0) = 0$. Defining $d_0 = -d_1 - d_2 - \cdots - d_k$, we then have $d_0 v_0 + d_1 v_1 + \cdots + d_k v_k = 0$ where not all d_i are zero, but the sum of all d_i is zero. So L is affinely dependent. $\qquad \square$

When using this result to detect affine independence of a finite set of vectors, we can list the elements of the set in any order. So, given a finite set $S \subseteq V$, we can test the affine independence of S by picking any convenient $x \in S$ and checking if the set $T = \{y - x : y \in S, y \ne x\}$ is linearly independent. For example, *if S is any set of linearly independent vectors in V, then $S \cup \{0_V\}$ is affinely independent.*

11.8 Affine Bases and Barycentric Coordinates

Let X be a nonempty affine subset of V. A list $L = (v_0, \ldots, v_k)$ is an *ordered affine basis* of X iff L is affinely independent and $X = \text{aff}(\{v_0, \ldots, v_k\})$. Similarly, a set S is an *affine basis* of X iff S is affinely independent and $X = \text{aff}(S)$.

We can use the direction subspace of X to find an affine basis for X. Specifically, write the given affine set X as $X = u + W$, where $u \in X$ and W is a linear subspace of V. Suppose $\dim(W) = k$, and let $(w_1, \ldots, w_k)$ be an ordered vector-space basis of W. We claim $L = (u, u + w_1, \ldots, u + w_k)$ is an ordered affine basis of X. Using the criterion from §11.7, we see that L is affinely independent, since subtracting u from every other vector on the list produces the linearly independent list $(w_1, \ldots, w_k)$. Next, $u \in X$ and each $u + w_i \in X$, so the set $S = \{u, u + w_1, \ldots, u + w_k\}$ is contained in the affine set X. Hence aff$(S) \subseteq X$. To prove the reverse inclusion, fix any $z \in X$. Then $z - u \in W$, so we can write $z - u = c_1 w_1 + \cdots + c_k w_k$ for some $c_i \in F$. Hence, z itself can be written

$$z = u + \sum_{i=1}^{k} c_i w_i = \left(1 - \sum_{i=1}^{k} c_i\right) u + \sum_{i=1}^{k} c_i(u + w_i),$$

which is an affine combination of the elements of L. We now see that X, which has affine dimension $k = \dim(W)$, has an ordered affine basis consisting of $k + 1$ vectors.

By a similar argument, we now show that any affine basis of X must have size $k+1$. We can rule out the possibility of infinite affine bases using the fact that $\dim(V) = n$ is finite. Now, let $L = (y_0, y_1, \ldots, y_s)$ be any ordered affine basis of X; we show $s = k$ (so the list L has size $k+1$). We know $L' = (y_1 - y_0, \ldots, y_s - y_0)$ is a linearly independent list. Moreover, $X = y_0 + W$, so each vector in L' belongs to W. Let us check that W is the linear span of L'. Given $w \in W$, we have $y_0 + w \in X$, so that $y_0 + w = c_0 y_0 + c_1 y_1 + \cdots + c_s y_s$ for some $c_0, c_1, \ldots, c_s \in F$ such that $c_0 + c_1 + \cdots + c_s = 1$. Solving for w gives

$$w = (-1 + c_0 + c_1 + \cdots + c_s)y_0 + c_1(y_1 - y_0) + \cdots + c_s(y_s - y_0).$$

The coefficient of y_0 is zero, so w is a linear combination of the vectors in L'. Thus, L' is an ordered basis of the k-dimensional space W, forcing $s = k$.

These proofs illustrate the technique of establishing facts about affine concepts involving X by looking at the corresponding linear concepts involving the direction subspace of X. As further examples of this method, you can prove that any affinely independent subset of X can be extended to an affine basis of X, and any affine spanning set for X contains a subset that is an affine basis of X (Exercise 28). Furthermore, the maximum size of an affinely independent set in an n-dimensional space V is $n + 1$.

Now let X be an affine set with ordered affine basis $L = (v_0, v_1, \ldots, v_k)$. We claim that for each $z \in X$, there exist unique scalars $c_0, c_1, \ldots, c_k \in F$ with $\sum_{i=0}^{k} c_i = 1$ and $\sum_{i=0}^{k} c_i v_i = z$. These scalars are called the *barycentric coordinates of z relative to L*. Existence of the c_i follows since $X = $ aff$(\{v_0, \ldots, v_k\})$. To see that the c_i are unique, suppose we also had $d_0, d_1, \ldots, d_k \in F$ with $\sum_{i=0}^{k} d_i = 1$ and $\sum_{i=0}^{k} d_i v_i = z$. Subtracting the two expressions for z gives $\sum_{i=0}^{k}(c_i - d_i)v_i = 0$, where the sum of the coefficients $c_i - d_i$ is $1 - 1 = 0$. By affine independence of L, $c_i - d_i = 0$ for all i, so $c_i = d_i$ for all i.

Let us work out an example of barycentric coordinates for an affine plane in $\mathbb{R}^3$. Let P be the set of points $(x, y, z) \in \mathbb{R}^3$ satisfying $x + 2y - 3z = 5$. By choosing values for y and z and calculating x, we obtain the three vectors $u = (5, 0, 0)$, $v = (3, 1, 0)$, and $w = (2, 0, -1)$ in P. You can check that (u, v, w) is an ordered affine basis for P. Consider a point $(-8, 2, -3)$ in P. We can express this point as the affine combination

$$(-8, 2, -3) = -4(5, 0, 0) + 2(3, 1, 0) + 3(2, 0, -1) = -4u + 2v + 3w.$$

So, the point in P with Cartesian coordinates $(-8, 2, -3)$ has barycentric coordinates $(-4, 2, 3)$ relative to the ordered affine basis (u, v, w).

11.9 Characterizations of Affine Sets

We defined affine sets to be those subsets of a vector space V that are closed under affine combinations. As in the case of linear subspaces, we can give several other characterizations of which nonempty subsets of V are affine.

Theorem Characterizing Affine Sets. Let X be a nonempty affine subset of an n-dimensional vector space V.

(a) X is a translate of a unique linear subspace of V (the direction subspace of X).

(b) X is the affine span of an ordered list of k affinely independent vectors, for some k between 1 and $n + 1$.

(c) If $V = F^n$ and $\dim(X) = n - m$, then X is the solution set of a system of m (possibly non-homogeneous) nonzero linear equations $Ax = b$ for some $A \in M_{m,n}(F)$ and $b \in F^m$.

(d) If $\dim(X) = n - m$, then X is an intersection of m affine hyperplanes in V.

Proof. We proved part (a) in §11.5 and part (b) in §11.8. For (c), let X be an $(n - m)$-dimensional affine set in F^n. Using (a), write $X = u + W$ for some $(n-m)$-dimensional linear subspace of F^n and some $u \in F^n$. By (b) of §11.3, there is a matrix $A \in M_{m,n}(F)$ such that $W = \{w \in F^n : Aw = 0\}$. We claim that for $b = Au$, we have $X = \{x \in F^n : Ax = b\}$. On one hand, if $x \in X$, then $x = u + w$ for some $w \in W$, so $Ax = A(u+w) = Au + Aw = b + 0 = b$. On the other hand, if $x \in F^n$ satisfies $Ax = b$, then $A(x - u) = Ax - Au = b - b = 0$, so $x - u \in W$ and $x \in u + W = X$. We prove (d) similarly: write $X = u + W$ as in (c), and invoke (c) of §11.3 to find linear hyperplanes $H_1, \ldots, H_m$ in V with $W = H_1 \cap \cdots \cap H_m$. Each translate $u + H_i$ is an affine hyperplane in V, and you may check that

$$X = u + W = u + (H_1 \cap \cdots \cap H_m) = (u + H_1) \cap \cdots \cap (u + H_m). \qquad \square$$

11.10 Affine Maps

Let V and W be two vector spaces over the field F. A *linear map* or *linear transformation* is a function $T : V \to W$ such that $T(x+y) = T(x) + T(y)$ and $T(cx) = cT(x)$ for all $x, y \in V$ and all $c \in F$. From this definition and induction, we see that a linear map T satisfies $T(c_1x_1 + \cdots + c_kx_k) = c_1T(x_1) + \cdots + c_kT(x_k)$ for all $k \in \mathbb{Z}_{\geq 0}$, all $c_i \in F$, and all $x_i \in V$. In other words, *linear maps preserve linear combinations.* By analogy, we define a map $U : V \to W$ to be an *affine map* or *affine transformation* iff U preserves affine combinations. In other words, U is an affine map iff $U(c_1x_1 + \cdots + c_kx_k) = c_1U(x_1) + \cdots + c_kU(x_k)$ for all $k \in \mathbb{Z}_{\geq 0}$, all $x_i \in V$, and all $c_i \in F$ satisfying $c_1 + \cdots + c_k = 1$. When $F = \mathbb{R}$, it suffices to check the condition for $k = 2$; see Exercise 51. More generally, we allow the domain (or codomain) of an affine map to be an affine subset of V (or W). An *affine isomorphism* is a bijective affine map.

Every linear map is an affine map, but the converse is not true. For example, given a fixed $u \in V$, the *translation map* $T_u : V \to V$ defined by $T_u(v) = u + v$ for $v \in V$ is an affine map. To check this, fix $k \geq 0$, $x_i \in V$, and $c_i \in F$ with $\sum_{i=1}^{k} c_i = 1$, and compute

$$T_u(c_1x_1 + \cdots + c_kx_k) = u + c_1x_1 + \cdots + c_kx_k = (c_1 + \cdots + c_k)u + c_1x_1 + \cdots + c_kx_k$$
$$= c_1(u + x_1) + \cdots + c_k(u + x_k) = c_1T_u(x_1) + \cdots + c_kT_u(x_k).$$

However, since linear maps must send 0_V to 0_V, T_u is a linear map only when $u = 0$. More generally, you can check that for $A \in M_{m,n}(F)$ and $b \in F^m$, the map $U : F^n \to F^m$ given by $U(x) = Ax + b$ for $x \in F^n$ is always affine, but is linear only when $b = 0$.

In fact, a map $U : V \to W$ is a linear map iff U is an affine map and $U(0_V) = 0_W$. The forward implication is immediate from the definitions. Conversely, assume U is an affine map with $U(0) = 0$. Given $x, y \in V$ and $c \in F$, compute:

$$U(x + y) = U(-1 \cdot 0 + 1x + 1y) = -1U(0) + 1U(x) + 1U(y) = U(x) + U(y);$$

$$U(cx) = U((1 - c) \cdot 0 + cx) = (1 - c)U(0) + cU(x) = cU(x).$$

So U is linear.

The following facts are routinely verified: the identity map on any affine set is an affine isomorphism; the composition of two affine maps (resp. affine isomorphisms) is an affine map (resp. affine isomorphism); and the inverse of an affine isomorphism is also an affine isomorphism. For example, every translation map T_u is an affine isomorphism with inverse $T_u^{-1} = T_{-u}$, which is also affine. You can also check that the direct image or preimage of an affine set under an affine map is an affine set.

We know that every linear map $T : F^n \to F^m$ has the form $T(x) = Ax$ for a unique matrix $A \in M_{m,n}(F)$ (namely, A is the matrix whose jth column is $T(\mathbf{e}_j)$ for $1 \le j \le n$). We use this to deduce a corresponding fact for affine maps.

Theorem on Affine Maps.
(a) Every affine map $U : F^n \to F^m$ has the form $U(x) = Ax + b$ for a unique $A \in M_{m,n}(F)$ and a unique $b \in F^m$.
(b) Let V and W be finite-dimensional vector spaces. Every affine map $U : V \to W$ has the form $U = T_b \circ S$ for a uniquely determined linear map $S : V \to W$ and a uniquely determined $b \in W$ (here, T_b is translation by b on W).

Proof. To prove (a), let $U : F^n \to F^m$ be an affine map. Choose $b = U(0) \in F^m$, and consider the map $T_b^{-1} \circ U : F^n \to F^m$, where T_b^{-1} is the inverse of translation by b in the space F^m. The composition $T_b^{-1} \circ U$ is an affine map sending zero to zero, so it is a linear map from F^n to F^m. Thus, there is a matrix A with $T_b^{-1} \circ U(x) = Ax$ for all $x \in F^n$. Solving for U, we see that $U(x) = T_b(Ax) = Ax + b$ for all $x \in F^n$. Now, b is uniquely determined by U since $U(x) = Ax + b$ forces $b = A0 + b = U(0)$ no matter what A is. Knowing this, A is uniquely determined because it is the unique matrix representing the linear map $T_b^{-1} \circ U$ relative to the standard bases of F^n and F^m. Part (b) can be proved by choosing ordered bases for V and W and invoking part (a). $\square$

The next theorem lets us build affine maps by specifying how the map should act on an affine basis.

Universal Mapping Property (UMP) for Affine Bases. Let V and W be finite-dimensional vector spaces. Suppose $X \subseteq V$ is an affine set with affine basis $S = \{v_0, \ldots, v_k\}$. For any $y_0, \ldots, y_k \in W$, there exists a unique affine map $U : X \to W$ with $U(v_i) = y_i$ for $0 \le i \le k$.

Proof. We can give an explicit formula for U by noting that each $x \in X$ has a unique expression as an affine combination $x = \sum_{i=0}^k c_i v_i$ with $c_i \in F$ and $\sum_{i=0}^k c_i = 1$. So if U exists with the required properties, U must be given by the formula $U(x) = \sum_{i=0}^k c_i U(v_i) = \sum_{i=0}^k c_i y_i$. A routine calculation confirms that this formula really does define an affine map sending each v_i to y_i. Thus, U does exist and is unique. $\square$

You can also prove the UMP by reducing to the direction subspace of X and appealing to the universal mapping property for the linear basis $\{v_1 - v_0, \ldots, v_k - v_0\}$ of this subspace (Exercise 50).

11.11   Convex Sets

For our study of convexity in the rest of this chapter, we shall consider only real vector spaces, taking V to be $\mathbb{R}^n$. To motivate the definition of a convex set, consider the line L shown in Figure 11.1. The entire line L through v and w is obtained by taking affine combinations $(1 - t)v + tw$ as t ranges over the entire real line $\mathbb{R}$. In contrast, if we only wanted to get the points of L on the line segment joining v to w, we would only take the affine combinations $(1-t)v+tw = v + t(w-v)$ with t ranging through real numbers in the closed interval $I = [0,1]$. (Taking $t = 0$ gives v, taking $t = 1$ gives w, and intermediate t's give precisely the points in the interior of the line segment.) In general, for any $v, w \in \mathbb{R}^n$, we define the *line segment joining v and w* to be the set $\{(1 - t)v + tw : t \in \mathbb{R}, 0 \le t \le 1\}$.

A subset C of $\mathbb{R}^n$ is *convex* iff for all $v, w \in C$ and all $t \in \mathbb{R}$ with $0 \le t \le 1$, $(1-t)v+tw$ belongs to C. Geometrically, a set C is convex iff the line segment joining any two points of C is always contained in C. All affine sets are convex, but the converse is not true. For example, any finite closed interval $[a, b]$ in $\mathbb{R}^1$ is readily seen to be convex but not affine.

Given finitely many points $v_1, \dots, v_k \in \mathbb{R}^n$, a *convex combination* of these points is a linear combination $c_1 v_1 + \cdots + c_k v_k$ with $c_1 + \cdots + c_k = 1$ and $c_i \ge 0$ for all i (and hence $c_i \in [0, 1]$ for all i). We now show that *a convex set $C \subseteq \mathbb{R}^n$ is closed under convex combinations*: for all positive integers k and all $v_1, \dots, v_k \in C$, every convex combination of $v_1, \dots, v_k$ belongs to C. The proof uses induction on k. When $k = 1$, we must have $c_1 = 1$. Given that $v_1 \in C$, we certainly have $c_1 v_1 = 1v_1 = v_1 \in C$. If $k = 2$, then we must have $c_1 = 1 - c_2$, so $c_1 v_1 + c_2 v_2 \in C$ by definition of a convex set. For the induction step, fix $k \ge 3$, and assume that convex combinations of $k-1$ or fewer points of C are already known to be in C. Fix $v_1, \dots, v_k \in C$ and $c_1, \dots, c_k \in [0, 1]$ with $\sum_{i=1}^{k} c_i = 1$. If $c_1 = 1$, then the conditions on the c_i force $c_2 = \cdots = c_k = 0$, so that $c_1 v_1 + \cdots + c_k v_k = v_1 \in C$. Otherwise, let $d_i = c_i/(1 - c_1)$ for $2 \le i \le k$, and note that

$$c_1 v_1 + c_2 v_2 + \cdots + c_k v_k = c_1 v_1 + (1 - c_1)(d_2 v_2 + \cdots + d_k v_k).$$

Since $c_2, \dots, c_k$ are nonnegative real numbers with sum $1 - c_1$, it follows that $d_2, \dots, d_k$ are nonnegative real numbers with sum 1. Then $w = d_2 v_2 + \cdots + d_k v_k$ is a convex combination of $k - 1$ elements of C, which is in C by the induction hypothesis. So $\sum_{i=1}^{k} c_i v_i = (1 - c_1)w + c_1 v_1 \in C$ by definition of a convex set. This completes the induction step.

11.12   Convex Hulls

Given $S \subseteq \mathbb{R}^n$, we have studied the *linear* span of S, which consists of all linear combinations of elements of S, and the *affine* span of S, which consists of all affine combinations of elements of S. By analogy, we define the *convex span* or *convex hull* of $S \subseteq \mathbb{R}^n$, denoted $\text{conv}(S)$, to be the set of all convex combinations of finitely many elements of S.

Theorem on Convex Hulls. For all $S \subseteq \mathbb{R}^n$, $\text{conv}(S)$ is a convex set containing S. If $S \subseteq T$ and T is a convex set, then $\text{conv}(S) \subseteq T$. The set $\text{conv}(S)$ of all convex combinations of vectors in S equals the intersection of all convex sets in $\mathbb{R}^n$ containing S.

Proof. To check the convexity of $\text{conv}(S)$, fix $x, y \in \text{conv}(S)$ and $t \in [0, 1]$; we must check that $z = tx + (1 - t)y \in \text{conv}(S)$. By definition, there are $v_1, \dots, v_k \in S$, $c_1, \dots, c_k \in [0, 1]$

with sum 1, and $d_1, \ldots, d_k \in [0,1]$ with sum 1, such that $x = \sum_{i=1}^{k} c_i v_i$ and $y = \sum_{i=1}^{k} d_i v_i$. (By taking some c_i and d_i to be zero, we can assume that the same elements $v_1, \ldots, v_k$ of S are used in the expressions for x and y.) We compute

$$z = tx + (1-t)y = \sum_{i=1}^{k} (tc_i + (1-t)d_i)v_i.$$

Each of c_i, d_i, t, $1-t$ is a nonnegative real number, so each coefficient $tc_i + (1-t)d_i$ is a nonnegative real number. The sum of these coefficients is

$$\sum_{i=1}^{k}(tc_i + (1-t)d_i) = t\sum_{i=1}^{k} c_i + (1-t)\sum_{i=1}^{k} d_i = t \cdot 1 + (1-t) \cdot 1 = 1.$$

So z is a convex combination of $v_1, \ldots, v_k \in S$, hence $z \in \operatorname{conv}(S)$ as required. Since any $v \in S$ is the convex combination $v = 1v$, we have $S \subseteq \operatorname{conv}(S)$. If T is any convex subset of $\mathbb{R}^n$ containing S, then $\operatorname{conv}(S) \subseteq T$ since T is closed under convex combinations of its elements. It follows, as in the proof of the Theorem on Affine Spans, that $\operatorname{conv}(S)$ is the intersection of all convex sets T in $\mathbb{R}^n$ containing S. $\qquad\square$

The convex hulls of some finite sets of points are illustrated in Figure 11.2. In general, the convex hull of a one-point set $\{v\}$ is $\{v\}$; the convex hull of $\{v, w\}$ is the line segment joining v and w; the convex hull of three non-collinear points $\{v, w, x\}$ is the triangle (including interior) with vertices v, w, x; and the convex hull of four non-coplanar points $\{v, w, x, y\}$ is the tetrahedron (including interior) with vertices v, w, x, y. For $k \geq 0$, we define a k-*dimensional simplex* to be a set $\operatorname{conv}(\{v_0, v_1, \ldots, v_k\})$, where $\{v_0, v_1, \ldots, v_k\}$ is an affinely independent subset of $\mathbb{R}^n$. Geometrically, simplices are k-dimensional analogs of triangles and tetrahedra. A specific example of a k-dimensional simplex is $\Delta_k = \operatorname{conv}(\{\mathbf{0}, \mathbf{e}_1, \ldots, \mathbf{e}_k\})$, where the $\mathbf{e}_i$ are standard basis vectors in $\mathbb{R}^k$. Expanding the definition, we see that Δ_k is the set of vectors $(c_1, \ldots, c_k) \in \mathbb{R}^k$ with $c_i \geq 0$ for all i and $\sum_{i=1}^{k} c_i \leq 1$. This sum can be less than 1, since the scalar c_0 disappears in the convex combination $c_0\mathbf{0} + c_1\mathbf{e}_1 + \cdots + c_k\mathbf{e}_k$.

11.13 Carathéodory's Theorem on Convex Hulls

Consider the six-point set $S = \{u, v, w, x, y, z\} \subseteq \mathbb{R}^2$ shown in the lower-right part of Figure 11.2. The convex hull $\operatorname{conv}(S)$ consists of all convex combinations of the six points of S. If we tried to create this convex hull by drawing line segments between every pair of points in S, the resulting set would not be convex (it would consist of the sides and diagonals of the shaded hexagonal region). On the other hand, suppose we took all convex combinations of three-element subsets of S. For each fixed subset of size 3, the set of convex combinations using these three elements is a triangle with interior. We see from the figure that the union of these triangles is all of $\operatorname{conv}(S)$. Thus, to obtain the convex hull in this example, it suffices to use only convex combinations of three or fewer points from S. The next theorem generalizes this remark to convex hulls in higher dimensions.

Carathéodory's Theorem on Convex Hulls. For any subset S of $\mathbb{R}^n$, $\operatorname{conv}(S)$ consists of all convex combinations of $n + 1$ or fewer points of S.

Proof. Fix $S \subseteq \mathbb{R}^n$, and pick any point $w \in \operatorname{conv}(S)$. By definition, we know there exist an integer $m \geq 0$, points $v_0, v_1, \ldots, v_m \in S$, and nonnegative real scalars $c_0, c_1, \ldots, c_m$ such

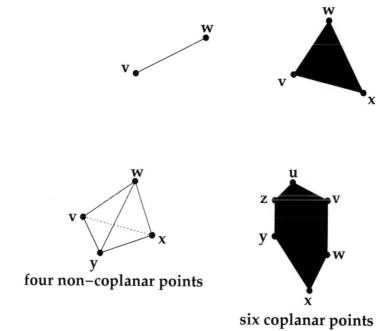

four non–coplanar points

six coplanar points

FIGURE 11.2
Examples of convex hulls.

that $c_0 + c_1 + \cdots + c_m = 1$ and $w = c_0 v_0 + c_1 v_1 + \cdots + c_m v_m$. If $m \leq n$, the claimed result holds for the point w. So assume $m > n$. We show how to find another expression for w as a convex combination of fewer than $m + 1$ points of S. Continuing this reduction process for at most $m - n$ steps, we eventually arrive at an expression for w as a convex combination of at most $n + 1$ points of S.

We can assume all v_i are distinct, since otherwise we can reduce m by using $cv + dv = (c + d)v$ to group together two terms in the convex combination. We can assume all $c_i > 0$, since otherwise we can reduce m by deleting the term $c_i v_i = 0$ from the sum. Since $m + 1 > n + 1$, the set $\{v_0, \ldots, v_m\}$ in $\mathbb{R}^n$ must be affinely dependent. So there exist $d_0, \ldots, d_m \in \mathbb{R}$ with some $d_i \neq 0$, $d_0 + \cdots + d_m = 0$, and $d_0 v_0 + \cdots + d_m v_m = 0$. At least one d_i is strictly positive, and we can reorder terms to ensure that $d_m > 0$ and $c_m/d_m \leq c_i/d_i$ for all $i < m$ such that $d_i > 0$. Define $b_i = c_i - (c_m/d_m)d_i$ for $0 \leq i < m$. We claim that each $b_i \geq 0$. If $d_i > 0$, the claim follows by multiplying $c_m/d_m \leq c_i/d_i$ by the positive quantity d_i and rearranging. If $d_i \leq 0$, the claim follows since $c_i > 0$, $c_m/d_m > 0$, hence $-(c_m/d_m)d_i \geq 0$ and $b_i \geq 0$. Next, $b_0 + b_1 + \cdots + b_{m-1} = 1$, because

$$\sum_{i=0}^{m-1} b_i = \sum_{i=0}^{m-1} c_i - (c_m/d_m) \sum_{i=0}^{m-1} d_i = (1 - c_m) - (c_m/d_m)(-d_m) = 1.$$

Finally, $w = \sum_{i=0}^{m-1} b_i v_i$, because

$$\sum_{i=0}^{m-1} b_i v_i = \sum_{i=0}^{m-1} c_i v_i - (c_m/d_m) \sum_{i=0}^{m-1} d_i v_i = (w - c_m v_m) - (c_m/d_m)(-d_m v_m) = w.$$

We have now expressed w as a convex combination of m points of S. $\qquad\qquad \square$

Exercise 77 uses Carathéodory's Theorem to prove that the convex hull of a closed and bounded (compact) subset of $\mathbb{R}^n$ is closed and bounded.

11.14 Hyperplanes and Half-Spaces in $\mathbb{R}^n$

For every affine hyperplane H in $\mathbb{R}^n$, there exist $a_1, \ldots, a_n, b \in \mathbb{R}$ such that

$$H = \{(x_1, \ldots, x_n) \in \mathbb{R}^n : a_1 x_1 + \cdots + a_n x_n = b\};$$

this follows from part (c) of the Theorem Characterizing Affine Sets. Recall that the *dot product* or *inner product* of two vectors $w = (w_1, \ldots, w_n)$ and $x = (x_1, \ldots, x_n)$ in $\mathbb{R}^n$ is defined by

$$w \bullet x = \langle w, x \rangle = w_1 x_1 + w_2 x_2 + \cdots + w_n x_n.$$

Letting $a = (a_1, \ldots, a_n) \neq 0$, we can write $H = \{x \in \mathbb{R}^n : a \bullet x = b\}$. Recall that the *Euclidean length* of the vector a is $||a|| = \sqrt{a \bullet a} = (a_1^2 + a_2^2 + \cdots + a_n^2)^{1/2}$. By multiplying both sides of the equation $a \bullet x = b$ by the nonzero scalar $1/||a||$, we obtain an equivalent equation $u \bullet x = c$ where u is a *unit vector* (i.e., $||u|| = 1$) and $c = ||a||^{-1} b \in \mathbb{R}$. By multiplying by -1 if needed, we can arrange that $c \geq 0$. You can check that when $c \neq 0$, the unit vector u and positive scalar c are uniquely determined by H. If $c = 0$, then there are exactly two possible unit vectors u that can be used in the equation defining H.

Given an affine hyperplane H, write $H = \{x \in \mathbb{R}^n : u \bullet x = c\}$ with $||u|| = 1$ and $c \geq 0$. If $c = 0$, then H consists of all vectors $x \in \mathbb{R}^n$ perpendicular to the unit vector u. If $c > 0$ and x_0 is any given point on H, then $x \in H$ iff $u \bullet x = u \bullet x_0$ iff $u \bullet (x - x_0) = 0$ iff $x - x_0$ is perpendicular to the unit vector u. Accordingly, u and its nonzero scalar multiples are called *normal vectors for H*.

The two *closed half-spaces* determined by H are the sets $\{x \in \mathbb{R}^n : u \bullet x \geq c\}$ and $\{x \in \mathbb{R}^n : u \bullet x \leq c\}$. Geometrically, the first set consists of points of H together with points outside of H on the same side of H as u (where u is a vector with tail drawn at some point $x_0 \in H$), and the second set consists of H together with the points on the opposite side of H from u. Similarly, the two *open half-spaces* determined by H are the sets $\{x \in \mathbb{R}^n : u \bullet x > c\}$ and $\{x \in \mathbb{R}^n : u \bullet x < c\}$. All four half-spaces determined by H, along with H itself, are convex. For example, suppose $x, y \in S = \{z \in \mathbb{R}^n : u \bullet z < c\}$ and $t \in [0, 1]$. We know $u \bullet x < c$ and $u \bullet y < c$. Since the inner product on $\mathbb{R}^n$ is linear in each variable and $0 \leq t \leq 1$, we have

$$u \bullet (tx + (1 - t)y) = t(u \bullet x) + (1 - t)(u \bullet y) < tc + (1 - t)c = c,$$

so $tx + (1 - t)y \in S$. A closed half-space is the solution set of a *linear inequality* $u_1 x_1 + \cdots + u_n x_n \leq c$, while an open half-space is the solution set of a *strict linear inequality*. You can check that the intersection of any family of convex subsets of $\mathbb{R}^n$ is also convex. Taking each convex set in the family to be a hyperplane or a closed half-space or an open half-space, we conclude: *the solution set of any family (possibly infinite) of linear equations, inequalities, and strict inequalities in n variables is a convex subset of $\mathbb{R}^n$*. Observe that each linear equation $u \bullet x = c$ appearing in the family could be replaced by the two linear inequalities $u \bullet x \leq c$ and $(-u) \bullet x \leq c$.

11.15   Closed Convex Sets

We might expect a description of convex sets as intersections of closed half-spaces, by analogy with our description of affine sets as intersections of affine hyperplanes. However, not all convex sets can be written in this form. To see why, recall that a subset C of $\mathbb{R}^n$ is *closed* iff for every sequence of vectors $v_1, \ldots, v_k, \ldots$ in C that converge to a vector $v \in \mathbb{R}^n$ (meaning $\lim_{k \to \infty} ||v_k - v|| = 0$), the limit vector v belongs to C. In other words, C is closed (in the topological sense) iff C is closed under taking limits of convergent sequences of vectors in C. Using this definition, you can check that the intersection of any family of closed subsets of $\mathbb{R}^n$ is also a closed set. Furthermore, every closed half-space is a closed set (as the name suggests). Combining these facts, we see that any intersection of closed half-spaces must be a closed and convex set. In this section, we prove the converse: every closed and convex subset of $\mathbb{R}^n$ is the intersection of all closed half-spaces containing it.

To obtain this result, we need a theorem that lets us separate certain subsets of $\mathbb{R}^n$ using an affine hyperplane. Recall that a set $D \subseteq \mathbb{R}^n$ is *bounded* iff there exists $M \in \mathbb{R}$ such that for all $x \in D$, $||x|| \leq M$.

Hyperplane Separation Theorem. For all disjoint, closed, convex sets C and D in $\mathbb{R}^n$ with C bounded, there is an affine hyperplane H in $\mathbb{R}^n$ (with associated open half-spaces H^+ and H^-) such that $C \subseteq H^+$ and $D \subseteq H^-$ or vice versa.

Proof. Let C and D be sets satisfying the hypotheses of the theorem. For points $x, y \in \mathbb{R}^n$, let $d(x, y) = ||x - y|| = \sqrt{(x - y) \bullet (x - y)}$ be the *distance* between x and y in $\mathbb{R}^n$. For a point $x \in \mathbb{R}^n$ and a nonempty set $B \subseteq \mathbb{R}^n$, let $d(x, B) = \inf_{y \in B} d(x, y)$ be the *distance between x and B*. For two nonempty sets A and B in $\mathbb{R}^n$, let $d(A, B) = \inf_{x \in A} d(x, B) = \inf_{x \in A, y \in B} d(x, y)$ be the *distance between A and B*. A theorem of real analysis states that a closed bounded subset C of $\mathbb{R}^n$ is *sequentially compact*, meaning that every sequence of points of C has a convergent subsequence. Using compactness of C, another theorem states that every continuous function $f : C \to \mathbb{R}$ attains its minimum value, meaning that there is $x \in C$ with $f(x) \leq f(w)$ for all $w \in C$. With these definitions and facts in hand, you can check that there exist $x \in C$ and $y \in D$ with $d(x, y) = d(C, D)$ (see Exercise 70). Set $r = d(x, y) = ||y - x||$; since C and D are disjoint, $x \neq y$ and $r > 0$.

Let u be the unit vector $r^{-1}(y - x)$ parallel to the line segment from x to y in $\mathbb{R}^n$. Let $c_1 = u \bullet x$ and $c_2 = u \bullet y$. Let $H_x = \{v \in \mathbb{R}^n : u \bullet v = c_1\}$ be the affine hyperplane through x perpendicular to u and $H_y = \{v \in \mathbb{R}^n : u \bullet v = c_2\}$ be the affine hyperplane through y perpendicular to u. We cannot have $c_1 = c_2$, since otherwise $u \bullet (y - x) = c_2 - c_1 = 0$ would give $r = r(u \bullet u) = u \bullet (ru) = 0$. We consider the case $c_1 < c_2$ below; the case $c_1 > c_2$ is handled similarly. We claim that $C \subseteq \{v \in \mathbb{R}^n : u \bullet v \leq c_1\}$ and that $D \subseteq \{v \in \mathbb{R}^n : u \bullet v \geq c_2\}$. The claim is illustrated by Figure 11.3, in which C is contained in the closed half-space weakly left of H_x, and D is contained in the closed half-space weakly right of H_y. The separation theorem follows from the claim by taking $H = \{z \in \mathbb{R}^n : u \bullet z = c_3\}$ for any choice of $c_3 \in \mathbb{R}$ with $c_1 < c_3 < c_2$.

We now prove the claim. Fix a point $z \in \mathbb{R}^n$ with $u \bullet z = c > c_1 = u \bullet x$; we show $z \notin C$. Suppose $z \in C$ to obtain a contradiction. Since $x, z \in C$ and C is convex, the entire line segment joining x and z is contained in C. Figure 11.3 suggests that we can drop an altitude from y to this line segment to find a point $w \in C$ with $d(w, y) < d(x, y) = d(C, D)$, giving a contradiction. To make this pictorial argument precise, we consider the function $g : [0, 1] \to \mathbb{R}$ such that $g(t) = d(y, x + t(z - x))^2$ for $t \in [0, 1]$. By definition of distance and

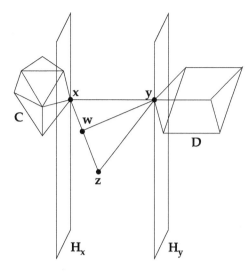

FIGURE 11.3
Finding a hyperplane to separate C and D.

the linearity of the dot product in each argument, we know

$$g(t) = [(y - x) - t(z - x)] \bullet [(y - x) - t(z - x)]$$
$$= (y - x) \bullet (y - x) - 2t(y - x) \bullet (z - x) + t^2(z - x) \bullet (z - x).$$

Recalling $r = d(x, y)$, $y - x = ru$, and setting $s = d(z, x)$, this becomes

$$g(t) = r^2 - 2t((ru) \bullet z - (ru) \bullet x) + s^2 t^2 = s^2 t^2 - 2r(c - c_1)t + r^2.$$

The graph of g in $\mathbb{R}^2$ is a parabola with slope $g'(0) = -2r(c - c_1) < 0$ at $t = 0$. So, for a sufficiently small positive $t \in [0, 1]$, $w = x + t(z - x)$ satisfies $d(w, y) = \sqrt{g(t)} < \sqrt{g(0)} = d(y, x)$. This contradiction shows $z \notin C$. A similar argument proves that no point $z \in \mathbb{R}^n$ with $u \bullet z < c_2$ can belong to D. $\qquad\square$

Theorem on Closed Convex Sets and Half-Spaces. Every closed convex set D in $\mathbb{R}^n$ is the intersection of all closed half-spaces S with $D \subseteq S$. D is also the intersection of all open half-spaces containing D.

Proof. Let $D \subseteq \mathbb{R}^n$ be closed and convex. Let E be the intersection of all closed half-spaces containing D. The inclusion $D \subseteq E$ is evident from the definition of E. For the reverse inclusion, suppose $x \in \mathbb{R}^n$ and $x \notin D$. Apply the Hyperplane Separation Theorem to the one-point set $C = \{x\}$ and the closed convex set D; note that C is closed, convex, bounded, and disjoint from D. We obtain a hyperplane H with $C \subseteq H^+$ (one of the open half-spaces for H) and $D \subseteq H^-$ (the other open half-space for H). Then $S = H^- \cup H$ is a closed half-space that contains D but not x. This is one of the half-spaces we intersect to obtain E, so $x \notin E$ as needed. The proof for open half-spaces is similar, taking $S = H^-$ this time. $\qquad\square$

11.16   Cones and Convex Cones

We would like to understand the structure of subsets of $\mathbb{R}^n$ formed by intersecting a finite number of closed half-spaces. Our ultimate goal is to prove that *a subset S of $\mathbb{R}^n$ is the convex hull of a finite set of points iff S is bounded and $S = H_1 \cap \cdots \cap H_k$ for finitely many closed half-spaces H_i.* This theorem is geometrically very plausible for subsets of $\mathbb{R}^2$ and $\mathbb{R}^3$ (cf. Figures 11.2 and 11.3), but it is quite tricky to prove in general dimensions. To prove this result, and to describe unbounded intersections of finitely many closed half-spaces, we must first develop some machinery involving cones.

A subset C of $\mathbb{R}^n$ is a *cone* iff $0 \in C$ and for all $x \in C$ and all $b > 0$ in $\mathbb{R}$, $bx \in C$. In other words, cones are sets containing 0 that are closed under multiplication by positive real scalars. Equivalently, cones are nonempty sets closed under multiplication by nonnegative scalars. We claim *a cone C is convex iff C is closed under addition.* On one hand, suppose C is a convex cone and $x, y \in C$. Then $z = (1/2)x + (1/2)y \in C$ by convexity, so $x + y = 2z \in C$ because C is a cone, so C is closed under addition. On the other hand, suppose a cone C is closed under addition. For $x, y \in C$ and $t \in [0,1]$, we know $tx \in C$ and $(1-t)y \in C$, since $t \geq 0$ and $1 - t \geq 0$. Then $tx + (1-t)y \in C$, so that C is convex.

A *positive combination* (or *conical combination*) of vectors $v_1, \ldots, v_k \in \mathbb{R}^n$ is a linear combination $c_1 v_1 + \cdots + c_k v_k$, where each c_i is in $\mathbb{R}_{\geq 0}$. By the remarks in the previous paragraph and induction on k, we see that *convex cones are closed under positive combinations of their elements.* Conversely, any subset of $\mathbb{R}^n$ closed under positive combinations is a convex cone (taking $k = 0$ shows that such a subset must contain zero).

Given $S \subseteq \mathbb{R}^n$, the *convex cone generated by S* is the set cone(S) consisting of all positive combinations of vectors $v_1, \ldots, v_k \in S$. Let us show that cone(S) is a convex cone containing S. Taking $k = 0$, we see that $0 \in$ cone(S). Fix $x, y \in$ cone(S) and $b > 0$ in $\mathbb{R}$. We can write $x = c_1 v_1 + \cdots + c_k v_k$ and $y = d_1 v_1 + \cdots + d_k v_k$ for some real scalars $c_i, d_i \geq 0$ and some $v_1, \ldots, v_k \in S$. Then $bx = \sum_{i=1}^{k}(bc_i)v_i \in$ cone(S) since each $bc_i \geq 0$, and $x + y = \sum_{i=1}^{k}(c_i + d_i)v_i \in$ cone(S) since each $c_i + d_i \geq 0$. So cone(S) is a convex cone. This cone contains S since each $v \in S$ is the positive combination $v = 1v$. Moreover, if T is any convex cone containing S, then cone(S) $\subseteq T$ since T is closed under positive combinations of its elements. We conclude that cone(S) can also be described as the intersection of all convex cones in $\mathbb{R}^n$ containing S.

All linear subspaces W of $\mathbb{R}^n$ are convex cones, as they are closed under all linear combinations of their elements (hence are closed under positive combinations). Define a *linear half-space* of $\mathbb{R}^n$ to be any closed half-space determined by a linear hyperplane in $\mathbb{R}^n$. Explicitly, linear half-spaces are sets of the form $\{x \in \mathbb{R}^n : u \bullet x \leq 0\}$ for some nonzero $u \in \mathbb{R}^n$. You can check that linear half-spaces are convex cones. Moreover, any intersection of convex cones is also a convex cone. In particular, intersections of linear half-spaces are convex cones.

By imposing finiteness conditions on the constructions in the two preceding paragraphs, we obtain two special classes of convex cones. First, a *V-cone* is a convex cone of the form cone(S), where S is a finite subset of $\mathbb{R}^n$. Explicitly, $C \subseteq \mathbb{R}^n$ is a V-cone iff there exist a positive integer k and $v_1, \ldots, v_k \in \mathbb{R}^n$ such that

$$C = \{c_1 v_1 + \cdots + c_k v_k : c_i \in \mathbb{R}, c_i \geq 0 \text{ for } 1 \leq i \leq k\}.$$

We call $v_1, \ldots, v_k$ *generators* of the cone C. Second, an *H-cone* is an intersection of finitely many linear half-spaces in $\mathbb{R}^n$. Explicitly, $D \subseteq \mathbb{R}^n$ is an H-cone iff there exist a positive integer k and $u_1, \ldots, u_k \in \mathbb{R}^n$ such that

$$D = \{x \in \mathbb{R}^n : u_i \bullet x \leq 0 \text{ for } 1 \leq i \leq k\}.$$

The next four sections are devoted to proving that *a subset of $\mathbb{R}^n$ is an H-cone iff it is a V-cone*. This theorem is the key to understanding the structure of finite intersections of general closed half-spaces.

11.17   Intersection Lemma for V-Cones

You can check that the intersection of any H-cone in $\mathbb{R}^n$ with any linear subspace W is also an H-cone. To help prove that H-cones are the same as V-cones, we first need to prove that V-cones satisfy a special case of this intersection property. Specifically, given a V-cone C in $\mathbb{R}^n$, let W be the linear subspace $\mathbb{R}^{n-1} \times \{0\}$, which is the solution set of the equation $x_n = 0$. We prove that $C \cap W$ *is also a V-cone*.

Since C is a V-cone, we have $C = \mathrm{cone}(\{v_1, \ldots, v_m\})$ for some positive integer m and $v_i \in \mathbb{R}^n$. Here and below, we write $v_i(n)$ for the nth coordinate of the vector v_i. We divide the generators v_i of C into classes based on the sign of their last coordinates. To do this, let $I = \{1, 2, \ldots, m\}$, and define

$$I_0 = \{i \in I : v_i(n) = 0\}; \quad I_+ = \{i \in I : v_i(n) > 0\}; \quad I_- = \{i \in I : v_i(n) < 0\}.$$

Next, define a V-cone

$$D = \mathrm{cone}(\{v_i : i \in I_0\} \cup \{v_i(n)v_j - v_j(n)v_i : i \in I_+, j \in I_-\}).$$

It suffices to show that $C \cap W = D$.

First, we check that each generator of D has nth coordinate zero, hence is in W. This is true by definition for the v_i with $i \in I_0$. Given $i \in I_+$ and $j \in I_-$, the nth coordinate of $v_i(n)v_j - v_j(n)v_i$ is $v_i(n)v_j(n) - v_j(n)v_i(n) = 0$. Since all generators of D belong to the subspace W, any positive combination of those generators is also in W. Hence, $D \subseteq W$. Next, note that each generator of D is a positive combination of generators of the convex cone C (e.g., for $i \in I_+$ and $j \in I_-$, $v_i(n) \geq 0$ and $-v_j(n) \geq 0$). Since all generators of D belong to the convex cone C, the convex cone D is contained in C. In summary, $D \subseteq C \cap W$.

For the reverse inclusion, fix $v \in C \cap W$. Since $v \in C$, we can write $v = c_1 v_1 + \cdots + c_m v_m$ for some real scalars $c_i \geq 0$. Since $v \in W$, $0 = v(n) = c_1 v_1(n) + \cdots + c_m v_m(n)$. Consider two cases. *Case 1:* for all $i \in I$, $c_i v_i(n) = 0$. Then for all $i \in I$, $c_i = 0$ or $v_i(n) = 0$, so that the only v_i appearing with nonzero coefficients in the expression for v are v_i with $i \in I_0$. So v is a positive combination of some of the generators of D, hence $v \in D$. *Case 2:* for some $i \in I$, $c_i v_i(n) \neq 0$. In the equation $\sum_{i \in I} c_i v_i(n) = 0$, drop terms indexed by $i \in I_0$, and move terms indexed by $i \in I_-$ to the right side. We obtain a strictly positive quantity

$$b = \sum_{i \in I_+} c_i v_i(n) = \sum_{j \in I_-} c_j(-v_j(n)) > 0.$$

Now, using both formulas for b, compute:

$$
\begin{aligned}
v &= \sum_{i \in I} c_i v_i = \sum_{i \in I_0} c_i v_i + \frac{1}{b} \sum_{i \in I_+} b c_i v_i + \frac{1}{b} \sum_{j \in I_-} b c_j v_j \\
&= \sum_{i \in I_0} c_i v_i + \frac{1}{b} \sum_{i \in I_+} \sum_{j \in I_-} c_j(-v_j(n)) c_i v_i + \frac{1}{b} \sum_{j \in I_-} \sum_{i \in I_+} c_i v_i(n) c_j v_j \\
&= \sum_{i \in I_0} c_i v_i + \sum_{i \in I_+} \sum_{j \in I_-} \frac{c_i c_j}{b} (v_i(n) v_j - v_j(n) v_i).
\end{aligned}
$$

Note the coefficients c_i and $c_i c_j/b$ are all nonnegative, so this expression is a positive combination of the generators for D. Therefore, $v \in D$ in Case 2, completing the proof that $C \cap W \subseteq D$.

We have now proved that for a V-cone C in $\mathbb{R}^n$, the intersection $C \cap (\mathbb{R}^{n-1} \times \{0\})$ is a V-cone in $\mathbb{R}^{n-1}$. Iterating this result, we deduce the *Intersection Lemma for V-Cones*: for any V-cone C in $\mathbb{R}^n$ and any $k \leq n$, $C \cap (\mathbb{R}^{n-k} \times \{0\}^k)$ is a V-cone in $\mathbb{R}^{n-k}$, where we identify $\mathbb{R}^{n-k}$ with $\mathbb{R}^{n-k} \times \{0\}^k \subseteq \mathbb{R}^n$.

11.18 All H-Cones Are V-Cones

We now use the Intersection Lemma for V-Cones to prove that *every H-cone is a V-cone*. Given an H-cone $C \subseteq \mathbb{R}^n$, choose a positive integer k and $u_1, \ldots, u_k \in \mathbb{R}^n$ with

$$C = \{x \in \mathbb{R}^n : u_i \bullet x \leq 0 \text{ for } 1 \leq i \leq k\}.$$

The first step is to convert this H-cone into a V-cone in the higher-dimensional space $\mathbb{R}^{n+k}$. To do this, define

$$D = \{(x, y) \in \mathbb{R}^n \times \mathbb{R}^k : u_i \bullet x \leq y(i) \text{ for } 1 \leq i \leq k\}, \tag{11.2}$$

where $y(i)$ denotes the ith component of the vector y. You can check that $0 \in D$, and D is closed under addition and positive scalar multiples, so D is a convex cone. For $1 \leq j \leq n$, let $\mathbf{e}_j$ be the jth standard basis vector in $\mathbb{R}^n$. Define $w_j = (u_1(j), u_2(j), \ldots, u_k(j)) \in \mathbb{R}^k$, so $w_j(i) = u_i(j)$ for $1 \leq i \leq k$. For $1 \leq i \leq k$, let $\mathbf{e}'_i$ be the ith standard basis vector in $\mathbb{R}^k$. We show that D is a V-cone by confirming that

$$D = \mathrm{cone}(\{(\mathbf{e}_j, w_j) : 1 \leq j \leq n\} \cup \{(-\mathbf{e}_j, -w_j) : 1 \leq j \leq n\} \cup \{(0, \mathbf{e}'_i) : 1 \leq i \leq k\}).$$

Call the convex cone on the right side D'. For $1 \leq i \leq k$ and $1 \leq j \leq n$, $u_i \bullet \mathbf{e}_j = u_i(j) = w_j(i) \leq w_j(i)$ and $u_i \bullet (-\mathbf{e}_j) = -u_i(j) = -w_j(i) \leq -w_j(i)$, so that $\pm(\mathbf{e}_j, w_j) \in D$. For $1 \leq i, r \leq k$, $u_i \bullet 0 = 0 \leq \mathbf{e}'_r(i)$ since $\mathbf{e}'_r(i)$ is 0 or 1. These inequalities prove that all generators of D' are in the convex cone D, so $D' \subseteq D$.

For the reverse inclusion, fix $(x, y) \in D$ satisfying the conditions in (11.2). For any real z, let $\mathrm{sgn}(z) = 1$ if $z \geq 0$, $\mathrm{sgn}(z) = -1$ if $z < 0$, so that $z = |z| \mathrm{sgn}(z)$ in all cases. Exercise 84, proves the identity

$$(x, y) = \sum_{j=1}^{n} |x(j)| (\mathrm{sgn}(x(j))\mathbf{e}_j, \mathrm{sgn}(x(j))w_j) + \sum_{i=1}^{k} (y(i) - u_i \bullet x)(0, \mathbf{e}'_i). \tag{11.3}$$

This identity expresses (x, y) as a linear combination of generators of D', where all the scalars $|x(j)|$ and $y(i) - u_i \bullet x$ are nonnegative by the assumed conditions on (x, y). It follows that $(x, y) \in D'$, completing the proof that $D = D'$ and D is a V-cone.

Comparing the original definitions of C and D, we see that $C \times \{0\}^k = D \cap (\mathbb{R}^n \times \{0\}^k)$. By invoking the Intersection Lemma for V-Cones, we see that $C \times \{0\}^k$ is a V-cone. Then C itself is also a V-cone.

11.19 Projection Lemma for H-Cones

You can check that the image of any V-cone in $\mathbb{R}^n$ under any linear map is also a V-cone. To help prove that V-cones are the same as H-cones, we first need to prove that H-cones satisfy a special case of this projection property. Specifically, consider the linear projection map $p : \mathbb{R}^n \to \mathbb{R}^{n-1}$ given by $p(x) = (x(1), \ldots, x(n-1))$ for $x \in \mathbb{R}^n$. Given any H-cone C in $\mathbb{R}^n$, we prove that $p[C] = \{p(x) : x \in C\}$ is also an H-cone. Geometrically, $p[C]$ is the projection of C onto $\mathbb{R}^{n-1}$ (identified with $\mathbb{R}^{n-1} \times \{0\}$ in $\mathbb{R}^n$).

By definition, there exist a positive integer k and $u_1, \ldots, u_k \in \mathbb{R}^n$ with

$$C = \{x \in \mathbb{R}^n : u_i \bullet x \le 0 \text{ for } 1 \le i \le k\}. \tag{11.4}$$

Also, $z \in \mathbb{R}^{n-1}$ belongs to $p[C]$ iff there exists $r \in \mathbb{R}$ with $(z, r) \in C$. We need to pass from the system of linear inequalities (11.4) defining C to a new system of linear inequalities (involving only the first $n-1$ variables) with solution set $p[C]$. The idea is to take carefully chosen positive combinations of the original linear inequalities to eliminate the variable x_n.

To implement this elimination process, write the index set $I = \{1, 2, \ldots, k\}$ for the original list of inequalities as the disjoint union of subsets

$$I_0 = \{i \in I : u_i(n) = 0\}; \quad I_+ = \{i \in I : u_i(n) > 0\}; \quad I_- = \{i \in I : u_i(n) < 0\}.$$

For $i \in I_0$, let $u_i' = (u_i(1), u_i(2), \ldots, u_i(n-1)) \in \mathbb{R}^{n-1}$. For $i \in I_+$ and $j \in I_-$, let $v_{ij} = u_i(n)u_j - u_j(n)u_i$. Each v_{ij} has nth component $u_i(n)u_j(n) - u_j(n)u_i(n) = 0$; let $v_{ij}' \in \mathbb{R}^{n-1}$ be v_{ij} with this zero deleted. Finally, let

$$D = \{z \in \mathbb{R}^{n-1} : u_i' \bullet z \le 0 \text{ for all } i \in I_0, \text{ and } v_{ij}' \bullet z \le 0 \text{ for all } i \in I_+, j \in I_-\}.$$

By definition, D is an H-cone in $\mathbb{R}^{n-1}$, so it suffices to prove that $p[C] = D$.

To prove $p[C] \subseteq D$, we fix $z \in p[C]$ and check that $z \in D$. There is $r \in \mathbb{R}$ with $x = (z, r) \in C$. On one hand, for $i \in I_0$, we know $u_i \bullet x \le 0$. Since $u_i(n) = 0$, the left side of this inequality does not involve r, and we get $u_i' \bullet z \le 0$. On the other hand, fix $i \in I_+$ and $j \in I_-$. Multiply the known inequality $u_j \bullet x \le 0$ by the scalar $u_i(n) \ge 0$, multiply the known inequality $u_i \bullet x \le 0$ by the scalar $-u_j(n) \ge 0$, and add the resulting inequalities. We thereby see that $v_{ij} \bullet x \le 0$. As noted above, $v_{ij}(n) = 0$ by construction, so we also have $v_{ij}' \bullet z \le 0$. This means that z satisfies all the inequalities defining D, so that $z \in D$.

To prove $D \subseteq p[C]$, we fix $z \in D$ and must prove the existence of $r \in \mathbb{R}$ with $x = (z, r) \in C$. On one hand, for $i \in I_0$, the known inequality $u_i' \bullet z \le 0$ guarantees the needed inequality $u_i \bullet x \le 0$ for any choice of r, since $u_i(n) = 0$. On the other hand, for any fixed $i \in I_+$, the needed inequality $u_i \bullet x \le 0$ holds for $x = (z, r)$ iff $\sum_{s=1}^{n-1} u_i(s)z(s) + u_i(n)r \le 0$ iff

$$r \le u_i(n)^{-1} \sum_{s=1}^{n-1} (-u_i(s)z(s))$$

(recall $u_i(n) > 0$). Similarly, for any fixed $j \in I_-$, $u_j \bullet x \le 0$ holds iff

$$r \ge u_j(n)^{-1} \sum_{s=1}^{n-1} (-u_j(s)z(s))$$

(note that the inequality is reversed since $u_j(n) < 0$). Combining these observations, we see that the remaining inequalities needed to ensure $x \in C$ hold for $x = (z, r)$ iff

$$u_j(n)^{-1} \sum_{s=1}^{n-1} (-u_j(s)z(s)) \le r \le u_i(n)^{-1} \sum_{s=1}^{n-1} (-u_i(s)z(s))$$

for all $i \in I_+$ and all $j \in I_-$. This collection of constraints on r is equivalent to the single condition

$$\max_{j \in I_-} \left[u_j(n)^{-1} \sum_{s=1}^{n-1} (-u_j(s)z(s)) \right] \leq r \leq \min_{i \in I_+} \left[u_i(n)^{-1} \sum_{s=1}^{n-1} (-u_i(s)z(s)) \right].$$

There exists an $r \in \mathbb{R}$ satisfying this condition iff

$$\max_{j \in I_-} \left[u_j(n)^{-1} \sum_{s=1}^{n-1} (-u_j(s)z(s)) \right] \leq \min_{i \in I_+} \left[u_i(n)^{-1} \sum_{s=1}^{n-1} (-u_i(s)z(s)) \right],$$

which holds iff

$$u_j(n)^{-1} \sum_{s=1}^{n-1} (-u_j(s)z(s)) \leq u_i(n)^{-1} \sum_{s=1}^{n-1} (-u_i(s)z(s))$$

for all $i \in I_+$ and all $j \in I_-$. Multiplying by the negative number $u_i(n)u_j(n)$, the inequality just written is equivalent to

$$u_i(n) \sum_{s=1}^{n-1} (-u_j(s)z(s)) \geq u_j(n) \sum_{s=1}^{n-1} (-u_i(s)z(s)),$$

which holds iff

$$\sum_{s=1}^{n-1} (u_i(n)u_j(s)z(s) - u_j(n)u_i(s)z(s)) \leq 0$$

iff $(u_i(n)u_j - u_j(n)u_i) \bullet (z,r) \leq 0$ iff $v'_{ij} \bullet z \leq 0$. All of these last inequalities are true, since $z \in D$, and we see therefore that $z \in p[C]$. So $D \subseteq p[C]$.

We have now proved that for an H-cone C in $\mathbb{R}^n$, the projection $p[C]$ onto $\mathbb{R}^{n-1}$ is an H-cone in $\mathbb{R}^{n-1}$. Iterating this result, we deduce the *Projection Lemma for H-Cones*: for an H-cone C in $\mathbb{R}^n$ and any $k \leq n$, the projection

$$p_k[C] = \{z \in \mathbb{R}^{n-k} : \exists r_1, \ldots, r_k \in \mathbb{R}, (z, r_1, \ldots, r_k) \in C\} = \{z \in \mathbb{R}^{n-k} : \exists y \in \mathbb{R}^k, (z,y) \in C\}$$

is an H-cone in $\mathbb{R}^{n-k}$.

11.20   All V-Cones Are H-Cones

We now use the Projection Lemma for H-Cones to prove that *every V-cone is an H-cone*. Given a V-cone $C \subseteq \mathbb{R}^n$, there exist a positive integer k and $v_1, \ldots, v_k \in \mathbb{R}^n$ with

$$C = \{c_1 v_1 + \cdots + c_k v_k : c_i \in \mathbb{R}, c_i \geq 0 \text{ for } 1 \leq i \leq k\}.$$

The key observation is that we can convert this V-cone into an H-cone in the higher-dimensional space $\mathbb{R}^{n+k}$. To do this, define

$$D = \left\{ (x,y) \in \mathbb{R}^{n+k} : x = \sum_{i=1}^{k} y(i)v_i \text{ and } y(i) \geq 0 \text{ for } 1 \leq i \leq k \right\}. \tag{11.5}$$

To see why D is an H-cone, let $\mathbf{e}_j$ and $\mathbf{e}'_i$ denote standard basis vectors in $\mathbb{R}^n$ and $\mathbb{R}^k$, respectively. Define $u_i = (0, -\mathbf{e}'_i)$ for $1 \leq i \leq k$ and $w_j = (\mathbf{e}_j, (-v_1(j), \ldots, -v_k(j)))$ for

$1 \leq j \leq n$. You can check that (x, y) satisfies the conditions in the definition of D iff $u_i \bullet (x, y) \leq 0$ for $1 \leq i \leq k$ and $w_j \bullet (x, y) \leq 0$ for $1 \leq j \leq n$ and $(-w_j) \bullet (x, y) \leq 0$ for $1 \leq j \leq n$. So D is the solution set of a finite system of homogeneous linear inequalities, hence D is an H-cone.

To finish the proof, we need only observe that $x \in \mathbb{R}^n$ belongs to C iff there exists $y \in \mathbb{R}^k$ with $(x, y) \in D$. By invoking the Projection Lemma for H-Cones, we see that C is an H-cone.

11.21   Finite Intersections of Closed Half-Spaces

Now that we know V-cones are the same as H-cones, we can prove the following result characterizing finite intersections of closed half-spaces.

Theorem on Finite Intersections of Closed Half-Spaces A subset C of $\mathbb{R}^n$ has the form $C = H_1 \cap H_2 \cap \cdots \cap H_s$ for finitely many closed half-spaces $H_1, \ldots, H_s$ iff there exist finitely many vectors $v_1, \ldots, v_k, w_1, \ldots, w_m \in \mathbb{R}^n$ with $C = \text{conv}(\{v_1, \ldots, v_k\}) + \text{cone}(\{w_1, \ldots, w_m\})$.

The plus symbol denotes the sum of sets in $\mathbb{R}^n$, namely $A + B = \{a + b : a \in A, b \in B\}$.

Proof. First assume that there are finitely many vectors v_i and w_j in $\mathbb{R}^n$ with

$$C = \text{conv}(\{v_1, \ldots, v_k\}) + \text{cone}(\{w_1, \ldots, w_m\}). \tag{11.6}$$

A typical point $v \in C$ looks like

$$v = c_1 v_1 + \cdots + c_k v_k + d_1 w_1 + \cdots + d_m w_m \tag{11.7}$$

where $c_i, d_j \geq 0$ are scalars such that $\sum_{i=1}^k c_i = 1$. Identifying $\mathbb{R}^n$ with $\mathbb{R}^n \times \{0\}$, we can regard each v_i and w_j as a vector in $\mathbb{R}^{n+1}$. In $\mathbb{R}^{n+1}$, define a V-cone

$$D = \text{cone}(v_1 + \mathbf{e}_{n+1}, \ldots, v_k + \mathbf{e}_{n+1}, w_1, \ldots, w_m).$$

D consists of all points $v \in \mathbb{R}^{n+1}$ that can be written in the form

$$v = c_1 v_1 + \cdots + c_k v_k + d_1 w_1 + \cdots + d_m w_m + \left(\sum_{i=1}^k c_i \right) \mathbf{e}_{n+1}, \tag{11.8}$$

for some scalars $c_i, d_j \geq 0$. Let H_0 denote the affine hyperplane $\mathbb{R}^n \times \{1\}$ in $\mathbb{R}^{n+1}$. We can obtain exactly those points in the intersection $D \cap H_0$ by choosing scalars $c_i, d_j \geq 0$ in (11.8) such that $\sum_{i=1}^k c_i = 1$. Since these are precisely the conditions imposed on the scalars in (11.7), we see that $D \cap H_0$ is the translate $C + \mathbf{e}_{n+1}$.

As shown in §11.20, D is an H-cone, so D is a finite intersection of certain linear half-spaces $H_1', \ldots, H_s'$ in $\mathbb{R}^{n+1}$. It follows that

$$C + \mathbf{e}_{n+1} = D \cap H_0 = (H_1' \cap H_0) \cap \cdots \cap (H_s' \cap H_0).$$

Translating back to $\mathbb{R}^n \times \{0\}$, we see that $C = \bigcap_{i=1}^s ((H_i' \cap H_0) - \mathbf{e}_{n+1})$, where each set in the intersection is readily seen to be a closed half-space in $\mathbb{R}^n$. So C has been expressed as a finite intersection of closed half-spaces.

Conversely, assume C is the intersection of finitely many closed half-spaces in $\mathbb{R}^n$, say

$$C = \{x \in \mathbb{R}^n : u_i \bullet x \le b_i \text{ for } 1 \le i \le p\}$$

for some integer $p \ge 0$ and $u_i \in \mathbb{R}^n \subseteq \mathbb{R}^{n+1}$. In $\mathbb{R}^{n+1}$, define an H-cone

$$D = \{z \in \mathbb{R}^{n+1} : (u_i - b_i e_{n+1}) \bullet z \le 0 \text{ for } 1 \le i \le p \text{ and } (-e_{n+1}) \bullet z \le 0\}.$$

The last linear inequality in this definition amounts to requiring that the last coordinate of each point in D be nonnegative. Observe that for points $z \in \mathbb{R}^{n+1}$ of the form $(x, 1)$, the other linear inequalities in the definition of D hold iff $u_i \bullet x \le b_i$ for each i. It follows from these remarks that $D \cap H_0 = C + e_{n+1}$, where (as before) H_0 denotes the affine hyperplane $\mathbb{R}^n \times \{1\}$ in $\mathbb{R}^{n+1}$.

As shown in §11.18, we can write $D = \text{cone}(\{z_1, \ldots, z_s\})$ for some $s \in \mathbb{Z}_{\ge 0}$ and $z_i \in \mathbb{R}^{n+1}$. By reordering the z_i and rescaling some z_i by positive scalars if needed, we can assume that $z_i(n+1) = 1$ for $1 \le i \le k$ and $z_i(n+1) = 0$ for $k < i \le s$. Then a typical element of D is a positive combination

$$\sum_{i=1}^{k} c_i z_i + \sum_{i=k+1}^{s} d_i z_i \qquad \text{for some } c_i, d_i \ge 0.$$

To find the points of $D \cap H_0$, we need to restrict the c_i and d_i to those nonnegative scalars that make the last coordinate equal to 1. By choice of the z_i, the restriction needed is precisely $\sum_{i=1}^{k} c_i = 1$. Since $D \cap H_0 = C + e_{n+1}$, we see that the points in C are exactly those points that can be written in the form

$$\sum_{i=1}^{k} c_i(z_i - e_{n+1}) + \sum_{i=k+1}^{s} d_i z_i \qquad \text{where } c_i, d_i \ge 0 \text{ and } \sum_{i=1}^{k} c_i = 1.$$

This means that

$$C = \text{conv}(\{z_1 - e_{n+1}, \ldots, z_k - e_{n+1}\}) + \text{cone}(\{z_{k+1}, \ldots, z_s\}),$$

where we view all generators as elements of $\mathbb{R}^n$. So C has been expressed in the form (11.6). $\qquad \square$

Theorem on Convex Hulls of Finite Sets. A set $C \subseteq \mathbb{R}^n$ is the convex hull of finitely many points iff C is a bounded intersection of finitely many closed half-spaces.

Proof. Assume $C = \text{conv}(\{v_1, \ldots, v_k\})$. This is the special case of (11.6) with $m = 0$, so C is an intersection of finitely many closed half-spaces. Moreover, any convex combination $w = \sum_{i=1}^{k} c_i v_i$ with $c_i \ge 0$ and $\sum_{i=1}^{k} c_i = 1$ satisfies

$$||w|| \le \sum_{i=1}^{k} |c_i| \cdot ||v_i|| \le \sum_{i=1}^{k} ||v_i||,$$

so C is bounded.

Conversely, assume C is a bounded set that is the intersection of finitely many closed half-spaces. We have proved that C can be written in the form (11.6), and we may assume no w_i is zero. If $m > 0$, then $\{dw_1 : d \ge 0\}$ would be an unbounded subset of C. So $m = 0$, and $C = \text{conv}(\{v_1, \ldots, v_k\})$ as needed. $\qquad \square$

11.22 Convex Functions

We conclude the chapter with a brief introduction to convex functions. Let C be a convex subset of $\mathbb{R}^n$. A function $f : C \to \mathbb{R}$ is called *convex* iff $f(cx + (1-c)y) \le cf(x) + (1-c)f(y)$ for all $x, y \in C$ and all $c \in [0,1]$. (Compare to the definition of an affine map, where equality was required to hold for all $c \in \mathbb{R}$.) To see how this definition is related to the idea of convexity for sets, define the *epigraph* of f to be the set $\text{epi}(f) = \{(x, z) \in \mathbb{R}^{n+1} : x \in C, z \ge f(x)\}$, which consists of all points in $\mathbb{R}^{n+1}$ above the graph of f. We claim that f is a convex function iff $\text{epi}(f)$ is a convex set.

To verify this, first assume $\text{epi}(f)$ is a convex set. Given $x, y \in C$ and $c \in [0,1]$, the points $(x, f(x))$ and $(y, f(y))$ are in $\text{epi}(f)$. By convexity of the epigraph, we know

$$c(x, f(x)) + (1-c)(y, f(y)) = (cx + (1-c)y, cf(x) + (1-c)f(y)) \in \text{epi}(f).$$

The definition of the epigraph now gives $f(cx + (1-c)y) \le cf(x) + (1-c)f(y)$, so f is a convex function. Conversely, assume f is a convex function. Fix $c \in [0,1]$ and two points $(x, z), (y, w) \in \text{epi}(f)$, where $x, y \in \mathbb{R}^n$ and $z, w \in \mathbb{R}$. We know $z \ge f(x)$ and $w \ge f(y)$. By convexity of f,

$$cz + (1-c)w \ge cf(x) + (1-c)f(y) \ge f(cx + (1-c)y),$$

and hence $c(x, z) + (1-c)(y, w) = (cx + (1-c)y, cz + (1-c)w) \in \text{epi}(f)$. So $\text{epi}(f)$ is a convex set. For $C \subseteq \mathbb{R}$, the convexity of f (or $\text{epi}(f)$) means that for any real $a < b$ in C, the graph of f on the interval $[a, b]$ is always weakly below the line segment joining $(a, f(a))$ to $(b, f(b))$.

Convexity is a rather strong condition to impose on a function. For instance, *for all real $c < d$, a convex function $f : [c, d] \to \mathbb{R}$ must be continuous on the open interval (c, d).* Figure 11.4 illustrates the proof. We verify continuity of f at an arbitrary point $x \in (c, d)$. Having fixed x, choose $y < x$ in (c, d) and $z > x$ in (c, d). Draw the line L_1 of slope m_1 through $(y, f(y))$ and $(x, f(x))$ and the line L_2 of slope m_2 through $(x, f(x))$ and $(z, f(z))$. Since $(x, f(x))$ is below the line L of slope m joining $(y, f(y))$ and $(z, f(z))$, it is geometrically evident that $m_1 \le m \le m_2$. So, in the part of the plane right of x, L_1 is below L_2. Now consider any sequence of points x_n in the open interval (x, z) such that (x_n) converges to x. To see that f is right-continuous at x, we must show $\lim_{n \to \infty} f(x_n) = f(x)$. By using

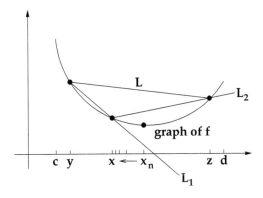

FIGURE 11.4
A convex function must be continuous on an open interval.

the remark at the end of the last paragraph with $a = x$ and $b = z$, we see that every point $(x_n, f(x_n))$ is below the line L_2. On the other hand, using the same remark with $a = y$ and $b = x_n$, we see that $(x, f(x))$ is below the line joining $(y, f(y))$ to $(x_n, f(x_n))$, and hence $(x_n, f(x_n))$ is above the line L_1. Each of the lines L_1 and L_2 is the graph of a continuous (in fact, affine) function whose limit as x_n approaches x is $f(x)$. By the Sandwich Theorem for Limits, (x_n) must also converge to x as n goes to infinity. Similar remarks prove the left-continuity of f at x. However, f need not be continuous at the endpoints c and d of its domain.

We have proved (§11.11) that convex sets are closed under convex combinations of their elements. Applying this remark to epi(f) for a convex function $f : C \to \mathbb{R}$, we obtain a result called *Jensen's Inequality*. Let $x_1, \ldots, x_k \in C$ and $c_1, \ldots, c_k \in \mathbb{R}$ with each $c_i \geq 0$ and $\sum_{i=1}^{k} c_i = 1$. Since $(x_i, f(x_i)) \in$ epi(f) for all i, the convex combination $(\sum_{i=1}^{k} c_i x_i, \sum_{i=1}^{k} c_i f(x_i))$ is in epi(f). Hence,

$$f\left(\sum_{i=1}^{k} c_i x_i\right) \leq \sum_{i=1}^{k} c_i f(x_i). \tag{11.9}$$

This is the discrete form of Jensen's Inequality. By applying this inequality to Riemann sums approximating a Riemann integral, we can obtain a version of *Jensen's Inequality for Integrals*. Specifically, suppose $f : (c, d) \to \mathbb{R}$ is convex and $g : [0, 1] \to \mathbb{R}$ is an integrable function taking values in (c, d). Then

$$f\left(\int_0^1 g(x)\, dx\right) \leq \int_0^1 f(g(x))\, dx.$$

This follows from (11.9) by setting $c_i = 1/k$ and $x_i = g(i/k)$ for $1 \leq i \leq k$, which causes the two sums to approximate the two integrals just written, and then taking the limit of the inequality as k goes to infinity. By continuity of f, the limit of the left side of (11.9) is $f(\int_0^1 g(x)\, dx)$. Similarly, the right side approaches $\int_0^1 f(g(x))\, dx$, where the integrability of $f \circ g$ follows since f is continuous. In the context of probability theory, sums involving convex combinations and integrals on $[0, 1]$ are special cases of expectations of random variables on a probability space. In this setting, Jensen's Inequality becomes $f(E[X]) \leq E[f(X)]$, where f is convex, X is a random variable, and E denotes expected value.

11.23 Derivative Tests for Convex Functions

We have not yet given any examples of convex functions. Every affine function $f : \mathbb{R}^n \to \mathbb{R}$ is certainly convex; when $n = 1$, the graph of such a function is a line. The absolute value function ($f(x) = |x|$ for $x \in \mathbb{R}$) is convex, since it is visually apparent that the epigraph of f is convex. More generally, suppose $x_0 < x_1 < x_2 < \cdots < x_n$ and $y_0, y_1, \ldots, y_n$ are given real numbers. Drawing the line segments from (x_{i-1}, y_{i-1}) to (x_i, y_i) for $1 \leq i \leq n$ gives the graph of a function $f : [x_0, x_n] \to \mathbb{R}$. This graph consists of a sequence of line segments having slopes $m_i = (y_i - y_{i-1})/(x_i - x_{i-1})$. You can check that f is convex iff $m_1 \leq m_2 \leq \cdots \leq m_n$, i.e., the slopes of successive line segments are weakly increasing.

This suggests that *a differentiable function $f : (c, d) \to \mathbb{R}$ is convex iff the first derivative $f' : (c, d) \to \mathbb{R}$ is an increasing function*. To explain this geometrically, assume f is convex, and fix $y < x$ in (c, d) as shown in Figure 11.4. The secant line L_1 joining $(y, f(y))$ to $(x, f(x))$ has some slope m_1. Convexity tells us that the graph of f between y and x is below this secant line, so it is geometrically evident that the tangent line to the graph of f at y has slope weakly less than m_1. In symbols, $f'(y) \leq m_1$. But, as we argued earlier, the graph of f to the right of x must be above the secant line L_1, so that the tangent line to the graph of f at x has slope weakly greater than m_1. Thus, $m_1 \leq f'(x)$, so $f'(y) \leq f'(x)$ and f' is increasing.

Conversely, assume f' is increasing. To verify convexity of f, it suffices to show that for all $y < x < z$ in (c, d), the point $(x, f(x))$ is weakly below the line segment L joining $(y, f(y))$ and $(z, f(z))$ (see Figure 11.4). Let the secant lines L_1 and L_2 shown in the figure have respective slopes m_1 and m_2. Then the point $(x, f(x))$ is weakly below L iff $m_1 \leq m_2$. By the Mean Value Theorem, there is a point y_1 with $y < y_1 < x$ and $f'(y_1) = m_1$, and there is z_1 with $x < z_1 < z$ and $f'(z_1) = m_2$. Since $y_1 < z_1$ and f' is increasing, $m_1 \leq m_2$ as needed.

By a theorem from calculus, we deduce that *a twice-differentiable function $f : (c, d) \to \mathbb{R}$ is convex iff the second derivative $f'' : (c, d) \to \mathbb{R}$ is nonnegative; i.e., $f''(x) \geq 0$ for all $x \in (c, d)$*. This provides a convenient criterion for testing convexity of many functions. For example, $f(x) = e^x$ for $x \in \mathbb{R}$, $g(x) = x^2$ for $x \in \mathbb{R}$, and $h(x) = -\ln x$ for $x > 0$, are all convex functions by the second derivative test. In elementary calculus, these functions are often described as being *concave up*. For convex $C \subseteq \mathbb{R}^n$, we say that a function $f : C \to \mathbb{R}$ is *concave* iff $-f$ is convex iff $f(cx + (1 - c)y) \geq cf(x) + (1 - c)f(y)$ for all $x, y \in C$ and all $c \in [0, 1]$. When $n = 1$, concave functions are called *concave down* in elementary calculus.

The second derivative test extends to multivariable functions as follows. Suppose C is an open convex subset of $\mathbb{R}^n$, and $f : C \to \mathbb{R}$ has continuous second-order partial derivatives. For each $z \in C$, define the *Hessian matrix* $H_z \in M_n(\mathbb{R})$ by letting $H_z(i, j) = f_{x_i, x_j}(z) = \frac{\partial^2 f}{\partial x_i \partial x_j}(z)$ for $1 \leq i, j \leq n$. Then f is convex on C iff for all $z \in C$, H_z is a positive semidefinite matrix (which means $v \bullet H_z v = \sum_{i,j} f_{x_i, x_j}(z) v_i v_j \geq 0$ for all $v \in \mathbb{R}^n$). The proof is outlined in Exercise 104; the idea is to reduce to the one-variable case by composing f with affine maps $g : (c, d) \to C$ and using the chain rule to compute $(f \circ g)''$.

11.24   Summary

Let F be a field in which $1_F + 1_F \neq 0_F$ and V be an n-dimensional F-vector space.

1. *Types of Linear Combinations.* A *linear* combination of $v_1, \ldots, v_k \in V$ is a vector of the form $c_1 v_1 + \cdots + c_k v_k$ with all $c_i \in F$. This vector is:
   (a) an *affine* combination iff $\sum_{i=1}^{k} c_i = 1_F$;
   (b) a *convex* combination iff $F = \mathbb{R}$, all $c_i \geq 0$, and $\sum_{i=1}^{k} c_i = 1$;
   (c) a *positive* combination iff $F = \mathbb{R}$ and all $c_i \geq 0$.

2. *Definitions of Special Sets and Maps.* Table 11.1 summarizes the definitions of the structured sets and maps studied in this chapter.

3. *Characterizations of Linear Subspaces.* Every k-dimensional linear subspace of F^n is: (a) $\mathrm{Sp}_F(S)$ for some linearly independent set S of size k;
   (b) the null space of some $(n - k) \times n$ matrix;
   (c) the range of some $n \times k$ matrix;
   (d) the intersection of $n - k$ linear hyperplanes in F^n;
   (e) the solution set of a system of $n - k$ homogeneous linear equations in n unknowns.

TABLE 11.1

Definitions of affine and convex concepts.

Concept	Definition	Additional Properties
linear subspace W	$0_V \in W$; $x, y \in W \Rightarrow x + y \in W$; $x \in W, c \in F \Rightarrow cx \in W$.	W is closed under linear combinations
affine set X	$x, y \in X, c \in F \Rightarrow$ $cx + (1 - c)y \in X$.	X is closed under affine combinations; $X = u + W$ for dir. subspace W
convex set C	$x, y \in C, t \in [0, 1] \Rightarrow$ $tx + (1 - t)y \in C$.	C is closed under convex combinations
cone C	$0 \in C$; $x \in C, t \geq 0 \Rightarrow tx \in C$	C is convex iff C is closed under $+$
convex cone C	$0 \in C$, $x, y \in C \Rightarrow x + y \in C$, $x \in C, t \geq 0 \Rightarrow tx \in C$.	C is closed under positive combinations
V-cone C	$C = \mathrm{cone}(S)$ for some finite $S \subseteq \mathbb{R}^n$	same as H-cone
H-cone C	$C = \{x \in \mathbb{R}^n : u_i \bullet x \leq 0$ for $u_1, \ldots, u_k \in \mathbb{R}^n\}$	same as V-cone
linear span $\mathrm{Sp}_F(S)$	$\{$lin. combs. of vectors in $S\}$	intersection of all subspaces $W \supseteq S$
affine span $\mathrm{aff}(S)$	$\{$aff. combs. of vectors in $S\}$	intersection of all affine sets $X \supseteq S$
convex hull $\mathrm{conv}(S)$	$\{$conv. combs. of points in $S\}$	intersection of all convex sets $C \supseteq S$
convex cone $\mathrm{cone}(S)$	$\{$pos. combs. of points in $S\}$	intersection of all convex cones $C \supseteq S$
linear hyperplane	$\{x \in \mathbb{R}^n : u \bullet x = 0\}$	2 choices for unit vector u
affine hyperplane	$\{x \in \mathbb{R}^n : u \bullet x = c\}$	unit vector u is unique when $c > 0$
closed half-space	$\{x \in \mathbb{R}^n : u \bullet x \leq c\}$	other half-space: $\{x \in \mathbb{R}^n : u \bullet x \geq c\}$
open half-space	$\{x \in \mathbb{R}^n : u \bullet x < c\}$	other half-space: $\{x \in \mathbb{R}^n : u \bullet x > c\}$
linear map T	$x, y \in V, c \in F \Rightarrow$ $T(x + y) = T(x) + T(y)$, $T(cx) = cT(x)$	T preserves linear combinations
affine map U	$x, y \in V, c \in F \Rightarrow$ $U(cx + (1 - c)y) =$ $cU(x) + (1 - c)U(y)$	U preserves affine combinations, $U(x) = T(x) + b$ (T linear)
convex function f (convex domain C)	$x, y \in C, t \in [0, 1] \Rightarrow$ $f(tx + (1 - t)y) \leq$ $tf(x) + (1 - t)f(y)$	f is continuous on open interval Jensen's Inequality holds
concave function g (convex domain C)	$x, y \in C, t \in [0, 1] \Rightarrow$ $g(tx + (1 - t)y) \geq$ $tg(x) + (1 - t)g(y)$	f is continuous on open interval $-g$ is convex

4. *Characterizations of Affine Sets.* Every nonempty affine subset X of F^n of affine dimension k is: (a) $\mathrm{aff}(S)$ for some affinely independent set S of size $k + 1$; (b) $u + W$ for a unique k-dimensional linear subspace W (the direction subspace of X) and all $u \in X$; (c) the solution set of a system of $n - k$ (possibly non-homogeneous) linear equations in n unknowns; (d) the intersection of $n - k$ affine hyperplanes in F^n.

5. *Affine Independence, Bases, and Dimension.* A list $L = (v_0, v_1, \ldots, v_k)$ of vectors in V is *affinely independent* iff for all $c_i \in F$ such that $\sum_{i=0}^{k} c_i = 0$, if $\sum_{i=0}^{k} c_i v_i = 0$ then all $c_i = 0$. L is affinely independent iff $(v_1 - v_0, \ldots, v_k - v_0)$ is linearly independent. An affine basis of an affine set X is an affinely independent set whose affine span is X. The dimension of X is the dimension of its direction subspace, which is one less than the size of any affine basis of X. Affinely

independent subsets of X can be extended to an affine basis; affine spanning sets for X contain an affine basis; no affinely independent set is larger than an affine basis for X; and any function defined on an affine basis of X extends uniquely to an affine map with domain X. Each point v in X has unique barycentric coordinates expressing v as an affine combination of an ordered affine basis of X.

6. *Carathéodory's Theorem on Convex Hulls.* For all $S \subseteq \mathbb{R}^n$, every element of $C = \text{conv}(S)$ is a convex combination of at most $n + 1$ elements of S. One consequence is that convex hulls of closed and bounded (compact) sets are also closed and bounded.

7. *Separation by Hyperplanes.* If C is a closed, bounded, convex subset of $\mathbb{R}^n$ and D is a closed, convex subset of $\mathbb{R}^n$ disjoint from C, there is an affine hyperplane H in $\mathbb{R}^n$ such that C and D are in opposite open half-spaces of H. So, closed convex sets are the intersection of all open (or all closed) half-spaces containing them.

8. *Theorems on Generation vs. Intersection.* For all subsets C of $\mathbb{R}^n$:

   (a) C has the form $\text{Sp}(S)$ for some set S iff C is an intersection of linear hyperplanes.

   (b) C has the form $\text{aff}(S)$ for some set S iff C is an intersection of affine hyperplanes.

   (c) C is closed and has the form $\text{conv}(S)$ for some set S iff C is an intersection of closed half-spaces.

   (d) C has the form $\text{conv}(S)$ for some finite set S iff C is a bounded intersection of finitely many closed half-spaces.

   (e) C has the form $\text{conv}(S) + \text{cone}(T)$ for some finite sets S and T iff C is an intersection of finitely many closed half-spaces.

   (f) C has the form $\text{cone}(T)$ for some finite set T iff C is an intersection of finitely many linear half-spaces (this says V-cones are the same as H-cones).

9. *Intersection Lemma for V-Cones.* Given a V-cone $C = \text{cone}(\{v_1, \ldots, v_m\}) \subseteq \mathbb{R}^n$, the intersection $C \cap (\mathbb{R}^{n-1} \times \{0\})$ is the V-cone

$$D = \text{cone}(\{v_i : v_i(n) = 0\} \cup \{v_i(n)v_j - v_j(n)v_i : v_i(n) > 0, v_j(n) < 0\}).$$

10. *Projection Lemma for H-Cones.* Given an H-cone

$$C = \{x \in \mathbb{R}^n : u_i \bullet x \leq 0 \text{ for } 1 \leq i \leq k\},$$

the projection $p[C] = \{z \in \mathbb{R}^{n-1} : \exists r \in \mathbb{R}, (z, r) \in C\}$ is the H-cone

$$D = \{z \in \mathbb{R}^{n-1} : u_i \bullet (z, 0) \leq 0 \text{ for all } i \text{ with } u_i(n) = 0 \text{ and}$$
$$(u_i(n)u_j - u_j(n)u_i) \bullet (z, 0) \leq 0 \text{ for all } i, j \text{ with } u_i(n) > 0, u_j(n) < 0\}.$$

11. *Theorems on Convex Functions.* A function f with convex domain is convex iff its epigraph $\text{epi}(f) = \{(x, y) : x \in C, y \geq f(x)\}$ is a convex set. For convex functions f and scalars $c_i \geq 0$ summing to 1, $f(\sum_i c_i x_i) \leq \sum_i c_i f(x_i)$ and $f(\int_0^1 g(x)\, dx) \leq \int_0^1 f(g(x))\, dx$ whenever these expressions are defined (Jensen's Inequality). A convex function whose domain is an open interval of $\mathbb{R}$ must be continuous. For one-variable functions f such that f' exists, f is convex iff f' is increasing. When f'' exists, f is convex iff $f'' \geq 0$. If $C \subseteq \mathbb{R}^n$ is open and convex and $f : C \to \mathbb{R}$ has continuous second partial derivatives on C, f is convex iff the Hessian matrix $H_z = (f_{x_i, x_j}(z))_{1 \leq i, j \leq n}$ is positive semidefinite for all z in C.

11.25   Exercises

In these exercises, assume V and W are vector spaces over a field F with $\dim(V) = n$ unless otherwise stated.

1. Prove that a linear subspace W of V is closed under linear combinations.

2. Prove: If $W \subseteq V$ is closed under linear combinations, then W is a subspace of V.

3. Let $A \in M_{m,n}(F)$.
   (a) Prove the range of A is a subspace of F^m.
   (b) Prove that $T : F^n \to F^m$, given by $T(x) = Ax$ for $x \in F^n$, is F-linear.
   (c) Confirm that $\ker(T)$ is the null space of A, and $\operatorname{img}(T)$ is the range of A.

4. Prove that the intersection of any collection of subspaces of V is a subspace.

5. Prove that the intersection of any collection of affine subsets of V is affine.

6. Prove that the intersection of any collection of convex subsets of $\mathbb{R}^n$ is convex.

7. Prove that the intersection of any collection of convex cones in $\mathbb{R}^n$ is a convex cone.

8. Given subsets $W_1, \ldots, W_k$ of V, the *sum* of these subsets is

$$W_1 + \cdots + W_k = \{x_1 + \cdots + x_k : x_1 \in W_1, \ldots, x_k \in W_k\}.$$

   Prove that the sum of linear subspaces is a linear subspace.

9. Prove that the sum of affine sets is affine.

10. Prove that the sum of convex sets is convex.

11. Prove that the sum of convex cones is a convex cone.

12. Prove or disprove: if X, Y, and $X \cup Y$ are all linear subspaces of V, then $X \subseteq Y$ or $Y \subseteq X$.

13. Prove or disprove: if X, Y, and $X \cup Y$ are all affine subsets of V, then $X \subseteq Y$ or $Y \subseteq X$.

14. Prove or disprove: if X, Y, and $X \cup Y$ are all convex subsets of $\mathbb{R}^n$, then $X \subseteq Y$ or $Y \subseteq X$.

15. Given vector spaces $V_1, \ldots, V_k$ and subsets $S_i \subseteq V_i$, the *product* of these subsets is $S_1 \times \cdots \times S_k = \{(x_1, \ldots, x_k) : x_i \in S_i\}$. Prove that the product of linear subspaces is a linear subspace.

16. Prove that the product of affine sets is an affine set.

17. Prove that the product of convex sets is a convex set.

18. Prove that the product of convex cones is a convex cone.

19. Let $S = \{v_1, \ldots, v_k\} \subseteq V$. Prove that $\operatorname{Sp}(S)$ is a linear subspace of V. Prove that $\operatorname{Sp}(S)$ is the intersection of all subspaces of V that contain S.

20. Prove every m-dimensional subspace of F^n is the range of some $A \in M_{n,m}(F)$.

21. Prove every k-dimensional subspace of V is the kernel of a linear map $T : V \to F^{n-k}$.

22. Prove every k-dimensional subspace of V is the image of a linear map $S : F^k \to V$.

23. Prove: for all $A \in M_{m,n}(F)$ and $b \in F^m$, the set $\{x \in F^n : Ax = b\}$ is an affine subset of F^n.

24. Prove: a nonempty $S \subseteq V$ is a linear subspace of V iff $S + S = S$ and $cS = S$ for all nonzero $c \in F$.

25. Prove: if $S \neq \emptyset$ is an affine set, then the direction subspace of S is $S + (-1)S$.

26. Prove: for all affine sets S and all $c \in F$, cS is an affine set.

27. Say that two nonempty affine sets $S, T \subseteq V$ are *parallel* iff S and T have the same direction subspace. (a) Show that two affine lines in $\mathbb{R}^2$ are parallel iff the lines are both vertical or both lines have the same slope. (b) Show that parallelism is an equivalence relation on the set of nonempty affine subsets of V, such that the equivalence class of S consists of all translates $u + S$ for $u \in V$.

28. Let X be an affine subset of V. (a) Prove: for all affinely independent $S \subseteq X$, there exists an affine basis of X containing S. (b) Prove: for all sets $T \subseteq X$ with $\mathrm{aff}(T) = X$, there exists an affine basis of X contained in T.

29. Show that any subset of an affinely independent set is also affinely independent.

30. Show that the maximum size of an affinely independent subset of the n-dimensional space V is $n + 1$.

31. Show that $\{v_0, v_1, \ldots, v_k\}$ is affinely dependent iff some v_i is an affine combination of the v_j with $j \neq i$.

32. Given $S = \{v_0, v_1, \ldots, v_k\} \subseteq \mathbb{R}^n$, let v'_k be an affine combination of $v_0, \ldots, v_k$ in which v_k appears with nonzero coefficient. Let S' be S with v_k replaced by v'_k. Show $\mathrm{aff}(S') = \mathrm{aff}(S)$. Show S' is affinely independent iff S is affinely independent.

33. Given $A \in M_{m,n}(F)$, let B be obtained from A by appending a column of 1s to the right end of A. Show that the rows of A are affinely independent in F^n iff the rows of B are linearly independent in F^{n+1}.

34. **Points in General Position.** Say that points $v_1, \ldots, v_k \in \mathbb{R}^n$ are *in general position* iff every subset of $\{v_1, \ldots, v_k\}$ of size $n+1$ or less is affinely independent. For $n = 2$, show that distinct points $v_1, \ldots, v_k \in \mathbb{R}^2$ are in general position iff no three v_i are collinear. State and prove an analogous result $n = 3$.

35. Let $c_1, \ldots, c_k$ be distinct real numbers. Set $v_i = (c_i^n, c_i^{n-1}, \ldots, c_i^2, c_i)$ for $1 \leq i \leq k$. Prove $v_1, \ldots, v_k \in \mathbb{R}^n$ are in general position. [*Hint:* Use the fact that the Vandermonde matrix (see Exercise 40 in Chapter 5) has nonzero determinant.]

36. Prove: for each positive integer n, it is impossible to write $\mathbb{R}^n$ as a union of finitely many affine hyperplanes.

37. Use the previous exercise and induction on k to prove that for all positive integers k and n, there exist k points in general position in $\mathbb{R}^n$ (see Exercise 34).

38. Assume $1_F + 1_F \neq 0_F$ in the field F. Prove $S \subseteq V$ is affine iff $cx + (1 - c)y \in S$ for all $x, y \in S$ and all $c \in F$. Give an example to show that this result may fail if $1_F + 1_F = 0_F$.

39. Prove: for any field F, $S \subseteq V$ is affine iff $ax + by + cz \in S$ for all $x, y, z \in S$ and all $a, b, c \in F$ with $a + b + c = 1_F$.

40. Prove: for all $A \in M_{m,n}(F)$ and all $b \in F^m$, the map $U : F^n \to F^m$ given by $U(x) = Ax + b$ is an affine map. Show that U is linear iff $b = 0$.

41. Prove the identity map on any affine set is an affine isomorphism.

42. Prove the composition of two affine maps (resp. affine isomorphisms) is an affine map (resp. affine isomorphism).

43. Prove the inverse of an affine isomorphism is also an affine isomorphism.

44. Let $U : X \to Y$ be an affine map between affine sets. Prove $U[S]$ is affine for all affine $S \subseteq X$, and $U^{-1}[T]$ is affine for all affine $T \subseteq Y$.

45. Let $U : \mathbb{R}^n \to \mathbb{R}^m$ be an affine map. Prove $U[S]$ is convex for all convex $S \subseteq X$, and $U^{-1}[T]$ is convex for all convex $T \subseteq Y$.

46. Let $U : V \to W$ be an affine map. Prove: for all $S \subseteq V$, $U[\mathrm{aff}(S)] = \mathrm{aff}(U[S])$. Deduce that $u + \mathrm{aff}(S) = \mathrm{aff}(u + S)$ for all $u \in V$ and $S \subseteq V$.

47. Prove: for every affine map $U : V \to W$ between vector spaces V and W, there exists a unique linear map $S : V \to W$ and a unique $b \in W$ with $U = T_b \circ S$, where $T_b : W \to W$ is translation by b.

48. Prove or disprove: for every affine map $U : V \to W$ between any vector spaces V and W, there exists a unique linear map $S : V \to W$ and a unique $c \in V$ with $U = S \circ T_c^{-1}$.

49. Check the map U defined at the end of §11.10 is affine and sends each v_i to y_i.

50. Prove the UMP for affine bases of X (see §11.10) by invoking the analogous UMP for a linear basis of the direction subspace of X (cf. (6.2)).

51. Assume $1_F + 1_F \neq 0_F$ in the field F. Prove $U : V \to W$ is an affine map iff

$$U(cx + (1 - c)y) = cU(x) + (1 - c)U(y)$$

for all $x, y \in V$ and all $c \in F$. Give an example to show that this result may fail if $1_F + 1_F = 0_F$.

52. Prove: for any field F, a map $U : V \to W$ is an affine map iff $U(ax + by + cz) = aU(x) + bU(y) + cU(z)$ for all $x, y, z \in V$ and all $a, b, c \in F$ with $a + b + c = 1_F$.

53. Let X and Y be affine sets, $C = (v_0, \ldots, v_k)$ be a list in X, $U : X \to Y$ be an affine map, and $C' = (U(v_0), \ldots, U(v_k))$. Prove:
    (a) If C is affinely independent and U is injective, then C' is affinely independent.
    (b) If $X = \mathrm{aff}(C)$ and U is surjective, then $Y = \mathrm{aff}(C')$.
    (c) If C is an affine basis of X and U is bijective, then C' is an affine basis of Y.

54. Let $B = (v_0, v_1, v_2) = ((1, 0, 1), (2, 3, 1), (1, 1, 2))$ and $X = \mathrm{aff}(B)$.
    (a) Prove B is an affinely independent ordered list.
    (b) Find the Cartesian coordinates of the point that has barycentric coordinates $(1/3, 1/2, 1/6)$ relative to B.
    (c) Find the barycentric coordinates relative to B of the point with Cartesian coordinates $(0, 0, 4)$.

55. Let $B = ((2, 0, 0), (1, 1, 0), (0, 1, 3), (-1, -1, 1))$, which is an affine basis of $\mathbb{R}^3$. Given $v = (x, y, z) \in \mathbb{R}^3$, find the barycentric coordinates of v relative to B.

56. Let $\Delta = \mathrm{conv}(\{\mathbf{e}_1, \mathbf{e}_2, \ldots, \mathbf{e}_{k+1}\}) \subseteq \mathbb{R}^{k+1}$. Show that Δ is a k-dimensional simplex such that Cartesian coordinates of points in Δ coincide with barycentric coordinates of points in Δ relative to the ordered affine basis $(\mathbf{e}_1, \ldots, \mathbf{e}_{k+1})$. Describe Δ as the solution set of a system of linear inequalities.

57. Show that for any two k-dimensional simplexes Δ_1 and Δ_2 in $\mathbb{R}^n$, there exists an affine isomorphism of $\mathbb{R}^n$ mapping Δ_1 onto Δ_2.

58. **Barycenter of a Simplex.** Let $\{v_0, \ldots, v_k\}$ be affinely independent in $\mathbb{R}^n$. The *barycenter* of the simplex $\Delta = \text{conv}(\{v_0, \ldots, v_k\})$ is the point in Δ all of whose barycentric coordinates (relative to the v_i) are $1/(k+1)$. (a) Show that when $k = 1$, the barycenter of Δ is the midpoint of the line segment with endpoints v_0 and v_1. (b) Find the barycenter of the triangle $\text{conv}(\{(1,2), (3,5), (2,-4)\})$, and illustrate in a sketch. (c) Find the barycenter of $\text{conv}(\{0, e_1, \ldots, e_n\})$ in $\mathbb{R}^n$.

59. Let $\{v_0, \ldots, v_k\}$ be affinely independent in $\mathbb{R}^n$. Let w be the barycenter of $\Delta = \text{conv}(\{v_0, \ldots, v_k\})$. For $0 \leq i \leq k$, let $\Delta_i = \text{conv}((\{v_0, \ldots, v_k\} \setminus \{v_i\}) \cup \{w\})$. Prove that $\Delta = \bigcup_{i=0}^{k} \Delta_i$.

60. The *graph* of a function $T : V \to W$ is the set

$$G(T) = \{(x, T(x)) : x \in V\} \subseteq V \times W.$$

(a) Show that the graph of a linear map is a linear subspace of $V \times W$.
(b) Show that the graph of an affine map is an affine subset of $V \times W$.
(c) Show that for every affine set S in F^n, there exist $m \leq n$, an affine map $U : F^{n-m} \to F^m$, and a linear isomorphism $P : F^n \to F^n$ that acts by permuting standard basis vectors, such that $P[S]$ is the graph of U.

61. (a) Show that for all linear hyperplanes H in $\mathbb{R}^n$, there exist exactly two unit vectors $u \in \mathbb{R}^n$ with $H = \{x \in \mathbb{R}^n : u \bullet x = 0\}$.
(b) Show that for all affine hyperplanes H in $\mathbb{R}^n$ not containing 0, there is exactly one unit vector $u \in \mathbb{R}^n$ and one $c > 0$ with $H = \{x \in \mathbb{R}^n : u \bullet x = c\}$.

62. In $\mathbb{R}^1$, explicitly describe all: (a) linear subspaces; (b) affine sets; (c) convex sets; (d) cones. Prove that the sets you describe, and no others, have each property.

63. Prove that $C = \{(x, y, z) \in \mathbb{R}^3 : z^2 = x^2 + y^2\}$ is a cone. Is C a convex cone?

64. Prove that $D = \{(x, y, z) \in \mathbb{R}^3 : z \geq \sqrt{x^2 + y^2}\}$ is a convex cone.

65. Is $E = \{(x, y, z) \in \mathbb{R}^3 : z^2 \geq x^2 + y^2\}$ a cone? Is E convex? Explain.

66. Fix $x_0 \in \mathbb{R}^n$ and $r \geq 0$.
(a) Prove the open ball $B = \{x \in \mathbb{R}^n : d(x, x_0) < r\}$ is convex.
(b) Is the closed ball $B' = \{x \in \mathbb{R}^n : d(x, x_0) \leq r\}$ convex? Explain.
(c) Is the sphere $S = B' \setminus B$ convex? Explain.

67. Let S and T be convex subsets of $\mathbb{R}^n$.
(a) Prove: for all $c \in \mathbb{R}$, cS is a convex set.
(b) Prove: if S and T are convex and $c \in [0, 1]$, then $cS + (1 - c)T$ is convex.

68. Let $S = \text{conv}(\{(\cos(2\pi k/6), \sin(2\pi k/6)) : 0 \leq k < 6\})$
and $T = \text{conv}(\{(\cos(\pi/2 + 2\pi k/3), \sin(\pi/2 + 2\pi k/3)) : 0 \leq k < 3\})$.
Sketch the sets $cS + (1 - c)T$ for $c \in \{0, 1/4, 1/2, 2/3, 1\}$.

69. Let $S \subseteq \mathbb{R}^n$ be a convex set.
(a) Prove the closure of S (the set of $x \in \mathbb{R}^n$ with $d(x, S) = 0$) is convex.
(b) Prove the interior of S (the union of all open balls of $\mathbb{R}^n$ contained in S) is convex.

70. Let C be a nonempty compact (closed and bounded) subset of $\mathbb{R}^n$. Given a subset $D \neq \emptyset$ in $\mathbb{R}^n$, define $f : C \to \mathbb{R}$ by $f(z) = d(z, D)$ for $z \in C$. Prove:
(a) f is continuous, and there is $x \in C$ with $d(x, D) = d(C, D)$.
(b) If D is compact, then there are $x \in C$ and $y \in D$ with $d(x, y) = d(C, D)$.
(c) The conclusion of (b) holds for all closed D, even if D is unbounded.

71. Give an example of two closed, unbounded sets $C, D \subseteq \mathbb{R}^2$ such that $d(C, D)$ is not of the form $d(v, w)$ for any $v \in C$, $w \in D$.

72. The Hyperplane Separation Theorem (§11.15) assumes that C is convex, closed, and bounded and that D is convex and closed. Give examples to show that the omission of any one of these five assumptions may cause the conclusion of the theorem to fail.

73. Complete the proof of the Hyperplane Separation Theorem by showing that $D \subseteq \{v \in \mathbb{R}^n : u \bullet v \geq c_2\}$ and indicating what adjustments are needed in the case $c_1 > c_2$.

74. Give an example of a convex subset of $\mathbb{R}^2$ that cannot be written as an intersection of any family of half-spaces (even allowing a mixture of open and closed half-spaces).

75. (a) Prove that the intersection of any family of closed subsets of $\mathbb{R}^n$ is closed.
    (b) Prove that every closed half-space is a closed set.
    (c) Prove that for fixed $x_0 \in \mathbb{R}^n$ and $r > 0$, the closed ball $\{x \in \mathbb{R}^n : d(x, x_0) \leq r\}$ is a closed set.

76. (a) Prove that every affine hyperplane is a closed set.
    (b) Prove that every affine set is a closed set.
    (c) Prove that every simplex is a closed set.

77. Prove that if $S \subseteq \mathbb{R}^n$ is closed and bounded, then $\text{conv}(S)$ is a closed convex subset of $\mathbb{R}^n$. (Use Carathéodory's Theorem and sequential compactness of S and $[0, 1]$.)

78. Give an example of a countably infinite set S such that $\text{conv}(S)$ is not closed.

79. If $S \subseteq \mathbb{R}^n$ is bounded but not closed, must $\text{conv}(S)$ be bounded? Prove or give a counterexample.

80. **Radon's Theorem.** Suppose $S = \{v_0, v_1, \ldots, v_{n+1}\}$ is a set of $n + 2$ vectors in $\mathbb{R}^n$. Prove there exist disjoint subsets T and U of S with $\text{conv}(T) \cap \text{conv}(U) \neq \emptyset$.

81. Fix $n, k \geq 0$. For any convex sets $C, D \subseteq \mathbb{R}^{n+k}$, let $C +_{n,k} D$ be the set

$$\{(x, y) \in \mathbb{R}^n \times \mathbb{R}^k : \exists u \in \mathbb{R}^k, \exists v \in \mathbb{R}^k, (x, u) \in C, (x, v) \in D, \text{ and } y = u + v\}.$$

Prove $C +_{n,k} D$ is convex. Prove $+_{n,k}$ is an associative, commutative binary operation on the set of convex subsets of $\mathbb{R}^{n+k}$.

82. Let $C, D \subseteq \mathbb{R}^n$ be convex sets. Define $C \# D = \bigcup_{t \in [0,1]} [(1-t)C \cap tD]$. Prove $C \# D$ is convex. [*Hint:* Study $\text{cone}(C + e_{n+1}) +_{n,1} \text{cone}(D + e_{n+1})$.]

83. (a) Prove: if C and D are convex cones, then $C \# D = C \cap D$.
    (b) Prove: if C and D are convex cones, then $C + D = \text{conv}(C \cup D)$.
    (c) Give examples to show that (a) and (b) can fail if C and D are convex sets that are not cones.

84. This exercise checks some details of the proof in §11.18.
    (a) Show that D defined in (11.2) is a convex cone.
    (b) Confirm identity (11.3).
    (c) Confirm that $C \times \{0\}^k = D \cap (\mathbb{R}^n \times \{0\}^k)$.
    (d) Show that since $C \times \{0\}^k$ is a V-cone, C is a V-cone.

85. For the set D defined in (11.5), confirm that $(x, y) \in D$ iff (x, y) satisfies the linear inequalities given in §11.20.

86. Let $S = \{(0, 1, 2), (0, -1, 1), (-1, 0, 2), (3, 0, 0)\} \subseteq \mathbb{R}^3$. Sketch S, $\text{Sp}(S)$, $\text{aff}(S)$, $\text{conv}(S)$, and $\text{cone}(S)$. Give a linear inequality defining the closed half-space of $\text{aff}(S)$ containing the origin. Give a description of $\text{cone}(S)$ as an H-cone.

87. Prove: for all $S \subseteq \mathbb{R}^n$, $\mathrm{conv}(S) \subseteq \mathrm{aff}(S) \cap \mathrm{cone}(S)$. Does equality always hold?

88. (a) Prove the intersection of an H-cone in $\mathbb{R}^n$ with any linear subspace of $\mathbb{R}^n$ is an H-cone. (b) Prove the image of a V-cone in $\mathbb{R}^n$ under a linear map is a V-cone. (c) Can you give a direct proof of (a) for V-cones, or (b) for H-cones, without using the theorem that V-cones and H-cones are the same?

89. Say $P \subseteq \mathbb{R}^n$ is a *V-polyhedron* iff $P = \mathrm{conv}(S)$ for a finite set $S \subseteq \mathbb{R}^n$. Prove:
    (a) The intersection of finitely many V-polyhedra is a V-polyhedron.
    (b) The intersection of a V-polyhedron with an affine set in $\mathbb{R}^n$ is a V-polyhedron.

90. Say $P \subseteq \mathbb{R}^n$ an *H-polyhedron* iff $P = H_1 \cap \cdots \cap H_s$ for finitely many closed half-spaces H_i. Prove:
    (a) The image of an H-polyhedron under an affine map is an H-polyhedron.
    (b) The sum of finitely many H-polyhedra is an H-polyhedron.

91. Let $S = \{(-2, 1), (-1, 4), (0, -1), (1, 1), (2, 3), (3, 0)\} \subseteq \mathbb{R}^2$. Sketch $\mathrm{conv}(S)$ in $\mathbb{R}^2$ and $\mathrm{cone}(S + \mathbf{e}_3)$ in $\mathbb{R}^3$. Express $\mathrm{cone}(S + \mathbf{e}_3)$ as a specific intersection of linear half-spaces. Express $\mathrm{conv}(S)$ as a specific intersection of closed half-planes.

92. Let S be the solution set in $\mathbb{R}^2$ of the system of linear inequalities

$$y - x \leq 3, \; x \leq 2, \; x + y \leq 3, \; x - 2y \leq 4, \; -y \leq 2, \; 2x + y \leq -2, \; -3x + y \leq 3.$$

Find a finite set T with $S = \mathrm{conv}(T)$.

93. Let S be the solution set in $\mathbb{R}^3$ of the linear inequalities $0 \leq z \leq 2$, $0 \leq y \leq z$, $y \leq x \leq (y+6)/3$, $x-y+z \leq 3$. Sketch S and find a finite set T with $S = \mathrm{conv}(T)$.

94. Let $S = \{(\pm 3, 0, 0), (1, \pm 1, 0), (-1, \pm 1, 0), (\pm 1/2, 0, 1), (0, \pm 1/2, 1)\}$. Draw $T = \mathrm{conv}(S)$ and find a system of linear inequalities with solution set T.

95. Let C be the V-cone $\mathrm{cone}(v_1, \ldots, v_5)$ in $\mathbb{R}^4$, where $v_1 = (3, 1, 1, 1)$, $v_2 = (0, 1, 2, 2)$, $v_3 = (1, 0, 2, 0)$, $v_4 = (-1, -1, 1, -1)$, $v_5 = (2, 2, 0, 3)$. Follow the proof in §11.17 to find a finite set S with $C \cap (\mathbb{R}^3 \times \{0\}) = \mathrm{cone}(S)$. Find a finite set T with $C \cap (\mathbb{R}^2 \times \{0\}^2) = \mathrm{cone}(T)$.

96. Let C be the H-cone in $\mathbb{R}^4 = \{(w, x, y, z) : w, x, y, z \in \mathbb{R}\}$ defined by the system of inequalities

$$2w - x + y \leq 0, \; w + x + 2y + 2z \leq 0, \; x - 3z \leq 0, \; w + y - z \leq 0, \; 2x + y + z \leq 0.$$

Follow the proof in §11.19 to find a system of inequalities whose solution set is the projection of C onto $\mathbb{R}^3 \times \{0\}$. Find a system of inequalities defining the projection of C onto $\mathbb{R}^2 \times \{0\}^2$. What is this projection?

97. Let C be the V-cone $\mathrm{cone}(v_1, \ldots, v_5)$ in $\mathbb{R}^3$, where

$$v_1 = (1, 2, 1), \; v_2 = (1, 4, 0), \; v_3 = (4, 1, 0), \; v_4 = (4, 3, -1), \; v_5 = (3, 4, -1).$$

Sketch C and find an explicit description of C as an H-cone.

98. Let C be the H-cone defined by the inequalities

$$x - 3y - z \leq 0, \; x + y - z \leq 0, \; -11x + 7y - z \leq 0, \; -x - 3y - z \leq 0.$$

Sketch C and find an explicit description of C as a V-cone.

99. In Figure 11.4, let the lines L, L_1, and L_2 have respective slopes m, m_1, and m_2. Carefully prove that $(x, f(x))$ is weakly below L iff $m_1 \leq m \leq m_2$. Use definitions to prove that for convex $C \subseteq \mathbb{R}$, $f : C \to \mathbb{R}$ is convex iff for all $y < x < z$ in C, $(x, f(x))$ is weakly below the line segment joining $(y, f(y))$ and $(z, f(z))$.

100. Determine whether each function below is convex, concave, or neither.
    (a) $f(x) = -2x + 5$ for $x \in \mathbb{R}$
    (b) $f(x) = x^3 - x$ for $x \in \mathbb{R}$
    (c) $f(x) = \sin^2 x$ for $x \in \mathbb{R}$
    (d) $f(x) = \cos x$ for $x \in [\pi/2, 3\pi/2]$
    (e) $f(x) = \arctan(x)$ for $x \geq 0$
    (f) $f(x) = \sqrt{1 - x^2}$ for $x \in [-1, 1]$
    (g) $f : [0, \infty) \to \mathbb{R}$ given by $f(x) = x^r$, for fixed $r \geq 1$
    (h) $f : [0, \infty) \to \mathbb{R}$ given by $f(x) = x^r$, for fixed $r \in [0, 1)$
    (i) $f : (0, \infty) \to \mathbb{R}$ given by $f(x) = x^r$, for fixed $r < 0$

101. Draw a diagram like Figure 11.4 and use it to prove that the convex function f is left-continuous at $x \in (c, d)$.

102. Give an example to show that a convex function $f : [c, d] \to \mathbb{R}$ may not be right-continuous at c or left-continuous at d.

103. Suppose $a < b$ in $\mathbb{R}$, $f : (c, d) \to \mathbb{R}$ is convex, and $g : [a, b] \to (c, d)$ is integrable. Prove $(b - a)f(\int_a^b g(x)\,dx) \leq \int_a^b f((b-a)g(x))\,dx$. Give an example to show that the inequality may fail if we omit $(b - a)$ from both sides.

104. Let C be an open convex subset of $\mathbb{R}^n$ and $f : C \to \mathbb{R}$ be a function with continuous second-order partial derivatives. Show f is convex on C iff for all $c < d$ in $\mathbb{R}$ and all affine maps $g : (c, d) \to C$, $f \circ g : (c, d) \to \mathbb{R}$ is a convex function. Let $g : (c, d) \to C$ have the formula $g(t) = x + tv$ for $t \in (c, d)$, where $x, v \in \mathbb{R}^n$ are fixed. Show that $f \circ g$ is convex iff $v \bullet H_y v \geq 0$ for all y in the image of g. Using this, prove f is convex iff H_y is positive semidefinite for all $y \in C$.

105. Prove $g : \mathbb{R}^n \to \mathbb{R}$, given by $g(x) = ||x|| = \sqrt{x \bullet x}$ for $x \in \mathbb{R}^n$, is convex.

106. Let $C = \{x \in \mathbb{R}^n : x(i) > 0 \text{ for } 1 \leq i \leq n\}$. Prove $f : C \to \mathbb{R}^n$, given by $f(x) = -(x(1)x(2)\cdots x(n))^{1/n}$ for $x \in C$, is convex.

107. Let $f : C \to \mathbb{R}$ be convex. Prove: for all $r \in \mathbb{R}$, $\{x \in C : f(x) < r\}$ and $\{x \in C : f(x) \leq r\}$ are convex sets. Must the sets $\{x \in C : f(x) > r\}$ and $\{x \in C : f(x) \geq r\}$ be convex? Prove or give a counterexample.

108. Prove: for all $k \geq 1$ and all real $z_1, \ldots, z_k > 0$, $(z_1 + z_2 + \cdots + z_k)/k \geq \sqrt[k]{z_1 z_2 \cdots z_k}$. (Apply Jensen's Inequality to a certain convex function.)

109. Let B and C be convex sets, $f : C \to \mathbb{R}$ and $g : \mathbb{R} \to \mathbb{R}$ be convex functions, and $T : B \to C$ be a linear map. (a) Prove: if g is increasing, then $g \circ f$ is convex. (b) Give an example to show that $g \circ f$ may not be convex if g is not increasing. (c) Prove: $f \circ T$ is convex. (d) If T is only an affine map, must (c) be true? Explain.

110. (a) Suppose $f_1, \ldots, f_k : C \to \mathbb{R}$ are convex functions. Prove every positive linear combination of these functions is convex. (b) Suppose $\{f_i : i \in I\}$ is a family of convex functions with domain C such that $g(x) = \sup_{i \in I} f_i(x)$ is finite for all $x \in C$. Prove $g : C \to \mathbb{R}$ is convex.

111. Prove: for all real $p, q > 1$ and $r, s > 0$, if $p^{-1} + q^{-1} = 1$, then $r^{1/p} s^{1/q} \leq r/p + s/q$.

12

Ruler and Compass Constructions

The ancient Greeks investigated and solved many construction problems in plane geometry. Here are some examples of geometric construction problems.

- *Altitudes:* Given a line ℓ and a point P, construct the line through P that meets ℓ at a right angle.

- *Parallels:* Given a line ℓ and a point P not on ℓ, construct the line through P that is parallel to ℓ.

- *Polygons:* Given a circle with center O and a point P on the circle, construct a regular n-sided polygon inscribed in the circle and having P as one vertex. For example, when $n = 3, 4, 5, 6, 7$, the goal is to construct equilateral triangles, squares, regular pentagons, regular hexagons, and regular heptagons.

- *Angle Bisection:* Given two lines ℓ_1, ℓ_2 that meet at a point P at an angle θ, draw a line through P that makes an angle $\theta/2$ with ℓ_1.

- *Angle Trisection:* Given two lines ℓ_1, ℓ_2 that meet at a point P at an angle θ, draw a line through P that makes an angle $\theta/3$ with ℓ_1.

- *Cube Duplication:* Given a line segment $\overline{PQ}$, construct a line segment $\overline{AB}$ such that a cube with $\overline{AB}$ as one side has twice the volume of a cube with $\overline{PQ}$ as one side.

- *Squaring the Circle:* Given a line segment $\overline{PQ}$, construct a line segment $\overline{AB}$ such that the square with $\overline{AB}$ as one side has the same area as the circle with $\overline{PQ}$ as radius.

To solve these construction problems, the Greeks only allowed themselves to use two basic tools: a *ruler*, which can draw the line passing through any two points; and a *compass*, which can draw the circle with a given center passing through a given point. The ruler cannot be used to measure distances, and the compass cannot be used to draw several circles with the same radius in different parts of the diagram. To emphasize these restrictions, the ruler is sometimes called a *straightedge*, and the compass is sometimes called a *collapsing compass*.

A great variety of construction problems can be solved using only a ruler and compass. For example, the constructions of altitudes, parallels, angle bisectors, and regular n-gons for $n = 3, 4, 5, 6, 8$ are all possible. On the other hand, some of the other problems on the preceding list are impossible to solve using only a ruler and compass. The most famous of these problems are squaring the circle, trisecting an arbitrary angle, and duplicating the cube. It is also impossible to construct a regular heptagon (7-sided polygon) and many other regular polygons. One of the amazing achievements of Gauss was the discovery that a regular 17-sided polygon can be constructed using a ruler and compass.

This chapter develops the mathematical tools needed to prove the unsolvability or solvability of constructions with ruler and compass. Remarkably, deciding the solvability of a given geometric construction can ultimately be translated into a linear algebra question involving the dimensions of certain vector spaces. To see how this occurs, we must first link the geometric operations occurring in ruler and compass constructions to various arithmetic

DOI: 10.1201/9781003484561-12

operations, including the extraction of square roots. We then use field theory to characterize the numbers obtainable by means of these arithmetic operations. Any field K with subfield F can be regarded as a vector space with scalars coming from F. The fact that each such vector space has a unique dimension is the key to proving the impossibility or possibility of various geometric constructions.

To implement this agenda, we begin by giving rigorous definitions of three different kinds of constructible numbers — one based on geometry, another based on arithmetic, and a third based on field theory. The core theorem of this subject asserts that the three notions of constructibility are all equivalent. Once we prove this theorem, we use it to analyze some of the famous construction problems mentioned above.

The main prerequisites for reading this chapter are the definitions of fields and vector spaces (see Chapter 1), facts about the dimension of vector spaces (see §1.8), an acquaintance with Euclidean geometry and analytic geometry, and some knowledge of irreducible polynomials (Chapter 3).

12.1   Geometric Constructibility

We first give an informal description of ruler and compass constructions, which we then translate into a rigorous definition. Our geometric constructions occur in a two-dimensional Euclidean plane. We use an x, y-coordinate system to identify this plane with $\mathbb{R}^2$, the set of ordered pairs (a, b) with $a, b \in \mathbb{R}$. Alternatively, we can identify the plane with the set $\mathbb{C}$ of complex numbers by letting the point (a, b) correspond to the complex number $a + ib$. The initial data for our geometric constructions consist of two points A and B in the plane corresponding to the complex numbers 0 and 1. Thus, A has coordinates $(0, 0)$, and B has coordinates $(1, 0)$.

We are allowed to construct new points, lines, and circles in $\mathbb{R}^2$ by applying a finite sequence of actions from the following list.

- If two different points P and Q have already been constructed, we can draw the unique line through P and Q, which we denote by $L(P, Q)$.

- If two different points P and Q have already been constructed, we can draw the unique circle with center P passing through Q, which we denote by $C(P; Q)$.

- If two unequal, non-parallel lines ℓ_1 and ℓ_2 have been constructed, we can locate the unique point where these two lines intersect.

- If a line ℓ and a circle C have been constructed, we can locate all intersection points of this line and this circle (if any).

- If two unequal circles C_1 and C_2 have been constructed, we can locate all intersection points of these two circles (if any).

For our proofs, we need a more formal definition of geometrically constructible points. We recursively define a set GC of complex numbers, called the *geometrically constructible* numbers, by the following rules.

G0. $0 \in \text{GC}$ and $1 \in \text{GC}$.

G1.   If $P, Q, R, S \in \text{GC}$, $L(P, Q)$ and $L(R, S)$ are unequal lines, and $L(P, Q) \cap L(R, S) = \{T\}$, then $T \in \text{GC}$.

G2. If $P, Q, R, S \in$ GC and $T \in L(P, Q) \cap C(R; S)$, then $T \in$ GC.

G3. If $P, Q, R, S \in$ GC, $C(P; Q)$ and $C(R; S)$ are unequal circles, and $T \in C(P; Q) \cap C(R; S)$, then $T \in$ GC.

G4. The only numbers in GC are those that can be obtained by applying rules G0, G1, G2, and G3 a finite number of times.

We can also rephrase this recursive definition in the following iterative fashion. A complex number Q is in GC iff there is a finite sequence of points

$$P_0 = 0, P_1 = 1, P_2, P_3, \ldots, P_k = Q,$$

where $k \in \mathbb{Z}_{\geq 0}$, and for all i with $2 \leq i \leq k$, there exist $r, s, t, u < i$ such that P_i is in $L(P_r, P_s) \cap L(P_t, P_u)$ or in $L(P_r, P_s) \cap C(P_t; P_u)$ or in $C(P_r; P_s) \cap C(P_t; P_u)$. (In the first and third alternatives, we require that the two lines or circles do not coincide.) We call the sequence $P_0, P_1, \ldots, P_k$ a *geometric construction sequence* for Q.

For example, the sequence $0, 1, -1, i\sqrt{3}, \sqrt{3}$ is a geometric construction sequence for $\sqrt{3}$, so $\sqrt{3} \in$ GC. To verify this, first note that -1 is a point in the intersection of the line $L(0, 1)$ (the x-axis) and the circle $C(0; 1)$ (the unit circle). Next, $i\sqrt{3}$ is one of the two intersection points of $C(1; -1)$ and $C(-1; 1)$, since these circles have radius 2 and the point $(0, \sqrt{3})$ has distance 2 from both $(-1, 0)$ and $(1, 0)$. Finally, $\sqrt{3}$ is in $C(0; i\sqrt{3}) \cap L(0, 1)$.

12.2   Arithmetic Constructibility

Next, we define a notion of constructibility involving arithmetic operations. We consider complex numbers that can be built up from the integers by performing the standard arithmetic operations (addition, subtraction, multiplication, and division in $\mathbb{C}$) together with the operation of extracting square roots. Such numbers appear naturally, for example, in connection with the quadratic formula. Recall that for $a, b, c \in \mathbb{C}$ with a nonzero, the roots of the equation $ax^2 + bx + c$ in $\mathbb{C}$ are

$$r_1 = \frac{-b + \sqrt{b^2 - 4ac}}{2a}, \qquad r_2 = \frac{-b - \sqrt{b^2 - 4ac}}{2a}.$$

We see that these roots can be built from the initial data a, b, c by performing arithmetic operations and taking square roots.

To formalize this idea, we recursively define a set AC of complex numbers, called the *arithmetically constructible* numbers, by the following rules.

A0. $0 \in$ AC and $1 \in$ AC.

A1. If $a, b \in$ AC, then $a + b \in$ AC.

A2. If $a \in$ AC, then $-a \in$ AC.

A3. If $a, b \in$ AC, then $a \cdot b \in$ AC.

A4. If $a \in$ AC is nonzero, then $a^{-1} \in$ AC.

A5. If $a \in$ AC and $b \in \mathbb{C}$ and $b^2 = a$, then $b \in$ AC.

A6. The only numbers in AC are those that can be obtained by applying rules A0 through A5 a finite number of times.

A less formal phrasing of rule A5 says that if $a \in AC$ then $\sqrt{a} \in AC$. Since every nonzero complex number has two complex square roots, the notation $\sqrt{a}$ used here is ambiguous. However, this is not a serious difficulty, since rule A2 guarantees that both square roots of a must belong to AC.

We can give an iterative formulation of the recursive definition of AC, as follows. A complex number z is in AC iff there is a finite sequence of numbers

$$x_0 = 0, x_1 = 1, x_2, x_3, \ldots, x_k = z,$$

where $k \in \mathbb{Z}_{\geq 0}$, and for all i with $2 \leq i \leq k$, either there exists $r < i$ with $x_i = -x_r$ or $x_i = x_r^{-1}$ ($x_r \neq 0$) or $x_i = \sqrt{x_r}$, or there exist $r, s < i$ with $x_i = x_r + x_s$ or $x_i = x_r \cdot x_s$. The sequence $x_0, x_1, \ldots, x_k$ is called an *arithmetic construction sequence* for z.

For example, an arithmetic construction sequence for $i\sqrt{3}$ is $0, 1, 2, 3, -3, i\sqrt{3}$, hence $i\sqrt{3} \in AC$. To build this sequence, we use rule A0 twice, then rule A1 twice, then rule A2, then rule A5. For a more elaborate illustration of arithmetic construction sequences, suppose $a, b, c \in AC$ with $a \neq 0$. Then the roots r_1, r_2 of the quadratic equation $ax^2 + bx + c$ are also in AC. For instance, we prove that $r_1 \in AC$ by producing an arithmetic construction sequence as follows. Begin the sequence with $0, 1, 2, 3, 4$, followed by the concatenation of arithmetic construction sequences for a, b, and c. Continue this sequence as follows:

$$b^2, ac, 4ac, -4ac, b^2 - 4ac, \sqrt{b^2 - 4ac}, -b, -b + \sqrt{b^2 - 4ac}, 2a, r_1.$$

You can check that each new term listed here is produced from one or two previous terms by invoking some rule A0 through A5.

12.3   Preliminaries on Field Extensions

To continue, we need some facts about field extensions. Let K be a field, as defined in §1.2. Recall (§1.4) that a *subfield* of K is a subset F of K such that $0_K \in F$, $1_K \in F$, and for all $a, b \in F$, we have $a + b \in F$, $-a \in F$, $a \cdot b \in F$, and if $a \neq 0_K$, then $a^{-1} \in F$. (Note the similarity of these closure conditions to rules A0 through A4.) A subfield F of K becomes a field by restricting the sum and product operations on K to the subset F. We say K is an *extension field* of F iff F is a subfield of K. A *chain of fields* is a sequence $F_0 \subseteq F_1 \subseteq F_2 \subseteq \cdots \subseteq F_n$ in which F_n is a field, and each F_i (for $0 \leq i < n$) is a subfield of F_n. In this chapter, the fields under consideration are subfields of $\mathbb{C}$, so the operations in these fields are ordinary addition and multiplication of complex numbers.

Suppose K is any field with subfield F. A key observation is that we can regard K as a vector space over the field F, by taking vector addition to be the field addition $+ : K \times K \to K$, and taking scalar multiplication $s : F \times K \to K$ to be the restriction of the field multiplication $\cdot : K \times K \to K$ to the domain $F \times K$. In other words, we multiply a scalar $c \in F$ by a vector $v \in K$ by forming the product $c \cdot v$ in the field K. The vector space axioms follow by comparing the axioms for the field K (and subfield F) to the axioms for an F-vector space. Like any other F-vector space, K has a unique dimension viewed as a vector space over F. This dimension is written $[K : F]$ and is called the *degree of K over F*. For example, $K = \mathbb{C}$ is a two-dimensional vector space over its subfield $F = \mathbb{R}$ (since $(1, i)$ is an ordered basis of the real vector space $\mathbb{C}$), so the degree $[\mathbb{C} : \mathbb{R}]$ is 2. On the other hand, $[\mathbb{R} : \mathbb{Q}] = \infty$, since $\mathbb{R}$ is uncountable but every finite-dimensional $\mathbb{Q}$-vector space is countable. (More precisely, $[\mathbb{R} : \mathbb{Q}]$ is a certain infinite cardinal giving the $\mathbb{Q}$-dimension of $\mathbb{R}$. However, for our purposes in this chapter, we do not need this extra precision. The notation

$[K : F] = \infty$ means that K is an infinite-dimensional F-vector space.) The following formula plays a critical role in all subsequent developments.

Degree Formula for Field Extensions. If $F \subseteq K \subseteq E$ is a chain of three fields, then

$$[E : F] = [E : K][K : F].$$

Proof. We assume in this proof that $[E : K] = m$ and $[K : F] = n$ for some positive integers m and n. It can be shown (Exercise 20) that if $[E : K] = \infty$ or $[K : F] = \infty$, then $[E : F] = \infty$. Since $[E : K] = m$, there exists an ordered basis $B = (y_1, y_2, \ldots, y_m)$ for the K-vector space E. We call B a K-*basis of* E to emphasize that scalars for this space come from K. Similarly, since $[K : F] = n$, there is an ordered F-basis $C = (z_1, \ldots, z_n)$ for the F-vector space K. Consider the list of all products $z_j y_i$:

$$D = (z_1 y_1, z_1 y_2, \ldots, z_1 y_m, \; z_2 y_1, \ldots, z_2 y_m, \; \ldots, \; z_n y_1, \ldots, z_n y_m).$$

D is a list of mn elements, which are not yet known to be distinct. We claim D is an ordered F-basis of E; the needed conclusion $[E : F] = mn$ follows from this claim.

First, we prove that D spans E as an F-vector space. Given any $w \in E$, we can write $w = \sum_{i=1}^{m} k_i y_i$ for certain scalars $k_i \in K$, since B is a K-basis of E. Since each $k_i \in K$, and since C is an F-basis of K, we can find further scalars $f_{ij} \in F$ such that $k_i = \sum_{j=1}^{n} f_{ij} z_j$ for $1 \leq i \leq m$. Inserting these expressions into the formula for w and simplifying, we obtain

$$w = \sum_{i=1}^{m} \left(\sum_{j=1}^{n} f_{ij} z_j \right) y_i = \sum_{i=1}^{m} \sum_{j=1}^{n} f_{ij}(z_j y_i).$$

We have expressed an arbitrary vector $w \in E$ as an F-linear combination of the elements of the list D, so this list does span the F-vector space E.

Next, we prove that the list D is F-linearly independent. Assume $c_{ij} \in F$ are scalars for which $\sum_{i=1}^{m} \sum_{j=1}^{n} c_{ij}(z_j y_i) = 0$. We must prove all c_{ij} are 0. Using the distributive laws to regroup terms, our assumption can be written

$$0 = \sum_{i=1}^{m} \left(\sum_{j=1}^{n} c_{ij} z_j \right) y_i.$$

Each term in parentheses belongs to the field K, since $z_j \in K$ and $c_{ij} \in F \subseteq K$. The list $B = (y_1, \ldots, y_m)$ is known to be K-linearly independent, so we deduce that $\sum_{j=1}^{n} c_{ij} z_j = 0$ for all i with $1 \leq i \leq m$. Now we can invoke the known F-linear independence of the list $C = (z_1, \ldots, z_n)$ to see that $c_{ij} = 0$ for all i, j with $1 \leq i \leq m$ and $1 \leq j \leq n$. $\quad \square$

Consider a chain of finitely many fields, say

$$F_0 \subseteq F_1 \subseteq F_2 \subseteq F_3 \subseteq \cdots \subseteq F_n.$$

Iteration of the previous result gives the *General Degree Formula*

$$[F_n : F_0] = [F_n : F_{n-1}] \cdots [F_3 : F_2][F_2 : F_1][F_1 : F_0] = \prod_{i=1}^{n} [F_i : F_{i-1}].$$

The last preliminary concept we need is the notion of a field extension generated by one element. Suppose K is a field with subfield F, and let $z \in K$. Define $F(z)$, the *field extension of F obtained by adjoining z*, to be the intersection of all subfields L of K such that $F \subseteq L$ and $z \in L$. You can check that $F(z)$ is a subfield of K with $F \subseteq F(z)$ and $z \in F(z)$; and for any subfield M of K with $F \subseteq M$ and $z \in M$, $F(z) \subseteq M$. You can also verify that $F(z)$ consists of all elements of K that can be written in the form $f(z)g(z)^{-1}$, for some polynomials f, g with coefficients in F such that $g(z) \neq 0$. In this chapter, we are mainly interested in the case where $K = \mathbb{C}$ and $F = \mathbb{Q}$, so $\mathbb{Q}(z)$ is the smallest subfield of the complex numbers containing the given complex number z.

12.4   Field-Theoretic Constructibility

Our third notion of constructibility involves field extensions of degree 2. We say that a complex number z *has a square root tower* iff there is a chain of fields

$$\mathbb{Q} = F_0 \subseteq F_1 \subseteq \cdots \subseteq F_n \subseteq \mathbb{C}$$

such that $z \in F_n$ and $[F_k : F_{k-1}] = 2$ for $1 \leq k \leq n$. Let SQC denote the set of all $z \in \mathbb{C}$ such that z has a square root tower. We claim:

$$\text{If } z \in \text{SQC, then } [\mathbb{Q}(z) : \mathbb{Q}] = 2^e \text{ for some } e \geq 0. \tag{12.1}$$

To prove this, first use the Degree Formula to see that $[F_n : \mathbb{Q}] = 2^n$. Since $z \in F_n$ and $\mathbb{Q} \subseteq F_n$, we have a chain of fields $\mathbb{Q} \subseteq \mathbb{Q}(z) \subseteq F_n$. It follows from the Degree Formula that $[\mathbb{Q}(z) : \mathbb{Q}]$ divides $[F_n : \mathbb{Q}] = 2^n$, so that $[\mathbb{Q}(z) : \mathbb{Q}]$ must be a power of 2. It can be shown that the converse of (12.1) is not true in general. However, the contrapositive of (12.1) is certainly true:

$$\text{If } [\mathbb{Q}(z) : \mathbb{Q}] \text{ is not a power of 2, then } z \notin \text{SQC.} \tag{12.2}$$

We can now precisely state the main goal of this chapter.

Constructibility Theorem. GC = AC = SQC. In other words, a complex number z is geometrically constructible iff z is arithmetically constructible iff z has a square root tower.

Once this theorem is known, we can use the criterion (12.2) to prove the unsolvability of various geometric construction problems. In fact, for such applications, it suffices to know the weaker result GC $\subseteq$ AC $\subseteq$ SQC. Similarly, we can demonstrate the solvability of certain geometric constructions by building appropriate square root towers. We illustrate this technique in §12.9 and §12.10, after we prove the Constructibility Theorem GC = AC = SQC in the intervening sections.

12.5   Proof that GC $\subseteq$ AC

In this section, we give the formal proof that GC $\subseteq$ AC. The details are somewhat lengthy, but the intuition behind the proof can be concisely summarized as follows. When computing intersections of lines and circles via analytic geometry, the coordinates of newly located points can be computed from coordinates of previous points by arithmetic operations and extractions of square roots. This is the key conceptual feature of the formulas derived here.

Our proof requires facts from plane analytic geometry, labeled AG1 through AG5.

AG1. For every line ℓ in the plane, there exist real numbers a, b, s such that

$$\ell = \{(x, y) \in \mathbb{R}^2 : ax + by = s\},$$

where a and b are not both 0. If $\ell = L(P, Q)$ where $P = (x_0, y_0)$ and $Q = (x_1, y_1)$, then we may take $a = y_1 - y_0$, $b = x_0 - x_1$, and $s = ax_0 + by_0 = ax_1 + by_1$. So if $x_0, y_0, x_1, y_1 \in$ AC, then we can choose a, b, s so that $a, b, s \in$ AC.

AG2. If C is a circle in $\mathbb{R}^2$ with center (c, d) and radius $r > 0$, then

$$C = \{(x, y) \in \mathbb{R}^2 : (x - c)^2 + (y - d)^2 = r^2\}.$$

If (e, f) is any point on C, then $r^2 = (e - c)^2 + (f - d)^2$. So if $c, d, e, f \in$ AC, then r^2 and r belong to AC.

AG3. Consider lines ℓ_1 and ℓ_2 with equations $ax + by = r$ and $cx + dy = s$. These lines are not parallel iff $ad - bc \neq 0$. In this case, the unique point (x_0, y_0) where ℓ_1 and ℓ_2 intersect is given by Cramer's Rule (see §5.12):

$$x_0 = \frac{rd - bs}{ad - bc}, \qquad y_0 = \frac{as - rc}{ad - bc}.$$

So if $a, b, c, d, r, s \in$ AC, then $x_0, y_0 \in$ AC.

AG4. Suppose a line ℓ has equation $ax + by = s$, and a circle C has equation $(x - c)^2 + (y - d)^2 = r^2$. A point (x, y) is on both ℓ and C iff both equations hold simultaneously. Suppose that b is nonzero. Substituting $y = (s - ax)/b$ into the equation of the circle, we see that x must satisfy the equation $(x - c)^2 + ((s - ax)/b - d)^2 = r^2$. This equation can be rewritten as $Ax^2 + Bx + D = 0$, where

$$A = 1 + \frac{a^2}{b^2}, \quad B = -2c + \frac{2ad}{b} - \frac{2as}{b^2}, \quad D = c^2 + d^2 - r^2 - \frac{2ds}{b} + \frac{s^2}{b^2}.$$

So x is a root of a linear or quadratic equation with the indicated coefficients, and $y = (s - ax)/b$ is determined by x. If $a, b, s, c, d, r \in$ AC, then we see from these formulas that $A, B, D \in$ AC and $x, y \in$ AC. A similar analysis holds when a is nonzero.

AG5. Suppose C_1 and C_2 are circles with equations $(x - a)^2 + (y - b)^2 = r^2$ and $(x - c)^2 + (y - d)^2 = s^2$. A point (x, y) is on both circles iff both equations hold simultaneously. Subtracting the second equation from the first, we obtain a third equation

$$(2c - 2a)x + (2d - 2b)y = c^2 + d^2 + r^2 - a^2 - b^2 - s^2.$$

Assuming that the circles are non-concentric, this is the equation of a line ℓ. The first and second equations have the same solution set as the second and third equations, as you can check. It follows that the intersection points of C_1 and C_2 are the same as the intersection points of ℓ and C_2. So we can find the coordinates of these points using the formulas from the preceding item. If $a, b, c, d, r, s \in$ AC, then the coefficients of the third equation are all in AC, so the coordinates x and y of the intersection point(s) also belong to AC.

We now prove that GC $\subseteq$ AC. Given $z = (x, y) = x + iy \in$ GC, we know there is a geometric construction sequence $z_0, z_1, \ldots, z_k$ with last term z. Write $z_r = x_r + iy_r$ with $x_r, y_r \in \mathbb{R}$ for $0 \leq r \leq k$. We show that $x = x_k$ and $y = y_k$ are in AC by strong induction on k. Since $i \in$ AC, it follows that $z = x + iy$ is in AC. If $k = 0$, then $z_0 = 0 = 0 + 0i$, and 0 is in AC. If $k = 1$, then $z_1 = 1 = 1 + 0i$, and $0, 1 \in$ AC. Fix $k > 1$. We make the induction hypothesis that for all $r < k$, $x_r \in$ AC and $y_r \in$ AC. Consider the three possible cases for z_k.

- *Case G1:* There exist $r, s, t, u < k$ such that $z_k = (x_k, y_k)$ is the unique point of intersection of the line ℓ_1 through (x_r, y_r) and (x_s, y_s) and the line ℓ_2 through (x_t, y_t) and (x_u, y_u). Since $x_r, y_r, x_s, y_s, x_t, y_t, x_u, y_u$ are all known to be in AC by the induction hypothesis, we see that x_k and y_k are in AC by AG1 and AG3.

- *Case G2:* There exist $r, s, t, u < k$ such that $z_k = (x_k, y_k)$ belongs to both the line ℓ through (x_r, y_r) and (x_s, y_s) and the circle C with center (x_t, y_t) passing through (x_u, y_u). Since $x_r, y_r, x_s, y_s, x_t, y_t, x_u, y_u$ are known to be in AC by the induction hypothesis, we see that x_k and y_k are in AC by AG1, AG2, and AG4.

- *Case G3:* There exist $r, s, t, u < k$ such that $z_k = (x_k, y_k)$ belongs to both the circle C_1 through (x_s, y_s) with center (x_r, y_r) and the circle C_2 through (x_u, y_u) with center (x_t, y_t). Since $x_r, y_r, x_s, y_s, x_t, y_t, x_u, y_u$ are known to be in AC by the induction hypothesis, we see that x_k and y_k are in AC by AG2 and AG5.

12.6   Proof that $\mathrm{AC} \subseteq \mathrm{GC}$

We proved that GC is contained in AC by showing that the coordinates of the intersections of lines and circles can always be found from previous coordinates by arithmetic operations and square root extractions. To prove the reverse containment, we must show the converse: arithmetic operations and square root extractions can be performed geometrically with a ruler and compass. We give the necessary geometric constructions (facts CR1 through CR12) followed by the proof that $\mathrm{AC} \subseteq \mathrm{GC}$. Each construction can be justified by the axioms and theorems of Euclidean geometry, but we omit the detailed justifications here.

CR1.   *Given distinct points A and B, we can construct an equilateral triangle with side $\overline{AB}$, the midpoint of $\overline{AB}$, and the perpendicular bisector of $\overline{AB}$.* First draw the line connecting A and B, then draw the circle centered at A passing through B, then draw the circle centered at B passing through A. The two circles must meet[1] in two points, say C and D. Then $\triangle ABC$ and $\triangle ABD$ are equilateral triangles, $\overleftrightarrow{CD}$ is the perpendicular bisector of $\overline{AB}$, and the intersection of $\overleftrightarrow{CD}$ and $\overleftrightarrow{AB}$ is the midpoint of $\overline{AB}$.

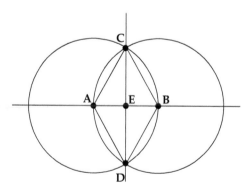

---

[1]A careful axiomatic development of Euclidean geometry would include a proof that the claimed intersection points really do exist. We omit this.

CR2.   *Given three points A, B, C with $A \neq B$, we can construct the unique altitude through C perpendicular to $\overleftrightarrow{AB}$.* Choose notation so that $C \neq A$. Draw $\overleftrightarrow{AB}$ and the circle with center C passing through A. If this circle meets $\overleftrightarrow{AB}$ at a second point $A' \neq A$, then the required altitude is the perpendicular bisector of $\overline{AA'}$, which can be built by CR1. If, instead, the circle is tangent to $\overleftrightarrow{AB}$ at A, the required altitude is $\overleftrightarrow{AC}$. The construction works whether or not C is on the line $\overleftrightarrow{AB}$.

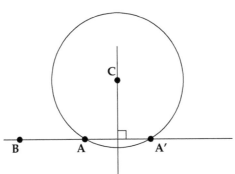

CR3.   *Given three points A, B, C with $A \neq B$, we can construct the unique line through C parallel to $\overleftrightarrow{AB}$.* Apply CR2 once to get a line $\overleftrightarrow{CD}$ perpendicular to $\overleftrightarrow{AB}$. Apply CR2 again to get a line $\overleftrightarrow{CE}$ perpendicular to $\overleftrightarrow{CD}$, hence parallel to $\overleftrightarrow{AB}$.

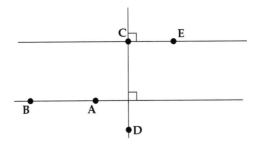

CR4.   *Given the initial points 0 and 1, we can draw the x-axis and the y-axis.* The x-axis is the line through 0 and 1. Now use CR2 to draw the y-axis.

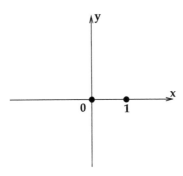

CR5.   *For all real x, y, $x + iy = (x, y)$ is in GC iff $x = (x, 0)$ is in GC and $y = (y, 0)$ is in GC.* If $x + iy = (x, y) \in$ GC, we can use CR2 to drop altitudes to the x-axis and the y-axis to find the points $(x, 0)$ and $(0, y)$. Drawing the circle centered at

0 and passing through $(0, y)$ allows us to locate $(y, 0)$. The converse statement is proved similarly, by reversing these steps.

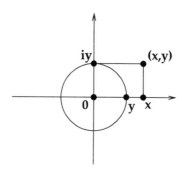

The next constructions show how to implement real arithmetic using a ruler and compass. We can then handle complex arithmetic by using CR5 to reduce to the real case.

CR6.   *For all real $x, y \in$ GC, $x + y \in$ GC and $x - y \in$ GC. Given x, y, locate the point $(x, y) = x + iy$ as in CR5. The circle with center x passing through $x + iy$ meets the real axis at the points $x + y$ and $x - y$. Hence $x + y$ and $x - y$ are in GC.*

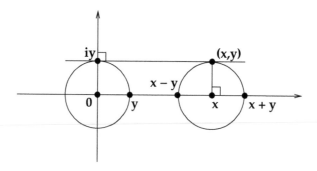

CR7.   *For all real $y \in$ GC, $-y \in$ GC. Take $x = 0$ in CR6.*

CR8.   *For all real $x, y \in$ GC, $xy \in$ GC. Drawing circles with center 0 through 1 and x, we can locate i and ix on the imaginary axis. Draw the line through i and y. Then use CR3 to draw the line through ix parallel to this line. Let z be the point where this line meets the real axis. By considering the two similar right triangles in the figure, we see that $y/1 = z/x$, so that $z = xy$. Our figure assumes that x and y are positive. The reader can adapt this construction to the case of negative variables (alternatively, this can be deduced from the positive case using CR7).*

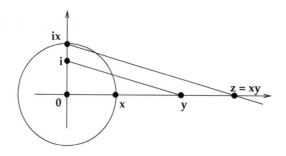

CR9. *For all nonzero real $x \in$ GC, $x^{-1} \in$ GC.* Draw i and the line through x and i. Then draw the line through 1 parallel to this line, which meets the imaginary axis in some point iz. Use a circle centered at 0 to find the real point z. Comparing similar right triangles again, we see that $z/1 = 1/x$, so $z = x^{-1}$ is in GC.

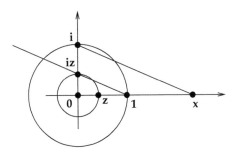

CR10. *For all positive real $x \in$ GC, $\sqrt{x} \in$ GC.* Draw -1, and then construct the midpoint P of the line segment joining -1 and x. Draw the circle through x with center P, and let iy be the point where this circle meets the imaginary axis. We can then construct y, which we claim is the positive square root of x. To see this, draw line segments from iy to -1 and from iy to x. The angle formed by these segments at iy is a right angle, since it is inscribed in a semicircle. It readily follows that the right triangle with vertices 0, -1, and iy is similar to the right triangle with vertices 0, iy, and x. So $1/y = y/x$, which shows that $y^2 = x$ and $y = \sqrt{x}$.

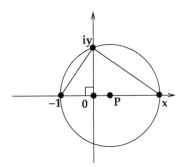

CR11. *For all complex $u, v \in$ GC, we have $u + v \in$ GC, $-u \in$ GC, $uv \in$ GC, and (if $u \neq 0$) $u^{-1} \in$ GC.* Write $u = a + ib$ and $v = c + id$ where $a, b, c, d \in \mathbb{R}$. By CR5, the real numbers a, b, c, d are all in GC. Now,

$$u + v = (a + c) + i(b + d), \quad -u = (-a) + i(-b),$$
$$uv = (ac - bd) + i(ad + bc), \quad u^{-1} = \left(\frac{a}{a^2 + b^2} \right) + i \left(\frac{-b}{a^2 + b^2} \right).$$

Using CR6, CR7, CR8, and CR9, we see that the real and imaginary parts of all these expressions are in GC. Hence, another application of CR5 shows that $u + v$, $-u$, uv, and u^{-1} are all in GC.

CR12. *For all complex $z \in$ GC, $\sqrt{z} \in$ GC.* Write $z = re^{i\theta}$ in polar form, where $r \geq 0$ and θ are real numbers. Drawing the circle through z with center 0, we see that $r = |z| \in$ GC. By CR10, we can construct the real square root of r. By CR1, we can construct the perpendicular bisector of the line segment joining z and r. Then the points w where this bisector intersects the circle with center 0 and radius $\sqrt{r}$ are in GC. You can check that $w = \pm\sqrt{r}e^{i\theta/2}$ in polar form, and these are the two complex square roots of z. Thus both square roots are in GC.

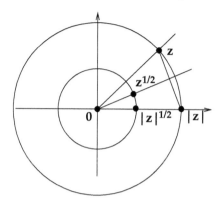

With these tools in hand, we prove AC $\subseteq$ GC as follows. Given $z \in$ AC, we know there is an arithmetic construction sequence $z_0, z_1, \ldots, z_k$ with last term z. We proceed by strong induction on k. If $k = 0$, then $z = 0$ is in GC. If $k = 1$, then $z = 1$ is in GC. Given $k > 1$, we may assume the induction hypothesis that $z_0, \ldots, z_{k-1}$ are all in GC. Now if $z_k = z_r + z_s$ for some $r, s < k$, then $z_k \in$ GC follows by CR11. The same conclusion holds if $z_k = -z_r$ or $z_k = z_r z_s$ or $z_k = 1/z_r$ where $r, s < k$. Finally, if $z_k = \sqrt{z_r}$ for some $r < k$, then $z_k \in$ GC by CR12. Thus, in all cases, $z = z_k$ is in GC.

12.7   Algebraic Elements and Minimal Polynomials

Before proving that AC = SQC, we need some more facts about field extensions of the form $F(z)$. Suppose K is a field, F is a subfield of K, and $z \in K$. We say that z is *algebraic over* F iff there exists a nonzero polynomial $g \in F[x]$ satisfying $g(z) = 0_K$.

Theorem on Minimal Polynomials. Suppose F is a subfield of K and $z \in K$ is algebraic over F. There exists a unique monic, irreducible polynomial $m \in F[x]$ such that $m(z) = 0_K$. No polynomial in $F[x]$ of smaller degree than m has z as a root.

The polynomial m is called the *minimal polynomial of z over F*.

Proof. Because $z \in K$ is algebraic over F, there exist nonzero polynomials in $F[x]$ having z as a root. Pick a nonzero polynomial $m \in F[x]$ of least degree such that $m(z) = 0$. Dividing by the leading coefficient if needed, we can assume that m is monic. We claim m must be an irreducible polynomial in $F[x]$, meaning there is no factorization $m = fh$ with $f, h \in F[x]$ each having lower degree than m. If such a factorization did exist, then evaluating both sides at z would give $0_K = m(z) = f(z)h(z)$. Since K is a field, this forces $f(z) = 0_K$ or $h(z) = 0_K$. Either possibility contradicts minimality of the degree of m.

Now we prove the uniqueness of m. To get a contradiction, assume $p \neq m$ is another monic, irreducible polynomial in $F[x]$ with $p(z) = 0_K$. Since p and m are non-associate irreducible polynomials, $\gcd(p, m) = 1$. It follows (see Chapter 3) that $pr + ms = 1$ for some $r, s \in F[x]$. Evaluating both sides at z and recalling that $p(z) = 0_K = m(z)$, we obtain $0_K = 1_K$. This contradicts the definition of a field. $\square$

We can use minimal polynomials to give an explicit basis for the field extension $F(z)$, viewed as an F-vector space, in the case where z is algebraic over F.

Theorem on a Basis for $F(z)$**.** Suppose K is a field with subfield F, and $z \in K$ is algebraic over F with minimal polynomial $m \in F[x]$ of degree d. Then $B = (1, z, z^2, \ldots, z^{d-1})$ is an ordered F-basis for $F(z)$, and $[F(z) : F] = d = \deg(m)$.

Proof. Let W be the subspace of the F-vector space K spanned by the vectors in the list B. W can also be described as the set of elements of the form $g(z)$, where $g \in F[x]$ is either 0 or has degree less than d. We first show that B is an F-linearly independent list. Suppose $c_0, c_1, \ldots, c_{d-1} \in F$ satisfy $c_0 1 + c_1 z + \cdots + c_{d-1} z^{d-1} = 0_K$. Then the polynomial $g = c_0 + c_1 x + \cdots + c_{d-1} x^{d-1}$ is in $F[x]$ and has z as a root. Since no polynomial of degree less than $d = \deg(m)$ has z as a root, it must be that g is the zero polynomial, so $c_0 = c_1 = \cdots = c_{d-1} = 0$. This shows that B is an ordered basis for the F-vector space W.

To finish the proof, we show that $W = F(z)$. On one hand, since $F(z)$ is a subfield of K containing z and F, closure under sums and products shows that $F(z)$ must contain all powers of z and all F-linear combinations of these powers. In particular, $W \subseteq F(z)$. To prove the reverse inclusion $F(z) \subseteq W$, it suffices to show that W is a subfield of K with $F \subseteq W$ and $z \in W$. Keeping in mind that W is the set of F-linear combinations of the powers of z in B, it is routine to check that W contains F (hence contains 0_K and 1_K), that $z \in W$, and that W is closed under addition, additive inverses, and scalar multiplication by elements of F.

We check that W is closed under multiplication. Consider two arbitrary elements $u, v \in W$, which have the form $u = g(z)$ and $v = h(z)$ for certain polynomials $g, h \in F[x]$ that are either 0 or have degree less than d. Note $uv = g(z)h(z) = (gh)(z)$, where the product polynomial $gh \in F[x]$ may have degree as high as $2d - 2$. We now show that for all $p \in F[x]$ of any degree, $p(z) \in W$; it follows that $uv = (gh)(z)$ is indeed in W. First we use induction on n to prove that $z^n \in W$ for all $n \geq 0$. This is true by definition of W for $0 \leq n < d$. Fix $n \geq d$, and assume as induction hypothesis that $z^0, z^1, \ldots, z^{n-1}$ belong to W. Let the minimal polynomial of z over F be $m = a_0 + a_1 x + \cdots + a_{d-1} x^{d-1} + x^d$, where all $a_i \in F$. Multiplying by x^{n-d} and evaluating at $x = z$, we find that

$$z^n = -a_0 z^{n-d} - a_1 z^{n-d+1} - \cdots - a_{d-1} z^{n-1}.$$

By induction hypothesis, each power of z on the right side is in W. So z^n, which is an F-linear combination of these powers, is also in W. It now follows that $p(z)$, which is an F-linear combination of powers of z, belongs to W.

To finish the proof that W is a subfield of K, we check that W is closed under multiplicative inverses. Fix a nonzero element $u \in W$, which has the form $u = g(z)$ for some nonzero $g \in F[x]$ of degree less than d. As m is irreducible and $\deg(g) < \deg(m)$, we must have $\gcd(g, m) = 1$, and so $1 = gr + ms$ for some $r, s \in F[x]$. Evaluating both sides at z gives $1_K = g(z)r(z) + m(z)s(z) = ur(z)$. Now, as seen in the previous paragraph, $r(z) \in W$; and $ur(z) = 1$ shows that $u^{-1} = r(z) \in W$. $\square$

Theorem on Algebraic Elements and Finite Degree Extensions. Let F be a subfield of K and $z \in K$. If z is algebraic over F, then the degree $[F(z) : F]$ is finite and equals the degree of the minimal polynomial of z over F. Conversely, if $d = [F(z) : F]$ is finite, then z is algebraic over F with minimal polynomial having degree d.

Proof. The first statement follows from the previous theorem. For the converse, assume $d = [F(z) : F]$ is finite. Consider the list of $d + 1$ vectors $(1, z, z^2, \ldots, z^d)$ in the F-vector space $F(z)$. Since $F(z)$ is d-dimensional, this list must be linearly dependent over F (see §1.8). So there exist scalars $c_0, c_1, \ldots, c_d \in F$, not all zero, with $\sum_{i=0}^{d} c_i z^i = 0_K$. The polynomial $\sum_{i=0}^{d} c_i x^i \in F[x]$ is nonzero and has z as a root. So z is algebraic over F. By the first part of the theorem, the minimal polynomial of z over F has degree d. $\square$

Here is an example needed in our subsequent discussion of SQC. Suppose F is a subfield of K, z is in K but not in F, and $y = z^2$ is in F. Then $[F(z) : F] = 2$ and $(1, z)$ is an ordered F-basis of the F-vector space $F(z)$. Indeed, since $x^2 - z^2 = x^2 - y$ is a nonzero polynomial in $F[x]$ having z as a root, we see that z is algebraic over F. The polynomial $x^2 - y = (x - z)(x + z)$ does not factor into linear factors in $F[x]$, since z is not in F. It follows that $x^2 - y$ is irreducible in $F[x]$. This polynomial is also monic and has z as a root, so $x^2 - y$ is the minimal polynomial of z over F. The stated conclusions now follow from the Theorem on a Basis for $F(z)$, since the minimal polynomial has degree 2. For instance, taking $F = \mathbb{Q}$ and $z = \sqrt{2}$, we see that $\mathbb{Q}(\sqrt{2})$ is a 2-dimensional $\mathbb{Q}$-vector space with ordered basis $(1, \sqrt{2})$. So, every element of the field $\mathbb{Q}(\sqrt{2})$ has a unique representation of the form $a + b\sqrt{2}$ for some $a, b \in \mathbb{Q}$.

12.8   Proof that $\mathrm{AC} = \mathrm{SQC}$

We now prove that $\mathrm{AC} \subseteq \mathrm{SQC}$, meaning that every arithmetically constructible complex number has a square root tower. Let $z \in \mathrm{AC}$ have an arithmetic construction sequence $z_0, z_1, \ldots, z_k$. We use this sequence to build a square root tower for z, as follows. Start with $\mathbb{Q} = K_0$, which contains $z_0 = 0$ and $z_1 = 1$. Suppose that, for a fixed $i \le k$, we already have a chain of fields

$$\mathbb{Q} = K_0 \subseteq K_1 \subseteq K_2 \subseteq \cdots \subseteq K_s \subseteq \mathbb{C}$$

such that each $[K_j : K_{j-1}] = 2$ and all the elements $z_0, z_1, \ldots, z_{i-1}$ belong to K_s. Consider the possible cases for z_i. If z_i is the sum or product of two preceding elements in the sequence, or if z_i is the additive inverse or multiplicative inverse of a preceding element, then z_i is in K_s because K_s is a subfield of $\mathbb{C}$. The only other possibility is that $z_i^2 = z_r$ for some $r < i$. If z_i happens to belong to K_s already, we can keep the current chain of fields. Otherwise, the example at the end of the last section applies to show that $K_{s+1} = K_s(z_i)$ is a field extension of K_s that contains z_i and satisfies $[K_{s+1} : K_s] = 2$. Thus, the recursive construction of the chain of fields can continue. Proceeding in this way taing $i = 2, 3, \ldots, k$, we eventually obtain a square root tower for z, so that $z \in \mathrm{SQC}$.

To prove $\mathrm{SQC} \subseteq \mathrm{AC}$, we need one final lemma about field extensions. Suppose $F \subseteq K$ are subfields of $\mathbb{C}$ such that $[K : F] = 2$. We assert that there is $w \in K$ such that $w \notin F$, $w^2 \in F$, $K = F(w)$, and $(1, w)$ is an ordered basis for the F-vector space K. Start by picking any $v \in K$ with $v \notin F$. We have a chain of fields $F \subseteq F(v) \subseteq K$, so the Degree Formula gives $2 = [K : F] = [K : F(v)][F(v) : F]$. Since $v \notin F$, the field $F(v)$ properly contains F, which forces $[F(v) : F] = 2$, $[K : F(v)] = 1$, and hence $K = F(v)$. The minimal polynomial of v exists and has degree $2 = [F(v) : F]$. Let this minimal polynomial be $x^2 + bx + c$ with $b, c \in F$, so $v^2 + bv + c = 0_K$. Completing the square gives $(v + b/2)^2 = b^2/4 - c \in F$. Let $w = v + b/2$. We have $w \in K$, $w^2 \in F$, and $w \notin F$, since otherwise $v = w - b/2$ would be in F. The same reasoning used for v shows that $K = F(w)$ and $[F(w) : F] = 2$. By the Theorem on a Basis of $F(z)$, $(1, w)$ is an F-basis of K, and every element of K can be written uniquely in the form $a + dw$ for some $a, d \in F$.

Now suppose $z \in \mathrm{SQC}$, and let

$$\mathbb{Q} = K_0 \subseteq K_1 \subseteq K_2 \subseteq \cdots \subseteq K_t \subseteq \mathbb{C}$$

be a square root tower for z. We show by induction on j that $K_j \subseteq \mathrm{AC}$ for $0 \le j \le t$, which implies $z \in \mathrm{AC}$ since $z \in K_t$. First consider $K_0 = \mathbb{Q}$. Since $0 \in \mathrm{AC}$, $1 \in \mathrm{AC}$, and AC is closed under addition, every positive integer n is in AC. So all integers are in AC by

A2, hence all rational numbers are in AC by A3 and A4. For the induction step, assume $0 \le j < t$ and $K_j \subseteq$ AC. Use the lemma to get $w \in K_{j+1}$ such that $w^2 \in K_j$ and every element of K_{j+1} has the form $a + dw$ for some $a, d \in K_j$. Now $w^2 \in K_j \subseteq$ AC, so A5 shows that $w \in$ AC. Then the induction hypothesis, A1, and A3 show that $a + dw \in$ AC for all $a, d \in K_j$. So $K_{j+1} \subseteq$ AC, completing the induction.

12.9 Impossibility of Geometric Construction Problems

In this section, we apply (12.2) to prove that the problems of duplicating a cube, squaring a circle, inscribing a regular heptagon in a circle, and trisecting a 60° angle are unsolvable with ruler and compass. To use (12.2), we compute $[\mathbb{Q}(z) : \mathbb{Q}]$ by finding the degree of the minimal polynomial of z over $\mathbb{Q}$. To check that a given polynomial $m \in \mathbb{Q}[x]$ is the minimal polynomial of a given $z \in \mathbb{C}$, it suffices to verify that m is monic, $m(z) = 0$, and m is irreducible in $\mathbb{Q}[x]$. The following results (proved in Chapter 3) can help establish the irreducibility of m.

First, given $m \in \mathbb{Q}[x]$, there is a nonzero $c \in \mathbb{Q}$ with $cm \in \mathbb{Z}[x]$, and m is irreducible in $\mathbb{Q}[x]$ iff cm is irreducible in $\mathbb{Q}[x]$. So we need only test irreducibility of polynomials with integer coefficients. If the new polynomial is $cm = a_n x^n + \cdots + a_1 x + a_0$ (where each $a_i \in \mathbb{Z}$), the Rational Root Theorem states that any rational root of cm (or of m) must have the form p/q, where p divides a_0 and q divides a_n in $\mathbb{Z}$. Thus, we can find all rational roots of the polynomial by checking finitely many possibilities. A polynomial of degree 2 or 3 in $\mathbb{Q}[x]$ with no rational roots is automatically irreducible in $\mathbb{Q}[x]$.

Another criterion for irreducibility in $\mathbb{Q}[x]$ involves reduction of the coefficients modulo a prime p. Suppose $cm = a_n x^n + \cdots + a_1 x + a_0 \in \mathbb{Z}[x]$ where a_n is not divisible by p. Consider the polynomial $q = (a_n \bmod p)x^n + \cdots + (a_1 \bmod p)x + (a_0 \bmod p) \in \mathbb{Z}_p[x]$ obtained by reducing each integer coefficient modulo p. If the polynomial q is irreducible in $\mathbb{Z}_p[x]$ for some prime p, then the original polynomial cm is irreducible in $\mathbb{Q}[x]$. Irreducibility in the polynomial ring $\mathbb{Z}_p[x]$ can be checked algorithmically since there are only finitely many possible factors of q. But this sufficient condition for irreducibility is not a necessary condition: there exist irreducible polynomials in $\mathbb{Q}[x]$ whose reductions modulo every prime are reducible in $\mathbb{Z}_p[x]$. There is an algorithm (due to Kronecker) for testing in finitely many steps whether an arbitrary polynomial in $\mathbb{Z}[x]$ is irreducible in $\mathbb{Q}[x]$; see §3.19.

We can now analyze the construction problems mentioned at the beginning of this section. First consider duplication of the cube. Given a cube with side length 1, the cube with twice the volume has side length $\sqrt[3]{2}$. You can check that if there exist $P, Q \in$ GC such that the line segment $\overline{PQ}$ has length $\sqrt[3]{2}$, then the real number $\sqrt[3]{2}$ must also belong to GC. The minimal polynomial of $\sqrt[3]{2}$ is $x^3 - 2 \in \mathbb{Q}[x]$. This polynomial is irreducible in $\mathbb{Q}[x]$ since it has degree 3 and has no rational roots (noting that ± 1, ± 2 are not roots). Observe that $[\mathbb{Q}(\sqrt[3]{2}) : \mathbb{Q}] = \deg(x^3 - 2) = 3$ is not a power of 2. By (12.2), $\sqrt[3]{2}$ is not in GC.

Next, consider the problem of trisecting an angle of measure $\theta = 60°$. We can assume that the angle in question is formed by the x-axis and the line $y = x\sqrt{3}$ passing through $(\cos 60°, \sin 60°)$. If this angle could be trisected, then $(\cos 20°, \sin 20°) \in$ GC, so (by CR5) the real number $\alpha = \cos 20°$ would be in GC. We compute the minimal polynomial of α in $\mathbb{Q}[x]$ as follows. First, repeated use of the trigonometric addition formulas leads to the identity $\cos(3t) = 4(\cos t)^3 - 3\cos t$. Letting $t = 20°$, we see that $1/2 = 4\alpha^3 - 3\alpha$. It follows that α is a root of the monic polynomial $x^3 - (3/4)x - 1/8 \in \mathbb{Q}[x]$, and α is a root of the rescaled polynomial $8x^3 - 6x - 1$ with integer coefficients. The only possible rational roots of these polynomials are ± 1, $\pm 1/2$, $\pm 1/4$, and $\pm 1/8$, but none of these are roots.

Since these polynomials have degree 3 with no roots in $\mathbb{Q}$, they are irreducible in $\mathbb{Q}[x]$. So $x^3 - (3/4)x - 1/8$ is the minimal polynomial of α over $\mathbb{Q}$, $[\mathbb{Q}(\alpha) : \mathbb{Q}] = 3$, and hence $\cos 20° \notin$ GC by (12.2).

Next, consider the problem of inscribing a regular heptagon (seven-sided polygon) in the unit circle. If one vertex of the heptagon is $(1, 0)$, an adjacent vertex is $(\cos(2\pi/7), \sin(2\pi/7)) = e^{2\pi i/7}$. So we must show that $\omega = e^{2\pi i/7} \notin$ GC. What is the minimal polynomial of ω in $\mathbb{Q}[x]$? Note that ω is a root of the polynomial $x^7 - 1$. But this polynomial is reducible, with factorization

$$x^7 - 1 = (x - 1)(x^6 + x^5 + x^4 + x^3 + x^2 + x + 1).$$

Since ω is not a root of $x - 1$, it must be a root of $m = x^6 + x^5 + \cdots + x + 1 \in \mathbb{Z}[x]$. You can check that the reduction of m in $\mathbb{Z}_3[x]$ is irreducible, and hence m is irreducible in $\mathbb{Q}[x]$. (More generally, for any prime p, it can be shown that $1 + x + x^2 + \cdots + x^{p-1}$ is irreducible in $\mathbb{Q}[x]$; see Exercise 83 in Chapter 3.) Thus, m is the minimal polynomial of ω. Since m has degree 6, $[\mathbb{Q}(\omega) : \mathbb{Q}] = 6$. This degree is not a power of 2, so ω is not geometrically constructible.

Finally, consider the notorious problem of squaring the circle. A circle with radius 1 has area π, so the square with the same area would have side length $\sqrt{\pi}$. If $\sqrt{\pi}$ were in GC, we would also have $\pi \in$ GC by CR8. But a hard theorem of Lindemann states that π is not algebraic over $\mathbb{Q}$. By the Theorem on Algebraic Elements and Finite Degree Extensions, we see that $[\mathbb{Q}(\pi) : \mathbb{Q}] = \infty$. Since this degree is not a finite power of 2, (12.2) applies to show that $\sqrt{\pi}$ and π are not in GC. We do not prove Lindemann's theorem here; the interested reader may find a proof in [32, Vol. 1, Chapter 4] or [23, §1.7].

12.10 Constructibility of a Regular 17-Sided Polygon

In this section, we show that a regular 17-sided polygon can be constructed using a ruler and compass. It suffices to show that $\omega = e^{2\pi i/17}$ is in GC, since once the line segment from 1 to ω is available, the other sides of the polygon can be readily constructed. Let $m = x^{16} + x^{15} + \cdots + x + 1 \in \mathbb{Q}[x]$. By Kronecker's Algorithm, or by reduction mod 3, or by Exercise 83 in Chapter 3, we can verify that m is irreducible in $\mathbb{Q}[x]$. Moreover, $x^{17} - 1 = (x - 1)m$, from which it follows that m is the minimal polynomial of ω over $\mathbb{Q}$. Thus, $[\mathbb{Q}(\omega) : \mathbb{Q}] = \deg(m) = 16 = 2^4$.

Unfortunately, this information is not enough by itself to conclude that $\omega \in$ GC, since the converse of (12.1) is not always true. So we adopt a different approach. Let $\theta = 2\pi/17$. We develop an explicit formula for $\cos \theta$ to show that $\cos \theta \in$ AC. Since AC = GC, we can construct $\cos \theta$ geometrically, and then we can construct $e^{2\pi i/17}$ by drawing the altitude to the real axis through the point $(\cos \theta, 0)$.

We make repeated use of the following observation. Suppose r_1 and r_2 are complex numbers with $r_1 + r_2 = b$ and $r_1 r_2 = c$. Then $(x - r_1)(x - r_2) = x^2 - bx + c$, and by the Quadratic Formula, $r_1, r_2 = (b \pm \sqrt{b^2 - 4c})/2$.

For each integer $j \geq 1$, we have $\omega^j = e^{2\pi i j/17} = \cos(j\theta) + i \sin(j\theta)$ and $\omega^{-j} = \cos(-j\theta) + i \sin(-j\theta) = \cos(j\theta) - i \sin(j\theta)$, hence $\omega^j + \omega^{-j} = 2\cos(j\theta)$. Consider the real numbers

$$x_1 = 2(\cos \theta + \cos(2\theta) + \cos(4\theta) + \cos(8\theta)), \quad x_2 = 2(\cos(3\theta) + \cos(5\theta) + \cos(6\theta) + \cos(7\theta)).$$

The previous remark shows that the sum $x_1 + x_2$ is

$$\omega + \omega^{-1} + \omega^2 + \omega^{-2} + \omega^4 + \omega^{-4} + \omega^8 + \omega^{-8} + \omega^3 + \omega^{-3} + \omega^5 + \omega^{-5} + \omega^6 + \omega^{-6} + \omega^7 + \omega^{-7}.$$

Since $\omega^{17} = 1$, we have $\omega^{-j} = \omega^{17-j}$, so we can also write the sum as

$$x_1 + x_2 = \sum_{i=1}^{16} \omega^i = m(\omega) - 1 = -1.$$

Next, recall the trigonometric identities $\cos(u+v)+\cos(u-v) = 2\cos u \cos v$ and $\cos(-u) = \cos u$. Also, for $k > 8$, $\cos(k\theta) = \cos(-k\theta) = \cos((17-k)\theta)$ since $17\theta = 2\pi$. Using these identities along with the distributive law, we find that

$$\begin{aligned}
x_1 x_2 = {} & 2(\cos(4\theta) + \cos(2\theta) + \cos(6\theta) + \cos(4\theta) + \cos(7\theta) + \cos(5\theta) + \cos(8\theta) + \cos(6\theta) \\
& + \cos(5\theta) + \cos(1\theta) + \cos(7\theta) + \cos(3\theta) + \cos(8\theta) + \cos(4\theta) + \cos(8\theta) + \cos(5\theta) \\
& + \cos(7\theta) + \cos(1\theta) + \cos(8\theta) + \cos(1\theta) + \cos(7\theta) + \cos(2\theta) + \cos(6\theta) + \cos(3\theta) \\
& + \cos(6\theta) + \cos(5\theta) + \cos(4\theta) + \cos(3\theta) + \cos(3\theta) + \cos(2\theta) + \cos(2\theta) + \cos(1\theta)).
\end{aligned}$$

Gathering terms, we discover that $x_1 x_2 = 4(x_1 + x_2) = -4$. By our initial observation, $x_1, x_2 = (-1 \pm \sqrt{17})/2$. Now, $x_1 \approx 1.56155 > 0$ and $x_2 \approx -2.56155 < 0$, so we conclude:

$$x_1 = \frac{-1 + \sqrt{17}}{2}, \qquad x_2 = \frac{-1 - \sqrt{17}}{2}.$$

The next step is to consider the real numbers

$$y_1 = 2(\cos\theta + \cos(4\theta)), \quad y_2 = 2(\cos(2\theta) + \cos(8\theta)),$$

$$y_3 = 2(\cos(3\theta) + \cos(5\theta)), \quad y_4 = 2(\cos(6\theta) + \cos(7\theta)).$$

We see that $y_1 + y_2 = x_1$ and, by the same trigonometric identities used above,

$$y_1 y_2 = 2(\cos(3\theta) + \cos(1\theta) + \cos(8\theta) + \cos(7\theta) + \cos(6\theta) + \cos(2\theta) + \cos(5\theta) + \cos(4\theta)).$$

So $y_1 y_2 = x_1 + x_2 = -1$. The initial observation applies again to give

$$y_1 = \frac{x_1 + \sqrt{x_1^2 + 4}}{2}, \quad y_2 = \frac{x_1 - \sqrt{x_1^2 + 4}}{2},$$

where the signs were found by comparison to decimal approximations of y_1 and y_2. Similarly, $y_3 + y_4 = x_2$ and

$$y_3 y_4 = 2(\cos(8\theta) + \cos(3\theta) + \cos(7\theta) + \cos(4\theta) + \cos(6\theta) + \cos(1\theta) + \cos(5\theta) + \cos(2\theta)).$$

So $y_3 y_4 = x_1 + x_2 = -1$, and hence

$$y_3 = \frac{x_2 + \sqrt{x_2^2 + 4}}{2}, \quad y_4 = \frac{x_2 - \sqrt{x_2^2 + 4}}{2}.$$

To finish, consider $z_1 = 2\cos\theta$ and $z_2 = 2\cos(4\theta)$. We have $z_1 + z_2 = y_1$ and $z_1 z_2 = 2\cos(5\theta) + 2\cos(3\theta) = y_3$. One final application of the initial observation gives

$$z_1 = \frac{y_1 + \sqrt{y_1^2 - 4y_3}}{2},$$

where the sign may be determined by noting that $z_1 > z_2 > 0$. We see by inspection of the formulas that $x_1, x_2 \in$ AC, hence each $y_i \in$ AC, hence $z_1 \in$ AC, hence finally $\cos(2\pi/17) = z_1/2 \in$ AC $=$ GC. More explicitly, if we substitute formulas for y_1, etc., into the formula for z_1 and simplify, we find that $\cos(2\pi/17)$ is exactly equal to

$$\frac{-1 + \sqrt{17} + \sqrt{34 - 2\sqrt{17}} + \sqrt{\left(-1 + \sqrt{17} + \sqrt{34 - 2\sqrt{17}}\right)^2 + 16\left(1 + \sqrt{17} - \sqrt{34 + 2\sqrt{17}}\right)}}{16},$$

and this expression is obviously in AC!

12.11   Overview of Solvability by Radicals

The recursive definition of AC allows us to construct new numbers by using addition, subtraction, multiplication, division, and the extraction of square roots. If we expand this definition to allow the extraction of nth roots for any $n \geq 2$, we can obtain even more numbers. This leads to the following recursive definition of RC, the set of *radically constructible* complex numbers.

R0. $0 \in \mathrm{RC}$ and $1 \in \mathrm{RC}$.

R1.   If $a, b \in \mathrm{RC}$ then $a + b \in \mathrm{RC}$.

R2.   If $a \in \mathrm{RC}$ then $-a \in \mathrm{RC}$.

R3.   If $a, b \in \mathrm{RC}$ then $a \cdot b \in \mathrm{RC}$.

R4.   If $a \in \mathrm{RC}$ is nonzero, then $a^{-1} \in \mathrm{RC}$.

R5.   If $a \in \mathrm{RC}$, $b \in \mathbb{C}$, $n \geq 2$ is an integer, and $b^n = a$, then $b \in \mathrm{RC}$.

R6.   The only numbers in RC are those that can be obtained by applying rules R0 through R5 a finite number of times.

Equivalently, $z \in \mathrm{RC}$ iff there is a finite sequence

$$z_0 = 0, z_1 = 1, z_2, \ldots, z_k = z,$$

where $k \geq 0$ and for all i with $2 \leq i \leq k$, either z_i is the sum or product of two preceding numbers in the sequence, or z_i is the additive inverse, multiplicative inverse, or nth root of some preceding number in the sequence. Such a sequence is called a *radical construction sequence* for z.

For example, consider a quadratic polynomial $ax^2 + bx + c$ with $a, b, c \in \mathrm{RC}$ and $a \neq 0$. The roots of this polynomial are $(-b \pm \sqrt{b^2 - 4ac})/(2a)$, which are also in RC. Next, consider a reduced cubic polynomial $y^3 + qy + r$ where $q, r \in \mathrm{RC}$ ("reduced" means the coefficient of y^2 is 0). The *Cubic Formula* states that the roots of this polynomial have the form $u - q/(3u)$ where u is any complex number satisfying $u^3 = (-r + \sqrt{r^2 + 4q^3/27})/2$. It follows that these roots are also in RC. Similarly, the *Quartic Formula* can be used to show that the roots of a fourth-degree polynomial with coefficients in RC are also in RC. However, there exist fifth-degree polynomials in $\mathbb{Z}[x]$ whose roots are not in RC. This is the famous theorem that quintic (degree 5) polynomials are not always solvable by radicals.

We only have space here to give an outline of the proof strategy. The first step is to give a field-theoretic rephrasing of the definition of RC, which is analogous to the result AC = SQC proved above. Specifically, for a prime p, let us say that a field extension $F \subseteq K$ is *pure* of type p if there exists $w \in K$ with $K = F(w)$ and $w^p \in F$. Informally, we obtain K from F by adjoining a pth root of an element of F. It can be shown that $z \in \mathrm{RC}$ iff there is a chain of fields

$$\mathbb{Q} = K_0 \subseteq K_1 \subseteq K_2 \subseteq \cdots \subseteq K_t \subseteq \mathbb{C}$$

where $z \in K_t$ and each K_j is a pure extension of K_{j-1}.

The second step is to use *Galois theory* to convert a chain of fields of the type just described to a chain of finite groups

$$G = G_0 \supseteq G_1 \supseteq G_2 \supseteq \cdots \supseteq G_t = \{e\}$$

in which G_i is a normal subgroup of G_{i-1} for each $i \geq 1$, and G_{i-1}/G_i is a cyclic group of size p. Not all finite groups G possess chains of subgroups satisfying these conditions; those groups that do are called *solvable*. Finally, with each polynomial in $\mathbb{Q}[x]$ we can associate a field extension of $\mathbb{Q}$ (the *splitting field* of the polynomial) and a finite group (the *Galois group* of the polynomial). We can construct degree 5 polynomials whose associated Galois groups are not solvable, and this can be combined with the preceding results to show that the roots of these polynomials cannot all belong to RC. For more information, consult the texts on Galois theory listed in the section on suggested further reading.

12.12 Summary

1. *Geometrically Constructible Numbers.* GC is the set of all points P in the complex plane for which there is a sequence $P_0 = 0, P_1 = 1, P_2, \ldots, P_k = P$ such that each P_j $(j \geq 2)$ is an intersection point of two lines and/or circles determined by preceding points in the sequence.

2. *Arithmetically Constructible Numbers.* AC is the set of all complex numbers z for which there is a sequence $z_0 = 0, z_1 = 1, z_2, \ldots, z_k = z$ such that each z_j $(j \geq 2)$ is the sum, additive inverse, product, multiplicative inverse, or square root of preceding number(s) in the sequence.

3. *Field Extensions.* Given a field K and subfield F, the degree $[K : F]$ is the dimension of K viewed as a vector space over F. For $z \in K$, the subfield $F(z)$ is the intersection of all subfields of K containing F and z. Given a chain of fields $F_0 \subseteq F_1 \subseteq F_2 \subseteq \cdots \subseteq F_n$, the Degree Formula says $[F_n : F_0] = \prod_{i=1}^{n}[F_i : F_{i-1}]$.

4. *Square Root Towers.* SQC is the set of all complex numbers z for which there is a chain of fields $\mathbb{Q} = K_0 \subseteq K_1 \subseteq \cdots \subseteq K_t \subseteq \mathbb{C}$ with $z \in K_t$ and $[K_i : K_{i-1}] = 2$ for all $i \geq 1$. We have $K_i = K_{i-1}(w)$ for some $w \in K_i$ such that $w^2 \in K_{i-1}$, and $K_i = \{a + bw : a, b \in K_{i-1}\}$. These facts are used to show SQC = AC.

5. *Constructibility Theorem.* GC = AC = SQC. For a point $z \in \mathbb{C}$ to be constructible by ruler and compass, it is necessary (but not always sufficient) that $[\mathbb{Q}(z) : \mathbb{Q}]$ be a finite power of 2. In particular, $\sqrt[3]{2}$, π, $\cos 20°$, and $e^{2\pi i/7}$ are not constructible, but $e^{2\pi i/17}$ is constructible.

6. *Algebraic Elements and Minimal Polynomials.* Given a field K, a subfield F, and $z \in K$, z is algebraic over F iff z is the root of some nonzero polynomial in $F[x]$; this holds iff $[F(z) : F]$ is finite. In this case, there exists a unique monic irreducible polynomial $m \in F[x]$ having z as a root, called the minimal polynomial of z over F. Letting $d = \deg(m)$, $(1, z, z^2, \ldots, z^{d-1})$ is an ordered F-basis of $F(z)$, so that $[F(z) : F] = d$, the degree of the minimal polynomial of z.

7. *Proof Idea for AC $\subseteq$ GC.* AC is contained in GC because it is possible to simulate each arithmetical operation $+, -, \times, \div, \sqrt{\ }$ by geometrical constructions. For example, if we draw the circle with center $(x-1)/2$ passing through $x \in \mathbb{R}_{>0}$, this circle hits the positive imaginary axis at the point $i\sqrt{x}$. Similar triangles can be used to implement real multiplication and division. Altitudes and parallels can be used to transfer distances and thereby perform real addition and subtraction. Complex arithmetic can be reduced to real arithmetic on real and imaginary parts, and complex square roots of $re^{i\theta}$ can be found by taking the square root of the real modulus r and bisecting the angle θ.

8. *Proof Idea for GC $\subseteq$ AC.* GC is contained in AC because the coordinates of the intersection points of two lines and/or circles can be calculated from the coordinates of points determining those lines and circles by arithmetic operations and square root extractions.

12.13   Exercises

1. Give explicit arithmetic construction sequences for each complex number.
   (a) 5   (b) $-2 + i$ (c) $\sqrt{7}$ (d) $2e^{5\pi i/6}$

2. Give explicit geometric construction sequences for each complex number.
   (a) 5   (b) $-2 + i$ (c) $\sqrt{7}$ (d) $2e^{5\pi i/6}$

3. Give explicit square root towers for each complex number.
   (a) 5   (b) $-2 + i$ (c) $\sqrt{7}$ (d) $2e^{5\pi i/6}$

4. Find all $z \in \mathbb{C}$ having geometric construction sequences of length at most 3.

5. Find all $z \in \mathbb{C}$ that have geometric construction sequences of length at most 4.

6. Find all $z \in \mathbb{C}$ having arithmetic construction sequences of length at most 4.

7. Find all $z \in \mathbb{C}$ that have arithmetic construction sequences of length at most 5.

8. Use the definition of AC to prove in detail that $\mathbb{Q} \subseteq$ AC.

9. Use the definition of AC to prove that AC is a subfield of $\mathbb{C}$.

10. Let n be a positive integer. Prove $0, 1, 2, \ldots, n$ is a geometric construction sequence for n. Prove if $P_0 = 0, P_1 = 1, P_2, \ldots, P_k = n$ is any geometric construction sequence for n, then $P_0, \ldots, P_k, 2n$ is a geometric construction sequence for $2n$. Prove if $P_0 = 0, P_1 = 1, P_2 = -1, \ldots, P_k = n$ is a geometric construction sequence for n, then $P_0, \ldots, P_k, 2n + 1$ is a geometric construction sequence for $2n + 1$.

11. Find a geometric construction sequence for 99 of length 9.

12. Prove: if $n \in \mathbb{Z}_{>0}$ is k bits long when written in base 2, then n has a geometric construction sequence of length $k + 2$.

13. Prove: for all positive integers n that are k bits long when written in base 2, n has an arithmetic construction sequence of length at most $2k$.

14. (a) Let $a, b \in$ AC have arithmetic construction sequences $x_0, \ldots, x_k = a$ and $y_0, \ldots, y_m = b$. Prove that $x_0, \ldots, x_k, y_0, \ldots, y_m, a + b$ is an arithmetic construction sequence for $a + b$. (b) Prove that the recursive definition of AC (using rules A0 through A6) is logically equivalent to the iterative definition of AC (using arithmetic construction sequences).

15. Let $\theta \in \mathbb{R}$. (a) Give a geometric proof that $\cos \theta \in$ GC iff $\sin \theta \in$ GC iff $e^{i\theta} \in$ GC.
    (b) Give an arithmetic proof that $\cos \theta \in$ AC iff $\sin \theta \in$ AC iff $e^{i\theta} \in$ AC.
    (c) Prove or disprove: for all θ, if $e^{i\theta} \in$ AC then $\theta \in$ AC.

16. Let F be a subfield of K. Prove that K is an F-vector space by writing the field axioms for K and comparing them to the axioms for a vector space over F.

17. (a) Let K be a field with subfield F. Prove that $F = K$ iff $[K : F] = 1$. (b) Give an example of a chain of fields $F \subseteq K \subseteq E$ where $[E : F] = [K : F]$, yet $K \neq E$.

18. Let $F = \{a + b\sqrt{3} + c\sqrt{5} + d\sqrt{15} : a, b, c, d \in \mathbb{Q}\}$. Prove F is a subfield of $\mathbb{C}$. Compute $[F : \mathbb{Q}]$, $[F : \mathbb{Q}(\sqrt{3})]$, and $[F : \mathbb{Q}(\sqrt{5})]$, and find ordered bases for each of these vector spaces. Find five different subfields of F, and prove carefully that they are all different.

19. Let $F \subseteq E \subseteq K$ be a chain of fields where $[K : F]$ is prime. Prove $E = F$ or $E = K$.

20. (a) Given a chain of fields $F \subseteq K \subseteq E$, prove that $[E : F]$ is finite iff $[E : K]$ is finite and $[K : F]$ is finite. (b) Generalize (a) to a chain of more than 3 fields.

21. **Field Extension Generated by a Set.** Given a field K, a subfield F, and a subset S of K, define $F(S)$ to be the intersection of all subfields of K containing $F \cup S$.
    (a) Prove $F(S)$ is a subfield of K, $F \subseteq F(S)$, and $S \subseteq F(S)$.
    (b) Prove: for any subfield M of K, $F(S) \subseteq M$ iff $F \subseteq M$ and $S \subseteq M$.
    (c) Prove: for all subsets S, T of K, $F(S)(T) = F(T)(S) = F(S \cup T)$.
    (d) Prove: for all subsets S, T of K, $F(S) = F(T)$ iff $S \subseteq F(T)$ and $T \subseteq F(S)$.

22. Give a specific example of a field K, a subfield F, and two disjoint nonempty subsets S and T of $K \setminus F$ with $|S| \neq |T|$ and yet $F(S) = F(T)$.

23. Let K be a field. (a) Prove that the set $\{n.1_K : n \in \mathbb{Z}\}$ of additive multiples of 1 in K is a subring of K that is isomorphic either to $\mathbb{Z}$ or to $\mathbb{Z}_p$ for some prime p.
    (b) Prove that the intersection of all subfields of K is a field that is isomorphic either to $\mathbb{Q}$ or to $\mathbb{Z}_p$ for some prime p.

24. Suppose K is a field, F is a subfield of K, and $z \in K$.
    (a) Prove $F(z) = \{f(z)/g(z) : f, g \in F[x] \text{ and } g(z) \neq 0\}$. (Call the right side E. To prove $F(z) \subseteq E$, show E is a subfield of K containing F and z. To prove $E \subseteq F(z)$, show every subfield M of K containing F and z must also contain E.)
    (b) Let $R = \{f(z) : f \in F[x]\}$. Prove that R is a subring of K that contains F and z. Prove that if z is algebraic over F, then $R = F(z)$.
    (c) Use the theorem that π is not algebraic over $\mathbb{Q}$ to prove that $\{f(\pi) : f \in \mathbb{Q}[x]\}$ is not a subfield of $\mathbb{C}$.

25. Describe a specific square root tower for $e^{2\pi i/17}$.

26. Suppose $x_0, y_0, x_1, y_1 \in \text{AC}$ satisfy $x_0 \neq x_1$. Let the unique line through (x_0, y_0) and (x_1, y_1) have slope m and y-intercept b for some $m, b \in \mathbb{R}$. Prove $m, b \in \text{AC}$.

27. Consider a line $y = mx + d$ and a parabola $y = ax^2 + bx + c$ where $a, b, c, d, m \in \text{AC}$. Prove that all intersection points of this line and this parabola are in AC.

28. In the proof of CR1, use theorems about triangles to confirm the assertions that E is the midpoint of $\overline{AB}$ and $\overline{CD}$ is perpendicular to $\overline{AB}$.

29. Use an actual ruler and compass to construct a regular hexagon inscribed in the unit circle with one vertex at $(1, 0)$.

30. Use an actual ruler and compass to construct a square inscribed in the unit circle with one vertex at $e^{\pi i/4}$.

31. Use an actual ruler and compass to construct: (a) the line $y = 3x + 1$; (b) the altitude to the line in (a) passing through $(2, 0)$; (c) the line parallel to the line in (a) passing through $(2, 0)$.

32. Use an actual ruler and compass to construct: (a) the real numbers $x = \sqrt{6}$ and $y = -3/2$; (b) $x + y$; (c) $x - y$; (d) xy; (e) $1/x$.

33. Use an actual ruler and compass to construct: (a) $u = 2 + i$ and $v = 3 - 2i$; (b) $u + v$; (c) $u - v$; (d) uv; (e) the two square roots of u; (f) $1/v$.

34. Prove geometrically: if P and Q are in GC and r is the length of the line segment $\overline{PQ}$, then the real number $r = (r, 0)$ is in GC. Conclude that the collapsing compass and ruler can simulate a real compass's ability to transfer a fixed distance from one part of a diagram to another.

35. Given an algebraic proof of the preceding exercise using the fact that GC = AC.

36. Prove geometrically: for all $\alpha, \beta \in \mathbb{R}$, if $e^{i\alpha} \in \mathrm{GC}$ and $e^{i\beta} \in \mathrm{GC}$, then $e^{i(\beta - \alpha)} \in \mathrm{GC}$. Conclude that when deciding the constructibility of regular n-gons inscribed in the unit circle, it suffices to consider n-gons with one vertex at $(1, 0)$.

37. Given an algebraic proof of the preceding exercise using the fact that $\mathrm{GC} = \mathrm{AC}$.

38. In CR11, we proved that for all complex $u, v \in \mathrm{GC}$, $u + v \in \mathrm{GC}$ by looking at real and imaginary parts. Give an alternate geometric proof of this result by using the parallelogram law for finding the vector sum of u and v in $\mathbb{R}^2$.

39. Prove geometrically that $u, v \in \mathrm{GC}$ implies $uv \in \mathrm{GC}$ by using the polar multiplication formula $(re^{i\alpha}) \cdot (se^{i\beta}) = (rs)e^{i(\alpha + \beta)}$.

40. Given two points P and Q, show how to use a ruler and compass to construct points R and S that trisect the line segment $\overline{PQ}$. Illustrate your construction with an actual ruler and compass.

41. Consider points $A = e^{\pi i/3}$, $B = 0$, and $C = 1$, so $\angle ABC$ is a 60° angle. As shown in the previous exercise, we can find points D and E that trisect the line segment $\overline{AC}$. Write D and E in the form $a + bi$ with $a, b \in \mathbb{R}$. Use this to approximate the measure of the angle $\angle EBC$ in degrees. Has this construction trisected $\angle ABC$?

42. Starting with the points 0 and 1, prove that a compass alone can be used to construct all points in the set $\{a + be^{2\pi i/6} : a, b \in \mathbb{Z}\}$. (Remarkably, it can be shown that *every point in* GC *can be constructed with only a compass*, starting from the two points 0 and 1.)

43. Find (with proof) the minimal polynomials over $\mathbb{Q}$ of each complex number.
    (a) $22/7$ (b) $2i$ (c) $\sqrt{11}$ (d) $\sqrt{3} + \sqrt{7}$ (e) $\sqrt[5]{2}$ (f) $\sqrt{2} + i$ (g) $e^{2\pi i/6}$ (h) $e^{2\pi i/12}$

44. Let F be a field and g be an irreducible polynomial in the polynomial ring $F[x]$. Let $I = F[x]g$ be the principal ideal in $F[x]$ generated by g. Let K be the quotient ring $F[x]/I$. (See Chapter 1 for the relevant definitions.)
    (a) Prove K is a field having a subfield $F_1 = \{c + I : c \in F\}$ isomorphic to F.
    (b) Use the isomorphism between F and F_1 to convert g to a polynomial $g_1 \in F_1[x]$. Prove $g_1(x + I) = 0 + I$, so the coset $x + I$ is a root of g_1 in the field K.
    (c) Suppose F is a subfield of $\mathbb{C}$, and $g \in F[x]$ is the minimal polynomial over F of some $z \in \mathbb{C}$. Prove that the field K in part (a) is isomorphic to the subfield $F(z)$ of $\mathbb{C}$. (Apply the Fundamental Homomorphism Theorem for Rings to an appropriate map.)

45. Construct a specific field K with 343 elements containing an element z with $z^3 = 2$ and containing a subfield isomorphic to $\mathbb{Z}_7$. (Use the previous exercise.)

46. Consider the polynomial $x^4 + x^3 + x^2 + x + 1 = (x^5 - 1)/(x - 1)$. (a) Find the complex roots of this polynomial in terms of cosines, sines, and/or complex exponentials. (b) Divide the polynomial by x^2 and let $y = x + x^{-1}$. Convert the quartic polynomial in x to a quadratic polynomial in y, and then use the Quadratic Formula to solve for y and then x algebraically. (c) Use (a) and (b) to prove $\cos 72° = \frac{-1 + \sqrt{5}}{4}$ and $\sin 72° = \frac{1}{2}\sqrt{\frac{5 + \sqrt{5}}{2}}$, and deduce that $e^{2\pi i/5} \in \mathrm{GC}$.

47. Use an actual ruler and compass to construct a regular pentagon inscribed in the unit circle.

48. Let z be the unique real root of the polynomial $x^3 + 2x + 1 \in \mathbb{Q}[x]$.
    (a) Find an ordered $\mathbb{Q}$-basis of $\mathbb{Q}(z)$ and compute $[\mathbb{Q}(z) : \mathbb{Q}]$.
    (b) For $-1 \le n \le 7$, express z^n as a $\mathbb{Q}$-linear combination of the basis in (a).
    (c) Compute $(z^2 - 3z + 2) \cdot (2z^2 + 5z - 1)$ in terms of the basis in (a).
    (d) Find $(z^2 - z + 1)^{-1}$ in terms of the basis in (a).

49. Let $z = e^{2\pi i/12}$.
    (a) Find an ordered $\mathbb{Q}$-basis of $\mathbb{Q}(z)$ and compute $[\mathbb{Q}(z) : \mathbb{Q}]$.
    (b) For all $n \in \mathbb{Z}$, express z^n as a $\mathbb{Q}$-linear combination of the basis in (a).
    (c) Compute $(z^3 + 2z - 5) \cdot (z^2 + 3z + 4)$ in terms of the basis in (a).
    (d) Find $(z^3 + 2z^2 + 3z + 4)^{-1}$ in terms of the basis in (a).

50. Suppose $z \in \mathbb{C}$ has a minimal polynomial over $\mathbb{Q}$ of odd degree. Prove that $x^2 - 2$ is the minimal polynomial of $\sqrt{2}$ over the subfield $\mathbb{Q}(z)$.

51. Let β be an irrational real number, and assume that for all integers $k \geq 1$ there exists a rational number $s_k = m_k/n_k$ (for some $m_k, n_k \in \mathbb{Z}$ with $n_k > 0$) satisfying $|\beta - s_k| < (kn_k^k)^{-1}$. Prove β is not algebraic over $\mathbb{Q}$, hence $\beta \notin GC$. Deduce that $\sum_{n \geq 1} 10^{-n!}$ is not algebraic over $\mathbb{Q}$.

52. (a) Prove: for all $z, w \in \mathbb{C}$, if z and w are algebraic over $\mathbb{Q}$, then $z + w$ and zw are algebraic over $\mathbb{Q}$. (Consider the chain of fields $\mathbb{Q} \subseteq \mathbb{Q}(z) \subseteq \mathbb{Q}(z)(w)$ and look at degrees.) (b) Let K be the set of all $z \in \mathbb{C}$ that are algebraic over $\mathbb{Q}$. Prove that K is a subfield of $\mathbb{C}$ containing $\mathbb{Q}$. (c) Is $[K : \mathbb{Q}]$ finite or infinite? Is K countable or uncountable?

53. Suppose K is a field with subfield F such that $[K : F] = 2$ and $1_K + 1_K \neq 0_K$. Prove there exists $w \in K$ with $K = F(w)$ and $w^2 \in F$.

54. Let $K = \{0, 1, 2, 3\}$, and define addition and multiplication in K as follows:

+	0	1	2	3		·	0	1	2	3
0	0	1	2	3		0	0	0	0	0
1	1	0	3	2		1	0	1	2	3
2	2	3	0	1		2	0	2	3	1
3	3	2	1	0		3	0	3	1	2

Prove that K is a field, $F = \{0, 1\}$ is a subfield of K, and $[K : F] = 2$. Prove there is no $w \in K$ with $K = F(w)$ and $w^2 \in F$.

55. For each n with $3 \leq n \leq 18$, decide (with proof) whether a regular n-sided polygon can be constructed with ruler and compass.

56. Prove: for all $n \geq 3$, a regular n-sided polygon is constructible with ruler and compass iff a regular $2n$-sided polygon is constructible.

57. Prove: if regular polygons with n sides and m sides are both constructible with ruler and compass and $k = \operatorname{lcm}(m, n)$, then a regular k-sided polygon is also constructible.

58. (a) Use the trigonometric addition formula for $\cos(\alpha + \beta)$ and other identities to derive the formula $\cos(3t) = 4(\cos t)^3 - 3\cos t$. (b) Similarly, prove that $\cos(u + v) + \cos(u - v) = 2\cos u \cos v$. (c) Reprove the formulas in (a) and (b) using $\cos t = (e^{it} + e^{-it})/2$ and facts about the exponential function.

59. Let $\theta \in (0, \pi/2)$ be the unique real number with $\cos \theta = 3/5$. Find the minimal polynomial of $\theta/3$. Explain how to construct $e^{i\theta}$ with a ruler and compass. Prove or disprove: the angle between the x-axis and the ray through $e^{i\theta}$ can be trisected with ruler and compass.

60. Let $\theta \in (0, \pi/2)$ be the unique real number with $\cos \theta = 11/16$. Find the minimal polynomial of $\theta/3$. Prove or disprove: the angle between the x-axis and the ray through $e^{i\theta}$ can be trisected with ruler and compass.

61. In §12.10, verify that the signs in the formulas for y_1, y_2, y_3, y_4 are correct by calculating decimal approximations for these quantities.

62. Given a cubic polynomial $x^3 + bx^2 + cx + d$, show that the substitution $y = x + b/3$ leads to a new reduced cubic $y^3 + qy + r$ where the coefficient of y^2 is 0.

63. Find a substitution of variables changing a monic degree n polynomial in x into a degree n polynomial in y where the coefficient of y^{n-1} is 0.

64. **Cubic Formula.** Given $q, r \in \mathbb{C}$, let $u \in \mathbb{C}_{\neq 0}$ satisfy $u^3 = (-r + \sqrt{r^2 + 4q^3/27})/2$. Verify by algebraic manipulations that $y = u - q/(3u)$ is a solution of the reduced cubic $y^3 + qy + r = 0$. What happens if $u = 0$?

65. **Quartic Formula.** Find a formula for the roots of a reduced quartic polynomial $y^4 + py^2 + qy + r$ as follows. Hypothesize a factorization of the form $(y^2 + ky + \ell)(y^2 - ky + m)$. Obtain equations for the unknown coefficients k, ℓ, m. Eliminate ℓ and m. Obtain a cubic equation in the variable k^2. Solve this with the Cubic Formula and deduce formulas for the roots of the original quartic. Conclude that if $p, q, r \in RC$, then all complex roots of the quartic polynomial are in RC.

66. Prove, as asserted in §12.11, that for all complex z, $z \in RC$ iff there is a chain of fields $\mathbb{Q} = K_0 \subseteq K_1 \subseteq K_2 \subseteq \cdots \subseteq K_t \subseteq \mathbb{C}$, where $z \in K_t$ and each K_j is a pure extension of K_{j-1}.

67. Prove that $\cos(2\pi/7) \in RC$ by finding an explicit algebraic formula for this number. (Use a substitution like the one in Exercise 46 to find the roots of $(x^7 - 1)/(x - 1)$.)

68. Find and justify an exact formula for $\cos 1°$ that is built up from integers using only arithmetical operations, square roots, and cube roots. (By the trigonometric addition formulas, this shows there are exact formulas for $\cos k°$ for any integer k. Curiously, grade school students are only made to memorize these formulas for a few select choices of k.)

69. True or false? Explain each answer.
    (a) The minimal polynomial of $e^{2\pi i/3}$ over $\mathbb{Q}$ is $x^3 - 1$.
    (b) GC is a subfield of $\mathbb{C}$.
    (c) AC = RC.
    (d) $e^{6\pi i/17} \in$ GC.
    (e) For all integers $n \geq 1$, there is a chain of fields $\mathbb{Q} \subseteq K \subseteq \mathbb{C}$ with $[K : \mathbb{Q}] = n$.
    (f) For all $w, z \in \mathbb{C}$, $\mathbb{Q}(w) = \mathbb{Q}(z)$ iff $w = z$.
    (g) $e^{\pi i/17} \in$ SQC.
    (h) For all fields K and subfields F with $[K : F] = 4$, there is at most one subfield E with $F \subsetneq E \subsetneq K$.
    (i) For every $k \geq 1$, there exists a sequence of distinct points $P_0, P_1, \ldots, P_k$ that is both an arithmetic construction sequence and a geometric construction sequence.
    (j) For any field K and any subfield F and any $z \in K$, if $[K : F]$ is finite then z is algebraic over F.
    (k) For all z in a field K with $\mathbb{Q} \subseteq K \subseteq \mathbb{C}$, if $[K : \mathbb{Q}] = 2$, then $z^2 \in \mathbb{Q}$.
    (l) For all real θ such that $0 < \theta < \pi/2$, if $e^{i\theta} \in$ GC then $e^{i\theta/3} \notin$ GC.

13

Dual Vector Spaces

A recurring theme in mathematics is the relationship between geometric spaces and structure-preserving functions defined on these spaces. Galois theory for field extensions, elementary algebraic geometry, and the theory of dual vector spaces can all be viewed as instances of this common theme. Given a space V and a set R of functions on V, the idea is to study the zero-set of a collection of functions in R, which is the set of points in V where all the functions in the collection are zero. Similarly, given a subset of V, we can consider the set of all functions in R that evaluate to zero at all points of the given subset. These two maps — one sending subsets of R to subsets of V, and the other sending subsets of V to subsets of R — may restrict to give a bijection (one-to-one correspondence) between certain distinguished subsets of V and certain distinguished subsets of R. In many cases, key aspects of the geometric structure of V are mirrored by the algebraic structure of R through this correspondence. This interplay between V and R allows us to obtain structural information about the properties of both V and R.

This chapter studies a fundamental and ubiquitous instance of this correspondence between spaces and functions, namely the theory of dual vector spaces. Here, the geometric space is a finite-dimensional vector space V over a field F. The collection R consists of all linear functions mapping V into F. We will see that R is also a vector space over F, called the *dual space* for V and denoted by V^*. The maps mentioned above restrict to inclusion-reversing bijections between the set of subspaces of V and the set of subspaces of V^*. Later, we discuss the connection between these results and the theory of bilinear pairings, real inner product spaces, and complex inner product spaces. After an interlude on Banach spaces, the chapter closes with a sketch of the ideal–variety correspondence, which is a cornerstone of algebraic geometry.

13.1 Vector Spaces of Linear Maps

Before beginning our study of dual spaces, we first give a general construction for manufacturing vector spaces. Let W be a vector space over a field F and S be any set. Let Z be the set of all functions $g : S \to W$. We introduce an F-vector space structure on Z as follows. Let g and h be two functions in Z. We define a new function $g + h : S \to W$ by letting $g + h$ send each x in S to $g(x) + h(x)$ in W. In symbols, $(g + h)(x) = g(x) + h(x)$ for all $x \in S$. Note that the plus symbol on the right side is the addition in the given vector space W, while the plus symbol on the left side is the new addition being defined on the set Z. It is routine to check that this new addition on Z, called *pointwise* addition of functions, satisfies the additive axioms for a vector space listeed in Table 1.2. The zero element of Z is the zero function that sends every $x \in S$ to 0_W, and the additive inverse of $f \in Z$ is the function that sends x to $-f(x)$ for all $x \in S$.

Given $g \in Z$ and $c \in F$, we define the scalar multiple $c \cdot g : S \to W$ by letting $(c \cdot g)(x) = c \cdot (g(x))$ for all $x \in S$. Here, the product on the right side is scalar multiplication in

DOI: 10.1201/9781003484561-13

the F-vector space W, and the product on the left side is the pointwise scalar multiplication being defined on Z. You can quickly verify the remaining vector space axioms for Z from Table 1.4 (Exercise 1).

Here are some special cases of this construction. If $S = \{1, 2, \ldots, n\}$ and W is the field F viewed as an F-vector space, then Z is the vector space F^n of all n-tuples of elements of F. If $S = \{1, 2, \ldots, m\} \times \{1, 2, \ldots, n\}$ and $W = F$, then Z is the vector space $M_{m,n}(F)$ of all $m \times n$ matrices with entries in F. See Chapter 4 for more details on these special cases. If the set S is itself an F-vector space V, then Z is the vector space $\text{Fun}(V, W)$ of all functions from V to W, using pointwise sum and scalar product operations.

Given F-vector spaces V and W, define $\text{Hom}_F(V, W)$ to be the set of all F-linear transformations (vector space homomorphisms) from V to W. A function $f : V \to W$ belongs to $\text{Hom}_F(V, W)$ iff $f(x + y) = f(x) + f(y)$ and $f(cx) = cf(x)$ for all $x, y \in V$ and all $c \in F$. We claim that $\text{Hom}_F(V, W)$ is an F-vector space under the pointwise operations on functions defined above. It suffices to show that $\text{Hom}_F(V, W)$ is a subspace of the vector space $\text{Fun}(V, W)$ of all functions from V to W. This follows since the zero function is F-linear, and sums and scalar multiples of F-linear functions are also F-linear (Exercise 2).

We define the *dual space* of an F-vector space V to be the vector space $V^* = \text{Hom}_F(V, F)$ of all F-linear transformations from V to F. Here, we are viewing the field F as an F-vector space where vector addition is addition in F, and scalar multiplication is the same as multiplication in the field F. We have $\dim(F) = 1$, and (1_F) is an ordered basis of F. Functions in V^* are called *linear functionals* on V.

13.2 Dual Bases

We often define linear maps on a vector space V by specifying where the map sends vectors in a given basis for V. The next theorem makes this process precise.

Universal Mapping Property (UMP) for Vector Spaces. Suppose V and W are F-vector spaces and $B = (x_1, \ldots, x_n)$ is an ordered basis for V. For any F-vector space W and any ordered list $(y_1, \ldots, y_n)$ of elements of W, there exists a unique F-linear map $T : V \to W$ such that $T(x_i) = y_i$ for all i between 1 and n.

The UMP also holds for infinite-dimensional vector spaces, with a similar proof, but we focus on the finite-dimensional case here.

Proof. Every $v \in V$ has a unique expansion as a linear combination of elements of B, say

$$v = c_1 x_1 + c_2 x_2 + \cdots + c_n x_n \qquad \text{for some } c_i \in F.$$

If $T : V \to W$ is any F-linear map sending x_i to y_i for all i, then by linearity, T must send v to $c_1 y_1 + \cdots + c_n y_n$. This observation proves that T is unique if it exists. To show existence of T, define $T(\sum_{i=1}^n c_i x_i) = \sum_{i=1}^n c_i y_i$ for any choice of $c_i \in F$, and check that T really is F-linear and sends each x_i to y_i (Exercise 6). Linear independence of B is needed to see that T is a well-defined function. This follows since $v \in V$ can be written in the form $\sum_{i=1}^n c_i x_i$ in exactly one way. $\qquad \square$

The universal mapping property can be expressed informally as follows. To define a linear map from V to W, we are free to send each basis element x_i to any element of W. After choosing the images of the basis elements, the entire map $T : V \to W$ is uniquely determined by linearity. Another way to say this is that each *function* $g : \{x_1, \ldots, x_n\} \to W$ "extends by linearity" to a unique *linear map* $T_g : V \to W$.

Define δ_{ij} be 1 for $i = j$ and 0 for $i \neq j$. The next theorem shows how to convert an ordered basis B for V to a related ordered basis B^d for V^*, called the *dual basis* of B.

Theorem on Dual Bases. Let V be a finite-dimensional F-vector space with ordered basis $B = (x_1, \ldots, x_n)$. There is a unique ordered basis $B^d = (f_1, \ldots, f_n)$ of V^* such that $f_i(x_j) = \delta_{ij}$ for all i, j between 1 and n. In particular, $\dim(V^*) = \dim(V) = n$.

Proof. For each fixed i between 1 and n, consider the function $g_i : \{x_1, \ldots, x_n\} \to F$ such that $g_i(x_i) = 1_F$ and $g_i(x_j) = 0_F$ for all $j \neq i$. This function g_i extends by linearity to give a linear map $f_i : V \to F$ satisfying

$$f_i(c_1 x_1 + \cdots + c_i x_i + \cdots + c_n x_n) = c_i \qquad \text{for all } c_1, \ldots, c_n \in F.$$

Each f_i is an element of $V^* = \operatorname{Hom}_F(V, F)$. The UMP shows that $B^d = (f_1, \ldots, f_n)$ is the unique ordered list in V^* such that $f_i(x_j) = \delta_{ij}$ for all i, j between 1 and n. To see that B^d is an ordered basis of V^*, we prove that B^d is a linearly independent list spanning V^*. To check linear independence, assume $d_1, \ldots, d_n \in F$ satisfy

$$d_1 f_1 + d_2 f_2 + \cdots + d_n f_n = 0. \tag{13.1}$$

Here, 0 is the zero function from V to F, and the operations are pointwise operations on functions. We need to prove that every $d_i = 0_F$. To do this, fix an index i, and evaluate both sides of (13.1) at the element x_i. We obtain

$$d_1 f_1(x_i) + \cdots + d_i f_i(x_i) + \cdots + d_n f_n(x_i) = 0(x_i).$$

By definition of f_j, the left side evaluates to d_i, while the right side is zero. Thus $d_i = 0_F$ for each i, so B^d is a linearly independent list.

Next, we check that B^d spans V^*. Given an arbitrary $h \in V^*$, we must express h as an F-linear combination of elements of B^d. We claim that, in the space V^*, we have

$$h = h(x_1) f_1 + h(x_2) f_2 + \cdots + h(x_n) f_n. \tag{13.2}$$

Here, each $h(x_i)$ is a scalar in F, since h maps V to F. Both sides of (13.2) are linear maps from V to F. Two linear functions with domain V are equal iff they have the same value at each basis element x_i in B. (This fact is a restatement of the uniqueness assertion in the UMP.) So it suffices to check whether the two sides of (13.2) agree at a fixed basis element x_i. The left side sends x_i to $h(x_i)$, and the right side sends x_i to

$$h(x_1) f_1(x_i) + \cdots + h(x_i) f_i(x_i) + \cdots + h(x_n) f_n(x_i) = h(x_i).$$

Thus (13.2) does hold, and B^d does span V.

Since the list B^d has the same length n as the list B, we get $\dim(V^*) = n = \dim(V)$. $\square$

Every vector space V has a basis, and any two bases of V have the same cardinality, namely the dimension $\dim(V)$. The Theorem on Dual Bases shows that in the finite-dimensional case, $\dim(V^*) = \dim(V)$. This dimension fact does not extend to infinite-dimensional vector spaces (see §13.12 and Exercise 12).

13.3   The Zero-Set Operator

Let V be an n-dimensional vector space over the field F. Our next task is to set up the correspondences between subsets of V^* and subsets of V mentioned in the introduction to this chapter. First we describe the map from subsets of V^* to subsets of V. Let S be any subset of V^*, so S is some collection of linear functionals mapping V into F. Define the *zero-set* of S to be

$$\mathcal{Z}(S) = \{x \in V : f(x) = 0_F \text{ for all } f \in S\}.$$

When $V = F^n$, we can think of $\mathcal{Z}(S)$ as the solution set of the system of homogeneous linear equations $f(x) = 0$ as f ranges over S (see Exercise 20).

Here are some properties of the zero-set operator $\mathcal{Z}$:

- For all subsets S of V^*, $\mathcal{Z}(S)$ is a subspace of V. *Proof:* First, $0_V \in \mathcal{Z}(S)$ because all maps $f \in S$ (being linear) must send 0_V to 0_F. Second, assume $x, y \in \mathcal{Z}(S)$. For any $f \in S$, $f(x+y) = f(x) + f(y) = 0 + 0 = 0$, so $x + y \in \mathcal{Z}(S)$. Third, assume $x \in \mathcal{Z}(S)$ and $c \in F$. For any $f \in S$, $f(cx) = cf(x) = c0 = 0$, so $cx \in \mathcal{Z}(S)$.

- $\mathcal{Z}$ is *inclusion-reversing:* $S \subseteq T$ in V^* implies $\mathcal{Z}(T) \subseteq \mathcal{Z}(S)$ in V. *Proof:* Assume $S \subseteq T$ and $x \in \mathcal{Z}(T)$. For all $f \in S$, f is also in T, and so $f(x) = 0$. This proves that $x \in \mathcal{Z}(S)$.

- $\mathcal{Z}(\emptyset) = V$. *Proof:* By definition, $\mathcal{Z}(\emptyset)$ is a subset of V. If it were a proper subset, then there would exist $x \in V$ with $x \notin \mathcal{Z}(\emptyset)$. This would imply the existence of $f \in \emptyset$ with $f(x) \neq 0$. This is impossible, since $\emptyset$ has no members.

- $\mathcal{Z}(V^*) = \{0_V\}$. *Proof:* First, $0_V \in \mathcal{Z}(V^*)$, since $\mathcal{Z}(V^*)$ is a subspace of V. Given $v \neq 0$ in V, we show that $v \notin \mathcal{Z}(V^*)$. We must find an $f \in V^*$ with $f(v) \neq 0$. For this, extend the one-element list (v) to an ordered basis $B = (v, v_2, \ldots, v_n)$ of V. The first function in the dual basis B^d is an element of V^* sending v to $1_F \neq 0_F$.

- If S is any subset of V^* and $W = \langle S \rangle$ is the subspace of V^* generated by S, then $\mathcal{Z}(S) = \mathcal{Z}(W)$. *Proof:* Since $S \subseteq W$, we already know $\mathcal{Z}(W) \subseteq \mathcal{Z}(S)$. Next, fix $x \in \mathcal{Z}(S)$. To prove $x \in \mathcal{Z}(W)$, fix $f \in W$. We can write $f = d_1 g_1 + \cdots + d_k g_k$ for some $d_i \in F$ and $g_i \in S$. Evaluating at x gives $f(x) = d_1 g_1(x) + \cdots + d_k g_k(x) = d_1 0 + \cdots + d_k 0 = 0$. So $x \in \mathcal{Z}(W)$.

13.4 The Annihilator Operator

Next, we describe the map from subsets of V to subsets of V^*. Let T be any subset of V, so T is a collection of vectors. Define the *annihilator* of T to be

$$\mathcal{A}(T) = \{g \in V^* : g(x) = 0_F \text{ for all } x \in T\}.$$

Here are some properties of the annihilator operator $\mathcal{A}$:

- For all subsets T of V, $\mathcal{A}(T)$ is a subspace of V^*. *Proof:* First, 0_{V^*} sends every element of V to 0_F. In particular, 0_{V^*} sends every element of T to 0_F, so $0_{V^*} \in \mathcal{A}(T)$. Second, assume $g, h \in \mathcal{A}(T)$. For any $x \in T$, $(g+h)(x) = g(x) + h(x) = 0 + 0 = 0$, so $g + h \in \mathcal{A}(T)$. Third, assume $g \in \mathcal{A}(T)$ and $c \in F$. For any $x \in T$, $(cg)(x) = c(g(x)) = c0 = 0$, so $cg \in \mathcal{A}(T)$.

- $\mathcal{A}$ is *inclusion-reversing:* $S \subseteq T$ in V implies $\mathcal{A}(T) \subseteq \mathcal{A}(S)$ in V^*. *Proof:* Assume $S \subseteq T$ and $g \in \mathcal{A}(T)$. For all $x \in S$, x is also in T, and so $g(x) = 0$. This proves that $g \in \mathcal{A}(S)$.

- $\mathcal{A}(\emptyset) = V^*$. The proof is analogous to the proof that $\mathcal{Z}(\emptyset) = V$ (Exercise 23).

- $\mathcal{A}(V) = \{0_{V^*}\}$. *Proof:* A linear functional $f : V \to F$ belongs to $\mathcal{A}(V)$ iff $f(x) = 0$ for all x in the domain V iff f is the zero function on V iff f is the zero functional in V^*.

- If T is any subset of V and $W = \langle T \rangle$ is the subspace of V generated by T, then $\mathcal{A}(T) = \mathcal{A}(W)$. *Proof:* Since $T \subseteq W$, we already know $\mathcal{A}(W) \subseteq \mathcal{A}(T)$. Next, fix $f \in \mathcal{A}(T)$. To prove $f \in \mathcal{A}(W)$, fix $x \in W$. We can write $x = c_1 x_1 + \cdots + c_k x_k$ for some $c_i \in F$ and $x_i \in T$. Linearity of f gives $f(x) = \sum_{i=1}^{k} c_i f(x_i) = \sum_{i=1}^{k} c_i 0 = 0$. So $f \in \mathcal{A}(W)$.

13.5   The Double Dual V^{**}

There is a striking similarity between the properties of $\mathcal{Z}$ and the properties of $\mathcal{A}$. This similarity is not a coincidence. To explain it, we introduce the double dual space for V. The *double dual* of V is the dual space of the vector space V^*, namely

$$V^{**} = (V^*)^* = \mathrm{Hom}_F(V^*, F) = \mathrm{Hom}_F(\mathrm{Hom}_F(V, F), F).$$

For finite-dimensional V, we know $\dim(V) = \dim(V^*) = \dim(V^{**})$.

Let us spell out the definition of V^{**} in more detail. An element of V^{**} is an F-linear function $E : V^* \to F$. The function E takes as input another F-linear function $g : V \to F$ and produces as output a scalar $E(g) \in F$. The F-linearity of E means that $E(g + h) = E(g) + E(h)$ and $E(dg) = dE(g)$ for all $g, h \in V^*$ and $d \in F$.

Elements of V^{**} may seem very abstract and difficult to visualize, since these elements are functions that operate on other functions. However, we can construct some concrete examples of elements of V^{**} as follows. Suppose x is a fixed vector in V. Define a map $E_x : V^* \to F$ by letting $E_x(g) = g(x)$ for all $g \in V^*$. In other words, E_x takes as input a linear functional $g : V \to F$ and returns as output the value of g at the particular point $x \in V$, namely $g(x)$. For this reason, the function E_x is called *evaluation at x*. To check that E_x really is F-linear, use the definitions of the pointwise operations in V^* to compute

$$E_x(g + h) = (g + h)(x) = g(x) + h(x) = E_x(g) + E_x(h) \text{ for all } g, h \in V^*;$$

$$E_x(dg) = (dg)(x) = d(g(x)) = d(E_x(g)) \text{ for all } g \in V^*, d \in F.$$

So every E_x really is an element of V^{**}.

We define a map $\mathrm{ev} : V \to V^{**}$ by setting $\mathrm{ev}(x) = E_x$ for all $x \in V$. So ev sends each vector x in V to the function "evaluation at x," which is in the double dual of V.

Theorem on Evaluation Maps. For any F-vector space V, $\mathrm{ev} : V \to V^{**}$ is a one-to-one F-linear map. If $\dim(V)$ is finite, then ev is a vector space isomorphism.

Proof. We first check that $\mathrm{ev} : V \to V^{**}$ is an F-linear map. (This is not the same as checking that $\mathrm{ev}(x) = E_x$ is F-linear for each $x \in V$, which we have already done. Do not confuse the function ev from V to V^{**} with one of its outputs $\mathrm{ev}(x)$, which is a function from V^* to F.) We first fix $x, y \in V$ and check that $\mathrm{ev}(x + y) = \mathrm{ev}(x) + \mathrm{ev}(y)$. In other words, we must prove that $E_{x+y} = E_x + E_y$ in V^{**}. These two functions from V^* to F are indeed equal, since for any $g \in V^*$,

$$E_{x+y}(g) = g(x + y) = g(x) + g(y) = E_x(g) + E_y(g) = (E_x + E_y)(g).$$

Next, fix $x \in V$ and $c \in F$. We must prove $\mathrm{ev}(cx) = c\,\mathrm{ev}(x)$, or equivalently $E_{cx} = cE_x$. For any $g \in V^*$,

$$E_{cx}(g) = g(cx) = cg(x) = c(E_x(g)) = (cE_x)(g).$$

This proves linearity of ev.

We show that the linear map ev is one-to-one by checking that for all $x \in V$, $\mathrm{ev}(x) = 0_{V^{**}}$ implies $x = 0_V$. We prove the contrapositive. Fix a nonzero $x \in V$. By extending $\{x\}$ to a basis of V and using the UMP, we can find $f \in V^*$ with $f(x) \neq 0$ (cf. Exercise 12). Then $E_x(f) \neq 0$, so $E_x = \mathrm{ev}(x)$ is not the zero function from V^* to F. Hence, $\mathrm{ev}(x) \neq 0$.

So far, we know ev $: V \to V^{**}$ is always F-linear and one-to-one. If V is finite-dimensional, then $\dim(V^{**}) = \dim(V)$ as remarked earlier. So the injective linear map ev must also be surjective since the domain and codomain have the same finite dimension. $\square$

The theorem shows that for each finite-dimensional F-vector space V, we have a natural vector space isomorphism ev $: V \to V^{**}$. The word *natural* is a technical mathematical adjective explained more fully later. Informally, the isomorphism is natural because it does not depend on any choices such as the chocie of an ordered basis. The fact that ev is a bijection means that for every $E \in V^{**}$, there exists a unique $x \in V$ such that $E = E_x$ (evaluation at x). Linearity of ev means $E_{x+y} = E_x + E_y$ and $E_{cx} = cE_x$ for all $x, y \in V$ and $c \in F$, as we confirmed earlier.

For the rest of this section, assume V is finite-dimensional. We often use the isomorphism ev to identify V^{**} with V, thereby blurring the distinction between an element $x \in V$ and the associated evaluation map $E_x \in V^{**}$. This identification may seem confusing at first, but it can clarify certain issues and make certain proofs less redundant. For instance, consider the relation between $\mathcal{Z}$ and $\mathcal{A}$ mentioned at the beginning of this subsection. Let $\mathcal{A}^*$ be the annihilator operator for the vector space V^*, which sends subsets of V^* to subsets of V^{**}. Given a subset S of V^*, the definition of $\mathcal{A}^*$ gives

$$\mathcal{A}^*(S) = \{E \in V^{**} : E(h) = 0 \text{ for all } h \in S\}.$$

But every $E \in V^{**}$ has the form E_x for a unique $x \in V$. Then $\mathrm{ev}(x) = E_x \in \mathcal{A}^*(S)$ iff $E_x(h) = 0$ for all $h \in S$ iff $h(x) = 0$ for all $h \in S$ iff $x \in \mathcal{Z}(S)$. This proves that $\mathcal{A}^*(S)$ is the image of $\mathcal{Z}(S)$ under the isomorphism ev. If we use ev to identify V with V^{**}, then the previous statement says that $\mathcal{A}^*(S) = \mathcal{Z}(S)$.

Similarly, letting $\mathcal{Z}^*$ be the zero-set operator for V^* (which maps subsets of V^{**} to subsets of V^*), we claim that $\mathcal{Z}^*(T) = \mathcal{A}(T)$ for all subsets T of V, under the identification of V^{**} with V. More precisely, we are asserting that $\mathcal{Z}^*(\mathrm{ev}[T]) = \mathcal{A}(T)$ for all $T \subseteq V$. The reason is that $g \in \mathcal{A}(T)$ iff $g(x) = 0$ for all $x \in T$ iff $E_x(g) = 0$ for all $x \in T$ iff g is sent to zero by all functions in $\mathrm{ev}[T]$ iff $g \in \mathcal{Z}^*(\mathrm{ev}[T])$.

With these facts in hand, we can see why the properties of the operators $\mathcal{Z}$ and $\mathcal{A}$ exactly mirror each other. For example, given that the annihilator operator has already been shown to be inclusion-reversing, it follows that the zero-set operator is also inclusion-reversing, because $S \subseteq T \subseteq V^*$ implies $\mathcal{Z}(T) = \mathcal{A}^*(T) \subseteq \mathcal{A}^*(S) = \mathcal{Z}(S)$.

As another example of this process, we prove:

- For all $T \subseteq V$, $T \subseteq \mathcal{Z}(\mathcal{A}(T))$;

- For all $S \subseteq V^*$, $S \subseteq \mathcal{A}(\mathcal{Z}(S))$.

For the first result, assume T is a subset of V, fix $x \in T$, and prove $x \in \mathcal{Z}(\mathcal{A}(T))$. We must show $g(x) = 0$ for all $g \in \mathcal{A}(T)$. But this follows from the very definition of $\mathcal{A}(T)$, since $x \in T$. The first result therefore holds for all vector spaces V. In particular, applying this result to the vector space V^* instead of V, we see that $S \subseteq \mathcal{Z}^*(\mathcal{A}^*(S))$ for all $S \subseteq V^*$. As seen above, this means $S \subseteq \mathcal{A}(\mathcal{Z}(S))$ for all $S \subseteq V^*$.

As a final example of working with the double dual, we show that the dual basis construction works in reverse. Suppose $B^* = (f_1, \ldots, f_n)$ is any ordered basis of V^*. We claim there exists a unique dual basis $(B^*)^D = (x_1, \ldots, x_n)$ of V such that $f_i(x_j) = \delta_{ij}$

for all i, j between 1 and n. To get this basis, first form the dual basis (in the old sense) $(B^*)^d = (E_1, \ldots, E_n)$, which is a basis of V^{**}. Then write $E_j = E_{x_j} = \mathrm{ev}(x_j)$ for a unique $x_j \in V$. Since ev is an isomorphism, $(x_1, \ldots, x_n)$ is an ordered basis for V. For each i, j, compute $f_i(x_j) = E_{x_j}(f_i) = E_j(f_i) = \delta_{ij}$. The uniqueness of the basis $(x_1, \ldots, x_n)$ follows from the uniqueness of $(E_1, \ldots, E_n)$ and the bijectivity of ev. We can also show that, for any ordered basis B of V, $B^{dD} = B$. To see why, write $B = (x_1, \ldots, x_n)$, $B^d = (f_1, \ldots, f_n)$, and $B^{dD} = (z_1, \ldots, z_n)$. We must have $x_i = z_i$ for each i, because $f_i(z_j) = \delta_{ij}$, and we know B is the unique ordered basis of V satisfying this condition.

13.6 Correspondence between Subspaces of V and V^*

Let V be a finite-dimensional vector space over a field F. Let $\mathcal{P}(V)$ be the set of all subsets of V and $\mathcal{S}(V)$ be the set of all subspaces of V. Define $\mathcal{P}(V^*)$ and $\mathcal{S}(V^*)$ similarly. In the previous subsections, we introduced two inclusion-reversing maps

$$\mathcal{Z} : \mathcal{P}(V^*) \to \mathcal{S}(V) \quad \text{and} \quad \mathcal{A} : \mathcal{P}(V) \to \mathcal{S}(V^*).$$

These maps are not one-to-one. For instance, if W is a subspace of V that is generated by a proper subset S, then $S \neq W$ but $\mathcal{A}(S) = \mathcal{A}(W)$. To get one-to-one correspondences, we restrict the operators $\mathcal{Z}$ and $\mathcal{A}$ to act only on subspaces of V^* and V, giving two inclusion-reversing maps

$$\mathcal{Z} : \mathcal{S}(V^*) \to \mathcal{S}(V) \quad \text{and} \quad \mathcal{A} : \mathcal{S}(V) \to \mathcal{S}(V^*).$$

Theorem on Correspondence Between Subspaces of V and V^*. Assume V is a finite-dimensional F-vector space with $\dim(V) = n$. The operators $\mathcal{Z}$ and $\mathcal{A}$ are mutually inverse inclusion-reversing bijections between $\mathcal{S}(V^*)$ and $\mathcal{S}(V)$ satisfying

$$\dim(W) + \dim(\mathcal{Z}(W)) = \dim(V) = \dim(Y) + \dim(\mathcal{A}(Y)) \quad \text{for all } W \in \mathcal{S}(V^*), \ Y \in \mathcal{S}(V).$$
$$(13.3)$$

The unrestricted $\mathcal{Z}$ and $\mathcal{A}$ operators satisfy

$$\mathcal{A}(\mathcal{Z}(S)) = \langle S \rangle \text{ and } \mathcal{Z}(\mathcal{A}(T)) = \langle T \rangle \quad \text{for all } S \in \mathcal{P}(V^*), \ T \in \mathcal{P}(V).$$

Proof. We first prove the dimension formula $n = \dim(V) = \dim(Y) + \dim(\mathcal{A}(Y))$, where Y is a fixed subspace of V. Let $(y_1, \ldots, y_k)$ be an ordered basis for Y. We extend this to an ordered basis $B = (y_1, \ldots, y_k, y_{k+1}, \ldots, y_n)$ for V. Let $B^d = (f_1, \ldots, f_k, f_{k+1}, \ldots, f_n)$ be the dual basis for V^*. We prove that $(f_{k+1}, \ldots, f_n)$ is an ordered basis for $\mathcal{A}(Y)$, which leads to the required conclusion $\dim(\mathcal{A}(Y)) = n - k = \dim(V) - \dim(Y)$. First, $(f_{k+1}, \ldots, f_n)$ is a linearly independent list in V^*, being a sublist of the ordered basis B^d. A generic element $f \in V^*$ can be written uniquely as

$$f = d_1 f_1 + \cdots + d_k f_k + d_{k+1} f_{k+1} + \cdots + d_n f_n \quad \text{for some } d_j \in F.$$

We claim $f \in \mathcal{A}(Y)$ iff $d_1 = \cdots = d_k = 0$. On one hand, $f \in \mathcal{A}(Y)$ iff $f \in \mathcal{A}(\{y_1, \ldots, y_k\})$, since these vectors generate the subspace Y. On the other hand, $f(y_i) = \sum_{j=1}^{n} d_j f_j(y_i) = d_i$ for all i. Therefore, $f \in \mathcal{A}(Y)$ iff $f(y_1) = \cdots = f(y_k) = 0$ iff $d_1 = \cdots = d_k = 0$. We now see that $\mathcal{A}(Y)$ consists precisely of all F-linear combinations of the linearly independent vectors $f_{k+1}, \ldots, f_n$. Thus $(f_{k+1}, \ldots, f_n)$ is an ordered basis for $\mathcal{A}(Y)$.

To complete the proof of (13.3), apply the result of the last paragraph to V^* instead of V. Using the identification of V^{**} with V, we get

$$\dim(V) = \dim(V^*) = \dim(W) + \dim(\mathcal{A}^*(W)) = \dim(W) + \dim(\mathcal{Z}(W)) \quad \text{for all } W \in \mathcal{S}(V^*).$$

Next, we show that the restricted $\mathcal{Z}$ and $\mathcal{A}$ operators are two-sided inverses of each other, so that $\mathcal{Z}$ and $\mathcal{A}$ are bijections. Suppose Y is any subspace of V. We have seen that $\mathcal{Z}(\mathcal{A}(Y))$ is a subspace of V containing Y. On the other hand, the dimension formulas in (13.3) show that

$$\dim(\mathcal{Z}(\mathcal{A}(Y))) = n - \dim(\mathcal{A}(Y)) = n - (n - \dim(Y)) = \dim(Y).$$

Since the dimensions are equal and finite, $\mathcal{Z}(\mathcal{A}(Y))$ cannot properly contain Y. So $\mathcal{Z}(\mathcal{A}(Y)) = Y$ holds. The same reasoning shows that $\mathcal{A}(\mathcal{Z}(W)) = W$ for all $W \in \mathcal{S}(V^*)$. Finally, for any subset S of V^*, we have

$$\mathcal{A}(\mathcal{Z}(S)) = \mathcal{A}(\mathcal{Z}(\langle S \rangle)) = \langle S \rangle.$$

Similarly, $\mathcal{Z}(\mathcal{A}(T)) = \langle T \rangle$ for all $T \subseteq V$. $\qquad \square$

To continue, we recall some definitions from the theory of posets (see the Appendix for more details). A *poset* (partially ordered set) consists of a set Z and a relation $\leq$ on Z that is reflexive, antisymmetric, and transitive. This means that for all $x, y, z \in Z$, $x \leq x$; if $x \leq y$ and $y \leq x$, then $x = y$; and if $x \leq y$ and $y \leq z$, then $x \leq z$. Given $S \subseteq Z$, an *upper bound* for S is an element $z \in Z$ such that for all $x \in S$, $x \leq z$. A *least upper bound* for S is an upper bound w for S such that $w \leq z$ for all upper bounds z for S. *Lower bounds* for S and the *greatest lower bound* for S are defined similarly. A subset S of a poset may or may not have a least upper bound. If the least upper bound exists, then it is unique and is written $\sup S$. Similar comments apply to the greatest lower bound of S, which is written $\inf S$ when it exists. The poset Z is called a *lattice* iff every two-element subset of Z has a least upper bound and a greatest lower bound. Z is a *complete lattice* iff every nonempty subset of Z has a least upper bound and a greatest lower bound.

Observe that $\mathcal{S}(V)$ and $\mathcal{S}(V^*)$ are both posets ordered by set inclusion: $X \leq Y$ means $X \subseteq Y$. You can check that if X and Y are subspaces of V, then in the poset $\mathcal{S}(V)$, $\inf(\{X, Y\}) = X \cap Y$ and $\sup(\{X, Y\}) = X + Y = \{x + y : x \in X, y \in Y\}$. So $\mathcal{S}(V)$ is a lattice. More generally, if $\{X_i : i \in I\}$ is any indexed collection of subspaces of V, then $\inf_{i \in I} X_i = \bigcap_{i \in I} X_i$ and $\sup_{i \in I} X_i = \sum_{i \in I} X_i$, where $\sum_{i \in I} X_i$ is the set of all finite sums $x_{i_1} + \cdots + x_{i_k}$ with $x_{i_j} \in X_{i_j}$. So $\mathcal{S}(V)$ is a complete lattice; similarly, $\mathcal{S}(V^*)$ is a complete lattice. We have shown that the maps $\mathcal{Z} : \mathcal{S}(V^*) \to \mathcal{S}(V)$ and $\mathcal{A} : \mathcal{S}(V) \to \mathcal{S}(V^*)$ are bijections that reverse inclusions. In poset terminology, we can say that these maps are *poset anti-isomorphisms*. It follows from this fact and the definitions of sup and inf that $\mathcal{Z}$ and $\mathcal{A}$ interchange least upper bounds and greatest lower bounds (Exercise 43). In other words,

$$\mathcal{Z}(X + Y) = \mathcal{Z}(X) \cap \mathcal{Z}(Y), \quad \mathcal{Z}(X \cap Y) = \mathcal{Z}(X) + \mathcal{Z}(Y) \quad \text{for all } X, Y \in \mathcal{S}(V^*); \quad (13.4)$$

$$\mathcal{A}(X + Y) = \mathcal{A}(X) \cap \mathcal{A}(Y), \quad \mathcal{A}(X \cap Y) = \mathcal{A}(X) + \mathcal{A}(Y) \quad \text{for all } X, Y \in \mathcal{S}(V); \quad (13.5)$$

and similar formulas hold for sums and intersections of indexed collections of subspaces.

13.7   Dual Maps

Suppose V and W are vector spaces over a field F, and $T : V \to W$ is a linear transformation. If $f : W \to F$ is any element of W^*, then the composition $f \circ T$ is a linear map from V to

F, so $f \circ T$ is an element of V^*. Thus, we can define a function $T^* : W^* \to V^*$ by setting $T^*(f) = f \circ T$ for all $f \in W^*$. The function T^* is itself a linear map, since

$$T^*(f + g) = (f + g) \circ T = (f \circ T) + (g \circ T) = T^*(f) + T^*(g) \quad \text{for all } f, g \in W^*;$$

$$T^*(cf) = (cf) \circ T = c(f \circ T) = cT^*(f) \quad \text{for all } c \in F, f \in W^*.$$

To verify the equality $(f + g) \circ T = (f \circ T) + (g \circ T)$, note that both sides are functions from V to F sending each $x \in V$ to $f(T(x)) + g(T(x))$. A similar calculation justifies the equality $(cf) \circ T = c(f \circ T)$. We call the linear map $T^* : W^* \to V^*$ the *dual map* to $T : V \to W$.

Here are some algebraic properties of dual maps. First, the function sending T to T^* is an F-linear map from $\operatorname{Hom}_F(V, W)$ into $\operatorname{Hom}_F(W^*, V^*)$. This means that for any linear maps $T, U : V \to W$ and scalar $c \in F$, we have $(T + U)^* = T^* + U^*$ and $(cT)^* = c(T^*)$. For example, the first equality holds because both sides are functions from W^* to V^* sending $g \in W^*$ to $g \circ (T + U) = (g \circ T) + (g \circ U)$. Similarly, both $(cT)^*$ and $c(T^*)$ send g to $c(g \circ T)$. Also, the mapping of T to T^* is *functorial*. This means that for any linear maps $T : V \to W$ and $S : W \to Z$, $(S \circ T)^* = T^* \circ S^*$; and for any identity map $\mathrm{id} : V \to V$, the dual map $\mathrm{id}^* : V^* \to V^*$ is the identity map on V^*. To check the first assertion, fix $h \in Z^*$ and compute

$$(S \circ T)^*(h) = h \circ (S \circ T) = (h \circ S) \circ T = (S^*(h)) \circ T = T^*(S^*(h)) = (T^* \circ S^*)(h).$$

The second assertion follows since $\mathrm{id}^*(f) = f \circ \mathrm{id} = f$ for $f \in V^*$.

Suppose $B = (v_1, \ldots, v_n)$ is an ordered basis for V, and $C = (w_1, \ldots, w_m)$ is an ordered basis for W. The matrix of T relative to input basis B and output basis C is the unique $m \times n$ matrix of scalars $A = [a_{ij}]$ such that $T(v_j) = \sum_{i=1}^{m} a_{ij} w_i$ for all j between 1 and n. We have seen that the dual basis $B^d = (f_1, \ldots, f_n)$ is an ordered basis for V^*, and the dual basis $C^d = (g_1, \ldots, g_m)$ is an ordered basis for W^*. Let us find the matrix of the linear map T^* relative to the bases C^d and B^d. By definition, this is the unique $n \times m$ matrix of scalars $A' = [a'_{rs}]$ such that $T^*(g_s) = \sum_{r=1}^{n} a'_{rs} f_r$ for all s between 1 and m. We claim that these equations hold if we take $a'_{rs} = a_{sr}$, meaning that A' is the *transpose* of the matrix A. To see this, fix s with $1 \le s \le m$, and check that the two functions $T^*(g_s)$ and $\sum_{r=1}^{n} a_{sr} f_r$ are equal. Both functions are linear maps from V to F, so it suffices to see that they take the same value at each basis element v_k, where $1 \le k \le n$. On one hand,

$$[T^*(g_s)](v_k) = (g_s \circ T)(v_k) = g_s(T(v_k)) = g_s\left(\sum_{i=1}^{m} a_{ik} w_i\right) = a_{sk}.$$

The last equality uses the fact that C^d is the dual basis for C. On the other hand,

$$\left[\sum_{r=1}^{n} a_{sr} f_r\right](v_k) = \sum_{r=1}^{n} a_{sr} f_r(v_k) = a_{sk}$$

since B^d is the dual basis for B. To summarize, we have shown that *if the matrix A represents a linear map T relative to certain bases, then the transpose of A represents the dual map T^* relative to the dual bases.* For this reason, the map T^* is sometimes called the *transpose* of the map T.

Given a linear map $T : V \to W$, we have the dual map $T^* : W^* \to V^*$. Iterating the construction produces the *double dual map* $T^{**} : V^{**} \to W^{**}$, which is defined by $T^{**}(E) = E \circ T^*$ for $E \in V^{**}$. Expanding this definition further, we see that $T^{**}(E)$ maps $g \in W^*$ to $E(T^*(g)) = E(g \circ T)$.

Now we can explain precisely what it means to say that the map ev $=$ ev$_V : V \to V^{**}$ is *natural*. Naturality means that for any vector spaces V and W and any linear map $T : V \to W$, the following diagram commutes:

$$\begin{array}{ccc} V & \xrightarrow{\;T\;} & W \\ {\scriptstyle \text{ev}_V}\downarrow & & \downarrow{\scriptstyle \text{ev}_W} \\ V^{**} & \xrightarrow{\;T^{**}\;} & W^{**} \end{array}$$

(that is, ev$_W \circ T = T^{**} \circ$ ev$_V$). To check this, fix $x \in V$, and ask whether the two functions ev$_W(T(x)) = E_{T(x)}$ and $T^{**}(\text{ev}_V(x)) = T^{**}(E_x)$ in W^{**} are equal. We fix $g \in W^*$ and apply each function to g. The first function produces $E_{T(x)}(g) = g(T(x))$. The second function gives

$$[T^{**}(E_x)](g) = (E_x \circ T^*)(g) = E_x(T^*(g)) = E_x(g \circ T) = (g \circ T)(x) = g(T(x)).$$

Thus, the functions are equal, and the diagram does commute.

In the finite-dimensional case, we say that V^{**} is *naturally* isomorphic to V under the map ev$_V$, since we can transfer linear maps $T : V \to W$ to linear maps $T^{**} : V^{**} \to W^{**}$. On the other hand, even though V and V^* are isomorphic, it can be shown that there is no *natural* isomorphism between V and V^* when $\dim(V) > 1$ (Exercise 46). As we will soon see, this situation can be partially remedied if V has appropriate additional structure (e.g., if V is an inner product space).

13.8   Bilinear Pairings of Vector Spaces

The rest of this chapter studies some variations and generalizations of the dual space construction for finite-dimensional vector spaces. Let V and W be finite-dimensional vector spaces over a field F. A function $B : W \times V \to F$ is called a *bilinear pairing* of W and V iff for all $x, x' \in W$, $y, y' \in V$, and $c \in F$,
(a) [linearity in first input] $B(x + x', y) = B(x, y) + B(x', y)$ and $B(cx, y) = cB(x, y)$;
(b) [linearity in second input] $B(x, y + y') = B(x, y) + B(x, y')$ and $B(x, cy) = cB(x, y)$.

Here are two examples of bilinear pairings. Given any finite-dimensional space V, let $W = V^*$ and $B(f, y) = f(y)$ for $f \in V^*$ and $y \in V$. Condition (a) holds by definition of the vector space operations in V^*. For instance, for $f, g \in V^*$ and $y \in V$, $B(f + g, y) = (f + g)(y) = f(y) + g(y) = B(f, y) + B(g, y)$. Condition (b) holds since each $f \in V^*$ is an F-linear map. For the second example, let $V = W = \mathbb{R}^n$ and $B(\mathbf{w}, \mathbf{v}) = \mathbf{w} \bullet \mathbf{v} = w_1 v_1 + \cdots + w_n v_n$, the dot product on $\mathbb{R}^n$. Conditions (a) and (b) are readily checked. When $W = V$, a bilinear pairing B is also called a *bilinear form on V*. We study bilinear forms in more detail in Chapter 14.

Return to the case of a general bilinear pairing $B : W \times V \to F$. For each fixed $y \in V$, define $R_y : W \to F$ by $R_y(x) = B(x, y)$ for $x \in W$. Condition (a) says that each R_y is a linear map from W to F, so R_y belongs to the dual space W^*. Similarly, for $x \in V$, define $L_x : V \to F$ by $L_x(y) = B(x, y)$ for $y \in V$. Condition (b) says that each L_x is a linear map from V to F, so $L_x \in V^*$. We call R_y *right-multiplication by y* and L_x *left-multiplication by x* (relative to the pairing B).

We can use these observations to define maps $L : W \to V^*$ and $R : V \to W^*$ determined by the pairing B. For $x \in W$ and $y \in V$, define $L(x) = L_x \in V^*$ and $R(y) = R_y \in W^*$. We already noted that each output $L(x)$ and $R(y)$ is an F-linear map. In fact, the functions L

and R are also F-linear maps. For example, given $x, x' \in W$, we have $L(x+x') = L(x)+L(x')$ because both sides are functions from V to F sending $y \in V$ to $B(x + x', y) = B(x, y) + B(x', y)$. For $x \in W$ and $c \in F$, $L(cx) = cL(x)$ because both sides are functions from V to F sending $y \in V$ to $B(cx, y) = cB(x, y)$. We similarly verify the linearity of R.

Next, we introduce analogs of the operators $\mathcal{Z}$ and $\mathcal{A}$ for bilinear pairings. Given $w \in W$ and $v \in V$, say that w is *orthogonal* to v and write $w \perp v$ to mean $B(w, v) = 0_F$. Given $S \subseteq W$ and $T \subseteq V$, say that S is *orthogonal* to T and write $S \perp T$ to mean $B(s, t) = 0_F$ for all $s \in S$ and all $t \in T$. When $S = \{s\}$ has a single element, we write $s \perp T$ to mean $\{s\} \perp T$; and $S \perp t$ means $S \perp \{t\}$. For any subset S of W, define the *orthogonal complement of S in V* to be

$$S^\perp = \{v \in V : S \perp v\} = \{v \in V : B(s, v) = 0 \text{ for all } s \in S\}.$$

By linearity of B in the second input, $S^\perp$ is always a subspace of V (Exercise 62). Similarly, for each subset T of V, define the *orthogonal complement of T in W* to be

$$^\perp T = \{w \in W : w \perp T\} = \{w \in W : B(w, t) = 0 \text{ for all } t \in T\}.$$

By linearity of B in the first input, $T^\perp$ is a subspace of W. Using these definitions, we readily check the following properties: for all $S_1, S_2, S \subseteq W$ and $T_1, T_2, T \subseteq V$, $S_1 \subseteq S_2$ implies $S_2^\perp \subseteq S_1^\perp$; $T_1 \subseteq T_2$ implies $^\perp T_2 \subseteq {}^\perp T_1$; $S \subseteq {}^\perp(S^\perp)$; and $T \subseteq ({}^\perp T)^\perp$.

We can now compute the kernels of the linear maps $L : W \to V^*$ and $R : V \to W^*$. For $w \in W$, the following conditions are equivalent: $w \in \ker(L)$; $L(w) = 0$, the zero map from V to F; $L(w)(v) = 0$ for all $v \in V$; $B(w, v) = 0$ for all $v \in V$; $w \in {}^\perp V$. So $\ker(L) = {}^\perp V$; we call this subspace of W the *left kernel of B*. Similarly, $\ker(R) = W^\perp$ is a subspace of V called the *right kernel of B*. We say that B is *left-nondegenerate* iff $\ker(L) = \{0\}$, in which case L is a one-to-one map of W into V^*. We say that B is *right-nondegenerate* iff $\ker(R) = \{0\}$, in which case R is a one-to-one map of V into W^*.

13.9 Theorems on Bilinear Pairings

Under appropriate nondegeneracy assumptions, bilinear pairings lead to isomorphisms involving dual vector spaces.

Theorem on Pairings Nondegenerate on Both Sides. Suppose W and V are finite-dimensional F-vector spaces and $B : W \times V \to F$ is a bilinear pairing that is both left-nondegenerate and right-nondegenerate. Then $L : W \to V^*$ and $R : V \to W^*$ are vector space isomorphisms, and $\dim(V) = \dim(W)$.

Proof. The nondegeneracy assumptions imply that L and R are both one-to-one, so

$$\dim(W) \leq \dim(V^*) = \dim(V) \leq \dim(W^*) = \dim(W).$$

Thus, the four spaces V, V^*, W, and W^* all have the same finite dimension n. Since L is a one-to-one map from W to V^*, where both spaces have dimension n, L must be onto. Similarly R is onto. So L and R are isomorphisms. $\square$

Bijectivity of L can be rephrased as follows: *for every linear map $g : V \to F$, there exists a unique vector $w \in W$ such that $g(v) = B(w, v)$ for all $v \in V$.* This w exists because L is onto, and w is unique because L is one-to-one. This gives a convenient concrete description of the functions in V^*: each such function is left-multiplication by w (relative to B) for some uniquely determined $w \in W$. Now $W \cong V^*$ implies $W^* \cong V^{**}$. Combining this isomorphism with R, we see that each function in V^{**} can be viewed as the element of W^* given by right-multiplication by v for some unique $v \in V$. Compare this to our earlier description of V^{**}, where every element of V^{**} was given by evaluation at some uniquely determined $v \in V$. For the pairing of V^* and V given by $B(f, x) = f(x)$, you can checkk that these two descriptions of elements of V^{**} coincide.

Theorem on Left-Nondegenerate Pairings. Suppose W and V are finite-dimensional vector spaces and $B : W \times V \to F$ is a left-nondegenerate bilinear pairing. For any subspace S of W, there are isomorphisms $V/S^\perp \cong S^*$ and $S \cong (V/S^\perp)^*$ induced by the maps L and R. We have $\dim(V) = \dim(S^\perp) + \dim(S)$ and $^\perp(S^\perp) = S$.

Taking $S = W$, we conclude in particular that $V/W^\perp \cong W^*$ and $W \cong (V/W^\perp)^*$. Analogous results hold for a right-nondegenerate pairing.

Proof. Let S be any subspace of W. Define $R' : V \to S^*$ by $R'(v)(s) = B(s, v)$ for $v \in V$, $s \in S$. Reasoning as we did earlier for R, the bilinearity of B shows that R' does map V into S^*, R' is F-linear, and $\ker(R') = S^\perp$. Passing to the quotient of V by this kernel, we get a well-defined one-to-one linear map $\overline{R'} : V/S^\perp \to S^*$.

On the other hand, let T be any subspace of V. Define a map $L' : {}^\perp T \to (V/T)^*$ by $L'(w)(v + T) = B(w, v)$ for $w \in {}^\perp T$, $v \in V$. To see that L' is well-defined, fix $w \in {}^\perp T$ and $v, v' \in V$ with $v + T = v' + T$. Then $v - v' \in T$, so $B(w, v - v') = 0$ because $w \in {}^\perp T$. This gives $B(w, v) - B(w, v') = 0$ and $B(w, v) = B(w, v')$. Bilinearity of B implies that $L'(w)$ does belong to $(V/T)^*$ and that L' is an F-linear map. Observe that for $w \in \ker(L')$, we have $B(w, v) = 0$ for all $v \in V$, so $w \in {}^\perp V = \{0\}$ because B is left-nondegenerate. So L' is a one-to-one linear map from ${}^\perp T$ into $(V/T)^*$. Apply this result to the subspace $T = S^\perp$ of V. Then $L' : {}^\perp(S^\perp) \to (V/S^\perp)^*$ is one-to-one and linear. We also have $S \subseteq {}^\perp(S^\perp)$. Combining all of this information gives

$$\dim(S) \le \dim({}^\perp(S^\perp)) \le \dim((V/S^\perp)^*) = \dim(V/S^\perp) \le \dim(S^*) = \dim(S).$$

So all the dimensions appearing here are equal and finite. It follows that $S = {}^\perp(S^\perp)$ and that we have isomorphisms $\overline{R'} : V/S^\perp \cong S^*$ and $L' : S \cong (V/S^\perp)^*$. From $\dim(V/S^\perp) = \dim(S)$, we deduce that $\dim(V) = \dim(S^\perp) + \dim(S)$. $\square$

For any vector space V, recall $\mathcal{S}(V)$ is the set of all subspaces of V, which is a lattice ordered by set inclusion. The next theorem shows how bilinear pairings induce correspondences between subspace lattices.

Theorem on Pairings and Subspace Lattices. Suppose W and V are finite-dimensional vector spaces and $B : W \times V \to F$ is a bilinear pairing. Define $F : \mathcal{S}(W) \to \mathcal{S}(V)$ by $F(S) = S^\perp$ for $S \in \mathcal{S}(W)$. Define $G : \mathcal{S}(V) \to \mathcal{S}(W)$ by $G(T) = {}^\perp T$ for $T \in \mathcal{S}(V)$. The maps F and G are inclusion-reversing. If B is left-nondegenerate, then $G(F(S)) = S$ and $\dim(V) = \dim(F(S)) + \dim(S)$ for all $S \in \mathcal{S}(W)$. If B is right-nondegenerate, then $F(G(T)) = T$ and $\dim(W) = \dim(G(T)) + \dim(T)$ for all $T \in \mathcal{S}(V)$. If B is nondegenerate on the left and right, then F and G are two-sided inverses of each other and are lattice anti-isomorphisms between $\mathcal{S}(W)$ and $\mathcal{S}(V)$.

Proof. All assertions follow from the previous theorem and its analog for right-nondegenerate pairings. $\square$

Theorem on General Bilinear Pairings. Suppose W and V are finite-dimensional vector spaces and $B : W \times V \to F$ is a bilinear pairing. Each space $W/^\perp V$ and $V/W^\perp$ is isomorphic to the dual of the other, so $\dim(W) - \dim(^\perp V) = \dim(V) - \dim(W^\perp)$. In particular, $\dim(V) = \dim(W)$ implies $\dim(^\perp V) = \dim(W^\perp)$. So when V and W have the same dimension, B is left-nondegenerate iff B is right-nondegenerate.

Proof. Define $B' : W/^\perp V \times V/W^\perp \to F$ by $B'(w + {}^\perp V, v + W^\perp) = B(w, v)$ for $w \in W$, $v \in V$. It is routine to check that B' is well-defined, bilinear, left-nondegenerate, and right-nondegenerate (Exercise 64). By the first theorem in this section, we get isomorphisms $W/^\perp V \cong (V/W^\perp)^*$ and $V/W^\perp \cong (W/^\perp V)^*$. The remaining assertions of the theorem follow from this. $\qquad\square$

We summarize our conclusions in the case where $\dim(V) = \dim(W) = n$ and the pairing $B : W \times V \to F$ is left-nondegenerate or right-nondegenerate. In this case, B is nondegenerate on both sides; $L : W \to V^*$ and $R : V \to W^*$ are isomorphisms; for any subspace S of W, $\dim(S) + \dim(S^\perp) = n$ and $^\perp(S^\perp) = S$; and for any subspace T of V, $\dim(T) + \dim(^\perp T) = n$ and $(^\perp T)^\perp = T$.

Theorem on Adjoints. Suppose W and V are n-dimensional F-vector spaces and $B : W \times V \to F$ is a nondegenerate bilinear pairing. For each linear operator $T : W \to W$, there exists a unique linear operator $T' : V \to V$ satisfying

$$B(T(x), y) = B(x, T'(y)) \quad \text{for all } x \in W,\ y \in V. \tag{13.6}$$

T' is called the *right adjoint* of T relative to B. A similar result holds for left adjoints.

Proof. Fix $y \in V$. The function sending $x \in W$ to $B(T(x), y) \in F$ is a linear map from W to F, as is readily checked. Since $R : V \to W^*$ is an isomorphism by our assumptions on B, there is a unique element of V, call it $T'(y)$, such that $B(T(x), y) = B(x, T'(y))$ holds for all $x \in W$. Thus there exists a unique function $T' : V \to V$ satisfying (13.6). It suffices to check that T' is linear. Given $y, z \in V$ and any $x \in W$, compute

$$B(T(x), y+z) = B(T(x), y) + B(T(x), z) = B(x, T'(y)) + B(x, T'(z)) = B(x, T'(y) + T'(z)).$$

On the other hand, $T'(y + z)$ is the unique vector in V satisfying $B(T(x), y + z) = B(x, T'(y + z))$ for all $x \in W$. So $T'(y+z) = T'(y) + T'(z)$. The proof that $T'(cy) = cT'(y)$ for $c \in F$ and $y \in V$ is similar. $\qquad\square$

Exercise 66 asks you to prove this theorem using the dual map $T^* : W^* \to W^*$. Specifically, you can check that $T' = R^{-1} \circ T^* \circ R$. Informally, using the isomorphism $R : V \cong W^*$, the right adjoint map $T' : V \to V$ can be identified with the dual map $T^* : W^* \to W^*$.

13.10   Real Inner Product Spaces

We can use the theory of dual spaces and bilinear pairings to derive some fundamental properties of inner product spaces. A *real inner product space* is a real vector space V and a bilinear form (pairing) $B : V \times V \to \mathbb{R}$ such that for all $x, y \in V$, $B(x, y) = B(y, x)$, and for all nonzero $v \in V$, $B(v, v)$ is a strictly positive real number. We often write $B(x, y) = \langle x, y \rangle$ and call B a *symmetric, positive definite inner product* on V. Such an inner product is automatically nondegenerate: if $x \in V$ satisfies $B(x, y) = 0$ for all $y \in V$, then $B(x, x) = 0$, so $x = 0$. Because B is symmetric, we need not distinguish left and right when discussing nondegeneracy or orthogonal complements. In particular, for any set $S \subseteq V$, we write $S^{\perp\perp}$

instead of $^\perp(S^\perp)$. The most familiar example of a real inner product space is the vector space $\mathbb{R}^n$ with the dot product.

Given a finite-dimensional real inner product space V, we can restate some results from the last section as follows. For each linear functional $f : V \to \mathbb{R}$, there is a unique $x \in V$ such that $f(y) = \langle y, x \rangle$ for all $y \in V$. For each linear map $T : V \to V$, there is a unique adjoint map $T' : V \to V$ characterized by $\langle T(x), y \rangle = \langle x, T'(y) \rangle$ for all $x, y \in V$. For any ordered basis $C = (x_1, \ldots, x_n)$ of V, there exists a unique *dual basis* $C' = (y_1, \ldots, y_n)$ for V such that $\langle x_i, y_j \rangle$ is 1 for $i = j$ and 0 otherwise. To see this, let $C^d = (f_1, \ldots, f_n)$ be the unique dual basis of C in V^*. Then let $y_1, \ldots, y_n$ be the unique elements of V such that f_j is right-multiplication by y_j. Note that $(y_1, \ldots, y_n)$ is an ordered basis of V, since it is the image of the ordered basis C^d under the isomorphism $R^{-1} : V^* \to V$. If T has matrix A relative to the ordered basis C of V, then the adjoint map T' has matrix A^{T} (the transpose of A) relative to the dual basis C^d (Exercise 67). For $\mathbb{R}^n$ with the dot product, the standard basis $C = (\mathbf{e}_1, \ldots, \mathbf{e}_n)$ is self-dual, meaning $C' = C$.

For any subspace U of V, the orthogonal complement of U relative to B is

$$U^\perp = \{y \in V : \langle u, y \rangle = 0 \text{ for all } u \in U\}.$$

The map F sending U to $U^\perp$ is an inclusion-reversing bijection on the lattice $\mathcal{S}(V)$ of subspaces of V, and F is its own inverse: $U^{\perp\perp} = U$ for all $U \in \mathcal{S}(V)$. Moreover, $\dim(U) + \dim(U^\perp) = \dim(V)$. Positive-definiteness shows that $U \cap U^\perp = \{0_V\}$, since any x belonging to both U and $U^\perp$ satisfies $B(x, x) = 0$, hence $x = 0$. Now $U + U^\perp$ is a subspace of V of dimension $\dim(U) + \dim(U^\perp) - \dim(U \cap U^\perp) = \dim(V)$, so $U + U^\perp$ must be all of V. We summarize the two conditions $U + U^\perp = V$ and $U \cap U^\perp = \{0\}$ by saying that V is the *direct sum* $U \oplus U^\perp$ of any subspace U and its orthogonal complement. Beware: if we replace positive definiteness by the weaker hypothesis that B is nondegenerate, then the conclusions $U \cap U^\perp = \{0\}$ and $V = U + U^\perp$ are not guaranteed to hold.

Orthogonal complements are closely related to the operators $\mathcal{A}$ and $\mathcal{Z}$ from our discussion of dual spaces. Given a finite-dimensional real inner product space V, nondegeneracy of the inner product means there is an isomorphism $R : V \to V^*$. You can check that for all $S \subseteq V$, $\mathcal{A}(S) = R[S^\perp]$; and for all $T \subseteq V^*$, $\mathcal{Z}(T) = R^{-1}[T]^\perp$ (Exercise 68). If we identify V with V^* using R, then these formulas can be written $\mathcal{A}(S) = S^\perp$ and $\mathcal{Z}(T) = T^\perp$. Informally, for a real inner product space, the $\mathcal{A}$ and $\mathcal{Z}$ operators coalesce into a single orthogonality operator acting on subsets of V.

13.11 Complex Inner Product Spaces

Recall that $\mathbb{C} = \{x + iy : x, y \in \mathbb{R}\}$ is the field of complex numbers, and complex conjugation is defined by $\overline{x + iy} = x - iy$ for $x, y \in \mathbb{R}$. A *complex inner product space* consists of a complex vector space V and a function $B : V \times V \to \mathbb{C}$ satisfying these conditions for all $v, v', w \in V$ and $c \in \mathbb{C}$:
(a) [linearity in first input] $B(v + v', w) = B(v, w) + B(v', w)$ and $B(cv, w) = cB(v, w)$;
(b) [conjugate symmetry] $B(w, v) = \overline{B(v, w)}$;
(c) [positive definiteness] if $v \neq 0$, then $B(v, v) \in \mathbb{R}_{>0}$.
These conditions readily imply a fourth condition:
(d) [conjugate-linearity in second input] $B(w, v + v') = B(w, v) + B(w, v')$ and $B(w, cv) = \overline{c}B(w, v)$.
We often write $B(v, w) = \langle v, w \rangle$ and call B a complex inner product. As before, $v \perp w$ means $B(v, w) = 0$, which is equivalent to $w \perp v$ since $\overline{0} = 0$. For $S \subseteq V$, the orthogonal

complement of S is $S^\perp = \{v \in V : S \perp v\}$. Some texts use the convention that a complex inner product is conjugate-linear in the first input and linear in the second input.

An example of a complex inner product space is $\mathbb{C}^n$ with the complex dot product, defined on inputs $\mathbf{v} = (v_1, \ldots, v_n)$ and $\mathbf{w} = (w_1, \ldots, w_n)$ in $\mathbb{C}^n$ by $\langle \mathbf{v}, \mathbf{w} \rangle = v_1 \overline{w_1} + v_2 \overline{w_2} + \cdots + v_n \overline{w_n}$. In particular, positive definiteness holds because $\langle \mathbf{v}, \mathbf{v} \rangle = |v_1|^2 + \cdots + |v_n|^2$, which is a positive real number for $\mathbf{v} \neq \mathbf{0}$. This fact would not hold without the conjugates in the complex dot product, which is the main reason why we require conjugate-symmetry rather than ordinary symmetry of B.

The main facts for a finite-dimensional complex inner product space V are similar to the real case. The map sending U to $U^\perp$ is a self-inverse, inclusion-reversing bijection on the lattice of subspaces of V; $U^{\perp\perp} = U$; $\dim(U) + \dim(U^\perp) = \dim(V)$; and $V = U \oplus U^\perp$.

To prove these facts using the general theory of bilinear pairings, we need a technical device to handle the conjugate-linearity in the second position. Define $\sigma : \mathbb{C} \to \mathbb{C}$ by $\sigma(z) = \overline{z}$ for $z \in \mathbb{C}$. The map σ is a ring isomorphism that is equal to its own inverse. For any complex vector space V, we define a new complex vector space V^σ that has the same underlying set and addition operation as V. Writing $\cdot$ for the scalar multiplication on V, define the scalar multiplication $\bullet$ on V^σ by $c \bullet v = \overline{c} \cdot v$ for all $c \in \mathbb{C}$, $v \in V$. It is routine to check that V^σ really is a complex vector space; a subset U of V is a subspace of V iff U is a subspace of V^σ; and a list of vectors in the space V spans U, or is linearly independent, or is an ordered basis of U, iff the list has the corresponding property in the space V^σ. A $\mathbb{C}$-linear map g from V^σ into another complex vector space Z is sometimes called a *semi-linear* map from V to Z. Such a map satisfies $g(v + w) = g(v) + g(w)$ and $g(c \cdot v) = g(\overline{c} \bullet v) = \overline{c} g(v)$ for all $v, w \in V$ and all $c \in \mathbb{C}$.

Suppose B is a complex inner product on V. The axioms for B show that $B : V \times V^\sigma \to \mathbb{C}$ is a bilinear pairing of $\mathbb{C}$-vector spaces. In particular, for $w \in V$, $v \in V^\sigma$, and $c \in \mathbb{C}$, axiom (d) gives $B(w, c \bullet v) = B(w, \overline{c} \cdot v) = \overline{\overline{c}} B(w, v) = c B(w, v)$. So B is $\mathbb{C}$-linear in both its first and second inputs. The pairing B is nondegenerate since $B(v, v) \neq 0$ for all nonzero $v \in V$. The facts claimed about orthogonal complements in the previous paragraph now follow from the general results in §13.9. We also see that V and V^σ are duals of each other via the isomorphisms $L : V \to (V^\sigma)^*$ and $R : V^\sigma \to V^*$. In particular, every $\mathbb{C}$-linear map $f : V \to \mathbb{C}$ has the form $f(x) = \langle x, y \rangle$ for a unique $y \in V$. Every semi-linear map $g : V \to \mathbb{C}$ has the form $g(y) = \langle x, y \rangle$ for a unique $x \in V$. For each ordered basis $C = (x_1, \ldots, x_n)$ of V, we obtain a unique dual basis $C' = (y_1, \ldots, y_n)$ of V^σ (which is also a basis of V), where $\langle x_i, y_j \rangle$ is 1 for $i = j$ and 0 for $i \neq j$. For each $\mathbb{C}$-linear map $T : V \to V$, there is a unique $\mathbb{C}$-linear adjoint map $T' : V^\sigma \to V^\sigma$ such that $\langle T(x), y \rangle = \langle x, T'(y) \rangle$ for all $x, y \in V$. In fact, T' is also a $\mathbb{C}$-linear (not semi-linear!) map from V to V, since

$$T'(c \cdot v) = T'(\overline{c} \bullet v) = \overline{c} \bullet T'(v) = c \cdot T'(v) \quad \text{for all } c \in \mathbb{C}, v \in V.$$

Suppose $A = [T]_C$ is the matrix of T relative to the ordered basis C. Viewing T' as a linear map on V^σ, the matrix of T relative to C' is the transpose of A. But viewing T' as a linear map on V, the matrix of T relative to C' is the *conjugate-transpose* of A, which has i, j-entry $\overline{A(j, i)}$ (Exercise 70). For $\mathbb{C}^n$ with the complex dot product, the standard basis $C = (\mathbf{e}_1, \ldots, \mathbf{e}_n)$ is self-dual, meaning $C' = C$.

13.12   Duality for Infinite-Dimensional Spaces

Although many of the results in this chapter apply to arbitrary vector spaces V, some theorems require the hypothesis that V be finite-dimensional. For example, we used finite-dimensionality in the proofs that $\dim(V) = \dim(V^*)$, that $\mathrm{ev} : V \to V^{**}$ is an isomorphism, and that $\mathcal{A}$ and $\mathcal{Z}$ are bijections between the subspace lattices of V and V^*. If we start with an infinite-dimensional space V, it can be shown that the dimension of the dual space V^* is a larger infinite cardinal than $\dim(V)$ (compare to Exercises 12 and 15). Then $\dim(V^{**})$ is larger still, so that V and V^{**} are never isomorphic in the infinite-dimensional case.

In order to obtain more satisfactory results for infinite-dimensional spaces, we can add extra structure to the vector space V and modify the definition of the dual space V^* to take into account this extra structure. We sketch how this is done for Banach spaces. In many real or complex vector spaces V, it is possible to define the *norm* or *length* of a vector, denoted $||x||$, satisfying these conditions: $||0|| = 0$; for all nonzero $x \in V$, $||x|| \in \mathbb{R}_{>0}$; $||cx|| = |c| \cdot ||x||$ for all scalars c and $x \in V$; and $||x + y|| \leq ||x|| + ||y||$ for all $x, y \in V$. The space V together with the norm function on V is called a *normed vector space*. Using the norm, we can define the *distance* between two vectors by setting $d(x, y) = ||x - y||$ for $x, y \in V$, and this turns V into a *metric space*. If this metric space is *complete* (meaning that every Cauchy sequence of vectors in V converges to a vector in V; see Chapter 15 for more details), then V is called a *Banach space*.

Given two Banach spaces V and W, we could study the vector space $\mathrm{Hom}_F(V, W)$ of all linear maps from V to W. However, it is more fruitful to restrict attention to the subspace of continuous linear maps from V to W. It can be shown that a linear map $T : V \to W$ is continuous iff T is continuous at 0_V iff there is a real constant $C > 0$ such that $||T(x)|| \leq C||x||$ for all $x \in V$. For this reason, continuous linear maps are also called *bounded* linear operators. In this setting, we adjust the definition of the dual space so that V^* consists of all continuous linear maps from V to the field of scalars ($\mathbb{R}$ or $\mathbb{C}$).

With this adjusted definition, various Banach spaces that occur in integration theory turn out to be duals of each other. For instance, for each fixed p with $1 < p < \infty$, there is a real Banach space L_p consisting of Lebesgue-measurable functions $f : \mathbb{R} \to \mathbb{R}$ such that $\int_{\mathbb{R}} |f(x)|^p \, dx$ is finite. Let $q \in \mathbb{R}_{>1}$ satisfy $1/p + 1/q = 1$. In advanced analysis, it is shown that L_q is isomorphic to L_p^* as a Banach space. The isomorphism sends $g \in L_q$ to the continuous linear functional that maps $f \in L_p$ to $\int_{\mathbb{R}} f(x)g(x) \, dx$. This is an instance of right-multiplication by g relative to a bilinear pairing; a key technical point is proving that the integral of the product fg is finite. Since $L_q \cong L_p^*$ and $L_q^* \cong L_p$, we deduce that $L_p^{**} \cong L_q^* \cong L_p$, as we might have expected from the finite-dimensional case. However, not all Banach spaces are isomorphic to their double duals. A Banach space V that is isomorphic to V^{**} is called *reflexive*.

Another subtlety that occurs in Banach spaces involves the $\mathcal{Z}$ and $\mathcal{A}$ operators. Recall that for $S \subseteq V^*$, $\mathcal{Z}(S)$ consists of all $x \in V$ such that $f(x) = 0$ for all $f \in S$. We can express this in symbols by writing $\mathcal{Z}(S) = \bigcap_{f \in S} f^{-1}[\{0\}]$. In the setting of Banach spaces, each $f \in S$ must be continuous, so that the preimage $f^{-1}[\{0\}]$ of the closed set $\{0\}$ is a closed subset of the metric space V. Then $\mathcal{Z}(S)$, being an intersection of closed sets, is always a closed set (as well as a subspace) in V. Similarly, it can be shown that V^* is a Banach space and $\mathcal{A}(T)$ is closed in V^* for all $T \subseteq V$. Thus, in order to turn the operators $\mathcal{Z}$ and $\mathcal{A}$ into bijective correspondences, it becomes necessary to restrict attention to the lattices of *closed* subspaces of V and V^*.

As the above discussion indicates, we often need to add topological ingredients to obtain an effective theory for infinite-dimensional vector spaces. In particular, restricting attention to *continuous* linear maps and *closed* linear subspaces is often essential. We shall see how this works in more detail in Chapter 15, which covers metric spaces and the basic elements of Hilbert space theory.

13.13 A Preview of Affine Algebraic Geometry

We end this chapter with a brief introduction to another correspondence between spaces and functions that appears in elementary algebraic geometry. We start with an algebraically closed field F such as the field of complex numbers. Let $F^n = \{(c_1, \ldots, c_n) : c_i \in F\}$ be the set of all n-tuples of elements of F. In this setting, F^n is called *affine n-space over F*. Next, let R be the polynomial ring $F[x_1, \ldots, x_n]$. Each formal polynomial $p \in R$ determines a polynomial function (also denoted p) from F^n to F. For instance, when $n = 3$ and $p = x_1^3 + 2x_2 x_3^2$, the polynomial function $p : F^3 \to F$ is given by $p(c_1, c_2, c_3) = c_1^3 + 2c_2 c_3^2$ for $c_1, c_2, c_3 \in F$.

As in the case of linear functionals, we can define a *zero-set operator* $\mathcal{Z}$ that maps subsets of R to subsets of F^n. Specifically, for all $S \subseteq R$, let

$$\mathcal{Z}(S) = \{(c_1, \ldots, c_n) \in F^n : p(c_1, \ldots, c_n) = 0 \text{ for all } p \in S\}.$$

We can regard $\mathcal{Z}(S)$ as the solution set of the (possibly infinite) system of polynomial equations $p(\mathbf{v}) = 0$ for $p \in S$. Any subset of F^n that has the form $\mathcal{Z}(S)$ for some $S \subseteq R$ is called an *affine variety* in F^n. Many texts write $\mathcal{V}(S)$ for $\mathcal{Z}(S)$ and call $\mathcal{Z}$ the *variety operator*.

We list without proof some properties of $\mathcal{Z}$. First, $\mathcal{Z}$ reverses inclusions: if $S \subseteq T \subseteq R$, then $\mathcal{Z}(T) \subseteq \mathcal{Z}(S)$. Each subset $S \subseteq R$ generates an ideal

$$I = \langle S \rangle = \{r_1 s_1 + \cdots + r_m s_m : m \in \mathbb{Z}_{\geq 0}, r_i \in R, s_i \in S\}$$

consisting of all finite R-linear combinations of polynomials in S. Define the *radical* of any ideal I of R to be $\sqrt{I} = \{p \in R : p^k \in I \text{ for some } k \in \mathbb{Z}_{>0}\}$. The second property of $\mathcal{Z}$ is $\mathcal{Z}(S) = \mathcal{Z}(I) = \mathcal{Z}(\sqrt{I})$ where $I = \langle S \rangle$. Third, $\mathcal{Z}(\{0\}) = F^n$ and $\mathcal{Z}(\{1\}) = \mathcal{Z}(R) = \emptyset$. Fourth, for a family of subsets $\{S_j : j \in J\}$ of R, $\mathcal{Z}(\bigcup_{j \in J} S_j) = \bigcap_{j \in J} \mathcal{Z}(S_j)$. Fifth, for a family of ideals $\{I_j : j \in J\}$ of R, $\mathcal{Z}(\sum_{j \in J} I_j) = \bigcap_{j \in J} \mathcal{Z}(I_j)$. Sixth, for ideals $I_1, I_2, \ldots, I_k$ in R, $\mathcal{Z}(I_1 I_2 \cdots I_k) = \mathcal{Z}(\bigcap_{j=1}^k I_j) = \bigcup_{j=1}^k \mathcal{Z}(I_j)$. The *Hilbert Basis Theorem* asserts that any ideal I in $R = F[x_1, \ldots, x_n]$ can be generated by a finite set S. Combining this theorem with the second property, we see that any affine variety can be defined as the solution set of a finite system of polynomial equations.

Next, we define an analog of the annihilator operator $\mathcal{A}$, which maps subsets of F^n to subsets of R. Given $S \subseteq F^n$, let

$$\mathcal{A}(S) = \{p \in R : p(c_1, \ldots, c_n) = 0 \text{ for all } (c_1, \ldots, c_n) \in S\}.$$

It can be checked that $\mathcal{A}(S)$ is always a *radical ideal* of R; this is an ideal I such that $I = \sqrt{I}$. Some texts write $\mathcal{I}(S)$ for $\mathcal{A}(S)$.

We state some properties of the annihilator operator and its relation to the zero-set operator. First, $\mathcal{A}(\emptyset) = R$ and $\mathcal{A}(F^n) = \{0_R\}$. Second, $\mathcal{A}$ reverses inclusions: if $S \subseteq T \subseteq F^n$, then $\mathcal{A}(T) \subseteq \mathcal{A}(S)$. Third, for any family of subsets $\{S_j : j \in J\}$ of F^n, $\mathcal{A}(\bigcup_{j \in J} S_j) = \bigcap_{j \in J} \mathcal{A}(S_j)$. Fourth, for any $S \subseteq F^n$, $\mathcal{Z}(\mathcal{A}(S))$ is the intersection of all varieties in F^n that contain S; this intersection is the smallest variety containing S. Fifth, for any subset S of R, $\mathcal{A}(\mathcal{Z}(S)) = \sqrt{\langle S \rangle}$. The fifth property, which requires algebraic closure of the field F, is one version of a difficult theorem of algebraic geometry called *Hilbert's Nullstellensatz*.

It follows from the results stated above that $\mathcal{Z}$ and $\mathcal{A}$ restrict to give poset anti-isomorphisms between the lattice of affine varieties in F^n and the lattice of radical ideals of $R = F[x_1, \ldots, x_n]$. This *ideal–variety correspondence* has many additional structural properties. For instance, maximal ideals of R correspond to individual points in F^n, prime ideals of R correspond to irreducible varieties in F^n, and so on. For more details, we refer the reader to the excellent text by Cox, Little, and O'Shea [12].

13.14 Summary

Let F be a field and V and W be finite-dimensional vector spaces over F. Table 13.1 summarizes definitions concerning dual spaces. We also recall the following results, some of which extend to infinite-dimensional spaces.

TABLE 13.1
Definitions related to dual spaces.

Concept	Definition
$\mathrm{Hom}_F(V, W)$	set of all F-linear maps $T: V \to W$ (a vector space under pointwise operations)
dual space V^*	$V^* = \mathrm{Hom}_F(V, F)$
dual basis of ordered basis $X = (x_1, \ldots, x_n)$ of V	unique ordered basis $X^d = (f_1, \ldots, f_n)$ of V^* with $f_i(x_j) = 1$ for $i = j$, $f_i(x_j) = 0$ for $i \neq j$
zero set $\mathcal{Z}(S)$ of $S \subseteq V^*$	subspace $\{x \in V : f(x) = 0 \text{ for all } f \in S\}$
annihilator $\mathcal{A}(T)$ of $T \subseteq V$	subspace $\{g \in V^* : g(y) = 0 \text{ for all } y \in T\}$
double dual V^{**}	$V^{**} = \mathrm{Hom}_F(V^*, F) = \mathrm{Hom}_F(\mathrm{Hom}_F(V, F), F)$
evaluation map E_x for $x \in V$	$E_x \in V^{**}$ sends $g \in V^*$ to $g(x) \in F$
$\mathrm{ev}: V \to V^{**}$	$\mathrm{ev}(x) = E_x$ for $x \in V$ (one-to-one linear map)
dual map of $T: V \to W$	$T^*: W^* \to V^*$ given by $T^*(g) = g \circ T$ for $g \in W^*$
bilinear pairing $B: W \times V \to F$	B is F-linear in first and second position
$\quad L: W \to V^*$	$L(w)$ sends $v \in V$ to $B(w, v)$
$\quad R: V \to W^*$	$R(v)$ sends $w \in W$ to $B(w, v)$
$\quad S \perp T$	$B(s, t) = 0$ for all $s \in S$, $t \in T$
$\quad S^{\perp}$ for $S \subseteq W$	subspace $\{v \in V : S \perp v\}$
$\quad {}^{\perp}T$ for $T \subseteq V$	subspace $\{w \in W : w \perp T\}$
$\quad$ left-nondegenerate pairing	${}^{\perp}V = \{0_W\}$, so L is one-to-one
$\quad$ right-nondegenerate pairing	$W^{\perp} = \{0_V\}$, so R is one-to-one
$\quad$ right adjoint of $T: W \to W$	unique linear map $T': V \to V$ such that $B(T(x), y) = B(x, T'(y))$ for $x \in W, y \in V$
real inner product	symmetric bilinear form on $\mathbb{R}$-vector space V with $\langle x, x \rangle > 0$ for all $x \neq 0$ in V
complex inner product	form on $\mathbb{C}$-vector space that is linear in input 1, $\langle v, w \rangle = \overline{\langle w, v \rangle}$, and $0 \neq v$ implies $\langle v, v \rangle \in \mathbb{R}_{>0}$

1. *Universal Mapping Property for Basis of a Vector Space.* For every ordered basis $B = (x_1, \ldots, x_n)$ of V and every list $(y_1, \ldots, y_n)$ of vectors in W, there exists a unique F-linear map $T: V \to W$ with $T(x_i) = y_i$ for $1 \leq i \leq n$.

2. *Dual Bases.* For every ordered basis $B = (x_1, \ldots, x_n)$ of V, there exists a unique dual basis $B^d = (f_1, \ldots, f_n)$ of V^* with $f_i(x_j) = \delta_{ij}$ for all i, j between 1 and n. For every ordered basis $C = (g_1, \ldots, g_n)$ of V^*, there is a unique ordered basis

$C^D = (z_1, \ldots, z_n)$ of V with $g_i(z_j) = \delta_{ij}$ for all i, j between 1 and n Since $\dim(V)$ is finite, $\dim(V) = \dim(V^*)$ and $V \cong V^*$. But this isomorphism is not natural; it depends on a choice of basis for V.

3. *Zero-Set Operator and Annihilator Operator.*
   (a) For $S \subseteq V^*$, $\mathcal{Z}(S)$ is the set of $x \in V$ such that $f(x) = 0$ for all $f \in S$.
   (b) For $T \subseteq V$, $\mathcal{A}(T)$ is the set of $g \in V^*$ such that $g(y) = 0$ for all $y \in T$.
   (c) The operators $\mathcal{Z}$ and $\mathcal{A}$ reverse inclusions, map subsets to subspaces, and satisfy $\mathcal{Z}(S) = \mathcal{Z}(\langle S \rangle)$, $\mathcal{A}(T) = \mathcal{A}(\langle T \rangle)$, $\mathcal{A}(\mathcal{Z}(S)) = \langle S \rangle$, and $\mathcal{Z}(\mathcal{A}(T)) = \langle T \rangle$.
   (d) Since $\dim(V)$ is finite, $\mathcal{Z}$ and $\mathcal{A}$ restrict to give poset anti-isomorphisms between the lattice of subspaces of V and the lattice of subspaces of V^*.
   (e) For subspaces U and W of V^*, $\dim(U) + \dim(\mathcal{Z}(U)) = \dim(V)$, $\mathcal{Z}(U + W) = \mathcal{Z}(U) \cap \mathcal{Z}(W)$, and $\mathcal{Z}(U \cap W) = \mathcal{Z}(U) + \mathcal{Z}(W)$; similar formulas hold for $\mathcal{A}$.

4. *Double Duals and Evaluation Maps.*
   (a) For $x \in V$, *evaluation at x* is the linear map $E_x : V^* \to F$ given by $E_x(g) = g(x)$ for $g \in V^*$. E_x belongs to the double dual space V^{**}.
   (b) The function $\mathrm{ev} : V \to V^{**}$ sending each $x \in V$ to E_x is linear and one-to-one.
   (c) Since $\dim(V)$ is finite, ev gives a natural vector-space isomorphism $V \cong V^{**}$. Naturality means that $T^{**} = \mathrm{ev}_W \circ T \circ \mathrm{ev}_V^{-1}$ for all linear maps $T : V \to W$.
   (d) Using ev, the zero-set and annihilator operators linking V^* and V^{**} can be identified with the annihilator and zero-set operators linking V and V^*.

5. *Dual Maps.*
   (a) For each linear map $T : V \to W$, there is a linear dual map $T^* : W^* \to V^*$ given by $T^*(g) = g \circ T$ for $g \in W^*$.
   (b) If A is the matrix of T relative to the ordered bases B for V and C for W, then the matrix of T^* relative to the dual bases C^d and B^d is the transpose of A.
   (c) For linear maps T and U and $r \in F$, the formulas $(T + U)^* = T^* + U^*$, $(rT)^* = r(T^*)$, $(\mathrm{id}_V)^* = \mathrm{id}_{V^*}$, and $(T \circ U)^* = U^* \circ T^*$ hold when defined.

6. *Bilinear Pairings.*
   (a) A bilinear pairing $B : W \times V \to F$ induces linear maps $L : W \to V^*$ and $R : V \to W^*$ with $\ker(L) = {}^{\perp}V$ and $\ker(R) = W^{\perp}$.
   (b) If B is left-nondegenerate and right-nondegenerate, then L and R are isomorphisms and $\dim(V) = \dim(W)$.
   (c) If B is left-nondegenerate and S is any subspace of W, then $V/S^{\perp} \cong S^*$, $S \cong (V/S^{\perp})^*$, $\dim(V) = \dim(S^{\perp}) + \dim(S)$, and ${}^{\perp}(S^{\perp}) = S$.
   (d) For general B, each space $W/{}^{\perp}V$ and $V/W^{\perp}$ is isomorphic to the dual of the other, and $\dim(V) = \dim(W)$ implies $\dim({}^{\perp}V) = \dim(W^{\perp})$.
   (e) When $\dim(V) = \dim(W)$, B is left-nondegenerate iff B is right-nondegenerate.

7. *Pairings and Subspace Lattices.* Suppose $\dim(V) = \dim(W) = n$ and the bilinear pairing $B : W \times V \to F$ is nondegenerate. The orthogonal complement operators (relative to B) define mutually inverse, inclusion-reversing poset anti-isomorphisms between the subspace lattice of W and the subspace lattice of V. For $S \in \mathcal{S}(W)$ and $T \in \mathcal{S}(V)$, $\dim(S) + \dim(S^{\perp}) = n = \dim(T) + \dim({}^{\perp}T)$.

8. *Adjoint Maps.* Suppose $\dim(V) = \dim(W) = n$ and the bilinear pairing $B : W \times V \to F$ is nondegenerate. For each linear map $T : W \to W$, there is a unique linear right adjoint $T' : V \to V$ such that $B(T(x), y) = B(x, T'(y))$ for all $x \in W$ and $y \in V$. A similar result holds for left adjoints.

9. *Real Inner Product Spaces.* For a real vector space V, an inner product is a symmetric bilinear form on V satisfying $\langle x, x \rangle > 0$ for all $x \neq 0$, which guarantees

nondegeneracy of the form. The map sending U to $U^\perp$ is a self-inverse, inclusion-reversing bijection on the lattice of subspaces of V. For all subspaces U of V, $U^{\perp\perp} = U$, $\dim(U) + \dim(U^\perp) = \dim(V)$, $U \cap U^\perp = \{0\}$, and $V = U \oplus U^\perp$.

10. *Complex Inner Product Spaces.* For a complex vector space V, a complex inner product is a map from $V \times V$ to $\mathbb{C}$ that is linear in the first input, conjugate-linear in the second input, conjugate symmetric, and positive definite. The subspace facts from the previous item hold. The matrix of an adjoint map (relative to a dual basis) is the conjugate-transpose of the matrix of the original map.

11. *Infinite-Dimensional Spaces.* For an infinite-dimensional F-vector space with topological structure (such as a Banach space where each vector has a norm), we redefine V^* to be the set of continuous linear maps from V to F. Here, V^{**} may or may not be isomorphic to V, and we must restrict $\mathcal{Z}$ and $\mathcal{A}$ to the lattices of closed linear subspaces to get lattice isomorphisms.

13.15 Exercises

In these exercises, assume F is a field and V, W are finite-dimensional F-vector spaces unless otherwise stated.

1. Let S be a set and Z be the set of functions $f : S \to W$.
   (a) Verify that Z is a commutative group under pointwise addition of functions.
   (b) Verify that Z is an F-vector space under pointwise operations.

2. Let V and W be F-vector spaces. Prove that $\operatorname{Hom}_F(V, W)$ is a subspace of the vector space of all functions from V to W (under pointwise operations).

3. Explain why the set V of all differentiable functions $f : (0, 3) \to \mathbb{R}$ is a subspace of the set of all functions from $(0, 3)$ to $\mathbb{R}$ (under pointwise operations).

4. Let V be the vector space of differentiable functions $f : (0, 3) \to \mathbb{R}$ under pointwise operations. Decide whether each operator below is a linear functional in V^*.
   (a) $D : V \to V$ given by $D(f) = f'$
   (b) $E : V \to \mathbb{R}$ given by $E(f) = f(1)$
   (c) $F : V \to \mathbb{R}$ given by $F(f) = f'(2)$
   (d) $G : V \to \mathbb{R}$ given by $G(f) = f(1)f(2)$
   (e) $H : V \to \mathbb{R}$ given by $H(f) = \int_1^2 f(x)\,dx$
   (f) $I : V \to \mathbb{R}$ given by $I(f) = \int_{1/2}^1 x f(x^2)\,dx$

5. For $n > 1$, decide which functions below (with domain $M_n(F)$) are in $M_n(F)^*$.
   (a) the map sending A to $A(1, 2)$
   (b) the trace map given by $\operatorname{tr}(A) = \sum_{i=1}^n A(i, i)$
   (c) the determinant function sending A to $\det(A)$
   (d) the transpose map sending A to A^{T}
   (e) the map sending A to the sum of all entries of A
   (f) for fixed $v \in M_{n,1}(F)$, the map sending A to $v^{\mathrm{T}} A v$
   (g) for $F = \mathbb{R}$, the map sending A to the number of zero entries of A

6. Check that the map T defined in the proof of the UMP (page 342) is F-linear and sends each x_i to y_i.

7. Let $B = \{E_{i,j} : 1 \leq i \leq m, 1 \leq j \leq n\}$ be the standard basis for $M_{m,n}(F)$, where $E_{i,j}$ is the matrix with 1 in the i,j-position and 0 elsewhere. Explicitly describe how each element in the dual basis B^d acts on an arbitrary matrix $A \in M_{m,n}(F)$.

8. Define a map $T : M_{1,n}(F) \to M_{n,1}(F)^*$ as follows. For a row vector $\mathbf{w}$, let $T(\mathbf{w})$ be the map sending each column vector $\mathbf{v}$ to the 1×1 matrix $\mathbf{wv}$, regarded as an element of F. Prove that T is a vector space isomorphism. We can use this isomorphism to identify the dual space of the vector space of $n \times 1$ column vectors with the vector space of $1 \times n$ row vectors.

9. Let $(E(1,1), E(2,1), \ldots, E(n,1))$ be the standard ordered basis of $M_{n,1}(F)$ and $(g_1, g_2, \ldots, g_n)$ be the dual basis of $M_{n,1}(F)^*$. Compute $T^{-1}(g_j)$ for each j, where T is the isomorphism in the previous exercise.

10. Find a dual basis for the ordered basis

$$B = \left(\begin{bmatrix} 2 \\ 1 \\ 1 \end{bmatrix}, \begin{bmatrix} 1 \\ 0 \\ 3 \end{bmatrix}, \begin{bmatrix} -3 \\ 4 \\ -1 \end{bmatrix} \right)$$

of $\mathbb{R}^3 = M_{3,1}(\mathbb{R})$, using Exercise 8 to describe the answer as a list of row vectors.

11. Given an ordered basis $(v_1, \ldots, v_n)$ of F^n (viewed as column vectors), show how to use matrix inversion to compute the dual basis of $(F^n)^*$ (viewed as row vectors).

12. Formulate and prove a version of the universal mapping property in §13.2 that applies to an infinite-dimensional vector space V with infinite basis B. Use the UMP to prove that the dual space V^* is isomorphic to the vector space of all functions from B to F with pointwise operations.

13. This exercise shows what can go wrong with the dual basis construction in the case of infinite-dimensional vector spaces. Recall (Chapter 3) that $F[x]$ is the vector space of formal polynomials with coefficients in F. The set $\{x^j : j \in \mathbb{Z}_{\geq 0}\}$ is a basis for $F[x]$. For each $k \in \mathbb{Z}_{\geq 0}$, define $C_k : F[x] \to F$ by $C_k(\sum_{i \geq 0} a_i x^i) = a_k$. Define $h : F[x] \to F$ by $h(\sum_{i=0}^n a_i x^i) = \sum_{i=0}^n a_i$.
    (a) Prove $C_k \in F[x]^*$, and compute $C_k(x^j)$ for all $j, k \in \mathbb{Z}_{\geq 0}$.
    (b) Prove $\{C_k : k \in \mathbb{Z}_{\geq 0}\}$ is F-linearly independent.
    (c) Prove $h \in F[x]^*$, but h is outside the span of $\{C_k : k \in \mathbb{Z}_{\geq 0}\}$.

14. Let $F[[x]]$ be the vector space of formal power series with coefficients in F (see §3.3). Define $C_k : F[[x]] \to F$ by $C_k(\sum_{i=0}^\infty a_i x^i) = a_k$. Is $\{C_k : k \in \mathbb{Z}_{\geq 0}\}$ a basis of $F[[x]]^*$?

15. Prove that $F[x]^*$ and $F[[x]]$ are isomorphic as F-vector spaces.

16. **Elementary Operations and Dual Bases.** Let $B = (x_1, \ldots, x_n)$ be an ordered basis of V with dual basis $B^d = (g_1, \ldots, g_n)$.
    (a) Let C be the ordered basis obtained from B by switching the positions of x_i and x_j. Describe (with proof) how C^d is related to B^d.
    (b) Let $C = (x_1, \ldots, dx_i, \ldots, x_n)$ where $d \in F$ is not 0. How is C^d related to B^d?
    (c) Let C be obtained from B by replacing x_i by $x_i + bx_j$ for some $i \neq j$ and some $b \in F$. How is C^d related to B^d?

17. Let Re and Im be the functions from $\mathbb{C}$ to $\mathbb{R}$ given by $\mathrm{Re}(a + ib) = a$ and $\mathrm{Im}(a + ib) = b$ for $a, b \in \mathbb{R}$. Find an ordered basis $B = (z_1, z_2)$ of $\mathbb{C}$ (viewed as a real vector space) such that $B^d = (\mathrm{Im}, \mathrm{Re})$.

18. Define $f, g \in (\mathbb{R}^2)^*$ by $f((x,y)) = x + y$ and $g((x,y)) = 2x - 3y$ for $x, y \in \mathbb{R}$. Find an ordered basis $B = (\mathbf{v}, \mathbf{w})$ of $\mathbb{R}^2$ such that $B^d = (f, g)$.

19. Given an ordered basis $C = (f_1, \ldots, f_n)$ of $(F^n)^*$ (viewed as row vectors), describe a matrix computation that finds an ordered basis $B = (\mathbf{v}_1, \ldots, \mathbf{v}_n)$ of F^n (viewed as column vectors) such that $C = B^d$.

20. Define functionals in $(\mathbb{R}^4)^*$ by $f((x_1, x_2, x_3, x_4)) = x_1 + 2x_2 - 3x_3 + x_4$, $g((x_1, x_2, x_3, x_4)) = 4x_1 - x_2 - x_4$, and $h((x_1, x_2, x_3, x_4)) = 3x_2 + 5x_3 - x_4$.
    (a) Find a basis for $\mathcal{Z}(\{f, g, h\})$.
    (b) Find a basis for $\mathcal{Z}(\{f, g\})$.
    (c) Find a basis for $\mathcal{Z}(\{f\})$.

21. Let V be the vector space of differentiable functions $f : (0, 3) \to \mathbb{R}$. Define $D, I \in V^*$ by $D(f) = f'(1/2)$ and $I(f) = \int_1^2 f(x)\, dx$ for $f \in V$. Describe geometrically which functions f belong to $\mathcal{Z}(\{D, I\})$.

22. Let $B = (x_1, \ldots, x_n)$ be an ordered basis of V with dual basis $B^d = (f_1, \ldots, f_n)$.
    (a) Prove: for all $v \in V$, $v = \sum_{i=1}^n f_i(v)x_i$.
    (b) For $1 \leq i \leq n$, explicitly describe $\mathcal{Z}(\{f_1, \ldots, f_i\})$ and $\mathcal{A}(\{x_1, \ldots, x_i\})$.

23. Suppose $T \subseteq V$ and $g \in V^*$. Negate the definition to give a useful completion of this sentence: "$g \notin \mathcal{A}(T)$ means..." Then prove that $\mathcal{A}(\emptyset) = V^*$.

24. For each subset of $\mathbb{R}^n$, describe $\mathcal{A}(T)$ as explicitly as possible.
    (a) $\{(2, 3)\}$
    (b) $\{(s, s, s) : s \in \mathbb{R}\}$
    (c) $\{(w, x, y, z) : w + x + y + z = 0 \text{ and } w + z = x + y\}$
    (d) $\{(x, y) : x^2 + y^2 = 1\}$
    (e) $\{(x, y, z) : x^2 + y^2 = 1, z = 0\}$

25. For each subset T in Exercise 24, describe $\mathcal{Z}(\mathcal{A}(T))$.

26. For each subset T of $M_n(\mathbb{R})$, describe $\mathcal{A}(T)$ in $M_n(\mathbb{R})^*$ as explicitly as possible.
    (a) $\{A \in M_n(\mathbb{R}) : \text{tr}(A) = 0\}$
    (b) $\{A \in M_n(\mathbb{R}) : \det(A) = 0\}$
    (c) $\{\text{upper-triangular } A \in M_n(\mathbb{R})\}$
    (d) $\{\text{diagonal } A \in M_n(\mathbb{R})\}$
    (e) $\{cI_n : c \in \mathbb{R}\}$

27. Prove: for an n-dimensional vector space V and all $T \subseteq V^*$, there exists a subset S of V^* of size at most n with $\mathcal{Z}(T) = \mathcal{Z}(S)$. Formulate and prove an analogous statement for annihilators.

28. Which of the five properties of zero-sets listed in §13.3 remain true for infinite-dimensional vector spaces V? Which of the five properties of annihilators listed in §13.4 remain true for infinite-dimensional vector spaces V?

29. Show directly from the definitions that $\mathcal{A}(\mathcal{Z}(S)) \supseteq S$ for all $S \subseteq V^*$. Use this and the identification of V and V^{**} to deduce $\mathcal{Z}(\mathcal{A}(T)) \supseteq T$ for all $T \subseteq V$.

30. Prove: for any family of subsets $\{S_i : i \in I\}$ of V^*, $\mathcal{Z}(\bigcup_{i \in I} S_i) = \bigcap_{i \in I} \mathcal{Z}(S_i)$. Formulate and prove an analogous statement for the $\mathcal{A}$ operator.

31. For general subsets $S, T \subseteq V^*$, must it be true that $\mathcal{Z}(S \cap T) = \mathcal{Z}(S) + \mathcal{Z}(T)$?

32. Use dual spaces to prove that every k-dimensional subspace W of F^n is the solution set of a system of $n - k$ homogeneous linear equations. Also prove that W is the intersection of $n - k$ linear hyperplanes (subspaces of dimension $n - 1$).

33. Let $F = \mathbb{Z}_2$ and $V = F^2$. Explicitly describe all elements of V^* by giving the domain, codomain, and values of each $f \in V^*$. Explicitly describe all elements $E \in V^{**}$. For each such E, find an $x \in V$ such that $E = E_x$ (evaluation at x). List all ordered bases of V. For each basis, find the associated dual basis of V^*.

34. Let $\mathbf{v}_1 = (1,2,3)$, $\mathbf{v}_2 = (0,1,-1)$, and $\mathbf{v}_3 = (0,0,1)$, so $B = (\mathbf{v}_1, \mathbf{v}_2, \mathbf{v}_3)$ is an ordered basis of $\mathbb{R}^3$. Let $B^d = (f_1, f_2, f_3)$ be the dual basis for B. Compute the matrix of each f_i relative to the standard ordered bases of $\mathbb{R}^3$ and $\mathbb{R}$. For $\mathbf{w} = (2,3,-1)$ and $i = 1, 2, 3$, find $E_\mathbf{w}(f_i)$. Use the UMP to obtain $E \in (\mathbb{R}^3)^{**}$ that sends f_1 to 4, f_2 to -1, and f_3 to 0. Find $\mathbf{x} \in \mathbb{R}^3$ such that $E = E_\mathbf{x}$.

35. Let $B = (x_1, \ldots, x_n)$ be an ordered basis of V and $B^d = (f_1, \ldots, f_n)$ be the dual basis of V^*. Show that $(E_{x_1}, \ldots, E_{x_n})$ is the dual basis B^{dd} of V^{**}.

36. Prove: if V is infinite-dimensional, then the map ev $: V \to V^{**}$ is not surjective. (Use Exercise 12.)

37. (a) Give a direct proof (using bases but not facts about $\mathcal{A}$) that $\dim(W) + \dim(\mathcal{Z}(W)) = \dim(V)$ for all subspaces W of V^*.
    (b) Use (a) to deduce $\dim(Y) + \dim(\mathcal{A}(Y)) = \dim(V)$ for all subspaces Y of V.

38. (a) In §13.6, give the details of the proof that $\mathcal{A}(\mathcal{Z}(W)) = W$ for all $W \in \mathcal{S}(V^*)$.
    (b) Explain why $\mathcal{Z}(\mathcal{A}(T)) = \langle T \rangle$ for all $T \subseteq V$.

39. Let $(\mathbf{e}_1, \mathbf{e}_2, \mathbf{e}_3)$ be the standard ordered basis of $\mathbb{R}^3$ and (f_1, f_2, f_3) be the dual basis of $(\mathbb{R}^3)^*$. Let $X = \langle f_1, f_2 \rangle$ and $Y = \langle f_2, f_3 \rangle$. Verify (13.4) by explicitly computing $X \cap Y$, $X + Y$, $\mathcal{Z}(X)$, $\mathcal{Z}(Y)$, $\mathcal{Z}(X) + \mathcal{Z}(Y)$, $\mathcal{Z}(X) \cap \mathcal{Z}(Y)$, $\mathcal{Z}(X \cap Y)$, and $\mathcal{Z}(X + Y)$.

40. Let $X = \langle (1,2,1) \rangle$ and $Y = \langle (2,-1,0) \rangle$ in $\mathbb{R}^3$. Verify (13.5) by explicitly computing $X \cap Y$, $X + Y$, $\mathcal{A}(X)$, $\mathcal{A}(Y)$, $\mathcal{A}(X) + \mathcal{A}(Y)$, $\mathcal{A}(X) \cap \mathcal{A}(Y)$, $\mathcal{A}(X \cap Y)$, and $\mathcal{A}(X + Y)$.

41. Let $T = \{(1,0,1), (2,2,2)\} \subseteq \mathbb{R}^3$. Find $\mathcal{Z}(\mathcal{A}(T))$ first by using a theorem, then again by using the definitions of $\mathcal{A}$ and $\mathcal{Z}$.

42. Let $\mathcal{S}(V)$ be the poset of subspaces of V, ordered by set inclusion.
    (a) Prove $\inf(X, Y) = X \cap Y$ for all $X, Y \in \mathcal{S}(V)$.
    (b) Prove $\sup(X, Y) = X + Y$ for all $X, Y \in \mathcal{S}(V)$.
    (c) Prove $\inf(X_i : i \in I) = \bigcap_{i \in I} X_i$ for all $X_i \in \mathcal{S}(V)$.
    (d) Prove $\sup(X_i : i \in I) = \sum_{i \in I} X_i$ for all $X_i \in \mathcal{S}(V)$.

43. Suppose U and V are lattices and $f : U \to V$ is a bijection such that for all $x, y \in U$, $x \le y$ iff $f(y) \le f(x)$. Prove $f(\inf(x,y)) = \sup(f(x), f(y))$ and $f(\sup(x,y)) = \inf(f(x), f(y))$ for all $x, y \in U$. Explain how (13.4) and (13.5) follow from this.

44. State and prove analogs of (13.4) and (13.5) that involve the greatest lower bound and least upper bound of an indexed collection of subspaces.

45. Let F be a finite field. Show that, for $0 \le k \le n$, the number of k-dimensional subspaces of an n-dimensional F-vector space V is the same as the number of $(n - k)$-dimensional subspaces of V.

46. Assume $\dim(V) = n > 1$. Prove there does not exist any isomorphism $f : V \to V^*$ such that $f \circ T = T^* \circ f$ for all linear maps $T : V \to V$.

47. Prove $(V \times W)^* \cong V^* \times W^*$. Show that $\mathcal{A}(V \times \{0\})$ maps to $\{0\} \times W^*$ under this isomorphism. Generalize to a direct product of k vector spaces.

48. If a linear map $T : V \to W$ has matrix A relative to an ordered basis B of V and an ordered basis C of W, what is the matrix of T^{**} using bases B^{dd} and C^{dd}?

49. Let $D : F[x] \to F[x]$ be the map given by $D(\sum_{k>0} a_k x^k) = \sum_{k \ge 1} k a_k x^{k-1}$. Describe how D^* acts on the vector space $F[x]^* \cong F[[x]]$ (see Exercise 15).

50. Prove: for all linear $T : V \to V$ and all integers $k \ge 0$, $(T^k)^* = (T^*)^k$. Prove this holds for all $k \in \mathbb{Z}$ if T is invertible.

51. Prove: if linear maps $T, U : V \to V$ are similar, then T^* and U^* are similar.

52. Let $T : V \to W$ be a linear map. Prove: for all $S \subseteq W$, $T^*[\mathcal{A}(S)] \subseteq \mathcal{A}(T^{-1}[S])$. Give an example where equality does not hold.

53. Let $T : V \to W$ be a linear map. Prove: for all $S \subseteq V$, $\mathcal{A}(T[S]) = (T^*)^{-1}[\mathcal{A}(S)]$.

54. Let $T : V \to W$ be a linear map. Prove $\ker(T^*) = \mathcal{A}(\operatorname{img}(T))$. Deduce that T^* is one-to-one iff T is onto.

55. Let $T : V \to W$ be a linear map. Prove $\operatorname{img}(T^*) = \mathcal{A}(\ker(T))$. Deduce that T^* is onto iff T is one-to-one.

56. Let V be any F-vector space, possibly infinite-dimensional. Prove: for any subspace U of V, $\mathcal{Z}(\mathcal{A}(U)) = U$, $\mathcal{A}(U) \cong (V/U)^*$, $\dim(U) = \dim(V^*/\mathcal{A}(U))$, and $\dim(\mathcal{A}(U)) = \dim(V/U)$.

57. Let V be any F-vector space, possibly infinite-dimensional. Let $\mathcal{S}_f(V^*)$ be the poset of finite-dimensional subspaces of V^*. Let $\mathcal{S}^f(V)$ be the poset of subspaces U of V such that $\dim(V/U)$ is finite. Prove $\mathcal{Z}$ and $\mathcal{A}$ restrict to give poset anti-isomorphisms between the lattices $\mathcal{S}_f(V^*)$ and $\mathcal{S}^f(V)$.

58. Let $R : V \to W^*$ be the map induced by a bilinear pairing $B : W \times V \to F$. Show R is F-linear with kernel $W^{\perp}$.

59. Define $B : M_{1,n}(F) \times M_{n,1}(F) \to F$ by letting $B(\mathbf{w}, \mathbf{v})$ be the sole entry of $\mathbf{wv}$. Show B is a bilinear pairing that is nondegenerate on both sides.

60. Suppose W has ordered basis $(w_1, \ldots, w_m)$ and V has ordered basis $(v_1, \ldots, v_n)$. Prove: for each $A \in M_{m,n}(F)$, there exists a unique bilinear pairing $B : W \times V \to F$ such that $B(w_i, v_j) = A(i, j)$ for $1 \le i \le m$ and $1 \le j \le n$.

61. Let B be related to $A \in M_{m,n}(F)$ as in the previous exercise. Prove B is left-degenerate iff $\mathbf{w}A = \mathbf{0}$ for some nonzero row vector $\mathbf{w} \in M_{1,m}(F)$. State and prove a similar condition for B to be right-degenerate. What can you say in the special case $m = n$?

62. Suppose $B : W \times V \to F$ is a bilinear pairing, $S \subseteq W$, and $T \subseteq V$. Prove $S^{\perp}$ is a subspace of V and $^{\perp}T$ is a subspace of W. Prove the orthogonal complement operators reverse inclusions, $S \subseteq {}^{\perp}(S^{\perp})$, and $T \subseteq ({}^{\perp}T)^{\perp}$.

63. State and prove an analog of the Theorem on Left-Nondegenerate Pairings for right-nondegenerate pairings $B : W \times V \to F$.

64. Check the assertions about B' in the proof of the Theorem on General Bilinear Pairings.

65. In the proof of the Theorem on Adjoints, check that $T'(cy) = cT'(y)$ for $c \in F$ and $y \in V$.

66. Prove the Theorem on Adjoints by checking that $T' = R^{-1} \circ T^* \circ R$.

67. Suppose V is a real inner product space, $T : V \to V$ is a linear map, T' is the adjoint of T, C is an ordered basis of V, and C' is the dual basis. Prove: if $[T]_C = A$, then $[T']_{C'} = A^{\mathrm{T}}$.

68. Let V be a real inner product space and $R : V \to V^*$ the isomorphism sending $y \in V$ to right-multiplication by y. Prove: for all $S \subseteq V$, $\mathcal{A}(S) = R[S^{\perp}]$; and for all $T \subseteq V^*$, $\mathcal{Z}(T) = R^{-1}[T]^{\perp}$.

69. Check the facts stated about V^{σ} in §13.11.

70. Suppose V is a complex inner product space, $T : V \to V$ is a linear map, T' is the adjoint of T, C is an ordered basis of V, C' is the dual basis, and $[T]_C = A$. Prove: if we view T' as a $\mathbb{C}$-linear map on V^σ, then $[T']_{C'} = A^{\mathrm{T}}$. If we view T' as a $\mathbb{C}$-linear map on V, then $[T']_{C'} = \overline{A^{\mathrm{T}}}$.

71. Prove results similar to Exercises 54 and 55 for the adjoint of a linear operator on a real or complex inner product space.

72. Give an example of a degenerate symmetric bilinear form B on $\mathbb{R}^3$ with $B(\mathbf{e}_i, \mathbf{e}_j) \neq 0$ for all $i, j \in \{1, 2, 3\}$. Find all $\mathbf{v} \in \mathbb{R}^3$ with $L(\mathbf{v}) = 0$.

73. Give an example of a nondegenerate bilinear form B on $\mathbb{R}^4$ with $B(\mathbf{v}, \mathbf{v}) = 0$ for all $\mathbf{v} \in \mathbb{R}^4$.

74. Suppose B is a symmetric bilinear form on a real vector space V such that $B(v, v) = 0$ for all $v \in V$. Prove $B = 0$.

75. Consider the following bilinear forms on $\mathbb{R}^3$:

$$B_1(\mathbf{x}, \mathbf{y}) = x_1 y_1 + x_2 y_2 + x_3 y_3;$$
$$B_2(\mathbf{x}, \mathbf{y}) = x_1 y_2 - x_2 y_1 + 2x_3 y_3;$$
$$B_3(\mathbf{x}, \mathbf{y}) = x_1 y_1 + x_1 y_2 + x_1 y_3 + x_2 y_2 + x_2 y_3 + x_3 y_3;$$
$$B_4(\mathbf{x}, \mathbf{y}) = x_1 y_1 + x_1 y_3 + 2x_2 y_2 + x_3 y_1 + x_3 y_3.$$

Which of these bilinear forms are symmetric? Which are nondegenerate? Let $f : \mathbb{R}^3 \to \mathbb{R}$ be the linear functional $f(\mathbf{x}) = 3x_1 - 2x_2 - x_3$ for $\mathbf{x} \in \mathbb{R}^3$. For $1 \leq i \leq 4$, find $\mathbf{y} \in \mathbb{R}^3$ such that $f(\mathbf{x}) = B_i(\mathbf{x}, \mathbf{y})$ for all $\mathbf{x} \in \mathbb{R}^3$, or explain why this cannot be done. For $1 \leq i \leq 4$, find $\mathbf{z} \in \mathbb{R}^3$ such that $f(\mathbf{x}) = B_i(\mathbf{z}, \mathbf{x})$ for all $\mathbf{x} \in \mathbb{R}^3$, or explain why this cannot be done.

76. Let V have ordered basis $X = (x_1, \ldots, x_n)$, let the dual basis of V^* be $X^d = (f_1, \ldots, f_n)$, let $A \in M_n(F)$, and let B be the unique bilinear form on V such that $B(x_i, x_j) = A(i, j)$ for i, j between 1 and n. What is the matrix of the linear map $R : V \to V^*$ relative to the basis X of V and the basis X^d of V^*? What is the matrix of the linear map $L : V \to V^*$ relative to the basis X of V and the basis X^d of V^*? Assume B is symmetric and nondegenerate, and $T : V \to V$ is a linear map with matrix C relative to X. What is the matrix of the adjoint map T' relative to X?

77. Let $T : \mathbb{R}^4 \to \mathbb{R}^4$ be the map $T(\mathbf{x}) = A\mathbf{x}$ for $\mathbf{x} \in \mathbb{R}^4$, where

$$A = \begin{bmatrix} 2 & 0 & -1 & 3 \\ 1 & 1 & -2 & -1 \\ 0 & 1 & 1 & 2 \\ 3 & 5 & -6 & 2 \end{bmatrix}.$$

Find the adjoint of T relative to the standard inner product on $\mathbb{R}^4$. Find the adjoint of T relative to the symmetric bilinear form $B(\mathbf{x}, \mathbf{y}) = x_1 y_1 + x_2 y_2 + x_3 y_3 - x_4 y_4$. Find the adjoint of T relative to the symmetric bilinear form $B(\mathbf{x}, \mathbf{y}) = x_1 y_2 + x_2 y_1 + 2x_3 y_4 + 2x_4 y_3$.

78. Let $U = \{(t, 3t, t) : t \in \mathbb{R}\}$ and $W = \{(x, y, z) \in \mathbb{R}^3 : 5x - 2y + 3z = 0\}$. Compute $U^\perp$ and $W^\perp$ relative to the standard inner product on $\mathbb{R}^3$.

79. For each linear operator T on V, let T' denote the adjoint operator relative to a fixed nondegenerate symmetric bilinear form B. Use the defining formula (13.6) to prove: for all $S, T \in \mathrm{Hom}_F(V, V)$ and $c \in F$, $(S + T)' = S' + T'$, $(cS)' = c(S')$, and $(S \circ T)' = T' \circ S'$. Prove these formulas again using the relation between T' and T^*.

80. Give an example of a vector space V, a symmetric nondegenerate bilinear form B on V, and a subspace W of V such that $W \cap W^\perp \neq \{0\}$ and $V \neq W \oplus W^\perp$. For your choice of W, verify that $W^{\perp\perp} = W$ and $\dim(W) + \dim(W^\perp) = \dim(V)$.

81. Show that for any ideal I in a commutative ring R, $\sqrt{I}$ (defined on page 357) is a radical ideal.

82. Prove the first five properties of the operator $\mathcal{Z}$ listed in §13.13.

83. Let I and J be ideals of a commutative ring R. Let IJ be the set of finite sums $i_1 j_1 + \cdots + i_m j_m$ where $i_1, \ldots, i_m \in I$ and $j_1, \ldots, j_m \in J$. Prove that IJ is an ideal of R, and $IJ \subseteq I \cap J$. Prove that $\mathcal{Z}(IJ) = \mathcal{Z}(I \cap J) = \mathcal{Z}(I) \cup \mathcal{Z}(J)$.

84. Prove the first three properties of the operator $\mathcal{A}$ listed in §13.13.

85. True or false? Explain each answer.
    (a) For all $x \in V$, there exists $f \in V^*$ with $f(x) \neq 0$.
    (b) For all vector spaces Z, $\dim(Z) = \dim(Z^*)$.
    (c) Given $\dim(V) = n$, every nonzero $f \in V^*$ satisfies $\dim(\ker(f)) = n - 1$.
    (d) For all vector spaces Z, $\mathrm{ev} : Z \to Z^{**}$ is an isomorphism.
    (e) For all $S, T \subseteq V^*$, if $\mathcal{Z}(S) = \mathcal{Z}(T)$ then $S = T$.
    (f) For all $x \neq y$ in V, there exists $f \in V^*$ with $f(x) \neq f(y)$.
    (g) For all linearly independent lists $(f_1, \ldots, f_k)$ in V^*, there exist $x_1, \ldots, x_k \in V$ with $f_i(x_j) = \delta_{ij}$ for $1 \leq i, j \leq k$.
    (h) For all subspaces U, Z of a real inner product space, if $U^\perp = Z^\perp$ then $U = Z$.
    (i) The map $D : \mathrm{Hom}_F(V, W) \to \mathrm{Hom}_F(W^*, V^*)$ given by $D(T) = T^*$ is F-linear.
    (j) For all linear maps $S, T : V \to V$, $(S \circ T)^* = S^* \circ T^*$.
    (k) For all $T \subseteq V^*$, $\mathcal{Z}(\mathcal{A}(\mathcal{Z}(T))) = \mathcal{Z}(T)$.

14

Bilinear Forms

The *dot product* on $\mathbb{R}^n$ is defined by $\mathbf{v} \bullet \mathbf{w} = v_1 w_1 + v_2 w_2 + \cdots + v_n w_n$ for all $\mathbf{v} = (v_1, v_2, \ldots, v_n)$ and $\mathbf{w} = (w_1, w_2, \ldots, w_n)$ in $\mathbb{R}^n$. The dot product supplies extra algebraic structure to the vector space $\mathbb{R}^n$, which allows us to define key geometric ideas: the length of one vector, the distance between two vectors, the angle between vectors, and perpendicularity of vectors. In more detail, the *Euclidean norm* of $\mathbf{v} \in \mathbb{R}^n$ is $||\mathbf{v}|| = \sqrt{\mathbf{v} \bullet \mathbf{v}} = \sqrt{v_1^2 + v_2^2 + \cdots + v_n^2}$. We regard $||\mathbf{v}||$ as the length of the arrow (directed line segment) from the origin of $\mathbb{R}^n$ to the point $(v_1, v_2, \ldots, v_n)$. More generally, the *Euclidean distance from* $\mathbf{v}$ *to* $\mathbf{w}$ is $d_2(\mathbf{v}, \mathbf{w}) = ||\mathbf{v} - \mathbf{w}|| = \sqrt{(\mathbf{v} - \mathbf{w}) \bullet (\mathbf{v} - \mathbf{w})}$. This distance is the length of the line segment joining the points $(v_1, \ldots, v_n)$ and $(w_1, \ldots, w_n)$. A *unit vector* is a vector $\mathbf{u} \in \mathbb{R}^n$ such that $||\mathbf{u}|| = 1$. The *angle* between nonzero vectors $\mathbf{v}, \mathbf{w} \in \mathbb{R}^n$ is the unique real θ in the range $0 \le \theta \le \pi$ such that $\cos(\theta) = \frac{\mathbf{v} \bullet \mathbf{w}}{||\mathbf{v}|| \cdot ||\mathbf{w}||}$. Vectors $\mathbf{v}, \mathbf{w} \in \mathbb{R}^n$ are called *orthogonal* iff $\mathbf{v} \bullet \mathbf{w} = 0$. Nonzero vectors $\mathbf{v}$ and $\mathbf{w}$ are orthogonal iff the angle between them is $\pi/2$ (a right angle), which means that the arrows representing $\mathbf{v}$ and $\mathbf{w}$ are perpendicular.

Using the Gram–Schmidt algorithm, you can show that every subspace W of $\mathbb{R}^n$ has an *orthonormal basis*, which is a list of unit vectors $(\mathbf{w}_1, \ldots, \mathbf{w}_k)$ spanning the subspace W such that $\mathbf{w}_i$ is orthogonal to $\mathbf{w}_j$ for all $i \ne j$. In other words, every $\mathbf{w} \in W$ has the form $\mathbf{w} = \sum_{i=1}^{k} c_i \mathbf{w}_i$ for some $c_i \in \mathbb{R}$, and $\mathbf{w}_i \bullet \mathbf{w}_j$ is 1 if $i = j$ and 0 if $i \ne j$. These conditions imply that $(\mathbf{w}_1, \ldots, \mathbf{w}_k)$ is linearly independent, so an orthonormal basis really is a basis of W. Every orthonormal list in W can be extended to an orthonormal basis of W.

Given any subspace W of $\mathbb{R}^n$, define the *orthogonal subspace* $W^\perp$ to be the set of $\mathbf{v} \in \mathbb{R}^n$ such that $\mathbf{w} \bullet \mathbf{v} = 0$ for all $\mathbf{w} \in W$. Using an orthonormal basis for W, you can check that $\mathbb{R}^n$ is the direct sum $W \oplus W^\perp$, meaning that every $\mathbf{z} \in \mathbb{R}^n$ can be written uniquely in the form $\mathbf{z} = \mathbf{w} + \mathbf{v}$, where $\mathbf{w} \in W$ and $\mathbf{v} \in W^\perp$. Among other facts, we have $W \cap W^\perp = \{0\}$, $(W^\perp)^\perp = W$, and $\dim(W) + \dim(W^\perp) = n = \dim(\mathbb{R}^n)$. The map sending W to $W^\perp$ is an inclusion-reversing bijection on the lattice of all subspaces of $\mathbb{R}^n$.

Our goal in this chapter is to study bilinear forms on general vector spaces V; these forms generalize the dot product on $\mathbb{R}^n$. General bilinear forms share some, but not all, of the properties of the dot product reviewed above. For example, the *Lorentz form* on $\mathbb{R}^4$ sends a pair of input vectors (t, x, y, z), (t', x', y', z') to the output $tt' - xx' - yy' - zz'$. Using this form, the product of a nonzero vector with itself can be negative or zero. The Lorentz form is important in the study of spacetime and the special theory of relativity.

We begin the chapter with the definition of bilinear forms and some basic examples. We show how to represent a bilinear form using a matrix by choosing an ordered basis for the underlying vector space V. We define congruent matrices, which represent the same bilinear form relative to different choices of bases. Next, we examine orthogonality, dual spaces, orthogonal complements, and the radical of a bilinear form. Using these concepts, we show that symmetric bilinear forms can be represented by diagonal matrices, while alternate bilinear forms have matrices with a simple block-diagonal structure. The chapter ends with more advanced structural results such as Witt's Decomposition Theorem, Witt's Cancellation Theorem, and the generation of orthogonal groups by reflections.

DOI: 10.1201/9781003484561-14

14.1 Definition of Bilinear Forms

Let V be a vector space over a field F. A *bilinear form* on V is a function $B : V \times V \to F$ such that, for all $u, v, w \in V$ and all $c \in F$:

- (linearity in first input) $B(u + v, w) = B(u, w) + B(v, w)$ and $B(cu, w) = cB(u, w)$;

- (linearity in second input) $B(w, u + v) = B(w, u) + B(w, v)$ and $B(w, cu) = cB(w, u)$.

In other words, for each fixed choice of $w \in V$, the function sending $u \in V$ to $B(u, w)$ is a linear map from V to F, and the function sending u to $B(w, u)$ is a linear map from V to F. Since linear maps preserve linear combinations, we deduce the following additional properties. For all $v_1, \ldots, v_k, w \in V$ and all $c_1, \ldots, c_k \in F$,

$$B\left(\sum_{i=1}^{k} c_i v_i, w\right) = \sum_{i=1}^{k} c_i B(v_i, w) \quad \text{and} \quad B\left(w, \sum_{i=1}^{k} c_i v_i\right) = \sum_{i=1}^{k} c_i B(w, v_i).$$

Combining these formulas, we get the *general bilinearity formula*

$$B\left(\sum_{i=1}^{k} c_i v_i, \sum_{j=1}^{m} d_j w_j\right) = \sum_{i=1}^{k} \sum_{j=1}^{m} c_i d_j B(v_i, w_j) \tag{14.1}$$

for all choices of v_i, w_j in V and c_i, d_j in F. Since linear maps send zero to zero, we have $B(v, 0_V) = 0_F$ and $B(0_V, w) = 0_F$ for all $v, w \in V$.

Here is terminology for some special types of bilinear forms. Let B be a bilinear form on V.

- B is *symmetric* iff $B(v, w) = B(w, v)$ for all $v, w \in V$.

- B is *antisymmetric* iff $B(v, w) = -B(w, v)$ for all $v, w \in V$.

- B is *alternate* iff $B(v, v) = 0$ for all $v \in V$.

- B is *degenerate* iff there is a nonzero $w \in V$ such that for all $v \in V$, $B(v, w) = 0$.

- B is *nondegenerate* iff for each nonzero $w \in V$, there is $v \in V$ (depending on w) with $B(v, w) \neq 0$. This means that 0_V is the *only* vector $w \in V$ satisfying $B(v, w) = 0$ for all $v \in V$.

- A *bilinear space* is a vector space V together with a specific bilinear form B on V. We may say that a bilinear space is symmetric, nondegenerate, etc., if the bilinear form B has these properties.

Antisymmetric bilinear forms and alternate bilinear forms are closely related. For any field F, define $2_F = 1_F + 1_F$. *Every alternate bilinear form is antisymmetric, and the converse holds if $2_F \neq 0_F$.* To prove this, first suppose B is an alternate bilinear form. For any v, w in V, we compute

$$0 = B(v + w, v + w) = B(v, v) + B(v, w) + B(w, v) + B(w, w) = 0 + B(v, w) + B(w, v) + 0,$$

so $B(v, w) = -B(w, v)$. Conversely, suppose B is an antisymmetric bilinear form and $2_F \neq 0_F$. For any v in V, we know $B(v, v) = -B(v, v)$, so $B(v, v) + B(v, v) = 0_F$, so $(1_F + 1_F)B(v, v) = 0_F$. Because F is a field, we can multiply both sides by the inverse of the nonzero element $2_F = 1_F + 1_F$ to conclude that $B(v, v) = 0_F$.

14.2   Examples of Bilinear Forms

The dot product on the real vector space $\mathbb{R}^n$ is the motivating example for the concept of a bilinear form. We can generalize the dot product to any field F as follows. Given n-tuples $\mathbf{v} = (v_1, \ldots, v_n)$ and $\mathbf{w} = (w_1, \ldots, w_n)$ in the F-vector space F^n, define $B(\mathbf{v}, \mathbf{w}) = v_1 w_1 + v_2 w_2 + \cdots + v_n w_n$. You can check from the field axioms that B is a symmetric, nondegenerate bilinear form on F^n. For example, given $\mathbf{v}, \mathbf{w} \in F^n$ and $c \in F$,

$$\begin{aligned} B(\mathbf{v}, c\mathbf{w}) &= v_1(cw_1) + v_2(cw_2) + \cdots + v_n(cw_n) \\ &= c(v_1 w_1) + c(v_2 w_2) + \cdots + c(v_n w_n) \\ &= c(v_1 w_1 + \cdots + v_n w_n) = cB(\mathbf{v}, \mathbf{w}) \end{aligned}$$

using associativity and commutativity of multiplication in F and the distributive law. To see that B is nondegenerate, suppose $\mathbf{w} = (w_1, \ldots, w_n) \in F^n$ is nonzero. Then some coordinate w_i is not 0_F. Letting $\mathbf{e}_i$ be the vector in F^n with 1_F in position i and 0_F elsewhere, we have $B(\mathbf{e}_i, \mathbf{w}) = w_i \neq 0_F$.

Based on your experience with the dot product in $\mathbb{R}^n$, you might think that $B(\mathbf{w}, \mathbf{w}) = w_1^2 + w_2^2 + \cdots + w_n^2$ must also be nonzero. But this need not be true for a general field F. For instance, if $F = \mathbb{Z}_3$ (the integers mod 3), $n = 3$, and $\mathbf{w} = (1, 1, 1) \in F^3$, then $B(\mathbf{w}, \mathbf{w}) = 1_F + 1_F + 1_F = 3_F = 0_F$. Even for the field $F = \mathbb{R}$, it is possible for a nondegenerate symmetric bilinear space to have vectors v with $B(v, v) = 0$, as we see in later examples. A bilinear form on a real vector space V is called *positive definite* iff for all nonzero $v \in V$, $B(v, v)$ is a strictly positive real number. The dot product on $\mathbb{R}^n$ is a positive definite form. Positive definiteness implies nondegeneracy, but the converse does not always hold.

Next, consider the F-vector space F^2. For $\mathbf{v} = (v_1, v_2)$ and $\mathbf{w} = (w_1, w_2)$ in F^2, let

$$B_1(\mathbf{v}, \mathbf{w}) = v_1 w_1 - v_2 w_2, \quad B_2(\mathbf{v}, \mathbf{w}) = v_1 w_2 + v_2 w_1, \quad B_3(\mathbf{v}, \mathbf{w}) = v_1 w_2 - v_2 w_1. \quad (14.2)$$

You can check that these three functions are all bilinear and nondegenerate, even though B_2 and B_3 satisfy $B(\mathbf{e}_1, \mathbf{e}_1) = 0 = B(\mathbf{e}_2, \mathbf{e}_2)$, and B_1 is not positive definite in the real case. Forms B_1 and B_2 are symmetric, while B_3 is alternate and antisymmetric. The bilinear space (F^2, B_2) is called a *hyperbolic plane* over the field F. Hyperbolic planes are important building blocks for analyzing the structure of symmetric bilinear spaces.

For any F-vector space V, the zero function $B_0 : V \times V \to F$, defined by $B_0(v, w) = 0_F$ for all $v, w \in V$, is a symmetric and alternate bilinear form. B_0 is degenerate except when V itself is the zero space. Next, let V be any 1-dimensional F-vector space, which is isomorphic to $F = F^1$. Given a nonzero $\mathbf{v} \in V$, there is a unique bilinear form $B_\mathbf{v}$ on V such that $B_\mathbf{v}(\mathbf{v}, \mathbf{v}) = 1_F$, given by $B_\mathbf{v}(c\mathbf{v}, d\mathbf{v}) = cd$ for $c, d \in F$. The form $B_\mathbf{v}$ is symmetric and nondegenerate.

We can obtain new bilinear spaces by taking products of given bilinear spaces, as follows. Suppose $(V_1, B_1), (V_2, B_2), \ldots, (V_k, B_k)$ are bilinear spaces over the same field F. Let $V = V_1 \times V_2 \times \cdots \times V_k$ be the direct product of the vector spaces $V_1, V_2, \ldots, V_k$. By definition, elements of V are k-tuples $(\mathbf{v}_1, \mathbf{v}_2, \ldots, \mathbf{v}_k)$ with each $\mathbf{v}_i \in V_i$, and the vector space operations on V are defined componentwise. We define $B : V \times V \to F$ by setting

$$B((\mathbf{v}_1, \mathbf{v}_2, \ldots, \mathbf{v}_k), (\mathbf{w}_1, \mathbf{w}_2, \ldots, \mathbf{w}_k)) = B_1(\mathbf{v}_1, \mathbf{w}_1) + B_2(\mathbf{v}_2, \mathbf{w}_2) + \cdots + B_k(\mathbf{v}_k, \mathbf{w}_k). \quad (14.3)$$

You can check that: B is a bilinear form on V; B is symmetric iff all B_i are symmetric; B is alternate iff all B_i are alternate; B is nondegenerate iff all B_i are nondegenerate. We call (V, B) the *direct product* or the *external direct sum* of the bilinear spaces (V_i, B_i).

For example, for each $s \in \mathbb{R}$, let $(\mathbb{R}, B_s)$ be the real vector space $\mathbb{R}$ with the unique bilinear form B such that $B_s(1, 1) = s$. Taking the direct product of n copies of $(\mathbb{R}, B_1)$ gives the space $\mathbb{R}^n$ with the dot product. For integers $a, b, c \geq 0$, taking the direct product of a copies of $(\mathbb{R}, B_0)$, b copies of $(\mathbb{R}, B_1)$, and c copies of $(\mathbb{R}, B_{-1})$ gives a real symmetric bilinear space of dimension $a + b + c$, which is nondegenerate iff $a = 0$. As we prove in §14.9, this is the most general example of a finite-dimensional real symmetric bilinear space. The Witt Decomposition Theorem generalizes this structural result to symmetric bilinear spaces over other fields.

As a final trio of examples, consider the following forms on the F-vector space F^{2n}. For $\mathbf{v} = (v_1, \ldots, v_{2n})$ and $\mathbf{w} = (w_1, \ldots, w_{2n})$ in F^{2n}, define

$$
\begin{aligned}
B_{1n}(\mathbf{v}, \mathbf{w}) &= v_1 w_1 + \cdots + v_n w_n - v_{n+1} w_{n+1} - \cdots - v_{2n} w_{2n}; \\
B_{2n}(\mathbf{v}, \mathbf{w}) &= \sum_{i=1}^{n} v_i w_{n+i} + \sum_{i=1}^{n} v_{n+i} w_i; \\
B_{3n}(\mathbf{v}, \mathbf{w}) &= \sum_{i=1}^{n} v_i w_{n+i} - \sum_{i=1}^{n} v_{n+i} w_i.
\end{aligned}
\tag{14.4}
$$

You can check that these are nondegenerate bilinear forms on F^{2n}, where the first two are symmetric and the third is alternate. For $j = 1, 2, 3$, we can think of the bilinear spaces (F^{2n}, B_{jn}) as direct sums of n copies of the spaces (F^2, B_j) considered in (14.2). Specifically, consider the subspaces $Z_i = F\mathbf{e}_i + F\mathbf{e}_{n+i}$ of F^{2n} for $i = 1, 2, \ldots, n$. Each Z_i is a copy of F^2, and B_{jn} acts on Z_i in the same way that B_j acts on F^2. For example, for all $a, b, c, d \in F$, $B_2((a, b), (c, d)) = ad + bc$, while $B_{2n}(a\mathbf{e}_i + b\mathbf{e}_{n+i}, c\mathbf{e}_i + d\mathbf{e}_{n+i}) = ad + bc$. Since F^{2n} is the (internal) direct sum of the subspaces $Z_1, \ldots, Z_n$, (F^{2n}, B_{jn}) is essentially the same as the (external) direct sum of n copies of (F^2, B_j).

To make this last comment more precise, we introduce the idea of isomorphic bilinear spaces. Given bilinear spaces (V, B) and (V', B') over the same field F, a *bilinear homomorphism* is an F-linear map $T : V \to V'$ such that $B'(T(x), T(y)) = B(x, y)$ for all $x, y \in V$. By linearity of T and bilinearity of B and B', it suffices to check the last condition for all x, y in any given basis for V (Exercise 10). A *bilinear isomorphism* is an invertible bilinear homomorphism. Continuing the previous example, take (V, B) to be the direct product of n copies of (F^2, B_j), and let $(V', B') = (F^{2n}, B_{jn})$. You can check that

$$
T(((v_1, x_1), (v_2, x_2), \ldots, (v_n, x_n))) = (v_1, v_2, \ldots, v_n, x_1, x_2, \ldots, x_n)
$$

is an isomorphism between these bilinear spaces for $j = 1, 2, 3$.

The apparently different bilinear spaces $(\mathbb{R}^2, B_1)$ and $(\mathbb{R}^2, B_2)$ defined in (14.2) are actually isomorphic. You can check that $T : \mathbb{R}^2 \to \mathbb{R}^2$ defined by $T(v_1, v_2) = \frac{1}{\sqrt{2}}(v_1 + v_2, v_1 - v_2)$ is an invertible linear map; in fact, $T^{-1} = T$. We verify that $B_2(T(x), T(y)) = B_1(x, y)$ by checking on the standard basis:

$$
B_2(T(\mathbf{e}_1), T(\mathbf{e}_1)) = \frac{1}{2} B_2((1, 1), (1, 1)) = 1 = B_1((1, 0), (1, 0)) = B_1(\mathbf{e}_1, \mathbf{e}_1);
$$

$$
B_2(T(\mathbf{e}_1), T(\mathbf{e}_2)) = \frac{1}{2} B_2((1, 1), (1, -1)) = 0 = B_1((1, 0), (0, 1)) = B_1(\mathbf{e}_1, \mathbf{e}_2);
$$

$$
B_2(T(\mathbf{e}_2), T(\mathbf{e}_1)) = \frac{1}{2} B_2((1, -1), (1, 1)) = 0 = B_1((0, 1), (1, 0)) = B_1(\mathbf{e}_2, \mathbf{e}_1);
$$

$$
B_2(T(\mathbf{e}_2), T(\mathbf{e}_2)) = \frac{1}{2} B_2((1, -1), (1, -1)) = -1 = B_1((0, 1), (0, 1)) = B_1(\mathbf{e}_2, \mathbf{e}_2).
$$

In fact, (F^2, B_1) and (F^2, B_2) are isomorphic for any field F such that $2_F \neq 0_F$, but a different proof is needed if 2 has no square root in F (Exercise 11).

14.3 Matrix of a Bilinear Form

Let V be an n-dimensional vector space over a field F. To define a general function B mapping the domain $V \times V$ into the codomain F, we must specify the scalar $B(v, w)$ in F for each ordered pair (v, w) of vectors v, w in V. When the function B happens to be a bilinear form, we can uniquely determine B using less information. Let $X = (x_1, x_2, \ldots, x_n)$ be any ordered basis for V. Suppose we know the values $B(x_i, x_j)$ for each ordered pair (x_i, x_j) of basis elements x_i, x_j in X. Then we can find $B(v, w)$ for any v, w in V as follows. We have expansions $v = \sum_{i=1}^{n} c_i x_i$ and $w = \sum_{j=1}^{n} d_j x_j$ for certain unique scalars c_i and d_j in F, since X is a basis of V. By the general bilinearity formula (14.1), we know

$$B(v, w) = \sum_{i=1}^{n} \sum_{j=1}^{n} c_i d_j B(x_i, x_j).$$

Introduce notation $[B]_X$ for the $n \times n$ matrix with entry $B(x_i, x_j)$ in row i, column j. We call $[B]_X$ *the matrix representing B relative to the ordered basis X.* We have just shown that each bilinear form $B : V \times V \to F$ is uniquely determined by its matrix $[B]_X$ relative to a given ordered basis X of V.

Conversely, suppose $A \in M_n(F)$ is any $n \times n$ matrix with entries in F, and let $X = (x_1, \ldots, x_n)$ be a given ordered basis of V. We can use the matrix A to define a bilinear form $B : V \times V \to F$, as follows. Given $v, w \in V$, write $v = \sum_i c_i x_i$ and $w = \sum_j d_j x_j$ for unique scalars $c_i, d_j \in F$, and set

$$B(v, w) = \sum_{i=1}^{n} \sum_{j=1}^{n} c_i d_j A(i, j). \tag{14.5}$$

You can check that this function B really is bilinear. We illustrate by verifying that $B(v, w + z) = B(v, w) + B(v, z)$ for all $v, w, z \in V$. Write $v = \sum_i c_i x_i$, $w = \sum_j d_j x_j$, $z = \sum_j e_j x_j$ with $c_i, d_j, e_j \in F$. Then $w + z = \sum_j (d_j + e_j) x_j$. Using the definition (14.5) of B and algebraic manipulation, we get

$$B(v, w + z) = \sum_{i=1}^{n} \sum_{j=1}^{n} c_i (d_j + e_j) A(i, j) = \sum_{i=1}^{n} \sum_{j=1}^{n} (c_i d_j A(i, j) + c_i e_j A(i, j))$$

$$= \sum_{i=1}^{n} \sum_{j=1}^{n} c_i d_j A(i, j) + \sum_{i=1}^{n} \sum_{j=1}^{n} c_i e_j A(i, j) = B(v, w) + B(v, z).$$

Also, $[B]_X$ is the matrix A we started with. To verify this, take $v = x_i$ and $w = x_j$ in the definition of B. Here, $c_i = 1$, $d_j = 1$, and all other c_k and d_k are 0. So the formula (14.5) for $B(x_i, x_j)$ simplifies to $1 \cdot 1 \cdot A(i, j) = A(i, j)$.

This means that if we start with a matrix A, use A and X to define a bilinear form B via (14.5), and form the matrix $[B]_X$, then $[B]_X = A$. On the other hand, if we start with any bilinear form B' on V, let A be the matrix $[B']_X$, and define a new bilinear form B by (14.5), then $B = B'$. We can rephrase these comments as follows. Let $\text{Bilin}(V)$ be the set of all bilinear forms on V and $M_n(F)$ be the set of all $n \times n$ matrices with entries in F. The map sending B to $[B]_X$ is a bijection from $\text{Bilin}(V)$ onto $M_n(F)$. The inverse of this bijection sends an input $A \in M_n(F)$ to the bilinear form B specified by (14.5).

In fact, the set $\text{Bilin}(V)$ of all bilinear forms on V can be made into an F-vector space using pointwise operations on functions. In more detail, given any bilinear forms B, B' in $\text{Bilin}(V)$ and $c \in F$, define $B + B' : V \times V \to F$ and $cB : V \times V \to F$ by

$$(B + B')(v, w) = B(v, w) + B'(v, w) \text{ and } (cB)(v, w) = c(B(v, w)) \quad \text{for all } v, w \in V.$$

You can check that $B + B'$ and cB really are bilinear forms on V, so that the set $\text{Bilin}(V)$ is closed under these operations of addition and scalar multiplication. Also, the zero function sending each pair (v, w) to 0_F is bilinear. So $\text{Bilin}(V)$ is a subspace of the vector space of all functions from $V \times V \to F$ (see §4.2) and is therefore a vector space, as claimed.

Using this vector space structure on $\text{Bilin}(V)$, the bijection from bilinear forms to matrices becomes an F-vector space isomorphism from $\text{Bilin}(V)$ to $M_n(F)$. In other words,

$$[B + B']_X = [B]_X + [B']_X \text{ and } [cB]_X = c[B]_X \quad \text{for all } B, B' \in \text{Bilin}(V) \text{ and all } c \in F.$$

To verify this, compare the i, j-entries of the matrices. Both $[B + B']_X$ and $[B]_X + [B']_X$ have i, j-entry $(B + B')(x_i, x_j) = B(x_i, x_j) + B'(x_i, x_j)$. Both $[cB]_X$ and $c[B]_X$ have i, j-entry $(cB)(x_i, x_j) = c(B(x_i, x_j))$. The following theorem summarizes our work so far.

Theorem on Matrix Representation of Bilinear Forms. Let V be an F-vector space of dimension n. For each ordered basis X of V, the map sending a bilinear form B to its representing matrix $[B]_X$ is a vector space isomorphism of $\text{Bilin}(V)$ onto $M_n(F)$.

Let us find the matrices representing some of the bilinear forms in §14.2. The dot product on $\mathbb{R}^n$ and the analogous form on F^n have matrix I_n (the $n \times n$ identity matrix) relative to the standard ordered basis $(\mathbf{e}_1, \ldots, \mathbf{e}_n)$. The zero bilinear form on any n-dimensional F-vector space V has matrix 0_n (the $n \times n$ zero matrix) relative to any basis of V. Relative to $X = (\mathbf{e}_1, \mathbf{e}_2)$, the forms B_1, B_2, B_3 defined in (14.2) have matrices

$$[B_1]_X = \begin{bmatrix} 1 & 0 \\ 0 & -1 \end{bmatrix}, \quad [B_2]_X = \begin{bmatrix} 0 & 1 \\ 1 & 0 \end{bmatrix}, \quad [B_3]_X = \begin{bmatrix} 0 & 1 \\ -1 & 0 \end{bmatrix}. \tag{14.6}$$

Note that $Z = ((1/\sqrt{2}, 1/\sqrt{2}), (1/\sqrt{2}, -1/\sqrt{2}))$ is an ordered basis for $\mathbb{R}^2$. The calculation at the end of §14.2 shows that $[B_2]_Z = \begin{bmatrix} 1 & 0 \\ 0 & -1 \end{bmatrix} = [B_1]_X$. In general, finite-dimensional bilinear spaces (V, B) and (V_*, B_*) are isomorphic iff V has an ordered basis X and V_* has an ordered basis X_* such that $[B]_X = [B_*]_{X_*}$ (Exercise 14).

Let Y be the standard ordered basis of F^{2n}. The forms B_{jn} defined in (14.4) have matrices given in block form by

$$[B_{1n}]_Y = \begin{bmatrix} I_n & 0_n \\ 0_n & -I_n \end{bmatrix}, \quad [B_{2n}]_Y = \begin{bmatrix} 0_n & I_n \\ I_n & 0_n \end{bmatrix}, \quad [B_{3n}]_Y = \begin{bmatrix} 0_n & I_n \\ -I_n & 0_n \end{bmatrix}.$$

Consider the reordered standard basis $Y' = (\mathbf{e}_1, \mathbf{e}_{n+1}, \mathbf{e}_2, \mathbf{e}_{n+2}, \ldots, \mathbf{e}_n, \mathbf{e}_{2n})$. You can check that for $j = 1, 2, 3$, the matrix $[B_{jn}]_{Y'}$ is block diagonal, consisting of n copies of the 2×2 matrices $[B_j]_X$ in (14.6). More generally, suppose $(V_1, B_1), \ldots, (V_k, B_k)$ are bilinear spaces with ordered bases $X_1, \ldots, X_k$. The external direct sum $V = V_1 \times \cdots \times V_k$ has an ordered basis X that can be regarded as the concatenation of $X_1, \ldots, X_k$. You can check that for the form B defined in (14.3), $[B]_X$ is block diagonal with diagonal blocks $[B_1]_{X_1}, \ldots, [B_k]_{X_k}$.

We return to the general case of a bilinear form B on V represented by the matrix $A = [B]_X$, where X is an ordered basis of V. Special properties of the form B translate to analogous properties of the matrix A, as follows.

(a) The bilinear form B is symmetric iff the matrix A is symmetric (meaning $A^{\mathrm{T}} = A$).

(b) The bilinear form B is antisymmetric iff the matrix A is skew (meaning $A^{\mathrm{T}} = -A$, and in particular $2_F A(i, i) = 0_F$ for all i).

(c) The bilinear form B is alternate iff $A^{\mathrm{T}} = -A$ and $A(i, i) = 0_F$ for all i.

(d) The bilinear form B is nondegenerate iff the matrix A is invertible.

We prove (c) and (d) here, asking you to prove (a) and (b) in the exercises. Write $X = (x_1, x_2, \ldots, x_n)$. First assume B is an alternate bilinear form. Then $A(i, i) = B(x_i, x_i) = 0$ for all i, and (since alternate forms are antisymmetric) $A(i, j) = B(x_i, x_j) = -B(x_j, x_i) = -A(j, i)$ for all i, j. Conversely, assume the entries of A satisfy these conditions. Fix arbitrary $v \in V$, and write $v = \sum_i c_i x_i$ with c_i in F. By general bilinearity, we compute

$$B(v, v) = B\left(\sum_{i=1}^{n} c_i x_i, \sum_{j=1}^{n} c_j x_j\right) = \sum_{i=1}^{n} \sum_{j=1}^{n} c_i c_j B(x_i, x_j).$$

Each summand indexed by (i, j) with $i = j$ is zero, since $B(x_i, x_i) = A(i, i) = 0$. The remaining summands, indexed by (i, j) with $i \neq j$, break up into pairs that cancel:

$$B(v, v) = \sum_{i<j} [c_i c_j B(x_i, x_j) + c_j c_i B(x_j, x_i)] = \sum_{i<j} c_i c_j (A(i, j) + A(j, i)) = \sum_{i<j} 0 = 0.$$

To prove (d), we translate the condition in the definition of degenerate bilinear forms into a condition on matrices and vectors. Suppose $w \in V$ satisfies $B(v, w) = 0$ for all $v \in V$. Write $w = \sum_j c_j x_j$ for scalars $c_j \in F$, and let $\mathbf{z}$ be the column vector in F^n with entries $(c_1, \ldots, c_n)$. Observe that $w = 0_V$ iff $\mathbf{z} = \mathbf{0}$, since X is linearly independent. For each i between 1 and n, we can take v to be x_i, the ith vector in the ordered basis X. We have

$$0 = B(x_i, w) = \sum_{j=1}^{n} c_j B(x_i, x_j) = \sum_{j=1}^{n} A(i, j) c_j \quad \text{for } 1 \leq i \leq n.$$

This says that $A\mathbf{z} = \mathbf{0}$. Conversely, $A\mathbf{z} = \mathbf{0}$ implies $B(x_i, w) = 0$ for all i, and hence (by linearity of B in the first position) $B(v, w) = 0$ for all $v \in V$. We now see that the following conditions are equivalent: B is degenerate; some nonzero $w \in V$ satisfies $B(v, w) = 0$ for all $v \in V$; some nonzero $\mathbf{z} \in F^n$ satisfies $A\mathbf{z} = \mathbf{0}$; A is not invertible.

14.4 Congruence of Matrices

Given an ordered basis X of V, there is a corresponding vector space isomorphism from $\mathrm{Bilin}(F)$ to $M_n(F)$, sending a bilinear form B to its matrix $[B]_X$ relative to X. We get one such isomorphism for each ordered basis of V. What happens to the matrix representing B if we change from X to a new ordered basis Y of V? In other words, how is the matrix $[B]_Y$ related to the matrix $[B]_X$?

To answer this question, we need some notation. Write $X = (x_1, x_2, \ldots, x_n)$, $Y = (y_1, y_2, \ldots, y_n)$, and let $p_{i,j} \in F$ be the unique scalars such that $y_j = \sum_{i=1}^{n} p_{i,j} x_i$ for $1 \leq i, j \leq n$. Define the matrix $P = [p_{i,j}]$ and $P^{\mathrm{T}} = [p_{j,i}]$. (Note that $P = {}_X[\mathrm{id}_V]_Y$ is the transition matrix from input basis Y to output basis X defined in §6.4, and P^{T} is the transpose of P.) We use general bilinearity to compute the i, j-entry of $[B]_Y$:

$$B(y_i, y_j) = B\left(\sum_{s=1}^{n} p_{s,i} x_s, \sum_{k=1}^{n} p_{k,j} x_k\right) = \sum_{s=1}^{n} \sum_{k=1}^{n} p_{s,i} p_{k,j} B(x_s, x_k)$$

$$= \sum_{s=1}^{n} \sum_{k=1}^{n} P^{\mathrm{T}}(i, s)[B]_X(s, k) P(k, j).$$

By definition of matrix multiplication, the right side equals the i, j-entry of the matrix product $P^{\mathrm{T}}[B]_X P$. To summarize:

$$[B]_Y = P^{\mathrm{T}}[B]_X P, \quad \text{where } P \text{ is the transition matrix from } Y \text{ to } X.$$

Compare this rule to the analogous rule for linear maps: if $T : V \to V$ is F-linear, then $[T]_Y = P^{-1}[T]_X P$, where P is the transition matrix from Y to X.

We may now consider two equivalence relations on the set $M_n(F)$ of $n \times n$ matrices with entries in F. Fix matrices $A, D \in M_n(F)$. First, A is *similar* to D iff $A = P^{-1}DP$ for some invertible matrix $P \in M_n(F)$. Second, A is *congruent* to D iff $A = P^{\mathrm{T}}DP$ for some invertible matrix $P \in M_n(F)$. Let us check that congruence really is an equivalence relation on $M_n(F)$. Fix $A, D, C \in M_n(F)$. First, $A = I_n^{\mathrm{T}} A I_n$ where I_n is the $n \times n$ identity matrix; so congruence is reflexive on $M_n(F)$. To check symmetry, assume A is congruent to D, say $A = P^{\mathrm{T}}DP$, where P is invertible with inverse Q. The inverse of P^{T} is Q^{T}, so $Q^{\mathrm{T}}AQ = Q^{\mathrm{T}}P^{\mathrm{T}}DPQ = D$, which means D is congruent to A. To check transitivity, assume A is congruent to D and D is congruent to C, say $A = P^{\mathrm{T}}DP$ and $D = R^{\mathrm{T}}CR$ for some invertible $P, R \in M_n(F)$. Then $A = P^{\mathrm{T}}R^{\mathrm{T}}CRP = (RP)^{\mathrm{T}}C(RP)$ where RP is invertible, so A is congruent to C.

For fixed $A \in M_n(F)$, the equivalence class of A relative to congruence is the set of matrices $P^{\mathrm{T}}AP$ as P varies through all invertible matrices in $M_n(F)$. Suppose $A = [B]_X$ for some bilinear form B on V and some ordered basis X of V. Then the congruence equivalence class of A (briefly, the *congruence class of A*) consists of exactly those matrices that represent the bilinear form B relative to some ordered basis of V. We can gain structural information about the bilinear form B by looking for matrices in the congruence class of A that have a simple form, such as diagonal matrices. This is the strategy for several classification theorems for bilinear forms, to be proven below.

As an example of congruence, the matrices $A = \begin{bmatrix} 1 & 0 \\ 0 & -1 \end{bmatrix}$ and $D = \begin{bmatrix} 0 & 1 \\ 1 & 0 \end{bmatrix}$ in $M_2(\mathbb{R})$ are congruent, since $A = P^{\mathrm{T}}DP$ holds with $P = \frac{1}{\sqrt{2}}\begin{bmatrix} 1 & 1 \\ 1 & -1 \end{bmatrix}$. The congruence of A and D also follows from (14.6) and the calculation below it, which shows that A and D both represent the same bilinear form B_2 relative to certain ordered bases of $\mathbb{R}^2$. The congruence relation can change if we change the field of scalars. For example, A is congruent to the identity matrix I_2 in $M_2(\mathbb{C})$, since $A = Q^{\mathrm{T}}I_2 Q$ holds with $Q = \begin{bmatrix} 1 & 0 \\ 0 & i \end{bmatrix}$. But A is not congruent to I_2 in $M_2(\mathbb{R})$, since $A = Q^{\mathrm{T}}I_2 Q$ with $Q \in M_2(\mathbb{R})$ implies $-1 = \det(A) = \det(Q^{\mathrm{T}})\det(Q) = \det(Q)^2$, which is impossible since $\det(Q)$ is real.

14.5   Orthogonality in Bilinear Spaces

We now generalize the idea of perpendicular vectors in $\mathbb{R}^n$ to the setting of a bilinear space (V, B) over a field F. Given vectors $v, w \in V$, we say v *is orthogonal to w* and write $v \perp w$ to mean that $B(v, w) = 0$. Given subsets S, T of V, we say S *is orthogonal to T* and write $S \perp T$ to mean that $B(s, t) = 0$ for all $s \in S$ and all $t \in T$. If S consists of a single vector v, we write $v \perp T$ to abbreviate $\{v\} \perp T$.

Geometric intuition tells us that orthogonality should be a symmetric relation on V, meaning that for all $v, w \in V$, $v \perp w$ implies $w \perp v$. If B is symmetric, antisymmetric, or alternate, then orthogonality is symmetric (see Exercise 33 for a converse). For example, in the antisymmetric case, $B(v, w) = 0$ implies $B(w, v) = -B(v, w) = -0 = 0$. In geometric settings, we focus mainly on bilinear forms that are either symmetric or alternate.

For an alternate bilinear form B, orthogonality is also reflexive on V: for all $v \in V$, $B(v, v) = 0$ and hence $v \perp v$. This runs counter to our intuition coming from Euclidean geometry, where $v \perp v$ is true only for $v = 0$. But this reflexivity is a key feature of symplectic geometry, which is an analog of Euclidean geometry based on alternate bilinear forms.

Given $v \in V$, we call v a *unit vector* iff $B(v, v) = 1$. Call a nonzero vector v *isotropic* iff $B(v, v) = 0$, and call v *non-isotropic* iff $B(v, v) \neq 0$. Suppose $L = (v_1, \ldots, v_k)$ is any finite list of vectors in V. We say L is an *orthogonal* list iff $v_i \perp v_j$ for all $i \neq j$ between 1 and k. We say L is an *orthonormal* list iff L is an orthogonal list of unit vectors.

Lemma on Orthogonality and Linear Independence. Every orthogonal list L of non-isotropic vectors in V is F-linearly independent.

Proof. Let $L = (v_1, \ldots, v_k)$ be an orthogonal list of non-isotropic vectors in V. Suppose $c_1, \ldots, c_k \in F$ are scalars such that $c_1 v_1 + c_2 v_2 + \cdots + c_k v_k = 0$. For each i between 1 and k,

$$0 = B(0, v_i) = B(c_1 v_1 + \cdots + c_i v_i + \cdots + c_k v_k, v_i)$$
$$= c_1 B(v_1, v_i) + \cdots + c_i B(v_i, v_i) + \cdots + c_k B(v_k, v_i) = c_i B(v_i, v_i).$$

Since $B(v_i, v_i)$ is a nonzero element of the field F, we conclude each c_i is 0, as needed. $\square$

Let (V, B) be any bilinear space. For any subset S of V, define the *orthogonal complement of S (on the right)* to be

$$S^\perp = \{w \in V : \text{ for all } s \in S, \ B(s, w) = 0\} = \{w \in V : S \perp w\}.$$

Similarly, the *orthogonal complement of S on the left* is

$$^\perp S = \{v \in V : \text{ for all } s \in S, \ B(v, s) = 0\} = \{v \in V : v \perp S\}.$$

When B is symmetric, antisymmetric, or alternate, $^\perp S = S^\perp$. For general B and all $S \subseteq V$, $S^\perp$ *and* $^\perp S$ *are subspaces of* V. We check this for $S^\perp$. First, $0 \in S^\perp$ since $B(s, 0) = 0$ for all $s \in S$. Second, fix $w, z \in S^\perp$ and $c \in F$. For each $s \in S$, we know $B(s, w) = B(s, z) = 0$, hence $B(s, w + z) = B(s, w) + B(s, z) = 0 + 0 = 0$ and $B(s, cw) = cB(s, w) = c0 = 0$. Thus, $w + z \in S^\perp$ and $cw \in S^\perp$.

For any $S \subseteq V$, we have $S \subseteq {}^\perp(S^\perp)$. To prove this, fix $s \in S$ and $w \in S^\perp$. Then $s \perp w$. This holds for all w in $S^\perp$, so $s \in {}^\perp(S^\perp)$. Similarly, $S \subseteq ({}^\perp S)^\perp$. In the main case of interest (where orthogonality is symmetric), we write $S^{\perp\perp}$ instead of $^\perp(S^\perp)$. Although $S \subseteq S^{\perp\perp}$ always holds, S could be a proper subset of $S^{\perp\perp}$. For instance, this always happens when S is a subset of V that is not a subspace. As another example, if B is the zero bilinear form on V, then $S^\perp = S^{\perp\perp} = V$ for every $S \subseteq V$.

In $\mathbb{R}^n$ with the dot product, we have $S \cap S^\perp = \{0\}$ for any subset S, since $v = 0$ is the only vector orthogonal to itself. But we may very well have $S \cap S^\perp \neq \{0\}$ for general bilinear forms. For example, if v is an isotropic vector in V and $S = Fv$, then $v \in S \cap S^\perp$ since $B(cv, v) = cB(v, v) = 0$ for all $c \in F$. In fact, $S \subseteq S^\perp$ in this example. By definition, B is a degenerate form iff there exists a nonzero $w \in V$ with $V \perp w$. For any subset S of V containing such a w, we have $w \in S \cap S^\perp$. Note that B *is nondegenerate iff* $V^\perp = \{0\}$.

Any subspace W of any bilinear space (V, B) becomes a bilinear space using the restriction of $B : V \times V \to F$ to the domain $W \times W$. The formal notation for this restricted function is $B|_{W \times W}$, but we often write $B|W$ or just B when W is clear from context. If

B is symmetric (or antisymmetric or alternate), then $B|W$ has the same property. But it is possible for B to be nondegenerate while $B|W$ is degenerate, or vice versa. For any $S \subseteq W$, the orthogonal complement of S in W is the intersection of W and the orthogonal complement of S in V. We may write this in symbols as $(S^\perp$ in $W) = W \cap (S^\perp$ in $V)$. Taking $S = W$ and applying the observation at the end of the last paragraph, we see that $B|W$ *is nondegenerate iff* $W \cap W^\perp = \{0\}$.

If V is an F-vector space with subspaces $W_1, W_2, \ldots, W_k$, the notation $V = W_1 \oplus W_2 \oplus \cdots \oplus W_k$ means that V is the *(internal) direct sum* of these subspaces. In other words, $V = W_1 + W_2 + \cdots + W_k$ and $W_i \cap (W_1 + \cdots + W_{i-1}) = \{0\}$ for $2 \leq i \leq k$. In this case, every $v \in V$ has a unique expression as $v = w_1 + w_2 + \cdots + w_k$ where $w_i \in W_i$. Concatenating ordered bases X_i for W_i gives an ordered basis X for V. Now suppose B is a bilinear form on V. We say V is the *orthogonal direct sum* of subspaces $W_1, \ldots, W_k$, writing $V = W_1 \boxplus W_2 \boxplus \cdots \boxplus W_k$, iff $V = W_1 \oplus W_2 \oplus \cdots \oplus W_k$ and $W_i \perp W_j$ for all $i \neq j$. In this situation, B is nondegenerate on V iff $B|W_i$ is nondegenerate for $1 \leq i \leq k$ (Exercise 36). Furthermore, (V, B) is isomorphic to the external direct sum $(W_1, B|W_1) \times \cdots \times (W_k, B|W_k)$ (Exercise 37). Using the ordered bases X_i and X mentioned above, $[B]_X$ is block-diagonal with blocks $[B|W_1]_{X_1}, \ldots, [B|W_k]_{X_k}$ (Exercise 38).

14.6   Bilinear Forms and Dual Spaces

This section describes the close connection between bilinear forms and dual vector spaces. For any F-vector space V, the *dual space* V^* is the set of all F-linear maps from V to F. The set V^* is an F-vector space under pointwise addition and scalar multiplication of maps. In more detail, let $g, h \in V^*$ and $c \in V$. The functions $g + h$ and cg are defined by $(g + h)(v) = g(v) + h(v)$ and $(cg)(v) = c(g(v))$ for all $v \in V$. You can check that $g + h \in V^*$, $cg \in V^*$, and the zero function from V to F is in V^*. So V^* is a subspace of the F-vector space of all functions from V to F. For finite-dimensional spaces V, we know $\dim(V^*) = \dim(V)$. This can be proved by considering matrices (see §6.7) or dual bases (see §13.2).

Let B be a fixed bilinear form on an F-vector space V with $\dim(V) = n$. Define a function $R : V \to V^*$ by letting $R(w)$ be the function from V to F that sends each input $v \in V$ to output $B(v, w) \in F$. More succinctly, we may write $R(w)(v) = B(v, w)$ for all $v, w \in V$. For fixed $w \in V$, note that $R(w)$ is an F-linear map, since B is F-linear in the first position. This means that the function R really does map V into the claimed codomain V^*. Next, R itself is an F-linear function, since B is F-linear in the second position. In more detail, for all $w, z \in V$, we have $R(w + z) = R(w) + R(z)$ since both sides send $v \in V$ to $B(v, w + z) = B(v, w) + B(v, z)$. For all $w \in V$ and $c \in F$, we have $R(cw) = cR(w)$ since both sides send $v \in V$ to $B(v, cw) = cB(v, w)$. So we have an F-linear map $R : V \to V^*$.

For $w \in V$, w is in the kernel of R iff $R(w)$ is the zero map on V iff $R(w)(v) = 0$ for all $v \in V$ iff $B(v, w) = 0$ for all $v \in V$. Thus, the kernel of R is $V^\perp = \{w \in V : v \perp w$ for all $v \in V\}$. The bilinear form B is nondegenerate iff $\ker(R) = \{0\}$ iff R is one-to-one. In this case, the image $R[V]$ is an n-dimensional subspace of V^*. Since $\dim(V^*) = \dim(V) = n$, we see that R must also be onto. To summarize, for any nondegenerate bilinear form B on V, we get an isomorphism $R : V \to V^*$ of F-vector spaces. Bijectivity of R means that for each linear functional $g \in V^*$, there exists a unique $w \in V$ such that for all $v \in V$, $g(v) = B(v, w)$. We say that g is *represented by* w *relative to* B.

For example, consider the real vector space $\mathbb{R}^n$ and take B to be the dot product. As a special case of the preceding discussion, for every linear functional $g : \mathbb{R}^n \to \mathbb{R}$, there is a unique vector $\mathbf{w} \in \mathbb{R}^n$ such that $g(\mathbf{v}) = \mathbf{v} \bullet \mathbf{w}$ for all $\mathbf{v} \in \mathbb{R}^n$. We can also verify this conclusion directly, as follows. Given $g \in (\mathbb{R}^n)^*$, let $w_i = g(\mathbf{e}_i)$ for $1 \le i \le n$, where $\mathbf{e}_i$ is the standard basis vector with 1 in position i and 0 in all other positions. Any $\mathbf{v} = (v_1, \ldots, v_n) \in \mathbb{R}^n$ can be written uniquely in the form $\mathbf{v} = \sum_{i=1}^n v_i \mathbf{e}_i$. By linearity of g,

$$g(\mathbf{v}) = \sum_{i=1}^n v_i g(\mathbf{e}_i) = \sum_{i=1}^n v_i w_i = \mathbf{v} \bullet \mathbf{w}.$$

This proves the existence of the vector $\mathbf{w}$ representing g. To see that $\mathbf{w}$ is unique, suppose $\mathbf{w}' = (w_1', \ldots, w_n')$ also represents g relative to the dot product. Then $w_i' = \mathbf{e}_i \bullet \mathbf{w}' = g(\mathbf{e}_i) = w_i$ for $1 \le i \le n$, so $\mathbf{w}' = \mathbf{w}$.

14.7   Theorem on Orthogonal Complements

View $\mathbb{R}^n$ as a bilinear space using the dot product. For any subspace W of $\mathbb{R}^n$, we have:
(i) $\mathbb{R}^n = W \boxplus W^\perp$; (ii) $n = \dim(\mathbb{R}^n) = \dim(W) + \dim(W^\perp)$; and (iii) $W^{\perp\perp} = W$.
Beware! These results do not always hold if we replace the dot product by a general bilinear form. Certain conclusions do follow with appropriate nondegeneracy hypotheses, as we see in the next theorem. Carefully memorize this theorem, as it will be used constantly.

Theorem on Orthogonal Complements. Let (V, B) be an n-dimensional bilinear space, where B is symmetric or alternate. Let W be any subspace of V.
(a) If $B|W$ is nondegenerate (meaning $W \cap W^\perp = \{0\}$), then $\dim(W) + \dim(W^\perp) = n = \dim(V)$ and $V = W \boxplus W^\perp$. But $W^{\perp\perp} = W$ is not guaranteed.
(b) If B is nondegenerate (meaning $V^\perp = \{0\}$), then $\dim(W) + \dim(W^\perp) = n = \dim(V)$ and $W^{\perp\perp} = W$. But $V = W \boxplus W^\perp$ is not guaranteed.

With minor modifications to the conclusions, the theorem holds without assuming B is symmetric or alternate (Exercise 29).

Proof. Step 1. We apply the Rank–Nullity Theorem to a variation of the map $R : V \to V^*$ considered in §14.6. Define $T : V \to W^*$ by letting $T(v)(w) = B(w, v)$ for all $v \in V$ and $w \in W$. In other words, $T(v)$ is the function from W to F sending an input $w \in W$ to the output $B(w, v)$. By the same proof used earlier for R, bilinearity of B implies that each function $T(v)$ is F-linear, and that T itself is F-linear. By the Rank–Nullity Theorem, $n = \dim(V) = \dim(\ker(T)) + \dim(\mathrm{img}(T))$.

Step 2. We compute the kernel of T. Fix $v \in V$. The following conditions on v are equivalent: $v \in \ker(T)$; $T(v) = 0$, the zero map from W to F; $B(w, v) = 0$ for all $w \in W$; $w \perp v$ for all $w \in W$; $v \in W^\perp$. Thus, $\ker(T) = W^\perp$.

Step 3. We prove part (a) of the theorem. Assume $W \cap W^\perp = \{0\}$. We use this to prove that $\mathrm{img}(T) = W^*$. Let $T' : W \to W^*$ be the restriction of the linear map T to the domain W. T' is a linear map with kernel $W \cap \ker(T) = W \cap W^\perp = \{0\}$. Thus, T' is one-to-one. As W and W^* have the same finite dimension, T' must also be onto. Since the restricted map T' maps W onto all of W^*, the original map T also maps V onto W^*. Using Steps 1 and 2,

$$n = \dim(\ker(T)) + \dim(\mathrm{img}(T)) = \dim(W^\perp) + \dim(W^*) = \dim(W^\perp) + \dim(W).$$

Since $W \cap W^\perp = \{0\}$, the sum of subspaces $W + W^\perp$ is an orthogonal direct sum $W \boxplus W^\perp$. This direct sum has dimension $\dim(W) + \dim(W^\perp) = n = \dim(V)$, so $W \boxplus W^\perp$ must be all of V. Exercise 24 asks you to find an example where $W^{\perp\perp} \ne W$.

Step 4. We prove part (b) of the theorem. Assume that B is nondegenerate. We use this to prove that $\text{img}(T) = W^*$. By §14.6, there is an F-linear isomorphism $R : V \to V^*$ given by $R(v)(x) = B(x, v)$ for all $v, x \in V$. Given $h \in W^*$, we can extend h to a linear map $g : V \to F$. For example, extend a basis of W to a basis of V, then let g be the linear map that acts as h does on basis vectors in W and sends the remaining basis vectors to 0. We know $g = R(v)$ for some $v \in V$ since R is onto. It follows that $h = T(v)$, since for each $w \in W$, $h(w) = g(w) = R(v)(w) = B(w, v) = T(v)(w)$. Because each $h \in W^*$ has the form $T(v)$ for some $v \in V$, T is onto. Using Steps 1 and 2,

$$\dim(V) = n = \dim(\ker(T)) + \dim(\text{img}(T)) = \dim(W^\perp) + \dim(W^*) = \dim(W^\perp) + \dim(W).$$

This conclusion holds for all subspaces W, so we also have $\dim(W^\perp) + \dim(W^{\perp\perp}) = n$. Because n is finite, we get $\dim(W^{\perp\perp}) = \dim(W)$. Since $W \subseteq W^{\perp\perp}$ and these spaces have the same finite dimension, $W = W^{\perp\perp}$ follows. Exercise 25 asks you to find an example where $V = W \boxplus W^\perp$ is not true. □

Here are three examples where part (a) of the theorem applies. First, suppose (V, B) is a symmetric bilinear space and $z \in V$ satisfies $B(z, z) \neq 0$. The 1-dimensional subspace $W = Fz$ has ordered basis (z), and the matrix of $B|W$ relative to this basis is $[B(z, z)]$. This matrix is invertible, so $B|W$ is nondegenerate, and the theorem gives $V = Fz \boxplus (Fz)^\perp$.

Second, suppose (V, B) is a symmetric bilinear space and (v, w) is a list of two vectors in V such that $B(v, v) = 0 = B(w, w)$ and $B(v, w) = 1 = B(w, v)$. We call such a list a *hyperbolic pair* in V, and the subspace $H = Fv + Fw$ of V is called a *hyperbolic plane*. It is routine to check that $X = (v, w)$ is an ordered basis of H, and $[B|H]_X = \begin{bmatrix} 0 & 1 \\ 1 & 0 \end{bmatrix}$. This matrix is invertible, so $B|H$ is nondegenerate, and the theorem gives $V = H \boxplus H^\perp$.

Third, suppose (V, B) is an alternate bilinear space and (v, w) is a list of two vectors in V such that $B(v, v) = 0 = B(w, w)$, $B(v, w) = 1$, and $B(w, v) = -1$. We call such a list a *symplectic pair* in V, and the subspace $S = Fv + Fw$ of V is called a *symplectic plane*. Again, $X = (v, w)$ is an ordered basis of S, and $[B|S]_X = \begin{bmatrix} 0 & 1 \\ -1 & 0 \end{bmatrix}$. This matrix is invertible, so $B|S$ is nondegenerate, and the theorem gives $V = S \boxplus S^\perp$.

14.8 Radical of a Bilinear Form

Let B be a symmetric, antisymmetric, or alternate bilinear form on an F-vector space V. The *radical of* (V, B) is the subspace $\text{Rad}(V) = V^\perp$, which is the set of all $z \in V$ such that $B(v, z) = 0$ for all $v \in V$. The form B is nondegenerate iff $\text{Rad}(V) = \{0\}$. Suppose W is any subspace that is complementary to $V^\perp$, meaning that V is the direct sum $V = V^\perp \oplus W$. Such subspaces W always exist. To find one, start with an ordered basis $(x_1, \ldots, x_k)$ for $V^\perp$, extend this list to an ordered basis $X = (x_1, \ldots, x_k, x_{k+1}, \ldots, x_n)$ for V, and let W be the subspace with basis $X' = (x_{k+1}, \ldots, x_n)$. There are often many possible choices of W, since there can be multiple ways of extending the basis of $V^\perp$. For any such W, the matrix $[B]_X$ has the block form

$$[B]_X = \left[\begin{array}{c|c} 0_{k \times k} & 0_{k \times (n-k)} \\ \hline 0_{(n-k) \times k} & [B|W]_{X'} \end{array} \right]. \tag{14.7}$$

The matrix has this form since for any i, j with $1 \leq i \leq k$ and $1 \leq j \leq n$, $B(x_i, x_j) = 0 = B(x_j, x_i)$. We claim $B|W$ is nondegenerate. Suppose $w \in W$ satisfies $B(y, w) = 0$ for all

$y \in W$; we show $w = 0$. Given any $v \in V$, we can write $v = z + y$ for some $z \in V^\perp$ and some $y \in W$. Then $B(v, w) = B(z, w) + B(y, w) = 0 + 0 = 0$ since z is orthogonal to everything in V and $B(y, w) = 0$. Since v was arbitrary, we conclude $w \in V^\perp$. Hence $w \in V^\perp \cap W = \{0\}$, so $w = 0$ as needed.

We often use this construction to reduce to the case of a nondegenerate bilinear form when proving a property for a general symmetric, antisymmetric, or alternate bilinear form.

We can also remove the radical using a quotient construction. Given (V, B) with radical $R = \mathrm{Rad}(V)$, let V/R be the quotient vector space (described in §1.6). Given cosets $x + R$ and $y + R$ in V/R, where $x, y \in V$, define $\overline{B}(x + R, y + R) = B(x, y)$. To see that $\overline{B}$ is well-defined, assume $x, x', y, y' \in V$ satisfy $x + R = x' + R$ and $y + R = y' + R$. Then $x' = x + r$ and $y' = y + s$ for some $r, s \in R$. We compute

$$B(x', y') = B(x + r, y + s) = B(x, y) + B(r, y) + B(x, s) + B(r, s) = B(x, y).$$

It is now routine to check that $(V/R, \overline{B})$ is a bilinear space, and $\overline{B}$ is symmetric, antisymmetric, or alternate if B is. We show $\overline{B}$ is nondegenerate. Assume $z + R \in V/R$ satisfies $\overline{B}(v + R, z + R) = 0$ for all $v \in V$. Then $B(v, z) = 0$ for all $v \in V$, which means $z \in R$ and $z + R = 0 + R$ in V/R. To compare this to the previous construction, let W be any subspace of V with $V = R \oplus W$. You can check that the map $p : W \to V/R$, given by $p(w) = w + R$ for $w \in W$, is a bilinear space isomorphism from $(W, B|W)$ to $(V/R, \overline{B})$. This approach shows that the bilinear space we get by removing the radical is (up to isomorphism) independent of the choice of the subspace W complementary to $\mathrm{Rad}(V)$.

14.9 Diagonalization of Symmetric Bilinear Forms

When studying a bilinear space (V, B), it is helpful to have an ordered basis $X = (x_1, \ldots, x_n)$ for V where the matrix $[B]_X$ is as simple as possible. Ideally, we would like $[B]_X$ to be a diagonal matrix, which means that $B(x_i, x_j) = 0$ for all $i \neq j$, or equivalently that the basis X is an orthogonal list of vectors. Since diagonal matrices are symmetric, a necessary condition for such a basis to exist is that the bilinear form B be symmetric. The next theorem shows that this necessary condition is also sufficient for most fields.

Theorem on Diagonalizing Symmetric Bilinear Forms. Assume $2_F \neq 0_F$. If B is a symmetric bilinear form on an n-dimensional F-vector space V, then there exists an ordered basis $X = (x_1, \ldots, x_n)$ for V such that $[B]_X$ is a diagonal matrix.

Proof. We prove the theorem by induction on $n = \dim(V)$. In the base case where $n \leq 1$, the result follows at once. Fix $n > 1$, and assume the theorem holds for all vector spaces of smaller dimension than n. First consider the case where B is degenerate. Choose any subspace W with $V = V^\perp \oplus W$. Since B is degenerate, $\dim(W) < n$, and we can apply the induction hypothesis to the symmetric bilinear space $(W, B|W)$. We obtain an ordered basis X' of W such that $[B|W]_{X'}$ is diagonal. Since $V = V^\perp \oplus W$, we can extend X' by prepending an ordered basis of $V^\perp$ to get an ordered basis X for V. By (14.7), $[B]_X$ is a diagonal matrix.

Next, consider the case where B is nondegenerate. We seek an $x \in V$ with $B(x, x) \neq 0$. Start with any nonzero v in V. If $B(v, v) \neq 0$, then we take $x = v$. Otherwise, use nondegeneracy of B to find $w \in V$ with $B(w, v) \neq 0$. If $B(w, w) \neq 0$, then we take $x = w$. If $B(v, v) = 0 = B(w, w)$, then we take $x = v + w$, noting that

$$B(x, x) = B(v + w, v + w) = B(v, v) + B(v, w) + B(w, v) + B(w, w) = 2_F B(w, v) \neq 0_F.$$

Let $W = Fx$ be the 1-dimensional subspace of V spanned by x. By the first example at the end of §14.7, $V = W \boxplus W^\perp$ and $\dim(W^\perp) = n - 1 < n$. By induction hypothesis, there is an ordered basis $X' = (x_2, \ldots, x_n)$ of $W^\perp$ such that $[B|W^\perp]_{X'}$ is diagonal. Since $x \perp x_i$ for $2 \leq i \leq n$, $X = (x, x_2, \ldots, x_n)$ is an ordered basis of V such that $[B]_X$ is diagonal. This completes the induction step of the proof. $\square$

We can prove a sharper version of this theorem when using a field F (such as $\mathbb{C}$) where every scalar in F has a square root in F.

Theorem on Diagonalizing Bilinear Forms over Fields with Square Roots. Let B be a symmetric bilinear form on an n-dimensional F-vector space V. Assume $2_F \neq 0_F$, and for all $c \in F$, there is $a \in F$ with $a^2 = c$. Then there is an ordered basis $X = (x_1, \ldots, x_n)$ for V such that $[B]_X$ is a diagonal matrix with diagonal entries consisting of k copies of 0_F followed by $n - k$ copies of 1_F for some k. We must have $k = \dim(V^\perp)$, so there is only one matrix of this form that represents B.

Proof. By the previous theorem, there is an ordered basis $X' = (x'_1, \ldots, x'_n)$ for V such that $[B]_{X'}$ is a diagonal matrix, say with diagonal entries $d_1, \ldots, d_n \in F$. We can choose the ordering of the list X' so that for some k between 0 and n, $d_i = 0_F$ for $1 \leq i \leq k$ and $d_i \neq 0_F$ for $k < i \leq n$. For each i in the range $k < i \leq n$, let $a_i \in F$ be a scalar (necessarily nonzero) with $a_i^2 = d_i$. For each i between 1 and k, let $a_i = 1_F$. Define $x_i = a_i^{-1} x'_i$ for $1 \leq i \leq n$. Let $X = (x_1, \ldots, x_n)$; X is an ordered basis for V since it comes from X' by rescaling basis vectors. We compute

$$B(x_i, x_i) = B(x'_i, x'_i) = 0 \quad \text{for } 1 \leq i \leq k;$$
$$B(x_i, x_i) = a_i^{-2} B(x'_i, x'_i) = d_i^{-1} d_i = 1_F \quad \text{for } k < i \leq n;$$
$$B(x_i, x_j) = a_i^{-1} a_j^{-1} B(x'_i, x'_j) = 0_F \quad \text{for all } i \neq j.$$

Thus $[B]_X$ has the structure described in the theorem statement.

To finish, we prove $k = \dim(V^\perp)$ for any ordered basis $X = (x_1, \ldots, x_n)$ such that $A = [B]_X$ is diagonal with k zeroes followed by $n-k$ ones on the diagonal. Let $v = \sum_{i=1}^n c_i x_i$ be any element of V, where all $c_i \in F$. Note that $B(x_i, v) = c_i A(i, i)$ for $1 \leq i \leq n$. We have $v \in V^\perp$ iff $B(x_i, v) = 0$ for $1 \leq i \leq n$ iff $c_i A(i, i) = 0$ for $1 \leq i \leq n$ iff $c_i = 0$ for $k < i \leq n$ (since $A(1, 1) = \cdots = A(k, k) = 0$). Thus, $(x_1, \ldots, x_k)$ spans $V^\perp$, and this list is F-linearly independent since it is a sublist of the basis X. So $(x_1, \ldots, x_k)$ is a basis of $V^\perp$, and $k = \dim(V^\perp)$. $\square$

We can prove a similar result for the real field $F = \mathbb{R}$, where all positive elements have square roots.

Theorem on Diagonalizing Real Symmetric Bilinear Forms. Let B be a symmetric bilinear form on an n-dimensional real vector space V. There is an ordered basis $X = (x_1, \ldots, x_n)$ for V such that $[B]_X$ is a diagonal matrix with diagonal consisting of k copies of 0, then s copies of 1, then t copies of -1, for some integers k, s, t that are uniquely determined by B.

Proof. We already know there is an ordered basis $X' = (x'_1, \ldots, x'_n)$ for V such that $[B]_{X'}$ is a diagonal matrix, say with diagonal entries $d_1, \ldots, d_n \in F$. Order the list X' so that the first k numbers $d_1, \ldots, d_k$ are 0, the next s numbers $d_{k+1}, \ldots, d_{k+s}$ are positive, and the last t numbers $d_{k+s+1}, \ldots, d_{k+s+t}$ are negative (where $k + s + t = n$). Now define $x_i = x'_i$ for $1 \leq i \leq k$, $x_i = \frac{1}{\sqrt{d_i}} x'_i$ for $k + 1 \leq i \leq k + s$, and $x_i = \frac{1}{\sqrt{|d_i|}} x'_i$ for $k + s + 1 \leq i \leq n$. You can check that $X = (x_1, \ldots, x_n)$ is an ordered basis of V such that $[B]_X$ has diagonal entries as described in the theorem. For example, for any i satisfying $k + s + 1 \leq i \leq n$, we have

$B(x_i, x_i) = \frac{1}{|d_i|} B(x_i', x_i') = d_i/|d_i| = -1$ since d_i is negative. Since each x_i is some scalar multiple of x_i', we have $B(x_i, x_j) = 0$ for all $i \neq j$ because $B(x_i, x_j)$ is a scalar multiple of $B(x_i', x_j') = 0$.

To prove uniqueness of k, s, and t, let $Y = (y_1, \ldots, y_n)$ be any ordered basis of V such that $A = [B]_Y$ is a diagonal matrix with k' copies of 0, then s' copies of 1, then t' copies of -1 on the diagonal, where $k' + s' + t' = n$. As in the proof of the previous theorem, we see that $k' = \dim(V^\perp) = k$. Let P be the subspace of V with basis $(x_{k+1}, \ldots, x_{k+s})$, and let N' be the subspace of V with basis $(y_1, \ldots, y_{k'}, y_{k'+s'+1}, \ldots, y_n)$. Given a nonzero $v \in P$, say $v = \sum_{i=k+1}^{k+s} c_i x_i$ with $c_i \in \mathbb{R}$ not all zero, we compute $B(v, v) = \sum_{i=k+1}^{k+s} c_i^2 > 0$. Similarly, any $v \in N'$ satisfies $B(v, v) \leq 0$. So we must have $P \cap N' = \{0\}$, and the subspace sum $P + N'$ is a direct sum $P \oplus N' \subseteq V$. Then

$$s + (k' + t') = \dim(P) + \dim(N') = \dim(P \oplus N') \leq \dim(V) = n = s' + k' + t'.$$

We conclude that $s \leq s'$. Interchanging the roles of X and Y, we see likewise that $s' \leq s$, so $s' = s$. Then $t' = n - s' - k' = n - s - k = t$. $\qquad\square$

14.10 Structure of Alternate Bilinear Forms

Any matrix representing an alternate bilinear form must have all diagonal entries 0. So, the only diagonalizable alternate bilinear form is the zero form. However, the next result shows that we can represent any alternate bilinear form with a block-diagonal matrix consisting of repetitions of a simple 2×2 block.

Theorem on Alternate Bilinear Forms. Let B be an alternate bilinear form on an n-dimensional F-vector space V. There is an ordered basis X of V such that $[B]_X$ is a block-diagonal matrix consisting of s diagonal blocks $C = \begin{bmatrix} 0 & 1 \\ -1 & 0 \end{bmatrix}$ followed by k diagonal entries equal to 0. The integers s and k are uniquely determined by B: namely, $k = \dim(V^\perp)$ and $s = (n - k)/2$.

An equivalent reformulation of the theorem says that any alternate bilinear space is the orthogonal direct sum of its radical and s symplectic planes for some $s \geq 0$.

Proof. We proceed by induction on $n = \dim(V)$. If $B = 0$, which must happen in the base cases $n = 0$ or $n = 1$, then $[B]_X = 0$ for any ordered basis X. The theorem holds for such B, taking $s = 0$ and $k = n$. Now fix $n > 1$, assume the theorem holds for all alternate bilinear forms on all vector spaces of dimension smaller than n, and assume B is nonzero. Then there exist $x, z \in V$ with $B(x, z) \neq 0$. We cannot have $z = cx$ with $c \in F$, since otherwise $B(x, z) = cB(x, x) = 0$. Let $y = B(x, z)^{-1} z$, so $B(x, y) = 1$, $B(y, x) = -1$, and $B(x, x) = B(y, y) = 0$. Then (x, y) is a symplectic pair spanning a 2-dimensional symplectic plane $S = Fx + Fy$ in V, and the matrix of $B|S$ relative to the basis (x, y) is C. By the third example at the end of §14.7, $V = S \boxplus S^\perp$. Since $\dim(S^\perp) = n - 2 < n$, the induction hypothesis says there is an ordered basis X' of $S^\perp$ such that $[B|S^\perp]_{X'}$ has the structure described in the theorem. Let X be the list X' with x, y added at the front. Then X is an ordered basis of V such that $[B]_X$ has the required structure.

For the uniqueness claim, suppose $Y = (x_1, y_1, x_2, y_2, \ldots, x_s, y_s, z_1, \ldots, z_k)$ is any ordered basis of V such that $[B]_Y$ is block-diagonal with s copies of C followed by k copies of 0 on the diagonal. A typical vector $v \in V$ has the form $\sum_{i=1}^s (a_i x_i + b_i y_i) + \sum_{j=1}^k c_j z_j$ where

$a_i, b_i, c_j \in F$. We have $v \in V^\perp$ iff $B(x_i, v) = B(y_i, v) = 0$ for $1 \le i \le s$ and $B(z_j, v) = 0$ for $1 \le j \le k$. By the assumed structure of $[B]_Y$, we see that $B(x_i, v) = b_i$, $B(y_i, v) = -a_i$, and $B(z_j, v) = 0$ for $1 \le i \le s$ and $1 \le j \le k$. Therefore, $v \in V^\perp$ iff each a_i and b_i is 0 iff v is in the subspace with basis $(z_1, \dots, z_k)$. We conclude that $k = \dim(V^\perp)$, which depends only on B. Then $s = (n - k)/2$ is also uniquely determined by B. $\square$

14.11 Totally Isotropic Subspaces

Let (V, B) be a bilinear space over a field F, where $2_F \ne 0_F$ and B is symmetric. We say that a subspace W of V *totally isotropic* iff $B(w, w) = 0$ for all $w \in W$. In this case, we have $B(w, z) = 0$ for all w, z in W, since

$$2B(w, z) = B(w, z) + B(z, w) = B(w + z, w + z) - B(w, w) - B(z, z) = 0 - 0 - 0 = 0$$

and 2 is invertible in F. Thus, a subspace W is totally isotropic iff $B|W$ is the zero bilinear form. Also, W is totally isotropic iff $W \subseteq W^\perp$. The next theorem gives a bound on the dimension of W when B is nondegenerate.

Dimension Theorem for Totally Isotropic Subspaces. Let (V, B) be an n-dimensional nondegenerate symmetric bilinear space over a field F, where $2_F \ne 0_F$. For any totally isotropic subspace W of V, $\dim(W) \le n/2$.

Proof. Let $k = \dim(W)$. Let $X = (x_1, \dots, x_n)$ be an ordered basis of V that extends an ordered basis $(x_1, \dots, x_k)$ for W. The matrix $A = [B]_X$ is an invertible $n \times n$ matrix that has a $k \times k$ zero submatrix in its upper-left corner. The list of all n columns of A is linearly independent in F^n, since A is invertible. So the sublist consisting of the first k columns of A is also linearly independent in F^n. We can regard the first k columns of A as column vectors in F^{n-k} by deleting all the zero entries in the first k rows. These k shortened columns must also be linearly independent in F^{n-k}, as is readily checked. Since F^{n-k} has dimension $n-k$, we conclude that $k \le n - k$ and hence $k \le n/2$. $\square$

With a little more work, we can prove a stronger version of this theorem that resembles the Structure Theorem for Alternate Bilinear Forms.

Embedding Theorem for Totally Isotropic Subspaces. Let (V, B) be an n-dimensional nondegenerate symmetric bilinear space over a field F, where $2_F \ne 0_F$. Let W be a k-dimensional totally isotropic subspace of V. There exist k hyperbolic planes $H_1, \dots, H_k$ in V such that $W \subseteq H_1 \boxplus \cdots \boxplus H_k$. More specifically, for any ordered basis $(w_1, \dots, w_k)$ of W, there is an ordered basis $X = (w_1, z_1, \dots, w_k, z_k, v_{2k+1}, \dots, v_n)$ of V such that $[B]_X$ is block-diagonal with k diagonal blocks equal to $\begin{bmatrix} 0 & 1 \\ 1 & 0 \end{bmatrix}$ followed by a block of size $(n - 2k) \times (n - 2k)$.

Proof. We prove the theorem by induction on $k = \dim(W)$. The conclusion is immediate in the base case $k = 0$. For the induction step, assume $k > 0$ and the theorem holds for smaller values of k. Let $(w_1, w_2, \dots, w_k)$ be any ordered basis of a k-dimensional totally isotropic subspace W of V. Let $Y = Fw_2 \oplus \cdots \oplus Fw_k$ be the $(k-1)$-dimensional subspace of W with ordered basis $(w_2, \dots, w_k)$. Since $Y \subseteq W$, we have $W^\perp \subseteq Y^\perp \subseteq V$. As B is nondegenerate, part (b) of the Theorem on Orthogonal Complements tells us that $\dim(W^\perp) = n - k$ and $\dim(Y^\perp) = n - k + 1$. So there exists z belonging to $Y^\perp$ but not to $W^\perp$.

Define $c = B(w_1, z) \in F$. We cannot have $c = 0$; otherwise z would be orthogonal to $w_1, w_2, \ldots, w_k$ and so z would belong to $W^\perp$. Replacing z by $c^{-1}z$, we can arrange that $c = 1$. For any $b \in F$, we have $B(w_1, bw_1 + z) = bB(w_1, w_1) + B(w_1, z) = 1$ and $B(bw_1 + z, bw_1 + z) = b^2 B(w_1, w_1) + 2bB(w_1, z) + B(z, z) = 2b + B(z, z)$. Define $z_1 = bw_1 + z$ where $b = -2^{-1}B(z, z)$. Then $B(z_1, z_1) = 0 = B(w_1, w_1)$ and $B(w_1, z_1) = 1$, so (w_1, z_1) is a hyperbolic pair spanning a hyperbolic plane H_1 in V.

Now $B|H_1$ is nondegenerate, since the matrix of this restricted form relative to the ordered basis (w_1, z_1) is the invertible matrix $\begin{bmatrix} 0 & 1 \\ 1 & 0 \end{bmatrix}$. By part (a) of the Theorem on Orthogonal Complements, $V = H_1 \boxplus H_1^\perp$. This decomposition shows that $B|H_1^\perp$ must be nondegenerate, since B is nondegenerate. Each basis vector $w_2, \ldots, w_k$ for Y is orthogonal to w_1 (since W is totally isotropic), orthogonal to z (since z was chosen in $Y^\perp$), and orthogonal to z_1. It follows that H_1 and Y are orthogonal subspaces, meaning that Y is a subspace of $H_1^\perp$. Y is a $(k-1)$-dimensional totally isotropic space. Applying the induction hypothesis with V replaced by $H_1^\perp$ and W replaced by Y, we get an ordered basis $(w_2, z_2, \ldots, w_k, z_k, v_{2k+1}, \ldots, v_n)$ of $H_1^\perp$ satisfying the conclusion of the theorem for the subspace Y of $H_1^\perp$. Since $V = H_1 \boxplus H_1^\perp$, $(w_1, z_1, w_2, z_2, \ldots, w_k, z_k, v_{2k+1}, \ldots, v_n)$ is an ordered basis of V satisfying the conclusion of the theorem for the subspace W of V. $\square$

We can also prove a structure theorem in the case where B might be degenerate. Call a subspace Y of a bilinear space (V, B) *non-isotropic* or *anisotropic* iff $B(y, y) \neq 0$ for all nonzero $y \in Y$.

Witt's Decomposition Theorem (Existence Part). Any symmetric bilinear space (V, B) has a decomposition $V = R \boxplus H \boxplus Y$, where R is totally isotropic, H is an orthogonal direct sum of k hyperbolic planes for some $k \geq 0$, and Y is non-isotropic.

Any such decomposition of V is called a *Witt decomposition* of V. For any such decomposition we have $\mathrm{Rad}(V) = \mathrm{Rad}(R) \boxplus \mathrm{Rad}(H) \boxplus \mathrm{Rad}(Y) = R$.

Proof. Let $R = V^\perp$, the radical of V. R is totally isotropic and orthogonal to all subspaces of V. Let V' be any complementary subspace to R (meaning $V = R \oplus R'$) and $B' = B|V'$. As we saw in §14.8, B' is nondegenerate. Next, let W be any maximal totally isotropic subspace of V', where *maximal* means that W is not properly contained in any totally isotropic subspace of V'. Since V' is finite-dimensional, such a subspace W must exist; we can take W to be a totally isotropic subspace of V' of maximum possible dimension k. By the Embedding Theorem, there exist k hyperbolic planes $H_1, \ldots, H_k$ in V' with $W \subseteq H_1 \boxplus \cdots \boxplus H_k$. Let $H = H_1 \boxplus \cdots \boxplus H_k$, and let $Y = H^\perp$, the orthogonal complement of H in V'. Since $B'|H$ is nondegenerate, we know $V' = H \boxplus Y$ by the Theorem on Orthogonal Complements. Thus $V = R \boxplus H \boxplus Y$. We need only check that Y is non-isotropic. If this were not true, then there would be a nonzero $y \in Y$ with $B(y, y) = 0$. On one hand, $y \notin W$, since $W \subseteq H$ and $H \cap Y = \{0\}$. On the other hand, since $y \in H^\perp$, $B(y, w) = 0$ for all $w \in W$. This implies $W + Fy$ is a totally isotropic subspace of V' properly containing W, contradicting the maximality of W. So Y is non-isotropic. $\square$

We have already observed that the subspace R in any Witt decomposition of V must equal the radical of V. Although H and Y do not satisfy such a strong uniqueness property, we prove in §14.16 that H and Y are uniquely determined up to bilinear isomorphisms. In particular, $\dim(H)$ is uniquely determined. On the other hand, the proof just given shows that $\dim(H) = 2k$, where k is the dimension of any maximal totally isotropic subspace W of V'. When B is nondegenerate, $V' = V$, and we see that *every maximal totally isotropic subspace of V has the same dimension k*. The number k is called the *Witt index* of (V, B).

14.12 Orthogonal Maps

A linear map $T : \mathbb{R}^n \to \mathbb{R}^n$ is called *orthogonal* iff T preserves dot products: $T(\mathbf{v}) \bullet T(\mathbf{w}) = \mathbf{v} \bullet \mathbf{w}$ for all $\mathbf{v}, \mathbf{w}$ in $\mathbb{R}^n$. It follows that an orthogonal map preserves the magnitude of vectors and the angle between nonzero vectors. Also, an orthogonal map on $\mathbb{R}^n$ is invertible, and the set of all orthogonal maps is a subgroup of the group of all invertible linear maps on $\mathbb{R}^n$. We can translate these facts into the language of matrices by fixing an orthonormal basis of $\mathbb{R}^n$ and taking the matrix of each linear map relative to this basis. Call a matrix $A \in M_n(\mathbb{R})$ *orthogonal* iff $A^{\mathrm{T}}A = I_n$, in which case $AA^{\mathrm{T}} = I_n$ and $A^{\mathrm{T}} = A^{-1}$ automatically follow. A linear map T is orthogonal iff the matrix representing T (relative to an orthonormal basis) is orthogonal. The set of orthogonal $n \times n$ real matrices forms a subgroup $\mathrm{O}_n(\mathbb{R})$ of the general linear group $\mathrm{GL}_n(\mathbb{R})$ of invertible $n \times n$ real matrices.

We now generalize these considerations by replacing the dot product on $\mathbb{R}^n$ by any bilinear form B on an n-dimensional F-vector space V. Define a linear map $T : V \to V$ to be *B-orthogonal* iff for all $v, w \in V$, $B(T(v), T(w)) = B(v, w)$. By linearity of T and bilinearity of B, it suffices to check this condition for v, w in a spanning set (or basis) of V. A B-orthogonal map is the same thing as a bilinear homomorphism from (V, B) to itself.

Here is a matrix formulation of B-orthogonality. Let $T : V \to V$ be a linear map. Given an ordered basis $X = (x_1, \ldots, x_n)$ of V, define $A = [B]_X$ and $C = [T]_X$, which means $A(i, j) = B(x_i, x_j)$ and $T(x_j) = \sum_{i=1}^n C(i, j)x_i$ for $1 \le i, j \le n$. We compute

$$
B(T(x_i), T(x_j)) = B\left(\sum_{r=1}^n C(r, i)x_r, \sum_{s=1}^n C(s, j)x_s \right) = \sum_{r=1}^n \sum_{s=1}^n C(r, i)B(x_r, x_s)C(s, j)
$$

$$
= \sum_{r=1}^n \sum_{s=1}^n C^{\mathrm{T}}(i, r)A(r, s)C(s, j) = [C^{\mathrm{T}}AC](i, j).
$$

So, the following conditions are equivalent (where i, j range from 1 to n): T is B-orthogonal; $B(T(x_i), T(x_j)) = B(x_i, x_j)$ for all i, j; $[C^{\mathrm{T}}AC](i, j) = A(i, j)$ for all i, j; $C^{\mathrm{T}}AC = A$. Thus, *B-orthogonality of T is equivalent to the matrix condition $A = C^{\mathrm{T}}AC$*, where $A = [B]_X$ and $C = [T]_X$. Our motivating example (the dot product in $\mathbb{R}^n$) has $A = I_n$ when X is the standard ordered basis. In this special case, we recover our original definition of an orthogonal real matrix: C is orthogonal iff $C^{\mathrm{T}}C = I_n$.

Let $\mathrm{O}_B(V)$ be the set of B-orthogonal linear maps on V. The identity map on V is B-orthogonal. Given B-orthogonal maps S and T, the composition $S \circ T$ is B-orthogonal, since $B(S(T(v)), S(T(w))) = B(T(v), T(w)) = B(v, w)$ for all $v, w \in V$. Similarly, the inverse of an invertible B-orthogonal map is B-orthogonal. It follows that the set $\mathrm{O}_B(V)^*$ of invertible B-orthogonal maps is a group under composition of functions. This group is a subgroup of $\mathrm{GL}(V)$, the general linear group of all invertible F-linear maps on V.

We claim that *if B is nondegenerate, then every B-orthogonal map is invertible*. To prove this, let $T : V \to V$ be B-orthogonal, X be any ordered basis of V, $A = [B]_X$, and $C = [T]_X$. We know A is invertible (since B is nondegenerate) and $A = C^{\mathrm{T}}AC$. Taking determinants shows that $\det(A) = \det(C^{\mathrm{T}})\det(A)\det(C) = \det(A)\det(C)^2$. Since $\det(A) \ne 0$, we have $\det(C)^2 = 1$, so $\det(C) = \pm 1_F \ne 0_F$. We conclude that T is an invertible map, and $\det(T)$ is 1_F or -1_F. (Recall that the determinant of a linear map is the determinant of any matrix representing T; all such matrices are similar and have the same determinant.) This proves the claim. So $\mathrm{O}_B(V)$ is a subgroup of $\mathrm{GL}(V)$ when B is nondegenerate. Similarly, for any fixed invertible matrix $A \in M_n(F)$, the set of *A-orthogonal matrices* $\mathrm{O}_n(A) = \{C \in M_n(F) : C^{\mathrm{T}}AC = A\}$ is a subgroup of $\mathrm{GL}_n(F)$, the group of all invertible matrices in $M_n(F)$.

Let us compute $O_B(V)$ explicitly for some low-dimensional bilinear spaces (V, B). It is routine to check that $O_B(V) = \{\mathrm{id}_V, -\mathrm{id}_V\}$ when (V, B) is a 1-dimensional nondegenerate bilinear space (Exercise 70). For a more interesting example, assume (V, B) is a hyperbolic plane spanned by a hyperbolic pair (v, w), so $B(v, v) = 0 = B(w, w)$ and $B(v, w) = 1 = B(w, v)$. Suppose $T : V \to V$ is any linear map. We have $T(v) = av + bw$ and $T(w) = cv + dw$ for some scalars $a, b, c, d \in F$. The following statements are equivalent:

(a) T is B-orthogonal;
(b) $B(T(v), T(v)) = B(v, v)$ and $B(T(w), T(w)) = B(w, w)$ and $B(T(v), T(w)) = B(v, w)$;
(c) $B(av + bw, av + bw) = 0$ and $B(cv + dw, cv + dw) = 0$ and $B(av + bw, cv + dw) = 1$;
(d) $2ab = 0$ and $2cd = 0$ and $ad + bc = 1$.

Let us assume $2_F \neq 0_F$; recall $1_F \neq 0_F$ holds in any field. To find all B-orthogonal maps T, consider various cases. In the case $a = 0$, we deduce $bc = 1$, $b \neq 0 \neq c$, $c = b^{-1}$, and $d = 0$, so T is the unique linear map such that $T(v) = bw$ and $T(w) = b^{-1}v$. In the case $b = 0$, we deduce $ad = 1$, $a \neq 0 \neq d$, $d = a^{-1}$, and $c = 0$, so T is the unique linear map such that $T(v) = av$ and $T(w) = a^{-1}w$. The group $O_B(V)$ consists of all maps given by these formulas, as a and b range over nonzero elements of F.

We can give a similar analysis of $O_B(V)$ when V is a symplectic plane spanned by a symplectic pair (v, w). Consider a linear map $T : V \to V$ and write $T(v) = av + bw$, $T(w) = cv + dw$ with $a, b, c, d \in F$. Since $B(u, u) = 0$ for all $u \in V$, we see that T is B-orthogonal iff $B(T(v), T(w)) = B(v, w)$ iff $B(av + bw, cv + dw) = 1$ iff $ad - bc = 1$ iff $\det(T) = 1$. In this case, $O_B(V)$ coincides with the special linear group $\mathrm{SL}(V)$ of linear maps on V with determinant 1.

As a final 2-dimensional example, consider the bilinear space $\mathbb{R}^2$ with the dot product. You can check that $O_B(\mathbb{R}^2)$ consists of all rotations around the origin, together with all reflections across lines through the origin (Exercise 62). Each rotation has determinant $+1$, and each reflection has determinant -1.

14.13 Reflections

Reflections play a central role in Euclidean geometry. Given a line ℓ in $\mathbb{R}^2$, the *geometric reflection across* ℓ is the map $s_\ell : \mathbb{R}^2 \to \mathbb{R}^2$ that sends each point P in the plane to its mirror image in ℓ. We give a formula for s_ℓ in the case where ℓ is the line through $(0, 0)$ and $\mathbf{v} = (v_1, v_2) \neq (0, 0)$. Let $\mathbf{z} = (v_2, -v_1)$, which is a vector orthogonal to $\mathbf{v}$. Each point P in $\mathbb{R}^2$ can be written as $P = a\mathbf{v} + b\mathbf{z}$ for unique scalars $a, b \in \mathbb{R}$. We define $s_\ell(P) = a\mathbf{v} - b\mathbf{z}$. Note that s_ℓ sends $\mathbf{z}$ to $-\mathbf{z}$ and fixes each point on the line $\ell = \mathbb{R}\mathbf{v}$ orthogonal to $\mathbf{z}$.

We now study reflections in the more general setting of a symmetric bilinear space (V, B) over F, where $2_F \neq 0_F$. For any $z \in V$ such that $B(z, z) \neq 0$, the *reflection determined by* z is the map $S_z : V \to V$ defined by

$$S_z(v) = v - \frac{2_F B(v, z)}{B(z, z)} z \quad \text{for } v \in V. \tag{14.8}$$

When S_z is defined, we have $V = Fz \boxplus (Fz)^\perp$ by the first example in §14.7. In particular, V has an ordered basis $(z, v_2, \ldots, v_n)$ such that $z \perp v_i$ for $2 \leq i \leq n$.

Properties of Reflections. Assume $B(z, z) \neq 0$, so that the reflection S_z is defined.
(a) S_z is an F-linear map.
(b) S_z is the unique F-linear map on V satisfying $S_z(z) = -z$ and $S_z(v) = v$ for all $v \in V$ such that $v \perp z$. Also $\det(S_z) = -1_F$.

(c) S_z is invertible, with $S_z^{-1} = S_z$.

(d) S_z is B-orthogonal.

(e) For any invertible B-orthogonal map T on V, $T \circ S_z \circ T^{-1} = S_{T(z)}$.

(f) If $v \in V$ satisfies $B(v,v) = B(z,z) \neq 0$ and $B(v-z, v-z) \neq 0$, then S_{v-z} interchanges v and z.

Proof. (a) Fix $v, w \in V$ and $c \in F$. Using bilinearity of B, compute

$$S_z(v+w) = (v+w) - \frac{2B(v+w, z)}{B(z,z)} z = v + w - \frac{2B(v,z) + 2B(w,z)}{B(z,z)} z$$

$$= v - \frac{2B(v,z)}{B(z,z)} z + w - \frac{2B(w,z)}{B(z,z)} z = S_z(v) + S_z(w);$$

$$S_z(cv) = cv - \frac{2B(cv,z)}{B(z,z)} z = cv - \frac{2cB(v,z)}{B(z,z)} z = cS_z(v).$$

(b) Taking $v = z$ in the definition gives $S_z(z) = z - \frac{2B(z,z)}{B(z,z)} z = z - 2z = -z$. For $v \in V$ such that $v \perp z$, we have $B(v,z) = 0$ and so $S_z(v) = v - 0z = v$. To prove uniqueness, suppose S is any linear map on V satisfying $S(z) = -z$ and $S(v) = v$ for all v orthogonal to z. Then $S = S_z$ since these linear maps agree on the ordered basis $(z, v_2, \ldots, v_n)$. The matrix of S_z relative to this ordered basis is diagonal with diagonal entries $-1, 1, \ldots, 1$. So $\det(S_z) = -1$.

(c) To show S_z is its own inverse, it suffices (by linearity) to show that $S_z(S_z(v)) = v$ for all v in the basis $(z, v_2, \ldots, v_n)$. By part (b), $S_z(S_z(z)) = S_z(-z) = -(-z) = z$, while $S_z(S_z(v_i)) = S_z(v_i) = v_i$ for $2 \leq i \leq n$.

(d) We prove that S_z is B-orthogonal by checking on the basis $(z, v_2, \ldots, v_n)$. Fix i, j between 2 and n. Using part (c), we compute:

$$B(S_z(z), S_z(z)) = B(-z, -z) = (-1)^2 B(z,z) = B(z,z);$$
$$B(S_z(z), S_z(v_i)) = B(-z, v_i) = -B(z, v_i) = -0 = 0 = B(z, v_i);$$
$$B(S_z(v_i), S_z(v_j)) = B(v_i, v_j).$$

(e) Let $T : V \to V$ be an invertible B-orthogonal map. Since $B(T(z), T(z)) = B(z,z) \neq 0$, the reflection $S_{T(z)}$ is defined. Define $S' = T \circ S_z \circ T^{-1}$, which is F-linear. We show $S' = S_{T(z)}$ by checking the conditions from part (b) that uniquely characterize $S_{T(z)}$. First,

$$S'(T(z)) = T(S_z(T^{-1}(T(z)))) = T(S_z(z)) = T(-z) = -T(z).$$

Second, suppose $v' \in V$ is orthogonal to $T(z)$. Since T^{-1} is an orthogonal map, $v = T^{-1}(v')$ is orthogonal to $T^{-1}(T(z)) = z$. So

$$S'(v') = T(S_z(T^{-1}(v'))) = T(S_z(v)) = T(v) = v'.$$

Thus S' is the reflection determined by $T(z)$.

(f) Assume $v \in V$ satisfies $B(v,v) = B(z,z) \neq 0$ and $B(v-z, v-z) \neq 0$. Then S_{v-z} is defined, $B(v-z, v-z) = B(v,v) - 2B(v,z) + B(z,z) = 2B(v,v) - 2B(v,z) = 2B(v, v-z)$, and hence

$$S_{v-z}(v) = v - \frac{2B(v, v-z)}{B(v-z, v-z)}(v-z) = v - (v-z) = z.$$

By part (c), $S_{v-z}(z) = v$. $\qquad\square$

As an example, consider a hyperbolic plane (V, B) spanned by a hyperbolic pair (v, w). We saw in §14.12 that for nonzero $b \in F$, the linear map T_b sending v to bw and w to $b^{-1}v$ is an orthogonal map on V. We claim that the map T_b equals the reflection S_{v-bw}. To verify this, first check that $(v - bw, v + bw)$ is a basis of V such that $B(v - bw, v - bw) = -2b \neq 0$ and $B(v - bw, v + bw) = 0$. Next, compute

$$T_b(v - bw) = T_b(v) - bT_b(w) = bw - b(b^{-1}v) = bw - v = -(v - bw) = S_{v-bw}(v - bw);$$
$$T_b(v + bw) = T_b(v) + bT_b(w) = bw + b(b^{-1}v) = v + bw = S_{v-bw}(v + bw).$$

Since the linear maps T_b and S_{v-bw} agree on a basis of V, they are equal. Taking $b = 1$, we have the special case $T_1 = S_{v-w}$, where T_1 interchanges v and w. Note that $T_1 \circ T_b$ is the linear map sending v to bv and w to $b^{-1}w$. We proved in §14.12 that $O_B(V)$ consists of all the maps T_b and $T_1 \circ T_b$, as b ranges over $F_{\neq 0}$. Since T_b and T_1 are reflections, we have now proved that *every orthogonal map on a hyperbolic plane is a composition of one or two reflections*. Similarly, every orthogonal map on $\mathbb{R}^2$ is a composition of one or two reflections (Exercise 62). We generalize these results in the next section.

14.14 Writing Orthogonal Maps as Compositions of Reflections

For the two-dimensional spaces considered in the last section, every orthogonal map is a product of at most two reflections. Cartan and Dieudonné generalized this result to higher-dimensional spaces. We give Artin's proof of their theorem, following the account in Jacobson's algebra text [32, §6.6].

Theorem on Orthogonal Maps and Reflections. Assume $2_F \neq 0_F$ and (V, B) is an n-dimensional symmetric nondegenerate bilinear space. Every B-orthogonal map $T : V \to V$ is the composition of at most n reflections.

Proof. We use induction on $n = \dim(V)$. The base case $n \leq 1$ is routine (Exercise 70), using the convention that id_V is the composition of zero reflections. Assume $n > 1$ and the theorem holds for all spaces of dimension less than n. Let T be an orthogonal map on an n-dimensional space (V, B). We consider four cases.

Case 1: There exists $x \in V$ with $B(x, x) \neq 0$ and $T(x) = x$. Let $V' = (Fx)^{\perp}$. By the first example at the end of §14.7, $V = Fx \boxplus V'$. Since $T : V \to V$ is a bilinear space isomorphism sending Fx onto itself, T maps V' onto V' (Exercise 23). We can therefore consider the restricted map $T' = T|V'$, which is an orthogonal map on the $(n - 1)$-dimensional bilinear space $(V', B|V')$. Since $V = Fx \boxplus V'$, $B|V'$ is nondegenerate. By induction hypothesis, for some $k \leq n - 1$ and some $z_1, \ldots, z_k \in V'$, we have $T' = S'_{z_1} \circ \cdots \circ S'_{z_k}$, where S'_z stands for the reflection on V' determined by z. Letting S_z be the reflection on V determined by z, we claim that $T = S_{z_1} \circ \cdots \circ S_{z_k}$. On one hand, both maps have the same effect on vectors in V'. On the other hand, both maps send x to itself, since $x \perp V'$. So the two linear maps are equal since they agree on Fx and V'. Thus we have written T as a product of k reflections in V, where $k < n$.

Case 2: There exists $x \in V$ with $B(x, x) \neq 0$ and $B(x - T(x), x - T(x)) \neq 0$. Note $B(T(x), T(x)) = B(x, x)$ since T is orthogonal. By property (f) of §14.13, the reflection $S_{x-T(x)}$ interchanges x and $T(x)$. Then $S_{x-T(x)} \circ T$ is orthogonal and sends x to itself, so Case 1 applies to this map. We get $S_{x-T(x)} \circ T = S_{z_1} \circ \cdots \circ S_{z_k}$ for some $z_1, \ldots, z_k \in V$ where $k < n$. Then $T = S_{x-T(x)} \circ S_{z_1} \circ \cdots \circ S_{z_k}$ is a product of at most n reflections.

Case 3: V is 2-dimensional. As a subcase, suppose every nonzero $v \in V$ satisfies $B(v,v) \neq 0$. Fix any nonzero $x \in V$. If $T(x) = x$, then Case 1 applies to T. If $T(x) \neq x$, then Case 2 applies to T. For the other subcase of Case 3, suppose $B(v,v) = 0$ for some nonzero $v \in V$. Since B is nondegenerate, there is $z \in V$ with $B(v,z) \neq 0$. By Exercise 72 (or by the Embedding Theorem for Totally Isotropic Spaces, applied to Fv), there is $w \in V$ such that (v,w) is a hyperbolic pair and $V = Fv \oplus Fw$ is a hyperbolic plane. Then the discussion at the end of §14.13 proves that T is a product of at most two reflections.

Case 4: V does not satisfy Case 1 or Case 2 or Case 3. This means that $n = \dim(V) \geq 3$ and for all $x \in V$, if $B(x,x) \neq 0$, then $T(x) \neq x$ and $B(x - T(x), x - T(x)) = 0$. Let $R : V \to V$ be the linear map given by $R(x) = x - T(x)$ for $x \in V$. We claim that $B(R(x), R(x)) = 0$ holds for all $x \in V$. By the assumption of Case 4, it suffices to check the claim for fixed nonzero $x \in V$ with $B(x,x) = 0$. Let $W = (Fx)^\perp$, which is a subspace of V containing Fx. By part (b) of the Theorem on Orthogonal Complements, $\dim(W) = n - 1$. Since $n \geq 3$, we have $n - 1 > n/2$, and so W cannot be totally isotropic (§14.11). Thus there is $w \in W$ with $B(w,w) \neq 0$. We know $B(w,x) = B(x,x) = 0$, which leads to $B(w + x, w + x) = B(w - x, w - x) = B(w,w) \neq 0$. By the assumption of Case 4, we get $B(R(w + x), R(w + x)) = B(R(w - x), R(w - x)) = B(R(w), R(w)) = 0$. Bilinearity of B gives $B(u + v, u + v) + B(u - v, u - v) = 2B(u,u) + 2B(v,v)$ for any $u, v \in V$ (Exercise 1). Applying this with $u = R(w)$ and $v = R(x)$ gives $0 + 0 = 2 \cdot 0 + 2B(R(x), R(x))$. So $B(R(x), R(x)) = 0$, proving the claim.

Let the linear map R have image I and kernel K. We prove $I = K$. The claim above shows that I is a totally isotropic subspace of V, which means that $I \subseteq I^\perp$. For any $x \in K$, we have $R(x) = 0$ and $T(x) = x$. By the contrapositive of the assumption of Case 4, we must have $B(x,x) = 0$. This means that K is also totally isotropic, so $K \subseteq K^\perp$. We next show that $I = K^\perp$. To prove $I \subseteq K^\perp$, fix $y \in I$. We can write $y = R(v) = v - T(v)$ for some $v \in V$. Fix $x \in K$, so $x = T(x)$. We compute $B(x,y) = B(x,v) - B(x, T(v)) = B(x,v) - B(T(x), T(v)) = 0$ using orthogonality of T. This shows $y \in K^\perp$. By part (b) of the Theorem on Orthogonal Complements and by the Rank–Nullity Theorem, $\dim(K^\perp) = n - \dim(K) = \dim(I)$. Since I is a subset of $K^\perp$ and both spaces have the same finite dimension, we get $I = K^\perp$. Taking the orthogonal complement of both sides gives $I^\perp = K^{\perp\perp} = K$. Then $K \subseteq K^\perp = I \subseteq I^\perp = K$, and therefore $K = I$. The fact that $I \subseteq K$ means that $R(R(u)) = 0$ for all u in V, or equivalently $(\mathrm{id}_V - T)^2$ is the zero map on V. By Exercise 75, $\det(T) = 1$ follows. Also, since $n = \dim(I) + \dim(K)$ and $K = I$, we see that n is even. To summarize, we have proved that if Case 4 occurs when analyzing a particular map T, then $n = \dim(V)$ must be even and $\det(T)$ must be 1.

To finish Case 4, pick a particular $u \in V$ such that $B(u,u) \neq 0$. Such u must exist, since otherwise V is totally isotropic and $B = 0$, contradicting nondegeneracy of B. Consider the map $T' = S_u \circ T$. This map is orthogonal, being the composition of orthogonal maps, and $\det(T') = \det(S_u)\det(T) = -1$. On one hand, this means that applying the proof to T' does not lead to Case 4. So we already know there exist $k \leq n$ and $z_1, \ldots, z_k \in V$ with $T' = S_{z_1} \circ \cdots \circ S_{z_k}$. On the other hand, since $\det(S_{z_i}) = -1$ for each i, $\det(T') = -1$ implies $(-1)^k = -1$ and k is odd. Since n is even in Case 4, we must have $k < n$. Finally, $T = S_u \circ S_{z_1} \circ \cdots \circ S_{z_k}$ is a product of $k + 1 \leq n$ reflections. $\square$

Exercise 77 outlines a simpler proof of the weaker result that every orthogonal map on an n-dimensional nondegenerate symmetric bilinear space can be written as a product of at most $2n$ reflections.

14.15   Witt's Cancellation Theorem

Suppose (V, B) and (V_*, B_*) are bilinear spaces, $f : V \to V_*$ is a bilinear space isomorphism, W is a subspace of V, and $W_* = f[W]$ is the image of W under f. It is routine to check that $f[W^\perp] = W_*^\perp$ in this situation (Exercise 23), so that f restricts to an isomorphism from $W^\perp$ onto $W_*^\perp$. The next theorem establishes a stronger result. We write $\cong$ to denote isomorphism of bilinear spaces.

Witt's Cancellation Theorem.   Assume $2_F \neq 0_F$, (V, B) and (V_*, B_*) are symmetric bilinear spaces, and $V \cong V_*$ via an isomorphism f. Suppose W is a nondegenerate subspace of V, W_* is a subspace of V_*, and $W \cong W_*$ via an isomorphism g. Then $W^\perp \cong W_*^\perp$.

Since W (and hence W_*) are nondegenerate subspaces, we know $V = W \boxplus W^\perp$ and $V_* = W_* \boxplus W_*^\perp$. Informally, the theorem says that if $W \boxplus W^\perp \cong W_* \boxplus W_*^\perp$ and $W \cong W_*$, then we can "cancel" the isomorphic summands W and W_* to conclude that $W^\perp \cong W_*^\perp$. The observation preceding the theorem is the special case where $g = f|W$. Some exercises explore whether analogous cancellation theorems hold for other algebraic structures.

Proof. Step 1. We prove the special case of the theorem where $V_* = V$, $B_* = B$, $f = \mathrm{id}_V$, and $\dim(W) = 1$. Write $W = Fx$ for some $x \in V$. Note $B(x, x) \neq 0$ since W is nondegenerate. Let $y = g(x)$, so $W_* = Fy$ and $B(y, y) = B(x, x) \neq 0$. By the Parallelogram Law (Exercise 1), $B(x+y, x+y) + B(x-y, x-y) = 2B(x, x) + 2B(y, y) = 2_F \cdot 2_F \cdot B(x, x) \neq 0_F$. Therefore $B(x + y, x + y) \neq 0$ or $B(x - y, x - y) \neq 0$. Define $z = y$ if $B(x - y, x - y) \neq 0$ and $z = -y$ otherwise. In both cases, we have $W_* = Fz$, $B(z, z) = B(x, x) \neq 0$, and $B(x - z, x - z) \neq 0$. By property (f) from §14.13, the reflection $S_{x-z} : V \to V$ is an orthogonal map interchanging x and z. Since S_{x-z} is a bilinear isomorphism of V sending $W = Fx$ onto $W_* = Fz$, S_{x-z} restricts to a bilinear isomorphism from $W^\perp$ onto $W_*^\perp$.

Step 2. We prove the special case of the theorem where $\dim(W) = 1$. Let $W_1 = f^{-1}[W_*]$, and note $f^{-1} \circ g : W \to W_1$ is a bilinear isomorphism. By Step 1, there is an isomorphism $h : W^\perp \to W_1^\perp$. Since f maps V onto V_* and W_1 onto W_*, f maps $W_1^\perp$ onto $W_*^\perp$. Then $f \circ h$ is an isomorphism from $W^\perp$ onto $W_*^\perp$.

Step 3. We prove the theorem by induction on $k = \dim(W)$. The case $k = 0$ is immediate, and Step 2 proves the case $k = 1$. Fix W with $\dim(W) = k > 1$, and assume the theorem holds for subspaces of dimension less than k. Since W is nondegenerate, there is $x \in W$ with $B(x, x) \neq 0$ (otherwise W would be totally isotropic and $B|W$ would be zero). Fix such an x, and let $Z = (Fx)^\perp$ be the orthogonal complement of Fx in W. We have $W = Fx \boxplus Z$ since $B|Fx$ is nondegenerate. Define $y = g(x) \in W_*$, and let $Z_* = (Fy)^\perp$ be the orthogonal complement of Fy in W_*. Then $W_* = Fy \boxplus Z_*$, where the isomorphism $g : W \to W_*$ maps Fx onto Fy and hence maps Z onto Z_*. We have

$$Fx \boxplus Z \boxplus W^\perp = V \cong V_* = Fy \boxplus Z_* \boxplus W_*^\perp.$$

The orthogonal complement of Fx in V is $Z \boxplus W^\perp$, and the orthogonal complement of Fy in V_* is $Z_* \boxplus W_*^\perp$. By Step 2, there is an isomorphism $f_1 : Z \boxplus W^\perp \to Z_* \boxplus W_*^\perp$. We also have an isomorphism $g_1 : Z \to Z_*$ (obtained by restricting g), and $\dim(Z) = k - 1$. Applying the induction hypothesis, we get an isomorphism $h : W^\perp \to W_*^\perp$. $\square$

14.16   Uniqueness Property of Witt Decompositions

This section proves a uniqueness property of Witt decompositions of a symmetric bilinear space. The proof uses Witt's Cancellation Theorem and the following analogous cancellation property for radicals.

Radical Cancellation Lemma. Suppose (V, B) is a symmetric bilinear space with radical R, (V_*, B_*) is a symmetric bilinear space with radical R_*, and $f : V \to V_*$ is a bilinear space isomorphism. If W and W_* are any subspaces such that $V = R \oplus W$ and $V_* = R_* \oplus W_*$, then $W \cong W_*$.

Proof. Recall from §14.8 the construction of the bilinear spaces $(V/R, \overline{B})$ and $(V_*/R_*, \overline{B_*})$. We saw in that section that $(W, B|W) \cong (V/R, \overline{B})$ and $(W_*, B_*|W_*) \cong (V_*/R_*, \overline{B_*})$. It is routine to check that f restricts to an isomorphism of R onto R_*, and that f induces an isomorphism $\overline{f} : V/R \to V_*/R_*$ given by $\overline{f}(x + R) = f(x) + R_*$ for $x \in V$. Combining this isomorphism with the ones just mentioned, we get $(W, B|W) \cong (W_*, B_*|W_*)$.   □

Witt's Decomposition Theorem (Uniqueness Part). Suppose (V, B) and (V_*, B_*) are isomorphic symmetric bilinear spaces, $V = R \boxplus H \boxplus Y$ is a Witt decomposition of V, and $V_* = R_* \boxplus H_* \boxplus Y_*$ is a Witt decomposition of V_*. Then $R \cong R_*$, $H \cong H_*$, and $Y \cong Y_*$.

Proof. Let $f : V \to V_*$ be a bilinear space isomorphism. As remarked in §14.11, $R = \mathrm{Rad}(V)$, $R_* = \mathrm{Rad}(V_*)$, and f restricts to an isomorphism $\mathrm{Rad}(V) \cong \mathrm{Rad}(V_*)$. By the Radical Cancellation Lemma, we get an isomorphism $g : H \boxplus Y \to H_* \boxplus Y_*$. Write $H = H_1 \boxplus \cdots \boxplus H_k$ and $H_* = H_{*1} \boxplus \cdots \boxplus H_{*k_*}$ for some hyperbolic planes H_i and H_{*j} and some $k, k_* \geq 0$. We can assume $k \leq k_*$ by interchanging the roles of V and V_* if needed. Any two hyperbolic planes over F are isomorphic and nondegenerate, so we can apply Witt's Cancellation Theorem k times to remove pairs of factors H_i and H_{*i} for $1 \leq i \leq k$. We are left with an isomorphism $h : Y \to H' \boxplus Y_*$ where H' is the orthogonal direct sum of $k_* - k$ hyperbolic planes. If $k_* - k > 0$, then we reach a contradiction since $H' \boxplus Y_*$ has a nonzero isotropic vector (coming from H_{*k_*}) but the isomorphic non-isotropic space Y has no such vector. Thus $k_* = k$, which shows that $H \cong H_*$, $H' = \{0\}$, and $Y \cong Y_*$ via h.   □

14.17   Summary

In this summary, we assume $2_F \neq 0_F$ and V is an n-dimensional vector space over F, although these assumptions are not always needed.

1. *Definitions.* Table 14.1 summarizes definitions concerning bilinear forms.

2. *Matrix Representation of Bilinear Forms.* For each ordered basis $X = (x_1, \ldots, x_n)$ of V, there is a vector space isomorphism from $\mathrm{Bilin}(V)$ onto $M_n(F)$ sending a bilinear form B to the matrix $A = [B]_X$ with i, j-entry $B(x_i, x_j)$. The form B is symmetric iff $A^{\mathrm{T}} = A$, and B is alternate iff $A^{\mathrm{T}} = -A$ and $A(i, i) = 0$ for all i. Changing from X to a new basis $Y = (y_1, \ldots, y_n)$ replaces $[B]_X$ by the congruent matrix $[B]_Y = P^{\mathrm{T}}[B]_X P$, where the transition matrix P is defined by $y_j = \sum_{i=1}^{n} P(i, j)x_i$ for $1 \leq i, j \leq n$.

TABLE 14.1

Definitions involving bilinear forms. Here, V is a finite-dimensional F-vector space.

Concept	Definition
bilinear form $B : V \times V \to F$	$B(x + y, z) = B(x, z) + B(y, z)$, $B(cx, y) = cB(x, y)$, $B(x, y + z) = B(x, y) + B(x, z)$, $B(x, cy) = cB(x, y)$ for all $x, y, z \in V$, $c \in F$
symmetric form	$B(x, y) = B(y, x)$ for all $x, y \in V$
antisymmetric form	$B(x, y) = -B(y, x)$ for all $x, y \in V$
alternate form	$B(x, x) = 0$ for all $x \in V$
nondegenerate form	if $B(v, w) = 0$ for all $v \in V$, then $w = 0$
matrix $[B]_X$ for basis $X = (x_1, \ldots, x_n)$	i, j-entry of $[B]_X$ is $B(x_i, x_j)$
matrix A is congruent to matrix D	$A = P^{\mathrm{T}} D P$ for some invertible matrix P
$\mathrm{Bilin}(V)$	vector space of bilinear forms on V
dual space V^*	vector space of linear functions from V to F
$v \perp w$ (where $v, w \in V$)	$B(v, w) = 0$
$S \perp T$ (where $S, T \subseteq V$)	$B(s, t) = 0$ for all $s \in S$, $t \in T$
$S^\perp$ (orthogonal complement)	$\{v \in V : B(s, v) = 0 \text{ for all } s \in S\}$
$^\perp S$ (left orthogonal complement)	$\{v \in V : B(v, s) = 0 \text{ for all } s \in S\}$
radical $\mathrm{Rad}(V)$	$V^\perp = \{w \in V : B(v, w) = 0 \text{ for all } v \in V\}$
$V = V_1 \oplus \cdots \oplus V_k$ (direct sum)	$V = V_1 + \cdots + V_k$ and $V_i \cap \sum_{j < i} V_j = \{0\}$ for all i
$V = V_1 \boxplus \cdots \boxplus V_k$ (orthog. direct sum)	$V = V_1 \oplus \cdots \oplus V_k$ and $V_i \perp V_j$ for all $i \neq j$
isotropic vector v	$B(v, v) = 0$ and $v \neq 0$
totally isotropic subspace W	$B(w, w) = 0$ for all $w \in W$ (so $B|W = 0$ and $W \subseteq W^\perp$)
non-isotropic subspace Y	$B(y, y) \neq 0$ for all nonzero $y \in Y$
Witt index of V	dimension of any maximal totally isotropic subspace of V
hyperbolic pair (v, w)	$B(v, v) = 0 = B(w, w)$ and $B(v, w) = 1 = B(w, v)$
hyperbolic plane	span of a hyperbolic pair
symplectic pair (v, w)	$B(v, v) = 0 = B(w, w)$ and $B(v, w) = 1 = -B(w, v)$
symplectic plane	span of a symplectic pair
Witt decomposition of V	$V = R \boxplus H_1 \boxplus \cdots \boxplus H_k \boxplus Y$ where R is totally isotropic, $H_1, \ldots, H_k$ are hyperbolic planes, and Y is non-isotropic
bilinear hom. $T : (V, B) \to (V_*, B_*)$	T is F-linear and $B_*(T(x), T(y)) = B(x, y)$ for $x, y \in V$
B-orthogonal map $T : V \to V$	linear map with $B(T(x), T(y)) = B(x, y)$ for $x, y \in V$
$O_B(V)$	set of B-orthogonal maps on V
reflection S_z (where $B(z, z) \neq 0$)	$S_z(v) = v - \frac{2B(v, z)}{B(z, z)} z$ for $v \in V$

3. *Basic Orthogonality Facts.* Every orthogonal list of non-isotropic vectors is linearly independent. For $S \subseteq V$, the orthogonal complements $S^\perp$ and $^\perp S$ are always subspaces of V such that $S \subseteq {}^\perp(S^\perp)$ and $S \subseteq ({}^\perp S)^\perp$. If $S \subseteq T$, then $T^\perp \subseteq S^\perp$. Orthogonality is a symmetric relation for symmetric and alternate bilinear forms. A bilinear form B on V is nondegenerate iff $V^\perp = \{0\}$, while the restriction of B to a subspace W is nondegenerate iff $W \cap W^\perp = \{0\}$. Given an orthogonal direct sum $V = W_1 \boxplus \cdots \boxplus W_k$, B is nondegenerate on V iff each $B|W_i$ is nondegenerate on W_i. If an ordered basis X for V is the concatenation of ordered bases X_i for W_i, then $[B]_X$ is block-diagonal with diagonal blocks $[B|W_i]_{X_i}$.

4. *Bilinear Forms and Dual Spaces.* If B is any nondegenerate bilinear form on V, there is a vector space isomorphism $R : V \to V^*$ given by $R(w)(v) = B(v, w)$ for all $v, w \in V$. This means that for every $g \in V^*$, there is a unique $w \in V$ such that $g(v) = B(v, w)$ for all $v \in V$.

5. *Theorem on Orthogonal Complements.* Let W be any subspace of V. If $B|W$ is nondegenerate, then $\dim(W) + \dim(W^\perp) = \dim(V) = \dim(W) + \dim({}^\perp W)$ and $V = W \oplus W^\perp = W \oplus {}^\perp W$. If B is nondegenerate, then $\dim(W) + \dim(W^\perp) = \dim(V) = \dim(W) + \dim({}^\perp W)$ and $^\perp(W^\perp) = W = ({}^\perp W)^\perp$.

6. *Diagonalization Theorems for Symmetric Bilinear Forms.* If B is symmetric, then there is an ordered basis X for V such that $[B]_X$ is diagonal. If square roots exist in F, then we can arrange that all diagonal entries of $[B]_X$ are in $\{0, 1\}$. If $F = \mathbb{R}$, then we can arrange that all diagonal entries of $[B]_X$ are in $\{-1, 0, 1\}$. In both cases, the number of -1s, 0s, and 1s is uniquely determined by B.

7. *Structure Theorem for Alternate Bilinear Forms.* An alternate bilinear space (V, B) is the orthogonal direct sum of its radical and s symplectic planes. So B can be represented by a block-diagonal matrix containins s copies of $C = \begin{bmatrix} 0 & 1 \\ -1 & 0 \end{bmatrix}$ and all other entries 0.

8. *Totally Isotropic Subspaces.* Let W be a totally isotropic subspace of a nondegenerate symmetric space V. Then $\dim(W) \leq \dim(V)/2$, and W is contained in an orthogonal direct sum of $\dim(W)$ hyperbolic planes.

9. *Witt Decompositions.* Any symmetric bilinear space has a Witt decomposition $V = R \boxplus H \boxplus Y$, where R is totally isotropic, H is an orthogonal direct sum of zero or more hyperbolic planes, and Y is non-isotropic. If $V = R_* \boxplus H_* \boxplus Y_*$ is a second Witt decomposition, then $R = R_* = \mathrm{Rad}(V)$, $H \cong H_*$, and $Y \cong Y_*$.

10. *Orthogonal Maps and Reflections.* For nondegenerate B, the set $O_B(V)$ of B-orthogonal maps is a subgroup of $\mathrm{GL}(V)$. For $z \in V$ with $B(z, z) \neq 0$, the reflection S_z is a self-inverse B-orthogonal linear map that sends z to $-z$ and sends each $v \in (Fz)^\perp$ to itself. If B is symmetric and nondegenerate, then every B-orthogonal map on V is the composition of at most $\dim(V)$ reflections.

11. *Witt's Cancellation Theorem.* If there are isomorphisms of symmetric bilinear spaces $W \boxplus W^\perp \cong W_* \boxplus W_*^\perp$ and $W \cong W_*$, then there is an isomorphism $W^\perp \cong W_*^\perp$.

14.18 Exercises

In these exercises, assume F is a field with $2_F \neq 0_F$, V is a finite-dimensional F-vector space, B is a bilinear form on V, and $X = (x_1, \ldots, x_n)$ is an ordered basis of V, unless otherwise stated.

1. **Parallelogram Law.** Prove: for all $u, v \in V$, $B(u+v, u+v) + B(u-v, u-v) = 2B(u, u) + 2B(v, v)$.

2. Assume B is symmetric. Define $Q : V \to V$ by $Q(v) = B(v, v)$. Prove: for all $c \in F$ and $v \in V$, $Q(cv) = c^2 Q(v)$. For $v, w \in V$, find a formula for $B(v, w)$ in terms of Q.

3. Define $B((x, y), (s, t)) = xs + 2yt$ for $x, y, s, t \in F$. Prove B is a bilinear form on F^2. Is B symmetric? alternate? nondegenerate? What is $[B]_X$ for $X = (\mathbf{e}_1, \mathbf{e}_2)$?

4. Define $B((x, y), (s, t)) = \det \begin{bmatrix} x & y \\ s & t \end{bmatrix}$ for $x, y, s, t \in F$. Is B a bilinear form on F^2? If so, is B symmetric? alternate? nondegenerate?

5. Define $B((x, y), (s, t)) = xy + st$ for $x, y, s, t \in F$. Is B a bilinear form on F^2? If so, is B symmetric? alternate? nondegenerate?

6. Define $B((x, y), (s, t)) = (x + y)(s - t)$ for $x, y, s, t \in F$. Is B a bilinear form on F^2? If so, is B symmetric? alternate? nondegenerate?

7. For $\mathbf{v}, \mathbf{w} \in \mathbb{R}^3$, define $B(\mathbf{v}, \mathbf{w}) = \mathbf{v} \times \mathbf{w}$ (the cross product of $\mathbf{v}$ and $\mathbf{w}$). Prove B is linear in the first and second position, and $B(\mathbf{v}, \mathbf{v}) = \mathbf{0}$ for all $\mathbf{v} \in \mathbb{R}^3$. But B is not a bilinear form on $\mathbb{R}^3$ — why not?

8. Check the claims about B_1, B_2, and B_3 in the sentence following equation (14.2) on page 369.

9. Check the claims about B in the sentence following equation (14.3) on page 369.

10. Suppose (V, B) and (V', B') are bilinear spaces, S spans V, and $T : V \to V'$ is a linear map. Prove T is a bilinear homomorphism iff for all $x, y \in S$, $B'(T(x), T(y)) = B(x, y)$.

11. Prove the bilinear spaces (F^2, B_1) and (F^2, B_2) defined in (14.2) are isomorphic. Does your proof use $2_F \neq 0_F$?

12. Let $F = \mathbb{Z}_2$, $V = F^2$, and define $B((a, b), (c, d)) = ad + bc$ for $a, b, c, d \in \mathbb{Z}_2$. Find all hyperbolic pairs in the hyperbolic plane (V, B). Is there an ordered basis X for V such that $[B]_X$ is diagonal?

13. Let $F = \mathbb{Z}_2$, $V = F^2$, and define $B((a, b), (c, d)) = ac - bd$ for $a, b, c, d \in \mathbb{Z}_2$. Is (V, B) a hyperbolic plane?

14. Let (V, B) and (V_*, B_*) be finite-dimensional bilinear spaces. Prove these spaces are isomorphic iff V has an ordered basis X and V_* has an ordered basis X_* such that $[B]_X = [B_*]_{X_*}$.

15. Suppose X_i is an ordered basis for the bilinear space (V_i, B_i) for $1 \leq i \leq k$, and let (V, B) be the product of these spaces. Give a precise formulation of the statement that the concatenation X of $X_1, \ldots, X_k$ is an ordered basis of V. Prove that $[B]_X$ is block diagonal with diagonal blocks $[B_1]_{X_1}, \ldots, [B_k]_{X_k}$.

16. Define $B'(v, w) = B(w, v)$ for $v, w \in V$. Prove B' is a bilinear form. How are the matrices $[B]_X$ and $[B']_X$ related?

17. Given a linear map $T : V \to V$, define $B'(v, w) = B(T(v), w)$ for $v, w \in V$. Prove B' is bilinear. Prove a formula relating $[B']_X$ to $[B]_X$ and $[T]_X$.

18. Given a linear map $S : V \to V$, define $B'(v, w) = B(v, S(w))$ for $v, w \in V$. Prove B' is bilinear. Prove a formula relating $[B']_X$ to $[B]_X$ and $[S]_X$.

19. Define $B_s(v, w) = B(v, w) + B(w, v)$ and $B_a(v, w) = B(v, w) - B(w, v)$ for $v, w \in V$. Prove B_s is a symmetric bilinear form. Prove B_a is an alternate bilinear form. How are $[B_s]_X$ and $[B_a]_X$ related to $[B]_X$?

20. Define $B(X, Y)$ to be the trace (sum of diagonal entries) of XY for $X, Y \in M_n(F)$. Prove B is a symmetric bilinear form on $M_n(F)$. Is B nondegenerate? Find the matrix of B relative to the ordered basis of matrix units $(E_{1,1}, E_{1,2}, \ldots, E_{n,n})$, where $E_{i,j} \in M_n(F)$ has i, j-entry 1_F and all other entries 0_F.

21. Let $A = [B]_X$. Prove B is symmetric iff $A^{\mathrm{T}} = A$. Prove B is antisymmetric iff $A^{\mathrm{T}} = -A$.

22. Suppose $v, w \in V$ have coordinates $[v]_X$ and $[w]_X$ (viewed as column vectors) relative to the ordered basis X. Prove $B(v, w) = [v]_X^{\mathrm{T}} [B]_X [w]_X$.

23. Suppose (V, B) and (V_*, B_*) are bilinear spaces, $f : V \to V_*$ is a bilinear space isomorphism, W is a subspace of V, and $W_* = f[W]$. Prove $f[W^\perp] = W_*^\perp$.

24. Give an example of a symmetric bilinear space (V, B) and a subspace $W \neq \{0\}$ where B is degenerate, $B|W$ is nondegenerate, and $W^{\perp\perp} \neq W$.

25. Give an example of an alternate bilinear space (V, B) and a subspace W where B is nondegenerate, $B|W$ is degenerate, and $V \neq W \boxplus W^\perp$.

26. Say that a bilinear form B on V is *degenerate on the left* if there exists $w \neq 0$ in V such that for all $v \in V$, $B(w, v) = 0$. If no such w exists, say B is *nondegenerate on the left*. Prove the following conditions are equivalent:
     (a) B is degenerate (as defined on page 368);
     (b) $[B]_X$ is invertible for some ordered basis X of V;
     (c) $[B]_X$ is invertible for every ordered basis X of V;
     (d) B is degenerate on the left.

27. Suppose the bilinear form B is nondegenerate on the left. Show that for each $g \in V^*$, there exists a unique $w \in V$ such that $g(v) = B(w, v)$ for all $v \in V$.

28. Prove: if B is nondegenerate, then for each ordered basis $(x_1, \ldots, x_n)$ of V, there is an ordered basis $(y_1, \ldots, y_n)$ for V such that for $1 \leq i, j \leq n$, $B(x_i, y_j)$ is 1 if $i = j$ and 0 if $i \neq j$.

29. Show that the Theorem on Orthogonal Complements, as stated in the chapter summary, holds for any bilinear form B.

30. Prove: if B is nondegenerate, then $f(S) = S^\perp$ and $g(S) = {}^\perp S$ define order-reversing, mutually inverse bijections on the lattice of subspaces of V.

31. Suppose $(V, B) \cong (V_*, B_*)$ via an isomorphism f, and let $R = \mathrm{Rad}(V)$, $R_* = \mathrm{Rad}(V_*)$. Prove that f restricts to an isomorphism of R onto R_*, and that f induces an isomorphism $\overline{f} : V/R \to V_*/R_*$ given by $\overline{f}(x + R) = f(x) + R_*$ for $x \in V$.

32. Prove: if (V, B) is a bilinear space containing a non-isotropic vector, then every isotropic vector in V is the difference of two non-isotropic vectors. (*Hint:* Use the Parallelogram Law.) What happens if $2_F = 0_F$?

33. Suppose (V, B) is a bilinear space such that orthogonality is symmetric, meaning: for all $x, y \in V$, $B(x, y) = 0$ implies $B(y, x) = 0$. Prove that B must be symmetric

or alternate, as follows. Let S be the set of $v \in V$ such that $B(v, w) = B(w, v)$ for all $w \in V$. Show S is a subspace of V containing all non-isotropic vectors in V. Use the previous exercise to complete the proof. Can you find a proof in the case $2_F = 0_F$?

34. Are $A = \begin{bmatrix} 1 & 0 \\ 0 & -1 \end{bmatrix}$ and $D = \begin{bmatrix} 0 & 1 \\ 1 & 0 \end{bmatrix}$ congruent in $M_2(\mathbb{Q})$?
    If so, find an invertible $P \in M_2(\mathbb{Q})$ with $A = P^T D P$.

35. Are $A = \begin{bmatrix} 3 & 0 \\ 0 & 4 \end{bmatrix}$ and $D = \begin{bmatrix} 2 & 0 \\ 0 & 3 \end{bmatrix}$ congruent in $M_2(\mathbb{Q})$?
    If so, find an invertible $P \in M_2(\mathbb{Q})$ with $A = P^T D P$.

36. Suppose $W_1, \ldots, W_k$ are subspaces of V with $V = W_1 \boxplus \cdots \boxplus W_k$. Prove $\mathrm{Rad}(V) = \mathrm{Rad}(W_1) \boxplus \cdots \boxplus \mathrm{Rad}(W_k)$. Prove that B is nondegenerate iff $B|W_i$ is nondegenerate for $1 \le i \le k$.

37. Suppose $W_1, \ldots, W_k$ are subspaces of V with $V = W_1 \boxplus \cdots \boxplus W_k$. Prove $f : W_1 \times \cdots \times W_k \to V$, given by $f((w_1, \ldots, w_k)) = w_1 + \cdots + w_k$ for $w_i \in W_i$, is an isomorphism of bilinear spaces.

38. Suppose $W_1, \ldots, W_k$ are subspaces of V with $V = W_1 \boxplus \cdots \boxplus W_k$. Let W_i have ordered basis X_i for $1 \le i \le k$ and X be the concatenation of $X_1, \ldots, X_k$. Prove $[B]_X$ is block-diagonal with diagonal blocks $[B|W_1]_{X_1}, \ldots, [B|W_k]_{X_k}$.

39. Give examples to show that the conclusions of the last three exercises do not always hold if we only assume $V = W_1 \oplus \cdots \oplus W_k$.

40. Suppose $W_1, \ldots, W_k$ are subspaces of V with $V = W_1 + \cdots + W_k$ and $W_i \perp W_j$ for all $i \ne j$. Does it always follow that $V = W_1 \boxplus \cdots \boxplus W_k$? Does the answer change if B is nondegenerate?

41. Let (V, B) be a bilinear space with radical R and W be any subspace of V such that $V = R \oplus W$. Prove that $p : W \to V/R$, given by $p(w) = w + R$ for $w \in W$, is a bilinear space isomorphism from $(W, B|W)$ to $(V/R, \overline{B})$.

42. Let B be a symmetric bilinear form on F^n where B has matrix A relative to the standard basis. Formulate an algorithm, similar to the Gram–Schmidt algorithm, that computes an ordered basis X of F^n such that $[B]_X$ is diagonal. Prove that your algorithm works.

43. For each matrix $A \in M_n(\mathbb{R})$, find a diagonal matrix D with entries in $\{0, 1, -1\}$ that is congruent to A.

    (a) $\begin{bmatrix} 4 & 4 & 4 \\ 4 & 4 & 4 \\ 4 & 4 & 4 \end{bmatrix}$ (b) $\begin{bmatrix} 0 & 1 & 0 & 1 \\ 1 & 0 & -1 & 0 \\ 0 & -1 & 0 & 1 \\ 1 & 0 & 1 & 0 \end{bmatrix}$ (c) $\begin{bmatrix} 1 & 2 & 0 \\ 2 & 0 & -1 \\ 0 & -1 & 3 \end{bmatrix}$

44. Define $A(i, j) = j - i$ for $1 \le i, j \le 5$. Find a matrix congruent to A in $M_5(\mathbb{R})$ that has the structure specified in the Theorem on Alternate Bilinear Forms.

45. How many congruence classes of symmetric matrices are there in $M_n(\mathbb{C})$?

46. How many congruence classes of symmetric matrices are there in $M_n(\mathbb{R})$?

47. How many congruence classes of skew-symmetric matrices are there in $M_n(F)$?

48. Let G be the multiplicative group of nonzero elements of the field F. Prove $G^2 = \{g^2 : g \in G\}$ is a subgroup of G. Prove that for all $A, C \in GL_n(F)$, if A and C are congruent, then $\det(A)G^2 = \det(C)G^2$ (equality of cosets in G/G^2). Prove that for a bilinear form B, the coset $\det([B]_X)G^2$ does not depend on the choice of ordered basis X. This coset is called the *discriminant of B*.

49. Give a specific infinite list of symmetric matrices in $M_2(\mathbb{Q})$ that are pairwise non-congruent.

50. Define $g \in (\mathbb{R}^2)^*$ by $g(x, y) = 3x - 5y$ for $x, y \in \mathbb{R}$. For each bilinear form B_j in (14.2), find $\mathbf{w} \in \mathbb{R}^2$ such that $g(\mathbf{v}) = B_j(\mathbf{v}, \mathbf{w})$ for all $\mathbf{v} \in \mathbb{R}^2$.

51. **Adjoints.** Given any function $T : V \to V$, a *right adjoint of T relative to B* is a function $S : V \to V$ such that for all $x, y \in V$, $B(T(x), y) = B(x, S(y))$. Suppose B is nondegenerate. Prove every F-linear map $T : V \to V$ has a unique right adjoint S, and S is F-linear. We write $S = T^*$.

52. Assume B is nondegenerate. For an F-linear map $T : V \to V$ with right adjoint T^*, what is the relation between the matrices $[T^*]_X$, $[T]_X$, and $[B]_X$?

53. Assume B is nondegenerate. Prove that the adjoint map (sending T to T^*) is F-linear. Prove: for all F-linear maps R, T, $(R \circ T)^* = T^* \circ R^*$.

54. Assume B is nondegenerate. Prove that a linear map $T : V \to V$ is B-orthogonal iff $T^* = T^{-1}$.

55. **Universal Forms.** A nondegenerate symmetric bilinear form B on V is called *universal* iff for all $c \in F$, there exists $v \in V$ with $B(v, v) = c$. Prove: if $B(x, x) = 0$ for some nonzero $x \in V$, then B is universal. Does your proof use $2_F \neq 0_F$?

56. Give an example of a field F, a nondegenerate symmetric universal bilinear form B on V, and a nondegenerate subspace W of V such that $B|W$ is not universal.

57. Prove: if $\dim(V) > 1$ and $|F| = 2k + 1$ for some integer k, then every symmetric nondegenerate bilinear form B on V is universal. (Reduce to showing that for fixed nonzero $a, b \in F$, for all $c \in F$, there exist $s, t \in F$ with $as^2 + bt^2 = c$. Study the images of the maps $g, h : F \to F$ given by $g(s) = as^2$ and $h(t) = c - bt^2$.)

58. Prove: if $\dim(V) > 1$, F is a finite field of odd size, and B is a nondegenerate symmetric bilinear form on V, then there is an ordered basis X of V and $d \in F$ such that $[B]_X$ is diagonal with diagonal entries $1, 1, \ldots, 1, d$.

59. Let $A \in M_n(F)$ be a fixed invertible matrix. Prove directly, without reference to bilinear forms, that $O_n(A) = \{C \in M_n(F) : C^{\mathrm{T}} A C = A\}$ is a group under matrix multiplication. Must this be true if A is not invertible?

60. For $\theta \in \mathbb{R}$, show that the matrix $R'_\theta = \begin{bmatrix} \cos\theta & \sin\theta \\ \sin\theta & -\cos\theta \end{bmatrix}$ is orthogonal. Show that the map $T(\mathbf{v}) = R'_\theta \mathbf{v}$ (for $\mathbf{v} \in \mathbb{R}^2$) is a reflection map $S_{\mathbf{z}}$.

61. For $\theta \in \mathbb{R}$, show that the matrix $R_\theta = \begin{bmatrix} \cos\theta & -\sin\theta \\ \sin\theta & \cos\theta \end{bmatrix}$ is orthogonal. Show that the map $T(\mathbf{v}) = R_\theta \mathbf{v}$ (for $\mathbf{v} \in \mathbb{R}^2$) rotates $\mathbf{v}$ counterclockwise θ radians around the origin. Show that T is a composition of two reflections.

62. Show that $O_2(\mathbb{R}) = \{R'_\theta : \theta \in \mathbb{R}\} \cup \{R_\theta : \theta \in \mathbb{R}\}$. Deduce that every orthogonal map T on $\mathbb{R}^2$ (with the dot product) is a rotation or a reflection and is a composition of at most two reflections.

63. Let $C(P; r)$ be the circle in $\mathbb{R}^2$ with center P and radius r. Prove that any two circles $C(P; r)$ and $C(Q; s)$ with unequal centers intersect in 0, 1, or 2 points. Prove that the number of intersection points depends only on r, s, and the distance from P to Q.

64. **Euclidean Isometries.** An *isometry of the Euclidean plane* is a function $F : \mathbb{R}^2 \to \mathbb{R}^2$ (not necessarily linear) that preserves Euclidean distance, meaning that $d_2(F(\mathbf{v}), F(\mathbf{w})) = d_2(\mathbf{v}, \mathbf{w})$ for all $\mathbf{v}, \mathbf{w} \in \mathbb{R}^2$. Prove that every isometry F

must be invertible. (Show F is one-to-one and onto. The previous exercise may help with the proof that F is onto.)

65. Show that the set of isometries of $\mathbb{R}^2$ is a group under function composition.

66. Show that a function $F : \mathbb{R}^2 \to \mathbb{R}^2$ is an orthogonal linear map iff F is an isometry such that $F(\mathbf{0}) = \mathbf{0}$.

67. For each $\mathbf{z} \in \mathbb{R}^2$, define *translation by* $\mathbf{z}$ to be the map $\tau_{\mathbf{z}}$ such that $\tau_{\mathbf{z}}(\mathbf{v}) = \mathbf{v} + \mathbf{z}$ for $\mathbf{v} \in \mathbb{R}^2$. Show that every isometry F of $\mathbb{R}^2$ can be written $F = \tau_{\mathbf{z}} \circ T$ for some orthogonal map T and some $\mathbf{z} \in \mathbb{R}^2$.

68. If ℓ is a line in $\mathbb{R}^2$ (not necessarily through the origin), let s_ℓ be the map on $\mathbb{R}^2$ that sends each point on ℓ to itself and sends each P not on ℓ to the point Q on the other side of ℓ such that ℓ is the perpendicular bisector of $\overline{PQ}$. We call s_ℓ *reflection through the line* ℓ. Prove that every translation $\tau_{\mathbf{z}}$ is a composition of two reflections. Prove that every isometry of $\mathbb{R}^2$ is a composition of reflections. How many reflections are needed at most?

69. Suppose B is a nondegenerate symmetric bilinear form on V, $z \in V$, and $B(z, z) \neq 0$. Prove: for all nonzero $c \in F$, S_{cz} is defined and $S_{cz} = S_z$.

70. Suppose B is a nondegenerate symmetric bilinear form on a 1-dimensional space V. Prove that $O_B(V) = \{\mathrm{id}_V, -\mathrm{id}_V\}$. Prove that for all nonzero $z \in V$, $S_z = -\mathrm{id}_V$.

71. Suppose B is a nondegenerate symmetric bilinear form on an n-dimensional space V. Prove that $-\mathrm{id}_V$ cannot be written as a product of fewer than n reflections.

72. Let B be a symmetric bilinear form on V. Prove: for all $v, z \in V$, if $B(v, v) = 0 \neq B(v, z)$, then there exists $w \in V$ such that (v, w) is a hyperbolic pair and $z \in Fv + Fw$.

73. **Quadratic Forms.** A function $Q : V \to V$ is called a *quadratic form on* V iff (a) $Q(cv) = c^2 Q(v)$ for all $c \in F$ and all $v \in V$; and (b) the function B given by $B(v, w) = (Q(v + w) - Q(v) - Q(w))/2_F$ for $v, w \in V$ is F-bilinear. Show that the map sending Q to B is a bijection from the set of quadratic forms on V onto the set of symmetric bilinear forms on V.

74. Given scalars $c_{ij} \in F$ for $1 \leq i, j \leq n$, define $Q : V \to V$ as follows. For $v = \sum_{i=1}^{n} a_i x_i \in V$ with $a_i \in F$, let $Q(v) = \sum_{i=1}^{n} \sum_{j=1}^{n} c_{ij} a_i a_j$. Show Q is a quadratic form, and find $[B]_X$ for the associated symmetric bilinear form B.

75. Suppose V is an n-dimensional vector space and $T : V \to V$ is a linear map such that $(T - \mathrm{id}_V)^2 = 0$. Prove that $\det(T) = 1$, as follows. Extend an ordered basis $(x_1, \ldots, x_k)$ of $\ker(T - \mathrm{id}_V)$ to an ordered basis $X = (x_1, \ldots, x_n)$ of V. Prove that the matrix of T relative to X has the block form $[T]_X = \begin{bmatrix} I_k & B \\ 0 & I_{n-k} \end{bmatrix}$ for some $k \times (n - k)$ matrix B.

76. Suppose V is an n-dimensional vector space (over any field F) and $T : V \to V$ is a linear map such that $(T - \mathrm{id}_V)^p = 0$ for some positive integer p. Prove $\det(T) = 1$. (One approach uses the Theorem Classifying Nilpotent Maps in §8.6. Alternatively, use induction on $\dim(V)$. Prove that T must have a fixed point $x \neq 0$, and study the action of T on the quotient space V/Fx.)

77. Let (V, B) be an n-dimensional nondegenerate symmetric bilinear space. Complete the following outline proving that every B-orthogonal map T on V is a composition of at most $2n$ reflections. Use induction on n. For the induction step, choose $x \in V$ with $B(x, x) \neq 0$. Prove there is a product U of one or

two reflections with $U(T(x)) = x$. Show $U \circ T$ maps $(Fx)^\perp$ into itself. Use the induction hypothesis to complete the proof.

78. Suppose V is an n-dimensional F-vector space and $H_1, \ldots, H_k$ are $(n-1)$-dimensional subspaces of V. Prove $H_1 \cap \cdots \cap H_k$ has dimension at least $n - k$.

79. Suppose (V, B) is a nondegenerate symmetric bilinear space and $T : V \to V$ is B-orthogonal. Prove:
    (a) If $\dim(V)$ is odd and $\det(T) = 1$, then T has a fixed point $x \neq 0$.
    (b) If $\dim(V)$ is even and $\det(T) = -1$, then T has a fixed point $x \neq 0$.

80. **Embedding Theorem for Degenerate Subspaces.** Suppose (V, B) and (V_*, B_*) are n-dimensional nondegenerate symmetric bilinear spaces with m-dimensional subspaces W and W_* such that $W \cong W_*$ as bilinear spaces. Let $k = \dim(\mathrm{Rad}(W))$. Show there exist nondegenerate subspaces U and U_* with $W \subseteq U \subseteq V$, $W_* \subseteq U_* \subseteq V_*$, $\dim(U) = m + k$, and $U \cong U_*$ as bilinear spaces.

81. **Witt's Extension Theorem.** Suppose (V, B) and (V_*, B_*) are n-dimensional nondegenerate symmetric bilinear spaces with m-dimensional subspaces W and W_* such that $V \cong V_*$ via an isomorphism f and $W \cong W_*$ via an isomorphism g. Show that g extends to an isomorphism from V to V_*.

82. Give an example to show that Witt's Cancellation Theorem need not hold if we omit the hypothesis that the subspace W is nondegenerate.

83. Suppose (V, B) and (V_*, B_*) are nondegenerate symmetric bilinear spaces, W is a subspace of V, W_* is a subspace of V_*, $V \cong V_*$ via f, and $W \cong W_*$ via g. Does $W^\perp \cong W_*^\perp$ necessarily follow?

84. (a) Prove: for all nonempty finite sets X, Y, Z, if $|X \times Y| = |X \times Z|$, then $|Y| = |Z|$.
    (b) Disprove: for all nonempty sets X, Y, Z, if $|X \times Y| = |X \times Z|$, then $|Y| = |Z|$.

85. Prove this cancellation theorem: for all finite-dimensional vector spaces X, Y, Z over a field F, if $X \times Y \cong X \times Z$, then $Y \cong Z$, where $\cong$ denotes isomorphism of vector spaces.

86. Disprove this statement: for all vector spaces X, Y, Z over a field F, if $X \times Y \cong X \times Z$, then $Y \cong Z$, where $\cong$ denotes isomorphism of vector spaces.

87. Prove: for all finite commutative groups X, Y, Z, if $X \times Y \cong X \times Z$, then $Y \cong Z$, where $\cong$ denotes isomorphism of groups. (You may need theorems from Chapter 16 to solve this.)

15

Metric Spaces and Hilbert Spaces

So far, our study of linear algebra has focused mostly on finite-dimensional vector spaces and the linear maps between such spaces. In this chapter, we introduce some ideas that are needed to do linear algebra in the infinite-dimensional setting. Infinite-dimensional vector spaces occur frequently in analysis, where we study vector spaces of functions. To understand such spaces in detail, we need not only algebraic concepts from linear algebra but also some tools from analysis and topology. In particular, the notions of the length of a vector, the distance between two vectors, and the convergence of a sequence of vectors are key ingredients in the study of infinite-dimensional spaces.

The first part of this chapter develops the necesssary analytic concepts in the setting of metric spaces. A metric space is a set in which we can define the distance between any two points. The distance function (also called a metric) satisfies a few basic axioms, from which many fundamental properties of convergent sequences can be derived. We use convergent sequences to define other topological concepts such as closed sets, open sets, continuous functions, compact sets, and complete spaces. Our discussion of metric spaces is far from comprehensive, but it does provide a self-contained account of the analytic material needed for our coverage of Hilbert spaces in the second part of the chapter.

Intuitively, a Hilbert space is a complex vector space (often infinite-dimensional) in which we can define both the length of a vector and the concept of orthogonality of vectors, which generalizes the geometric idea of perpendicular vectors. Orthogonality is defined via a scalar product similar to the dot product in $\mathbb{R}^3$, but using complex scalars. A crucial technical condition in the definition of a Hilbert space is the requirement that it be complete as a metric space. Informally, this means that if the terms of a given sequence in the Hlbert space get arbitrarily close to each other, then the sequence must converge to some point in the space.

This completeness assumption provides an analytic substitute for the dimension-counting arguments that we often need when proving facts about finite-dimensional inner product spaces. For example, given any subspace W of a finite-dimensional inner product space V, the theorem that $V = W \oplus W^{\perp}$ (see §13.10)is proved using dimension counting. The corresponding result in Hilbert spaces requires an additional hypothesis (the assumption that the subspace W be a closed set), and the proof uses completeness in an essential way.

Orthonormal bases play a central role in the theory of finite-dimensional inner product spaces. The analogous concept for infinite-dimensional Hilbert spaces is the idea of a maximal orthonormal set. We show that every Hilbert space has a maximal orthonormal set X, and every vector in the space can be expressed as an infinite linear combination of the vectors in X. (We needs analytic ideas to give a precise definition of what this means.) These results lead to a classification theorem showing that every abstract Hilbert space is isomorphic to a concrete Hilbert space consisting of square-summable functions defined on X. After proving these results, the chapter closes with a discussion of spaces of continuous linear maps, the identification of a Hilbert space with its dual space, and adjoint operators.

DOI: 10.1201/9781003484561-15

15.1   Metric Spaces

Metric spaces were discussed briefly in §10.5; we repeat the relevant definitions here to keep this chapter self-contained. A *metric space* is a set X together with a function (called a *metric* or *distance function*) that measures the distance between any two points in the set X. The distance function $d : X \times X \to \mathbb{R}$ must satisfy the following conditions for all $x, y, z \in X$. First, $0 \le d(x, y) < \infty$, and $d(x, y) = 0$ iff $x = y$. Second, $d(x, y) = d(y, x)$. Third, $d(x, z) \le d(x, y) + d(y, z)$. The second axiom is called *symmetry*; the third axiom is called the *Triangle Inequality*. If several metric spaces are being considered, we sometimes write d_X for the metric defined on X.

The set $\mathbb{R}$ of real numbers is a metric space, with distance function $d(x, y) = |x - y|$ for all $x, y \in \mathbb{R}$. More generally, for each positive integer m, $\mathbb{R}^m$ is a metric space under the *Euclidean distance function*, defined by setting $d_2(x, y) = \sqrt{\sum_{i=1}^{m} |x_i - y_i|^2}$ for all $x = (x_1, \ldots, x_m)$ and $y = (y_1, \ldots, y_m)$ in $\mathbb{R}^m$. Analogous formulas define metrics on the spaces $\mathbb{C}$ and $\mathbb{C}^m$. We verify the metric space axioms for these examples later in the chapter, after we introduce Hilbert spaces.

Here are a few more abstract examples of metric spaces. Given any set X and $x, y \in X$, define $d(x, y) = 0$ if $x = y$, and $d(x, y) = 1$ if $x \ne y$. The first two axioms for a metric are immediately verified. To see that the triangle inequality must hold, suppose that it failed. Then there would exist $x, y, z \in X$ with $d(x, z) > d(x, y) + d(y, z)$. Since the distance function only takes values 0 and 1, the inequality just written can only occur if $d(x, z) = 1$ and $d(x, y) = d(y, z) = 0$. But then the definition of d gives $x = y = z$, hence $x = z$, which contradicts $d(x, z) = 1$. This distance function d is called the *discrete metric* on X.

Given any metric space (X, d) and any subset S of X, the set S becomes a metric space by restricting the domain of the distance function d from $X \times X$ to $S \times S$. We call S a *subspace* of the metric space X.

Next, suppose X and Y are metric spaces with metrics d_X and d_Y. Consider the *product space* $Z = X \times Y$, which is the set of ordered pairs (x, y) with $x \in X$ and $y \in Y$. We define the *product metric* $d = d_Z : Z \times Z \to \mathbb{R}$ by setting

$$d((x_1, y_1), (x_2, y_2)) = d_X(x_1, x_2) + d_Y(y_1, y_2) \quad \text{for all } (x_1, y_1), (x_2, y_2) \in Z.$$

Let us verify the metric space axioms for Z and d. Fix $z_1 = (x_1, y_1)$, $z_2 = (x_2, y_2)$, and $z_3 = (x_3, y_3)$ in Z, where $x_i \in X$ and $y_i \in Y$ for $i = 1, 2, 3$. First, since $0 \le d_X(x_1, x_2) < \infty$ and $0 \le d_Y(y_1, y_2) < \infty$, we have $0 \le d(z_1, z_2) = d_X(x_1, x_2) + d_Y(y_1, y_2) < \infty$. Furthermore, since the sum of two nonnegative real numbers is zero iff both summands are zero, we have $d(z_1, z_2) = 0$ iff $d_X(x_1, x_2) = 0$ and $d_Y(y_1, y_2) = 0$ iff $x_1 = x_2$ and $y_1 = y_2$ iff $(x_1, y_1) = (x_2, y_2)$ iff $z_1 = z_2$. Second, $d(z_1, z_2) = d_X(x_1, x_2) + d_Y(y_1, y_2) = d_X(x_2, x_1) + d_Y(y_2, y_1) = d(z_2, z_1)$ by symmetry of d_X and d_Y. Third, using the Triangle Inequality for d_X and d_Y, we compute

$$d(z_1, z_3) = d_X(x_1, x_3) + d_Y(y_1, y_3) \le d_X(x_1, x_2) + d_X(x_2, x_3) + d_Y(y_1, y_2) + d_Y(y_2, y_3)$$
$$= d_X(x_1, x_2) + d_Y(y_1, y_2) + d_X(x_2, x_3) + d_Y(y_2, y_3) = d_Z(z_1, z_2) + d_Z(z_2, z_3).$$

Next, define $d' : Z \times Z \to \mathbb{R}$ by $d'((x_1, y_1), (x_2, y_2)) = \max(d_X(x_1, x_2), d_Y(y_1, y_2))$. You can check that d' also satisfies the metric space axioms. The metrics d and d' on the set Z are not equal, but we see in Exercise 12 that they are equivalent for many purposes.

We can iterate the definition of d (or d') to define metrics on products of finitely many metric spaces $X_1, X_2, \ldots, X_m$. Specifically, let $X = X_1 \times X_2 \times \cdots \times X_m$ and $x_i, y_i \in X_i$ for

$1 \leq i \leq m$. The formulas

$$d_1((x_1,\ldots,x_m),(y_1,\ldots,y_m)) = d_{X_1}(x_1,y_1) + d_{X_2}(x_2,y_2) + \cdots + d_{X_m}(x_m,y_m),$$
$$d_\infty((x_1,\ldots,x_m),(y_1,\ldots,y_m)) = \max(d_{X_1}(x_1,y_1), d_{X_2}(x_2,y_2),\ldots,d_{X_m}(x_m,y_m))$$

both define metrics on the product space X. In particular, these constructions provide two additional metrics on the spaces $\mathbb{R}^m$ and $\mathbb{C}^m$ that are not equal to the Euclidean metric given earlier. Although all three metrics are equivalent in some respects (Exercise 7), we see later that the Euclidean metric d_2 has additional geometric structure compared to d_1 and d_∞.

As a final remark, note that the constructions given here do not automatically generalize to products of infinitely many metric spaces. The reason is that the sum or maximum of an infinite sequence of positive real numbers may be $+\infty$, which cannot be a value of the distance function, by the first metric space axiom.

15.2 Convergent Sequences

Formally, a *sequence* of points in a set X is a function $\mathbf{x} : \mathbb{Z}_{\geq 0} \to X$. We often present such a sequence as an infinite list $\mathbf{x} = (x_0, x_1, x_2, \ldots, x_n, \ldots) = (x_n : n \in \mathbb{Z}_{\geq 0})$, where $x_n = \mathbf{x}(n) \in X$ is the *nth term* of the sequence. We sometimes index a sequence starting at x_1 instead of x_0, or starting at x_i for any fixed $i \in \mathbb{Z}$. A *subsequence* of $(x_0, x_1, x_2, \ldots, x_n, \ldots)$ is a sequence of the form $(x_{k_0}, x_{k_1}, x_{k_2}, \ldots, x_{k_n}, \ldots)$ where $0 \leq k_0 < k_1 < k_2 < \cdots < k_n < \cdots$ is a strictly increasing sequence of integers.

Given a metric space (X, d), a sequence (x_n) of points in X, and a point $y \in X$, we say that the sequence (x_n) *converges to* y in the metric space iff for all $\epsilon > 0$, there exists $n_0 \in \mathbb{Z}_{\geq 0}$ such that for all $n \geq n_0$, $d(x_n, y) < \epsilon$. When this condition holds, we write $x_n \to y$ or $\lim_{n \to \infty} x_n = y$ and call y the *limit* of the sequence (x_n). Intuitively, the formal definition means that for any given positive distance, no matter how small, the sequence (x_n) eventually comes closer than that distance to the limit y and remains closer than that distance forever. (The phrasing of this intuitive description arises by viewing the subscript n as a time index, so that x_n is the location of the sequence at time n.)

Not every sequence (x_n) in a metric space (X, d) converges. Those sequences that do converge to some point $y \in X$ are called *convergent* sequences. We note that the limit of a sequence, when it exists, must be unique. For suppose a particular sequence (x_n) converged to both y and z in X. We show that $y = z$. For any fixed $\epsilon > 0$, we can choose $n_1 \in \mathbb{Z}_{\geq 0}$ such that $n \geq n_1$ implies $d(x_n, y) < \epsilon/2$. Similarly, there exists $n_2 \in \mathbb{Z}_{\geq 0}$ such that $n \geq n_2$ implies $d(x_n, z) < \epsilon/2$. Picking $n = \max(n_1, n_2)$, we conclude that $d(y, z) \leq d(y, x_n) + d(x_n, z) = d(x_n, y) + d(x_n, z) < \epsilon/2 + \epsilon/2 = \epsilon$. Thus the fixed number $d(y, z)$ is less than every positive number ϵ, forcing $d(y, z) = 0$ and $y = z$.

Here are some examples of sequences in metric spaces. In any metric space, the constant sequence $(y, y, y, \ldots)$ converges to y. In the metric space $\mathbb{R}$, the sequences $(1/n : n \in \mathbb{Z}_{>0})$, $(1/n^2 : n \in \mathbb{Z}_{>0})$, and $(1/2^n : n \in \mathbb{Z}_{\geq 0})$ all converge to zero. The sequence $(1, -1, 1, -1, 1, -1, \ldots) = ((-1)^n : n \geq 0)$ does not converge. However, the subsequence of odd-indexed terms $(-1, -1, -1, \ldots)$ converges to -1, and the subsequence of even-indexed terms $(1, 1, 1, \ldots)$ converges to 1. We see that a subsequence of a non-convergent sequence may converge, and different subsequences might converge to different limits. On the other hand, the sequence $(n : n \in \mathbb{Z}_{\geq 0})$ in $\mathbb{R}$ is non-convergent and has no convergent subsequence ($+\infty$ is not allowed as a limit, since it is not in the set $\mathbb{R}$).

Suppose $(x_n : n \geq 0)$ is a convergent sequence in a metric space (X, d) with limit y. We now show that *all subsequences of* (x_n) *also converge to* y. Fix such a subsequence $(x_{k_0}, x_{k_1}, \ldots)$. Given $\epsilon > 0$, there exists $m_0 \in \mathbb{Z}_{\geq 0}$ such that for all $m \geq m_0$, $d(x_m, y) < \epsilon$. Since $k_0 < k_1 < \cdots$ is a strictly increasing sequence of integers, we can find $n_0 \in \mathbb{Z}_{\geq 0}$ such that $k_n \geq m_0$ for all $n \geq n_0$. (It suffices to take $n_0 = m_0$, although a smaller n_0 may also work.) Now, for any $n \geq n_0$, $d(x_{k_n}, y) < \epsilon$, confirming that $\lim_{n \to \infty} x_{k_n} = y$. We develop further connections between the limit of a sequence (if any) and the limits of its subsequences when we discuss compactness and completeness.

15.3 Closed Sets

Let (X, d) be any metric space. A subset C of X is called a *closed set* (relative to the metric d) iff for every sequence $(x_n : n \geq 0)$ with all terms x_n belonging to C, if x_n converges to some point $y \in X$, then y also belongs to C. Intuitively, the set is called "closed" because we can never go outside the set by passing to the limit of a convergent sequence of points all of which are in the set. In §1.4, we saw that many algebraic subsystems (subgroups, ideals, etc.) can be defined in terms of certain closure conditions, where performing various operations on elements of the set produces another object in that same set. Here, we can say that C is a closed set (in the sense just defined) iff C is closed under the operation of taking the limit of a convergent sequence of points in C. Negating the definition, note that C is *not closed* means there exists a sequence (x_n) with all $x_n \in C$ such that $x_n \to y$ for some $y \in X$, but $y \notin C$.

Here are some examples of closed sets. First, for any fixed $y \in X$, the one-point set $C = \{y\}$ is closed. To see why, note that the only sequence of points in C is the constant sequence $(y, y, y, \ldots)$. This sequence converges to the unique limit y, which belongs to C, and so C is indeed closed. Second, we claim the empty set $\emptyset$ is a closed subset of X. Otherwise, there would be a convergent sequence $(x_n : n \geq 0)$ with all $x_n \in \emptyset$ converging to a limit $y \notin \emptyset$. But the condition $x_n \in \emptyset$ is impossible, so there can be no such sequence. Third, the entire space X is a closed subset of X. Proof: given points $x_n \in X$ such that $x_n \to y \in X$, we certainly have $y \in X$, so that X is closed.

We continue by giving examples of closed sets and non-closed sets in the metric space $\mathbb{R}$. For any fixed $a \leq b$, the *closed interval* $[a, b] = \{x \in \mathbb{R} : a \leq x \leq b\}$ is a closed set, as the name suggests. We prove this by contradiction. If $[a, b]$ is not closed, choose a sequence (x_n) with all $x_n \in [a, b]$, such that (x_n) converges to some real number $y \notin [a, b]$. In the case $y < a$, note that $\epsilon = a - y > 0$. For this ϵ, there is $n_0 \geq 0$ such that $n \geq n_0$ implies $d(x_n, y) < \epsilon$. In particular, x_{n_0} must satisfy $y - \epsilon < x_{n_0} < y + \epsilon = a$, contradicting $x_{n_0} \in [a, b]$. Similarly, the case $y > b$ leads to a contradiction. So $[a, b]$ is closed. On the other hand, for fixed $a < b$, the open interval $(a, b) = \{x \in \mathbb{R} : a < x < b\}$ is not closed. You can check that the sequence defined by $x_n = a + (b - a)/(2n)$ for $n \geq 1$ has all terms x_n in (a, b) and converges to $a \notin (a, b)$. Similarly, the half-open intervals $(a, b]$ and $[a, b)$ are not closed. The set $\mathbb{Z}$ of integers is a closed subset of $\mathbb{R}$. We again prove this by contradiction. If $\mathbb{Z}$ is not closed, then we can choose a convergent sequence (x_n) of integers with $x_n \to y$, where $y \in \mathbb{R}$ is not an integer. We know y is between two consecutive integers, say $k < y < k + 1$. Take $\epsilon = \min(y - k, k + 1 - y) > 0$. For large enough n, we must have $y - \epsilon < x_n < y + \epsilon$. But no real number in this range is an integer, which contradicts $x_n \in \mathbb{Z}$.

Let us return to the case of a general metric space (X, d). Given any (possibly infinite) collection $\{C_i : i \in I\}$ of closed sets in X, we claim the intersection $C = \bigcap_{i \in I} C_i$ is closed. Proof: suppose (x_n) is a sequence in C converging to some $y \in X$; we must prove $y \in C$.

Fix an index $i \in I$. Since all x_n belong to C, we know all x_n are in C_i. Because C_i is a closed set, it follows that the limit y belongs to C_i. This holds for every i, so y belongs to the intersection C of all the C_i.

Next, we show that if C and D are closed sets in X, then the union $C \cup D$ is also closed. Let (x_n) be a sequence of points in $C \cup D$ converging to a limit $y \in X$; we must prove $y \in C \cup D$. Consider two cases. Case 1: there are only finitely many indices n with $x_n \in C$. Let n_0 be the largest index with $x_n \in C$, or $n_0 = 0$ if every x_n is in D. The subsequence $(x_{n_0+1}, x_{n_0+2}, \dots)$ of the original sequence still converges to y, and all points in this subsequence belong to D. As D is closed, we must have $y \in D$, so that $y \in C \cup D$ as well. Case 2: there are infinitely many indices $k_0 < k_1 < k_2 < \cdots$ with $x_{k_n} \in C$. Then the subsequence $(x_{k_0}, x_{k_1}, \dots)$ of the original sequence still converges to y, and all points in this subsequence belong to C. As C is closed, we must have $y \in C$, so that $y \in C \cup D$.

It follows by induction that if m is any positive integer and $C_1, C_2, \dots, C_m$ are closed sets in (X, d), then $C_1 \cup C_2 \cup \cdots \cup C_m$ is also closed. In particular, since one-point sets are closed, we see that *all finite subsets of X are closed*. However, the union of an infinite collection of closed subsets may or may not be closed. On one hand, we saw that $\mathbb{Z} = \bigcup_{n \in \mathbb{Z}} \{n\}$ is a closed set in $\mathbb{R}$. On the other hand, the union C of the one-point sets $\{1/n\}$ for $n \in \mathbb{Z}_{>0}$ is not closed in $\mathbb{R}$, since $(1/n : n > 0)$ is a sequence in C converging to the point $0 \notin C$.

To summarize: *in any metric space X, $\emptyset$ and X are closed. Finite subsets of X are closed. The union of finitely many closed sets is closed. The intersection of arbitrarily many closed sets is closed.*

15.4 Open Sets

Given a point x in a metric space (X, d) and a real $r > 0$, the *open ball of radius r and center x* is $B(x; r) = \{y \in X : d(x, y) < r\}$. For example, in $\mathbb{R}$, $B(x; r)$ is the open interval $(x - r, x + r)$. In $\mathbb{C}$ or $\mathbb{R}^2$ with the Euclidean metric, $B(x; r)$ is the interior of a circle with center x and radius r. In a discrete metric space X, $B(x; r) = \{x\}$ for all $r \leq 1$, whereas $B(x; r) = X$ for all $r > 1$.

A subset U of a metric space (X, d) is called an *open set* (relative to the metric d) iff for every $x \in U$, there exists $\epsilon > 0$ (depending on x) such that $B(x; \epsilon) \subseteq U$. Intuitively, the set U is called open because all points sufficiently close to a point in U are also in U.

Here are some examples of open sets in general metric spaces. First, *every open ball is an open set*, as the name suggests. To prove this, consider an open ball $U = B(y; r)$ and fix $x \in U$. We know $d(y, x) < r$, so the number $\epsilon = r - d(y, x)$ is strictly positive. We show $B(x; \epsilon) \subseteq U$. We fix $z \in B(x; \epsilon)$ and check $z \in U$. We know $d(x, z) < \epsilon$, so the Triangle Inequality gives $d(y, z) \leq d(y, x) + d(x, z) < d(y, x) + \epsilon = r$. This shows that $z \in B(y; r) = U$, as needed.

Second, *the entire space X is an open subset of X*. Proof: given $x \in X$, we can take $\epsilon = 1$ and note that $B(x; \epsilon)$ is a subset of X by definition.

Third, *the empty set $\emptyset$ is an open subset of X*. Proof: if $\emptyset$ were not open, then there exists $x \in \emptyset$ such that for all $\epsilon > 0$, $B(x; \epsilon)$ is not a subset of $\emptyset$. But the existence of $x \in \emptyset$ is impossible.

Fourth, *given any collection $\{U_i : i \in I\}$ of open subsets of X, the union $U = \bigcup_{i \in I} U_i$ is also open in X*. To prove this, fix x in the union U of the U_i. We know $x \in U_i$ for some $i \in I$. Since U_i is open, there exists $\epsilon > 0$ with $B(x; \epsilon) \subseteq U_i$. As U_i is a subset of U, we also have $B(x; \epsilon) \subseteq U$, so U is open.

Fifth, *if U and V are open subsets of X, then $U \cap V$ is open in X.* For the proof, fix $x \in U \cap V$. Since $x \in U$, there is $\epsilon_1 > 0$ with $B(x; \epsilon_1) \subseteq U$. Since $x \in V$, there is $\epsilon_2 > 0$ with $B(x; \epsilon_2) \subseteq V$. For $\epsilon = \min(\epsilon_1, \epsilon_2) > 0$, we see that $B(x; \epsilon)$ is contained in both U and V, hence is a subset of $U \cap V$. Thus, $U \cap V$ is open. By induction, it follows that if m is any positive integer and $U_1, \ldots, U_m$ are open subsets of X, then $U_1 \cap \cdots \cap U_m$ is also open. However, the intersection of infinitely many open subsets of X need not be open. For instance, all the sets $B(0; 1/n) = (-1/n, 1/n)$ are open subsets of $\mathbb{R}$ (being open balls). But their intersection, namely $\{0\}$, is not open in $\mathbb{R}$, since for every $\epsilon > 0$, $B(0; \epsilon) = (-\epsilon, \epsilon)$ is not a subset of $\{0\}$.

Open sets and closed sets are related in the following way: *a set U is open in X iff the complement $C = X \setminus U$ is closed in X.* We prove the contrapositive in both directions. First assume $C = X \setminus U$ is not closed in X. Then there is a sequence (x_n) of points of C and a point $x \in X$ such that $x_n \to x$ but $x \notin C$. Note x is a point of $X \setminus C = U$. For fixed $\epsilon > 0$, there is $n_0 \geq 0$ such that for all $n \geq n_0$, $d(x_n, x) < \epsilon$. Each such x_n is in both $C = X \setminus U$ and $B(x; \epsilon)$, which shows that $B(x; \epsilon)$ cannot be a subset of U. As ϵ was arbitrary, we see that U is not open.

Conversely, assume U is not open; we prove $C = X \setminus U$ is not closed. There is a point $x \in U$ such that for all $\epsilon > 0$, $B(x; \epsilon)$ is not a subset of U. Taking $\epsilon = 1/n$ for each positive integer n, we obtain points x_n with $x_n \in B(x; 1/n)$ but $x_n \notin U$. Thus, (x_n) is a sequence of points in C. Furthermore, we claim x_n converges to x in X. To see this, fix $\epsilon > 0$, and choose a positive integer n_0 with $1/n_0 < \epsilon$. For all $n \geq n_0$, $d(x_n, x) < 1/n \leq 1/n_0 < \epsilon$, as needed. Since the limit of (x_n) is $x \notin C$, we see that C is not closed.

Note carefully that "S is not an open set" does not mean the same thing as "S is a closed set." Most subsets of metric spaces are neither open nor closed. For instance, half-open intervals $(a, b]$ in $\mathbb{R}$ are neither open nor closed. It is possible for a subset of a metric space to be both open and closed; consider $\emptyset$ and X, for example.

15.5   Continuous Functions

The concept of continuity plays a central role in calculus. This concept can be generalized to the setting of metric spaces as follows. Let (X, d_X) and (Y, d_Y) be two metric spaces. A function $f : X \to Y$ is *continuous on X* iff whenever (x_n) is a sequence of points in X converging to some $x \in X$, the sequence $(f(x_n))$ in Y converges to $f(x)$. Briefly, continuity of f means that whenever $x_n \to x$ in X, $f(x_n) \to f(x)$ in Y. Using limit notation, continuity of f means

$$f\left(\lim_{n \to \infty} x_n\right) = \lim_{n \to \infty} f(x_n),$$

whenever the sequence (x_n) has a limit, so that f commutes with the operation of taking the limit of a convergent sequence. For a fixed element x^* in X, we say f is *continuous at the point x^** iff whenever $x_n \to x^*$ in X, $f(x_n) \to f(x^*)$ in Y.

You can check that a constant function ($f(x) = y_0$ for all $x \in X$) is continuous, as is the identity function $\mathrm{id}_X : X \to X$. We show later that addition and multiplication (viewed as functions from the product metric space $\mathbb{R} \times \mathbb{R}$ to $\mathbb{R}$) are continuous. Let us check that $f : \mathbb{R} \to \mathbb{R}$, given by $f(x) = -3x$ for $x \in \mathbb{R}$, is continuous. Suppose $x_n \to x$ in $\mathbb{R}$; we prove $-3x_n \to -3x$. Fix $\epsilon > 0$, and choose n_0 so that $n \geq n_0$ implies $d(x_n, x) < \epsilon/3$. Now notice that $d(-3x_n, -3x) = |(-3x_n) - (-3x)| = 3|x_n - x| = 3d(x_n, x)$. So, $n \geq n_0$ implies $d(f(x_n), f(x)) < \epsilon$, as needed.

A fundamental fact about continuity is that *compositions of continuous functions are continuous.* In detail, let $f : X \to Y$ and $g : Y \to Z$ be continuous functions between metric spaces X, Y, and Z; we show $g \circ f : X \to Z$ is continuous. To do so, let (x_n) be a sequence in X converging to $x \in X$. By continuity of f, the sequence $(f(x_n))$ converges to $f(x)$ in Y. By continuity of g, the sequence $(g(f(x_n)))$ converges to $g(f(x))$ in Z. So, the sequence $((g \circ f)(x_n))$ converges to $(g \circ f)(x)$ in Z, proving continuity of $g \circ f$.

A more subtle, but equally fundamental, property of continuity is that $f : X \to Y$ is *continuous iff for every closed set D in Y, the preimage $f^{-1}[D] = \{x \in X : f(x) \in D\}$ is closed in X.* To prove the forward direction, assume f is continuous, fix a closed set D in Y, and consider a sequence (x_n) of points in $f^{-1}[D]$ that converge to some limit $x^* \in X$. To see that $f^{-1}[D]$ is closed, we must prove $x^* \in f^{-1}[D]$. Now, each x_n is in $f^{-1}[D]$, so $f(x_n) \in D$ for all $n \geq 0$. As x_n converges to x^*, continuity of f tells us that $f(x_n)$ converges to $f(x^*)$. Since D is closed in Y, we deduce $f(x^*) \in D$, and hence $x^* \in f^{-1}[D]$, as needed.

For the other direction, assume that $f : X \to Y$ is not continuous. So there is a sequence (x_n) in X converging to a point $x \in X$, such that $f(x_n)$ does not converge to $y = f(x)$ in Y. This means that there exists $\epsilon > 0$ such that for every integer $n_0 \geq 0$ there exists $n \geq n_0$ with $d(f(x_n), y) \geq \epsilon$. It follows that we can find a subsequence (x'_n) of (x_n), consisting of all terms x_n such that $d(f(x_n), f(x)) \geq \epsilon$. We know (x'_n) still converges to x. Note that $D = \{z \in Y : d_Y(z, y) \geq \epsilon\}$ is a closed set in Y, being the complement of the open ball $B(y; \epsilon)$. To complete the proof, we show $f^{-1}[D]$ is not a closed set in X. Note that each term x'_n is in $f^{-1}[D]$, since $f(x'_n) \in D$ by construction. But the limit x of the sequence (x'_n) is not in $f^{-1}[D]$, because $f(x) = y$ satisfies $d_Y(y, y) = 0$, so that $f(x) \notin D$.

Since open sets are complements of closed sets, we readily deduce the following characterization of continuous functions: $f : X \to Y$ *is continuous iff for every open set V in Y, the preimage $f^{-1}[V]$ is open in X.* To prove this, suppose f is continuous and V is open in Y. Then $Y \setminus V$ is closed in Y, so $f^{-1}[Y \setminus V]$ is closed in X. From set theory, we know $f^{-1}[Y \setminus V] = f^{-1}[Y] \setminus f^{-1}[V] = X \setminus f^{-1}[V]$. Therefore, $f^{-1}[V]$ is the complement of a closed set in X, so it is open in X. The converse is proved similarly. This characterization of continuity is closely related to the ϵ-δ definition often given in calculus. Specifically, $f : X \to Y$ is continuous iff for all $x_0 \in X$ and $\epsilon > 0$, there exists $\delta > 0$ such that for all $x \in X$, if $d_X(x, x_0) < \delta$, then $d_Y(f(x), f(x_0)) < \epsilon$. We ask the reader to prove the equivalence of this condition to the condition involving open sets in Exercise 50.

15.6 Compact Sets

We know that not every sequence in a general metric space is convergent. It would be convenient if we could take a non-convergent sequence (x_n) and extract a convergent subsequence from it. However, even this is not always possible. For example, in $\mathbb{R}$, no subsequence of the sequence $(n : n \geq 0)$ converges. On the other hand, if we insist that all terms of the sequence (x_n) come from a closed interval $[a, b]$, it can be shown that (x_n) must have a convergent subsequence (Exercise 58). This suggests that we may have to restrict our attention to sequences coming from a sufficiently nice subset of the metric space. In a general metric space (X, d), a subset K of X is called *(sequentially) compact* iff for every sequence (x_n) of points in K, there exists a subsequence (x_{k_n}) and a point y belonging to K such that $x_{k_n} \to y$.

Every finite subset of X is compact. To prove this, suppose $K = \{y_1, \ldots, y_m\}$ is nonempty and finite, and (x_n) is any sequence of points in K. At least one y_j must occur infinitely often in the sequence (x_n). So the sequence (x_n) has a constant subsequence

$(y_j, y_j, y_j, \ldots)$, which converges to the point $y_j \in K$. Also, $\emptyset$ is compact, since the definition of compactness is vacuously satisfied (there are no sequences with values in $\emptyset$).

Every compact subset of X must be closed in X. To prove this, suppose $K \subseteq X$ is compact, (x_n) is a sequence with all $x_n \in K$, and x_n converges to a point x in X. On one hand, every subsequence of (x_n) converges to the unique limit x. On the other hand, the definition of compactness shows that some subsequence of (x_n) converges to a point of K. Therefore, x must belong to K, proving that K is closed. You can show that *unions of finitely many compact sets are compact, whereas intersections of arbitrarily many compact sets are compact.* The proof imitates the proofs of the analogous properties of closed sets. Similarly, *a closed subset of a compact set is compact.*

We say that a subset S of a nonempty metric space (X, d) is *bounded* iff there exists $z \in X$ and $M \in \mathbb{R}$ such that for all $x \in S$, $d(z, x) \leq M$. *Every compact subset of X must be bounded.* To prove this, suppose $S \subseteq X$ is not bounded. We construct a sequence (x_n) of points in S that has no convergent subsequence. Fix $z \in X$. For each $n \in \mathbb{Z}_{\geq 0}$, we can find $x_n \in S$ such that $d(z, x_n) > n$. To get a contradiction, suppose some subsequence (x_{k_n}) converges to some point $y \in X$. Choose an integer n_0 so that $n \geq n_0$ implies $d(y, x_{k_n}) < 1$. Then choose an integer $n \geq n_0$ with $k_n \geq d(z, y) + 1$. For this n, $d(z, x_{k_n}) \leq d(z, y) + d(y, x_{k_n}) < d(z, y) + 1 \leq k_n$, which contradicts the choice of x_{k_n}.

So far, we have seen that *every compact subset of a metric space is closed and bounded.* In $\mathbb{R}^m$ and $\mathbb{C}^m$ with any of the metrics d_1, d_2, and d_∞ discussed earlier, the converse also holds: *a subset K of $\mathbb{R}^m$ or $\mathbb{C}^m$ is compact iff K is closed and bounded.* The proof of this, which is rather difficult, is sketched in the exercises and can be found in texts on advanced calculus. On the other hand, *in general metric spaces, there can exist closed and bounded sets that are not compact.* For instance, consider $X = \mathbb{Z}$ with the discrete metric. The entire space X is bounded, since $d(0, x) \leq 1$ for all $x \in \mathbb{Z}$. X is also closed in X. But X is not compact, since $(n : n \geq 0)$ is a sequence in X with no convergent subsequence.

Continuous functions preserve compact sets, in the following sense. *If $f : X \to Y$ is a continuous function between two metric spaces and $K \subseteq X$ is compact in X, then the direct image $f[K] = \{f(x) : x \in K\}$ is compact in Y.* To prove this, assume $f : X \to Y$ is continuous and K is a compact subset of X. To prove $f[K]$ is compact, let (y_n) be any sequence of points in $f[K]$. Each y_n has the form $y_n = f(x_n)$ for some $x_n \in K$. Now (x_n) is a sequence of points in the compact set K, so there is a subsequence (x_{k_n}) converging to some point $x^* \in K$. By continuity, $(y_{k_n}) = (f(x_{k_n}))$ is a subsequence of (y_n) converging to $f(x^*) \in f[K]$. So $f[K]$ is indeed compact.

Compactness can also be defined in terms of open sets. Given any set S in a metric space (X, d), an *open cover* of S is a collection $\{U_i : i \in I\}$ of open subsets of X such that $S \subseteq \bigcup_{i \in I} U_i$. A *finite subcover* is a finite subcollection $\{U_{i_1}, \ldots, U_{i_m}\}$ of the given open cover such that $S \subseteq \bigcup_{j=1}^m U_{i_j}$. Such a finite subcover need not exist. We say that a subset K of X is *(topologically) compact* iff for each open cover of K, there does exist a finite subcover. It can be shown that *in metric spaces, sequential compactness is equivalent to topological compactness.* The topological definition applies in more general situations, but our study of Hilbert spaces only requires the more intuitive definition in terms of subsequences.

15.7   Completeness

A *Cauchy sequence* in a metric space (X, d) is a sequence $(x_n : n \geq 0)$ with the following property: for all $\epsilon > 0$, there exists an integer n_0 such that for all $m, n \geq n_0$, $d(x_n, x_m) < \epsilon$. This definition says that the terms of a Cauchy sequence get arbitrarily close to each other if we go far enough out in the sequence. In contrast, for a convergent sequence with limit x, the terms of the sequence get arbitrarily close to the limit value x. Let us explore the relationship between these concepts.

On one hand, *every convergent sequence in a metric space is a Cauchy sequence.* To prove this, suppose (x_n) is a sequence in (X, d) converging to $x \in X$. We show (x_n) is a Cauchy sequence. Fix $\epsilon > 0$, and choose an integer n_0 so that for all $n \geq n_0$, $d(x_n, x) < \epsilon/2$. For all $n, m \geq n_0$, the Triangle Inequality gives $d(x_n, x_m) \leq d(x_n, x) + d(x, x_m) < \epsilon/2 + \epsilon/2 = \epsilon$.

On the other hand, *a Cauchy sequence in a general metric space may not converge.* For example, consider the set $\mathbb{R}_{>0}$ of strictly positive real numbers with the metric $d(x, y) = |x - y|$ for $x, y \in \mathbb{R}_{>0}$. The sequence $(1/n : n \geq 1)$ converges in the larger metric space $\mathbb{R}$ to the unique limit 0, so this sequence is a Cauchy sequence (in $\mathbb{R}$ and in $\mathbb{R}_{>0}$). But the sequence does not converge to any point of the set $\mathbb{R}_{>0}$. For another example, consider the metric space $\mathbb{Q}$ of rational numbers with $d(x, y) = |x - y|$ for $x, y \in \mathbb{Q}$. We can find a sequence of rational numbers $(x_n : n \geq 0)$ converging to the irrational real number $\sqrt{2}$ (for instance, let x_n consist of the decimal expansion of $\sqrt{2}$ truncated n places after the decimal). This sequence is a Cauchy sequence in $\mathbb{Q}$ that does not converge in $\mathbb{Q}$.

A metric space (X, d) is called *complete* iff every Cauchy sequence (x_n) in X does converge to some point in the space X. The preceding examples show that $\mathbb{R}_{>0}$ and $\mathbb{Q}$ are not complete. On the other hand, *any compact metric space is complete.* To prove this, suppose (x_n) is a Cauchy sequence in a compact metric space (X, d). By compactness, this sequence has a subsequence (x_{k_n}) converging to some $x \in X$. We now show that the full sequence must also converge to x. Given $\epsilon > 0$, choose an integer n_0 so that for all $n \geq n_0$, $d(x_{k_n}, x) < \epsilon/2$. Also choose an integer n_1 so that for all $i, j \geq n_1$, $d(x_i, x_j) < \epsilon/2$. Fix any $i \geq n_1$. There is an integer $j \geq n_1$ such that $j = k_n$ for some $n \geq n_0$. Using this j, we see that $d(x_i, x) \leq d(x_i, x_j) + d(x_j, x) < \epsilon/2 + d(x_{k_n}, x) < \epsilon$. This proves that (x_n) converges to x.

A subset of a complete metric space is complete iff it is closed. To prove this, suppose (X, d) is complete, $C \subseteq X$ is closed, and (x_n) is a Cauchy sequence with all $x_n \in C$. Then (x_n) is a Cauchy sequence in X, hence converges to some $x \in X$. Since C is closed, the limit x must belong to C. So C is complete. Conversely, if $C \subseteq X$ is not closed, there exists a sequence (x_n) with all $x_n \in C$ converging to some $x \in X \setminus C$. The convergent sequence (x_n) is a Cauchy sequence, and x is its unique limit. Since this limit does not belong to C, (x_n) is a Cauchy sequence in C that does not converge in C. So C is not complete.

Each closed interval $[a, b]$ is a compact, hence complete, subset of $\mathbb{R}$ (see Exercise 58). We use this fact to show that $\mathbb{R}$ *is complete.* The proof requires the lemma that *a Cauchy sequence in any metric space is bounded* (Exercise 16). Given a Cauchy sequence (x_n) in $\mathbb{R}$, we can therefore choose $M \in \mathbb{R}$ such that every term x_n is in the closed interval $[-M, M]$. The completeness of this interval ensures that the given Cauchy sequence converges to some real number.

More generally, $\mathbb{R}^k$ *(and similarly $\mathbb{C}^k$) with the Euclidean metric is complete.* To prove this, let (x_n) be a Cauchy sequence in $\mathbb{R}^k$, where $x_n = (x_n(1), x_n(2), \ldots, x_n(k))$ for certain real numbers $x_n(i)$. For $1 \leq i \leq k$, the inequality $|x_n(i) - x_m(i)| \leq d_2(x_n, x_m)$ shows that $(x_n(i) : n \geq 0)$ is a Cauchy sequence in $\mathbb{R}$. By completeness of $\mathbb{R}$, this sequence converges to some real number $x(i)$. Let $x = (x(1), x(2), \ldots, x(k)) \in \mathbb{R}^k$. Given $\epsilon > 0$, choose integers $n_1, \ldots, n_k$ such that $n \geq n_i$ implies $|x_n(i) - x_i| < \epsilon/\sqrt{k}$. Then for $n \geq \max(n_1, \ldots, n_k)$, we have

$$d_2(x_n, x) = \sqrt{\sum_{i=1}^{k} |x_n(i) - x(i)|^2} \leq \sqrt{\sum_{i=1}^{k} \epsilon^2/k} = \epsilon.$$

So the given Cauchy sequence (x_n) in $\mathbb{R}^k$ converges to x.

15.8 Definition of a Hilbert Space

Having covered the necessary background on metric spaces, we are now ready to define Hilbert spaces. Briefly, a *Hilbert space* is a complex inner product space that is complete relative to the metric induced by the inner product. Let us spell this definition out in more detail.

We begin with the algebraic ingredients of a Hilbert space. The Hilbert space consists of a set H of vectors, together with operations of vector addition $+ : H \times H \to H$ and scalar multiplication $\cdot : \mathbb{C} \times H \to H$ satisfying the vector space axioms listed in Table 1.4. There is also defined on H a *complex inner product* $B : H \times H \to H$, denoted $B(v, w) = \langle v, w \rangle$, satisfying these axioms for all $v, w, z \in H$ and $c \in \mathbb{C}$:

(1) $\langle v + w, z \rangle = \langle v, z \rangle + \langle w, z \rangle$;

(2) $\langle cv, z \rangle = c \langle v, z \rangle$;

(3) $\langle w, v \rangle = \overline{\langle v, w \rangle}$, where the bar denotes complex conjugation;

(4) if $v \neq 0$, then $\langle v, v \rangle$ is a strictly positive real number.

It follows (as in §13.11) that the inner product is linear in the first input and conjugate-linear in the second input. In other words, for all $c_1, \ldots, c_m \in \mathbb{C}$ and $v_1, \ldots, v_m, w \in H$,

$$\langle c_1 v_1 + \cdots + c_m v_m, w \rangle = c_1 \langle v_1, w \rangle + \cdots + c_m \langle v_m, w \rangle;$$

$$\langle w, c_1 v_1 + \cdots + c_m v_m \rangle = \overline{c_1} \langle w, v_1 \rangle + \cdots + \overline{c_m} \langle w, v_m \rangle.$$

Note that $\langle 0, w \rangle = 0 = \langle w, 0 \rangle$ for all $w \in H$. We say that $v, w \in H$ are *orthogonal* iff $\langle v, w \rangle = 0$, which holds iff $\langle w, v \rangle = 0$. The inner product in a Hilbert space generalizes the dot product from $\mathbb{R}^2$ and $\mathbb{R}^3$, and orthogonality generalizes the geometric concept of perpendicularity in $\mathbb{R}^2$ and $\mathbb{R}^3$.

We use the inner product to define the analytic ingredients of the Hilbert space H, namely the length (or norm) of a vector and the distance between two vectors. For all $v \in H$, define the *norm* of v by setting $||v|| = \sqrt{\langle v, v \rangle}$. For all $v, w \in H$ and $c \in \mathbb{C}$, the following properties hold:

(a) $||v|| \in \mathbb{R}_{\geq 0}$; and $||v|| = 0$ iff $v = 0$;

(b) $||cv|| = |c| \cdot ||v||$;

(c) $|\langle v, w \rangle| \leq ||v|| \cdot ||w||$ (the *Cauchy–Schwarz Inequality*);

(d) $||v + w|| \leq ||v|| + ||w||$ (the *Triangle Inequality* for norms).

Property (a) is true $\langle v, v \rangle$ is either a positive real number (when $v \neq 0$) or zero (when $v = 0$). Property (b) is true because

$$||cv|| = \sqrt{\langle cv, cv \rangle} = \sqrt{c \overline{c} \langle v, v \rangle} = \sqrt{|c|^2 \langle v, v \rangle} = |c| \sqrt{\langle v, v \rangle} = |c| \cdot ||v||.$$

The Cauchy–Schwarz Inequality has a more subtle proof. The inequality holds if $\langle v, w \rangle = 0$, so we may assume $\langle v, w \rangle \neq 0$, hence $v \neq 0$ and $w \neq 0$. We first prove the inequality in the case where $||v|| = ||w|| = 1$ and $\langle v, w \rangle$ is a positive real number. Then $\langle v, v \rangle = 1 = \langle w, w \rangle$ and $|\langle v, w \rangle| = \langle v, w \rangle = \langle w, v \rangle$. Using the axioms for the complex inner product, we compute

$$0 \leq \langle v - w, v - w \rangle = \langle v, v \rangle - \langle v, w \rangle - \langle w, v \rangle + \langle w, w \rangle. \tag{15.1}$$

Using our current assumptions on v and w, this inequality becomes $0 \leq 1 - 2|\langle v, w \rangle| + 1$, which rearranges to $|\langle v, w \rangle| \leq 1 = ||v|| \cdot ||w||$. Still assuming $||v|| = ||w|| = 1$, we next prove the case where $\langle v, w \rangle$ is an arbitrary nonzero complex scalar. In polar form, $\langle v, w \rangle = r e^{i\theta}$ for some real $r > 0$ and some $\theta \in [0, 2\pi)$. Let $v_0 = e^{-i\theta} v$. Then $||v_0|| = |e^{-i\theta}| \cdot ||v|| = ||v|| = 1$,

$|\langle v_0, w \rangle| = |e^{-i\theta} \langle v, w \rangle| = |\langle v, w \rangle|$, and $\langle v_0, w \rangle = e^{-i\theta} \langle v, w \rangle = r$ is a positive real number. By the case already proved, $|\langle v_0, w \rangle| \leq ||v_0|| \cdot ||w||$, and therefore $|\langle v, w \rangle| \leq ||v|| \cdot ||w||$. Finally, we drop the assumption that $||v|| = ||w|| = 1$. Write $v = cv_1$ and $w = dw_1$, where $c = ||v||$, $v_1 = c^{-1}v$, $d = ||w||$, and $w_1 = d^{-1}w$. We see that $||v_1|| = c^{-1}||v|| = 1$ and $||w_1|| = d^{-1}||w|| = 1$. On one hand, $0 \neq |\langle v, w \rangle| = |\langle cv_1, dw_1 \rangle| = |c\bar{d}\langle v_1, w_1 \rangle| = cd|\langle v_1, w_1 \rangle|$. On the other hand, the cases already proved now give $|\langle v_1, w_1 \rangle| \leq ||v_1|| \cdot ||w_1|| = 1$. Therefore,

$$|\langle v, w \rangle| = cd|\langle v_1, w_1 \rangle| \leq cd = ||v|| \cdot ||w||,$$

completing the proof of the Cauchy–Schwarz Inequality.

We can now deduce the Triangle Inequality for norms from the Cauchy–Schwarz Inequality. Given $v, w \in H$, compute

$$||v+w||^2 = \langle v+w, v+w \rangle = \langle v, v \rangle + \langle v, w \rangle + \langle w, v \rangle + \langle w, w \rangle = ||v||^2 + \langle v, w \rangle + \overline{\langle v, w \rangle} + ||w||^2.$$

The two middle terms add up to twice the real part of $\langle v, w \rangle$, which is at most $2|\langle v, w \rangle| \leq 2||v|| \cdot ||w||$. Hence,

$$||v + w||^2 \leq ||v||^2 + 2||v|| \cdot ||w|| + ||w||^2 = (||v|| + ||w||)^2.$$

Taking the positive square root of both sides gives $||v + w|| \leq ||v|| + ||w||$, as needed.

We use the norm to define the metric space structure of H. For $v, w \in H$, define the distance between v and w by $d(v, w) = ||v - w||$. The properties of the norm derived above imply the required axioms for a metric space (as we saw in §10.5). In particular, the Triangle Inequality for the metric follows from the Triangle Inequality for the norm, because

$$d(v, z) = ||v - z|| = ||(v - w) + (w - z)|| \leq ||v - w|| + ||w - z|| = d(v, w) + d(w, z)$$

for all $v, w, z \in H$. Like any metric induced from a norm, the metric on a Hilbert space is compatible with translations and dilations, meaning that $d(v + w, z + w) = d(v, z)$ and $d(cv, cw) = |c|d(v, w)$ for all $v, w, z \in H$ and $c \in \mathbb{C}$.

The final topological ingredient in the definition of a Hilbert space is the assumption that H is complete relative to the metric just defined. The definition requires that every Cauchy sequence in H converges to a point of H. Writing what this means, we see that whenever $(x_n : n \geq 0)$ is a sequence of vectors in H such that $\lim_{m,n\to\infty} ||x_m - x_n|| = 0$, there exists a (necessarily unique) $x \in H$ such that $\lim_{n\to\infty} ||x_n - x|| = 0$.

We can define a *real Hilbert space* by the same conditions discussed above, restricting all scalars to come from $\mathbb{R}$ and using a real-valued inner product. In this case, $\langle v, w \rangle = \langle w, v \rangle$, and the inner product is $\mathbb{R}$-linear (as opposed to conjugate-linear) in both the first and second positions. Henceforth in this chapter, we continue to consider only complex Hilbert spaces, letting the reader make the required modifications to obtain analogous results for real Hilbert spaces.

15.9 Examples of Hilbert Spaces

A basic example of a Hilbert space is the space $\mathbb{C}^n$ of n-tuples $v = (v_1, \ldots, v_n)$, where all $v_k \in \mathbb{C}$. The inner product is defined by $\langle v, w \rangle = \sum_{k=1}^{n} v_k \overline{w_k}$, which can be written $\langle v, w \rangle = w^*v$ if we think of v and w as column vectors. The norm of v is $||v|| = \sqrt{\sum_{k=1}^{n} |v_k|^2}$, and the distance between v and w is the Euclidean distance $d_2(v, w)$ discussed in §15.1. The completeness of $\mathbb{C}^n$ under this metric was proved in §15.7. The other axioms of a

Hilbert space (namely, that $\mathbb{C}^n$ is a complex vector space and inner product space) may be routinely verified. In $\mathbb{C}^n$, the Cauchy–Schwarz Inequality $|\langle v, w \rangle| \leq ||v|| \cdot ||w||$ and the Triangle Inequality $||v + w|| \leq ||v|| + ||w||$ for norms translate to the following facts about sums of complex numbers:

$$\left| \sum_{k=1}^{n} v_k \overline{w_k} \right| \leq \sqrt{\sum_{k=1}^{n} |v_k|^2} \cdot \sqrt{\sum_{k=1}^{n} |w_k|^2}; \tag{15.2}$$

$$\sqrt{\sum_{k=1}^{n} |v_k + w_k|^2} \leq \sqrt{\sum_{k=1}^{n} |v_k|^2} + \sqrt{\sum_{k=1}^{n} |w_k|^2}. \tag{15.3}$$

An example of an infinite-dimensional Hilbert space is the space ℓ_2 of all infinite sequences $v = (v_1, v_2, \ldots, v_k, \ldots) = (v_k : k \geq 1)$ such that $v_k \in \mathbb{C}$ and $\sum_{k=1}^{\infty} |v_k|^2 < \infty$. The inner product is defined by $\langle v, w \rangle = \sum_{k=1}^{\infty} v_k \overline{w_k}$. We prove the Hilbert space axioms for this example in §15.10.

Both of the previous examples are special cases of a general construction to be described shortly. First, we need a technical digression on general infinite summations. Let X be any set, which could be finite, countably infinite, or uncountable. Let $\mathcal{F}(X)$ be the set of all finite subsets of X. Given a nonnegative real number p_k for each $k \in X$, the sum $\sum_{k \in X} p_k$ is defined to be the least upper bound (possibly ∞) of all sums $\sum_{k \in X'} p_k$, as X' ranges over all finite subsets of X. In symbols,

$$\sum_{k \in X} p_k = \sup \left\{ \sum_{k \in X'} p_k : X' \in \mathcal{F}(X) \right\}.$$

Given real numbers r_k for $k \in X$, let $p_k = r_k$ if $r_k \geq 0$ and $p_k = 0$ otherwise; and let $n_k = |r_k|$ if $r_k < 0$ and $n_k = 0$ otherwise. Define $\sum_{k \in X} r_k$ to be $\sum_{k \in X} p_k - \sum_{k \in X} n_k$ if at most one of the latter sums is ∞. Finally, given complex numbers z_k for $k \in X$, write $z_k = x_k + iy_k$ for $x_k, y_k \in \mathbb{R}$. Define $\sum_{k \in X} z_k = \sum_{k \in X} x_k + i \sum_{k \in X} y_k$ if both sums on the right side are finite. Properties of summations over the set X are analogous to properties of infinite series. Some of these properties are covered in the exercises.

We now describe the general construction for producing Hilbert spaces. Fix a set X, and let $\ell_2(X)$ be the set of all functions $f : X \to \mathbb{C}$ such that $\sum_{x \in X} |f(x)|^2 < \infty$. We sometimes think of such a function as a generalized sequence or X-tuple $(f(x) : x \in X) = (f_x : x \in X)$. The set $\ell_2(X)$ becomes a Hilbert space under the following operations. Given $f, g \in \ell_2(X)$ and $c \in \mathbb{C}$, define $f + g$ and cf by setting $(f + g)(x) = f(x) + g(x)$ and $(cf)(x) = c(f(x))$ for all $x \in X$. Define $\langle f, g \rangle = \sum_{x \in X} f(x)\overline{g(x)}$. The norm and metric in this Hilbert space are given by $||f|| = \sqrt{\sum_{x \in X} |f(x)|^2} < \infty$ and $d(f, g) = \sqrt{\sum_{x \in X} |f(x) - g(x)|^2}$. The verification of the Hilbert space axioms, which is somewhat technical, is given in §15.10. Note that both previous examples are special cases of this construction, since $\mathbb{C}^n$ is $\ell_2(\{1, 2, \ldots, n\})$, and ℓ_2 is $\ell_2(\mathbb{Z}_{>0})$.

The spaces $\ell_2(X)$ are themselves special cases of Hilbert spaces that occur in the theory of Lebesgue integration. It is beyond the scope of this book to discuss this topic in detail, but we allude to a few facts for readers familiar with this theory. Given any measure space X with measure μ, we let $L_2(X, \mu)$ be the set of measurable functions $f : X \to \mathbb{C}$ such that $\int_X |f|^2 \, d\mu < \infty$. This is a complex vector space under pointwise operations on functions, and the inner product is defined by $\langle f, g \rangle = \int_X f\overline{g} \, d\mu$ for $f, g \in L_2(X, \mu)$. It can be proved that $L_2(X, \mu)$ is a Hilbert space with these operations; the fact that this space is complete is a difficult theorem. By taking μ to be counting measure on an arbitrary set X, we obtain the examples $\ell_2(X)$ discussed above.

Finally, given any Hilbert space H, we can form new examples of Hilbert spaces by considering subspaces of H. An arbitrary vector subspace W of H (a subset closed under zero, addition, and scalar multiplication) automatically satisfies all the Hilbert space axioms except possibly completeness. Since H is complete, we know from §15.7 that the subspace W is complete iff W is closed. It follows that *closed subspaces of a Hilbert space are also Hilbert spaces, but non-closed subspaces are not complete.* This example illustrates a general theme: to obtain satisfactory results in infinite-dimensional settings, we often need to impose a topological condition (in this case, being closed relative to the metric) in addition to algebraic conditions. We remark that *every finite-dimensional subspace of H is automatically closed*; see Exercise 75. But there are examples of infinite-dimensional subspaces that are not closed (Exercises 70 and 71).

15.10 Proof of the Hilbert Space Axioms for $\ell_2(X)$

Let X be any set. This section gives the proof that $\ell_2(X)$, with the operations defined above, does satisfy all the axioms for a Hilbert space. We divide the proof into three parts: checking the vector space axioms, checking the inner product axioms, and verifying completeness of the metric.

Vector Space Axioms. Let $^X\mathbb{C}$ denote the set of all functions $f : X \to \mathbb{C}$. The space $\ell_2(X)$ is the subset of $^X\mathbb{C}$ consisting of those f satisfying $\sum_{x \in X} |f(x)|^2 < \infty$. We saw in §4.2 that $^X\mathbb{C}$ is a complex vector space under pointwise operations on functions. Hence, to see that $\ell_2(X)$ is a vector space under the same operations, it suffices to check that $\ell_2(X)$ is a subspace of $^X\mathbb{C}$. First, the zero vector in $^X\mathbb{C}$, which is the function sending every $x \in X$ to 0, belongs to $\ell_2(X)$ because $\sum_{x \in X} |0|^2 = 0 < \infty$. Second, for $f \in \ell_2(X)$ and $c \in \mathbb{C}$, it follows readily from the definition of summation over X (Exercise 76) that $\sum_{x \in X} |(cf)(x)|^2 = \sum_{x \in X} |c(f(x))|^2 = |c|^2 \sum_{x \in X} |f(x)|^2 < \infty$, so that $cf \in \ell_2(X)$. Third, fix $f, g \in \ell_2(X)$; we must show $f + g \in \ell_2(X)$. Let $X' = \{x_1, \ldots, x_n\} \in \mathcal{F}(X)$ be any finite subset of X. By definition of $\ell_2(X)$, we know that $||f||^2 = \sum_{x \in X} |f(x)|^2$ and $||g||^2 = \sum_{x \in X} |g(x)|^2$ are finite. Using the known version (15.3) of the Triangle Inequality for finite lists of complex numbers, we see that

$$\sum_{k=1}^{n} |f(x_k) + g(x_k)|^2 \leq \left(\sqrt{\sum_{k=1}^{n} |f(x_k)|^2} + \sqrt{\sum_{k=1}^{n} |g(x_k)|^2} \right)^2.$$

Now, since the square root function is increasing, the definition of summation over X shows that $\sqrt{\sum_{k=1}^{n} |f(x_k)|^2} \leq \sqrt{\sum_{x \in X} |f(x)|^2} = ||f||$, and similarly $\sqrt{\sum_{k=1}^{n} |g(x_k)|^2} \leq ||g||$. So

$$\sum_{x \in X'} |f(x) + g(x)|^2 = \sum_{k=1}^{n} |f(x_k) + g(x_k)|^2 \leq (||f|| + ||g||)^2.$$

This calculation shows that the finite number $(||f|| + ||g||)^2$, which does not depend on X', is an upper bound in $\mathbb{R}$ for all the finite sums $\sum_{x \in X'} |f(x) + g(x)|^2$ as X' ranges over $\mathcal{F}(X)$. Thus, the least upper bound of the set of these sums is finite, proving that $f + g \in \ell_2(X)$. In fact, our calculation shows that $||f + g||^2 \leq (||f|| + ||g||)^2$, giving a direct proof of the Triangle Inequality for norms in $\ell_2(X)$.

Inner Product Axioms. The main technical point to be checked when verifying the inner product axioms for $\ell_2(X)$ is that $\langle f, g \rangle = \sum_{x \in X} f(x)\overline{g(x)}$ is a well-defined complex number for all $f, g \in \ell_2(X)$. We first prove this under the additional assumption that $f(x)$ and $g(x)$ are nonnegative real numbers for all $x \in X$. Given any $X' = \{x_1, \ldots, x_n\} \in \mathcal{F}(X)$, the Cauchy–Schwarz Inequality for n-tuples (see (15.2)) tells us that

$$\sum_{k=1}^{n} f(x_k)g(x_k) \leq \sqrt{\sum_{k=1}^{n} f(x_k)^2} \cdot \sqrt{\sum_{k=1}^{n} g(x_k)^2}.$$

The right side is at most $||f|| \cdot ||g||$, which is a finite upper bound on $\sum_{x \in X'} f(x)g(x)$ that is independent of X'. So $\sum_{x \in X} f(x)g(x) < \infty$ in this case; in fact, the sum is bounded by $||f|| \cdot ||g||$.

We next consider the case where $f, g \in \ell_2(X)$ are real-valued. Note that $|f|$ has the same squared norm as f, namely $\sum_{x \in X} |f(x)|^2 < \infty$, so $|f|$ (and similarly $|g|$) are in $\ell_2(X)$. By the case already considered, $\sum_{x \in X} |f(x)g(x)| < \infty$. Now, using a general property of sums over X (Exercise 76), we can conclude that $\sum_{x \in X} f(x)g(x)$ is a finite real number, and in fact, $|\sum_{x \in X} f(x)g(x)| \leq \sum_{x \in X} |f(x)g(x)| \leq ||f|| \cdot ||g|| < \infty$. Finally, for complex-valued $f, g \in \ell_2(X)$, we can write $f = t + iu$ and $g = v + iw$ where $t, u, v, w : X \to \mathbb{R}$ give the real and imaginary parts of f and g. Since $|t(x)| \leq |f(x)|$ for all $x \in X$, we see that t (and similarly u, v, w) are real-valued functions in $\ell_2(X)$. Also, $f(x)\overline{g(x)} = [t(x)v(x) + u(x)w(x)] + i[u(x)v(x) - t(x)w(x)]$ for all $x \in X$. By cases already considered, $\sum_{x \in X} t(x)v(x)$, $\sum_{x \in X} u(x)w(x)$, $\sum_{x \in X} u(x)v(x)$, and $\sum_{x \in X} t(x)w(x)$ are all finite. It follows that $\sum_{x \in X} f(x)\overline{g(x)}$ is well-defined. In fact, using Exercise 76, we have the bound $|\sum_{x \in X} f(x)\overline{g(x)}| \leq \sum_{x \in X} |f(x)g(x)| \leq ||f|| \cdot ||g||$. This gives a direct proof of the Cauchy–Schwarz Inequality for $\ell_2(X)$.

Knowing that $\langle f, g \rangle$ is a well-defined complex number, we can prove the axioms for the inner product without difficulty. We prove $\langle f + g, h \rangle = \langle f, h \rangle + \langle g, h \rangle$ for all $f, g, h \in \ell_2(X)$, leaving the remaining axioms as exercises. Using the general property $\sum_{x \in X} (a(x) + b(x)) = \sum_{x \in X} a(x) + \sum_{x \in X} b(x)$ (see Exercise 76), we compute

$$\begin{aligned}
\langle f + g, h \rangle &= \sum_{x \in X} (f + g)(x)\overline{h(x)} = \sum_{x \in X} (f(x) + g(x))\overline{h(x)} = \sum_{x \in X} [f(x)\overline{h(x)} + g(x)\overline{h(x)}] \\
&= \sum_{x \in X} f(x)\overline{h(x)} + \sum_{x \in X} g(x)\overline{h(x)} = \langle f, h \rangle + \langle g, h \rangle.
\end{aligned}$$

Completeness of the Metric Space $\ell_2(X)$. Let $(f_0, f_1, f_2, \ldots) = (f_n : n \geq 0)$ be a Cauchy sequence in $\ell_2(X)$; we must prove that this sequence converges to some $g \in \ell_2(X)$. To find g, fix $x_0 \in X$ and consider the sequence of complex numbers $(f_n(x_0) : n \geq 0)$. Given $\epsilon > 0$, there is an integer m_0 so that for all $m, n \geq m_0$, $d(f_m, f_n) = \sqrt{\sum_{x \in X} |f_m(x) - f_n(x)|^2} < \epsilon$. Then for all $m, n \geq m_0$, $|f_m(x_0) - f_n(x_0)| = \sqrt{|f_m(x_0) - f_n(x_0)|^2} \leq d(f_m, f_n) < \epsilon$. So $(f_n(x_0) : n \geq 0)$ is a Cauchy sequence in the complete metric space $\mathbb{C}$. Therefore, there is a unique complex number $g(x_0)$ such that $\lim_{n \to \infty} f_n(x_0) = g(x_0)$. This holds for each $x_0 \in X$, so we have a function $g : X \to \mathbb{C}$ such that f_n converges to g pointwise. It remains to show that $g \in \ell_2(X)$ and $\lim_{n \to \infty} f_n = g$ in the metric space $\ell_2(X)$.

To see that $g \in \ell_2(X)$, fix $m_0 \in \mathbb{Z}_{\geq 0}$ (corresponding to $\epsilon = 1$) so that $m, n \geq m_0$ implies $d(f_m, f_n) < 1$. We show that $(||f_{m_0}|| + 2)^2$ is a finite upper bound for all sums $\sum_{x \in X'} |g(x)|^2$ as X' ranges over $\mathcal{F}(X)$, so that $\sum_{x \in X} |g(x)|^2 \leq (||f_{m_0}|| + 2)^2 < \infty$. Fix $X' = \{x_1, \ldots, x_N\} \in \mathcal{F}(X)$. Since $f_n(x_k) \to g(x_k)$ for $k = 1, 2, \ldots, N$, we can choose $m_1, \ldots, m_N$ such that for all $n \geq m_k$, $|f_n(x_k) - g(x_k)| < 1/\sqrt{N}$. Choose an n larger than all of $m_0, m_1, \ldots, m_N$. Writing $g(x_k) = (g(x_k) - f_n(x_k)) + (f_n(x_k) - f_{m_0}(x_k)) + f_{m_0}(x_k)$

and using (15.3) (extended to a sum of three terms), compute

$$\sum_{x \in X'} |g(x)|^2 \leq \left(\sqrt{\sum_{k=1}^N |g(x_k) - f_n(x_k)|^2} + \sqrt{\sum_{k=1}^N |f_n(x_k) - f_{m_0}(x_k)|^2} + \sqrt{\sum_{k=1}^N |f_{m_0}(x_k)|^2} \right)^2$$

$$\leq \left(\sqrt{\sum_{k=1}^N \frac{1}{N}} + ||f_n - f_{m_0}|| + ||f_{m_0}|| \right)^2 \leq (||f_{m_0}|| + 2)^2.$$

The proof that $f_n \to g$ in $\ell_2(X)$ requires a similar computation. Fix $\epsilon > 0$, and choose an integer m_0 so that $m, n \geq m_0$ implies $d(f_m, f_n) < \epsilon/2$. We fix $n \geq m_0$ and show that $||f_n - g|| \leq \epsilon$. To do so, pick $X' = \{x_1, \ldots, x_N\}$ in $\mathcal{F}(X)$, and choose $m_1, \ldots, m_N$ so that $m \geq m_k$ implies $|f_m(x_k) - g(x_k)| < \epsilon/2\sqrt{N}$. Choose an m larger than all of $m_0, m_1, \ldots, m_N$. Writing $f_n(x_k) - g(x_k) = (f_n(x_k) - f_m(x_k)) + (f_m(x_k) - g(x_k))$, we compute (using (15.3))

$$\sum_{x \in X'} |f_n(x) - g(x)|^2 \leq \left(\sqrt{\sum_{k=1}^N |f_n(x_k) - f_m(x_k)|^2} + \sqrt{\sum_{k=1}^N |f_m(x_k) - g(x_k)|^2} \right)^2$$

$$< \left(||f_n - f_m|| + \sqrt{\sum_{k=1}^N \frac{\epsilon^2}{4N}} \right)^2 < (\epsilon/2 + \epsilon/2)^2 = \epsilon^2.$$

The upper bound of ϵ^2 holds for all X', so $\sum_{x \in X} |f_n(x) - g(x)|^2 \leq \epsilon^2$, and hence $||f_n - g|| \leq \epsilon$, as needed.

15.11 Basic Properties of Hilbert Spaces

In this section, we derive some basic algebraic and analytic properties of Hilbert spaces that do not involve the axiom of completeness. Let H be any Hilbert space. For all $x, y \in H$, we have already derived the Triangle Inequality for norms: $||x + y|| \leq ||x|| + ||y||$. We can obtain a sharper result called the *Pythagorean Theorem* when x and y are orthogonal vectors. This theorem says that if $\langle x, y \rangle = 0$, then $||x + y||^2 = ||x||^2 + ||y||^2$. Geometrically, the square of the length of the hypotenuse of a right triangle equals the sum of the squares of the two legs. To prove the Pythagorean Theorem, we compute (cf. (15.1))

$$||x + y||^2 = \langle x + y, x + y \rangle = \langle x, x \rangle + \langle x, y \rangle + \langle y, x \rangle + \langle y, y \rangle = ||x||^2 + ||y||^2,$$

where $\langle x, y \rangle = 0$ by hypothesis, and $\langle y, x \rangle = \overline{\langle x, y \rangle} = 0$. More generally, we say that a list $x_1, \ldots, x_n$ of vectors in H is *orthogonal* iff $\langle x_i, x_j \rangle = 0$ for all $i \neq j$ in $[n]$. By induction on n, we see that for any orthogonal list $x_1, \ldots, x_n$ in H, $||x_1 + x_2 + \cdots + x_n||^2 = ||x_1||^2 + ||x_2||^2 + \cdots + ||x_n||^2$. More generally, if $x_1, \ldots, x_n$ is an orthogonal list and $c_1, \ldots, c_n \in \mathbb{C}$ are any scalars, then

$$||c_1 x_1 + c_2 x_2 + \cdots + c_n x_n||^2 = |c_1|^2 ||x_1||^2 + |c_2|^2 ||x_2||^2 + \cdots + |c_n|^2 ||x_n||^2.$$

The key to the induction proof is that $c_n x_n$ is orthogonal to any linear combination $c_1 x_1 + \cdots + c_{n-1} x_{n-1}$, so that the Pythagorean Theorem for the sum of two vectors can be applied.

Here are two more identities resembling the Pythagorean Theorem. First, the *Parallelogram Law* states that for any x, y in a Hilbert space H,

$$||x + y||^2 + ||x - y||^2 = 2||x||^2 + 2||y||^2. \tag{15.4}$$

Geometrically, the sum of the squares of the lengths of the two diagonals of any parallelogram equals the sum of the squares of the lengths of the four sides of the parallelogram. To prove this, compute

$$\begin{aligned}
||x + y||^2 + ||x - y||^2 &= \langle x + y, x + y \rangle + \langle x - y, x - y \rangle \\
&= [\langle x, x \rangle + \langle x, y \rangle + \langle y, x \rangle + \langle y, y \rangle] + [\langle x, x \rangle - \langle x, y \rangle - \langle y, x \rangle + \langle y, y \rangle] \\
&= 2||x||^2 + 2||y||^2.
\end{aligned}$$

Second, the *Polarization Identity* states that for all x, y in a Hilbert space H,

$$||x + y||^2 + i||x + iy||^2 - ||x - y||^2 - i||x - iy||^2 = 4\langle x, y \rangle. \tag{15.5}$$

This identity is proved by a calculation similar to the one just given (Exercise 78). You can use the Polarization Identity to prove that any complex normed vector space (as defined in §10.4) that is complete and whose norm satisfies the Parallelogram Law must be a Hilbert space (Exercise 81).

Next, we discuss some continuity properties of the operations appearing in the definition of a Hilbert space. First, *addition is continuous*: if $x_n \to x$ and $y_n \to y$ in a Hilbert space H, then $x_n + y_n \to x + y$. To prove this, fix $\epsilon > 0$ and let n_0, n_1 be integers such that $n \geq n_0$ implies $||x_n - x|| < \epsilon/2$, while $n \geq n_1$ implies $||y_n - y|| < \epsilon/2$. Then for $n \geq \max(n_0, n_1)$,

$$||(x_n + y_n) - (x + y)|| = ||(x_n - x) + (y_n - y)|| \leq ||x_n - x|| + ||y_n - y|| < \epsilon.$$

Second, *scalar multiplication is continuous*: if $x_n \to x$ in H and $c_n \to c$ in $\mathbb{C}$, then $c_n x_n \to cx$ in H. The convergent sequence (c_n) must be bounded in $\mathbb{C}$. Let M be a positive constant such that $|c_n| \leq M$ for all $n \in \mathbb{Z}_{\geq 0}$. Now, given $\epsilon > 0$, choose integers n_0, n_1 so that $n \geq n_0$ implies $||x_n - x|| < \epsilon/(2M)$, while $n \geq n_1$ implies $|c_n - c| < \epsilon/(2(1 + ||x||))$. Then $n \geq \max(n_0, n_1)$ implies

$$||c_n x_n - cx|| = ||c_n(x_n - x) + (c_n - c)x|| \leq |c_n| \cdot ||x_n - x|| + |c_n - c| \cdot ||x|| < M(\epsilon/(2M)) + \epsilon/2 = \epsilon.$$

Third, *the inner product on H is continuous*: if $x_n \to x$ and $y_n \to y$ in H, then $\langle x_n, y_n \rangle \to \langle x, y \rangle$ in $\mathbb{C}$. As before, there is a single constant $M > 0$ such that $||x_n||, ||x||, ||y_n||$, and $||y||$ are all bounded above by M. Given $\epsilon > 0$, choose integers n_0, n_1 so that $n \geq n_0$ implies $||x_n - x|| < \epsilon/(2M)$, while $n \geq n_1$ implies $||y_n - y|| < \epsilon/(2M)$. Then $n \geq \max(n_0, n_1)$ implies

$$|\langle x_n, y_n \rangle - \langle x, y \rangle| = |\langle x_n - x, y_n \rangle + \langle x, y_n - y \rangle| \leq |\langle x_n - x, y_n \rangle| + |\langle x, y_n - y \rangle|.$$

By the Cauchy–Schwarz inequality in H,

$$|\langle x_n - x, y_n \rangle| + |\langle x, y_n - y \rangle| \leq ||x_n - x|| \cdot ||y_n|| + ||x|| \cdot ||y_n - y|| < (\epsilon/(2M))M + M(\epsilon/(2M)) = \epsilon.$$

Fourth, *the norm on H is continuous*: if $x_n \to x$ in H, then $||x_n|| \to ||x||$ in $\mathbb{R}$. To prove this, note $||x_n|| = ||x_n - x + x|| \leq ||x_n - x|| + ||x||$, so $||x_n|| - ||x|| \leq ||x_n - x||$. Similarly, $||x|| - ||x_n|| \leq ||x - x_n|| = ||x_n - x||$, so the absolute value of $||x_n|| - ||x||$ is at most $||x_n - x||$. Given $\epsilon > 0$, choose n_0 so $n \geq n_0$ implies $||x_n - x|| < \epsilon$. Then $n \geq n_0$ implies $|\,||x_n|| - ||x||\,| < \epsilon$, so that $||x_n|| \to ||x||$ in $\mathbb{R}$.

15.12   Closed Convex Sets in Hilbert Spaces

Recall from Chapter 11 that a subset C of a (real or complex) vector space V is called *convex* iff for all $x, y \in C$ and all real $t \in [0, 1]$, $tx + (1 - t)y \in C$. Geometrically, this condition says that a convex set must contain the line segment joining any two of its points. Every subspace W of V is a convex set. If C is a convex subset of V and $z \in V$, the translate $z + C = \{z + w : w \in C\}$ is convex. The empty set is also convex.

The following geometric lemma is a key technical tool for studying Hilbert spaces. *For every nonempty closed convex set C in a Hilbert space H, there exists a unique $x \in C$ of minimum norm; so $||x|| < ||z||$ for all $z \neq x$ in C.* We prove uniqueness first. Suppose $x, y \in C$ are two elements such that $r = ||x|| = ||y|| \leq ||z||$ for all $z \in C$; we must prove $x = y$. Consider $z = (1/2)x + (1/2)y$. On one hand, by convexity of C, $z \in C$ and hence $r \leq ||z||$. On the other hand, applying the Parallelogram Law to $x/2$ and $y/2$ shows that

$$||(x/2) + (y/2)||^2 + ||(x/2) - (y/2)||^2 = 2||(x/2)||^2 + 2||(y/2)||^2, \qquad (15.6)$$

which simplifies to $||z||^2 + (1/4)||x - y||^2 = (2/4)r^2 + (2/4)r^2 = r^2$. Then $||x - y||^2 = 4r^2 - 4||z||^2 \leq 4r^2 - 4r^2 = 0$, forcing $||x - y|| = 0$ and $x = y$.

Turning to the existence proof, let s be the greatest lower bound in $\mathbb{R}$ of the set $\{||z|| : z \in C\}$; this set is nonempty and bounded below by zero, so s does exist. By definition of greatest lower bound, for each positive integer n we can find $x_n \in C$ such that $s \leq ||x_n|| < s + 1/n$. Observe that $\lim_{n \to \infty} ||x_n|| = s$. We first show that $(x_n : n > 0)$ is a Cauchy sequence in H. Fix $\epsilon > 0$; we must find an integer m_0 such that $m, n \geq m_0$ implies $d(x_n, x_m) = ||x_n - x_m|| < \epsilon$. For any positive integers m, n, we have $(1/2)x_m + (1/2)x_n \in C$ by convexity of C, so that $s \leq ||(x_m/2) + (x_n/2)||$ by definition of s. Applying the Parallelogram Law to $x_n/2$ and $x_m/2$ gives

$$||(x_n/2) - (x_m/2)||^2 = 2||x_n/2||^2 + 2||x_m/2||^2 - ||(x_m/2) + (x_n/2)||^2$$
$$\leq \frac{(s + 1/n)^2}{2} + \frac{(s + 1/m)^2}{2} - s^2,$$

which rearranges to

$$||x_n - x_m|| \leq 2\sqrt{s/n + s/m + 1/(2n^2) + 1/(2m^2)}.$$

Since s is fixed, each term inside the square root approaches zero as n and m increase to infinity. So we can choose m_0 so that $m, n \geq m_0$ implies $||x_n - x_m|| < \epsilon$.

Now we know $(x_n : n > 0)$ is a Cauchy sequence. By completeness of the Hilbert space H, this sequence must converge to some point $y \in H$. Because C is a closed subset of H, we must have $y \in C$. By continuity of the norm, $s = \lim_{n \to \infty} ||x_n|| = ||\lim_{n \to \infty} x_n|| = ||y||$. So $||y||$ is a lower bound of all the norms $||x||$ for $x \in C$, completing the existence proof.

Here is a slight generalization of the lemma just proved. *For every nonempty, closed, convex set C in H and all $w \in H$, there exists a unique $x \in C$ minimizing $d(x, w)$;* we call x the point of C *closest to* w. To prove this, consider the translate $C' = (-w) + C$, which is readily verified to be nonempty, closed, and convex (Exercise 84). The map $x \mapsto x - w$ is a bijection between C and C'. Moreover, $d(x, w) = ||x - w||$. Thus, $x \in C$ minimizes $d(x, w)$ iff $x - w \in C'$ has minimum norm among all elements of C'. By the lemma already proved, there exists a unique element of C' with the latter property. Therefore, there exists a unique $x \in C$ minimizing $d(x, w)$, as needed.

15.13   Orthogonal Complements

In our study of finite-dimensional inner product spaces in Chapter 13, we introduced the *orthogonal complement* of a subspace W, denoted $W^\perp$. For each subspace W of the inner product space V, $W^\perp$ is the subspace consisting of all vectors in V that are orthogonal to every vector in W. We showed that the map $W \mapsto W^\perp$ is an inclusion-reversing bijection on the lattice of subspaces of V, and for every subspace W, $W^{\perp\perp} = W$ and $V = W \oplus W^\perp$. A key ingredient in proving these results was the dimension formula $\dim(V) = \dim(W) + \dim(W^\perp)$, whose proof required V to be finite-dimensional.

Our goal is to extend these results to the case of a general Hilbert space H, which may be infinite-dimensional. To begin, we define the *orthogonal complement* of an arbitrary subset S of H to be

$$S^\perp = \{v \in H : \langle v, w \rangle = 0 \text{ for all } w \in S\}.$$

We claim that $S^\perp$ *is always a closed subspace of* H. First, $0_H \in S^\perp$ since $\langle 0, w \rangle = 0$ for all $w \in S$. Second, given $u, v \in S^\perp$, we know $\langle u, w \rangle = 0 = \langle v, w \rangle$ for all $w \in S$. So $\langle u+v, w \rangle = \langle u, w \rangle + \langle v, w \rangle = 0+0 = 0$ for all $w \in S$, proving $u+v \in S^\perp$. Third, given $u \in S^\perp$ and $c \in \mathbb{C}$, we find that $\langle cu, w \rangle = c\langle u, w \rangle = c0 = 0$ for all $w \in S$, so $cu \in S^\perp$. Fourth, to see that $S^\perp$ is closed, define a map $R_w : H \to \mathbb{C}$ (for each $w \in H$) by letting $R_w(x) = \langle x, w \rangle$ for all $x \in H$. If $x_n \to x$ in H, then (as seen in §15.11) $R_w(x_n) = \langle x_n, w \rangle \to \langle x, w \rangle = R_w(x)$. So, R_w is a continuous map from H to $\mathbb{C}$. In particular, since the one-point set $\{0\}$ is closed in $\mathbb{C}$, the inverse image $R_w^{-1}[\{0\}] = \{x \in H : \langle x, w \rangle = 0\}$ is a closed subset of H. By definition of $S^\perp$, we have $S^\perp = \bigcap_{w \in S} R_w^{-1}[\{0\}]$. This is an intersection of a family of closed sets, so $S^\perp$ is closed. We remark that each R_w is a $\mathbb{C}$-linear map (by the inner product axioms), and $R_w^{-1}[\{0\}]$ is precisely the kernel of R_w. This gives another way to see that $S^\perp$ is a subspace, since the kernel of a linear map is a subspace of the domain, and the intersection of a family of subspaces is also a subspace.

To obtain a bijective correspondence $W \mapsto W^\perp$ in the setting of Hilbert spaces, we must restrict attention to the set of closed subspaces W, since applying the orthogonal complement operator always produces a closed subspace. When proving the next theorem, dimension-counting arguments are no longer available. Instead we invoke the geometric lemma of §15.12, whose proof made critical use of the completeness of H.

Theorem on Orthogonal Complements. Suppose H is a Hilbert space. For any closed subspace W of H, $H = W \oplus W^\perp$ and $W^{\perp\perp} = W$.

Proof. Given $x \in H$, we must prove there exist unique $y \in W$ and $z \in W^\perp$ with $x = y + z$. We prove uniqueness first: assume $y_1, y_2 \in W$ and $z_1, z_2 \in W^\perp$ satisfy $x = y_1 + z_1 = y_2 + z_2$. Let $u = y_1 - y_2 = z_2 - z_1$. Since W is a subspace, $u = y_1 - y_2 \in W$. Since $W^\perp$ is a subspace, $u = z_2 - z_1 \in W^\perp$. Then $\langle u, u \rangle = 0$, forcing $u = 0$ by the inner product axioms. So $y_1 = y_2$ and $z_1 = z_2$.

To prove existence of y and z, we can gain intuition from the case where W is a two-dimensional subspace of $\mathbb{R}^3$ and $W^\perp$ is the line through 0 perpendicular to W. In this case, we could find y given x by dropping an altitude from x to the plane W. This altitude meets W at the point y on that plane closest to x, and then $x = y + (x - y)$ where the vector $x - y$ is parallel to the altitude and hence is in $W^\perp$. This suggests that in the general case, we could define y to be the unique point in W closest to x, and let $z = x - y$. The point y does exist, since W is a nonempty closed convex subset of H. It is evident that $x = y + z$, but we must still check that $z \in W^\perp$.

Fix $w \in W$; we must show $\langle z, w \rangle = 0$. The conclusion holds for $w = 0$, so assume $w \neq 0$. Write $w = cu$, where $c = ||w||$ and $u = c^{-1}w \in W$ satisfies $||u|| = 1$. For any $s \in \mathbb{C}$, the vector $y - su$ is in the subspace W. Since y is the closest point in W to x, we have $||z|| = ||x - y|| \leq ||x - (y - su)|| = ||z + su||$ for all $s \in \mathbb{C}$. Squaring this inequality and rewriting using scalar products,

$$\langle z, z \rangle \leq \langle z, z \rangle + \overline{s}\langle z, u \rangle + s\langle u, z \rangle + |s|^2\langle u, u \rangle.$$

Since $\langle u, u \rangle = 1$ and $\langle u, z \rangle = \overline{\langle z, u \rangle}$, the inequality becomes $0 \leq \overline{s}\langle z, u \rangle + s\overline{\langle z, u \rangle} + |s|^2$ for all $s \in \mathbb{C}$. Choose $s = -\langle z, u \rangle$ to get

$$0 \leq -|\langle z, u \rangle|^2 - |\langle z, u \rangle|^2 + |\langle z, u \rangle|^2 = -|\langle z, u \rangle|^2,$$

which forces $\langle z, u \rangle = 0$. Then $\langle z, w \rangle = c\langle z, u \rangle = 0$, as needed.

We have now proved $H = W \oplus W^\perp$ for any closed subspace W; we use this to prove $W^{\perp\perp} = W$. Recalling that $\langle x, y \rangle = 0$ iff $\langle y, x \rangle = 0$, we see from the definitions that $W \subseteq W^{\perp\perp}$ without any hypothesis on the subset W. The result just proved shows that $H = W \oplus W^\perp$ and also $H = W^\perp \oplus W^{\perp\perp}$, since $W^\perp$ is a closed subspace. We use this to prove $W^{\perp\perp} \subseteq W$, as follows. Fix $x \in W^{\perp\perp}$. There exist unique $y \in W$ and $z \in W^\perp$ with $x = y + z$. Similarly, x can be written in exactly one way as the sum of a vector in $W^{\perp\perp}$ and a vector in $W^\perp$. Since $y \in W^{\perp\perp}$, one such sum is $x = y + z$. On the other hand, since $x \in W^{\perp\perp}$ and $0 \in W^\perp$, another such sum is $x = x + 0$. By uniqueness, this forces $y = x$ and $z = 0$, so $x = y$ is in W, as needed. □

Let $\mathcal{L}$ be the set of all closed subspaces of the given Hilbert space H. You can check that $\mathcal{L}$ (ordered by set inclusion) is a complete lattice. We have shown that $f : \mathcal{L} \to \mathcal{L}$, given by $f(W) = W^\perp$ for $W \in \mathcal{L}$, does map into the codomain $\mathcal{L}$ and satisfies $f(f(W)) = W$. Therefore, f is a bijection on $\mathcal{L}$ with $f^{-1} = f$. You can check that for $W, X \in \mathcal{L}$, $W \subseteq X$ implies $f(W) \supseteq f(X)$, so f is order-reversing. To summarize, $W \mapsto W^\perp$ *is a lattice anti-isomorphism of the lattice of all closed subspaces of H.*

15.14   Orthonormal Sets

Orthonormal bases play a central role in finite-dimensional inner product spaces. Every such space has an orthonormal basis, and every vector in the space can be written as a (finite) linear combination of these basis elements. In the setting of Hilbert spaces, we develop an analytic version of orthonormal bases in which infinite linear combinations of basis elements are allowed. To prepare for this, we first study finite orthonormal sets in a general Hilbert space H.

A subset X of H is called *orthonormal* iff $||x||^2 = \langle x, x \rangle = 1$ for all $x \in X$, and $\langle x, y \rangle = 0$ for all $x \neq y$ in X. An orthonormal set X is automatically linearly independent. To prove this, suppose $\{x_1, \ldots, x_N\}$ is any finite subset of X and $c_1, \ldots, c_N \in \mathbb{C}$ satisfy $c_1 x_1 + \cdots + c_N x_N = 0$. Then, for all j between 1 and N,

$$0 = \langle 0, x_j \rangle = \langle c_1 x_1 + \cdots + c_N x_N, x_j \rangle = c_j\langle x_j, x_j \rangle + \sum_{k \neq j} c_k\langle x_k, x_j \rangle = c_j.$$

Now suppose $X = \{x_1, \ldots, x_N\}$ is a finite orthonormal subset of H. Let W be the subspace of H spanned by X. We give a direct argument to show that $H = W \oplus W^\perp$. (This also follows from previous results and the fact that the finite-dimensional subspace W must be closed.) It is routine to check that $W \cap W^\perp = \{0\}$, since 0 is the only vector orthogonal to itself. Next, we show how to write any $x \in H$ in the form $x = y + z$, where $y \in W$ and

$z \in W^\perp$. Define $y = \sum_{k=1}^{N} \langle x, x_k \rangle x_k \in W$ and $z = x - y$. To prove $z \in W^\perp$, it suffices (by the inner product axioms) to show that $\langle z, x_j \rangle = 0$ for $1 \le j \le N$. We compute

$$\langle z, x_j \rangle = \langle x, x_j \rangle - \langle y, x_j \rangle = \langle x, x_j \rangle - \sum_{k=1}^{N} \langle x, x_k \rangle \langle x_k, x_j \rangle = \langle x, x_j \rangle - \langle x, x_j \rangle = 0.$$

Thus $H = W \oplus W^\perp$. We already know $H = W^{\perp\perp} \oplus W^\perp$, so the argument used at the end of §15.13 shows that $W = W^{\perp\perp}$, and hence W is a closed subspace. It now follows from the proof in §15.13 that $y = \sum_{k=1}^{N} \langle x, x_k \rangle x_k$ is the closest element of W to x. We call y the *orthogonal projection of x onto W*.

Since y is orthogonal to z and the x_k are orthogonal to each other, the Pythagorean Theorem shows that $||x||^2 = ||y||^2 + ||z||^2 = \sum_{k=1}^{N} |\langle x, x_k \rangle|^2 + ||z||^2$. Discarding the error term $||z||^2$, we obtain the inequality

$$\sum_{k=1}^{N} |\langle x, x_k \rangle|^2 \le ||x||^2,$$

which is the finite version of *Bessel's Inequality*. This inequality can be viewed as an approximate version of the Pythagorean Theorem: the sum of the squared norms of the components of x in certain orthogonal directions is at most the squared length of x itself.

Next, let X be any orthonormal set (possibly infinite) in a Hilbert space H. Given $w \in H$, define a function $f_w : X \to \mathbb{C}$ by setting $f_w(x) = \langle w, x \rangle$ for all $x \in X$. The complex scalars $\langle w, x \rangle$ are called the *Fourier coefficients* of w relative to the orthonormal set X. We claim that for all $w \in H$, f_w is in the space $\ell_2(X)$ of square-summable sequences indexed by X (see §15.9). In other words, $\sum_{x \in X} |f_w(x)|^2 = \sum_{x \in X} |\langle w, x \rangle|^2 < \infty$.

To verify the claim, fix any finite subset $X' = \{x_1, \ldots, x_N\}$ of X, which is also orthonormal. By the finite version of Bessel's Inequality,

$$\sum_{x \in X'} |f_w(x)|^2 = \sum_{k=1}^{N} |\langle w, x_k \rangle|^2 \le ||w||^2.$$

Thus $||w||^2$ is a finite upper bound for all these sums as X' ranges over $\mathcal{F}(X)$. Thus we obtain the *general version of Bessel's Inequality*, namely

$$\sum_{x \in X} |\langle w, x \rangle|^2 \le ||w||^2$$

for all $w \in H$ and all orthonormal sets $X \subseteq H$.

15.15 Maximal Orthonormal Sets

A *maximal orthonormal set* in a Hilbert space H is an orthonormal set X such that for any set Y properly containing X, Y is not orthonormal. Some texts refer to maximal orthonormal sets as *complete* orthonormal sets, but we avoid this term to prevent confusion with the notion of a complete metric space. By appealing to Zorn's Lemma (Exercise 89), you can show that *maximal orthonormal sets exist in any Hilbert space.*

Let X be a maximal orthonormal set in a Hilbert space H. We prove that in this case, equality holds in Bessel's Inequality, namely,

$$\sum_{x \in X} |\langle w, x \rangle|^2 = ||w||^2 \tag{15.7}$$

for all $w \in H$. To get a contradiction, assume this equality fails for some $w \in H$. Let $r = \sum_{x \in X} |\langle w, x \rangle|^2 < ||w||^2$. For each positive integer n, we can find a finite subset X_n of X such that $r - 1/n < \sum_{x \in X_n} |\langle w, x \rangle|^2 \leq r$. By replacing each X_n by $X_1 \cup X_2 \cup \cdots \cup X_n$ (which is still finite), we can arrange that $X_1 \subseteq X_2 \subseteq \cdots \subseteq X_n \subseteq \cdots$. Now define x_n and y_n in H by setting $x_n = \sum_{x \in X_n} \langle w, x \rangle x$ and $y_n = w - x_n$ for each positive integer n.

We claim (x_n) is a Cauchy sequence in H. Given $\epsilon > 0$, choose an integer m_0 so that $1/m_0 < \epsilon$. For $m \geq n \geq m_0$, the Pythagorean Theorem gives

$$||x_m - x_n||^2 = \left\| \sum_{x \in X_m \setminus X_n} \langle w, x \rangle x \right\|^2 = \sum_{x \in X_m} |\langle w, x \rangle|^2 - \sum_{x \in X_n} |\langle w, x \rangle|^2 < r - (r - 1/n) < \epsilon.$$

So (x_n) is Cauchy, hence (x_n) converges to a point $z \in H$ by completeness of H. Letting $y = w - z$, we have $y_n = (w - x_n) \to (w - z) = y$ as n goes to infinity.

Now, $||z|| = \lim_{n \to \infty} ||x_n|| = \lim_{n \to \infty} \sqrt{\sum_{x \in X_n} |\langle w, x \rangle|^2} = \sqrt{r} < ||w||$. It follows that $z \neq w$ and $y \neq 0$. We next claim that $\langle y, x \rangle = 0$ for all $x \in X$. Once this claim is proved, we can deduce that $y \notin X$, $y/||y|| \notin X$, yet $X \cup \{y/||y||\}$ is orthonormal, contradicting the maximality of the orthonormal set X. To prove the claim, fix $x \in X$ and note that $\langle y, x \rangle = \lim_{n \to \infty} \langle y_n, x \rangle = \langle w, x \rangle - \lim_{n \to \infty} \langle x_n, x \rangle$. If $x \in X_{n_0}$ for some n_0, then for all $n \geq n_0$, x_n is the sum of $\langle w, x \rangle x$ plus other vectors orthogonal to x, so $\lim_{n \to \infty} \langle x_n, x \rangle = \langle w, x \rangle$ and $\langle y, x \rangle = 0$. On the other hand, if $x \notin X_{n_0}$ for all n_0, then $\langle x_n, x \rangle = 0$ for all n. We show that $\langle w, x \rangle = 0$ in this case. If $\langle w, x \rangle \neq 0$, choose n so that $|\langle w, x \rangle|^2 > 1/n$. Then $|\langle w, x \rangle|^2 + \sum_{u \in X_n} |\langle w, u \rangle|^2 > 1/n + (r - 1/n) = r$, which contradicts the definition of r. This completes the proof of (15.7). Recalling that $f_w(x) = \langle w, x \rangle$ for all $x \in X$, we have $||f_w||^2 = \sum_{x \in X} |\langle w, x \rangle|^2 = ||w||^2$ for all $w \in H$.

Now we can define the concept of an infinite linear combination of vectors in an orthonormal set. We show that for any maximal orthonormal set X and all $w \in H$, $w = \sum_{x \in X} \langle w, x \rangle x$. By definition, this means that given any $\epsilon > 0$, there exists a finite subset $X' \in \mathcal{F}(X)$ such that for every finite subset $Y \in \mathcal{F}(X)$ containing X', $||w - \sum_{x \in Y} \langle w, x \rangle x|| < \epsilon$. Given $w \in H$ and $\epsilon > 0$, define the sets X_n as in the proof above (taking $r = ||w||^2$ here). Choose n with $1/n < \epsilon^2$, and take X' to be the finite set X_n. For any finite subset Y containing X_n, we know from the calculation in §15.14 that

$$\left\| w - \sum_{x \in Y} \langle w, x \rangle x \right\|^2 = ||w||^2 - \sum_{x \in Y} |\langle w, x \rangle|^2 \leq ||w||^2 - \sum_{x \in X_n} |\langle w, x \rangle|^2 < 1/n < \epsilon^2,$$

as needed.

15.16 Isomorphism of H and $\ell_2(X)$

Recall that a *vector space isomorphism* is a bijection $f : V \to W$ between F-vector spaces V and W such that $f(x + y) = f(x) + f(y)$ and $f(cx) = cf(x)$ for all $x, y \in V$ and all $c \in F$. Given two metric spaces X and Y, an *isometry* from X to Y is a function $f : X \to Y$ that preserves distances, meaning that $d_Y(f(u), f(v)) = d_X(u, v)$ for all $u, v \in X$. An isometry is necessarily one-to-one, since if $u, v \in X$ satisfy $f(u) = f(v)$, then $d_X(u, v) = d_Y(f(u), f(v)) = 0$ and hence $u = v$. Similarly, you can check that *an isometry must be continuous* (Exercise 43). Given Hilbert spaces H_1 and H_2, a *Hilbert space isomorphism* (also called an *isometric isomorphism*) is a bijection $f : H_1 \to H_2$ that is both a vector space isomorphism and an isometry.

A Hilbert space isomorphism preserves norms, meaning that $||f(u)|| = ||u||$ for all $u \in H_1$. This holds because $||f(u)|| = d(f(u), 0) = d(f(u), f(0)) = d(u, 0) = ||u||$. It follows from the Polarization Identity that a Hilbert space isomorphism preserves inner products, meaning that $\langle f(u), f(v) \rangle = \langle u, v \rangle$ for all $u, v \in H_1$. To see why, use (15.5) in H_2 and in H_1 to compute

$$\langle f(u), f(v) \rangle = \frac{1}{4}(||f(u) + f(v)||^2 - ||f(u) - f(v)||^2 + i||f(u) + if(v)||^2 - i||f(u) - if(v)||^2)$$

$$= \frac{1}{4}(||f(u + v)||^2 - ||f(u - v)||^2 + i||f(u + iv)||^2 - i||f(u - iv)||^2)$$

$$= \frac{1}{4}(||u + v||^2 - ||u - v||^2 + i||u + iv||^2 - i||u - iv||^2) = \langle u, v \rangle.$$

Conversely, if a given linear map $f : H_1 \to H_2$ preserves inner products, we see by taking $u = v$ that f preserves norms. By linearity, f must be an isometry.

Theorem Classifying Hilbert Spaces. Every Hilbert space H is isomorphic to a Hilbert space of the form $\ell_2(X)$. We can take X to be any maximal orthonormal subset of H.

Although we omit the proof, it is also true that for all sets X and Y, the Hilbert spaces $\ell_2(X)$ and $\ell_2(Y)$ are isometrically isomorphic if and only if $|X| = |Y|$ (meaning that there is a bijection from X onto Y).

Proof. Let H be any Hilbert space and X be any maximal orthonormal subset of H. Such subsets do exist, by Exercise 89. We define a map $f : H \to \ell_2(X)$ by letting $f(w) = f_w$ for all $w \in H$. Recall that $f_w : X \to \mathbb{C}$ gives the Fourier coefficients of w relative to X, namely $f_w(x) = \langle w, x \rangle$ for $x \in X$. Also recall that f_w does belong to $\ell_2(X)$, by Bessel's Inequality. We prove the theorem by showing that f is an isometric isomorphism.

First, we check $\mathbb{C}$-linearity of f. Fix $w, y \in H$ and $c \in \mathbb{C}$. On one hand, $f(w + y) = f_{w+y}$ is the function sending $x \in X$ to $\langle w + y, x \rangle = \langle w, x \rangle + \langle y, x \rangle$. On the other hand, $f(w) + f(y) = f_w + f_y$ is the function sending $x \in X$ to $f_w(x) + f_y(x) = \langle w, x \rangle + \langle y, x \rangle$. These functions are equal, so $f(w + y) = f(w) + f(y)$. Similarly, both functions $f(cw)$ and $cf(w)$ send each $x \in X$ to $\langle cw, x \rangle = c\langle w, x \rangle$, so $f(cw) = cf(w)$. Next, we observe that f is an isometry, since (15.7) says

$$||w|| = \sqrt{\sum_{x \in X} |\langle w, x \rangle|^2} = \sqrt{\sum_{x \in X} |f_w(x)|^2} = ||f_w||$$

for all $w \in H$. We deduce that f is one-to-one. It also follows that f preserves inner products, so that

$$\langle f_w, f_z \rangle = \sum_{x \in X} \langle w, x \rangle \overline{\langle z, x \rangle} = \langle w, z \rangle$$

for all $w, z \in H$. This formula is called *Parseval's Identity*.

The only thing left to prove is that f is surjective. Fix $g \in \ell_2(X)$; we must find $w \in H$ with $f(w) = g$. To do so, we first build a sequence of partial sums $g_n \in \ell_2(X)$ such that $\lim_{n \to \infty} g_n = g$. We know $r = \sum_{x \in X} |g(x)|^2 < \infty$. So, for each positive integer n, there is a finite subset X_n of X such that $r - 1/n < \sum_{x \in X_n} |g(x)|^2 \le r$. Define $g_n(x) = g(x)$ for $x \in X_n$, and $g_n(x) = 0$ for $x \in X \setminus X_n$. Evidently, $g_n \in \ell_2(X)$, and $||g - g_n||^2 = \sum_{x \in X \setminus X_n} |g(x)|^2 < 1/n$. It readily follows that $g_n \to g$ in the Hilbert space $\ell_2(X)$. In particular, the convergent sequence (g_n) is also a Cauchy sequence.

For each positive integer n, define $w_n \in H$ by $w_n = \sum_{x \in X_n} g(x)x$. By orthonormality of X, $\langle w_n, x \rangle = g(x) = g_n(x)$ for $x \in X_n$, and $\langle w_n, x \rangle = 0 = g_n(x)$ for $x \in X \setminus X_n$. Therefore, $f(w_n) = g_n$ for all $n > 0$. Since f is an isometry and (g_n) is a Cauchy sequence, (w_n) must also be a Cauchy sequence (Exercise 45). By completeness of H, w_n converges to some point $w \in H$. Now f is continuous (being an isometry), so $w_n \to w$ in H implies $g_n = f(w_n) \to f(w)$ in $\ell_2(X)$. On the other hand, by construction, $g_n \to g$ in $\ell_2(X)$. Since limits are unique, $g = f(w)$, as needed. $\qquad\square$

Compare the classification of Hilbert spaces in this section to the analogous classification of F-vector spaces based on dimension (cardinality of a basis). Note carefully that an infinite maximal orthonormal set X in a Hilbert space H is not the same as a basis B for the complex vector space H. The reason is that each $z \in H$ must be expressible as a finite linear combination of basis vectors in B, as opposed to the infinite linear combinations of the orthonormal vectors in X.

15.17 Continuous Linear Maps

We intend to study operators and linear functionals on Hilbert spaces. Before doing so, we establish some facts about continuous linear maps in the more general setting of normed vector spaces. Recall from Chapter 10 that a *normed vector space* consists of a real or complex vector space V and a norm function $|| \cdot || : V \to \mathbb{R}$ satisfying these axioms: for all $x, y \in V$ and all scalars c,
(a) $||x|| \in \mathbb{R}_{\geq 0}$; and $||x|| = 0$ iff $x = 0$;
(b) $||x + y|| \leq ||x|| + ||y||$ (the Triangle Inequality);
(c) $||cx|| = |c| \cdot ||x||$.

Any normed vector space is a metric space with distance function $d(x, y) = ||x - y||$ for $x, y \in V$. A *Banach space* is a normed vector space that is complete using this metric.

Consider a map $T : V \to W$ between two normed vector spaces with norms $|| \cdot ||_V$ and $|| \cdot ||_W$. Recall that T is *continuous* iff whenever $v_n \to v$ in V, $T(v_n) \to T(v)$ in W. To check the continuity of a linear map T, it suffices to check that for all sequences (z_n) of points in V such that $z_n \to 0_V$ in V, $T(z_n) \to 0_W$ in W. To prove this, suppose (v_n) is a sequence in V converging to $v \in V$. Then the sequence $(v_n - v)$ converges to $v - v = 0$, so the stated condition on T implies that $T(v_n - v)$ converges to 0_W. By linearity, $T(v_n) - T(v)$ converges to 0_W, so $T(v_n) \to T(v)$, and T is continuous.

A linear map $T : V \to W$ is called *bounded* iff there is a finite real constant $M \geq 0$ such that $||T(x)||_W \leq M||x||_V$ for all $x \in V$. We now prove that *a linear map T is continuous iff it is bounded*. On one hand, suppose T is bounded and $x_n \to 0$ in V. If $M = 0$, then $T(x) = 0_W$ for all $x \in V$, so T is continuous. Otherwise, given $\epsilon > 0$, choose n_0 so that $n \geq n_0$ implies $||x_n|| < \epsilon/M$. Then for $n \geq n_0$, $||T(x_n)||_W \leq M||x_n||_V < \epsilon$. So $T(x_n) \to 0$, and T is continuous. On the other hand, suppose T is not bounded. Then for each integer $n > 0$, there is $x_n \in V$ (necessarily nonzero) with $||T(x_n)||_W > n||x_n||_V$. By linearity of T and properties of norms, this inequality still holds if we replace x_n by any nonzero scalar multiple of itself. Picking an appropriate scalar multiple, we can arrange that $||x_n||_V = 1/n$ for each positive integer n. Then $x_n \to 0$ in V, but $||T(x_n)||_W > 1$ for all n, so $T(x_n)$ does not converge to zero in W, and T is not continuous.

Given two normed vector spaces V and W, let $B(V, W)$ be the set of all bounded (i.e., continuous) linear maps from V to W. You can check that $B(V, W)$ is a subspace of the vector space of all linear maps from V to W under pointwise operations on functions (see §4.2). We now show that $B(V, W)$ is itself a normed vector space using the *operator norm* defined by

$$||T|| = \sup\{||T(x)||_W : x \in V, ||x||_V = 1\}$$

for $T \in B(V, W)$. (In the special case $V = \{0\}$, define $||0_{B(V,W)}|| = 0$.) First, since the continuous map T is bounded, we know there is a finite upper bound M for the set in the definition, so that $||T||$ is a finite nonnegative real number. Second, if $||T|| = 0$, then $||T(x)||_W = 0$ for all $x \in V$ with $||x||_V = 1$. Multiplying by a scalar, it follows that $||T(y)||_W = 0$ for all $y \in V$, so $T(y) = 0_W$ for all $y \in V$, so T is the zero map, which is the zero element of $B(V, W)$. The axiom $||cT|| = |c| \cdot ||T||$ follows readily from the fact that $||(cT)(x)||_W = ||c(T(x))||_W = |c| \cdot ||T(x)||_W$. Finally, for $S, T \in B(V, W)$ and any $x \in V$ of norm 1,

$$||(S+T)(x)||_W = ||S(x) + T(x)||_W \leq ||S(x)||_W + ||T(x)||_W \leq ||S|| + ||T||,$$

so that $||S + T|| \leq ||S|| + ||T||$.

You can check the following additional properties of the operator norm. Here, V, W, Z are normed linear spaces and $T \in B(V, W)$, $U \in B(W, Z)$ are arbitrary.

(a) $||T(y)||_W \leq ||T|| \cdot ||y||_V$ for all $y \in V$.

(b) $||T|| = \sup\{||T(x)||_W / ||x||_V : x \in V, x \neq 0\}$.

(c) $||T|| = \inf\{M \in \mathbb{R}_{>0} : ||T(y)||_W \leq M||y||_V \text{ for all } y \in V\}$.

(d) $U \circ T \in B(V, Z)$ and $||U \circ T|| \leq ||U|| \cdot ||T||$.

We conclude this section by showing that *if W is a Banach space, then $B(V, W)$ is a Banach space.* In other words, completeness of the metric space W implies completeness of $B(V, W)$. Let $(T_n : n \geq 0)$ be a Cauchy sequence in $B(V, W)$. For each fixed $x \in V$, $||T_n(x) - T_m(x)||_W = ||(T_n - T_m)(x)||_W \leq ||T_n - T_m|| \cdot ||x||_V$. It follows that $(T_n(x))$ is a Cauchy sequence in W for each fixed $x \in V$. By completeness of W, for each $x \in V$ there exists a unique $y \in W$ (denoted $T(x)$) so that $\lim_{n \to \infty} T_n(x) = y = T(x)$.

We now have a function $T : V \to W$; we must check that T is linear, T is bounded, and $T_n \to T$ in $B(V, W)$. For linearity, fix $x, z \in V$ and a scalar c. By continuity and linearity of each T_n,

$$T(x + z) = \lim_n T_n(x + z) = \lim_n [T_n(x) + T_n(z)] = \lim_n T_n(x) + \lim_n T_n(z) = T(x) + T(z);$$

$$T(cx) = \lim_n T_n(cx) = \lim_n cT_n(x) = c \lim_n T_n(x) = cT(x).$$

Next, since (T_n) is a Cauchy sequence, $\{T_n : n \geq 0\}$ is a bounded subset of $B(V, W)$. So there exists a finite constant $M \in \mathbb{R}_{>0}$ with $||T_n|| \leq M$ for all $n \geq 0$. For any $x \in V$, $||T(x)||_W = ||\lim_n T_n(x)||_W = \lim_n ||T_n(x)||_W$, where $||T_n(x)||_W \leq ||T_n|| \cdot ||x||_V \leq M||x||_V$ for all n. So $||T(x)||_W \leq M||x||_V$ for all $x \in V$, proving that T is bounded. Finally, we show $T_n \to T$ in $B(V, W)$. Given $\epsilon > 0$, we find n_0 so that $n \geq n_0$ implies $||T_n - T|| < \epsilon$. Since (T_n) is a Cauchy sequence, we can choose n_0 so that $m, n \geq n_0$ implies $||T_n - T_m|| < \epsilon/4$. Next, given $x \in V$ with $||x||_V = 1$, choose $m \geq n_0$ (depending on x) so that $||T_m(x) - T(x)||_W < \epsilon/4$. Now, for $n \geq n_0$,

$$||(T_n - T)(x)||_W = ||T_n(x) - T_m(x) + T_m(x) - T(x)||_W$$
$$\leq ||(T_n - T_m)(x)||_W + ||T_m(x) - T(x)||_W < \epsilon/4 + \epsilon/4 = \epsilon/2.$$

Note that n_0 is independent of x here, so we see that $||T_n - T|| \leq \epsilon/2 < \epsilon$, completing the proof.

15.18   Dual Space of a Hilbert Space

In Chapter 13, we studied the *dual space* V^* of an F-vector space V, which is the vector space of all linear maps from V to F. For finite-dimensional V, we saw that V and V^* are isomorphic. In the case of a finite-dimensional real inner product space V, we could realize this isomorphism by mapping $y \in V$ to the linear functional $R_y \in V^*$ given by $R_y(x) = \langle x, y \rangle$ for $x \in V$. Similar results hold for finite-dimensional complex inner product spaces, but here we must distinguish between linear maps and semi-linear maps.

Now let H be a Hilbert space. We define the *dual space* H^* to be the set $B(H, \mathbb{C})$ of all continuous linear maps from H to $\mathbb{C}$. A function $f : H \to \mathbb{C}$ is in H^* iff for all $x, y, z_n, z \in H$ and all $c \in \mathbb{C}$, $f(x+y) = f(x) + f(y)$, $f(cx) = cf(x)$, and $z_n \to z$ in H implies $f(z_n) \to f(z)$ in $\mathbb{C}$. For example, given $y \in H$, consider the map $R_y : H \to \mathbb{C}$ defined by $R_y(x) = \langle x, y \rangle$ for all $x \in H$. As noted in §15.13, each R_y is linear and continuous, so $R_y \in H^*$ for all $y \in H$.

Since $\mathbb{C}$ is complete, it follows from the theorem proved in §15.17 that $H^* = B(H, \mathbb{C})$ is a Banach space with norm $||f|| = \sup\{|f(x)| : x \in H, ||x|| = 1\}$ for $f \in H^*$. Our goal here is to define a semi-linear bijective isometry $R : H \to H^*$, which shows that the Hilbert space H and the normed vector space H^* are essentially isomorphic (up to conjugation of scalars). We can use this semi-isomorphism to define an inner product on H^* that makes H^* a Hilbert space.

The map $R : H \to H^*$ is given by $R(y) = R_y$ for all $y \in H$, where $R_y(x) = \langle x, y \rangle$ for all $x \in H$. You can check from the inner product axioms that R is a semi-linear map, meaning $R(y + y') = R(y) + R(y')$ and $R(cy) = \overline{c}R(y)$ for all $y, y' \in H$ and all $c \in \mathbb{C}$. Next, we prove that R is a bijection; in other words, *for every $f \in H^*$, there exists a unique $y \in H$ with $f = R_y$*. Fix $f \in H^*$. To prove uniqueness of y, assume $f = R_w = R_y$ for some $w, y \in H$; we show $w = y$. Compute

$$||w - y||^2 = \langle w - y, w - y \rangle = \langle w - y, w \rangle - \langle w - y, y \rangle = R_w(w - y) - R_y(w - y) = 0,$$

so $w - y = 0$ and $w = y$. To prove existence of y, consider the null space W of f, namely $W = \{x \in H : f(x) = 0\} = f^{-1}[\{0\}]$. W is a subspace of H, because it is the kernel of the linear map f. W is closed, because it is the preimage of the closed set $\{0\}$ under the continuous map f. So we can write $H = W \oplus W^\perp$. If $W = H$, then f is the zero map, and we may take $y = 0$. If $W \neq H$, then there exists a nonzero $z \in W^\perp$. Using the Fundamental Homomorphism Theorem for Vector Spaces, we see that $W^\perp$ is the 1-dimensional subspace spanned by z. Let $y = cz$, where $c = ||z||^{-2}\overline{f(z)}$. We claim $f = R_y$. To prove this, fix $x \in H$, and write $x = w + dz$ for some $w \in W$ and $d \in \mathbb{C}$. On one hand, $f(x) = f(w + dz) = f(w) + df(z) = df(z)$. On the other hand, $R_y(x) = \langle w + dz, cz \rangle = \langle w, cz \rangle + \langle dz, cz \rangle = 0 + d\overline{c}||z||^2 = df(z)$. So the claim holds, and R is a bijection.

To finish, we check that R is an isometry. Fix $x, y \in H$ with $||x|| = 1$ and $y \neq 0$. On one hand, by the Cauchy–Schwarz Inequality,

$$|R_y(x)| = |\langle x, y \rangle| \leq ||y|| \cdot ||x|| = ||y||,$$

so that $||R_y|| \leq ||y||$. On the other hand, letting $u = ||y||^{-1}y$, we have $||u|| = 1$ and

$$|R_y(u)| = |\langle y, y \rangle|/||y|| = ||y||,$$

so that $||R_y|| \geq ||y||$. Thus, $||R_y|| = ||y||$, and this equality also holds for $y = 0$. So R is indeed an isometry. You can now check that H^* is a Hilbert space with inner product $\langle R_x, R_y \rangle = \langle y, x \rangle$ for $x, y \in H$; we reverse the order of x and y because R is semi-linear.

15.19   Adjoints

Next, we discuss adjoints of operators on Hilbert spaces. We can approach this topic through the dual space $H^\star$ (as we did in §13.11), or as follows. An *operator* on a Hilbert space H is a continuous linear map $T : H \to H$. We write $B(H) = B(H, H)$ for the set of all such operators. We have seen that $B(H)$ is a Banach space (complete normed vector space). Given any $T \in B(H)$, we claim there exists a unique operator T^* on H, called the *adjoint of T*, such that $\langle T(x), y \rangle = \langle x, T^*(y) \rangle$ for all $x, y \in H$. Fix $y \in H$. The map f sending $x \in H$ to $\langle T(x), y \rangle$ is $\mathbb{C}$-linear, as you can check. It is also continuous, since $x_n \to x$ implies $T(x_n) \to T(x)$, hence $f(x_n) = \langle T(x_n), y \rangle \to \langle T(x), y \rangle = f(x)$. As we saw in §15.18, there exists a unique $w \in H$ with $f(x) = R_w(x) = \langle x, w \rangle$ for all $x \in H$. Writing $T^*(y) = w$ for each y, we obtain a unique function $T^* : H \to H$ satisfying $\langle T(x), y \rangle = \langle x, T^*(y) \rangle$ for all $x, y \in H$.

Recall that the map sending $u \in H$ to R_u is one-to-one. It follows that, for $u, v \in H$, we can show that $u = v$ by checking $\langle x, u \rangle = \langle x, v \rangle$ for all $x \in H$. We use this to prove that T^* is a linear map. Given $y, z \in H$, we show $T^*(y + z) = T^*(y) + T^*(z)$ by fixing $x \in H$ and computing

$$\langle x, T^*(y + z) \rangle = \langle T(x), y + z \rangle = \langle T(x), y \rangle + \langle T(x), z \rangle = \langle x, T^*(y) \rangle + \langle x, T^*(z) \rangle$$
$$= \langle x, T^*(y) + T^*(z) \rangle.$$

Similarly, for $y \in H$ and $c \in \mathbb{C}$, $T^*(cy) = cT^*(y)$ holds because for each $x \in H$,

$$\langle x, T^*(cy) \rangle = \langle T(x), cy \rangle = \bar{c}\langle T(x), y \rangle = \bar{c}\langle x, T^*(y) \rangle = \langle x, cT^*(y) \rangle.$$

To finish, we check continuity of T^*. Fix $y \in H$, and use the Cauchy–Schwarz Inequality and the boundedness of T to compute

$$||T^*(y)||^2 = \langle T^*(y), T^*(y) \rangle = \langle T(T^*(y)), y \rangle \leq ||T(T^*(y))|| \cdot ||y|| \leq ||T|| \cdot ||T^*(y)|| \cdot ||y||.$$

If $||T^*(y)|| > 0$, we divide by $||T^*(y)||$ to see that $||T^*(y)|| \leq ||T|| \cdot ||y||$; and this inequality also holds if $||T^*(y)|| = 0$. It follows that T^* is bounded, hence continuous. In fact, the proof shows that $||T^*|| \leq ||T||$.

Here are some properties of adjoint operators: for all $S, T \in B(H)$ and all $c \in \mathbb{C}$, $(S + T)^* = S^* + T^*$, $(cS)^* = \bar{c}(S^*)$, $S^{**} = S$, $(S \circ T)^* = T^* \circ S^*$, and $||T^*|| = ||T||$. We can prove these by the method used above. For instance, $S^{**} = S$ since for all $x, y \in H$,

$$\langle x, S^{**}(y) \rangle = \langle S^*(x), y \rangle = \overline{\langle y, S^*(x) \rangle} = \overline{\langle S(y), x \rangle} = \langle x, S(y) \rangle.$$

Similarly, $(S \circ T)^* = (T^* \circ S^*)$ because for all $x, y \in H$,

$$\langle x, (S \circ T)^*(y) \rangle = \langle (S \circ T)(x), y \rangle = \langle S(T(x)), y \rangle = \langle T(x), S^*(y) \rangle = \langle x, T^*(S^*(y)) \rangle$$
$$= \langle x, (T^* \circ S^*)(y) \rangle.$$

We showed earlier that $||T^*|| \leq ||T||$ for all $T \in B(H)$. Replacing T by T^* gives $||T|| = ||T^{**}|| \leq ||T^*||$, so $||T^*|| = ||T||$. We leave semi-linearity of the map $S \mapsto S^*$ as an exercise.

Now that we have the concept of an adjoint operator, we can generalize the special types of matrices studied in Chapter 7 to the setting of Hilbert spaces. An operator T on a Hilbert space H is called *self-adjoint* iff $T^* = T$; *positive* iff $T^* = T$ and $\langle T(x), x \rangle \in \mathbb{R}_{\geq 0}$ for all $x \in H$; *normal* iff $T \circ T^* = T^* \circ T$; and *unitary* iff $T \circ T^* = \mathrm{id}_H = T^* \circ T$. We indicate some basic properties of these operators in the exercises, but we must refer the reader to more advanced texts for a detailed account of the structure theory of normal operators on Hilbert spaces.

15.20 Summary

1. *Definitions for Metric Spaces.* Table 15.1 summarizes definitions of concepts related to metric spaces, sequences, and continuous functions.

2. *Examples of Metric Spaces.* $\mathbb{R}^m$ and $\mathbb{C}^m$ are metric spaces with the Euclidean metric $d_2(x, y) = \sqrt{\sum_{k=1}^m |x_k - y_k|^2}$. Any set X has the discrete metric defined by $d(x, y) = 0$ for $x = y$, $d(x, y) = 1$ for $x \neq y$. A product $X = X_1 \times \cdots \times X_m$ of metric spaces X_k has metrics $d_1(x, y) = \sum_{k=1}^m d_{X_k}(x_k, y_k)$ and $d_\infty(x, y) = \max\{d_{X_k}(x_k, y_k) : 1 \leq k \leq m\}$.

3. *Convergent Sequences and Cauchy Sequences.* The limit of a convergent sequence is unique. Convergent sequences are Cauchy sequences; the converse holds in complete spaces. Convergent sequences and Cauchy sequences are bounded. Every subsequence of a convergent sequence converges to the same limit as the full sequence. If one subsequence of a Cauchy sequence converges, then the full sequence converges to the same limit.

4. *Closed Sets.* In any metric space X, $\emptyset$ and X are closed. Finite subsets of X are closed. The union of finitely many closed sets is closed. The intersection of arbitrarily many closed sets is closed. C is closed iff $X \setminus C$ is open. Closed intervals $[a, b]$ in $\mathbb{R}$ are closed sets. If $C_1, \ldots, C_m$ are closed in $X_1, \ldots, X_m$ respectively, then $C_1 \times \cdots \times C_m$ is closed in the product metric space $X_1 \times \cdots \times X_m$ (with metric d_1 or d_∞).

5. *Open Sets.* In any metric space X, $\emptyset$ and X are open. The union of arbitrarily many open sets is open. The intersection of finitely many open sets is open. U is open iff $X \setminus U$ is closed. Open balls $B(x; r)$ are open sets. Open intervals (a, b) in $\mathbb{R}$ are open sets, and every open set in $\mathbb{R}$ is a disjoint union of countably many open

TABLE 15.1
Definitions of concepts for metric spaces.

Concept	Definition
metric space (X, d)	$\forall x, y \in X, 0 \leq d(x, y) < \infty; \ d(x, y) = 0 \Leftrightarrow x = y$ $\forall x, y \in X, d(x, y) = d(y, x)$ (symmetry) $\forall x, y, z \in X, d(x, z) \leq d(x, y) + d(y, z)$ (Triangle Ineq.)
convergent sequence (x_n)	$x_n \to x$ means $\forall \epsilon > 0, \exists n_0, \forall n \geq n_0, d(x_n, x) < \epsilon$.
Cauchy sequence (x_n)	$\forall \epsilon > 0, \exists n_0, \forall m, n \geq n_0, d(x_n, x_m) < \epsilon$.
closed set C	If $x_n \to x$ and all $x_n \in C$, then $x \in C$.
open set U	$\forall x \in U, \exists r > 0, \forall y \in X, d(x, y) < r \Rightarrow y \in U$.
bounded set S	$\exists x \in X, \exists M \in \mathbb{R}, \forall y \in S, d(x, y) \leq M$.
totally bounded set S (Exc. 62)	$\forall \epsilon > 0, \exists m < \infty, \exists x_1, \ldots, x_m \in X, S \subseteq \bigcup_{k=1}^m B(x_k; \epsilon)$.
sequentially compact set K	Any sequence (x_n) in K has a subsequence converging to a point of K.
topologically compact set K	Whenever $K \subseteq \bigcup_{i \in I} U_i$ with all U_i open, there is a finite $F \subseteq I$ with $K \subseteq \bigcup_{i \in F} U_i$.
complete set K	Every Cauchy sequence in K converges to a point of K.
continuous $f : X \to Y$	Whenever $x_n \to x$ in X, $f(x_n) \to f(x)$ in Y.

intervals. If $U_1, \ldots, U_m$ are open in $X_1, \ldots, X_m$ respectively, then $U_1 \times \cdots \times U_m$ is open in the product metric space $X_1 \times \cdots \times X_m$ (with metric d_1 or d_∞).

6. *Continuous Functions.* The continuity of $f : X \to Y$ is equivalent to each of these conditions: $x_n \to x$ in X implies $f(x_n) \to f(x)$ in Y; for all open $V \subseteq Y$, $f^{-1}[V]$ is open in X; for all closed $D \subseteq Y$, $f^{-1}[D]$ is closed in X; for all $\epsilon > 0$ and $x \in X$, there exists $\delta > 0$ such that for all $z \in X$ with $d_X(x, z) < \delta$, $d_Y(f(x), f(z)) < \epsilon$. If K is compact in X and f is continuous, then $f[K]$ is compact in Y. The image of an open (or closed) subset of X under a continuous map need not be open (or closed) in Y.

7. *Compact Sets.* A subset K of a metric space X is sequentially compact iff every sequence of points in K has a subsequence converging to a point of K. In metric spaces, sequential compactness is equivalent to topological compactness (every open cover of K has a finite subcover). Finite subsets of X are compact. Finite unions and arbitrary intersections of compact subsets are compact. Compact subsets of X must be closed in X and bounded, but the converse does not hold in general. However, in $\mathbb{R}^m$ and $\mathbb{C}^m$ with the metrics d_1, d_2, and d_∞, a subset K is compact iff K is closed and bounded. A closed subset of a compact set is compact. The image of a compact set under a continuous function is compact; but the preimage of a compact set may not be compact. K is compact iff K is complete and totally bounded.

8. *Complete Spaces.* A metric space X is complete iff every Cauchy sequence in X converges to a point of X. Compactness implies completeness, but not conversely. $\mathbb{R}^m$ and $\mathbb{C}^m$ with the metrics d_1, d_2, and d_∞ are complete but not compact. A subset S of a complete space is complete iff S is closed.

9. *Hilbert Spaces.* A Hilbert space is a complex inner product space, with norm $\|x\| = \sqrt{\langle x, x \rangle}$ and metric $d(x, y) = \|x - y\|$, that is complete as a metric space. $\mathbb{C}^m$ is an m-dimensional Hilbert space. For any set X, the set $\ell_2(X)$ of functions $f : X \to \mathbb{C}$ with $\sum_{x \in X} |f(x)|^2 < \infty$ is a Hilbert space with inner product $\langle f, g \rangle = \sum_{x \in X} f(x)\overline{g(x)}$ and norm $\|f\| = \sqrt{\sum_{x \in X} |f(x)|^2}$. The set of integrable functions on a measure space (X, μ) is a Hilbert space with inner product $\langle f, g \rangle = \int_X f(x)\overline{g(x)} \, d\mu$ and norm $\|f\| = \sqrt{\int_X |f(x)|^2 \, d\mu}$.

10. *Properties of Hilbert Spaces.* All x, y in a Hilbert space H satisfy:
    (a) *Cauchy–Schwarz Inequality*: $|\langle x, y \rangle| \leq \|x\| \cdot \|y\|$.
    (b) *Parallelogram Law*: $\|x + y\|^2 + \|x - y\|^2 = 2\|x\|^2 + 2\|y\|^2$.
    (c) *Polarization Identity*: $\|x+y\|^2 - \|x-y\|^2 + i\|x+iy\|^2 - i\|x-iy\|^2 = 4\langle x, y \rangle$.
    (d) *Pythagorean Theorem*: If $\langle x, y \rangle = 0$, then $\|x \pm y\|^2 = \|x\|^2 + \|y\|^2$.
    If $x_1, \ldots, x_m \in H$ is an orthogonal list ($\langle x_i, x_j \rangle = 0$ for all $i \neq j$) and $c_1, \ldots, c_m \in \mathbb{C}$, then $\|c_1 x_1 + \cdots + c_m x_m\|^2 = |c_1|^2 \|x_1\|^2 + \cdots + |c_m|^2 \|x_m\|^2$. The Hilbert space operations are continuous: if $x_n \to x$ and $y_n \to y$ in H and $c_n \to c$ in $\mathbb{C}$, then $x_n + y_n \to x + y$, $c_n x_n \to cx$, $\langle x_n, y_n \rangle \to \langle x, y \rangle$, and $\|x_n\| \to \|x\|$.

11. *Closed Convex Sets.* For any nonempty closed convex subset C of a Hilbert space H, C has a unique element of minimum norm. For each $w \in H$, there exists a unique point of C closest to w.

12. *Orthogonal Complements.* For each subset S of a Hilbert space H, the orthogonal complement $S^\perp = \{v \in H : \langle v, w \rangle = 0 \text{ for all } w \in S\}$ is a closed subspace of H. The map $W \mapsto W^\perp$ is an order-reversing bijection on the lattice of all closed subspaces of H. For each closed subspace W, $H = W \oplus W^\perp$, $W^{\perp\perp} = W$, and in the unique expression of $x \in H$ as a sum of $y \in W$ and $z \in W^\perp$, y (resp. z) is

the closest point to x in W (resp. $W^\perp$). For any subset S, $S^{\perp\perp}$ is the smallest closed subspace containing S.

13. *Orthonormal Sets.* A subset X of a Hilbert space H is orthonormal iff $\langle x, x \rangle = 1$ and $\langle x, y \rangle = 0$ for all $x \neq y$ in X. For X orthonormal and $w \in H$, Bessel's Inequality states that $\sum_{x \in X} |\langle w, x \rangle|^2 \leq ||w||^2$. Equality holds for all $w \in H$ iff X is a maximal orthonormal set.

14. *The Isomorphism $H \cong \ell_2(X)$.* By Zorn's Lemma, every Hilbert space H has a maximal orthonormal subset X. The map $f : H \to \ell_2(X)$ sending $w \in H$ to the function $f_w : X \to \mathbb{C}$, given by $f_w(x) = \langle w, x \rangle$ for $x \in X$, is a bijective continuous linear isometry (Hilbert space isomorphism). So $\sum_{x \in X} |\langle w, x \rangle|^2 = ||w||^2$ and $\sum_{x \in X} \langle w, x \rangle \overline{\langle z, x \rangle} = \langle w, z \rangle$ for all $w, z \in H$ (Parseval's Identity).

15. *Continuous Linear Maps.* A linear map $T : V \to W$ between two normed vector spaces is continuous ($v_n \to v$ in V implies $T(v_n) \to T(v)$ in W) iff T is continuous at zero ($v_n \to 0$ in V implies $T(v_n) \to 0$ in W) iff T is bounded (for some finite M, $||T(v)|| \leq M||v||$ for all $v \in V$). The set $B(V, W)$ of continuous linear maps from V to W is a normed vector space with norm $||T|| = \sup\{||T(x)|| : x \in V, ||x|| = 1\}$. If W is complete, then $B(V, W)$ is complete.

16. *Dual of a Hilbert Space.* Given a Hilbert space H, the dual space $H^\star$ is $B(H, \mathbb{C})$, the set of bounded (continuous) linear maps from H to $\mathbb{C}$. For every $f \in H^\star$, there exists a unique $y \in H$ such that $f = R_y$, where $R_y(x) = \langle x, y \rangle$ for $x \in H$. The map $R : H \to H^\star$ is a bijective semi-linear isometry. So, every Hilbert space is semi-isomorphic to its dual space.

17. *Adjoint Operators.* For each operator $T \in B(H) = B(H, H)$, there exists a unique adjoint operator $T^* \in B(H)$ satisfying $\langle T(x), y \rangle = \langle x, T^*(y) \rangle$ for all $x, y \in H$. For $S, T \in B(H)$ and $c \in \mathbb{C}$, $(S + T)^* = S^* + T^*$, $(cS)^* = \bar{c}(S^*)$, $S^{**} = S$, $(S \circ T)^* = T^* \circ S^*$, and $||T^*|| = ||T||$. The operator T is *self-adjoint* iff $T = T^*$; *positive* iff $T = T^*$ and $\langle T(x), x \rangle \geq 0$ for all $x \in H$; *normal* iff $T \circ T^* = T^* \circ T$; and *unitary* iff $T \circ T^* = \mathrm{id}_H = T^* \circ T$.

15.21   Exercises

Unless otherwise specified, assume X and Y are arbitrary metric spaces and H is an arbitrary Hilbert space in these exercises.

1. Which of the following functions define metrics on $\mathbb{R}^2$? Explain.
   (a) $d((x, y), (u, v)) = |x - u|$.
   (b) $d((x, y), (u, v)) = (x - u)^2 + (y - v)^2$.
   (c) $d((x, y), (u, v)) = \sqrt[3]{(x - u)^3 + (y - v)^3}$.
   (d) $d((x, y), (u, v)) = |x| + |y|$ if $(x, y) \neq (u, v)$, and 0 otherwise.
   (e) $d((x, y), (u, v)) = 3$ if $(x, y) \neq (u, v)$, and 0 otherwise.

2. Let X be the set of all bounded functions $f : [0, 1] \to \mathbb{R}$. For $f, g \in X$, let $d(f, g) = \sup\{|f(x) - g(x)| : x \in [0, 1]\}$ Prove d is a metric on X. Find $d(x^2, x^3)$.

3. Let Y be the set of all continuous functions $f : [0, 1] \to \mathbb{R}$. For $f, g \in Y$, let $d(f, g) = \int_0^1 |f(x) - g(x)|\, dx$. Prove d is a metric on Y. Find $d(x^2, x^3)$. Is d a metric on the set Z of all Riemann-integrable functions $f : [0, 1] \to \mathbb{R}$?

4. Let p be a fixed prime integer. Define $d_p : \mathbb{Q} \times \mathbb{Q} \to \mathbb{R}$ by letting $d_p(x,y) = 0$ if $x = y$, $d_p(x,y) = 1/p^k$ if $x - y$ is a nonzero integer and p^k is the largest power of p dividing $x - y$, and $d_p(x,y) = 1$ otherwise. Prove: for all $x, y, z \in \mathbb{Q}$, $d_p(x,z) \leq \max(d_p(x,y), d_p(y,z))$. Show that $(\mathbb{Q}, d_p)$ is a metric space.

5. **Equivalent Metrics.** We say that two metrics d and d' defined on the same set X are *equivalent* iff they have the same open sets, i.e., for all $U \subseteq X$, U is open in the metric space (X, d) iff U is open in the metric space (X, d'). Prove that equivalent metrics have the same closed sets, the same compact sets, and the same convergent sequences. Prove that $f : X \to Y$ is continuous relative to d iff f is continuous relative to d'. Show that equivalence of metrics is an equivalence relation on the set of all metrics on a fixed set X.

6. Let (X, d) be a metric space. Fix $m \in \mathbb{R}_{>0}$, and define $d' : X \times X \to \mathbb{R}$ by $d'(x,y) = \min(m, d(x,y))$ for $x, y \in X$. Prove d' is a metric on X. Prove d' is equivalent to d (see Exercise 5). Show that all subsets of X are bounded relative to d'. Construct an example of a closed, bounded, non-compact subset of a metric space.

7. **Comparable Metrics.** We say that two metrics d and d' defined on the same set X are *comparable* iff there exist finite positive constants M, N such that for all $x, y \in X$, $d(x,y) \leq M d'(x,y)$ and $d'(x,y) \leq N d(x,y)$.
   (a) Prove: if d and d' are comparable, then they are equivalent (Exercise 5).
   (b) Prove d_2 and d_∞ are comparable metrics on $\mathbb{R}^n$ and $\mathbb{C}^n$.
   (c) Prove d_1 and d_∞ are comparable metrics on $\mathbb{R}^n$ and $\mathbb{C}^n$.
   (d) Prove comparability is an equivalence relation on the set of metrics for X.
   (e) Give an example of two equivalent metrics on $\mathbb{R}$ that are not comparable.

8. Let X have metric d, Z be any set, and $f : X \to Z$ be a bijection. Show there exists a unique metric on Z such that f is an isometry.

9. Give an example of two equivalent metrics on $\mathbb{R}$ (Exercise 5) such that $\mathbb{R}$ is complete with respect to one of the metrics but not the other. (Study $f : (-\pi/2, \pi/2) \to \mathbb{R}$ given by $f(x) = \tan x$.)

10. Given two comparable metrics d_1 and d_2 on a set X (Exercise 7), show that (X, d_1) is complete iff (X, d_2) is complete.

11. Verify the metric space axioms for d', d_1, and d_∞ (defined in §15.1).

12. Given metric spaces $X_1, \ldots, X_m$, prove the metrics d_1 and d_∞ on $X_1 \times \cdots \times X_m$ are comparable.

13. Negate the definition of convergent sequence to obtain the formal definition of a non-convergent sequence. Use this to prove carefully that the sequence $(n : n \geq 0)$ is non-convergent in the metric space $\mathbb{R}$.

14. A sequence $(x_n : n \geq 0)$ is called *eventually constant* iff there exists an integer n_0 such that for all $n \geq n_0$, $x_n = x_{n_0}$. Prove every eventually constant sequence in (X, d) converges to x_{n_0}. If d is the discrete metric on X, prove that every convergent sequence is eventually constant.

15. Fix $k \in \mathbb{Z}_{\geq 0}$. Show that a sequence $(x_n : n \geq 0)$ converges to x iff the *tail sequence* $(x_{k+n} : n \geq 0)$ converges to x. Show that $(x_n : n \geq 0)$ is Cauchy iff $(x_{k+n} : n \geq 0)$ is Cauchy. Show that $(x_n : n \geq 0)$ is bounded iff $(x_{k+n} : n \geq 0)$ is bounded.

16. Prove: if $(x_n : n \geq 0)$ converges in X, then $\{x_n : n \geq 0\}$ is bounded.

17. Prove: if $(x_n : n \geq 0)$ is a Cauchy sequence, then $\{x_n : n \geq 0\}$ is bounded.

18. Let $f : X \to Y$ be a continuous map between metric spaces and (x_n) be a sequence in X. If (x_n) is convergent, or Cauchy, or bounded, must the same be true of the sequence $(f(x_n))$? If $(f(x_n))$ is convergent, or Cauchy, or bounded, must the same be true of (x_n)?

19. Let $(x_n) = ((x_n(1), x_n(2), \ldots, x_n(m)) : n \geq 0)$ be a sequence in a product metric space $X = X_1 \times X_2 \times \cdots \times X_m$ with metric d_1. Prove that (x_n) converges to $y = (y_1, \ldots, y_m)$ iff $(x_n(k))$ converges to y_k for all k between 1 and m. Prove (x_n) is Cauchy iff $(x_n(k))$ is Cauchy for all k between 1 and m.

20. Repeat Exercise 19 using the metric d_∞ on X.

21. Repeat Exercise 19 taking X to be $\mathbb{R}^m$ or $\mathbb{C}^m$ using the Euclidean metric d_2.

22. Let $(x_n : n \geq 0)$ be a sequence in (X, d). Show that the set S of all limits of subsequences of $(x_n : n \geq 0)$ is a closed subset of X.

23. Give an example of a sequence $(x_n : n \geq 0)$ in $\mathbb{R}$ such that every $k \in \mathbb{Z}$ is a limit of some subsequence of (x_n).

24. Give an example of a sequence $(x_n : n \geq 0)$ in $\mathbb{R}$ such that every $r \in [0, 1]$ is a limit of some subsequence of (x_n).

25. Show that every subset of a discrete metric space is both open and closed.

26. Decide (with proof) whether each subset of $(\mathbb{R}^2, d_2)$ is closed.
    (a) $\{(x, y) \in \mathbb{R}^2 : y \geq 0\}$ (b) $\{(x, y) \in \mathbb{R}^2 : x^2 + y^2 < 1\}$ (c) $\mathbb{Z} \times \mathbb{Z}$
    (d) the graph $\{(x, f(x)) : x \in \mathbb{R}\}$, where $f : \mathbb{R} \to \mathbb{R}$ is continuous

27. Let $Z \subseteq Y \subseteq X$, where the metric on Y is the restriction of the metric on X.
    (a) Prove: if Z is closed in Y and Y is closed in X, then Z is closed in X.
    (b) Give an example to show that (a) can fail if Y is not closed in X.
    (c) Prove: if Z is open in Y and Y is open in X, then Z is open in X.
    (d) Give an example to show that (c) can fail if Y is not open in X.

28. **Closure of a Set.** Given a subset S of a metric space (X, d), the *closure of S in X*, denoted $\overline{S}$, is the intersection of all closed subsets of X containing S. Show that $S \subseteq \overline{S}$, $\overline{S}$ is a closed set, and for any closed set T, if $S \subseteq T$ then $\overline{S} \subseteq T$. Show that $\overline{S}$ equals the set S' of all $y \in X$ such that there exists a sequence (x_n) with all $x_n \in S$ and $\lim_{n \to \infty} x_n = y$. (First show S' is closed.)

29. In $(\mathbb{R}, d_2)$, find the closure of these sets: the interval (a, b), $\mathbb{Q}$, $\mathbb{R}$, and $\mathbb{Z}$.

30. **Boundary of a Set.** Given a subset S of the metric space (X, d), the *boundary of S* is the set $\mathrm{Bdy}(S)$ consisting of all $x \in X$ such that for all $\epsilon > 0$, $B(x; \epsilon)$ contains at least one point in S and at least one point not in S. Prove $\mathrm{Bdy}(S) = \overline{S} \cap \overline{X \setminus S}$ (see Exercise 28). Prove $\mathrm{Bdy}(S)$ is a closed set.

31. In $(\mathbb{R}, d_2)$, find the boundary of these sets: the interval (a, b), $\mathbb{Q}$, $\mathbb{R}$, and $\mathbb{Z}$.

32. In $(\mathbb{R}^2, d_2)$, find the boundary of $\mathbb{R} \times \{0\}$, $\{(x, y) : x^2 + y^2 < 1\}$, and $[0, 1] \times \mathbb{Q}$.

33. Consider a product metric space $X = X_1 \times \cdots \times X_m$ with metric d_1.
    (a) Prove: if S_i is closed in X_i for all i, then $S_1 \times \cdots \times S_m$ is closed in X.
    (b) Prove: if S_i is open in X_i for all i, then $S_1 \times \cdots \times S_m$ is open in X.
    (c) Prove: if S_i is compact in X_i for all i, then $S_1 \times \cdots \times S_m$ is compact in X.

34. Which results in the previous exercise are true if we use the metric d_∞ on X? Which results hold when X is $(\mathbb{R}^m, d_2)$ and each X_i is $(\mathbb{R}, d_2)$?

35. Let W be a subspace of a normed vector space V. Show that the closure $\overline{W}$ (Exercise 28) is a closed subspace of V.

36. Sketch the open ball $B((2,1);1)$ in $\mathbb{R}^2$ for each of the metrics d_1, d_2, and d_∞.

37. *Interior of a Set.* Given a subset S of the metric space (X,d), the *interior of S in X*, denoted $\text{Int}(S)$ or S°, is the union of all open subsets of X contained in S. Show that $\text{Int}(S) \subseteq S$, $\text{Int}(S)$ is an open set, and for any open set U, if $U \subseteq S$ then $U \subseteq \text{Int}(S)$. Show that $\text{Int}(S) = X \sim \overline{X \sim S}$.

38. In $(\mathbb{R}, d_2)$, give an example of a set S with $\text{Int}(S) = \emptyset$ and $\overline{S} = \mathbb{R}$.

39. For $x \in X$ and $r > 0$, define the *closed ball* $B[x;r] = \{y \in X : d(x,y) \le r\}$.
    (a) Prove $B[x;r]$ is a closed subset of X.
    (b) Show that open and closed balls in normed vector spaces are convex.
    (c) Give an example to show that $\overline{B(x;r)}$ can be a proper subset of $B[x;r]$.

40. Given $S \subseteq X$ and $r > 0$, let $N_r(S) = \{y \in X : d(x,y) < r$ for some $x \in S\}$. Show that $N_r(S)$ is always an open set.

41. Let $N_r[S] = \{y \in X : d(x,y) \le r$ for some $x \in S\}$. Must $N_r[S]$ be a closed set? Explain.

42. Prove: every open set in $\mathbb{R}$ is the union of countably many disjoint open intervals.

43. Prove that every isometry is continuous. Deduce that id_X is continuous.

44. A *contraction* is a function $f : X \to Y$ such that $d_Y(f(x_1), f(x_2)) \le d_X(x_1, x_2)$ for all $x_1, x_2 \in X$. Prove that every contraction is continuous. Deduce that constant functions are continuous.

45. Let $f : X \to Y$ be an isometry. Prove: for all $x_n \in X$, (x_n) is a Cauchy sequence in X iff $(f(x_n))$ is a Cauchy sequence in Y. What can you say about Cauchy sequences if f is a contraction?

46. Let $X = X_1 \times \cdots \times X_m$ be a product metric space using the metric d_1 defined in §15.1. For $1 \le i \le m$, define $p_i : X \to X_i$ by $p_i((x_1, \ldots, x_m)) = x_i$.
    (a) Prove each p_i is continuous.
    (b) Prove: $g : Y \to X$ is continuous iff $p_i \circ g$ is continuous for $1 \le i \le m$.
    (c) Explain why (a) and (b) also hold for (X, d_∞) and $(\mathbb{C}^m, d_2)$.

47. Let $Z = \{z \in \mathbb{C} : |z| = 1\}$ with the Euclidean metric. Define $f : [0, 2\pi) \to Z$ by $f(x) = (\cos x, \sin x)$ for $x \in [0, 2\pi)$. Verify that f is a continuous bijection, but f^{-1} is not continuous.

48. Prove that if X has the discrete metric, then every function $f : X \to Y$ is continuous. Use this to give an example of a continuous bijection whose inverse is not continuous.

49. Let V, W, Z be normed vector spaces. Suppose (T_n) is a sequence in $B(V, W)$ converging to T, and (U_n) is a sequence in $B(W, Z)$ converging to U. Prove $U_n \circ T_n \to U \circ T$ in $B(V, Z)$.

50. Prove $f : X \to Y$ is continuous iff for all $x_0 \in X$ and all $\epsilon > 0$, there exists $\delta > 0$ such that for all $x \in X$, if $d_X(x, x_0) < \delta$, then $d_Y(f(x), f(x_0)) < \epsilon$.

51. **Uniform Continuity.** A function $f : X \to Y$ is called *uniformly continuous* iff for all $\epsilon > 0$, there exists $\delta > 0$ such that for all $x_1, x_2 \in X$, if $d_X(x_1, x_2) < \delta$, then $d_Y(f(x_1), f(x_2)) < \epsilon$. Note that the δ appearing here depends on ϵ but not on x_1 and x_2, whereas the δ in Exercise 50 depends on both ϵ and x_0. Prove:
    (a) Contractions and isometries are uniformly continuous.
    (b) Bounded linear maps on normed vector spaces are uniformly continuous.
    (c) Addition (viewed as a map from $(H \times H, d_1)$ to H) is uniformly continuous.
    (d) The norm on a normed vector space is uniformly continuous.

52. Prove $f : \mathbb{R} \to \mathbb{R}$, given by $f(x) = x^2$ for $x \in \mathbb{R}$, is continuous but not uniformly continuous.

53. Prove that the union of a finite collection of compact subsets of X is compact.

54. Show that a subset S of a compact set is compact iff S is closed.

55. Prove that the intersection of any collection of compact subsets of X is compact.

56. Suppose X is a discrete metric space. Under what conditions is X compact? Under what conditions is X complete?

57. Let $f : X \to Y$ be continuous. Answer each question with full explanation.
    (a) If U is open in X, must $f[U]$ be open in Y?
    (b) If C is closed in X, must $f[C]$ be closed in Y?
    (c) If K is compact in Y, must $f^{-1}[K]$ be compact in X?

58. **Theorem:** *For all $a < b$ in $\mathbb{R}$, the closed interval $[a, b]$ is (sequentially) compact.* Prove this theorem by completing the following outline. Let $\mathbf{x} = (x_n : n \in \mathbb{Z}_{\geq 0})$ be a sequence of points in $[a, b]$. We construct a sequence of nested closed intervals $I_0, I_1, \ldots, I_k, \ldots$, where $I_0 = [a, b]$, $I_{k+1} \subseteq I_k$ for all $k \geq 0$, and the length of I_k is $(b - a)/2^k$ for all $k \geq 0$. I_{k+1} is either the left half or the right half of I_k. We also construct a list of sequences $\mathbf{x}^0 = \mathbf{x}, \mathbf{x}^1, \mathbf{x}^2, \ldots$, such that each sequence $\mathbf{x}^{k+1}$ is a subsequence of the previous sequence $\mathbf{x}^k$, and for all k, all terms of $\mathbf{x}^k$ belong to I_k. The construction proceeds recursively, with I_0 and $\mathbf{x}^0$ given above. Suppose that I_k and $\mathbf{x}^k$ have been constructed with the stated properties. Each term in $\mathbf{x}^k$ belongs to either the left half or the right half of I_k. So one of these two halves must contain infinitely many terms of the sequence $\mathbf{x}^k$. Let I_{k+1} be the left half of I_k if this half contains infinitely many terms of $\mathbf{x}^k$; otherwise, let I_{k+1} be the right half of I_k. Define $\mathbf{x}^{k+1}$ to be the subsequence of $\mathbf{x}^k$ consisting of all terms of $\mathbf{x}^k$ that are in I_{k+1}. (a) Verify by induction that $\mathbf{x}^k$ and I_k have the properties stated above. (b) Use the fact that every bounded subset of $\mathbb{R}$ has a greatest lower bound and least upper bound to see that $\bigcap_{k \geq 0} I_k$ must consist of a single point $x^* \in \mathbb{R}$. (c) Consider the diagonal sequence (y_k), where $y_k = \mathbf{x}^k_k$. Check that this sequence is a subsequence of the original sequence (x_n), and $y_k \in I_k$ for all k. (d) Show that (y_k) converges to x^*.

59. Use Exercise 58 and Exercise 33 to prove that a subset K of $\mathbb{R}^m$ (using any of the metrics d_1, d_2, or d_∞) is compact iff K is closed and bounded.

60. **Extreme Value Theorem.** Prove: if X is a compact metric space and $f : X \to \mathbb{R}$ is continuous, then there exist $x_1, x_2 \in X$ such that for all $x \in X$, $f(x_1) \leq f(x) \leq f(x_2)$. In other words, *a continuous real-valued function on a compact domain attains its maximum and minimum value somewhere on that domain.*

61. Let c_n be positive real constants with $\sum_{n=1}^{\infty} c_n^2 < \infty$. Let $K = \{f \in \ell_2(\mathbb{Z}_{>0}) : 0 \leq |f(n)| \leq c_n \text{ for all } n\}$. Prove K is a compact subset of the Hilbert space $\ell_2(\mathbb{Z}_{>0})$. (Use ideas from the proof in Exercise 58.)

62. Call a subset S of X *totally bounded* iff for all $\epsilon > 0$, there exist finitely many points $x_1, \ldots, x_m \in X$ such that $S \subseteq \bigcup_{i=1}^{m} B(x_i; \epsilon)$. Prove that S is (sequentially) compact iff S is complete and totally bounded.

63. Prove that a topologically compact metric space X is sequentially compact. (If X is not sequentially compact, fix a sequence (x_n) in X with no convergent subsequence. Without loss of generality, assume no two x_n are equal. For each $x \in X$, prove that there is an open ball $U_x = B(x; \epsilon(x))$ containing x such that

at most one x_n is in U_x. Show that $\{U_x : x \in X\}$ is an open cover of X with no finite subcover.)

64. A metric space X is called *separable* iff there exists a countable set $S = \{x_n : n \in \mathbb{Z}_{>0}\}$ with $\overline{S} = X$. (a) Prove that sequentially compact metric spaces are separable. (b) Prove that given any open cover $\{U_i : i \in I\}$ of a separable metric space X, there exists a countable subcover $\{U_{i_m} : m \in \mathbb{Z}_{\geq 0}\}$ with $X = \bigcup_{m=0}^{\infty} U_{i_m}$. (Fix a set U_{i_0} in the open cover. Given S as above, for each $n, k \in \mathbb{Z}_{>0}$ let $V_{n,k}$ be one particular set U_i in the open cover such that $B(x_n; 1/k) \subseteq U_i$, if such a set exists; otherwise let $V_{n,k} = U_{i_0}$. Show that the set of all $V_{n,k}$ gives a countable subcover.)

65. Prove that a sequentially compact metric space X is topologically compact. (By Exercise 64, reduce to showing that every countably infinite open cover $\{V_n : n \in \mathbb{Z}_{>0}\}$ of X has a finite subcover. If no such finite subcover exists, then all of the closed sets $C_n = X \setminus (V_1 \cup \cdots \cup V_n)$ are nonempty. Pick $x_n \in C_n$ and study the sequence (x_n).)

66. Let X and Y be complete metric spaces with subsets $X_1 \subseteq X$ and $Y_1 \subseteq Y$. Assume $f_1 : X_1 \to Y_1$ is an isometry, $\overline{X_1} = X$, and $\overline{Y_1} = Y$. Show that there exists a unique extension of f_1 to an isometry $f : X \to Y$.

67. Find necessary and sufficient conditions on $v, w \in H$ for equality to hold in the Cauchy–Schwarz Inequality.

68. Find necessary and sufficient conditions on $v, w \in H$ for equality to hold in the Triangle Inequality for norms.

69. Show that every finite-dimensional complex inner product space is complete relative to the metric induced from the inner product. (Use orthonormal bases to show that such a space is isometrically isomorphic to $\mathbb{C}^n$ with the Euclidean metric.)

70. Let W be the subset of $H = \ell_2(\mathbb{Z}_{\geq 0})$ consisting of all sequences with only finitely many nonzero terms. Show that W is a subspace of H that is not closed.

71. Let H be the Hilbert space of (Lebesgue-measurable) functions $f : [0, 1] \to \mathbb{C}$ such that $\int_0^1 |f(x)|^2 \, dx < \infty$. Show that the set of continuous $f \in H$ is a subspace of H that is not closed.

72. Let S be any subset of H. (a) Let $\langle S \rangle$ be the span of S, consisting of all finite $\mathbb{C}$-linear combinations of elements of S. Prove $S^\perp = \langle S \rangle^\perp$. (b) Let $\overline{S}$ be the closure of S (see Exercise 28). Prove $S^\perp = (\overline{S})^\perp$. (c) Show that $S^{\perp\perp} = \overline{\langle S \rangle}$, which is the smallest closed subspace of H containing S. (d) Show that for any subspace W of H, $W^{\perp\perp} = \overline{W}$.

73. **Gram–Schmidt Algorithm in a Hilbert Space.** Let $(x_1, x_2, \ldots, x_m, \ldots)$ be an infinite sequence of linearly independent vectors in H. Give a constructive procedure for computing a sequence $(z_1, z_2, \ldots, z_m, \ldots)$ of orthonormal vectors in H such that for all $m \geq 1$, $(x_1, \ldots, x_m)$ and $(z_1, \ldots, z_m)$ span the same subspace. Conclude that every infinite-dimensional Hilbert space has an infinite orthonormal set.

74. (a) Verify that the set L of closed subspaces of H is a complete lattice. (b) Explain why the map $W \mapsto W^\perp$ for $W \in L$ is a lattice anti-isomorphism. (c) Prove or disprove: given $\{W_i : i \in I\} \subseteq L$, the least upper bound of this collection must be the sum of subspaces $\sum_{i \in I} W_i$.

75. Show that if W and Z are closed subspaces of H such that $\langle w, z \rangle = 0$ for all $w \in W$ and $z \in Z$, then $W + Z$ is a closed subspace. Show that every 1-dimensional subspace of H is closed. Use these facts and the existence of orthonormal bases to show that every finite-dimensional subspace of H must be closed.

76. Let X be a set, $w_k, z_k \in \mathbb{C}$ for each $k \in X$, and $c \in \mathbb{C}$.
    (a) Prove $\sum_{k \in X} w_k + \sum_{k \in X} z_k = \sum_{k \in X}(w_k + z_k)$ if both sums on the left side are finite. (Proceed in three steps, assuming first that all $w_k, z_k \geq 0$, next that all w_k, z_k are real, and then handling the complex case.)
    (b) Prove $c \sum_{k \in X} w_k = \sum_{k \in X}(c w_k)$ when the sum on the left is finite.
    (c) Prove: if $\sum_{k \in X} |w_k| < \infty$, then $\sum_{k \in X} w_k < \infty$ and $|\sum_{k \in X} w_k| \leq \sum_{k \in X} |w_k|$.
    (d) Prove: if $0 \leq w_k \leq z_k$ for each $k \in X$, then $\sum_{k \in X} w_k \leq \sum_{k \in X} z_k$.

77. In §15.10, prove the remaining axioms for the inner product.

78. Prove the Polarization Identity (15.5).

79. Prove: For all $n \in \mathbb{Z}_{\geq 3}$, for all $x, y \in H$, $\sum_{k=0}^{n-1} e^{2\pi i k/n} ||x + e^{2\pi i k/n} y||^2 = n\langle x, y \rangle$.

80. For fixed $x, y \in H$, evaluate the integral $\int_0^{2\pi} e^{it} ||x + e^{it} y||^2 \, dt$.

81. Let V be a complex Banach space in which the norm satisfies the Parallelogram Law (15.4). Take the Polarization Identity (15.5) as the definition of $\langle x, y \rangle$ for $x, y \in V$. Show that the axioms for a complex inner product hold, and show that $\langle x, x \rangle = ||x||^2$, where $||x||$ is the given norm on V. (Informally, this says that Hilbert spaces are the Banach spaces where the Parallelogram Law holds.)

82. Give an example to show that the Parallelogram Law is not true in $\mathbb{C}^n$ if we use the 1-norm or the sup-norm (see §10.4). Conclude that these normed vector spaces do not have the structure of a Hilbert space.

83. If C and D are closed subsets of H, must $C + D = \{x + y : x \in C, y \in D\}$ be closed in H?

84. For fixed $z \in H$, define $T_z : H \to H$ by $T_z(x) = z + x$ for $x \in H$. Prove T_z is an isometry with inverse T_{-z}. Prove T_z maps closed sets to closed sets and convex sets to convex sets.

85. Given a closed subspace W of H and $x \in H$, we proved that $x = y + z$ for unique $y \in W$ and $z \in W^\perp$. Prove that z is the closest point to x in $W^\perp$.

86. Given a closed subspace W of H, define maps $P : H \to H$ and $Q : H \to H$ as follows. Given $x \in H$, we know $x = y + z$ for unique $y \in W$ and $z \in W^\perp$. Define $P(x) = y$ and $Q(x) = z$. Prove that P and Q are linear maps such that $P \circ P = P$, $Q \circ Q = Q$, and $P \circ Q = 0 = Q \circ P$. Prove that P and Q are continuous maps. Find the image and kernel of P and Q.

87. Let H be the Hilbert space of Lebesgue integrable functions with domain $[0, 2\pi]$. For $n \in \mathbb{Z}$, define $u_n \in H$ by $u_n(t) = e^{int}/\sqrt{2\pi}$ for $t \in [0, 2\pi]$. Prove that $\{u_n : n \in \mathbb{Z}\}$ is an orthonormal set in H. (This orthonormal set, which can be shown to be maximal, plays a central role in the theory of Fourier series.)

88. Let $S = \{u_n : n \in \mathbb{Z}_{>0}\}$ be an infinite orthonormal set in H.
    (a) Compute $||u_n - u_m||$ for all $n \neq m$.
    (b) Prove that a sequence (x_n) with all $x_n \in S$ converges in H iff the sequence is eventually constant, i.e., for some integer n_0, $x_n = x_{n_0}$ for all $n \geq n_0$.
    (c) Using (b), show that S is closed and bounded in H, but not compact.
    (d) Prove: if C is a closed set in H that contains $B(0; \epsilon)$ for some $\epsilon > 0$, then C is not compact. (This shows infinite-dimensional Hilbert spaces are not locally compact.)

89. Use Zorn's Lemma (see §17.13) to prove that in any Hilbert space H, there exists a maximal orthonormal set.

90. Let X be an orthonormal subset of H. Prove:
    (a) If $||w||^2 = \sum_{x \in X} |\langle w, x \rangle|^2$ for all $w \in H$, then X is a maximal orthonormal set.
    (b) If $\sum_{x \in X} \langle w, x \rangle \overline{\langle z, x \rangle} = \langle w, z \rangle$ for all $w, z \in H$, then X is maximal.

91. Prove: if $h : X \to Y$ is a bijection, then there is a Hilbert space isomorphism from $\ell_2(X)$ to $\ell_2(Y)$ sending $f \in \ell_2(X)$ to $f \circ h^{-1} \in \ell_2(Y)$.

92. Prove: if $i : X \to Y$ is an injection, then i induces a linear isometry mapping $\ell_2(X)$ onto a closed subspace of $\ell_2(Y)$.

93. Prove: for any two Hilbert spaces H_1 and H_2, H_1 is isometrically isomorphic to a closed subspace of H_2, or vice versa.

94. Let V, W, Z be normed vector spaces, $T \in B(V, W)$, and $U \in B(W, Z)$. Prove:
    (a) $||T(y)|| \leq ||T|| \cdot ||y||$ for all $y \in V$.
    (b) $||T|| = \sup\{||T(x)||/||x|| : 0 \neq x \in V\}$.
    (c) $||T|| = \inf\{M \in \mathbb{R}_{>0} : ||T(y)|| \leq M||y|| \text{ for all } y \in V\}$.
    (d) $U \circ T \in B(V, Z)$ and $||U \circ T|| \leq ||U|| \cdot ||T||$.

95. Given a nonzero $f \in H^\star$, let $W = \ker(f)$. Prove $\dim(W^\perp) = 1$.

96. Check that $H^\star$ is a Hilbert space with the inner product $\langle R_x, R_y \rangle_{H^\star} = \langle y, x \rangle_H$ for $x, y \in H$.

97. Define $E : H \to H^{\star\star}$ by letting $E(z)$ be the evaluation map $E_z : H^\star \to \mathbb{C}$ given by $E_z(f) = f(z)$ for $f \in H^\star$ and $z \in H$. In §15.18, we defined a semi-linear bijective isometry $R_H : H \to H^\star$. Prove that $E = R_{H^\star} \circ R_H$. Conclude that E is an isometric isomorphism, so that $H \cong H^{\star\star}$ as Hilbert spaces.

98. Prove: for all $S, T \in B(H)$ and $c \in \mathbb{C}$, $(S + T)^* = S^* + T^*$ and $(cS)^* = \overline{c}(S^*)$.

99. Prove: for all $T \in B(H)$, $||T^* \circ T|| = ||T||^2 = ||T \circ T^*||$ (calculate $||T(x)||^2$).

100. Given $T \in B(H)$, show that the *dual map* $T^\star : H^\star \to H^\star$, given by $T^\star(f) = f \circ T$ for $f \in H^\star$, is in $B(H^\star)$. Prove that dual maps satisfy properties analogous to the properties of adjoint maps listed in §15.19. Let $R : H \to H^\star$ be the semi-isomorphism from §15.18. Prove: for all $T \in B(H)$, $T^* = R^{-1} \circ T^\star \circ R$.

101. Prove: the set of self-adjoint operators on H is a closed *real* subspace of $B(H)$.

102. If $S, T \in B(H)$ are self-adjoint, must $S \circ T$ be self-adjoint?

103. (a) Prove: if $T \in B(H)$ satisfies $\langle T(x), x \rangle = 0$ for all $x \in H$, then $T = 0$.
     (b) Prove: if $S, T \in B(H)$ satisfy $\langle T(x), x \rangle = \langle S(x), x \rangle$ for all $x \in H$, then $S = T$.
     (c) Prove: $T \in B(H)$ is self-adjoint iff $\langle T(x), x \rangle$ is real for all $x \in H$.

104. Let Z be the set of normal operators in $B(H)$. Is Z closed under addition? scalar multiplication? composition? Is Z a closed set? Prove: $T \in Z$ iff $||T^*(x)|| = ||T(x)||$ for all $x \in H$. Prove: if $T \in Z$, then $||T^2|| = ||T||^2$.

105. Define $T : \ell_2(\mathbb{Z}_{>0}) \to \ell_2(\mathbb{Z}_{>0})$ by setting $T((c_1, c_2, \ldots)) = (0, c_1, c_2, \ldots)$ for $c_n \in \mathbb{C}$. Prove $T \in B(\ell_2(\mathbb{Z}_{>0}))$ and compute $||T||$. Find an explicit formula for the map T^*. Show that $T^* \circ T = \text{id}$, but $T \circ T^* \neq \text{id}$. Is T self-adjoint? normal? unitary?

106. Prove that $T \in B(H)$ is unitary iff $T : H \to H$ is an isometric isomorphism.

Part V

Modules and Classification Theorems

16

Finitely Generated Commutative Groups

A major goal of abstract algebra is the classification of algebraic structures. An example of such a classification is the theorem of linear algebra stating that every finite-dimensional real vector space V is isomorphic to $\mathbb{R}^n$ for some integer $n \geq 0$. Moreover, the number n is uniquely determined by V and is the dimension of V as a vector space. This classification theorem is useful because, in principle, we can use it to reduce the study of abstract vector spaces such as V to the specific concrete vector spaces $\mathbb{R}^n$.

We could hope for similar classification theorems in the theory of groups. A complete classification of all groups seems far too difficult to ever be achieved. However, much more can be said about special classes of groups, such as simple groups or commutative groups. For example, here is a structure theorem for finite commutative groups.

Theorem Classifying Finite Commutative Groups. Every finite commutative group G is isomorphic to a direct product of cyclic groups, where each cyclic group has size p^e for some prime p and some integer $e > 0$. The list of prime power sizes of the cyclic groups (written in decreasing order) is uniquely determined by G.

This chapter proves a more general classification theorem describing the structure of all finitely generated commutative groups. The techniques used to obtain this structural result for groups can also be applied in a linear-algebraic context to derive canonical forms for matrices and linear transformations. Further abstraction of these arguments eventually leads to a structure theorem classifying finitely generated modules over principal ideal domains, which is a cornerstone of abstract algebra. These topics (modules over PIDs and canonical forms) are covered in Chapter 18, which can be read independently from this chapter.

To obtain the structure theorem for finitely generated commutative groups, we first develop the theory of free commutative groups. Free commutative groups are the analogs in group theory of the vector spaces that appear so prominently in linear algebra. Thus, our development of free commutative groups has a distinctively linear-algebraic flavor. In particular, integer-valued matrices emerge as a key tool for understanding group homomorphisms between two free commutative groups. As you study this chapter, look for analogies between the material presented here and the parallel theory of vector spaces, linear transformations, and matrices.

16.1 Commutative Groups

We begin by reviewing some basic facts about commutative groups (see also Chapter 1). A *commutative group* consists of a set G together with a binary operation on G (denoted by the symbol $+$), satisfying these five axioms:
(i) *Closure:* for all $g, h \in G$, $g + h$ belongs to the set G.
(ii) *Associativity:* for all $g, h, k \in G$, $(g + h) + k = g + (h + k)$.
(iii) *Commutativity:* for all $g, h \in G$, $g + h = h + g$.

DOI: 10.1201/9781003484561-16

(iv) *Additive Identity:* there exists $0 \in G$ so that for all $g \in G$, $g + 0 = g = 0 + g$.
(v) *Additive Inverses:* for all $g \in G$, there exists $-g \in G$ such that $g + (-g) = 0 = (-g) + g$.

Some examples of commutative groups are the number systems $\mathbb{Z}$, $\mathbb{Q}$, $\mathbb{R}$, and $\mathbb{C}$, where the group operation is ordinary addition of numbers.

Let H be a subset of a commutative group G. H is called a *subgroup* of G iff the following closure conditions hold:

(a) 0 (the additive identity of G) is in the set H.
(b) For all $h, k \in H$, $h + k$ is in H.
(c) For all $h \in H$, $-h$ is in H.

In this case, the set H is a commutative group using the addition operation $+$ from G, restricted to inputs in H.

Given a subgroup H of G, we can form the *quotient group* G/H whose elements are the cosets $x + H = \{x + h : h \in H\}$, with x ranging through G. The Coset Equality Theorem states that for all $x, y \in G$, $x + H = y + H$ if and only if $x - y \in H$. Here, $x - y$ is defined to mean $x + (-y)$. The group operation in G/H is given by

$$(x + H) + (y + H) = (x + y) + H \qquad \text{for all } x, y \in G.$$

See §1.6 for a fuller discussion of quotient groups.

Recall $\mathbb{Z}$ is the commutative group of integers under addition. In our later work, we need to know all the subgroups of $\mathbb{Z}$. The next theorem classifies these subgroups.

Theorem on Subgroups of $\mathbb{Z}$. Every subgroup of $\mathbb{Z}$ has the form $n\mathbb{Z} = \{nk : k \in \mathbb{Z}\}$ for some uniquely determined integer $n \geq 0$.

Proof. It is routine to verify that for each $n \in \mathbb{Z}_{\geq 0}$, the subset $n\mathbb{Z}$ is indeed a subgroup of $\mathbb{Z}$, and different values of n produce unequal subgroups. Given an arbitrary subgroup H of $\mathbb{Z}$, we show that $H = n\mathbb{Z}$ for some $n \in \mathbb{Z}_{\geq 0}$. If H consists of 0 alone, then $H = 0\mathbb{Z}$. Otherwise, there must exist a nonzero element $h \in H$. Since $h \in H$ implies $-h \in H$, we see that H contains at least one positive integer. Using the Least Natural Number Axiom, let n be the smallest positive integer belonging to H. We claim that $n\mathbb{Z} = H$. You can check the inclusion $n\mathbb{Z} = \{nk : k \in \mathbb{Z}\} \subseteq H$ by induction on k, since $n \in H$ and H is closed under addition and inverses. We now prove the reverse inclusion $H \subseteq n\mathbb{Z}$. Fix $h \in H$. Dividing h by the nonzero integer n, we can write $h = nq + r$ for some integers q, r such that $0 \leq r < n$. Now, since $n\mathbb{Z} \subseteq H$, nq is in H. It follows that $r = h + -(nq)$ is in H as well, since H is closed under addition and inverses. Because n is the least positive integer in H and $r < n$, we must have $r = 0$. This means $h = nq$ is in $n\mathbb{Z}$. So $H \subseteq n\mathbb{Z}$, and hence $H = n\mathbb{Z}$. $\qquad \square$

For each integer $n \geq 1$, the quotient group $\mathbb{Z}/n\mathbb{Z}$ gives one way of defining the commutative group of *integers modulo* n. Using integer division with remainder, you can check that $\mathbb{Z}/n\mathbb{Z}$ consists of the n distinct cosets

$$\bar{0} = 0 + n\mathbb{Z}, \ \bar{1} = 1 + n\mathbb{Z}, \ \ldots, \ \overline{n-1} = (n-1) + n\mathbb{Z}.$$

Furthermore, this quotient group is isomorphic to the group $\mathbb{Z}_n = \{0, 1, 2, \ldots, n-1\}$ with operation $\oplus$ (addition modulo n), as defined in §1.1. You can check that the quotient group $\mathbb{Z}/0\mathbb{Z}$ is isomorphic to $\mathbb{Z}$ itself.

Suppose $G_1, \ldots, G_k$ are commutative groups. Consider the product set

$$G_1 \times G_2 \times \cdots \times G_k = \{(x_1, x_2, \ldots, x_k) : x_i \in G_i \text{ for all } i \text{ between 1 and } k\}.$$

This product set becomes a commutative group under the operation

$$(x_1, x_2, \ldots, x_k) + (y_1, y_2, \ldots, y_k) = (x_1 + y_1, x_2 + y_2, \ldots, x_k + y_k) \qquad \text{for all } x_i, y_i \in G_i,$$

as you can verify. We call the group $G_1 \times G_2 \times \cdots \times G_k$ the *direct product* of $G_1, \ldots, G_k$. When working with commutative groups, the direct product is also written $G_1 \oplus G_2 \oplus \cdots \oplus G_k$ and called the *(external) direct sum* of the G_i. If every G_i equals the same group G, we may write G^k instead of $G_1 \times \cdots \times G_k$.

16.2   Generating Sets for Commutative Groups

Suppose G is a commutative group, $x \in G$, and $n \in \mathbb{Z}$. Define the *integer multiple* nx by

$$
nx = \begin{cases}
x + x + \cdots + x & \text{(sum of n copies of x)} & \text{if } n > 0; \\
0 & \text{(the identity of G)} & \text{if } n = 0; \\
-x + -x + \cdots + -x & \text{(sum of n copies of $-x$)} & \text{if } n < 0.
\end{cases}
$$

For all $x, y \in G$ and all $m, n \in \mathbb{Z}$, the following rules hold:

$$(m+n)x = (mx) + (nx), \quad m(nx) = (mn)x, \quad 1x = x, \quad -(mx) = (-m)x, \tag{16.1}$$

$$m(x+y) = (mx) + (my).$$

The first four rules are the Laws of Exponents, translated from multiplicative notation (involving powers x^n) to additive notation (involving multiples nx). These laws hold in all groups, not just commutative groups. The fifth rule requires commutativity of G. For positive m, we can justify this rule informally by writing

$$
\begin{aligned}
m(x+y) &= (x+y) + (x+y) + \cdots + (x+y) \quad \text{(m copies of $x+y$)} \\
&= \underbrace{(x + x + \cdots + x)}_{m \text{ copies of } x} + \underbrace{(y + y + \cdots + y)}_{m \text{ copies of } y} \quad \text{(since G is commutative)} \\
&= (mx) + (my).
\end{aligned}
$$

A more formal verification of this rule (see Exercise 7) uses induction to establish its validity for $m \geq 0$, followed by a separate argument for negative m.

Next, suppose $v_1, \ldots, v_k$ are elements of a commutative group G. A $\mathbb{Z}$-*linear combination* of these elements is an element of the form

$$c_1 v_1 + c_2 v_2 + \cdots + c_k v_k = \sum_{i=1}^{k} c_i v_i,$$

where each c_i is an integer. The set H of all $\mathbb{Z}$-linear combinations of $v_1, \ldots, v_k$ is a subgroup of G, denoted by $\langle v_1, v_2, \ldots, v_k \rangle$ or $\mathbb{Z}v_1 + \cdots + \mathbb{Z}v_k$. This follows from the rules in (16.1) and commutativity of addition in G. For instance, H is closed under addition since

$$\sum_{i=1}^{k} c_i v_i + \sum_{i=1}^{k} d_i v_i = \sum_{i=1}^{k} (c_i v_i + d_i v_i) = \sum_{i=1}^{k} (c_i + d_i) v_i$$

for all $c_i, d_i \in \mathbb{Z}$. We call $\langle v_1, v_2, \ldots, v_k \rangle$ the *subgroup of G generated by $v_1, \ldots, v_k$*. If there exist finitely many elements $v_1, \ldots, v_k$ that generate G, then G is called a *finitely generated commutative group*. If G can be generated by a single element v_1, then G is called a *cyclic group*. Given any commutative group G and any element $x \in G$, the set of multiples $\langle x \rangle = \{nx : n \in \mathbb{Z}\} = \mathbb{Z}x$ is a cyclic subgroup of G.

Every finite commutative group G is finitely generated, since we can take all the elements of G as a generating set. Of course, there are often other generating sets that are much smaller. The groups $\mathbb{Z}$ and $\mathbb{Z}/n\mathbb{Z}$ (for $n \geq 1$) are finitely generated groups. Indeed, they are cyclic groups since $\mathbb{Z} = \langle 1 \rangle$ and $\mathbb{Z}/n\mathbb{Z} = \langle 1 + n\mathbb{Z} \rangle$. Generators of cyclic groups are usually not unique; for instance, $\mathbb{Z} = \langle 1 \rangle = \langle -1 \rangle$. As another example, for each prime p, $\mathbb{Z}_p$ is generated by each of its elements other than 0 (Exercise 11). *Some finitely generated groups are not finite groups.* For example, $\mathbb{Z}$ is generated by 1 element but is infinite. More examples appear in the next paragraph.

Let $G_1, \ldots, G_k$ be finitely generated commutative groups, say $G_i = \langle v_{i,1}, \ldots, v_{i,n_i} \rangle$ for some integers $n_i \in \mathbb{Z}_{>0}$ and $v_{i,j} \in G_i$. Given the product group $G = G_1 \times \cdots \times G_k$, we can associate with each $v_{i,j} \in G_i$ the k-tuple $(0, 0, \ldots, v_{i,j}, \ldots, 0) \in G$, where $v_{i,j}$ appears in position i. You can check that the finite list of all such k-tuples generates G. Thus, *the direct product of finitely many finitely generated groups is finitely generated.* For example, since $\mathbb{Z} = \langle 1 \rangle$, $\mathbb{Z}^k$ is generated by the k elements $e_i = (0, 0, \ldots, 1, \ldots, 0)$, where the 1 occurs in position i. Note that e_i is the image in $\mathbb{Z}^k$ of the generator 1 of the ith copy of $\mathbb{Z}$. For each $k > 0$, $\mathbb{Z}^k$ is an infinite commutative group that is finitely generated.

Not every commutative group is finitely generated. For example, consider the additive group $\mathbb{Q}$ of rational numbers. To prove that $\mathbb{Q}$ is not finitely generated, we use proof by contradiction. Assume that $v_1, \ldots, v_k \in \mathbb{Q}$ is a finite generating set for $\mathbb{Q}$. Write $v_i = a_i/b_i$ for some integers a_i, b_i with $b_i > 0$. By finding a common denominator, we can change notation so that $v_i = c_i/d$ for some integers c_i, d with $d > 0$. For instance, let $d = b_1 b_2 \cdots b_k$ and $c_i = da_i/b_i \in \mathbb{Z}$. Now consider an arbitrary $\mathbb{Z}$-linear combination of $v_1, \ldots, v_k$:

$$x = n_1 v_1 + \cdots + n_k v_k = \frac{n_1 c_1 + \cdots + n_k c_k}{d} \qquad \text{where each } n_i \in \mathbb{Z}.$$

Since the numerator $n_1 c_1 + \cdots + n_k c_k$ is an integer, any such x is either 0 or has absolute value at least $1/d$. But the rational number $1/(2d)$ cannot be expressed in this form, because its absolute value is too small.

Although this chapter is concerned mainly with finitely generated commutative groups, it is possible to define the notion of an infinite generating set. Let S be an arbitrary subset (possibly infinite) of a commutative group G. By definition, a $\mathbb{Z}$-*linear combination of elements of S* is a $\mathbb{Z}$-linear combination of some finite subset S' of S. We often write $\sum_{x \in S} n_x x$ for such a linear combination, understanding that all but a finite number of the coefficients $n_x \in \mathbb{Z}$ must be zero. You can check that the set of $\mathbb{Z}$-linear combinations of elements of S is a subgroup of G, written $\langle S \rangle$ or $\sum_{s \in S} \mathbb{Z}s$. We say S *generates* G iff $G = \langle S \rangle$. For example, using the Fundamental Theorem of Arithmetic, you can verify that $S = \{1/p^e : p \text{ is prime}, e \geq 1\}$ generates the commutative group $\mathbb{Q}$.

Our main goal in this chapter is to prove the following theorem.

Theorem Classifying Finitely Generated Commutative Groups. Every finitely generated commutative group G is isomorphic to a direct product of cyclic groups, specifically $G \cong \mathbb{Z}^b \times \mathbb{Z}_{a_1} \times \mathbb{Z}_{a_2} \times \cdots \times \mathbb{Z}_{a_s}$, where $b \geq 0$, $s \geq 0$, $a_1 \geq a_2 \geq \cdots \geq a_s > 1$, and every a_i is a power of a prime. The nonnegative integer b and the prime powers $a_1, \ldots, a_s$ are uniquely determined by G.

16.3 $\mathbb{Z}$-Independence and $\mathbb{Z}$-Bases

A generating set for a commutative group resembles a spanning set for a vector space. The only difference is that the scalars multiplying the group elements are integers rather than field elements. The analogy to linear algebra motivates the following definitions. Let $(v_1, \ldots, v_k)$ be a finite list of elements in a commutative group G. We call this list $\mathbb{Z}$-*linearly independent* (or $\mathbb{Z}$-*independent*) iff for all integers $c_1, \ldots, c_k$,

$$\text{if } c_1 v_1 + c_2 v_2 + \cdots + c_k v_k = 0_G, \text{ then } c_1 = c_2 = \cdots = c_k = 0.$$

This means that no linear combination of $v_1, \ldots, v_k$ produces 0_G except the one where all coefficients are 0. If S is a subset of G (possibly infinite), we say that S is $\mathbb{Z}$-*linearly independent* iff every finite nonempty list of distinct elements of S is $\mathbb{Z}$-independent.

An *ordered* $\mathbb{Z}$-*basis* of G is a list $B = (v_1, \ldots, v_k)$ of elements $v_i \in G$ such that B is $\mathbb{Z}$-independent and $G = \langle v_1, \ldots, v_k \rangle$. A subset S of G (finite or not) is called a $\mathbb{Z}$-*basis* of G iff $G = \langle S \rangle$ and S is $\mathbb{Z}$-linearly independent. If G has a basis, then G is called a *free* commutative group. If G has a k-element basis for some $k \in \mathbb{Z}_{\geq 0}$, then G is called a *finitely generated free commutative group* with *dimension* (or *rank*) k. We prove later that the dimension of such a group is uniquely determined.

Not every commutative group is free. For instance, suppose G is a finite group with size at least 2. Consider any list $(v_1, \ldots, v_k)$ of elements of G. By group theory, there exists an integer $n > 0$ with $n v_1 = 0$; for instance, $n = |G|$ has this property (see Exercise 11 in Chapter 1 for a proof). Then the relation $n v_1 + 0 v_2 + \cdots + 0 v_k = 0$ shows that the given list must be $\mathbb{Z}$-linearly dependent. On the other hand, if $G = \{0\}$ is a 1-element group, then the empty set is a $\mathbb{Z}$-basis of G. This follows from the convention $\langle \emptyset \rangle = \{0\}$ and the fact that the empty set is $\mathbb{Z}$-independent, which is a logical consequence of the definition.

For more interesting examples of bases, consider the group $\mathbb{Z}^k$ of all k-tuples of integers, with group operation

$$(a_1, a_2, \ldots, a_k) + (b_1, b_2, \ldots, b_k) = (a_1 + b_1, a_2 + b_2, \ldots, a_k + b_k) \qquad \text{for } a_i, b_i \in \mathbb{Z}.$$

For $1 \leq i \leq k$, let $e_i = (0, \ldots, 0, 1, 0, \ldots, 0)$, where the 1 occurs in position i. Let us verify explicitly that the list $B = (e_1, e_2, \ldots, e_k)$ is an ordered basis of $\mathbb{Z}^k$; B is called the *standard ordered basis of* $\mathbb{Z}^k$. First, B spans $\mathbb{Z}^k$, because for any $(a_1, \ldots, a_k) \in \mathbb{Z}^k$, we have $(a_1, \ldots, a_k) = \sum_{i=1}^k a_i e_i$. Second, B is $\mathbb{Z}$-independent, because the relation $\sum_{i=1}^k b_i e_i = \mathbf{0}$ (where $b_i \in \mathbb{Z}$) means that $(b_1, \ldots, b_k) = (0, \ldots, 0)$, which implies that each b_i is 0 by equating corresponding components. In summary, we have proved that $\mathbb{Z}^k$ *is a* k-*dimensional free commutative group with basis* B.

16.4   Elementary Operations on $\mathbb{Z}$-Bases

As in the case of vector spaces, most free commutative groups have many different bases. There are three *elementary operations* that produce new ordered bases from given ordered bases. We now define how each of these operations affects a given ordered $\mathbb{Z}$-basis $B = (v_1, \ldots, v_k)$ of a finitely generated free commutative group G.

(B1) For any $i \neq j$, we can interchange the position of v_i and v_j in the ordered list B. Using commutativity, we see (Exercise 28) that the new ordered list still generates G and is still $\mathbb{Z}$-independent, so this new list is a $\mathbb{Z}$-basis of G.

(B2) For any i, we can replace v_i in B by $-v_i$. Since $c_i v_i = (-c_i)(-v_i)$, it readily follows (Exercise 29) that the new ordered list is still a $\mathbb{Z}$-basis of G.

(B3) For any $i \neq j$ and any integer c, we can replace v_i in B by $w_i = v_i + cv_j$. Let us check that this gives another $\mathbb{Z}$-basis of G. Using (B1), it suffices to consider the case $i = 1$ and $j = 2$. Write $B = (v_1, v_2, \ldots, v_k)$ and $B' = (v_1 + cv_2, v_2, \ldots, v_k)$. First, does B' generate G? Given any $g \in G$, we can write

$$g = n_1 v_1 + n_2 v_2 + \cdots + n_k v_k$$

for some $n_i \in \mathbb{Z}$, because B is known to generate G. Manipulating this expression gives

$$g = n_1(v_1 + cv_2) + (n_2 - cn_1)v_2 + n_3 v_3 + \cdots + n_k v_k$$

where all coefficients are integers. Thus, g is a linear combination of the elements in the list B'. Second, is B' a $\mathbb{Z}$-independent list? Assume that $d_1, \ldots, d_k$ are fixed integers such that

$$d_1(v_1 + cv_2) + d_2 v_2 + \cdots + d_k v_k = 0;$$

we must prove every $d_i = 0$. Rewriting the assumption gives

$$d_1 v_1 + (cd_1 + d_2)v_2 + d_3 v_3 + \cdots + d_k v_k = 0.$$

By the known $\mathbb{Z}$-independence of B, we conclude that $d_1 = cd_1 + d_2 = d_3 = \cdots = d_k = 0$. Then $d_2 = (cd_1 + d_2) - cd_1 = 0 - c0 = 0$. So every d_i is 0.

For example, starting with the standard ordered basis (e_1, e_2, e_3) of $\mathbb{Z}^3$, we can apply a sequence of elementary operations to produce new ordered bases of this group. As a specific illustration, you can check that the list $(e_2 - 5e_3, -e_3, e_1 + 2e_2 + 3e_3) = ((0, 1, -5), (0, 0, -1), (1, 2, 3))$ can be obtained from (e_1, e_2, e_3) by appropriate elementary operations. Hence, this list is an ordered $\mathbb{Z}$-basis of $\mathbb{Z}^3$.

We remark that similar elementary operations can be applied to ordered bases of vector spaces over a field F. In operation (B3), we replace the integer c by a scalar $c \in F$. In operation (B2), we may now select any nonzero scalar $c \in F$ and replace v_i by cv_i. You can check that the corresponding operation for commutative groups only produces a $\mathbb{Z}$-basis when $c = \pm 1$. The reason is that $+1$ and -1 are the only integers whose multiplicative inverses are also integers.

16.5 Coordinates and $\mathbb{Z}$-Linear Maps

The next few sections derive some fundamental facts about bases and free commutative groups. We focus on the case of finitely generated groups, but all the proofs can be extended to the case of groups with infinite bases.

Let G be a free commutative group with ordered basis $B = (v_1, \ldots, v_k)$. We first prove: *for every $g \in G$, there exist unique integers $n_1, \ldots, n_k$ such that $g = n_1 v_1 + \cdots + n_k v_k$.* We call $(n_1, \ldots, n_k)$ the *coordinates of g relative to B.* The existence of the integers n_i follows immediately from the fact that the list $v_1, \ldots, v_k$ generates G. To prove uniqueness, suppose $g \in G$ and

$$g = n_1 v_1 + \cdots + n_k v_k = m_1 v_1 + \cdots + m_k v_k$$

for some $n_i, m_i \in \mathbb{Z}$. Subtracting the equations and using the rules in (16.1), we get

$$(n_1 - m_1)v_1 + \cdots + (n_k - m_k)v_k = 0.$$

By $\mathbb{Z}$-independence of B, it follows that $n_i - m_i = 0$ for all i. Thus, $n_i = m_i$ for all i, proving that the coordinates n_i are uniquely determined by g and B.

To continue our study of free commutative groups, we need the idea of $\mathbb{Z}$-linear maps. A $\mathbb{Z}$-*linear map* is a group homomorphism $T : G \to H$ between two commutative groups G and H. To say that T is a homomorphism means that $T(x + y) = T(x) + T(y)$ for all $x, y \in G$. It follows by induction that $T(nx) = nT(x)$ for all $n \in \mathbb{Z}$ and all $x \in G$. w The analogy to the condition $T(cx) = cT(x)$ (for linear transformations on vector spaces) explains the terminology "$\mathbb{Z}$-linear." Another induction proof shows that *any $\mathbb{Z}$-linear map $T : G \to H$ preserves $\mathbb{Z}$-linear combinations:*

$$T\left(\sum_{i=1}^{k} n_i v_i\right) = \sum_{i=1}^{k} n_i T(v_i) \qquad \text{for all } k > 0, \, n_i \in \mathbb{Z}, \, v_i \in G.$$

This continues the analogy to linear transformations between F-vector spaces; such maps preserve F-linear combinations. The *kernel* of a $\mathbb{Z}$-linear map $T : G \to H$ is $\ker(T) = \{x \in G : T(x) = 0_H\}$. The *image* of T is $\text{img}(T) = \{T(x) : x \in G\}$. A $\mathbb{Z}$-linear map T is injective iff $\ker(T) = \{0_G\}$. A $\mathbb{Z}$-linear map T is surjective iff $\text{img}(T) = H$. The *Fundamental Homomorphism Theorem for Commutative Groups* states that any $\mathbb{Z}$-linear map $T : G \to H$ with kernel K and image I induces a $\mathbb{Z}$-linear isomorphism $T' : G/K \to I$ given by $T'(x + K) = T(x)$ for all $x \in G$. See §1.7 for a proof of this theorem.

We pause to establish two basic facts about $\mathbb{Z}$-linear maps. Let $T : G \to H$ be a $\mathbb{Z}$-linear map between two commutative groups. Assume H is generated by elements $w_1, \ldots, w_m$. The first fact says that *the image of T is all of H iff every generator w_j is in the image of T*. The forward implication is immediate. Conversely, suppose each $w_j = T(x_j)$ for some $x_j \in G$. Given $h \in H$, h can be written (not necessarily uniquely) in the form $h = n_1 w_1 + \cdots + n_m w_m$ where $n_i \in \mathbb{Z}$. Choosing $x = n_1 x_1 + \cdots + n_m x_m \in G$, $\mathbb{Z}$-linearity implies that $T(x) = h$. To state the second fact, let $S : G \to H$ be another $\mathbb{Z}$-linear map, and assume G is generated by elements $v_1, \ldots, v_k$. The second fact says $T = S$ iff $T(v_i) = S(v_i)$ for all i with $1 \le i \le k$. In other words, *two $\mathbb{Z}$-linear maps are equal iff they agree on a generating set for the domain*. The forward implication is immediate. To prove the converse, suppose $T(v_i) = S(v_i)$ for all i, and let $g \in G$. We can write $g = p_1 v_1 + \cdots + p_k v_k$ for some integers p_i. Using $\mathbb{Z}$-linearity and the hypothesis on T and S, we see that

$$T(g) = \sum_{i=1}^{k} p_i T(v_i) = \sum_{i=1}^{k} p_i S(v_i) = S(g).$$

16.6 UMP for Free Commutative Groups

The next theorem states a fundamental property of free commutative groups.

Universal Mapping Property (UMP) for Free Commutative Groups. Suppose $X = \{v_1, \ldots, v_k\}$ is a basis of a free commutative group G. For every commutative group H and every function $f : X \to H$, there exists a unique $\mathbb{Z}$-linear map $T_f : G \to H$ such that $T_f(v_i) = f(v_i)$ for all $v_i \in X$.

We call T_f the $\mathbb{Z}$-*linear extension of f* from X to G.

Proof. First we prove existence of T_f. To define the value of T_f at a given $g \in G$, write g in the form

$$g = n_1 v_1 + \cdots + n_k v_k \qquad \text{with all } n_j \in \mathbb{Z}.$$

We already proved that the integers n_j in this expression are uniquely determined by g. Therefore, the formula

$$T_f(g) = n_1 f(v_1) + \cdots + n_k f(v_k)$$

gives a well-defined element of H for each $g \in G$. We must check that the function T_f is a group homomorphism extending f. First, we prove $T_f(g + g') = T_f(g) + T_f(g')$ for all $g, g' \in G$. Given $g, g' \in G$, write $g = \sum_i n_i v_i$ and $g' = \sum_i m_i v_i$ where $n_i, m_i \in \mathbb{Z}$. By the definition of T_f,

$$T_f(g) = n_1 f(v_1) + \cdots + n_k f(v_k);$$

$$T_f(g') = m_1 f(v_1) + \cdots + m_k f(v_k).$$

On the other hand, note that $g + g' = \sum_i (n_i + m_i) v_i$ is the unique expression for $g + g'$ as a linear combination of $v_1, \ldots, v_k$. Applying the definition of T_f gives

$$T_f(g + g') = (n_1 + m_1) f(v_1) + \cdots + (n_k + m_k) f(v_k).$$

Comparing to the previous formulas, we see that $T_f(g+g') = T_f(g) + T_f(g')$. Next, we show T_f extends f. Fix an index j between 1 and k. Observe that $v_j = 0v_1 + \cdots + 1v_j + \cdots + 0v_k$ is the unique expansion of v_j in terms of $v_1, \ldots, v_k$. Therefore, by definition,

$$T_f(v_j) = 0f(v_1) + \cdots + 1f(v_j) + \cdots + 0f(v_k) = f(v_j).$$

This completes the existence proof.

To prove uniqueness of T_f, suppose $S : G \to H$ is any $\mathbb{Z}$-linear map extending f. Then $T_f(v_j) = f(v_j) = S(v_j)$ for all j between 1 and k. This says T_f and S agree on a generating set for G, so $T_f = S$. $\qquad \square$

The UMP holds (with an analogous proof) for all free commutative groups, not just finitely generated ones.

Next, we use the UMP to prove that *every k-dimensional free commutative group G is isomorphic to $\mathbb{Z}^k$.* More precisely, let $B = (v_1, \ldots, v_k)$ be an ordered basis of G. Define a function $f : \{v_1, \ldots, v_k\} \to \mathbb{Z}^k$ by setting $f(v_i) = e_i$, where e_i is the k-tuple with a 1 in position i and 0s elsewhere. By the UMP, there is a unique $\mathbb{Z}$-linear map $T : G \to \mathbb{Z}^k$ extending f. Now, T is surjective because the image of T contains the generating set $\{e_1, \ldots, e_k\}$ of $\mathbb{Z}^k$. To see that T is injective, recall the formula defining T:

$$T(n_1 v_1 + \cdots + n_k v_k) = n_1 f(v_1) + \cdots + n_k f(v_k) = n_1 e_1 + \cdots + n_k e_k \qquad \text{where } n_i \in \mathbb{Z}.$$

If T sends $x = n_1 v_1 + \cdots + n_k v_k \in G$ to $\mathbf{0}$, then $n_1 = \cdots = n_k = 0$ by the $\mathbb{Z}$-independence of $e_1, \ldots, e_k$. Hence, $x = 0$, proving that T has kernel $\{0\}$. It follows that T is a bijective $\mathbb{Z}$-linear map, so T is an isomorphism from G to $\mathbb{Z}^k$.

16.7 Quotient Groups of Free Commutative Groups

Our goal in this chapter is to classify all finitely generated commutative groups. More precisely, part of our goal is to show that every finitely generated commutative group is isomorphic to a direct product of cyclic groups (namely $\mathbb{Z}$ or $\mathbb{Z}/n\mathbb{Z} \cong \mathbb{Z}_n$). We achieved part of this goal at the end of the last section, by showing that every finitely generated *free* commutative group is isomorphic to one of the groups $\mathbb{Z}^k$. This fact is analogous to the linear algebra theorem stating that every finite-dimensional vector space over a field F is isomorphic to F^k for some $k \geq 0$. However, in the case of commutative groups, there is more work to do since not all commutative groups are free. (We must also eventually address the question of the uniqueness of k.)

The next step toward our goal is to prove that *every finitely generated commutative group is isomorphic to a quotient group F/P, for some finitely generated free commutative group F and some subgroup P.* (This statement remains true, with the same proof, if the two occurrences of "finitely generated" are deleted.) Let H be a commutative group generated by $w_1, \ldots, w_k$. Let $\mathbb{Z}^k$ be the free commutative group with standard ordered basis $(e_1, \ldots, e_k)$. Define a map $f : \{e_1, \ldots, e_k\} \to H$ by setting $f(e_i) = w_i$ for $1 \leq i \leq k$. Next, use the UMP to obtain a $\mathbb{Z}$-linear extension $T : \mathbb{Z}^k \to H$. The image of T is all of H, since all the generators w_j of H are in the image of T. Let P be the kernel of T. Applying the Fundamental Homomorphism Theorem to T, we see that T induces a group isomorphism $T' : \mathbb{Z}^k/P \to H$. Thus, H is isomorphic to a quotient group of a free commutative group whose dimension is the same as the size of the given generating set for H.

To see why this result helps us in the classification of finitely generated commutative groups, consider the special case where the subgroup P of $\mathbb{Z}^k$ has the particular form

$$P = n_1\mathbb{Z} \times n_2\mathbb{Z} \times \cdots \times n_k\mathbb{Z}, \tag{16.2}$$

for some integers $n_1, \ldots, n_k \geq 0$. Define a map $S : \mathbb{Z}^k \to (\mathbb{Z}/n_1\mathbb{Z}) \times \cdots \times (\mathbb{Z}/n_k\mathbb{Z})$ by setting $S((a_1, \ldots, a_k)) = (a_1 + n_1\mathbb{Z}, \ldots, a_k + n_k\mathbb{Z})$ for $a_i \in \mathbb{Z}$. You can check that S is a surjective group homomorphism with kernel P. Hence, by the Fundamental Homomorphism Theorem, we obtain an isomorphism

$$H \cong \mathbb{Z}^k/P \cong \frac{\mathbb{Z}}{n_1\mathbb{Z}} \times \frac{\mathbb{Z}}{n_2\mathbb{Z}} \times \cdots \times \frac{\mathbb{Z}}{n_k\mathbb{Z}} \cong \mathbb{Z}_{n_1} \times \mathbb{Z}_{n_2} \times \cdots \times \mathbb{Z}_{n_k}.$$

Thus, in the case where P is a subgroup of the form (16.2), we have succeeded in writing H as a direct product of cyclic groups.

Unfortunately, when $k > 1$, not every subgroup of $\mathbb{Z}^k$ has the form given in (16.2). For example, when $k = 2$, consider the subgroup $P = \{(m, 2m) : m \in \mathbb{Z}\} \subseteq \mathbb{Z} \times \mathbb{Z}$. You can check that the set P is not of the form $A \times B$ for any choice of $A, B \subseteq \mathbb{Z}$. In general, the subgroup structure of the free commutative groups $\mathbb{Z}^k$ is rich and subtle. A deeper study of these subgroups is needed to achieve our goal of classifying all finitely generated commutative groups.

Before continuing our investigation of the subgroups of $\mathbb{Z}^k$, let us revisit the analogy to vector spaces to gain some intuition. The analog (for vector spaces) of the statement "not every P has the form (16.2)" is the statement "not every subspace of $\mathbb{R}^k$ is spanned by a subset of the standard basis vectors e_i." The analog of the subgroup $P = \{(n, 2n) : n \in \mathbb{Z}\}$ is the subspace $\{(t, 2t) : t \in \mathbb{R}\}$ of $\mathbb{R}^2$. This 1-dimensional subspace of $\mathbb{R}^2$ is not spanned by either of the vectors $(1, 0)$ or $(0, 1)$. However, it is spanned by the vector $(1, 2)$. We know from linear algebra that the one-element set $\{(1, 2)\}$ can be extended to a basis of $\mathbb{R}^2$. This suggests the possibility of changing the basis of $\mathbb{Z}^k$ to force the subgroup P to assume the nice form given in (16.2). We pursue this idea in the next few sections.

16.8 Subgroups of Free Commutative Groups

To begin our more detailed study of subgroups of $\mathbb{Z}^k$, we prove a key technical point about subgroups of free commutative groups. We show that *if G is a k-dimensional free* commutative group and H is a subgroup of G, then H is also a free commutative group with dimension at most k. Since we know G is isomorphic to $\mathbb{Z}^k$, we can prove this result for the particular group $G = \mathbb{Z}^k$ without loss of generality. The proof uses induction on k.

The case $k = 0$ is immediate. Suppose $k = 1$. In §16.1, we saw that every subgroup of $\mathbb{Z}^1$ has the form $n\mathbb{Z}$ for some integer $n \geq 0$. If $n = 0$, then $0\mathbb{Z}$ is a zero-dimensional free commutative group with an empty basis. For $n > 0$, you can check that $n\mathbb{Z} \cong \mathbb{Z}$ is a 1-dimensional free commutative group with basis $\{n\}$. (We observe in passing that, for $n > 1$, this basis of the subgroup $n\mathbb{Z}$ cannot be extended to a basis of $\mathbb{Z}$. This reveals one notable difference between free commutative groups and vector spaces.)

For the induction step, fix $k > 1$ and assume the theorem holds for all free commutative groups of dimension less than k. Let H be a fixed subgroup of $\mathbb{Z}^k$. To apply our induction hypothesis, we need a subgroup of $\mathbb{Z}^{k-1}$. To obtain such a subgroup, let $H' = H \cap (\mathbb{Z}^{k-1} \times \{0\})$ be the set of all elements of H with last coordinate 0. Note that H' is a subgroup of $\mathbb{Z}^{k-1} \times \{0\}$, which is a $(k-1)$-dimensional free commutative group isomorphic to $\mathbb{Z}^{k-1}$. Applying the induction hypothesis, we conclude that H' is free and has some ordered basis $(v_1, \ldots, v_{m-1})$ where $m - 1 \leq k - 1$. We must somehow pass from this basis to a basis of the full subgroup H.

Toward this end, consider the projection map $P : \mathbb{Z}^k \to \mathbb{Z}$ given by $P((a_1, \ldots, a_k)) = a_k$. It is routine to check that P is a group homomorphism. Therefore, $P[H] = \{P(h) : h \in H\}$ is a subgroup of $\mathbb{Z}$. By our earlier classification of the subgroups of $\mathbb{Z}$, we know there is some integer $q \geq 0$ such that $P[H] = q\mathbb{Z}$. Let v_m be any fixed element of H such that $P(v_m) = q$. So $v_m = (a_1, \ldots, a_{k-1}, q) \in H$ for some integers a_i.

Now consider two cases. First, suppose $q = 0$. Then $P[H] = \{0\}$, which means that every element of H has last coordinate 0. Then $H = H'$, and we already know that H' is a free commutative group of dimension $m - 1 \leq k - 1 < k$.

The second case is that $q > 0$. We claim that $X = (v_1, \ldots, v_m)$ is an ordered basis of H, so that H is free with dimension $m \leq k$. To prove the claim, we first check the $\mathbb{Z}$-independence of X. Suppose $c_1 v_1 + \cdots + c_m v_m = 0$ for some integers c_i. Apply the $\mathbb{Z}$-linear projection map P to this relation to obtain $c_1 P(v_1) + \cdots + c_{m-1} P(v_{m-1}) + c_m P(v_m) = P(0) = 0$. For $i < m$, $P(v_i) = 0$ since $v_i \in H'$. On the other hand, $P(v_m) = q$ by choice of v_m. So the relation reduces to $c_m q = 0$, which implies $c_m = 0$ because $q > 0$ and $\mathbb{Z}$ has no zero divisors. However, once we know that $c_m = 0$, the original relation becomes $c_1 v_1 + \cdots + c_{m-1} v_{m-1} = 0$. We can now conclude that $c_1 = \cdots = c_{m-1} = 0$ because $v_1, \ldots, v_{m-1}$ are already known to be $\mathbb{Z}$-linearly independent. To finish the proof, we must show that X generates H. Fix $h = (b_1, \ldots, b_k) \in H$, where $b_i \in \mathbb{Z}$. We have $P(h) = b_k \in P[H] = q\mathbb{Z}$, so $b_k = tq$ for some integer t. Note that

$$h - tv_m = (b_1, \ldots, b_k) - t(a_1, \ldots, a_{k-1}, q) = (b_1 - ta_1, \ldots, b_{k-1} - ta_{k-1}, 0).$$

So $h - tv_m \in H$ and $h - tv_m$ has last coordinate 0, meaning that $h - tv_m \in H'$. Since $(v_1, \ldots, v_{m-1})$ is known to generate H', we have $h - tv_m = d_1 v_1 + \cdots + d_{m-1} v_{m-1}$ for certain integers d_i. Then $h = d_1 v_1 + \cdots + d_{m-1} v_{m-1} + tv_m$, showing that h is a $\mathbb{Z}$-linear combination of $v_1, \ldots, v_m$.

More generally, *every subgroup of an infinite-dimensional free commutative group is also free*. We omit the proof, which is similar to the preceding proof but requires transfinite induction.

16.9 ℤ-Linear Maps and Integer Matrices

We have not yet finished our analysis of the subgroups of $\mathbb{Z}^k$. However, to complete our work in this area, we must first develop some machinery for understanding general $\mathbb{Z}$-linear maps. Eventually, we apply this material to the $\mathbb{Z}$-linear inclusion map of a subgroup into $\mathbb{Z}^k$ to gain information about that subgroup.

Suppose $T : G \to H$ is a $\mathbb{Z}$-linear map between two finitely generated free commutative groups. Let $X = (v_1, \ldots, v_n)$ be an ordered basis of G and $Y = (w_1, \ldots, w_m)$ be an ordered basis of H. We first show how to use the ordered bases X and Y to represent T by an $m \times n$ matrix with integer entries. The following construction is exactly analogous to the procedure used in linear algebra associating a matrix of scalars with a given linear transformation between two vector spaces (see Chapter 6).

We first remark that the $\mathbb{Z}$-linear map T is completely determined by its effect on the generators $v_1, \ldots, v_n$ of G. Thus, to specify T, we need only record the n elements $T(v_1), \ldots, T(v_n)$. Next, for each j, we know that $T(v_j)$ can be expressed uniquely as a $\mathbb{Z}$-linear combination of $w_1, \ldots, w_m$. In other words, for all j with $1 \leq j \leq n$, we can write

$$T(v_j) = \sum_{i=1}^{m} a_{ij} w_i \qquad (16.3)$$

for certain uniquely determined integers a_{ij}. The $m \times n$ matrix $A = [a_{ij}]$ is called the *matrix of T relative to the bases X and Y*. As long as X and Y are fixed and known, the passage from T to A is completely reversible. In other words, we could start with the matrix A and use (16.3) as the definition of T on the generators v_j. By the UMP for G, this definition extends uniquely to a $\mathbb{Z}$-linear map from G to H.

For example, consider the matrix

$$A = \begin{bmatrix} 7 & 2 & -1 & 5 \\ 0 & 4 & 0 & 2 \\ 3 & 3 & 1 & 0 \end{bmatrix}.$$

Take $G = \mathbb{Z}^4$, $H = \mathbb{Z}^3$, $v_1 = (1,0,0,0)$, $v_2 = (0,1,0,0)$, $v_3 = (0,0,1,0)$, $v_4 = (0,0,0,1)$, $w_1 = (1,0,0)$, $w_2 = (0,1,0)$, and $w_3 = (0,0,1)$, so that X and Y are the standard ordered bases of $\mathbb{Z}^4$ and $\mathbb{Z}^3$, respectively. Given this data, we obtain a $\mathbb{Z}$-linear map $T : \mathbb{Z}^4 \to \mathbb{Z}^3$ defined on basis elements by

$$
\begin{aligned}
T(v_1) &= 7w_1 + 0w_2 + 3w_3 = (7,0,3); \\
T(v_2) &= 2w_1 + 4w_2 + 3w_3 = (2,4,3); \\
T(v_3) &= -1w_1 + 0w_2 + 1w_3 = (-1,0,1); \\
T(v_4) &= 5w_1 + 2w_2 + 0w_3 = (5,2,0).
\end{aligned}
$$

Note that the jth column of A contains the coordinates of the image of the jth input basis element relative to the given output basis. We can use $\mathbb{Z}$-linearity to compute explicitly the image of an arbitrary element $(a,b,c,d) \in \mathbb{Z}^4$. In detail, for any $a,b,c,d \in \mathbb{Z}$,

$$
\begin{aligned}
T((a,b,c,d)) &= T(av_1 + bv_2 + cv_3 + dv_4) \\
&= aT(v_1) + bT(v_2) + cT(v_3) + dT(v_4) \\
&= a(7,0,3) + b(2,4,3) + c(-1,0,1) + d(5,2,0) \\
&= (7a + 2b - c + 5d, 4b + 2d, 3a + 3b + c).
\end{aligned}
$$

The same answer can be found by computing the following matrix-vector product:

$$
\begin{bmatrix} 7 & 2 & -1 & 5 \\ 0 & 4 & 0 & 2 \\ 3 & 3 & 1 & 0 \end{bmatrix}
\begin{bmatrix} a \\ b \\ c \\ d \end{bmatrix}
=
\begin{bmatrix} 7a + 2b - c + 5d \\ 4b + 2d \\ 3a + 3b + c \end{bmatrix}.
$$

We can interpret matrix addition and matrix multiplication in terms of $\mathbb{Z}$-linear maps. Returning to the general setup, suppose $T, U : G \to H$ are two $\mathbb{Z}$-linear maps. Let $A = [a_{ij}]$ and $B = [b_{ij}]$ be the integer matrices representing T and U relative to the ordered bases X and Y. Then, by definition, we must have

$$
T(v_j) = \sum_{i=1}^{m} a_{ij} w_i \quad \text{for } 1 \le j \le n;
$$

$$
U(v_j) = \sum_{i=1}^{m} b_{ij} w_i \quad \text{for } 1 \le j \le n.
$$

The *sum* of T and U is the map $T + U : G \to H$ defined by $(T + U)(x) = T(x) + U(x)$ for all $x \in G$. We see at once that $T + U$ is $\mathbb{Z}$-linear. What is the matrix of $T + U$ relative to X and Y? To find the jth column of this matrix, we must find the coordinates of $(T + U)(v_j)$ relative to Y. We find that

$$
(T + U)(v_j) = T(v_j) + U(v_j) = \sum_{i=1}^{m} a_{ij} w_i + \sum_{i=1}^{m} b_{ij} w_i = \sum_{i=1}^{m} (a_{ij} + b_{ij}) w_i.
$$

Thus, $T + U$ is represented by the $m \times n$ matrix $[a_{ij} + b_{ij}] = A + B$. This shows that *addition of matrices corresponds to addition of $\mathbb{Z}$-linear maps.*

Products are a bit more subtle. Take T and A as above, and suppose $V : H \to K$ is a $\mathbb{Z}$-linear map from H into a third free commutative group K with ordered basis $Z = (z_1, \ldots, z_p)$. There are unique integers c_{ik} such that

$$
V(w_k) = \sum_{i=1}^{p} c_{ik} z_i \quad \text{for } 1 \le k \le m;
$$

here, $C = [c_{ik}]$ is the $p \times m$ matrix of V relative to the ordered bases Y and Z. We know that the composite map $V \circ T : G \to K$ is $\mathbb{Z}$-linear. What is the matrix of this map relative to the ordered bases X and Z? As before, we discover the answer by applying this map to a general basis vector $v_j \in X$. Using $\mathbb{Z}$-linearity of the maps, commutativity of addition, and the distributive law, we calculate:

$$
\begin{aligned}
(V \circ T)(v_j) &= V(T(v_j)) = V\left(\sum_{k=1}^{m} a_{kj} w_k \right) \\
&= \sum_{k=1}^{m} a_{kj} V(w_k) = \sum_{k=1}^{m} a_{kj} \left(\sum_{i=1}^{p} c_{ik} z_i \right) \\
&= \sum_{k=1}^{m} \sum_{i=1}^{p} a_{kj} c_{ik} z_i = \sum_{i=1}^{p} \sum_{k=1}^{m} c_{ik} a_{kj} z_i \\
&= \sum_{i=1}^{p} \left(\sum_{k=1}^{m} c_{ik} a_{kj} \right) z_i.
\end{aligned}
$$

So, the ij-entry of the matrix of $V \circ T$ is $\sum_{k=1}^{m} c_{ik} a_{kj}$, which is precisely the ij-entry of the matrix product CA. We thereby see that CA is the matrix of $V \circ T$ relative to the ordered bases X and Z. So, *matrix multiplication corresponds to composition of* $\mathbb{Z}$*-linear maps*, as long as the matrices are found using the same ordered basis Y for the middle group H.

16.10 Elementary Operations and Change of Basis

Let $T : G \to H$ be a $\mathbb{Z}$-linear map between two finitely generated free commutative groups G and H. Once we fix an ordered basis X for G and an ordered basis Y for H, we obtain a unique integer-valued matrix A representing T. A key point is that this matrix depends on the ordered bases X and Y as well as the map T.

This raises the possibility of changing the matrix of T by replacing X and Y by other ordered bases. Our eventual goal is to select ordered bases for G and H judiciously, so that the matrix of T takes an especially simple form. Before pursuing this objective, we need to understand precisely how modifications of the ordered bases X and Y affect the matrix A.

Recall from §16.4 that there are three elementary operations on ordered bases: (B1) interchanges the positions of two basis elements; (B2) multiplies a basis element by -1; and (B3) adds an integer multiple of one basis element to a different basis element. We shall soon see that these operations on ordered bases correspond to analogous elementary operations on the rows and columns of A. In particular, we consider the following elementary row operations on an integer-valued matrix (analogous to the row operations used to solve linear equations via Gaussian elimination): (R1) interchanges two rows of the matrix; (R2) multiplies some row of the matrix by -1; (R3) adds an integer multiple of one row to a different row. There are similar elementary operations (C1), (C2), and (C3) that act on the columns of the matrix. The critical question is how performing the operations (B1), (B2), (B3) on Y or X causes an associated row or column operation on the matrix representing T.

The rules are most readily understood by considering a concrete example. Consider once again the matrix

$$A = \begin{bmatrix} 7 & 2 & -1 & 5 \\ 0 & 4 & 0 & 2 \\ 3 & 3 & 1 & 0 \end{bmatrix},$$

which is the matrix of a $\mathbb{Z}$-linear map $T : \mathbb{Z}^4 \to \mathbb{Z}^3$ relative to the standard ordered bases $X = (v_1, v_2, v_3, v_4)$ and $Y = (w_1, w_2, w_3)$ of $\mathbb{Z}^4$ and $\mathbb{Z}^3$. Let us first study the effect of applying one of the basis operations (B1), (B2), and (B3) to the input basis X.

In the case of (B1), let us find the matrix A_1 of T relative to the ordered bases (v_1, v_4, v_3, v_2) and (w_1, w_2, w_3). Here, we have switched the second and fourth basis elements in X. Since $T(v_1) = 7w_1 + 0w_2 + 3w_3$, the first column of A_1 has entries $7, 0, 3$. Since $T(v_4) = 5w_1 + 2w_2 + 0w_3$, the second column of A_1 has entries $5, 2, 0$. Continuing in this way, we thereby obtain

$$A_1 = \begin{bmatrix} 7 & 5 & -1 & 2 \\ 0 & 2 & 0 & 4 \\ 3 & 0 & 1 & 3 \end{bmatrix}.$$

Note that A_1 is obtained from A by interchanging columns 2 and 4. You can check that this holds in general: *if we modify the input basis by switching the vectors in positions i and j, then the new matrix is obtained by interchanging columns i and j.*

In the case of (B2), let us find the matrix of T relative to the ordered bases $(v_1, v_2, -v_3, v_4)$ and (w_1, w_2, w_3). Here, we have multiplied the third basis element in X by -1. Evidently, columns 1, 2, and 4 of the matrix are unchanged by this modification. To find the new column 3, compute

$$T(-v_3) = -T(v_3) = w_1 + 0w_2 - w_3.$$

So the new matrix is

$$A_2 = \begin{bmatrix} 7 & 2 & 1 & 5 \\ 0 & 4 & 0 & 2 \\ 3 & 3 & -1 & 0 \end{bmatrix},$$

which was obtained from A by multiplying column 3 by -1. This remark holds in general: *if we modify the input basis by negating the vector in position i, then the new matrix is obtained by negating column i.*

In the case of (B3), let us find the matrix of T relative to the ordered bases $(v_1, v_2, v_3 + 2v_4, v_4)$ and (w_1, w_2, w_3). As before, columns 1, 2, and 4 of the matrix are the same as in the original matrix A. To find the new column 3, compute

$$T(v_3 + 2v_4) = T(v_3) + 2T(v_4) = (-w_1 + w_3) + 2(5w_1 + 2w_2) = 9w_1 + 4w_2 + w_3.$$

So the new matrix is

$$A_3 = \begin{bmatrix} 7 & 2 & 9 & 5 \\ 0 & 4 & 4 & 2 \\ 3 & 3 & 1 & 0 \end{bmatrix},$$

which was obtained from A by adding two times column 4 to column 3. You can verify that this holds in general: *if we modify the ith input basis vector by adding c times the jth basis vector to it, then the new matrix is obtained by adding c times column j to column i.*

To summarize, performing elementary operations on the input basis causes elementary column operations on the matrix of T. Next, we show that performing elementary operations on the output basis causes elementary row operations on the matrix of T.

In the case of (B1), let us find the matrix of T relative to the ordered bases (v_1, v_2, v_3, v_4) and (w_1, w_3, w_2). Taking into account the new ordering of the output basis, we have

$$T(v_1) = 7w_1 + 0w_2 + 3w_3 = 7w_1 + 3w_3 + 0w_2.$$

So the first column of the new matrix has entries $7, 3, 0$ (in this order) instead of $7, 0, 3$. The other columns are affected similarly. So the new matrix is

$$A_4 = \begin{bmatrix} 7 & 2 & -1 & 5 \\ 3 & 3 & 1 & 0 \\ 0 & 4 & 0 & 2 \end{bmatrix},$$

which is obtained from A by interchanging rows 2 and 3. This holds in general: *if we modify the output basis by switching the elements in positions i and j, then the associated matrix is found by interchanging rows i and j.*

In the case of (B2), let us find the matrix of T relative to the ordered bases (v_1, v_2, v_3, v_4) and $(-w_1, w_2, w_3)$. First,

$$T(v_1) = 7w_1 + 0w_2 + 3w_3 = (-7)(-w_1) + 0w_2 + 3w_3,$$

so the first column of the new matrix has entries $-7, 0, 3$. Second,

$$T(v_2) = 2w_1 + 4w_2 + 3w_3 = (-2)(-w_1) + 4w_2 + 3w_3,$$

so the second column of the new matrix has entries $-2, 4, 3$. Continuing similarly, we obtain the new matrix

$$A_5 = \begin{bmatrix} -7 & -2 & 1 & -5 \\ 0 & 4 & 0 & 2 \\ 3 & 3 & 1 & 0 \end{bmatrix},$$

which is obtained from A by multiplying the first row by -1. This holds in general: *if we modify the output basis by negating the ith element, then the associated matrix is found by negating the ith row.*

In the case of (B3), let us find the matrix of T relative to the ordered bases (v_1, v_2, v_3, v_4) and $(w_1 + 2w_3, w_2, w_3)$. Compute

$$\begin{aligned} T(v_1) &= 7w_1 + 0w_2 + 3w_3 = 7(w_1 + 2w_3) + 0w_2 + (3 - 7 \cdot 2)w_3; \\ T(v_2) &= 2w_1 + 4w_2 + 3w_3 = 2(w_1 + 2w_3) + 4w_2 + (3 - 2 \cdot 2)w_3; \\ T(v_3) &= -1w_1 + 0w_2 + 1w_3 = -1(w_1 + 2w_3) + 0w_2 + (1 - (-1) \cdot 2)w_3; \\ T(v_4) &= 5w_1 + 2w_2 + 0w_3 = 5(w_1 + 2w_3) + 2w_2 + (0 - 5 \cdot 2)w_3. \end{aligned}$$

This leads to the new matrix

$$A_6 = \begin{bmatrix} 7 & 2 & -1 & 5 \\ 0 & 4 & 0 & 2 \\ -11 & -1 & 3 & -10 \end{bmatrix},$$

which is obtained from A by adding -2 times row 1 to row 3. This result, which may not be what you were expecting, generalizes as follows: *if we modify the output basis by replacing w_i by $w_i + cw_j$, then the associated matrix is found by adding $-c$ times row i to row j.* Here is the proof of the general case, where we take $i = 1$ and $j = 2$ solely for notational convenience. Relative to the original input basis $(v_1, \ldots, v_n)$ and output basis $(w_1, \ldots, w_m)$,

$$T(v_k) = a_{1k}w_1 + a_{2k}w_2 + \sum_{i=3}^{m} a_{ik}w_i \qquad \text{for } 1 \le k \le n.$$

We now replace w_1 by $w_1 + cw_2$. To maintain equality, we must subtract the term $ca_{1k}w_2$. Regrouping terms, we get

$$T(v_k) = a_{1k}(w_1 + cw_2) + (a_{2k} - ca_{1k})w_2 + \sum_{i=3}^{m} a_{ik}w_i.$$

Thus, for all k, the kth column of the new matrix has entries $a_{1k}, a_{2k} - ca_{1k}, a_{3k}, \ldots, a_{mk}$. So, the new matrix is indeed found by adding $-c$ times row 1 to row 2.

16.11 Reduction Theorem for Integer Matrices

Now that we understand the connection between $\mathbb{Z}$-linear maps and integer matrices, we prove a theorem describing how much an integer matrix can be simplified by row and column operations. This theorem helps us uncover structural properties of $\mathbb{Z}$-linear maps between finitely generated free commutative groups.

Reduction Theorem for Integer Matrices. Let A be an $m \times n$ matrix with integer entries. There is a finite sequence of elementary row and column operations that reduces A to a matrix

$$\begin{bmatrix} a_1 & 0 & 0 & \dots & 0 \\ 0 & a_2 & 0 & \dots & 0 \\ 0 & 0 & a_3 & \dots & 0 \\ \vdots & & & \ddots & \end{bmatrix}, \tag{16.4}$$

where there are r positive integers $a_1, \dots, a_r$ on the main diagonal, a_i divides a_{i+1} for $1 \le i < r$, and all other entries in the matrix are 0.

By Exercise 100, the matrix (16.4) satisfying the stated conditions is uniquely determined by A. This reduced matrix is sometimes called the *Smith normal form* of A.

Proof. We prove the theorem by induction. This proof can be translated into an explicit recursive algorithm for reducing a given input matrix A to the required form. The base cases of the induction occur when $m = 0$ or $n = 0$ or when every entry of A is 0. In these cases, A already has the required form, so there is nothing to do.

For the induction step, assume that the reduction theorem is known to hold when the total number of rows and columns of A is less than $m + n$. Our strategy in this step is to transform A to a matrix of the form

$$\begin{bmatrix} a_1 & 0 & \cdots & 0 \\ 0 & & & \\ \vdots & & A' & \\ 0 & & & \end{bmatrix}, \tag{16.5}$$

where $a_1 > 0$ and A' is an $(m-1) \times (n-1)$ integer-valued matrix all of whose entries are divisible by a_1. Assuming that this has been done, we can use the induction hypothesis to continue to reduce A' to a matrix with some positive entries $a_2, \dots, a_r$ on its diagonal, such that a_i divides a_{i+1} for $2 \le i < r$, and with zeroes elsewhere. The operations used to reduce A' do not affect the zeroes in row 1 and column 1 of the overall matrix. Furthermore, you can check that if an integer a_1 divides every entry of a matrix, and if we perform an elementary row or column operation on that matrix, then a_1 still divides every entry of the new matrix (Exercise 63). So, as we continue to reduce A', a_1 always divides every entry of all the matrices obtained along the way. In particular, at the end, a_1 divides a_2, and A has been reduced to a matrix of the required form.

To summarize, we need only find a way of reducing A to a matrix of the form (16.5). We assume that A is not a zero matrix, since that situation was handled in the base cases. There are two possibilities to consider. *Case 1:* There exists an entry $d = a_{ij}$ in A such that d divides all the entries of A. In this case, switch rows 1 and i, and then switch columns 1 and j, to bring d into the $1, 1$-position. Since d divides every other entry in its column, we can subtract appropriate integer multiples of row 1 from the other rows to produce zeroes below d in column 1. Similarly, we can use column operations to produce zeroes in row 1 to the right of d. By the remark in the last paragraph, d continues to divide all the entries in the matrix as we perform these various operations on rows and columns. Finally, we can multiply row 1 by -1 to make d positive, if necessary. We now have a matrix of the form (16.5), which completes the proof in this case.

Case 2: There does not exist an entry of A that divides all the entries of A. When this case occurs, we adopt the following strategy. Let $m(A)$ be the smallest of the integers $|a_{ij}|$ as a_{ij} ranges through the nonzero entries of A. We aim to reduce A to a new matrix A_2 such that $m(A) > m(A_2)$. We then repeat the whole reduction algorithm on A_2. If A_2

satisfies Case 1, then we can finish reducing as above. On the other hand, if A_2 satisfies Case 2, we reduce A_2 to a new matrix A_3 such that $m(A_2) > m(A_3)$. Continuing in this way, either Case 1 eventually occurs (in which case the reduction succeeds), or Case 2 occurs indefinitely. But in the latter situation, we have an infinite strictly decreasing sequence of positive integers

$$m(A) > m(A_2) > m(A_3) > \cdots,$$

which violates the Least Natural Number Axiom for $\mathbb{Z}_{>0}$. So the reduction of A must always terminate after a finite number of steps.

We still need to explain how to obtain the matrix A_2 such that $m(A) > m(A_2)$. For this, let $e = a_{ij}$ be a nonzero entry of A with minimum absolute value (so that $m(A) = |e|$). By definition of Case 2, there exist entries in the matrix that are not divisible by e. We consider various subcases.

Case 2a: There exists $k \neq j$ such that e does not divide a_{ik}. In other words, there is an entry in the same row as e that is not divisible by e. Dividing a_{ik} by e, we then have $a_{ik} = qe + r$ where $0 < r < |e|$. Subtracting q times column j from column k produces a matrix A_2 with an r in the i, k-position. Now $m(A_2) \leq r < |e|$, so we have achieved our goal in this case.

Case 2b: There exists $k \neq i$ such that e does not divide a_{kj}. In other words, there is an entry in the same column as e that is not divisible by e. Dividing a_{kj} by e, we then have $a_{kj} = qe + r$ where $0 < r < |e|$. Subtracting q times row i from row k produces a matrix A_2 with an r in the k, j-position. As before, $m(A_2) \leq r < |e|$, so we have achieved our goal in this case.

Case 2c: e divides every entry in its row and column, but for some $k \neq i$ and $t \neq j$, e does not divide a_{kt}. So, for some $u, v \in \mathbb{Z}$, the matrix A looks like this (where we only show the four relevant entries in rows i, k and columns j, t):

$$\begin{bmatrix} & & \cdots & & \\ & e & \cdots & ue & \\ & \vdots & \vdots & \vdots & \vdots \\ & ve & \cdots & a_{kt} & \\ & & \cdots & & \end{bmatrix}.$$

Now, add $(1 - v)$ times row i to row k, obtaining:

$$\begin{bmatrix} & & \cdots & & \\ & e & \cdots & ue & \\ & \vdots & \vdots & \vdots & \vdots \\ & e & \cdots & a_{kt} + (1 - v)ue & \\ & & \cdots & & \end{bmatrix}.$$

The new k, t-entry is not divisible by e, since otherwise e would divide a_{kt}. So we can proceed as in Case 2a, subtracting an appropriate multiple of column j from column t to get a new k, t-entry $r < |e|$. The new matrix A_2 satisfies $m(A_2) \leq r < |e|$. The case analysis is finally complete. $\square$

An example of the reduction process for a specific integer-valued matrix appears in §16.14.

16.12   Structure of $\mathbb{Z}$-Linear Maps of Free Commutative Groups

We now apply the Reduction Theorem for Integer Matrices to study the structure of $\mathbb{Z}$-linear maps and finitely generated commutative groups. First, suppose $T : G \to H$ is a $\mathbb{Z}$-linear map between two finitely generated free commutative groups. Start with any ordered basis X for G and any ordered basis Y for H, and let A be the matrix of T relative to X and Y. Use row and column operations to reduce A to the form given in the theorem, modifying the bases X and Y accordingly. At the end, we have a new ordered basis $X' = (x_1, \ldots, x_n)$ for G and a new ordered basis $Y' = (y_1, \ldots, y_m)$ for H such that the matrix of T relative to X' and Y' has the form

$$\begin{bmatrix} a_1 & 0 & 0 & \ldots & 0 \\ 0 & a_2 & 0 & \ldots & 0 \\ 0 & 0 & a_3 & \ldots & 0 \\ & & & \ddots & \end{bmatrix},$$

where $a_1, \ldots, a_r$ are positive integers such that a_i divides a_{i+1} for $1 \leq i < r$. Inspection of the columns of this matrix tells us how T acts on each input x_i. Specifically, $T(x_i) = a_i y_i$ for $1 \leq i \leq r$ and $T(x_i) = 0$ for $r < i \leq n$. We call the matrix displayed above the *Smith normal form* for the $\mathbb{Z}$-linear map T; it is uniquely determined by T (cf. Exercise 100).

Compare these results to the corresponding fact about linear transformations of vector spaces over fields. In that setting, we are allowed to multiply rows and columns by arbitrary nonzero scalars. The net effect of this extra ability is that we can ensure that each a_i in the final matrix equals 1 (Exercise 66). The reduction process is also easier because we can use any nonzero scalar to create zeroes in all the other entries in its row and column — no integer division is needed. So, if $T : V \to W$ is a linear map between vector spaces, there exist ordered bases $X = (x_1, \ldots, x_n)$ for V and $Y = (y_1, \ldots, y_m)$ for W such that $T(x_i) = y_i$ for $1 \leq i \leq r$, and $T(x_i) = 0$ for $r < i \leq n$. The number r is called the *rank* of the linear map T. We can also reach this result without reducing any matrices. Start with a basis $(x_{r+1}, \ldots, x_n)$ for the null space of T, and extend it to a basis of $(x_1, \ldots, x_n)$ of V. You can check that the list $T(x_1), \ldots, T(x_r)$ is linearly independent, so this list can be extended to a basis of the target space Y. We used this argument to prove the Rank–Nullity Theorem in §1.8.

16.13   Structure of Finitely Generated Commutative Groups

Returning to commutative groups, we are now ready to prove the existence part of the Theorem Classifying Finitely Generated Commutative Groups. Suppose H is a commutative group generated by m elements. We have seen (§16.7) that H is isomorphic to a quotient group $\mathbb{Z}^m/P$, where P is some subgroup of the free commutative group $\mathbb{Z}^m$. We have also seen (§16.8) that the subgroup P must also be free, with a basis of size $n \leq m$.

Consider the inclusion map $I : P \to \mathbb{Z}^m$, given by $I(x) = x$ for $x \in P$. I is certainly $\mathbb{Z}$-linear, and it is a map between two finitely generated free commutative groups. So our structural result for such maps can be applied. We see, therefore, that there is a basis $(x_1, \ldots, x_n)$ for P, a basis $(y_1, \ldots, y_m)$ of $\mathbb{Z}^m$, an integer $r \geq 0$, and positive integers $a_1, \ldots, a_r$ with a_i dividing a_{i+1} for $1 \leq i < r$, such that $I(x_i) = a_i y_i$ for $1 \leq i \leq r$, and

$I(x_i) = 0$ for $r < i \leq n$. Since I is an inclusion map, this says that $x_i = a_i y_i$ for $1 \leq i \leq r$, and $x_i = 0$ for $r < i \leq n$. But basis elements are never 0 (by $\mathbb{Z}$-independence), so we deduce that $r = n \leq m$. To summarize, $H \cong \mathbb{Z}^m / P$, where $\mathbb{Z}^m$ has some ordered basis $(y_1, \ldots, y_n, \ldots, y_m)$ and P has an ordered basis $(a_1 y_1, \ldots, a_n y_n)$.

We now apply an isomorphism of $\mathbb{Z}^m$ with itself that forces the subgroup P to assume the special form given in (16.2). Consider the function $f : \{y_1, \ldots, y_m\} \to \mathbb{Z}^m$ such that $f(y_i) = e_i$ (the standard basis vector) for $1 \leq i \leq m$. By the UMP for free commutative groups, f extends to a $\mathbb{Z}$-linear map $T : \mathbb{Z}^m \to \mathbb{Z}^m$. As seen at the end of §16.6, T is an isomorphism. The isomorphism T maps P to a new subgroup P_1 of $\mathbb{Z}^m$ with ordered basis $(a_1 e_1, \ldots, a_n e_n)$. You can check that T induces an isomorphism from the quotient group $\mathbb{Z}^m / P$ to the quotient group $\mathbb{Z}^m / P_1$. Now, we can write P_1 as the Cartesian product

$$a_1 \mathbb{Z} \times a_2 \mathbb{Z} \times \cdots \times a_n \mathbb{Z} \times 0\mathbb{Z} \times \cdots \times 0\mathbb{Z},$$

where there are $m - n$ factors equal to $\{0\}$. Applying the Fundamental Homomorphism Theorem (see the discussion below (16.2)), we conclude that

$$H \cong \mathbb{Z}^m / P \cong \mathbb{Z}^m / P_1 \cong \mathbb{Z}_{a_1} \oplus \cdots \oplus \mathbb{Z}_{a_n} \oplus \mathbb{Z}^{m-n}.$$

Thus, *every finitely generated commutative group is isomorphic to a direct sum of finitely many cyclic groups, where the sizes of the finite cyclic summands (if any) successively divide each other.* Note that if some of the a_i are equal to 1, we can omit these factors from the product, since $\mathbb{Z}_1$ is the one-element group.

We now derive another version of this result, obtained by splitting apart the cyclic groups $\mathbb{Z}_{a_i}$ based on the prime factorizations of each a_i. First we need a group-theoretic lemma. Suppose $a > 1$ is an integer with prime factorization $a = p_1^{e_1} p_2^{e_2} \cdots p_s^{e_s}$, where each $e_i \geq 1$ and $p_1, \ldots, p_s$ are distinct primes. We claim that $\mathbb{Z}_a \cong \mathbb{Z}_{p_1^{e_1}} \oplus \mathbb{Z}_{p_2^{e_2}} \oplus \cdots \oplus \mathbb{Z}_{p_s^{e_s}}$, or equivalently,

$$\mathbb{Z}/a\mathbb{Z} \cong (\mathbb{Z}/p_1^{e_1}\mathbb{Z}) \oplus (\mathbb{Z}/p_2^{e_2}\mathbb{Z}) \oplus \cdots \oplus (\mathbb{Z}/p_s^{e_s}\mathbb{Z}). \tag{16.6}$$

To prove this, call the product group on the right side K, and consider the map sending the integer 1 to the s-tuple of cosets $(1 + p_1^{e_1}\mathbb{Z}, \ldots, 1 + p_s^{e_s}\mathbb{Z}) \in K$. Since $\mathbb{Z}$ is free with basis $\{1\}$, the UMP furnishes a $\mathbb{Z}$-linear extension $T : \mathbb{Z} \to K$ such that

$$T(n) = nT(1) = (n + p_1^{e_1}\mathbb{Z}, \ldots, n + p_s^{e_s}\mathbb{Z}) \qquad \text{for all } n \in \mathbb{Z}.$$

Let us find the kernel of T. By the formula just written, n belongs to the kernel iff all cosets $n + p_i^{e_i}\mathbb{Z}$ are zero iff $n \in p_i^{e_i}\mathbb{Z}$ for all i iff $p_i^{e_i}$ divides n for all i iff $\operatorname{lcm}(p_i^{e_i} : 1 \leq i \leq s)$ divides n iff $a = p_1^{e_1} \cdots p_s^{e_s}$ divides n (since $p_1, \ldots, p_s$ are distinct primes). In other words, the kernel of T is $a\mathbb{Z}$. By the Fundamental Homomorphism Theorem, the quotient group $\mathbb{Z}/a\mathbb{Z} \cong \mathbb{Z}_a$ is isomorphic to the image of T in K. But the size of the product group K is $p_1^{e_1} p_2^{e_2} \cdots p_s^{e_s} = a$, which is the same size as $\mathbb{Z}/a\mathbb{Z}$. It follows that the isomorphic copy of $\mathbb{Z}/a\mathbb{Z}$ in K must be all of K, completing the proof of the claim.

Apply this result to each of the integers a_i in the preceding decomposition of H. We conclude that *every finitely generated commutative group is isomorphic to a direct sum of finitely many cyclic groups, each of which is either infinite or has size equal to a prime power.* This concludes the proof of the existence part of the Theorem Classifying Finitely Generated Commutative Groups. We must still address the question of the uniqueness of the two decompositions of H found above. We consider this issue shortly, but first we give a concrete example illustrating the reduction algorithm and the ideas in the proofs just given.

16.14   Example of the Reduction Algorithm

Let P be the subgroup of $\mathbb{Z}^4$ generated by $v_1 = (10, 0, -8, 4)$, $v_2 = (12, 6, -6, -6)$, and $v_3 = (20, 48, 14, -82)$. We use the ideas in the last few sections to determine the structure of the quotient group $H = \mathbb{Z}^4/P$.

You can verify that v_1, v_2, v_3 are $\mathbb{Z}$-independent. (This also follows from the calculations below; see Exercise 69.) So we can consider the matrix of the inclusion map $I : P \to \mathbb{Z}^4$ relative to the ordered basis $X = (v_1, v_2, v_3)$ of P and the standard ordered basis $Y = (e_1, e_2, e_3, e_4)$ of $\mathbb{Z}^4$. The jth column of this matrix gives the coordinates of v_j relative to the standard ordered basis, so the matrix is

$$A = \begin{bmatrix} 10 & 12 & 20 \\ 0 & 6 & 48 \\ -8 & -6 & 14 \\ 4 & -6 & -82 \end{bmatrix}.$$

We proceed to reduce this matrix. Inspection reveals that no entry of A divides every other entry. So our first goal (following the proof of Case 2 of the reduction theorem) is to reduce the magnitude of the smallest nonzero entry of A. This entry is 4, which fails to divide the entry 10 in its column. As prescribed by Case 2b of the reduction proof, we replace row 1 by row 1 minus 2 times row 4, obtaining

$$A_1 = \begin{bmatrix} 2 & 24 & 184 \\ 0 & 6 & 48 \\ -8 & -6 & 14 \\ 4 & -6 & -82 \end{bmatrix}.$$

This is the matrix of I relative to the bases (v_1, v_2, v_3) and $(e_1, e_2, e_3, e_4 + 2e_1)$.

The new $1, 1$-entry, namely 2, does divide every entry of the matrix. So we are in Case 1 of the reduction proof. First, we use two column operations to produce zero entries in the rest of row 1:

$$A_2 = \begin{bmatrix} 2 & 0 & 0 \\ 0 & 6 & 48 \\ -8 & 90 & 750 \\ 4 & -54 & -450 \end{bmatrix}.$$

This is the matrix of I relative to the bases $(v_1, v_2 - 12v_1, v_3 - 92v_1)$ and $(e_1, e_2, e_3, e_4 + 2e_1)$. Second, we use two row operations to produce zero entries in the rest of column 1:

$$A_3 = \begin{bmatrix} 2 & 0 & 0 \\ 0 & 6 & 48 \\ 0 & 90 & 750 \\ 0 & -54 & -450 \end{bmatrix}.$$

This is the matrix of I relative to the input basis $(v_1, v_2 - 12v_1, v_3 - 92v_1)$ and output basis $(e_1 - 4e_3 + 2(e_4 + 2e_1), e_2, e_3, e_4 + 2e_1) = (5e_1 - 4e_3 + 2e_4, e_2, e_3, e_4 + 2e_1)$.

Now, we proceed to reduce the 3×2 submatrix in the lower-right corner. The upper-left entry of this submatrix (namely 6) already divides every other entry of this submatrix. Adding -8 times column 2 to column 3 gives

$$A_4 = \begin{bmatrix} 2 & 0 & 0 \\ 0 & 6 & 0 \\ 0 & 90 & 30 \\ 0 & -54 & -18 \end{bmatrix};$$

the new bases are $(v_1, v_2 - 12v_1, v_3 - 92v_1 - 8(v_2 - 12v_1))$ and $(5e_1 - 4e_3 + 2e_4, e_2, e_3, e_4 + 2e_1)$.
Next, two row operations produce

$$A_5 = \begin{bmatrix} 2 & 0 & 0 \\ 0 & 6 & 0 \\ 0 & 0 & 30 \\ 0 & 0 & -18 \end{bmatrix};$$

the new bases are $(v_1, v_2 - 12v_1, 4v_1 - 8v_2 + v_3)$ and

$$(5e_1 - 4e_3 + 2e_4, e_2 + 15e_3 - 9(e_4 + 2e_1), e_3, e_4 + 2e_1).$$

To continue, we must reduce the 2×1 submatrix with entries 30 and -18. We are in Case 2b again. Adding 2 times row 4 to row 3 gives

$$A_6 = \begin{bmatrix} 2 & 0 & 0 \\ 0 & 6 & 0 \\ 0 & 0 & -6 \\ 0 & 0 & -18 \end{bmatrix};$$

the new bases are $(v_1, v_2 - 12v_1, 4v_1 - 8v_2 + v_3)$ and

$$(5e_1 - 4e_3 + 2e_4, -18e_1 + e_2 + 15e_3 - 9e_4, e_3, e_4 + 2e_1 - 2e_3).$$

Next, multiply row 3 by -1, obtaining the matrix

$$A_7 = \begin{bmatrix} 2 & 0 & 0 \\ 0 & 6 & 0 \\ 0 & 0 & 6 \\ 0 & 0 & -18 \end{bmatrix}$$

and bases $(v_1, v_2 - 12v_1, 4v_1 - 8v_2 + v_3)$ and

$$(5e_1 - 4e_3 + 2e_4, -18e_1 + e_2 + 15e_3 - 9e_4, -e_3, e_4 + 2e_1 - 2e_3).$$

Finally, add 3 times row 3 to row 4 to get the reduced matrix

$$A_8 = \begin{bmatrix} 2 & 0 & 0 \\ 0 & 6 & 0 \\ 0 & 0 & 6 \\ 0 & 0 & 0 \end{bmatrix}$$

and bases $(v_1, v_2 - 12v_1, 4v_1 - 8v_2 + v_3)$ and

$$(5e_1 - 4e_3 + 2e_4, -18e_1 + e_2 + 15e_3 - 9e_4, -e_3 - 3(e_4 + 2e_1 - 2e_3), e_4 + 2e_1 - 2e_3).$$

To check our work, note that the final ordered input basis for P is (x_1, x_2, x_3), where:

$$\begin{aligned} x_1 &= v_1 = (10, 0, -8, 4); \\ x_2 &= v_2 - 12v_1 = (-108, 6, 90, -54); \\ x_3 &= 4v_1 - 8v_2 + v_3 = (-36, 0, 30, -18); \end{aligned}$$

and the final ordered output basis for $\mathbb{Z}^4$ is (y_1, y_2, y_3, y_4), where:

$$\begin{aligned} y_1 &= 5e_1 - 4e_3 + 2e_4 = (5, 0, -4, 2); \\ y_2 &= -18e_1 + e_2 + 15e_3 - 9e_4 = (-18, 1, 15, -9); \\ y_3 &= -6e_1 + 5e_3 - 3e_4 = (-6, 0, 5, -3); \\ y_4 &= 2e_1 - 2e_3 + e_4 = (2, 0, -2, 1). \end{aligned}$$

As predicted by the proof, we have $x_1 = 2y_1$, $x_2 = 6y_2$ and $x_3 = 6y_3$. Therefore,

$$H \cong \frac{\mathbb{Z}^4}{P} \cong \frac{\mathbb{Z} \times \mathbb{Z} \times \mathbb{Z} \times \mathbb{Z}}{2\mathbb{Z} \times 6\mathbb{Z} \times 6\mathbb{Z} \times 0\mathbb{Z}} \cong \mathbb{Z}_2 \oplus \mathbb{Z}_6 \oplus \mathbb{Z}_6 \oplus \mathbb{Z}.$$

Using the prime factorization $6 = 2 \cdot 3$, we also have

$$H \cong \mathbb{Z}_2 \oplus \mathbb{Z}_2 \oplus \mathbb{Z}_2 \oplus \mathbb{Z}_3 \oplus \mathbb{Z}_3 \oplus \mathbb{Z}.$$

16.15   Some Special Subgroups

To finish our classification of finitely generated commutative groups, we prove the following two uniqueness theorems.

Theorem on Uniqueness of Invariant Factors. Suppose G is a commutative group such that

$$\mathbb{Z}^b \oplus \mathbb{Z}_{a_1} \oplus \cdots \oplus \mathbb{Z}_{a_r} \cong G \cong \mathbb{Z}^d \oplus \mathbb{Z}_{c_1} \oplus \cdots \oplus \mathbb{Z}_{c_t}, \tag{16.7}$$

where $b, d, r, t \in \mathbb{Z}_{\geq 0}$, all a_i and c_j are in $\mathbb{Z}_{>1}$, a_i divides a_{i+1} for $1 \leq i < r$, and c_j divides c_{j+1} for $1 \leq j < t$. Then $b = d$ and $r = t$ and $a_i = c_i$ for $1 \leq i \leq r$.

We call b the *Betti number* of G, and we call $a_1, \ldots, a_r$ the *invariant factors* of G. The theorem says these integers are isomorphism invariants of G.

Theorem on Uniqueness of Elementary Divisors. Suppose G is a commutative group such that

$$\mathbb{Z}^b \oplus \mathbb{Z}_{a_1} \oplus \cdots \oplus \mathbb{Z}_{a_r} \cong G \cong \mathbb{Z}^d \oplus \mathbb{Z}_{c_1} \oplus \cdots \oplus \mathbb{Z}_{c_t}, \tag{16.8}$$

where $b, d, r, t \in \mathbb{Z}_{\geq 0}$, $a_1 \geq a_2 \geq \cdots \geq a_r > 1$, $c_1 \geq c_2 \geq \cdots \geq c_t > 1$, and every a_i and every c_j is a prime power. Then $b = d$ and $r = t$ and $a_i = c_i$ for $1 \leq i \leq r$.

The integer b is the Betti number of G, and the prime powers $a_1, \ldots, a_r$ are called *elementary divisors* of G. The theorem says these integers are isomorphism invariants of G.

We prove these results in stages, first considering the cases where $r = t = 0$ (which means G is free) and where $b = d = 0$ (which means G is finite). To aid our proofs, we must first introduce some special subgroups of G that are invariant under group isomorphisms. Let G be an arbitrary commutative group and n be any fixed integer. Consider the map $M_n : G \to G$ given by $M_n(g) = ng$ for $g \in G$. Because G is commutative, M_n is a group homomorphism: for each $g, h \in G$,

$$M_n(g + h) = n(g + h) = ng + nh = M_n(g) + M_n(h).$$

(This result need not hold for non-commutative groups.) Define $nG = \{ng : g \in G\}$ and $G[n] = \{g \in G : ng = 0\}$. The sets nG and $G[n]$ are subgroups of G, since nG is the image of the homomorphism M_n, and $G[n]$ is the kernel of M_n.

We assert that these subgroups are preserved by group isomorphisms. More precisely, suppose $f : G \to H$ is a group isomorphism. We claim the restriction of f to $G[n] \subseteq G$ is a bijection of $G[n]$ onto $H[n]$. To prove this, suppose $x \in G[n]$. Then $nx = 0$, so $f(nx) = 0$, so $nf(x) = 0$, so $f(x) \in H[n]$. So the restriction of f to the domain $G[n]$ does map into the codomain $H[n]$. Applying the same argument to the inverse isomorphism f^{-1}, we see that f^{-1} maps $H[n]$ into $G[n]$. Hence, $G[n] \cong H[n]$ via the restriction of f to this domain and codomain. Similar reasoning proves that f restricts to a group isomorphism

$nG \cong nH$. Here, the key point is that for $x \in nG$, we have $x = ng$ for some $g \in G$, hence $f(x) = f(ng) = nf(g)$ where $f(g) \in H$, hence $f(x) \in nH$. So f maps nG into nH, and likewise f^{-1} maps nH into nG. We can conclude that f induces isomorphisms of quotient groups $G/G[n] \cong H/H[n]$ and $G/nG \cong H/nH$. For instance, the first of these isomorphisms follows by applying the Fundamental Homomorphism Theorem to the homomorphism from G to $H/H[n]$ sending $g \in G$ to $f(g) + H[n]$, which is a surjective homomorphism with kernel $G[n]$.

Another special subgroup of G is the set $\text{tor}(G) = \bigcup_{n \geq 1} G[n]$, which consists of all elements of G of finite order: $g \in \text{tor}(G)$ iff $g \in G$ and $ng = 0$ for some $n \in \mathbb{Z}_{>0}$. We call $\text{tor}(G)$ the *torsion subgroup of G*. To see that $\text{tor}(G)$ is a subgroup, first note that $0_G \in \text{tor}(G)$ since $1(0_G) = 0$. Given $g \in \text{tor}(G)$, we know $ng = 0$ for some $n \in \mathbb{Z}_{>0}$. Then $n(-g) = -(ng) = -0 = 0$, so $-g \in \text{tor}(G)$. To check closure under addition, suppose $g, h \in \text{tor}(G)$, so that $ng = 0 = mh$ for some $n, m > 0$. Because G is commutative,

$$nm(g + h) = nm(g) + nm(h) = m(ng) + n(mh) = m0 + n0 = 0.$$

So $g + h \in \text{tor}(G)$, and $\text{tor}(G)$ is indeed a subgroup. (This result is false in some infinite non-commutative groups.) As with the previous subgroups, the torsion subgroup is an isomorphism invariant: if $f : G \to H$ is an isomorphism, then f restricts to an isomorphism $\text{tor}(G) \cong \text{tor}(H)$. So f induces an isomorphism of quotient groups $G/\text{tor}(G) \cong H/\text{tor}(H)$.

16.16 Uniqueness Proof: Free Case

To see how the subgroups introduced in the preceding section can be relevant, let us prove the uniqueness theorem for finitely generated free commutative groups. Suppose G is a free commutative group with an n-element basis X and an m-element basis Y. We prove that $m = n$. This shows that the dimension (Betti number) of a finitely generated free commutative group is well-defined. The assumptions on G imply that $G \cong \mathbb{Z}^n$ and $G \cong \mathbb{Z}^m$ (§16.6). Combining these isomorphisms gives an isomorphism $f : \mathbb{Z}^n \to \mathbb{Z}^m$. We show that the existence of f forces $n = m$.

We know from the previous section that the isomorphism f induces an isomorphism $f' : \mathbb{Z}^n/2\mathbb{Z}^n \to \mathbb{Z}^m/2\mathbb{Z}^m$. Now, $2\mathbb{Z}^n = \{2v : v \in \mathbb{Z}^n\} = \{2(a_1, \ldots, a_n) : a_i \in \mathbb{Z}\} = \{(2a_1, \ldots, 2a_n) : a_i \in \mathbb{Z}\} = 2\mathbb{Z} \oplus 2\mathbb{Z} \oplus \cdots \oplus 2\mathbb{Z}$. This is a subgroup of $\mathbb{Z}^n$ of the form (16.2). So, as shown below (16.2),

$$\frac{\mathbb{Z}^n}{2\mathbb{Z}^n} = \frac{\mathbb{Z} \oplus \cdots \oplus \mathbb{Z}}{2\mathbb{Z} \oplus \cdots \oplus 2\mathbb{Z}} \cong \frac{\mathbb{Z}}{2\mathbb{Z}} \oplus \cdots \oplus \frac{\mathbb{Z}}{2\mathbb{Z}} \cong \mathbb{Z}_2^n.$$

Similarly, $\mathbb{Z}^m/2\mathbb{Z}^m \cong \mathbb{Z}_2^m$. So we obtain an isomorphism $\mathbb{Z}_2^n \cong \mathbb{Z}_2^m$. Now, the product group $\mathbb{Z}_2^n$ has 2^n elements, while $\mathbb{Z}_2^m$ has 2^m elements. Since isomorphic groups have the same size, we deduce $2^n = 2^m$, which in turn implies $n = m$.

An analogous result holds in linear algebra: any two bases of a vector space have the same cardinality (see Chapter 17 for a proof). In particular, if the real vector spaces $\mathbb{R}^n$ and $\mathbb{R}^m$ are isomorphic, then $m = n$. However, we cannot prove this result by the counting argument used above. Remarkably, the *commutative groups* $\mathbb{R}^n$ and $\mathbb{R}^m$ *are* isomorphic for all positive integers m and n (see Exercise 101). Intuition suggests that the *topological spaces* $\mathbb{R}^n$ and $\mathbb{R}^m$ should be homeomorphic iff $m = n$. This is true, but it is quite difficult to prove when $n, m > 1$. We need the tools of algebraic topology to establish that $\mathbb{R}^n$ and $\mathbb{R}^m$ are not homeomorphic when $m \neq n$. See the texts of Munkres [46] or Rotman [53] for details. We remark that algebraic topology makes heavy use of the classification theorems for commutative groups proved in this chapter.

16.17 Uniqueness Proof: Prime Power Case

As the next step in the uniqueness proof, we prove the following result about commutative groups whose size is a prime power. *Suppose p is prime, and there are integers $a_1 \geq a_2 \geq \cdots \geq a_r > 0$ and $c_1 \geq c_2 \geq \cdots \geq c_t > 0$ such that*

$$\mathbb{Z}_{p^{a_1}} \oplus \mathbb{Z}_{p^{a_2}} \oplus \cdots \oplus \mathbb{Z}_{p^{a_r}} \cong \mathbb{Z}_{p^{c_1}} \oplus \mathbb{Z}_{p^{c_2}} \oplus \cdots \oplus \mathbb{Z}_{p^{c_t}}.$$

Then $r = t$ and $a_i = c_i$ for $1 \leq i \leq r$.

The proof is greatly clarified by introducing the notion of an integer partition (cf. Chapter 8). A *partition* of an integer n is a weakly decreasing sequence $(a_1 \geq a_2 \geq \cdots \geq a_r)$ of positive integers that sum to n. For example, the seven partitions of $n = 5$ are:

$$(5), \quad (4,1), \quad (3,2), \quad (3,1,1), \quad (2,2,1), \quad (2,1,1,1), \quad (1,1,1,1,1).$$

We can visualize a partition by drawing a collection of n squares such that there are a_i squares in row i. This picture is called the *diagram* of the partition. For example, the diagrams of the seven partitions of 5 are shown here:

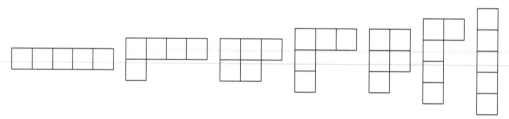

Each partition encodes a possible commutative group of size p^n (where p is any fixed prime). For example, the seven partitions above correspond to the commutative groups listed here:

$$\mathbb{Z}_{p^5}, \quad \mathbb{Z}_{p^4} \oplus \mathbb{Z}_p, \quad \mathbb{Z}_{p^3} \oplus \mathbb{Z}_{p^2}, \quad \mathbb{Z}_{p^3} \oplus \mathbb{Z}_p \oplus \mathbb{Z}_p,$$

$$\mathbb{Z}_{p^2} \oplus \mathbb{Z}_{p^2} \oplus \mathbb{Z}_p, \quad \mathbb{Z}_{p^2} \oplus \mathbb{Z}_p \oplus \mathbb{Z}_p \oplus \mathbb{Z}_p, \quad \mathbb{Z}_p \oplus \mathbb{Z}_p \oplus \mathbb{Z}_p \oplus \mathbb{Z}_p \oplus \mathbb{Z}_p.$$

The existence part of the classification theorem says that every commutative group of size p^5 is isomorphic to one of these seven groups. The uniqueness part of the theorem (to be proved momentarily) says that no two of these seven groups are isomorphic.

Before beginning the proof, let us see how to use partition diagrams to gain algebraic information about the associated commutative group. For definiteness, let us consider the commutative group

$$G = \mathbb{Z}_{7^4} \oplus \mathbb{Z}_{7^4} \oplus \mathbb{Z}_{7^2} \oplus \mathbb{Z}_{7^2} \oplus \mathbb{Z}_{7^2} \oplus \mathbb{Z}_7,$$

which corresponds to $p = 7$ and the partition $\mu = (4, 4, 2, 2, 2, 1)$. A typical element of G is a 6-tuple

$$x = (n_1, n_2, n_3, n_4, n_5, n_6)$$

where $0 \leq n_1 < 7^4$, $0 \leq n_2 < 7^4$, $0 \leq n_3 < 7^2$, $0 \leq n_4 < 7^2$, $0 \leq n_5 < 7^2$, and $0 \leq n_6 < 7$. Suppose we write n_1 as a 4-digit number in base 7:

$$n_1 = d_3 7^3 + d_2 7^2 + d_1 7^1 + d_0 7^0 \qquad \text{with } 0 \leq d_i < 7.$$

Writing similar expressions for $n_2, \ldots, n_6$, we can represent the group element x by filling the squares of the partition diagram of μ with arbitrary digits in the range $\{0, 1, 2, 3, 4, 5, 6\}$. For example, the element $x = (3600, 0250, 55, 41, 30, 0)$ (where we write all entries of x in base 7 with leading zero digits, if needed) is represented by the following filled diagram:

$$x = \begin{array}{|c|c|c|c|}
\hline
3 & 6 & 0 & 0 \\
\hline
0 & 2 & 5 & 0 \\
\hline
\end{array}
\begin{array}{|c|c|}
\hline
5 & 5 \\
\hline
4 & 1 \\
\hline
3 & 0 \\
\hline
\end{array}
\begin{array}{|c|}
\hline
0 \\
\hline
\end{array} \quad .$$

To multiply an integer written in base 7 by 7, we shift all the digits one place left and append a zero. In the example above, this multiplication produces

$$(36000, 02500, 550, 410, 300, 00).$$

However, to obtain $7x$ in the group G, we must now reduce the first two entries modulo 7^4, the next three entries modulo 7^2, and the last entry modulo 7. The effect of all these reductions is to make the leading digit of each entry disappear, leaving us with

$$7x = (6000, 2500, 50, 10, 00, 0).$$

The associated filled diagram is

$$7x = \begin{array}{|c|c|c|c|}
\hline
6 & 0 & 0 & 0 \\
\hline
2 & 5 & 0 & 0 \\
\hline
\end{array}
\begin{array}{|c|c|}
\hline
5 & 0 \\
\hline
1 & 0 \\
\hline
0 & 0 \\
\hline
\end{array}
\begin{array}{|c|}
\hline
0 \\
\hline
\end{array} \quad .$$

This diagram arises from the diagram for x by shifting the entries of each row one step left, erasing the entries that fall off the left end and bringing in zeroes at the right end. Similarly, we can compute $7^2 x$ from the original filled diagram for x by shifting all the digits two places to the left, filling in zeroes on the right side:

$$7^2 x = \begin{array}{|c|c|c|c|}
\hline
0 & 0 & 0 & 0 \\
\hline
5 & 0 & 0 & 0 \\
\hline
\end{array}
\begin{array}{|c|c|}
\hline
0 & 0 \\
\hline
0 & 0 \\
\hline
0 & 0 \\
\hline
\end{array}
\begin{array}{|c|}
\hline
0 \\
\hline
\end{array} \quad .$$

Inspection of the diagram shows that $7^3 x = 0$, since the only nonzero digits in the diagram for x occur in the first three columns. More generally, for any $y \in G$, $7^i y = 0$ iff all the nonzero digits in the filled diagram for y occur in the first i columns of μ. This observation allows us to determine the size of the various subgroups $G[7^i] = \{y \in G : 7^i y = 0\}$. For example, how large is $G[7^2]$? Consider the following schematic diagram:

$\star$	$\star$	0	0
$\star$	$\star$	0	0
$\star$	$\star$		
$\star$	$\star$		
$\star$	$\star$		
$\star$			

By the previous observation, we obtain a typical element of $G[7^2]$ by choosing an arbitrary base-7 digit for each of the starred positions (possibly zero), and filling in the remaining squares of the diagram with zeroes. There are seven choices for each star, leading to the conclusion that

$$|G[7^2]| = 7^{11}.$$

These observations generalize to arbitrary diagrams. Suppose p is prime, n is a positive integer, $\mu = (a_1 \geq a_2 \geq \cdots \geq a_r)$ is a partition of n, and $G = \mathbb{Z}_{p^{a_1}} \oplus \cdots \oplus \mathbb{Z}_{p^{a_r}}$. Let $a_1' \geq a_2' \geq \cdots \geq a_s'$ denote the number of squares in each column of the diagram of μ, reading from left to right. As a convention, set $a_k = 0$ for $k > r$ and $a_k' = 0$ for $k > s$. As in the example, for each $j \in \mathbb{Z}_{\geq 0}$, we obtain a typical element of $G[p^j]$ by filling the squares in the first j columns with arbitrary base-p digits, and filling the remaining squares with zeroes. By the Product Rule from combinatorics, we see that

$$|G[p^j]| = p^{a_1' + a_2' + \cdots + a_j'} \qquad \text{for all } j \in \mathbb{Z}_{\geq 0}.$$

We finally have the tools needed to attack the uniqueness proof for commutative groups of prime power size. Fix a prime p, and assume

$$G = \mathbb{Z}_{p^{a_1}} \oplus \cdots \oplus \mathbb{Z}_{p^{a_r}} \cong H = \mathbb{Z}_{p^{c_1}} \oplus \cdots \oplus \mathbb{Z}_{p^{c_t}},$$

where $a_1 \geq \cdots \geq a_r > 0$ and $c_1 \geq \cdots \geq c_t > 0$. We need to show that the two partitions $(a_i : i \geq 1)$ and $(c_i : i \geq 1)$ are the same. Let $(a_j' : j \geq 1)$ and $(c_j' : j \geq 1)$ be the column lengths of the associated partition diagrams. To show that the row lengths a_i and c_i are the same for all i, it suffices to show that the column lengths a_j' and c_j' are the same for all j. Equivalently, it suffices to show that $a_1' + \cdots + a_j' = c_1' + \cdots + c_j'$ for all $j \geq 1$. To prove these equalities, note on the one hand that

$$|G[p^j]| = p^{a_1' + \cdots + a_j'},$$

while on the other hand

$$|H[p^j]| = p^{c_1' + \cdots + c_j'}.$$

Now $G \cong H$ implies that $G[p^j] \cong H[p^j]$, so that

$$p^{a_1' + \cdots + a_j'} = p^{c_1' + \cdots + c_j'}.$$

Since $p > 1$, this finally leads to $a_1' + \cdots + a_j' = c_1' + \cdots + c_j'$ for all j, completing the proof.

16.18 Uniqueness of Elementary Divisors

The next step is to prove the uniqueness theorems for finite commutative groups. Suppose

$$\mathbb{Z}_{a_1} \oplus \cdots \oplus \mathbb{Z}_{a_r} \cong G \cong \mathbb{Z}_{c_1} \oplus \cdots \oplus \mathbb{Z}_{c_t}, \tag{16.9}$$

where $a_1 \geq \cdots \geq a_r > 1$ are prime powers and $c_1 \geq \cdots \geq c_t > 1$ are prime powers. We must prove $r = t$ and $a_i = c_i$ for $1 \leq i \leq r$.

The idea here is to somehow isolate those a_i and c_j that are powers of a given prime p, so that we can apply the uniqueness result from the previous section. To see how this can be done, suppose p and q are distinct primes and $d, e \geq 1$. For $d \leq e$, every element x of the cyclic group $\mathbb{Z}_{p^d}$ satisfies $p^e x = p^{e-d}(p^d x) = 0$, so $\mathbb{Z}_{p^d}[p^e] = \mathbb{Z}_{p^d}$. On the other hand, for arbitrary d, $\mathbb{Z}_{q^d}[p^e] = \{0\}$. To see why, suppose $y \in \mathbb{Z}_{q^d}$ satisfies $p^e y = 0$. Then q^d must divide the product $p^e y$ computed in $\mathbb{Z}$, so q^d divides y (since p and q are distinct primes). As $0 \leq y < q^d$, it follows that $y = 0$. Next, you can check that for any integer b,

$$(G_1 \oplus \cdots \oplus G_k)[b] = (G_1[b]) \oplus \cdots \oplus (G_k[b]).$$

Now assume the situation in (16.9). Let p be any fixed prime and e be the highest power of p dividing $|G|$. Compute $G[p^e]$ in two ways, using the two isomorphic versions of G as direct sums of cyclic groups appearing in (16.9). On one hand, by the remarks in the previous paragraph, we obtain a direct sum of the form $H_1 \oplus \cdots \oplus H_r$, where $H_i = \mathbb{Z}_{a_i}$ if a_i is a power of p, and $H_i = \{0\}$ if a_i is a power of another prime. Deleting zero factors and rearranging, we see that this direct sum is isomorphic to the direct sum involving precisely those a_i that are powers of p, arranged in decreasing order. On the other hand, applying the same reasoning to the other representation of G, we see that $G[p^e]$ is also isomorphic to the direct sum involving those c_j that are powers of p, arranged in decreasing order. Since $G[p^e]$ is an isomorphism invariant, the two direct sums just mentioned are isomorphic. By the results in the previous section, those a_i and c_j that are powers of the given prime p must be the same (counting multiplicities). Since p was arbitrary and every a_i and c_j is a power of some prime, we see that all the a_i and c_j must match.

16.19 Uniqueness of Invariant Factors

We can use the last result to deduce the other uniqueness theorem for finite commutative groups. Suppose now that

$$\mathbb{Z}_{a_1} \oplus \cdots \oplus \mathbb{Z}_{a_r} \cong G \cong \mathbb{Z}_{c_1} \oplus \cdots \oplus \mathbb{Z}_{c_t} \tag{16.10}$$

where the a_i and c_j are integers larger than 1 such that a_i divides a_{i+1} and c_j divides c_{j+1} for $1 \leq i < r$ and $1 \leq j < t$. Our goal is to prove that $r = t$ and $a_i = c_i$ for all i. Recall (§16.13) that we can split and merge cyclic groups based on their prime factorizations, using the isomorphism

$$\mathbb{Z}_{p_1^{e_1} \cdots p_s^{e_s}} \cong \mathbb{Z}_{p_1^{e_1}} \oplus \cdots \oplus \mathbb{Z}_{p_s^{e_s}} \quad \text{(for distinct primes } p_k\text{)}.$$

We use this fact to split each of the cyclic groups of size a_i and c_j in (16.10) into direct sums of cyclic groups of prime power size. Let $|G|$ have prime factorization $p_1^{e_1} \cdots p_s^{e_s}$. Write

$a_i = \prod_{k=1}^{s} p_k^{f_{ki}}$ and $c_j = \prod_{k=1}^{s} p_k^{g_{kj}}$ for some integers $f_{ki}, g_{kj} \geq 0$. Then

$$\bigoplus_{k=1}^{s} \bigoplus_{i=1}^{r} \mathbb{Z}_{p_k^{f_{ki}}} \cong G \cong \bigoplus_{k=1}^{s} \bigoplus_{j=1}^{t} \mathbb{Z}_{p_k^{g_{kj}}}.$$

By the previously proved uniqueness result, for each fixed k between 1 and s, the list of numbers $p_k^{f_{ki}}$ with $f_{ki} > 0$ (with multiplicities) equals the list of numbers $p_k^{g_{kj}}$ with $g_{kj} > 0$ (with multiplicities). So it suffices to show that the list $a_1, \ldots, a_r$ is uniquely determined by the list of powers $p_k^{f_{ki}}$ together with the divisibility conditions linking successive a_i (and similarly, the list $c_1, \ldots, c_t$ is uniquely determined by the list of $p_k^{g_{kj}}$).

We describe an algorithm for reconstructing $a_1, \ldots, a_r$ from the list of all prime powers $p_k^{f_{ki}}$ that exceed 1. Construct an $s \times r$ matrix whose kth row contains the powers $p_k^{f_{ki}}$ arranged in increasing order, padded on the left with 1s so that there are exactly r entries in the row. Note that s is known, being the number of distinct prime factors of $|G|$. Furthermore, letting n_k be the number of positive powers of p_k that occur in the given list, r can be calculated as the maximum of $n_1, \ldots, n_s$. This follows since $a_1 > 1$, which means that at least one prime p_k divides all r integers $a_1, \ldots, a_r$. Every prime power in the matrix arises by splitting the prime factorization of some a_i into prime powers. Taking the divisibility relations among the a_i into account, it follows that a_r must be the product of the largest possible power of each prime. So, a_r is the product of the prime powers in the rth (rightmost) column of the matrix. Proceeding inductively from right to left, it follows similarly that a_{r-1} must be the product of the prime powers in column $r-1$ of the matrix, and so on. Thus we can recover $a_r, a_{r-1}, \ldots, a_1$ uniquely from the given matrix of prime powers.

The following example of the reconstruction algorithm may clarify the preceding argument. Suppose we are given the list of prime powers:

$$[2, 2^4, 2^4, 2^4, 2^7, 3^2, 3^2, 3^2, 7, 7, 7^5, 7^8].$$

We see that $s = 3$ and $r = 5$; the matrix of prime powers is

$$\begin{bmatrix} 2 & 2^4 & 2^4 & 2^4 & 2^7 \\ 1 & 1 & 3^2 & 3^2 & 3^2 \\ 1 & 7 & 7 & 7^5 & 7^8 \end{bmatrix}.$$

Multiplying the entries in each column, we get

$$a_1 = 2, \quad a_2 = 2^4 \cdot 7, \quad a_3 = 2^4 \cdot 3^2 \cdot 7, \quad a_4 = 2^4 \cdot 3^2 \cdot 7^5, \quad a_5 = 2^7 \cdot 3^2 \cdot 7^8.$$

Evidently, we do have $a_1|a_2$, $a_2|a_3$, $a_3|a_4$, $a_4|a_5$, and splitting each a_i into prime power divisors does produce the given list of prime powers. Furthermore, you can check in this example that the choice of a_5 (then a_4, etc.) really is forced by these conditions.

16.20   Uniqueness Proof: General Case

To finish proving the uniqueness assertion of the classification theorem in full generality, we use torsion subgroups to separate the torsion part and the free part of a finitely generated commutative group. More precisely, consider a group $G = \mathbb{Z}^b \times H$, where $b \geq 0$ and H

is a (possibly empty) direct sum of finitely many finite cyclic groups. You can check that $\mathrm{tor}(G) = \{0\} \times H \cong H$, from which it follows that

$$G/\mathrm{tor}(G) = \frac{\mathbb{Z}^b \times H}{\{0\} \times H} \cong \frac{\mathbb{Z}^b}{\{0\}} \times \frac{H}{H} \cong \mathbb{Z}^b.$$

Let us begin the uniqueness proof. Suppose $G_1 = \mathbb{Z}^b \times H$ is isomorphic to $G_2 = \mathbb{Z}^d \times K$, where $H = \mathbb{Z}_{a_1} \times \cdots \times \mathbb{Z}_{a_r}$, $K = \mathbb{Z}_{c_1} \times \cdots \times \mathbb{Z}_{c_t}$, $b, d, r, t \in \mathbb{Z}_{\geq 0}$, every a_i and c_j is in $\mathbb{Z}_{>1}$, and either (i) a_i divides a_{i+1} and c_i divides c_{i+1} for all applicable i, or (ii) both lists $a_1, \ldots, a_r$ and $c_1, \ldots, c_t$ consist of prime powers in weakly decreasing order. We must show $b = d$ and $r = t$ and $a_i = c_i$ for all i.

Since $G_1 \cong G_2$, we know $\mathrm{tor}(G_1) \cong \mathrm{tor}(G_2)$ and $G_1/\mathrm{tor}(G_1) \cong G_2/\mathrm{tor}(G_2)$ (§16.15). Using the preceding remarks, this means that $H \cong K$ and $\mathbb{Z}^b \cong \mathbb{Z}^d$. The isomorphism between H and K guarantees that $r = t$ and $a_i = c_i$ for all i (as shown in §16.18 and §16.19). The isomorphism between $\mathbb{Z}^b$ and $\mathbb{Z}^d$ guarantees that $b = d$ (by §16.16). This completes the proof of the Theorem Classifying Finitely Generated Commutative Groups.

16.21   Summary

We now review the facts about commutative groups established in this chapter.

1. *Definitions.* A *commutative group* is a set G closed under a commutative, associative binary operation (written as addition) that has an identity element and additive inverses. A map $T : G \to H$ between commutative groups is $\mathbb{Z}$-*linear* iff $T(x + y) = T(x) + T(y)$ for all $x, y \in G$, which automatically implies $T(nx) = nT(x)$ for all $x \in G$ and $n \in \mathbb{Z}$. A $\mathbb{Z}$-*linear combination* of elements $v_1, \ldots, v_k \in G$ is an element of the form $n_1 v_1 + \cdots + n_k v_k$ where $n_1, \ldots, n_k$ are integers. G is *generated* by $v_1, \ldots, v_k$ iff every $g \in G$ is a $\mathbb{Z}$-linear combination of $v_1, \ldots, v_k$. The list $v_1, \ldots, v_k$ is $\mathbb{Z}$-*independent* iff for every $n_1, \ldots, n_k \in \mathbb{Z}$, $n_1 v_1 + \cdots + n_k v_k = 0$ implies $n_1 = \cdots = n_k = 0$. The list $v_1, \ldots, v_k$ is an *ordered* $\mathbb{Z}$-*basis* of G iff the list generates G and is $\mathbb{Z}$-independent. Equivalently, $v_1, \ldots, v_k$ is a $\mathbb{Z}$-basis of G iff for each $g \in G$ there exist unique integers n_i with $g = \sum_{i=1}^{k} n_i v_i$. A commutative group is *finitely generated* iff it has a finite generating set; the group is *free* iff it has a basis; the group is k-*dimensional* iff it has a basis with k elements.

2. *Properties of Generating Sets.* Two $\mathbb{Z}$-linear maps $S, T : G \to H$ that agree on a generating set of G must be equal. A $\mathbb{Z}$-linear map $T : G \to H$ is surjective iff its image contains a generating set for H.

3. *Universal Mapping Property for Free Commutative Groups.* Suppose X is a $\mathbb{Z}$-basis of a commutative group G. Given any commutative group H and any function $f : X \to H$, there exists a unique $\mathbb{Z}$-linear extension $T_f : G \to H$ such that $T_f(v) = f(v)$ for all $v \in X$.

4. *Consequences of the Universal Mapping Property.* Every k-dimensional free commutative group G is isomorphic to $\mathbb{Z}^k$. Different isomorphisms can be obtained by choosing an ordered basis $(v_1, \ldots, v_k)$ for G and sending $n_1 v_1 + \cdots + n_k v_k \in G$ to the k-tuple of coordinates $(n_1, \ldots, n_k) \in \mathbb{Z}^k$. Every commutative group with a k-element generating set is isomorphic to a quotient group of $\mathbb{Z}^k$.

5. *Matrix Representation of* $\mathbb{Z}$*-Linear Maps.* Given a $\mathbb{Z}$-linear map $T : G \to H$, an ordered basis $X = (v_1, \ldots, v_n)$ of G, and an ordered basis $Y = (w_1, \ldots, w_m)$ of H, the *matrix of* T *relative to the input basis* X *and output basis* Y is the unique $m \times n$ integer-valued matrix $A = [a_{ij}]$ such that

$$T(v_j) = \sum_{i=1}^{m} a_{ij} w_i \qquad \text{for } 1 \leq j \leq n.$$

Matrix addition corresponds to pointwise addition of $\mathbb{Z}$-linear maps, while matrix multiplication corresponds to composition of linear maps.

6. *Elementary Operations on Bases, Rows, and Columns.* Given an ordered $\mathbb{Z}$-basis of a free commutative group G, we can create new ordered bases of G by the following operations: switch two basis vectors; negate a basis vector; add an integer multiple of one basis vector to another basis vector. Similar operations can be performed on the rows and columns of integer matrices. Column operations on the matrix of a $\mathbb{Z}$-linear map T correspond to changes in the input basis, while row operations correspond to changes in the output basis.

7. *Reduction of Integer Matrices.* Using row and column operations, we can reduce any $m \times n$ integer-valued matrix A to a new matrix B such that the nonzero entries of B (if any) occupy the first r positions on the main diagonal of B, and these nonzero entries are positive integers each of which divides the next one. B is uniquely determined by A and is called the *Smith normal form* of A.

8. *Canonical Form for* $\mathbb{Z}$*-Linear Maps.* Suppose $T : G \to H$ is a $\mathbb{Z}$-linear map between two finite-dimensional free commutative groups. There exist an ordered basis $X = (x_1, \ldots, x_n)$ for G, an ordered basis $Y = (y_1, \ldots, y_m)$ for H, an integer $r \geq 0$, and positive integers $a_1, \ldots, a_r$ such that a_i divides a_{i+1} for all i with $1 \leq i < r$, $T(x_i) = a_i y_i$ for $1 \leq i \leq r$, and $T(x_i) = 0$ for $r < i \leq n$.

9. *Subgroups of Finitely Generated Free Commutative Groups.* Any subgroup P of a k-dimensional free commutative group G is also free with dimension at most k. By choosing appropriate bases for P and G, we can find an isomorphism $G \cong \mathbb{Z}^k$ such that P maps under this isomorphism to a subgroup of the form

$$a_1 \mathbb{Z} \oplus \cdots \oplus a_m \mathbb{Z} \oplus \{0\} \oplus \cdots \oplus \{0\},$$

where each a_i divides the next one. It follows that G/P is isomorphic to a direct sum of cyclic groups.

10. *Invariant Subgroups.* For any commutative group G and integer n, the subsets $G[n] = \{g \in G : ng = 0\}$, $nG = \{ng : g \in G\}$, and $\text{tor}(G) = \bigcup_{n \geq 1} G[n]$ are subgroups of G. Any isomorphism $f : G \to H$ restricts to give isomorphisms $G[n] \cong H[n]$, $nG \cong nH$, and $\text{tor}(G) \cong \text{tor}(H)$, and f induces isomorphisms of quotient groups $G/G[n] \cong H/H[n]$, $G/nG \cong H/nH$, and $G/\text{tor}(G) \cong H/\text{tor}(H)$.

11. *Theorem Classifying Finitely Generated Commutative Groups (Version 1).* Every finitely generated commutative group G is isomorphic to a direct sum of cyclic groups

$$\mathbb{Z}^b \oplus \mathbb{Z}_{a_1} \oplus \cdots \oplus \mathbb{Z}_{a_r},$$

where $b, r \geq 0$, every $a_i > 1$, and a_i divides a_{i+1} for $1 \leq i < r$. The integers b, r, and $a_1, \ldots, a_r$ (satisfying these conditions) are uniquely determined by G.

12. *Theorem Classifying Finitely Generated Commutative Groups (Version 2).* Every finitely generated commutative group G is isomorphic to a direct sum of cyclic groups

$$\mathbb{Z}^b \oplus \mathbb{Z}_{a_1} \oplus \cdots \oplus \mathbb{Z}_{a_r},$$

where $b, r \geq 0$, $a_1 \geq \cdots \geq a_r > 1$, and every a_i is a prime power. The integers b, r, and $a_1, \ldots, a_r$ (satisfying these conditions) are uniquely determined by G.

13. *Restatement of Uniqueness in the Classification Theorem.* Suppose

$$G \cong \mathbb{Z}^b \oplus \mathbb{Z}_{a_1} \oplus \cdots \oplus \mathbb{Z}_{a_r} \cong \mathbb{Z}^d \oplus \mathbb{Z}_{c_1} \oplus \cdots \oplus \mathbb{Z}_{c_t},$$

where $b, d, r, t \in \mathbb{Z}_{\geq 0}$, all a_i and c_j are in $\mathbb{Z}_{>1}$, and either: (i) both lists $a_1, \ldots, a_r$ and $c_1, \ldots, c_t$ consist of prime powers in weakly decreasing order; or (ii) a_i divides a_{i+1} for $1 \leq i < r$ and c_j divides c_{j+1} for $1 \leq j < t$. Then $b = d$ and $r = t$ and $a_i = c_i$ for $1 \leq i \leq r$. In particular (when $r = t = 0$), this says that the dimension (or *Betti number*) b of a finite-dimensional free commutative group is well-defined. In case (i), $a_1, \ldots, a_r$ are called the *elementary divisors* of G. In case (ii), $a_1, \ldots, a_r$ are called the *invariant factors* of G.

14. *Remarks for Non-Finitely Generated Commutative Groups.* A subset S of a commutative group G *generates* G iff every $g \in G$ is a $\mathbb{Z}$-linear combination of some finite subset of S; S is $\mathbb{Z}$-*independent* iff every finite list of distinct elements of S is $\mathbb{Z}$-independent; S is a $\mathbb{Z}$-*basis of* G iff S is $\mathbb{Z}$-independent and generates G. A commutative group G is called *free* iff it has a $\mathbb{Z}$-basis. The additive group $\mathbb{Q}$ is an example of a commutative group that is not finitely generated. The Universal Mapping Property is valid for free commutative groups. Every free commutative group G is isomorphic to a direct sum of copies of $\mathbb{Z}$, where the number of factors in the direct sum is the cardinality of a basis of G. Every commutative group is isomorphic to a quotient group of a free commutative group.

16.22 Exercises

Unless otherwise specified, assume $(G, +)$ and $(H, +)$ are commutative groups in these exercises.

1. (a) For fixed $n \in \mathbb{Z}$, verify that $n\mathbb{Z} = \{ni : i \in \mathbb{Z}\}$ is a subgroup of $\mathbb{Z}$.
   (b) Find all $m, n \in \mathbb{Z}$ with $m\mathbb{Z} = n\mathbb{Z}$.

2. Let H be a subgroup of G. Prove: for all $x \in G$, $\langle x \rangle \subseteq H$ iff $x \in H$.

3. Fix $n \in \mathbb{Z}_{>0}$. Prove that every element of $\mathbb{Z}/n\mathbb{Z}$ equals one of the cosets $0 + n\mathbb{Z}$, $1 + n\mathbb{Z}$, ..., $(n-1) + n\mathbb{Z}$, and show that these cosets are all distinct. Then prove that the groups $\mathbb{Z}/n\mathbb{Z}$ and $(\mathbb{Z}_n, \oplus)$ are isomorphic.

4. Prove that $\mathbb{Z}/0\mathbb{Z} \cong \mathbb{Z}$.

5. In the text, we constructed a cyclic group of size n by forming the quotient group $\mathbb{Z}/n\mathbb{Z}$. For each group, find a specific subgroup of that group that is isomorphic to $\mathbb{Z}/n\mathbb{Z}$. (a) $(S_n, \circ)$ (b) $\mathbb{C}_{\neq 0}$ under multiplication (c) $\mathrm{GL}_n(\mathbb{R})$

6. Let $G_1, \ldots, G_k$ be commutative groups. Show $G = G_1 \times \cdots \times G_k$ is a commutative group under componentwise addition. For $1 \leq i \leq k$, let $j_i : G_i \to G$ be given by $j_i(x_i) = (0, \ldots, x_i, \ldots, 0)$ for $x_i \in G_i$, where the x_i occurs in position i. Show each

j_i is an injective group homomorphism. Are these results still true for arbitrary groups $G_1, \ldots, G_k$?

7. Let $n \in \mathbb{Z}_{\geq 0}$ and $x \in G$. The definition of nx in §16.2 can be stated more precisely in the following recursive way: $0x = 0$, and $(n+1)x = (nx) + x$. Use this recursive definition and induction to prove the following laws of multiples from (16.1) assuming $m, n \in \mathbb{Z}_{\geq 0}$ and $x, y \in G$ (a separate argument is needed when m or n is negative). (a) $(m+n)x = (mx) + (nx)$ (b) $m(nx) = (mn)x$ (c) $1x = x$ (d) $m(x+y) = (mx) + (my)$

8. Give a specific example of a non-commutative group $(G, +)$, $x, y \in G$, and $m \in \mathbb{Z}_{>0}$ with $m(x+y) \neq mx + my$.

9. Let $v_1, \ldots, v_k \in G$ and $H = \{c_1 v_1 + \cdots + c_k v_k : c_i \in \mathbb{Z}\}$.
   (a) Prove that $0_G \in H$ and that H is closed under inverses.
   (b) If $(G, +)$ is not commutative, must H be closed under the group operation?
   (c) Let K be the set of all group elements of the form $e_1 w_1 + e_2 w_2 + \cdots + e_s w_s$, where $s \in \mathbb{Z}_{\geq 0}$, each $e_i \in \{1, -1\}$, and each w_i is some v_j (with repeats allowed). Prove: for all G (commutative or not), K is a subgroup of G, and prove $K = H$ when G is commutative.

10. Find all generators of each of these groups. (a) $\mathbb{Z}_8$ (b) $\mathbb{Z}_{12}$ (c) $\mathbb{Z}_{30}$ (d) $\mathbb{Z}_3 \times \mathbb{Z}_5$

11. (a) Prove that for prime p and all nonzero a in $\mathbb{Z}_p$, $\mathbb{Z}_p = \langle a \rangle$.
    (b) For prime p and $e \in \mathbb{Z}_{>0}$, how many $a \in \mathbb{Z}_{p^e}$ satisfy $\mathbb{Z}_{p^e} = \langle a \rangle$?

12. Given $n \in \mathbb{Z}_{>0}$ and $k \in \mathbb{Z}_n$, find and prove a criterion for when $\mathbb{Z}_n = \langle k \rangle$.

13. (a) Suppose G_1 and G_2 are commutative groups with $G_1 = \langle v_1, \ldots, v_m \rangle$ and $G_2 = \langle w_1, \ldots, w_n \rangle$. Prove $G_1 \times G_2 = \langle (v_1, 0), \ldots, (v_m, 0), (0, w_1), \ldots, (0, w_n) \rangle$.
    (b) Generalize (a) to direct products $G_1 \times G_2 \times \cdots \times G_k$ where $k \in \mathbb{Z}_{>0}$.

14. (a) Describe an explicit generating set for the product group $G = \mathbb{Z} \times \cdots \times \mathbb{Z} \times \mathbb{Z}_{a_1} \times \cdots \times \mathbb{Z}_{a_s}$, where there are b copies of $\mathbb{Z}$ and every $a_i = p_i^{e_i}$ is a prime power.
    (b) How many generating sets S does G have where $|S| = b + s$ and every element of S has exactly one nonzero component?

15. Let $S \subseteq G$ and H be the set of $\mathbb{Z}$-linear combinations of elements of S. Prove H is a subgroup of G.

16. Prove that $S = \{1/p^e : p \text{ is prime}, e \in \mathbb{Z}_{>0}\}$ generates $(\mathbb{Q}, +)$.

17. Prove or disprove: the set $T = \{1/p : p \text{ is prime}\}$ generates $(\mathbb{Q}, +)$.

18. Find a generating set for $\mathbb{Q}_{>0}$, the group of positive rational numbers under multiplication.

19. Show that $\mathbb{Z}[x]$ (polynomials with integer coefficients, under addition) is a commutative group that is not finitely generated. Find an infinite generating set for this group.

20. Give an example of a commutative group $(G, +)$ that is not finitely generated and is not isomorphic to $\mathbb{Z}[x]$ or to $(\mathbb{Q}, +)$.

21. Prove or disprove: the commutative groups $\mathbb{Z}[x]$ and $(\mathbb{Q}, +)$ are isomorphic.

22. Prove or disprove: the commutative groups $\mathbb{Z}[x]$ and $\mathbb{Q}_{>0}$ (positive rationals under multiplication) are isomorphic.

23. Let $B = (v_1, \ldots, v_k)$ be a list of elements in G. Say what it means for B to be $\mathbb{Z}$-dependent, by negating the definition of $\mathbb{Z}$-linear independence. Say what it means for a subset S of G (possibly infinite) to be $\mathbb{Z}$-dependent. Then explain why $\emptyset$ is $\mathbb{Z}$-independent relative to G.

24. Which of the following commutative groups are free? Explain.
    (a) $3\mathbb{Z}$ (a subgroup of $\mathbb{Z}$) (b) $\mathbb{Z} \times \mathbb{Z}_5$ (c) $\mathbb{Z}[x]$ under addition
    (d) $\{a + bi : a \in 2\mathbb{Z}, 2b \in \mathbb{Z}\}$ under complex addition (e) $\mathbb{Q}$ under addition

25. Let V be a $\mathbb{Q}$-vector space. Prove that a list $(v_1, \ldots, v_k)$ of elements of V is $\mathbb{Z}$-linearly independent iff the list is $\mathbb{Q}$-linearly independent.

26. Show that for all $a \in \mathbb{Z}_{>0}$ with a not a perfect square, $(1, \sqrt{a})$ is a $\mathbb{Z}$-linearly independent list in $\mathbb{R}$.

27. Show that $(1, \sqrt{2}, \sqrt{3}, \sqrt{6})$ is $\mathbb{Z}$-linearly independent.

28. Let $B = (v_1, \ldots, v_k)$ be a list of elements in G. Suppose C is obtained from B by switching v_i and v_j, for some $i \neq j$. Prove $G = \langle B \rangle$ iff $G = \langle C \rangle$. Prove B is $\mathbb{Z}$-independent iff C is $\mathbb{Z}$-independent. Prove B is an ordered basis of G iff C is an ordered basis of G.

29. Repeat Exercise 28, but now assume C is obtained from B by replacing some v_i by $-v_i$. Which implications in Exercise 28 are true if C is obtained from B by replacing some v_i by nv_i, where $n \notin \{-1, 0, 1\}$ is a fixed integer?

30. Check that $B = ((0, 1, -5), (0, 0, -1), (1, 2, 3))$ is an ordered basis of $\mathbb{Z}^3$: (a) by proving from the definitions that B generates $\mathbb{Z}^3$ and is $\mathbb{Z}$-linearly independent; (b) by showing how to obtain B from the known ordered basis (e_1, e_2, e_3) by a sequence of elementary operations.

31. Repeat Exercise 30 for $B = ((12, -7, -2), (8, 5, 3), (1, 2, 1))$.

32. (a) Find and prove necessary and sufficient conditions on $a, b, c, d \in \mathbb{Z}$ so that $((a, b), (c, d))$ is an ordered basis of $\mathbb{Z}^2$. (b) Can you generalize your answer to (a) to characterize ordered bases of $\mathbb{Z}^k$ for all $k \geq 1$?

33. Suppose F is a field and $B = (v_1, \ldots, v_n)$ is an ordered basis for an F-vector space V. Prove that the three elementary operations in §16.4 (as modified in the last paragraph of that section) send the basis B to another ordered basis of V.

34. Prove or disprove: for all $k \geq 1$, every ordered basis of the free commutative group $\mathbb{Z}^k$ can be obtained from the standard ordered basis $(e_1, e_2, \ldots, e_k)$ by a finite sequence of elementary operations (B1), (B2), and (B3).

35. Give a justified example of a finitely generated free commutative group G and a $\mathbb{Z}$-linearly independent list $(v_1, \ldots, v_k)$ in G that cannot be extended to an ordered $\mathbb{Z}$-basis $(v_1, \ldots, v_k, \ldots, v_s)$ of G.

36. Give a justified example of a finitely generated free commutative group G and a generating set $\{v_1, \ldots, v_k\}$ of G such that no subset of this generating set is a $\mathbb{Z}$-basis of G.

37. Let $T : G \to H$ be homomorphism of commutative groups.
    (a) Prove by induction on n that $T(nx) = nT(x)$ for all $x \in G$ and all $n \in \mathbb{Z}_{\geq 0}$.
    (b) Prove $T(nx) = nT(x)$ for all $x \in G$ and all $n \in \mathbb{Z}$.
    (c) Prove by induction on k that $T(\sum_{i=1}^{k} n_i v_i) = \sum_{i=1}^{k} n_i T(v_i)$ for all $k \in \mathbb{Z}_{\geq 0}$, $n_i \in \mathbb{Z}$ and $v_i \in G$.

38. Find the coordinates of $(1, 2, 3)$ and $(4, -1, 1)$ relative to each ordered basis for $\mathbb{Z}^3$.
    (a) (e_1, e_2, e_3) (b) $((0, 1, -5), (0, 0, -1), (1, 2, 3))$ (c) $((-12, 7, 2), (8, 5, 3), (1, 2, 1))$

39. Let $T : G \to H$ be a $\mathbb{Z}$-linear map and $X \subseteq G$.
    (a) Prove: If X generates G, then $\{T(x) : x \in X\}$ generates $\text{img}(T)$.
    (b) Prove or disprove: If X generates G, then $X \cap \ker(T)$ generates $\ker(T)$.

40. Let G be a free commutative group with ordered basis $B = (v_1, v_2, \ldots, v_k)$. Use the UMP for free commutative groups to construct an isomorphism $T : G \to \mathbb{Z}v_1 \oplus \mathbb{Z}v_2 \oplus \cdots \oplus \mathbb{Z}v_k$. Show that $\mathbb{Z}v_i \cong \mathbb{Z}$, preferably by using the UMP, and conclude that $G \cong \mathbb{Z}^k$.

41. Let G and H be finitely generated free commutative groups of dimensions n and m, respectively. Use the UMP for free commutative groups to prove:
    (a) If $n \le m$, then there exists an injective $\mathbb{Z}$-linear map $S : G \to H$.
    (b) If $n \ge m$, then there exists a surjective $\mathbb{Z}$-linear map $T : G \to H$.
    (c) If $n = m$, then there exists a bijective $\mathbb{Z}$-linear map $U : G \to H$.

42. Give a justified example of a free commutative group G, a commutative group H, a finite spanning set X for G, and a function $f : X \to H$ that has no extension to a $\mathbb{Z}$-linear map with domain G.

43. Give a justified example of a free commutative group G, a commutative group H, a $\mathbb{Z}$-independent subset X of G, and a function $f : X \to H$ that has infinitely many extensions to $\mathbb{Z}$-linear maps with domain G.

44. Let G be a free commutative group with infinite basis X. (a) Prove: for every $g \in G$, there exist unique integers $\{n_x : x \in X\}$ such that $n_x \neq 0$ for only finitely many $x \in X$ and $g = \sum_{x \in X} n_x x$. (b) Prove: if $S, T : G \to H$ are $\mathbb{Z}$-linear maps that agree on X, then $S = T$. (c) Prove: for every commutative group H and every function $f : X \to H$, there exists a unique $\mathbb{Z}$-linear map $T_f : G \to H$ extending f. (d) For any set Y, let $\mathbb{Z}^{(Y)}$ be the set of all functions $f : Y \to \mathbb{Z}$ such that $f(y) = 0$ for all but finitely many $y \in Y$. Check $\mathbb{Z}^{(Y)}$ is a free commutative group under pointwise addition of functions, with a basis in bijective correspondence with the set Y. (e) Prove G is isomorphic to the group $\mathbb{Z}^{(X)}$. (f) Prove that every commutative group is isomorphic to a quotient group of a free commutative group.

45. Fix $k, n \in \mathbb{Z}_{>0}$, and let $X = \{e_1, \ldots, e_k\}$ where each $e_i = (0, \ldots, 1, \ldots, 0)$ is viewed as an element of $G = \mathbb{Z}_n^k$. Prove that G and X satisfy the following UMP: for all commutative groups H such that $ny = 0$ for all $y \in H$ and for all functions $f : X \to H$, there exists a unique $\mathbb{Z}$-linear map $T_f : G \to H$ extending f.

46. Use the UMP for the free commutative group $\mathbb{Z}$, together with the Fundamental Homomorphism Theorem for Groups, to show that every cyclic group is isomorphic to $\mathbb{Z}$ or to $\mathbb{Z}/n\mathbb{Z}$ for some $n \in \mathbb{Z}_{>0}$.

47. Prove that S defined below (16.2) is a surjective $\mathbb{Z}$-linear map with kernel P.

48. Assume H_i is a normal subgroup of a group G_i for $1 \le i \le k$. Prove

$$(G_1 \times \cdots \times G_k)/(H_1 \times \cdots \times H_k) \cong (G_1/H_1) \times \cdots \times (G_k/H_k).$$

49. Suppose G is a free commutative group, A and B are commutative groups, $f : A \to B$ is a surjective group homomorphism, and $g : G \to B$ is a group homomorphism. Prove there exists a group homomorphism $h : G \to A$ with $f \circ h = g$.

50. In §16.8, we proved that every subgroup of a finitely generated free commutative group is free. (a) Trace through the construction in that proof, applied to the subgroup $H = \{(t, 2t) : t \in \mathbb{Z}\}$ of $\mathbb{Z}^2$, to construct an ordered basis for this subgroup. (b) Similarly, use the proof to find an ordered basis for $H = \langle (2, 4, 15), (4, 6, 6) \rangle$, which is a subgroup of $\mathbb{Z}^3$.

51. Let F be a field. Modify the proof in §16.8 to prove that for all $k \in \mathbb{Z}_{\geq 0}$, every subspace of F^k has an ordered basis of size at most k. Use only the definitions and induction, avoiding any theorems whose conclusions involve the existence of a basis.

52. Let $e_1 = (1,0)$, $e_2 = (0,1)$, $f_1 = (2,5)$, $f_2 = (1,3)$. Let $T : \mathbb{Z}^2 \to \mathbb{Z}^2$ be the $\mathbb{Z}$-linear map given by $T((a,b)) = (4a - b, 2a + 3b)$ for $a, b \in \mathbb{Z}$. Find the matrix of T relative to each pair of ordered bases.
    (a) input basis (e_1, e_2), output basis (e_1, e_2)
    (b) input basis (e_1, e_2), output basis (f_1, f_2)
    (c) input basis (f_1, f_2), output basis (e_1, e_2)
    (d) input basis (f_1, f_2), output basis (f_1, f_2)

53. Let $v_1 = (7,2,2)$, $v_2 = (2,-1,0)$, and $v_3 = (3,1,1)$. Let $I : \mathbb{Z}^3 \to \mathbb{Z}^3$ be the identity map, $X = (e_1, e_2, e_3)$, and $Y = (v_1, v_2, v_3)$. Check that Y is an ordered $\mathbb{Z}$-basis of $\mathbb{Z}^3$. Then find the matrix of I relative to each pair of ordered bases.
    (a) input basis X, output basis X (b) input basis Y, output basis X
    (c) input basis X, output basis Y (d) input basis Y, output basis Y

54. Let $T, U : \mathbb{Z}^3 \to \mathbb{Z}^3$ be the $\mathbb{Z}$-linear maps whose matrices (using the standard ordered basis (e_1, e_2, e_3) as both input and output basis) are

$$ A = \begin{bmatrix} 2 & 0 & -1 \\ 4 & 4 & 1 \\ 0 & 3 & -2 \end{bmatrix} \text{ and } B = \begin{bmatrix} -1 & 5 & 7 \\ 2 & 0 & -2 \\ -3 & 1 & 4 \end{bmatrix}. $$

    (a) Find $T((a,b,c))$ and $U((a,b,c))$ for all $a, b, c \in \mathbb{Z}$.
    (b) Find the matrix of $T + U$ relative to the standard ordered basis.
    (c) Find the matrix of TU relative to the standard ordered basis.
    (d) Find the matrix of UT relative to the standard ordered basis.

55. Let $T : \mathbb{Z}^n \to \mathbb{Z}^m$ be a $\mathbb{Z}$-linear map with matrix A relative to the standard ordered bases of $\mathbb{Z}^n$ and $\mathbb{Z}^m$. Prove: for $v \in \mathbb{Z}^n$, $T(v) = Av$ (the product of the matrix A and the column vector v).

56. Suppose $T : G \to H$ is $\mathbb{Z}$-linear, $X = (x_1, \ldots, x_n)$ is an ordered basis of G, $Y = (y_1, \ldots, y_m)$ is an ordered basis of H, and A is the matrix of T relative to these bases. Show that if $g \in G$ has coordinates $v \in \mathbb{Z}^n$ relative to X, then $Av \in \mathbb{Z}^m$ gives the coordinates of $T(g)$ relative to Y. (View v and Av as column vectors.)

57. Let G and H be finitely generated free commutative groups with ordered bases X and Y, respectively. Suppose $T : G \to H$ is a $\mathbb{Z}$-linear map whose matrix relative to X and Y is A. Given $c \in \mathbb{Z}$, show that $cT : G \to H$, defined by $(cT)(x) = cT(x)$ for $x \in G$, is $\mathbb{Z}$-linear. Find the matrix of cT relative to X and Y.

58. **Dual Groups.** For any commutative group G, define the *dual group* of G, written G^* or $\mathrm{Hom}_{\mathbb{Z}}(G, \mathbb{Z})$, to be the set of all $\mathbb{Z}$-linear maps $f : G \to \mathbb{Z}$.
    (a) Prove G^* is a commutative group under pointwise addition of functions.
    (b) Prove $\mathbb{Z}^* \cong \mathbb{Z}$.
    (c) For commutative groups $G_1, \ldots, G_k$, prove $(G_1 \times \cdots \times G_k)^* \cong (G_1^*) \times \cdots \times (G_k^*)$.
    (d) Prove: if G is a free commutative group of dimension k, then G^* is a free commutative group of dimension k.

59. (a) Suppose G is free with ordered basis $X = (v_1, \ldots, v_n)$. Show there exists a unique *dual basis* $X^* = (v_1^*, \ldots, v_n^*)$ of the dual group G^* (defined in Exercise 58)

satisfying $v_i^*(v_j) = 1$ if $i = j$ and 0 if $i \neq j$. (b) Suppose H is also free with ordered basis $Y = (w_1, \ldots, w_m)$. Given a $\mathbb{Z}$-linear map $T : G \to H$, show the map $T^* : H^* \to G^*$ given by $T^*(f) = f \circ T$ for $f \in H^*$ is $\mathbb{Z}$-linear. (c) How is the matrix of T^* relative to the bases Y^* and X^* related to the matrix A of T relative to the bases X and Y?

60. Assume the setup in Exercise 54. For each input basis X and output basis Y, find the matrix of T relative to these bases.
    (a) $X = (e_1 + 4e_3, e_2, e_3)$, $Y = (e_1, e_2, e_3)$ (b) $X = (e_1, e_2, -e_3)$, $Y = (e_1, e_2, e_3)$
    (c) $X = (e_1, e_2, e_3)$, $Y = (e_1, e_3, e_2)$ (d) $X = (e_1, e_2, e_3)$, $Y = (-e_1, -e_2, e_3)$
    (e) $X = (e_2, e_1, e_3)$, $Y = (e_1, e_2, e_3)$ (f) $X = (e_1, e_2, e_3)$, $Y = (e_1, e_2 - 3e_3, e_3)$
    (g) $X = (e_1, e_2 - 2e_1, e_3 + e_1)$, $Y = (e_1 + e_3, e_2 - 2e_3, e_3)$

61. Suppose G is a free commutative group with ordered basis (v_1, v_2, v_3, v_4), H is a free commutative group with ordered basis (w_1, w_2, w_3), and $T : G \to H$ is a $\mathbb{Z}$-linear map whose matrix (relative to these bases) is

$$A = \begin{bmatrix} 2 & -4 & -3 & 0 \\ 1 & -2 & 3 & 3 \\ 4 & -3 & -2 & 4 \end{bmatrix}.$$

Compute $T(3v_2 - v_3 + 2v_4)$. Find the matrix of T relative to each pair of bases.
    (a) input basis (v_3, v_1, v_2, v_4), output basis (w_3, w_2, w_1)
    (b) input basis $(v_1, -v_2, v_3, v_4)$, output basis $(-w_1, w_2, -w_3)$
    (c) input basis $(v_1, v_2 - v_1, v_3 + 3v_1, v_4 + v_1)$, output basis (w_1, w_2, w_3)
    (d) input basis (v_1, v_2, v_3, v_4), output basis $(w_1, w_1 + w_2 + w_3, 2w_2 + w_3)$

62. Carefully prove the italicized statements in §16.10, which indicate how elementary operations on input and output bases affect the matrix of a $\mathbb{Z}$-linear map.

63. Suppose A is in $M_{m,n}(\mathbb{Z})$ and an integer b divides every entry of A. Let C be obtained from A by applying a single elementary row or column operation. Show that b divides every entry of C. Show that the gcd of all entries of A equals the gcd of all entries of C (interpret the gcd as zero in the case of a zero matrix). Show that the $1, 1$-entry of the Smith normal form of A is the gcd of all entries of A. (For a generalization, see Exercise 100.)

64. Use elementary row and column operations in $\mathbb{Z}$ to reduce each integer matrix to its Smith normal form (16.4).
    (a) $\begin{bmatrix} 9 & 18 \\ 36 & 6 \end{bmatrix}$ (b) $\begin{bmatrix} 9 & 8 & 7 \\ 6 & 5 & 4 \\ 3 & 2 & 1 \end{bmatrix}$ (c) $\begin{bmatrix} 12 & 0 & 9 & 15 \\ -21 & -9 & 27 & 18 \\ 0 & 15 & 33 & -21 \end{bmatrix}$

65. Let $T : \mathbb{Z}^2 \to \mathbb{Z}^2$ be the $\mathbb{Z}$-linear map $T((a, b)) = (4a - b, 2a + 3b)$. Find $\mathbb{Z}$-bases $X = (g_1, g_2)$ and $Y = (h_1, h_2)$ of $\mathbb{Z} \times \mathbb{Z}$ such that the matrix of T with respect to the input basis X and the output basis Y has the form $\begin{bmatrix} c & 0 \\ 0 & d \end{bmatrix}$, where $c, d \in \mathbb{Z}_{\geq 0}$ and c divides d. (Reduce the matrix found in Exercise 52(a), keeping track of how each operation changes the input or output basis.)

66. Let F be a field. Prove: For any matrix $A \in M_{m,n}(F)$, we can perform finitely many elementary row and column operations on A to obtain a matrix B such that, for some r with $0 \leq r \leq \min(m, n)$, $B(i, i) = 1_F$ for $1 \leq i \leq r$ and all other entries of B are 0_F. (Imitate the first part of the proof of the Reduction Theorem for Integer Matrices.)

67. Write a computer program that takes as input a matrix $A \in M_{m,n}(\mathbb{Z})$ and returns as output the Smith normal form (16.4) of A.

68. Prove that a $\mathbb{Z}$-linear map between two finitely generated free commutative groups is invertible iff the Smith normal form for the map is an identity matrix.

69. Prove that the columns of $A \in M_{m,n}(\mathbb{Z})$ are $\mathbb{Z}$-linearly dependent iff the Smith normal form of A has at least one column of zeroes.

70. Let M be the subgroup of $\mathbb{Z}^3$ generated by $v_1 = (6, 6, 9)$ and $v_2 = (12, 6, 6)$. (a) Explain why (v_1, v_2) is an ordered $\mathbb{Z}$-basis of M. (b) Define $T : M \to \mathbb{Z}^3$ by $T(x) = x$ for $x \in M$. Find the matrix of T relative to the input basis (v_1, v_2) and output basis (e_1, e_2, e_3). (c) Use matrix reduction to find a new basis (x_1, x_2) for M, a new basis (y_1, y_2, y_3) for $\mathbb{Z}^3$, and positive integers d_1, d_2 (where d_1 divides d_2) such that $x_1 = d_1 y_1$ and $x_2 = d_2 y_2$. (d) Find a product of cyclic groups (satisfying the conclusions of the classification theorem for finitely generated commutative groups) that is isomorphic to $\mathbb{Z}^3/M$.

71. In the proof in §16.13, show that the map T extending the function f (sending each y_i to e_i) is an isomorphism. Prove $P_1 = T[P]$ has ordered basis $(a_1 e_1, \ldots, a_n e_n)$. Prove T induces a group isomorphism from $\mathbb{Z}^m/P$ to $\mathbb{Z}^m/P_1$.

72. Let m and n be relatively prime positive integers. (a) Use (16.6) to prove that $\mathbb{Z}/(mn\mathbb{Z}) \cong (\mathbb{Z}/m\mathbb{Z}) \times (\mathbb{Z}/n\mathbb{Z})$. (b) Imitate the proof of (16.6) to prove that $\mathbb{Z}/(mn\mathbb{Z}) \cong (\mathbb{Z}/m\mathbb{Z}) \times (\mathbb{Z}/n\mathbb{Z})$. (c) Which steps in the proof in (b) fail when $\gcd(m, n) > 1$?

73. Assume the setup in Exercise 54. Use matrix reduction to find new bases X and Y for $\mathbb{Z}^3$ such that the matrix of T relative to X and Y is in Smith normal form.

74. Assume the setup in Exercise 54. Use matrix reduction to find new bases X and Y for $\mathbb{Z}^3$ such that the matrix of U relative to X and Y is in Smith normal form.

75. A certain $\mathbb{Z}$-linear map $T : \mathbb{Z}^4 \to \mathbb{Z}^3$ is represented by the following matrix relative to the standard ordered bases of $\mathbb{Z}^4$ and $\mathbb{Z}^3$:

$$\begin{bmatrix} 15 & 0 & -10 & 20 \\ 30 & -20 & 20 & 10 \\ 25 & -15 & 55 & 40 \end{bmatrix}.$$

Give a formula for $T((i, j, k, p))$, where $i, j, k, p \in \mathbb{Z}$. Use matrix reduction to find an ordered basis $X = (v_1, v_2, v_3, v_4)$ of $\mathbb{Z}^4$, an ordered basis $Y = (w_1, w_2, w_3)$ of $\mathbb{Z}^3$, and a matrix B in Smith normal form such that B is the matrix of T relative to X and Y.

76. Use matrix reduction to determine the Betti numbers, elementary divisors, and invariant factors for each of the following quotient groups.
(a) $\mathbb{Z}^2/\langle(-4, 4), (-8, -4)\rangle$
(b) $\mathbb{Z}^3/\langle(-255, -12, -60), (-114, -6, -27)\rangle$
(c) $\mathbb{Z}^4/\langle(50, 160, 70, 210), (69, 213, 81, 282), (29, 88, 31, 117)\rangle$

77. Give a specific example of a group $(G, \star)$ and a positive integer n such that the map $M_n : G \to G$ given by $M_n(g) = g^n$ for $g \in G$ is not a group homomorphism, the image of M_n is not a subgroup of G, and the set $\{g \in G : M_n(g) = e_G\}$ is not a subgroup of G.

78. Suppose $f : G \to H$ is a group homomorphism.
(a) Prove: for all $n \in \mathbb{Z}_{>0}$, $f[G[n]] \subseteq H[n]$.
(b) Give an example where strict inclusion holds in (a).

(c) If f is injective, must equality hold in (a)? Explain.

(d) If f is surjective, must equality hold in (a)? Explain.

79. Suppose $f : G \to H$ is a group homomorphism.

(a) Prove: for all $n \in \mathbb{Z}_{>0}$, $f[nG] \subseteq nH$.

(b) Give an example where strict inclusion holds in (a).

(c) If f is injective, must equality hold in (a)? Explain.

(d) If f is surjective, must equality hold in (a)? Explain.

80. Let $|G| = m$. Prove: for all $n \in \mathbb{Z}$ with $\gcd(m, n) = 1$, $nG = G$ and $G[n] = \{0\}$.

81. Suppose $f : G \to H$ is a group homomorphism.

(a) Prove $f[\mathrm{tor}(G)] \subseteq \mathrm{tor}(H)$.

(b) Give an example where strict inclusion holds in (a).

(c) Prove: for an isomorphism f, f restricts to an isomorphism $\mathrm{tor}(G) \cong \mathrm{tor}(H)$.

(d) Deduce from (c) that for an isomorphism f, we get an induced isomorphism $G/\mathrm{tor}(G) \cong H/\mathrm{tor}(H)$.

82. In §16.16, we proved that $\mathbb{Z}^n \cong \mathbb{Z}^m$ implies $n = m$. Where does this proof break down if we try to use it to show that $\mathbb{R}^n \cong \mathbb{R}^m$ (isomorphism of additive groups) implies $n = m$?

83. For $1 \leq n \leq 4$, list all integer partitions of n and draw their diagrams.

84. Let p be a fixed prime. Use partitions to make a complete list of all non-isomorphic commutative groups of size p^6.

85. Let G be the group $\mathbb{Z}_{5^5} \times \mathbb{Z}_{5^5} \times \mathbb{Z}_{5^5} \times \mathbb{Z}_{5^3} \times \mathbb{Z}_{5^3} \times \mathbb{Z}_{5^2} \times \mathbb{Z}_{5^2}$.

Draw pictures of partition diagrams to help answer the following questions.

(a) For each $i \geq 1$, find the size of the subgroup $G[5^i]$.

(b) For each $i \geq 1$, find the size of the subgroup $5^i G$.

(c) Find the size of $G[125] \cap 25G$ (explain).

86. Suppose p is prime and $G = \mathbb{Z}_{p^{a_1}} \times \mathbb{Z}_{p^{a_2}} \times \cdots \times \mathbb{Z}_{p^{a_k}}$, where $(a_1 \geq a_2 \geq \cdots \geq a_k)$ is a partition. What is the size of the subgroup pG? Describe how to use partition diagrams to find the size of $p^i G$ for all $i \geq 1$. Describe how to use partition diagrams to find the size of $p^i G \cap G[p^j]$ for all $i, j \in \mathbb{Z}_{\geq 0}$.

87. Prove that two integer partitions $(a_i : i \geq 1)$ and $(c_i : i \geq 1)$ are equal iff for all $j \geq 1$, the column sums $a_1' + \cdots + a_j'$ and $c_1' + \cdots + c_j'$ are equal.

88. Let $G_1, \ldots, G_k$ be commutative groups and $b \in \mathbb{Z}_{>0}$.

(a) Prove $(G_1 \oplus \cdots \oplus G_k)[b] = (G_1[b]) \oplus \cdots \oplus (G_k[b])$.

(b) Prove $b(G_1 \oplus \cdots \oplus G_k) = (bG_1) \oplus \cdots \oplus (bG_k)$.

(c) Prove $\mathrm{tor}(G_1 \oplus \cdots \oplus G_k) = \mathrm{tor}(G_1) \oplus \cdots \oplus \mathrm{tor}(G_k)$.

89. For each n below, make a complete list of all non-isomorphic commutative groups of size n. Use decompositions that display the elementary divisors of each group.

(a) 400   (b) 1001   (c) 666   (d) $p^2 q^3$, where p and q are distinct primes.

90. For each n below, make a complete list of all non-isomorphic commutative groups of size n. Use decompositions that display the invariant factors of each group.

(a) 24   (b) 300   (c) 32   (d) $p^3 q^3$, where p and q are distinct primes.

91. For each commutative group, find its elementary divisors.

(a) $\mathbb{Z}_{9900}$ (b) $\mathbb{Z}_{60} \times \mathbb{Z}_{100} \times \mathbb{Z}_{80}$ (c) $\mathbb{Z}_{48} \times \mathbb{Z}_{111} \times \mathbb{Z}_{99} \times \mathbb{Z}_{1001}$

92. For each commutative group, find its invariant factors.

(a) $\mathbb{Z}_{32} \times \mathbb{Z}_{16} \times \mathbb{Z}_4 \times \mathbb{Z}_4 \times \mathbb{Z}_9 \times \mathbb{Z}_9 \times \mathbb{Z}_3$ (b) $\mathbb{Z}_{60} \times \mathbb{Z}_{100} \times \mathbb{Z}_{80}$ (c) $\mathbb{Z}_{48} \times \mathbb{Z}_{111} \times \mathbb{Z}_{99} \times \mathbb{Z}_{1001}$

93. Let p and q be distinct primes. How many non-isomorphic commutative groups of size $p^5 q^3$ are there?

94. Let P be a logical property such that every cyclic group has property P; if a group G has property P and $G \cong H$, then H has property P; and whenever groups G and H have property P, the product group $G \times H$ has property P. Prove that all finitely generated commutative groups have property P.

95. Use the classification of finite commutative groups to characterize all positive integers n such that every commutative group of size n is cyclic. Give two proofs, one based on elementary divisors and one based on invariant factors.

96. Use the classification of finite commutative groups to prove that for all n-element commutative groups G and all positive divisors d of n, G has a subgroup of size d. Can you prove this without using the classification results from this chapter?

97. Prove that if G is an n-element commutative group that has at most one subgroup of size d for each positive divisor d of n, then G must be cyclic.

98. Suppose A, B, and C are finitely generated commutative groups such that $A \times C \cong B \times C$. Prove that $A \cong B$. Give an example to show this result can fail if C is not finitely generated.

99. (a) Let G be a commutative group of size $p_1^{e_1} p_2^{e_2} \cdots p_k^{e_k}$, where $p_1, \ldots, p_k$ are distinct primes and $e_1, \ldots, e_k > 0$. Prove: for $1 \le i \le k$, G has a unique subgroup P_i of size $p_i^{e_i}$. (b) Give an example to show that (a) can fail if G is not commutative. (c) Show that if the conclusion in (a) holds, then $G \cong P_1 \times P_2 \times \cdots \times P_k$ even if G is not commutative. (Use Exercise 12 from Chapter 1.)

100. Let $A \in M_{m,n}(\mathbb{Z})$. For $1 \le k \le \min(m,n)$, a *minor* of A of order k is the determinant of some $k \times k$ submatrix of A obtained by looking at the entries in k fixed rows and k fixed columns of A. Let $G_k(A)$ be the gcd of all kth order minors of A, with the convention that $\gcd(0,0,\ldots,0) = 0$. (a) Show that performing one elementary row or column operation on A does not change any of the integers $G_k(A)$. (First show that if $c \in \mathbb{Z}$ divides all kth order minors of A, then c divides all kth order minors of the new matrix.) (b) Show that if B is any Smith normal form of A, then $G_k(B) = G_k(A)$. (c) Show that if B is any matrix in Smith normal form, then $G_k(B) = \prod_{i=1}^{k} B(i,i)$ for $1 \le k \le \min(m,n)$. (d) Use (b) and (c) to prove that the Smith normal form of a matrix is unique. (e) Use (b) and (c) to compute the Smith normal form of

$$A = \begin{bmatrix} 10 & 8 & 0 \\ -4 & 8 & 6 \\ 0 & 12 & -8 \end{bmatrix}.$$

101. Let B be a basis for $\mathbb{R}$, viewed as a $\mathbb{Q}$-vector space. (a) For $m \in \mathbb{Z}_{>0}$, use B to describe a basis B_m for the $\mathbb{Q}$-vector space $\mathbb{R}^m$. (b) Show that $|B| = |B_m|$ for all $m \in \mathbb{Z}_{>0}$. Conclude that all of the commutative groups $(\mathbb{R}^m, +)$ (for $m = 1, 2, 3, \ldots$) are isomorphic to the commutative group $(\mathbb{R}, +)$.

102. Give justified examples of each of the following.
    (a) an infinite commutative group that is not free
    (b) a free commutative group that is not infinite
    (c) a $\mathbb{Z}$-independent list in $\mathbb{R}^2$ that is not $\mathbb{R}$-independent
    (d) a free commutative group with an infinite basis

(e) an infinite group G such that for all $x \in G$, $nx = 0_G$ for some $n > 0$

(f) a commutative group G and a proper subgroup H with $G \cong H$

(g) a group G in which $\text{tor}(G)$ is not a subgroup of G

(h) a commutative group G with isomorphic subgroups A and B such that G/A is not isomorphic to G/B

103. Prove: for all finite commutative groups G and H, if $G \times G \cong H \times H$, then $G \cong H$.

104. True or false? Explain each answer.

(a) Every subgroup of a finitely generated free commutative group is free.

(b) Every quotient group of a finitely generated free commutative group is free.

(c) A product of finitely many finitely generated free commutative groups is free.

(d) If (x_1, x_2, x_3) is an ordered basis for a commutative group G, then $(x_3, x_2 + x_3, x_1 + x_2 + x_3)$ is always an ordered basis for G.

(e) If $(x_1, \ldots, x_n)$ is any $\mathbb{Z}$-independent list in a commutative group G and c is a nonzero integer, then $(cx_1, x_2, \ldots, x_n)$ must also be $\mathbb{Z}$-independent.

(f) For all commutative groups G, $(G \sim \text{tor}(G)) \cup \{e_G\}$ is a subgroup of G.

(g) For all commutative groups G and all $n \in \mathbb{Z}_{>0}$, $G/G[n] \cong nG$.

(h) Every finite commutative group is isomorphic to a subgroup of $\mathbb{Z}_n$ for some $n > 0$.

(i) Every finite commutative group is isomorphic to a quotient group of $\mathbb{Z}_n$ for some $n > 0$.

(j) Every finite commutative group is isomorphic to a subgroup of S_n for some $n > 0$.

(k) Every finitely generated commutative group is isomorphic to a subgroup of $\mathbb{Z}^n$ for some $n > 0$.

(l) Every finitely generated commutative group is isomorphic to a quotient group of $\mathbb{Z}^n$ for some $n > 0$.

(m) Every finitely generated commutative group G with $\text{tor}(G) = \{0\}$ is free.

(n) Every k-element generating set for $\mathbb{Z}^k$ is a $\mathbb{Z}$-basis for $\mathbb{Z}^k$.

(o) Every k-element $\mathbb{Z}$-linearly independent subset of $\mathbb{Z}^k$ generates $\mathbb{Z}^k$.

Introduction to Modules

Introductions to linear algebra often cover the following concepts involving vector spaces over a field F: axioms for a vector space, subspaces, quotient spaces, direct sums, linear independence, spanning sets, bases, and linear transformations. The goal of this chapter is to cover the same material in a more general context. We replace the field F by a general ring R and consider vector spaces over this ring, which are now called *R-modules*. Intuitively, an R-module is a set of vectors on which we define operations of vector addition and scalar multiplication satisfying certain axioms. In this setting, the scalars come from the ring R rather than a field. As we will see, much of the theory of vector spaces extends without change to this more general situation, although the terminology used is a bit different.

However, we warn the reader that certain aspects of the theory of R-modules are quite different from what might be expected based on experience with vector spaces. The most glaring example of this phenomenon is that not every R-module has a basis. Those modules that do have a basis are called *free* R-modules. Even if an R-module is free, it may possess bases with different cardinalities, something which does not happen for vector spaces over a field. Fortunately, for commutative rings R (and certain other classes of rings), the cardinality of a basis is invariant for free R-modules.

Before we can even define modules, we must point out another complication that occurs when R is non-commutative. Suppose F is a field, V is an F-vector space, $c, d \in F$, and $v \in V$. We have the associative axiom for scalar multiplication, $(cd)v = c(dv)$, which we might also write as $v(cd) = (vc)d$. The point is that, although we often write scalars to the left of the vectors on which they act, we could have written the scalars on the right instead. However, if we replace F by a non-commutative ring R, so that we now have $c, d \in R$, then the two versions of the associative law just written are no longer equivalent. So we must distinguish between the concepts of scalar multiplication on the left and scalar multiplication on the right. This distinction leads to the concepts of *left* and *right* R-modules.

In the rest of this chapter, we define left and right R-modules and discuss fundamental constructions for R-modules, including submodules, direct products, direct sums, and quotient modules. We also discuss R-module homomorphisms, generating sets for a module, and independent sets in a module, which generalize the concepts of linear transformations, spanning sets, and linearly independent sets in vector spaces. We conclude by studying free R-modules, bases, the Jordan–Hölder Theorem, and the length of a module.

17.1   Module Axioms

Let R be an arbitrary ring (see §1.2 for the definition of a ring; recall our convention that every ring has a multiplicative identity element, denoted 1_R). We now present the axioms defining a *left R-module*, which are motivated by the corresponding axioms for a vector space (see Table 1.4). A left R-module consists of a set M, an *addition* operation $+ : M \times M \to M$, and a *scalar multiplication* operation $\cdot : R \times M \to M$, often denoted by juxtaposition. The addition operation on M must satisfy the following axioms:

(A1)   For all $a, b \in M$, $a + b$ belongs to the set M (closure under addition).

(A2)   For all $a, b, c \in M$, $(a + b) + c = a + (b + c)$ (associativity).

(A3)   For all $a, b \in M$, $a + b = b + a$ (commutativity).

(A4)   There exists an element $0 \in M$ (necessarily unique) such that $a + 0 = a = 0 + a$ for all $a \in M$ (additive identity).

(A5)   For each $a \in M$, there exists an element $-a \in M$ (necessarily unique) such that $a + (-a) = 0 = (-a) + a$ (additive inverses).

In other words, $(M, +)$ is a commutative group (see §1.1). The scalar multiplication operation must satisfy the following axioms:

(M1)   For all $r \in R$ and $m \in M$, $r \cdot m$ belongs to the set M (closure under scalar multiplication).

(M2)   For all $m \in M$, $1_R \cdot m = m$ (identity law).

(M3)   For all $r, s \in R$ and $m \in M$, $(rs) \cdot m = r \cdot (s \cdot m)$ (left associativity).

Also, addition and scalar multiplication are linked by the following distributive laws:

(D1)   For all $r, s \in R$ and $m \in M$, $(r + s) \cdot m = r \cdot m + s \cdot m$ (distributive law for ring addition).

(D2)   For all $r \in R$ and $m, n \in M$, $r \cdot (m + n) = r \cdot m + r \cdot n$ (distributive law for module addition).

We define *subtraction* by setting $a - b = a + (-b)$ for $a, b \in M$. Elements of M may be called *vectors*, and elements of R *scalars*, although this terminology is more often used when discussing vector spaces over a field.

Now we give the corresponding definition of a *right R-module*. A right R-module consists of a set N, an *addition* operation $+ : N \times N \to N$, and a *scalar multiplication* operation $\star : N \times R \to N$. We require $(N, +)$ to be a commutative group (so axioms (A1) through (A5) must hold with M replaced by N). The multiplication operation must satisfy the following axioms:

(M1′)   For all $n \in N$ and $r \in R$, $n \star r$ belongs to the set N (closure under scalar multiplication).

(M2′)   For all $n \in N$, $n \star 1_R = n$ (identity law).

(M3′)   For all $n \in N$ and $r, s \in R$, $n \star (rs) = (n \star r) \star s$ (right associativity).

Also, addition and scalar multiplication are linked by the following distributive laws:

(D1′)   For all $n \in N$ and $r, s \in R$, $n \star (r + s) = n \star r + n \star s$ (distributive law for ring addition).

(D2′)   For all $m, n \in N$ and $r \in R$, $(m + n) \star r = m \star r + n \star r$ (distributive law for module addition).

Subtraction is defined as before.

When R is commutative, there is no essential difference between left R-modules and right R-modules. To see why, let $(M, +)$ be a fixed commutative group. We can set up a one-to-one correspondence between left scalar multiplications $\cdot : R \times M \to M$ and right scalar multiplications $\star : M \times R \to M$, given by $m \star r = r \cdot m$ for all $r \in R$ and $m \in M$. For any ring R, you can verify that the left module axioms (M1), (M2), (D1), and (D2) hold for $\cdot$ iff the corresponding right module axioms (M1'), (M2'), (D1'), and (D2') hold for $\star$. If R is commutative, then (M3) holds for $\cdot$ iff (M3') holds for $\star$. Thus, in the commutative case, it makes little difference whether we multiply by scalars on the left or on the right. Later in this chapter, the unqualified term "module" always means *left* R-module. All definitions and results for left R-modules have analogs for right R-modules.

Next, we introduce the concept of *R-module homomorphisms*, which are analogous to linear transformations of vector spaces. Let M and N be left R-modules. A map $f : M \to N$ is a *left R-module homomorphism* iff $f(x+y) = f(x)+f(y)$ and $f(rx) = rf(x)$ for all $x, y \in M$ and $r \in R$. The definition of a *right R-module homomorphism* between right R-modules M and N is analogous: we require $f(x+y) = f(x)+f(y)$ and $f(xr) = f(x)r$ for all $x, y \in M$ and $r \in R$. Homomorphisms of R-modules (left or right) are also called *R-maps* or *R-linear maps*. The following terminology is sometimes used for R-maps satisfying additional conditions: a module homomorphism $f : M \to N$ is called a *monomorphism* iff f is injective, an *epimorphism* iff f is surjective, an *isomorphism* iff f is bijective, an *endomorphism* iff $M = N$, and an *automorphism* iff $M = N$ and f is bijective. Modules M and N are called *isomorphic* or *R-isomorphic* iff there exists an R-module isomorphism $g : M \to N$; in this case, we write $M \cong N$. You can verify that the following facts about linear transformations and group homomorphisms are also true for R-maps: the identity map on M is a bijective R-map; the composition of R-maps is an R-map (similarly for monomorphisms, epimorphisms, etc.); the inverse of an R-isomorphism is an R-isomorphism; the relation "M is isomorphic to N as an R-module" is an equivalence relation on any collection of left R-modules; and for an R-map $f : M \to N$, $f(0_M) = 0_N$ and $f(-x) = -f(x)$ for all $x \in M$. You can also check that $0_R x = 0_M$ for all $x \in M$; $r0_M = 0_M$ for all $r \in R$; and $r(-x) = (-r)x = -(rx)$ for all $r \in R$ and all $x \in M$.

17.2   Examples of Modules

We now discuss four fundamental examples of R-modules.

First, any ring R is a left R-module, if we take addition and scalar multiplication to be the given addition and multiplication in the ring R. In this case, the module axioms reduce to the axioms in the definition of a ring (see §1.2). Similarly, every ring R is a right R-module. Suppose $f : R \to R$ is a function satisfying $f(x + y) = f(x) + f(y)$ for all $x, y \in R$. Then f is a ring homomorphism iff $f(1_R) = 1_R$ and $f(xy) = f(x)f(y)$ for all $x, y \in R$. On the other hand, f is a left R-module homomorphism iff $f(xy) = xf(y)$ for all $x, y \in R$, while f is a right R-module homomorphism iff $f(xy) = f(x)y$ for all $x, y \in R$.

Second, generalizing the first example, let R be a subring of a ring S. (Recall from §1.4 that this means R is a subset of S containing 0_S and 1_S and closed under addition, subtraction, and multiplication; R itself is a ring under the operations inherited from S.) The ring S is a left R-module, if we take addition to be the addition in S and scalar multiplication $\cdot : R \times S \to S$ to be the restriction of the multiplication $\cdot : S \times S \to S$ in the ring S. Here, the conditions in the module axioms hold because they are a subset of the conditions in the ring axioms for S. Similarly, S is a right R-module.

Third, let F be a field. Comparing the module axioms to the axioms defining a vector space over F (see Table 1.4), we see that a left F-module V is exactly the same as an F-vector space. An F-module homomorphism $T : V \to W$ is exactly the same as an F-linear map from V to W.

Fourth, let $(M, +)$ be any commutative group, and consider the ring $R = \mathbb{Z}$. We make M into a left $\mathbb{Z}$-module by defining $0 \cdot x = 0_M$, $n \cdot x = x + x + \cdots + x$ (n copies of x), and $-n \cdot x = -(x + x + \cdots + x)$ (n copies of x) for all $n > 0$ and $x \in M$. The module axioms involving multiplication are the additive versions of the laws of exponents for commutative groups. Moreover, you can verify (using axioms (M2) and (D1) and induction) that $\cdot$ is the unique scalar multiplication map from $\mathbb{Z} \times M$ into M that makes $(M, +)$ into a left $\mathbb{Z}$-module. In other words, the $\mathbb{Z}$-module structure of M is completely determined by the structure of M as an additive group. Next, consider a function $f : M \to N$ between two $\mathbb{Z}$-modules. On one hand, if f is a $\mathbb{Z}$-module homomorphism, then it is in particular a homomorphism of commutative groups (meaning $f(x + y) = f(x) + f(y)$ for all $x, y \in M$). Conversely, if f is a group homomorphism, then we know from group theory that $f(n \cdot x) = n \cdot f(x)$ for all $x \in M$ and $n \in \mathbb{Z}$, so that f is automatically a $\mathbb{Z}$-module homomorphism. These remarks show that $\mathbb{Z}$-modules and $\mathbb{Z}$-module homomorphisms are essentially the same as additive commutative groups and group homomorphisms.

The examples illustrate one of the advantages of module theory as a unifying notational tool: facts about vector spaces, commutative groups, and rings can all be formulated and proved in the general framework of module theory.

17.3   Submodules

The next few sections discuss some general constructions for manufacturing new modules from old ones: submodules, direct products, direct sums, Hom modules, quotient modules, and changing the ring of scalars. Chapter 20 discusses more advanced constructions for modules, such as tensor products, tensor powers, exterior powers, and symmetric powers.

Let N be a subset of a left R-module M. We say N is an R-*submodule* of M iff N is a subgroup of the additive group $(M, +)$ such that for all $r \in R$ and $x \in N$, $r \cdot x$ belongs to the set N. We may omit R from the term "R-submodule" when R is understood. A submodule of M is a subset of M containing 0_M and closed under addition, additive inverses, and left multiplication by scalars from R. Using the fact that $-1 \in R$, we see that closure under inverses follows from the other closure conditions, since $-n = -(1_R \cdot n) = (-1_R) \cdot n$ for $n \in N$. If we restrict the addition and scalar multiplication operations for M to the domains $N \times N$ and $R \times N$ (respectively), then N becomes a left R-module. The axioms (A1) and (M1) hold for N by definition of a submodule. The other axioms for N are special cases of the corresponding axioms for M, keeping in mind that $0_M \in N$ and $-x \in N$ whenever $x \in N$.

The definition of a submodule N of a right R-module M is similar: we require N to be an additive subgroup of M such that $x \star r \in N$ for all $x \in N$ and $r \in R$.

Consider the examples in the preceding section. Regarding R as a left R-module, the definition of a left R-submodule of R is exactly the definition of a *left ideal* of R (see §1.4). Regarding R as a right R-module, the definition of a right R-submodule of R is exactly the definition of a *right ideal* of R. If R is commutative, the R-submodules of R (viewed as a left or right R-module) are precisely the *ideals* of the ring R. Next, if V is a vector space over a field F (so V is an F-module), the F-submodules of V are precisely the *subspaces* of the vector space V. Finally, if M is a commutative group and hence a $\mathbb{Z}$-module, the $\mathbb{Z}$-submodules of M are precisely the *subgroups* of M. On one hand, $\mathbb{Z}$-submodules of M are subgroups of M by definition. On the other hand, if H is a subgroup, $x \in H$, and $n \in \mathbb{Z}$, then $n \cdot x \in H$ follows by induction from the fact that H is closed under addition and additive inverses.

We now discuss intersections and sums of submodules. Let $\{M_i : i \in I\}$ be a nonempty collection of submodules of an R-module M. The *intersection* $N = \bigcap_{i \in I} M_i$ is a submodule of M, hence an R-module. *Proof:* First, $0_M \in N$ since 0_M belongs to every M_i. Given $x, y \in N$, we know $x, y \in M_i$ for all $i \in I$, so $x + y \in M_i$ for all $i \in I$, so $x + y \in N$. Given $x \in N$ and $r \in R$, we have $x \in M_i$ for all $i \in I$, so $r \cdot x \in M_i$ for all $i \in I$, so $r \cdot x \in N$. The submodule N is the largest submodule of M contained in every M_i.

Next, let M and N be submodules of an R-module P. You can verify that the set $M + N = \{m + n : m \in M, n \in N\}$ is a submodule of P, called the *sum* of M and N. Sums of finitely many submodules of P are defined analogously. For the general case, let $\{M_i : i \in I\}$ be a family of submodules of P. The *sum of submodules* $\sum_{i \in I} M_i$ is the set of all finite sums $m_{i_1} + \cdots + m_{i_s}$ where $i_j \in I$, $s \in \mathbb{Z}_{\geq 0}$, and $m_{i_j} \in M_{i_j}$ for $1 \leq j \leq s$. You can check that $\sum_{i \in I} M_i$ is an R-submodule of P, and this is the smallest submodule of P containing every M_i. If I is empty, we consider the vacuous sum $\sum_{i \in I} M_i$ to be the submodule $\{0_M\}$ consisting of zero alone.

A *partially ordered set* is a set X and a relation $\leq$ on X that is *reflexive* (for all $x \in X$, $x \leq x$), *antisymmetric* (for all $x, y \in X$, if $x \leq y$ and $y \leq x$, then $x = y$), and *transitive* (for all $x, y, z \in X$, if $x \leq y$ and $y \leq z$, then $x \leq z$). A *lattice* is a partially ordered set in which any two elements have a least upper bound and a greatest lower bound (see the Appendix for more detailed definitions). A *complete lattice* is a partially ordered set $(X, \leq)$ in which any nonempty subset of X has a least upper bound and a greatest lower bound. In more detail, given any nonempty $S \subseteq X$, the set of upper bounds $\{x \in X : s \leq x \text{ for all } s \in S\}$ must be nonempty and have a least element, and similarly for the set of lower bounds of S. For any left R-module M, the set X of all submodules of M is a partially ordered set under the ordering defined by $N \leq P$ iff $N \subseteq P$ (for $N, P \in X$). Our results on intersections and sums of submodules show that *the poset X of submodules of a left R-module is a complete lattice*, since every nonempty family of submodules $\{M_i : i \in I\}$ has a greatest lower bound (the intersection of the M_i) and a least upper bound (the sum of the M_i).

A left R-module M is called a *simple* module iff $M \neq \{0\}$ and the only R-submodules of M are $\{0\}$ and M. The adjective "simple" really describes the submodule lattice of M, which is a poset with only two elements. For example, $\mathbb{Z}_2$, $\mathbb{Z}_3$, and $\mathbb{Z}_5$ are simple $\mathbb{Z}$-modules, but $\mathbb{Z}_1$ and $\mathbb{Z}_4$ are not simple $\mathbb{Z}$-modules. You can check that a commutative ring R is simple (as a left R-module) iff R is a field.

17.4 Submodule Generated by a Subset

Let S be any subset of a left R-module M. Our next goal is to construct a submodule of M containing S that is as small as possible. To do this, let $\{M_i : i \in I\}$ be the family of all submodules of M that contain S. This family is nonempty, since M itself is a submodule of M containing S. Let $N = \bigcap_{i \in I} M_i$. Then N is a submodule, N contains S, and N is contained in any submodule of M that contains S (since that submodule must be one of the M_i). We write $N = \langle S \rangle$, and call N the *submodule generated by S*. Elements of S are called *generators* for N. In the case of vector spaces, we say that the subspace N is *spanned* by the vectors in S, and that S forms a *spanning set* for N.

We can give a more explicit description of $\langle S \rangle$ that shows how elements of this submodule are built up from elements of S. Let N' be the set of all finite sums $r_1 s_1 + \cdots + r_n s_n$, where $r_i \in R$, $s_i \in S$, and $n \in \mathbb{Z}_{\geq 0}$. Such a sum is called an R-*linear combination* of elements of S. If $n = 0$, the sum is defined to be zero, so 0_M is always in N'. We show that $N' = N = \langle S \rangle$. First, you can check that N' is an R-submodule of M. Since R has an identity, each $s \in S$

can be written as $1_R \cdot s$, which implies that N' contains S. Therefore, N' is one of the submodules M_i appearing in the definition of N, and hence $N' \supseteq N$. To show the reverse inclusion, consider an arbitrary submodule M_i that contains S. Since M_i is a submodule, it must contain $r_1 s_1 + \cdots + r_n s_n$, which is the typical element of N'. Thus, $N' \subseteq M_i$ for every M_i, and hence $N' \subseteq N$.

We draw attention to some special cases of this result. If $S = \{s_1, \ldots, s_k\}$ is finite, then $\langle S \rangle = \{\sum_{i=1}^{k} r_i s_i : r_i \in R\}$. If $S = \{s\}$ has one element, then $\langle S \rangle = \{rs : r \in R\}$. If S is empty, then $\langle S \rangle = \{0_M\}$. Given any R-module M, we say M is *finitely generated* iff there exists a finite subset S of M such that $M = \langle S \rangle$. We say M is *cyclic* iff there exists a one-element set S such that $M = \langle S \rangle$. If $S = \{x\}$, we often write $M = \langle x \rangle$ or $M = Rx$ instead of $M = \langle\{x\}\rangle$. You can check that for any $x_1, \ldots, x_n$ in an R-module M, the sum of submodules $Rx_1 + \cdots + Rx_n$ equals the R-submodule $\langle x_1, \ldots, x_n \rangle$ generated by $\{x_1, \ldots, x_n\}$. More generally, for any $S \subseteq M$, $\sum_{x \in S} Rx = \langle S \rangle$.

One property of generating sets is that *module homomorphisms that agree on a generating set must be equal.* More formally, let M be an R-module generated by S, and let $f, g : M \to P$ be R-module homomorphisms such that $f(s) = g(s)$ for all $s \in S$. Then $f = g$. To prove this, take any nonzero element x of M and write it as $x = r_1 s_1 + \cdots + r_k s_k$ with $r_i \in R$ and $s_i \in S$. Since f and g are module homomorphisms agreeing on S, $f(x) = \sum_{i=1}^{k} r_i f(s_i) = \sum_{i=1}^{k} r_i g(s_i) = g(x)$. Also, $f(0_M) = 0_P = g(0_M)$, so that $f = g$.

Let us use the ideas of cyclic submodules and generating sets to determine all the $\mathbb{Z}$-submodules of the $\mathbb{Z}$-module $M = \mathbb{Z}_4 \times \mathbb{Z}_2$. M is an 8-element commutative group, and its submodules are precisely the subgroups of M. It is known from group theory (Lagrange's Theorem) that the size of each such subgroup must be a divisor of 8, namely 1 or 2 or 4 or 8. We can start finding subgroups by looking at the cyclic submodules generated by each element of M. We discover the following submodules:

$$
\begin{aligned}
A &= \mathbb{Z}(0,0) &=\quad &\{(0,0)\} \\
B &= \mathbb{Z}(0,1) &=\quad &\{(0,0),(0,1)\} \\
C &= \mathbb{Z}(1,0) &=\quad &\{(0,0),(1,0),(2,0),(3,0)\} = \mathbb{Z}(3,0) \\
D &= \mathbb{Z}(2,0) &=\quad &\{(0,0),(2,0)\} \\
E &= \mathbb{Z}(2,1) &=\quad &\{(0,0),(2,1)\} \\
F &= \mathbb{Z}(1,1) &=\quad &\{(0,0),(1,1),(2,0),(3,1)\} = \mathbb{Z}(3,1).
\end{aligned}
$$

We can build further submodules by taking sums of the cyclic submodules found above. This produces two new submodules, namely $G = B + D = \{(0,0),(0,1),(2,0),(2,1)\}$ and $B + C = \mathbb{Z}_4 \times \mathbb{Z}_2 = M$. Figure 17.1 displays the lattice of submodules of the $\mathbb{Z}$-module M. In the figure, we abbreviate (x,y) as xy, and we draw a thick line from a submodule U up to a submodule V whenever $U \subseteq V$ and there is no submodule properly between U and V. This figure is called the *Haase diagram* of the poset of submodules of M.

17.5 Direct Products and Direct Sums

Let I be any set, and suppose we have a left R-module M_i for each $i \in I$. Let N be the set of all functions $f : I \to \bigcup_{i \in I} M_i$ such that $f(i) \in M_i$ for all $i \in I$. We may regard such a function as an I-*tuple* $(f(i) : i \in I)$, particularly when I is a finite set such as $\{1, 2, \ldots, n\}$. Given $f, g \in N$, define the *sum* $f + g$ by the rule $(f + g)(i) = f(i) + g(i) \in M_i$ for $i \in I$.

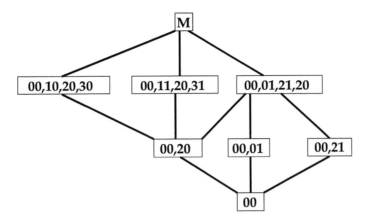

FIGURE 17.1
The lattice of submodules of the $\mathbb{Z}$-Module $\mathbb{Z}_4 \times \mathbb{Z}_2$.

You can check that N becomes a commutative group with this addition operation. Next, for $f \in N$ and $r \in R$, define the *scalar multiple* rf by the rule $(rf)(i) = r \cdot [f(i)] \in M_i$ for $i \in I$. You can verify that this scalar multiplication makes N into a left R-module. The axioms for N follow from the corresponding axioms for the M_i. For instance, to check (D1), fix $r, s \in R$ and $f \in N$, and compare the effects of the functions $(r+s)f$ and $rf + sf$ on a generic index $i \in I$:

$$\begin{aligned}
[(r+s)f](i) &= (r+s) \cdot [f(i)] = r \cdot [f(i)] + s \cdot [f(i)] \quad \text{(by (D1) in } M_i) \\
&= (rf)(i) + (sf)(i) = (rf + sf)(i).
\end{aligned}$$

We write $N = \prod_{i \in I} M_i$ and call N the *direct product* of the modules M_i. If $I = \{1, 2, \ldots, n\}$, we write $N = M_1 \times M_2 \times \cdots \times M_n$. In this case, elements of N are often regarded as n-tuples instead of functions. Using n-tuple notation for elements of N, the module operations are

$$(f_1, \ldots, f_n) + (g_1, \ldots, g_n) = (f_1 + g_1, \ldots, f_n + g_n) \qquad \text{for all } f_i, g_i \in M_i;$$

$$r(f_1, \ldots, f_n) = (rf_1, \ldots, rf_n) \qquad \text{for all } f_i \in M_i, r \in R,$$

where we have written f_i instead of $f(i)$.

As a special case of the direct product, we can take every M_i equal to a given module M, and we write $N = M^I$ in this case. If $I = \{1, 2, \ldots, n\}$, we write $N = M^n$. For instance, the vector spaces F^n (for F a field) are special cases of this construction. Taking $R = \mathbb{Z}$, we see that direct products of commutative groups are also special cases. There is a similar construction for producing the direct product of an indexed set of right R-modules.

Consider a general direct product $N = \prod_{i \in I} M_i$ of left R-modules. We say that a function $f \in N$ has *finite support* iff the set $\{i \in I : f(i) \neq 0_{M_i}\}$ is finite. Let N_0 consist of all functions $f \in N$ with finite support. You can check that N_0 contains 0_N and is closed under addition and scalar multiplication, so that N_0 is an R-submodule of N. We call N_0 the *direct sum* of the R-modules M_i, written $N_0 = \bigoplus_{i \in I} M_i$. If I is finite, the direct sum coincides with the direct product. For example, $M_1 \times M_2 \times \cdots \times M_n = M_1 \oplus M_2 \oplus \cdots \oplus M_n$.

Let F be a field. We show that *every F-vector space V is isomorphic to a direct sum of copies of F*. The proof requires the well-known fact (proved in §17.14) that every vector space has a basis. Given V, let X be a fixed basis of V and W be the direct sum of copies of F indexed by X. Thus, a typical element of W is a function $g : X \to F$ with finite support. To get a vector space isomorphism $T : W \to V$, define $T(g) = \sum_{x \in X} g(x)x \in V$ for $g \in W$. The sum is well-defined because, for each fixed $g \in W$, there are only finitely many nonzero summands. To describe the inverse map $S : V \to W$, recall that each $v \in V$ can be uniquely written as a finite linear combination $v = c_1 x_1 + \cdots + c_n x_n$ for some $n \geq 0$, nonzero $c_i \in F$, and distinct $x_i \in X$. Define $S(v)$ to be the function $g : X \to F$ such that $g(x_i) = c_i$ for $1 \leq i \leq n$, and $g(x) = 0$ for all other $x \in X$. You can verify that S and T are linear maps that are inverses of each other. This result for vector spaces does not extend to general R-modules, because (as we see later) not every R-module has a basis.

17.6 Homomorphism Modules

Let M and N be left R-modules. Define $\mathrm{Hom}(M, N)$ to be the set of all group homomorphisms $f : M \to N$. Note that $\mathrm{Hom}(M, N)$ is a subset of the direct product $N^M = \prod_{x \in M} N$, which is a left R-module. We check that the subset $\mathrm{Hom}(M, N)$ is an R-submodule of N^M. The zero element of N^M is the zero map $z : M \to N$ such that $z(x) = 0_N$ for all $x \in M$. This map is a group homomorphism, so $0_{N^M} \in \mathrm{Hom}(M, N)$. Next, fix $f, g \in \mathrm{Hom}(M, N)$. Then $f + g \in \mathrm{Hom}(M, N)$ because for all $x, y \in M$,

$$
\begin{aligned}
(f + g)(x + y) &= f(x + y) + g(x + y) = f(x) + f(y) + g(x) + g(y) \\
&= f(x) + g(x) + f(y) + g(y) \quad \text{(since + in } N \text{ is commutative)} \\
&= (f + g)(x) + (f + g)(y).
\end{aligned}
$$

Suppose $f \in \mathrm{Hom}(M, N)$ and $r \in R$. Then $rf \in \mathrm{Hom}(M, N)$ because for all $x, y \in M$,

$$
(rf)(x + y) = r \cdot [f(x + y)] = r \cdot [f(x) + f(y)] = r \cdot f(x) + r \cdot f(y) = (rf)(x) + (rf)(y).
$$

Since $\mathrm{Hom}(M, N)$ is an R-submodule of N^M, it is a left R-module under the pointwise operations on functions inherited from N^M.

Define $\mathrm{Hom}_R(M, N)$ to be the set of all left R-module homomorphisms $g : M \to N$. The set $\mathrm{Hom}_R(M, N)$ is a subset of N^M and also a subset of $\mathrm{Hom}(M, N)$. We show that, *if R is commutative, then $\mathrm{Hom}_R(M, N)$ is a submodule of N^M and is therefore a left R-module.* You can check that $\mathrm{Hom}_R(M, N)$ is an additive subgroup of N^M (for any ring R). Fix $r \in R$, $f \in \mathrm{Hom}_R(M, N)$, $s \in R$, and $x \in M$. For R commutative, we have

$$
\begin{aligned}
(rf)(s \cdot x) &= r \cdot [f(s \cdot x)] = r \cdot [s \cdot f(x)] = (rs) \cdot [f(x)] \\
&= (sr) \cdot [f(x)] \quad \text{(by commutativity of } R) \\
&= s \cdot (r \cdot [f(x)]) = s \cdot ((rf)(x)).
\end{aligned}
$$

The other condition for being an R-map, namely $(rf)(x + y) = (rf)(x) + (rf)(y)$ for $x, y \in M$, was already checked above. Therefore, $rf \in \mathrm{Hom}_R(M, N)$, as needed.

As a special case of this construction, consider F-vector spaces V and W. The set of all linear transformations from V to W, sometimes denoted by $L(V, W)$, becomes an F-vector space under pointwise operations on functions. This follows by noting that $L(V, W) = \mathrm{Hom}_F(V, W)$ and F is commutative.

17.7 Quotient Modules

In §1.6, we discussed the construction of the quotient group of a commutative group by a subgroup, and the formation of the quotient space of a vector space by a subspace. These constructions are special cases of quotient modules, which are defined as follows.

Let N be a submodule of a left R-module M. For each $x \in M$, we have the *coset* $x + N = \{x + n : n \in N\}$. The *quotient set* $M/N = \{x + N : x \in M\}$ is the collection of all cosets of N in M. The Coset Equality Theorem (§1.6) states that for all $x, z \in M$, $x + N = z + N$ iff $x - z \in N$. We also proved that M is the disjoint union of the distinct cosets of N, meaning that every $u \in M$ belongs to exactly one coset in M/N. Since N is a subgroup of $(M, +)$, we can define an addition operation on M/N by setting $(x + N) + (y + N) = (x + y) + N$ for all $x, y \in M$. We proved in §1.6 that this binary operation is well-defined and satisfies the axioms for a commutative group. The identity element of this quotient group is $0_{M/N} = 0_M + N = \{0 + n : n \in N\} = N$, and the additive inverse of the coset $x + N$ is the coset $(-x) + N$, for each $x \in M$.

To complete the definition of the quotient R-module M/N, we introduce a scalar multiplication operation $\cdot : R \times M/N \to M/N$. Given $r \in R$ and $x \in M$, define $r \cdot (x + N) = (rx) + N$. We must check that this operation is well-defined. Suppose $r \in R$ and $x, z \in M$ satisfy $x + N = z + N$. We prove that $r \cdot (x + N) = r \cdot (z + N)$. By the Coset Equality Theorem, $x - z \in N$. Since N is an R-submodule, $r(x - z) \in N$. So $rx - rz \in N$, and hence $(rx) + N = (rz) + N$. This shows that $r \cdot (x + N) = r \cdot (z + N)$. Now that we know scalar multiplication is well-defined, it is routine to check that M/N is a left R-module. The following calculation verifies axiom (D1): for all $x \in M$ and $r, s \in R$,

$$
\begin{aligned}
(r + s) \cdot (x + N) &= ((r + s)x + N) \\
&= ((rx + sx) + N) \quad \text{(by (D1) in } M\text{)} \\
&= (rx + N) + (sx + N) = [r \cdot (x + N)] + [s \cdot (x + N)].
\end{aligned}
$$

You can prove the other axioms for scalar multiplication similarly.

For any left R-module M and submodule N, there is a surjection $p : M \to M/N$ defined by $p(x) = x + N$ for $x \in M$. This map is called the *natural map* from M to M/N, the *canonical map* from M to M/N, or the *projection* of M onto M/N. The map p is R-linear, since

$$p(x + y) = (x + y) + N = (x + N) + (y + N) = p(x) + p(y),$$
$$\text{and } p(rx) = (rx) + N = r \cdot (x + N) = r \cdot p(x)$$

for all $x, y \in M$ and $r \in R$.

Here are some special cases and extensions of the quotient module construction. For $R = \mathbb{Z}$, the quotient $\mathbb{Z}$-module M/N is identical to the quotient group M/N (since, as we have seen, the $\mathbb{Z}$-module structure on M/N is uniquely determined by the addition operation on this set). If R is a field F, then a submodule N is a vector subspace of the F-vector space M, and M/N is the quotient vector space of M by N. Finally, suppose R is any ring, and M is R viewed as a left R-module. Given any left ideal I of R, R/I is a left R-module. However, more can be said if I is a (two-sided) ideal of R (meaning that I is both a left ideal and a right ideal, which always happens for commutative rings R). By analogy with the definition of scalar multiplication $\cdot : R \times R/I \to R/I$, we can define a multiplication operation $\star : R/I \times R/I \to R/I$ by setting $(a + I) \star (b + I) = (ab) + I$ for all $a, b \in R$. To check that $\star$ is well-defined, fix $a, b, a', b' \in R$ and assume $a + I = a' + I$ and $b + I = b' + I$. Using the Coset Equality Theorem, we see that $(ab) + I = (a'b') + I$ because

$$a - a' \in I, \quad b - b' \in I, \quad ab - a'b' = a(b - b') + (a - a')b',$$

and the last expression is in I because I is a two-sided ideal. You can check that $(R/I, +, \star)$ is a ring, which is commutative if R is commutative. R/I is called the *quotient ring of R by the ideal I.*

One nice feature of modules is that the quotient of a module by any submodule is always defined, unlike the situation for general groups (where the subgroup must be normal) or rings (where the subset must be a two-sided ideal).

Suppose S is a generating set for the R-module M, and N is a submodule of M. Then the image of S in M/N, namely $p[S] = \{s + N : s \in S\}$, is a generating set for M/N. To prove this, note that any coset in M/N has the form $x + N$ for some $x \in M$. Write $x = \sum_i a_i s_i$ with $a_i \in R$ and $s_i \in S$. Then $x + N = (\sum_i a_i s_i) + N = \sum_i ((a_i s_i) + N) = \sum_i a_i (s_i + N)$.

17.8 Changing the Ring of Scalars

This section describes two constructions for changing the ring of scalars for a given module. First, let M be a left R-module, and suppose S is a subring of R. Restricting the scalar multiplication $\cdot : R \times M \to M$ to the domain $S \times M$, we get a scalar multiplication of S on M that satisfies the module axioms. Hence, we can regard M as a left S-module. More generally, suppose S is any ring and $f : S \to R$ is a ring homomorphism. You can verify that the scalar multiplication $\star : S \times M \to M$, defined by $s \star x = f(s) \cdot x$ for $s \in S$ and $x \in M$, turns M into a left S-module. The situation where S is a subring is the special case where $f : S \to R$ is the inclusion map given by $f(s) = s$ for all $s \in S$.

Second, let M be a left R-module. Suppose I is an ideal of R, and let S be the quotient ring R/I. Assume that I *annihilates* M, meaning that $i \cdot x = 0_M$ for all $i \in I$ and $x \in M$. We claim that M is an S-module (i.e., an R/I-module) with scalar multiplication defined by $(r + I) \bullet x = r \cdot x$ for $r \in R$ and $x \in M$. To check that $\bullet$ is well-defined, fix $r, r' \in R$ and $x \in M$ with $r + I = r' + I$. We know $r - r' \in I$, so $(r - r') \cdot x = 0$ (by the annihilation condition), hence $r \cdot x = r' \cdot x$ and $(r + I) \bullet x = (r' + I) \bullet x$. The S-module axioms now follow routinely from the corresponding R-module axioms. For instance, (M2) is true since $1_{R/I} \bullet x = (1_R + I) \bullet x = 1_R \cdot x = x$ for all $x \in M$.

Let N be any R-module, not necessarily annihilated by the ideal I. Define the subset IN of N to be the set of all finite sums of terms $i \cdot x$ with $i \in I$ and $x \in N$. You can check that IN is an R-submodule of N, so that the quotient module $M = N/IN$ is defined. Now, M is annihilated by I, because $i \cdot (x + IN) = (i \cdot x) + IN = 0_N + IN = 0_M$ for all $i \in I$ and $x \in N$ (since $i \cdot x - 0_N \in IN$). Therefore, M is an S-module. We restate this result for emphasis: *given any R-module N and any ideal I of R, N/IN is an R/I-module via the rule $(r + I) \bullet (x + IN) = (r \cdot x) + IN$ for $r \in R$ and $x \in N$.*

Here is one situation in which this result can be helpful. Let R be a nonzero commutative ring, N an R-module, and I a *maximal* ideal of R (meaning that $I \neq R$ and there are no ideals J with $I \subsetneq J \subsetneq R$). It can be shown that such maximal ideals exist in R (see §17.13) and that maximality of I implies $F = R/I$ is a field (see Exercise 53). Hence, we can pass from the general module N over the ring R to an associated vector space over the field F, namely N/IN. As we see later, this association lets us use known theorems about vector spaces over a field to deduce facts about modules over a commutative ring.

17.9 Fundamental Homomorphism Theorem for Modules

The next two sections describe some key theorems concerning R-module homomorphisms, which have analogs for homomorphisms of groups, rings, and vector spaces (cf. §1.7).

Let $f : M \to N$ be a homomorphism of left R-modules. The *kernel* of f, written $\ker(f)$, is $\{x \in M : f(x) = 0_N\}$. The *image* of f, written $\operatorname{img}(f)$, is $\{y \in N : y = f(x) \text{ for some } x \in M\}$. You can check that $\ker(f)$ is a submodule of M, and $\operatorname{img}(f)$ is a submodule of N. More generally, if M' is any submodule of M, then the image $f[M'] = \{f(x) : x \in M'\}$ is a submodule of N. The image of f is the special case where $M' = M$. If N' is any submodule of N, then the preimage $f^{-1}[N'] = \{x \in M : f(x) \in N'\}$ is a submodule of M. The kernel of f is the special case obtained by taking $N' = \{0_N\}$. For any submodule N' of N, $f^{-1}[N']$ contains $\ker(f)$.

It is immediate from the definitions that f is surjective iff $\operatorname{img}(f) = N$. We now prove that f is injective iff $\ker(f) = \{0_M\}$. First suppose the kernel is $\{0\}$. Assume $x, z \in M$ satisfy $f(x) = f(z)$. Then $f(x - z) = f(x) - f(z) = 0$, so that $x - z \in \ker(f) = \{0\}$. Therefore, $x = z$, proving f is injective. Conversely, assume f is injective. On one hand, $0_M \in \ker(f)$ since $f(0_M) = 0_N$. On the other hand, for any $x \in M$ unequal to 0, injectivity of f gives $f(x) \neq f(0) = 0$, so $x \notin \ker(f)$. Thus, $\ker(f) = \{0_M\}$.

Using kernels, images, and quotient modules, the next theorem builds an R-module isomorphism from any given R-module homomorphism.

Fundamental Homomorphism Theorem for Modules. Let $f : M \to N$ be a homomorphism of left R-modules. There is an R-module isomorphism $f' : M/\ker(f) \to \operatorname{img}(f)$ given by $f'(x + \ker(f)) = f(x)$ for $x \in M$.

Proof. Although this result can be deduced quickly from the Fundamental Homomorphism Theorem for Groups (see §1.7), we prove the full result here for emphasis. The crucial first step is to check that f' is well-defined. Let $K = \ker(f)$, and suppose $x + K = z + K$ for fixed $x, z \in M$. Then $z = x + k$ for some $k \in K$, and

$$f'(z + K) = f(z) = f(x + k) = f(x) + f(k) = f(x) + 0 = f(x) = f'(x + K),$$

so that f' is well-defined. Second, let us check that f' is an R-module homomorphism. For fixed $x, y \in M$ and $r \in R$, compute:

$$f'((x + K) + (y + K)) = f'((x + y) + K) = f(x + y) = f(x) + f(y) = f'(x + K) + f'(y + K);$$

$$f'(r(x + K)) = f'((rx) + K) = f(rx) = rf(x) = rf'(x + K).$$

Third, f' is injective, since for $x, y \in M$, $f'(x + K) = f'(y + K)$ implies $f(x) = f(y)$, hence $f(x - y) = 0$, hence $x - y \in K$, hence $x + K = y + K$. Fourth, the image of f' is $\{f'(x + K) : x \in M\} = \{f(x) : x \in M\}$, which is the image of f, so that f' is surjective as a mapping into the codomain $\operatorname{img}(f)$. Thus, f' is an R-module isomorphism. $\qquad\square$

By modifying the proof, we obtain the following more general result (Exercise 59).

Generalized Fundamental Homomorphism Theorem for Modules. Let $f : M \to N$ be an R-module homomorphism and H be a submodule of M such that $H \subseteq \ker(f)$. There is a well-defined R-module homomorphism $f' : M/H \to N$ given by $f'(x + H) = f(x)$ for $x \in M$. We have $\operatorname{img}(f') = \operatorname{img}(f)$ and $\ker(f') = \ker(f)/H$.

The generalized theorem can also be restated as a universal mapping property (UMP) for quotient modules.

Universal Mapping Property (UMP) for Quotient Modules. Let $f : M \to N$ be an R-module homomorphism, H be a submodule of M with $H \subseteq \ker(f)$, and $p : M \to M/H$ be the canonical projection. There exists a unique R-module homomorphism $f' : M/H \to N$ such that $f = f' \circ p$. Also, $\mathrm{img}(f') = \mathrm{img}(f)$ and $\ker(f') = \ker(f)/H$.

The setup for the UMP is displayed in the following diagram:

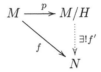

Existence of f' follows from the generalized theorem, together with the observation that $(f' \circ p)(x) = f'(p(x)) = f'(x + H) = f(x)$ for all $x \in M$. Similarly, uniqueness of f' holds since the requirement $f = f' \circ p$ forces us to define $f'(x + H) = f'(p(x)) = (f' \circ p)(x) = f(x)$ for all $x \in M$.

17.10 More Module Isomorphism Theorems

We now use the Fundamental Homomorphism Theorem for Modules to deduce some further isomorphism theorems that play a prominent role in module theory.

Diamond Isomorphism Theorem. Let M and N be submodules of a left R-module P. The quotient modules $M/(M \cap N)$ and $(M + N)/N$ are isomorphic, via the R-map $g : M/(M \cap N) \to (M + N)/N$ given by $g(m + M \cap N) = m + N$ for $m \in M$.

Figure 17.2, showing part of the submodule lattice of P, explains the name "Diamond Isomorphism Theorem" and can aid in remembering the theorem. The edges marked by double lines indicate which two quotient modules are isomorphic. By interchanging M and N, the theorem also yields an R-module isomorphism $(M + N)/M \cong N/(M \cap N)$.

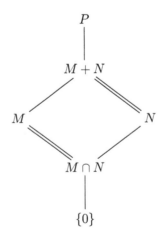

FIGURE 17.2
Picture of the Diamond Isomorphism Theorem.

Proof. Define $f : M \to (M + N)/N$ by $f(m) = m + N$ for $m \in M$. The map f is the composition of an inclusion $M \to M+N$ and a canonical map $M+N \to (M+N)/N$, hence is a module homomorphism. For any m in the domain M of f, $f(m) = 0$ in $(M + N)/N$ iff $m+N = 0+N$ iff $m \in N$. So the kernel of f equals $M \cap N$. We claim that the image of f is all of $(M+N)/N$. A typical element of the set $(M+N)/N$ is a coset $(m+n)+N$, where $m \in M$ and $n \in N$. But $(m+n)+N = (m+N)+(n+N) = (m+N)+0_{(M+N)/N} = m+N = f(m)$, so that this coset is in the image of f. Applying the Fundamental Homomorphism Theorem to f provides an isomorphism g given by the stated formula. $\qquad\square$

Nested Quotient Isomorphism Theorem. Suppose M is a left R-module, with submodules H and N such that $H \subseteq N$. There is an R-module isomorphism $g : (M/H)/(N/H) \to (M/N)$ given by $g((x + H) + (N/H)) = x + N$ for $x \in M$.

Proof. Consider the projection homomorphism $p : M \to M/N$. Since $H \subseteq \ker(p) = N$ by assumption, applying the Generalized Fundamental Homomorphism Theorem to p produces a well-defined R-map $f : M/H \to M/N$ given by $f(x+H) = x+N$ for $x \in M$. That theorem also tells us that $\mathrm{img}(f) = \mathrm{img}(p) = M/N$ and $\ker(f) = \ker(p)/H = N/H$. Applying the Fundamental Homomorphism Theorem to f, we get the isomorphism g given by the stated formula. $\qquad\square$

Correspondence Theorem for Modules. Let N be a fixed submodule of a left R-module M and $p : M \to M/N$ be the projection map. Let $\mathcal{A}$ be the collection of all submodules U of M such that $N \subseteq U \subseteq M$. Let $\mathcal{B}$ be the collection of all submodules of M/N. There are inclusion-preserving, mutually inverse bijections $T : \mathcal{A} \to \mathcal{B}$ and $S : \mathcal{B} \to \mathcal{A}$ given by $T(U) = p[U] = U/N$ (the image of U under p) for $U \in \mathcal{A}$ and $S(V) = p^{-1}[V]$ (the preimage of V under p) for $V \in \mathcal{B}$.

The statement that T *preserves inclusions* means that for all $U_1, U_2 \in \mathcal{A}$, if $U_1 \subseteq U_2$ then $T(U_1) \subseteq T(U_2)$; similarly for S. Note that square brackets denote the image or preimage of a subset under p, whereas round parentheses denote evaluation of a function at an input. Each $U \in \mathcal{A}$ is an element of the domain of T, while U is a subset of the domain of p. Similarly, each $V \in \mathcal{B}$ is an element of the domain of S, while V is a subset of the codomain of p. In particular, the function T is not the same as the function p. Similarly, $S \neq p^{-1}$; in fact, the inverse function p^{-1} only exists when $N = \{0\}$.

Proof. The proof of the Correspondence Theorem consists of a sequence of routine but lengthy verifications, some details of which appear in Exercises 55 and 61. First, T does map $\mathcal{A}$ into $\mathcal{B}$, since the image $p[U]$ is a submodule of M/N for any submodule U of M (whether or not U contains N). Second, S does map $\mathcal{B}$ into $\mathcal{A}$, since the preimage $p^{-1}[V]$ is a submodule of M containing $\ker(p) = N$ for any submodule V of M/N. Third, for any subset W of M/N, $p[p^{-1}[W]] = W$ follows from the surjectivity of the function $p : M \to M/N$. Consequently, taking W to be any submodule V of M/N, we see that $T(S(V)) = V = \mathrm{id}_{\mathcal{B}}(V)$ for all $V \in \mathcal{B}$, and hence $T \circ S = \mathrm{id}_{\mathcal{B}}$. Fourth, we prove that $S(T(U)) = U = \mathrm{id}_{\mathcal{A}}(U)$ for any $U \in \mathcal{A}$ (hence $S \circ T = \mathrm{id}_{\mathcal{A}}$). It follows from the definition of images and preimages that $S(T(U)) = p^{-1}[p[U]] \supseteq U$. To check the reverse inclusion, recall that U is a submodule of M containing N. Let $x \in p^{-1}[p[U]]$. Then $p(x) = x + N \in p[U] = U/N$, so there exists $z \in U$ with $x + N = z + N$. This means there exists $n \in N$ with $x = z + n$. Since $N \subseteq U$ and U is closed under addition, we have $x \in U$. This establishes the inclusion $p^{-1}[p[U]] \subseteq U$. Fifth, you can show that inclusions are preserved when we take images or preimages of subsets under any function. This fact (applied to p) implies that T and S preserve inclusions. $\qquad\square$

The Correspondence Theorem shows that T and S are lattice isomorphisms between the lattice $\mathcal{A}$ of submodules of M containing N and the lattice $\mathcal{B}$ of all submodules of M/N. (A *lattice isomorphism* is a bijection f between two lattices such that f and its inverse preserve the order relation, which in this situation is set inclusion.) This fact provides some retroactive motivation for the quotient module construction: if we are studying the submodule lattice of M, we can focus attention on the part of the lattice above the submodule N by passing to the submodule lattice of M/N. See Figure 17.3 for a picture of the relevant lattices.

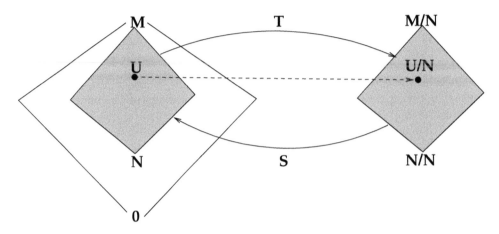

FIGURE 17.3
The lattice isomorphisms in the Correspondence Theorem for Modules.

Recognition Theorem for Direct Products. Suppose N and P are submodules of a left R-module M such that $N + P = M$ and $N \cap P = \{0_M\}$. There is an R-module isomorphism $g : N \times P \to M$ given by $g((x, y)) = x + y$ for $x \in N$ and $y \in P$.

Proof. We check that g is a one-to-one, onto, R-linear map. To verify R-linearity, fix $x_1, x_2 \in N$ and $y_1, y_2 \in P$, and compute

$$g((x_1, y_1) + (x_2, y_2)) = g((x_1 + x_2, y_1 + y_2)) = (x_1 + x_2) + (y_1 + y_2)$$
$$= (x_1 + y_1) + (x_2 + y_2) = g((x_1, y_1)) + g((x_2, y_2)).$$

Given $x \in N$, $y \in P$, and $r \in R$, we compute

$$g(r(x, y)) = g((rx, ry)) = rx + ry = r(x + y) = rg((x, y)).$$

To prove g is one-to-one, we show that $\ker(g) = \{(0, 0)\}$. For any (x, y) in $\ker(g)$, $0 = g((x, y)) = x + y$, so $y = -x$. We know $y \in P$, and since $y = -x$, y is also in the submodule N. Thus $y \in N \cap P = \{0_M\}$, hence $y = 0$. Then $x = -y = 0$ and $(x, y) = (0, 0)$, as needed. To prove g is onto, let $z \in M$ be arbitrary. Since $N + P = M$, there exist $x \in N$ and $y \in P$ with $z = x + y = g((x, y))$. $\qquad\square$

Exercise 79 generalizes the Recognition Theorem to products of more than two factors.

17.11   Free Modules

The concepts of spanning sets, linearly independent sets, and bases play a key role in the theory of vector spaces over a field. Now we consider the analogous concepts for a left R-module M where R is any ring. For $S \subseteq M$, we say S *spans* M and M *is generated by* S to mean that every element of M can be written as a finite sum $a_1 s_1 + \cdots + a_k s_k$ for some $k \in \mathbb{Z}_{\geq 0}$, $a_1, \ldots, a_k \in R$, and $s_1, \ldots, s_k \in S$. Such a sum is called an R-*linear combination of elements of* S. An ordered list $(s_1, \ldots, s_k)$ of elements of M is called R-*independent* (or R-*linearly independent*) iff for all $a_1, \ldots, a_k \in R$, the relation $a_1 s_1 + \cdots + a_k s_k = 0_M$ implies $a_i = 0_R$ for all i. Negating this definition, we see that the list $(s_1, \ldots, s_k)$ is R-*dependent* or R-*linearly dependent* iff there exist $a_1, \ldots, a_k \in R$ with $a_1 s_1 + \cdots + a_k s_k = 0_M$ and some $a_j \neq 0_R$. The list $(s_1, \ldots, s_k)$ is an *ordered basis* of the R-module M iff the list is R-linearly independent and $\{s_1, \ldots, s_k\}$ spans M.

A subset S of M (finite or not) is R-*independent* iff every finite list of distinct elements of S is linearly independent; otherwise, S is R-*dependent*. If there exists an R-independent generating set S for the module M, then M is called a *free* R-*module*, and S is called an R-*basis* or *basis* of M.

Certain results about generating sets, independence, and free R-modules are exactly like the corresponding results for vector spaces, but other familiar theorems about vector spaces and bases do not always hold for general left R-modules. Here are the key facts in the module case.

1. *If M is a free left R-module with basis S, then every $y \in M$ can be uniquely expressed as a finite R-linear combination of elements of S.* The existence of such an expression follows from the fact that S generates M. For uniqueness, suppose that $y = a_1 s_1 + \cdots + a_k s_k = b_1 s_1 + \cdots + b_k s_k$ where $k \in \mathbb{Z}_{\geq 0}$, $a_i, b_i \in R$, and $s_1, \ldots, s_k$ are distinct elements of S. (We can assume that the same elements of S are used in both expressions, by adding new terms with zero coefficients if needed.) Subtracting gives $0 = \sum_{i=1}^{k} (a_i - b_i) s_i$. By the R-independence of S, we deduce $a_i - b_i = 0$ for all i. So $a_i = b_i$ for all i.

2. **Universal Mapping Property (UMP) for Free R-modules.** Let M be a free left R-module with basis S. Given any left R-module N and any function $g : S \to N$, there exists a unique R-module homomorphism $g' : M \to N$ whose restriction to S equals g.

$$
\begin{array}{ccc}
S & \overset{\subseteq}{\longrightarrow} & M \\
 & {\scriptstyle g} \searrow & \big\downarrow {\scriptstyle \exists! g'} \\
 & & N
\end{array}
$$

    To prove existence, take any $y \in M$, and write y uniquely as $y = \sum_{x \in S} a_x x$ where $a_x \in R$ and all but finitely many a_x are 0. Define $g'(y) = \sum_{x \in S} a_x g(x) \in N$. You can check that g' is R-linear and that $g'(x) = g(x)$ for $x \in S$. For uniqueness, suppose g'' is another R-linear map such that g'' extends g. Then $g' = g''$ follows because two R-linear maps that agree on the generating set S must be equal.

3. *Every free left R-module M is isomorphic to a direct sum of copies of R.* We imitate the proof of the corresponding vector space result (see §17.5). Let S be an R-basis for the free module M. Let $N = \bigoplus_{x \in S} R$. Each element of N is a function $g : S \to R$ that is nonzero for only finitely many elements of its domain

S. Define a map $T : N \to M$ by setting $T(g) = \sum_{x \in S} g(x) \cdot x$, which is well-defined because the sum has only a finite number of nonzero terms. Define a map $T' : M \to N$ as follows. Given $y \in M$, write y uniquely as $y = \sum_{x \in S} c_x x$ where only finitely many $c_x \in R$ are nonzero. Define $T'(y)$ to be the function given by $g(x) = c_x$ for $x \in S$. You can verify that T and T' are R-linear maps that are inverses of each other. We call the S-tuple $T'(y) = (c_x : x \in S)$ the *coordinates of y relative to the basis S*.

4. *Given any set S and any nonzero ring R, there exists a free left R-module with basis S.* We let $M = \bigoplus_{x \in S} R$, the direct sum of copies of R indexed by S. Elements of M are functions $g : S \to R$ with finite support. For each $s \in S$, we have an associated function $e_s : S \to R$ such that $e_s(x) = 0_R$ for all $x \neq s$ in S, and $e_s(s) = 1_R$. (Note that $0_R \neq 1_R$ in the nonzero ring R.) Each e_s belongs to M; we show that $\{e_s : s \in S\}$ is a basis for M. Given any nonzero $g \in M$, let $\{s_1, \ldots, s_k\}$ be the elements of S for which $g(s_i) \neq 0$. Then $g = \sum_{i=1}^{k} g(s_i) e_{s_i}$, as can be seen by evaluating both sides at each $s \in S$. Next, to show R-independence, suppose $\sum_{t \in S} a_t e_t = 0$ for some $a_t \in R$ (with all but finitely many a_t equal to zero). Evaluating this function at $s \in S$, we see that each $a_s = 0$. Finally, we change notation by replacing each element $e_s \in M$ by the corresponding element $s \in S$. Then M is a free R-module with basis S (before the notation change, we had a basis $\{e_s\}$ in bijective correspondence with S).

5. *For some rings R, not every left R-module is free.* For example, consider $R = \mathbb{Z}$. A free R-module M is isomorphic to a direct sum of copies of $\mathbb{Z}$. Hence, either M is the zero module or M is infinite. On the other hand, any finite commutative group G is a $\mathbb{Z}$-module. Therefore, if G is not the zero group, then G is a non-free $\mathbb{Z}$-module. More generally, suppose R is an infinite ring containing a left ideal I such that R/I is finite and nonzero. Then R/I is a left R-module that cannot be isomorphic to a direct sum of copies of R; hence R/I is a non-free R-module.

6. *For some rings R, every left R-module is free.* For example, if R is a field, then every R-module is free because every vector space has a basis. Later in this chapter, we prove the more general result that all modules over a division ring R are free.

7. *For some rings R, not every left R-module is isomorphic to a submodule of a free R-module.* For example, consider $R = \mathbb{Z}$. If G is a nonzero finite commutative group, then G is not isomorphic to a submodule of any free R-module M. This holds since M must be a direct sum of copies of $\mathbb{Z}$, which implies that every nonzero submodule of M is infinite.

8. *Every left R-module is isomorphic to a quotient module of a free left R-module.* Let M be an arbitrary left R-module. Let F be a free left R-module having the set M as a basis. Consider the identity map $\mathrm{id}_M : M \to M$, where the domain is viewed as a subset of F, and the codomain is viewed as a left R-module. By the UMP, this map extends uniquely to an R-module homomorphism $g : F \to M$, which is evidently surjective. By the Fundamental Homomorphism Theorem, $F/\ker(g) \cong M$. More generally, if S is any generating set for M, then we could take F to be a free left R-module with S as a basis. The R-map $g : F \to M$ that extends the inclusion map $i : S \to M$ must be surjective, since the image of g is a submodule of M containing the generating set S of M. This shows that *every finitely generated left R-module M is isomorphic to a quotient module of a finitely generated free left R-module F.*

17.12   Finitely Generated Modules over a Division Ring

A *division ring* is a ring D such that $1_D \neq 0_D$ and for all nonzero $x \in D$, there exists $y \in D$ with $xy = 1_D = yx$. This condition says that every nonzero element of D has a two-sided multiplicative inverse. A *field* is the same thing as a division ring where multiplication is commutative. The quaternion ring $\mathbb{H}$ (see Exercise 40 of Chapter 1) is an example of a division ring that is not a field.

Let D be a division ring. A left D-module V is also called a *left vector space over D*. Our next goal is to show that facts about bases and dimensions of vector spaces over fields extend to the setting of modules over division rings. These facts are summarized in the next theorem.

Theorem on Modules over Division Rings.
Suppose D is a division ring and M is a left D-module.
(a) There exists a D-basis of M.
(b) Any D-independent set in M can be extended to a basis of M.
(c) Any generating set for M contains a basis of M.
(d) For all $S, T \subseteq M$, if S is D-independent and T generates M, then $|S| \leq |T|$.
(e) Any two bases of M have the same cardinality, called the *dimension* of M.

We prove this theorem for finitely generated D-modules here. The proofs in the general case require Zorn's Lemma and are covered later. Part (a) says that *every module over a division ring is free*.

Lemma on Extending an Independent List. Suppose D is a division ring, M is a left D-module, $x_1, \ldots, x_k$ is a D-independent list in M, and $y \in M$. The list $x_1, \ldots, x_k, y$ is D-independent iff $y \notin \langle x_1, \ldots, x_k \rangle$.

Proof. We give a contrapositive proof of both implications. First, we assume $y \in \langle x_1, \ldots, x_k \rangle$ and prove that $x_1, \ldots, x_k, y$ is a D-linearly dependent list. We know there are scalars $d_1, \ldots, d_k \in D$ with $y = d_1 x_1 + \cdots + d_k x_k$. Then $d_1 x_1 + \cdots + d_k x_k + (-1_D)y = 0_D$ expresses 0_D as a D-linear combination of the list $x_1, \ldots, x_k, y$ where the coefficient of y is $-1_D \neq 0_D$. So this list is linearly dependent.

To prove the converse, assume $x_1, \ldots, x_k, y$ is linearly dependent. So there exist $d_1, \ldots, d_k, d \in D$ with $d_1 x_1 + \cdots + d_k x_k + dy = 0_D$ and at least one of the scalars $d_1, \ldots, d_k, d$ is nonzero. If $d = 0$, then $d_1 x_1 + \cdots + d_k x_k = 0$ with some $d_i \neq 0$. This contradicts the assumed linear independence of $x_1, \ldots, x_k$. Therefore $d \neq 0$, which means d^{-1} exists in the division ring D. Rearranging the given relation, we get $dy = -d_1 x_1 - \cdots - d_k x_k$ and then $y = -d^{-1} d_1 x_1 - \cdots - d^{-1} d_k x_k$. So $y \in \langle x_1, \ldots, x_k \rangle$, as needed.   □

A similar proof establishes the following variation of the lemma (Exercise 62).

Lemma on Extending an Independent Set. Suppose D is a division ring, M is a left D-module, and S is a D-linearly independent subset of M. For all $y \in M \setminus S$, $S \cup \{y\}$ is D-linearly independent iff $y \notin \langle S \rangle$.

Lemma on Building an Ordered Basis. Suppose D is a division ring, M is a finitely generated left D-module, $x_1, \ldots, x_k$ is a D-linearly independent list in M, and $y_1, \ldots, y_m$ is a list generating M (where k and m are finite). Then M has an ordered D-basis of the form $x_1, \ldots, x_k, y_{i_1}, \ldots, y_{i_s}$ for some $\{i_1, \ldots, i_s\} \subseteq \{1, 2, \ldots, m\}$.

Informally, the lemma says we can extend any given independent list to a basis of M by appending suitable vectors from any generating list for M.

Proof. Let $N = \langle x_1, \ldots, x_k \rangle$, which is a submodule of M. If every $y_i \in N$, then $N = M$, since N contains a generating set for M. In this case, the list $x_1, \ldots, x_k$ is already a D-basis for M. The other possibility is that some y_i is not in N. Let i_1 be the least index i with this property, and note $x_1, \ldots, x_k, y_{i_1}$ is D-independent by the Lemma on Extending an Independent List.

We now apply the same reasoning to the submodule $N_1 = \langle x_1, \ldots, x_k, y_{i_1} \rangle$ generated by the new D-independent list. If $N_1 = M$, then we have found a basis for M. Otherwise, some y_i is not in N_1. Let i_2 be the least index i with this property, and note $x_1, \ldots, x_k, y_{i_1}, y_{i_2}$ is D-independent by the lemma. Continuing in this way, the extension process must eventually terminate with a D-basis of M, since there are only finitely many y_i available. □

Assuming M is a finitely generated D-module, we can now prove parts (a), (b), and (c) of the main theorem with sets replaced by finite lists. Let $y_1, \ldots, y_m$ be any finite list generating M. The lemma shows that any finite D-independent list L in M can be extended to a basis of M, proving (b). Taking L to be the empty list, we see that the given generating list contains an ordered basis of M, proving (a) and (c).

Comparison Lemma for Finite Independent Sets and Spanning Sets. Suppose D is a division ring, M is a finitely generated left D-module, $S = \{x_1, \ldots, x_k\}$ is a k-element D-linearly independent subset of M, and $T = \{y_1, \ldots, y_m\}$ is an m-element generating set for M, where k and m are finite. Then $k \leq m$.

Proof. We prove the lemma by induction on $\ell = |S \setminus T|$, the number of elements in S but not in T. The base case is $\ell = 0$, which occurs iff $S \subseteq T$. In this case, $k \leq m$ certainly holds since S is a subset of T.

For the induction step, suppose $|S \setminus T| = \ell > 0$. Assume this induction hypothesis: for all finite independent sets S_1 in M and all finite generating sets T_1 of M such that $|S_1 \setminus T_1| < \ell$, we have $|S_1| \leq |T_1|$. We must prove $|S| \leq |T|$.

Take a fixed $s \in S$ with $s \notin T$. Since $s \in M$ and T generates M, we have $s = d_1 y_1 + \cdots + d_m y_m$ for some $d_1, \ldots, d_m \in D$. Now $s \neq 0_M$, since otherwise S would be D-linearly dependent. Thus, at least one d_j is nonzero. Keeping only the nonzero d_j, we have $s = d_{i_1} y_{i_1} + \cdots + d_{i_p} y_{i_p}$ for some $p > 0$. We know $y_{i_1}, \ldots, y_{i_p}$ all belong to T, while s is not in T. If all of $y_{i_1}, \ldots, y_{i_p}$ belong to S, then the dependence relation $d_{i_1} y_{i_1} + \cdots + d_{i_p} y_{i_p} + (-1_D)s = 0$ contradicts the linear independence of S. So some y_{i_j} belongs to T but not S. By changing notation for the elements of T, we can assume y_{i_j} is y_1.

To summarize the situation so far, we have $s \in S \setminus T$, $y_1 \in T \setminus S$, and $s = d_1 y_1 + \cdots + d_m y_m$ with $d_1 \neq 0$. Consider $T_1 = \{s, y_2, \ldots, y_m\}$, which is the set obtained from T by exchanging s for y_1. Since $y_1 = d_1^{-1}(s - d_2 y_2 - \cdots - d_m y_m)$, T_1 still generates M. Also, T_1 still has size m, since s (which is not in T) does not equal y_j for any j between 2 and m. The crucial point is that $S \setminus T_1 = (S \setminus T) \setminus \{s\}$, so that $|S \setminus T_1| = \ell - 1 < \ell$. Applying the induction hypothesis to the sets S and T_1, we conclude that $k \leq m$, as needed. □

This lemma proves part (d) of the main theorem for finite subsets S, T of M. Assuming M is finitely generated, there exists a generating set T for M of finite size m. In this situation, an infinite subset S of M cannot be linearly independent. Otherwise, S would have a subset of size $m + 1$ that is necessarily linearly independent, but this contradicts the lemma. So (in the finitely generated case), part (d) is now proved for all subsets S, T of M.

We can now prove part (e) of the theorem, still assuming M is finitely generated. Let S and T be any two bases of M. We just proved that S and T must both be finite (since they are linearly independent subsets of M). Since S is independent and T generates M, $|S| \leq |T|$. Since T is independent and S generates M, $|T| \leq |S|$. Thus, $|S| = |T|$.

17.13   Zorn's Lemma

To extend the results of the previous section to modules that are not finitely generated, we need an axiom of set theory called Zorn's Lemma. This axiom is equivalent to the Axiom of Choice, which has many other equivalent formulations. For algebraic applications of the Axiom of Choice, it is often most convenient to use Zorn's Lemma, so that is the version of the axiom presented here. For more details on the Axiom of Choice and its variants, the reader may consult the set theory texts by Halmos [27] or Monk [42].

Recall that a partially ordered set (or *poset*) is a structure $(Z, \leq)$ where Z is a set and $\leq$ is a relation on Z that is reflexive, antisymmetric, and transitive. For example, if X is any set and Z is the set of all subsets of X, then Z with the relation $\subseteq$ (set inclusion) is a poset. More generally, for any collection Z of subsets of X, $(Z, \subseteq)$ is a poset.

For any poset $(Z, \leq)$, $x \in Z$ is a *maximal element* of Z iff for all $y \in Z$, $x \leq y$ implies $x = y$. This definition says that no element of the poset is strictly larger than x under the given ordering. Given any subset Y of Z, we say $z \in Z$ is an *upper bound* for Y iff $y \leq z$ for all $y \in Y$. A subset Y of Z is a *chain* iff for all $x, y \in Y$, $x \leq y$ or $y \leq x$.

Zorn's Lemma. For any poset $(Z, \leq)$, if every chain $Y \subseteq Z$ has an upper bound in Z, then Z has a maximal element.

We adopt this statement as an axiom, so it does not require proof (although, as mentioned above, it can be proved as a consequence of the Axiom of Choice).

We frequently apply Zorn's Lemma to posets $(Z, \subseteq)$, where Z is some collection of subsets of a given set X. To verify the hypothesis of Zorn's Lemma in this setting, we consider an arbitrary chain Y of elements of Z, which we can write as an indexed set $Y = \{S_i : i \in I\}$. Note that every S_i is a subset of X. The assumption that Y is a chain means that for all $i, j \in I$, either $S_i \subseteq S_j$ or $S_j \subseteq S_i$. If Z is the collection of all subsets of X, we see immediately that $S = \bigcup_{i \in I} S_i$ is an upper bound for the chain Y in Z, since $S_i \subseteq S$ for all $i \in I$. But in general, Z does not consist of all subsets of X, so the fact that S belongs to Z requires proof. In fact, for some choices of Z, the union of the S_i is not in Z, so some other subset of X must be used as an upper bound for Y in the poset Z. In any case, once we have found an upper bound for the chain Y that does belong to Z, we can conclude via Zorn's Lemma that Z has a maximal element. By definition, this is a subset M of X such that no subset of X properly containing M belongs to Z.

One other subtlety that occurs when using Zorn's Lemma is the fact that the empty subset of $(Z, \leq)$ is always a chain. Any element of Z is an upper bound for this chain. Checking that the hypothesis of Zorn's Lemma does hold for this special chain amounts to proving that the set Z is nonempty. Depending on how Z is defined, it may be a nontrivial task to verify this assertion. Once this verification has been done, we can assume (when checking the hypothesis of Zorn's Lemma) that all chains being considered are nonempty.

Before using Zorn's Lemma to prove theorems about bases of left vector spaces, we prove a useful result in ring theory to illustrate the application of Zorn's Lemma. By definition, a *maximal ideal* in a commutative ring R is an ideal M such that $M \neq R$ and there is no ideal J of R with $M \subsetneq J \subsetneq R$.

Theorem on Existence of Maximal Ideals. Every nonzero commutative ring R contains a maximal ideal.

Proof. To apply Zorn's Lemma, we introduce the poset $(Z, \subseteq)$, where Z is the collection of all ideals I of R with $I \neq R$. Since R is a nonzero ring, the set $\{0_R\}$ is an ideal of R that does not equal R, so Z is nonempty. To check the hypothesis of Zorn's Lemma, let

$\{J_t : t \in T\}$ be any nonempty chain of elements of Z, so each J_t is an ideal of R with $J_t \neq R$. Our candidate for an upper bound for this chain is $J = \bigcup_{t \in T} J_t$. We must check that J does belong to Z, meaning that J is an ideal of R and $J \neq R$. On one hand, 1_R does not belong to any J_t for $t \in T$, since otherwise $J_t = R$. Then 1_R does not belong to the union J of the J_t, so that J is a proper subset of R. On the other hand, let us check that J is an ideal of R. First, $0_R \in J$ since $0_R \in J_t$ for every J_t (this uses the fact that $T \neq \emptyset$). Next, fix $x, y \in J$ and $r \in R$. We have $x \in J_s$ and $y \in J_t$ for some $s, t \in T$. Then $-x \in J_s$ (since J_s is an ideal), hence $-x \in J$. Similarly, $rx \in J_s$ and so $rx \in J$. To finish, we prove $x + y \in J$. Since we are looking at a chain of ideals, either $J_s \subseteq J_t$ or $J_t \subseteq J_s$. In the first case, x and y both belong to J_t, so $x + y \in J_t \subseteq J$. In the second case, $x, y \in J_s$, so $x + y \in J_s \subseteq J$. Thus, J is an ideal. We have now proved that J is an upper bound for the given chain that belongs to the poset Z.

Having checked the hypothesis of Zorn's Lemma, we deduce from the lemma that the poset Z has a maximal element. By definition, this is an ideal M of R such that $M \neq R$ and there does not exist $N \in Z$ with $M \subsetneq N$. In other words, there is no ideal N of R with $M \subsetneq N \subsetneq R$, which is exactly what it means for M to be a maximal ideal of R. $\qquad\square$

17.14   Existence of Bases for Modules over Division Rings

Zorn's Lemma is exactly the tool needed to prove the following fundamental result on existence of bases for left vector spaces over division rings.

Theorem on Building Bases. Suppose D is a division ring, M is a left D-module, S is a D-linearly independent subset of M, and T is a generating set for M. There exists a basis B for M with $S \subseteq B \subseteq S \cup T$.

As special cases of this theorem, we see that M has a D-basis (take $S = \emptyset$ and $T = M$); any D-independent set $S \subseteq M$ can be extended to a basis (take $T = M$); and any generating set T of M contains a basis (take $S = \emptyset$).

Proof. Consider the poset $(Z, \subseteq)$, where Z is the collection of all D-linearly independent subsets C such that $S \subseteq C \subseteq S \cup T$. We know Z is nonempty since the D-linearly independent set S belongs to Z. To check the hypothesis of Zorn's Lemma, let $\{C_i : i \in I\}$ be any nonempty chain of elements of Z. Each C_i is a D-linearly independent set such that $S \subseteq C_i \subseteq S \cup T$. Define $C = \bigcup_{i \in I} C_i$. We check that C is an upper bound for the given chain that belongs to Z. Evidently, $C_i \subseteq C$ for all $i \in I$, and $S \subseteq C \subseteq S \cup T$. To finish showing $C \in Z$, we need only confirm that C is D-linearly independent.

Assume, to get a contradiction, that C is linearly dependent. Then there exist a finite list of distinct elements $x_1, \ldots, x_n \in C$ and nonzero scalars $d_1, \ldots, d_n \in D$ such that $d_1 x_1 + \cdots + d_n x_n = 0_M$. Because C is the union of the sets C_i, each x_j belongs to C_{i_j} for some $i_j \in I$. We prove by induction on n that there exists a single index $i \in I$ such that $x_1, \ldots, x_n$ all belong to C_i. This certainly holds if $n = 1$. If $n > 1$, we can assume by induction that $x_1, \ldots, x_{n-1}$ all belong to $C_{i'}$ for some $i' \in I$. Now, $C_{i'} \subseteq C_{i_n}$ or $C_{i_n} \subseteq C_{i'}$ since $\{C_i : i \in I\}$ is a chain. In the first case, $x_1, \ldots, x_n$ all belong to C_{i_n}. In the second case, $x_1, \ldots, x_n$ all belong to $C_{i'}$. This completes the induction proof. Now $x_1, \ldots, x_n$ is a finite list of distinct elements of C_i that satisfy a nontrivial dependence relation. This contradicts the assumed linear independence of C_i.

The hypothesis of Zorn's Lemma has now been checked. By that lemma, there exists a maximal element B in the poset Z. Since $B \in M$, we know $S \subseteq B \subseteq S \cup T$ and B is

D-linearly independent. We claim B is a D-basis of M. It suffices to show that every y in the generating set T for M is in the submodule $\langle B \rangle$ generated by B. To get a contradiction, assume some $y \in T$ does not belong to $\langle B \rangle$. Let $B' = B \cup \{y\}$, so $S \subseteq B \subsetneq B' \subseteq S \cup T$. By the Lemma on Extending an Independent Set, B' is D-linearly independent. So B' belongs to Z but properly contains B. This contradicts the maximality of B in Z. Thus B is a D-basis of M, as needed. $\qquad\square$

17.15 Basis Invariance for Modules over Division Rings

As our next application of Zorn's Lemma, we prove a result that says (informally) that an independent set cannot be larger than a spanning set in a left vector space. First we need another technical lemma.

Exchange Lemma. Suppose D is a division ring, M is a left D-module, X is a D-linearly independent subset of M, and Y is a generating set for M. For any $x \in X \setminus Y$, there exists $y \in Y \setminus X$ such that $(X \setminus \{x\}) \cup \{y\}$ is D-linearly independent.

Proof. Fix $x \in X \setminus Y$, and let $X' = X \setminus \{x\}$. We must have $x \neq 0$, since otherwise the relation $1x = 0$ shows that X is linearly dependent. Since Y generates M, there exist a positive integer n, elements $y_1, \ldots, y_n \in Y$, and nonzero scalars $d_1, \ldots, d_n \in D$ with $x = d_1 y_1 + \cdots + d_n y_n$. If $y_1, \ldots, y_n$ all belong to the submodule $\langle X' \rangle$, then x belongs to $\langle X' \rangle$ also. But then x would be a linear combination of other elements of X, violating the linear independence of X. So some y_i, say y_1, is not in $\langle X' \rangle$. By the Lemma on Extending an Independent Set, $X' \cup \{y_1\}$ is linearly independent. We also see that $y_1 \notin X'$ and $y_1 \neq x$ (since $x \notin Y$), so that $y_1 \in Y \setminus X$, as needed. $\qquad\square$

Comparison Theorem for Independent Sets and Spanning Sets. Suppose D is a division ring, M is a left D-module, S is a D-linearly independent subset of M, and T is a generating set for M. Then $|S| \leq |T|$.

Proof. By definition, $|S| \leq |T|$ means there exists an injection (one-to-one function) $h : S \to T$. The idea of the proof is to use Zorn's Lemma to assemble partial injections (each mapping a subset of S into T) to get an injection with the largest possible domain. The details are rather intricate, so we proceed in steps.

Step 1: We define the poset. Let Z be the set of triples (A, C, f), where A and C are sets, $S \cap T \subseteq A \subseteq S$, $S \cap T \subseteq C \subseteq T$, $f : A \to C$ is a bijection, $f(z) = z$ for all $z \in S \cap T$, and $(S \setminus A) \cup C$ is D-linearly independent. Partially order Z by defining $(A, C, f) \leq (A_1, C_1, f_1)$ to mean $A \subseteq A_1$, $C \subseteq C_1$, and $f \subseteq f_1$. (Here, we view the functions f and f_1 as sets of ordered pairs. For example, $f = \{(a, f(a)) : a \in A\}$. The condition $f \subseteq f_1$ means $f_1(a) = f(a)$ for all $a \in A \subseteq A_1$, which says that f_1 extends f.) The poset axioms for $(Z, \leq)$ follow routinely from the known reflexivity, antisymmetry, and transitivity of $\subseteq$.

Step 2: We check the hypothesis of Zorn's Lemma. First, the poset Z is nonempty, because $(S \cap T, S \cap T, \mathrm{id}_{S \cap T})$ is in Z (note that $(S \setminus (S \cap T)) \cup (S \cap T) = S$ is independent). Second, given a nonempty chain $\{(A_i, C_i, f_i) : i \in I\} \subseteq Z$, we must find an upper bound (A, C, f) for this chain that belongs to Z. Let $A = \bigcup_{i \in I} A_i$, $C = \bigcup_{i \in I} C_i$, and $f = \bigcup_{i \in I} f_i$. We certainly have $A_i \subseteq A$, $C_i \subseteq C$, and $f_i \subseteq f$ for all $i \in I$; we need only show that $(A, C, f) \in Z$. We must first check that f, which is a certain set of ordered pairs, really is a (single-valued) function with domain A. In more detail, we must show that for all $a \in A$, there exists a unique c with $c \in C$ and $(a,c) \in f$. Fix $a \in A$. Now $a \in A_i$ for some $i \in I$, so there is $c \in C_i \subseteq C$ with $(a,c) \in f_i \subseteq f$, namely $c = f_i(a)$. Suppose we also have $(a,d) \in f$ for some d; we must prove $d = c$. Note $(a,d) \in f$ means $f_j(a) = d$ for some $j \in I$. We are working with a chain, so $(A_i, C_i, f_i) \leq (A_j, C_j, f_j)$ or $(A_j, C_j, f_j) \leq (A_i, C_i, f_i)$. In the first case, f_j extends f_i, so $d = f_j(a) = f_i(a) = c$; similarly in the other case. Analogous reasoning shows that f maps A one-to-one onto C, so that we have a well-defined bijection $f : A \to C$. Since f extends every f_i, we also have $f(z) = z$ for all $z \in S \cap T$. Note $S \cap T \subseteq A \subseteq S$ and $S \cap T \subseteq C \subseteq T$ since these set inclusions hold for every A_i and C_i.

The key point still to be checked is that $(S \setminus A) \cup C$ is D-linearly independent. By definition, we must check that every finite list of distinct elements of $(S \setminus A) \cup C$ is D-independent. Let $s_1, \ldots, s_m, c_1, \ldots, c_n$ be such a list, where each $s_j \in S \setminus A$ and each $c_k \in C$. For each k, there is an index $i(k) \in I$ with $c_k \in C_{i(k)}$. Since we are working with a chain and n is finite, there is a single index $i_0 \in I$ with $c_k \in C_{i_0}$ for all k between 1 and n. We also have $s_j \in (S \setminus A) \subseteq (S \setminus A_{i_0})$ for all j between 1 and m. Hence, $s_1, \ldots, s_m, c_1, \ldots, c_n$ is a finite list of distinct elements of the set $(S \setminus A_{i_0}) \cup C_{i_0}$. This set is known to be independent (since $(A_{i_0}, C_{i_0}, f_{i_0}) \in Z$), so the list is independent, as needed. We have now proved that $(A, C, f) \in Z$, so that the given chain has an upper bound in Z.

Step 3: We analyze a maximal element of Z. By Zorn's Lemma, there is a maximal element (A, C, f) in the poset Z. If $A = S$, then f is a bijection from S onto the subset C of T. By enlarging the codomain from C to T, we get an injection from S into T, as needed.

To finish the proof, we must rule out the case where $A \subsetneq S$. In this case, there exists $x \in S \setminus A$. Since A contains $S \cap T$, $x \notin T$ but x does belong to the set $X = (S \setminus A) \cup C$. We can apply the Exchange Lemma to the linearly independent set X and the generating set T. By the lemma, there exists $y \in T \setminus X$ such that $(X \setminus \{x\}) \cup \{y\}$ is linearly independent. Define $A' = A \cup \{x\}$, $C' = C \cup \{y\}$, and $f' = f \cup \{(x,y)\}$. The conditions on x and y ensure that $S \cap T \subseteq A \subsetneq A' \subseteq S$, $S \cap T \subseteq C \subsetneq C' \subseteq T$, $f' : A' \to C'$ is a bijection extending f, $f'(z) = z$ for all $z \in S \cap T$, and $(S \setminus A') \cup C' = (X \setminus \{x\}) \cup \{y\}$ is linearly independent. Thus, (A', C', f') belongs to Z and is strictly larger than (A, C, f) relative to the ordering. This contradicts the maximality of (A, C, f). □

Basis Cardinality Theorem for Left Vector Spaces. Suppose D is a division ring and M is a left D-module. Any two bases of M have the same cardinality.

Proof. Let B and C be two bases for M. Since B is linearly independent and C generates M, $|B| \leq |C|$. Since C is linearly independent and B generates M, $|C| \leq |B|$. By the Schröder–Bernstein Theorem from set theory, $|B| \leq |C|$ and $|C| \leq |B|$ imply $|B| = |C|$, as needed. □

17.16 Basis Invariance for Free Modules over Commutative Rings

We have just proved that any two bases of a module over a division ring have the same size. The next theorem leverages this result to show that any two bases of a free module over a commutative ring have the same size. Later, we give an example to show that this property need not hold for free modules over non-commutative rings.

Basis Cardinality Theorem for Free Modules over Commutative Rings. Suppose R is a nonzero commutative ring and N is a free R-module. For any two bases X and Y of N, $|X| = |Y|$.

Proof. First consider the case where X and Y are finite ordered bases of N, say $X = (x_1, \ldots, x_n)$ and $Y = (y_1, \ldots, y_m)$. Let I be a maximal ideal of R (see §17.13) and note that $F = R/I$ is a field (Exercise 53). We know that $V = N/IN$ is an R/I-module, i.e., an F-vector space. Let $X' = (x_1 + IN, \ldots, x_n + IN)$ be the list of images of elements of X in N/IN. We claim that X' is an ordered basis for the F-vector space V. You can check that X' generates V, since X generates N. To show that X' is F-linearly independent, suppose $\sum_{i=1}^{n} (s_i + I)(x_i + IN) = 0$ for fixed $s_i \in R$. We must show that each $s_i + I = 0$ in R/I, i.e., that $s_i \in I$ for all i. From the given relation and the definition of the operations in N/IN, we deduce $(\sum_{i=1}^{n} s_i x_i) + IN = 0$. Therefore, the element $x = \sum_{i=1}^{n} s_i x_i$ is in IN. By definition of IN, we can write $x = \sum_{j=1}^{p} a_j z_j$ where $a_j \in I$ and $z_j \in N$. Writing $z_j = \sum_{i=1}^{n} c_{i,j} x_i$ (for some $c_{i,j} \in R$) and substituting, we see that $x = \sum_{i=1}^{n} (\sum_{j=1}^{p} a_j c_{i,j}) x_i$. By the R-linear independence of the list X in N, we conclude that $s_i = \sum_{j=1}^{p} a_j c_{i,j}$ for all i. Since each a_j is in the ideal I, we have $s_i \in I$, as needed. So X' is an ordered F-basis for V. By the same argument, $Y' = (y_1 + IN, \ldots, y_m + IN)$ is an ordered F-basis for V. Applying the Basis Cardinality Theorem for Vector Spaces, we deduce $n = m$.

Now consider the general case, where X and Y are sets (possibly infinite). Let $p : N \to N/IN$ be the canonical map, and set $X' = p[X]$, $Y' = p[Y]$. As before, X' and Y' generate the F-vector space N/IN. The proof in the preceding paragraph, applied to each finite list of distinct elements of X, shows that X' is F-linearly independent. Similarly, Y' is F-linearly independent. Therefore, $|X'| = |Y'|$ by the Basis Cardinality Theorem for Vector Spaces. To finish, it suffices to show that the restriction of p to X is injective, so that $|X| = |X'|$ (and similarly $|Y| = |Y'|$). Injectivity follows from the argument used to show linear independence of X'. To see why, fix $x_1 \neq x_2$ in X. The list (x_1, x_2) is R-independent in N, so that $(x_1 + IN, x_2 + IN)$ is F-independent in N/IN. In particular, $p(x_1) = x_1 + IN \neq x_2 + IN = p(x_2)$ in N/IN. $\square$

Here is an example of a non-commutative ring R and a free R-module that has R-bases with different cardinalities. Let F be any field, $X = \{x_n : n \geq 0\}$ be a countably infinite set, and V be an F-vector space with basis X. Let $R = \mathrm{Hom}_F(V, V)$ be the set of linear transformations from V to itself. We have seen that R is an F-vector space. In fact, R is also a non-commutative ring, if we define multiplication of elements $f, g \in R$ to be composition of functions. The ring axioms may be routinely verified. In particular, the distributive laws follow since addition of functions is defined pointwise.

Like any ring, R is a left R-module. For each integer $k \geq 1$, we produce an R-basis for R consisting of k elements. Fix a positive integer k. We define certain elements f_j and g_j in R (for $0 \leq j < k$) by specifying how these linear maps operate on the basis X of V. Recall (by integer division) that every integer $n \geq 0$ can be written uniquely in the form $n = ki + j$, for some integers i, j with $0 \leq j < k$. Let f_j send x_{ki+j} to x_i for all $i \geq 0$; let f_j send all other elements of X to 0. Let g_j send x_i to x_{ki+j} for all $i \geq 0$. We have $f_j g_j = \mathrm{id}_V$ for all j, since both sides have the same effect on the basis X. For the same reason, $f_{j'} g_j = 0_R$ for $j \neq j'$ between 0 and $k - 1$, and

$$g_0 f_0 + g_1 f_1 + \cdots + g_{k-1} f_{k-1} = \mathrm{id}_V . \tag{17.1}$$

The last identity follows because, given any $x_n \in X$, we can write $n = ki + j$ for a unique j between 0 and $k - 1$. Then $g_j f_j(x_n) = x_n$ for this j, while $g_{j'} f_{j'}(x_n) = 0$ for all $j' \neq j$.

We can now show that $B_k = (f_0, \ldots, f_{k-1})$ is a k-element ordered R-basis for the left R-module R. Suppose f is any element of R. Multiplying (17.1) on the left by f, we get $(fg_0)f_0 + (fg_1)f_1 + \cdots + (fg_{k-1})f_{k-1} = f$ where $fg_j \in R$. This identity shows that B_k is a generating set for the left R-module R. Next, to test R-independence, suppose $h_0 f_0 + \cdots + h_{k-1}f_{k-1} = 0_R$ for some $h_j \in R$. For each j_0 between 0 and $k-1$, multiply this equation on the right by g_{j_0} to obtain $h_{j_0} = 0_R$ (using the relations above to simplify products $f_j g_{j_0}$). Thus, B_k is an R-independent list. Note how prominently the non-commutativity of R is used in this proof.

17.17   Jordan–Hölder Theorem for Modules

If M is a left vector space over a division ring, or a free module over a commutative ring, then the dimension of M (the cardinality of any basis) is uniquely determined by M. The *Jordan–Hölder Theorem* is a more general uniqueness result regarding decompositions of a module into building blocks that are simple modules. To see how these decompositions occur, we must first discuss chains of submodules.

Suppose R is a ring and M is a left R-module. A *chain of submodules* of M is a list $(M_0, M_1, \ldots, M_m)$ where each M_i is a submodule of M and $M_0 \supsetneq M_1 \supsetneq \cdots \supsetneq M_m$. We say that this chain has *length* m. Chains of submodules $(M_0, M_1, \ldots, M_n, \ldots)$ of infinite length are defined similarly. A chain of finite length is called a *maximal chain* iff it cannot be extended to a longer chain by adding another submodule to the beginning, middle, or end of the list. The chain $(M_0, \ldots, M_m)$ is maximal iff $M_0 = M$ and $M_m = \{0\}$ and for all i between 1 and m, there exists no submodule P of M with $M_{i-1} \supsetneq P \supsetneq M_i$. By the Correspondence Theorem, the last condition is equivalent to saying that the only submodules of the quotient module M_{i-1}/M_i are $\{0\} = M_i/M_i$ and M_{i-1}/M_i. In other words, the chain of submodules $(M_0, \ldots, M_m)$ is maximal iff $M = M_0$ and $M_m = \{0\}$ and M_{i-1}/M_i is a simple module for all i between 1 and m.

Jordan–Hölder Theorem for Modules. Suppose R is a ring and M is an R-module such that there exists a maximal chain of submodules $(M_0, M_1, \ldots, M_m)$ of finite length m. (a) Any chain $(N_0, N_1, \ldots)$ of submodules of M has finite length $n \leq m$. (b) Any maximal chain $(N_0, \ldots, N_n)$ of submodules of M satisfies $n = m$ and, for some permutation f of $\{1, 2, \ldots, m\}$, $M_{i-1}/M_i \cong N_{f(i)-1}/N_{f(i)}$ for all i between 1 and m.

Proof. We prove the theorem by induction on m, the length of the given maximal chain. If $m = 0$, then $M = M_0 = M_m = \{0\}$, and the conclusions of the theorem are evident. If $m = 1$, then $M \cong M/\{0\} = M_0/M_1$ must be a simple module, and again the needed conclusions follow immediately. Now assume $m > 1$ and that the theorem is already known for all R-modules having a maximal chain of length less than m. Fix a chain $(N_0, N_1, \ldots)$ of submodules of M. We first prove (a).

Case 1: $N_1 \subseteq M_1$. We can apply the induction hypothesis to the R-module M_1, which has a maximal chain $(M_1, \ldots, M_m)$ of length $m - 1$. The chain $(N_1, N_2, \ldots)$ must therefore have some finite length $n - 1 \leq m - 1$, so that $(N_0, N_1, \ldots, N_n)$ has finite length $n \leq m$.

Case 2: N_1 is not a subset of M_1. Note $M \neq N_1 \neq M_1$, but M and M_1 are the only two submodules of M containing M_1. So M_1 is not a subset of N_1, and $P = M_1 \cap N_1$ is a proper submodule of both M_1 and N_1. Moreover, $M_1 + N_1$ is a submodule properly containing M_1, so $M = M_1 + N_1$. By the Diamond Isomorphism Theorem, $N_1/P = N_1/(N_1 \cap M_1) \cong (N_1 + M_1)/M_1 = M/M_1$ is a simple module. Starting with the chain (M_1, P), we can build longer and longer chains of submodules of M_1 by repeatedly inserting a new submodule somewhere in the existing chain, as long as this can be done. By the induction hypothesis applied to M_1, this insertion process must terminate in finitely many steps. When it does terminate, we have by definition a maximal chain

$(P_1 = M_1, P_2, \ldots, P_m)$ of submodules of M_1, which must have length $m - 1$ by induction applied to the module M_1. We know $M_1 \cap N_1 = P = P_i$ for some i with $2 \leq i \leq m$. Since N_1/P is simple, $(N_1, P = P_i, P_{i+1}, \ldots, P_m)$ is a maximal chain of submodules of N_1 of finite length $m - i + 1 \leq m - 1$. So the induction hypothesis is applicable to the R-module N_1, and we conclude that the chain $(N_1, N_2, \ldots)$ of submodules of N_1 has finite length at most $m - i + 1 \leq m - 1$. So $(N_0, N_1, \ldots, N_n)$ has finite length $n \leq m$, proving (a) for the module M.

To prove (b), we now assume that $(N_0, N_1, \ldots, N_n)$ is a maximal chain of submodules of M. Part (a) says $n \leq m$. Using part (a) with the roles of the two maximal chains reversed, we also get $m \leq n$ and hence $n = m$. To obtain the further conclusion about the isomorphism of quotient modules, we again consider two cases.

Case 1: $N_1 \subseteq M_1$. Since $M_1 \subsetneq M$, maximality of the chain $(N_0, N_1, \ldots, N_n)$ forces $M_1 = N_1$. Then $M_0/M_1 = N_0/N_1$. We can match up the rest of the quotient modules by applying part (b) of the induction hypothesis to the module $M_1 = N_1$ and the maximal chains $(M_1, \ldots, M_m)$ and $(N_1, \ldots, N_n)$ within this module.

Case 2: N_1 is not a subset of M_1. As before, we let $P = M_1 \cap N_1$ and note that $M = M_1 + N_1$. By the Diamond Isomorphism Theorem and the assumed maximality of the two given chains, $N_1/P \cong M_0/M_1$ and $M_1/P \cong N_0/N_1$ are simple modules. As before, we can extend the chain (M_1, P) to a maximal chain $(M_1, P, P_3, \ldots, P_m)$ of submodules of M_1. We now have four maximal chains of submodules of M that look like this:

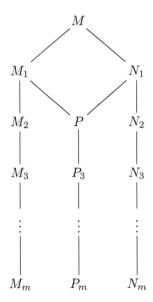

Note that $(N_1, P, P_3, \ldots, P_m)$ is also a maximal chain of submodules of N_1. Applying the induction hypothesis to M_1, we know that the modules $M_0/M_1, M_1/M_2, \ldots, M_{m-1}/M_m$ are isomorphic (in some order) to the modules $M/M_1, M_1/P, P/P_3, \ldots, P_{m-1}/P_m$. By the diamond isomorphisms at the top of the figure, these modules (switching the first two) are isomorphic to $M/N_1, N_1/P, P/P_3, \ldots, P_{m-1}/P_m$. Applying the induction hypothesis to N_1, we know that the modules just listed are isomorphic (in some order) to the modules $N_0/N_1, N_1/N_2, \ldots, N_{m-1}/N_m$. Combining all these steps, the modules M_{i-1}/M_i (for $1 \leq i \leq m$) are isomorphic in some order to the modules N_{i-1}/N_i. $\square$

17.18   Modules of Finite Length

We say that an R-module M has *finite length* iff there exists a maximal chain of submodules of M of finite length. In this case, we have proved that all maximal chains of submodules of M have the same length m, and we define the *length of the module* M to be $\operatorname{len}(M) = m$. If M does not have finite length, set $\operatorname{len}(M) = \infty$. The length of a module is a generalization of the dimension of a vector space. You can check that if M is a finitely generated module over a division ring R, then $\operatorname{len}(M) = \dim(M)$ (Exercise 63).

We have $\operatorname{len}(M) = 0$ iff M is the zero module, while $\operatorname{len}(M) = 1$ iff M is a simple module. You can check that if $M \cong M'$, then $\operatorname{len}(M) = \operatorname{len}(M')$. Next, suppose M is any R-module with a submodule N. We claim $\operatorname{len}(M) < \infty$ iff $\operatorname{len}(N) < \infty$ and $\operatorname{len}(M/N) < \infty$, in which case $\operatorname{len}(M) = \operatorname{len}(N) + \operatorname{len}(M/N)$. On one hand, suppose $\operatorname{len}(M) = m < \infty$. Any chain of submodules of N is also a chain of submodules of M, so has length at most m. Hence N has finite length. On the other hand, by the Correspondence Theorem, any chain of submodules of M/N has the form $M_0/N \supsetneq M_1/N \supsetneq \cdots \supsetneq M_k/N$ for some submodules $M_0 \supsetneq M_1 \supsetneq \cdots \supsetneq M_k$ of M containing N. So the given chain in M/N has length at most m. Conversely, suppose $\operatorname{len}(N) = n < \infty$ and $\operatorname{len}(M/N) = k < \infty$. Fix a maximal chain of submodules of N, say $N = N_0 \supsetneq N_1 \supsetneq \cdots \supsetneq N_n = \{0\}$. Similarly, fix a maximal chain of submodules of M/N, which must have the form $M/N = M_0/N \supsetneq M_1/N \supsetneq \cdots \supsetneq M_k/N = N/N$ for certain submodules $M = M_0 \supsetneq M_1 \supsetneq \cdots \supsetneq M_k = N$. Splicing these chains together, we get a chain of submodules

$$M = M_0 \supsetneq M_1 \supsetneq \cdots \supsetneq M_k = N = N_0 \supsetneq N_1 \supsetneq \cdots \supsetneq N_n = \{0\}$$

of length $k + n$. This chain must be maximal, since otherwise the chain for N or the chain for M/N would not be maximal. This proves $\operatorname{len}(M) = k + n = \operatorname{len}(M/N) + \operatorname{len}(N) < \infty$.

Given R-modules $M_1, \ldots, M_k$, we show by induction that the product module $M = M_1 \times \cdots \times M_k$ has finite length iff all M_i have finite length, in which case $\operatorname{len}(M) = \operatorname{len}(M_1) + \cdots + \operatorname{len}(M_k)$. This is evident if $k = 1$. Assume $k > 1$ and the result is known for products of $k - 1$ modules. On one hand, if $\operatorname{len}(M) < \infty$, then each submodule $\{0\} \times \cdots \times M_i \times \cdots \times \{0\}$ has finite length. As M_i is isomorphic to this submodule, $\operatorname{len}(M_i) < \infty$. On the other hand, suppose each M_i has finite length. Note $M = N \times M_k$ where $N = M_1 \times \cdots \times M_{k-1}$. By the induction hypothesis, $\operatorname{len}(N \times \{0\}) = \operatorname{len}(N) = \operatorname{len}(M_1) + \cdots + \operatorname{len}(M_{k-1}) < \infty$. We have $M/(N \times \{0\}) \cong M_k$, which has finite length. By the result in the previous paragraph, $\operatorname{len}(M) = \operatorname{len}(N) + \operatorname{len}(M_k) = \sum_{i=1}^{k} \operatorname{len}(M_i) < \infty$.

17.19   Summary

Let R be a ring. Here, we summarize the definitions and results for R-modules, module homomorphisms, and free modules that were covered in this chapter.

Definitions

1. A *left R-module* is an additive commutative group M and a scalar multiplication $\cdot : R \times M \to M$ satisfying closure, left associativity, the two distributive laws, and the identity axiom.

2. *Right R-modules* are defined like left modules, using a scalar multiplication $\cdot : M \times R \to M$ obeying right associativity. The two types of modules are equivalent for commutative rings.

3. A *homomorphism* of R-modules (left or right) is a map between modules that preserves addition and scalar multiplication. Homomorphisms of R-modules are also called *R-maps* or *R-linear maps*.

4. For fields F, F-modules and F-maps are the same as F-vector spaces and linear transformations. For $R = \mathbb{Z}$, $\mathbb{Z}$-modules and $\mathbb{Z}$-maps are the same as commutative groups and group homomorphisms.

5. A *submodule* of a module is a subset containing 0 and closed under addition and scalar multiplication (and hence under additive inverses).

6. Let $f : M \to N$ be an R-module homomorphism. The set of all $x \in M$ such that $f(x) = 0_N$ is the *kernel* of f. The set of all $y \in N$ of the form $y = f(x)$ for some $x \in M$ is the *image* of f. The kernel is a submodule of M, and the image is a submodule of N.

7. If S is a subset of a module M, an *R-linear combination* of elements of S is a finite sum $\sum_i a_i s_i$ where $a_i \in R$ and $s_i \in S$. The set of all such R-linear combinations is $\langle S \rangle$, the submodule *generated by* S. If $\langle S \rangle = M$, we say that S *spans* M or S *generates* M. If M is generated by some finite set S, then we say M is *finitely generated*.

8. A left R-module M is *cyclic* iff there exists $x \in M$ with $M = \langle x \rangle = Rx = \{rx : r \in R\}$. M is a *simple R-module* iff $M \neq \{0\}$ and the only R-submodules of M are $\{0\}$ and M.

9. A *division ring* is a ring R with $1_R \neq 0_R$ such that for all nonzero $x \in R$, there exists $y \in R$ with $xy = 1_R = yx$. A *field* is a commutative division ring. A *left vector space* is a left R-module over a division ring R.

10. A subset S of a module M is called *R-independent* iff for all $k \in \mathbb{Z}_{>0}$, $a_i \in R$, and distinct $s_i \in S$, $\sum_{i=1}^{k} a_i s_i = 0_M$ implies every $a_i = 0_R$.

11. An R-independent generating set of an R-module M is called an *R-basis* for M. M is called a *free R-module* iff M has an R-basis. If R is a division ring or a commutative ring, then the size of any R-basis of a free R-module M is the *dimension* of M.

12. The *length* $\text{len}(M)$ of an R-module M is the maximum n such that there is a chain of submodules $M_0 \supsetneq M_1 \supsetneq \cdots \supsetneq M_n$ of M, or ∞ if there is no such n.

Module Constructions

1. *Submodules:* Every submodule of an R-module is itself an R-module. Intersections and sums of submodules are submodules. The image or preimage of a submodule under an R-linear map is a submodule.

2. *Submodule Generated by a Set:* If S is any subset of an R-module M, there exists a smallest submodule $\langle S \rangle$ of M containing S. This submodule can be defined as the intersection of all submodules of M containing S, or as the set of all R-linear combinations of elements of S (including 0_M).

3. *Direct Products and Direct Sums:* If M_i is a left R-module for each i in a set I, then the set of functions f with domain I such that $f(i) \in M_i$ for all $i \in I$ is a left R-module under pointwise operations on functions. This module is the

direct product of the M_i, written $\prod_{i \in I} M_i$. The subset consisting of functions with $f(i) = 0_{M_i}$ for all but finitely many $i \in I$ is a submodule called the *direct sum* of the M_i, written $\bigoplus_{i \in I} M_i$. A special case is R^n, the module of n-tuples of elements of R, which is a free R-module having an n-element R-basis.

4. *Hom Modules:* Let M and N be left R-modules. The set $\text{Hom}(M, N)$ of group homomorphisms from M to N is a left R-module. For commutative R, the set $\text{Hom}_R(M, N)$ of R-module homomorphisms from M to N is a left R-module.

5. *Quotient Modules:* Let N be a submodule of an R-module M. The *quotient module* M/N consists of all cosets $x + N$ with $x \in M$. For all $x, z \in M$, $x + N = z + N$ iff $x - z \in N$. The operations in the quotient module are defined by $(x + N) + (y + N) = (x + y) + N$ and $r \cdot (x + N) = (rx) + N$ for $x, y \in M$ and $r \in R$. The canonical projection $p : M \to M/N$, given by $p(x) = x + N$ for $x \in M$, is a surjective R-map with kernel N. If S generates M, then $p[S]$ generates M/N.

6. *Change of Scalars:* If T is a subring of R, any R-module can be regarded as a T-module. If $f : S \to R$ is a ring homomorphism, then an R-module M becomes an S-module via $s \star x = f(s) \cdot x$ for $s \in S$ and $x \in M$. If I is an ideal of R annihilating an R-module M (meaning $ix = 0$ for all $i \in I$ and $x \in M$), then M can be regarded as an R/I-module via $(r + I) \bullet x = r \cdot x$ for $r \in R$ and $x \in M$. In particular, for any R-module N and any ideal I of R, N/IN is an R/I-module. Taking I to be a maximal ideal in a commutative ring R, so that R/I is a field, we can convert R-modules to R/I-vector spaces.

Results about Module Homomorphisms and Submodules

Let M and N be left R-modules.

1. If $f, g : M \to N$ are two R-maps agreeing on a generating set for M, then $f = g$.

2. *Fundamental Homomorphism Theorem for R-modules:* Let $f : M \to N$ be an R-module homomorphism. There is an R-module isomorphism $f' : M/\ker(f) \to \text{img}(f)$ given by $f'(x + \ker(f)) = f(x)$ for $x \in M$.

3. *Universal Mapping Property for the Quotient Module M/H:* Let $f : M \to N$ be an R-module homomorphism, H be a submodule of M, and $p : M \to M/H$ be the canonical map. If $H \subseteq \ker(f)$, then there exists a unique R-module homomorphism $f' : M/H \to N$ such that $f = f' \circ p$. Moreover, $\text{img}(f') = \text{img}(f)$ and $\ker(f') = \ker(f)/H$.

4. *Diamond Isomorphism Theorem:* Let M and N be submodules of an R-module P. The function $g : M/(M \cap N) \to (M + N)/N$ given by $g(m + M \cap N) = m + N$ (for $m \in M$) is an R-module isomorphism.

5. *Nested Quotient Isomorphism Theorem:* Given $H \subseteq N \subseteq M$ with H and N submodules of M, there is an R-module isomorphism $g : (M/H)/(N/H) \to (M/N)$ given by $g((x + H) + (N/H)) = x + N$ for $x \in M$.

6. *Correspondence Theorem for Modules:* Assume N is a submodule of M, and let $p : M \to M/N$ be the canonical map. Let $\mathcal{A}$ be the collection of all submodules U of M such that $N \subseteq U \subseteq M$. Let $\mathcal{B}$ be the collection of all submodules of M/N. There are inclusion-preserving, mutually inverse bijections $T : \mathcal{A} \to \mathcal{B}$ and $S : \mathcal{B} \to \mathcal{A}$ given by $T(U) = p[U] = U/N$ (the image of U under p) and $S(V) = p^{-1}[V]$ (the preimage of V under p).

7. *Recognition Theorem for Direct Products:* Let N and P be submodules of M such that $N + P = M$ and $N \cap P = \{0_M\}$. There is an R-module isomorphism $g : N \times P \to M$ given by $g((x,y)) = x + y$ for $x \in N$ and $y \in P$.

8. *Jordan–Hölder Theorem for Modules:* If M has one maximal chain of submodules of finite length m, then every chain of submodules of M has length at most m, and every maximal chain has length m. Given two maximal chains $M_0 \supsetneq M_1 \supsetneq \cdots \supsetneq M_m$ and $N_0 \supsetneq N_1 \supsetneq \cdots \supsetneq N_m$ of M, the quotient modules $M_0/M_1, M_1/M_2, \ldots, M_{m-1}/M_m$ are simple modules and are isomorphic (in some order) to $N_0/N_1, N_1/N_2, \ldots, N_{m-1}/N_m$.

9. *Results for Finite Length Modules:* $M = \{0\}$ iff $\operatorname{len}(M) = 0$. M is simple iff $\operatorname{len}(M) = 1$. Isomorphic modules have the same length. For a submodule N of M, M has finite length iff N and M/N have finite length, and then $\operatorname{len}(M) = \operatorname{len}(N) + \operatorname{len}(M/N)$. For $M = M_1 \times \cdots \times M_k$, M has finite length iff all M_i have finite length, and then $\operatorname{len}(M) = \operatorname{len}(M_1) + \cdots + \operatorname{len}(M_k)$.

Results about Free R-modules

1. If M is a free R-module with basis S, then every $y \in M$ can be uniquely expressed as a (finite) R-linear combination of elements of S.

2. *Universal Mapping Property for Free R-Modules:* Let M be a free R-module with basis X. Given any R-module N and any function $g : X \to N$, there exists a unique R-module homomorphism $g' : M \to N$ whose restriction to X equals g.

3. Every free R-module M is isomorphic to a direct sum of copies of R. Conversely, any such direct sum is a free R-module.

4. Given any set X and any nonzero ring R, there exists a free R-module with R-basis X.

5. Suppose R is a division ring. Every R-module M is free, and every R-basis of M has the same cardinality. For all R-independent sets S in M and all generating sets T for M: $|S| \le |T|$; S can be extended to a basis of M; and T contains a basis of M.

6. For some rings R, not every R-module is free, and not every R-module is isomorphic to a submodule of a free R-module. For all rings R, every R-module M is isomorphic to a quotient module of a free R-module F. If M is finitely generated, then F can be chosen to be finitely generated with the same number of generators as M.

7. If R is a nonzero commutative ring and N is a free R-module, then any two bases for N have the same cardinality. However, there exists a non-commutative ring R and a free R-module that has an R-basis of size k for every positive integer k.

17.20 Exercises

Unless otherwise specified, assume R is an arbitrary ring in these exercises.

1. Let R be any nonzero ring. For each commutative group $(M, +)$ below, show that M is not a left R-module under the indicated scalar multiplication $\cdot : R \times M \to M$ by pointing out one or more module axioms that fail to hold.

(a) $M = R$, $r \cdot m = 0_M$ for all $r \in R$ and $m \in M$.

(b) $M = R$, $r \cdot m = m$ for all $r \in R$ and $m \in M$.

(c) $M = R$, $r \cdot m = r$ for all $r \in R$ and $m \in M$.

(d) $M = R^2$, $r \cdot (m_1, m_2) = (rm_1, m_2)$ for all $r, m_1, m_2 \in R$.

(e) $M = R$, $r \cdot m = mr$ for all $r, m \in R$ (assume R is non-commutative).

2. Fix a positive integer n. Show that the additive group R^n (viewed as a set of column vectors) is a left $M_n(R)$-module if scalar multiplication $A \cdot v$ (for $A \in M_n(R)$ and $v \in R^n$) is defined to be the matrix-vector product Av. If we define $v \star A = Av$, do we get a right $M_n(R)$-module structure on R^n?

3. Given $n \in \mathbb{Z}_{>0}$, show that R^n (viewed as a set of row vectors) is a right $M_n(R)$-module if scalar multiplication $w \cdot A$ (for $A \in M_n(R)$ and $w \in R^n$) is defined to be the vector-matrix product wA.

4. Let V be a vector space over a field F and R be the ring of all F-linear transformations $T : V \to V$. Show that $(V, +)$ is a left R-module if we define $T \cdot v = T(v)$ for $T \in R$ and $v \in V$.

5. Let $(M, +)$ be a commutative group and $\cdot : R \times M \to M$ be a function. Define $\star : M \times R \to M$ by $m \star r = r \cdot m$ for all $r \in R$ and $m \in M$. Prove that each axiom (M1), (M2), (D1), and (D2) in §17.1 holds for $\cdot$ iff the analogous axiom (M1'), (M2'), (D1'), (D2') holds for $\star$. Prove that, if R is commutative, then axiom (M3) holds for $\cdot$ iff axiom (M3') holds for $\star$.

6. Let M, N be left R-modules and assume $f : M \to N$ is an R-linear map. Prove the following facts, which were stated in §17.1.

(a) The identity map $\mathrm{id}_M : M \to M$ is an automorphism.

(b) The composition of R-maps is an R-map (similarly for monomorphisms, epimorphisms, isomorphisms, endomorphisms, and automorphisms).

(c) If f is an isomorphism, then f^{-1} is an isomorphism.

(d) Given a set X of left R-modules, the relation given by $M \cong N$ iff M and N are isomorphic R-modules (for $M, N \in X$) is an equivalence relation on X.

7. Let $f : M \to N$ be an R-linear map between two R-modules. Prove $f(0_M) = 0_N$ and $f(-x) = -f(x)$ for all $x \in M$.

8. Let M be any R-module.

(a) Prove $0_R \cdot x = 0_M$ for all $x \in M$.

(b) Prove $r \cdot 0_M = 0_M$ for all $r \in R$.

(c) Prove $r \cdot (-x) = (-r) \cdot x = -(r \cdot x)$ for all $r \in R$ and all $x \in M$.

9. The complex number system $\mathbb{C}$ is a ring, a left $\mathbb{C}$-module, and a left $\mathbb{R}$-module.

(a) Show that $f : \mathbb{C} \to \mathbb{C}$ defined by $f(a + ib) = a - ib$ for $a, b \in \mathbb{R}$ is a ring homomorphism and an $\mathbb{R}$-linear map, but not a $\mathbb{C}$-linear map.

(b) Show that $g : \mathbb{C} \to \mathbb{C}$ defined by $g(z) = iz$ for $z \in \mathbb{C}$ is a $\mathbb{C}$-linear map, but not a ring homomorphism.

10. Prove or disprove: there exists exactly one function $\cdot : \mathbb{C} \times \mathbb{C} \to \mathbb{C}$ such that $(\mathbb{C}, +, \cdot)$ is a left $\mathbb{C}$-module, where $+$ denotes complex addition.

11. **Opposite Rings.** Let $(R, +, \bullet)$ be a ring. Define an operation $\star : R \times R \to R$ by setting $a \star b = b \bullet a$ for all $a, b \in R$.

(a) Prove $(R, +, \star)$ is a ring, called the *opposite ring* R^{op} of R.

(b) Let $(M, +)$ be a commutative group. Prove $\cdot : M \times R \to M$ satisfies the axioms for a right R-module iff $* : R^{\mathrm{op}} \times M \to M$, defined by $r * m = m \cdot r$ for $r \in R$ and $m \in M$, satisfies the axioms for a left R^{op}-module. (This result reduces the study of right modules to the study of left modules over the opposite ring.)

12. Define a left R-module structure on the set of matrices $M_{m,n}(R)$. Is this a free R-module? If so, describe an R-basis.

13. Let $(M,+)$ be a commutative group. Prove there is at most one scalar multiplication $\cdot : \mathbb{Z} \times M \to M$ that makes $(M,+,\cdot)$ a left $\mathbb{Z}$-module. Prove that the scalar multiplication $\cdot : \mathbb{Z} \times M \to M$, defined in §17.2, does satisfy the axioms for a left $\mathbb{Z}$-module.

14. Let $(M,+)$ be a commutative group and $n \in \mathbb{Z}_{>0}$. Prove there is at most one scalar multiplication $\cdot : \mathbb{Z}_n \times M \to M$ that makes $(M,+,\cdot)$ a left $\mathbb{Z}_n$-module. Find and prove a condition on $(M,+)$ that is necessary and sufficient for there to exist a left $\mathbb{Z}_n$-module with underlying additive group $(M,+)$.

15. Given a commutative group $(M,+)$, recall (Exercise 67 in Chapter 1) that the *endomorphism ring* $\operatorname{End}(M)$ is the set of all group homomorphisms $f : M \to M$. We add and multiply $f, g \in \operatorname{End}(M)$ via the rules $(f+g)(x) = f(x) + g(x)$ and $(f \circ g)(x) = f(g(x))$ for all $x \in M$. Prove: for any subring S of $\operatorname{End}(M)$, M is a left S-module under the scalar multiplication $f \cdot x = f(x)$ for $f \in S$ and $x \in M$.

16. Let $(M,+)$ be a fixed commutative group with endomorphism ring $\operatorname{End}(M)$.
    (a) Given a scalar multiplication $\cdot : R \times M \to M$ satisfying the axioms for a left R-module, define a map $L_r : M \to M$ (for each $r \in R$) by setting $L_r(x) = r \cdot x$ for $x \in M$. The map L_r is called *left multiplication by r*. Confirm that $L_r \in \operatorname{End}(M)$ for each $r \in R$. Then show that the map $L : R \to \operatorname{End}(M)$, defined by $L(r) = L_r$ for $r \in R$, is a ring homomorphism.
    (b) Conversely, suppose $T : R \to \operatorname{End}(M)$ is a given ring homomorphism. Define $\star : R \times M \to M$ by $r \star x = T(r)(x)$ for all $r \in R$ and all $x \in M$. Show that this scalar multiplication turns M into a left R-module.
    (c) Let $\mathcal{X}$ be the set of all scalar multiplication functions $\cdot : R \times M \to M$ satisfying the axioms for a left R-module and $\mathcal{Y}$ be the set of all ring homomorphisms $L : R \to \operatorname{End}(M)$. The constructions in (a) and (b) define maps $\phi : \mathcal{X} \to \mathcal{Y}$ and $\psi : \mathcal{Y} \to \mathcal{X}$. Check that $\phi \circ \psi = \operatorname{id}_\mathcal{Y}$ and $\psi \circ \phi = \operatorname{id}_\mathcal{X}$, so that ϕ and ψ are bijections. (This exercise shows that left R-module structures on a given commutative group $(M,+)$ correspond bijectively with ring homomorphisms of R into the endomorphism ring $\operatorname{End}(M)$.)

17. Prove that for any ring S, there exists exactly one ring homomorphism $f : \mathbb{Z} \to S$. Deduce from this and Exercise 16 that for every commutative group $(M,+)$, there exists a unique scalar multiplication that turns M into a left $\mathbb{Z}$-module.

18. Show that N is a submodule of an R-module M iff N is a nonempty subset of M closed under subtraction and left multiplication by scalars in R.

19. Prove that every additive subgroup of a $\mathbb{Z}$-module M is a $\mathbb{Z}$-submodule.

20. Give a specific example of a ring R, a left R-module M, and an additive subgroup N of M that is not an R-submodule of M.

21. Prove or disprove: for each positive integer n, every additive subgroup of any $\mathbb{Z}_n$-module M is a $\mathbb{Z}_n$-submodule.

22. Prove or disprove: if M and N are submodules of a left R-module P, then $M \cup N$ is always a submodule of P.

23. Prove: if $\{M_i : i \in I\}$ is an indexed family of submodules of a left R-module P such that $I \neq \emptyset$ and for all $i, j \in I$, either $M_i \subseteq M_j$ or $M_j \subseteq M_i$, then $N = \bigcup_{i \in I} M_i$ is a submodule of P.

24. Given submodules M and N of a left R-module P, confirm that $M + N$ is a submodule of P.

25. Given a family $\{M_i : i \in I\}$ of submodules of an R-module P, confirm that $\sum_{i \in I} M_i$ is a submodule of P. Prove that if N is any R-submodule of P containing every M_i, then $\sum_{i \in I} M_i \subseteq N$. (So the sum of the M_i is the smallest submodule containing every M_i).

26. Let S be any set and $\mathcal{X}$ be the set of all subsets of S. $\mathcal{X}$ is a partially ordered set where $U \leq V$ means $U \subseteq V$ for $U, V \in \mathcal{X}$. Prove that $\mathcal{X}$ is a complete lattice.

27. Prove that every simple R-module is cyclic. Give an example to show that the converse is not true in general.

28. Show that a $\mathbb{Z}$-module M is simple iff $|M|$ is prime. (You may need Lagrange's Theorem from group theory.)

29. Show: for all fields F, an F-module M is simple iff $\dim_F(M) = 1$.

30. Show: for all commutative rings R, the left R-module R is simple iff R is a field.

31. Give an example (with proof) of an infinite ring R and an infinite cyclic R-module M such that there exists a unique $x \in M$ with $M = Rx$.

32. For any left ideal I of R, prove R/I is a cyclic left R-module. Conversely, prove that every cyclic left R-module M is isomorphic to a module R/I for some left ideal I of R. (Use the Fundamental Homomorphism Theorem.)

33. A subset I of R is called a *maximal left ideal* iff I is a submodule of the left R-module R such that $I \neq R$ and for any submodule J with $I \subseteq J \subseteq R$, either $J = I$ or $J = R$. Prove that if I is a maximal left ideal of R, then R/I is a simple left R-module. (Use the Correspondence Theorem.) Conversely, prove that every simple left R-module M is isomorphic to a module R/I for some maximal left ideal I of R.

34. The set $M_2(\mathbb{R})$ of 2×2 real matrices is an additive group, a ring, a left $\mathbb{Z}$-module, a left $\mathbb{R}$-module, and a left $M_2(\mathbb{R})$-module. Let N be the set of matrices of the form $\begin{bmatrix} 0 & a \\ 0 & b \end{bmatrix}$ for some $a, b \in \mathbb{R}$. Show that N is an $\mathbb{R}$-submodule of $M_2(\mathbb{R})$ and an $M_2(\mathbb{R})$-submodule of $M_2(\mathbb{R})$, but N is not an ideal of $M_2(\mathbb{R})$. Show that N is not a simple $\mathbb{R}$-module, but N is a simple $M_2(\mathbb{R})$-module. (Prove any nonzero $M_2(\mathbb{R})$-submodule P of N must be equal to N.)

35. Give an example of submodules A, B, C of the $\mathbb{R}$-module $\mathbb{R} \times \mathbb{R}$ such that the distributive law $A \cap (B + C) = (A \cap B) + (A \cap C)$ is not true.

36. Let A, B, and C be submodules of a left R-module M. Prove: if $A \subseteq C$, then $A + (B \cap C) = (A + B) \cap C$.

37. Let S be a subset of a left R-module M.
    (a) Show that the set N' of R-linear combinations of elements of S is an R-submodule of M by checking the closure conditions in the definition.
    (b) Show that N' is a submodule of M by verifying that $N' = \sum_{s \in S} Rs$.

38. Let N be a subset of a left R-module M. Let I be the set of all $r \in R$ such that $r \cdot x = 0_M$ for all $x \in N$. Prove that I is a submodule of the left R-module R. Prove that if N is a submodule of M, then I is a two-sided ideal of the ring R.

39. Assuming $R \neq \{0\}$, prove that $R[x]$ is not a finitely generated R-module.

40. Given an index set I and left R-modules M_i for each $i \in I$, verify that $\prod_{i \in I} M_i$ satisfies all the module axioms.

41. Given an index set I and left R-modules M_i for each $i \in I$, verify that $\bigoplus_{i \in I} M_i$ is a submodule of $\prod_{i \in I} M_i$.

42. Verify that the maps S and T in §17.5 are R-linear maps with $S = T^{-1}$.

43. Let M and N be left R-modules. Prove that $\operatorname{Hom}_R(M, N)$ is always an additive subgroup of N^M.

44. **Bimodules.** Given rings R and S, an R, S-*bimodule* is a commutative group $(M, +)$ that has both a left R-module structure, given by $\cdot : R \times M \to M$, and a right S-module structure, given by $\star : M \times S \to M$, that are connected by the axiom $(r \cdot x) \star s = r \cdot (x \star s)$ for all $r \in R$, $s \in S$, and $x \in M$.
    (a) Prove that any ring R is an R, R-bimodule if we take $\cdot$ and $\star$ to be the multiplication of R.
    (b) Prove that R^n (viewed as column vectors) is an $M_n(R), R$-bimodule using the natural action of matrices and scalars on column vectors.

45. Let M be a left R-module and N an R, S-bimodule. Show that the commutative group $\operatorname{Hom}_R(M, N)$ of R-maps from M to N is a right S-module if we define $f \cdot s$ (for $f \in \operatorname{Hom}_R(M, N)$ and $s \in S$) to be the function from M to N sending $x \in M$ to $f(x) \star s$.

46. Let M be an R, S-bimodule and N a left R-module. Show that the commutative group $\operatorname{Hom}_R(M, N)$ of R-maps from M to N is a left S-module if we define $s \cdot f$ (for $f \in \operatorname{Hom}_R(M, N)$ and $s \in S$) to be the function from M to N sending $x \in M$ to $f(x \star s)$.

47. Let M be a left R-module with submodule N. Define a relation $\equiv$ on M by setting $x \equiv y$ iff $x - y \in N$ (for all $x, y \in M$). Prove $\equiv$ is an equivalence relation, and prove the equivalence class of x is the coset $x + N$.

48. Prove that $R[x]$ is a free left R-module by finding an explicit basis for this module.

49. Assume R is an integral domain and $g \in R[x]$ is monic of degree $n > 0$. Let $I = R[x]g$. Prove that $R[x]/I$ is a free R-module with ordered basis $(1 + I, x + I, x^2 + I, \ldots, x^{n-1} + I)$.

50. Suppose S is a ring, $f : S \to R$ is a ring homomorphism, and M is a left R-module. (a) Prove that M is a left S-module via $s \star x = f(s) \cdot x$ for $s \in S$ and $x \in M$ by checking the S-module axioms. (b) Give a short proof that M is a left S-module using the results of Exercise 16.

51. Assume M is a left R-module annihilated by an ideal I of R. Complete the verification (from §17.8) that M is a left R/I-module via $(r + I) \bullet m = r \cdot m$ for $r \in R$ and $m \in M$.

52. Given a left R-module N and an ideal I of R, show IN is a submodule of N.

53. Suppose R is a commutative ring with maximal ideal M. Prove R/M is a field. (Given a nonzero $x + M \in R/M$ with $x \in R$, consider the ideal $M + Rx$.)

54. Let $N = \left\{ \begin{bmatrix} 0 & b \\ 0 & d \end{bmatrix} : b, d \in \mathbb{R} \right\}$, which is an $M_2(\mathbb{R})$-submodule of $M_2(\mathbb{R})$ by Exercise 34. Use the Fundamental Homomorphism Theorem to prove that there is an $M_2(\mathbb{R})$-module isomorphism $M_2(\mathbb{R})/N \cong N$.

55. Let $f : M \to N$ be an R-map between left R-modules M and N. Prove: for any submodule M' of M, $f[M']$ is a submodule of N, and this submodule is contained in $\operatorname{img}(f)$. Prove: for any submodule N' of N, $f^{-1}[N']$ is a submodule of M, and this submodule contains $\ker(f)$.

56. Let M be a left R-module. Prove $M/\{0_M\}$ is isomorphic to M. Prove M/M is isomorphic to $\{0_M\}$.

57. Let M and N be left R-modules. Prove $\frac{M \times N}{M \times \{0_N\}} \cong N$.

58. Given an index set I, left R-modules M_i for $i \in I$, and a submodule N_i of M_i for each $i \in I$, prove $\frac{\prod_{i \in I} M_i}{\prod_{i \in I} N_i} \cong \prod_{i \in I} (M_i/N_i)$.

59. Suppose $f : M \to N$ is a homomorphism of left R-modules, and H is a submodule of M. Try to define $f' : M/H \to N$ by $f'(x + H) = f(x)$ for all $x \in M$.
    (a) Prove the Generalized Fundamental Homomorphism Theorem stated in §17.9.
    (b) Prove: If H is not contained in $\ker(f)$, then f' is not a well-defined function.

60. Assume H and N are submodules of a left R-module M with $H \subseteq N$. Prove: for all $x \in M$, $x + H \in N/H$ iff $x \in N$.

61. This exercise proves some results from set theory that are needed in the proof of the Correspondence Theorem for Modules. Let $f : X \to Y$ be any function.
    (a) Prove: for all $W \subseteq Y$, $f[f^{-1}[W]] \subseteq W$.
    (b) Give an example to show equality need not hold in (a).
    (c) Prove: if f is surjective, then equality does hold in (a).
    (d) Prove: for all $U \subseteq X$, $f^{-1}[f[U]] \supseteq U$.
    (e) Give an example to show equality need not hold in (d).
    (f) Prove: if f is one-to-one, then equality does hold in (d).
    (g) Prove: for all U_1, U_2 with $U_1 \subseteq U_2 \subseteq X$, $f[U_1] \subseteq f[U_2]$.
    (h) Prove: for all W_1, W_2 with $W_1 \subseteq W_2 \subseteq Y$, $f^{-1}[W_1] \subseteq f^{-1}[W_2]$.

62. Prove the Lemma on Extending an Independent Set in §17.12.

63. Prove: if M is a finitely generated module over a division ring R, then M has finite length and $\mathrm{len}(M) = \dim(M)$.

64. Prove that each structure $(Z, \leq)$ is a poset.
    (a) $(Z, \subseteq)$, where Z is a set of subsets of a fixed set
    (b) $(\mathbb{Z}_{>0}, |)$, where $a|b$ means a divides b
    (c) $(\mathbb{Z}_{>0}, \preceq)$, where $a \preceq b$ means b divides a
    (d) Z is the set of functions $f : \mathbb{R} \to \mathbb{R}$; $f \leq g$ means $f(x) \leq g(x)$ for all $x \in \mathbb{R}$
    (e) Z is the set of functions $f : D \to \mathbb{R}$ for some $D \subseteq \mathbb{R}$, and for $f : D \to \mathbb{R}$ and $g : E \to \mathbb{R}$ in Z, $f \leq g$ means $D \subseteq E$ and $f(x) = g(x)$ for all $x \in D$

65. For each poset in (b) through (e) of Exercise 64, describe all maximal elements of the poset, or explain why none exist.

66. For each poset in (b) through (e) of Exercise 64, give an example of an infinite chain in that poset, and state whether the chain has an upper bound in the poset.

67. Let R be a commutative ring, I be an ideal of R, and S be a nonempty subset of $R \setminus I$. Use Zorn's Lemma to prove that the set of ideals J of R with $I \subseteq J$ and $J \cap S = \emptyset$ has a maximal element (relative to set inclusion).

68. An ideal P in a nonzero commutative ring R is called *prime* iff $P \neq R$ and for all $x, y \in R$, $xy \in P$ implies $x \in P$ or $y \in P$. Use Zorn's Lemma to prove that for a given prime ideal P of R, the set of prime ideals Q contained in P has a minimal element relative to set inclusion.

69. Give an example of a commutative group G with no maximal proper subgroups. Suppose we try to use Zorn's Lemma to prove that every commutative group has a maximal proper subgroup, by adapting the proof that maximal ideals exist in commutative rings. Exactly where does the proof break down?

70. Use Zorn's Lemma to prove that any poset $(Z, \leq)$ has a maximal chain (which is a chain $C \subseteq Z$ such that any set properly containing C is not a chain).

71. For any set X, let $\mathcal{P}'(X)$ be the set of nonempty subsets of X. Use Zorn's Lemma to prove this version of the Axiom of Choice: for any set X, there exists a function (a set of ordered pairs) $f : \mathcal{P}'(X) \to X$ with $f(S) \in S$ for all $S \in \mathcal{P}'(X)$.

72. Assume M is a finite left R-module.
    (a) Given a submodule C of M, explain why $|M| = |C| \cdot |M/C|$.
    (b) Let A and B be submodules of M. Use an appropriate isomorphism theorem to prove that $|A + B| = |A| \cdot |B|/|A \cap B|$.

73. Use Figure 17.1 to find all maximal chains of submodules of the $\mathbb{Z}$-module $\mathbb{Z}_4 \times \mathbb{Z}_2$. Confirm that the conclusions of the Jordan–Hölder Theorem are true for these chains.

74. We know every subgroup of $\mathbb{Z}$ has the form $\mathbb{Z}m$ for a unique $m \geq 0$. Also, for all $a, b \in \mathbb{Z}$, $\mathbb{Z}a \subseteq \mathbb{Z}b$ iff b divides a. Use this information and the Correspondence Theorem to draw the submodule lattices of the following quotient $\mathbb{Z}$-modules: $\mathbb{Z}/8\mathbb{Z}$, $\mathbb{Z}/35\mathbb{Z}$, and $\mathbb{Z}/60\mathbb{Z}$. Verify by inspection of the drawings that all maximal chains of submodules have the same length.

75. Draw the lattice of all $\mathbb{Z}_2$-submodules of the $\mathbb{Z}_2$-module $\mathbb{Z}_2 \times \mathbb{Z}_2 \times \mathbb{Z}_2$. Confirm that the conclusions of the Jordan–Hölder Theorem are true for this module.

76. Let R be the set of upper-triangular matrices in $M_2(\mathbb{Z}_2)$.
    (a) Prove that R is a subring of $M_2(\mathbb{Z}_2)$.
    (b) Draw the submodule lattice of R, viewed as a left R-module.
    (c) Draw the submodule lattice of R, viewed as a right R-module.
    (d) Draw the lattice of two-sided ideals of the ring R.

77. Solve this exercise without using the Jordan–Hölder Theorem.
    (a) Suppose A is a submodule of an R-module M with the property that A and M/A are simple R-modules. Suppose B is a submodule of M different from $\{0\}$, A, and M. Prove that B and M/B are simple R-modules. (Draw a picture of what you know about the submodule lattice of M.)
    (b) Give an example of R, A, B, M satisfying the conditions in (a), such that A and B are non-isomorphic R-modules.

78. Fix $m, n \in \mathbb{Z}_{>0}$ with $\gcd(m, n) = 1$. Use the Recognition Theorem for Direct Products to show that the $\mathbb{Z}$-module $\mathbb{Z}_{mn} = \{0, 1, \ldots, mn - 1\}$ (with operation addition mod mn) is isomorphic to the direct product of its submodules $\mathbb{Z}n$ and $\mathbb{Z}m$. Conclude that $\mathbb{Z}_{mn} \cong \mathbb{Z}_m \times \mathbb{Z}_n$ as $\mathbb{Z}$-modules.

79. **Recognition Theorem for Finite Direct Products.** Suppose $N_1, N_2, \ldots, N_k$ are submodules of a left R-module M such that $M = N_1 + N_2 + \cdots + N_k$ and $N_i \cap (N_1 + N_2 + \cdots + N_{i-1}) = \{0\}$ for all i between 2 and k. Prove that $g : N_1 \times N_2 \times \cdots \times N_k \to M$, defined by $g(x_1, \ldots, x_k) = x_1 + \cdots + x_k$ for $x_i \in N_i$, is an R-module isomorphism.

80. **Recognition Theorem for Direct Sums.** Suppose I is an index set, and $\{N_i : i \in I\}$ is a collection of submodules of a left R-module M such that $M = \sum_{i \in I} N_i$ and $N_i \cap \sum_{j \in I: \, j \neq i} N_j = \{0_M\}$ for all $i \in I$. Prove that $M \cong \bigoplus_{i \in I} N_i$.

81. Find the length of the following $\mathbb{Z}$-modules.
    (a) $\mathbb{Z}_{32}$ (b) $\mathbb{Z}_{60}$ (c) $\mathbb{Z}_{12} \times \mathbb{Z}_{15}$ (d) $\mathbb{Z}$ (e) $\mathbb{Z}_p^n$, where p is prime and $n \geq 1$

82. Given a positive integer n, apply the Jordan–Hölder Theorem to the $\mathbb{Z}$-module $\mathbb{Z}_n$ to prove that n can be factored into a product of prime integers that are uniquely determined by n (up to reordering).

83. Let V be a vector space over a field F.
    (a) Prove: if V has an n-element basis for some integer n, then there is a maximal chain of subspaces of V of length n.
    (b) Prove: if $\dim(V) = \infty$, then V has an infinitely long chain of subspaces.
    (c) Use the Jordan–Hölder Theorem to prove that if V is finite-dimensional, then all bases of V have the same (finite) size.

84. Let $(x_1, \ldots, x_n)$ be a list of elements in a left R-module M, where R is a nonzero ring. Show that this list is R-linearly dependent in each of the following situations.
    (a) $x_i = 0_M$ for some i; (b) $x_i = x_j$ for some $i \neq j$; (c) $x_1, \ldots, x_{n-1}$ generate M.

85. Let $v_1 = (2, 3)$ and $v_2 = (1, 5)$.
    (a) Show that (v_1, v_2) is an ordered basis of the $\mathbb{R}$-module $\mathbb{R} \times \mathbb{R}$.
    (b) Show that (v_1, v_2) is not an ordered basis of the $\mathbb{Z}$-module $\mathbb{Z} \times \mathbb{Z}$.
    (c) Is (v_1, v_2) an ordered basis of the $(\mathbb{R} \times \mathbb{R})$-module $\mathbb{R} \times \mathbb{R}$? Explain.

86. Let $f : N \to P$ be an R-linear map. Prove: if f is surjective and the list $(x_1, \ldots, x_n)$ generates N, then $(f(x_1), \ldots, f(x_n))$ generates P.

87. Let $f : N \to P$ be an R-linear map. Prove: if f is injective and the list $(x_1, \ldots, x_n)$ is R-linearly independent in N, then $(f(x_1), \ldots, f(x_n))$ is R-linearly independent in P.

88. Check in detail that the map g' (from the UMP in §17.11) is an R-linear map and $g'(x) = g(x)$ for all $x \in S$.

89. Let I be an ideal of a commutative ring R such that $\{0\} \neq I \neq R$. Show that the R-module R/I has no finite basis. Indicate where you use the hypotheses $\{0\} \neq I$ and $I \neq R$. Does the R/I-module R/I have a basis? Explain.

90. Prove: for all $x, y \in R$, the list (x, y) is an R-basis of the left R-module R iff there exist $r, s \in R$ with $xr = 1 = ys$, $yr = 0 = xs$, and $rx + sy = 1$.

91. Let A, B, C be submodules of a left R-module M such that $A \subseteq B$, $A + C = B + C$, and $A \cap C = B \cap C$. Prove that $A = B$.

92. Give an example of an $\mathbb{R}$-module M and submodules A, B, C such that $A + C = B + C$, $A \cap C = B \cap C$, and yet $A \neq B$.

93. Give a justified example of each of the following:
    (a) a ring R and an R-module that is not finitely generated
    (b) a ring R and an infinite non-cyclic finitely generated R-module
    (c) a ring R and a finitely generated R-module that has no basis
    (d) a list (v_1, v_2) in the $\mathbb{Z}$-module $\mathbb{Z} \times \mathbb{Z}$ that is $\mathbb{Z}$-linearly independent but does not generate $\mathbb{Z} \times \mathbb{Z}$
    (e) a ring R, a commutative group $(M, +)$, and a function $\bullet : R \times M \to M$ satisfying every module axiom except $1_R \bullet x = x$ for all $x \in M$
    (f) a simple left $M_3(\mathbb{Q})$-module.

94. True or false? Explain each answer.
    (a) The set of all subgroups of a fixed R-module M, ordered by set inclusion, is always a complete lattice.
    (b) Every submodule of a product R-module $M \times N$ must have the form $A \times B$, where A is a submodule of M and B is a submodule of N.
    (c) For any submodules A and B of a left R-module M, there is an R-module isomorphism $(A + B)/A \cong A/(A \cap B)$.
    (d) A left R-module M is simple iff M has exactly two submodules.

(e) There exists exactly one function $\cdot : \mathbb{Z} \times (\mathbb{Z} \times \mathbb{Z}) \to \mathbb{Z} \times \mathbb{Z}$ that turns the additive group $\mathbb{Z} \times \mathbb{Z}$ into a left $\mathbb{Z}$-module.

(f) There exists exactly one function $\cdot : (\mathbb{Z} \times \mathbb{Z}) \times \mathbb{Z} \to \mathbb{Z}$ that turns the additive group $\mathbb{Z}$ into a left $(\mathbb{Z} \times \mathbb{Z})$-module.

(g) Every left R-module M is a sum of cyclic submodules.

95. Suppose we are given ten left R-modules and thirteen R-linear maps as shown in the following diagram:

$$
\begin{array}{ccccccccc}
A & \xrightarrow{f} & B & \xrightarrow{g} & C & \xrightarrow{h} & D & \xrightarrow{k} & E \\
\downarrow{\alpha} & & \downarrow{\beta} & & \downarrow{\gamma} & & \downarrow{\delta} & & \downarrow{\epsilon} \\
A' & \xrightarrow{f'} & B' & \xrightarrow{g'} & C' & \xrightarrow{h'} & D' & \xrightarrow{k'} & E'
\end{array}
$$

(This means $f : A \to B$, $\alpha : A \to A'$, etc.) Assume that:

(1) $\operatorname{img}(f) = \ker(g)$
(2) $\operatorname{img}(g) = \ker(h)$
(3) $\operatorname{img}(h) = \ker(k)$
(4) $\operatorname{img}(f') = \ker(g')$
(5) $\operatorname{img}(g') = \ker(h')$
(6) $\operatorname{img}(h') = \ker(k')$
(7) $f' \circ \alpha = \beta \circ f$
(8) $g' \circ \beta = \gamma \circ g$
(9) $h' \circ \gamma = \delta \circ h$
(10) $k' \circ \delta = \epsilon \circ k$
(11) β is one-to-one
(12) δ is one-to-one
(13) α is onto

(a) Prove that γ is one-to-one. Indicate which of the 13 hypotheses is needed in each step of the proof.

(b) Keep assumptions (1) through (10), but replace (11), (12), (13) by:

(11') β is onto
(12') δ is onto
(13') ϵ is one-to-one

Prove that γ is onto. Indicate which of the 13 hypotheses is needed in each step of the proof.

(c) Deduce that, if (1) through (10) hold and α, β, δ, and ϵ are isomorphisms, then γ is an isomorphism.

18

Principal Ideal Domains, Modules over PIDs, and Canonical Forms

Given a field F, we know that every finite-dimensional F-vector space V is isomorphic to F^n for some integer $n \geq 0$, where F^n is the direct product of n copies of the vector space F. Similarly, given any finitely generated commutative group G, we proved in Chapter 16 that G is isomorphic to a direct product of finitely many cyclic groups. Now, F-vector spaces are the same thing as F-modules, and commutative groups are the same thing as $\mathbb{Z}$-modules. So, the two theorems just mentioned provide a classification of all finitely generated R-modules, where R is either a field or the ring $\mathbb{Z}$.

In this chapter, we prove a more general classification theorem that includes both of the previous results as special cases. To obtain this theorem, we must isolate the key properties of fields and the ring $\mathbb{Z}$ that made the previous theorems work. This leads us to study principal ideal domains (abbreviated PIDs), which are integral domains where every ideal can be generated by a single element. We will see that $\mathbb{Z}$ is a PID, and so is the polynomial ring $F[x]$ in one variable with coefficients in a field F. As in the cases of $\mathbb{Z}$ and $F[x]$, we will see that elements in general PIDs have unique factorizations into irreducible elements (which are analogous to prime integers or irreducible polynomials).

The key theorem in this chapter asserts that for any PID R, every finitely generated R-module M is isomorphic to a direct product of cyclic R-modules

$$R/Ra_1 \times R/Ra_2 \times \cdots \times R/Ra_k$$

for some $k \in \mathbb{Z}_{>0}$ and some $a_i \in R$. As in the case of $\mathbb{Z}$-modules, we can arrange that the a_i satisfy certain additional conditions (for instance, that each a_i divide a_{i+1}, or that every nonzero a_i be a prime power in R). When we impose appropriate conditions of this kind, the ideals $Ra_1, \ldots, Ra_k$ appearing in the decomposition of M are uniquely determined by M.

The proof of this theorem mimics the proof of the classification theorem for commutative groups, which occupies most of Chapter 16. The current chapter can be read independently of that one and yields the main results of that chapter as a special case. However, the reader is urged to study that chapter first, to get used to the essential ideas of the proof in the very concrete setting of integer-valued matrices. This chapter does assume knowledge of definitions and basic facts about R-modules and free R-modules, which we cover in Chapter 17. We also need some material on one-variable polynomials over a field from Chapter 3.

As an application of the Classification Theorem for Modules over PIDs, we prove the Rational Canonical Form Theorem for Linear Operators. This theorem shows that every linear map $T : V \to V$ defined on a finite-dimensional F-vector space V can be represented (relative to an appropriate ordered basis) by a matrix with an especially simple structure. To obtain this matrix and to prove its uniqueness, we first use T to make V into a finitely generated $F[x]$-module, then apply the structure theorems of this chapter to this module. For fields F satisfying appropriate hypotheses, we use similar techniques to give another

DOI: 10.1201/9781003484561-18

derivation of the Jordan Canonical Form Theorem from Chapter 8. We also discuss Smith normal forms, rational canonical forms, and Jordan forms for matrices.

18.1 Principal Ideal Domains

We begin by spelling out the definition of a principal ideal domain in more detail. Recall from §1.2 that an *integral domain* is a commutative ring $(R, +, \cdot)$ such that $1_R \neq 0_R$ and R has no zero divisors other than 0. The last condition means that for all $a, b \in R$, if $a \cdot b = 0_R$, then $a = 0_R$ or $b = 0_R$. The following *Cancellation Law* holds in integral domains R: for all $x, y, z \in R$ with $x \neq 0_R$, if $xy = xz$ then $y = z$. To prove this, rewrite $xy = xz$ as $xy - xz = 0$ and then as $x(y - z) = 0$. Since $x \neq 0$, we get $y - z = 0$ and $y = z$.

Next, recall from §1.4 that an *ideal* of a commutative ring R is a subset I of R satisfying these closure conditions: $0_R \in I$; for all $x, y \in I$, $x + y \in I$; for all $x \in I$, $-x \in I$; and for all $x \in I$ and $r \in R$, $r \cdot x \in I$. You can check that for any commutative ring R and any $a \in R$, the set $Ra = \{r \cdot a : r \in R\}$ is an ideal of R containing a. This ideal is called the *principal ideal generated by a*. An ideal I of R is a *principal ideal* iff there exists $a \in R$ with $I = Ra$. A *principal ideal domain (PID)* is an integral domain R such that every ideal of R is a principal ideal.

We know that the ring $\mathbb{Z}$ is an integral domain, since the product of any two nonzero integers is nonzero. To prove that $\mathbb{Z}$ is a PID, consider any ideal I of $\mathbb{Z}$. By the first three closure conditions in the definition of an ideal, we see that I is an additive subgroup of the group $(\mathbb{Z}, +)$. In §16.1, we used integer division with remainder to prove that such a subgroup must have the form $\mathbb{Z}n = \{kn : k \in \mathbb{Z}\}$ for some integer $n \geq 0$. So I is a principal ideal.

You can check (Exercise 2) that every field F is a PID. Next, we show that *a one-variable polynomial ring $F[x]$ with coefficients in a field F is a PID*. In §3.4, we used degree arguments to see that $F[x]$ is an integral domain. To show that every ideal I of $F[x]$ is principal, we use the division theorem for one-variable polynomials (§3.6). On one hand, if $I = \{0\}$, then $I = F[x]0$ is a principal ideal. On the other hand, if I is a nonzero ideal, we can choose a nonzero $g \in I$ of minimum possible degree. Because I is an ideal and $g \in I$, I contains the principal ideal $F[x]g = \{pg : p \in F[x]\}$. We now prove that $I \subseteq F[x]g$, which shows that $I = F[x]g$ is principal. Fix any $f \in I$. Dividing f by g produces a unique quotient $q \in F[x]$ and remainder $r \in F[x]$ with $f = qg + r$ and $r = 0$ or $\deg(r) < \deg(g)$. If $r = 0$, then $f = qg \in F[x]g$ as needed. If $r \neq 0$, then $r = f + (-q)g \in I$ since $f, g \in I$ and I is an ideal. But then $\deg(r) < \deg(g)$ contradicts minimality of the degree of g, so $r \neq 0$ cannot occur.

You can check (Exercises 3 and 4 that $\mathbb{Z}[x]$ and $F[x_1, \ldots, x_n]$ (where $n > 1$) are examples of integral domains that are not PIDs.

18.2 Divisibility in Commutative Rings

In any commutative ring R, we can define concepts related to divisibility by analogy with the familiar definitions for integers and polynomials. Given $a, b \in R$, we say *a divides b in R* and write $a|b$ iff there exists $c \in R$ with $b = ac$. When $a|b$, we also say b is a *multiple* of a and a is a *divisor* of b. You can check that: for all $a \in R$, $a|a$ (reflexivity); for all $a, b, c \in R$, if $a|b$ and $b|c$ then $a|c$ (transitivity); and for all $a \in R$, $1|a$ and $a|0$. A *unit* of the ring R is an element $u \in R$ such that $u|1$. This means that $1 = uv = vu$ for some $v \in R$, so that the units of R are precisely the invertible elements of R relative to the multiplication in R. We let R^* be the set of units of R.

The set R under the divisibility relation $|$ is almost a poset, since $|$ is reflexive and transitive on R. However, for almost all commutative rings R, antisymmetry does not hold for $|$. In other words, there can exist $x \neq y$ in R with $x|y$ and $y|x$. For all $x, y \in R$, we define x and y to be *associates in R*, denoted $x \sim y$, iff $x|y$ and $y|x$. You can check that $\sim$ is an equivalence relation on the set R. Associate ring elements behave identically with respect to divisibility; more precisely, given $a, a', b, c \in R$ with $a \sim a'$, $a|b$ iff $a'|b$, and $c|a$ iff $c|a'$. Similarly, $u \in R$ is a unit of R iff $u \sim 1$.

If R is an integral domain, then there is an alternate description of when two elements $x, y \in R$ are associates: $x \sim y$ iff $y = ux$ for some unit $u \in R^*$. To prove one direction, assume $y = ux$ for some unit u of R. There is $v \in R$ with $vu = 1$, so $vy = vux = 1x = x$. Since $y = ux$ and $x = vy$, we see that $x|y$ and $y|x$ in R, hence $x \sim y$. Conversely, assume x and y are associates in R. Then $y = ax$ and $x = by$ for some $a, b \in R$. Combining these, we get $1x = by = (ba)x$ and $1y = ax = aby = (ba)y$. If x or y is nonzero, then the Cancellation Law for integral domains gives $ba = 1$, so that $a \in R^*$ and $y = ax$ as needed. If $x = 0 = y$, then $y = ux$ holds for the unit $u = 1_R$.

For example, in $\mathbb{Z}$ the units are $+1$ and -1, so $x \sim y$ in $\mathbb{Z}$ iff $y = \pm x$. In $F[x]$ where F is a field, the units are the nonzero constant polynomials, so $p \sim q$ in $F[x]$ iff $q = cp$ for some nonzero $c \in F$. It follows that each equivalence class of $\sim$ in $\mathbb{Z}$ contains exactly one nonnegative integer, and each equivalence class of $\sim$ in $F[x]$ (other than $\{0\}$) contains exactly one monic polynomial.

Next, we define common divisors, common multiples, gcds, and lcms. Let $a_1, \ldots, a_k$ be fixed elements of a commutative ring R. We say $d \in R$ is a *common divisor* of $a_1, \ldots, a_k$ iff $d|a_i$ for $1 \leq i \leq k$. We say $e \in R$ is a *common multiple* of $a_1, \ldots, a_k$ iff $a_i|e$ for $1 \leq i \leq k$. We say $d \in R$ is a *greatest common divisor (gcd)* of $a_1, \ldots, a_k$ iff d is a common divisor of $a_1, \ldots, a_k$ such that for all common divisors c of $a_1, \ldots, a_k$, $c|d$. We say $e \in R$ is a *least common multiple (lcm)* of $a_1, \ldots, a_k$ iff e is a common multiple of $a_1, \ldots, a_k$ such that for all common multiples c of $a_1, \ldots, a_k$, $e|c$. (Compare these definitions to the definitions of lower bounds, upper bounds, greatest lower bounds, and least upper bounds in a poset, given in the Appendix.)

We warn the reader that greatest common divisors of $a_1, \ldots, a_k$ need not exist in general. Even if gcds do exist, they may not be unique. Indeed, given associate ring elements $d, d' \in R$, it follows from the definitions that d is a gcd of $a_1, \ldots, a_k$ iff d' is a gcd of $a_1, \ldots, a_k$. On the other hand, if $c, d \in R$ are any two gcds of $a_1, \ldots, a_k$, then $c \sim d$. Similar results hold for lcms. In $\mathbb{Z}$ and $F[x]$, we can remedy the non-uniqueness by using nonnegative integers and monic polynomials as our gcds and lcms.

18.3 Divisibility and Ideals

Early in the development of abstract algebra, it was realized that *divisibility of elements* in a commutative ring R can be conveniently studied by instead looking at *containment of principal ideals*. To explain this, we need the fundamental observation that *for all $a, b \in R$, $a|b$ in R iff $Rb \subseteq Ra$.* Fix $a, b \in R$. On one hand, assume $a|b$ in R, say $b = ac$ for some $c \in R$. To prove $Rb \subseteq Ra$, fix $x \in Rb$. Then $x = rb$ for some $r \in R$, hence $x = r(ac) = (rc)a \in Ra$. On the other hand, assume $Rb \subseteq Ra$. Then $b = 1b \in Rb$, so $b \in Ra$, so $b = sa$ for some

$s \in R$, so $a|b$. When using this result, we must remember that the smaller element a in the divisibility relation $a|b$ corresponds to the larger ideal in the containment relation $Rb \subseteq Ra$.

You can check that the set $\mathcal{X}$ of all ideals of R is a poset ordered by set inclusion. The subset $\mathcal{Z}$ of $\mathcal{X}$ consisting of the principal ideals of R is also a poset ordered by $\subseteq$. Let us translate some of the definitions in the last section into statements about ideal containment. First, $a, b \in R$ are associates in R iff a and b generate the same principal ideal in R, because: $a \sim b$ iff $a|b$ and $b|a$ iff $Rb \subseteq Ra$ and $Ra \subseteq Rb$ iff $Ra = Rb$. Second, u is a unit of R iff $Ru = R$, because: $u \in R^*$ iff $u \sim 1$ iff $Ru = R1 = R$. Third, given $d, a_1, \ldots, a_k \in R$, d is a common divisor of $a_1, \ldots, a_k$ iff $d|a_i$ for all i iff $Ra_i \subseteq Rd$ for all i iff the ideal Rd is an upper bound for the set of ideals $\{Ra_1, \ldots, Ra_k\}$ in the poset $\mathcal{Z}$. Fourth, the element d is a common multiple of the elements a_i iff the ideal Rd is a lower bound for $\{Ra_1, \ldots, Ra_k\}$ in $\mathcal{Z}$. Fifth, d is a gcd of the a_i iff Rd is the least upper bound of $\{Ra_1, \ldots, Ra_k\}$ in $\mathcal{Z}$ Sixth, d is an lcm of the a_i iff Rd is the greatest lower bound of $\{Ra_1, \ldots, Ra_k\}$ in $\mathcal{Z}$. We illustrate these ideas by proving the following result.

Theorem on GCDs and LCMs in PIDs. Let R be a PID. For any $a_1, \ldots, a_k$ in R, $d = \gcd(a_1, \ldots, a_k)$ exists in R and $L = \operatorname{lcm}(a_1, \ldots, a_k)$ exists in R. Furthermore, there exist $r_1, \ldots, r_k \in R$ with $d = r_1 a_1 + \cdots + r_k a_k$.

The last statement says that in a PID R, the gcd of $a_1, \ldots, a_k$ is an R-linear combination of $a_1, \ldots, a_k$. This property of gcds need not hold in more general integral domains (see Exercise 42).

Proof. To prove that the gcd and lcm of $a_1, \ldots, a_k$ exist, it suffices to find a least upper bound and a greatest lower bound for the set of ideals $\{Ra_1, Ra_2, \ldots, Ra_k\}$ in the poset $\mathcal{Z}$. Recall from §17.3 that the poset of all submodules of a fixed left R-module M is a lattice. More specifically (taking $M = R$), the least upper bound in $\mathcal{X}$ of a set of ideals $\{I_1, \ldots, I_k\}$ is the ideal sum $I_1 + \cdots + I_k$. The greatest lower bound in $\mathcal{X}$ of $\{I_1, \ldots, I_k\}$ is the intersection $I_1 \cap \cdots \cap I_k$. The crucial point is that R is a PID, so $\mathcal{Z}$ and $\mathcal{X}$ are the same poset.

On one hand, $I = Ra_1 + Ra_2 + \cdots + Ra_k = \{r_1 a_1 + r_2 a_2 + \cdots + r_k a_k : r_1, \ldots, r_k \in R\}$ is an ideal of R that is the least upper bound for $\{Ra_1, \ldots, Ra_k\}$ in the poset $\mathcal{X} = \mathcal{Z}$. We know I is principal, so $I = Rd$ for some $d \in R$. Any generator d for I is a gcd of $a_1, \ldots, a_k$. Every such generator has the form $d = r_1 a_1 + r_2 a_2 + \cdots + r_k a_k$ for some $r_i \in R$, by definition of I. On the other hand, $J = Ra_1 \cap Ra_2 \cap \cdots \cap Ra_k$ is an ideal of R that is the greatest lower bound for $\{Ra_1, \ldots, Ra_k\}$ in the poset $\mathcal{X} = \mathcal{Z}$. We know J is principal, so $J = Re$ for some $e \in R$, and any generator e for J is an lcm of $a_1, \ldots, a_k$. $\qquad \square$

18.4 Prime and Irreducible Elements

To continue our discussion of factorization theory in a commutative ring R, we need to generalize the definitions of prime integers and irreducible polynomials. Two different generalizations are possible, but we will see that these generalizations coincide when R is a PID. Let $p \in R$ be a nonzero element of R that is not a unit of R. We say p is a *prime in R* iff for all $f, g \in R$, $p|(fg)$ implies $p|f$ or $p|g$. We say p is *irreducible in R* iff for all $q \in R$, $q|p$ implies $q \sim p$ or $q \in R^*$. Irreducibility of p means that the only divisors of p in R are the units of R (which divide everything) and the associates of p.

In any integral domain R, every prime p must be irreducible. To see why, let p be prime in R and assume $q \in R$ divides p. Then $p = qg$ for some $g \in R$. By primeness of p, either $p|q$ or $p|g$. If $p|q$, then (since also $q|p$) we see that $q \sim p$. On the other hand, if $p|g$, write

$g = rp$ for some $r \in R$. Then $1p = qg = (qr)p$. As $p \neq 0$ in the integral domain R, we can cancel p to get $qr = 1$, so that $q \in R^*$. Since $q \sim p$ or $q \in R^*$, p is irreducible as claimed. On the other hand, there exist integral domains R and irreducible elements $p \in R$ that are not prime in R (see Exercise 31).

Let us translate the definitions of prime and irreducible elements into statements about principal ideals. First, the assumption that p is a nonzero non-unit means that $Rp \neq \{0\}$ and $Rp \neq R$, so that Rp is a proper nonzero ideal of R. Rewriting the definition of prime element, we see that p is prime in R iff for all $f, g \in R$, $R(fg) \subseteq Rp$ implies $Rf \subseteq Rp$ or $Rg \subseteq Rp$. Now, $R(fg) \subseteq Rp$ iff $fg \in Rp$, and similarly for Rf and Rg. So we can also say that p is prime in R iff for all $f, g \in R$, $fg \in Rp$ implies $f \in Rp$ or $g \in Rp$. In ring theory, an ideal I of a commutative ring R is called a *prime ideal* iff $I \neq R$ and for all $f, g \in R$, $fg \in I$ implies $f \in I$ or $g \in I$. So we have shown that p *is a prime element in* R *iff* Rp *is a nonzero prime ideal of* R.

Next, rewriting the definition of irreducible element, we get that p is irreducible in R iff for all $q \in R$, $Rp \subseteq Rq$ implies $Rq = Rp$ or $Rq = R$. (Recall that p still satisfies $\{0\} \neq Rp \neq R$.) In other words, *irreducibility of a nonzero non-unit* p *means that the principal ideal* Rp *is a maximal element in the poset of all proper, principal ideals of* R. In ring theory, an ideal I of a commutative ring R is called a *maximal ideal* iff $I \neq R$ and for all ideals J with $I \subseteq J$, either $I = J$ or $I = R$. In other words, maximal ideals (as just defined) are maximal elements in the poset of all proper ideals of R. In the case of a PID R, the poset of proper principal ideals of R is the same as the poset of proper ideals of R. We conclude that, *in a PID* R, p *is irreducible iff* Rp *is a nonzero maximal ideal of* R.

You can prove that *every maximal ideal in any commutative ring* R *is a prime ideal* (Exercise 17). By the remarks in the last two paragraphs, we conclude that every irreducible element in a PID is also a prime element. Since the converse holds in all integral domains, *irreducible elements and prime elements coincide in a PID*.

18.5   Irreducible Factorizations in PIDs

A *unique factorization domain (UFD)* is an integral domain R such that for each nonzero non-unit $f \in R$: (a) there exist irreducible elements $p_1, \ldots, p_k \in R$ with $f = p_1 p_2 \cdots p_k$; (b) for any two factorizations $f = up_1 p_2 \cdots p_k = vq_1 q_2 \cdots q_m$ with all p_i and q_j irreducible in R and $u, v \in R^*$, we must have $k = m$ and, after reordering the q_j, $p_i \sim q_i$ for $1 \leq i \leq k$.

$\mathbb{Z}$ is a UFD, and we saw in Chapter 3 that $F[x]$ is a UFD for any field F. The next theorem includes these two results as special cases.

Theorem: Every PID is a UFD.

Proof. We need a preliminary lemma about chains of ideals in a PID. Suppose R is a PID and we have an infinite sequence of ideals

$$I_1 \subseteq I_2 \subseteq I_3 \subseteq \cdots \subseteq I_k \subseteq \cdots .$$

Then there exists k_0 such that $I_k = I_{k_0}$ for all $k \geq k_0$. Informally, we say that *every ascending chain of ideals in a PID must stabilize*. To prove this, let $I = \bigcup_{k=1}^{\infty} I_k$ be the union of all the ideals in the sequence. You can verify that I is an ideal of R, and $I_k \subseteq I$ for all $k \geq 1$. Since R is a PID, there exists $a \in R$ with $I = Ra$. Now $a \in I$, so $a \in I_{k_0}$ for some fixed index k_0. Then, for any $k \geq k_0$, $I = Ra \subseteq I_{k_0} \subseteq I_k \subseteq I$, which proves that $I = I_k = I_{k_0}$ for all such k.

We prove (a) holds for a PID R by contradiction. Assuming (a) fails, we can find a nonzero non-unit $f_0 \in R$ that cannot be factored into a product of irreducible elements in R. In particular, f_0 itself cannot be irreducible, so we can write $f_0 = gh$ where $g, h \in R$ are not zero, are not units of R, and are not associates of f_0. In terms of ideals, these conditions mean that $Rf_0 \subsetneq Rg \subsetneq R$ and $Rf_0 \subsetneq Rh \subsetneq R$. Now, if g and h could both be factored into products of irreducible elements, then f could be so factored as well by combining the two factorizations. It follows that g or h must also be a counterexample to (a). Let $f_1 = g$ if g is a counterexample, and $f_1 = h$ otherwise. Now $Rf_0 \subsetneq Rf_1 \subsetneq R$ and f_1 is a counterexample to (a). Then we can repeat the argument to produce another counterexample f_2 with $Rf_0 \subsetneq Rf_1 \subsetneq Rf_2 \subsetneq R$. This process can be continued indefinitely (using the Axiom of Choice), ultimately producing an infinite strictly ascending chain of ideals in the PID R. But this contradicts the lemma proved in the previous paragraph. So (a) does hold for R.

To prove (b), assume $f = up_1p_2 \cdots p_k = vq_1q_2 \cdots q_m$, where $u, v \in R^*$, $k, m \in \mathbb{Z}_{>0}$, and all p_i and q_j are irreducible in the PID R. Because R is a PID, all p_i and q_j are prime. In particular, the prime element p_1 divides $f = vq_1q_2 \cdots q_m$, so p_1 must divide some q_s. (Note p_1 cannot divide the unit v, or p_1 would also be a unit of R.) Reordering the q_j if needed, assume that p_1 divides q_1. As p_1 is a non-unit and q_1 is irreducible, we obtain $p_1 \sim q_1$. Write $q_1 = wp_1$ for some $w \in R^*$. Then $up_1p_2 \cdots p_k = (vw)p_1q_2 \cdots q_m$ where u and $v' = vw$ are units of R. We are in an integral domain, so we can cancel p_1 to obtain $up_2 \cdots p_k = v'q_2 \cdots q_m$. Now we repeat the argument to get $p_2 \sim q_j$ for some $j \geq 2$. We can reorder to ensure $j = 2$ and then modify the unit v' to replace q_2 by its associate p_2. Then cancel p_2 from both sides and continue until all p_i have been matched with appropriate q_j. Note that $k < m$ is impossible, since otherwise we would obtain $u = v^*q_{k+1} \cdots q_m$ after k cancellation steps, contradicting the fact that q_m is not a unit. Similarly, $k > m$ is impossible, so $k = m$, and $p_i \sim q_i$ for $1 \leq i \leq k$ after reordering the q_j. $\square$

18.6   Free Modules over a PID

We now begin our study of finitely generated modules over a fixed PID R. As in the case of $\mathbb{Z}$-modules (Chapter 16), the first step is to look at properties of finitely generated free R-modules. See §17.11 for proofs of the general facts about free modules reviewed here. An R-module M is *free and finitely generated* (abbreviated f.g. free) iff M has a finite ordered basis $\mathcal{B} = (v_1, \ldots, v_n)$, which is an R-linearly independent list of vectors that spans the R-module M. In more detail, R-*independence of $\mathcal{B}$* means that for all $c_1, \ldots, c_n \in R$, if $c_1v_1 + \cdots + c_nv_n = 0_M$, then $c_1 = \cdots = c_n = 0_R$. Saying that $\mathcal{B}$ *spans M* means that for all $w \in M$, there exist $d_1, \ldots, d_n \in R$ such that $w = d_1v_1 + \cdots + d_nv_n$. Because $\mathcal{B}$ is R-independent, the list of scalars $(d_1, \ldots, d_n)$ is uniquely determined by w. We call $(d_1, \ldots, d_n)$ the *coordinates of w relative to the ordered basis $\mathcal{B}$*. The map $T : M \to R^n$ such that $T(w) = (d_1, \ldots, d_n)$ is an R-module isomorphism. Hence, every f.g. free R-module M with an n-element basis is isomorphic to the free R-module R^n whose elements are n-tuples of scalars in R.

The R-module M with ordered basis $\mathcal{B} = (v_1, \ldots, v_n)$ satisfies the following *universal mapping property* (UMP): for every R-module N and every list $w_1, \ldots, w_n \in N$, there exists a unique R-linear map $U : M \to N$ with $U(v_i) = w_i$ for $1 \leq i \leq n$, namely $U(\sum_{i=1}^{n} d_iv_i) = \sum_{i=1}^{n} d_iw_i$. In §17.11, we used the UMP to show that for any (possibly non-free) finitely generated R-module N generated by n elements, there is a surjective R-linear map $U : R^n \to N$, and hence an R-module isomorphism $N \cong R^n / \ker(U)$. In other words, *every finitely generated R-module is isomorphic to a quotient module of a f.g. free R-module by some submodule.* This explains why the study of f.g. free R-modules helps us understand the structure of all finitely generated R-modules. Submodules of f.g. free modules also play a role, as seen in the next theorem.

Theorem on Submodules of a Free Module over a PID. Let R be a PID. Any submodule P of any f.g. free R-module M is also f.g. free. If M has a k-element basis, then P has a d-element basis for some $d \le k$.

Proof. We imitate the proof in §16.8 for the special case $R = \mathbb{Z}$. Let M be a f.g. free module over a PID R. We know $M \cong R^k$ for some integer $k \ge 0$, so we can assume $M = R^k$ without loss of generality. The proof uses induction on k. If $k = 0$, then P and M must be the zero module, which is f.g. free with an empty basis. Suppose $k = 1$; the assumption that P is a submodule of R^1 means that P is an ideal of the ring R. Because R is a PID, there exists $a \in R$ with $P = Ra$. If $a = 0$, then $P = \{0\}$ is f.g. free with a basis of size zero. Otherwise, $a \ne 0$, and $\mathcal{B} = (a)$ is a generating list for P of size 1. Is this list R-linearly independent? Given $c \in R$ with $ca = 0$, we see that $c = 0$ since $a \ne 0$ and R is an integral domain. So $\mathcal{B}$ is a basis for P of size 1, and P is f.g. free. Note how the conditions in the definition of a PID were exactly what we needed to make this base case work.

Proceeding to the induction step, fix $k > 1$ and assume the theorem is known for all f.g. free R-modules having bases of size less than k. Let P be a fixed submodule of R^k. Define $Q = P \cap (R^{k-1} \times \{0\})$, which is a submodule of the free R-module $R^{k-1} \times \{0\} \cong R^{k-1}$. By induction, Q is f.g. free with some ordered basis $(v_1, \ldots, v_{d-1})$ where $d - 1 \le k - 1$. Consider the projection map $T : R^k \to R$ given by $T((r_1, \ldots, r_k)) = r_k$ for $r_i \in R$. T is R-linear, so $T[P] = \{T(x) : x \in P\}$ is an R-submodule of R. Since R is a PID, $T[P] = Ry$ for some $y \in R$. Fix an element $v_d \in P$ with $T(v_d) = y$, so v_d is a k-tuple of elements of R with last coordinate y.

If $y = 0$, then $P = Q$ is f.g. free with a basis of size $d - 1 < k$. Assuming $y \ne 0$, we now show that $\mathcal{B} = (v_1, \ldots, v_d)$ is an ordered R-basis of P, so that P is f.g. free with a basis of size $d \le k$. First, we show $\mathcal{B}$ is R-linearly independent. Assume $c_1, \ldots, c_d \in R$ satisfy $c_1 v_1 + \cdots + c_d v_d = 0$. Applying the R-linear map T and noting that $T(v_i) = 0$ for each $i < d$ (since $v_1, \ldots, v_{d-1} \in Q$), we get $0 = c_1 T(v_1) + \cdots + c_d T(v_d) = c_d y$. As $y \ne 0$ and R is an integral domain, $c_d = 0$ follows. Now, since $c_1 v_1 + \cdots + c_{d-1} v_{d-1} = 0$, the known R-linear independence of $v_1, \ldots, v_{d-1}$ gives $c_1 = \cdots = c_{d-1} = 0$. So $\mathcal{B}$ is R-linearly independent.

Second, we check that $\mathcal{B}$ spans the R-module P. Fix $z = (z_1, \ldots, z_k) \in P$. Since $z_k = T(z) \in T[P] = Ry$, we have $z_k = ry$ for some $r \in R$. Then $z - r v_d$ is in the R-submodule P and has last coordinate $z_k - ry = 0$, so $z - r v_d \in Q$. Therefore $z - r v_d = e_1 v_1 + \cdots + e_{d-1} v_{d-1}$ for some $e_i \in R$, and we see that z itself is an R-linear combination of $v_1, \ldots, v_d$. This completes the induction step. $\qquad \square$

18.7   Operations on Bases

Assume R is a PID and M is a f.g. free R-module with ordered basis $\mathcal{X} = (v_1, \ldots, v_n)$. We can perform various transformations on $\mathcal{X}$ that produce new ordered bases for M. For example, by analogy with §16.4, there are three *elementary operations* we could apply to $\mathcal{X}$. Operation (B1) interchanges v_i and v_j for some $i \ne j$; operation (B2) replaces v_i by uv_i for some i and some unit $u \in R^*$; and operation (B3) replaces v_i by $v_i + bv_j$ for some $i \ne j$ and some $b \in R$. You can check that applying any finite sequence of such operations to $\mathcal{X}$ produces a new ordered basis of M, and each elementary operation is reversible.

We require an even more general operation on bases that includes (B1), (B2), and (B3) as special cases. Suppose we are given a, b, c, d in a PID R such that $u = ad - bc$ is a unit of R. Operation (B4) acts on $\mathcal{X}$ by replacing v_i with $v'_i = av_i + bv_j$ and replacing v_j with $v'_j = cv_i + dv_j$ for some $i \neq j$ in $\{1, 2, \ldots, n\}$. We can also write this as

$$\begin{bmatrix} v'_i \\ v'_j \end{bmatrix} = \begin{bmatrix} a & b \\ c & d \end{bmatrix} \begin{bmatrix} v_i \\ v_j \end{bmatrix}. \tag{18.1}$$

We claim the new list $\mathcal{X}' = (v_1, \ldots, v'_i, \ldots, v'_j, \ldots, v_n)$ is another ordered basis for M, and $\mathcal{X}$ can be recovered from $\mathcal{X}'$ by another operation of type (B4).

By inverting the 2×2 matrix in (18.1), we get

$$\begin{bmatrix} v_i \\ v_j \end{bmatrix} = \begin{bmatrix} u^{-1}d & -u^{-1}b \\ -u^{-1}c & u^{-1}a \end{bmatrix} \begin{bmatrix} v'_i \\ v'_j \end{bmatrix}. \tag{18.2}$$

where $(u^{-1}d)(u^{-1}a) - (-u^{-1}b)(-u^{-1}c) = u^{-2}(da - bc) = u^{-1}$ is a unit in R since $u \in R^*$. This shows that we can go from $\mathcal{X}'$ back to $\mathcal{X}$ by an operation of type (B4).

Let us check that $\mathcal{X}'$ spans M and is R-linearly independent. Given $w \in M$, write $w = \sum_{k=1}^{n} d_k v_k$ for scalars $d_k \in R$. Using (18.2), we can replace $d_i v_i$ and $d_j v_j$ in this expression by R-linear combinations of v'_i and v'_j. So w is in the span of $\mathcal{X}'$. Next, assume $0 = e_i v'_i + e_j v'_j + \sum_{k \neq i,j} e_k v_k$ for scalars $e_k \in R$. Using (18.1), this equation becomes

$$\begin{aligned}
0 &= e_i(av_i + bv_j) + e_j(cv_i + dv_j) + \sum_{k \neq i,j} e_k v_k \\
&= (e_i a + e_j c)v_i + (e_i b + e_j d)v_j + \sum_{k \neq i,j} e_k v_k.
\end{aligned}$$

By the known linear independence of $\mathcal{X}$, it follows that $e_k = 0$ for all $k \neq i, j$ and $e_i a + e_j c = e_i b + e_j d = 0$. In matrix notation, $[e_i \ e_j] \begin{bmatrix} a & b \\ c & d \end{bmatrix} = [0 \ 0]$. Right-multiplying by the inverse matrix gives $e_i = e_j = 0$. So $\mathcal{X}'$ is linearly independent.

Note that the elementary operations (B1), (B2), and (B3) really are special cases of operation (B4): for (B1), take $a = d = 0$ and $b = c = 1$; for (B2), take $a = u$, $d = 1$ (for any j), and $b = c = 0$; for (B3), take $a = d = 1$, $c = 0$ and any b. You can check that (B2) can still be applied to replace v_1 by uv_1 (for some $u \in R^*$) in the case $n = 1$.

18.8 Matrices of Linear Maps between Free Modules

Our next step is to study the matrices that represent linear maps between f.g. free modules. Assume R is a PID, M is a f.g. free R-module with ordered basis $\mathcal{X} = (v_1, \ldots, v_n)$, N is a f.g. free R-module with ordered basis $\mathcal{Y} = (w_1, \ldots, w_m)$, and $T : M \to N$ is a fixed R-linear map. For $1 \leq j \leq n$, we can write $T(v_j) = \sum_{i=1}^{m} A(i, j)w_i$ for unique scalars $A(i, j) \in R$ (these scalars are the coordinates of $T(v_j)$ relative to the basis $\mathcal{Y}$). The $m \times n$ matrix $A = [A(i, j)]$ is called the *matrix of T relative to the bases $\mathcal{X}$ and $\mathcal{Y}$*. By linearity of T, T is uniquely determined by the matrix A when $\mathcal{X}$ and $\mathcal{Y}$ are fixed and known. As in previously studied cases (when $R = \mathbb{Z}$ or R is a field), you can show that matrix addition corresponds to pointwise addition of linear maps, and matrix multiplication corresponds to composition of linear maps.

Changing the input basis $\mathcal{X}$ or the output basis $\mathcal{Y}$ changes the matrix A that represents the given linear map T. Let us investigate how an application of basis operation (B4) to $\mathcal{X}$ or to $\mathcal{Y}$ affects the matrix A. Suppose $a, b, c, d \in R$ satisfy $u = ad - bc \in R^*$. Define $a' = u^{-1}d$, $b' = -u^{-1}b$, $c' = -u^{-1}c$, and $d' = u^{-1}a$, which are the entries of the inverse matrix shown in (18.2). Recall that $A^{[k]}$ denotes column k of A, and $A_{[k]}$ denotes row k of A. We make two claims:

1. Suppose $\mathcal{X}'$ is obtained from $\mathcal{X}$ by replacing v_i and v_j by $v_i' = av_i + bv_j$ and $v_j' = cv_i + dv_j$. Then the matrix B of T relative to the bases $\mathcal{X}'$ and $\mathcal{Y}$ satisfies $B^{[i]} = aA^{[i]} + bA^{[j]}$, $B^{[j]} = cA^{[i]} + dA^{[j]}$, and $B^{[k]} = A^{[k]}$ for all $k \neq i, j$. We say B is obtained from A by a *type 4 column operation* on columns i and j.

2. Suppose $\mathcal{Y}'$ is obtained from $\mathcal{Y}$ by replacing w_i and w_j by $w_i' = aw_i + bw_j$ and $w_j' = cw_i + dw_j$. Then the matrix C of T relative to the bases $\mathcal{X}$ and $\mathcal{Y}'$ satisfies $C_{[i]} = A_{[i]}a' + A_{[j]}c'$, $C_{[j]} = A_{[i]}b' + A_{[j]}d'$, and $C_{[k]} = A_{[k]}$ for all $k \neq i, j$. We say C is obtained from A by a *type 4 row operation* on rows i and j.

We prove the second claim, leaving the first claim as Exercise 36. For fixed $p \in \{1, \ldots, n\}$, we must find the unique scalars $C(s, p) \in R$ satisfying

$$T(v_p) = C(i, p)w_i' + C(j, p)w_j' + \sum_{k \neq i, j} C(k, p)w_k.$$

We know

$$T(v_p) = A(i, p)w_i + A(j, p)w_j + \sum_{k \neq i, j} A(k, p)w_k.$$

Recalling from (18.2) that $w_i = a'w_i' + b'w_j'$ and $w_j = c'w_i' + d'w_j'$, we get

$$T(v_p) = (A(i, p)a' + A(j, p)c')w_i' + (A(i, p)b' + A(j, p)d')w_j' + \sum_{k \neq i, j} A(k, p)w_k.$$

Since $\mathcal{Y}'$ is R-independent, we can equate coefficients to conclude that $C(i, p) = A(i, p)a' + A(j, p)c'$, $C(j, p) = A(i, p)b' + A(j, p)d'$, and $C(k, p) = A(k, p)$ for all $k \neq i, j$. This holds for all p, so the rows of C are related to the rows of A as stated in the claim.

As special cases of the result for (B4), we can deduce how elementary column and row operations on the matrix A correspond to elementary operations of types (B1), (B2), and (B3) on the input basis $\mathcal{X}$ and the output basis $\mathcal{Y}$. Specifically, you can check that:

3. Switching columns i and j in A corresponds to switching v_i and v_j in the input basis $\mathcal{X}$; and switching rows i and j of A corresponds to switching w_i and w_j in the output basis $\mathcal{Y}$.

4. Multiplying column i of A by a unit $u \in R^*$ corresponds to replacing v_i by uv_i in the input basis $\mathcal{X}$; and multiplying row i of A by $u \in R^*$ corresponds to replacing w_i by $u^{-1}w_i$ in the output basis $\mathcal{Y}$.

5. Adding b times column j of A to column i of A (where $b \in R$) corresponds to replacing v_i by $v_i + bv_j$ in the input basis $\mathcal{X}$. Adding b times row j of A to row i of A corresponds to replacing w_j by $w_j - bw_i$ in the output basis $\mathcal{Y}$.

We also need the following properties, which you are asked to prove in Exercise 36:

6. Suppose $e \in R$ divides every entry of a matrix $A \in M_{m,n}(R)$. If we apply any sequence of type 4 row and column operations to A, then e still divides every entry of the new matrix.

7. Suppose B is obtained from A by applying a type 4 column operation as in Claim 1. Then $B = AV$, where $V \in M_n(R)$ is an invertible matrix with entries $V(i,i) = a$, $V(j,i) = b$, $V(i,j) = c$, $V(j,j) = d$, $V(k,k) = 1_R$ for all $k \neq i,j$, and all other entries of V are zero (cf. §4.7 and §4.9).

8. Suppose C is obtained from A by applying a type 4 row operation as in Claim 2. Then $C = UA$, where $U \in M_m(R)$ is an invertible matrix with entries $U(i,i) = a'$, $U(i,j) = c'$, $U(j,i) = b'$, $U(j,j) = d'$, $U(k,k) = 1_R$ for all $k \neq i,j$, and all other entries of U are zero (cf. §4.8 and §4.9).

18.9 Reduction Theorem for Matrices over a PID

We now have all the tools needed to prove the following structural result.

Matrix Reduction Theorem for PIDs. . Let A be an $m \times n$ matrix with entries in a PID R. There is a finite sequence of type 4 row and column operations on A that reduces A to a new matrix

$$B = \begin{bmatrix} a_1 & 0 & 0 & \cdots & 0 \\ 0 & a_2 & 0 & \cdots & 0 \\ 0 & 0 & a_3 & \cdots & 0 \\ \vdots & & & \ddots & \end{bmatrix}, \tag{18.3}$$

in which there are $s \geq 0$ nonzero elements $a_1, \ldots, a_s \in R$ on the main diagonal, a_i divides a_{i+1} in R for $1 \leq i < s$, and all other entries of B are zero.

For brevity, we write $B = \mathrm{diag}(a_1, \ldots, a_s)_{m \times n}$, omitting the $m \times n$ when $s = m = n$. In terms of ideals, the divisibility conditions on the a_j are equivalent to $Ra_1 \supseteq Ra_2 \supseteq Ra_3 \supseteq \cdots \supseteq Ra_s$. Later, we prove that s and the ideals satisfying this containment condition are uniquely determined by A, and hence the a_j are unique up to associates in R. So the matrix B is unique in this sense; it is called a *Smith normal form* of A. By repeated use of properties 7 and 8 from §18.8, we see that $B = PAQ$ for some invertible $P \in M_m(R)$ and $Q \in M_n(R)$, where P (resp. Q) is the product of all the matrices U (resp. V) used to accomplish the type 4 row (resp. column) operations needed to reduce A to B.

We proved the reduction theorem for $R = \mathbb{Z}$ in §16.11. If we try to repeat that proof in the setting of general PIDs, a problem emerges. The old proof made critical use of integer division with remainder, as well as the fact that there is no infinite strictly decreasing sequence of positive integers. These proof ingredients can be generalized to a class of rings called *Euclidean domains* (defined in Exercise 6), but they are not available in all PIDs. To execute the proof at this level of generality, a new trick is needed.

Recall that all PIDs are UFDs, so that every nonzero non-unit $a \in R$ has a factorization $a = p_1 p_2 \cdots p_s$ into irreducible elements p_i in R; and any other irreducible factorization $a = q_1 q_2 \cdots q_t$ has $s = t$ and $p_i \sim q_i$ after appropriate reordering. Define the *length of a* in R to be $\mathrm{len}(a) = s$, the number of factors appearing in any irreducible factorization of a. For any unit u of R, let $\mathrm{len}(u) = 0$; $\mathrm{len}(0_R)$ is undefined. For any nonzero matrix A with entries in R, let $\mathrm{len}(A)$ be the minimum length of all the nonzero entries of A. Given nonzero $a, b, d \in R$, you can check that: $\mathrm{len}(ab) = \mathrm{len}(a) + \mathrm{len}(b)$; if d divides a in R, then $\mathrm{len}(d) \leq \mathrm{len}(a)$ with equality iff $d \sim a$; and if $d = \gcd(a,b)$ where a does not divide b, then $\mathrm{len}(d) < \mathrm{len}(a)$.

We now begin the proof of the Matrix Reduction Theorem for a PID R. The theorem holds when $A = 0$ or $m = 0$ or $n = 0$, so we assume $m, n > 0$ and $A \neq 0$ throughout the rest of the proof. Using induction on m (the number of rows), we can assume the theorem is known to hold for all matrices with fewer than m rows.

Step 1: We show we can apply finitely many type 4 row and column operations to A to produce a matrix A_1 such that some nonzero entry in A_1 divides all entries of A_1. The proof uses induction on $\mathrm{len}(A)$. If $\mathrm{len}(A) = 0$, then some entry of A is a unit of R, which divides all elements of R and hence divides all entries of A. Next, assume $\mathrm{len}(A) = \ell > 0$ and the result of Step 1 is known to hold for all matrices of length less than ℓ. Let $e = A(i,j)$ be a nonzero entry of A with $\mathrm{len}(e) = \ell$. If e happens to divide all entries of A, then the conclusion of Step 1 already holds for the matrix A. Suppose instead that there exists at least one entry of A not divisible by $e = A(i,j)$.

Case 1: There is $k \neq j$ such that e does not divide $f = A(i,k)$. Since R is a PID, we know $g = \gcd(e,f)$ exists in R, and $g = ae + bf$ for some $a, b \in R$. We have $e = gd$ and $f = gc$ for some $c, d \in R$. Cancelling $g \neq 0$ in $g = a(gd) + b(gc)$ gives $1_R = ad + bc$. So, we can apply a type 4 column operation to A that replaces $A^{[j]}$ by $aA^{[j]} + bA^{[k]}$ and $A^{[k]}$ by $-cA^{[j]} + dA^{[k]}$. The new matrix A' has i,j-entry $ae + bf = g$, and $\mathrm{len}(g) < \mathrm{len}(e)$ since e does not divide f. So $\mathrm{len}(A') \leq \mathrm{len}(g) < \mathrm{len}(A)$. By induction, we can apply further reduction steps to A' to achieve the conclusion of Step 1.

Case 2: There is $k \neq i$ such that $e = A(i,j)$ does not divide $f = A(k,j)$. We argue as in Case 1, but this time we use a type 4 row operation to replace e by $g = \gcd(e,f)$, which lowers the length of the matrix.

Case 3: $e = A(i,j)$ divides everything in row i and column j, but for some $i_1 \neq i$ and $j_1 \neq j$, e does not divide $f = A(i_1, j_1)$. Pictorially, rows i, i_1 and columns j, j_1 look like this for some $u, v \in R$:

$$\begin{bmatrix} e & \cdots & ue \\ \vdots & & \vdots \\ ve & \cdots & f \end{bmatrix}.$$

Adding $(1 - v)$ times row i to row i_1 produces:

$$\begin{bmatrix} e & \cdots & ue \\ \vdots & & \vdots \\ e & \cdots & f + (1-v)ue \end{bmatrix}.$$

If this new matrix has lower length than A, we are done by induction. Otherwise, note e cannot divide $f + (1 - v)ue$ in R, lest e divide f. So Case 1 now applies to row i_1, and we can complete Step 1 as in that case.

Step 2: Let $a_1 = A_1(i,j)$ be a nonzero entry in A_1 dividing all entries of A_1. We show A_1 can be further reduced to the form

$$A_2 = \begin{bmatrix} a_1 & 0 & \cdots & 0 \\ 0 & & & \\ \vdots & & A' & \\ 0 & & & \end{bmatrix}, \tag{18.4}$$

where $A' \in M_{m-1,n-1}(R)$, and a_1 divides all entries of A' in R. To prove this, recall that applying type 4 row and column operations to A_1 never changes the property that a_1 divides every entry of the matrix. To begin, bring a_1 into the $1,1$-position by switching row 1 and row i and switching column 1 and column j. Since a_1 divides every entry in row 1, we can subtract appropriate multiples of column 1 from each later column to make the other entries in row 1 become zero. Similarly, we can use row operations to produce zeroes below a_1 in the first column. The current matrix now looks like (18.4). Since a_1 still divides all entries of the full matrix, a_1 divides every entry of A'.

Step 3: We show how A_2 can be reduced to the normal form (18.3). Since A' has $m - 1 < m$ rows, the induction hypothesis shows that we can apply type 4 row and column operations to this $(m - 1) \times (n - 1)$ matrix to obtain a matrix in normal form with entries $a_2, \ldots, a_s$ on the main diagonal, all other entries zero, and $a_i | a_{i+1}$ in R for $2 \le i < s$. We can apply the same type 4 operations to the full matrix A_2, and these operations do not disturb the zeroes we have already created in row 1 and column 1. Furthermore, a_1 continues to divide all entries of the matrix throughout the reduction of A_2. In particular, at the end, a_1 divides a_2, and we have reached the required normal form for A.

18.10   Structure Theorems for Linear Maps and Modules

The Matrix Reduction Theorem for PIDs translates into the following result on linear maps.

Structure Theorem for Linear Maps between Free Modules over PIDs. Let R be a PID, N and M be f.g. free R-modules, and $T : N \to M$ be an R-linear map. There exist an ordered basis $\mathcal{X} = (x_1, \ldots, x_n)$ for N and an ordered basis $\mathcal{Y} = (y_1, \ldots, y_m)$ for M such that the matrix of T relative to $\mathcal{X}$ and $\mathcal{Y}$ is in Smith normal form (18.3). So there exist $s \in \mathbb{Z}_{\geq 0}$ and nonzero $a_1, \ldots, a_s \in R$ with $a_i | a_{i+1}$ for $1 \le i < s$, $T(x_i) = a_i y_i$ for $1 \le i \le s$, and $T(x_i) = 0_M$ for $s < i \le n$.

Later, we prove that s and the ideals Ra_i (satisfying $Ra_1 \supseteq Ra_2 \supseteq \cdots \supseteq Ra_s \neq \{0\}$) are uniquely determined by T.

Proof. To prove existence, start with any ordered bases $\mathcal{X}_0$ for N and $\mathcal{Y}_0$ for M, and let A be the matrix of T relative to $\mathcal{X}_0$ and $\mathcal{Y}_0$. Use a finite sequence of type 4 operations to bring A into Smith normal form. To ensure that each new matrix still represents T, we perform the appropriate basis operation (B4) on the input basis (when we do a column operation on A) or on the output basis (when we do a row operation on A), using the rules explained in §18.8. At the end, we get new ordered bases $\mathcal{X}$ for N and $\mathcal{Y}$ for M satisfying the required properties.   □

Next, we prove the existence part of our main classification result.

Classification Theorem for Finitely Generated Modules over PIDs (Version 1). For any finitely generated module M over a PID R, there exist $d \in \mathbb{Z}_{\geq 0}$ and $b_1, \ldots, b_d \in R$ with $R \neq Rb_1 \supseteq Rb_2 \supseteq \cdots \supseteq Rb_d \supseteq \{0\}$ and

$$M \cong R/Rb_1 \times R/Rb_2 \times \cdots \times R/Rb_d. \tag{18.5}$$

Note that some b_j could be zero, in which case $R/Rb_j \cong R$. Each ideal Rb_j is called an *invariant factor* of M. The generators b_j of these ideals are also called invariant factors. In §18.14, we prove that d and the sequence of ideals $(Rb_1, Rb_2, \ldots, Rb_d)$ satisfying the stated conditions are uniquely determined by M.

Proof. To prove the existence part of the theorem, suppose M is an R-module generated by m elements. Recall (§18.6) that $M \cong R^m / P$ for some submodule P of R^m, where P is a free R-module with a basis of size $n \le m$. The inclusion map $T : P \to R^m$, given by $T(x) = x$ for all $x \in P$, is an R-linear map between free R-modules. So there exist an ordered basis $\mathcal{X} = (x_1, \ldots, x_n)$ of P, an ordered basis $\mathcal{Y} = (y_1, \ldots, y_m)$ for R^m, $s \in \mathbb{Z}_{\geq 0}$, and nonzero

$a_1, \ldots, a_s \in R$ with $Ra_1 \supseteq \cdots \supseteq Ra_s \neq \{0\}$, $T(x_i) = a_i y_i$ for $1 \leq i \leq s$, and $T(x_i) = 0$ for $s < i \leq n$. Now $T(x_i) = x_i \neq 0$ for all i (since $\mathcal{X}$ is a basis), so we must have $s = n$ and $x_i = a_i y_i$ for $1 \leq i \leq n$. Define $a_i = 0$ for $n < i \leq m$, so $Ra_1 \supseteq \cdots \supseteq Ra_m \supseteq \{0\}$. As in §16.13, the UMP for free R-modules provides an isomorphism $R^m \cong R^m$ sending y_i to $\mathbf{e}_i$ (the standard basis vector) for $1 \leq i \leq m$, and this isomorphism sends P to $P_1 = Ra_1 \times Ra_2 \times \cdots \times Ra_m$. So

$$M \cong R^m/P \cong R^m/P_1 \cong (R/Ra_1) \times (R/Ra_2) \cdots \times (R/Ra_m),$$

where the last step uses the Fundamental Homomorphism Theorem for Modules (see Exercise 58 of Chapter 17). To finish, we delete any initial factors R/Ra_i that are equal to zero. Such a factor occurs iff $Ra_i = R$ iff a_i is a unit of R.   □

Continuing to imitate §16.13, we now derive a prime power version of the structure theorem for modules.

Lemma on Splitting into Prime Powers. Suppose R is a PID and $a \in R$ has irreducible factorization $a = p_1^{e_1} \cdots p_k^{e_k}$ where $p_1, \ldots, p_k$ are non-associate irreducible elements of R and $e_1, \ldots, e_k \in \mathbb{Z}_{>0}$. There is an R-module isomorphism

$$R/Ra \cong (R/Rp_1^{e_1}) \times \cdots \times (R/Rp_k^{e_k}).$$

Proof. Define $T : R \to \prod_{i=1}^k (R/Rp_i^{e_i})$ by $T(x) = (x + Rp_1^{e_1}, \ldots, x + Rp_k^{e_k})$ for $x \in R$. The map T is R-linear, and $x \in \ker(T)$ iff $x + Rp_i^{e_i} = 0$ for all i iff $p_i^{e_i}|x$ for all i iff $a = \mathrm{lcm}(p_1^{e_1}, \ldots, p_k^{e_k})|x$ iff $x \in Ra$. So T induces an isomorphism $T' : R/Ra \to \mathrm{img}(T)$. It suffices to show T is onto, which is accomplished by showing that each generator $(0, \ldots, 1 + Rp_i^{e_i}, \ldots, 0)$ of $\prod_{i=1}^k (R/Rp_i^{e_i})$ is in the image of T. Note $r = p_i^{e_i}$ and $s = \prod_{j \neq i} p_j^{e_j}$ have gcd 1_R (by comparing unique prime factorizations), so there exist $b, c \in R$ with $br + cs = 1_R$. Consider $T(cs)$. For any $k \neq i$, the coset $cs + Rp_k^{e_k}$ is zero since $p_k^{e_k}$ divides s. On the other hand, $cs + Rp_i^{e_i} = (1 - br) + Rp_i^{e_i} = 1 + Rp_i^{e_i}$ since $-br + Rp_i^{e_i}$ is the zero coset. Thus, $T(cs) = (0, \ldots, 1 + Rp_i^{e_i}, \ldots, 0)$ as needed.   □

Applying this lemma to each nonzero b_j in (18.5), we obtain the following structural result.

Classification Theorem for Finitely Generated Modules over PIDs (Version 2). For any finitely generated module M over a PID R, there exist $k \in \mathbb{Z}_{\geq 0}$ and $q_1, \ldots, q_k \in R$ such that each q_i is either zero or $p_i^{e_i}$ for some irreducible $p_i \in R$ and $e_i \in \mathbb{Z}_{>0}$, and

$$M \cong R/Rq_1 \times R/Rq_2 \times \cdots \times R/Rq_k. \tag{18.6}$$

The ideals Rq_j (as well as the generators q_j of these ideals) are called the *elementary divisors* of the R-module M. In §18.15, we prove that these ideals (counted with multiplicity) are uniquely determined by the module M.

18.11   Minors and Matrix Invariants

Our next goal is to prove the uniqueness of the Smith normal form of a matrix $A \in M_{m,n}(R)$ or a linear map between f.g. free R-modules. Before doing so, we need some preliminary results on *matrix invariants*, which are quantities depending on A that do not change when we multiply A on the left or right by an invertible matrix with entries in R.

Let A be an $m \times n$ matrix with entries in a PID R. We consider submatrices of A formed by keeping only certain rows and columns of A. More precisely, given a subset $I = \{i_1 < i_2 < \cdots < i_k\}$ of $[m] = \{1, 2, \ldots, m\}$ and a subset $J = \{j_1 < j_2 < \cdots < j_\ell\}$ of $[n] = \{1, 2, \ldots, n\}$, the *submatrix of A with rows in I and columns in J* is the matrix $A_{I,J}$ with r, s-entry $A_{I,J}(r, s) = A(i_r, j_s)$ for $1 \leq r \leq k$ and $1 \leq s \leq \ell$. For example, $A_{[m],\{2,3,5\}}$ is the submatrix obtained by keeping all m rows of A and columns 2, 3, and 5 of A. Using this notation, the Cauchy–Binet Formula (proved in §5.14) can be stated as follows: Given $U \in M_{k,m}(R)$ and $V \in M_{m,k}(R)$ with $k \leq m$,

$$\det(UV) = \sum_{L \subseteq [m], |L| = k} \det(U_{[k],L}) \det(V_{L,[k]}).$$

Fix k with $1 \leq k \leq \min(m, n)$. A $k \times k$ *submatrix of A* is a matrix $A_{I,J}$ with $|I| = |J| = k$. For each choice of I and J of size k, let $d_{I,J} = \det(A_{I,J}) \in R$; $d_{I,J}$ is called the *order k minor of A indexed by I and J*. Let $g_k(A)$ be a gcd in R of all the determinants $d_{I,J}$ with $|I| = |J| = k$. We claim that for all invertible $P \in M_m(R)$ and all invertible $Q \in M_n(R)$, $g_k(A) \sim g_k(PAQ)$ in R. In terms of ideals, this says that $Rg_k(A) = Rg_k(PAQ)$, so that the *ideal generated by any gcd of all the order k minors of A is a matrix invariant of A.*

First, we use the Cauchy–Binet Formula to prove that $g_k(A) \sim g_k(PA)$. Fix $I = \{i_1 < \cdots < i_k\} \subseteq [m]$ and $J = \{j_1 < \cdots < j_k\} \subseteq [n]$ of size k. Note that $(PA)_{I,J} = P_{I,[m]} A_{[m],J}$, since the r, s-entry of both sides is $(PA)(i_r, j_s) = \sum_{t=1}^{m} P(i_r, t) A(t, j_s)$. Since $k \leq m$, we can apply the Cauchy–Binet Formula to the $k \times m$ matrix $U = P_{I,[m]}$ and the $m \times k$ matrix $V = A_{[m],J}$. We obtain

$$\det((PA)_{I,J}) = \det(UV) = \sum_{\substack{L \subseteq [m], \\ |L| = k}} \det(U_{[k],L}) \det(V_{L,[k]}) = \sum_{\substack{L \subseteq [m], \\ |L| = k}} \det(P_{I,L}) \det(A_{L,J}).$$

This formula shows that the minor $\det((PA)_{I,J})$ of the matrix PA is an R-linear combination of various order k minors $d_{L,J}$ of the matrix A. Therefore, if $e \in R$ is a common divisor of all the order k minors of A, then e divides each order k minor of PA. In particular, $e = g_k(A)$ divides all minors $\det(A_{I,J})$, so $g_k(A)$ is a common divisor of all minors $\det((PA)_{I,J})$, so $g_k(A)$ divides $g_k(PA) = \gcd\{\det((PA)_{I,J}) : |I| = |J| = k\}$. We have now proved $g_k(A)$ divides $g_k(PA)$ for all $A \in M_{m,n}(R)$ and all invertible $P \in M_m(R)$. Applying this result with A replaced by PA and P replaced by P^{-1}, we see that $g_k(PA)$ divides $g_k(P^{-1}(PA)) = g_k(A)$. Hence, $g_k(A) \sim g_k(PA)$ as needed.

By a similar argument (Exercise 59), we can use the Cauchy–Binet Formula to show $g_k(A) \sim g_k(AQ)$ for any $A \in M_{m,n}(R)$ and any invertible $Q \in M_n(R)$. So $g_k(A) \sim g_k(PAQ)$ when P and Q are invertible over R. (You can also prove $g_k(PA) \sim g_k(A) \sim g_k(AQ)$ without appealing to the Cauchy–Binet Formula, but instead invoking multilinearity properties of determinants — see Exercise 60.)

18.12 Uniqueness of Smith Normal Form

Let R be a PID and A be an $m \times n$ matrix with entries in R. In §18.9, we proved that there exist invertible matrices $P \in M_m(R)$ and $Q \in M_n(R)$ such that $B = PAQ$ has the form

$$B = \operatorname{diag}(a_1, a_2, \ldots, a_s)_{m \times n},$$

where $s \geq 0$, $a_1, \ldots, a_s \in R$, and $Ra_1 \supseteq Ra_2 \supseteq \cdots \supseteq Ra_s \neq \{0\}$.

Our goal here is to prove the uniqueness of s and the ideals $Ra_1, \ldots, Ra_s$ satisfying the properties just stated. More specifically, we show that for any invertible $P' \in M_m(R)$ and $Q' \in M_n(R)$ such that

$$B' = P'AQ' = \mathrm{diag}(b_1, b_2, \ldots, b_t)_{m \times n}$$

for some $t \geq 0$ and $b_1, \ldots, b_t \in R$ with $Rb_1 \supseteq Rb_2 \supseteq \cdots \supseteq Rb_t \neq \{0\}$, we must have $s = t$ and $Ra_i = Rb_i$ for $1 \leq i \leq s$ (equivalently, a_i and b_i are associates in R for all i). The ideals Ra_i (and their generators a_i) are called the *invariant factors* of A, and s is called the *rank* of A.

Define $a_i = 0$ for $s < i \leq \min(m, n)$ and $b_j = 0$ for $t < j \leq \min(m, n)$. Fix k with $1 \leq k \leq \min(m, n)$. On one hand, we have seen that

$$g_k(B) = g_k(PAQ) \sim g_k(A) \sim g_k(P'AQ') = g_k(B'). \tag{18.7}$$

On the other hand, we can use the special form of B to compute $g_k(B)$ directly from the definition. If $k > s$, every $k \times k$ submatrix $B_{I,J}$ must have a row and column of zeroes, so every order k minor $\det(B_{I,J})$ is zero. Then $g_k(B) = 0$ since this is the gcd of a list of zeroes. Now suppose $k \leq s$. You can check that every order k minor $\det(B_{I,J})$ is either zero or is some product of the form $a_{i_1} a_{i_2} \cdots a_{i_k} \neq 0$, where $1 \leq i_1 < i_2 < \cdots < i_k \leq s$. Furthermore, one of these minors is $\det(B_{[k],[k]}) = a_1 a_2 \cdots a_k \neq 0$. Since $a_i | a_j$ for all $1 \leq i \leq j \leq s$, we see that $a_1 a_2 \cdots a_k$ divides all the order k minors of B. So this ring element is a gcd of all of these minors, and we can therefore take $g_k(B) = a_1 a_2 \cdots a_k \neq 0$ for $1 \leq k \leq s$. Letting $g_0(B) = 1_R$, we see that $a_k = g_k(B)/g_{k-1}(B)$ for $1 \leq k \leq s$. (More precisely, a_k is the unique x in the integral domain R solving $g_{k-1}(B)x = g_k(B)$.) Replacing $g_k(B)$ or $g_{k-1}(B)$ by associate ring elements replaces a_k by an associate of a_k in R.

Applying the same reasoning to B', we see that $g_k(B') = 0$ for all $k > t$, $g_k(B') = b_1 b_2 \cdots b_k \neq 0$ for $0 \leq k \leq t$, and $b_k = g_k(B')/g_{k-1}(B')$ for $1 \leq k \leq t$. Returning to (18.7), we now see that $s = t = \max\{k : g_k(A) \neq 0_R\}$ and $a_k \sim g_k(A)/g_{k-1}(A) \sim b_k$ for $1 \leq k \leq s$. This completes the uniqueness proof.

We can deduce a similar uniqueness result for R-linear maps between f.g. free R-modules. *Given a PID R, f.g. free R-modules N and M, and an R-linear map $T : N \to M$, there exist **unique** $s \geq 0$ and ideals $Ra_1 \supseteq Ra_2 \supseteq \cdots \supseteq Ra_s \neq \{0\}$ such that for some ordered bases $\mathcal{X} = (x_1, \ldots, x_n)$ for N and $\mathcal{Y} = (y_1, \ldots, y_m)$ for M, $T(x_i) = a_i y_i$ for $1 \leq i \leq s$ and $T(x_i) = 0_M$ for $s < i \leq n$.* Existence of one choice of $s, a_1, \ldots, a_s, \mathcal{X}, \mathcal{Y}$ was shown in §18.10. To prove uniqueness, suppose $s', Ra_1', \ldots, Ra_{s'}', \mathcal{X}', \mathcal{Y}'$ also satisfies all of the conclusions above. Let A be the matrix of T relative to the bases $\mathcal{X}$ and $\mathcal{Y}$. Then A is in Smith normal form (18.3). Similarly, the matrix A' of T relative to $\mathcal{X}'$ and $\mathcal{Y}'$ is in Smith normal form with the elements a_j' on its diagonal. Considering transition matrices between the bases $\mathcal{X}$ and $\mathcal{X}'$ and the bases $\mathcal{Y}$ and $\mathcal{Y}'$, you can check that $A' = PAQ$ for some invertible $P \in M_m(R)$ and $Q \in M_n(R)$ (Exercise 40). Applying the uniqueness result for matrices proved above, we obtain $s = s'$ and $a_i \sim a_i'$ (hence $Ra_i = Ra_i'$) for $1 \leq i \leq s$.

18.13   Torsion Submodules

Our next task is to prove the uniqueness of the ideals Rb_i and Rq_j appearing in the decompositions (18.5) and (18.6). We begin in this section by showing how to split off the free part of a finitely generated module over a PID.

Given an integral domain R and any R-module M, the *torsion submodule* of M is

$$\text{tor}(M) = \{x \in M : \text{for some } r \in R, r \neq 0 \text{ and } r \cdot x = 0\}.$$

To see that $\text{tor}(M)$ really is a submodule, first note $1_R \neq 0_R$ and $1_R \cdot 0_M = 0_M$, so $0_M \in \text{tor}(M)$. Next, fix $x, y \in \text{tor}(M)$ and $t \in R$. Choose nonzero $r, s \in R$ with $rx = 0 = sy$. Then $rs \neq 0$ since R is an integral domain, and $(rs) \cdot (x + y) = (rs) \cdot x + (rs) \cdot y = s \cdot (r \cdot x) + r \cdot (s \cdot y) = s0 + r0 = 0$, so $x + y \in \text{tor}(M)$. Also $r \cdot (t \cdot x) = t \cdot (r \cdot x) = t \cdot 0 = 0$ since R is commutative, so $t \cdot x \in \text{tor}(M)$.

Now suppose M and N are R-modules and $f : M \to N$ is an R-module isomorphism. You can check that $f[\text{tor}(M)] = \text{tor}(N)$, so that f restricts to an isomorphism from $\text{tor}(M)$ to $\text{tor}(N)$. It follows from this and the Fundamental Homomorphism Theorem that f induces a module isomorphism $f' : M/\text{tor}(M) \to N/\text{tor}(N)$ given by $f'(x + \text{tor}(M)) = f(x) + \text{tor}(N)$ for $x \in M$. To summarize: if $M \cong N$, then $\text{tor}(M) \cong \text{tor}(N)$ and $M/\text{tor}(M) \cong N/\text{tor}(N)$.

For example, consider an R-module

$$P = R/Ra_1 \times R/Ra_2 \times \cdots \times R/Ra_k \times R^d,$$

where $a_1, \ldots, a_k$ are nonzero elements of R and $d \geq 0$. We claim

$$\text{tor}(P) = R/Ra_1 \times \cdots \times R/Ra_k \times \{0_{R^d}\}.$$

A typical element of the right side is $z = (x_1 + Ra_1, \ldots, x_k + Ra_k, 0)$ where each $x_i \in R$. Let $r = a_1 a_2 \cdots a_k \neq 0$; note rx_i is divisible by a_i, so $r(x_i + Ra_i) = rx_i + Ra_i = 0 + Ra_i$ for $1 \leq i \leq k$, so that $rz = (0, \ldots, 0, 0)$ and hence $z \in \text{tor}(P)$. On the other hand, consider $y = (x_1 + Ra_1, \ldots, x_k + Ra_k, (r_1, \ldots, r_d)) \in P$ with some $r_j \neq 0_R$. Multiplying y by any nonzero $s \in R$ produces $sy = (sx_1 + Ra_1, \ldots, sx_k + Ra_k, (sr_1, \ldots, sr_d))$, where $sr_j \neq 0_R$. So sy cannot be 0, hence $y \notin \text{tor}(P)$. This proves the claim. From the claim, we readily deduce that $P/\text{tor}(P) \cong R^d$.

The preceding remarks imply the following result that lets us break apart the free piece and the torsion piece of a finitely generated module.

Splitting Theorem for Modules over a PID. Suppose R is a PID and M is a finitely generated R-module such that

$$R/Ra_1 \times \cdots \times R/Ra_k \times R^d \cong M \cong R/Rb_1 \times \cdots \times R/Rb_\ell \times R^e,$$

where every Ra_i and Rb_j is a nonzero ideal of R and $d, e \geq 0$. Then

$$R/Ra_1 \times \cdots \times R/Ra_k \cong \text{tor}(M) \cong R/Rb_1 \times \cdots \times R/Rb_\ell \text{ and } R^d \cong M/\text{tor}(M) \cong R^e,$$

and hence $d = e$.

To see why $d = e$ follows, note $M/\text{tor}(M)$ is a free R-module having a basis of size d (since this module is isomorphic to R^d) and a basis of size e (since this module is isomorphic to R^e). Since the PID R is commutative, $d = e$ follows from the theorem proved in §17.16. We call d the *Betti number* of M.

18.14   Uniqueness of Invariant Factors

We are now ready to prove the uniqueness of the sequence of ideals (the invariant factors) appearing in the decomposition (18.5). Using the Splitting Theorem to separate out all factors of the form $R/R0 \cong R$, it suffices to prove the following statement.

Theorem on Uniqueness of Invariant Factors. Let R be a PID. Assume $a_1, \ldots, a_k$, $b_1, \ldots, b_\ell \in R$ satisfy

$$R \neq Ra_k \supseteq \cdots \supseteq Ra_1 \neq \{0\}, \qquad R \neq Rb_\ell \supseteq \cdots \supseteq Rb_1 \neq \{0\},$$

$$\text{and } R/Ra_k \times \cdots \times R/Ra_1 \cong R/Rb_\ell \times \cdots \times R/Rb_1. \tag{18.8}$$

Then $k = \ell$ and $Ra_i = Rb_i$ for $1 \leq i \leq k$.

We have reversed the indexing order of the a_i and b_j to simplify notation in the induction proof below.

Proof. All quotient modules appearing in the theorem statement are nonzero (as $Ra_i \neq R \neq Rb_j$), so $k = 0$ iff $\ell = 0$. Assume $k, \ell > 0$. We first prove that $Ra_1 = Rb_1$ using the following ideas. Given any R-module M, the *annihilator of M* is

$$\mathrm{ann}_R(M) = \{r \in R : \text{for all } x \in M, rx = 0\}.$$

You can check that for any commutative ring R, $\mathrm{ann}_R(M)$ is an ideal of R. Also, $M \cong M'$ implies $\mathrm{ann}_R(M) = \mathrm{ann}_R(M')$, so that isomorphic R-modules have equal annihilators.

Given $M = R/Ra_k \times \cdots \times R/Ra_1$ as above, we show that $\mathrm{ann}_R(M) = Ra_1$. A typical element of M is a k-tuple of cosets $x = (x_k + Ra_k, \ldots, x_1 + Ra_1)$ with all $x_i \in R$. Multiplying x by $ra_1 \in Ra_1$ (where $r \in R$) produces $(ra_1)x = (ra_1 x_k + Ra_k, \ldots, ra_1 x_1 + Ra_1)$. Every $ra_1 x_i$ is in Ra_1, which is contained in all the other ideals Ra_i by assumption. So $ra_1 x_i + Ra_i = 0 + Ra_i$ for all i between 1 and k, proving that $(ra_1)x = 0$. This means that $Ra_1 \subseteq \mathrm{ann}_R(M)$. To prove the reverse inclusion, fix $s \in \mathrm{ann}_R(M)$. Then $s \cdot (0, \ldots, 0, 1 + Ra_1) = 0_M$, so that $s + Ra_1 = 0 + Ra_1$, so that $s \in Ra_1$. The same reasoning shows that $\mathrm{ann}_R(R/Rb_\ell \times \cdots \times R/Rb_1)$ is Rb_1. By the result in the last paragraph, $Ra_1 = Rb_1$ follows.

Fix i with $1 \leq i - 1 \leq \min(k, \ell)$, and make the induction hypothesis that $Ra_1 = Rb_1$, $Ra_2 = Rb_2, \ldots, Ra_{i-1} = Rb_{i-1}$. We now prove that $k \geq i$ iff $\ell \geq i$, in which case $Ra_i = Rb_i$. Assume $k \geq i$; we show $\ell \geq i$ and $b_i | a_i$ in R. The proof requires facts about the length of a module proved in §17.18. Let $P = R/Ra_{i-1} \times \cdots \times R/Ra_1$, which appears in both of the product modules (18.8) by the induction hypothesis. You can check that $\mathrm{len}(R/Ra) = \mathrm{len}(a)$ for any nonzero a in a PID R (Exercise 63), so that $\mathrm{len}(P) = \mathrm{len}(a_1) + \cdots + \mathrm{len}(a_{i-1}) < \infty$. If we had $\ell = i - 1$, then (18.8) says

$$[R/Ra_k \times \cdots \times R/Ra_i] \times P \cong P,$$

where the term in brackets is a nonzero module Q. On one hand, the isomorphic modules $Q \times P$ and P have the same finite length. On the other hand, $\mathrm{len}(Q \times P) > \mathrm{len}(P)$ since $Q \neq \{0\}$. This contradiction shows $\ell \geq i$.

Note that for any R-module N and any $c \in R$ (where R is a commutative ring), $cN = \{c \cdot n : n \in N\}$ is a submodule of N; and if $N \cong N'$ are isomorphic R-modules, then $cN \cong cN'$. Furthermore, for a direct product $N = N_1 \times N_2 \times \cdots \times N_k$, we have $cN = (cN_1) \times (cN_2) \times \cdots \times (cN_k)$. Taking $c = a_i$ and applying these remarks to (18.8), we get an isomorphism

$$[a_i(R/Ra_k) \times \cdots \times a_i(R/Ra_i)] \times a_i P \cong [a_i(R/Rb_\ell) \times \cdots \times a_i(R/Rb_i)] \times a_i P. \tag{18.9}$$

We know a_i belongs to all the ideals $Ra_i \subseteq Ra_{i+1} \subseteq \cdots \subseteq Ra_k$. It follows that the product in brackets on the left side of (18.9) is the zero module. Comparing lengths of both sides (noting that $\mathrm{len}(a_i P) \leq \mathrm{len}(P) < \infty$), we conclude that the product in brackets on the right side must also be the zero module. In particular, $a_i(R/Rb_i) = \{0\}$, so $a_i \cdot (1+Rb_i) = 0+Rb_i$. This means $a_i \in Rb_i$, so $b_i|a_i$ in R.

By interchanging the roles of the two product modules in (18.8), we prove similarly that $\ell \geq i$ implies $k \geq i$ and $a_i|b_i$ in R. So $k \geq i$ iff $\ell \geq i$, in which case $b_i|a_i$ and $a_i|b_i$, hence $a_i \sim b_i$, hence $Ra_i = Rb_i$. This completes the induction step. Taking $i = \min(k, \ell) + 1$, we see that $k = \ell$ and $Ra_j = Rb_j$ for $1 \leq j \leq k$. $\qquad\square$

18.15 Uniqueness of Elementary Divisors

Next, we prove the uniqueness (up to reordering) of the elementary divisors appearing in the decomposition (18.6). By invoking the Splitting Theorem from §18.13 to remove all factors of the form $R/R0 \cong R$, it suffices to prove the following statement.

Theorem on Uniqueness of Elementary Divisors. Let R be a PID. Assume $q_1, \ldots, q_k$, $r_1, \ldots, r_\ell$ are positive powers of irreducible elements in R such that

$$R/Rq_1 \times \cdots \times R/Rq_k \cong R/Rr_1 \times \cdots \times R/Rr_\ell. \tag{18.10}$$

Then $k = \ell$ and the list of ideals $(Rq_1, \ldots, Rq_k)$ is a rearrangement of the list $(Rr_1, \ldots, Rr_\ell)$.

To simplify the proof, let $M = [Rq_1, \ldots, Rq_k]$ be the set of all rearrangements of the list $(Rq_1, \ldots, Rq_k)$; we call M a *multiset* of ideals. This word indicates that the order in which we list the ideals is unimportant, but the number of times each ideal occurs is significant. Let X be the set of all such multisets arising from lists of finitely many ideals Rq_j with each $q_j = p_j^{e_j}$ for some irreducible $p_j \in R$ and $e_j \in \mathbb{Z}_{>0}$. Let Y be the set of all finite lists of ideals $(Ra_1, \ldots, Ra_m)$ with $R \neq Ra_1 \supseteq Ra_2 \supseteq \cdots \supseteq Ra_m \neq \{0\}$. The idea of the proof is to define bijections $f : X \to Y$ and $g : Y \to X$ that let us use the known uniqueness of the invariant factors of a module. We saw this idea in the simpler setting of commutative groups in §16.19.

Let Z be a fixed set of irreducible elements in R such that no two elements of Z are associates, but every irreducible element in R is associate to some element of Z. The map g acts on $L = (Ra_1, \ldots, Ra_m) \in Y$ as follows. We know each a_i factors in R into a product $u \prod_{j=1}^{n_i} p_{ij}^{e_{ij}}$ where $u \in R^*$, $p_{i1}, p_{i2}, \ldots, p_{in_i}$ are distinct irreducible elements in Z and $e_{ij} \in \mathbb{Z}_{>0}$. Define $g(L)$ to be the multiset in X consisting of all ideals $Rp_{ij}^{e_{ij}}$ for $1 \leq i \leq m$ and $1 \leq j \leq n_i$. Using the Lemma on Splitting into Prime Powers, note that $\prod_{i=1}^{m} R/Ra_i \cong \prod_{i=1}^{m} \prod_{j=1}^{n_i} R/Rp_{ij}^{e_{ij}}$, no matter what order we list the terms in the direct product on the right side.

The map f acts on $M = [Rq_1, \ldots, Rq_k] \in X$ as follows. Let $p_1, \ldots, p_n \in Z$ be the distinct irreducible elements such that each $q_i \sim p_j^{e_j}$ for some j between 1 and n. Place the elements of M in a matrix such that row j contains all the ideals Rq_i with $q_i \sim p_j^{e_j}$, listed with multiplicities so that the exponents e_j weakly increase reading from left to right. Suppose the longest row in the matrix has length m. Pad all shorter rows with copies of 1_R on the left so that all rows have length m. Define $f(M) = (Ra_1, Ra_2, \ldots, Ra_m)$, where a_k is the product of all q_i appearing in column k. You can check that $f(M) \in Y$, since the construction ensures that $a_i|a_{i+1}$ for all $i < m$. Since splitting the a_k back into prime powers recovers the q_i in some order, we see that $g(f(M)) = M$ for all $M \in X$, and $\prod_{i=1}^{k} R/Rq_i \cong \prod_{j=1}^{m} R/Ra_j$. You can also check that $f(g(L)) = L$ for all $L \in Y$, though we do not need this fact below.

To begin the uniqueness proof, assume we have an isomorphism as in (18.10). Let $M_1 = [Rq_1, \ldots, Rq_k]$, $M_2 = [Rr_1, \ldots, Rr_\ell]$, $L_1 = f(M_1) = [Ra_1, \ldots, Ra_m]$, and $L_2 = f(M_2) = [Rb_1, \ldots, Rb_n]$. We have seen that

$$\prod_{j=1}^{m} R/Ra_j \cong \prod_{i=1}^{k} R/Rq_i \cong \prod_{i=1}^{\ell} R/Rr_i \cong \prod_{j=1}^{n} R/Rb_j.$$

Since $L_1, L_2 \in Y$, the uniqueness result for invariant factors shows that $L_1 = L_2$. Then $M_1 = g(f(M_1)) = g(L_1) = g(L_2) = g(f(M_2)) = M_2$, which proves the required uniqueness result for elementary divisors.

18.16 $F[x]$-Module Defined by a Linear Operator

In the rest of this chapter, we apply the structure theorems for finitely generated modules over PIDs to derive results on canonical forms of matrices and linear operators on a vector space. Throughout, we let F be a field and V be an n-dimensional vector space over F. We also fix an F-linear map $T : V \to V$. We will define a collection of matrices in $M_n(F)$ called *rational canonical forms* and show that each T is represented (relative to an appropriate ordered basis of V) by exactly one of these matrices.

To obtain this result from the preceding theory, we use T to turn the vector space (F-module) V into an $F[x]$-module. The addition in the $F[x]$-module V is the given addition in the vector space V. For $v \in V$ and $p = \sum_{i=0}^{d} p_i x^i \in F[x]$, define scalar multiplication by $p \cdot v = \sum_{i=0}^{d} p_i T^i(v)$, where $T^0 = \mathrm{id}_V$ and T^i denotes the composition of i copies of T.

Let us check the $F[x]$-module axioms. The five additive axioms are already known to hold. For $p \in F[x]$ and $v \in V$ as above, $p \cdot v$ is in V, since T and each T^i map V to V and V is closed under addition and multiplication by scalars in F. Next, $1_{F[x]} \cdot v = 1T^0(v) = v$. Given $q = \sum_{i \geq 0} q_i x^i \in F[x]$, note $qp = \sum_{i \geq 0} \left(\sum_{k=0}^{i} q_k p_{i-k} \right) x^i$. Using linearity of T and its powers, we compute:

$$
\begin{aligned}
(qp) \cdot v &= \sum_{i \geq 0} \left(\sum_{k=0}^{i} q_k p_{i-k} \right) T^i(v) = \sum_{i \geq 0} \sum_{k=0}^{i} q_k T^k(p_{i-k} T^{i-k}(v)) = \sum_{k \geq 0} \sum_{j \geq 0} q_k T^k(p_j T^j(v)) \\
&= \sum_{k \geq 0} q_k T^k \left(\sum_{j \geq 0} p_j T^j(v) \right) = q \cdot \left(\sum_{j \geq 0} p_j T^j(v) \right) = q \cdot (p \cdot v).
\end{aligned}
$$

Next,

$$(p+q) \cdot v = \sum_{i \geq 0} (p_i + q_i) T^i(v) = \sum_{i \geq 0} p_i T^i(v) + \sum_{i \geq 0} q_i T^i(v) = p \cdot v + q \cdot v.$$

Finally, given $w \in V$,

$$p \cdot (v+w) = \sum_{i \geq 0} p_i T^i(v+w) = \sum_{i \geq 0} p_i [T^i(v) + T^i(w)] = \sum_{i \geq 0} p_i T^i(v) + \sum_{i \geq 0} p_i T^i(w) = p \cdot v + p \cdot w.$$

Let us spell out some definitions from module theory in the setting of the particular $F[x]$-module V determined by the linear operator T. First, what is an $F[x]$-submodule of

V? This is an additive subgroup W of V such that $p \cdot w \in W$ for all $p \in F[x]$ and all $w \in W$. Taking p to be a constant polynomial, we see that a submodule W must be closed under multiplication by scalars in F. Taking $p = x$, we see that a submodule W must satisfy $x \cdot w = T(w) \in W$ for all $w \in W$. Conversely, suppose W is a subspace such that $T[W] \subseteq W$; a subspace satisfying this condition is called a *T-invariant* subspace of V. By induction on i, we see that $T^i(w) \in W$ for all $w \in W$ and all $i \geq 0$. Since W is a subspace, we then see that $p \cdot w = \sum_{i \geq 0} p_i T^i(w) \in W$ for all $w \in W$ and all $p \in F[x]$. So, *submodules of the $F[x]$-module V are the same thing as T-invariant subspaces of V*.

Second, is V *finitely generated* as an $F[x]$-module? We know V is finitely generated as an F-module, since the vector space V has an n-element basis $\mathcal{B} = (v_1, \ldots, v_n)$. We claim $\{v_1, \ldots, v_n\}$ generates the $F[x]$-module V. Given $v \in V$, write $v = c_1 v_1 + \cdots + c_n v_n$ for some $c_1, \ldots, c_n \in F$. Each c_i is also a constant polynomial in $F[x]$, so v has been expressed as an $F[x]$-linear combination of the v_i. (V might also be generated, as an $F[x]$-module, by a proper subset of the v_i, since we could also act on each v_i by non-constant polynomials.)

Third, what does a *cyclic* $F[x]$-submodule of V look like? Recall this is a submodule of the form $W = F[x]z$ for some fixed $z \in W$; z is called a *generator* of the submodule. Such a submodule is also called a *T-cyclic subspace* of V. Define a map $g : F[x] \to W$ by $g(p) = p \cdot z$ for all $p \in F[x]$. You can check that g is a surjective $F[x]$-module homomorphism, which induces an $F[x]$-module isomorphism $g' : F[x]/\ker(g) \to W$. The kernel of g is a submodule (ideal) of $F[x]$, called the *T-annihilator* of z. Since $F[x]$ is a PID, $\ker(g) = F[x]h$ for some $h \in F[x]$. Now h cannot be zero, since otherwise W would be isomorphic to $F[x]$ as an $F[x]$-module, hence also isomorphic to $F[x]$ as an F-module. But $F[x]$ is an infinite-dimensional F-vector space and W is finite-dimensional. So $h \neq 0$, and we can take h to be the unique monic generator of $\ker(g)$. Write $h = h_0 + h_1 x + \cdots + h_d x^d$, where $d \geq 0$, each $h_i \in F$, and $h_d = 1$.

So far, we know that $F[x]/F[x]h$ and $W = F[x]z$ are isomorphic (both as $F[x]$-modules and F-modules) via the map g' sending $p + F[x]h$ to $p \cdot z$ for all $p \in F[x]$. Now, using polynomial division with remainder (cf. §3.20), you can check that

$$(1 + F[x]h, x + F[x]h, x^2 + F[x]h, \ldots, x^{d-1} + F[x]h)$$

is an ordered basis for the F-vector space $F[x]/F[x]h$. Applying the F-isomorphism g' to this basis, we conclude that

$$\mathcal{B}_z = (1 \cdot z, x \cdot z, x^2 \cdot z, \ldots, x^{d-1} \cdot z) = (z, T(z), T^2(z), \ldots, T^{d-1}(z))$$

is an ordered F-basis for the subspace W of V.

Since W is T-invariant, we know T restricts to a linear map $T|_W : W \to W$. What is the matrix of $T|_W$ relative to the ordered basis $\mathcal{B}_z$? Note $T|_W(z) = T(z)$, which has coordinates $(0, 1, 0, \ldots, 0)$ relative to $\mathcal{B}_z$. Note $T|_W(T(z)) = T(T(z)) = T^2(z)$, which has coordinates $(0, 0, 1, 0, \ldots, 0)$ relative to the basis $\mathcal{B}_z$. Similarly, $T|_W(T^j(z)) = T^{j+1}(z)$ for $0 \leq j < d-1$. But, when we apply $T|_W$ to the final element in $\mathcal{B}_z$, $T|_W(T^{d-1}(z)) = T^d(z)$ is not in the basis. As $h \in \ker(g)$, we know $0 = h \cdot z = \sum_{i=0}^{d-1} h_i T^i(z) + T^d(z)$, so the coordinates of $T^d(z)$ relative to $\mathcal{B}_z$ must be $(-h_0, -h_1, \ldots, -h_{d-1})$. In conclusion, the matrix we want is

$$[T|_W]_{\mathcal{B}_z} = \begin{bmatrix} 0 & 0 & 0 & \cdots & 0 & -h_0 \\ 1 & 0 & 0 & \cdots & 0 & -h_1 \\ 0 & 1 & 0 & \cdots & 0 & -h_2 \\ 0 & 0 & 1 & \cdots & 0 & -h_3 \\ & \cdots & & \cdots & & \cdots \\ 0 & 0 & 0 & \cdots & 1 & -h_{d-1} \end{bmatrix}_{d \times d}. \tag{18.11}$$

This matrix is called the *companion matrix* of the monic polynomial h and is written C_h. Conversely, let W be any T-invariant subspace such that for some $z \in W$, $\mathcal{B}_z = (z, T(z), \ldots, T^{d-1}(z))$ is an F-basis of W and $[T|_W]_{\mathcal{B}_z} = C_h$. You can check that $W = F[x]z$ is a T-cyclic subspace isomorphic to $F[x]/F[x]h$ (Exercise 74).

18.17 Rational Canonical Form of a Linear Map

As in the last section, let V be an n-dimensional vector space over a field F and $T : V \to V$ be a fixed linear map. Make V into an $F[x]$-module via T, as described above. We know $F[x]$ is a PID and V is a finitely generated $F[x]$-module, so version 1 of the Classification Theorem for Modules over a PID (see (18.5)) gives us an $F[x]$-module isomorphism

$$\phi : V \to F[x]/F[x]h_1 \times F[x]/F[x]h_2 \times \cdots \times F[x]/F[x]h_k \qquad (18.12)$$

for uniquely determined ideals

$$F[x] \neq F[x]h_1 \supseteq F[x]h_2 \supseteq \cdots \supseteq F[x]h_k \supseteq \{0\}$$

generated by $h_1, \ldots, h_k \in F[x]$. Since V is finite-dimensional as an F-module, none of the ideals $F[x]h_j$ can be zero. So we can assume the generators $h_1, \ldots, h_k$ of these ideals are monic with respective degrees $d_1, \ldots, d_k > 0$. These polynomials satisfy $h_i | h_{i+1}$ in $F[x]$ for $1 \leq i < k$.

Call the product module on the right side of (18.12) V'. For $1 \leq i \leq k$, let W_i' be the submodule $\{0\} \times \cdots \times F[x]/F[x]h_i \times \cdots \times \{0\}$ of V'. Note W_i' is a cyclic $F[x]$-module generated by $z_i' = (0, \ldots, 1 + F[x]h_i, \ldots, 0)$. Letting $W_i = \phi^{-1}[W_i']$ and $z_i = \phi^{-1}(z_i')$ for each i, we obtain cyclic submodules $W_i = F[x]z_i$ of V such that $W_i \cong W_i' \cong F[x]/F[x]h_i$. You can check that h_i is the monic polynomial of least degree sending the coset $1 + F[x]h_i$ to $0 + F[x]h_i$, and hence $F[x]h_i$ is the T-annihilator of z_i for all i.

You can also check that we get an ordered basis for the F-vector space V' by concatenating ordered bases for the submodules W_i' corresponding to each factor in the direct product. Applying ϕ^{-1}, we get an ordered basis $\mathcal{B}$ for the F-vector space V by concatenating ordered bases for $W_1, \ldots, W_k$. Using the bases $\mathcal{B}_{z_1}, \ldots, \mathcal{B}_{z_k}$ constructed in §18.16, we get an ordered basis

$$\mathcal{B} = (z_1, T(z_1), \ldots, T^{d_1-1}(z_1), z_2, T(z_2), \ldots, T^{d_2-1}(z_2), \ldots, z_k, T(z_k), \ldots, T^{d_k-1}(z_k))$$

for V. The matrix of T relative to the basis $\mathcal{B}$ is the block-diagonal matrix

$$[T]_{\mathcal{B}} = \begin{bmatrix} C_{h_1} & 0 & \cdots & 0 \\ 0 & C_{h_2} & \cdots & 0 \\ \vdots & & \ddots & \vdots \\ 0 & 0 & \cdots & C_{h_k} \end{bmatrix}, \qquad (18.13)$$

where C_{h_i} is the companion matrix of h_i.

In general, given any square matrices $A_1, \ldots, A_k$, we write blk-diag$(A_1, \ldots, A_k)$ for the block-diagonal matrix with diagonal blocks $A_1, \ldots, A_k$. A *rational canonical form* is a matrix $A \in M_n(F)$ of the form $A = $ blk-diag$(C_{h_1}, \ldots, C_{h_k})$, where: $k > 0$, $h_1, \ldots, h_k \in F[x]$ are monic, non-constant polynomials, and $h_i | h_{i+1}$ in $F[x]$ for $1 \leq i < k$. The matrix in (18.13) is called *the rational canonical form of* T, and the polynomials h_i are called the *invariant factors of* T. We have now proved the existence part of the following theorem.

Theorem on Rational Canonical Forms. Let V be a finite-dimensional F-vector space. Every linear map $T : V \to V$ is represented by exactly one matrix in rational canonical form (18.13).

To prove the uniqueness of the rational canonical form, suppose there were another ordered basis $\mathcal{B}^*$ of V such that $[T]_{\mathcal{B}^*} = \text{blk-diag}(C_{g_1}, \ldots, C_{g_\ell})$, where $g_1, \ldots, g_\ell \in F[x]$ are monic polynomials in $F[x]$ with $g_i | g_{i+1}$ for $1 \le i < \ell$ and $\deg(g_i) = d_i^* > 0$. Let W_1^* be the subspace of V generated by the first d_1^* vectors in $\mathcal{B}^*$, W_2^* the subspace generated by the next d_2^* vectors in $\mathcal{B}^*$, and so on. The form of the matrix $[T]_{\mathcal{B}^*}$ shows that each W_i^* is a cyclic $F[x]$-submodule of V annihilated by g_i, namely $W_i^* = F[x]z_i^* \cong F[x]/F[x]g_i$, where z_i^* is the first basis vector in the generating list for W_i^*. You can check (Exercise 74) that

$$V \cong W_1^* \times \cdots \times W_\ell^* \cong F[x]/F[x]g_1 \times \cdots \times F[x]/F[x]g_\ell$$

as F-modules and $F[x]$-modules. The uniqueness result proved in §18.14 now gives $\ell = k$ and $F[x]h_i = F[x]g_i$ for all i, hence $h_i = g_i$ for all i since h_i and g_i are monic.

18.18   Jordan Canonical Form of a Linear Map

Assume $T : V \to V$ is a fixed linear map on an n-dimensional F-vector space V. We know (using (18.6)) that there is an $F[x]$-module isomorphism

$$\psi : V \cong F[x]/F[x]q_1 \times \cdots \times F[x]/F[x]q_s$$

where each q_i is $p_i^{e_i}$ for some monic irreducible $p_i \in F[x]$ and some $e_i > 0$. We saw in §18.15 that s and the multiset of q_i are uniquely determined by these conditions. The q_i are called *elementary divisors* of the linear map T.

By exactly the same argument used to derive (18.13) from (18.12), we see that V is the direct sum of T-cyclic subspaces $W_1, \ldots, W_s$ with $W_i = F[x]y_i \cong F[x]/F[x]q_i$ for some $y_i \in V$, and V has an ordered basis $\mathcal{B}$ such that $[T]_{\mathcal{B}} = \text{blk-diag}(C_{q_1}, \ldots, C_{q_s})$. To obtain further structural results, we now impose the additional hypothesis that every p_i has degree 1, say $p_i = x - c_i$ where $c_i \in F$. For example, this hypothesis automatically holds when $F = \mathbb{C}$, or when F is any algebraically closed field (which means all irreducible polynomials in $F[x]$ have degree 1).

Our goal is to change the basis of each T-cyclic subspace $W_i = F[x]y_i$ to obtain an even nicer matrix than the companion matrix $C_{q_i} = C_{(x-c_i)^{e_i}}$. Fix $c \in F$ and a positive integer e, and consider any T-cyclic subspace $W = F[x]y$ of V with ordered basis $\mathcal{B}_y = (y, T(y), \ldots, T^{e-1}(y))$, such that $[T|_W]_{\mathcal{B}_y} = C_{(x-c)^e}$. For $0 \le k \le e$, define $y_k = (x - c)^k \cdot y$, which is a vector in W. Since $(x - c)^k = (x - c)(x - c)^{k-1}$, we have $y_0 = y$, $y_e = 0$, and for $1 \le k \le e$, $y_k = (x - c) \cdot y_{k-1} = T(y_{k-1}) - cy_{k-1}$, hence $T(y_{k-1}) = y_k + cy_{k-1}$. Using these facts, you can show by induction that the list $(y_0, \ldots, y_k)$ spans the same F-subspace as the list $(y, T(y), \ldots, T^k(y))$ for $0 \le k < e$, and therefore $\mathcal{B} = (y_{e-1}, \ldots, y_2, y_1, y_0)$ is another ordered F-basis for W.

Let us compute the matrix $[T|_W]_{\mathcal{B}}$. First, $T(y_{e-1}) = y_e + cy_{e-1} = cy_{e-1}$, which has coordinates $(c, 0, \ldots, 0)$ relative to $\mathcal{B}$. Next, for $1 < j \le e$, $T(y_{e-j}) = 1y_{e-j+1} + cy_{e-j}$, so that column j of the matrix has a 1 in row $j - 1$, a c in row j, and zeroes elsewhere. In other words, $[T|_W]_\mathcal{B}$ is the *Jordan block* defined by

$$J(c; e) = \begin{bmatrix} c & 1 & 0 & \cdots & 0 \\ 0 & c & 1 & \cdots & 0 \\ 0 & 0 & c & \cdots & 0 \\ 0 & 0 & 0 & \cdots & 1 \\ 0 & 0 & 0 & \cdots & c \end{bmatrix}.$$

(In the case $e = 1$, this is a 1×1 matrix with sole entry c.)

By concatenating bases of the form $(y_{e-1}, \ldots, y_0)$ for all the T-cyclic subspaces W_i, we obtain an ordered basis for V such that the matrix of T relative to this basis is

$$\mathbf{J} = \begin{bmatrix} J(c_1; e_1) & 0 & \cdots & 0 \\ 0 & J(c_2; e_2) & \cdots & 0 \\ \vdots & & \ddots & \vdots \\ 0 & 0 & \cdots & J(c_s; e_s) \end{bmatrix}. \tag{18.14}$$

Any block-diagonal matrix in $M_n(F)$ with Jordan blocks on the diagonal is called a *Jordan canonical form*; the particular matrix $\mathbf{J}$ is called a *Jordan canonical form of the linear map* T. The Jordan canonical form for T may not be unique, since we can obtain other Jordan canonical forms by reordering the multiset of elementary divisors of T, which leads to a permutation of the Jordan blocks of $\mathbf{J}$. But, arguing as in §18.17, we can prove that these are the only possible Jordan canonical forms for T. The only new detail is showing that for any subspace W of V such that $[T|_W]_\mathcal{B} = J(c; e)$ for some basis $\mathcal{B}$ of W, W is a T-cyclic subspace with $W = F[x]y \cong F[x]/F[x](x-c)^e$ for some $y \in W$ (Exercise 78). We summarize these results as follows.

Theorem on Jordan Canonical Forms. Let V be a finite-dimensional F-vector space. Assume $T : V \to V$ is a linear map such that every elementary divisor of T has the form $(x - c)^e$ for some $c \in F$ and $e \in \mathbb{Z}_{>0}$ (which always holds for $F = \mathbb{C}$). Then there is an ordered basis $\mathcal{B}$ for V such that the matrix $\mathbf{J} = [T]_\mathcal{B}$ is a Jordan canonical form. All Jordan canonical forms for T are obtained from $\mathbf{J}$ by reordering the Jordan blocks.

18.19　Canonical Forms of Matrices

Let $A \in M_n(F)$ be an $n \times n$ matrix with entries in a field F. We use A to define a linear map $T : F^n \to F^n$ such that $T(v) = Av$ for all $v \in F^n$. Then F^n becomes an $F[x]$-module via T, where the action is given by $p \cdot v = \sum_{i=0}^{d} p_i A^i v$ for $p = \sum_{i=0}^{d} p_i x^i \in F[x]$ and $v \in F^n$. The matrix of T relative to the standard ordered basis of F^n is A. We know (Chapter 6) that the matrix C of T relative to any other ordered basis of F^n has the form $C = S^{-1}AS$ for some invertible $S \in M_n(F)$; in other words, C is *similar* to A. Conversely, every matrix similar to A is the matrix of T relative to some ordered basis of F^n. We can rephrase our previous theorems on canonical forms of linear maps as statements about canonical forms of matrices.

Theorem on Canonical Forms of Matrices. Let F be a field.

(a) Every matrix $A \in M_n(F)$ is similar to exactly one matrix of the form blk-diag$(C_{h_1}, \ldots, C_{h_k})$, where $h_1, \ldots, h_k \in F[x]$ are monic non-constant polynomials with $h_i | h_{i+1}$ for $1 \leq i < k$. This matrix is called the *rational canonical form of A*, and the polynomials $h_1, \ldots, h_k$ are the *invariant factors* of A.

(b) If F is algebraically closed, then every matrix $A \in M_n(F)$ is similar to a matrix of the form blk-diag$(J(c_1; e_1), \ldots, J(c_s; e_s))$, and the only matrices of this form that are similar to A are those obtained by reordering the Jordan blocks. These matrices are called *Jordan canonical forms of A*.

(c) Two matrices in $M_n(F)$ are similar iff they have the same rational canonical form. For algebraically closed F, two matrices are similar iff they have the same Jordan canonical forms.

Recall that the *minimal polynomial* of $A \in M_n(F)$ is the unique monic polynomial $m_A \in F[x]$ of minimum degree such that $m_A(A) = 0$, and the *characteristic polynomial* of A is $\chi_A = \det(xI_n - A) \in F[x]$. You can check that similar matrices have the same minimal polynomial and the same characteristic polynomial. For monic $h \in F[x]$, it follows directly from these definitions that $m_{C_h} = \chi_{C_h} = h$ (Exercise 79). For any block-diagonal matrix $B = $ blk-diag$(B_1, \ldots, B_k)$, we have $m_B = \text{lcm}(m_{B_1}, \ldots, m_{B_k})$ and $\chi_B = \prod_{i=1}^{k} \chi_{B_i}$. In the special case where B is the rational canonical form of A, we see that $\chi_A = \chi_B = \prod_{i=1}^{k} h_i$ and $m_A = m_B = h_k$. These observations reprove the *Cayley–Hamilton Theorem*, which states that $m_A | \chi_A$ in $F[x]$, or equivalently $\chi_A(A) = 0$.

Theorem on Invariant Factors of a Matrix. Let F be a field. Given $A \in M_n(F)$, the invariant factors of A are precisely the invariant factors of positive degree in the Smith normal form of the matrix $xI_n - A \in M_n(F[x])$.

We can use this theorem and the matrix reduction algorithm in §18.9 (applied to the matrix $xI_n - A$) to calculate the rational canonical form of a given matrix A. Alternatively, we can invoke the results in §18.12 to give specific formulas for each invariant factor (involving quotients of gcds of appropriate minors of $xI_n - A$) and to show once again that the invariant factors of A are uniquely determined by A.

Proof. To prove the theorem, first consider the case where $A = C_h$ is the companion matrix of a monic polynomial $h \in F[x]$ of degree n. Using elementary row and column operations, the matrix $xI_n - C_h \in M_n(F[x])$ can be reduced to the diagonal matrix diag$(1, \ldots, 1, h)_{n \times n}$ (Exercise 79). This matrix is a Smith normal form of $xI_n - C_h$. For a general matrix A, we know there is an invertible matrix $S \in M_n(F)$ such that $S^{-1}AS$ is the unique rational canonical form of A. Suppose $S^{-1}AS = $ blk-diag$(C_{h_1}, \ldots, C_{h_k})$, where $h_1, \ldots, h_k$ are the invariant factors of A. Write $\deg(h_i) = d_i > 0$ for $1 \leq i \leq k$. It follows that $S^{-1}(xI_n - A)S = xI_n - S^{-1}AS = $ blk-diag$(xI_{d_1} - C_{h_1}, \ldots, xI_{d_k} - C_{h_k})$. By the observation above, for each block $xI_{d_i} - C_{h_i}$ there are invertible $P_i, Q_i \in M_n(F[x])$ such that $P_i(xI_{d_i} - C_{h_i})Q_i = $ diag$(1, \ldots, 1, h_i)_{d_i \times d_i}$. Letting $P = $ blk-diag$(P_1, \ldots, P_k)$ and $Q = $ blk-diag$(Q_1, \ldots, Q_k)$, $PS^{-1}(xI_n - A)SQ$ is a diagonal matrix where the diagonal entries (in some order) are $h_1, \ldots, h_k$, and $n - k$ copies of 1. By performing row and column interchanges, we can arrange that the diagonal consists of all the 1s followed by $h_1, \ldots, h_k$. So we have found invertible $P', Q' \in M_n(F[x])$ such that $P'(xI_n - A)Q' = $ diag$(1, \ldots, 1, h_1, \ldots, h_k)_{n \times n}$. This matrix is a Smith normal form of $xI_n - A$. We see that the non-constant invariant factors in this normal form are exactly $h_1, \ldots, h_k$, which are the invariant factors of the matrix A. $\square$

18.20 Summary

Let R be a commutative ring.

1. *Divisibility Definitions.* Given x, y in R:
   - x *divides* y (written $x|y$) iff $y = rx$ for some $r \in R$ iff $Ry \subseteq Rx$;
   - x is an *associate* of y (written $x \sim y$) iff $x|y$ and $y|x$ iff $Rx = Ry$;
   - x is a *unit* of R (written $x \in R^*$) iff $xy = 1_R$ for some $y \in R$ iff $x \sim 1$;
   - x is a *zero divisor* of R iff $x \neq 0$ and there is $y \neq 0$ in R with $xy = 0_R$.
   - When R is an integral domain, $x \sim y$ iff there exists $u \in R^*$ with $y = ux$.
   - Given $a_1, \ldots, a_k, d, e \in R$, d is a *gcd* of $a_1, \ldots, a_k$ iff d divides all a_i, and every common divisor of the a_i divides d; e is an *lcm* of $a_1, \ldots, a_k$ iff all a_i divide e, and e divides every common multiple of the a_i.

2. *Types of Ideals.* Given an ideal I in R:
   - I is *principal* iff there exists $c \in R$ with $I = Rc = \{rc : r \in R\}$;
   - I is *prime* iff $I \neq R$ and for all $x, y \in R$, $xy \in I$ implies $x \in I$ or $y \in I$;
   - I is *maximal* iff $I \neq R$ and for all ideals J with $I \subseteq J \subseteq R$, $J = I$ or $J = R$.
   All maximal ideals are prime, but not all prime ideals are maximal in general.

3. *Prime and Irreducible Elements.* Given a nonzero non-unit x in R:
   - x is *prime in* R iff for all $y, z \in R$, $x|(yz)$ implies $x|y$ or $x|z$;
   - x is *irreducible in* R iff for all $w \in R$, $w|x$ implies $w \sim x$ or w is a unit.
   - x is prime iff Rx is a nonzero prime ideal of R;
   - x is irreducible iff Rx is a nonzero ideal that is maximal in the poset of proper, principal ideals of R.
   - In an integral domain, every prime element is irreducible.
   - In PIDs and UFDs (but not for general R), every irreducible element is prime.

4. *Types of Rings.*
   - R is an *integral domain* iff $0_R \neq 1_R$ and R has no (nonzero) zero divisors.
   - R is a *principal ideal domain (PID)* iff R is an integral domain in which every ideal I has the form $I = Rc$ for some $c \in R$.
   - R is a *unique factorization domain (UFD)* iff R is an integral domain in which every $r \neq 0$ in R factors as $r = up_1 \cdots p_k$ for some $u \in R^*$ and irreducible $p_i \in R$; and whenever $up_1 \cdots p_k = vq_1 \cdots q_t$ with $u, v \in R^*$ and all p_i, q_j irreducible in R, $k = t$ and $p_i \sim q_i$ for $1 \leq i \leq k$ after reordering the q_j.

5. *Theorems about PIDs and UFDs.*
   - $\mathbb{Z}$ and $F[x]$ (for any field F) are PIDs.
   - For all x in a PID or UFD R, x is prime in R iff x is irreducible in R.
   - Every PID is a UFD.
   - Given $a_1, \ldots, a_k$ in a PID R, there exist gcds and lcms of $a_1, \ldots, a_k$, and each gcd is an R-linear combination of the a_j. Specifically, d is a gcd of $a_1, \ldots, a_k$ iff $Rd = Ra_1 + \cdots + Ra_k$, and e is an lcm of $a_1, \ldots, a_k$ iff $Re = Ra_1 \cap \cdots \cap Ra_k$.
   - Every ascending chain of ideals in a PID must stabilize.
   - Every nonzero prime ideal in a PID is maximal.
   - In a UFD R, gcds and lcms of $a_1, \ldots, a_k$ exist, but the gcd may not be an R-linear combination of the a_i (if it is in all cases, then R must be a PID).

6. *Module Lemmas.* Assume M and N are isomorphic R-modules.
   When R is a general commutative ring:
   - for any $c \in R$, cM is a submodule of M with $cM \cong cN$ and $M/cM \cong N/cN$;
   - every basis of an f.g. free R-module has the same size;
   - $\operatorname{ann}_R(M) = \{r \in R : \forall x \in M, rx = 0\} = \operatorname{ann}_R(N)$ is an ideal of R.

   When R is an integral domain:
   - $\operatorname{tor}(M) = \{x \in M : \exists r \in R, r \neq 0 \text{ and } rx = 0\}$ is a submodule of M;
   - $\operatorname{tor}(M) \cong \operatorname{tor}(N)$ and $M/\operatorname{tor}(M) \cong M/\operatorname{tor}(N)$.

   When R is a PID:
   - every submodule of an f.g. free R-module is also f.g. free;
   - given $a = p_1^{e_1} \cdots p_k^{e_k} \in R$ where the p_i are non-associate irreducible elements of R, $R/Ra \cong \prod_{i=1}^{k} R/Rp_i^{e_i}$.

7. *Smith Normal Form of a Matrix.* For a PID R and any matrix $A \in M_{m,n}(R)$, there exist unique ideals $Ra_1 \supseteq Ra_2 \supseteq \cdots \supseteq Ra_s \neq \{0\}$ (where $s \geq 0$) such that for some invertible matrices $P \in M_m(R)$ and $Q \in M_n(R)$, $PAQ = \operatorname{diag}(a_1, \ldots, a_s)_{m \times n}$. Note $a_i | a_{i+1}$ for $1 \leq i < s$. We say s is the *rank* of A, the ideals Ra_i are the *invariant factors* of A, and the matrix PAQ is a *Smith normal form* of A. We can pass from A to PAQ using a sequence of type 4 row and column operations. For $1 \leq k \leq s$, $a_1 a_2 \cdots a_k$ is a gcd of all the order k minors of A (which are determinants of $k \times k$ submatrices obtained by keeping any k rows and any k columns of A), and s is the largest k for which some order k minor of A is nonzero.

8. *Structure Theorem for Linear Maps between Free Modules.* Let N and M be f.g. free R-modules where R is a PID. For every R-linear map $T : N \to M$, there exist unique ideals $Ra_1 \supseteq Ra_2 \supseteq \cdots \supseteq Ra_s \neq \{0\}$ (where $s \geq 0$) such that for some ordered basis $\mathcal{X} = (x_1, \ldots, x_n)$ of N and some ordered basis $\mathcal{Y} = (y_1, \ldots, y_m)$ of M, $T(x_i) = a_i y_i$ for $1 \leq i \leq s$ and $T(x_i) = 0_M$ for $s < i \leq n$.

9. *Structure of Finitely Generated Modules over PIDs.* For any finitely generated module M over a PID R, there exist unique $k, d \geq 0$ and a unique sequence of ideals $R \neq Ra_1 \supseteq Ra_2 \supseteq \cdots \supseteq Ra_k \neq \{0\}$ such that $M \cong R/Ra_1 \times R/Ra_2 \times \cdots \times R/Ra_k \times R^d$. There also exist unique $m, d \geq 0$ and a unique multiset of ideals $[Rq_1, \ldots, Rq_m]$, with every q_i a power of an irreducible in R, such that $M \cong R/Rq_1 \times \cdots \times R/Rq_m \times R^d$. We call d the *Betti number* of M, $Ra_1, \ldots, Ra_k$ the *invariant factors* of M, and $Rq_1, \ldots, Rq_m$ the *elementary divisors* of M.

10. *Rational Canonical Forms.* Let F be any field. Given $h = x^d + h_{d-1}x^{d-1} + \cdots + h_0 \in F[x]$, the *companion matrix* C_h has $C_h(i+1, i) = 1$ for $1 \leq i < d$, $C_h(i, d) = -h_{i-1}$ for $1 \leq i \leq d$, and all other entries zero. A *rational canonical form* is a matrix of the form $\operatorname{blk-diag}(C_{h_1}, \ldots, C_{h_k})$, where $h_1, \ldots, h_k \in F[x]$ are monic and non-constant and $h_i | h_{i+1}$ for $1 \leq i < k$. For every n-dimensional vector space V and every linear map $T : V \to V$, there exists a unique rational canonical form $B \in M_n(F)$ such that $[T]_{\mathcal{X}} = B$ for some ordered basis $\mathcal{X}$ of V. For any matrix $A \in M_n(F)$, there exists a unique rational canonical form $B \in M_n(F)$ similar to A (i.e., $B = P^{-1}AP$ for some invertible $P \in M_n(F)$). In this case, T and A and B have minimal polynomial h_k and characteristic polynomial $h_1 \cdots h_k$, so $m_A | \chi_A$. The invariant factors $h_1, \ldots, h_k$ in the rational canonical form of $A \in M_n(F)$ coincide with the non-constant monic invariant factors in the Smith normal form of $xI_n - A \in M_n(F[x])$.

11. *Jordan Canonical Forms.* Let F be an algebraically closed field (such as $\mathbb{C}$). For $c \in F$, the *Jordan block* $J(c; e)$ is the $e \times e$ matrix with cs on the main diagonal, 1s on the next higher diagonal, and zeroes elsewhere. A *Jordan canonical form* is a matrix of the form blk-diag$(J(c_1; e_1), \ldots, J(c_s; e_s))$. For every n-dimensional F-vector space V and every linear map $T : V \to V$, there exists a Jordan canonical form $C \in M_n(F)$ such that $[T]_{\mathcal{X}} = C$ for some ordered basis $\mathcal{X}$ of V. For any matrix $A \in M_n(F)$, there exists a Jordan canonical form $C \in M_n(F)$ similar to A. The only other Jordan forms for T (or A) are obtained by reordering the Jordan blocks of C.

12. *$F[x]$-Module of a Linear Operator.* Given a field F, a finite-dimensional F-vector space V, and a linear map $T : V \to V$, V becomes an $F[x]$-module via $(\sum_{i=0}^{d} p_i x^i) \cdot v = \sum_{i=0}^{d} p_i T^i(v)$ for $p_i \in F$ and $v \in V$. $F[x]$-submodules of V are the *T-invariant subspaces* (subspaces W with $T[W] \subseteq W$). A cyclic $F[x]$-submodule W has an F-basis of the form $\mathcal{B}_z = (z, T(z), T^2(z), \ldots, T^{d-1}(z))$ for some $z \in V$ and $d \geq 0$. Such a cyclic submodule is isomorphic to $F[x]/F[x]h$, where $h \in F[x]$ is the monic polynomial of least degree d with $h \cdot z = 0$. The matrix of $[T|_W]_{\mathcal{B}_z}$ is the companion matrix C_h.

18.21   Exercises

1. (a) Prove: for all commutative rings R and all $a \in R$, Ra is an ideal of R.
   (b) Give an example to show (a) can be false if R is not commutative.

2. Prove that every field F is a PID.

3. Let $R = \mathbb{Z}[x]$. Prove that $I = \{2f + xg : f, g \in R\}$ is an ideal of R that is not principal. Conclude that R is not a PID.

4. For any integral domain R and any integer $n \geq 2$, prove $S = R[x_1, \ldots, x_n]$ is not a PID.

5. Let $R = \{a + bi : a, b \in \mathbb{Z}\}$. Prove R is a subring of $\mathbb{C}$ and hence an integral domain. Prove: for all $f, g \in R$ with $g \neq 0$, there exist $q, r \in R$ with $f = qg + r$ and $r = 0$ or $|r| < |g|$, where $|a + bi| = \sqrt{a^2 + b^2}$. (First modify the division algorithm in $\mathbb{Z}$ to see that for all $u, v \in \mathbb{Z}$ with $v \neq 0$, there exist $q, r \in \mathbb{Z}$ with $u = qv + r$ and $|r| \leq v/2$.) Deduce that R is a PID.

6. A *Euclidean domain* is a ring R satisfying the following hypotheses. First, R is an integral domain. Second, R has a *degree function* $\deg : R_{\neq 0} \to \mathbb{Z}_{\geq 0}$ such that for all nonzero $f, g \in R$, $\deg(f) \leq \deg(fg)$. Third, R satisfies a *Division Theorem* relative to the degree function: for all $f, g \in R$ with $g \neq 0$, there exist $q, r \in R$ with $f = qg + r$ and $r = 0$ or $\deg(r) < \deg(g)$. Prove that every Euclidean domain is a PID. Explain why it follows that fields F, polynomial rings $F[x]$, and the ring $\mathbb{Z}$ are PIDs. (It can be shown [14] that not all PIDs are Euclidean domains; the standard example is the ring $R = \{a + bz : a, b \in \mathbb{Z}\}$, where $z = (1 + i\sqrt{19})/2$.)

7. Let F be a field. Prove $F[[x]]^*$ consists of all formal power series $\sum_{i \geq 0} p_i x^i$ with $p_0 \neq 0$. Prove $F[[x]]$ is a PID. (Show that every nonzero ideal of $F[[x]]$ has the form $F[[x]]x^j$ for some $j \in \mathbb{Z}_{\geq 0}$.)

8. Let R be a commutative ring. Prove: for all $a, a', b, c \in R$:
    (a) $a|a$;
    (b) if $a|b$ and $b|c$ then $a|c$;
    (c) $1_R|a$ and $a|0_R$;
    (d) $0|a$ iff $a = 0$;
    (e) $\sim$ (defined by $a \sim b$ iff $a|b$ and $b|a$) is an equivalence relation on R;
    (f) if $a \sim a'$, then: $a|b$ iff $a'|b$; and $c|a$ iff $c|a'$;
    (g) $a \in R^*$ iff $a \sim 1$.

9. Let R be a commutative ring. Prove: for all $k > 0$ and all $r_1, \ldots, r_k, a_1, \ldots, a_k,$ $c \in R$, if c divides every r_i, then $c|(a_1 r_1 + \cdots + a_k r_k)$. Translate this result into a statement about principal ideals.

10. Suppose R is a commutative ring and $d \in R$ is a gcd of $a_1, \ldots, a_k \in R$. Prove: the set of all gcds of $a_1, \ldots, a_k$ in R equals the set of all $d' \in R$ with $d \sim d'$. Prove a similar result for lcms.

11. Let $R = \{a + bi : a, b \in \mathbb{Z}\}$, which is a subring of $\mathbb{C}$.
    (a) Find all units of R.
    (b) What are the associates of $3 - 4i$ in R?
    (c) How do the answers to (a) and (b) change if we replace R by $\mathbb{C}$?

12. Given $d, a_1, \ldots, a_k$ in a commutative ring R, carefully prove that d is an lcm of $a_1, \ldots, a_k$ iff Rd is the greatest lower bound of $\{Ra_1, \ldots, Ra_k\}$ in the poset of principal ideals of R ordered by $\subseteq$.

13. Suppose R is a commutative ring and $a, b \in R$ are associates.
    (a) Prove a is irreducible in R iff b is irreducible in R using divisibility definitions.
    (b) Prove the result in (a) by considering principal ideals.
    (c) Prove a is prime in R iff b is prime in R using divisibility definitions.
    (d) Prove the result in (c) by considering principal ideals.

14. Suppose p is a prime element in a commutative ring R. Prove by induction: for all $k \geq 1$ and all $a_1, \ldots, a_k \in R$, if $p|(a_1 a_2 \cdots a_k)$ then p divides some a_i.

15. Let p be irreducible in an integral domain R. Prove: for all $k \geq 1$ and all $b_1, \ldots, b_k$ in R, if $p \sim b_1 b_2 \cdots b_k$, then for some i, $p \sim b_i$ and all other b_j are units of R.

16. Let R be a commutative ring.
    (a) Prove an ideal P of R is a prime ideal iff R/P is an integral domain.
    (b) Prove an ideal M of R is a maximal ideal iff R/M is a field.
    (c) Deduce from (a) and (b) that every maximal ideal is a prime ideal.

17. Let R be a commutative ring with a maximal ideal M. Without using quotient rings, prove that M is a prime ideal. (Argue by contradiction, and assume $x, y \in R$ satisfy $x, y \notin M$ but $xy \in M$. Consider the ideals $M + Rx$ and $M + Ry$ to deduce $1_R \in M$, which cannot occur.)

18. Suppose R is a PID and $I = Ra$ is an ideal of R. Under what conditions on a is R/I a PID? Explain.

19. Consider the commutative ring $R = \mathbb{Z}_{12}$, which is not an integral domain.
    (a) Find all units of $\mathbb{Z}_{12}$.
    (b) Find all prime elements in $\mathbb{Z}_{12}$.
    (c) Find all irreducible elements in $\mathbb{Z}_{12}$.
    (d) Describe the equivalence classes of $\sim$ in $\mathbb{Z}_{12}$.
    (e) Is it true that for all $a, b \in \mathbb{Z}_{12}$, $a \sim b$ iff $a = ub$ for some $u \in \mathbb{Z}_{12}^*$?

20. Let R and S be rings. Suppose K is any ideal in the product ring $R \times S$. Prove there exists an ideal I in R and an ideal J in S such that $K = I \times J$.

21. Consider the product ring $T = \mathbb{Z}_4 \times \mathbb{Z}_2$. Find all ideals in T. Draw the ideal lattice of T (compare to Figure 17.1). Decide, with explanation, whether each ideal is prime and whether each ideal is maximal. Decide, with explanation, whether each nonzero element of T is a unit, a zero divisor, a prime element, or an irreducible element (indicate all that apply).

22. Repeat the previous exercise for the product ring $T = \mathbb{Z}_2 \times \mathbb{Z}_2 \times \mathbb{Z}_2$. (Compare the ideal lattice to the one in Exercise 75 of Chapter 17.)

23. Let R be a PID that is not a field.
    (a) Explain why $\{0\}$ is a prime ideal of R that is not a maximal ideal.
    (b) Show that every nonzero prime ideal of R is a maximal ideal.

24. Let R be an integral domain in which every ascending chain of principal ideals stabilizes. Show R satisfies condition (a) in the definition of a UFD (existence of irreducible factorizations).

25. Let R be an integral domain in which every irreducible Show R satisfies condition (b) in the definition of a UFD (uniqueness of irreducible factorizations).

26. Suppose R is a UFD. Prove $p \in R$ is irreducible in R iff p is prime in R.

27. Let R be a UFD and $\{p_i : i \in I\}$ be a set of irreducible elements of R such that every irreducible q in R is an associate of exactly one p_i. Prove: for all nonzero $r \in R$, there exist unique $u \in R^*$ and exponents $e_i \geq 0$ such that all but finitely many e_i are zero and $r = u \prod_{i \in I} p_i^{e_i}$. Write $u = u(r)$ and $e_i = e_i(r)$ to indicate that these parameters are functions of r.

28. Let R be a UFD. With the notation of the previous exercise, prove:
    (a) for $a, b, c \in R$, $c = ab$ iff $u(c) = u(a)u(b)$ and $e_i(c) = e_i(a) + e_i(b)$ for all $i \in I$;
    (b) for all $r, s \in R_{\neq 0}$, $r|s$ iff $e_i(r) \leq e_i(s)$ for all $i \in I$;
    (c) for all $d, a_1, \ldots, a_k \in R_{\neq 0}$, d is a gcd of $a_1, \ldots, a_k$ iff $e_i(d) = \min_{1 \leq j \leq k} e_i(a_j)$ for all $i \in I$. Deduce that *gcds of finite lists of elements in a UFD always exist.*

29. State and prove a formula characterizing lcms of $a_1, \ldots, a_k$ in a UFD R, similar to part (c) of the previous exercise.

30. Suppose R is a Euclidean domain (Exercise 6) for which there is an algorithm to compute the quotient and remainder (relative to deg) when f is divided by g. Describe an algorithm that takes as input $f, g \in R$ and produces as output $d, a, b \in R$ such that $d = af + bg$ and d is a gcd of f and g. (Imitate the proof in §3.8.)

31. Let $R = \{a + bi\sqrt{5} : a, b \in \mathbb{Z}\}$, which is a subring of $\mathbb{C}$. Define $N : R \to \mathbb{Z}$ by $N(a + bi\sqrt{5}) = a^2 + 5b^2$ for $a, b \in \mathbb{Z}$. Prove $N(rs) = N(r)N(s)$ for all $r, s \in R$. Prove: for all $r \in R$, r is a unit of R iff $N(r) = \pm 1$. Find all units of R. Show that 2, 3, $1 + i\sqrt{5}$, and $1 - i\sqrt{5}$ are irreducible in R but not prime in R. Is R a UFD? Is R a PID? Why?

32. Let $R = \mathbb{Z}_6$. Show that every ideal of R is principal, but give an example of a f.g. free R-module M and a submodule N of M that is not free. Explain exactly why the proof in §18.6 fails for this ring.

33. Give a specific example of an integral domain R, a f.g. free R-module M, and a submodule N of M that is not free. Why does the proof in §18.6 fail here?

34. Prove or disprove: for every f.g. free module M over a PID R and every submodule N of M, there is a submodule P of M with $P + N = M$ and $P \cap N = \{0\}$.

35. Without using operation (B4), give direct proofs that applying the operations (B1), (B2), and (B3) in §18.7 to an ordered basis of M produces a new ordered basis of M.

36. Prove item 1 and items 3—8 in §18.8.

37. Let $R = \mathbb{Q}[x]$, let $\mathcal{X} = \mathcal{Y} = (\mathbf{e}_1, \mathbf{e}_2)$ be the standard ordered basis of the R-module R^2, and let $T : R^2 \to R^2$ be the R-linear map with matrix
$$A = \begin{bmatrix} x - 3 & x^2 + 1 \\ x^3 - x & 5x + 2 \end{bmatrix}$$ relative to the bases $\mathcal{X}$ and $\mathcal{Y}$.
(a) Compute $T((3x - 1, x^2 + x - 1))$.
(b) Let $\mathcal{Z} = ((x - 1, 2x), (x/2, x + 1))$. Show $\mathcal{Z}$ is obtained from $\mathcal{X}$ by operation (B4), so is an ordered basis of R^2.
(c) What is the matrix of T relative to input basis $\mathcal{Z}$ and output basis $\mathcal{Y}$? How is this matrix related to A?
(d) What is the matrix of T relative to input basis $\mathcal{X}$ and output basis $\mathcal{Z}$? How is this matrix related to A?
(e) What is the matrix of T relative to input basis $\mathcal{Z}$ and output basis $\mathcal{Z}$? How is this matrix related to A?

38. Continuing the previous exercise:
(a) Find one type 4 column operation on A making the $1, 1$-entry become 1. Find an ordered basis $\mathcal{X}'$ such that the new matrix represents T relative to $\mathcal{X}'$ and $\mathcal{Y}$.
(b) Find one type 4 row operation on A making the $1, 1$-entry become 1. Find an ordered basis $\mathcal{Y}'$ such that the new matrix represents T relative to $\mathcal{X}$ and $\mathcal{Y}'$.

39. Let M be a f.g. free module over a PID R, $\mathcal{X} = (x_1, \ldots, x_m)$ be an ordered basis of M, and $P \in M_m(R)$ be an invertible matrix. Consider the operation that maps $\mathcal{X}$ to $\mathcal{X}' = (x_1', \ldots, x_m')$, where $x_i' = \sum_{j=1}^{m} P(j, i)x_j$ for $1 \le i \le m$. Prove $\mathcal{X}'$ is an ordered basis of M. Prove that for any ordered basis $\mathcal{Y}$ of M, we can choose $P \in M_m(R)$ such that $\mathcal{X}' = \mathcal{Y}$. Show that operation (B4) is a special case of the operation considered here.

40. Let R be a PID and $T : N \to M$ be an R-linear map between two f.g. free R-modules. Let $\mathcal{X} = (x_1, \ldots, x_n)$ be an ordered basis for N, $\mathcal{Y} = (y_1, \ldots, y_m)$ be an ordered basis for M, and A be the matrix of T relative to X and Y.
(a) Given an invertible $P \in M_n(R)$, define $\mathcal{X}' = \mathcal{X}P$ as in Exercise 39, and let A' be the matrix of T relative to $\mathcal{X}'$ and $\mathcal{Y}$. How is A' related to A and P?
(b) Given an invertible $Q \in M_m(R)$, let $\mathcal{Y}' = \mathcal{Y}Q$ as in Exercise 39, and let A'' be the matrix of T relative to $\mathcal{X}$ and $\mathcal{Y}'$. How is A'' related to A and Q?
(c) Suppose $\mathcal{X}'$ and $\mathcal{Y}'$ are any ordered bases for N and M (respectively), and B is the matrix of T relative to $\mathcal{X}'$ and $\mathcal{Y}'$. Show that $B = UAV$ for some invertible $U \in M_m(R)$ and $V \in M_n(R)$.

41. Let a, b, d be nonzero elements in a PID (or UFD) R. Prove:
(a) $\operatorname{len}(ab) = \operatorname{len}(a) + \operatorname{len}(b)$;
(b) if $d|a$ in R, then $\operatorname{len}(d) \le \operatorname{len}(a)$, and equality holds iff $d \sim a$;
(c) if $d = \gcd(a, b)$ and a does not divide b, then $\operatorname{len}(d) < \operatorname{len}(a)$.

42. Suppose R is a UFD such that for all $x, y \in R$ and every gcd d of x and y in R, there exist $a, b \in R$ with $d = ax + by$. Prove that R must be a PID. (Given any nonzero ideal I of R, show I is generated by any $x \in I$ of minimum length.) This exercise shows that in UFDs that are not PIDs, the gcd of a list of elements is not always an R-linear combination of those elements.

43. Prove the Matrix Reduction Theorem in §18.9 for Euclidean domains R (see Exercise 6) without using unique factorization or length, by imitating the proof for $R = \mathbb{Z}$ in §16.11. Also show that the matrix reduction never requires general type 4 operations, but can be achieved using only elementary operations (B1), (B2), and (B3).

44. Let R be the PID $\{a + bi : a, b \in \mathbb{Z}\}$ (see Exercise 5). Use matrix reduction to find a Smith normal form of these matrices:

    (a) $\begin{bmatrix} 8 - 21i & 12 + 51i \\ 49 + 32i & -136 + 2i \end{bmatrix}$
    (b) $\begin{bmatrix} 4 + i & 3 & 1 + i \\ 16 + 5i & 10 - 7i & 1 - 3i \\ 1 + 10i & 8 - 3i & 3 - 7i \end{bmatrix}$

45. Solve Exercise 44 again, but find the rank and invariant factors by computing gcds of order k minors.

46. Let $R = \mathbb{Q}[x]$. Use matrix reduction to find a Smith normal form of these matrices:

    (a) $\begin{bmatrix} x^3 - x^2 - 3x + 2 & x^2 - 4 \\ x^3 - 6x + 4 & x^2 + x - 6 \end{bmatrix}$

    (b) $\begin{bmatrix} x^4 - x^2 + x & x^4 + x^3 - x^2 & x^3 - x \\ x^5 - x^4 - x^3 + 2x^2 - 2x & x^5 - 2x^3 + x^2 - x & x^4 - x^3 - x^2 + x \end{bmatrix}$

    (c) $\begin{bmatrix} x + 1 & x^2 + x - 1 \\ 1 & x \\ x^2 + x & x^3 + x^2 - x \end{bmatrix}$

47. Solve Exercise 46 again, but find the rank and invariant factors by computing gcds of order k minors.

48. Let R be a PID and $a \in R$. What is a Smith normal form of an $m \times n$ matrix all of whose entries equal a?

49. Given $c \in \mathbb{C}$, what is a Jordan canonical form of an $n \times n$ complex matrix all of whose entries equal c?

50. Given $c \in \mathbb{Q}$, what is the rational canonical form of the matrix in $M_n(\mathbb{Q})$ all of whose entries equal c?

51. For prime p, what is the rational canonical form of the matrix in $M_p(\mathbb{Z}_p)$ all of whose entries equal 1?

52. Write a computer program that finds the Smith normal form of a matrix $A \in M_{m,n}(\mathbb{Q}[x])$ via the matrix reduction algorithm described in §18.9.

53. Write a computer program to find the Smith normal form of A using the formulas for the invariant factors as gcds of minors of A (see §18.12).

54. Comment on the relative efficiency of the programs in the previous two exercises for large m and n.

55. Let $R = \mathbb{Q}[x]$ and $T : R^3 \to R^2$ be the R-linear map

    $$T((f, g, h)) = (x^3 f + (x^2 + x)g + x^2 h, (x^4 - x^3)f + x^2 g + x^3 h) \text{ for } f, g, h \in R.$$

    (a) What is the matrix of T relative to the standard ordered bases of R^3 and R^2?
    (b) Find a Smith normal form B of T and ordered bases $\mathcal{X}$ of R^3 and $\mathcal{Y}$ of R^2 such that B is the matrix of T relative to $\mathcal{X}$ and $\mathcal{Y}$.

56. Find the Betti number, invariant factors, and elementary divisors of each R-module M.

(a) $R = \mathbb{Q}[x]$, $M = R^4/(Rv_1 + Rv_2 + Rv_3)$ where

$$v_1 = (x^4 - 2x^3 + 2x^2, -2x^2, -x^2, x^2),$$
$$v_2 = (x^3 - x^2, x^2, x^2, 0),$$
$$v_3 = (x^5 - 2x^3 + 2x^2, x^4 - 2x^2, x^4 - x^2, x^2).$$

(b) $R = \mathbb{Q}[x]$, $M = R^3/(Rw_1 + Rw_2 + Rw_3)$ where $w_1 = (x^2, x^3, x^3)$, $w_2 = (x^3, x^3, x^2)$, and $w_3 = (x^2, x^3, x^4)$.

(c) $R = \{a + bi : a, b \in \mathbb{Z}\}$, $M = R^2/(Rz_1 + Rz_2)$ where $z_1 = (4 + 3i, 12 + 8i)$ and $z_2 = (5 - 7i, 7 - 4i)$.

57. Let R be a PID. Prove: if $a, b \in R$ satisfy $\gcd(a, b) = 1$, then R/Rab is isomorphic to $R/Ra \times R/Rb$ both as R-modules and as rings.

58. Show that for $R = \mathbb{Q}[x, y]$, R/Rxy and $R/Rx \times R/Ry$ are not isomorphic as R-modules (compare to the previous exercise).

59. Let A be an $m \times n$ matrix with entries in a PID R. Use the Cauchy–Binet Formula to prove: for all k between 1 and $\min(m, n)$ and all invertible $Q \in M_n(R)$, $g_k(AQ) \sim g_k(A)$. If Q is not invertible, is there any relation between $g_k(AQ)$ and $g_k(A)$?

60. Let A be an $m \times n$ matrix with entries in a PID R. Without using the Cauchy–Binet Formula, prove: for all k between 1 and $\min(m, n)$ and all invertible $P \in M_m(R)$ and all invertible $Q \in M_n(R)$, $g_k(PA) \sim g_k(A) \sim g_k(AQ)$. (Use ideas from §4.8 to show that each row of $(PA)_{I,J}$ is an R-linear combination of certain rows of $A_{[m],J}$. Then use multilinearity of the determinant as a function of the rows of a matrix (§5.6) to show that $\det((PA)_{I,J})$ is some R-linear combination of order k minors of A. The Cauchy–Binet Formula shows us explicitly what this linear combination is.)

61. Give an example of a commutative ring R and an R-module M such that $\text{tor}(M)$ is not a submodule of M.

62. Suppose R is an integral domain and $f : M \to N$ is an R-module isomorphism. Carefully check that $f[\text{tor}(M)] = \text{tor}(N)$ and that f induces an isomorphism $f' : M/\text{tor}(M) \to N/\text{tor}(N)$.

63. Let a be a nonzero element in a PID R. Prove that $\text{len}(R/Ra)$ [as defined in §17.18] equals $\text{len}(a)$ [as defined in §18.9].

64. Let N and N' be isomorphic R-modules, where R is a commutative ring, and fix $c \in R$. Prove cN is a submodule of N. Prove $cN \cong cN'$ and $N/cN \cong N'/cN'$. If $N = N_1 \times \cdots \times N_k$, prove $cN = cN_1 \times \cdots \times cN_k$.

65. Define $T : \mathbb{R}^4 \to \mathbb{R}^4$ by $T(v) = Av$ for $v \in \mathbb{R}^4$, where $A = \text{blk-diag}(J(3; 2), J(5; 2))$ is a Jordan canonical form. What is the rational canonical form of T? Use this and results about submodule lattices to find all T-invariant subspaces of $\mathbb{R}^4$ and a specific generator for each T-cyclic subspace.

66. Let $A = \text{blk-diag}(J(4; 2), J(4; 2))$ and $T(v) = Av$ for $v \in \mathbb{R}^4$. What is the rational canonical form of T? Show that $\mathbb{R}^4$ has infinitely many 1-dimensional T-invariant subspaces and infinitely many 2-dimensional T-invariant subspaces.

67. A matrix in $M_n(\mathbb{Q})$ has invariant factors $(x - 1, x^3 - 3x + 2, x^5 + x^4 - 5x^3 - x^2 + 8x - 4)$. Find n and the elementary divisors of the matrix.

68. A matrix in $M_m(\mathbb{Q})$ has elementary divisors

    $$[x-1, x-1, x-1, (x-1)^3, (x-1)^4, (x^2-2)^2, (x^2-2)^2, (x^2-2)^3, x^2+1, x^2+1].$$

    Find m and the invariant factors of the matrix.

69. Let M be a finitely generated module over a PID R. Give an alternate proof of the uniqueness of the elementary divisors of M by imitating the arguments for $\mathbb{Z}$-modules in §16.17 and §16.18.

70. (a) In §18.15, prove $f(g(L)) = L$ for all $L \in Y$.
    (b) Let R be a PID. Assume we have proved the uniqueness of the elementary divisors of a finitely generated R-module M (see Exercise 69). Use this result and (a) to prove the uniqueness of the invariant factors of M (cf. §16.19).

71. In §18.16, we proved that V was an $F[x]$-module by checking all the module axioms. Give a more conceptual proof of this fact by using Exercise 16 in Chapter 17 and the Universal Mapping Property for $F[x]$.

72. Let $T, S : V \to V$ be linear operators on an n-dimensional vector space V over a field F with $S \circ T = T \circ S$. Show that there exists a unique $F[x, y]$-module action $\cdot : F[x, y] \times V \to V$ such that for all $v \in V$, $x \cdot v = T(v)$, $y \cdot v = S(v)$, and for all $c \in F$, $c \cdot v$ is the given scalar multiplication in V.

73. Let F be a field and $h \in F[x]$ have degree $d > 0$. Show that the F-vector space $F[x]/F[x]h$ has ordered basis $(1 + F[x]h, x + F[x]h, \ldots, x^{d-1} + F[x]h)$.

74. Let T be a linear map on an n-dimensional F-vector space V. Let W be any T-invariant subspace of V such that for some $z \in W$, $\mathcal{B}_z = (z, T(z), \ldots, T^{d-1}(z))$ is an F-basis of W and $[T|_W]_{\mathcal{B}_z} = C_h$. Prove that $W = F[x]z$ is a T-cyclic subspace isomorphic to $F[x]/F[x]h$.

75. Define $T : \mathbb{R}^6 \to \mathbb{R}^6$ by $T((x_1, x_2, x_3, x_4, x_5, x_6)) = (x_2, x_3, x_4, x_5, x_6, x_1)$.
    (a) For $1 \le i \le 6$, find the T-annihilator of $\mathbf{e}_i$ in $\mathbb{R}[x]$.
    (b) What is the T-annihilator of $\mathbf{e}_1 + \mathbf{e}_3 + \mathbf{e}_5$?
    (c) What is the T-annihilator of $\mathbf{e}_1 + \mathbf{e}_4$?
    (d) If possible, find $v \in \mathbb{R}^6$ whose T-annihilator is $\mathbb{R}[x](x^2 - x + 1)$.
    (e) If possible, find $w \in \mathbb{R}^6$ whose T-annihilator is $\mathbb{R}[x](x^3 + 2x^2 + 2x + 1)$.

76. In the last paragraph of §18.17, confirm that each W_i^* is a cyclic $F[x]$-submodule of V generated by z_i^*, and $V \cong W_1^* \times \cdots \times W_\ell^* \cong \prod_{i=1}^{\ell} F[x]/F[x]g_i$.

77. Check the claim in §18.18 that $\mathcal{B} = (y_{e-1}, \ldots, y_2, y_1, y_0)$ is an ordered F-basis for W.

78. In §18.18, prove that if W is a T-invariant subspace of V such that $[T|_W]_{\mathcal{B}} = J(c; e)$ for some ordered basis $\mathcal{B}$ of W, then W is a T-cyclic subspace with $W = F[x]y \cong F[x]/F[x](x-c)^e$ for some $y \in W$. Use this to show that all Jordan canonical forms of T are obtained from (18.14) by reordering the Jordan blocks.

79. Let $h \in F[x]$ be monic of degree $n > 0$. Show the minimal polynomial of C_h is h. Show that $xI_n - C_h \in M_n(F[x])$ can be reduced by elementary row and column operations to $\mathrm{diag}(1, \ldots, 1, h)$. (Start by adding $-x$ times row i to row $i-1$, for $i = n, n-1, \ldots, 2$.) Show that $\chi_{C_h} = h$ by calculating $\det(xI_n - C_h)$.

80. Prove: If $A, B \in M_n(F)$ are similar matrices, then $m_A = m_B$ and $\chi_A = \chi_B$.

81. Show that for a block-diagonal matrix $B = \mathrm{blk\text{-}diag}(B_1, \ldots, B_k) \in M_n(F)$, $m_B = \mathrm{lcm}(m_{B_1}, \ldots, m_{B_k})$ and $\chi_B = \prod_{i=1}^{k} \chi_{B_i}$.

82. What are the minimal polynomial and characteristic polynomial of a Jordan block $J(c; e)$? What are the minimal polynomial and characteristic polynomial of a Jordan canonical form $\text{blk-diag}(J(c_1; e_1), \ldots, J(c_s, e_s))$?

83. Let F be a field. What are the invariant factors in the rational canonical form of the matrix bI_n? What are the invariant factors for a diagonal matrix $\text{diag}(a_1, \ldots, a_n)_{n \times n}$ where all $a_i \in F$ are distinct? Given that $a, b, c, d \in F$ are distinct, what are the invariant factors of $\text{diag}(a, a, a, a, b, b, b, b, c, c, d, d)$?

84. Let F be a field. Find (with proof) all matrices in $M_n(F)$ whose rational canonical form is diagonal.

85. Use the Jordan canonical form to prove that a matrix $A \in M_n(\mathbb{C})$ is diagonalizable iff the minimal polynomial of A in $\mathbb{C}[x]$ has no repeated roots.

86. For any field F, give a simple characterization of the rational canonical forms of diagonalizable matrices in $M_n(F)$.

87. Find the rational canonical form of each matrix in $M_n(\mathbb{Q})$:

(a) $\begin{bmatrix} 1 & 2 & 3 \\ 4 & 5 & 6 \\ 7 & 8 & 9 \end{bmatrix}$ (b) $\begin{bmatrix} 0 & 1 & 1 \\ 1 & 0 & 1 \\ 1 & 1 & 0 \end{bmatrix}$ (c) $\begin{bmatrix} 1 & 0 & 0 & 0 \\ 0 & -1 & -1 & 0 \\ 0 & 4 & 3 & 0 \\ 1 & -4 & -2 & 1 \end{bmatrix}$

(d) $\begin{bmatrix} 12 & 15 & 5 & -30 \\ 1 & 7/2 & 1/2 & -3 \\ 13 & 39/2 & 17/2 & -39 \\ 6 & 9 & 3 & -16 \end{bmatrix}$

88. Find a Jordan canonical form in $M_n(\mathbb{C})$ for each matrix in Exercise 87.

89. Prove: for any PID R and any $B \in M_{m,n}(R)$, B and B^{T} have the same rank and invariant factors.

90. Prove: for any field F and any $A \in M_n(F)$, A is similar to A^{T}.

91. Let F be a subfield of K. Prove: for all $A \in M_n(F)$, the rational canonical form of A in $M_n(F)$ is the same as the rational canonical form of A in $M_n(K)$.

Part VI

Universal Mapping Properties and Multilinear Algebra

19

Introduction to Universal Mapping Properties

The concept of a *Universal Mapping Property* (abbreviated UMP) occurs frequently in linear and multilinear algebra. Indeed, this concept pervades every branch of abstract algebra and occurs in many other parts of mathematics as well. To introduce this fundamental notion, we start by considering a well-known result about vector spaces over a field F.

Theorem on Bases and Linear Maps. Let V be an n-dimensional F-vector space with basis $X = \{x_1, \dots, x_n\}$. For any F-vector space W and any function $f : X \to W$, there exists a unique F-linear map $T : V \to W$ extending f (i.e., $T(x) = f(x)$ for all $x \in X$).

Proof of uniqueness of T: Any $v \in V$ can be uniquely expressed as $v = \sum_{i=1}^n c_i x_i$ for scalars $c_i \in F$. If T is to be an F-linear map extending f, then we must have

$$T(v) = T\left(\sum_{i=1}^n c_i x_i\right) = \sum_{i=1}^n c_i T(x_i) = \sum_{i=1}^n c_i f(x_i). \tag{19.1}$$

This shows that the map T is uniquely determined by f, if T exists at all.

Proof of existence of T: To prove existence of T, we define $T(\sum_{i=1}^n c_i x_i) = \sum_{i=1}^n c_i f(x_i)$ for all $c_i \in F$, as in formula (19.1). Note that T is a well-defined map from V into W, since every vector $v \in V$ has a unique expansion in terms of the basis X. Taking $v = x_j = 1x_j + \sum_{i \neq j} 0x_i$, the formula shows that $T(x_j) = 1f(x_j) + \sum_{i \neq j} 0f(x_i) = f(x_j)$ for all $x_j \in X$, so that T does extend f. To check F-linearity of T, fix $v, w \in V$ and $a \in F$. Write $v = \sum_{i=1}^n c_i x_i$ and $w = \sum_{i=1}^n d_i x_i$ for some $c_i, d_i \in F$. Then $v + w = \sum_{i=1}^n (c_i + d_i) x_i$ and $av = \sum_{i=1}^n (ac_i) x_i$. Using the definition of T several times, we get

$$T(v + w) = \sum_{i=1}^n (c_i + d_i) f(x_i) = \sum_{i=1}^n c_i f(x_i) + \sum_{i=1}^n d_i f(x_i) = T(v) + T(w);$$

$$T(av) = \sum_{i=1}^n (ac_i) f(x_i) = a \sum_{i=1}^n c_i f(x_i) = aT(v).$$

Therefore, T is an F-linear map. $\qquad\square$

We now restate the Theorem on Bases and Linear Maps in three equivalent ways. Throughout, we fix the field F, the vector space V, and the basis X. Let $i : X \to V$ be the inclusion map given by $i(x) = x$ for all $x \in X$. Note that T extends f iff $T(x) = f(x)$ for all $x \in X$ iff $T(i(x)) = f(x)$ for all $x \in X$ iff $(T \circ i)(x) = f(x)$ for all $x \in X$ iff the two functions $T \circ i : X \to W$ and $f : X \to W$ are equal.

1. *Diagram Completion Property.* For any F-vector space W and any function $f : X \to W$, there exists a unique F-linear map $T : V \to W$ such that the following diagram commutes.

DOI: 10.1201/9781003484561-19

(Saying the *diagram commutes* means, by definition, that $f = T \circ i$.)

2. *Unique Factorization Property.* For any F-vector space W and any function $f : X \to W$, there exists a unique F-linear map $T : V \to W$ such that the factorization $f = T \circ i$ holds.

3. *Bijection between Collections of Functions.* For any F-vector space W, there is a bijection from the collection

$$A = \{F\text{-linear maps } T : V \to W\}$$

onto the collection

$$B = \{\text{arbitrary maps } f : X \to W\}.$$

The bijection sends $T \in A$ to $T \circ i \in B$. The inverse bijection sends $f \in B$ to the unique F-linear map $T : V \to W$ that extends f.

The first two restatements follow immediately from the original theorem and the fact that T extends f iff $f = T \circ i$ iff the given diagram commutes. To prove the third restatement, define a map ϕ with domain A by $\phi(T) = T \circ i$ for $T \in A$. For any such T, $T \circ i$ is a function from X to W, so that ϕ does map A into the codomain B. The unique factorization property amounts to the statement that $\phi : A \to B$ is a bijection. In more detail, given any $f \in B$, that property says there exists a unique $T \in A$ with $f = T \circ i = \phi(T)$. Existence of such a T means that ϕ is onto, and uniqueness of T means that ϕ is one-to-one. So, ϕ is a bijection. The relation $f = \phi(T)$ now implies that $\phi^{-1}(f) = T$, where T is the unique F-linear map extending f.

The theorem and its three restatements are all referred to as the *universal mapping property* for the basis X of the vector space V. More precisely, we might refer to this result as the UMP for the inclusion map $i : X \to V$. Note that composition with i establishes a correspondence (bijection) between F-*linear* functions from V to another vector space W and *arbitrary* functions from X to W. The term "universal" indicates that the same map $i : X \to V$ works for all possible target vector spaces W.

The rest of this chapter discusses many examples of UMPs that occur in linear and abstract algebra. Even more UMPs are discussed in the following chapter on multilinear algebra. All of these universal mapping properties involve variations of the setup illustrated by the preceding example. In that example, we are interested in understanding special types of functions (namely, F-linear maps) from the fixed vector space V into arbitrary vector spaces W. The universal mapping property helps us understand these functions by setting up a bijection (for each fixed W) between these linear maps and another, simpler set of maps (namely, the set of all functions from the finite set X into W). The central idea of studying a linear map by computing its matrix relative to an ordered basis is really a manifestation of this universal mapping property (the columns of the matrix give the coordinates of $f(x_j)$ for each $x_j \in X$). Other UMPs have a similar purpose: roughly speaking, composition with a universal map induces a bijection from collections of functions having one kind of structure to collections of functions having another kind of structure. The bijection is valuable since one of the two kinds of functions might be much easier to understand and analyze than the other. For instance, in the next chapter, we study UMPs that convert multilinear maps (and their variations) into linear maps, thus permitting us to use linear algebra in the study of the more difficult subject of multilinear algebra.

19.1 Bases of Free R-Modules

The rest of this chapter assumes familiarity with the material on modules covered in Chapter 17. Our introductory example, involving bases of finite-dimensional vector spaces, readily generalizes to free modules over an arbitrary ring. Let R be a ring and M be a free left R-module with basis X. Recall this means that for each $v \in M$, there exists a unique expression $v = \sum_{x \in X} c_x x$ in which $c_x \in R$ and all but finitely many scalars c_x are zero. Let $i : X \to M$ be the inclusion map given by $i(x) = x$ for all $x \in X$. We have the following equivalent versions of the Universal Mapping Property for $i : X \to M$.

1. *Diagram Completion Property.* For any left R-module N and any function $f : X \to N$, there exists a unique R-module homomorphism $T : M \to N$ such that the following diagram commutes.

2. *Unique Factorization Property.* For any left R-module N and any function $f : X \to N$, there exists a unique R-module homomorphism $T : M \to N$ such that the factorization $f = T \circ i$ holds.

3. *Bijection between Collections of Functions.* For any left R-module N, there is a bijection from the collection

$$A = \{R\text{-module homomorphisms } T : M \to N\}$$

onto the collection

$$B = \{\text{arbitrary maps } f : X \to N\}.$$

The bijection sends $T \in A$ to $T \circ i \in B$. The inverse bijection sends $f \in B$ to the unique R-module homomorphism $T : M \to N$ that extends f.

The proof proceeds as before. Given N and $f : X \to N$, the only map $T : M \to N$ that could possibly be an R-module homomorphism extending f must be given by the formula

$$T\left(\sum_{x \in X} c_x x\right) = \sum_{x \in X} c_x f(x), \qquad \text{where all } c_x \in R.$$

So T is unique if it exists. To prove existence of T, use the preceding formula as the definition of T. Then T is a well-defined function from M into N (the definition is unambiguous, since M is free with basis X). You can check, as we did for vector spaces, that T is an R-module homomorphism extending f. This proves the first version of the UMP, and the other two versions follow as before.

19.2 Homomorphisms out of Quotient Modules

Let R be a ring and M be any left R-module with submodule N. Let $p : M \to M/N$ be the R-homomorphism given by $p(x) = x + N$ for all $x \in M$. The following universal mapping property of p helps explain the significance of the construction of the quotient module M/N. As before, we can state the UMP in several equivalent ways.

1. *UMP for Projection onto M/N (Diagram Completion Formulation):* For every left R-module Q and every R-linear map $f : M \to Q$ such that $f(x) = 0$ for all $x \in N$, there exists a unique R-module homomorphism $f' : M/N \to Q$ satisfying $f = f' \circ p$, meaning that $f'(x+N) = f(x)$ for all $x \in M$. (Furthermore, $\mathrm{img}(f') = \mathrm{img}(f)$ and $\ker(f') = \ker(f)/N$.)

2. *UMP for Projection onto M/N (Bijective Formulation):* For any left R-module Q, there is a bijection from the collection

$$A = \{\text{all } R\text{-module homomorphisms } f' : M/N \to Q\}$$

onto the collection

$$B = \{R\text{-module homomorphisms } f : M \to Q \text{ such that } f(x) = 0 \text{ for all } x \in N\}.$$

The bijection sends $f' \in A$ to $f' \circ p \in B$. The inverse bijection sends $f \in B$ to the unique R-module homomorphism $f' : M/N \to Q$ such that $f' \circ p = f$.

To prove the diagram completion property, let $f : M \to Q$ satisfy $f(x) = 0$ for all $x \in N$. Uniqueness of $f' : M/N \to Q$ follows from the requirement that $f = f' \circ p$. This requirement means that $f'(x + N) = f'(p(x)) = (f' \circ p)(x) = f(x)$ for all $x \in M$, and every element of the set M/N has the form $x+N$ for some $x \in M$. Thus, if f' exists at all, it must be defined by the formula $f'(x + N) = f(x)$ for $x \in M$. Naturally, then, this is the formula we must use in the proof that f' does exist. To see that the formula defines a single-valued function, suppose $x, y \in M$ are such that $x + N = y + N$. Then $x - y \in N$, so that $f(x - y) = 0$ by assumption on f. Since f is a homomorphism, we have $f(x) - f(y) = f(x - y) = 0$ and hence $f(x) = f(y)$. Therefore, $f'(x + N) = f(x) = f(y) = f'(y + N)$, proving that $f' : M/N \to Q$ is a well-defined (single-valued) function. Knowing this, we can now check that f' is a R-module homomorphism. For all $x, z \in M$ and $r \in R$, compute

$$f'((x+N)+(z+N)) = f'((x+z)+N) = f(x+z) = f(x)+f(z) = f'(x+N)+f'(z+N);$$

$$f'(r(x + N)) = f'((rx) + N) = f(rx) = rf(x) = rf'(x + N).$$

As seen in the uniqueness proof, the very definition of f' guarantees that $f = f' \circ p$.

To verify the bijective version of the UMP, consider the function ϕ with domain A given by $\phi(f') = f' \circ p$ for $f' \in A$. We check ϕ does map into the claimed codomain B. Given $f' \in A$, $f' \circ p$ is a R-homomorphism from M to Q, being a composition of R-homomorphisms. Also, $f' \circ p$ does send every $x \in N$ to 0_Q, since $(f' \circ p)(x) = f'(x + N) = f'(0 + N) = 0_Q$. The existence and uniqueness assertions proved above show that ϕ is a bijection.

To finish, we prove the parenthetical remark about the image and kernel of f'. Note

$$\mathrm{img}(f') = \{f'(w) : w \in M/N\} = \{f'(x + N) : x \in M\} = \{f(x) : x \in M\} = \mathrm{img}(f).$$

Next, $N \subseteq \ker(f)$ holds by our assumption on f, so the quotient module $\ker(f)/N$ exists. For each $x \in M$, we have $x + N \in \ker(f')$ iff $f'(x + N) = 0$ iff $f(x) = 0$ iff $x \in \ker(f)$ iff $x + N \in \ker(f)/N$.

19.3 Direct Product of Two Modules

Let R be a ring and M and N be left R-modules. We can form the direct product $M \times N = M \oplus N = \{(m,n) : m \in M, n \in N\}$, which is a left R-module under componentwise operations (see §17.5). We define four R-module homomorphisms associated with the module $M \times N$:

$$
\begin{array}{llll}
p: M \times N \to M & \text{given by} & p((m,n)) = m & \text{for } m \in M,\, n \in N; \\
q: M \times N \to N & \text{given by} & q((m,n)) = n & \text{for } m \in M,\, n \in N; \\
i: M \to M \oplus N & \text{given by} & i(m) = (m,0) & \text{for } m \in M; \\
j: N \to M \oplus N & \text{given by} & j(n) = (0,n) & \text{for } n \in N.
\end{array}
$$

We call p and q the *canonical projections*, and we call i and j the *canonical injections*. These maps satisfy the following identities:

$$p \circ i = \mathrm{id}_M, \quad q \circ j = \mathrm{id}_N, \quad q \circ i = 0, \quad p \circ j = 0, \quad i \circ p + j \circ q = \mathrm{id}_{M \oplus N}.$$

In this section and the next, we discuss two different universal mapping properties for $M \times N$, one involving the canonical projections, and the other involving the canonical injections.

1. *UMP for Projections (Diagram Completion Formulation):* Suppose Q is any left R-module, and we are given two R-homomorphisms $f: Q \to M$ and $g: Q \to N$. There exists a unique R-homomorphism $h: Q \to M \times N$ satisfying $f = p \circ h$ and $g = q \circ h$, meaning that this diagram commutes:

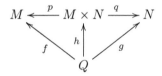

2. *UMP for Projections (Bijective Formulation):* For any left R-module Q, there is a bijection from the collection

   $$A = \{\text{all } R\text{-module homomorphisms } h: Q \to M \times N\} = \mathrm{Hom}_R(Q, M \times N)$$

   onto the collection

   $$
   \begin{aligned}
   B &= \{\text{pairs } (f,g) \text{ of } R\text{-homomorphisms } f: Q \to M,\ g: Q \to N\} \\
   &= \mathrm{Hom}_R(Q, M) \times \mathrm{Hom}_R(Q, N).
   \end{aligned}
   $$

   The bijection sends $h \in A$ to the pair $(p \circ h, q \circ h) \in B$. The inverse bijection sends $(f,g) \in B$ to the unique R-module homomorphism $h: Q \to M \times N$ such that $(f,g) = (p \circ h, q \circ h)$.

To verify the first version of the UMP, we start by proving the uniqueness of $h: Q \to M \times N$. If the function h exists at all, it must have the form $h(x) = (h_1(x), h_2(x))$ for all $x \in Q$, where $h_1: Q \to M$ and $h_2: Q \to N$ are certain functions. The requirement on h that $p \circ h = f$ forces $f(x) = p(h(x)) = p((h_1(x), h_2(x))) = h_1(x)$ for all $x \in Q$, so that $h_1 = f$. Similarly, the requirement $q \circ h = g$ forces $h_2 = g$. Therefore, h (if it exists at all) must be given by the formula $h(x) = (f(x), g(x))$ for $x \in Q$, so that uniqueness of h is proved.

To prove existence of h, define h (as we must) by setting $h(x) = (f(x), g(x))$ for $x \in Q$. By a computation similar to the one in the last paragraph, we have $p \circ h = f$ and $q \circ h = g$. We must still prove that h is an R-module homomorphism. Fix $x, y \in Q$ and $r \in R$, and calculate:

$$h(x + y) = (f(x + y), g(x + y)) = (f(x) + f(y), g(x) + g(y))$$
$$= (f(x), g(x)) + (f(y), g(y)) = h(x) + h(y);$$
$$h(rx) = (f(rx), g(rx)) = (rf(x), rg(x)) = r(f(x), g(x)) = rh(x).$$

To verify the bijective version of the UMP, consider the function ϕ with domain A given by $\phi(h) = (p \circ h, q \circ h)$ for $h \in A$. Note ϕ does map A into the set B, since for any $h \in A$, $p \circ h$ is an R-homomorphism from Q to M, and $q \circ h$ is an R-homomorphism from Q to N. The existence and uniqueness assertions proved above show that ϕ is a bijection.

19.4   Direct Sum of Two Modules

Let R be a ring, let M and N be left R-modules, let $M \oplus N = \{(m, n) : m \in M, n \in N\}$, and let i and j be the canonical injections defined in §19.3. We now consider the universal mapping property satisfied by these maps.

1. *UMP for Injections (Diagram Completion Formulation):* Suppose Q is any left R-module, and we are given two R-homomorphisms $f : M \to Q$ and $g : N \to Q$. There exists a unique R-homomorphism $h : M \oplus N \to Q$ satisfying $f = h \circ i$ and $g = h \circ j$, meaning that this diagram commutes:

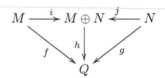

2. *UMP for Injections (Bijective Formulation):* For any left R-module Q, there is a bijection from the collection

    $$A = \{\text{all } R\text{-module homomorphisms } h : M \oplus N \to Q\} = \text{Hom}_R(M \oplus N, Q)$$

    onto the collection

    $$B = \{\text{pairs } (f, g) \text{ of } R\text{-homomorphisms } f : M \to Q, \; g : N \to Q\}$$
    $$= \text{Hom}_R(M, Q) \times \text{Hom}_R(N, Q).$$

    The bijection sends $h \in A$ to the pair $(h \circ i, h \circ j) \in B$. The inverse bijection sends $(f, g) \in B$ to the unique R-module homomorphism $h : M \oplus N \to Q$ such that $(f, g) = (h \circ i, h \circ j)$.

To verify the first version of the UMP, we start by proving the uniqueness of $h : M \oplus N \to Q$. The requirement $h \circ i = f$ means that $h((m, 0)) = h(i(m)) = f(m)$ for all $m \in M$. The requirement $h \circ j = g$ means that $h((0, n)) = h(j(n)) = g(n)$ for all $n \in N$. Now, h is also required to be an R-homomorphism. So if h exists at all, we must have

$$h((m, n)) = h((m, 0) + (0, n)) = h((m, 0)) + h((0, n)) = f(m) + g(n)$$

for all $m \in M$ and $n \in N$. This proves uniqueness of h.

To prove existence of $h : M \oplus N \to Q$, define h (as we must) by setting $h((m, n)) = f(m) + g(n)$ for all $m \in M$ and all $n \in N$. For each $m \in M$, we have $(h \circ i)(m) = h(i(m)) = h((m, 0)) = f(m) + g(0) = f(m) + 0 = f(m)$, so that $h \circ i = f$. Similarly, $f(0) = 0$ implies that $h \circ j = g$. To finish, we check that h is an R-homomorphism. Let $m, m' \in M$, $n, n' \in N$, $r \in R$, and calculate:

$$h((m, n) + (m', n')) = h((m + m', n + n')) = f(m + m') + g(n + n')$$
$$= f(m) + f(m') + g(n) + g(n') = f(m) + g(n) + f(m') + g(n')$$
$$= h((m, n)) + h((m', n')); \quad (19.2)$$

$$h(r(m, n)) = h((rm, rn)) = f(rm) + g(rn) = rf(m) + rg(n) = r(f(m) + g(n)) = rh((m, n)).$$

Note that commutativity of addition in Q was needed for the fourth equality in (19.2).

To verify the bijective version of the UMP, consider the function ϕ with domain A given by $\phi(h) = (h \circ i, h \circ j)$ for $h \in A$. Note ϕ does map into the set B, because for any $h \in A$, $h \circ i$ is a R-homomorphism from M to Q, and $h \circ j$ is a R-homomorphism from N to Q. The existence and uniqueness assertions proved above show that ϕ is a bijection.

You can check (Exercises 11 and 12) that all the constructions and results in this section and Section 19.3 extend to modules $M_1 \times \cdots \times M_n = M_1 \oplus \cdots \oplus M_n$ where there are only finitely many factors. In the next two sections, we generalize the universal mapping properties even further, discussing arbitrary direct products and direct sums of R-modules. When there are infinitely many nonzero factors, the direct product $\prod_{i \in I} M_i$ is distinct from the direct sum $\bigoplus_{i \in I} M_i$, so that the two universal mapping properties (one for projections, one for injections) involve different R-modules.

19.5 Direct Products of Arbitrary Families of R-Modules

Let R be a ring, I be an index set, M_i be a left R-module for each $i \in I$, and $M = \prod_{i \in I} M_i$ be the direct product of the M_i. Recall from §17.5 that elements $x \in M$ are functions $x : I \to \bigcup_{i \in I} M_i$ such that $x(i) \in M_i$ for all $i \in I$. We often visualize these functions as I-tuples $(x_i : i \in I)$, particularly when $I = \{1, 2, \ldots, n\}$. Module operations in M are defined pointwise: given $x, y \in M$, we have $(x + y)(i) = x(i) + y(i) \in M_i$ and $(rx)(i) = r(x(i)) \in M_i$ for all $i \in I$.

For each $i \in I$, we have the *canonical projection map* $p_i : M \to M_i$, which sends $x \in M$ to $x(i) \in M_i$. Each p_i is an R-homomorphism, since for $x, y \in M$ and $r \in R$,

$$p_i(x + y) = (x + y)(i) = x(i) + y(i) = p_i(x) + p_i(y);$$

$$p_i(rx) = (rx)(i) = r(x(i)) = rp_i(x).$$

The family of projection maps $\{p_i : i \in I\}$ satisfies the following universal mapping property.

1. *UMP for Direct Products of Modules (Diagram Completion Formulation):* Suppose Q is any left R-module, and for each $i \in I$ we have an R-homomorphism $f_i : Q \to M_i$. There exists a unique R-homomorphism $f : Q \to \prod_{i \in I} M_i$ satisfying $f_i = p_i \circ f$ for all $i \in I$, meaning that these diagrams commute for all $i \in I$:

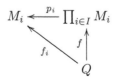

2. *UMP for Direct Products of Modules (Bijective Formulation):* For any left R-module Q, there is a bijection from the collection

$$A = \operatorname{Hom}_R\left(Q, \prod_{i\in I} M_i\right) = \left\{R\text{-module homomorphisms } f : Q \to \prod_{i\in I} M_i\right\}$$

onto the collection

$$B = \prod_{i\in I} \operatorname{Hom}_R(Q, M_i) = \{I\text{-tuples } (f_i : i \in I) \text{ of } R\text{-homomorphisms } f_i : Q \to M_i\}.$$

The bijection sends $f \in A$ to $(p_i \circ f : i \in I) \in B$. The inverse bijection sends $(f_i : i \in I) \in B$ to the unique R-module homomorphism $f : Q \to M$ such that $f_i = p_i \circ f$ for all $i \in I$.

The proof of this UMP is similar to the one given earlier for $M \times N$. Consider the diagram completion property for fixed Q and fixed R-maps $f_i : Q \to M_i$ ($i \in I$). If $f : Q \to M$ exists, then $f(x)$ is a function from I to $\bigcup_{i\in I} M_i$ for each $x \in Q$. Furthermore, the requirement that $f_i = p_i \circ f$ means that $f_i(x) = p_i(f(x))$ for all $x \in Q$. By definition of p_i, $p_i(f(x)) = f(x)(i)$ is the value of the function $f(x)$ at the point $i \in I$. Therefore, f is completely determined by the requirements $f_i = p_i \circ f$ for $i \in I$: we must have $f(x)(i) = f_i(x)$ for all $i \in I$ and all $x \in Q$. To say the same thing in I-tuple notation, we must have

$$f(x) = (f_i(x) : i \in I) \qquad \text{for all } x \in Q.$$

This proves uniqueness of f.

Proceeding to the existence proof, we must use the formula just written as the definition of $f : Q \to M$. Since $f(x)(i) = f_i(x) \in M_i$ for all $i \in I$, $f(x)$ is a well-defined element of $M = \prod_{i\in I} M_i$, and it is true that $f_i = p_i \circ f$ for all $i \in I$. We need only check that $f : Q \to M$ is an R-module homomorphism. Let $x, y \in Q$ and $r \in R$, and calculate:

$$
\begin{aligned}
f(x+y) &= (f_i(x+y) : i \in I) = (f_i(x) + f_i(y) : i \in I)\\
&= (f_i(x) : i \in I) + (f_i(y) : i \in I) = f(x) + f(y);\\
f(rx) &= (f_i(rx) : i \in I) = (rf_i(x) : i \in I)\\
&= r(f_i(x) : i \in I) = rf(x).
\end{aligned}
$$

For the second version of the UMP, define a function ϕ with domain A by setting $\phi(f) = (p_i \circ f : i \in I)$ for $f \in A$. For each $i \in I$, $p_i \circ f$ is an R-homomorphism from Q to M_i, as required. So ϕ does map into B, and the first version of the UMP shows that ϕ is a bijection.

19.6 Direct Sums of Arbitrary Families of R-Modules

Let R be a ring, I be an index set, M_i be a left R-module for each $i \in I$, and $M = \bigoplus_{i\in I} M_i$ be the direct sum of the M_i. Recall that M is the submodule of the direct product $\prod_{i\in I} M_i$ consisting of those functions $x : I \to \bigcup_{i\in I} M_i$ such that $x(i) \neq 0_{M_i}$ for only finitely many indices i. If the index set I is finite, then $\bigoplus_{i\in I} M_i = \prod_{i\in I} M_i$, but these modules are distinct when I is infinite and all M_i are nonzero. The module $\bigoplus_{i\in I} M_i$ is also referred to as the *coproduct* of the R-modules M_i.

For each $i \in I$, we have the *canonical injection map* $j_i : M_i \to M$, which sends $m \in M_i$ to the function $x \in M$ such that $x(i) = m$ and $x(k) = 0_{M_k}$ for all $k \neq i$ in I. Informally, $j_i(m)$ is the I-tuple with m in position i and zeroes in all other positions. For all $m \in M_i$ and $i, k \in I$, let $\delta_{k,i}m$ denote m if $k = i$ and 0_{M_k} if $k \neq i$. With this notation, we can write $j_i(m) = (\delta_{k,i}m : k \in I)$ for $m \in M_i$. Each j_i is an R-homomorphism, since for $m, n \in M_i$ and $r \in R$,

$$j_i(m + n) = (\delta_{k,i}(m + n) : k \in I) = (\delta_{k,i}m : k \in I) + (\delta_{k,i}n : k \in I) = j_i(m) + j_i(n);$$

$$j_i(rm) = (\delta_{k,i}(rm) : k \in I) = r(\delta_{k,i}m : k \in I) = rj_i(m).$$

Next, we claim that any $x \in M = \bigoplus_{i \in I} M_i$ can be written as follows:

$$x = \sum_{i \in I} j_i(x(i)).$$

The right side is a sum of elements of M, in which all but finitely many summands $j_i(x(i))$ are zero, since all but finitely many of the elements $x(i)$ are zero. To verify the claim that the two functions x and $\sum_{i \in I} j_i(x(i))$ are equal, we evaluate each of them at an arbitrary $k \in I$:

$$\left[\sum_{i \in I} j_i(x(i)) \right](k) = \sum_{i \in I} j_i(x(i))(k) = \sum_{i \in I} \delta_{k,i}x(i) = x(k).$$

Now we are ready to state the UMP satisfied by the family of injection maps $\{j_i : i \in I\}$.

1. *UMP for Direct Sums of Modules (Diagram Completion Formulation):* Suppose Q is any left R-module, and for each $i \in I$ we are given an R-homomorphism $f_i : M_i \to Q$. There exists a unique R-homomorphism $f : \bigoplus_{i \in I} M_i \to Q$ satisfying $f_i = f \circ j_i$ for all i, meaning that these diagrams commute for all $i \in I$:

$$M_i \xrightarrow{\ j_i\ } \bigoplus_{i \in I} M_i$$

2. *UMP for Direct Sums of Modules (Bijective Formulation):* For any left R-module Q, there is a bijection from the collection

$$A = \operatorname{Hom}_R \left(\bigoplus_{i \in I} M_i, Q \right) = \left\{ R\text{-module homomorphisms } f : \bigoplus_{i \in I} M_i \to Q \right\}$$

onto the collection

$$B = \prod_{i \in I} \operatorname{Hom}_R(M_i, Q) = \{I\text{-tuples } (f_i : i \in I) \text{ of } R\text{-homomorphisms } f_i : M_i \to Q\}.$$

The bijection sends $f \in A$ to $(f \circ j_i : i \in I) \in B$. The inverse bijection sends $(f_i : i \in I) \in B$ to the unique R-module homomorphism $f : \bigoplus_{i \in I} M_i \to Q$ such that $f_i = f \circ j_i$ for all $i \in I$.

We begin by proving uniqueness of f in the diagram completion version of the UMP. Fix Q and the R-maps $f_i : M_i \to Q$ for $i \in I$. Suppose $f : M = \bigoplus_{i \in I} M_i \to Q$ is any

R-homomorphism such that $f \circ j_i = f_i$ for all i. Take any function $x \in M$, and write $x = \sum_{i \in I} j_i(x(i))$, as above. We must have

$$
\begin{aligned}
f(x) &= f\left(\sum_{i \in I} j_i(x(i))\right) = \sum_{i \in I} f(j_i(x(i))) \\
&= \sum_{i \in I} (f \circ j_i)(x(i)) = \sum_{i \in I} f_i(x(i)).
\end{aligned}
$$

This proves that f is uniquely determined on M by the f_i, if f exists at all. (Note that all sums written here and below are really finite sums, since we disregard all zero summands. The calculations in this proof would not make sense for infinite direct products.)

To prove existence of f, define $f(x) = \sum_{i \in I} f_i(x(i))$ for all $x \in M$. This sum is a finite sum of elements of Q, since x is an element of the direct sum of the M_i, so $f : M \to Q$ is a well-defined function. To confirm that $f \circ j_k = f_k$ for fixed $k \in I$, let us check that these functions agree at each $y \in M_k$. First, $(f \circ j_k)(y) = f(j_k(y)) = \sum_{i \in I} f_i(j_k(y)(i))$. Now, $j_k(y)(i) = 0$ if $i \neq k$, while $j_k(y)(k) = y$. Therefore, the sum has at most one nonzero summand, corresponding to $i = k$, and $(f \circ j_k)(y) = f_k(j_k(y)(k)) = f_k(y)$. It remains to check that $f : M \to Q$ is an R-module homomorphism. Let $x, y \in M$ and $r \in R$, and calculate:

$$
\begin{aligned}
f(x+y) &= \sum_{i \in I} f_i((x+y)(i)) = \sum_{i \in I} f_i(x(i) + y(i)) \\
&= \sum_{i \in I} [f_i(x(i)) + f_i(y(i))] = \sum_{i \in I} f_i(x(i)) + \sum_{i \in I} f_i(y(i)) = f(x) + f(y); \\
f(rx) &= \sum_{i \in I} f_i((rx)(i)) = \sum_{i \in I} f_i(r(x(i))) \\
&= \sum_{i \in I} r f_i(x(i)) = r \sum_{i \in I} f_i(x(i)) = r f(x).
\end{aligned}
$$

For the second version of the UMP, define a function ϕ with domain A by setting $\phi(f) = (f \circ j_i : i \in I)$ for $f \in A$. For each $i \in I$, $f \circ j_i$ is an R-homomorphism from M_i to Q, as required. So ϕ does map into B, and the first version of the UMP shows that ϕ is a bijection.

Let I and K be index sets and M_i (for $i \in I$) and N_k (for $k \in K$) be left R-modules. By combining the bijections discussed in this section and the previous one, we obtain a bijection

$$
\mathrm{Hom}_R\left(\bigoplus_{i \in I} M_i, \prod_{k \in K} N_k\right) \to \prod_{i \in I} \prod_{k \in K} \mathrm{Hom}_R(M_i, N_k)
$$

that maps an R-module homomorphism $g : \bigoplus_{i \in I} M_i \to \prod_{k \in K} N_k$ to the tuple

$$
(p_k \circ g \circ j_i : i \in I, k \in K),
$$

where the p_k are the canonical projections of $\prod_k N_k$ and the j_i are the canonical injections of $\bigoplus_{i \in I} M_i$. Note that $p_k \circ g \circ j_i$ is an R-map from M_i to N_k. Given R-maps $g_{i,k} : M_i \to N_k$ for all $i \in I$ and $k \in K$, the inverse bijection maps the tuple $(g_{i,k} : i \in I, k \in K)$ to the R-homomorphism from $\bigoplus_{i \in I} M_i$ to $\prod_{k \in K} N_k$ such that

$$
(x_i : i \in I) \text{ maps to } \left(\sum_{i \in I} g_{i,k}(x_i) : k \in K\right).
$$

This function can also be written

$$\left(x \mapsto \left(\sum_{i \in I} g_{i,k}(p_i'(x)) : k \in K \right) : x \in \bigoplus_{i \in I} M_i \right),$$

where p_i' denotes the canonical projection of $\bigoplus_{j \in I} M_j$ onto M_i.

19.7 Solving Universal Mapping Problems

In this chapter, we have analyzed some basic algebraic constructions for modules and discovered the universal mapping properties (UMPs) of these constructions. The next chapter adopts the opposite point of view: we start by specifying some *universal mapping problem* (also abbreviated UMP), and we then seek to construct a new object and maps that solve this problem. If we succeed in finding a solution to the UMP, then we may inquire to what extent our solution is unique.

For example, our discussion of direct sums (coproducts) of R-modules suggests the analogous universal mapping problem for sets:

Problem (Coproducts for Sets). Given a family of sets $\{S_i : i \in I\}$, construct a set S and maps $j_i : S_i \to S$ satisfying the following UMP: for any set T and any collection of functions $g_i : S_i \to T$, there exists a unique function $g : S \to T$ with $g_i = g \circ j_i$ for all $i \in I$.

An equivalent formulation of this problem is to construct a set S and maps $j_i : S_i \to S$ such that, for any set T, there is a bijection from the set

$$A = \{\text{all functions } g : S \to T\}$$

onto the set

$$B = \{\text{families of functions } (g_i : i \in I) \text{ where } g_i : S_i \to T\}$$

given by $g \mapsto (g \circ j_i : i \in I)$.

Note that we cannot form the direct sum of the S_i, since the S_i are sets, not R-modules. So a modification of our construction for R-modules is required. Here is one possible solution.

Construction of Solution to the UMP. Let S be the disjoint union of the S_i. More formally, define

$$S = \{(i, x) : i \in I, \ x \in S_i\}.$$

For $i \in I$, define the function $j_i : S_i \to S$ by $j_i(x) = (i, x)$ for all $x \in S_i$. We must verify that S and the maps j_i do have the necessary universal mapping property. Assume T and $g_i : S_i \to T$ are given. To prove uniqueness, consider any function $g : S \to T$ satisfying $g_i = g \circ j_i$ for all $i \in I$. For $i \in I$ and $x \in S_i$, we then have $g((i, x)) = g(j_i(x)) = g_i(x)$. Thus, the value of g at every $(i, x) \in S$ is completely determined by the given g_i. So g is unique if it exists at all.

To prove existence, we must define $g((i,x)) = g_i(x)$ for all $(i,x) \in S$. It is immediate that $g : S \to T$ is a well-defined function such that $g_i = g \circ j_i$ for all $i \in I$. The bijective version of the UMP follows, as in earlier proofs, once we note that the function $g \mapsto (g \circ j_i : i \in I)$ does indeed map the domain A into the claimed codomain B.

We have now solved the UMP posed above. But is our solution unique? Certainly not — we can always change notation to obtain superficially different solutions to the UMP. For instance, we could have defined $S = \{(x,i) : i \in I, \ x \in S_i\}$ and $j_i(x) = (x,i)$ for $i \in I$ and $x \in S_i$. On the other hand, we claim our solution is unique up to a unique isomorphism compatible with the UMP. In the case at hand, this means that for any solution (S', j_i') to the UMP, there exists a unique bijection $g : S \to S'$ such that $j_i' = g \circ j_i$ for all $i \in I$. Briefly, although ungrammatically, we say that the solution (S, j_i) to the UMP is *essentially unique*.

Proof of Essential Uniqueness of Solution. Suppose (S', j_i') also solves the UMP. Applying the universal mapping property of (S, j_i) to the set $T = S'$ and the family of maps $g_i = j_i'$, we conclude that there exists a unique function $g : S \to S'$ with $j_i' = g \circ j_i$ for all $i \in I$, meaning that the following diagram commutes for all $i \in I$:

$$
\begin{array}{ccc}
S_i & \xrightarrow{\ j_i\ } & S \\
 & {\scriptstyle j_i'} \searrow & \downarrow {\scriptstyle g} \\
 & & S'
\end{array}
$$

To complete the proof of essential uniqueness, we need only show that g is a bijection. To do so, we can use the universal mapping property of (S', j_i') to construct a candidate for the inverse of g. Specifically, let $T = S$ and $g_i = j_i$ in the UMP for (S', j_i'). The UMP says that there exists a unique function $g' : S' \to S$ with $j_i = g' \circ j_i'$ for all $i \in I$, so that this diagram commutes for all $i \in I$:

$$
\begin{array}{ccc}
S_i & \xrightarrow{\ j_i'\ } & S' \\
 & {\scriptstyle j_i} \searrow & \downarrow {\scriptstyle g'} \\
 & & S
\end{array}
$$

Now, $\mathrm{id}_S \circ j_i = j_i = g' \circ j_i' = (g' \circ g) \circ j_i$. Thus, $h = \mathrm{id}_S$ and $h = g' \circ g$ are two functions from S to S with the property that $h \circ j_i = j_i$ for all $i \in I$. But according to the UMP for (S, j_i) (with $T = S$ and $g_i = j_i$), there is a *unique* map $h : S \to S$ with this property. Therefore, $g' \circ g = \mathrm{id}_S$. Visually, we are invoking the fact that exactly one map h makes the following diagram commute for all $i \in I$:

$$
\begin{array}{ccc}
S_i & \xrightarrow{\ j_i\ } & S \\
 & {\scriptstyle j_i} \searrow & \downarrow {\scriptstyle h} \\
 & & S
\end{array}
$$

Similarly, $\mathrm{id}_{S'} \circ j_i' = j_i' = g \circ j_i = (g \circ g') \circ j_i'$. So, $h = \mathrm{id}_{S'}$ and $h = g \circ g'$ are two functions from S' to S' such that $h \circ j_i' = j_i'$ for all $i \in I$ (see the diagram below). But according to the UMP for (S', j_i') (with $T = S'$ and $g_i = j_i'$), there is a *unique* map $h : S' \to S'$ with this property.

$$
\begin{array}{ccc}
S_i & \xrightarrow{\ j_i'\ } & S' \\
 & {\scriptstyle j_i'} \searrow & \downarrow {\scriptstyle h} \\
 & & S'
\end{array}
$$

Therefore, $g \circ g' = \mathrm{id}_{S'}$. We now see that g' is the two-sided inverse of g, so both functions are bijections. $\qquad\square$

The next chapter further develops the ideas presented here by posing and solving some universal mapping problems that appear at the foundations of multilinear algebra. For each UMP, we give an explicit construction showing that a solution to the UMP does exist. In each case, a proof completely analogous to the one just given proves that our solution is essentially unique, up to a unique isomorphism compatible with the universal maps. Once the universal mapping properties are available, we can use them to derive the fundamental facts about the algebraic structures occurring in multilinear algebra.

19.8 Summary

Here, we summarize the universal mapping properties discussed in this chapter. We state each result as a diagram completion property and as a bijection between appropriate collections of functions.

1. *UMP for Basis of a Finite-Dimensional Vector Space.* Let $X = \{x_1, \ldots, x_n\}$ be a basis of the vector space V over the field F and $i : X \to V$ be the inclusion map. For each F-vector space W, composition with i gives a bijection from the set $\mathrm{Hom}_F(V, W)$ of all F-linear maps $T : V \to W$ onto the set of all functions $f : X \to W$. So, for each function $f : X \to W$ there exists a unique F-linear map $T : V \to W$ extending f (meaning $f = T \circ i$):

Explicitly, $T(\sum_{i=1}^{n} c_i x_i) = \sum_{i=1}^{n} c_i f(x_i)$ for $c_i \in F$.

2. *UMP for Basis of a Free Module.* Let R be a ring, M be a free left R-module with basis X, and $i : X \to M$ be the inclusion map. For each left R-module N, composition with i gives a bijection from the set $\mathrm{Hom}_R(M, N)$ of all R-linear maps $T : M \to N$ onto the set of all functions $f : X \to N$. So, for each function $f : X \to N$ there exists a unique R-linear map $T : M \to N$ extending f (meaning $f = T \circ i$):

Explicitly, $T(\sum_{x \in X} c_x x) = \sum_{x \in X} c_x f(x)$ for all $c_x \in R$ such that only finitely many c_x are nonzero.

3. *UMP for Quotient Modules.* Let R be a ring, M be a left R-module, N be a submodule of M, and $p : M \to M/N$ be the canonical projection map. For each left R-module Q, composition with p gives a bijection from the set $\mathrm{Hom}_R(M/N, Q)$ of all R-linear maps $f' : M/N \to Q$ onto the set of those R-linear maps $f : M \to Q$ satisfying $f(z) = 0$ for all $z \in N$. So, for each R-linear

map f on M that sends all of N to zero, there exists a unique lifting of f to an R-linear map f' on M/N (meaning $f = f' \circ p$):

Explicitly, $f'(x + N) = f(x)$ for all $x \in M$. Moreover, $\mathrm{img}(f') = \mathrm{img}(f)$ and $\ker(f') = \ker(f)/N$.

4. *UMP for Direct Product of Two Modules.* Let R be a ring, M and N be left R-modules, and $p : M \times N \to M$ and $q : M \times N \to N$ be the canonical projections. For each left R-module Q, the map $h \mapsto (p \circ h, q \circ h)$ is a bijection from $\mathrm{Hom}_R(Q, M \times N)$ onto $\mathrm{Hom}_R(Q, M) \times \mathrm{Hom}_R(Q, N)$. So, for each pair of R-linear maps (f, g) with $f : Q \to M$ and $g : Q \to N$, there exists a unique R-linear $h : Q \to M \times N$ with $f = p \circ h$ and $g = q \circ h$:

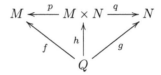

Explicitly, $h(x) = (f(x), g(x))$ for $x \in Q$.

5. *UMP for Direct Sum of Two Modules.* Let R be a ring, M and N be left R-modules, and $i : M \to M \oplus N$ and $j : N \to M \oplus N$ be the canonical injections. For each left R-module Q, the map $h \mapsto (h \circ i, h \circ j)$ is a bijection from $\mathrm{Hom}_R(M \oplus N, Q)$ onto $\mathrm{Hom}_R(M, Q) \times \mathrm{Hom}_R(N, Q)$. So, for each pair of R-linear maps (f, g) with $f : M \to Q$ and $g : N \to Q$, there exists a unique R-linear $h : M \oplus N \to Q$ with $f = h \circ i$ and $g = h \circ j$:

Explicitly, $h((m, n)) = f(m) + g(n)$ for $m \in M$ and $n \in N$.

6. *UMP for Direct Product of a Family of Modules.* Let R be a ring, I be an index set, M_i be a left R-module for each $i \in I$, and $p_i : \prod_{j \in I} M_j \to M_i$ be the canonical projections. For each left R-module Q, the map $f \mapsto (p_i \circ f : i \in I)$ is a bijection from $\mathrm{Hom}_R(Q, \prod_{j \in I} M_j)$ onto $\prod_{j \in I} \mathrm{Hom}_R(Q, M_j)$. So, for each family of R-linear maps $(f_i : i \in I)$ with $f_i : Q \to M_i$ for all $i \in I$, there exists a unique R-linear $f : Q \to \prod_{j \in I} M_j$ with $f_i = p_i \circ f$ for all $i \in I$:

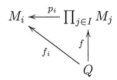

Explicitly, $f(x) = (f_i(x) : i \in I)$ for $x \in Q$.

7. *UMP for Direct Sum of a Family of Modules.* Let R be a ring, I be an index set, M_i be a left R-module for each $i \in I$, and $j_i : M_i \to \bigoplus_{k \in I} M_k$ be the canonical injections. For each left R-module Q, the map $f \mapsto (f \circ j_i : i \in I)$ is a bijection from $\operatorname{Hom}_R(\bigoplus_{k \in I} M_k, Q)$ onto $\prod_{k \in I} \operatorname{Hom}_R(M_k, Q)$. So, for each family of R-linear maps $(f_i : i \in I)$ with $f_i : M_i \to Q$ for all $i \in I$, there exists a unique R-linear $f : \bigoplus_{k \in I} M_k \to Q$ with $f_i = f \circ j_i$ for all $i \in I$:

$$M_i \xrightarrow{\ j_i\ } \bigoplus_{k \in I} M_k$$

with maps f_i and f to Q.

Explicitly, $f(x) = \sum_{i \in I} f_i(x_i)$ for $x = (x_i : i \in I) \in \bigoplus_{k \in I} M_k$.

8. Combining the last two items, we have a bijection

$$\operatorname{Hom}_R\left(\bigoplus_{i \in I} M_i, \prod_{k \in K} N_k\right) \longrightarrow \prod_{i \in I} \prod_{k \in K} \operatorname{Hom}_R(M_i, N_k)$$

that sends an R-map $g : \bigoplus_{i \in I} M_i \to \prod_{k \in K} N_k$ to $(p_k \circ g \circ j_i : i \in I, k \in K)$.

9. *UMP for Coproduct of Sets.* Let I be an index set and S_i be a set for each $i \in I$. Define $S = \{(i, x) : i \in I, x \in S_i\}$ and define $j_i : S_i \to S$ by $j_i(x) = (i, x)$ for $i \in I$ and $x \in S_i$. For each set T, the map $f \mapsto (f \circ j_i : i \in I)$ is a bijection from the set of functions from S to T to the set of families $(f_i : i \in I)$ where $f_i : S_i \to T$ for $i \in I$. So, for each family $(f_i : i \in I)$ of functions from S_i to T, there exists a unique $f : S \to T$ with $f_i = f \circ j_i$ for all $i \in I$:

$$S_i \xrightarrow{\ j_i\ } S$$

with maps f_i and f to T.

Explicitly, $f((i, x)) = f_i(x)$ for $i \in I$ and $x \in S_i$.

The solution of a universal mapping problem is essentially unique, meaning that for any two solutions to a UMP, there is a unique bijection between the underlying sets that respects the associated universal maps.

19.9 Exercises

In these exercises, assume R is a ring unless otherwise stated.

1. Let V and W be nonzero finite-dimensional vector spaces over a field F. Let $X = \{x_1, \ldots, x_n\}$ be a subset of V. Assume X spans V but is linearly dependent over F. (a) Prove or disprove: for every function $f : X \to W$, there exists an F-linear map $T : V \to W$ extending f. (b) Prove or disprove: for all $f : X \to W$, there is at most one F-linear map $T : V \to W$ extending f.

2. Repeat Exercise 1, but now assume X is a linearly independent subset of V that does not span V.

3. Give complete details of the proof of the UMP for bases of a free R-module stated in §19.1. Do not assume the basis X is finite.

4. **UMP for Quotient Groups.** Let $(G, \star)$ be a group with normal subgroup N. Let $p : G \to G/N$ be the projection $p(x) = x \star N$ for $x \in G$. (a) Prove: for every group L and every group homomorphism $f : G \to L$ such that $f(x) = e_L$ for all $x \in N$, there exists a unique group homomorphism $f' : G/N \to L$ such that $f = f' \circ p$. What are $\text{img}(f')$ and $\ker(f')$? (b) Restate (a) in terms of a bijection between two collections of functions.

5. **UMP for Quotient Rings.** Let $(R, +, \cdot)$ be a ring with ideal I. Let $p : R \to R/I$ be the projection $p(x) = x+I$ for $x \in R$. Formulate and prove a universal mapping property characterizing the quotient ring R/I and the map $p : R \to R/I$.

6. Let M and N be left R-modules. Check carefully that the canonical projections and injections for $M \times N = M \oplus N$ are R-linear and satisfy the following identities:

$$p \circ i = \text{id}_M, \quad q \circ j = \text{id}_N, \quad q \circ i = 0, \quad p \circ j = 0, \quad i \circ p + j \circ q = \text{id}_{M \oplus N}.$$

7. Assume R is a commutative ring. Prove that the bijection

$$\phi : \text{Hom}_R(Q, M \times N) \to \text{Hom}_R(Q, M) \times \text{Hom}_R(Q, N)$$

constructed in §19.3 is an R-linear map.

8. Assume R is commutative. Prove that the bijection

$$\phi : \text{Hom}_R(M \times N, Q) \to \text{Hom}_R(M, Q) \times \text{Hom}_R(N, Q)$$

constructed in §19.4 is an R-linear map.

9. Assume R is commutative. Prove $\text{Hom}_R(Q, \prod_{i \in I} M_i)$ and $\prod_{i \in I} \text{Hom}_R(Q, M_i)$ are isomorphic R-modules, via the bijection in §19.5.

10. Assume R is commutative. Prove $\text{Hom}_R(\bigoplus_{i \in I} M_i, Q)$ and $\prod_{i \in I} \text{Hom}_R(M_i, Q)$ are isomorphic R-modules, via the bijection in §19.6.

11. For fixed $k > 0$, let $M_1, M_2, \ldots, M_k$ be left R-modules. Prove that for all left R-modules Q, there is a bijection from $\text{Hom}_R(Q, M_1 \times M_2 \times \cdots \times M_k)$ to $\text{Hom}_R(Q, M_1) \times \text{Hom}_R(Q, M_2) \times \cdots \times \text{Hom}_R(Q, M_k)$
(a) by using induction on k and the bijections in §19.3;
(b) by imitating the construction in §19.3.

12. For fixed $k > 0$, let $M_1, M_2, \ldots, M_k$ be left R-modules. Prove that for all left R-modules Q, there is a bijection from $\text{Hom}_R(M_1 \oplus M_2 \oplus \cdots \oplus M_k, Q)$ to $\text{Hom}_R(M_1, Q) \times \text{Hom}_R(M_2, Q) \times \cdots \times \text{Hom}_R(M_k, Q)$
(a) by using induction on k and the bijections in §19.4;
(b) by imitating the construction in §19.4.

13. Assume R is a commutative ring. With the setup in §19.1, show that the bijection $T \mapsto T \circ i$ is an R-module isomorphism between the R-module $\text{Hom}_R(M, N)$ and the product R-module N^X.

14. With the setup in §19.5, let $f : Q \to \prod_{i \in I} M_i$ be the R-map corresponding to a given family of R-maps $f_i : Q \to M_i$. Prove $\ker(f) = \bigcap_{i \in I} \ker(f_i)$.

15. With the setup in §19.6, let $f : \bigoplus_{i \in I} M_i \to Q$ be the R-map corresponding to a given family of R-maps $f_i : M_i \to Q$. Prove $\text{img}(f) = \sum_{i \in I} \text{img}(f_i)$.

16. **UMP for Products of Sets.** Given an index set I and a set S_i for each $i \in I$, formulate and prove a universal mapping property satisfied by the Cartesian product set $S = \prod_{i \in I} S_i$ and the canonical projection functions $p_i : S \to S_i$ given by $p_i((x_k : k \in I)) = x_i$ for each $i \in I$.

17. **UMP for Products of Groups.** Given a family of groups $\{G_i : i \in I\}$ (not necessarily commutative), construct a group G and group homomorphisms $p_i : G \to G_i$ (for all $i \in I$) such that, for any group K, there is a bijection from the set A of all group homomorphisms $f : K \to G$ onto the set B of all families of group homomorphisms $(f_i : i \in I)$ with $f_i : K \to G_i$ for all $i \in I$, given by $f \mapsto (p_i \circ f : i \in I)$ for $f \in A$.

18. **UMP for Products of Rings.** Repeat the previous exercise, replacing groups and group homomorphisms by rings and ring homomorphisms.

19. In §19.4, we showed that for any two left R-modules M and N, the direct sum $M \oplus N$ and the canonical injections $i : M \to M \oplus N$ and $j : N \to M \oplus N$ satisfy the UMP for the coproduct of two modules. (a) Does this construction of the coproduct still work if we assume M and N are groups (possibly non-commutative), and demand that all maps be group homomorphisms? (b) Does this construction of the coproduct still work if we assume M and N are rings and demand that all maps be ring homomorphisms? What if we only allow commutative rings?

20. Let M be a left R-module with submodule N, and suppose there is a left R-module Z and a map $q : M \to Z$ such that $q[N] = \{0_Z\}$, and Z and q satisfy the UMP for quotient modules from §19.2. Prove that there exists a unique R-module isomorphism $g : M/N \to Z$ with $q = g \circ p$, where $p : M \to M/N$ is the canonical projection. (Imitate the essential uniqueness proof in §19.7.)

21. Give a specific example using $\mathbb{Z}$-modules to show that the result of Exercise 20 might fail without the hypothesis that $q[N] = \{0_Z\}$.

22. Let $\{M_i : i \in I\}$ be an indexed family of left R-modules. Carefully state what it means to say that the direct product $\prod_{i \in I} M_i$ and the associated projection maps are *essentially unique*, and then prove it.

23. Let $\{M_i : i \in I\}$ be an indexed family of left R-modules. Carefully state what it means to say that the direct sum $\bigoplus_{i \in I} M_i$ and the associated injection maps are *essentially unique*, and then prove it.

24. Given left R-modules M and N, prove that $M \times N \cong N \times M$ using the fact that both modules solve the same universal mapping problem. Then find a formula for the unique isomorphism compatible with the canonical injections.

25. We are given an index set I, left R-modules M_i and N_i for each $i \in I$, and an R-map $f_i : M_i \to N_i$ for each $i \in I$. Let $p_i : \prod_{k \in I} M_k \to M_i$ and $q_i : \prod_{k \in I} N_k \to N_i$ be the canonical projection maps, for each $i \in I$.
    (a) Use the UMP for direct products to show there exists a unique R-module homomorphism $F : \prod_{k \in I} M_k \to \prod_{k \in I} N_k$ such that $q_i \circ F = f_i \circ p_i$ for all $i \in I$.
    (b) Find an explicit formula for F, and describe $\ker(F)$ and $\mathrm{img}(F)$.
    (c) Write $F = F(f_i : i \in I) = F(f_i)$ to indicate the dependence of F on the given maps f_i. Suppose P_i is a left R-module and $g_i : N_i \to P_i$ is an R-map, for all $i \in I$. Let $r_i : \prod_{k \in I} P_k \to P_i$ be the canonical projection maps. Prove that $F(g_i \circ f_i) = F(g_i) \circ F(f_i)$ in two ways: using the explicit formula in (b), and using the uniqueness of F proved in (a).

26. With the setup in Exercise 25, state and prove results analogous to (a), (b), and
   (c) of that exercise for a map $G : \bigoplus_{k \in I} M_k \to \bigoplus_{k \in I} N_k$ induced by the f_i and
   compatible with the canonical injections.

27. Let M, N, and P be free left R-modules with bases X, Y, and Z and inclusion
   maps $i : X \to M$, $j : Y \to N$, and $k : Z \to P$.
   (a) Show that for each function $f : X \to Y$, there is a unique R-homomorphism
   $F(f) : M \to N$ with $F(f) \circ i = j \circ f$. Do not find a formula for $F(f)$.
   (b) Use (a) (and similar results for functions from Y to Z, etc.) to show that for
   all functions $f : X \to Y$ and $g : Y \to Z$, $F(g \circ f) = F(g) \circ F(f)$.
   (c) Use (a) to show that $F(\mathrm{id}_X) = \mathrm{id}_M$. [In the language of category theory, F is a
   functor from the category of sets and functions to the category of left R-modules
   and R-maps.]

28. **UMP for Quotient Sets.** Let X be a set, $\sim$ be an equivalence relation on
   X, and $X/\sim$ be the set of all equivalence classes of $\sim$. Call a function f with
   domain X *compatible with* $\sim$ iff for all $x, y \in X$, $x \sim y$ implies $f(x) = f(y)$.
   Let $p : X \to X/\sim$ be the map that sends each $x \in X$ to its equivalence class
   $[x]$ relative to $\sim$, given by $[x] = \{z \in X : x \sim z\}$. Note that p is surjective and
   compatible with $\sim$. (a) Prove that for every set Z, the map $h \mapsto h \circ p$ defines
   a bijection from the set A of all functions $h : X/\sim \to Z$ to the set B of all
   functions $f : X \to Z$ that are compatible with $\sim$. (b) Prove that $X/\sim$ (and the
   map p, compatible with $\sim$) is the essentially unique solution to the UMP in (a).
   (c) Explain how quotient modules are a special case of the construction in this
   problem (cf. §19.2).

29. **UMP for Quotient Topologies.** A *topological space* is a set X and a family
   of subsets of X, called *open sets*, such that $\emptyset$ and X are open; the union of any
   collection of open sets is open; and the intersection of finitely many open sets is
   open. A function $f : X \to Y$ between two such spaces is *continuous* iff for all open
   subsets V of Y, $f^{-1}[V]$ is an open subset of X. Let X be a topological space and $\sim$
   be an equivalence relation on X. Define $X/\sim$ and $p : X \to X/\sim$ as in Exercise 28.
   Define $V \subseteq X/\sim$ to be an open set iff $p^{-1}[V] \subseteq X$ is open in the given topological
   space X. (a) Show that this definition makes $X/\sim$ into a topological space and
   p into a continuous surjective map. (b) Prove that for every topological space Z,
   the map $h \mapsto h \circ p$ defines a bijection from the set of all continuous functions
   $h : X/\sim \to Z$ to the set of all continuous functions $f : X \to Z$ that are compatible
   with $\sim$. (Explain why it suffices, using Exercise 28(a), to show that h is continuous
   if and only if $h \circ p$ is continuous.)

30. **Products of Topological Spaces.** Let $\{X_i : i \in I\}$ be a family of topological
   spaces. Construct a *product* of the spaces X_i satisfying a universal mapping
   property analogous to the product of modules (all functions are required to be
   continuous here).

31. **Coproducts of Topological Spaces.** Let $\{X_i : i \in I\}$ be a family of topological
   spaces. Construct a *coproduct* of the spaces X_i satisfying a universal mapping
   property analogous to direct sums of modules and coproducts of sets (all functions
   considered must be continuous here).

32. **UMP for Polynomial Rings.** Let R be a commutative ring and $i : R \to R[x]$
   be given by $i(r) = (r, 0, 0, \ldots)$ for r in R (meaning that i maps r to the constant
   polynomial with constant term r). Prove: for each commutative ring S, the map
   $H \mapsto (H \circ i, H(x))$ is a bijection from the set of ring homomorphisms $H : R[x] \to$

S to the set of pairs (h, c), where $h : R \to S$ is a ring homomorphism and $c \in S$ (cf. §3.5).

33. Fix a positive integer m. Generalize the previous exercise to obtain a bijective formulation of the UMP for the polynomial ring $R[x_1, \ldots, x_m]$ (cf. §3.21).

34. **Localization of a Commutative Ring.** For any commutative ring T, we say $t \in T$ is a *unit of* T iff there exists $u \in T$ with $tu = 1_T = ut$. Let T^* be the set of units of T. The goal of this exercise is to solve the following universal mapping problem. Let R be a commutative ring and S be a subset of R that contains 1_R and is closed under the multiplication of R. Construct a commutative ring L and a ring homomorphism $i : R \to L$ such that $i(s) \in L^*$ for all $s \in S$; and for all commutative rings T, the map $g \mapsto g \circ i$ defines a bijection from the set of all ring homomorphisms $g : L \to T$ onto the set of all ring homomorphisms $f : R \to T$ such that $f(s) \in T^*$ for all $s \in S$. The diagram below visualizes what is needed:

$$
\begin{array}{ccc}
R \xrightarrow{\ i\ } L & & (i[S] \subseteq L^*) \\
 \searrow_{f} \ \big\downarrow{g} & & \\
 T & & (f[S] \subseteq T^*)
\end{array}
$$

Intuitively, we need to build a ring L in which all elements of S become invertible, such that L is as close to R as possible. The idea is to use *fractions* r/s, with $r \in R$ and $s \in S$, as the elements of L. Proceed formally as follows.

(a) Let $X = R \times S = \{(r, s) : r \in R, s \in S\}$. Define a binary relation $\sim$ on X by setting $(r, s) \sim (r', s')$ iff there exists $t \in S$ with $t(rs' - sr') = 0$. Check that $\sim$ is an equivalence relation on X.

(b) Let L be the set of equivalence classes of $\sim$ on X and r/s be the equivalence class of $(r, s) \in X$. For r/s and u/v in L, define $r/s + u/v = (rv + su)/(sv)$ and $(r/s) \cdot (u/v) = (ru)/(sv)$. Check that these two operations are well-defined.

(c) Verify that $(L, +, \cdot)$ is a commutative ring.

(d) Define $i : R \to L$ by setting $i(r) = r/1_R$ for all $r \in R$. Check that i is a ring homomorphism, and $i(s) \in L^*$ for all $s \in S$.

(e) Verify that L and i solve the UMP described above.

(f) Explain why L and i are essentially unique.

(g) Explain why the construction of $\mathbb{Q}$ from $\mathbb{Z}$ is a special case of the construction in this exercise.

35. Define an *inverse chain* to be a collection of left R-modules $(M_n : n \geq 0)$ and R-maps $f_n : M_n \to M_{n-1}$ for $n > 0$:

$$
M_0 \xleftarrow{f_1} M_1 \xleftarrow{f_2} M_2 \xleftarrow{f_3} \cdots \xleftarrow{f_{n-1}} M_{n-1} \xleftarrow{f_n} M_n \xleftarrow{f_{n+1}} \cdots .
$$

Solve the following universal mapping problem: construct a left R-module L and R-maps $p_n : L \to M_n$ for $n \geq 0$ with $f_n \circ p_n = p_{n-1}$ for all $n > 0$, such that for any left R-module P and R-maps $g_n : P \to M_n$ (for $n \geq 0$) with $f_n \circ g_n = g_{n-1}$ for all $n > 0$, there exists a unique R-map $h : P \to L$ with $p_n \circ h = g_n$ for all $n \geq 0$. L is called the *inverse limit* of the inverse chain. (Define L to be a certain submodule of the direct product $\prod_{n \geq 0} M_n$.)

36. Define a *direct chain* to be a collection of left R-modules $(M_n : n \geq 0)$ and R-maps $f_n : M_n \to M_{n+1}$ for $n \geq 0$:

$$
M_0 \xrightarrow{f_0} M_1 \xrightarrow{f_1} M_2 \xrightarrow{f_2} \cdots \xrightarrow{f_{n-1}} M_n \xrightarrow{f_n} M_{n+1} \xrightarrow{f_{n+1}} \cdots .
$$

Solve the following universal mapping problem: construct a left R-module D and R-maps $j_n : M_n \to D$ for $n \geq 0$ with $j_{n+1} \circ f_n = j_n$ for all $n \geq 0$, such that for any left R-module P and R-maps $g_n : M_n \to P$ $(n \geq 0)$ with $g_{n+1} \circ f_n = g_n$ for all $n \geq 0$, there exists a unique R-map $h : D \to P$ with $h \circ j_n = g_n$ for all $n \geq 0$. D is called the *direct limit* of the direct chain. (Define D to be a certain quotient module of the direct sum $\bigoplus_{n \geq 0} M_n$. Use known UMPs for direct sums and quotient modules.)

37. **Free Monoid Generated by a Set.** A *monoid* is a pair $(M, \star)$ satisfying the first three group axioms in Table 1.1 (closure, associativity, and identity). Given monoids M and N, a *monoid homomorphism* is a map $f : M \to N$ such that $f(xy) = f(x)f(y)$ for all $x, y \in M$, and $f(1_M) = 1_N$. Given any set X, our goal is to solve the following universal mapping problem (cf. §19.1): construct a monoid M and a function $i : X \to M$ such that for any monoid N and any function $f : X \to N$, there exists a unique monoid homomorphism $T : M \to N$ with $f = T \circ i$. (a) Let M be the set of all finite sequences $w = w_1 w_2 \cdots w_k$ (called *words*) where $k \geq 0$ and each $w_i \in X$. Note that the empty sequence, denoted ϵ, is in M. Given $w = w_1 w_2 \cdots w_k$ and $y = y_1 y_2 \cdots y_m$ in M, define $w \star y$ to be the concatenation $w_1 w_2 \cdots w_k y_1 y_2 \cdots y_m$. Show that $(M, \star)$ is a monoid with identity ϵ. (b) Define $i : X \to M$ by letting $i(x)$ be the word (sequence) of length 1 with sole entry x, for $x \in X$. Prove that M and i solve the universal mapping problem posed above.

38. **Coproduct of Two Groups.** Given groups M and N, our goal is to construct a group $M * N$ and group homomorphisms $i : M \to M * N$ and $j : N \to M * N$ satisfying an analog of the universal mapping problem in §19.4 (with modules and R-maps replaced by groups and group homomorphisms). Let $M' = M \sim \{e_M\}$ be the set of non-identity elements in M; let $N' = N \sim \{e_N\}$. If needed, change notation so that M' and N' are disjoint sets. Let the set $M * N$ consist of all words $w_1 w_2 \cdots w_k$ with $k \geq 0$, $w_i \in M' \cup N'$ for all i, and (for all $i < k$) $w_i \in M'$ iff $w_{i+1} \in N'$ (meaning that letters alternate between M' and N'). Given $w = w_1 w_2 \cdots w_k$ and $y = y_1 y_2 \cdots y_m$ in $M * N$, define $w \star y$ by the following recursive rules. If $w = \epsilon$ (where ϵ is the empty word), then $w \star y = y$. If $y = \epsilon$, then $w \star y = w$. If $w_k \in M'$ and $y_1 \in N'$, or if $w_k \in N'$ and $y_1 \in M'$, then $w \star y$ is the concatenation $w_1 w_2 \cdots w_k y_1 y_2 \cdots y_m$. If $w_k, y_1 \in M'$ and $w_k y_1 = z \neq e_M$, then $w \star y = w_1 \cdots w_{k-1} z y_2 \cdots y_m$. If $w_k, y_1 \in M'$ and $w_k y_1 = e_M$, we recursively define $w \star y = (w_1 \cdots w_{k-1}) \star (y_2 \cdots y_m)$. If $w_k, y_1 \in N'$ and $w_k y_1 = z \neq e_N$, then $w \star y = w_1 \cdots w_{k-1} z y_2 \cdots y_m$. If $w_k, y_1 \in N'$ and $w_k y_1 = e_N$, we recursively define $w \star y = (w_1 \cdots w_{k-1}) \star (y_2 \cdots y_m)$.
(a) Prove $(M * N, \star)$ is a group. (The verification of associativity is tricky.)
(b) Define injective group homomorphisms $i : M \to M * N$ and $j : N \to M * N$, and prove that $M * N$ with these maps solves the UMP posed above.

39. Generalize the construction in the previous exercise by defining a coproduct of a family of groups $\{G_i : i \in I\}$ satisfying a UMP analogous to the one in §19.6.

40. **Free Group Generated by a Set X.** Given any set X, construct a group $(G, \star)$ and a function $i : X \to G$ such that for any group K and any function $f : X \to K$, there exists a unique group homomorphism $T : G \to K$ with $f = T \circ i$. (Ideas from the previous three exercises can help here. Let G consist of certain words in the alphabet $X \cup X'$, where X' is a set with $|X| = |X'|$ and $X \cap X' = \emptyset$; the elements of X' represent formal inverses of elements of X.)

20

Universal Mapping Problems in Multilinear Algebra

This chapter introduces a subject called *multilinear algebra*, which studies functions of several variables that are linear in each variable. After defining these multilinear maps, as well as alternating maps and symmetric maps, we formulate and solve universal mapping problems that convert these less familiar maps to linear maps. The modules that arise in these constructions are called tensor products, exterior powers, and symmetric powers.

After building the tensor product, we use its universal mapping property to prove some isomorphisms and other general facts about tensor products. We show that linear maps between modules induce associated linear maps on tensor products, exterior powers, and symmetric powers. In the case of free modules, we find bases for these new modules in terms of bases for the original modules. This leads to a discussion of tensor products of matrices and the relation between determinants and exterior powers. The chapter ends with the construction of the tensor algebra of a module. Throughout the whole development, we stress the use of universal mapping properties as a means of organizing, motivating, and proving the basic results of multilinear algebra.

To read this chapter, you need to know facts about modules and universal mapping properties covered in Chapters 17 and 19, as well as properties of permutations from Chapter 2.

20.1 Multilinear Maps

Throughout this chapter, let R be a fixed commutative ring. Given R-modules $M_1, \ldots, M_n$, consider the product R-module $M = M_1 \times \cdots \times M_n$. We distinguish two special types of maps from M to another R-module N. On one hand, recall that $f : M \to N$ is an *R-module homomorphism* or an *R-linear map* iff $f(m + m') = f(m) + f(m')$ and $f(rm) = rf(m)$ for all $m, m' \in M$ and all $r \in R$. Writing this condition in terms of components, R-linearity of f means that for all $m_k, m'_k \in M_k$ and all $r \in R$,

$$f(m_1 + m'_1, m_2 + m'_2, \ldots, m_n + m'_n) = f(m_1, m_2, \ldots, m_n) + f(m'_1, m'_2, \ldots, m'_n)$$

$$\text{and } f(rm_1, rm_2, \ldots, rm_n) = rf(m_1, m_2, \ldots, m_n).$$

On the other hand, define $f : M_1 \times \cdots \times M_n \to N$ to be *R-multilinear* iff f is R-linear in each of its n inputs separately. More precisely, for each i between 1 and n and each fixed choice of $m_1, \ldots, m_{i-1}, m_{i+1}, \ldots, m_n$, we require that

$$f(m_1, \ldots, m_{i-1}, m_i + m'_i, m_{i+1}, \ldots, m_n) = f(m_1, \ldots, m_i, \ldots, m_n) + f(m_1, \ldots, m'_i, \ldots, m_n), \tag{20.1}$$

$$f(m_1, \ldots, m_{i-1}, rm_i, m_{i+1}, \ldots, m_n) = rf(m_1, \ldots, m_i, \ldots, m_n) \tag{20.2}$$

for all $m_i, m'_i \in M_i$ and all $r \in R$. When $n = 2$, we say f is *R-bilinear*; when $n = 3$, we say f is *R-trilinear*; when R is understood, we may speak of *n-linear* maps.

Let $f : M_1 \times \cdots \times M_n \to N$ be R-multilinear. By induction on s, we see that

$$f\left(m_1, \ldots, m_{i-1}, \sum_{j=1}^{s} r_j x_j, m_{i+1}, \ldots, m_n\right) = \sum_{j=1}^{s} r_j f(m_1, \ldots, x_j, \ldots, m_n) \qquad (20.3)$$

for each i, where $m_k \in M_k$, $r_1, \ldots, r_s \in R$, and $x_1, \ldots, x_s \in M_i$. Iterating this formula, we obtain

$$f\left(\sum_{j_1=1}^{s_1} r_{1,j_1} x_{1,j_1}, \sum_{j_2=1}^{s_2} r_{2,j_2} x_{2,j_2}, \ldots, \sum_{j_n=1}^{s_n} r_{n,j_n} x_{n,j_n}\right)$$

$$= \sum_{j_1=1}^{s_1} \sum_{j_2=1}^{s_2} \cdots \sum_{j_n=1}^{s_n} r_{1,j_1} r_{2,j_2} \cdots r_{n,j_n} f(x_{1,j_1}, \ldots, x_{n,j_n}) \qquad (20.4)$$

for all choices of $r_{ij} \in R$ and $x_{ij} \in M_i$. Furthermore, if P is an R-module and $g : N \to P$ is any R-linear map, then $g \circ f : M_1 \times \cdots \times M_n \to P$ is R-multilinear. This follows by applying g to each side of (20.1) and (20.2).

20.2 Alternating Maps

For R-modules M and P, define a map $f : M^n \to P$ to be *alternating* iff f is R-multilinear and $f(m_1, \ldots, m_n) = 0$ whenever $m_i = m_j$ for some $i \neq j$. The alternating condition is related to the following *anti-commutativity* conditions on an R-multilinear map f:

(AC1) $f(m_1, \ldots, m_i, m_{i+1}, \ldots, m_n) = -f(m_1, \ldots, m_{i+1}, m_i, \ldots, m_n)$ for $1 \leq i < n$ and all $m_k \in M$. In other words, interchanging two adjacent inputs of f multiplies the output by -1.

(AC2) $f(m_1, \ldots, m_i, \ldots, m_j, \ldots, m_n) = -f(m_1, \ldots, m_j, \ldots, m_i, \ldots, m_n)$ for all $i < j$ and all $m_k \in M$. In other words, interchanging any two inputs of f multiplies the output by -1.

(AC3) $f(m_{w(1)}, \ldots, m_{w(n)}) = \mathrm{sgn}(w) f(m_1, \ldots, m_n)$ for all permutations $w \in S_n$ and all $m_k \in M$. In other words, rearranging the inputs of f according to the permutation w multiplies the output by $\mathrm{sgn}(w)$.

We make the following claims regarding these conditions.

Claim 1: The conditions (AC1), (AC2), and (AC3) are equivalent. *Proof:* By letting w be the transposition (i, j), which satisfies $\mathrm{sgn}((i,j)) = -1$, we see that (AC3) implies (AC2). Evidently (AC2) implies (AC1). To see that (AC1) implies (AC3), recall from Chapter 2 that the list of inputs $(m_{w(1)}, \ldots, m_{w(n)})$ can be sorted into the list $(m_1, \ldots, m_n)$ using exactly $\mathrm{inv}(w(1), \ldots, w(n)) = \mathrm{inv}(w)$ basic transposition moves, where a basic transposition move switches two adjacent elements in a list. According to (AC1), each such move multiplies the value of f by -1. Therefore,

$$f(m_1, \ldots, m_n) = (-1)^{\mathrm{inv}(w)} f(m_{w(1)}, \ldots, m_{w(n)}) = \mathrm{sgn}(w) f(m_{w(1)}, \ldots, m_{w(n)}).$$

Since $\mathrm{sgn}(w)$ is ± 1, this relation is equivalent to (AC3). We say that an R-multilinear map $f : M^n \to P$ is *anti-commutative* iff the equivalent conditions (AC1), (AC2), and (AC3) hold for f.

Claim 2: The alternating condition implies all the anti-commutativity conditions. *Proof:* Assume f is alternating; we show that condition (AC1) holds. Fix $i < n$, and fix the inputs of f at positions different from i and $i+1$. For any $x, y \in M$, the alternating property gives

$$f(\ldots, x+y, x+y, \ldots) = f(\ldots, x, x, \ldots) = f(\ldots, y, y, \ldots) = 0,$$

where the displayed inputs occur at positions i and $i+1$. On the other hand, linearity of f in input i and in input $i+1$ shows that

$$f(\ldots, x+y, x+y, \ldots) = f(\ldots, x, x+y, \ldots) + f(\ldots, y, x+y, \ldots)$$
$$= f(\ldots, x, x, \ldots) + f(\ldots, x, y, \ldots) + f(\ldots, y, x, \ldots) + f(\ldots, y, y, \ldots). \quad (20.5)$$

Substituting zero in three places and rearranging, we get $f(\ldots, x, y, \ldots) = -f(\ldots, y, x, \ldots)$, as needed.

Claim 3: For rings R such that $1_R + 1_R$ is not zero and not a zero divisor, any of the anti-commutativity conditions implies the alternating condition. (For instance, the result holds when R is a field or integral domain such that $1_R + 1_R \neq 0_R$.) *Proof:* We deduce the alternating condition from condition (AC2). Suppose $(m_1, \ldots, m_n) \in M^n$ is such that $m_i = m_j$ where $i < j$. By (AC2),

$$f(m_1, \ldots, m_i, \ldots, m_j, \ldots, m_n) = -f(m_1, \ldots, m_j, \ldots, m_i, \ldots, m_n).$$

Since $m_i = m_j$, this relation gives

$$f(m_1, \ldots, m_i, \ldots, m_j, \ldots, m_n) = -f(m_1, \ldots, m_i, \ldots, m_j, \ldots, m_n).$$

Grouping terms, $(1_R + 1_R)f(m_1, \ldots, m_i, \ldots, m_j, \ldots, m_n) = 0$ in R. By hypothesis on R, it follows that $f(m_1, \ldots, m_n) = 0$.

Claim 4: If $g : P \to Q$ is R-linear and $f : M^n \to P$ is alternating (resp. anti-commutative), then $g \circ f : M^n \to Q$ is alternating (resp. anti-commutative). *Proof:* This follows by applying g to each side of the identities defining the alternating or anti-commutative properties.

20.3 Symmetric Maps

For R-modules M and P, we define a map $f : M^n \to P$ to be *symmetric* iff f is R-multilinear and $f(m_1, \ldots, m_n) = f(m'_1, \ldots, m'_n)$ whenever the list $(m_1, \ldots, m_n) \in M^n$ is a rearrangement of the list $(m'_1, \ldots, m'_n)$. It is equivalent to require that

$$f(m_1, \ldots, m_i, \ldots, m_j, \ldots, m_n) = f(m_1, \ldots, m_j, \ldots, m_i, \ldots, m_n)$$

for all $i < j$; i.e., the value of f is unchanged whenever two distinct inputs of f are interchanged. It is also equivalent to require that

$$f(m_1, \ldots, m_i, m_{i+1}, \ldots, m_n) = f(m_1, \ldots, m_{i+1}, m_i, \ldots, m_n)$$

for $1 \leq i < n$; i.e., the value of f is unchanged whenever two adjacent inputs of f are interchanged. The equivalence of the conditions follows from the fact that an arbitrary rearrangement of the list $(m_1, \ldots, m_n)$ can be accomplished by a finite sequence of interchanges of two adjacent inputs (see §2.6).

If $g : P \to Q$ is R-linear and $f : M^n \to P$ is symmetric, then $g \circ f : M^n \to Q$ is symmetric. This follows by applying g to each side of the identities $f(m_1, \ldots, m_n) = f(m'_1, \ldots, m'_n)$ in the definition of a symmetric map.

20.4   Tensor Product of Modules

Suppose P and $M_1, M_2, \ldots, M_n$ are R-modules, and let $X = M_1 \times \cdots \times M_n$. Recall the distinction between R-*linear maps* from X to P and R-*multilinear maps* from X to P (§20.1). It would be convenient if we could somehow reduce the study of multilinear maps to the study of R-linear maps. This raises the following universal mapping problem.

Problem (UMP for Tensor Products). Given a commutative ring R and R-modules $M_1, \ldots, M_n$, construct an R-module N and an R-multilinear map $j : M_1 \times \cdots \times M_n \to N$ satisfying the following UMP: for any R-module P, there is a bijection from the set

$$A = \{R\text{-linear maps } g : N \to P\}$$

onto the set

$$B = \{R\text{-multilinear maps } f : M_1 \times \cdots \times M_n \to P\}$$

sending g to $g \circ j$ for all $g \in A$. In other words, for each R-multilinear map $f : M_1 \times \cdots \times M_n \to P$, there exists a unique R-linear map $g : N \to P$ with $f = g \circ j$:

$$M_1 \times \cdots \times M_n \xrightarrow{\;j\;} N$$
$$f \searrow \quad \downarrow g$$
$$P$$

Construction of Solution to the UMP. We will define N to be the quotient of a certain free R-module F by a certain submodule K. The main idea of the construction is to fit together the diagrams describing two previously solved universal mapping problems, as shown here and in Figure 20.1:

$$X \xrightarrow{\;i\;} F \xrightarrow{\;\nu\;} N$$
$$f \searrow \quad \downarrow h \quad \swarrow g$$
$$P$$

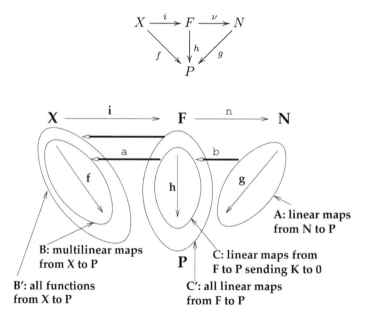

FIGURE 20.1
Bijections between sets of maps used to construct tensor products.

To begin the construction, let X be the set $M_1 \times \cdots \times M_n$, F be a free R-module with basis X (see §17.11), and $i : X \to F$ be the inclusion map. Recall that each element of F can be written uniquely as a finite R-linear combination $c_1 x_1 + \cdots + c_k x_k$, where $c_j \in R$ and $x_j \in X$. Also recall the universal mapping property of i (§19.1): for any R-module P, there is a bijection α from the set

$$C' = \{\text{all } R\text{-linear maps } h : F \to P\}$$

onto the set

$$B' = \{\text{all functions } f : X \to P\},$$

given by $\alpha(h) = h \circ i$ for all $h \in C'$.

Next, let K be the R-submodule of F generated by all elements of F of the form

$$1_R(m_1, \ldots, m_{k-1}, m_k + m_k', m_{k+1}, \ldots, m_n)$$
$$- 1_R(m_1, \ldots, m_k, \ldots, m_n) - 1_R(m_1, \ldots, m_k', \ldots, m_n), \quad (20.6)$$

$$1_R(m_1, \ldots, m_{k-1}, r m_k, m_{k+1}, \ldots, m_n) - r(m_1, \ldots, m_k, \ldots, m_n), \quad (20.7)$$

where $1 \leq k \leq n$, $m_k, m_k' \in M_k$, $m_s \in M_s$ for $s \neq k$, and $r \in R$. Let N be the R-module F/K and $\nu : F \to N$ be the projection map given by $\nu(z) = z + K$ for $z \in F$. Recall the universal mapping property of ν (§19.2): for any R-module P, there is a bijection β from the set

$$A = \{\text{all } R\text{-linear maps } g : N = F/K \to P\}$$

onto the set

$$C = \{\text{all } R\text{-linear maps } h : F \to P \text{ such that } h(z) = 0 \text{ for all } z \in K\},$$

given by $\beta(g) = g \circ \nu$ for all $g \in A$.

Note that $C \subseteq C'$. We claim that $\alpha[C] = B$, the set of all R-multilinear maps from $X = M_1 \times \cdots \times M_n$ to P. *Proof:* An R-linear map $h : F \to P$ in C' belongs to C iff $h[K] = \{0\}$ iff h maps every generator of the submodule K to zero iff

$$h(m_1, \ldots, m_{k-1}, m_k + m_k', m_{k+1}, \ldots, m_n)$$
$$- h(m_1, \ldots, m_k, \ldots, m_n) - h(m_1, \ldots, m_k', \ldots, m_n) = 0$$

and $h(m_1, \ldots, m_{k-1}, r m_k, m_{k+1}, \ldots, m_n) - r h(m_1, \ldots, m_k, \ldots, m_n) = 0$

for all choices of the variables iff

$$(h \circ i)(m_1, \ldots, m_k + m_k', \ldots, m_n) = (h \circ i)(m_1, \ldots, m_k, \ldots, m_n) + (h \circ i)(m_1, \ldots, m_k', \ldots, m_n)$$

and $(h \circ i)(m_1, \ldots, r m_k, \ldots, m_n) = r(h \circ i)(m_1, \ldots, m_k, \ldots, m_n)$

for all choices of the variables iff $\alpha(h) = h \circ i : X \to P$ is R-multilinear iff $\alpha(h) \in B$. By the claim, the restriction $\alpha|_C : C \to B$ is a bijection sending h to $h \circ i$ for $h \in C$. We also have the bijection $\beta : A \to C$ given by $\beta(g) = g \circ \nu$ for $g \in A$. Composing these bijections, we obtain a bijection γ from A to B given by $\gamma(g) = g \circ (\nu \circ i)$ for $g \in A$. Letting $j = \nu \circ i : X \to N$, we see that $\gamma : A \to B$ is given by composition with j. See Figure 20.1.

To summarize the construction, we have $N = F/K$, where F is the free R-module with basis $X = M_1 \times \cdots \times M_n$ and K is the R-submodule generated by all elements of the form (20.6) and (20.7). The map $j : X \to N$ sends $(m_1, \ldots, m_n) \in X$ to the coset $(m_1, \ldots, m_n) + K$ in N. By the Coset Equality Theorem, we know that

$$(m_1, \ldots, m_k + m_k', \ldots, m_n) + K = [(m_1, \ldots, m_k, \ldots, m_n) + K] + [(m_1, \ldots, m_k', \ldots, m_n) + K],$$

$$(m_1, \ldots, r m_k, \ldots, m_n) + K = r[(m_1, \ldots, m_k, \ldots, m_n) + K],$$

and these observations show that $j : X \to N$ is an R-multilinear map from X to N, as needed. We call N the *tensor product over R of the modules $M_1, \ldots, M_n$* and write

$$N = M_1 \otimes_R M_2 \otimes_R \cdots \otimes_R M_n.$$

Given $m_k \in M_k$, we introduce the *tensor notation* $m_1 \otimes m_2 \otimes \cdots \otimes m_n = j(m_1, \ldots, m_n) \in N$. In this notation, R-multilinearity of j translates into the identities

$$m_1 \otimes \cdots \otimes (m_k + m'_k) \otimes \cdots \otimes m_n = m_1 \otimes \cdots \otimes m_k \otimes \cdots \otimes m_n + m_1 \otimes \cdots \otimes m'_k \otimes \cdots \otimes m_n; \quad (20.8)$$

$$m_1 \otimes \cdots \otimes (r m_k) \otimes \cdots \otimes m_n = r(m_1 \otimes \cdots \otimes m_k \otimes \cdots \otimes m_n), \quad (20.9)$$

valid for all $m_k, m'_k \in M_k$, $m_s \in M_s$, and $r \in R$.

Uniqueness of Solution to the UMP. To justify our new notation for N and j, we show that the solution (N, j) to our universal mapping problem is unique up to a unique isomorphism compatible with the universal map j. Suppose (N', j') is another solution to the UMP. The proof involves the following four diagrams of sets and mappings:

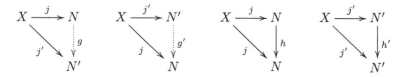

Since $j' : X \to N'$ is R-multilinear and (N, j) solves the UMP, we get a unique R-linear map $g : N \to N'$ with $j' = g \circ j$ (see the first diagram above). It now suffices to show that g is an isomorphism. Since $j : X \to N$ is R-multilinear and (N', j') solves the UMP, we get a unique R-linear map $g' : N' \to N$ with $j = g' \circ j'$ (see the second diagram above). It follows that $\mathrm{id}_N \circ j = j = (g' \circ g) \circ j$. By the uniqueness assertion in the UMP for (N, j), there is only one R-linear map $h : N \to N$ with $j = h \circ j$ (see the third diagram above). Therefore, $g' \circ g = \mathrm{id}_N$. Similarly, using the uniqueness of h' in the fourth diagram, we see that $g \circ g' = \mathrm{id}_{N'}$. So g' is the two-sided inverse of g, hence both maps are isomorphisms. Note that this uniqueness proof is essentially identical to the earlier uniqueness proof given for the coproduct of a family of sets (§19.7). In general, this same proof template can be used repeatedly to establish the uniqueness of solutions to various universal mapping problems (up to a unique isomorphism compatible with the universal maps). For future universal mapping problems, we omit the details of this uniqueness proof, asking the reader to verify the applicability of the proof template used here.

Let M be an R-module and n a positive integer. Consider the Cartesian product $M^n = M \times M \times \cdots \times M$, where there are n copies of M. By letting each $M_k = M$ in the tensor product construction, we obtain the *nth tensor power of M*, denoted

$$\bigotimes^n M = M^{\otimes n} = M \otimes_R M \otimes_R \cdots \otimes_R M.$$

The associated universal map j sends $(m_1, \ldots, m_n) \in M^n$ to $m_1 \otimes \cdots \otimes m_n \in M^{\otimes n}$. The UMP for tensor products says that there is a bijection γ from the set of R-linear maps (R-module homomorphisms) $g : M^{\otimes n} \to P$ onto the set of R-multilinear maps $f : M^n \to P$, given by $\gamma(g) = g \circ j$. In the next two sections, we use the tensor power $M^{\otimes n}$ to solve universal mapping problems for alternating maps and symmetric maps.

20.5 Exterior Powers of a Module

In the last section, we solved a universal mapping problem that converted R-multilinear maps into R-linear maps. Here, we formulate and solve a similar problem that converts alternating maps into R-linear maps.

Problem (UMP for Exterior Powers). Given a commutative ring R, an R-module M, and a positive integer n, construct an R-module N and an alternating map $i : M^n \to N$ satisfying the following UMP: for any R-module P, there is a bijection from the set

$$A = \{R\text{-linear maps } g : N \to P\}$$

onto the set

$$B = \{\text{alternating maps } f : M^n \to P\}$$

that sends $g \in A$ to $g \circ i \in B$. In other words, for each alternating map $f : M^n \to P$, there exists a unique R-linear map $g : N \to P$ with $f = g \circ i$:

Construction of Solution to the UMP. As before, the idea is to fit together two previously solved universal mapping problems, as shown in the following diagram and in Figure 20.2:

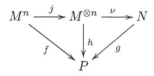

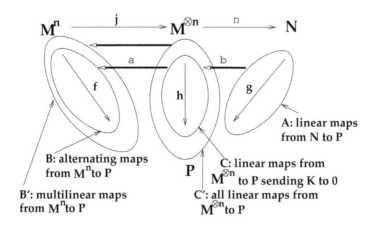

FIGURE 20.2
Bijections between sets of maps used to construct exterior powers.

To explain this, first recall the universal mapping property for $M^{\otimes n}$ and $j : M^n \to M^{\otimes n}$ (§20.4): for any R-module P, there is a bijection α from the set

$$C' = \{\text{all } R\text{-linear maps } h : M^{\otimes n} \to P\}$$

onto the set

$$B' = \{\text{all } R\text{-multilinear maps } f : M^n \to P\},$$

given by $\alpha(h) = h \circ j$ for all $h \in C'$.

Next, let K be the R-submodule of $M^{\otimes n}$ generated by all elements $m_1 \otimes m_2 \otimes \cdots \otimes m_n \in M^{\otimes n}$ such that $m_k = m_\ell$ for some $k \neq \ell$. Let N be the R-module $M^{\otimes n}/K$ and $\nu : M^{\otimes n} \to N$ be the projection map given by $\nu(z) = z + K$ for $z \in M^{\otimes n}$. Recall the universal mapping property of ν (§19.2): for any R-module P, there is a bijection β from the set

$$A = \{\text{all } R\text{-linear maps } g : N = M^{\otimes n}/K \to P\}$$

onto the set

$$C = \{\text{all } R\text{-linear maps } h : M^{\otimes n} \to P \text{ such that } h(z) = 0 \text{ for all } z \in K\},$$

given by $\beta(g) = g \circ \nu$ for all $g \in A$.

Note that $C \subseteq C'$. We claim that $\alpha[C] = B$, the set of all alternating maps from M^n to P. *Proof:* An R-linear map $h : M^{\otimes n} \to P$ in C' belongs to C iff $h[K] = \{0\}$ iff h maps every generator of the submodule K to zero iff $h(m_1 \otimes \cdots \otimes m_n) = 0$ whenever $m_k = m_\ell$ for some $k \neq \ell$ iff $(h \circ j)(m_1, \ldots, m_n) = 0$ whenever $m_k = m_\ell$ for some $k \neq \ell$ iff $\alpha(h) = h \circ j : M^n \to P$ is alternating iff $\alpha(h) \in B$. By the claim, α restricts to a bijection $\alpha|_C : C \to B$ sending $h \in C$ to $h \circ j \in B$. We also have the bijection $\beta : A \to C$ sending $g \in A$ to $g \circ \nu \in C$. Composing these bijections, we obtain a bijection $\gamma : A \to B$ given by $\gamma(g) = g \circ (\nu \circ j)$ for $g \in A$. Letting $i = \nu \circ j : M^n \to N$, we see that the bijection from A to B is given by composition with i. Since j is R-multilinear and ν is R-linear, the composite map i is R-multilinear. See Figure 20.2.

The map $i : M^n \to N$ sends $(m_1, \ldots, m_n) \in M^n$ to the coset $(m_1 \otimes \cdots \otimes m_n) + K$ in N. By definition of K, $i(m_1, \ldots, m_n) = 0 + K = 0_N$ whenever $m_k = m_\ell$ for some $k \neq \ell$. Therefore, i is alternating. The standard argument proves that the solution (N, i) to the UMP is unique up to a unique R-isomorphism.

We call N the *nth exterior power of M* and write $N = \bigwedge^n M$. We also write

$$m_1 \wedge m_2 \wedge \cdots \wedge m_n = i(m_1, \ldots, m_n) = m_1 \otimes \cdots \otimes m_n + K \in N$$

and call this element the *wedge product of $m_1, \ldots, m_n$*. In this notation, the R-multilinearity and alternating properties of i translate into the identities:

$$m_1 \wedge \cdots \wedge (m_k + m_k') \wedge \cdots \wedge m_n = m_1 \wedge \cdots \wedge m_k \wedge \cdots \wedge m_n + m_1 \wedge \cdots \wedge m_k' \wedge \cdots \wedge m_n;$$

$$m_1 \wedge \cdots \wedge (rm_k) \wedge \cdots \wedge m_n = r(m_1 \wedge \cdots \wedge m_k \wedge \cdots \wedge m_n);$$

$$m_1 \wedge \cdots \wedge m_n = 0 \text{ whenever } m_k = m_\ell \text{ for some } k \neq \ell.$$

The anti-commutativity of i, which follows from the alternating property, translates into the following facts:

$$m_1 \wedge \cdots \wedge m_k \wedge m_{k+1} \wedge \cdots \wedge m_n = -m_1 \wedge \cdots \wedge m_{k+1} \wedge m_k \wedge \cdots \wedge m_n;$$

$$m_1 \wedge \cdots \wedge m_k \wedge \cdots \wedge m_\ell \wedge \cdots \wedge m_n = -m_1 \wedge \cdots \wedge m_\ell \wedge \cdots \wedge m_k \wedge \cdots \wedge m_n;$$

$$m_{w(1)} \wedge \cdots \wedge m_{w(n)} = (\text{sgn}(w))m_1 \wedge \cdots \wedge m_n \text{ for all } w \in S_n.$$

20.6 Symmetric Powers of a Module

Next, we give a universal construction for converting symmetric maps into R-linear maps.

Problem (UMP for Symmetric Powers). Given a commutative ring R, an R-module M, and a positive integer n, construct an R-module N and a symmetric map $i : M^n \to N$ satisfying the following UMP: for any R-module P, there is a bijection from the set

$$A = \{R\text{-linear maps } g : N \to P\}$$

onto the set

$$B = \{\text{symmetric maps } f : M^n \to P\}$$

sending $g \in A$ to $g \circ i \in B$. In other words, for each symmetric map $f : M^n \to P$, there exists a unique R-linear map $g : N \to P$ with $f = g \circ i$:

Construction of Solution to the UMP. The proof is nearly identical to what we did for alternating maps (see the diagram below and Figure 20.3).

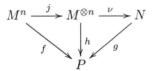

To start, recall once again the universal mapping property for $M^{\otimes n}$ and $j : M^n \to M^{\otimes n}$ (§20.4): for any R-module P, there is a bijection α from the set

$$C' = \{\text{all } R\text{-linear maps } h : M^{\otimes n} \to P\}$$

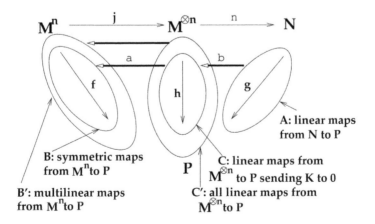

FIGURE 20.3
Bijections between sets of maps used to construct symmetric powers.

onto the set
$$B' = \{\text{all } R\text{-multilinear maps } f : M^n \to P\},$$
given by $\alpha(h) = h \circ j$ for all $h \in C'$.

Next, let K be the R-submodule of $M^{\otimes n}$ generated by all elements of the form

$$m_1 \otimes m_2 \otimes \cdots \otimes m_n - m_1' \otimes m_2' \otimes \cdots \otimes m_n',$$

where the list $(m_1', \ldots, m_n')$ is a rearrangement of the list $(m_1, \ldots, m_n)$. Let N be the R-module $M^{\otimes n}/K$ and $\nu : M^{\otimes n} \to N$ be the projection map given by $\nu(z) = z + K$ for $z \in M^{\otimes n}$. Recall once again the universal mapping property of ν (§19.2): for any R-module P, there is a bijection β from the set

$$A = \{\text{all } R\text{-linear maps } g : N = M^{\otimes n}/K \to P\}$$

onto the set

$$C = \{\text{all } R\text{-linear maps } h : M^{\otimes n} \to P \text{ such that } h(z) = 0 \text{ for all } z \in K\},$$

given by $\beta(g) = g \circ \nu$ for all $g \in A$.

Note that $C \subseteq C'$. We claim that $\alpha[C] = B$, the set of all symmetric maps from M^n to P. *Proof:* An R-linear map $h : M^{\otimes n} \to P$ in C' is in C iff $h[K] = \{0\}$ iff h maps every generator of the submodule K to zero iff $h(m_1 \otimes \cdots \otimes m_n) - h(m_1' \otimes \cdots \otimes m_n') = 0$ whenever $(m_1', \ldots, m_n')$ is a rearrangement of $(m_1, \ldots, m_n)$ iff $(h \circ j)(m_1, \ldots, m_n) = (h \circ j)(m_1', \ldots, m_n')$ whenever $(m_1', \ldots, m_n')$ is a rearrangement of $(m_1, \ldots, m_n)$ iff $\alpha(h) = h \circ j : M^n \to P$ is symmetric iff $\alpha(h) \in B$. By the claim, α restricts to a bijection $\alpha|_C : C \to B$ sending $h \in C$ to $h \circ j \in B$. We also have the bijection $\beta : A \to C$ sending $g \in A$ to $g \circ \nu \in C$. Composing these bijections, we obtain a bijection $\gamma : A \to B$ given by $\gamma(g) = g \circ (\nu \circ j)$ for $g \in A$. Letting $i = \nu \circ j : M^n \to N$, we see that the bijection from A to B is given by composition with i. Since j is R-multilinear and ν is R-linear, the composite map i is R-multilinear. See Figure 20.3.

The map $i : M^n \to N$ sends $(m_1, \ldots, m_n) \in M^n$ to the coset $(m_1 \otimes \cdots \otimes m_n) + K$ in N. By definition of K, $i(m_1, \ldots, m_n) = i(m_1', \ldots, m_n')$ in N whenever $(m_1', \ldots, m_n')$ is a rearrangement of $(m_1, \ldots, m_n)$. Therefore, i is symmetric. The standard argument proves that the solution (N, i) to the UMP is unique up to a unique R-isomorphism.

We call N the *nth symmetric power of M* and write $N = \operatorname{Sym}^n M$. We also write

$$m_1 m_2 \cdots m_n = i(m_1, \ldots, m_n) = m_1 \otimes \cdots \otimes m_n + K \in N$$

and call this element the *symmetric product of* $m_1, \ldots, m_n$. In this notation, the R-multilinearity and symmetric properties of i translate into the identities:

$$m_1 m_2 \cdots (m_k + m_k') \cdots m_n = m_1 m_2 \cdots m_k \cdots m_n + m_1 m_2 \cdots m_k' \cdots m_n;$$

$$m_1 m_2 \cdots (r m_k) \cdots m_n = r(m_1 m_2 \cdots m_k \cdots m_n);$$

$$m_1 m_2 \cdots m_n = m_1' m_2' \cdots m_n'$$

whenever $(m_1', \ldots, m_n')$ is a rearrangement of $(m_1, \ldots, m_n)$.

20.7 Myths about Tensor Products

The tensor product construction is more subtle than other module constructions such as the direct product. To help the reader avoid common errors, we discuss some misconceptions or "myths" about tensor products in this section. For simplicity, let us consider the tensor product $M \otimes_R N$ of two R-modules M and N.

Myth 1: "Every element of $M \otimes_R N$ has the form $x \otimes y$ for some $x \in M$ and $y \in N$." This myth arises from the faulty assumption that the underlying set of the module $M \otimes_R N$ is the Cartesian product $M \times N = \{(x, y) : x \in M, y \in N\}$. However, recalling the construction in §20.4, we see that $M \otimes_R N$ was built by starting with a free module F having the set $M \times N$ as a basis, and then taking the quotient by a certain submodule K. For any basis element (x, y) of F, we wrote $x \otimes y$ for the coset $(x, y) + K$. Since every element of F is a finite R-linear combination of basis elements, it *is* true that every element of $M \otimes_R N$ is a finite R-linear combination of the form $\sum_{i=1}^{k} r_i(x_i \otimes y_i)$, where $k \in \mathbb{Z}_{\geq 0}$, $r_i \in R$, $x_i \in M$, and $y_i \in N$. In fact, we can use (20.9) to write $r_i(x_i \otimes y_i) = (r_i x_i) \otimes y_i$ for each i. This means that *every element of $M \otimes_R N$ can be written as a finite sum of "basic" tensors $x \otimes y$ with $x \in M$ and $y \in N$.*

Myth 2: "The set $\{x \otimes y : x \in M, y \in N\}$ is a basis for the R-module $M \otimes_R N$." In fact, the set of all ordered pairs (x, y) with $x \in M$ and $y \in N$ is a basis for the free R-module F used in the construction of $M \otimes_R N$. However, after passing to the quotient module F/K, the set of cosets $(x, y) + K$ is almost never a basis for F/K. For instance, using (20.8), we have $(x_1 + x_2) \otimes y - (x_1 \otimes y) - (x_2 \otimes y) = 0$ for all $x_1, x_2 \in M$ and $y \in N$, which gives a dependence relation among three basic tensors. On the other hand, as we saw above, it is true that the set of all tensors $x \otimes y$ forms a *generating set* for the R-module $M \otimes_R N$. In the coming sections, we will find bases for tensor products, exterior powers, and symmetric powers of free modules.

Myth 3: "All we need to do to define a function with domain $M \otimes_R N$ is to declare $f(x \otimes y)$ to be any formula involving x and y." To illustrate the problems that can arise here, consider the following proposed proof that the R-modules $M \otimes_R N$ and $N \otimes_R M$ are isomorphic: "Define $f : M \otimes_R N \to N \otimes_R M$ by $f(x \otimes y) = y \otimes x$ for all $x \in M$ and $y \in N$. The map f is evidently an R-linear bijection, so $M \otimes_R N \cong N \otimes_R M$." The first difficulty is that the stated formula does not define f on the entire domain $M \otimes_R N$ (see Myth 1). Since most maps of interest preserve addition, we could try to get around this by extending f additively, i.e., by setting $f(\sum_{i=1}^{k} x_i \otimes y_i) = \sum_{i=1}^{k} y_i \otimes x_i$.

However, there is still a problem with this extended definition. When computing $f(z)$ for any $z \in M \otimes_R N$, the output appears to depend on the particular representation of z as a sum of tensors $x_i \otimes y_i$. Usually, z has many such representations; how do we know that the formula for f gives the same answer no matter which representation we use? For similar reasons, it is not at all clear that f must be one-to-one.

To resolve these difficulties in defining f, we must return to the universal mapping property characterizing $M \otimes_R N$. We have seen that there is a bijection between the set of R-linear maps from $M \otimes_R N$ to a given R-module P and the set of R-bilinear maps from $M \times N$ to P. In this case, P is the R-module $N \otimes_R M$. To define the required R-linear map $f : M \otimes_R N \to N \otimes_R M$, we instead define a function $g : M \times N \to N \otimes_R M$ on the product set $M \times N$, by letting $g(x, y) = y \otimes x$ for $x \in M$ and $y \in N$. Note that this definition has none of the problems that we encountered earlier, since every element of the Cartesian product $M \times N$ can be written uniquely as (x, y) for some $x \in M$ and $y \in N$. Furthermore, using (20.8) and (20.9) in $N \otimes_R M$, we see that g is R-bilinear: for all $x, x_1, x_2 \in M$ and

$y, y_1, y_2 \in N$ and $r \in R$,

$$g(x_1 + x_2, y) = y \otimes (x_1 + x_2) = (y \otimes x_1) + (y \otimes x_2) = g(x_1, y) + g(x_2, y);$$

$$g(x, y_1 + y_2) = (y_1 + y_2) \otimes x = (y_1 \otimes x) + (y_2 \otimes x) = g(x, y_1) + g(x, y_2);$$

$$g(rx, y) = y \otimes (rx) = r(y \otimes x) = rg(x, y); \qquad g(x, ry) = (ry) \otimes x = r(y \otimes x) = rg(x, y).$$

The UMP applies and gives us a unique (and well-defined) R-linear map $f : M \otimes_R N \to N \otimes_R M$ with $g = f \circ j$; i.e., $f(x \otimes y) = f(j(x, y)) = g(x, y) = y \otimes x$. This is our original definition of f on generators of $M \otimes_R N$, but now we are sure that f is well-defined and R-linear. In most situations, this is the method we must use to define maps whose domain is a tensor product, exterior power, or symmetric power.

Why is f a bijection? Seeing that f is onto is not too hard, but checking the injectivity of f directly can be difficult. Instead, we show that f is bijective by exhibiting a two-sided inverse map. Start with the map $g_1 : N \times M \to M \otimes_R N$ given by $g_1(y, x) = x \otimes y$ for all $y \in N$ and $x \in M$. As above, we see that g_1 is R-bilinear, so the UMP furnishes a unique R-linear map $f_1 : N \otimes_R M \to M \otimes_R N$ given on generators by $f_1(y \otimes x) = x \otimes y$. To check that $f \circ f_1 = \mathrm{id}_{N \otimes_R M}$, observe that $f \circ f_1(y \otimes x) = f(x \otimes y) = y \otimes x$ for every generator $y \otimes x$ of the R-module $N \otimes_R M$. The R-linear map $\mathrm{id}_{N \otimes_R M}$ also sends $y \otimes x$ to $y \otimes x$ for all $y \in N$ and $x \in M$. We know that two R-linear maps agreeing on a set of generators for their common domain must be equal. So $f \circ f_1 = \mathrm{id}_{N \otimes_R M}$, and similarly $f_1 \circ f = \mathrm{id}_{M \otimes_R N}$. Finally, we have a rigorous proof of the R-module isomorphism $M \otimes_R N \cong N \otimes_R M$.

20.8   Tensor Product Isomorphisms

In this section, we give further illustrations of the technique used above to prove that $M \otimes_R N$ is isomorphic to $N \otimes_R M$. As a first example, we prove that *for every R-module M, $R \otimes_R M \cong M$.*

Given an R-module M, let $r \star m$ denote the action of a scalar $r \in R$ on a module element $m \in M$. Define a map $g : R \times M \to M$ by $g(r, m) = r \star m$ for $r \in R$ and $m \in M$. Now, g is R-bilinear, since for all $r, s \in R$ and $m, n \in M$:

$$
\begin{aligned}
g(r + s, m) &= (r + s) \star m = (r \star m) + (s \star m) = g(r, m) + g(s, m); \\
g(rs, m) &= (rs) \star m = r \star (s \star m) = rg(s, m); \\
g(r, m + n) &= r \star (m + n) = (r \star m) + (r \star n) = g(r, m) + g(r, n); \\
g(r, sm) &= r \star (sm) = (rs) \star m = (sr) \star m = s \star (r \star m) = sg(r, m).
\end{aligned}
$$

Note that commutativity of R is needed to justify the last equality. We therefore get a unique R-linear map $f : R \otimes_R M \to M$ defined on generators by $f(r \otimes m) = r \star m$ for $r \in R$ and $m \in M$.

Next, we define $h : M \to R \otimes_R M$ by $h(m) = 1_R \otimes m$ for $m \in M$. The map h is R-linear because $h(m + n) = 1 \otimes (m + n) = (1 \otimes m) + (1 \otimes n) = h(m) + h(n)$ for all $m, n \in M$; and $h(rm) = 1 \otimes (rm) = r(1 \otimes m) = rh(m)$ for all $r \in R$ and $m \in M$. We claim h is the two-sided inverse of f, so that both maps are R-module isomorphisms. On one hand, for any $m \in M$, $f(h(m)) = f(1 \otimes m) = 1 \star m = m = \mathrm{id}_M(m)$, so $f \circ h = \mathrm{id}_M$. On the other hand, for any $r \in R$ and $m \in M$,

$$h(f(r \otimes m)) = h(r \star m) = 1 \otimes (rm) = r(1 \otimes m) = (r1) \otimes m = r \otimes m.$$

Thus, the R-linear maps $h \circ f$ and $\mathrm{id}_{R \otimes_R M}$ have the same effect on all generators of $R \otimes_R M$, so these maps are equal. This completes the proof that $R \otimes_R M \cong M$ as R-modules.

For our next result, fix a ring R and R-modules M, N, and P. We prove that

$$(M \oplus N) \otimes_R P \cong (M \otimes_R P) \oplus (N \otimes_R P)$$

as R-modules. Recall that $M \oplus N = M \times N$ is the set of ordered pairs (m, n) with $m \in M$ and $n \in N$, which is an R-module under componentwise operations.

First, define a map $g : (M \oplus N) \times P \to (M \otimes_R P) \oplus (N \otimes_R P)$ by $g((m, n), p) = (m \otimes p, n \otimes p)$ for all $m \in M$, $n \in N$, and $p \in P$. You can check that g is R-bilinear; for example, if $u = (m_1, n_1)$ and $v = (m_2, n_2)$ are in $M \oplus N$, we compute:

$$
\begin{aligned}
g(u + v, p) &= g((m_1 + m_2, n_1 + n_2), p) = ((m_1 + m_2) \otimes p, (n_1 + n_2) \otimes p) \\
&= (m_1 \otimes p + m_2 \otimes p, n_1 \otimes p + n_2 \otimes p) \\
&= (m_1 \otimes p, n_1 \otimes p) + (m_2 \otimes p, n_2 \otimes p) \\
&= g((m_1, n_1), p) + g((m_2, n_2), p) = g(u, p) + g(v, p).
\end{aligned}
$$

By the UMP, there is a unique R-linear map $f : (M \oplus N) \otimes_R P \to (M \otimes_R P) \oplus (N \otimes_R P)$ given on generators by

$$f((m, n) \otimes p) = (m \otimes p, n \otimes p).$$

To construct the inverse of f, we first define maps $g_1 : M \times P \to (M \oplus N) \otimes_R P$ and $g_2 : N \times P \to (M \oplus N) \otimes_R P$, by setting $g_1(m, p) = (m, 0) \otimes p$ and $g_2(n, p) = (0, n) \otimes p$ for all $m \in M$, $n \in N$, and $p \in P$. You can verify that g_1 and g_2 are R-bilinear, so the UMP provides R-linear maps $h_1 : M \otimes_R P \to (M \oplus N) \otimes_R P$ and $h_2 : N \otimes_R P \to (M \oplus N) \otimes_R P$ defined on generators by

$$h_1(m \otimes p) = (m, 0) \otimes p, \qquad h_2(n \otimes p) = (0, n) \otimes p.$$

Now, using the UMP for the direct sum of two R-modules (§19.4), we can combine h_1 and h_2 to get an R-linear map $h : (M \otimes_R P) \oplus (N \otimes_R P) \to (M \oplus N) \otimes_R P$ given by $h(y, z) = h_1(y) + h_2(z)$ for $y \in M \otimes_R P$ and $z \in N \otimes_R P$.

To finish the proof, we prove that $f \circ h$ and $h \circ f$ are identity maps by checking that each map sends all relevant generators to themselves. A typical generator of $(M \oplus N) \otimes_R P$ is $w = (m, n) \otimes p$ with $m \in M$, $n \in N$, and $p \in P$. We compute

$$
\begin{aligned}
h(f(w)) = h(m \otimes p, n \otimes p) &= h_1(m \otimes p) + h_2(n \otimes p) \\
&= (m, 0) \otimes p + (0, n) \otimes p = ((m, 0) + (0, n)) \otimes p = (m, n) \otimes p = w.
\end{aligned}
$$

On the other hand, you can check that elements of the form $(m \otimes p, 0)$ and $(0, n \otimes p)$ (with $m \in M$, $n \in N$, and $p \in P$) generate the R-module $(M \otimes_R P) \oplus (N \otimes_R P)$. For generators of the first type,

$$f(h(m \otimes p, 0)) = f(h_1(m \otimes p)) = f((m, 0) \otimes p) = (m \otimes p, 0 \otimes p) = (m \otimes p, 0).$$

For generators of the second type,

$$f(h(0, n \otimes p)) = f(h_2(n \otimes p)) = f((0, n) \otimes p) = (0 \otimes p, n \otimes p) = (0, n \otimes p).$$

These calculations use the fact that $0 \otimes p = 0$ for any $p \in P$. We conclude that h is the inverse of f, so both maps are R-module isomorphisms.

An analogous proof shows that $P \otimes_R (M \oplus N) \cong (P \otimes_R M) \oplus (P \otimes_R N)$. More generally, for all R-modules M_i and P_j, the same techniques (Exercise 36) prove that

$$\left(\bigoplus_{i \in I} M_i \right) \otimes_R P_2 \otimes_R \cdots \otimes_R P_k \cong \bigoplus_{i \in I} (M_i \otimes_R P_2 \otimes_R \cdots \otimes_R P_k). \qquad (20.10)$$

Even more generally, there is an R-module isomorphism

$$\left(\bigoplus_{i_1 \in I_1} M_{i_1} \right) \otimes_R \cdots \otimes_R \left(\bigoplus_{i_k \in I_k} M_{i_k} \right) \cong \bigoplus_{(i_1,\ldots,i_k) \in I_1 \times \cdots \times I_k} (M_{i_1} \otimes_R M_{i_2} \otimes_R \cdots \otimes_R M_{i_k}). \qquad (20.11)$$

20.9 Associativity of Tensor Products

Next, we prove an associativity result for tensor products. Fix an R-module M and positive integers k and m. We show that $M^{\otimes k} \otimes_R M^{\otimes m} \cong M^{\otimes(k+m)}$ as R-modules by constructing isomorphisms in both directions. First, define $g : M^{k+m} \to M^{\otimes k} \otimes_R M^{\otimes m}$ by

$$g(x_1, \ldots, x_k, x_{k+1}, \ldots, x_{k+m}) = (x_1 \otimes \cdots \otimes x_k) \otimes (x_{k+1} \otimes \cdots \otimes x_{k+m})$$

for $x_i \in M$. It is routine to check (by repeated use of (20.8) and (20.9)) that g is an R-multilinear map. Hence g induces a unique R-linear map $f : M^{\otimes(k+m)} \to M^{\otimes k} \otimes_R M^{\otimes m}$ defined on generators by

$$f(x_1 \otimes \cdots \otimes x_k \otimes x_{k+1} \otimes \cdots \otimes x_{k+m}) = (x_1 \otimes \cdots \otimes x_k) \otimes (x_{k+1} \otimes \cdots \otimes x_{k+m}).$$

Constructing f^{-1} is somewhat tricky, since the domain of f^{-1} involves three tensor products. To begin, fix $z = (x_{k+1}, \ldots, x_{k+m}) \in M^m$ and define a map $p'_z : M^k \to M^{\otimes(k+m)}$ by

$$p'_z(x_1, \ldots, x_k) = x_1 \otimes \cdots \otimes x_k \otimes x_{k+1} \otimes \cdots \otimes x_{k+m}$$

for all $(x_1, \ldots, x_k) \in M^k$. You can check that p'_z is R-multilinear. Invoking the UMP for $M^{\otimes k}$, we get an R-linear map $p_z : M^{\otimes k} \to M^{\otimes(k+m)}$ given on generators by

$$p_z(x_1 \otimes \cdots \otimes x_k) = x_1 \otimes \cdots \otimes x_k \otimes x_{k+1} \otimes \cdots \otimes x_{k+m}.$$

Next, for each fixed $y \in M^{\otimes k}$, we define a map $q'_y : M^m \to M^{\otimes(k+m)}$ by $q'_y(z) = p_z(y)$ for all $z \in M^m$. You can check, using the formula for p_z on generators, that q'_y is R-multilinear. Invoking the UMP for $M^{\otimes m}$, we get an R-linear map $q_y : M^{\otimes m} \to M^{\otimes(k+m)}$ given on generators by

$$q_y(x_{k+1} \otimes \cdots \otimes x_{k+m}) = p_{(x_{k+1},\ldots,x_{k+m})}(y).$$

Finally, define a map $t' : M^{\otimes k} \times M^{\otimes m} \to M^{\otimes(k+m)}$ by setting $t'(y, w) = q_y(w)$ for all $y \in M^{\otimes k}$ and all $w \in M^{\otimes m}$. You can check that t' is R-bilinear, so we finally get an R-linear map $t : M^{\otimes k} \otimes_R M^{\otimes m} \to M^{\otimes(k+m)}$ given on generators by $t(y \otimes w) = q_y(w)$. Tracing through all the definitions, we find that

$$t((x_1 \otimes \cdots \otimes x_k) \otimes (x_{k+1} \otimes \cdots \otimes x_{k+m})) = x_1 \otimes \cdots \otimes x_k \otimes x_{k+1} \otimes \cdots \otimes x_{k+m}$$

for all $x_j \in M$. Hence, $t \circ f = \mathrm{id}_{M^{\otimes(k+m)}}$ since these two R-linear maps agree on a generating set. You can see similarly that $f \circ t$ is an identity map, after checking that elements of the form $(x_1 \otimes \cdots \otimes x_k) \otimes (x_{k+1} \otimes \cdots \otimes x_{k+m})$ generate the R-module $M^{\otimes k} \otimes_R M^{\otimes m}$.

By an analogous proof, you can show that

$$(M_1 \otimes_R \cdots \otimes_R M_k) \otimes_R (M_{k+1} \otimes_R \cdots \otimes_R M_{k+m}) \cong (M_1 \otimes_R \cdots \otimes_R M_{k+m}) \qquad (20.12)$$

for any R-modules $M_1, \ldots, M_{k+m}$ (Exercise 40). More generally, no matter how we insert parentheses into $M_1 \otimes_R \cdots \otimes_R M_{k+m}$ to indicate nested tensor product constructions, we always obtain a module isomorphic to the original tensor product. Another approach to proving these isomorphisms is to show that both modules solve the same universal mapping problem.

20.10   Tensor Product of Maps

Let $M_1, \ldots, M_n$ and $P_1, \ldots, P_n$ be R-modules and $f_k : M_k \to P_k$ be an R-linear map for each k between 1 and n. We show that there exists a unique R-linear map $g : M_1 \otimes_R \cdots \otimes_R M_n \to P_1 \otimes_R \cdots \otimes_R P_n$ given on generators by

$$g(x_1 \otimes x_2 \otimes \cdots \otimes x_n) = f_1(x_1) \otimes f_2(x_2) \otimes \cdots \otimes f_n(x_n) \qquad (20.13)$$

for all $x_k \in M_k$. The map g is written $f_1 \otimes f_2 \otimes \cdots \otimes f_n$ or $\bigotimes_{k=1}^n f_k : \bigotimes_{k=1}^n M_k \to \bigotimes_{k=1}^n P_k$ and called the *tensor product of the maps* f_k.

As in previous sections, to obtain the R-linear map g we must invoke the UMP for tensor products. Define a map $h : M_1 \times \cdots \times M_n \to P_1 \otimes_R \cdots \otimes_R P_n$ by

$$h(x_1, x_2, \ldots, x_n) = f_1(x_1) \otimes f_2(x_2) \otimes \cdots \otimes f_n(x_n)$$

for all $x_k \in M_k$. The function h is R-multilinear, since

$$
\begin{aligned}
h(x_1, \ldots, x_k + x_k', \ldots, x_n) &= f_1(x_1) \otimes \cdots \otimes f_k(x_k + x_k') \otimes \cdots \otimes f_n(x_n) \\
&= f_1(x_1) \otimes \cdots \otimes (f_k(x_k) + f_k(x_k')) \otimes \cdots \otimes f_n(x_n) \\
&= f_1(x_1) \otimes \cdots \otimes f_k(x_k) \otimes \cdots \otimes f_n(x_n) \\
&\quad + f_1(x_1) \otimes \cdots \otimes f_k(x_k') \otimes \cdots \otimes f_n(x_n) \\
&= h(x_1, \ldots, x_k, \ldots, x_n) + h(x_1, \ldots, x_k', \ldots, x_n);
\end{aligned}
$$

and, for all $r \in R$,

$$
\begin{aligned}
h(x_1, \ldots, rx_k, \ldots, x_n) &= f_1(x_1) \otimes \cdots \otimes f_k(rx_k) \otimes \cdots \otimes f_n(x_n) \\
&= f_1(x_1) \otimes \cdots \otimes r f_k(x_k) \otimes \cdots \otimes f_n(x_n) \\
&= r(f_1(x_1) \otimes \cdots \otimes f_k(x_k) \otimes \cdots \otimes f_n(x_n)) = rh(x_1, \ldots, x_k, \ldots, x_n).
\end{aligned}
$$

Applying the UMP for tensor products to the R-multilinear map h, we obtain a unique R-linear map g satisfying (20.13).

With the same setup as above, suppose also that we have R-modules $Q_1, \ldots, Q_n$ and R-linear maps $g_k : P_k \to Q_k$ for k between 1 and n. Then we have R-linear maps $\bigotimes_{k=1}^n f_k : \bigotimes_{k=1}^n M_k \to \bigotimes_{k=1}^n P_k$, $\bigotimes_{k=1}^n g_k : \bigotimes_{k=1}^n P_k \to \bigotimes_{k=1}^n Q_k$, and (for each k) $g_k \circ f_k : M_k \to Q_k$. We claim that

$$\left(\bigotimes_{k=1}^n g_k \right) \circ \left(\bigotimes_{k=1}^n f_k \right) = \bigotimes_{k=1}^n (g_k \circ f_k). \qquad (20.14)$$

Both sides are R-linear maps from $\bigotimes_{k=1}^{n} M_k$ to $\bigotimes_{k=1}^{n} Q_k$, so it suffices to check that these functions have the same effect on all generators of $\bigotimes_{k=1}^{n} M_k$. To check this, note that for all $x_k \in M_k$,

$$(g_1 \otimes \cdots \otimes g_k) \circ (f_1 \otimes \cdots \otimes f_k)(x_1 \otimes \cdots \otimes x_k) = (g_1 \otimes \cdots \otimes g_k)(f_1(x_1) \otimes \cdots \otimes f_k(x_k))$$
$$= g_1(f_1(x_1)) \otimes \cdots \otimes g_k(f_k(x_k)) = ((g_1 \circ f_1) \otimes \cdots \otimes (g_k \circ f_k))(x_1 \otimes \cdots \otimes x_k).$$

Similarly, we have $\bigotimes_{k=1}^{n} \mathrm{id}_{M_k} = \mathrm{id}_{\bigotimes_{k=1}^{n} M_k}$ because both sides are R-linear and

$$(\mathrm{id}_{M_1} \otimes \cdots \otimes \mathrm{id}_{M_n})(x_1 \otimes \cdots \otimes x_n) = \mathrm{id}_{M_1}(x_1) \otimes \cdots \otimes \mathrm{id}_{M_n}(x_n)$$
$$= x_1 \otimes \cdots \otimes x_n = \mathrm{id}_{\bigotimes_{k=1}^{n} M_k}(x_1 \otimes \cdots \otimes x_n).$$

We can use a similar construction to define linear maps on exterior and symmetric powers of R-modules. Suppose $f : M \to P$ is an R-linear map between R-modules M and P, and $n \in \mathbb{Z}_{>0}$. We claim there exists a unique R-linear map $\bigwedge^n f : \bigwedge^n M \to \bigwedge^n P$ given on generators by

$$\left(\bigwedge^n f \right) (x_1 \wedge x_2 \wedge \cdots \wedge x_n) = f(x_1) \wedge f(x_2) \wedge \cdots \wedge f(x_n)$$

for all $x_i \in M$. This map is called the *nth exterior power of f*. To obtain this map, define $h : M^n \to \bigwedge^n P$ by $h(x_1, \ldots, x_n) = f(x_1) \wedge \cdots \wedge f(x_n)$. Using R-linearity of f, you can check that h is R-multilinear. The map h is alternating as well, since $x_i = x_j$ for $i < j$ gives $f(x_i) = f(x_j)$, hence $f(x_1) \wedge \cdots \wedge f(x_i) \wedge \cdots \wedge f(x_j) \wedge \cdots \wedge f(x_n) = 0$. Applying the UMP for exterior powers to h, we obtain a unique R-linear map $\bigwedge^n f$ satisfying the formula above. If Q is another R-module and $g : P \to Q$ is another R-linear map, you can prove that $(\bigwedge^n g) \circ (\bigwedge^n f) = \bigwedge^n (g \circ f)$ by a calculation on generators analogous to the one used to prove (20.14). Similarly, $\bigwedge^n \mathrm{id}_M = \mathrm{id}_{\bigwedge^n M}$.

With the same setup as the previous paragraph, the same method proves the existence of a unique R-linear map $\mathrm{Sym}^n f : \mathrm{Sym}^n M \to \mathrm{Sym}^n P$ given on generators by

$$(\mathrm{Sym}^n f)(x_1 x_2 \cdots x_n) = f(x_1) f(x_2) \cdots f(x_n)$$

for $x_i \in M$. This map is called the *nth symmetric power of f*. You can check that $(\mathrm{Sym}^n g) \circ (\mathrm{Sym}^n f) = \mathrm{Sym}^n(g \circ f)$ and $\mathrm{Sym}^n \mathrm{id}_M = \mathrm{id}_{\mathrm{Sym}^n M}$.

20.11 Bases and Multilinear Maps

In this section, we study a universal mapping property involving multilinear maps defined on a product of free R-modules. We show that each such multilinear map is uniquely determined by its effect on the product of bases of the given modules. By comparing this result to the UMP for tensor products, we obtain an explicit basis for a tensor product of free R-modules.

To begin, assume that $M_1, \ldots, M_n$ are free R-modules with respective bases $X_1, \ldots, X_n$. Let $M = M_1 \times \cdots \times M_n$, $X = X_1 \times \cdots \times X_n$, and $i : X \to M$ be the inclusion map. Then the following universal mapping property holds.

UMP for Multilinear Maps on Free Modules. For any R-module N and any function $f : X \to N$, there exists a unique R-multilinear map $T : M \to N$ such that $f = T \circ i$.

$$
\begin{array}{ccc}
X & \xrightarrow{\ i\ } & M \\
 & {\scriptstyle f}\searrow & \downarrow{\scriptstyle T} \\
 & & N
\end{array}
$$

Equivalently: for any R-module N, there is a bijection from the set

$$A = \{R\text{-multilinear maps } T : M \to N\}$$

onto the set

$$B = \{\text{all functions } f : X \to N\},$$

which sends each $T \in A$ to $T \circ i \in B$. Informally, we say that each function $f : X \to N$ *extends by multilinearity* to a unique multilinear map $T : M \to N$.

To prove the UMP, fix the module N and the map $f : X \to N$. Suppose $T : M \to N$ is R-multilinear and extends f. Since each M_k is free with basis X_k, each element $m_k \in M_k$ can be written uniquely as a finite R-linear combination of elements of X_k, say $m_k = \sum_{j_k=1}^{s_k} c_{j_k,k} x_{j_k,k}$ with each $x_{j_k,k} \in X_k$ and each $c_{j_k,k} \in R$. Using (20.4), we see that $T(m_1, m_2, \ldots, m_n)$ must be given by the formula

$$
\sum_{j_1=1}^{s_1} \sum_{j_2=1}^{s_2} \cdots \sum_{j_n=1}^{s_n} c_{j_1,1} c_{j_2,2} \cdots c_{j_n,n} f(x_{j_1,1}, x_{j_2,2}, \ldots, x_{j_n,n}).
$$

This shows that T, if it exists at all, is uniquely determined by f. To prove existence, use the previous formula as the definition of $T(m_1, \ldots, m_n)$. It is routine to check that T extends f. To see that T is multilinear, fix an index k, a scalar $r \in R$, and $m_k' = \sum_{j_k=1}^{s_k} d_{j_k,k} x_{j_k,k} \in M_k$. On one hand, since $r m_k = \sum_{j_k=1}^{s_k} (r c_{j_k,k}) x_{j_k,k}$, we get

$$
\begin{aligned}
T(m_1, \ldots, r m_k, \ldots, r_n) &= \sum_{j_1=1}^{s_1} \cdots \sum_{j_n=1}^{s_n} c_{j_1,1} \cdots (r c_{j_k,k}) \cdots c_{j_n,n} f(x_{j_1,1}, \ldots, x_{j_n,n}) \\
&= r \sum_{j_1=1}^{s_1} \cdots \sum_{j_n=1}^{s_n} c_{j_1,1} \cdots c_{j_k,k} \cdots c_{j_n,n} f(x_{j_1,1}, \ldots, x_{j_n,n}) \\
&= r T(m_1, \ldots, m_k, \ldots, r_n),
\end{aligned}
$$

where the second equality needs commutativity of R. On the other hand, since $m_k + m_k' = \sum_{j_k=1}^{s_k} (c_{j_k,k} + d_{j_k,k}) x_{j_k,k}$, we get

$$
\begin{aligned}
T(m_1, \ldots, m_k + m_k', \ldots, m_n) &= \sum_{j_1=1}^{s_1} \cdots \sum_{j_n=1}^{s_n} c_{j_1,1} \cdots (c_{j_k,k} + d_{j_k,k}) \cdots c_{j_n,n} f(x_{j_1,1}, \ldots, x_{j_n,n}) \\
&= \sum_{j_1=1}^{s_1} \cdots \sum_{j_n=1}^{s_n} c_{j_1,1} \cdots c_{j_k,k} \cdots c_{j_n,n} f(x_{j_1,1}, \ldots, x_{j_n,n}) \\
&\quad + \sum_{j_1=1}^{s_1} \cdots \sum_{j_n=1}^{s_n} c_{j_1,1} \cdots d_{j_k,k} \cdots c_{j_n,n} f(x_{j_1,1}, \ldots, x_{j_n,n}) \\
&= T(m_1, \ldots, m_k, \ldots, m_n) + T(m_1, \ldots, m_k', \ldots, m_n).
\end{aligned}
$$

So T is multilinear, completing the proof of the UMP.

20.12   Bases for Tensor Products of Free Modules

We continue to assume that $M_1, \ldots, M_n$ are free R-modules with respective bases $X_1, \ldots, X_n$. We use the UMP proved in §20.11 to construct a new solution to the universal mapping problem for multilinear maps from §20.4. Our ultimate goal is to show that $\{x_1 \otimes x_2 \otimes \cdots \otimes x_n : x_k \in X_k\}$ is a basis of $M_1 \otimes_R M_2 \otimes_R \cdots \otimes_R M_n$.

We have already constructed an R-module $N = M_1 \otimes_R \otimes \cdots \otimes_R M_n$ and a multilinear map $j : M_1 \times \cdots \times M_n \to N$ such that (N, j) solves the UMP from §20.4. To build the second solution, let N' be a free R-module with basis $X = X_1 \times \cdots \times X_n$ (see §17.11). Define a multilinear map $j' : M_1 \times \cdots \times M_n \to N'$ as follows. For each $x = (x_1, \ldots, x_n) \in X$, define $j'(x) = x \in N'$. Using the UMP proved in §20.11, j' extends by multilinearity to a unique multilinear map with domain $M_1 \times \cdots \times M_n$. Explicitly, we have

$$j'\left(\sum_{k_1 \geq 1} c_{k_1,1} x_{k_1,1}, \ldots, \sum_{k_n \geq 1} c_{k_n,n} x_{k_n,n} \right) = \sum_{k_1 \geq 1} \cdots \sum_{k_n \geq 1} c_{k_1,1} \cdots c_{k_n,n} (x_{k_1,1}, \ldots, x_{k_n,n}) \in N'.$$

To prove that (N', j') solves the UMP, suppose P is any R-module and $f : M_1 \times \cdots \times M_n \to P$ is any multilinear map. We must show there exists a unique R-linear map $g' : N' \to P$ with $f = g' \circ j'$.

$$
\begin{array}{ccc}
M_1 \times \cdots \times M_n & \xrightarrow{\ j'\ } & N' \\
& f \searrow & \downarrow g' \\
& & P
\end{array}
$$

To see that g' exists, define $g'(x) = f(x)$ for all $x \in X$ and extend g' by linearity to an R-linear map $g' : N' \to P$ (using the UMP for free R-modules). Now f and $g' \circ j'$ are two multilinear maps from $M_1 \times \cdots \times M_n$ to P that agree on $X = X_1 \times \cdots \times X_n$, since $f(x) = g'(x) = g'(j'(x)) = (g' \circ j')(x)$ for all $x \in X$. By the uniqueness property in the UMP from §20.11, $f = g' \circ j'$ as needed. To see that g' is unique, suppose we also had $f = h' \circ j'$ for some R-linear $h' : N' \to P$. Then $h'(x) = h'(j'(x)) = f(x) = g'(x)$ for all $x \in X$. Two R-linear maps that agree on a basis of N' must be equal, so $g' = h'$.

Now we know that (N', j') and (N, j) both solve the same UMP. Thus there exists a unique R-module isomorphism $g' : N' \to N$ such that $j = g' \circ j'$.

$$
\begin{array}{ccc}
M_1 \times \cdots \times M_n & \xrightarrow{\ j'\ } & N' \\
& j \searrow & \downarrow g' \\
& & M_1 \otimes_R \cdots \otimes_R M_n
\end{array}
$$

The isomorphism g' sends the R-basis $X = X_1 \times \cdots \times X_n$ of N' onto an R-basis of the tensor product $M_1 \otimes_R \cdots \otimes_R M_n$. Explicitly, g' sends $x = (x_1, \ldots, x_n) \in X$ to $g'(x) = g'(j'(x)) = j(x) = x_1 \otimes \cdots \otimes x_n$. We have now proved that

$$\{x_1 \otimes x_2 \otimes \cdots \otimes x_n : x_k \in X_k\}$$

is a basis for the R-module $M_1 \otimes_R M_2 \otimes_R \cdots \otimes_R M_n$. In particular, if $\dim(M_k) = d_k < \infty$ for each k, we see that $\dim(\bigotimes_{k=1}^{n} M_k) = d_1 d_2 \cdots d_n$.

20.13 Bases and Alternating Maps

Let M be a free R-module with basis X. We use X to build a basis for the exterior power $\bigwedge^n M$. First, we need to establish a universal mapping property for alternating maps. Fix a total ordering $<$ on X. For instance, if $X = \{x_1, x_2, \ldots, x_m\}$ is finite, we can use the ordering $x_1 < x_2 < \cdots < x_m$. Let $X^n = \{(z_1, \ldots, z_n) : z_i \in X\}$ and

$$X^n_< = \{(z_1, \ldots, z_n) \in X^n : z_1 < z_2 < \cdots < z_n\}$$

be the set of strictly increasing sequences of n basis elements. If $n > |X|$, then $X^n_<$ is empty. Let $i : X^n_< \to M^n$ be the inclusion mapping of $X^n_<$ into the product module M^n.

UMP for Alternating Maps on Free Modules. For any R-module N and any function $f : X^n_< \to N$, there exists a unique alternating map $T : M^n \to N$ such that $f = T \circ i$.

Equivalently: for any R-module N, there is a bijection from the set

$$A = \{\text{alternating maps } T : M^n \to N\}$$

onto the set

$$B = \{\text{all functions } f : X^n_< \to N\},$$

which sends each $T \in A$ to $T \circ i \in B$. Intuitively, the UMP says that we can build alternating maps by deciding where to send each strictly increasing list of n basis elements, and the alternating map is uniquely determined by these decisions.

To prove the UMP, fix the module N and the map $f : X^n_< \to N$. We first extend f to a map $g : X^n \to N$. Given $z = (z_1, z_2, \ldots, z_n) \in X^n$, consider two cases. If $z_i = z_j$ for some $i \neq j$, let $g(z) = 0_N$. If all z_i are distinct, let $\mathrm{sort}(z) \in X^n_<$ be the unique sequence obtained by rearranging the entries of z into increasing order. We can write $\mathrm{sort}(z) = (z_{w(1)}, z_{w(2)}, \ldots, z_{w(n)})$ for a unique $w \in S_n$. In this case, define $g(z) = \mathrm{sgn}(w) f(\mathrm{sort}(z)) \in N$. Recall from Chapter 2 that $\mathrm{sgn}(w) = (-1)^{\mathrm{inv}(w)}$, where $\mathrm{inv}(w)$ is the number of interchanges of adjacent elements needed to pass from $\mathrm{sort}(z)$ to z or vice versa. It follows from this that if z and z' differ by interchanging two adjacent elements, then $g(z) = -g(z')$. In turn, we deduce that if z and z' differ by interchanging any two elements, then $g(z) = -g(z')$.

From §20.11, we know that $g : X^n \to N$ extends uniquely by multilinearity to give a multilinear map $T : M^n \to N$. Since T extends g, T also extends f, so $f = T \circ i$. We must show that T is alternating. Fix $(m_1, \ldots, m_n) \in M^n$ with $m_i = m_j$ for some $i \neq j$. Write $m_k = \sum_{z \in X} r(k, z) z$ for some $r(k, z) \in R$. Since $m_i = m_j$, $r(i, z) = r(j, z)$ for all $z \in X$. Since T is multilinear,

$$\begin{aligned}
T(m_1, \ldots, m_n) &= \sum_{z_1 \in X} \cdots \sum_{z_n \in X} r(1, z_1) \cdots r(n, z_n) T(z_1, \ldots, z_n) \\
&= \sum_{z = (z_1, \ldots, z_n) \in X^n} r(1, z_1) \cdots r(i, z_i) \cdots r(j, z_j) \cdots r(n, z_n) g(z_1, \ldots, z_n).
\end{aligned}$$

By definition of g, we can drop all terms in this sum in which two entries of z are equal. The other terms can be split into pairs $z = (z_1, \ldots, z_i, \ldots, z_j, \ldots, z_n)$ and $z' = (z_1, \ldots, z_j, \ldots, z_i, \ldots, z_n)$ by switching the basis elements in positions i and j. By the observations above, $g(z') = -g(z)$. The term indexed by z is

$$r(1, z_1) \cdots r(i, z_i) \cdots r(i, z_j) \cdots r(n, z_n) g(z),$$

and the term indexed by z' is

$$r(1, z_1) \cdots r(i, z_j) \cdots r(i, z_i) \cdots r(n, z_n) g(z').$$

Since R is commutative, the sum of these two terms is zero. Thus, adding all these pairs, the total effect is that $T(m_1, \ldots, m_i, \ldots, m_i, \ldots, m_n) = 0_N$.

We prove uniqueness of T. Suppose $T' : M^n \to N$ is another alternating map that extends $f : X_<^n \to N$. If we can show that T' extends $g : X^n \to N$, we can conclude that $T = T'$ using the known uniqueness property from the UMP in §20.11. Suppose $z = (z_1, \ldots, z_n) \in X^n$ has two repeated entries. Then $T'(z) = 0 = g(z)$ since T' is alternating. Otherwise, when all entries of z are distinct, let $\text{sort}(z) = (z_{w(1)}, \ldots, z_{w(n)})$ for some $w \in S_n$. Property (AC3) in §20.2 holds for the alternating map T', so

$$T'(z) = \text{sgn}(w) T'(\text{sort}(z)) = \text{sgn}(w) f(\text{sort}(z)) = g(z).$$

Thus T and T' both extend g, hence $T = T'$.

20.14   Bases for Exterior Powers of Free Modules

Assume that M is a free R-module with a basis X totally ordered by $<$. Following the pattern of §20.12, we prove that

$$\{z_1 \wedge z_2 \wedge \cdots \wedge z_n : z_i \in X, z_1 < z_2 < \cdots < z_n\}$$

is a basis of the R-module $\bigwedge^n M$.

We have already constructed an R-module $N = \bigwedge^n M$ and an alternating map $i : M^n \to N$ such that (N, i) solves the UMP from §20.5. We construct a second solution to this UMP by letting N' be a free R-module with basis $X_<^n = \{(z_1, \ldots, z_n) \in X^n : z_1 < \cdots < z_n\}$. Noting that $X_<^n$ is a subset of both M^n and N', we can define an alternating map $i' : M^n \to N'$ by letting $i'(z) = z$ for each $z \in X_<^n$ and extending i' to M^n by the UMP in §20.13. To prove that (N', i') solves the UMP for exterior powers, suppose P is any R-module and $f : M^n \to P$ is any alternating map. We must show there exists a unique R-linear map $g' : N' \to P$ with $f = g' \circ i'$.

To see that g' exists, define $g'(x) = f(x)$ for all $x \in X_<^n$ and extend g' by linearity to an R-linear map $g' : N' \to P$ (using the UMP for free R-modules). Note f and $g' \circ i'$ are two alternating maps from M^n to P that agree on $X_<^n$, since $f(x) = g'(x) = g'(i'(x)) = (g' \circ i')(x)$ for all $x \in X_<^n$. By the uniqueness property in the UMP from §20.13, $f = g' \circ i'$ as needed.

To see that g' is unique, suppose we also have $f = h' \circ i'$ for some R-linear $h' : N' \to P$. Then $h'(x) = h'(i'(x)) = f(x) = g'(x)$ for all $x \in X_<^n$. Two R-linear maps that agree on a basis of N' must be equal, so $g' = h'$.

Now we know that (N', i') and (N, i) both solve the same UMP. Thus there exists a unique R-module isomorphism $g' : N' \to N$ such that $i = g' \circ i'$.

$$M^n \xrightarrow{\ i'\ } N'$$

with i going diagonally down to $\bigwedge^n M$ and g' going down from N' to $\bigwedge^n M$.

The isomorphism g' sends the R-basis $X_<^n$ of N' onto an R-basis of the exterior power $\bigwedge^n M$. Explicitly, g' sends $z = (z_1, \ldots, z_n) \in X_<^n$ to $g'(z) = g'(i'(z)) = i(z) = z_1 \wedge \cdots \wedge z_n$. We have now proved that

$$i[X_<^n] = \{z_1 \wedge z_2 \wedge \cdots \wedge z_n : z_i \in X, z_1 < \cdots < z_n\}$$

is a basis for the R-module $\bigwedge^n M$. In particular, if $\dim(M) = d < \infty$, we see that $\dim(\bigwedge^n M) = \binom{d}{n} = \frac{d!}{n!(d-n)!}$ for $0 \leq n \leq d$, and $\dim(\bigwedge^n M) = 0$ for $n > d$.

20.15   Bases for Symmetric Powers of Free Modules

Assume M is a free R-module with totally ordered basis X. Let $X_\leq^n$ be the set of all sequences $z = (z_1, \ldots, z_n) \in X^n$ with $z_1 \leq z_2 \leq \cdots \leq z_n$ relative to the total ordering $<$ on X. Let $i : X_\leq^n \to M^n$ be the inclusion map given by $i(z) = z$ for $z \in X_\leq^n$.

UMP for Symmetric Maps on Free Modules. For any R-module N and any function $f : X_\leq^n \to N$, there exists a unique symmetric map $T : M^n \to N$ such that $f = T \circ i$.

Equivalently: for any R-module N, there is a bijection from the set

$$A = \{\text{symmetric maps } T : M^n \to N\}$$

onto the set

$$B = \{\text{all functions } f : X_\leq^n \to N\},$$

which sends each $T \in A$ to $T \circ i \in B$. So, we can build symmetric maps uniquely by specifying where to send each weakly increasing list of n basis elements.

The proof is similar to the one in §20.13, so we leave certain details as exercises. Fix the module N and the map $f : X_\leq^n \to N$. Define $g : X^n \to N$ by $g(z) = f(\text{sort}(z))$ for all $z \in X^n$, where $\text{sort}(z)$ is the unique weakly increasing sequence of basis elements that can be obtained by sorting the entries of z. From §20.11, we know that $g : X^n \to N$ extends uniquely by multilinearity to give a multilinear map $T : M^n \to N$. Since T extends g, T also extends f, so $f = T \circ i$. Using multilinearity and the definition of g, one checks that T

is a symmetric map. For uniqueness of T, suppose $T' : M^n \to N$ is another symmetric map that extends $f : X^n_{\leq} \to N$. Using symmetry and the definition of g, check that T and T' must both extend $g : X^n \to N$. Hence, $T = T'$ follows from the known uniqueness property in the UMP from §20.11.

Next, we show that

$$i[X^n_{\leq}] = \{z_1 z_2 \cdots z_n : z_i \in X, z_1 \leq z_2 \leq \cdots \leq z_n\}$$

is a basis of the R-module $\text{Sym}^n M$. We have already constructed an R-module $N = \text{Sym}^n M$ and a symmetric map $i : M^n \to N$ such that (N, i) solves the UMP from §20.6. We construct a second solution to this UMP by letting N' be a free R-module with basis $X^n_{\leq}$. Define a symmetric map $i' : M^n \to N'$ by sending z to z for each $z \in X^n_{\leq}$ and extending to M^n by the UMP shown above. You can now verify, as in the case of exterior powers, that (N', i') solves the same UMP that (N, i) does. So there is a unique isomorphism between N' and N compatible with the universal maps, and this isomorphism maps the R-basis $X^n_{\leq}$ of N' to the claimed basis of $N = \text{Sym}^n M$.

In particular, if M is a free module of dimension $d < \infty$, then for all $n \geq 0$, $\dim(\text{Sym}^n M)$ is the number of weakly increasing sequences of length n drawn from a d-letter totally ordered alphabet. A counting argument (Exercise 48) shows that $\dim(\text{Sym}^n M) = \binom{d+n-1}{n}$.

20.16   Tensor Product of Matrices

Suppose M is a free R-module with ordered basis $X = (x_1, \ldots, x_m)$ and N is a free R-module with ordered basis $Y = (y_1, \ldots, y_n)$. We have seen that the list of mn basic tensors

$$Z = (x_1 \otimes y_1, x_2 \otimes y_1, \ldots, x_m \otimes y_1, x_1 \otimes y_2, x_2 \otimes y_2, \ldots x_m \otimes y_2, \ldots, x_m \otimes y_n)$$

is an ordered basis of $M \otimes_R N$. Now suppose $f : M \to M$ and $g : N \to N$ are R-linear maps. Let $A \in M_m(R)$ be the matrix of f relative to the basis X and $B \in M_n(R)$ be the matrix of g relative to the basis Y. What is the matrix C of the R-linear map $f \otimes g : M \otimes_R N \to M \otimes_R N$ relative to the basis Z?

To answer this question, first recall that $f(x_j) = \sum_{i=1}^m A(i, j) x_i$ for $1 \leq j \leq m$ and $g(y_j) = \sum_{i=1}^n B(i, j) y_i$ for $1 \leq j \leq n$. We label the rows and columns of the matrix C with the elements of Z in the order they appear above. To compute the entries in the column of C labeled by $x_i \otimes y_j$, we apply $f \otimes g$ to this element, obtaining

$$
\begin{aligned}
(f \otimes g)(x_i \otimes y_j) &= f(x_i) \otimes g(y_j) = \left(\sum_{k=1}^m A(k, i) x_k \right) \otimes \left(\sum_{\ell=1}^n B(\ell, j) y_\ell \right) \\
&= \sum_{k=1}^m \sum_{\ell=1}^n A(k, i) B(\ell, j)(x_k \otimes y_\ell).
\end{aligned}
$$

So, *the entry of C in the row labeled $(x_k \otimes y_\ell)$ and the column labeled $(x_i \otimes y_j)$ is $A(k, i)B(\ell, j)$.* We write $C = A \otimes B$ and call C the *tensor product of the matrices A and B.*

Note that C is an $mn \times mn$ matrix, where each entry is a scalar in R. We can also think of C as an $n \times n$ block matrix where each block is itself an $m \times m$ matrix. For $1 \leq i, j \leq n$, the i, j-block of C has rows labeled $x_1 \otimes y_i, x_2 \otimes y_i, \ldots, x_m \otimes y_i$ and columns labeled $x_1 \otimes y_j, x_2 \otimes y_j, \ldots, x_m \otimes y_j$. The above calculation shows that for $1 \leq r, s \leq m$, the r, s-entry of the i, j-block of C is $A(r, s)B(i, j)$. Thus, the entire i, j-block of C is found by multiplying the entire matrix A by the scalar $b_{ij} = B(i,j)$. Pictorially, the block matrix C is

$$C = A \otimes B = \begin{bmatrix} Ab_{11} & Ab_{12} & \cdots & Ab_{1n} \\ Ab_{21} & Ab_{22} & \cdots & Ab_{2n} \\ \vdots & \vdots & \vdots & \vdots \\ Ab_{n1} & Ab_{n2} & \cdots & Ab_{nn} \end{bmatrix}.$$

Suppose that $f_1 : M \to M$ and $g_1 : N \to N$ are also R-linear maps, represented (relative to the bases X and Y) by matrices A_1 and B_1. We have seen that $(f_1 \circ f) \otimes (g_1 \circ g) = (f_1 \otimes g_1) \circ (f \otimes g)$. Passing to matrices, this yields the matrix identity

$$(A_1 A) \otimes (B_1 B) = (A_1 \otimes B_1)(A \otimes B).$$

You can check that $(f_1 + f) \otimes g = (f_1 \otimes g) + (f \otimes g)$ and $(rf) \otimes g = r(f \otimes g)$ for all $r \in R$. We therefore see that

$$(A_1 + A) \otimes B = (A_1 \otimes B) + (A \otimes B), \qquad (rA) \otimes B = r(A \otimes B),$$

and similar identities hold in the second input.

20.17   Determinants and Exterior Powers

Let M be a free R-module with ordered basis $X = (x_1, \ldots, x_n)$. Given an R-linear map $f : M \to M$, let $A \in M_n(R)$ be the matrix of f relative to X. For each k with $0 \le k \le n$, there is an induced R-linear map $\bigwedge^k f : \bigwedge^k M \to \bigwedge^k M$. What is the matrix of this map relative to the basis $i[X_{<}^k]$ of $\bigwedge^k M$?

Before answering this question in general, consider the special case $k = n$. Here, $\bigwedge^n M$ is a one-dimensional free R-module, since its basis $i[X_{<}^n]$ consists of the single wedge product $x^* = x_1 \wedge x_2 \wedge \cdots \wedge x_n$ in which all elements of X appear in order. So the matrix of $\bigwedge^n f$ has just one entry; we find it by applying $\bigwedge^n f$ to x^* and seeing which multiple of x^* results. Recall $f(x_i) = \sum_{k=1}^n A(k,i)x_k$ for $1 \le i \le n$. Using this, we compute:

$$\begin{aligned} \left(\bigwedge^n f\right)(x^*) &= f(x_1) \wedge f(x_2) \wedge \cdots \wedge f(x_n) \\ &= \left(\sum_{k_1=1}^n A(k_1, 1)x_{k_1}\right) \wedge \left(\sum_{k_2=1}^n A(k_2, 2)x_{k_2}\right) \wedge \cdots \wedge \left(\sum_{k_n=1}^n A(k_n, n)x_{k_n}\right) \\ &= \sum_{k_1=1}^n \sum_{k_2=1}^n \cdots \sum_{k_n=1}^n A(k_1, 1)A(k_2, 2) \cdots A(k_n, n)(x_{k_1} \wedge x_{k_2} \wedge \cdots \wedge x_{k_n}). \end{aligned}$$

The last step uses the multilinearity of wedge products. To continue simplifying, recall that any wedge product with a repeated term is zero. So instead of summing over all sequences $(k_1, \ldots, k_n)$, it suffices to sum only over the permutations $k = (k_1, \ldots, k_n) \in S_n$. Using the last identity in §20.5, we obtain

$$\left(\bigwedge^n f\right)(x^*) = \left(\sum_{k \in S_n} \operatorname{sgn}(k)A(k_1, 1)A(k_2, 2) \cdots A(k_n, n)\right)(x_1 \wedge x_2 \wedge \cdots \wedge x_n) = \det(A)x^*.$$

We see that $\bigwedge^n f$ has matrix $[\det(A)]$ relative to the basis (x^*) of $\bigwedge^n M$. This computation shows how the mysterious definition of $\det(A)$ (first given in Chapter 5, equation (5.1)) occurs naturally in the theory of exterior powers.

We can now give a one-sentence proof of the product formula for determinants (see §5.13). Take maps $f, g : M \to M$ whose matrices relative to X are A and B, respectively; since $\bigwedge^n(f \circ g) = (\bigwedge^n f) \circ (\bigwedge^n g)$, applying the preceding result to the maps $f \circ g$, f, and g gives $\det(AB) = \det(A)\det(B)$.

We return to the question of computing the matrix C of $\bigwedge^k f$ relative to the basis $i[X_<^k]$, for any k between 0 and n. Each element of $i[X_<^k]$ has the form $x_{i_1} \wedge x_{i_2} \wedge \cdots \wedge x_{i_k}$ for a unique k-element subset $I = \{i_1 < i_2 < \cdots < i_k\}$ of $\{1, 2, \ldots, n\}$. Let us label the rows and columns of C with these subsets. To find the entries of C in the column labeled $J = \{j_1 < j_2 < \cdots < j_k\}$, apply $\bigwedge^k f$ to $x_{j_1} \wedge x_{j_2} \wedge \cdots \wedge x_{j_k}$. We obtain

$$f(x_{j_1}) \wedge f(x_{j_2}) \wedge \cdots \wedge f(x_{j_k})$$

$$= \left(\sum_{w_1=1}^{n} A(w_1, j_1) x_{w_1} \right) \wedge \left(\sum_{w_2=1}^{n} A(w_2, j_2) x_{w_2} \right) \wedge \cdots \wedge \left(\sum_{w_k=1}^{n} A(w_k, j_k) x_{w_k} \right)$$

$$= \sum_{w_1=1}^{n} \cdots \sum_{w_k=1}^{n} A(w_1, j_1) \cdots A(w_k, j_k) x_{w_1} \wedge \cdots \wedge x_{w_k}.$$

As before, we can discard zero terms to reduce to a sum over $w = (w_1, \ldots, w_k)$ with all entries distinct. For each k-element subset $I = \{i_1 < i_2 < \cdots < i_k\}$, there are $k!$ terms in the sum indexed by words w that are rearrangements of the entries of I. If a permutation $f \in S_k$ rearranges I into w, then

$$A(w_1, j_1) \cdots A(w_k, j_k) x_{w_1} \wedge \cdots \wedge x_{w_k} = \mathrm{sgn}(f) A(i_{f(1)}, j_1) \cdots A(i_{f(k)}, j_k) x_{i_1} \wedge \cdots \wedge x_{i_k}.$$

These $k!$ terms in the sum, and no others, contribute to the coefficient of $x_{i_1} \wedge \cdots \wedge x_{i_k}$. It follows that *the entry of C in the row labeled I and the column labeled J is*

$$\sum_{f \in S_k} \mathrm{sgn}(f) A(i_{f(1)}, j_1) \cdots A(i_{f(k)}, j_k) = \det(A_{I,J}),$$

where $\det(A_{I,J})$ denotes the determinant of the $k \times k$ submatrix of A obtained by keeping only the k rows in I and the k columns in J. Thus, the entries in the matrix for $\bigwedge^k f$ are precisely the kth order minors of A, which we studied in §18.11.

20.18   From Modules to Algebras

An *R-algebra* is a ring $(A, +, \star)$ that is also an R-module, such that $c(x \star y) = (cx) \star y = x \star (cy)$ for all $x, y \in A$ and all $c \in R$. An *R-algebra homomorphism* is a map between R-algebras that is both a ring homomorphism and an R-linear map. Given any R-module M, our goal is to build an R-algebra $T(M)$, called the *tensor algebra of M*, solving the following universal mapping problem.

UMP for Tensor Algebras. Given an R-module M, construct an R-algebra $T(M)$ and an R-linear map $i : M \to T(M)$ such that, for any R-algebra A and any R-linear map $f : M \to A$, there exists a unique R-algebra homomorphism $g : T(M) \to A$ with $f = g \circ i$.

Construction of $T(M)$. Let $T(M)$ be the (external) direct sum $\bigoplus_{k=0}^{\infty} M^{\otimes k}$, where $M^{\otimes 0} = R$. By definition, an element of $T(M)$ is an infinite sequence $z = (z_0, z_1, \ldots, z_k, \ldots)$, where $z_k \in M^{\otimes k}$ and all but finitely many z_k are zero. We already know $T(M)$ is an R-module, and (using Exercise 15) there is an injective R-linear map $i : M \to T(M)$ given by $i(x) = (0, x, 0, 0, \ldots)$ for $x \in M$. We must define an algebra structure on $T(M)$ by specifying the ring multiplication $\star : T(M) \times T(M) \to T(M)$.

In §20.9, we constructed (for each $m, k \in \mathbb{Z}_{>0}$) an R-linear map $\nu_{k,m} : M^{\otimes k} \otimes_R M^{\otimes m} \to M^{\otimes(k+m)}$ given on generators by

$$\nu_{k,m}((x_1 \otimes \cdots \otimes x_k) \otimes (x_{k+1} \otimes \cdots \otimes x_{k+m})) = x_1 \otimes \cdots \otimes x_k \otimes x_{k+1} \otimes \cdots \otimes x_{k+m}.$$

Recall that $\nu_{k,m}$ is induced (via the UMP for tensor products) from an R-bilinear map $\mu_{k,m} : M^{\otimes k} \times M^{\otimes m} \to M^{\otimes(k+m)}$, which acts on generators by

$$\mu_{k,m}(x_1 \otimes \cdots \otimes x_k, x_{k+1} \otimes \cdots \otimes x_{k+m}) = x_1 \otimes \cdots \otimes x_k \otimes x_{k+1} \otimes \cdots \otimes x_{k+m}.$$

Similarly, we have R-bilinear maps $\mu_{0,m}$ and $\mu_{k,0}$ such that (for $r \in R$ and $x_i \in M$)

$$\mu_{0,m}(r, x_1 \otimes \cdots \otimes x_m) = (rx_1) \otimes \cdots \otimes x_m; \qquad \mu_{k,0}(x_1 \otimes \cdots \otimes x_k, r) = (rx_1) \otimes \cdots \otimes x_k.$$

We can assemble all the maps $\mu_{k,m}$ to obtain a map $\star : T(M) \times T(M) \to T(M)$, as follows. Given $y = (y_k : k \geq 0)$ and $z = (z_k : k \geq 0)$ in $T(M)$, define

$$y \star z = \left(\sum_{k+m=n} \mu_{k,m}(y_k, z_m) : n \geq 0 \right).$$

It is tedious but routine to confirm that $(T(M), +, \star)$ satisfies the axioms for a ring and R-algebra. In particular, the left and right distributive laws for $\star$ follow from the preceding definition and bilinearity of each $\mu_{k,m}$. Associativity of $\star$ follows from the fact that

$$\mu_{k+m,p}(\mu_{k,m}(x_1 \otimes \cdots \otimes x_k, y_1 \otimes \cdots \otimes y_m), z_1 \otimes \cdots \otimes z_p)$$
$$= \mu_{k,m+p}(x_1 \otimes \cdots \otimes x_k, \mu_{m,p}(y_1 \otimes \cdots \otimes y_m, z_1 \otimes \cdots \otimes z_p)),$$

which holds since both sides equal

$$x_1 \otimes \cdots \otimes x_k \otimes y_1 \otimes \cdots \otimes y_m \otimes z_1 \otimes \cdots \otimes z_p.$$

The multiplicative identity of $T(M)$ is $(1_R, 0, 0, \ldots)$.

Next, we verify that $(T(M), i)$ solves the UMP. Let $(A, +, *)$ be any R-algebra and $f : M \to A$ be an R-linear map. We need to build an R-algebra map $g : T(M) \to A$ with $f = g \circ i$. Define $g_0 : R \to A$ by $g_0(r) = r \cdot 1_A$ for all $r \in R$. For each $k > 0$, define $f_k : M^k \to A$ by $f_k(x_1, x_2, \ldots, x_k) = f(x_1) * f(x_2) * \cdots * f(x_k)$ for all $x_j \in M$. Since A is an R-algebra and f is R-linear, f_k is R-multilinear. So we get an induced R-linear map $g_k : M^{\otimes k} \to A$ given on generators by

$$g_k(x_1 \otimes x_2 \otimes \cdots \otimes x_k) = f(x_1) * f(x_2) * \cdots * f(x_k).$$

The UMP for direct sums (§19.6) combines all the maps g_k to give an R-linear map $g : T(M) \to A$ such that $g(z_0, z_1, \ldots, z_k, \ldots) = \sum_{k \geq 0} g_k(z_k)$. Note that $g \circ i = f$, since for all $z_1 \in M$, $g(i(z_1)) = g(0, z_1, 0, \ldots) = g_1(z_1) = f(z_1)$. Since g is already known to be R-linear, we need only check that $g(y \star z) = g(y) * g(z)$ for all $y, z \in T(M)$. Using the definition of $\star$ and the distributive laws, this verification reduces to the fact that $g_{k+m}(\mu_{k,m}(y, z)) = g_k(y) * g_m(z)$ for all $y \in M^{\otimes k}$ and all $z \in M^{\otimes m}$. This fact holds because

$$g_{k+m}(\mu_{k,m}(y_1 \otimes \cdots \otimes y_k, z_1 \otimes \cdots \otimes z_m)) = f(y_1) * f(y_2) * \cdots * f(y_k) * f(z_1) * \cdots * f(z_m)$$
$$= g_k(y_1 \otimes \cdots \otimes y_k) * g_m(z_1 \otimes \cdots \otimes z_m),$$

$\mu_{k,m}$ is R-bilinear, and g_k, g_m, g_{k+m} are R-linear.

We must show that g is unique. Suppose $h : T(M) \to A$ is also an R-algebra homomorphism with $h \circ i = f$. By the uniqueness property in the UMP for direct sums, it suffices to check that $h \circ j_k = g \circ j_k$ for all $k \geq 0$, where j_k is the injection of $M^{\otimes k}$ into $T(M) = \bigoplus_{s \geq 0} M^{\otimes s}$. When $k = 0$, we have

$$h \circ j_0(r) = rh(1_{T(M)}) = r1_A = rg_0(1_{T(M)}) = g \circ j_0(r)$$

for all $r \in R$. When $k = 1$, we have

$$h \circ j_1(x) = h \circ i(x) = f(x) = g \circ i(x) = g \circ j_1(x)$$

for all $x \in M$. When $k \geq 2$, note that $x_1 \otimes x_2 \otimes \cdots \otimes x_k = x_1 \star x_2 \star \cdots \star x_k$ for all $x_j \in M$. Since g and h preserve multiplication and coincide with f on M, we get

$$h \circ j_k(x_1 \otimes \cdots \otimes x_k) = h(x_1 \star \cdots \star x_k) = f(x_1) * f(x_2) * \cdots * f(x_k)$$
$$= g(x_1 \star \cdots \star x_k) = g \circ j_k(x_1 \otimes \cdots \otimes x_k).$$

Thus $g = h$, completing the proof of the uniqueness assertion in the UMP.

In general, the algebra $T(M)$ is not commutative. By replacing tensor powers of M by symmetric powers of M throughout the preceding construction, we can build a commutative algebra $\mathrm{Sym}(M)$ such that any R-linear map from M into a commutative algebra A uniquely extends to an R-algebra map from $\mathrm{Sym}(M)$ to A (Exercise 59). $\mathrm{Sym}(M)$ is called the *symmetric algebra of M*. A similar construction using exterior powers produces an algebra $\bigwedge(M)$ solving an appropriate UMP (Exercise 60). $\bigwedge(M)$ is called the *exterior algebra of M*.

20.19 Summary

In this summary, assume R is a commutative ring and all other capital letters are R-modules unless otherwise stated.

1. *Special Maps.* A map $f : M \to N$ is *R-linear* iff $f(m + m') = f(m) + f(m')$ and $f(rm) = rf(m)$ for all $m, m' \in M$ and $r \in R$. A map $f : M_1 \times \cdots \times M_n \to N$ is *R-multilinear* iff for $1 \leq i \leq n$, f is R-linear in position i when all other inputs are held fixed. An R-multilinear map $f : M^n \to N$ is *alternating* iff f has value zero when any two inputs are equal. An R-multilinear map $f : M^n \to N$ is *anti-commutative* iff $f(m_{w(1)}, \ldots, m_{w(n)}) = \text{sgn}(w)f(m_1, \ldots, m_n)$ for all $m_k \in M$ and all $w \in S_n$. To prove anti-commutativity, it suffices to check that f changes sign when any two adjacent inputs are switched. Alternating maps are anti-commutative, but the converse only holds when $1_R + 1_R$ is not a zero divisor. An R-multilinear map $f : M^n \to N$ is *symmetric* iff $f(m_{w(1)}, \ldots, m_{w(n)}) = f(m_1, \ldots, m_n)$ for all $m_k \in M$ and all $w \in S_n$. To prove symmetry, it suffices to check that f is unchanged when any two adjacent inputs are switched.

2. *Generators for $\bigotimes_{k=1}^n M_k$, $\bigwedge^n M$, and $\text{Sym}^n M$.* Every element of $M_1 \otimes_R \cdots \otimes_R M_n$ is a finite sum of basic tensors $x_1 \otimes \cdots \otimes x_n$ with $x_k \in M_k$. Every element of $\bigwedge^n M$ is a finite sum of elements $x_1 \wedge \cdots \wedge x_n$ with $x_k \in M$. Every element of $\text{Sym}^n M$ is a finite sum of elements $x_1 \cdots x_n$ with $x_k \in M$. If X_k generates M_k, then $\{x_1 \otimes \cdots \otimes x_n : x_k \in X_k\}$ generates $\bigotimes_{k=1}^n M_k$; similarly for $\bigwedge^n M$ and $\text{Sym}^n M$.

3. *Bases for $\bigotimes_{k=1}^n M_k$, $\bigwedge^n M$, and $\text{Sym}^n M$.* If each M_k is free with basis X_k, then $\bigotimes_{k=1}^n M_k$ is free with basis $\{x_1 \otimes \cdots \otimes x_n : x_k \in X_k\}$. If M is free with basis X totally ordered by $<$, then $\bigwedge^n M$ is free with basis $\{x_1 \wedge \cdots \wedge x_n : x_k \in X, x_1 < \cdots < x_n\}$, and $\text{Sym}^n M$ is free with basis $\{x_1 \cdots x_n : x_k \in X, x_1 \leq \cdots \leq x_n\}$. In the case where $\dim(M_k) = d_k < \infty$ and $\dim(M) = d < \infty$, we have $\dim(M_1 \otimes_R \cdots \otimes_R M_n) = d_1 d_2 \cdots d_n$, $\dim(\bigwedge^n M) = \binom{d}{n}$ for $0 \leq n \leq d$, $\dim(\bigwedge^n M) = 0$ for $n > d$, and $\dim(\text{Sym}^n M) = \binom{d+n-1}{n}$.

4. *Myths about Tensor Products.* It is false that every element of $M_1 \otimes_R \cdots \otimes_R M_n$ must have the form $x_1 \otimes \cdots \otimes x_n$ for some $x_k \in M_k$. It is false that $\{x_1 \otimes \cdots \otimes x_n : x_k \in M_k\}$ must be a basis for $M_1 \otimes_R \cdots \otimes_R M_n$. It is false that we can define a function on $M_1 \otimes_R \cdots \otimes_R M_n$ (with no further work) by specifying where the function sends basic tensors. It is false that a tensor product of nonzero modules must be nonzero.

5. *Tensor Product Isomorphisms.* Five R-module isomorphisms (defined on generators) are:

   (a) *Commutativity:* $M \otimes_R N \cong N \otimes_R M$ via the map $m \otimes n \mapsto n \otimes m$.

   (b) *Associativity:* $(M \otimes_R N) \otimes_R P \cong M \otimes_R (N \otimes_R P)$ via the map $(m \otimes n) \otimes p \mapsto m \otimes (n \otimes p)$.

   (c) *Left Distributivity:* $(M \oplus N) \otimes_R P \cong (M \otimes_R P) \oplus (N \otimes_R P)$ via the map $(m, n) \otimes p \mapsto (m \otimes p, n \otimes p)$.

   (d) *Right Distributivity:* $P \otimes_R (M \oplus N) \cong (P \otimes_R M) \oplus (P \otimes_R N)$ via the map $p \otimes (m, n) \mapsto (p \otimes m, p \otimes n)$.

   (e) *Identity for $\otimes_R$:* $R \otimes_R M \cong M$ via the map $r \otimes m \mapsto rm$; similarly, $M \otimes_R R \cong M$.

More generally:

$$M_{w(1)} \otimes_R \cdots \otimes_R M_{w(n)} \cong M_1 \otimes_R \cdots \otimes_R M_n \text{ for all } w \in S_n;$$

$$(M_1 \otimes_R \cdots \otimes_R M_k) \otimes_R (M_{k+1} \otimes_R \cdots \otimes_R M_n) \cong M_1 \otimes_R \cdots \otimes_R M_n;$$

$$\left(\bigoplus_{i_1 \in I_1} M_{i_1} \right) \otimes_R \cdots \otimes_R \left(\bigoplus_{i_n \in I_n} M_{i_n} \right) \cong \bigoplus_{(i_1, \ldots, i_n) \in I_1 \times \cdots \times I_n} M_{i_1} \otimes_R \cdots \otimes_R M_{i_n};$$

and all factors of R can be deleted in a tensor product to give an isomorphic tensor product.

6. *Tensor Product of Linear Maps.* Given R-linear maps $f_k : M_k \to N_k$, there is an induced R-linear map $\bigotimes_{k=1}^n f_k : \bigotimes_{k=1}^n M_k \to \bigotimes_{k=1}^n N_k$ given on generators by $(f_1 \otimes \cdots \otimes f_n)(x_1 \otimes \cdots \otimes x_n) = f_1(x_1) \otimes \cdots \otimes f_n(x_n)$. Given R-linear maps $g_k : N_k \to P_k$, we have $\bigotimes_{k=1}^n (g_k \circ f_k) = (\bigotimes_{k=1}^n g_k) \circ (\bigotimes_{k=1}^n f_k)$.

7. *Exterior Powers and Symmetric Powers of Linear Maps.* Given an R-linear map $f : M \to N$, there are induced R-linear maps $\bigwedge^n f : \bigwedge^n M \to \bigwedge^n N$ and $\operatorname{Sym}^n f : \operatorname{Sym}^n M \to \operatorname{Sym}^n N$ that act on generators by

$$\left(\bigwedge^n f \right)(x_1 \wedge \cdots \wedge x_n) = f(x_1) \wedge \cdots \wedge f(x_n),$$

$$(\operatorname{Sym}^n f)(x_1 \cdots x_n) = f(x_1) \cdots f(x_n).$$

For an R-linear map $g : N \to P$, we have $\bigwedge^n (g \circ f) = (\bigwedge^n g) \circ (\bigwedge^n f)$ and $\operatorname{Sym}^n(g \circ f) = (\operatorname{Sym}^n g) \circ (\operatorname{Sym}^n f)$.

8. *Tensor Product of Matrices.* Given $A \in M_m(R)$ and $B \in M_n(R)$, $A \otimes B \in M_{mn}(R)$ is the $n \times n$ block matrix whose i, j-block is $B(i,j)A$. If A is the matrix of f and B is the matrix of g relative to certain bases, $A \otimes B$ is the matrix of $f \otimes g$ relative to the tensor product of these bases. For $C \in M_m(R)$ and $D \in M_n(R)$ and $r \in R$, $(CA) \otimes (DB) = (C \otimes D)(A \otimes B)$, $(C + A) \otimes B = (C \otimes B) + (A \otimes B)$, $A \otimes (B + D) = (A \otimes B) + (A \otimes D)$, and $(rA) \otimes B = r(A \otimes B) = A \otimes (rB)$.

9. *Exterior Powers and Determinants.* Suppose X is an n-element basis for M and $f : M \to M$ is R-linear. If A is the matrix of f relative to X, then $[\det(A)]$ is the matrix of $\bigwedge^n f$ relative to the basis $i[X_<^n]$. Similarly, the matrix of $\bigwedge^k f$ relative to the basis $i[X_<^k]$ has entries $\det(A_{I,J})$, where $I, J \subseteq \{1, 2, \ldots, n\}$ have size k.

Summary of Universal Mapping Properties

1. *UMP for Tensor Products.* Let $j : M_1 \times \cdots \times M_n \to M_1 \otimes_R \cdots \otimes_R M_n$ send $(m_1, \ldots, m_n)$ to $m_1 \otimes \cdots \otimes m_n$. For every R-module P and every R-multilinear map $f : M_1 \times \cdots \times M_n \to P$, there exists a unique R-linear map $g : M_1 \otimes_R \cdots \otimes_R M_n \to P$ with $f = g \circ j$.

$$M_1 \times \cdots \times M_n \xrightarrow{\ j\ } M_1 \otimes_R \cdots \otimes_R M_n$$

So *tensor products convert R-multilinear maps to R-linear maps.*

2. *UMP for Exterior Powers.* Let $i : M^n \to \bigwedge^n M$ send $(m_1, \ldots, m_n)$ to $m_1 \wedge \cdots \wedge m_n$. For every R-module P and every alternating map $f : M^n \to P$, there exists a unique R-linear map $g : \bigwedge^n M \to P$ with $f = g \circ i$.

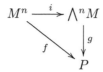

So *exterior powers convert alternating maps to R-linear maps.*

3. *UMP for Symmetric Powers.* Let $i : M^n \to \operatorname{Sym}^n M$ send $(m_1, \ldots, m_n)$ to $m_1 \cdots m_n$. For every R-module P and every symmetric map $f : M^n \to P$, there exists a unique R-linear map $g : \operatorname{Sym}^n M \to P$ with $f = g \circ i$.

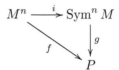

So *symmetric powers convert symmetric maps to R-linear maps.*

4. *UMP for Multilinear Maps on Free Modules.* Let $M_1, \ldots, M_n$ be free R-modules with respective bases $X_1, \ldots, X_n$. Let $i : X_1 \times \cdots \times X_n \to M_1 \times \cdots \times M_n$ be the inclusion map. For any R-module N and any function $f : X_1 \times \cdots \times X_n \to N$, there exists a unique R-multilinear map $T : M_1 \times \cdots \times M_n \to N$ such that $f = T \circ i$.

$$X_1 \times \cdots \times X_n \xrightarrow{\ i\ } M_1 \times \cdots \times M_n$$
$$\diagdown f \qquad \downarrow T$$
$$N$$

So *arbitrary functions on the product of bases extend uniquely to multilinear maps.*

5. *UMP for Alternating Maps on Free Modules.* Let M be a free R-module with a basis X totally ordered by $<$. Let $X_<^n$ be the set of strictly increasing sequences of n elements of X and $i : X_<^n \to M^n$ be the inclusion map. For any R-module N and any function $f : X_<^n \to N$, there exists a unique alternating map $T : M^n \to N$ such that $f = T \circ i$.

So *arbitrary functions on $X_<^n$ extend uniquely to alternating maps.*

6. *UMP for Symmetric Maps on Free Modules.* Let M be a free R-module with a basis X totally ordered by $<$. Let $X_\leq^n$ be the set of weakly increasing sequences of n elements of X and $i : X_\leq^n \to M^n$ be the inclusion map. For any R-module N and any function $f : X_\leq^n \to N$, there exists a unique symmetric map $T : M^n \to N$ such that $f = T \circ i$.

So *arbitrary functions on $X_\leq^n$ extend uniquely to symmetric maps.*

7. *UMP for Tensor Algebras.* For every R-module M, there is an R-algebra $T(M)$ and an R-linear injection $i : M \to T(M)$ such that for every R-algebra A and every R-linear map $f : M \to A$, there exists a unique R-algebra homomorphism $g : T(M) \to A$ with $f = g \circ i$.

So *tensor algebras convert R-linear maps (with codomain an R-algebra) into R-algebra maps.*

8. *UMP for Symmetric Algebras.* For every R-module M, there is a commutative R-algebra $\mathrm{Sym}(M)$ and an R-linear injection $i : M \to \mathrm{Sym}(M)$ such that for every commutative R-algebra A and every R-linear map $f : M \to A$, there exists a unique R-algebra homomorphism $g : \mathrm{Sym}(M) \to A$ with $f = g \circ i$.

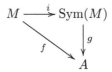

So *symmetric algebras convert R-linear maps (with codomain a commutative R-algebra) into R-algebra maps.*

9. *UMP for Exterior Algebras.* For every R-module M, there is an R-algebra $\bigwedge(M)$ with $z \star z = 0$ for all $z \in \bigwedge(M)$ and an R-linear injection $i : M \to \bigwedge(M)$ such that for every R-algebra A and every R-linear map $f : M \to A$ such that $f(x) \star f(x) = 0$ for all $x \in M$, there exists a unique R-algebra homomorphism $g : \bigwedge(M) \to A$ with $f = g \circ i$.

So *exterior algebras convert R-linear maps (where all images square to zero) into R-algebra maps.*

20.20   Exercises

Unless otherwise stated, assume R is a commutative ring and M, N, P, M_i are R-modules in these exercises.

1. Define $f : R^n \to R$ by $f(r_1, r_2, \ldots, r_n) = r_1 r_2 \cdots r_n$. Prove that f is a symmetric map. Justify each step using the ring axioms.

2. Decide (with proof) whether each map below is $\mathbb{R}$-linear, $\mathbb{R}$-bilinear, or neither. For the bilinear maps, say whether the map is symmetric or alternating.

    (a) $f : \mathbb{R}^2 \to \mathbb{R}$ given by $f(x, y) = x + y$.

    (b) $f : \mathbb{R}^2 \to \mathbb{R}$ given by $f(x, y) = 5xy$.

    (c) $f : \mathbb{R}^2 \to \mathbb{R}$ given by $f(x, y) = x^2 - y^2$.

    (d) $f : M_3(\mathbb{R}) \times M_3(\mathbb{R}) \to M_3(\mathbb{R})$ given by $f(A, B) = AB^{\mathrm{T}}$.

    (e) $f : \mathbb{R}^2 \times \mathbb{R}^2 \to \mathbb{R}$ given by $f((a, b), (c, d)) = ad - bc$ for $a, b, c, d \in \mathbb{R}$.

    (f) $I(g, h) = \int_0^1 t g(t) h(t)\, dt$ where $g, h : [0, 1] \to \mathbb{R}$ are continuous functions.

    (g) $C(g, h) = g \circ h - h \circ g$, where $g, h : \mathbb{R}^n \to \mathbb{R}^n$ are linear maps.

3. Define $f : \mathbb{R}^n \times \mathbb{R}^n \to \mathbb{R}$ by $f(\mathbf{v}, \mathbf{w}) = \mathbf{v}^{\mathrm{T}} A \mathbf{w}$, where $A \in M_n(\mathbb{R})$ is a fixed matrix and $\mathbf{v}, \mathbf{w}$ are column vectors. Prove that f is $\mathbb{R}$-bilinear. For which matrices A is f alternating? symmetric? $\mathbb{R}$-linear?

4. Prove that every $\mathbb{R}$-bilinear map $f : \mathbb{R}^n \times \mathbb{R}^n \to \mathbb{R}$ has the form $f(\mathbf{v}, \mathbf{w}) = \mathbf{v}^{\mathrm{T}} A \mathbf{w}$ for some $A \in M_n(\mathbb{R})$.

5. Prove (20.3) by induction on s. Deduce (20.4) from (20.3).

6. Give an example of an anti-commutative bilinear map that is not alternating.

7. Give an example of a nonzero bilinear map that is symmetric and alternating.

8. Find conditions on n and R so that every symmetric alternating map $f : M^n \to P$ must be zero.

9. Prove: for all multilinear $f : M_1 \times \cdots \times M_n \to P$, if $m_i = 0$ for some i, then $f(m_1, \ldots, m_n) = 0$. Deduce that if $m_i = 0$ for some i, then $m_1 \otimes \cdots \otimes m_n = 0$ in $M_1 \otimes_R \cdots \otimes_R M_n$.

10. Prove that $2 \otimes 3 = 0$ in $\mathbb{Z}_6 \otimes_{\mathbb{Z}} \mathbb{Z}_7$.

11. Prove or disprove: if $f : M \times N \to P$ is R-linear and R-bilinear, then f must be the zero map.

12. Suppose R is any ring, possibly non-commutative. Define R-bilinearity as in the text. Prove: if M and N are left R-modules and $f : M \times N \to R$ is an R-bilinear map such that $f(x, y)$ is nonzero and not a zero divisor for some $x \in M$ and $y \in N$, then R is commutative.

13. Show that if (N', i') solves the UMP in §20.5, then there exists a unique R-module isomorphism $g : \bigwedge^n M \to N'$ with $g \circ i = i'$ (where i is the map from M^n to $\bigwedge^n M$ defined in §20.5).

14. State and prove a result similar to the previous exercise for $\mathrm{Sym}^n M$.

15. Prove: for all R-modules M, $M \cong M^{\otimes 1} \cong \bigwedge^1 M \cong \mathrm{Sym}^1 M$.

16. Formulate and solve a UMP that converts anti-commutative maps to R-linear maps.

17. Assume $1_R + 1_R$ is not zero and not a zero divisor. Let K_1 be the submodule of $M^{\otimes n}$ generated by all elements of the form

$$m_1 \otimes \cdots \otimes m_i \otimes \cdots \otimes m_j \otimes \cdots \otimes m_n + m_1 \otimes \cdots \otimes m_j \otimes \cdots \otimes m_i \otimes \cdots \otimes m_n$$

for all $m_k \in M$ and all $i < j$. Prove that $M^{\otimes n}/K_1 \cong \bigwedge^n M$.

18. Assume $1_R + 1_R$ is not zero and not a zero divisor. Let K_2 be the submodule of $M^{\otimes n}$ generated by all elements of the form

$$m_1 \otimes \cdots \otimes m_i \otimes m_{i+1} \otimes \cdots \otimes m_n + m_1 \otimes \cdots \otimes m_{i+1} \otimes m_i \otimes \cdots \otimes m_n$$

for all $m_k \in M$ and all $i < n$. Prove that $M^{\otimes n}/K_2 \cong \bigwedge^n M$.

19. Let K_3 be the submodule of $M^{\otimes n}$ generated by all elements of the form

$$m_1 \otimes \cdots \otimes m_i \otimes m_{i+1} \otimes \cdots \otimes m_n - m_1 \otimes \cdots \otimes m_{i+1} \otimes m_i \otimes \cdots \otimes m_n$$

for all $m_k \in M$ and all $i < n$. Prove that $M^{\otimes n}/K_3 \cong \operatorname{Sym}^n M$.

20. Assume $n! \cdot 1_R$ is invertible in R. Let L be the submodule of $M^{\otimes n}$ generated by elements of the form

$$(n!)^{-1} \sum_{w \in S_n} m_{w(1)} \otimes m_{w(2)} \otimes \cdots \otimes m_{w(n)},$$

where all $m_k \in M$. Prove $L \cong \operatorname{Sym}^n M$.

21. Assume $n! \cdot 1_R$ is invertible in R. Let L' be the submodule of $M^{\otimes n}$ generated by elements of the form

$$(n!)^{-1} \sum_{w \in S_n} \operatorname{sgn}(w) m_{w(1)} \otimes m_{w(2)} \otimes \cdots \otimes m_{w(n)},$$

where all $m_k \in M$. Prove $L' \cong \bigwedge^n M$.

22. Assume $X_1, \ldots, X_n$ are generating sets for the R-modules $M_1, \ldots, M_n$ (respectively). Prove that $X = \{x_1 \otimes \cdots \otimes x_n : x_i \in X_i\}$ generates the R-module $M_1 \otimes_R \cdots \otimes_R M_n$.

23. Assume X generates the R-module M. (a) Prove $\{x_1 \wedge \cdots \wedge x_n : x_i \in X\}$ generates the R-module $\bigwedge^n M$. (b) Suppose $<$ is a total ordering on X. Prove $\{x_1 \wedge \cdots \wedge x_n : x_i \in X, x_1 < \cdots < x_n\}$ generates $\bigwedge^n M$.

24. State and prove results similar to (a) and (b) in Exercise 23 for $\operatorname{Sym}^n M$.

25. Let V be the free $\mathbb{Z}_2$-module $\mathbb{Z}_2^2$, which has basis $X = \{(1,0), (0,1)\}$.
    (a) Use X to find bases for $V^{\otimes 2}$, $\bigwedge^2 V$, and $\operatorname{Sym}^2 V$.
    (b) List all elements $z \in V \otimes_{\mathbb{Z}_2} V$. For each z, express z in all possible ways as a basic tensor $u \otimes v$ with $u, v \in V$, or explain why this cannot be done.
    (c) List all $w \in \bigwedge^2 V$. For each w, express w in all possible ways in the form $u \wedge v$, or explain why this cannot be done.
    (d) List all $y \in \operatorname{Sym}^2 V$. For each y, express y in all possible ways in the form uv, or explain why this cannot be done.

26. **Myth:** "The tensor product of two nonzero modules must be nonzero." Disprove this myth by showing that for all $a, b \in \mathbb{Z}_{>0}$ with $\gcd(a, b) = 1$, $\mathbb{Z}_a \otimes_{\mathbb{Z}} \mathbb{Z}_b = \{0\}$.

27. **Myth:** "If $M \neq \{0\}$, then $M^{\otimes 2} \neq \{0\}$." Disprove this myth by considering the $\mathbb{Z}$-module $M = \mathbb{Q}/\mathbb{Z}$.

28. In the proof that $M \otimes_R N \cong N \otimes_R M$ in §20.7, show directly that f is surjective.

29. **Commutativity Isomorphisms for Tensor Products.** Prove: for all R-modules $M_1, \ldots, M_n$ and all $w \in S_n$, $M_1 \otimes_R \cdots \otimes_R M_n \cong M_{w(1)} \otimes_R \cdots \otimes_R M_{w(n)}$.

30. Prove $M \otimes_R R \cong M$: (a) by using previously proved isomorphisms; (b) by defining specific isomorphisms in both directions.

31. Suppose $I \subseteq \{1, 2, \ldots, n\}$ and $M_i = R$ for all i between 1 and n with $i \notin I$. Prove $M_1 \otimes_R \cdots \otimes_R M_n \cong \bigotimes_{i \in I} M_i$.

32. Prove or disprove: for all commutative rings R, $\bigwedge^2 R \cong R$ as R-modules.

33. Prove or disprove: for all commutative rings R, $\operatorname{Sym}^2 R \cong R$ as R-modules.

34. In the proof of $(M \oplus N) \otimes_R P \cong (M \otimes_R P) \oplus (N \otimes_R P)$ in §20.8, check that g, g_1, and g_2 are R-bilinear; and check that elements of the form $(m \otimes p, 0)$ and $(0, n \otimes p)$ generate $(M \otimes_R P) \oplus (N \otimes_R P)$.

35. Prove that $P \otimes_R (M \oplus N) \cong (P \otimes_R M) \oplus (P \otimes_R N)$ by showing both sides solve the same UMP.

36. Prove (20.10).

37. Prove (20.11).

38. Use the isomorphisms in §20.8 to give a new proof that the tensor product of free R-modules is free, and that (in the finite-dimensional case) $\dim(M_1 \otimes_R \cdots \otimes_R M_n) = \prod_{k=1}^{n} \dim(M_k)$.

39. In the proof in §20.9, check that p'_z is R-multilinear; q'_y is R-multilinear; t' is R-bilinear; and the elements $(x_1 \otimes \cdots \otimes x_k) \otimes (x_{k+1} \otimes \cdots \otimes x_{k+m})$ generate $M^{\otimes k} \otimes_R M^{\otimes m}$.

40. Prove (20.12) by imitating the proof in §20.9.

41. Prove (20.12) by showing both sides solve the same UMP.

42. **Extension of Scalars.** Suppose M is a free R-module with basis X, and assume R is a subring of a commutative ring S.
    (a) Prove $M \otimes_R S$ is a free S-module with basis $X \otimes 1_S = \{x \otimes 1_S : x \in X\}$.
    (b) Suppose N is a free R-module with basis Y and $T : M \to N$ is an R-linear map with matrix A relative to the bases X and Y. Prove $T \otimes \mathrm{id}_S : M \otimes_R S \to N \otimes_R S$ is an S-linear map with matrix A relative to the bases $X \otimes 1_S$ and $Y \otimes 1_S$.

43. Given an R-linear map $f : M \to P$, give the details of the construction of the induced map $\mathrm{Sym}^n f : \mathrm{Sym}^n M \to \mathrm{Sym}^n P$. Prove: if $g : P \to Q$ is also R-linear, then $(\mathrm{Sym}^n g) \circ (\mathrm{Sym}^n f) = \mathrm{Sym}^n (g \circ f)$ and $\mathrm{Sym}^n \mathrm{id}_M = \mathrm{id}_{\mathrm{Sym}^n M}$.

44. The $\mathbb{Z}_5$-module $V = \mathbb{Z}_5^2$ has a $\mathbb{Z}_5$-basis $X = \{\mathbf{e}_1, \mathbf{e}_2\}$, where $\mathbf{e}_1 = (1, 0)$ and $\mathbf{e}_2 = (0, 1)$. Define $f : X \times X \to \mathbb{Z}_5$ by $f(\mathbf{e}_1, \mathbf{e}_1) = 3$, $f(\mathbf{e}_1, \mathbf{e}_2) = 1$, $f(\mathbf{e}_2, \mathbf{e}_1) = 4$, and $f(\mathbf{e}_2, \mathbf{e}_2) = 0$.
    (a) Extend f by multilinearity to $T : V \times V \to \mathbb{Z}_5$. Compute $T((2, 3), (4, 1))$.
    (b) T induces a $\mathbb{Z}_5$-linear map $S : V \otimes_{\mathbb{Z}_5} V \to \mathbb{Z}_5$. Describe the kernel of S.

45. The $\mathbb{R}$-module $V = \mathbb{R}^4$ has the standard ordered basis $X = (\mathbf{e}_1, \mathbf{e}_2, \mathbf{e}_3, \mathbf{e}_4)$.
    (a) Give an $\mathbb{R}$-basis for $V^{\otimes 2}$.
    (b) For all $k \geq 0$, give an $\mathbb{R}$-basis for $\bigwedge^k V$.
    (c) Give an $\mathbb{R}$-basis for $\mathrm{Sym}^3 V$.

46. Let $X = (\mathbf{e}_1, \mathbf{e}_2, \mathbf{e}_3)$ be the standard ordered basis of $V = \mathbb{R}^3$. Define $f : X_<^2 \to \mathbb{R}$ by $f(\mathbf{e}_1, \mathbf{e}_2) = 5$, $f(\mathbf{e}_1, \mathbf{e}_3) = -2$, and $f(\mathbf{e}_2, \mathbf{e}_3) = 1$. Let $T : \mathbb{R}^3 \times \mathbb{R}^3 \to \mathbb{R}$ be the unique alternating map induced from f. Compute $T((a, b, c), (x, y, z))$ for all $a, b, c, x, y, z \in \mathbb{R}$.

47. In §20.15: check that T is a symmetric map; prove that $T = T'$; and show that (N', i') solves the same UMP that (N, i) does.

48. Define a bijection from the set X of weakly increasing sequences of length n using entries in $\{1, 2, \ldots, d\}$ to the set Y of strictly increasing sequences of length n using entries in $\{1, 2, \ldots, d + n - 1\}$. Conclude that $|X| = |Y| = \binom{d+n-1}{n}$.

49. Let $A = \begin{bmatrix} 2 & 1 \\ 0 & -1 \end{bmatrix}$ and $B = \begin{bmatrix} -1 & -2 & -1 \\ 2 & 0 & 2 \\ 1 & 0 & 1 \end{bmatrix}$. Compute $A \otimes B$ and $B \otimes A$.

50. Let $A = \begin{bmatrix} 1 & 1 \\ 0 & 1 \end{bmatrix}$. Compute $A^{\otimes k}$ (an iterated tensor product of matrices) for all $k \geq 0$. (Label the rows and columns of $A^{\otimes k}$ by subsets of $\{1, 2, \ldots, k\}$.) Compute the inverses of the matrices $A^{\otimes k}$.

51. By computing the general entry on each side, prove: for all $A, C \in M_m(R)$ and $B, D \in M_n(R)$, $(CA) \otimes (DB) = (C \otimes D)(A \otimes B)$.

52. Fix $m, n \in \mathbb{Z}_{>0}$. Show that the map $p : M_m(R) \times M_n(R) \to M_{mn}(R)$ given by $p(A, B) = A \otimes B$ (tensor product of matrices) is R-bilinear. Deduce the existence of an R-isomorphism $M_m(R) \otimes_R M_n(R) \cong M_{mn}(R)$.

53. Suppose $M_1, \ldots, M_n$ are free R-modules with respective bases $X_1, \ldots, X_n$ of sizes $d_1, \ldots, d_n$. Suppose $f_k : M_k \to M_k$ is an R-linear map represented by the matrix $A_k \in M_{d_k}(R)$, for $1 \leq k \leq n$. Let A be the matrix of the map $\bigotimes_{k=1}^n f_k$ relative to the basis $X = \{x_1 \otimes \cdots \otimes x_n : x_k \in X_k\}$ of $\bigotimes_{k=1}^n M_k$. What is the size of A? Describe the entries of A in terms of the entries of each A_k.

54. Let $f : \mathbb{R}^3 \to \mathbb{R}^3$ have matrix $\begin{bmatrix} 2 & 2 & -1 \\ 0 & 1 & 3 \\ 0 & -1 & 1 \end{bmatrix}$ relative to the standard ordered basis. Compute the matrix of $\bigwedge^k f$ for $0 \leq k \leq 3$.

55. Let M and N be free R-modules, and suppose an R-linear map $f : M \to N$ has matrix A relative to ordered bases X for M and Y for N. Compute the entries in the matrix of $\bigwedge^k f$ relative to the bases derived from $X_<^k$ and $Y_<^k$.

56. Use Exercise 55 and the relation $\bigwedge^k (g \circ f) = (\bigwedge^k g) \circ (\bigwedge^k f)$ to deduce a formula relating the kth order minors of rectangular matrices A, B, and AB.

57. Deduce the Cauchy–Binet formula (§5.14) from the previous exercise.

58. In §20.18: check that the structure $(T(M), +, \star)$ is a ring and an R-algebra; check that f_k is R-multilinear; and fill in the missing details in the proof that g is a ring homomorphism.

59. Given an R-module M, construct a commutative R-algebra $\mathrm{Sym}(M)$ and an R-linear map $i : M \to \mathrm{Sym}(M)$ solving this UMP: for any commutative R-algebra $(A, +, *)$ and any R-linear map $f : M \to A$, there exists a unique R-algebra homomorphism $g : \mathrm{Sym}(M) \to A$ with $f = g \circ i$. (You can solve this problem in two ways: either imitate the construction in §20.18, or take the quotient of the tensor algebra $T(M)$ by an appropriate ideal.)

60. Given an R-module M, construct an R-algebra $\bigwedge(M)$ such that $z \star z = 0$ for all $z \in \bigwedge(M)$ and an R-linear map $i : M \to \bigwedge(M)$ solving this UMP: for any R-algebra $(A, +, *)$ and any R-linear map $f : M \to A$ such that $f(x) * f(x) = 0_A$ for all $x \in M$, there exists a unique R-algebra homomorphism $g : \bigwedge(M) \to A$ with $f = g \circ i$.

61. Suppose M, N, and P are R-modules, and $f : M \to N$, $g : N \to P$ are R-linear maps. Show that f extends to a unique R-algebra homomorphism $T(f) : T(M) \to T(N)$. For $y = (y_k : k \geq 0) \in T(M)$, give a formula for $T(f)(y)$. Show $T(g \circ f) = T(g) \circ T(f)$ and $T(\mathrm{id}_M) = \mathrm{id}_{T(M)}$.

62. Prove results for $\mathrm{Sym}(M)$ and $\bigwedge(M)$ analogous to the results in the previous exercise.

63. **Tensor Product for Bimodules.** Let R, S, and T be rings, possibly non-commutative. We use the notation $_RM_S$ to indicate that M is an R, S-*bimodule*, which is a left R-module and a right S-module such that $(rm)s = r(ms)$ for all $r \in$

$R, m \in M$, and $s \in S$. A *bimodule homomorphism* between two R, S-bimodules is a map that is both R-linear and S-linear. Fix bimodules $_R M_S$, $_S N_T$, and $_R P_T$. Say that a map $g : M \times N \to P$ is S-*biadditive* iff $g(m + m', n) = g(m, n) + g(m', n)$, $g(m, n + n') = g(m, n) + g(m, n')$, and $g(ms, n) = g(m, sn)$ for all $m, m' \in M$, $n, n' \in N$, and $s \in S$. Call the map g an R, S, T-*map* iff g is S-biadditive and $g(rm, n) = rg(m, n)$ and $g(m, nt) = g(m, n)t$ for all $m \in M$, $n \in N$, $r \in R$, and $t \in T$. Construct a bimodule $_R(M \otimes_S N)_T$ (called the *tensor product of M and N over S*) and an R, S, T-map $i : M \times N \to M \otimes_S N$, solving the following UMP: for each bimodule $_R P_T$, there is a bijection from the set of group homomorphisms $f : M \otimes_S N \to P$ onto the set of S-biadditive maps $g : M \times N \to P$, that maps f to $f \circ i$. Furthermore, show that f is an R, T-bimodule homomorphism iff $g = f \circ i$ is an R, S, T-map. (Define the commutative group $M \otimes_S N$ to be a certain quotient of the free commutative group with basis $M \times N$. Verify the UMP for S-biadditive maps. Aided by this, carefully define a left R-action and a right T-action on $M \otimes_S N$, verify the bimodule axioms, and prove that bimodule homomorphisms correspond to R, S, T-maps.)

64. Establish tensor product isomorphisms (analogous to those proved in §20.8) for tensor products of bimodules.

65. Given rings R, S, T, U and bimodules $_R M_S$, $_S N_T$, $_T P_U$, prove there is an R, U-bimodule isomorphism $(M \otimes_S N) \otimes_T P \cong M \otimes_S (N \otimes_T P)$.

66. Let $(A, +, \star)$ be an R-algebra. (a) Show there is a well-defined R-linear map $m : A \otimes_R A \to A$ given on generators by $m(x \otimes y) = x \star y$ for $x, y \in A$. (b) Identifying $(A \otimes_R A) \otimes_R A$ and $A \otimes_R (A \otimes_R A)$ with $A^{\otimes 3}$, show that $m \circ (m \otimes \mathrm{id}_A) = m \circ (\mathrm{id}_A \otimes m)$. (c) Let $c : A \otimes_R A \to A \otimes_R A$ be the isomorphism given by $c(x \otimes y) = y \otimes x$ for $x, y \in A$. Show: if A is commutative, then $m \circ c = m$. (d) Define an R-linear map $e : R \to A$ by $e(r) = r 1_A$ for $r \in R$. Let $g : R \otimes_R A \to A$ and $h : A \otimes_R R \to A$ be the canonical isomorphisms. Show $g = m \circ (e \otimes \mathrm{id}_A)$ and $h = m \circ (\mathrm{id}_A \otimes e)$.

67. Let A be an R-module. Suppose $m : A \otimes_R A \to A$ and $e : R \to A$ are R-linear maps satisfying the equations in (b) and (d) of Exercise 66. Define $x \star y = m(x \otimes y)$ for $x, y \in A$. Show that $(A, +, \star)$ is an R-algebra with identity $1_A = e(1_R)$. Also show that if m satisfies the equation in (c) of Exercise 66, then A is commutative.

68. Given R-modules M and P, call a function $f : M^4 \to P$ *peculiar* iff f is R-multilinear and for all $w, x, y, z \in M$, $f(x, x, y, z) = 0$, $f(x, y, z, z) = 0$, and

$$f(w, x, y, z) = f(y, x, w, z) + f(w, y, x, z)$$
$$= f(z, x, y, w) + f(w, z, y, x)$$
$$= f(y, z, w, x).$$

Define an R-module N and a peculiar function $i : M^4 \to N$ solving the following UMP: for each R-module P and each peculiar $f : M^4 \to P$, there exists a unique R-linear map $g : N \to P$ such that $f = g \circ i$.

69. Suppose M is a free R-module with ordered basis $X = (x_1 < x_2 < \cdots < x_n)$. Show that the R-module N in Exercise 68 is free with basis consisting of all elements $i(x_a, x_b, x_c, x_d)$ where $a, b, c, d \in \{1, 2, \ldots, n\}$, $a < b$, $c < d$, $a \leq c$, and $b \leq d$.

Appendix: Basic Definitions

This appendix records some general mathematical definitions and notations that occur throughout the text. See [40] for a more detailed exposition of this material. The word *iff* is defined to mean "if and only if."

Sets

We first review some definitions from set theory. All capital letters used here denote sets.

- *Set Membership:* $x \in S$ means x is a member of the set S.

- *Set Non-membership:* $x \notin S$ means x is not a member of the set S.

- *Subsets:* $A \subseteq B$ means for all x, if $x \in A$, then $x \in B$.

- *Binary Union:* For all x, $x \in A \cup B$ iff $x \in A$ or $x \in B$.

- *Binary Intersection:* For all x, $x \in A \cap B$ iff $x \in A$ and $x \in B$.

- *Set Difference:* For all x, $x \in A \setminus B$ iff $x \in A$ and $x \notin B$.

- *Empty Set:* For all x, $x \notin \emptyset$.

- *Indexed Unions:* For all x, $x \in \bigcup_{i \in I} A_i$ iff there exists $i \in I$ with $x \in A_i$.

- *Indexed Intersections:* For all x, $x \in \bigcap_{i \in I} A_i$ iff for all $i \in I$, $x \in A_i$.

- *Cartesian Products:* $A \times B$ is the set of all ordered pairs (a, b) with $a \in A$ and $b \in B$. $A_1 \times \cdots \times A_n$ is the set of all ordered n-tuples $(a_1, \ldots, a_n)$ with $a_i \in A_i$ for $1 \leq i \leq n$.

- *Number Systems:* We write $\mathbb{Z}_{\geq 0} = \{0, 1, 2, 3, \ldots\}$ for the set of natural numbers, $\mathbb{Z}_{>0} = \{1, 2, 3, \ldots\}$ for the set of positive integers, $\mathbb{Z}$ for the set of integers, $\mathbb{Q}$ for the set of rational numbers, $\mathbb{R}$ for the set of real numbers, and $\mathbb{C}$ for the set of complex numbers. $\mathbb{Q}_{>0}$ denotes the set of positive rational numbers. $\mathbb{R}_{>0}$ denotes the set of positive real numbers. For each positive integer n, we write $[n]$ to denote the finite set $\{1, 2, \ldots, n\}$.

Functions

Formally, a *function* is an ordered triple $f = (X, Y, G)$, where X is a set called the *domain* of f, Y is a set called the *codomain* of f, and $G \subseteq X \times Y$ is a set called the *graph* of f, which is required to satisfy this condition: for all $x \in X$, there exists a unique $y \in Y$ with $(x, y) \in G$. For all $x \in X$, we write $y = f(x)$ iff $(x, y) \in G$.

The notation $f : X \to Y$ means that f is a function with domain X and codomain Y. We often introduce a new function by a phrase such as: "Let $f : X \to Y$ be given by $f(x) = \cdots$," where $\cdots$ is some formula involving x. We must check that for each fixed $x \in X$, this formula always does produce exactly one output, and that this output belongs to the claimed codomain Y. By our definition, two functions f and g are *equal* iff they have the same domain and the same codomain and the same graph. To check equality of the graphs, we must check that $f(x) = g(x)$ for all x in the common domain of f and g.

Given functions $f : X \to Y$ and $g : Y \to Z$, the *composite function* $g \circ f$ is the function with domain X, codomain Z, and graph $\{(x, g(f(x))) : x \in X\}$. Thus, $g \circ f : X \to Z$ satisfies $(g \circ f)(x) = g(f(x))$ for all $x \in X$.

Let $f : X \to Y$ be any function. We say f is *one-to-one* (or *injective*, or an *injection*) iff for all $x_1, x_2 \in X$, if $f(x_1) = f(x_2)$ then $x_1 = x_2$. We say f is *onto* (or *surjective*, or a *surjection*) iff for each $y \in Y$, there exists $x \in X$ with $y = f(x)$. We say f is *bijective* (or a *bijection*) iff f is one-to-one and onto iff for each $y \in Y$, there exists a unique $x \in X$ with $y = f(x)$. The composition of two injections is an injection. The composition of two surjections is a surjection. The composition of two bijections is a bijection.

The *identity function* on any set X is the function $\mathrm{id}_X : X \to X$ given by $\mathrm{id}_X(x) = x$ for all $x \in X$. Given $f : X \to Y$, we say that a function $g : Y \to X$ is *the inverse* of f iff $f \circ g = \mathrm{id}_Y$ and $g \circ f = \mathrm{id}_X$, in which case we write $g = f^{-1}$. We can show that the inverse of f is unique when it exists. Also, f^{-1} exists iff f is a bijection, in which case f^{-1} is also a bijection and $(f^{-1})^{-1} = f$.

Suppose $f : X \to Y$ is a function and $Z \subseteq X$. We obtain a new function $g : Z \to Y$ with domain Z by setting $g(z) = f(z)$ for all $z \in Z$. We call g the *restriction of f to Z*, denoted $g = f|Z$ or $f|_Z$.

Suppose $f : X \to Y$ is any function. For all $A \subseteq X$, the *image* of A under f is the set $f[A] = \{f(a) : a \in A\} \subseteq Y$. For all $B \subseteq Y$, the *preimage* of B under f is the set $f^{-1}[B] = \{x \in X : f(x) \in B\} \subseteq X$. This notation does not mean that the inverse function f^{-1} must exist. But, when f^{-1} does exist, the preimage of B under f coincides with the image of B under f^{-1}, so the notation $f^{-1}[B]$ is not ambiguous. The *image* of the function f is the set $f[X]$; f is a surjection iff $f[X] = Y$. We use square brackets for images and preimages to prevent ambiguity. More precisely, if A is both a member of X and a subset of X, then $f(A)$ is the value of f at the point A in its domain, whereas $f[A]$ is the image under f of the subset A of the domain.

Relations

A *relation from X to Y* is a subset R of $X \times Y$. For $x \in X$ and $y \in Y$, xRy means $(x, y) \in R$. A *relation on a set X* is a relation R from X to X. R is called *reflexive on X* iff for all $x \in X$, xRx. R is called *symmetric* iff for all x, y, if xRy, then yRx. R is called *antisymmetric* iff for all x, y, if xRy and yRx, then $x = y$. R is called *transitive* iff for all x, y, z, if xRy and yRz, then xRz. R is called an *equivalence relation on X* iff R is reflexive on X, symmetric, and transitive.

Suppose R is an equivalence relation on a set X. For $x \in X$, the *equivalence class of x relative to R* is the set $[x]_R = \{y \in X : xRy\}$. A given equivalence class typically has many names; more precisely, for all $x, z \in R$, $[x]_R = [z]_R$ iff xRz. The *quotient set X modulo R* is the set of all equivalence classes of R, namely $X/R = \{[x]_R : x \in X\}$.

A *set partition* of a given set X is a collection P of nonempty subsets of X such that for all $x \in X$, there exists a unique $S \in P$ with $x \in S$. For every equivalence relation R

on a fixed set X, the quotient set X/R is a set partition of X consisting of the equivalence classes $[x]_R$ for $x \in X$. Conversely, given any set partition P of X, the relation R defined by "xRy iff there exists $S \in P$ with $x \in S$ and $y \in S$" is an equivalence relation on X with $X/R = P$. Formally, letting EQ_X be the set of all equivalence relations on X and SP_X be the set of all set partitions on X, the map $f : EQ_X \to SP_X$ given by $f(R) = X/R$ for $R \in EQ_X$ is a bijection.

Partially Ordered Sets

A *partial ordering* on a set X is a relation $\leq$ on X that is reflexive on X, antisymmetric, and transitive. A *partially ordered set* or *poset* is a pair $(X, \leq)$ where $\leq$ is a partial ordering on X. A poset $(X, \leq)$ is *totally ordered* iff for all $x, y \in X$, $x \leq y$ or $y \leq x$. A subset Y of a poset $(X, \leq)$ is called a *chain* iff for all $x, y \in Y$, $x \leq y$ or $y \leq x$. By definition, $y \geq x$ means $x \leq y$; $x < y$ means $x \leq y$ and $x \neq y$; and $x > y$ means $x \geq y$ and $x \neq y$.

Let S be a subset of a poset $(X, \leq)$. An *upper bound* for S is an element $x \in X$ such that for all $y \in S$, $y \leq x$. A *greatest element* of S is an element $x \in S$ such that for all $y \in S$, $y \leq x$; x is unique if it exists. A *lower bound* for S is an element $x \in X$ such that for all $y \in S$, $x \leq y$. A *least element* of S is an element $x \in S$ such that for all $y \in S$, $x \leq y$; x is unique if it exists.

We say $x \in X$ is a *least upper bound* for S iff x is the least element of the set of upper bounds of S in X. In detail, this means $y \leq x$ for all $y \in S$; and for any $z \in X$ such that $y \leq z$ for all $y \in S$, $x \leq z$. The least upper bound of S is unique if it exists; we write $x = \sup S$ in this case. For $S = \{y_1, \ldots, y_n\}$, the notation $y_1 \vee y_2 \vee \cdots \vee y_n$ is also used to denote $\sup S$.

We say $x \in X$ is a *greatest lower bound* for S iff x is the greatest element of the set of lower bounds of S in X. In detail, this means $x \leq y$ for all $y \in S$; and for any $z \in X$ such that $z \leq y$ for all $y \in S$, $z \leq x$. The greatest lower bound of S is unique if it exists; we write $x = \inf S$ in this case. For $S = \{y_1, \ldots, y_n\}$, the notation $y_1 \wedge y_2 \wedge \cdots \wedge y_n$ is also used to denote $\inf S$.

A *lattice* is a poset $(X, \leq)$ such that for all $a, b \in X$, the least upper bound $a \vee b$ and the greatest lower bound $a \wedge b$ exist in X. A *complete lattice* is a poset $(X, \leq)$ such that for every nonempty subset S of X, $\sup S$ and $\inf S$ exist in X.

A *maximal element* in a poset $(X, \leq)$ is an element $x \in X$ such that for all $y \in X$, if $x \leq y$ then $y = x$. A *minimal element* of X is an element $x \in X$ such that for all $y \in X$, if $y \leq x$ then $y = x$. *Zorn's Lemma* states that if $(X, \leq)$ is a poset in which every chain $Y \subseteq X$ has an upper bound in X, then X has a maximal element. Zorn's Lemma is discussed in detail in §17.13.

Further Reading

Chapter 1. There are many introductory accounts of modern algebra, including the texts by Durbin [15], Fraleigh [16], Gallian [17], and Rotman [51]. For more advanced treatments of modern algebra, you may consult the textbooks by Dummit and Foote [14], Hungerford [31], Jacobson [32], and Rotman [50]. Introductions to linear algebra at various levels abound; among many others, we mention the books by Larson and Falvo [35], Lay [36], and Strang [59]. Two more advanced linear algebra books that are similar, in some respects, to the present volume are the texts by Halmos [25] and Hoffman and Kunze [29].

Chapter 2. For basic facts on permutations, you may consult any of the abstract algebra texts mentioned above. There is a vast literature on permutations and the symmetric group; we direct the reader to the texts by Bona [7], Rotman [54], and Sagan [57] for more information.

Chapter 3. Thorough algebraic treatments of polynomials may be found in most texts on abstract algebra, such as those by Dummit and Foote [14] or Hungerford [31]. For more details on formal power series, you may consult [38, Chpt. 7]. The matrix reduction algorithm in §3.10 for computing gcds of polynomials (or integers) comes from an article by W. Blankinship [6]. Cox, Little, and O'Shea have written an excellent book [12] on multivariable polynomials and their role in computational algebraic geometry.

Chapter 4. Three classic texts on matrix theory are the books by Gantmacher [19], Horn and Johnson [30], and Lancaster [34].

Chapter 5. There is a vast mathematical literature on the subject of determinants. Lacking the space to give tribute to all of these, we only mention the text by Turnbull [62], the book of Aitken [2], and the treatise of Muir [44]. Muir has also written an extensive four-volume work chronicling the historical development of the theory of determinants [43].

Chapter 6. This chapter developed a "dictionary" linking abstract concepts defined for vector spaces and linear maps to concrete concepts defined for column vectors and matrices. Many texts on matrix theory, such as Horn and Johnson [30], heavily favor matrix-based descriptions and proofs. Other texts, most notably Bourbaki [8, Chpt. II], prefer a very abstract development that makes almost no mention of matrices. We think it is advisable to gain facility with both languages for discussing linear algebra. For alternative developments of this material, see Halmos [25] and Hoffman and Kunze [29].

Chapter 7. The analogy between complex numbers and complex matrices, including the theorems on the polar decomposition of a matrix, is based on the exposition in Halmos [25]. A wealth of additional material on properties of Hermitian, unitary, positive definite, and normal matrices may be found in Horn and Johnson's text [30].

Chapter 8. We can derive the Jordan canonical form theorem in many different ways. In abstract algebra, this theorem can be deduced from the rational canonical form theorem [29], which in turn is derivable from the classification of finitely generated modules over principal ideal domains. See Chapter 18 or [32] for this approach. Matrix theorists might prefer a more algorithmic construction that triangularizes a complex matrix and then gradually reduces it to Jordan form [13, 30]. Various elementary derivations can be found in [9, 18, 20, 24, 63].

Chapter 9. Further information on QR factorizations can be found in Chapter 5 of Golub and van Loan [21], part II of Trefethen and Bau [61], and §5.3 of Kincaid and Cheney [33]. For LU factorizations, see [21, Chpt. 3] or [30, Sec. 3.5]. These references also contain a wealth of information on the numerical stability properties of matrix factorizations and the associated algorithms.

Chapter 10. Our treatment of iterative algorithms for solving linear systems and computing eigenvalues is similar to that found in §4.6 and §5.1 of Kincaid and Cheney [33]. For more information on this topic, you may consult the numerical analysis texts authored by Ackleh, Allen, Kearfott, and Seshaiyer [1, §3.4, Chpt. 5], Cheney and Kincaid [10, §8.2, §8.4], Trefethen and Bau [61, Chpt. VI], and Golub and Van Loan [21].

Chapter 11. For a very detailed treatment of convex sets and convex functions, the reader may consult Rockafellar's text [49]. A wealth of material on convex polytopes can be found in Grünbaum's encyclopedic work [22]. The presentation in §11.17 through §11.21 is similar to [66, Lecture 1].

Chapter 12. Our exposition of ruler and compass constructions is similar to the accounts found in [32, Vol. 1, Chpt. 4] and [52, App. C]. Treatments of Galois theory may be found in these two texts, as well as in the books by Cox [11], Dummit and Foote [14], and Hungerford [31]. See Tignol's book [60] for a very nice historical account of the development of Galois theory. Another good reference for geometric constructions and other problems in field theory is Hadlock [23].

Chapter 13. Another treatment of dual spaces and their relation to complex inner product spaces appears in Halmos [25]. A good discussion of dual spaces in the context of Banach spaces is given in Simmons [58, Chpt. 9]. The book of Cox, Little, and O'Shea [12] contains an excellent exposition of the ideal-variety correspondence and other aspects of affine algebraic geometry.

Chapter 14. Our coverage of bilinear forms closely follows Chapter 6 of Volume 1 of Jacobson's algebra text [32]. That chapter contains additional information about the structure of orthogonal groups. Another superb reference for this material is Artin's classic monograph [4].

Chapter 15. Two other introductions to Hilbert spaces at a level similar to ours can be found in Simmons [58, Chpt. 10] and Rudin [56, Chpt. 4]. Halmos' book [26] contains an abundance of problems on Hilbert spaces. For more on metric spaces, see Simmons [58] or Munkres [45].

Chapter 16. A nice treatment of commutative groups, finitely generated or not, appears in Chapter 10 of Rotman's group theory text [54]. Another exposition of the reduction algorithm for integer matrices and its connection to classifying finitely generated commutative groups is given by Munkres [46, §11].

Chapter 17. Four excellent accounts of module theory appear in Anderson and Fuller's text [3], Atiyah and Macdonald's book [5], Jacobson's *Basic Algebra 1 and 2* [32] (especially the third chapter in each volume), and Rotman's homological algebra book [55]. Bourbaki [8, Chpt. II] provides a very thorough and general, but rather difficult, treatment of modules.

Chapter 18. The classification of finitely generated modules over principal ideal domains is a standard topic covered in advanced abstract algebra texts such as [14, 32]. I hope that the coverage here may be more quickly accessible to readers with a little less background in group theory and ring theory. The book by Hartley and Hawkes [28] is an excellent, concise reference that covers the classification theorem, the necessary background on rings and modules, and applications to canonical forms. There are two approaches to proving the rational canonical form for square matrices over a field. The approach adopted here deduces this result from the general theory for PIDs. The other approach avoids the abstraction of PIDs by proving all necessary results at the level of finite-dimensional vector spaces, T-invariant subspaces, and T-cyclic subspaces. See [29] for such a treatment. The author's opinion is that proving the special case of the classification theorem for torsion $F[x]$-modules is not much simpler than proving the full theorem for all finitely generated modules over all PIDs. In fact, because of all the extra structure of the ring $F[x]$, focusing on this special case might even give the reader less intuition for what the proof is doing. To help the reader build intuition, we chose to cover the much more concrete case of $\mathbb{Z}$-modules in an earlier chapter.

Chapter 19. Two sources that give due emphasis to the central role of universal mapping properties in abstract algebra are Jacobson's two-volume algebra text [32] and Rotman's homological algebra book [55]. The appropriate general context for understanding UMP's is category theory, the basic elements of which are covered in the two references just cited. A more comprehensive introduction to category theory is given in Mac Lane's book [41].

Chapter 20. A nice introduction to multilinear algebra is the text by Northcott [47]. A very thorough account of the subject, including detailed discussions of tensor algebras, exterior algebras, and symmetric algebras, appears in [8, Chpt. III].

Bibliography

[1] Azmy Ackleh, Edward J. Allen, Ralph Kearfott, and Padmanabhan Seshaiyer, *Classical and Modern Numerical Analysis: Theory, Methods, and Practice*, Chapman and Hall/CRC Press, Boca Raton, FL (2010).

[2] A. C. Aitken, *Determinants and Matrices* (eighth ed.), Oliver and Boyd Ltd., Edinburgh (1954).

[3] Frank W. Anderson and Kent R. Fuller, *Rings and Categories of Modules* (Graduate Texts in Mathematics, Vol. 13, second ed.), Springer-Verlag, New York (1992).

[4] Emil Artin, *Geometric Algebra*, Dover Publications, Mineola, NY (2016).

[5] M. F. Atiyah and I. G. Macdonald, *Introduction to Commutative Algebra*, Addison-Wesley, Reading, MA (1969).

[6] W. A. Blankinship, "A new version of the Euclidean algorithm," *Amer. Math. Monthly* **70** #7 (1963), 742–745.

[7] Miklòs Bòna, *Combinatorics of Permutations*, Chapman and Hall/CRC, Boca Raton, FL (2004).

[8] Nicolas Bourbaki, *Algebra 1*, Springer-Verlag, New York (1989).

[9] R. Brualdi, "The Jordan canonical form: an old proof," *Amer. Math. Monthly* **94** #3 (1987), 257–267.

[10] E. Ward Cheney and David R. Kincaid, *Numerical Mathematics and Computing* (sixth ed.), Brooks/Cole, Pacific Grove, CA (2007).

[11] David A. Cox, *Galois Theory* (second ed.), John Wiley and Sons, New York (2012).

[12] David A. Cox, John Little, and Donal O'Shea, *Ideals, Varieties, and Algorithms: An Introduction to Computational Algebraic Geometry and Commutative Algebra* (third ed.), Springer-Verlag, New York (2010).

[13] R. Fletcher and D. Sorenson, "An algorithmic derivation of the Jordan canonical form," *Amer. Math. Monthly* **90** #1 (1983), 12–16.

[14] David S. Dummit and Richard M. Foote, *Abstract Algebra* (third ed.), John Wiley and Sons, New York (2003).

[15] John R. Durbin, *Modern Algebra: An Introduction* (sixth ed.), John Wiley and Sons, New York (2008).

[16] John B. Fraleigh, *A First Course in Abstract Algebra* (seventh ed.), Addison Wesley, Reading (2002).

[17] Joseph A. Gallian, *Contemporary Abstract Algebra* (fifth ed.), Houghton Mifflin, Boston (2001).

[18] A. Galperin and Z. Waksman, "An elementary approach to Jordan theory," *Amer. Math. Monthly* **87** #9 (1980), 728–732.

[19] F. R. Gantmacher, *The Theory of Matrices* (two volumes), Chelsea Publishing Co., New York (1960).

[20] I. Gohberg and S. Goldberg, "A simple proof of the Jordan decomposition theorem for matrices," *Amer. Math. Monthly* **103** #2 (1996), 157–159.

[21] Gene Golub and Charles Van Loan, *Matrix Computations* (third ed.), The Johns Hopkins University Press, Baltimore (1996).

[22] Branko Grünbaum, *Convex Polytopes* (Graduate Texts in Mathematics, Vol. 221, second ed.), Springer-Verlag, New York (2003).

[23] Charles R. Hadlock, *Field Theory and Its Classical Problems*, Carus Mathematical Monograph no. 19, Mathematical Association of America, Washington, D.C., (1978).

[24] J. Hall, "Another elementary approach to the Jordan form," *Amer. Math. Monthly* **98** #4 (1991), 336–340.

[25] Paul R. Halmos, *Finite-Dimensional Vector Spaces*, Springer-Verlag, New York (1974).

[26] Paul R. Halmos, *A Hilbert Space Problem Book* (Graduate Texts in Mathematics, Vol. 19, second ed.), Springer-Verlag, New York (1982).

[27] Paul R. Halmos, *Naive Set Theory*, Springer-Verlag, New York (1998).

[28] Brian Hartley and Trevor Hawkes, *Rings, Modules, and Linear Algebra*. Chapman and Hall, London (1970).

[29] Kenneth Hoffman and Ray Kunze, *Linear Algebra* (second ed.), Prentice Hall, Upper Saddle River, NJ (1971).

[30] Roger Horn and Charles Johnson, *Matrix Analysis* (second ed.), Cambridge University Press, Cambridge (2012).

[31] Thomas W. Hungerford, *Algebra* (Graduate Texts in Mathematics, Vol. 73), Springer-Verlag, New York (1980).

[32] Nathan Jacobson, *Basic Algebra I and II* (second ed.), Dover Publications, Mineola, NY (2009).

[33] David Kincaid and Ward Cheney, *Numerical Analysis: Mathematics of Scientific Computing* (second ed.), Brooks/Cole, Pacific Grove, CA (1996).

[34] Peter Lancaster, *Theory of Matrices*, Academic Press, New York (1969).

[35] Ron Larson and David Falvo, *Elementary Linear Algebra* (sixth ed.), Brooks Cole, Belmont, CA (2009).

[36] David C. Lay, *Linear Algebra and Its Applications* (fourth ed.), Addison Wesley, Reading, MA (2011).

[37] Hans Liebeck, "A proof of the equality of column and row rank of a matrix," *Amer. Math. Monthly* **73** #10 (1966), 1114.

[38] Nicholas A. Loehr, *Bijective Combinatorics*, Chapman and Hall/CRC, Boca Raton, FL (2011).

[39] Nicholas A. Loehr, "A direct proof that row rank equals column rank," *College Math. J.* **38** #4 (2007), 300–301.

[40] Nicholas A. Loehr, *An Introduction to Mathematical Proofs*, CRC Press, Boca Raton, FL (2020).

[41] Saunders Mac Lane, *Categories for the Working Mathematician* (Graduate Texts in Mathematics, Vol. 5, second ed.), Springer-Verlag, New York (1998).

[42] J. Donald Monk, *Introduction to Set Theory*, McGraw-Hill, New York (1969).

[43] Thomas Muir, *The Theory of Determinants in the Historical Order of Development* (four volumes), Dover Publications, New York (1960).

[44] Thomas Muir, *A Treatise on the Theory of Determinants*, revised and enlarged by William Metzler, Dover Publications, New York (1960).

[45] James R. Munkres, *Topology* (second ed.), Prentice Hall, Upper Saddle River, NJ (2000).

[46] James R. Munkres, *Elements of Algebraic Topology*, Perseus Publishing, Cambridge, MA (1984).

[47] D. G. Northcott, *Multilinear Algebra*, Cambridge University Press, Cambridge (1984).

[48] James G. Oxley, *Matroid Theory* (second ed.), Oxford University Press, Oxford (2011).

[49] R. Tyrrell Rockafellar, *Convex Analysis*, Princeton University Press, Princeton, NJ (1972).

[50] Joseph J. Rotman, *Advanced Modern Algebra* (second ed.), American Mathematical Society, Providence, RI (2010).

[51] Joseph J. Rotman, *A First Course in Abstract Algebra* (third ed.), Prentice Hall, Upper Saddle River, NJ (2005).

[52] Joseph J. Rotman, *Galois Theory* (second ed.), Springer-Verlag, New York (1998).

[53] Joseph J. Rotman, *An Introduction to Algebraic Topology* (Graduate Texts in Mathematics, Vol. 119), Springer-Verlag, New York (1988).

[54] Joseph J. Rotman, *An Introduction to the Theory of Groups* (fourth ed.), Springer-Verlag, New York (1994).

[55] Joseph J. Rotman, *Notes on Homological Algebra*, Van Nostrand Reinhold, New York (1970).

[56] Walter Rudin, *Real and Complex Analysis* (third ed.), McGraw-Hill, Boston (1987).

[57] Bruce E. Sagan, *The Symmetric Group: Representations, Combinatorial Algorithms, and Symmetric Functions* (second ed.), Springer-Verlag, New York (2001).

[58] George F. Simmons, *Introduction to Topology and Modern Analysis*, Krieger Publishing Co., Malabar, FL (2003).

[59] Gilbert Strang, *Introduction to Linear Algebra* (fourth ed.), Wellesley Cambridge Press, Wellesley, MA (2009).

[60] Jean-Pierre Tignol, *Galois' Theory of Algebraic Equations*, World Scientific Publishing, Singapore (2001).

[61] Lloyd N. Trefethen and David Bau III, *Numerical Linear Algebra*, SIAM, Philadelphia (1997).

[62] Herbert W. Turnbull, *The Theory of Determinants, Matrices, and Invariants* (second ed.), Blackie and Son Ltd., London (1945).

[63] H. Valiaho, "An elementary approach to the Jordan form of a matrix," *Amer. Math. Monthly* **93** #9 (1986), 711–714.

[64] D. J. Welsh, *Matroid Theory*, Academic Press, New York (1976).

[65] Douglas B. West, *Introduction to Graph Theory* (second ed.), Prentice Hall, Upper Saddle River, NJ (2001).

[66] Günter M. Ziegler, *Lectures on Polytopes* (Graduate Texts in Mathematics, Vol. 152), Springer-Verlag, New York (1995).

Index